Electrons, Atoms, and Molecules in Inorganic Chemistry

Electrons, Atoms, and Molecules in Inorganic Chemistry

A Worked Examples Approach

Joseph J. Stephanos

Anthony W. Addison

ELSEVIER

ACADEMIC PRESS

An imprint of Elsevier

Academic Press is an imprint of Elsevier
125 London Wall, London EC2Y 5AS, United Kingdom
525 B Street, Suite 1800, San Diego, CA 92101-4495, United States
50 Hampshire Street, 5th Floor, Cambridge, MA 02139, United States
The Boulevard, Langford Lane, Kidlington, Oxford OX5 1GB, United Kingdom

Library of Congress Cataloging-in-Publication Data
A catalog record for this book is available from the Library of Congress

British Library Cataloguing-in-Publication Data
A catalogue record for this book is available from the British Library

ISBN: 978-0-12-811048-5

Publisher: John Fedor
Acquisition Editor: Emily McCloskey
Editorial Project Manager: Katerina Zaliva
Production Project Manager: Paul Prasad Chandramohan
Cover Designer: Mathew Limbert

Typeset by SPi Global, India

To our Students in Chemistry

Contents

Preface

This book presents the chemical concepts that govern the chemistry of molecular construction. The emphasis is on the building up of an understanding of essential principles and on familiarization with basic inorganic concepts. Necessary background information is introduced to comprehend the field from both chemical and practical areas. The book explains and details the fundamentals that serve as a source of numerous basic concepts of methods and applications. The combination of the basic concepts, methods and applications with example exercises yields a more positive outcome for students and teachers.

Many inorganic textbooks that are available cover too much material and do not go into the depth needed for fundamental principles. Most of these are seem to be either fairly elementary or very advanced. Students might be displeased by the current selection available, and there is a great need for an inorganic principles of chemical bonding text. In our book, we bridge and integrate both elementary and more advanced principles. A student should be able to become familiar with the topics presented with this one book, rather than learn the basics in one and use another for the more advanced aspects of the material.

Given the complex and abstract nature of the subject, the book is easy to follow. The text is carefully thought through and laid out. The approach of developing the material by answering questions and problems relevant to inorganic chemistry in extensive mathematical detail is unique and makes the book attractive especially for university students, as a study material source for examinations. Every mathematical step in the book is elaborated with close attention to every detail. That necessary mathematical foundation is found in a separate supplement, as an entirely non-mathematical approach will be of little value for the purpose. The bullet point approach of answering common questions rather writing a narrative and starting each chapter with the circle scheme describing the sections to be discussed are intended to be attention-catching and could attract certain students and aid them as they study.

The content and the style of the book should capture readers from both traditional and modern schools. It is useful as a reference and text for specialized and graduate courses in physical inorganic or advanced organic chemistry. The book is intended for advanced undergraduates and for postgraduates taking courses in chemistry, students studying atomic structure and molecule formation in chemical engineering and material science. It should also be of value to research workers in other fields, who might need an introduction to essential inorganic principles. This book is very suitable for self-study; the range covered is so extensive that this book can be student's companion throughout his or her university career. At the same time, teachers can turn to it for ideas and inspiration.

This book is divided into 11 chapters, and covers a full range of topics in inorganic chemistry: wave-particle duality, electrons in atoms, chemical bonding, molecular symmetry, theories of bonding, valence bond theory, VSEPR theory, orbital hybridization, molecular orbital theory, crystal field theory, ligand field theory, vibrational, rotational, and electronic spectroscopy, magnetism and finally, a mathematics supplement outlining the necessary methods.

In the beginning of this book we develop and provide an understanding of the dual wave-particle nature of electrons, photons, and other particles of small mass.

Schrödinger's method is linked for exploring the modern theory of atomic structure, and to establish the formal mathematical framework. This framework is employed to set up the final real solution for the full orbital wave function and identify the four quantum numbers n, l, m, and s, also to compute the most probable radius, mean radius of an orbital, and the boundary surfaces of s, p and d orbitals. The orbital wave equations are used as the key feature in order to explain the orbital and spin angular momenta, as well the electronic configurations of many-electron atoms, and spin-orbital coupling. We examine how to identify the term symbols of the ground state and the different terms of the excited microstates of polyelectronic atoms, and how many subterms arise when spin-orbit coupling is taken into consideration. The splitting of Russell–Saunders terms into microstates in an external magnetic field and their energies are identified. The term wave functions and the corresponding single electron wave functions are described in order to understand the effect of the ligand field.

Then we begin with a review of basic electron accounting procedures for different types of bond formation and proceed to a model for predicting three-dimensional molecular structure. The basic concepts of metallic structure that describe the bonding, define the role of the free valence electrons, and relate the physical properties and theories of metallic bonding are explored. Emphasis is also placed on ionic bonding and the relationships among the lattice energy, thermodynamic parameters and covalency. We review the grounds of the ionic crystal structure, in which radius ratios govern the geometrical arrangements, and review the factors that influence the solubilities. The foundation and basics of coordination chemistry are laid out: firstly, characterization, formulation, formation and stability including hard and soft acid/base interactions, the chelate and macrocycle effects, cavity size, solvation enthalpy, donor atom basicity, solvent competition, steric effects, metal oxidation state, and metal ionization potential. An extensive discussion is given of intermolecular forces, exploring their rôles, consequences and significances. Considerations are given to the structural and chemical nature of the covalent networks in giant molecules such diamond, graphite, fullerenes, graphene, nanotubes and asbestos.

In order to deal with molecular structures, where many energy levels of atoms are involved, symmetry concepts are extensively invoked. Thus, it is appropriate to explain how to establish a proper system for sorting molecules according to their structures. One purpose of this sorting is to introduce some ideas and mathematical techniques that are essential for understanding the structure and properties of molecules and crystals. Matrix representations of symmetry operations, point group, translation, rotation motions, and atomic orbital are thoroughly examined. This leads to presentation of the character tables, and finally shows how that the symmetry representations for the atomic orbitals form bases for the molecular wave functions.

The valence bond theory concept is explained, then we investigate how to predict the shapes and geometries of simple molecules using the valence shell electron-pair repulsion method. The process of predicting the molecule's structure is reviewed. As well, the relationships between the chemical bonds in molecules and its geometry using orbital hybridization theory are addressed. Special attention is devoted to the angles between the bonds formed by a given atom, also to multiple bonding and σ/π hybridization of atomic orbitals. Symmetry-adapted linear combination of atomic wave functions (SALC's) are composed and detailed, then used to compute the contribution of each atomic orbital to the hybrid orbitals.

A brief representation of molecular orbital theory is elaborated. Understanding the electronic distribution of some elected small molecules, and approaches to the relative energies of the molecular orbitals are reviewed. We then explain how the electron distribution changes upon going to some low-lying excited electronic states. The theory is employed to estimate energy changes in chemical reactions, to study stability & reactivity, to find the delocalization energy, electron density, formal charge, bond order, ionization energy, equilibrium constant, and configuration interaction. The orbital combination introduces the band theory concept that makes it possible to rationalize conductivity, insulation and semiconductivity.

Having now available the valence orbital's wave functions of the central ion in their real forms, it is possible to explore the impact of various distributions of ligand atoms around the central ion upon its valence orbitals. A quantitative basis of this effect in the case of a purely ionic model of coordination, and other degrees of mixing are addressed. In this part, we indicate why there is a need for both crystal field theory and ligand field theory. The effect of a cubic crystal field on *d*- and *f*-electrons is introduced, then the expressions of the Hamiltonian to find the crystal field potential experienced by electrons in octahedral, square planar, tetragonally distorted octahedral, and tetrahedral ligand arrangements are computed. The perturbation theory for degenerate systems is used to explain how the crystal field potential of the surrounding ligands perturbs the degeneracy of *d* orbitals of the central ion. The energies of the perturbed *d*-orbitals are calculated by solving the secular determinant. The obtained energies are fed back into secular equations that are derived from the secular determinant to yield wave functions appropriate for the presence of the potential. The variation in the potential energy of each *d* electron due to the crystal field is determined and the splitting of *d*-orbitals so deduced in octahedral, tetrahedral, tetragonally distorted D_{4h} geometries in terms of D_q, D_t, and D_s. Problems and the required approximations are discussed for the free ion in weak crystal field. Then, we study the influence of weak field on polyelectronic configuration of free ion terms, and find the splitting in each term and the wave function for each state. In the strong field situation, the first concern was how the strong field differs from the weak field approach; define the determinant, symmetry and the energy of each state. We compute the appropriate Hamiltonian and the diagonal and off-diagonal interelectronic repulsion in terms of the Racah parameters A, B, and C.

We firstly examine how it is possible to use symmetry and group theory to find what states will be obtained when an ion is placed into a crystalline environment of definite symmetry. Secondly, the relative energies of these states will be investigated. Thirdly, we show how the energies of the various states into which the free ion term are split depend on the strength of the interaction of the ion with its environment. The relationship between the energy of the excited states and D_q are discussed using correlation, Orgel, and Tanabe-Sugano diagrams.

The vibrational spectra of diatomic molecules establish most of the essential principles that are used for complicated polyatomic molecules. Since infrared radiation will excite not only molecular vibration but also rotation, there is a need to comprehend both rotation and vibration of diatomic molecules in order to analyze their spectra. Molecular vibrations are explained by classical mechanics using a simple ball and spring model, whereas vibrational energy levels and transitions between them are concepts taken from quantum mechanics. The quantum mechanics of the translation, vibration and rotation motions are explored in detail. As well, the expression for the vibration-rotation energies of diatomic molecule for the harmonic and anharmonic oscillator models is introduced. Then, we identify the Schrödinger equation for the vibration system of n-atom molecules. We elucidate how to obtain, monitor, and explain the vibrational–rotational excitations, find the quantum mechanical expression for the vibrational and rotational energy levels, predict the frequency of the bands, and compare harmonic *vs.* anharmonic oscillators and between rigid and nonrigid rotor models for the possible excitations. Lagrange's equation is used to show the change in the amplitude of displacement with time. We examine also how to calculate the relative amplitudes of motion and the kinetic and potential energies for the vibrational motions of an n-atom molecule. Calculation of the force constants using the GF-matrix method are discussed. The general steps to determine the normal modes of vibration, and the symmetry representation of these modes are outlined. We examine the relationship and the differences among the cartesian, internal, and normal coordinates used to characterize the stretching vibrations. Then we show why the normal coordinates are used to calculate the vibrational energy of polyatomic molecules. In this part we also focus on how the molecules interact with the radiation and the chemical information obtainable by measurement of the infrared and Raman spectra. Only the radiation electric field interacts significantly with molecules and is important in explaining infrared absorption and Raman scattering. The requirements and the selection rules for the allowed vibrational and rotational excitations are explored. We investigate the relationship between the center of symmetry and the mutual exclusion rule, how to distinguish among isomers and ligand binding modes, and define the forms of the normal modes of vibration and which of these modes are infrared and/or Raman active.

We then focus on the chemical information obtained from electronic spectra in the visible and ultraviolet regions. We examine the relationship between the solute concentration and light absorbance, as well the correlation among the molar extinction coefficient, integrated intensity and dipole strength. The significance of the Born-Oppenheimer approximation is considered, and symmetry considerations are elaborated with respect to allowedness of electronic transitions. The consequences of spin, orbital and vibrational constraints are investigated to explore the basis of the electronic absorption selection rules. The electronic excitation of functional groups, donor–acceptor complexes, and porphyrins are spectroscopically characterized. The study outlines the roles of vibronic coupling, configuration interaction, and π-bonding. Factors that affect the bandwidth, band intensity, and intense colors of certain metal complexes are also identified. The text addresses the effects of Jahn–Teller distortion, temperature, and reduced symmetry, and elaborates the spectrochemical series. Comparative studies of octahedral versus tetrahedral, low-spin versus high-spin, and d^n versus d^{10-n} configuration are conducted. We illustrate how to evaluate D_q and β from the positions of the absorption peaks, and discuss the unexpectedly weak absorbances, simultaneous pair excitations, and the absorption of unpolarized light in general. The role of the magnetic dipole moment on the absorbance intensity is investigated, using circular dichroism spectroscopy and the Kuhn anisotropy factor to examine the effects of lower symmetry, absolute configuration, and the energy levels within the molecule

Finally we investigate the types of magnetic behaviors, giving key definitions and concepts leading to the relationships relevant to magnetochemistry. These provide a bridge to understand spin and orbital contributions to magnetic moments. Then, thermal spreading using the Boltzmann distribution is employed to investigate and estimate the magnetic moments and susceptibilities. The subsequent section deals with the van Vleck treatment and the second order Zeeman Effect to link the spin and orbital contributions to the magnetic susceptibility. Requirements and conditions for nonzero orbital contribution are discussed. The following section is devoted to the effect of imposition of a ligand field on spin-orbital coupling in A, E, and T ground terms. The Curie law, deviations, and data presentation modes are shown. Spin-crossover and the effects of thermal distortion are discussed, followed by the behavior of dinuclear systems with exchange coupling. Finally, Gouy's, Faraday's, Quincke's, and NMR methods for susceptibility are described.

Joseph J. Stephanos
Anthony W. Addison

Chapter 1

Particle Wave Duality

In this chapter we shall develop an understanding of the dual wave-particle nature of electrons, photons, and other particles of small mass (Scheme 1.1).

The particle nature of the electrons had been confirmed by cathode rays, Millikan's capacitor, and Thomson's experimentation. Models for atomic structures were proposed by Thomson and Rutherford.

Studies of black-body radiation shows that energy emits in a small, specific quantity called quanta. Also, hydrogen emission spectrum indicates that electrons in atoms exist only in very specific energy states. Quantum has been concluded as the smallest amount of energy that can be lost or gained by an atom. Bohr's and Bohr-Sommerfeld's atomic models are presented and discussed. Bohr in his atomic models used quantum theory not quantum mechanics, he did not recognize the wave nature of electrons.

Wave interference and diffraction are used as evidences to confirm the wave properties of the electrons. Einstein's relationships are explored to describe and clarify the interdependence of mass and energy.

Furthermore, the corpuscular nature of light is revealed in the photoelectric effect and the Compton effect. de Broglie shows that the dual wave-particle nature is true not only for photon, but for any other material particle as well.

Heisenberg, in his uncertainty principle, points out that only the probability of finding an electron in a particular volume of space can be determined. This probability of finding an electron is proportional to the square of the absolute value of the wave function.

Subatomic particles are examined and classified as fundamental particles (fermions) and force particle (bosons) that mediate interactions among fermions.

In the following outline, we shall try to understand:

- 1.1: Cathode and anode rays
- 1.2: Charge of the electron
- 1.3: Mass of the electron and proton
- 1.4: Rutherford's atomic model
- 1.5: Quantum of energy
- 1.6: The hydrogen-atom line-emission spectra
- 1.7: Bohr's quantum theory of the hydrogen atom
- 1.8: The Bohr-Sommerfeld model
- 1.9: The corpuscular nature of electrons, photons, and particles of very small mass
- 1.10: Relativity theory: mass and energy
- 1.11: The corpuscular nature of electromagnetic waves: the photoelectric effect and the Compton effect
- 1.12: de Broglie's considerations
- 1.13: Werner Heisenberg's uncertainty principle, or the principle of indeterminacy
- 1.14: The probability of finding an electron and the wave function
- 1.15: Atomic and subatomic particles
- Suggestions for further reading

1.1 CATHODE AND ANODE RAYS

How do the cathode and anode rays expose and characterize the subatomic particles?

- The discovery of subatomic particles resulted from investigations into the relationship between electricity and matter.
- When electric current was passed through various gases at low pressures (cathode-tube), the surface of the tube directly opposite the cathode glows (Fig. 1.1).

Electrons, Atoms, and Molecules in Inorganic Chemistry. http://dx.doi.org/10.1016/B978-0-12-811048-5.00001-8

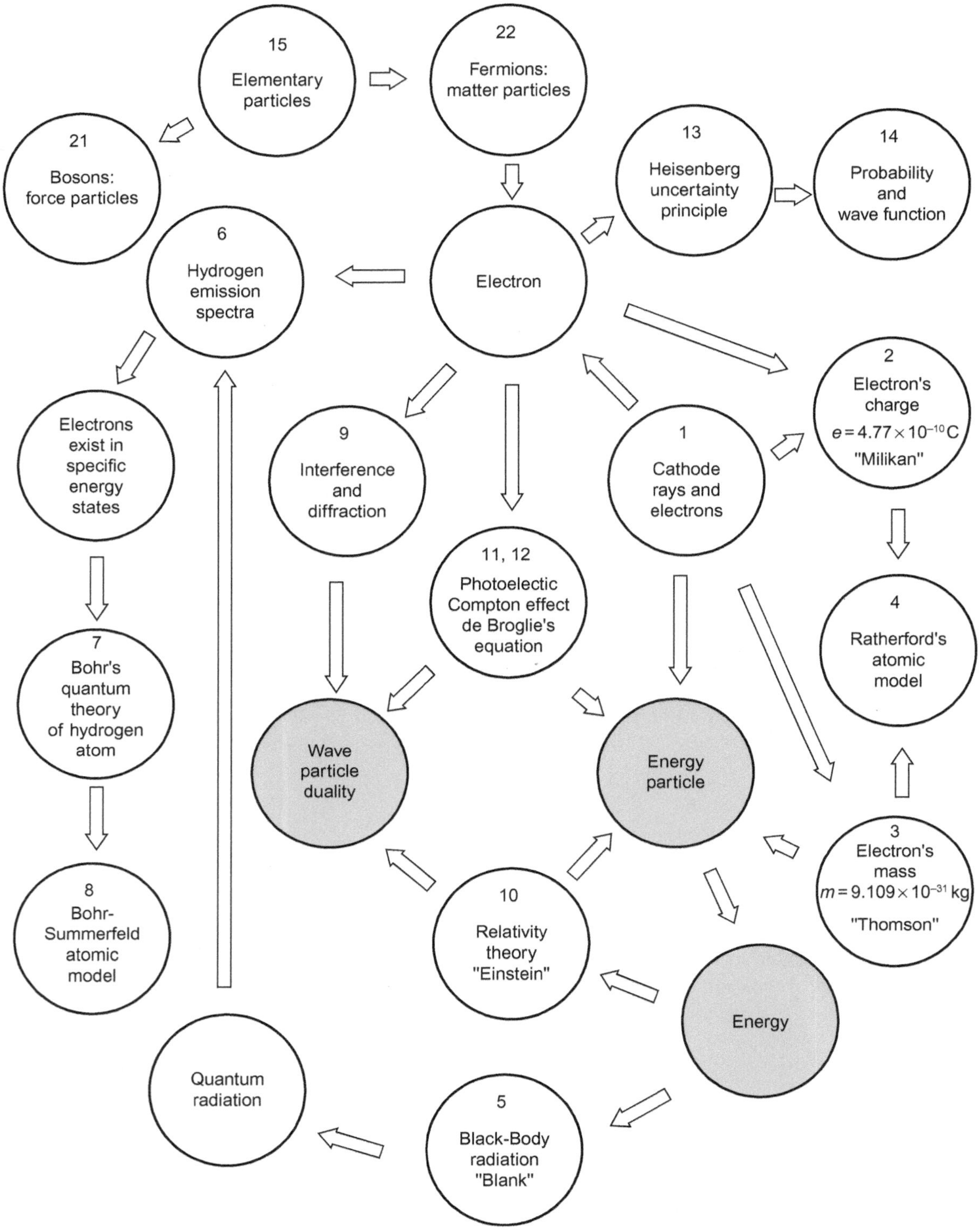

SCHEME 1.1 Schematic chart presents the development of the dual wave-particle nature. Note that mass and energy are entirely different properties of matter. The only conclusion shows that mass of material bodies depends on their motion.

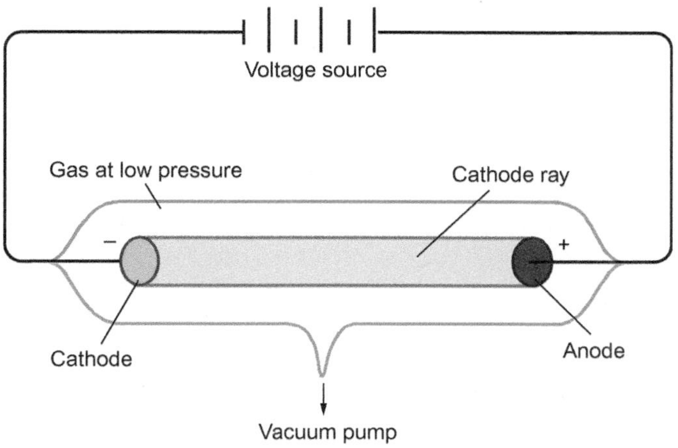

FIG. 1.1 Cathode-tube.

- It has been hypothesized that the glow was caused by a stream of particles, called a cathode ray.
- The ray travels from the cathode to the anode when current is applied.
- The following observations are revealed:
  - If an object placed between the cathode and the anode, it will cause a shadow on the glass. This supports the existence of a cathode ray.
  - If a paddle wheel placed on rails between the electrodes, it will roll along the rails from the cathode toward the anode (Fig. 1.2A). This shows that a cathode ray *had sufficient mass* to set the wheel in motion.
  - The rays were deflected away from a negatively charged object.
  - Cathode rays were deflected by a magnetic field in the same manner as a wire carrying electric current, which was known to have negative charge.
- Because cathode rays have the same properties regardless of the element used to produce them, it was concluded that:
  - The cathode rays are composed of previously unknown negatively charged particles, which were later called electrons.
  - Electrons are present in atoms of all elements.
- Because atoms are electrically neutral, they must have a positive charge to balance each negative electron.
- If one electron is removed from a neutral atom or molecule, the resulting residue has a positive charge equal to the sum of the negative charges of the electron removed.
- Positive ions are formed in the gas discharged tube when electrons from the cathode collide with gaseous atoms (Fig. 1.2B).
- The positive ions move toward the cathode, while the negatively charged electrons of the cathode rays move in the opposite direction.

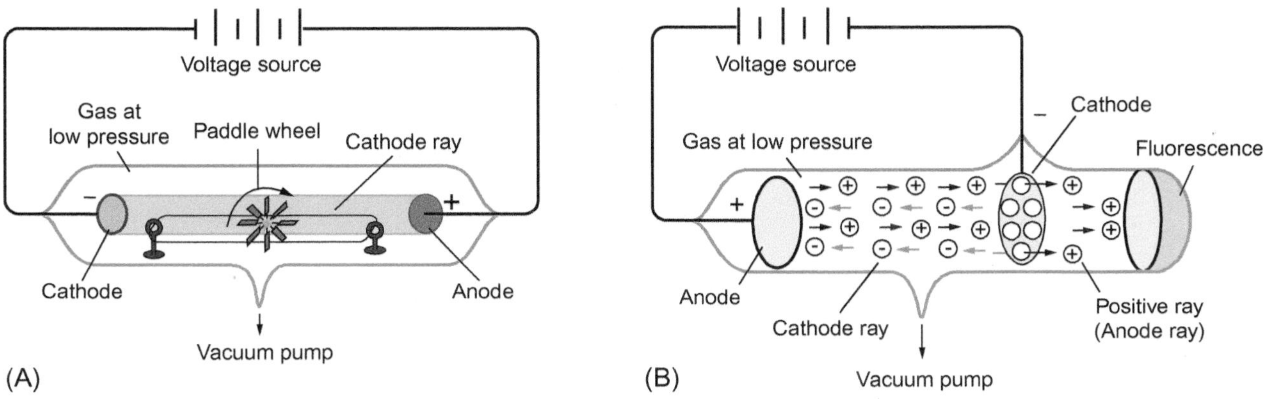

FIG. 1.2 (A) Cathode ray had sufficient mass to set the wheel in motion. (B) Anode ray produced in gas discharged.

- If canals have been bored in this electrode, the positive ions pass through them and cause fluorescence when they strike the end of the tube.
- When different gases are used in the discharged tube, different types of positive ions are produced.
- The deflections of positive rays in the electrical and magnetic fields were studied.

1.2 CHARGE OF THE ELECTRON

Describe Millikan's oil drop apparatus. How could he calculate the charge of the electron?

- The precise determination of the charge of the electron was first computed by Millikan.
- The most significant part of Millikan's apparatus was an electric capacitor inside a thermostated metal chamber (Fig. 1.3).
- A fog of small oil droplets was formed in the chamber by an atomizer.
- The droplets flow through an aperture in the upper plate of the capacitor.
- The movement of the droplets between the plates of the capacitor could be scanned with an eyepiece.
- The droplets were ionized by exposure to X-rays emitted from a radiation tube.
- By altering the voltage, V, across the plates of the capacitor, it was possible to reach a specific voltage at which the electric field strength was balanced by the force of the gravity of the charge droplet, e_d.
  - As a result:

$$mg = e_d E \tag{1.2.1}$$

where
m is the mass of the droplet
g is the acceleration due gravity
E is the strength of the electrical field is the electric force per unit charge at a particular location

  - However, the electric force between the plates of the capacitor:

$$E = \frac{V}{d} \tag{1.2.2}$$

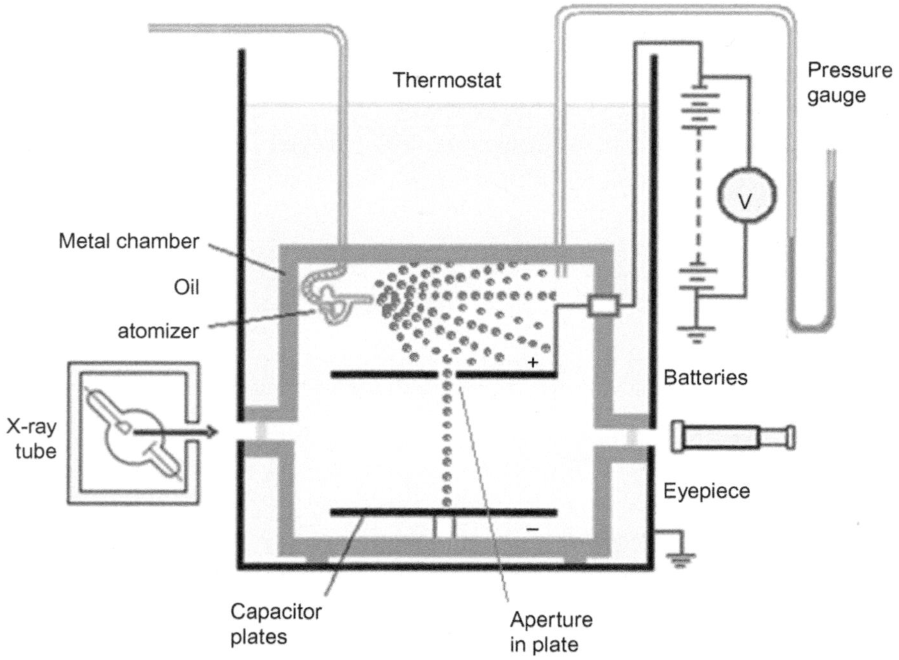

FIG. 1.3 Millikan apparatus that used to measure the charge on the electron.

where

V is the voltage applied to the plates

d is the distance between plates

By substituting in Eq. (1.2.1):

$$\therefore mg = e_d E = e_d \frac{V}{d}, \quad \text{then}$$

$$e_d = \frac{mgd}{V} \tag{1.2.3}$$

- The value of e_d can be found if the mass of the droplet, m, is known.

- The charged droplets could be forced to go up or down by varying the voltage across the plates.
- The mass of the droplet, m, can be evaluated from its falling velocity in absence of the electrical field.
- Initially the oil drops are allowed to fall between the plates while the electric field is turned off. The droplets reach a terminal velocity due to the air friction.
- The field is then turned on; if the voltage is large enough, some of the charged drops will start to go up. (This is because the upwards electric force F_{el} is greater than the downwards gravitation force, $W = mg$.)
- A perfectly spherical droplet is selected and held in the middle of the field of view by varying the voltage while all the other drops have fallen. The experiment is then continued with this one drop.
- The drop is allowed to fall in the absence of an electric field and its terminal velocity u_1 is calculated. The force of air friction for falling drop can then be estimated using Stokes' Law:

$$F_d = 6\pi r \eta u_1 \tag{1.2.4}$$

where u_1 is the terminal velocity of the falling drop (i.e., velocity in the absence of an electric field), η is the viscosity of the air, and r is the radius of the drop.

- The weight W' of the drop is the volume V_d multiplied by the density ρ and the acceleration due to gravity g.

$$W' = V_d g \rho \tag{1.2.5}$$

- However, the apparent weight in air is the true weight minus the weight of air displaced by the oil drop.

$$\therefore W = V_d g \rho - V_d g \rho_{air} = V_d g(\rho - \rho_{air})$$

ρ_{air} is the density of the air

- For a perfectly spherical droplet, the apparent weight can be written as:

$$\because V_d = \frac{4}{3}\pi r^3$$

$$\therefore W = mg = \frac{4}{3}\pi r^3 g(\rho - \rho_{air}) \tag{1.2.6}$$

- The oil drop is not accelerating at terminal velocity. Therefore, the total force acting on it must be zero and the two forces F and W must cancel one another out (i.e., $F_d = W$, Eqs 1.2.4, 1.2.6). This implies

$$6\pi r \eta u_1 = \frac{4}{3}\pi r^3 g(\rho - \rho_{air})$$

$$r^2 = \frac{9\eta u_1}{2g(\rho - \rho_{air})} \tag{1.2.7}$$

- Once r is calculated, W, m, and e_d can easily be computed (Eqs. 1.2.3, 1.2.6).
- In practice this is very hard to accomplish precisely. Estimating F_d is complicated because the mass of the oil drop is difficult to determine without the use of Stokes' Law.
- A more practical approach is to turn V up slightly, so that the oil drop goes up with a new terminal velocity u_2 Then

$$\text{if}: 6\pi r \eta = \frac{W}{u_1}, \text{ then}:$$

$$e_d E - W = 6\pi r \eta u_2 = \frac{W u_2}{u_1}$$

- If the mass of the droplet is known, e_d can be found.

- Millikan noticed that the charge on the droplets were always multiples of a certain value of e, the smallest charge experimentally found. This could be explained by the fact that a droplet can possess whole numbers of electrons, but never a fraction, because an electron is undividable.
- Millikan obtained the following value: $e = 4.803 \times 10^{-10}$ esu $= 1.60206 \times 10^{-19}$ coulomb.

1.3 MASS OF ELECTRON AND PROTON

How could Thomson determine the mass of the electron and proton, and what is his atomic model?

- Thomson concluded that all cathode rays are composed of indistinguishable negatively charged particles, which were named electrons.
- The ratio between the charge of the electron to its mass, e/m_e, was first established by Thomson.
- This ratio is based on the deflection of a beam of electrons in electric and magnetic fields (Fig. 1.4).
- Consider a beam of electrons passing between the plates of capacitor (Fig. 1.5); the force, f_{el}, that acts on the electron in the electric field is equal to

FIG. 1.4 Cathode ray tube for determining the value of e/m for electron.

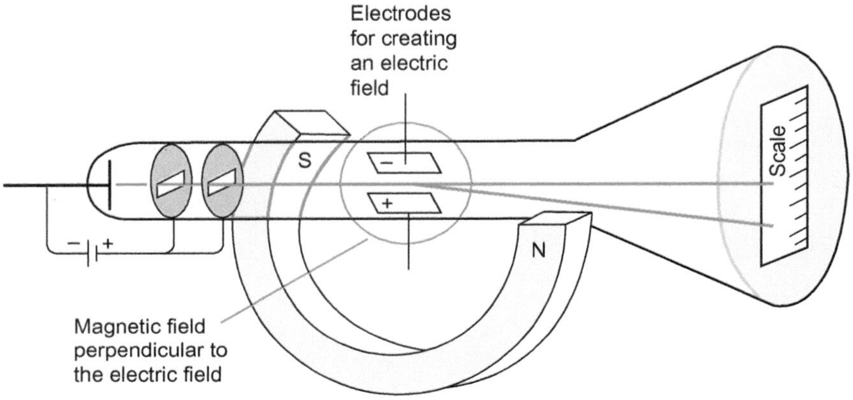

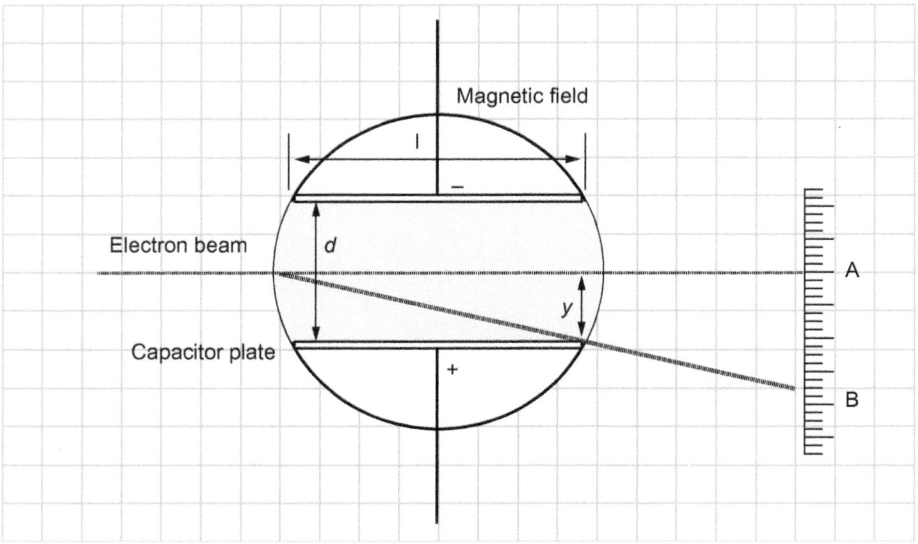

FIG. 1.5 Lay out for determining e/m_e.

$$f_{el} = eE = e\frac{V}{d} \tag{1.3.1}$$

where
- e is the charge of electron
- E is the field strength in the capacitor
- V is the voltage across the plates
- d is the distance between the plates

- This force accelerates the electrons in the direction perpendicular to the original direction of the electron beam.

$$f_{el} = m_e a \tag{1.3.2}$$

where
- a is the acceleration of electron
- m_e is the mass of electron

Then

$$\therefore f_{el} = e\frac{V}{d} = m_e a \tag{1.3.3}$$

$$\therefore a = \frac{e}{m_e}\frac{V}{d} \tag{1.3.4}$$

- For a period of time t during which the electron is between the plates, the beam is displaced by a distance y. The value of y is determined by

$$y = \frac{1}{2}at^2, \quad \text{where } t = \frac{l}{u}$$

where l is the length of the plates and u is the velocity of the electron.

Substituting for a, using Eq. (1.3.4), therefore,

$$y = \frac{1}{2}\frac{e}{m_e} \cdot \frac{V}{d} \cdot \frac{l^2}{u^2} \tag{1.3.5}$$

The distance y can be found from the distance AB on the screen (Fig. 1.5).

The velocity of the electron u can be evaluated from the deviation of the electron in the magnetic field.

- When the magnetic field compensates for the deviation of the electron in the electric field, the direction of the electron beam remains unchanged, and the magnetic and electric fields are equal:

$$f_{el} = f_{mag.} \tag{1.3.6}$$

where $f_{mag.}$ is the force of the magnetic field acting on the electron current and $i = eu$.

- According to electrodynamics, the magnetic field will act on the electron moving perpendicular to the field with a force:

$$f_{mag.} = iH = euH \tag{1.3.7}$$

where H is the intensity of the magnetic field.

Therefore, from Eq. (1.3.6),

$$f_{el} = e\frac{V}{d} = euH, \quad \text{and}$$

$$u = \frac{V}{dH} \tag{1.3.8}$$

- By substituting in Eq. (1.3.5):

$$y = \frac{1}{2}\frac{e}{m_e} \cdot \frac{V}{d} \cdot \frac{l^2}{u^2} \tag{1.3.5}$$

The only unknown is the value of $\frac{e}{m_e}$, which now can be determined:

$$\therefore \frac{e}{m_e} = \frac{2y}{l^2}\frac{V}{dH^2} \qquad (1.3.9)$$

- It was found that

$$\frac{e}{m_e} = 5.273 \times 10^{17}\,\text{esu/g}$$

- The mass of the electron can be calculated if e/m_e and e are known ($e = 4.803 \times 10^{-10}$ esu).
- This shows that the mass of the electron is 9.109×10^{-31} kg (about 1/1837 the mass of the hydrogen atom).
- Thomson showed that electron is a particle with
  - mass
  - energy
  - momentum

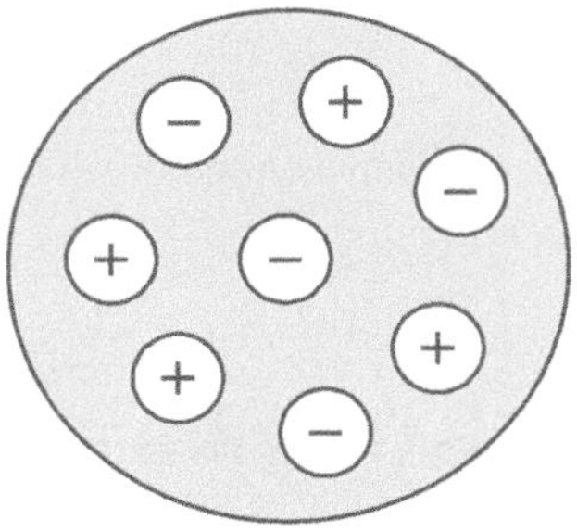

Thomson atom

- Because electrons have so much less mass than atoms, atoms must contain other particles that account for most of their mass.
- The values of e/m were estimated for the positive ions by using basically the same technique employed in the study of cathode ray.
- When hydrogen gas is used, a positive particle results that has the smallest mass (the largest e/m value) of any positive ion observed:

$$\frac{e}{m} = +9.5791 \times 10^4\ \text{C/g}$$

- These particles:
  - are known as protons
  - have equal charge to that of the electron, but are opposite in sign:

$$e = +1.6022 \times 10^{-19}\,\text{C}$$

  - the mass of the proton:

$$m = \frac{e}{e/m} = \frac{+1.6022 \times 10^{-19}\,\text{C}}{+9.5791 \times 10^4\ \text{C/g}} = 1.6726 \times 10^{-24}\,\text{g}$$

  which is about 1837 times heavier than the mass of electron.
  - It is assumed to be a component of all atoms.

1.4 RUTHERFORD'S ATOMIC MODEL

What are the experimental observations and the atomic model that concluded by Rutherford, and what are the weaknesses of this model?

- Rutherford, Geiger, and Marsden firstly assumed that mass and charge are uniformly distributed throughout the atoms, in order to experiment that, they bombarded a thin gold foil with fast moving alpha particles (Fig. 1.6).
- They predicted that alpha particles will pass through with only a slight deflection. However, only 1 in 8000 of the alpha particles had been redirected back toward the source (Fig. 1.7).

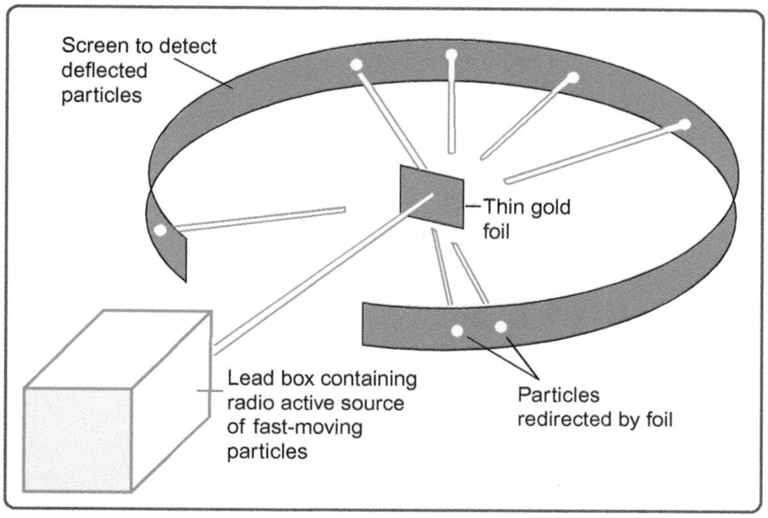

FIG. 1.6 Rutherford bombarded a thin gold foil with fast-moving alpha particles.

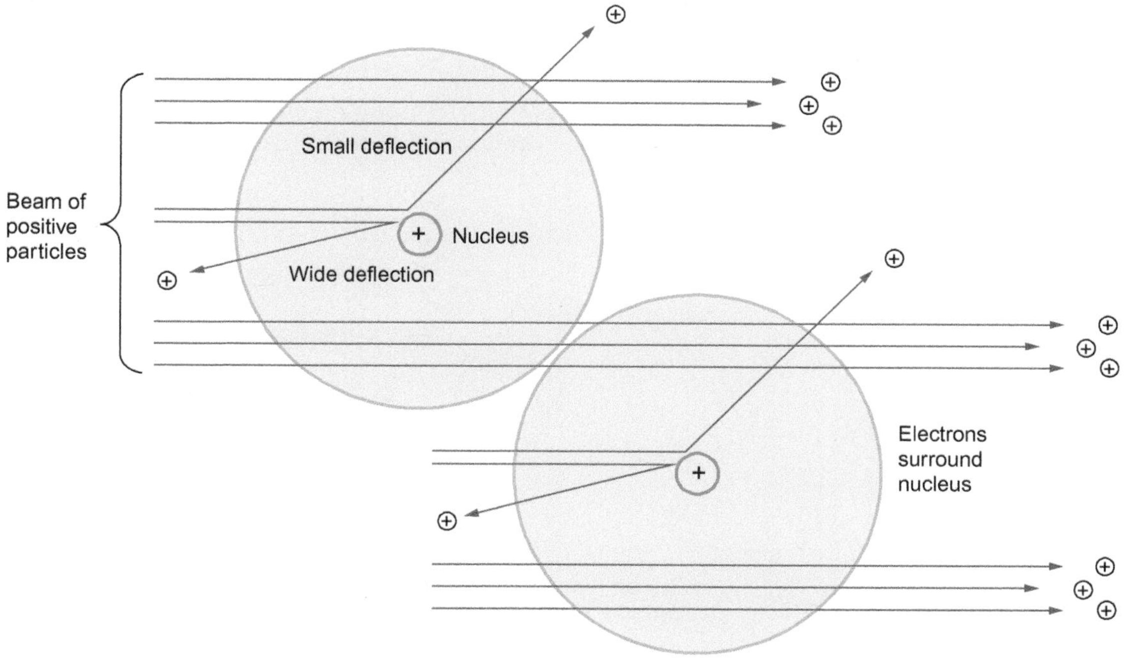

FIG. 1.7 Few fractions of the alpha particles had been redirected back toward the source.

- Rutherford had proposed the following:
  - The volume of the nucleus is very small compared with the volume of an atom.
  - The electrons encircle the positively charged nucleus like planets around the sun.
  - Because atoms are electrically neutral, a given atom must have as many electrons as protons.
  - To account for the total masses of atoms, he proposed the existence of an uncharged particle, which now called neutron.

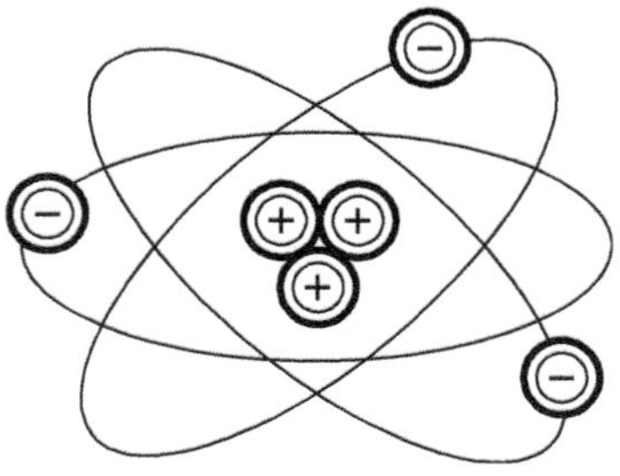

Rutherford atom

- The weaknesses of this model: according to electromagnetic theory, the atom of Rutherford cannot be real. The electrons circling around the nucleus are accelerating charged particles; therefore, they should emit radiation, spend energy, and execute descending spirals until they collapse into the positive nucleus.

1.5 QUANTUM OF ENERGY

Describe the ideal black-body radiator. What is the expected relationship between the energy and the frequency of the emitted radiation?

How did Max Planck explain that the intensity of radiation rises with increasing frequency to a maximum and then falls steeply (the ultraviolet catastrophe)?

- For any object to be in equilibrium with its surroundings; the absorbed radiation must be equivalent in wave length and energy to the emitted radiation.
- Consider a red hot block of a metal with a spherical cavity of which the interior is blackened by iron oxide or a mixture of chromium, nickel and cobalt oxides, black-body. The density of the emitted radiant light can be observed and monitored through a small hole (Fig. 1.8).

FIG. 1.8 Black-body cavity and emitted radiation.

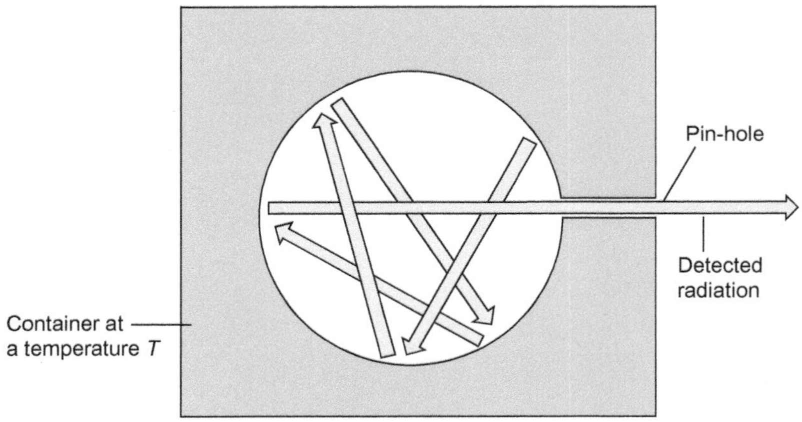

- An ideal Black-body is a physical body that absorbs all incident electromagnetic radiation (no reflected energy), regardless of frequency or angle of incident.
- As a consequence of the equipartition principle, all frequencies of this emitted radiation should have the same average energy.
- The energy density of the radiated light, $E(\nu)\partial\nu$, is simply the number of the oscillators, ∂n that can occur per unit volume between ν and $\nu + \partial\nu$ times the average energy of an oscillator, $\bar{\varepsilon}$. For electromagnetic radiation, there is an extra factor of two, because both magnetic and electric fields are oscillating.

  If:

$$\partial n = \frac{4\pi V}{u^3}\nu^2 \partial\nu$$

where V is the volume, u is the phase velocity, and ν is the frequency of the electromagnetic wave:

$$\because \nu = \frac{u}{\lambda}$$

$$\therefore E(\nu)\partial\nu = \partial n \cdot 2\bar{\varepsilon} = \left(\frac{4\pi V}{u^3}\nu^2 \cdot 2\bar{\varepsilon}\right)\partial\nu \tag{1.5.1}$$

Classic theory predicts that

$$\bar{\varepsilon} = kT \tag{1.5.2}$$

where k is the Boltzmann constant and T is the temperature.

If $V = 1$, then

$$\therefore E(\nu)\partial\nu = \partial n \cdot 2\bar{\varepsilon} = \left(\frac{8\pi k}{u^3}\nu^3 T\right)\partial\nu \tag{1.5.3}$$

- Accordingly, the intensity of black-body radiation (or cavity radiation) should increase endlessly with rising frequency.
- But the experimental data reveal that the intensity of radiation rises to a maximum, then falls steeply with increasing frequency (the ultraviolet catastrophe) (Fig. 1.9).
- When Max Planck was studying the emission of light by a hot object, he proposed that a hot object does not emit electromagnetic energy continuously.

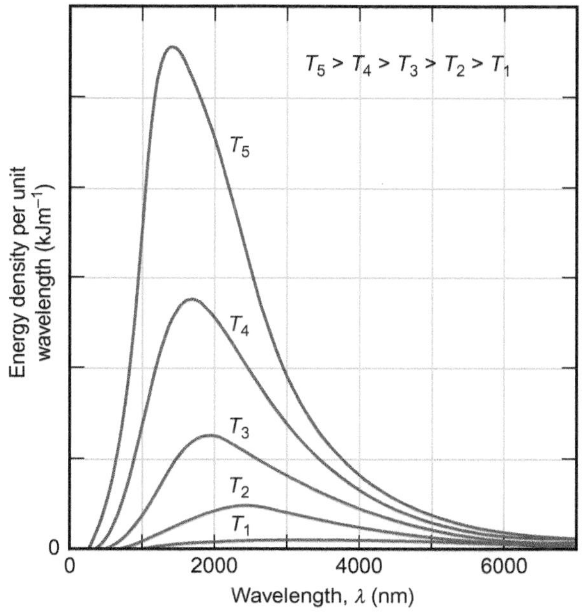

FIG. 1.9 The energy density per unit wavelength in black-body cavity recorded at different temperatures. The energy density increases and the peak shifts to shorter wavelength as the temperature is increased. The total energy density (the area under the curve) increases as temperature is raised.

- Planck proposed that the object emits energy in a small, specific quantity called quanta. A quantum is the smallest amount of energy that can be lost or gained by an atom.
- Planck suggested the following relationship between a quantum of energy and the frequency of radiation:

$$E = h\nu \tag{1.5.4}$$

where
E is the energy in joules, of a quantum radiation
ν is the frequency of radiation emitted
h is the Planck constant; $h = 6.626 \times 10^{-34}\,\mathrm{J\,s}$

- Consider a set of N oscillators having a fundamental vibration frequency, ν.
- If these can take up energy only in multiples of $h\nu$, then the allowed energies are: $0h\nu$, $2h\nu$, $3h\nu$, etc.
- According to the Boltzmann formula, if N_o is the number oscillators in the lowest energy state:

$$N_i = N_o e^{-\frac{\varepsilon_i}{kT}} \tag{1.5.5}$$

where N_i is the number of oscillators having energy ε_i, and

$$N_1 = N_o e^{-\frac{h\nu}{kT}},\quad N_2 = N_o e^{-\frac{2h\nu}{kT}},\quad N_3 = N_o e^{-\frac{3h\nu}{kT}},\ldots$$

- Then the total number of the oscillators is

$$N = N_o + N_o e^{-\frac{h\nu}{kT}} + N_o e^{-\frac{2h\nu}{kT}} + N_o e^{-\frac{3h\nu}{kT}} + \cdots = N_o \sum_{i=0}^{\infty} e^{-\frac{ih\nu}{kT}} \tag{1.5.6}$$

and the total energy E is

$$E = 0N_o + h\nu N_o e^{-\frac{h\nu}{kT}} + 2h\nu N_o e^{-\frac{2h\nu}{kT}} + 3h\nu N_o e^{-\frac{3h\nu}{kT}} + \cdots = ih\nu N_o \sum_{i=0}^{\infty} e^{-\frac{ih\nu}{kT}} \tag{1.5.7}$$

- The average energy of an oscillator, $\overline{\varepsilon}$, is

$$\overline{\varepsilon} = \frac{E}{N} = h\nu \frac{\sum_{i=0}^{\infty} i e^{-\frac{ih\nu}{kT}}}{\sum_{i=0}^{\infty} e^{-\frac{ih\nu}{kT}}} \tag{1.5.8}$$

$$\text{if}:\ x = \frac{h\nu}{kT} \ \text{and}\ y = e^{-x}$$

Then

$$\sum y^i = 1 + y + y^2 + \cdots = \frac{1}{1-y},\ (y < 1),\ \text{and}$$

$$\sum iy^i = y\left(1 + 2y + 3y^2 + \cdots\right) = \frac{y}{(1-y)^2},\ (y < 1)$$

$$\therefore \frac{\sum iy^i}{\sum y^i} = \frac{y}{1-y}$$

Accordingly,

$$\overline{\varepsilon} = \frac{E}{N} = \frac{h\nu e^{-\frac{h\nu}{kT}}}{1 - e^{-\frac{h\nu}{kT}}} = \frac{h\nu}{e^{\frac{h\nu}{kT}} - 1} \tag{1.5.9}$$

$$\text{when}\ \ h\nu \ll kT,\ e^{\frac{h\nu}{kT}} \cong 1 + \frac{h\nu}{kT}\ \ \text{then}\ \ \overline{\varepsilon} \cong kT$$

- By substituting $E(\nu)\,\partial\nu$, in the energy density,

$$\because E(\nu)\partial\nu = \partial n \cdot 2\overline{\varepsilon} = \left(\frac{4\pi V}{u^3}\nu^2 \cdot 2\,\overline{\varepsilon}\right)\partial\nu \tag{1.5.1}$$

$$\therefore E(\nu)\partial\nu = \partial n \cdot 2\bar{\varepsilon} = \left(\frac{4\pi V}{u^3}\nu^2 \cdot \frac{2h\nu}{e^{\frac{h\nu}{kT}} - 1}\right)\partial\nu, \text{ if } V = 1$$

Then

$$E(\nu)\partial\nu = \frac{8\pi h\nu^3}{u^3}\frac{\partial\nu}{e^{\frac{h\nu}{kT}} - 1} \quad (1.5.10)$$

- This expression exactly matches the experimental curve at all wavelengths, and quantum theory had achieved its first great success.

1.6 HYDROGEN ATOM LINE-EMISSION SPECTRA; ELECTRONS IN ATOMS EXIST ONLY IN VERY SPECIFIC ENERGY STATES

Why do excited hydrogen atoms emit only specific frequencies of electromagnetic radiation and not a continuous range of frequencies?

What are the proposed mathematic relationships between wave-numbers of the emitted frequencies and the energy state of the excited electrons?

- Classical theory expected that the hydrogen atoms would be excited by any quantity of supplied energy.
- Consequently, scientists expected to detect a continuous spectrum (Fig. 1.10), which is the emission of a continuous range of frequencies of electromagnetic radiation.
- When researchers passed electric current through a vacuum tube containing hydrogen gas at low pressure, they observed the emission of a characteristic pinkish glow (Fig. 1.11).
- When a narrow beam of the emitted pinkish light was shone through a prism, it was separated into a series of specific frequencies, known as the line-emission spectrum.

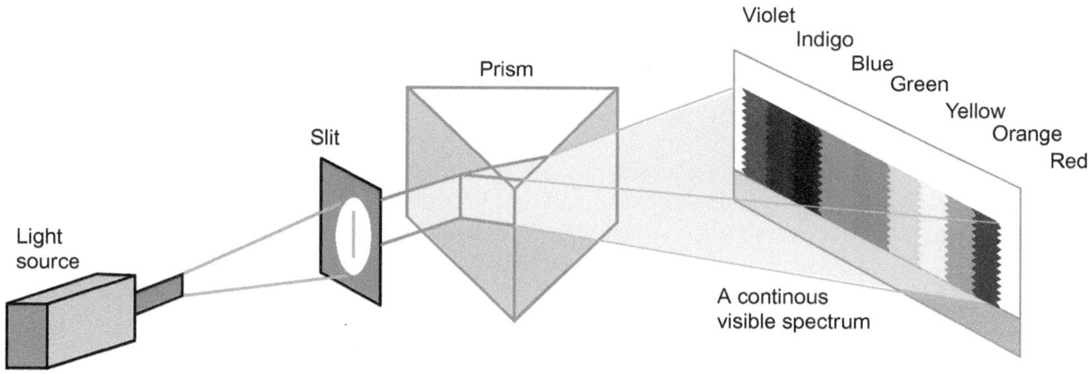

FIG. 1.10 A continuous visible spectrum is produced when a narrow beam of white light is passed through a prism.

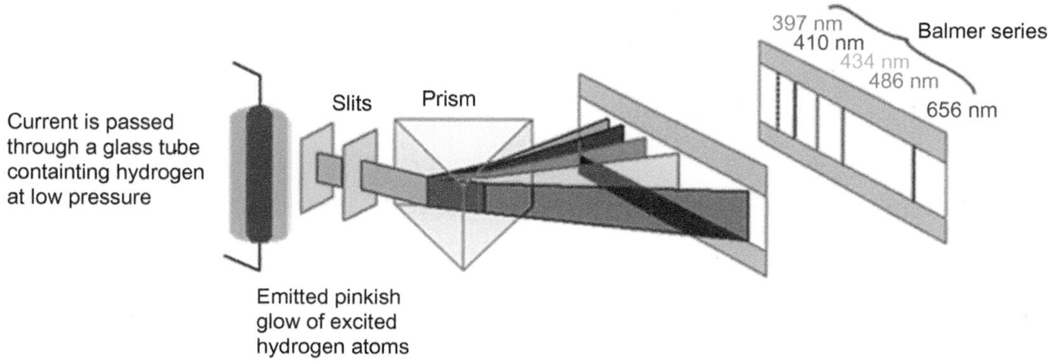

FIG. 1.11 A series of specific wavelengths of emitted light makes up hydrogen's line-emission spectra.

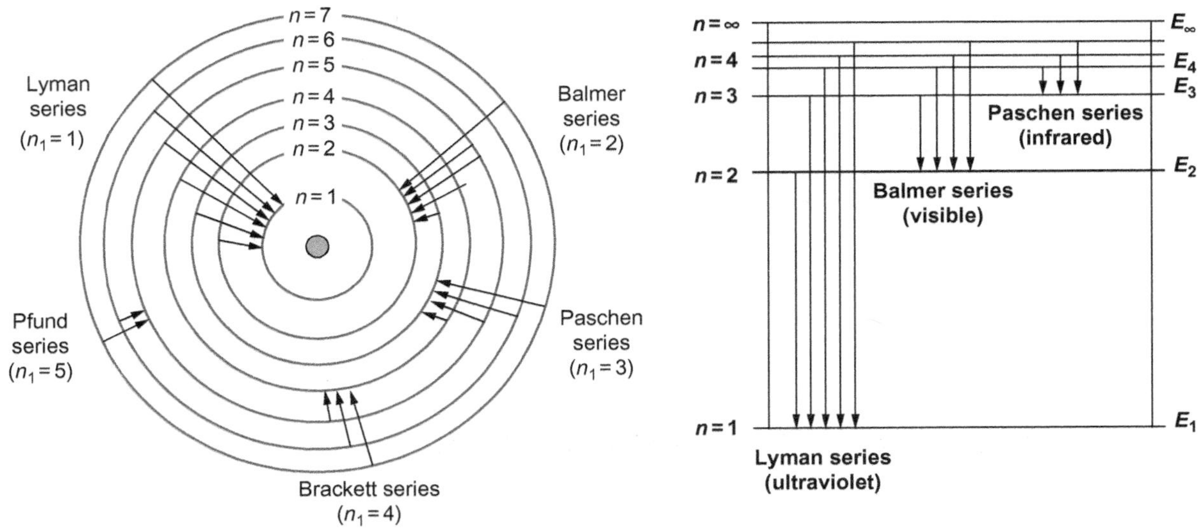

FIG. 1.12 Electron energy-level diagram for hydrogen and the energy transitions for the Lyman, Balmer, and Paschen spectral series.

- Additional series of lines were discovered in the ultraviolet and infrared regions of hydrogen's line-emission spectrum (Fig. 1.12); they are known as the Lyman, Balmer, and Paschen series.
- The fact that hydrogen atoms emit only specific frequencies of light indicated that the energy differences between the atom' energy states were fixed.
- This suggested that the electron of hydrogen atom exist only in very specific energy states.
- Balmer discovered a relationship between the frequencies of the atomic hydrogen line in the visible region of the spectrum; the wave numbers, $\bar{\nu}$, are given by

$$\bar{\nu} = R\left(\frac{1}{2^2} - \frac{1}{n^2}\right) \tag{1.6.1}$$

with $n=3, 4, 5$, etc.

The bright red line at $\lambda = 656.28$ nm corresponds to $n=3$, the blue line at $\lambda = 486.13$ nm to $n=4$, etc. (Figs. 1.11 and 1.12). The constant R is called the Rydberg constant and has the value 109 677.581 cm^{-1} (1.10×10^5 cm^{-1}).

- Other hydrogen series were discovered later, which obeyed the more general formula:

$$\bar{\nu} = R\left(\frac{1}{n_1^2} - \frac{1}{n_2^2}\right) \tag{1.6.2}$$

- Lyman found the series with $n_1 = 1$ in the far ultraviolet and other were found in the infrared.

$$\text{Paschen}: \ \bar{\nu} = R\left(\frac{1}{3^2} - \frac{1}{n^2}\right) \quad n = 4, 5, 6, 7, \ldots$$

$$\text{Brackett}: \ \bar{\nu} = R\left(\frac{1}{4^2} - \frac{1}{n^2}\right) \quad n = 5, 6, 7, 8, \ldots$$

$$\text{Pfund}: \ \bar{\nu} = R\left(\frac{1}{5^2} - \frac{1}{n^2}\right) \quad n = 6, 7, 8, 9, \ldots$$

In what zone of the electromagnetic spectrum would you expect a spectral line resulting from the electronic transition from the fifth to the tenth level of the hydrogen atom?

$$\because \bar{\nu} = R\left(\frac{1}{n_1^2} - \frac{1}{n_2^2}\right) \tag{1.6.2}$$

$$\because R = 1.10 \times 10^5\,\text{cm}^{-1}, \quad n_1 = 5, \quad \text{and} \quad n_2 = 10$$

$$\because \bar{\nu} = 1.10 \times 10^5 \left(\frac{1}{5^2} - \frac{1}{10^2}\right) = 3.3 \times 10^3 \text{cm}^{-1}$$

This line would be seen in the infrared zone of the electromagnetic spectrum; it is a member of the Pfund series.

What would be the maximum number of emission lines that you would expect to see in a spectroscope for atomic hydrogen if only seven electronic energy levels were involved (Fig. 1.12)?

- Emission spectrum lines result by electronic transitions from high electronic energy levels to a lower one.
- The number of the emission line for hydrogen = 6 + 5 + 4 + 3 + 2 + 1 = 21 emission lines.
- In general, the maximum number of emission lines is given by = ½ $(n)(n-1)$.

$$= \frac{1}{2}(7)(7-1) = 21$$

1.7 BOHR'S QUANTUM THEORY OF THE HYDROGEN ATOM

What did Bohr conclude in his atomic model, and what are the weaknesses of this model?
How could Bohr drive the radius of the electronic orbits?
What are the radius and the ionization potential of the hydrogen atom?
How could Bohr explain hydrogen's line-emission spectrum?

- The modern theories of atomic structure began with the ideas of Niels Bohr, thus these are essential to be considered and explored.
- Bohr combined Planck's and Rydberg's equations, and suggested a mechanism for discreet emission from excited atomic vapor. He concluded with the following in his atomic model:
  - The electron moves about the nucleus in a circular orbit.
  - Only orbits in which the electron's angular momentum, L, is an integral multiple of $\hbar$ are allowed.
  - The electron does not radiate energy when it is in an allowed orbit.
  - The electron can absorb energy and move to a higher orbit (or it can loss energy and move to a lower orbit) (Fig. 1.13).
- Bohr used quantum theory not quantum mechanics; he did not recognize the wave nature of electrons.
- He superimposed the quantum condition on the angular momentum but he assumed that the electron was a particle with a definite position.
- The idea of an electron circles the nucleus in a well-defined orbit (Bohr orbits) gained wide acceptance.
- One difficulty was the fact that an atom in a magnetic field has a more complicated emission spectrum than the same atom in the absence of a magnetic field. This phenomenon is known as the Zeeman Effect, and is not explainable by simple Bohr Theory.

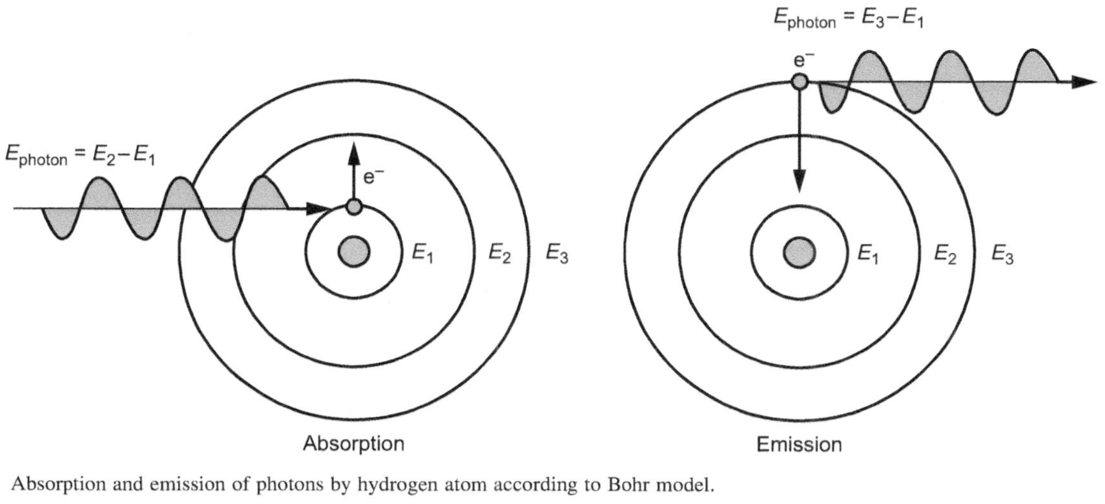

FIG. 1.13 Absorption and emission of photons by hydrogen atom according to Bohr model.

- For an electron to remain in a specific orbit, the electrostatic attraction between the electron and the nucleus must be equal to the centrifugal force. For an electron of mass m, moving with a velocity u in an orbit of radius r:

$$\text{Centrifuge force} = \frac{mu^2}{r} \tag{1.7.1}$$

- If the charge on the electron is e, the number of charges on the nucleus Z, and the permittivity of a vacuum ε_o, then:

$$\text{Columbic attractive force} = \frac{Ze^2}{4\pi\varepsilon_o r^2} \tag{1.7.2}$$

$$\because \text{Centrifuge force} = \text{Coulombic attractive force}$$

$$\therefore \frac{mu^2}{r} = \frac{Ze^2}{4\pi\varepsilon_o r^2}$$

$$\therefore u^2 = \frac{Ze^2}{4\pi\varepsilon_o m r} \tag{1.7.3}$$

- In accordance with Planck's quantum theory, energy is not continuous but is quantized. This means that energy occurs in "packets" called quanta, of magnitude $h/2\pi$, where h is Planck's constant.
- Bohr assumed that the energy of an electron in an orbit, that is, its angular momentum equals mur, must be equal to a whole number n of quanta ($nh/2\pi$):

$$mur = \frac{nh}{2\pi} \quad \text{"Bohr assumption"} \tag{1.7.4}$$

The integral n is called the principle quantum number.

$$\therefore u = \frac{nh}{2\pi mr}$$

$$\therefore u^2 = \frac{n^2h^2}{4\pi^2m^2r^2}$$

$$\because u^2 = \frac{Ze^2}{4\pi\varepsilon_o m r} \tag{1.7.3}$$

$$\therefore \frac{Ze^2}{4\pi\varepsilon_o m r} = \frac{n^2h^2}{4\pi^2m^2r^2}$$

$$\therefore r = \frac{\varepsilon_o n^2h^2}{\pi m e^2 Z} \tag{1.7.4}$$

$$r = n^2\frac{\varepsilon_o h^2}{\pi m e^2 Z} = n^2 \times 0.05292\text{nm}$$

- For hydrogen, the charge on the nucleus $Z=1$, and if:
  - $n=1$ this gives a value $r=1^2 \times 0.0529$ nm
  - $n=2$ this gives a value $r=2^2 \times 0.0529$ nm
  - $n=3$ this gives a value $r=3^2 \times 0.0529$ nm
- The kinetic energy of an electron is $\frac{1}{2}mu^2$:

$$\because u^2 = \frac{Ze^2}{4\pi\varepsilon_o m r} \tag{1.7.3}$$

$$E_k = \frac{1}{2}mu^2 = \frac{Ze^2}{8\pi\varepsilon_o r}$$

We substitute for r, using

$$r = \frac{\varepsilon_o n^2h^2}{\pi m e^2 Z} \tag{1.7.4}$$

$$E_k = \frac{Z^2 e^4 m}{8\varepsilon_o^2 n^2 h^2} \tag{1.7.5}$$

- The total energy of any state is the sum of the kinetic energy, E_k, and potential energy E_P:

$$E_p = -\frac{Ze^2}{4\pi\varepsilon_o r}$$

$$E = E_k + E_p = \frac{1}{2}mu^2 - \frac{Ze^2}{4\pi\varepsilon_o r} = \frac{Ze^2}{8\pi\varepsilon_o r} - \frac{Ze^2}{4\pi\varepsilon_o r} = -\frac{Ze^2}{4\pi\varepsilon_o r}$$

$$E = -\frac{Z^2 e^4 m}{8\varepsilon_o^2 n^2 h^2} \tag{1.7.6}$$

- The difference in energy between the series limit and ground level is called ionization potential. In atoms containing more than one electron, there will be first, second, third ionization potentials, etc.
- This gives a picture of the hydrogen atom where an electron moves in circular orbits of radius proportional to 1^2, 2^2, 3^2... The atom will only radiate energy when the electron jumps from one orbit to another.
- If an electron jumps from an initial orbit i to a final orbit f, the change in energy ΔE is

$$\Delta E = \left(-\frac{Z^2 e^4 m}{8\varepsilon_o^2 n_i^2 h^2}\right) - \left(-\frac{Z^2 e^4 m}{8\varepsilon_o^2 n_f^2 h^2}\right)$$

$$\Delta E = \frac{Z^2 e^4 m}{8\varepsilon_o^2 h^2}\left(\frac{1}{n_f^2} - \frac{1}{n_i^2}\right) \tag{1.7.7}$$

- Energy is related to wavelength ($E = hc\overline{\nu}$, $\overline{\nu} = 1/\lambda$), so this equation is of the same form as the Rydberg equation:

$$\overline{\nu} = \frac{Z^2 e^4 m}{8\varepsilon_o^2 h^3 u}\left(\frac{1}{n_f^2} - \frac{1}{n_i^2}\right) \tag{1.7.8}$$

$$\overline{\nu} = R\left(\frac{1}{n_f^2} - \frac{1}{n_i^2}\right) \text{ (Rydberg equation)}$$

- Thus the Rydberg constant is as follows:

$$R = \frac{Z^2 e^4 m}{8\varepsilon_o^2 h^3 u}$$

  - The experimental value of R is $1.097373 \times 10^7\ \text{m}^{-1}$, in good agreement with the theoretical value of $1.096776 \times 10^7\ \text{m}^{-1}$.
  - Bohr theory provides an explanation of the atomic spectrum of hydrogen. The different series of spectral lines can be obtained by varying the values of n_i and n_f.
    - Thus with $n_f = 1$ and $n_i = 2, 3, 4,\ldots$, we obtain the Lyman series of lines in the UV region.
    - With $n_f = 2$ and $n_i = 3, 4, 5,\ldots$, we get the Balmer series of lines in the visible spectrum.
    - Similarly, $n_f = 3$ and $n_i = 4, 5, 6,\ldots$, give the Paschen series.
    - With $n_f = 4$ and $n_i = 5, 6, 7,\ldots$, we have the Brackett series.
    - For $n_f = 6$ and $n_i = 7, 8, 9\ldots$, we have the Pfund series. The various transitions that are possible between orbits are shown in Fig. 1.12.

Calculate the radius of the hydrogen atom

- The radius, r:

$$r = n^2 \frac{\varepsilon_o h^2}{\pi m e^2 Z}$$

for hydrogen atom: $n = 1$, and $z = 1$

$$r = \frac{\varepsilon_0 h^2}{\pi m e^2} = \frac{(8.85419 \times 10^{-12}\,\text{C}^2\text{N}^{-1})(6.626 \times 10^{-34}\,\text{Js})^2}{(^{22}/_7)(9.109 \times 10^{-31}\,\text{kg})(1.6022 \times 10^{-19}\,\text{C})^2} = 5.292 \times 10^{-11}\,\text{m} = 0.05292\,\text{nm}$$

If the ionization energy of the hydrogen atom is 13.598 eV/atom (1312.0 kJ/mol.), what is the second ionization energy of the He^{+}?

$$\because E = -\frac{Z^2 e^4 m}{8\varepsilon_0^2 n^2 h^2} \tag{1.7.6}$$

- the ionization energy:

$$I = \frac{Z^2 e^4 m}{8\varepsilon_0^2 n^2 h^2}$$

For hydrogen atom, $Z=1$, $n=1$, then

$$I_{\text{H}} = \frac{e^4 m}{8\varepsilon_0^2 h^2} = 13.598\,\text{eV/atom} = 1312.0\,\text{kJ/mol}$$

For the second ionization of He^{+}, $Z=2$:

$$I_{\text{He}^+} = 4 \times 13.598 = 54.392\ \text{eV/atom} = 5248.0\,\text{kJ/mol}$$

1.8 THE BOHR-SOMMERFELD MODEL

What did Sommerfeld propose to modify Bohr's atomic model? What are the weaknesses of this model?

- Sommerfeld proposed elliptical orbits as well as circular orbits for electron.
- For the orbital closest to the nucleus, the principle quantum number $n=1$, and only a circular orbit is possible. For the next orbital, the principle quantum number $n=2$, both circular and elliptical orbits are possible.
- To characterize the elliptical orbit, a second quantum number k is needed. The shape of the ellipse is defined by the ratio of the lengths of the major and minor axes.

  Thus:

$$\frac{\text{Major axis}}{\text{Minor axis}} = \frac{n}{k}$$

  where k is called the azimuthal or subsidiary quantum number, and may have values from 1, 2,...,n.

  Thus, for $n=2$, n/k may have the values 2/2 (circular orbit) and 2/1 (elliptical orbit).

  For the principal quantum number $n=3$, n/k may have values 3/3 (circular), 3/2 (ellipse), and 3/1 (narrower ellipse).
- The existence of these additional orbits, which have slightly different energies from each other, explains the extra lines in the spectrum revealed under high resolution.
- The combined Bohr-Sommerfeld theory rationalized the Zeeman Effect very nicely, but failed to explain the spectral details of atoms that have several electrons.

1.9 THE CORPUSCULAR NATURE OF ELECTRONS, PHOTONS, AND PARTICLES OF VERY SMALL MASS

What are the experimental observations that establish the wave nature of electrons?

- The particle nature of the electrons had previously been confirmed by cathode rays, Millikan's capacitor, and Thomson's experimentation.
- Two properties characterize waves;
  - wave interference
  - wave diffraction.
- Firstly, wave interference is the effect produced when a train of waves is reinforced or weakened by another one.
  - Constructive interference is when identical waves meet in phase, so that their high and low points coincide, reinforcing each other (Fig. 1.14).

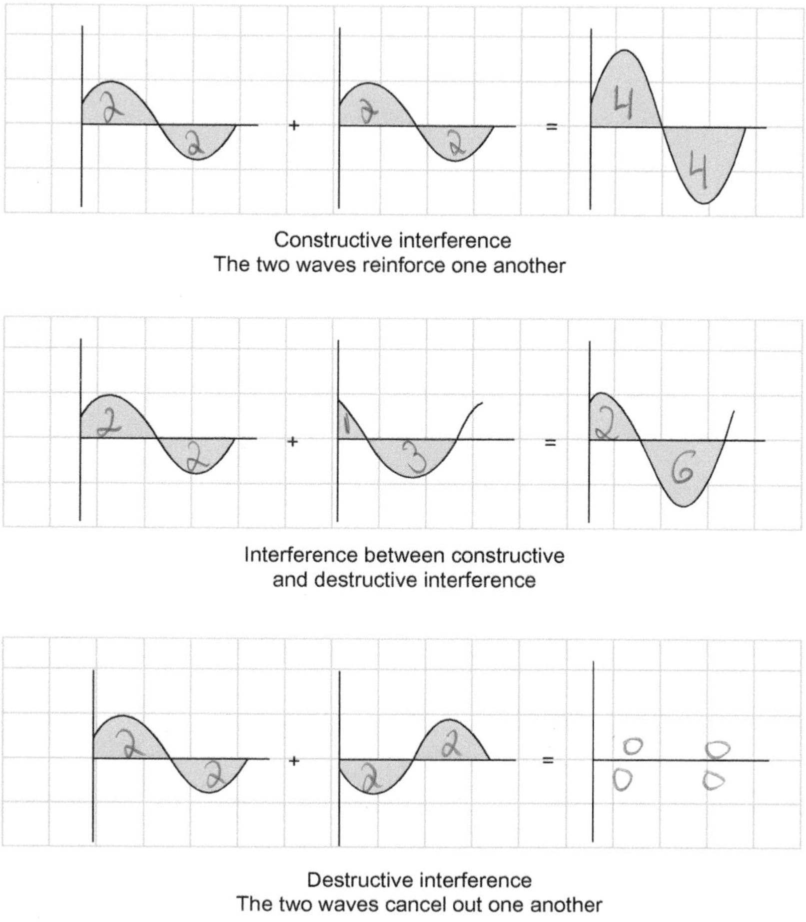

FIG. 1.14 Constructive and destructive interferences.

  - Destructive interference is when identical waves meet out of phase, so that the high points of one exactly match the low points of the other. The two waves then cancel out one another (Fig. 1.14).
- Secondly, diffraction is the spreading of waves as they pass obstacles or openings comparable in size to their wavelength. It is due to the separation of the wave into several groups of waves that interfere with each other.
  - When light passes through two slits, the diffracted light waves interfere with each other and produce a pattern of light and dark (Fig. 1.15).
  - The obstacle may be a diffraction grating (Fig. 1.16), with a great number of clear slits equidistant from each other and of a size that is of the same order as that of the wavelength.
  - Each slit becomes an independent source of waves that spread in all directions and interfere with each other.
  - As seen in Fig. 1.17, using simple geometry, when the waves are in phase it defined by:

$$\sin\theta = \frac{n\lambda}{\Delta y}, \quad \text{where } n = 1, 2, 3, \tag{1.9.1}$$

    where

    θ is the angle between the given direction and the line perpendicular to the grating

    λ is the wavelength

    Δy is the distance between the clear slits of the grating
  - Intensification of the wave occurs in the directions that satisfy this relation.
  - In contrast, similar relation containing half-integer values of n ($n = 1/2$, $3/2$, $5/2, \ldots$), when the waves are out-of-phase, at which extinction of the waves occurs.

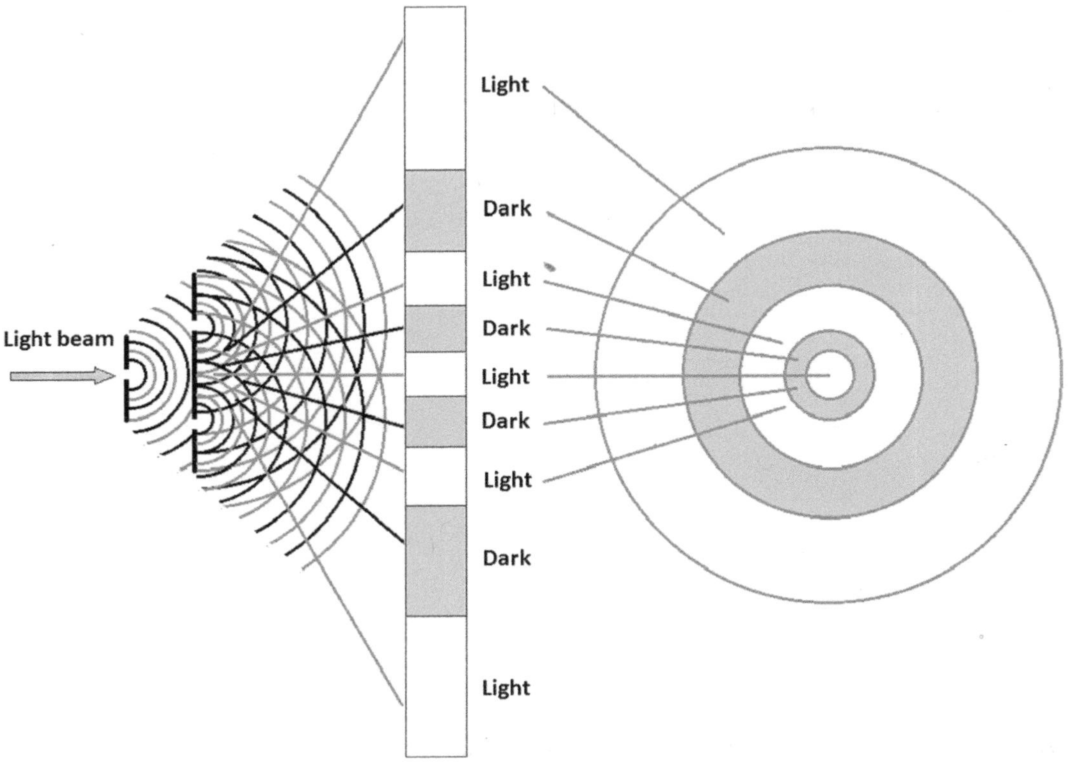

FIG. 1.15 A pattern of light and dark is produced by interference of light waves that have passed through two or more opening close in size to the wavelength of light. Similar diffraction pattern is produced when a beam of electrons passed through a crystal.

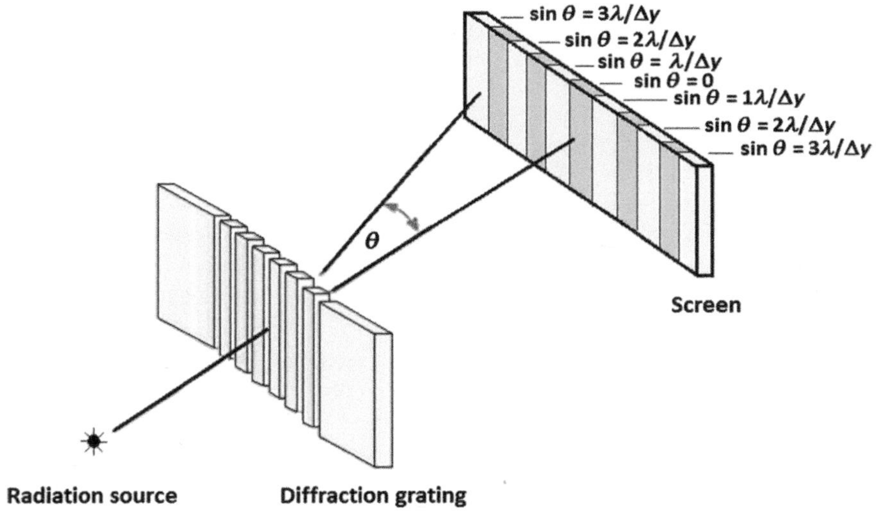

FIG. 1.16 Formation of diffraction pattern when radiation passes through a diffraction grating.

- Diffraction patterns are also obtained by passing beams of electrons through thin gold foil or a thin film of chromium, confirming the wave properties of the electrons.
- Additionally, diffraction patterns have been emitted from crystals placed in beams of neutrons or hydrogen atoms, although these more massive particles also exhibit wave properties.
- When interference and diffraction are observed, they are interpreted as evidence that the passage of waves has taken place.

1.10 RELATIVITY THEORY: MASS AND ENERGY, MOMENTUM, AND WAVELENGTH INTERDEPENDENCE

What are Einstein's relationships that describe the interdependence of mass and energy? What are the accepted mass of a photon moving with velocity of light and interdependence of the momentum and the wavelength?

- Einstein established that the mass of a body in motion exceeds its mass at rest, according to the equation:

$$m = \frac{m_o}{\sqrt{1 - \frac{u^2}{c^2}}} \quad (1.10.1)$$

where

m is the mass of the body in motion
m_o is the mass of the body at rest
u is the velocity of the body
c is the velocity of light in vacuum

An increase in the velocity of the body causes a consequent increase in its energy, results in an increase in its mass.

- Einstein also showed that the mass of a body is related to its energy according to

$$E = mc^2 \quad (1.10.2)$$

This equation shows the interdependence of the changes in Δm and ΔE in any process:

$$\Delta E = \Delta mc^2 \quad (1.10.3)$$

- It cannot be concluded that mass can be converted to energy or vice versa. Mass and energy are entirely different properties of matter. Eq. (1.10.3) only shows that mass of material bodies depends on their motion.
- Albert Einstein established that light energy is not distributed continuously in space, but is quantized in small bundles called photon.
- When a photon is at rest (does not exist), the mass, m, and energy, E, of a photon would be infinitely large.
- A photon has no rest mass because it moves with velocity of light; thus a motionless photon does not exist. Therefore, all mass of a photon is dynamic, and according to Planck's formula:

$$E = h\nu = h\frac{c}{\lambda}$$

$$\because E = mc^2 \quad (1.10.2)$$

$$\therefore mc^2 = h\frac{c}{\lambda}$$

$$\lambda = \frac{h}{mc} = \frac{h}{p} \quad (1.10.4)$$

where p is the momentum (impulse) of a photon. The momentum is a vector quantity; its direction coincides with that of the velocity. Eq. (1.10.4) expresses the interdependence of the momentum of the photon mc and the wavelength of light.

1.11 THE CORPUSCULAR NATURE OF ELECTROMAGNETIC WAVES

How could the dual wave-particle nature of electromagnetic waves or photons be confirmed?

What is the most precise technique for determining Planck's constant?

What are the difficulties in explaining the photoelectric effect and the Compton effect, and how could these problems be solved?

How could the corpuscular nature of light be revealed in the photoelectric effect and the Compton effect?

- The interference and diffraction experimentally confirm that light consists of transverse electromagnetic vibrations. The occurrence of interference and diffraction is a characteristic of any wave process.
- The corpuscular properties of light (that light made up of small particles, which move in straight lines with finite velocity and have kinetic energy) are obviously revealed in two phenomena: the photoelectric effect and the Compton effect.

The Photoelectric Effect

- Several materials, such as potassium (and semiconductors), emit electrons when irradiated with visible light, or in certain cases with ultraviolet light.
- The emission of electrons from metal when exposed to light is known as the photoelectric effect.
- Fig. 1.18 shows the circuit to monitor and study the photoelectric effect. When light is focused on the clean metal surface of the cathode, electrons are emitted. If some of these electrons hit the anode, there is a current in the external circuit.
- The number of the emitted electrons reaching the cathode can be increased or decreased by making the anode positive or negative with respect to the cathode.
- At a specific large enough potential, V, nearly all the emitted electrons get to the anode and current attains its maximum (saturation) value (Fig. 1.19). Any increase in V does not affect the current, the maximum current is directly proportional to light intensity, and doubling the energy per unit time incident upon the cathode should double the number of electrons emitted.
- Those electrons that are emitted without losing any of their energy to the atoms of the metal surface will have the maximum energy, $(E_e)_{max}$.
- The maximum energy of the emitted electrons was computed by applying an external electrical field at which the photoelectric current is stopped.

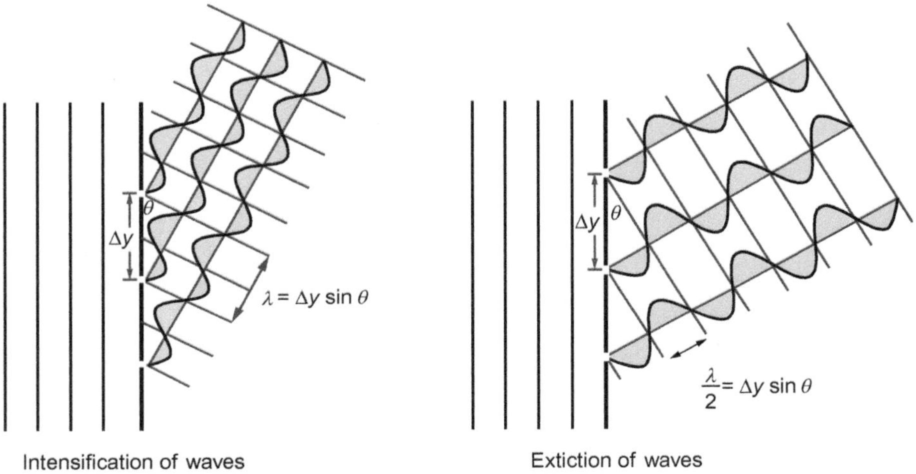

FIG. 1.17 Interference of waves on passing through a diffraction grating.

FIG. 1.18 Schematic drawing of photoelectric effect circuit, the anode can be made positive or negative with respect to the cathode, to attract or repel the emitted electrons.

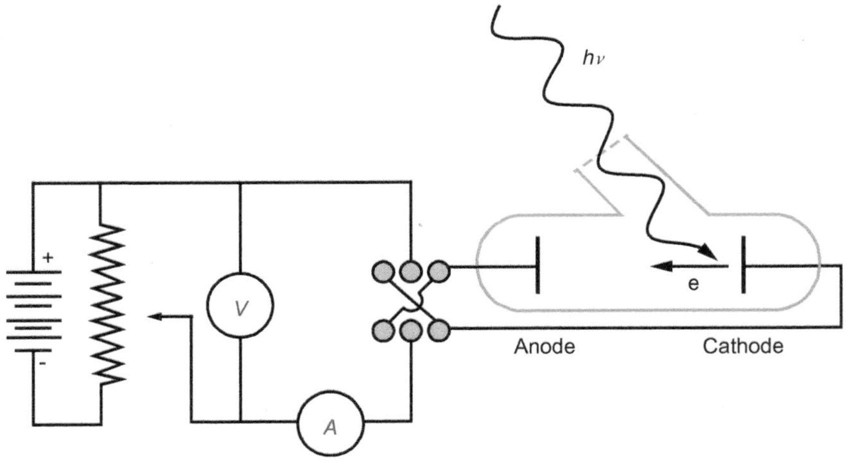

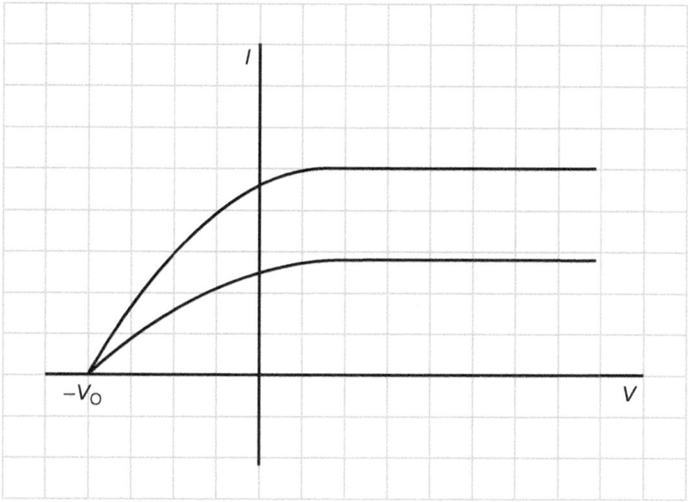

FIG. 1.19 Plots of photoelectric current versus voltage V for two values of light intensity.

- In Fig. 1.19, if V is less than V_o, no electrons reach the anode. The potential V_o is called the stopping potential. The experimental results indicate that V_o is independent of the intensity of the incident light.
- The stopping potential, V_o, is related to the maximum kinetic energy of the emitted electrons by

$$(E_e)_{max} = \left(\frac{1}{2}m_e u^2\right)_{max} = eV_o \tag{1.11a.1}$$

  where m_e, e, and u are the mass, charge, and velocity of the electron, respectively. In fact, increasing the rate of the energy falling on the cathode does not increase the maximum kinetic energy of the electrons emitted, $(E_e)_{max}$.
- According to wave theory, the energy E_e of the electrons emitted by the metal should be proportional to the intensity of the incident light. It was, however, established that:
  - E_e does not depend on the intensity of the light. The increase in the intensity of light only causes a greater number of electrons to be emitted from the metal
  - E_e depends on its frequency of the light, ν. E_e increases with ν
- Einstein showed that the photoelectric effect could be explained very simply if:
  - light were regarded as a stream of particles, called photons
  - a photon is a particle of electromagnetic radiation having zero mass and carries a quantum of energy
  - the energy of particular photon depends on the frequency of the radiation (ν)

$$E_{photon} = h\nu$$

  where h is Planck's constant.
- In order for an electron to be forced out from a metal surface, the electron must be bumped by a single photon carrying at least the minimum energy required to expel the electron.
- Einstein rationalized the photoelectric effect by suggesting that
  - electromagnetic radiation is absorbed by matter only in whole numbers of photons
  - when the intensity of the light of a given frequency is increased, more photons fall on the surface per unit time, but the energy absorbed by each electron is unchanged
  - for a given metal, no electrons were emitted if the light frequency was below a certain minimum
  - clearly, the energy of such electrons is equal to the difference between the energy of the photon $h\nu$ and the work required to overcome the force holding the electron in the metal
  - if a photon's frequency is below the minimum, then the electron remains bound to the metal surface
  - electrons in different metals are bound more or less tightly, so different metals require different minimum frequencies to exhibit the photoelectric effect.
- If Φ is the energy necessary to remove an electron from the surface of a metal, the maximum kinetic energy of the electrons emitted will be

$$\left(\frac{1}{2}m_e u^2\right)_{max} = e V_o = h\nu - \Phi \qquad (1.11a.2)$$

where

m_e, e, u are the mass, charge, and velocity of the electron, respectively

Φ is called the work function, the equation is known as Einstein's photoelectric equation

- Millikan showed that the Einstein equation is correct, and measurements of h agreed with the value found by Planck. Fig. 1.20 represents the dependence of the voltage at which the photoelectric current is discontinued on the frequency of the incident light. The data fall on a straight line which has a slope h/e. This method is one of the most precise techniques for determining Planck's constant. By placing V_o equal to zero in

$$\left(\frac{1}{2}m_e u^2\right)_{max} = e V_o = h\nu - \phi$$

when $\nu \rightarrow \nu_i$, then

$$\Phi = h\nu_i = \frac{hu}{\lambda_i} \qquad (1.11a.3)$$

Photons of frequency less than ν_i do not have enough energy to eject electrons from the metal.

Calculate the kinetic energy and the velocity of ejected electrons, when a light with wavelength of 300 nm is incident on a potassium surface for which the work function Φ is 2.26 eV.

- If the kinetic energy, E_e:

$$E_e = \left(\frac{1}{2}m_e u^2\right)_{max} = eV_o = h\nu - \Phi \qquad (1.11a.2)$$

$$\because h\nu = \frac{hc}{\lambda}$$

$$\therefore E_e = \frac{hc}{\lambda} - \Phi$$

$$E_e = \frac{(6.626 \times 10^{-34}\,\text{Js})(2.998 \times 10^8\,\text{m/s})}{300 \times 10^{-9}\,\text{m}} - (2.26\text{eV})(1.602 \times 10^{-19}\,\text{J/ev})$$

$$E_e = 6.62 \times 10^{-19}\,\text{J} - 3.62 \times 10^{-19}\,\text{J} = 3 \times 10^{-19}\,\text{J}$$

- The velocity, u, of the ejected electron is as follows:

$$u = \left(\frac{2 \cdot E_e}{m}\right)^{1/2} = \left(\frac{2 \times 3 \times 10^{-19}\,\text{J}}{9.109 \times 10^{-31}\,\text{kg}}\right)^{1/2} = 8.12 \times 10^5\,\text{m/s}$$

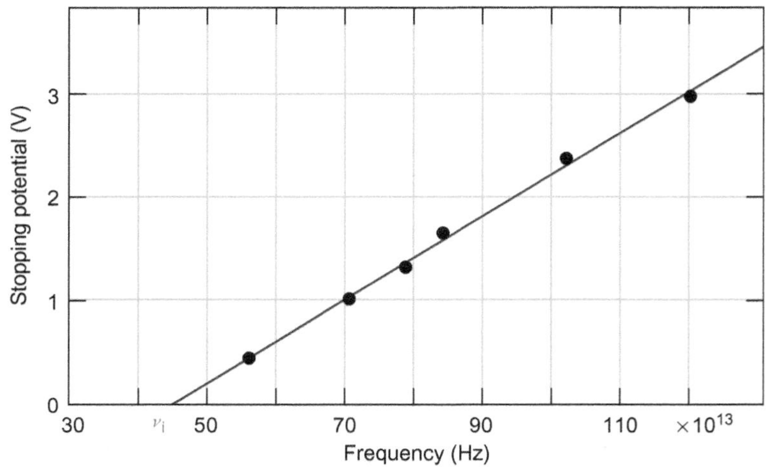

FIG. 1.20 Dependence of the voltage at which the photoelectric current is discontinued on the frequency of the incident light.

The Compton Effect

- When a photon collides with an electron, it discharges part of its energy to the electron. As a result, the radiation is scattered and its wavelength increased (Fig. 1.21).
- Compton noticed that when a variety of substances are exposed to X-rays, the wavelength of the scattered radiation is greater than the initial wavelength.
- The variation in the wavelength, $\Delta\lambda$, does not depend on the nature of the substance or the wavelength of the original radiation. It always has a definite value that is determined by the scattering angle ϕ (the angle between the directions of the scattered and original radiation).
- The accurate expression of the Compton effect could be determined if the hitting of photon and an electron could be considered as an elastic collision of two particles, in which the laws of conservation of energy and impulse are observed.
- Let us assume that a photon of energy $h\nu$ collides with an electron (Fig. 1.21). The energy and the momentum of the electron are taken to be zero.
- After collision:
  - the energy of the photon become $h\nu'$ and scattered by an angle ϕ to the original direction
  - the kinetic energy of the electron changes and moves in the direction of an angle θ to the direction of the original photon.
- According to the law of conservation of energy, the kinetic energy of the electron, T, is given by

$$T = h\nu - h\nu' = -h(\nu' - \nu) = -h\Delta\nu$$

$$T = \frac{1}{2}m_e u^2 \quad \text{and} \quad p_e = m_e u$$

where m_e, u, and p_e are the mass, velocity, and momentum of the electron, therefore,

$$T = \frac{p_e^2}{2m_e} = -h\Delta\nu$$

$$p_e^2 = -2m_e h\Delta\nu \tag{1.11b.1}$$

- According to the law of conservation of momentum, the sum of the vectors of the scattered photon and recoiled electron is equal the momentum of the original photon:

$$p_e = p_{ph_1} - p_{ph_2} \tag{1.11b.2}$$

$$p_e^2 = p_{ph_1}^2 + p_{ph_2}^2 - 2p_{ph_1}p_{ph_2}\cos\phi$$

where p_{ph_1} and p_{ph_2} are the magnitudes of the impulses of the original and scattered photons. The difference between these magnitudes is negligible, therefore,

$$\therefore p_{ph_1}^2 \approx p_{ph_2}^2$$

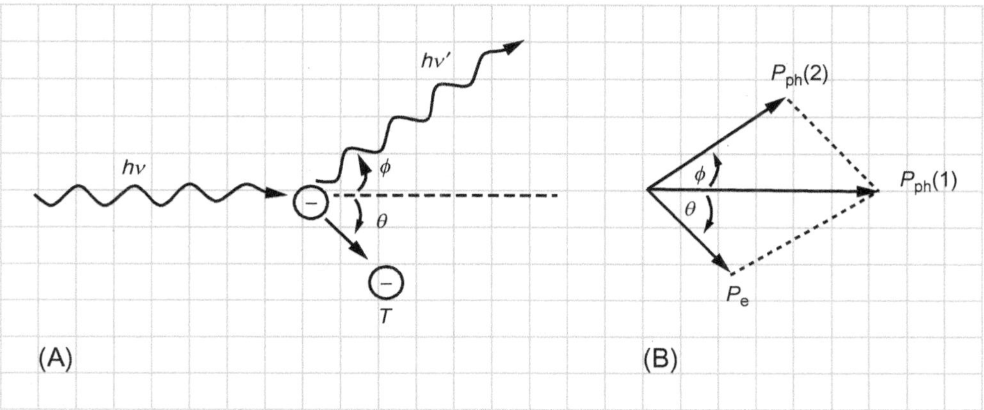

FIG. 1.21 (A) Schematic representation of the collision of the photon and electron. (B) Vector addition of momenta of the recoil electron and scattered photon.

$$\therefore p_e^2 = 2p_{ph_1}^2 - 2p_{ph_1}^2 \cos\phi$$

$$p_e^2 = 2p_{ph_1}^2(1 - \cos\phi) \quad (1.11b.3)$$

If: $(1 - \cos\phi) = 2\sin^2\frac{\phi}{2}$

$$\therefore p_e^2 = 4p_{ph_1}^2 \sin^2\frac{\phi}{2} \quad (1.11b.4)$$

$$\because p = \frac{h}{\lambda} = \frac{h\nu}{u} \quad (1.10.4)$$

$$p_e^2 = 4\frac{h^2\nu^2}{u^2}\sin^2\frac{\phi}{2} \quad (1.11b.5)$$

$$\because p_e{}^2 = -2m_e h\Delta\nu \quad (1.11b.1)$$

$$-2m_e h\Delta\nu = 4\frac{h\nu^2}{u^2}\sin^2\frac{\phi}{2}$$

$$\because \nu = \frac{u}{\lambda}$$

$$\therefore \partial\nu = -\frac{u}{\lambda^2}\partial\lambda$$

$$\therefore \Delta\nu = -\frac{u}{\lambda^2}\Delta\lambda \quad (1.11b.6)$$

$$\therefore -2m_e h\left(-\frac{u}{\lambda^2}\Delta\lambda\right) = 4\frac{h\nu^2}{u^2}\sin^2\frac{\phi}{2}, \text{ and } \because \frac{1}{\lambda^2} = \frac{\nu^2}{u^2}$$

$$\therefore \Delta\lambda = 2\frac{h}{m_e u}\sin^2\frac{\phi}{2} \quad (1.11b.7)$$

- The value $h/m_e u$ has the dimension of length and is equal to 0.0242 Å, and is called the Compton wavelength of the electron.
- Investigation has shown that this equation is in good agreement with the experimental data.

1.12 DE BROGLIE'S CONSIDERATIONS

How did de Broglie's show the following?

(a) The dual wave-particle nature is true not only for photon but for any other material particle as well.

(b)

$$mur = \frac{nh}{2\pi}, \text{ Bohr assumption for a stable orbital} \quad (1.7.4, \text{ this chapter page 16})$$

- Max Planck proposed that light wave have particle properties.
- Since the particle has wave properties, two fundamental equations need to be obeyed:
  $E = h\nu$, where ν is the frequency
  $E = mu^2$, where u is the velocity

$$\therefore h\nu = mu^2$$

$$\therefore \frac{h}{mu} = \frac{u}{\nu}$$

$$\because \lambda = \frac{u}{\nu}$$

The de Broglie equation is as follows:

$$\lambda = \frac{h}{mu} = \frac{h}{p} \quad (1.12.1)$$

This is a fundamental relation between the momentum of the electron (as a particle) and the wavelength (de Broglie wave).

- Every Bohr orbit is associated with a number of waves:

$$\text{Circumference} = 2\pi r = n\lambda$$

$$\lambda = \frac{2\pi r}{n}$$

When $n=1$, the number of waves equals 1, etc.

$$\lambda = \frac{h}{mu} = \frac{2\pi r}{n}$$

$$mur = \frac{nh}{2\pi} \quad (1.7.4)$$

which is simply the original Bohr condition for a stable orbital.

If an electron in a field of 1 volt/cm has a velocity of 6×10^7 cm/s, find λ.

$$\lambda = \frac{h}{mu} = \frac{6.63\times 10^{-27}\ \text{ergs}}{(2.1\times 10^{-20}\,\text{g})\times(6\times 10^7\,\text{cm/s})} = 12.0\times 10^{-8}\,\text{cm} = 12\,\text{Å}$$

Such an electron has a wavelength equivalent to an X-ray.

- An X-ray undergoes diffraction, as all waves do.
- Electron beams undergo diffraction. Therefore electrons have wave properties. Previously, Thomson showed that electron is a particle with mass, energy, and momentum.

1.13 WERNER HEISENBERG'S UNCERTAINTY PRINCIPLE, OR THE PRINCIPLE OF INDETERMINACY

If a particle has the wave nature, is it possible to determine its position and its velocity at the same time?

How can you verify the uncertainty principle by diffraction of a beam of particles at slit?

Why can we not precisely define the position of the electron without sacrifice information about its momentum (or velocity), and what is the significance of this finding?

- Assume a particle of mass m moves along x direction with a speed u (Fig. 1.22).The momentum of this particle is in x-direction, p_x. Consequently, the y-component of momentum p_y is zero. At a point x_o, the particle is allowed to be diffracted by passing through a slit of width Δy. The diffracted beam has momentum p with components in both x and y directions.

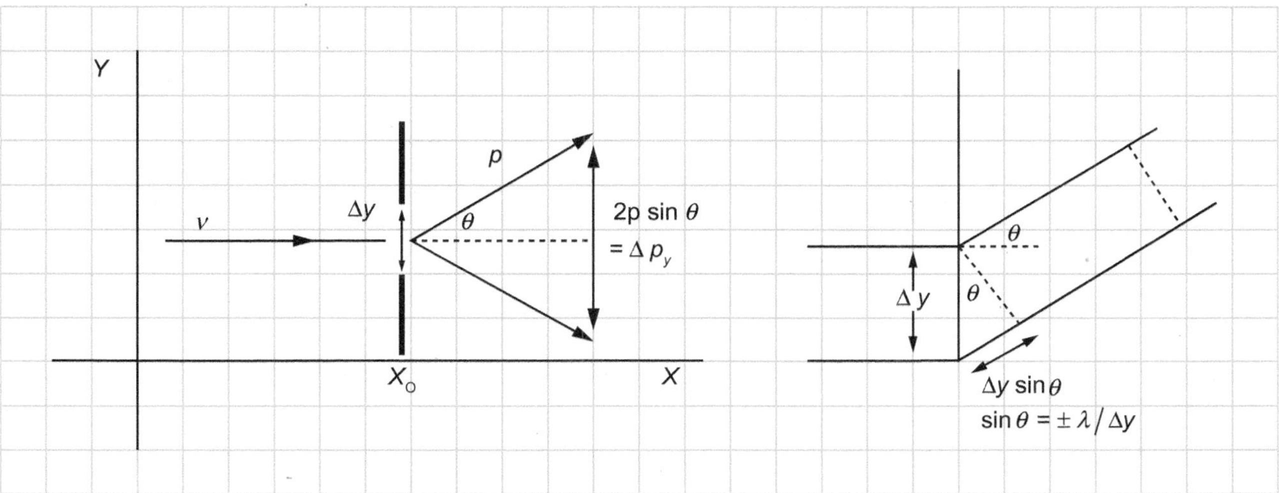

FIG. 1.22 Illustration of the path difference and uncertainty principle by diffraction of a beam of particles at the slit.

- The pattern of the diffracted intensity is exposed on the screen placed beyond x_o. The necessity for reinforcement is

$$\sin\theta = \frac{n\lambda}{\Delta y} \tag{1.9.1}$$

where, $n = 1, 2, 3\ldots$

The difference in the path for the two waves must be an integral number, n, of the wavelength, λ. The path difference becomes (Figs. 1.17 and 1.22)

$$\Delta y \sin\theta = n\lambda$$

- For the first two minima in the diffracted pattern, $n = 1$:

$$\Delta y \sin\theta = \lambda, \quad \text{then: } \sin\theta = \frac{\lambda}{\Delta y}$$

- The new direction of the momentum can be defined only within an angular-spread of $\pm\theta$ (Fig. 1.22), so that

$$\Delta p_y = 2p\sin\theta = \frac{2p\lambda}{\Delta y} \tag{1.13.1}$$

$$\because \lambda = \frac{h}{mu} = \frac{h}{p}$$

$$\Delta p_y \cdot \Delta y \simeq 2p\lambda = 2h \tag{1.13.2}$$

Therefore, the product of uncertainty in momentum Δp_y times the uncertainty in its conjugated coordinate Δy is of the order of h. A more accurate derivation shows the following:

$$\Delta y \cdot \Delta p_y \geq \frac{h}{4\pi} = \frac{1}{2}\hbar \tag{1.13.3}$$

- If we attempt to locate exactly the position of a particle, we must ignore the information about its momentum (or velocity).
- Heisenberg suggested that we cannot simultaneously know the exact position, x, and the exact momentum, p, of a particle:

$$(dx)(dp) \geq \frac{h}{4\pi}$$

- The uncertainty principle suggests that we cannot assume that the electron is moving around from point to point, with exact momentum at each point. Rather, we have to assume that the electron has only a specific probability of being found at each fixed point in space.
- The concept of an electron following a specific orbit, where its position and velocity are identified exactly, must therefore be replaced by the probability of finding an electron in a certain position or in a particular volume of space.

1.14 THE PROBABILITY OF FINDING AN ELECTRON AND THE WAVE FUNCTION

How can the electron's location be determined?

- In quantum mechanics we do not ask: what is the location of a particle? However, we may ask: what is the probability of having a particle in an extremely small volume?
- The probability of finding an electron in any volume element in space is proportional to the square of the absolute value of the wave function, $|\psi|^2$, integrated over that volume of space. This is the physical significance of the wave function.

$$\text{Probability}(x, y, z) \propto |\psi(x, y, z)|^2$$

$|\psi(x,y,z)|^2$ is equal to the probability that a particle, whose wave function is ψ, is located in a volume element ∂V at x, y, z, provided that

$$N^2 \int^{\text{All space}} \psi^2 \partial\tau = 1$$

where the normalized condition is $\partial\tau$ and the volume element $= \partial x \partial y \partial z$.

This interpretation is suggested by the theory of electromagnetic radiation, where intensity $\equiv \psi^2$.

- Acceptable solutions to the wave equation must be physically possible, and have certain properties:
  - ψ must be continuous.
  - ψ must be finite.
  - ψ must be single valued.
  - The probability of finding the electron over all the space from plus infinity to minus infinity must be equal to one.

1.15 ATOMIC AND SUBATOMIC PARTICLES

Elementary Particles

What do we mean by elementary particles?

- Elementary particles:
  - are the fundamental units that compose matter
  - exhibit wave-particle duality, displaying particle-like behavior under certain experimental conditions and wave like behavior in others
  - have neither known structure nor size
  - are often formed in high-energy collisions between known particles at particle accelerators
  - when formed, tend to be unstable and have short half-lives ($t_{1/2}$: $10^{-6} - 10^{-23}$ s)
  - have their dynamics governed by quantum mechanics.

What are the differences between fermions and bosons?
Name 12 leptons and 12 quarks.
What are the chemical characteristic features of leptons, hadrons, baryons, and mesons?

- Elementary particles are classified as fundamental particles (fermions) and mediating field particles (bosons).
- Pairs from each classification are grouped together to form a generation, with corresponding particles exhibiting similar physical behavior:
  - Charged particles of the first generation do not decay such electrons.
  - Charged particles of the second and third generations decay with very short half-lives, and are observed in very high-energy environments.
- Elementary fermions: include leptons and hadrons.
  - Leptons:
    There are six known leptons: electron, muon, tau, and their three associated neutrinos.

Lepton (negatively charged)	e^-	μ^-	τ^-
Neutrino	ν_e	ν_μ	ν_τ
Generation	1st	2nd	3rd

Each of these six leptons has an antiparticles: e^+ (positron), $\overline{\nu}_e$, μ^+, $\overline{\nu}_\mu$, τ^+, and $\overline{\nu}_\tau$.
These leptons do not break down into small units, and have no measurable size or internal structure.

  - Hadrons include baryons and mesons.
    (a) Baryons: protons and neutrons are common examples. Baryons have masses equal to or greater than proton.
    (b) Mesons are unstable.
  - Baryons are heavier than mesons, and both are heavier than leptons.
  - All hadrons are composed of two or three fundamental particles, which are known as quarks.
  - There are six quarks that fit together in pairs: up and down, charm and strange, and top and bottom.

Charge −1/3 quark	d: dawn quark	s: strange quark	b: bottom quark
Charge +2/3 quark	u: up quark	c: charm quark	t: top quark
Generation	1st	2nd	3rd

Associated with each quark is an antiquark of opposite charge: $\overline{d}$, $\overline{u}$, $\overline{s}$, $\overline{c}$, $\overline{b}$, and $\overline{t}$

  - All quarks have an associated fractional electrical charge $\left(-\frac{1}{3}e \text{ or } +\frac{2}{3}e\right)$.

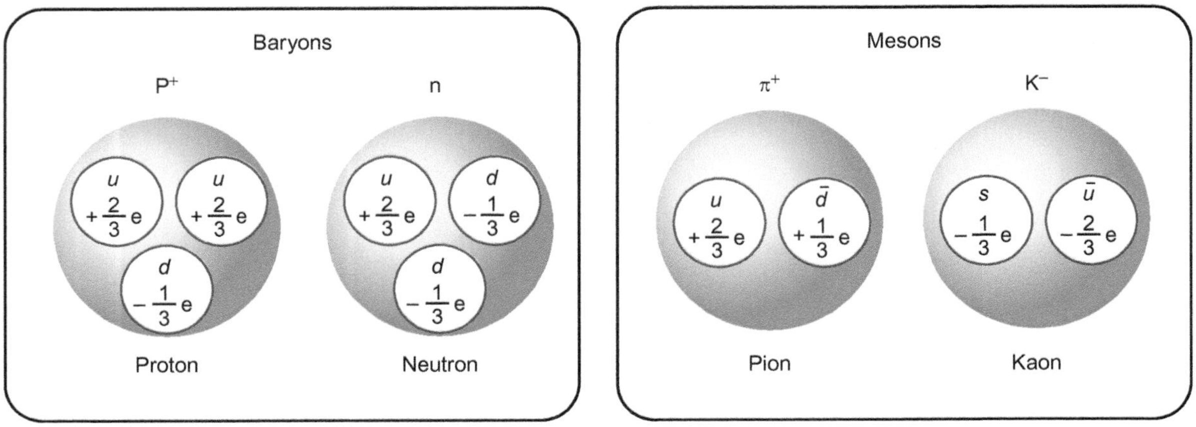

FIG. 1.23 Examples of baryons and mesons.

- Baryons and mesons are distinguished by the composition of their internal structure (Fig. 1.23):
  meson: one quark + one antiquark, such as pion and kaon
  baryon: three quarks, like proton and neutron
  antibaryon: three antiquarks
- The charge of the hadrons is either a multiple of e or zero.
- The total charge of a hadron is equal to the sum of the charges of its constituent's quarks (Fig. 1.23).
- Quarks come in six categories (names): d, u, s, c, b, and t. Quarks have an additionally character, which is whimsically called color; however, they are of course not really colored. Quark "colors" are red (r), blue (b), and green (g). Antiquarks have "anticolors": antired ($\overline{r}$), antiblue ($\overline{b}$), and antigreen ($\overline{g}$).
- It is believed that six leptons and six quarks (and their antiparticles) are the fundamental particles. Fermions are the building blocks of the matter we see around us.

What are the main particles that intermediate fundamental interaction?

- Elementary particles that intermediate fundamental interaction are known as bosons.
- Types of bosons include: gluon, photon, $W^{\pm}$, Z boson, graviton, Gauge boson, and Higgs boson.
  - These force carrier particles to intermediate the fundamental interaction among fermions.
  - Such forces result from matter particles exchanging other particles, and are generally referred to as force-intermediating particles.
  - The exchange bosons can mediate an interaction only over a range that is inversely proportional to its mass.
  - Gluon
    - Strong interaction is mediated by exchange particles called gluons.
    - Gluon is responsible for the binding of protons and neutrons into nuclei, and holds the nucleons together.
    - Force mediated by gluons is the strongest of all the fundamental interactions; it has a very short range (the range of force is about 10^{-15} m).
    - A gluon carries one color quark and one anticolor quark.
    - Totally color-neutral gluon such as r$\overline{r}$, b$\overline{b}$, and g$\overline{g}$ are not possible. This is similar to electric charges: like colors repel, and opposite colors attract.
  - Photon
    - A photon is the exchange boson that mediates the electromagnetic interaction between electrically charged particles.
    - It is massless and has charge of zero.
    - The electromagnetic interaction is responsible for
      (1) the attraction of unlike charges and the repulsion of the like charges
      (2) binding of atoms and molecules
    - Fig. 1.24 represent two electrons repelling each other through the exchange of a photon. Because the momentum is conserved, the two electrons change directions.
    - The electromagnetic interaction is a long-range interaction and proportional to $1/r^2$.

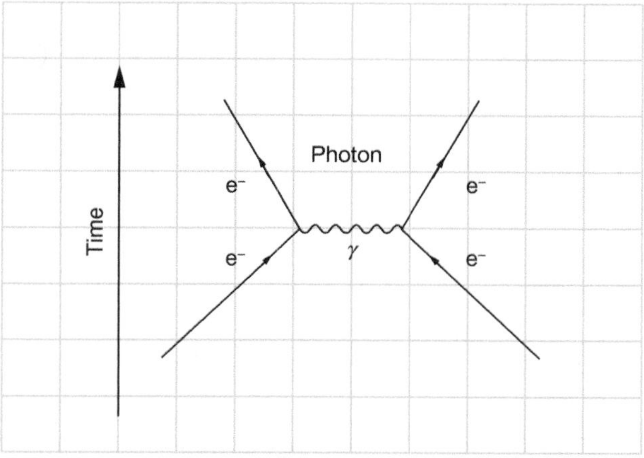

FIG. 1.24 One electron emits a photon and the other electron absorbs a photon.

  - It is about 10^{-2} times the strength of the strong interaction at the nuclei.
- Weak interaction, $W^{\pm}$
  - Quarks and leptons can interact by a so-called weak interaction.
  - This weak interaction is mediated by particles called W^{+} or W^{-}.
  - $W^{\pm}$ carries an electric charge of +1 and −1 and couples to electromagnetic interaction.
  - In this interaction, one quark type is changed into another by the emission of a lepton and neutrino:

$$\underbrace{d}_{-\frac{1}{3}e} \longrightarrow \underbrace{u}_{+\frac{2}{3}e} + \underbrace{e^{-} + \overline{\nu}_{e}}_{W^{-}}$$

$$\underbrace{n}_{0e} \longrightarrow \underbrace{p}_{+1e} + \underbrace{e^{-} + \overline{\nu}_{e}}_{W^{-}}$$

In the above reactions, d and n emit W^{-} boson, which decays to an electron, e^{-}, and an antielectron-neutrino, $\overline{\nu}_{e}$.

  - It is a short-range nuclear interaction (with a range of force of about 10^{-18}m).
  - It is involved in β-decay.
- Z boson
  - The Z boson permits uncharged elementary particles to participate in weak interactions.
  - The mediator of this interaction needs to be a zero-charged boson, the Z boson.
  - The Z boson decays into an electron, e^{-}, and a positron, e^{+}. This provides a clear signature of the Z boson.
- Higgs boson
  - The Higgs boson couples to all elementary particles to give them mass. This includes the masses of the $W^{\pm}$ and Z boson, and the masses of fermions, i.e., the quark and leptons.
- Graviton
  - The gravitational interaction is mediated by gravitons.
  - The gravitational force has infinite range.
  - The exchange bosons have a mass of zero.

What are the main laws that must hold for any reaction or decay?

- All physical process have to obey the conservation laws of:
  - charge
  - energy
  - momentum
  - angular momentum

- In addition to all these, two fundamental conservation laws must hold:
  - The total lepton number, l_{net}:
    Number of lepton minus the number of antilepton is constant in time.

$$l_{net} = n_l - n_{\bar{l}} = \text{constant}$$

For example,

$$n \rightarrow p + \underbrace{\underbrace{\bar{e}}}_{1\ \text{lepton}} + \underbrace{\bar{\nu}_e}_{1\ \text{antilepton}}$$

$$\therefore l_{net} = 1\ \text{lepton}(\bar{e}) - 1\ \text{antilepton}(\bar{\nu}_e) = 0$$

  - The net quark number, Q_{net}:
    The difference between the total number of quarks and the total number of antiquarks is constant in time.

$$Q_{net} = n_q - n_{\bar{q}} = \text{constant}$$

The conservation law of a net quark number is more often expressed as the law of baryon number conservation.

  - The law of baryon number conservation:
    The difference between the number of baryons and the number of antibaryons is constant in time.

SUGGESTIONS FOR FURTHER READING

General

R.S. Berry, V—Atomic orbitals, J. Chem. Educ. 43 (1966) 283.

Cathode Rays and Atomic Constituents

E. Goldstein, Preliminary communications on electric discharges in rarefied gases, Monthly Reports of the Royal Prussian Academy of Science in Berlin, 1876, p. 279.

J.F. Keithley, The Story of Electrical and Magnetic Measurements: From 500 B.C. to the 1940s, John Wiley and Sons, New York, NY, 1999. ISBN: 0-7803-1193-0, p. 205.

M. Faraday, VIII. Experimental researches in electricity. Thirteenth series, Philos. Trans. R. Soc. Lond. 128 (1838) 125.

J.J. Thomson, On bodies smaller than atoms, PSM (August) (1901) 323.

Charge and Mass of the Electron

R.A. Millikan, F. Harvey, Elektrizitätsmengen, Phys. Zeit. 10 (1910) 308.

R.A. Millikan, On the elementary electric charge and the Avogadro constant, Phys. Rev. 2 (2) (1913) 109.

F.P. Michael, Remembering the oil drop experiment, Phys. Today 60 (5) (2007) 56.

F. Harvey, My work with millikan on the oil-drop experiment, Phys. Today 43 (June) (1982).

A. Franklin, Millikan's oil-drop experiments, Chem. Educ. 2 (1) (1997) 1.

D. Goodstein, In defense of Robert Andrews Millikan, Eng. Sci. 63 (4) (2000) 30.

J.J. Thomson, Cathode rays, Philos. Mag. 44 (1897) 293.

J.J. Thomson, Rays of Positive Electricity, Proc. R. Soc. A 89 (1913) 1.

H.A. Boorse, L. Motz, The World of the Atom, vol. 1, Basic Books, New York, NY, 1966.

Rutherford's Atomic Model

E.E. Salpeter, Models and modelers of hydrogen, in: L. Akhlesh (Ed.), Am. J. Phys., 65, 9, 1996, ISBN: 981-02-2302-1, p. 933.

E. Rutherford, The scattering of α and β particles by matter and the structure of the atom, Philos. Mag. 21 (1911) Series 6.

Quantum Radiation

M. Planck, On the Theory of the Energy Distribution Law of the Normal Spectrum, Verhandl. Deut. Ges. Phys. 2 (1900) 237.

A. Einstein, Does the Inertia of a Body Depend on its Energy Content? Ann. Phys. 17 (1905) 132.

J.K. Robert, A.R. Miller, Heat and Thermodynamics, fifth ed., Interscience Publishers, New York, NY, 1960. p. 526.

E. Schrödinger, Quantum Mechanics and the Magnetic Moment of Atoms, Ann. Phys. 81 (1926) 109.

T.S. Kuhn, Black-Body Theory and Quantum Discontinuity, 1894–1912, Oxford University Press, New York, NY, 1978.

The Hydrogen Atom Line-Emission Spectra

E. Whittaker, History of Aether and Electricity, Vol. 2, Th. Nelson and Sons, London, 1953.

Bohr's Theory of Hydrogen Atom

N. Bohr, On the constitution of atoms and molecules, part I, Philos. Mag. 26 (151) (1913a) 1.

N. Bohr, On the constitution of atoms and molecules, part II systems containing only a single nucleus, Philos. Mag. 26 (153) (1913b) 476.

N. Bohr, On the constitution of atoms and molecules, part III systems containing several nuclei, Philos. Mag. 26 (1913c) 857.

N. Bohr, The spectra of helium and hydrogen, Nature 92 (2295) (1914) 231.

N. Bohr, Atomic structure, Nature 107 (2682) (1921) 104.

A. Pais, Niels Bohr's Times: in Physics, Philosophy, and Polity, Clarendon Press, Oxford, 1991.

Bohr-Sommerfeld Model

L. Pauling, E.B. Wilson, Introduction to Quantum Mechanics, McGraw-Hill Book Company, New York, NY, 1935 (Chapter 2).

Electrons as Waves: Interference and Diffraction

C.J. Davission, L.H. Germer, The Scattering of Electrons by a Single Crystal of Nickel, Nature 119 (1927) 558.

R.P. Feynman, The Feynman Lectures on Physics, vol. I, Addison-Wesley, Reading, MA, 1963. p. 16.

L.A. Bendersky, F.W. Gayle, Electron diffraction using transmission electron microscopy, J. Res. Natl. Inst. Stand. Technol. 106 (2001) 997.

G. Thomas, M.J. Goringe, Transmission Electron Microscopy of Materials, John Wiley, New York, NY, 1979. ISBN: 0-471-12244-0.

Relativity Theory: Mass and Energy

A. Einstein, Relativity: The Special and General Theory, H. Holt and Company, New York, NY, 1916.

E. Freundlich, The Foundations of Einstein's Theory of Gravitation (translation by H. L. Brose) The University Press, Cambridge, 1920.

M.N. Roy, Discussion on the theory of relativity, Astron. Soc. IXXX (2) (1919) 96.

The Corpuscular Nature of Electromagnetic Waves: The Photoelectric Effect and Compton Effect

A.H. Compton, A Quantum Theory of the Scattering of X-rays by Light Elements, Phys. Rev. 21 (1923) 483.

D.A. Skoog, R.C. Stanley, F.J. Holler, Principles of Instrumental Analysis, Thomson Brooks, Belmont, CA, 2007. ISBN: 0-495-01201-7.

R.A. Serway, Physics for Scientists & Engineers, Saunders, Philadelphia, 1990. ISBN: 0030302587. p. 1150.

J.D. Dana, E.S. Dana, Photoelectric Effect, Am. J. Sci. (1880) 234.

P. Christillin, Nuclear compton scattering, J. Phys. G: Nucl. Phys. 12 (9) (1986) 837.

J.R. Taylor, C.D. Zafiratos, M.A. Dubson, Modern Physics for Scientists and Engineers, second ed., Prentice-Hall, Englewood Cliffs, NJ, 2004. ISBN: 0-13-805715-X p. 136.

M. Cooper, X-ray Compton Scattering, Oxford University Press, Oxford, 2004. ISBN: 978-0-19-850168-8 14 October.

de Broglie's Considerations

L. de Broglie, Researches on the quantum theory, Ann. De. Phys. 3 (1925) 22.

L. de Broglie, Matter and Light, Dover Publications, New York, NY, 1946 (first ed., W. W. Norton Co.).

Werner Heisenberg's Uncertainty Principle, the Uncertainty Principle or the Principle of Indeterminacy

W. Heisenberg, Uber den anschauclichen Inhalt der quantentheoretischen Kinematik und Mechanik, Z. Phys. 43 (1927) 172.

N. Bohr, Atomic Physics and Human Knowledge, John Wiley and Sons, New York, NY, 1958. p. 38ff.

W. Kauzmann, Quantum Chemistry, Academic Press, New York, NY, 1957 (Chapter 7).

W. Sherwin, Introduction to Quantum Mechanics, Holt, Rinehart & Winston, New York, NY, 1959. p. 130.

D.C. Cassidy, Uncertainty: The Life and Science of Werner Heisenberg, W. H. Freeman & Co, New York, NY, 1992.

Subatomic Particles

S. Braibant, G. Giacomelli, M. Spurio, Particles and Fundamental Interaction: An Introduction to Particle Physics, Springer, New York, NY, 2009. ISBN: 978-94-007-2463-1.

K. Nakamura, Review of Particle Physics, J. Phys. G: Nucl. Phys. 37 (7A) (2010) 075021 (1422pp).

W. Bauer, G.D. Westfall, University Physics With Modern Physics, McGraw-Hill Companies, New York, NY, 2011. ISBN: 978-007-131366-7.

R.A. Serway, J.S. Faughn, Holt Physics, Holt, Rinehart and Winston, Austin, 2002. ISBN: 0-03-056544-8.

F. Close, Particle Physics: A Very Short Introduction, Oxford University Press, Oxford, 2004. ISBN: 0-19-280434-0.

D.J. Griffiths, Introduction to Elementary Particles, John Wiley & Sons, Weinheim, 1987. ISBN: 0-471-60386-4.

G.L. Kane, Modern Elementary Particle Physics, Perseus Books, Reading, MA, United States, 1987. ISBN: 0-201-11749-5.

O. Boyarkin, Advanced Particle Physics Two-Volume Set, CRC Press, Taylor & Francis Group, Boca, Raton, London, New York, 2011. ISBN: 978-1-4398-0412-4.

Chapter 2

Electrons in Atoms

In the previous chapter, the following physical observations were discussed to support the idea that particle energies are quantized:

- The electrons in atoms can have only strictly definite values of energy, which are characterized by a series of integers.
- Energy intensity is increased by a specific small quantity called quanta.

We then addressed the physical significance of the particle-wave behavior, and how classic mechanics can be employed to explain and predict the wave-particle duality of microparticles. Such duality reflects itself in classic mechanics when we examined the occurrence of interference, diffraction, photoelectric effect, Compton Effect, and de Broglie waves. We saw that the dual wave-particle nature is true not only for photon but for any other material particle as well.

The mechanics of microparticles is called quantum mechanics or wave mechanics. In quantum mechanics, the laws of motion of microparticles are described by the Schrödinger equation, which is of the same significance for microparticles as Newton's laws are for classic mechanics. The mechanics based on Newton's laws are applicable to the motion of ordinary bodies, and are called classic mechanics. Quantum mechanics give the probability of finding a particle. Concepts such as coordinates and velocity at a given moment mean nothing in quantum mechanics. However, the concepts of mass, energy, and angular momentum of the particle retain their importance.

Heisenberg and Schrödinger independently proposed two alternative routes to describe quantum mechanics. It was found that both routes lead to identical results; however, Schrödinger's method proved to be more convenient for investigating atomic and molecular structures.

First, in this chapter we present how Schrödinger drove his wave mechanics equation that competent to deal with the wave-particle duality of matter and energy, and presented the main postulates of this equation. The appropriate considerations for Schrödinger equation of the hydrogen atom and transformation of the Cartesian form of the equation into the corresponding spherical polar form are detailed. The spherical wave function factored into Θ- and Φ-angular and R-radial wave functions. The solutions of these equations have been meticulously analyzed using *Legendre* and *Laguerre* polynomial differential equations. As results, the physical meaning of the four quantum number: n, l, m, and s, relation among angular momentum, moment of inertia, and energy of the electron, the final solution for the full wave function, and how it relates to the atomic orbital are clarified. The real wave functions that represent the atomic orbitals have been constructed in polar and Cartesian coordinates. The orthonormal properties are used to verify the reality of the wave functions. The wave functions of s, p, and d orbital are used to draw graphically the boundary surface of the corresponding orbitals and to evaluate atomic radius, kinetic, potential energy, and the orbital angular momentum. Then we demonstrate how to calculate most probable radius and the mean radius, and to find out the number radial nodes of the orbitals.

Secondly, we examine the taken procedures in order to apply Schrödinger equations for many-electron atoms. The Pauli Exclusion Principle, fermions and bosons exchange, and Slater determinant are used to express acceptable wave functions. Orbital degeneracy in each shell and why it has been removed are laid out. Slater's empirical rules for electron shielding are considered and used to calculate the ionization potential and the energy of atomic electrons. The building-up principle and the empirical rules concerning the order of filling the orbitals are discussed.

Thirdly, we examine how to identify the term symbols of the ground state and the different terms of the excited microstates of polyelectronic atom, and how many subterms arise when spin-orbit coupling is taken into consideration. The term wave functions and the corresponding single electron wave functions are described in order to understand the effect of the ligand field on term of the free electron in the following chapters.

Finally, we investigate the splitting and the energy of the Russell–Saunders terms due to the interaction between the spin and orbital angular momentums. The energy difference between two adjacent spin-orbital states is also defined. The coupling of the orbital and the spin angular momentums with external magnetic field are examined, and the potential energy is assigned. The effect of external magnetic field and the splitting each Russell–Saunders term into microstates and the energy of the microstate are identified.

Electrons, Atoms, and Molecules in Inorganic Chemistry. http://dx.doi.org/10.1016/B978-0-12-811048-5.00002-X

In this chapter, Schrödinger's method was applied for exploring the modern theory of the atomic structure (Scheme 2.1). In order to do that the following will be investigated:

- 2.1: The wave function (the Schrödinger Equation)
- 2.2: Properties of the wave function
- 2.3: Schrödinger equation of the hydrogen atom
- 2.4: Transformation of the Schrödinger equation from Cartesians to spherical polar coordinates
- 2.5: The angular equation
- 2.6: The Φ-equation
- 2.7: The Θ-equation
- 2.8: The radial equation, R

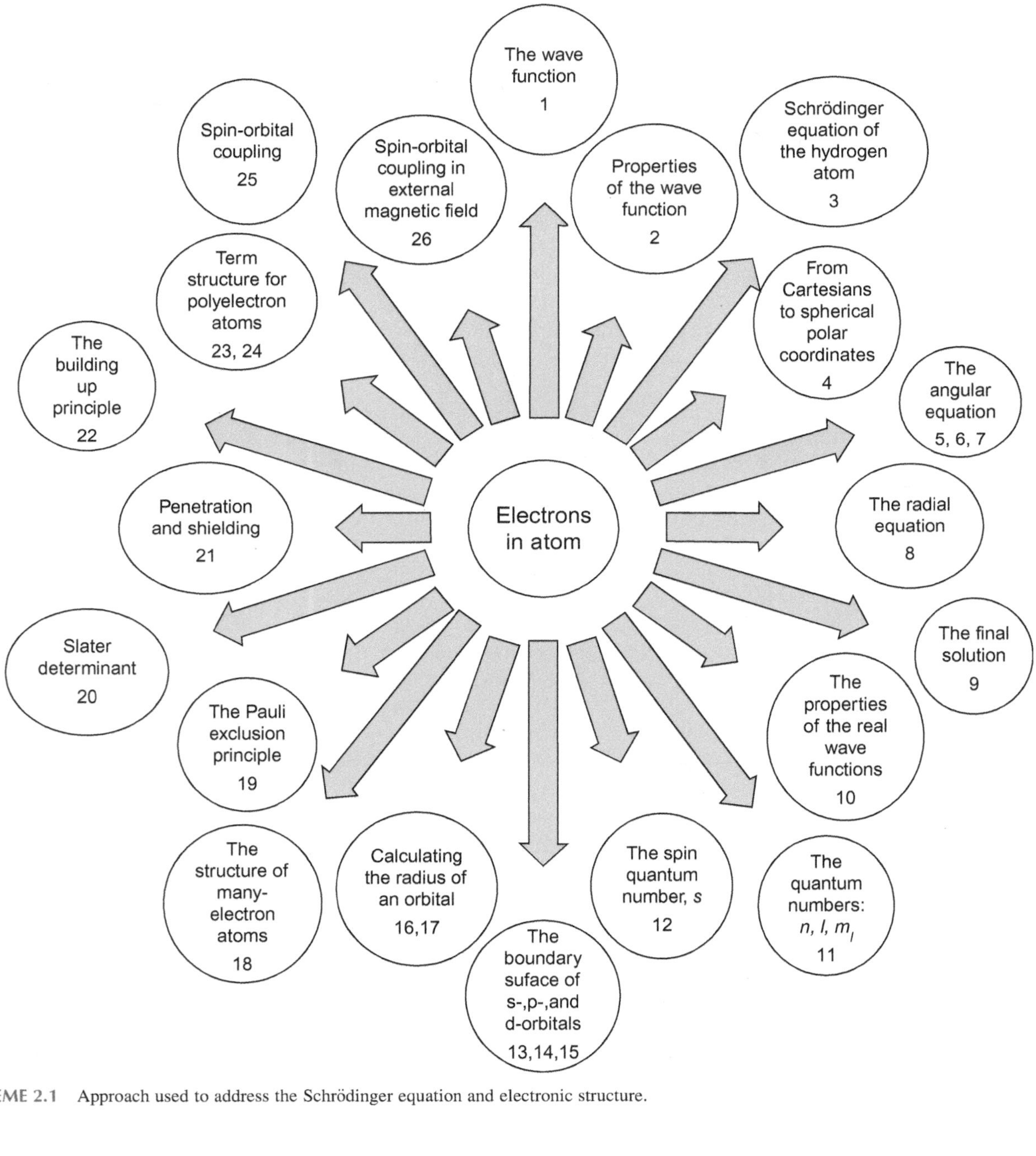

SCHEME 2.1 Approach used to address the Schrödinger equation and electronic structure.

2.1 THE WAVE FUNCTION (THE SCHRÖDINGER EQUATION)

How did Schrödinger derive his wave mechanics equation that competent to deal with the wave-particle duality of matter and energy?

- Erwin Schrödinger presented the equation relating the energy of a system to the wave motion.
- He started by assuming that particle could be described by the general wave (Eq. (2.1.1)):

$$\psi(x,t) = a_o e^{2\pi i\left(\frac{x}{\lambda} - \nu t\right)} \tag{2.1.1}$$

  where a_o is constant, λ and ν are wavelength and frequency of the wave, which describe the behavior of the particle, and $i^2 = -1$.
- The wave function ψ contains all of the information about the particle that we can know: its position, momentum, energy, etc.
- The differential equation of the wave function ψ was derived, if

$$y = ae^{bx}, \quad \text{then}$$

$$\frac{\partial y}{\partial x} = abe^{bx} = by, \quad \text{and}$$

$$\frac{\partial^2 y}{\partial x^2} = ab^2 e^{bx} = b^2 y, \quad \text{consequently,}$$

$$\frac{\partial^2 \psi(x,t)}{\partial x^2} = -\left(\frac{2\pi}{\lambda}\right)^2 \psi(x,t) \tag{2.1.2}$$

- If momentum: $p = mu$, where m and u are the mass and the velocity of the electron, and potential energy: $V(x, t)$, and

$$\because \lambda = \frac{h}{p} \rightarrow \therefore \frac{1}{\lambda} = \frac{p}{h}, \hbar = \frac{h}{2\pi}, \quad \text{and}$$

$$\because p^2 = 2mK, \quad K \rightarrow \left(\begin{matrix}\text{kinetic}\\ \text{energy}\end{matrix} = \frac{1}{2}mu^2\right)$$

$$\therefore \frac{\partial^2 \psi(x,t)}{dx^2} = -\left(\frac{p^2}{\hbar^2}\right)\psi(x,t) = -\left(\frac{2mK}{\hbar^2}\right)\psi(x,t) \tag{2.1.3}$$

$$\because E = K + V(x,t) \rightarrow \therefore K = E - V(x,t)$$

$$\therefore \frac{\partial^2 \psi(x,t)}{\partial x^2} = -\left(\frac{2mE}{\hbar^2}\right)\psi(x,t) + \left(\frac{2mV}{\hbar^2}\right)\psi(x,t) \tag{2.1.4}$$

Therefore,

$$-\frac{\hbar^2}{2m}\frac{\partial^2 \psi(x,t)}{\partial x^2} + V(x,t)\psi(x,t) = E\psi(x,t) \tag{2.1.5}$$

- This equation can be extended to three dimensions:

$$-\frac{\hbar^2}{2m}\left(\frac{\partial^2}{\partial x^2} + \frac{\partial^2}{\partial y^2} + \frac{\partial^2}{\partial z^2}\right)\psi(x,y,z,t) + V(x,y,z,t)\psi(x,y,z,t) = E\psi(x,y,z,t)$$

or

$$\left(-\frac{\hbar^2}{2m}\nabla^2 + V(x,y,z,t)\right)\psi(x,y,z,t) = E\psi(x,y,z,t) \tag{2.1.6}$$

where $\nabla^2 = \frac{\partial^2}{\partial x^2} + \frac{\partial^2}{\partial y^2} + \frac{\partial^2}{\partial z^2}$.

- The Schrödinger wave equation is commonly written in the form:

$$\hat{H}\psi = E\psi \tag{2.1.7}$$

where $\hat{H} = \left(-\frac{\hbar^2}{2m}\nabla^2 + V(x,y,z,t)\right) \rightarrow \begin{pmatrix}\text{Hamiltonian}\\ \text{operator}\end{pmatrix}$.

  The function, ψ, is called the eigenfunction.
  E is called the eigenvalue of the Hamiltonian operator.

- Also, if

$$\psi(x,y,z,t) = a_0 e^{2\pi i\left(\frac{(x,y,z)}{\lambda} - \nu t\right)} \tag{2.1.1}$$

then

$$\frac{\partial\psi(x,y,z,t)}{\partial t} = -2\pi i\nu\, a_0 e^{2\pi i\left(\frac{(x,y,z)}{\lambda} - \nu t\right)} = -2\pi i\nu\,\psi(x,y,z,t)$$

$$E = h\nu \rightarrow \nu = \frac{E}{2\pi\hbar}$$

$$\therefore \frac{\partial\psi(x,y,z,t)}{\partial t} = -\left(\frac{iE}{\hbar}\right)\psi(x,y,z,t)$$

$$\therefore -\left(\frac{\hbar}{i}\right)\frac{\partial\psi(x,y,z,t)}{\partial t} = E\psi(x,y,z,t)$$

$$\therefore i\hbar\frac{\partial}{\partial t}\psi(x,y,z,t) = E\psi(x,y,z,t) \tag{2.1.8}$$

The Schrödinger wave equation is written in the form:

$$\hat{E}\psi = E\psi \tag{2.1.7}$$

where

$$\hat{E} = i\hbar\frac{\partial}{\partial t} \rightarrow \begin{pmatrix}\text{Hamiltonian}\\ \text{operator}\end{pmatrix}$$

Note that:

$$\left(-\frac{\hbar^2}{2m}\nabla^2 + V(x, y, z, t)\right)\psi = \left(i\hbar\frac{\partial}{\partial t}\right)\psi$$

- The function, ψ, is the eigenfunction.
  E is the eigenvalue of the Hamiltonian operator.

Factor the wave function:

$$-\frac{\hbar^2}{2m}\frac{\partial^2\psi(x,t)}{\partial x^2} + V(x,t)\psi(x,t) = E\psi(x,t)$$

into separate functions of position and time.

- The wave function can be factored into separate functions of position and time:

$$\because e^{a+b} = e^a \cdot e^b$$

$$\therefore \psi(x,t) = a_0 e^{2\pi i\left(\frac{x}{\lambda} - \nu t\right)} = a_0 e^{\frac{2\pi i x}{\lambda}} e^{-2\pi i \nu t} = \psi(x)\phi(t) \quad (2.1.1)$$

$$E = h\nu = \frac{2\pi h\nu}{2\pi} = 2\pi\hbar\nu$$

$$\nu = \frac{E}{2\pi\hbar} \quad (2.1.9)$$

$$e^{-2\pi i \nu t} = e^{-\frac{2itE}{\hbar}} = \phi(t)$$

$$\therefore \psi(x,t) = a_0 e^{2\pi i\left(\frac{x}{\lambda} - \nu t\right)} = a_0 e^{\frac{2\pi i x}{\lambda}} e^{-\frac{2itE}{\hbar}} = \psi(x)\phi(t) \quad (2.1.10)$$

$$\because -\frac{\hbar^2}{2m}\frac{\partial^2\psi(x,t)}{\partial x^2} + V(x,t)\psi(x,t) = E\psi(x,t) \quad (2.1.5)$$

$$\therefore -\frac{\hbar^2}{2m}\frac{\partial^2\psi(x)\,\phi(t)}{\partial x^2} + V(x,t)\psi(x)\phi(t) = E\psi(x)\,\phi(t)$$

and

$$-\frac{\hbar^2}{2m}\frac{\partial^2\psi(x)}{\partial x^2} + V(x,t)\psi(x) = E\psi(x) \quad (2.1.11)$$

Or,

$$\hat{H}\psi = E\psi \quad (2.1.7)$$

2.2 PROPERTIES OF THE WAVE FUNCTION

What are the main postulates of the Schrödinger wave equation?

- The physical state of a particle at time t is fully expressed by a complex wave function ψ.
- The wave function ψ and its first and second derivatives (ψ' and ψ'') must be:
  - finite
  - single valued
  - continuous.
- Any quantity that is physically observable can be recalled in quantum mechanics by the Hamiltonian operator.
- The Hamiltonian operator corresponding to a physical observation is generated by writing down the classical expression in terms of the variable x, p_x, t, E, and exchange that expression to operator by using the following (Table 2.1):

TABLE 2.1 Classical Variables and Corresponding Operators

Classical Variable	x	p_x	t	E
Quantum mechanical operator	$\hat{x}$	$\hat{p}_x$	$\hat{t}$	$\hat{E}$
Expression for operator	x	$\frac{\hbar}{i}\frac{\partial}{\partial x}$	t	$-\frac{\hbar}{i}\frac{\partial}{\partial t}$
Operation	Multiply by x	Take derivative with respect to x and multiply by $\frac{\hbar}{i}$	Multiply by t	Take derivative with respect to t and multiply by $-\frac{\hbar}{i}$

- Only acceptable quantities of a physical observation represented by $\hat{F}$ are any of the eigenvalues f_i.
  If ψ_i is an eigenfunction of $\hat{F}$ with an eigenvalue f_i,

$$\hat{F}\psi_i = f_i\psi_i \tag{2.1.7}$$

  then a measurement of F is certain to yield the value f_i.
- The expectation or average value f of any observable f, which relates to an operator $\hat{F}$, is the estimated form:

$$\bar{f} = F = \psi^*|\hat{F}|\psi = \int_{-\infty}^{\infty} \psi^* \hat{F}\, \psi\, \partial x \tag{2.2.1}$$

  Assuming that the wave function is normalized:

$$\int_{-\infty}^{\infty} \psi^* \psi \partial x = 1 \tag{2.2.2}$$

  where ψ^* is the complex conjugate of ψ, formed by placing i with $-i$ where it occurs in ψ function.
- The Hamiltonian operator, $\hat{F}$, is a linear operator that satisfies the condition,

$$\int \psi_1{}^* \hat{F} \psi_2\, \partial x = \int \psi_2 \left(\hat{F}\psi_1\right)^* \partial x \tag{2.2.3}$$

  for any pair of functions ψ_1 and ψ_2 that represent physical states of the particle.
- In this chapter, we shall consider solution of Schrödinger equation for the hydrogen atom (motion of a particle in columbic force field).

2.3 SCHRÖDINGER EQUATION OF THE HYDROGEN ATOM

What are the appropriate considerations for the Schrödinger equation of the hydrogen atom?

- The translation motion of the atom as a whole is neglected.
- The problem of the hydrogen atom is reduced to that of single electron of mass m_e in the columbic field.
- The motion of the nucleus can be considered by using the reduced mass μ instead of m_e, where $\mu = \frac{m_e m_p}{m_e + m_p}$ and m_p is the mass of the proton.
- The problem now is a particle in spherical symmetry of gradual rise in the potential with distance from the nucleus: at $r = \infty$, $V = 0$; at $r = 0$, $V = -\infty$.
- The potential energy in the field of the nucleus of charge Ze is given by

$$V = -\frac{Ze^2}{r}$$

Why are the polar coordinates invoked to solve Schrödinger equation?

- The Schrödinger equation is a complicated differential equation, and cannot be solved analytically unless some simplifying features can be exploited.

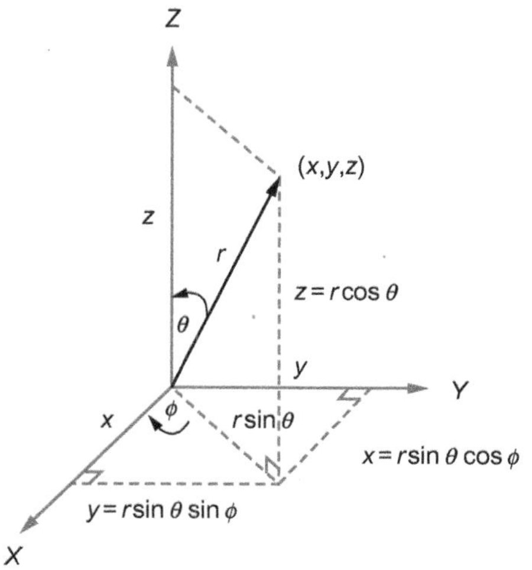

FIG. 2.1 Spherical polar coordinates.

- The spherical symmetry of the potential-energy field implies that the appropriate coordinate system for such a problem is not the Cartesian x, y, and z, but the spherical polar coordinator r, θ, and ϕ. The coordinate r measures the radial distance from the origin, θ is a colatitude, ϕ a longitude as shown in Fig. 2.1:

2.4 TRANSFORMATION OF THE SCHRÖDINGER EQUATION FROM CARTESIANS TO SPHERICAL POLAR COORDINATES

Show in detail how you can convert the Cartesian form of the Schrödinger equation:

$$\left(-\frac{\hbar^2}{2\mu}\right)\nabla^2\psi + V(x,y,z)\,\psi = E\psi$$

where

$$\nabla^2 = \frac{\partial^2}{\partial x^2} + \frac{\partial^2}{\partial y^2} + \frac{\partial^2}{\partial z^2}$$

into the corresponding spherical polar form.

$$\frac{-\hbar^2}{2\mu}\left[\frac{1}{r^2}\frac{\partial}{\partial r}\left(r^2\frac{\partial}{\partial r}\right) + \frac{1}{r^2\sin\theta}\frac{\partial}{\partial\theta}\left(\sin\theta\frac{\partial}{\partial\theta}\right) + \frac{1}{r^2\sin^2\theta}\frac{\partial^2}{\partial\phi^2}\right]\psi + V(r)\psi = \psi E$$

- The Laplacian operator, ∇^2, has its simplest form in Cartesian coordinates:

$$\nabla^2 = \frac{\partial^2}{\partial x^2} + \frac{\partial^2}{\partial y^2} + \frac{\partial^2}{\partial z^2}$$

or

$$\nabla^2 = \frac{\partial}{\partial x}\frac{\partial}{\partial x} + \frac{\partial}{\partial y}\frac{\partial}{\partial y} + \frac{\partial}{\partial z}\frac{\partial}{\partial z} \tag{2.4.1}$$

- Using the chain rule:

$$\frac{\partial}{\partial x} = \frac{\partial r}{\partial x}\frac{\partial}{\partial r} + \frac{\partial\theta}{\partial x}\frac{\partial}{\partial\theta} + \frac{\partial\phi}{\partial x}\frac{\partial}{\partial\phi} \tag{2.4.2}$$

$$\frac{\partial}{\partial y}=\frac{\partial r}{\partial y}\frac{\partial}{\partial r}+\frac{\partial \theta}{\partial y}\frac{\partial}{\partial \theta}+\frac{\partial \phi}{\partial y}\frac{\partial}{\partial \phi} \tag{2.4.3}$$

and

$$\frac{\partial}{\partial z}=\frac{\partial r}{\partial z}\frac{\partial}{\partial r}+\frac{\partial \theta}{\partial z}\frac{\partial}{\partial \theta}+\frac{\partial \phi}{\partial z}\frac{\partial}{\partial \phi} \tag{2.4.4}$$

- We must now determine the transformation derivatives:

$$\frac{\partial r}{\partial x}, \frac{\partial r}{\partial y}, \frac{\partial r}{\partial z} \tag{2.4.5}$$

$$\frac{\partial \theta}{\partial x}, \frac{\partial \theta}{\partial y}, \frac{\partial \theta}{\partial z} \tag{2.4.6}$$

and

$$\frac{\partial \phi}{\partial x}, \frac{\partial \phi}{\partial y}, \frac{\partial \phi}{\partial z} \tag{2.4.7}$$

- The correspondence values of x, y, and z in polar coordinates system (Fig. 2.1) are simply:

$$x = r \sin\theta \cos\phi$$

$$y = r \sin\theta \sin\phi$$

$$z = r \cos\theta$$

where $r=(x^2+y^2+z^2)^{1/2}$,

$$\cos\theta=\frac{z}{(x^2+y^2+z^2)^{1/2}} \quad \text{and} \quad \tan\phi=\frac{y}{x}$$

If,

$$r=(x^2+y^2+z^2)^{1/2},$$

and

$$\because \frac{\partial(u^n)}{\partial x}=nu^{n-1}\frac{\partial u}{\partial x},$$

then

$$\frac{\partial r}{\partial x}=\frac{1}{2}(x^2+y^2+z^2)^{-1/2}(2x)=\frac{x}{r}=\sin\theta\cos\phi \tag{2.4.8}$$

Likewise,

$$\frac{\partial r}{\partial y}=\frac{1}{2}(x^2+y^2+z^2)^{-1/2}(2y)=\frac{y}{r}=\sin\theta\sin\phi \tag{2.4.9}$$

$$\frac{\partial r}{\partial z}=\frac{1}{2}(x^2+y^2+z^2)^{-1/2}(2z)=\frac{z}{r}=\cos\theta \tag{2.4.10}$$

- A simple way to find $\frac{\partial \theta}{\partial x}$ is to differentiate $\cos\theta$ directly:

$$\frac{\partial \cos\theta}{\partial \theta}=-\sin\theta$$

$$\partial \cos\theta=-\sin\theta\,\partial\theta$$

$$\therefore \frac{\partial \cos\theta}{\partial x}=-\sin\theta\frac{\partial \theta}{\partial x}$$

$$\because \cos\theta = \frac{z}{r}$$

$$\therefore \frac{\partial \cos\theta}{\partial x} = \frac{\partial}{\partial x}\left(\frac{z}{r}\right) = \frac{\partial}{\partial x}\frac{z}{(x^2+y^2+z^2)^{1/2}},$$

using $\frac{\partial(u^n)}{\partial x} = nu^{n-1}\frac{\partial u}{\partial x}$, then

$$\frac{\partial \cos\theta}{\partial x} = -z\left(\frac{1}{2}\right)\left(x^2+y^2+z^2\right)^{-3/2}(2x) = -\frac{zx}{r^3}$$

$$\frac{\partial \cos\theta}{\partial x} = \cancel{-\sin\theta}\frac{\partial\theta}{\partial x} = -\frac{(r\cos\theta)(r\cancel{\sin\theta}\cos\phi)}{r^3}$$

$$\therefore \frac{\partial\theta}{\partial x} = \frac{\cos\theta\cos\phi}{r} \tag{2.4.11}$$

In the same way:

$$\frac{\partial \cos\theta}{\partial y} = \frac{\partial}{\partial y}\left(\frac{z}{r}\right) = \frac{\partial}{\partial y}\frac{z}{(x^2+y^2+z^2)^{1/2}}$$

$$\cancel{-\sin\theta}\frac{\partial\theta}{\partial y} = -z\left(\frac{1}{2}\right)(x^2+y^2+z^2)^{-3/2}(2y) = -\frac{zy}{r^3} = -\frac{(r\cos\theta)(r\cancel{\sin\theta}\sin\phi)}{r^3}$$

$$\therefore \frac{\partial\theta}{\partial y} = \frac{\sin\phi\cos\theta}{r} \tag{2.4.12}$$

Similarly:

$$-\sin\theta\frac{\partial\theta}{\partial z} = \frac{\partial}{\partial z}\frac{z}{(x^2+y^2+z^2)^{1/2}} = -z\left(\frac{1}{2}\right)\left(x^2+y^2+z^2\right)^{-3/2}(2z) + \left(x^2+y^2+z^2\right)^{-1/2}$$
$$= -\frac{z^2}{r^3} + \frac{1}{r} = \frac{r^2}{r^3} - \frac{z^2}{r^3}$$

$$\cancel{-\sin\theta}\frac{\partial\theta}{\partial z} = r^2\left(\frac{1-\cos^2\theta}{r^3}\right) = \frac{\sin^2\theta}{r} = \frac{\sin\theta\cancel{\sin\theta}}{r}$$

$$\therefore \frac{\partial\theta}{\partial z} = -\frac{\sin\theta}{r} \tag{2.4.13}$$

- To find $\frac{\partial\phi}{\partial x}$, we differentiate the $\tan\phi$ directly:

$$\tan\phi = \frac{y}{x}$$

$$\frac{\partial\tan\phi}{\partial\phi} = \sec^2\phi = \frac{1}{\cos^2\phi}$$

$$\frac{\partial\tan\phi}{\partial x} = \sec^2\phi\frac{\partial\phi}{\partial x} = \frac{1}{\cos^2\phi}\cdot\frac{\partial\phi}{\partial x} = \frac{\partial}{\partial x}\left(\frac{y}{x}\right) = -\frac{y}{x^2} = -\frac{r\sin\theta\sin\phi}{r^2\sin^2\theta\cos^2\phi}$$

$$\therefore \frac{\partial\phi}{\partial x} = -\frac{\sin\phi}{r\sin\theta} \tag{2.4.14}$$

Likewise,

$$\frac{\partial \tan\phi}{\partial y}=\frac{1}{\cos^2\phi}\cdot\frac{\partial\phi}{\partial y}=\frac{\partial}{\partial y}\left(\frac{y}{x}\right)=\frac{1}{x}=\frac{1}{r\sin\theta\cos\phi}$$

$$\therefore\frac{\partial\phi}{\partial y}=\frac{\cos\phi}{r\sin\theta} \tag{2.4.15}$$

Finally,

$$\therefore\frac{\partial\phi}{\partial z}=0 \tag{2.4.16}$$

- The transformation equations for the first derivatives are found using the chain rule:

$$\frac{\partial}{\partial x}=\frac{\partial r}{\partial x}\frac{\partial}{\partial r}+\frac{\partial\theta}{\partial x}\frac{\partial}{\partial\theta}+\frac{\partial\phi}{\partial x}\frac{\partial}{\partial\phi} \tag{2.4.2}$$

From Eqs. (2.4.8), (2.4.11), and (2.4.14):

$$\therefore\frac{\partial}{\partial x}=\sin\theta\cos\phi\frac{\partial}{\partial r}+\frac{\cos\theta\cos\phi}{r}\frac{\partial}{\partial\theta}-\frac{\sin\phi}{r\sin\theta}\frac{\partial}{\partial\phi} \tag{2.4.17}$$

$$\frac{\partial}{\partial y}=\frac{\partial r}{\partial y}\frac{\partial}{\partial r}+\frac{\partial\theta}{\partial y}\frac{\partial}{\partial\theta}+\frac{\partial\phi}{\partial y}\frac{\partial}{\partial\phi} \tag{2.4.3}$$

From Eqs. (2.4.9), (2.4.12), and (2.4.15):

$$\therefore\frac{\partial}{\partial y}=\sin\theta\sin\phi\frac{\partial}{\partial r}+\frac{\cos\theta\sin\phi}{r}\frac{\partial}{\partial\theta}+\frac{\cos\phi}{r\sin\theta}\frac{\partial}{\partial\phi} \tag{2.4.18}$$

$$\frac{\partial}{\partial z}=\frac{\partial r}{\partial z}\frac{\partial}{\partial r}+\frac{\partial\theta}{\partial z}\frac{\partial}{\partial\theta}+\frac{\partial\phi}{\partial z}\frac{\partial}{\partial\phi} \tag{2.4.4}$$

From Eqs. (2.4.10), (2.4.13), and (2.4.16):

$$\therefore\frac{\partial}{\partial z}=\cos\theta\frac{\partial}{\partial r}-\frac{\sin\theta}{r}\frac{\partial}{\partial\theta} \tag{2.4.19}$$

- To find the second derivatives of these operators (Eqs. (2.4.17), (2.4.18), and (2.4.19)):

$$\nabla^2=\frac{\partial}{\partial x}\frac{\partial}{\partial x}+\frac{\partial}{\partial y}\frac{\partial}{\partial y}+\frac{\partial}{\partial z}\frac{\partial}{\partial z} \tag{2.4.1}$$

We must operate each operator on itself. Remember that when $\sin\theta\cos\phi\frac{\partial}{\partial r}$ operates on $\frac{\cos\theta\cos\phi}{r}\frac{\partial}{\partial\theta}$:

i. Cos θ cos ϕ passes through the $\frac{\partial}{\partial r}$ operator; and

ii. the term $\frac{1}{r}\frac{\partial}{\partial\theta}$ must be differentiated as a product.

$$\frac{\partial}{\partial x}\frac{\partial}{\partial x}=\frac{\partial^2}{\partial x^2}=\sin\theta\cos\phi\frac{\partial}{\partial r}\left(\sin\theta\cos\phi\frac{\partial}{\partial r}+\frac{\cos\theta\cos\phi}{r}\frac{\partial}{\partial\theta}-\frac{\sin\phi}{r\sin\theta}\frac{\partial}{\partial\phi}\right)$$

$$+\frac{\cos\theta\cos\phi}{r}\frac{\partial}{\partial\theta}\left(\sin\theta\cos\phi\frac{\partial}{\partial r}+\frac{\cos\theta\cos\phi}{r}\frac{\partial}{\partial\theta}-\frac{\sin\phi}{r\sin\theta}\frac{\partial}{\partial\phi}\right)$$

$$-\frac{\sin\phi}{r\sin\theta}\frac{\partial}{\partial\phi}\left(\sin\theta\cos\phi\frac{\partial}{\partial r}+\frac{\cos\theta\cos\phi}{r}\frac{\partial}{\partial\theta}-\frac{\sin\phi}{r\sin\theta}\frac{\partial}{\partial\phi}\right)$$

$$\frac{\partial}{\partial x}\frac{\partial}{\partial x}=\frac{\partial^2}{\partial x^2}=\sin^2\theta\cos^2\phi\frac{\partial^2}{\partial r^2}+\sin\theta\cos\phi\left(\frac{\partial}{\partial r}\sin\theta\cos\phi\right)\frac{\partial}{\partial r}$$
$$+\frac{(\sin\theta\cos\phi)(\cos\theta\cos\phi)}{r}\frac{\partial^2}{\partial r\,\partial\theta}+\sin\theta\cos\phi\left(\frac{\partial}{\partial r}\frac{\cos\theta\cos\phi}{r}\right)\frac{\partial}{\partial\theta}$$
$$-\frac{(\sin\theta\cos\phi)(\sin\phi)}{r\sin\theta}\frac{\partial^2}{\partial r\,\partial\phi}-\sin\theta\cos\phi\left(\frac{\partial}{\partial r}\frac{\sin\phi}{r\sin\theta}\right)\frac{\partial}{\partial\phi}+\cdots+\cdots$$

$$\begin{aligned}\therefore\frac{\partial^2}{\partial x^2}=&\sin^2\theta\cos^2\phi\frac{\partial^2}{\partial r^2}+\frac{\sin\theta\cos\theta\cos^2\phi}{r}\frac{\partial^2}{\partial r\,\partial\theta}-\frac{\sin\theta\cos\theta\cos^2\phi}{r^2}\frac{\partial}{\partial\theta}\\&-\frac{\cos\phi\sin\phi}{r}\frac{\partial^2}{\partial r\,\partial\phi}+\frac{\cos\phi\sin\phi}{r^2}\frac{\partial}{\partial\phi}+\frac{\sin\theta\cos\theta\cos^2\phi}{r}\frac{\partial^2}{\partial r\,\partial\theta}\\&-\frac{\sin\theta\cos\theta\cos^2\phi}{r^2}\frac{\partial}{\partial\theta}+\frac{\cos^2\theta\cos^2\phi}{r^2}\frac{\partial^2}{\partial\theta^2}+\frac{\cos^2\theta\cos^2\phi}{r}\frac{\partial}{\partial r}\\&-\frac{\cos\theta\cos\phi\sin\phi}{r^2\sin\theta}\frac{\partial^2}{\partial\phi\,\partial\theta}-\frac{\cos\phi\sin\phi}{r}\frac{\partial^2}{\partial r\,\partial\phi}+\frac{\sin\phi\cos\phi\cos^2\theta}{r^2\sin^2\theta}\frac{\partial}{\partial\phi}\\&+\frac{\sin^2\phi}{r}\frac{\partial}{\partial r}-\frac{\sin\phi\cos\phi\cos\theta}{r^2\sin\theta}\frac{\partial^2}{\partial\phi\,\partial r}+\frac{\sin^2\phi\cos\theta}{r^2\sin^2\theta}\frac{\partial}{\partial\theta}\\&+\frac{\sin^2\phi}{r^2\sin^2\theta}\frac{\partial^2}{\partial\phi^2}+\frac{\sin\phi\cos\phi}{r^2\sin^2\theta}\frac{\partial}{\partial\phi}\end{aligned}\tag{2.4.20}$$

$$\because\frac{\partial^2}{\partial y^2}=\frac{\partial}{\partial y}\frac{\partial}{\partial y}=\sin\theta\sin\phi\frac{\partial}{\partial r}\left(\sin\theta\sin\phi\frac{\partial}{\partial r}+\frac{\cos\theta\sin\phi}{r}\frac{\partial}{\partial\theta}+\frac{\cos\phi}{r\sin\theta}\frac{\partial}{\partial\phi}\right)$$
$$+\frac{\cos\theta\sin\phi}{r}\frac{\partial}{\partial\theta}\left(\sin\theta\sin\phi\frac{\partial}{\partial r}+\frac{\cos\theta\sin\phi}{r}\frac{\partial}{\partial\theta}+\frac{\cos\phi}{r\sin\theta}\frac{\partial}{\partial\phi}\right)$$
$$+\frac{\cos\phi}{r\sin\theta}\frac{\partial}{\partial\phi}\left(\sin\theta\sin\phi\frac{\partial}{\partial r}+\frac{\cos\theta\sin\phi}{r}\frac{\partial}{\partial\theta}+\frac{\cos\phi}{r\sin\theta}\frac{\partial}{\partial\phi}\right)$$

$$\begin{aligned}\therefore\frac{\partial^2}{\partial y^2}=\frac{\partial}{\partial y}\frac{\partial}{\partial y}=&\sin^2\theta\sin^2\phi\frac{\partial^2}{\partial r^2}+\frac{\sin\theta\cos\theta\sin^2\phi}{r}\frac{\partial^2}{\partial r\,\partial\theta}-\frac{\sin\theta\cos\theta\sin^2\phi}{r^2}\frac{\partial}{\partial\theta}\\&+\frac{\sin\phi\cos\phi}{r}\frac{\partial^2}{\partial r\,\partial\phi}-\frac{\sin\phi\cos\phi}{r^2}\frac{\partial}{\partial\phi}+\frac{\cos\theta\sin\theta\sin^2\phi}{r}\frac{\partial^2}{\partial r\,\partial\theta}\\&-\frac{\sin\theta\cos\theta\sin^2\phi}{r^2}\frac{\partial}{\partial\theta}+\frac{\cos^2\theta\sin^2\phi}{r^2}\frac{\partial^2}{\partial\theta^2}+\frac{\cos^2\theta\sin^2\phi}{r}\frac{\partial}{\partial r}\\&+\frac{\sin\theta\cos\phi\sin\phi}{r^2\sin\theta}\frac{\partial^2}{\partial\phi\,\partial\theta}+\frac{\sin\phi\cos\phi}{r}\frac{\partial^2}{\partial r\,\partial\phi}-\frac{\sin\phi\cos\phi\cos^2\theta}{r^2\sin^2\theta}\frac{\partial}{\partial\phi}\\&+\frac{\cos^2\phi}{r}\frac{\partial}{\partial r}+\frac{\sin\phi\cos\phi\cos\theta}{r^2\sin\theta}\frac{\partial^2}{\partial\phi\,\partial r}+\frac{\cos^2\phi\cos\theta}{r^2\sin\theta}\frac{\partial}{\partial\theta}\\&+\frac{\cos^2\phi}{r^2\sin^2\theta}\frac{\partial^2}{\partial\phi^2}-\frac{\sin\phi\cos\phi}{r^2\sin^2\theta}\frac{\partial}{\partial\phi}\end{aligned}\tag{2.4.21}$$

$$\because \frac{\partial^2}{\partial z^2} = \frac{\partial}{\partial z}\frac{\partial}{\partial z} = \cos\theta \frac{\partial}{\partial r}\left(\cos\theta \frac{\partial}{\partial r} - \frac{\sin\theta}{r}\frac{\partial}{\partial \theta}\right) - \frac{\sin\theta}{r}\frac{\partial}{\partial \theta}\left(\cos\theta \frac{\partial}{\partial r} - \frac{\sin\theta}{r}\frac{\partial}{\partial \theta}\right)$$

$$\therefore \frac{\partial^2}{\partial z^2} = \frac{\partial}{\partial z}\frac{\partial}{\partial z} = \cos^2\theta \frac{\partial^2}{\partial r^2} - \frac{\sin\theta\cos\theta}{r}\frac{\partial^2}{\partial r\,\partial\theta} + \frac{\sin\theta\cos\theta}{r^2}\frac{\partial}{\partial\theta} - \frac{\sin\theta\cos\theta}{r}\frac{\partial^2}{\partial r\,\partial\theta} + \frac{\sin^2\theta}{r}\frac{\partial}{\partial r} + \frac{\sin^2\theta}{r^2}\frac{\partial^2}{\partial\theta^2} + \frac{\sin\theta\cos\theta}{r^2}\frac{\partial}{\partial\theta} \quad (2.4.22)$$

- Adding the second derivatives (Eqs. 2.4.20–2.4.22):

$$\frac{\partial^2}{\partial x^2} + \frac{\partial^2}{\partial y^2} + \frac{\partial^2}{\partial z^2} = \left(\sin^2\theta\cos^2\phi + \sin^2\theta\sin^2\phi + \cos^2\theta\right)\frac{\partial^2}{\partial r^2}$$
$$+ \frac{\left(\cos^2\theta\cos^2\phi + \sin^2\phi + \cos^2\theta\sin^2\phi + \cos^2\phi + \sin^2\theta\right)}{r}\frac{\partial}{\partial r} + (\cdots\cdots)$$
$$\frac{1}{r^2}\frac{\partial^2}{\partial\theta^2} + (\cdots\cdots)\frac{\cos\theta}{r^2\sin\theta}\frac{\partial}{\partial\theta} + (\cdots\cdots)\frac{1}{r^2\sin^2\theta}\frac{\partial^2}{\partial\phi^2}$$

$$\therefore \nabla^2 = \frac{\partial^2}{\partial x^2} + \frac{\partial^2}{\partial y^2} + \frac{\partial^2}{\partial z^2} = \frac{\partial^2}{\partial r^2} + \frac{2}{r}\frac{\partial}{\partial r} + \frac{1}{r^2}\frac{\partial^2}{\partial\theta^2} + \frac{\cos\theta}{r^2\sin\theta}\frac{\partial}{\partial\theta} + \frac{1}{r^2\sin^2\theta}\frac{\partial^2}{\partial\phi^2}$$

or as it is usually expressed,

$$\nabla^2 = \frac{1}{r^2}\frac{\partial}{\partial r}\left(r^2\frac{\partial}{\partial r}\right) + \frac{1}{r^2\sin\theta}\frac{\partial}{\partial\theta}\left(\sin\theta\frac{\partial}{\partial\theta}\right) + \frac{1}{r^2\sin^2\theta}\frac{\partial^2}{\partial\phi^2} \quad (2.4.23)$$

- The time-independent wave equation can be written in the spherical coordinate system as:

$$\left(-\frac{\hbar^2}{2\mu}\right)\nabla^2\psi + V(x, y, z)\,\psi = E\psi$$

where $\psi = \psi(x, y, z) = \psi(r, \theta, \phi)$

$$\nabla^2 = \left(\frac{\partial^2}{\partial x^2}\right) + \left(\frac{\partial^2}{\partial y^2}\right) + \left(\frac{\partial^2}{\partial z^2}\right)$$

The equation can also be

$$\nabla^2\psi + \left(\frac{2\mu}{\hbar^2}\right)[E - V(r)]\,\psi = 0$$

In polar coordinate:

$$\because \nabla^2 = \frac{1}{r^2}\frac{\partial}{\partial r}\left(r^2\frac{\partial}{\partial r}\right) + \frac{1}{r^2\sin\theta}\frac{\partial}{\partial\theta}\left(\sin\theta\frac{\partial}{\partial\theta}\right) + \frac{1}{r^2\sin^2\theta}\frac{\partial^2}{\partial\phi^2} \quad (2.4.24)$$

Then

$$\therefore \frac{-\hbar^2}{2\mu}\left[\frac{1}{r^2}\frac{\partial}{\partial r}\left(r^2\frac{\partial}{\partial r}\right) + \frac{1}{r^2\sin\theta}\frac{\partial}{\partial\theta}\left(\sin\theta\frac{\partial}{\partial\theta}\right) + \frac{1}{r^2\sin^2\theta}\frac{\partial^2}{\partial\phi^2}\right]\psi + V(r)\psi = \psi E \quad (2.4.25)$$

Factor the spherical wave function:

$$\boldsymbol{\frac{-\hbar^2}{2\mu}\left[\frac{1}{r^2}\frac{\partial}{\partial r}\left(r^2\frac{\partial}{\partial r}\right) + \frac{1}{r^2\sin\theta}\frac{\partial}{\partial\theta}\left(\sin\theta\frac{\partial}{\partial\theta}\right) + \frac{1}{r^2\sin^2\theta}\frac{\partial^2}{\partial\phi^2}\right]\psi + V(r)\psi = \psi E} \quad (2.4.25)$$

into separate functions of radial, $R(r)$, and angular, $Y(\Phi,\Theta)$.

- The variables in this equation can be separated because the potential is a function of r alone:

$$\text{If}: \psi(r,\theta,\phi) = R(r)\,Y(\theta,\phi) = RY$$

Substitute Eq. (2.4.25):

$$\left[\frac{\partial}{\partial r}\left(r^2\frac{\partial}{\partial r}\right) + \frac{1}{\sin\theta}\frac{\partial}{\partial\theta}\left(\sin\theta\frac{\partial}{\partial\theta}\right) + \frac{1}{\sin^2\theta}\frac{\partial^2}{\partial\phi^2}\right]RY = -\frac{2\mu r^2}{\hbar^2}(E - V(r))RY \tag{2.4.25}$$

$$Y\frac{\partial}{\partial r}\left(r^2\frac{\partial R}{\partial r}\right) + R\left[\frac{1}{\sin\theta}\frac{\partial}{\partial\theta}\left(\sin\theta\frac{\partial Y}{\partial\theta}\right) + \frac{1}{\sin^2\theta}\frac{\partial^2 Y}{\partial\phi^2}\right] = -\frac{2\mu r^2}{\hbar^2}(E - V(r))RY$$

Then divide by RY, then

$$\frac{1}{R}\frac{\partial}{\partial r}\left(r^2\frac{\partial R}{\partial r}\right) + \frac{2\mu r^2}{\hbar^2}(E - V(r)) = -\frac{1}{Y}\left[\frac{1}{\sin\theta}\frac{\partial}{\partial\theta}\left(\sin\theta\frac{\partial Y}{\partial\theta}\right) + \frac{1}{\sin^2\theta}\frac{\partial^2 Y}{\partial\phi^2}\right]$$

- The left-hand side of the equation is only a function of r, radial equation.
- The right-hand side of the equation is only a function of θ and ϕ, angular equation.
- Since the two sides are equal and each is dependent upon independent coordinate, each side has to be equal to a constant (λ).
  - Then the angular part of the Schrödinger equation:

$$\frac{1}{Y}\left[\frac{1}{\sin\theta}\frac{\partial}{\partial\theta}\left(\sin\theta\frac{\partial Y}{\partial\theta}\right) + \frac{1}{\sin^2\theta}\frac{\partial^2 Y}{\partial\phi^2}\right] = \lambda$$

$$\therefore \frac{1}{\sin\theta}\frac{\partial}{\partial\theta}\left(\sin\theta\frac{\partial Y}{\partial\theta}\right) + \frac{1}{\sin^2\theta}\frac{\partial^2 Y}{\partial\phi^2} + \lambda Y = 0 \tag{2.4.26}$$

  - And the radial part of the Schrödinger equation:

$$\frac{1}{R}\frac{\partial}{\partial r}\left(r^2\frac{\partial R}{\partial r}\right) + \frac{2\mu r^2}{\hbar^2}(E - V(r)) = \lambda$$

$$\therefore \frac{1}{r^2}\frac{\partial}{\partial r}\left(r^2\frac{\partial R}{\partial r}\right) + \left\{\frac{2\mu}{\hbar^2}(E - V(r)) - \frac{\lambda}{r^2}\right\}R = 0 \tag{2.4.27}$$

2.5 THE ANGULAR EQUATION

Factor the angular wave function:

$$\mathbf{\frac{1}{\sin\theta}\frac{\partial}{\partial\theta}\left(\sin\theta\frac{\partial Y}{\partial\theta}\right) + \frac{1}{\sin^2\theta}\frac{\partial^2 Y}{\partial\Phi^2} + \lambda Y = 0} \tag{2.4.26}$$

into separate functions $\Phi(\phi)$ and $\Theta(\theta)$.

- The angular equation can be separated further into Θ and Φ equations using the following substitution:

$$Y(\theta,\phi) = \Theta(\theta)\Phi(\phi) = \Theta\Phi$$

- Making this substitution in

$$\frac{1}{\sin\theta}\frac{\partial}{\partial\theta}\left(\sin\theta\frac{\partial Y}{\partial\theta}\right) + \frac{1}{\sin^2\theta}\frac{\partial^2 Y}{\partial\phi^2} + \lambda Y = 0 \tag{2.4.26}$$

$$\frac{\Phi}{\sin\theta}\frac{\partial}{\partial\theta}\left(\sin\theta\frac{\partial\Theta}{\partial\theta}\right) + \frac{\Theta}{\sin^2\theta}\frac{\partial^2\Phi}{\partial\phi^2} + \lambda\Theta\Phi = 0$$

$$\frac{\Phi}{\sin\theta}\frac{\partial}{\partial\theta}\left(\sin\theta\frac{\partial\Theta}{\partial\theta}\right) + \lambda\Phi\Theta = -\frac{\Theta}{\sin^2\theta}\frac{\partial^2\Phi}{\partial\phi^2}$$

multiplies by $\dfrac{\sin^2\theta}{\Theta\Phi}$

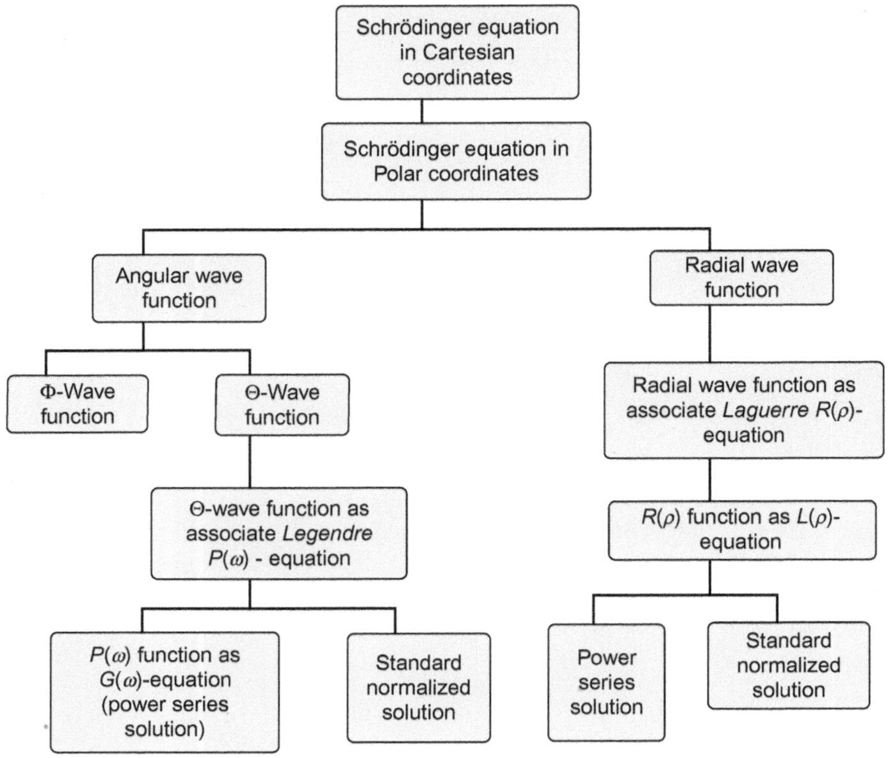

SCHEME 2.2 A schematic description to obtain the final solution for the full wave function.

$$\frac{1}{\Theta}\sin\theta\frac{\partial}{\partial\theta}\left(\sin\theta\frac{\partial\Theta}{\partial\theta}\right)+\lambda\sin^2\theta=-\frac{1}{\Phi}\frac{\partial^2\Phi}{\partial\phi^2}$$

- Using the argument used before, and after using the constant υ:

$$\frac{1}{\Theta}\sin\theta\frac{\partial}{\partial\theta}\left(\sin\theta\frac{\partial\Theta}{\partial\theta}\right)+\lambda\sin^2\theta=-\frac{1}{\Phi}\frac{\partial^2\Phi}{\partial\phi^2}=\upsilon$$

$$\therefore\Phi-\text{equation}:\frac{\partial^2\Phi}{\partial\phi^2}+\Phi\upsilon=0 \tag{2.5.1}$$

and

$$\frac{1}{\Theta}\sin\theta\frac{\partial}{\partial\theta}\left(\sin\theta\frac{\partial\Theta}{\partial\theta}\right)+\lambda\sin^2\theta=\upsilon$$

multiplies by $\frac{\Theta}{\sin^2\theta}$

$$\therefore\Theta-\text{equation}:\frac{1}{\sin\theta}\frac{\partial}{\partial\theta}\left(\sin\theta\frac{\partial\Theta}{\partial\theta}\right)+\left(\lambda-\frac{\upsilon}{\sin^2\theta}\right)\Theta=0 \tag{2.5.2}$$

- Scheme 2.2 describes the steps to reach the final solution for the full wave function.

2.6 THE Φ-EQUATION

If the Φ-equation:

$$\frac{\partial^2\Phi}{\partial\phi^2}+\Phi\upsilon=0 \tag{2.5.1}$$

where υ is a constant, prove the following:

a. **υ is equal to a square of integral donated m ($\upsilon = m^2$)**

b. $\Phi = \frac{1}{\sqrt{2\pi}} e^{im\phi}$

c. **$m = 0, \pm 1, \pm 2, \pm 3$ ......; and show the physical meaning of m**

d. $m = \pm \frac{(2IE)^{\frac{1}{2}}}{h}$, **where I is the moment of inertia.**

- The Φ-equation has a simple solution:
  If: $y = ae^{bx}$, then

$$\frac{dy}{dx} = abe^{bx} = by \text{ and } \frac{d^2y}{dx^2} = b^2 y$$

In the same respect:

$$\text{If} : \Phi = e^{am\phi}$$

$$\frac{\partial \Phi}{\partial \phi} = ame^{am\phi} = am\Phi$$

$$\frac{\partial^2 \Phi}{\partial \phi^2} = a^2 m^2 e^{am\phi} = a^2 m^2 \Phi$$

$$\because \frac{\partial^2 \Phi}{\partial \phi^2} = -\upsilon \Phi \tag{2.5.1}$$

$$\therefore \frac{\partial^2 \Phi}{\partial \phi} = -\upsilon \Phi = a^2 m^2 \Phi$$

$$\therefore -\upsilon = a^2 m^2$$

This required that υ be square of an integral donated m:

$$-\upsilon = (am)^2 = (im)^2, \quad a = i,$$

then

$$\upsilon = m^2 \tag{2.6.1}$$

Then

$$\Phi_{\text{unnormalized}} = e^{im\phi}$$

to normalize the Φ function:

$$N^2 \int_0^{2\pi} \Phi^* \Phi \partial \phi = 1$$

$$N^2 \int_0^{2\pi} \left(e^{-im\phi}\right)\left(e^{im\phi}\right) \partial \phi = 1$$

$$N^2 \int_0^{2\pi} \partial \phi = 1$$

$$N^2 (2\pi) = 1, \quad N = \frac{1}{\sqrt{2\pi}}$$

The general normalized solution to Φ is

$$\Phi = \frac{1}{\sqrt{2\pi}} e^{im\phi} \tag{2.6.2}$$

- The wave function must satisfy the cyclic boundary condition:

$$\Phi_m(\phi) = \Phi_m(\phi + 2\pi)$$

The requirement that Φ and $\partial\Phi/\partial\phi$ be continuous throughout the range of Φ from 0 to 2π:

$$\Phi_m(\phi + 2\pi) = \frac{1}{\sqrt{2\pi}} e^{im(\phi + 2\pi)} = \frac{1}{\sqrt{2\pi}} e^{im\phi} e^{2\pi im}$$
$$= e^{2\pi im}\Phi_m(\phi) = (-1)^{2m}\Phi_m(\phi) = \Phi_m(\phi)$$

where $e^{\pi i} = -1$, and the value of 2 m is an even number; therefore, in order to satisfy the cyclic boundary condition:

$$\therefore m = 0, \pm 1, \pm 2, \pm 3 \tag{2.6.3}$$

- The relation between the angular momentum about z axis and the m quantum number:

$$\because \Phi_m(\phi) = \Phi_m(\phi + 2\pi), \text{ and } \Phi = \frac{1}{\sqrt{2\pi}} e^{im\phi}$$

- This implies a standing wave about the z-axis that follows:

$$[\cos(m\phi) \pm i \sin(m\phi)]$$

- Such a wave repeats itself in $\frac{2\pi r}{m}$ radians.
- Therefore, the wavelength is (Fig. 2.2)

$$\therefore \lambda = \frac{2\pi r}{m}, \tag{2.6.4}$$

$$\frac{1}{\lambda} = \frac{m}{2\pi r}$$

- From the de Broglie relationship:

$$p = \frac{h}{\lambda} \tag{1.12.1}$$

The angular momentum around the z-axis is:

$$\therefore p_\phi = \frac{h}{\lambda} r = \frac{hmr}{2\pi r} = m\hbar \tag{2.6.5}$$

$$(m = 0, \pm 1, \pm 2, \pm 3.....) \tag{2.6.6}$$

- Thus, the angular momentum about the z-axis is given by m, in units of $\hbar$.
- A positive m value implies a clockwise motion for the electron, while a negative m value implies a counterclockwise motion.
- The total energy is equal to the kinetic energy, because the potential energy is zero everywhere.

$$\therefore E = \frac{p^2}{2m}$$

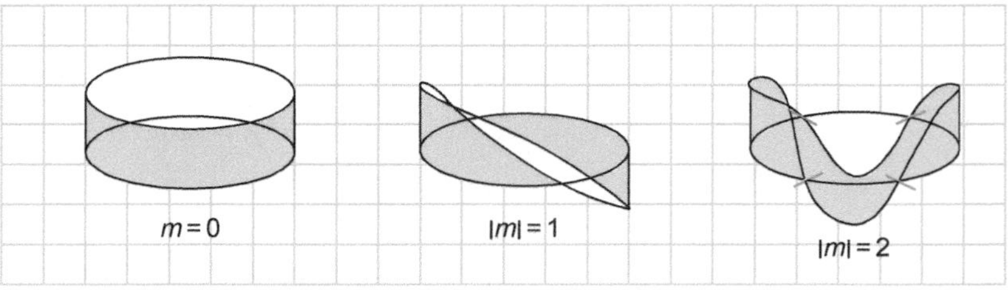

FIG. 2.2 As shorter wavelengths are reached, the angular momentum increases in steps of $\hbar$.

And if the moment of inertia, $I = mr^2$, then

$$E = \frac{p^2\,r^2}{2mr^2} = \frac{p_\phi{}^2}{2I} = m^2\frac{\hbar^2}{2I}$$

$$\therefore m = \pm\frac{(2IE)^{\frac{1}{2}}}{\hbar} \tag{2.6.7}$$

2.7 THE Θ-EQUATION

The resulting solution of the Θ-equation:

$$\frac{1}{\sin\theta}\frac{\partial}{\partial\theta}\left(\sin\theta\frac{\partial\Theta}{\partial\theta}\right)+\left(\lambda-\frac{\upsilon}{\sin^2\theta}\right)\Theta=0 \tag{2.5.2}$$

is based on the well-known associate Legendre polynomials equation. Substitute $\Theta(\theta)$ by the polynomial $P(\omega)$, where $\omega = \cos\theta$, and $\upsilon = m^2$, to generate the *Legendre* ω-polynomial wave function:

$$\frac{\partial}{\partial\omega}\left((1-\omega^2)\frac{\partial P}{\partial\omega}\right)+\left(\lambda-\frac{m^2}{1-\omega^2}\right)P=0$$

- The Θ-equation:

$$\frac{1}{\sin\theta}\frac{\partial}{\partial\theta}\left(\sin\theta\frac{\partial\Theta}{\partial\theta}\right)+\left(\lambda-\frac{\upsilon}{\sin^2\theta}\right)\Theta=0 \tag{2.5.2}$$

$$\because\frac{\partial}{\partial\theta}\left(\sin\theta\frac{\partial\Theta}{\partial\theta}\right)=\sin\theta\frac{\partial^2\Theta}{\partial\theta^2}+\left(\frac{\cos\theta\,\partial\Theta}{\sin\theta\;\partial\theta}\right)$$

$$\therefore\frac{\partial^2\Theta}{\partial\theta^2}+\left(\frac{\cos\theta\,\partial\Theta}{\sin\theta\;\partial\theta}\right)+\left(\lambda-\frac{\upsilon}{\sin^2\theta}\right)\Theta=0$$

or

$$\frac{\partial^2\Theta}{\partial\theta^2}+\left(\cot\theta\frac{\partial\Theta}{\partial\theta}\right)+\left(\lambda-\frac{\upsilon}{\sin^2\theta}\right)\Theta=0$$

- We substitute $\Theta(\theta)$ by the polynomial $P(\omega)$, $\cos\theta = \omega$, and $\upsilon = m^2$:

$$\Theta(\theta)=P(\omega)=P$$

$$\sin^2\theta=1-\cos^2\theta=1-\omega^2$$

$$\frac{\partial^2 P}{\partial\theta^2}+\left(\frac{\cos\theta\,\partial P}{\sin\theta\;\partial\theta}\right)+\left(\lambda-\frac{m^2}{1-\omega^2}\right)P=0 \tag{2.7.1}$$

$$\frac{\partial\omega}{\partial\theta}=\frac{\partial\cos\theta}{\partial\theta}=-\sin\theta$$

$$\frac{\partial\Theta}{\partial\theta}=\frac{\partial P}{\partial\theta}=\frac{\partial P}{\partial\omega}\frac{\partial\omega}{\partial\theta}=-\sin\theta\frac{\partial P}{\partial\omega}=-(1-\omega^2)^{\frac{1}{2}}\frac{\partial P}{\partial\omega}$$

$$\begin{aligned}\frac{\partial^2 P}{\partial\theta^2}&=(1-\omega^2)^{\frac{1}{2}}\frac{\partial P}{\partial\omega}(1-\omega^2)^{\frac{1}{2}}\frac{\partial P}{\partial\omega}\\&=(1-\omega^2)\frac{\partial^2 P}{\partial\omega^2}+(1-\omega^2)^{\frac{1}{2}}\left(\frac{1}{2}\right)(1-\omega^2)^{-\frac{1}{2}}(-2\omega)\frac{\partial P}{\partial\omega}\\&=(1-\omega^2)\frac{\partial^2 P}{\partial\omega^2}-\omega\frac{\partial P}{\partial\omega}\end{aligned}$$

- Then

$$\frac{\partial^2 P}{\partial \theta^2} + \left(\frac{\cos\theta}{\sin\theta}\frac{\partial P}{\partial \theta}\right) + \left(\lambda - \frac{m^2}{1-\omega^2}\right)P = 0 \tag{2.7.1}$$

becomes

$$(1-\omega^2)\frac{\partial^2 P}{\partial \omega^2} - \omega\frac{\partial P}{\partial \omega} - \frac{\omega}{(1-\omega^2)^{\frac{1}{2}}}\left((1-\omega^2)^{\frac{1}{2}}\frac{\partial P}{\partial \omega}\right) + \left(\lambda - \frac{m^2}{1-\omega^2}\right)P = 0$$

$$(1-\omega^2)\frac{\partial^2 P}{\partial \omega^2} - 2\omega\frac{\partial P}{\partial \omega} + \left(\lambda - \frac{m^2}{1-\omega^2}\right)P = 0 \tag{2.7.2}$$

or

$$\frac{\partial}{\partial \omega}\left((1-\omega^2)\frac{\partial P}{\partial \omega}\right) + \left(\lambda - \frac{m^2}{1-\omega^2}\right)P = 0 \tag{2.7.3}$$

Write the proposed solution of the associated Legendre ω-polynomial function of the Θ-equation:

$$\frac{\partial}{\partial \omega}\left((1-\omega^2)\frac{\partial P}{\partial \omega}\right) + \left(\lambda - \frac{m^2}{1-\omega^2}\right)P = 0 \tag{2.7.3}$$

What are the requirements in order that the solution does not go to infinity at $\omega = \pm 1$?

- This is a second-order differential equation with well-known solutions, the Legendre polynomial function. The solutions go to infinity as $\omega \rightarrow = 1$, whenever $m_l \neq 0$. The equation has physical solutions that do not go to infinity only when $\lambda = l(l+1)$.
- To obtain a two-term recursion relation when we try a power-series solution, we make the following change of dependent variable:

$$P(\omega) = (1-\omega^2)^{\frac{|m|}{2}} G(\omega) \tag{2.7.4}$$

where

$$G(\omega) = \sum_j a_j \omega^j \tag{2.7.5}$$

If the general solution of the Legendre ω-polynomial function of the Θ-equation is

$$\frac{\partial}{\partial \omega}\left((1-\omega^2)\frac{\partial P}{\partial \omega}\right) + \left(\lambda - \frac{m^2}{1-\omega^2}\right)P = 0 \tag{2.7.3}$$

is

$$P = P(\omega) = (1-\omega^2)^{\frac{|m|}{2}} G(\omega) \tag{2.7.4}$$

where

$$G(\omega) = \sum_j a_j \omega^j \tag{2.7.5}$$

Show that the Legendre G-polynomial function of the Θ-equation is

$$(1-\omega^2)G''(\omega) - 2\omega G'(\omega)[|m|+1] + G(\omega)(\lambda - |m|(|m|+1)) = 0$$

- The Legendre ω-polynomial function:

$$\frac{\partial}{\partial \omega}\left((1-\omega^2)\frac{\partial P}{\partial \omega}\right) + \left(\lambda - \frac{m^2}{1-\omega^2}\right)P = 0 \tag{2.7.3}$$

$$\therefore\ (1-\omega^2)\frac{\partial^2 P}{\partial \omega^2}-2\omega\frac{\partial P}{\partial \omega}+\left(\lambda-\frac{m^2}{1-\omega^2}\right)P=0 \tag{2.7.5}$$

- $$\text{If}: P(\omega)=(1-\omega^2)^{\frac{|m|}{2}}G(\omega) \tag{2.7.4}$$

Then

$$P'(\omega)=(1-\omega^2)^{\frac{|m|}{2}}G'(\omega)+\frac{|m|}{2}(1-\omega^2)^{\frac{|m|}{2}-1}(-2\omega)\,G(\omega)$$

$$\therefore\ P'(\omega)=(1-\omega^2)^{\frac{|m|}{2}}G'(\omega)-\omega|m|(1-\omega^2)^{\frac{|m|}{2}-1}G(\omega)$$

$$\therefore\ -2\omega P'(\omega)=-2\omega(1-\omega^2)^{\frac{|m|}{2}}G'(\omega)+2\omega^2|m|(1-\omega^2)^{\frac{|m|}{2}-1}G(\omega) \tag{2.7.5}$$

$$\because P''(\omega)=(1-\omega^2)^{\frac{|m|}{2}}G''(\omega)-\omega|m|(1-\omega^2)^{\frac{|m|}{2}-1}G'(\omega)$$
$$-\omega|m|(1-\omega^2)^{\frac{|m|}{2}-1}G'(\omega)-|m|(1-\omega^2)^{\frac{|m|}{2}-1}G(\omega)$$
$$+2\omega^2|m|\left(\frac{|m|}{2}-1\right)(1-\omega^2)^{\frac{|m|}{2}-2}G(\omega)$$

$$\therefore (1-\omega^2)P''(\omega)=(1-\omega^2)(1-\omega^2)^{\frac{|m|}{2}}G''(\omega)-\omega|m|(1-\omega^2)^{\frac{|m|}{2}}G'(\omega)$$
$$-\omega|m|(1-\omega^2)^{\frac{|m|}{2}}G'(\omega)-|m|(1-\omega^2)^{\frac{|m|}{2}}G(\omega) \tag{2.7.6}$$
$$+2\omega^2|m|\left(\frac{|m|}{2}-1\right)(1-\omega^2)^{\frac{|m|}{2}-1}G(\omega)$$

- By substituting Eqs. (2.7.5) and (2.7.6) in

$$(1-\omega^2)\frac{\partial^2 P}{\partial \omega^2}-2\omega\frac{\partial P}{\partial \omega}+\left(\lambda-\frac{m^2}{1-\omega^2}\right)P=0 \tag{2.7.5}$$

$$(1-\omega^2)(1-\omega^2)^{\frac{|m|}{2}}G''(\omega)-2\omega\,(1-\omega^2)^{\frac{|m|}{2}}G'(\omega)[|m|+1]$$
$$+(1-\omega^2)^{\frac{|m|}{2}}G(\omega)\left\{\lambda-\frac{m^2}{1-\omega^2}+\frac{2\omega^2|m|}{1-\omega^2}-|m|+\frac{2\omega^2|m|\left(\frac{|m|}{2}-1\right)}{1-\omega^2}\right\}=0$$

divided by $(1-\omega^2)^{\frac{|m|}{2}}$

$$(1-\omega^2)G''(\omega)-2\omega G'(\omega)[|m|+1]$$
$$+G(\omega)\left(\lambda-|m|\left(\frac{|m|}{1-\omega^2}-\frac{2\omega^2}{1-\omega^2}+1-\frac{\omega^2|m|-2\omega^2}{1-\omega^2}\right)\right)=0$$
$$(1-\omega^2)G''(\omega)-2\omega G'(\omega)[|m|+1]$$
$$+G(\omega)\left(\lambda-|m|\left(\frac{|m|-2\omega^2+1-\omega^2-\omega^2|m|+2\omega^2}{1-\omega^2}\right)\right)=0$$

Then

$$(1-\omega^2)G''(\omega)-2\omega G'(\omega)[|m|+1]+G(\omega)(\lambda-|m|(|m|+1))=0 \tag{2.7.7}$$

In the Legendre G-polynomial function of the Θ-equation

$$(1-\omega^2)G''(\omega)-2\omega G'(\omega)[|m|+1]+G(\omega)(\lambda-|m|(|m|+1))=0 \tag{2.7.7}$$

$$\text{if}: G(\omega)=\sum_j a_j\omega^j, \text{ and } \lambda=l(l+1)$$

Prove the following:

a. $a_{j+2}=\dfrac{(j+|m|)(j+|m|+1)-(l(l+1))}{(j+1)(j+2)}a_j$

b. $G(\omega)=\sum_{\substack{j=0,2\ldots\\ j=1,3\ldots}}^{l-|m|} a_j\omega^j$

c. $l=|m|+j$

d. $\Theta_{l,m}(\theta)=P(\omega)=\sin^{|m|}\theta\sum_{\substack{j=0,2\ldots\\ j=1,3\ldots}}^{l-|m|} a_j\cos^j\theta$

e. $m=\pm l, \pm(l-1), \pm(l-2), \ldots\ldots, 0$

f. $Y_l^m=\dfrac{1}{\sqrt{2\pi}}P(\omega)e^{im\phi}$

- $$\text{If}: G(\omega)=\sum_j a_j\omega^j \tag{2.7.5}$$

$$G'(\omega)=\sum_j j\,a_j\omega^{j-1}$$

$$G''(\omega)=\sum_j j(j-1)a_j\omega^{j-2}=\sum_j (j+2)(j+1)\,a_{j+2}\,\omega^j$$

- Substitution these power series in Eq. (2.7.7):

$$(1-\omega^2)G''(\omega)-2\omega G'(\omega)[|m|+1]+G(\omega)(\lambda-|m|(|m|+1))=0 \tag{2.7.7}$$

$$G''(\omega)-\omega^2G''(\omega)-2\omega G'(\omega)|m|-2\omega G'(\omega)+G(\omega)(\lambda-|m|(|m|+1))=0$$

$$\sum_j (j+2)(j+1)\,a_{j+2}\,\omega^j-\sum_j j(j-1)\,a_j\omega^j-2|m|\sum_j j\,a_j\omega^j$$
$$-2\sum_j j\,a_j\omega^j+\sum_j a_j\omega^j(\lambda-|m|(|m|+1))=0$$

$$\sum_j\left[(j+2)(j+1)\,a_{j+2}+\left(-j^2-j-2|m|j+\lambda-|m|^2-|m|\right)a_j\right]\omega^j=0$$

- Assuming $\omega^j=0$, and $\lambda=l(l+1)$,

$$a_{j+2}=\frac{\left(j^2+j+|m|j+|m|j-\lambda+|m|^2+|m|\right)}{(j+1)(j+2)}a_j$$

$$a_{j+2}=\frac{(j(j+|m|+1)+|m|(j+|m|+1)-(l(l+1)))}{(j+1)(j+2)}a_j$$

$$\therefore a_{j+2}=\frac{(j+|m|)(j+|m|+1)-(l(l+1))}{(j+1)(j+2)}a_j \tag{2.7.8}$$

- The series P will terminate when

$$\begin{aligned}(\boldsymbol{j}+|\boldsymbol{m}|)(\boldsymbol{j}+|\boldsymbol{m}|+\mathbf{1})-(\boldsymbol{l}(\boldsymbol{l}+\mathbf{1}))&=\mathbf{0}\\ (\boldsymbol{j}+|\boldsymbol{m}|)(\boldsymbol{j}+|\boldsymbol{m}|+\mathbf{1})&=(\boldsymbol{l}(\boldsymbol{l}+\mathbf{1}))\\ \boldsymbol{j}+|\boldsymbol{m}|&=\boldsymbol{l}\\ \therefore\ \boldsymbol{j}&=\boldsymbol{l}-|\boldsymbol{m}|\end{aligned} \tag{2.7.9}$$

- Consequently, the general solution is a linear combination of a series of even powers ($a_o, a_2 \ldots, a_{l-m)}$ or a series of odd powers ($a_1, a_3 \ldots., a_{l-m}$).

$$\therefore G(\omega)=\sum_{\substack{j=0,\,2\ldots\\ j=1,\,3\ldots}}^{l-|m|} a_j\omega^j \tag{2.7.10}$$

$$l=|m|+j$$

where the sum is over even or odd values of j, depending on whether $l-m$ is even or odd, and $\Theta(\theta)$:

$$\therefore \Theta_{l,\,m}(\theta)=P(\omega)=\sin^{|m|}\theta\sum_{\substack{j=0,\,2\ldots\\ j=1,\,3\ldots}}^{l-|m|} a_j\cos^j\theta \tag{2.7.11}$$

- When

$$\begin{aligned}j+|m|&=l\\ |m|&=l-j\\ m&=\pm(l-j)\end{aligned}$$

at: $j=0, 1, 2, \ldots\ldots.., l$

$$\therefore m=\pm l, \pm(l-1), \pm(l-2),\ldots\ldots,0 \tag{2.7.12}$$

and there are $2\,l+1$ values for m

If $Y(\theta, \phi)=\Theta(\theta)\,\Phi(\phi)$, then

$$\therefore Y_l^m=\Theta_{l,\,m}(\theta)\Phi(\phi)=\frac{1}{\sqrt{2\pi}}P(\omega)e^{im\phi} \tag{2.7.13}$$

- The functions of $P(\omega)$ are well known in mathematics, and are associated Legendre multiplied by the normalization constant.

Write the standard solution and the normalized solutions of $\Theta_{l,\,m}$-associated Legendre polynomials equation:

$$\frac{\mathbf{1}}{\sin\boldsymbol{\theta}}\frac{\partial}{\partial\boldsymbol{\theta}}\left(\sin\boldsymbol{\theta}\frac{\partial \boldsymbol{P}_l^m}{\partial\boldsymbol{\theta}}\right)+\left(\boldsymbol{l}(\boldsymbol{l}+\mathbf{1})-\frac{\boldsymbol{m}^2}{\sin^2\boldsymbol{\theta}}\right)\boldsymbol{P}_l^m=\mathbf{0} \tag{2.5.7}$$

where $\boldsymbol{\lambda}=\boldsymbol{l}(\boldsymbol{l}+\mathbf{1})$

Find all possible values of $\Theta_{l,m}(\theta)$ when $l=0, 1, 2$

- The Legendre polynomials satisfy the equation:

$$\because(1-\omega^2)\frac{\partial^2 P}{\partial\omega^2}-2\omega\frac{\partial P}{\partial\omega}+\left(\lambda-\frac{m^2}{1-\omega^2}\right)P=0 \tag{2.7.2}$$

or

$$\frac{1}{\sin\theta}\frac{\partial}{\partial\theta}\left(\sin\theta\frac{\partial P_l^m}{\partial\theta}\right)+\left(l(l+1)-\frac{m^2}{\sin^2\theta}\right)P_l^m=0$$

where

$$P_l^{|m|}(\cos\theta)=\frac{(1-\cos^2\theta)^{|m|/2}}{2^l l!}\frac{\partial^{l+|m|}}{\partial\cos^{l+|m|}\theta}(\cos^2\theta-1)^l \tag{2.7.14}$$

$$:-1\le\cos\theta\le 1$$

or

$$P_l^{|m|}(\omega) = \frac{(1-\omega^2)^{|m|/2}}{2^l l!} \frac{\partial^{l+|m|}}{\partial \omega^{l+|m|}} (\omega^2 - 1)^l$$
$$\text{where}: l = 0, 1, 2, \ldots; \quad |m| = 0, 1, 2, \ldots\ldots, l \tag{2.7.15}$$
$$\text{if} \quad \frac{\partial(u^n)}{\partial x} = nu^{n-1}\frac{\partial u}{\partial x}, \quad \text{then}$$
$$\frac{\partial}{\partial \omega}(\omega^2 - 1)^n = n(\omega^2 - 1)^{n-1}\frac{\partial(\omega^2 - 1)}{\partial \omega}, \quad \text{then}$$

- When: $l = 0$ and $m = 0$,

$$P_0^0(\omega) = 1$$

- When $l = 1$, $m = 0, \pm 1$,

$$P_1^0(\omega) = \omega$$
$$P_1^1(\omega) = (1 - \omega^2)^{1/2}$$

- When $l = 2$, $m = 0, \pm 1, \pm 2$.

$$P_2^0(\omega) = \frac{1}{2}(3\omega^2 - 1)$$
$$P_2^1(\omega) = 3\omega(1 - \omega^2)^{1/2}$$
$$P_2^2(\omega) = 3 - 3\omega^2$$

- The normalization relation for Legendre polynomials satisfies the following normalization:

$$\int_{-1}^{1} \left[P_l^{|m|}(\omega)\right]^2 \partial\omega = \frac{2}{(2l+1)} \frac{(l+|m|)!}{(l-|m|)!} = \frac{1}{N^2}$$

Consequentially,

$$\Theta(\theta) = \left[\frac{(2l+1)}{2} \frac{(l-|m|)!}{(l+|m|)!}\right]^{1/2} P_l^{|m|}(\cos\theta) \tag{2.7.16}$$

Write the angular functions: $Y_l^m(\theta,\phi)$ when $l = 0, 1, 2$ in polar and Cartesian coordinates.

- The function $Y_l^m(\theta, \phi)$, which represents the angular part of the particle wave function, is called the spherical harmonic and has the following form:

$$\because Y_l^m(\theta, \phi) = N_{ml} P_l^m(\cos\theta)\Phi_m(\phi) \tag{2.7.17}$$

$$\therefore Y_l^m = (-1)^m \left[\frac{(2l+1)}{4\pi} \frac{(l-|m|)!}{(l+|m|)!}\right]^{1/2} P_l^{|m|}(\cos\theta) e^{im\phi} \; : m \geq 0 \tag{2.7.18}$$

$$P_l^{|m|}(\omega) = \frac{(1-\omega^2)^{|m|/2}}{2^l l!} \frac{\partial^{l+|m|}}{\partial \omega^{l+|m|}} (\omega^2 - 1)^l \tag{2.7.19}$$

$$Y_l^{-m} = (-1)^m Y^*{}_{ml}(\theta, \phi) \tag{2.7.20}$$

- Table 2.2 represents the spherical harmonic function in polar and Cartesian coordinates. In order to convert from polar to Cartesian, the correspondence values of x, y, and z in the polar coordinates system are simply used:

$$x = r \sin\theta \cos\phi$$

TABLE 2.2 The Spherical Harmonic Function in Polar and Cartesian Coordinates

Harmonic	Polar Coordinates	Cartesian Coordinates
Y_0^0	$\frac{1}{\sqrt{2}}\frac{1}{\sqrt{2\pi}}$	$\frac{1}{\sqrt{2}}\frac{1}{\sqrt{2\pi}}$
Y_1^0	$\sqrt{\left(\frac{3}{2}\right)}\left(\frac{1}{\sqrt{2\pi}}\right)\cos\theta$	$\sqrt{\left(\frac{3}{2}\right)}\left(\frac{1}{\sqrt{2\pi}}\right)\frac{z}{r}$
$Y_1^{\pm 1}$	$\pm\sqrt{\left(\frac{3}{4}\right)}\left(\frac{1}{\sqrt{2\pi}}\right)\sin\theta\, e^{\pm i\phi}$	$\pm\sqrt{\left(\frac{3}{4}\right)}\left(\frac{1}{\sqrt{2\pi}}\right)\frac{(x\pm iy)}{r}$
Y_2^0	$\sqrt{\left(\frac{5}{8}\right)}\left(\frac{1}{\sqrt{2\pi}}\right)(3\cos^2\theta-1)$	$\sqrt{\left(\frac{5}{8}\right)}\left(\frac{1}{\sqrt{2\pi}}\right)\left(\frac{1}{2}\right)\frac{(3z^2-r^2)}{r^2}$
$Y_2^{\pm 1}$	$\pm\sqrt{\left(\frac{15}{4}\right)}\left(\frac{1}{\sqrt{2\pi}}\right)\sin\theta\cos\theta\, e^{\pm i\phi}$	$\pm\sqrt{\left(\frac{5}{2}\right)}\left(\frac{1}{\sqrt{2\pi}}\right)\sqrt{\left(\frac{3}{2}\right)}\frac{z(x\pm iy)}{r^2}$
$Y_2^{\pm 2}$	$\sqrt{\left(\frac{15}{16}\right)}\left(\frac{1}{\sqrt{2\pi}}\right)\sin^2\theta\, e^{\pm 2i\phi}$	$\sqrt{\left(\frac{5}{2}\right)}\left(\frac{1}{\sqrt{2\pi}}\right)\sqrt{\left(\frac{3}{8}\right)}\frac{(x\pm iy)^2}{r^2}$

$$y = r\sin\theta\sin\phi$$

$$z = r\cos\theta$$

$$e^{\pm im\phi} = \cos(m\phi) \pm i\sin(m\phi)$$

2.8 THE RADIAL EQUATION

The resulting solution of the radial wave function:

$$\frac{1}{r^2}\frac{\partial}{\partial r}\left(r^2\frac{\partial R}{\partial r}\right)+\left\{\frac{2\mu}{\hbar^2}(E-V(r))-\frac{\lambda}{r^2}\right\}R=0 \qquad (2.4.27)$$

is based on the well-known associate Laguerre polynomial function. Substitute:

- **λ by $l(l+1)$,**

$$V(r) by -\frac{Ze^2}{(4\pi\epsilon_o)r},$$

$$\frac{8\mu r^2|E|}{\hbar^2} by\, \rho^2 \text{ and } \frac{2\mu Ze^2}{(4\pi\epsilon_o)\alpha\hbar^2} \text{ by } \bar{\lambda}, \text{ where } \rho^2=\alpha^2 r^2$$

Rewrite the radial equation to match the standard mathematic equation described by the associated Laguerre:

$$R''+\frac{2}{\rho}R'+\left[\frac{\bar{\lambda}}{\rho}-\frac{1}{4}-\frac{l(l+1)}{\rho^2}\right]R=0$$

- The general radial equation is

$$\frac{1}{r^2}\frac{\partial}{\partial r}\left(r^2\frac{\partial R}{\partial r}\right)+\left\{\frac{2\mu}{\hbar^2}(E-V(r))-\frac{\lambda}{r^2}\right\}R=0 \qquad (2.4.27)$$

The solution of the radial equation in the electron-proton system is straight forward and required to replace

$$V(r) = -\frac{Ze^2}{(4\pi\epsilon_o)r}$$

and λ by $l(l+1)$.

Note that we are interested in the bound state energies, where the eigenvalue of E is negative (attraction):

$$\frac{1}{r^2}\frac{\partial}{\partial r}\left(r^2\frac{\partial R}{\partial r}\right) + \left\{\frac{2\mu}{\hbar^2}\left[-|E| + \frac{Ze^2}{(4\pi\epsilon_o)r}\right] - \frac{l(l+1)}{r^2}\right\}R = 0$$

Multiplies by $\frac{\hbar^2}{2\mu}$

$$\frac{\hbar^2}{2\mu r^2}\frac{1}{}\frac{\partial}{\partial r}\left(r^2\frac{\partial R}{\partial r}\right) - |E|R + \frac{Ze^2}{(4\pi\epsilon_o)r}R - \left(\frac{l(l+1)\hbar^2}{2\mu r^2}\right)R = 0$$

and divide by 4 $|E|$

$$\frac{\hbar^2}{8\mu|E|}\frac{1}{r^2}\frac{\partial}{\partial r}\left(r^2\frac{\partial R}{\partial r}\right) + \frac{Ze^2}{(16\pi\epsilon_o)r|E|}R - \left(\frac{l(l+1)\hbar^2}{8\mu r^2}\right)\frac{R}{|E|} - \frac{R}{4} = 0 \quad (2.8.1)$$

- This equation can be rewritten to match the standard mathematic equation described by the associated Laguerre polynomials.
  - We start with introduction of a dimensionless variable
    $\rho = \alpha r$
    with the choice

$$\alpha^2 = \frac{8\mu|E|}{\hbar^2} \quad (2.8.2)$$

$$\rho^2 = \frac{8\mu r^2|E|}{\hbar^2} \quad (2.8.3)$$

$$\rho = \frac{2\sqrt{2}r(\mu|E|)^{1/2}}{\hbar}$$

We also write

$$\bar{\lambda} = \frac{2\mu Ze^2}{(4\pi\epsilon_o)\alpha\hbar^2} = \frac{Ze^2}{(4\pi\epsilon_o)\hbar}\left(\frac{\mu}{2|E|}\right)^{1/2} \quad (2.8.4)$$

$$\frac{\bar{\lambda}}{\rho} = \frac{Ze^2}{(16\pi\epsilon_o)Er}$$

Substituting Eq. (2.8.1):

$$\frac{1}{\rho^2}\frac{\partial}{\partial\rho}\left(\rho^2\frac{\partial R}{\partial\rho}\right) + \left[\frac{\bar{\lambda}}{\rho} - \frac{1}{4} - \frac{l(l+1)}{\rho^2}\right]R = 0$$

$$\frac{1}{\rho^2}\rho^2 R'' + \frac{2\rho}{\rho^2}R' + \left[\frac{\bar{\lambda}}{\rho} - \frac{1}{4} - \frac{l(l+1)}{\rho^2}\right]R = 0$$

$$R'' + \frac{2}{\rho}R' + \left[\frac{\bar{\lambda}}{\rho} - \frac{1}{4} - \frac{l(l+1)}{\rho^2}\right]R = 0 \quad (2.8.5)$$

Write the proposed solution of the associated Laguerre ρ-polynomial of the radial function:

$$R'' + \frac{2}{\rho}R' + \left[\frac{\lambda}{\rho} - \frac{1}{4} - \frac{l(l+1)}{\rho^2}\right]R = 0 \tag{2.8.5}$$

- We look to a solution of the form

$$R(\rho) = \rho^s L(\rho) e^{-\rho/2}, \quad s \geq 0 \tag{2.8.6}$$

where

$$L(\rho) = a_0 + a_1\rho + a_2\rho^2 + \ldots\ldots, a_0 \neq 0 \tag{2.8.7}$$

If the proposed solution of the associated Laguerre ρ-polynomial of radial function:

$$R'' + \frac{2}{\rho}R' + \left[\frac{\lambda}{\rho} - \frac{1}{4} - \frac{l(l+1)}{\rho^2}\right]R = 0 \tag{2.8.5}$$

is

$$R(\rho) = \rho^s L(\rho) e^{-\rho/2}, \quad s \geq 0$$

where

$$L(\rho) = a_0 + a_1\rho + a_2\rho^2 + \ldots\ldots, a_0 \neq 0$$

Show that:

a. $s = l$

b.

$$\rho L'' + [2(l+1) - \rho]L' + [\lambda - l - 1]L = 0$$

- If:

$$R(\rho) = \rho^s L e^{-\rho/2} \tag{2.8.6}$$

$$R'(\rho) = -\frac{1}{2}(\rho^s L) e^{-\rho/2} + \rho^s L' e^{-\rho/2} + s\rho^{s-1} L e^{-\rho/2}$$

$$R'(\rho) = -\frac{1}{2}(\rho^s L) e^{-\rho/2} + [\rho^s L' + s\rho^{s-1} L] e^{-\rho/2}$$

$$R''(\rho) = -\frac{1}{2}\left\{-\frac{1}{2}(\rho^s L) e^{-\rho/2} + [\rho^s L' + s\rho^{s-1} L] e^{-\rho/2}\right\}$$
$$+\left\{-\frac{1}{2}(\rho^s L') e^{-\rho/2} + [\rho^s L'' + s\rho^{s-1} L'] e^{-\rho/2}\right\}$$
$$+s\left\{-\frac{1}{2}(\rho^{s-1} L) e^{-\rho/2} + [\rho^{s-1} L' + (s-1)\rho^{s-2} L] e^{-\rho/2}\right\}$$

- The radial Eq. (2.8.5) is now

$$e^{-\rho/2}\left\{\begin{array}{c}\rho^s L'' + \left[\frac{2}{\rho}\rho^s - \frac{1}{2}\rho^s - \frac{1}{2}\rho^s + s\rho^{s-1} + s\rho^{s-1}\right]L' \\ + \left[-\frac{1}{2}\left(\frac{2}{\rho}\right)\rho^s + \frac{2}{\rho}s\rho^{s-1} + \frac{1}{4}\rho^s - \frac{1}{2}s\rho^{s-1} + s(s-1)\rho^{s-2} + \left(\frac{\bar{\lambda}}{\rho} - \frac{1}{4} - \frac{l(l+1)}{\rho^2}\right)\rho^s\right]L\end{array}\right\} = 0$$

$$\text{Divide by } e^{-\rho/2}\rho^s$$

$$L'' + \left[\frac{2}{\rho} - \frac{1}{2} - \frac{1}{2} + \frac{s}{\rho} + \frac{s}{\rho}\right]L' + \left[-\frac{1}{\rho} + \frac{2s}{\rho^2} + \frac{1}{4} - \frac{s}{2\rho} - \frac{s}{2\rho} + \frac{s(s-1)}{\rho^2} + \frac{\bar{\lambda}}{\rho} - \frac{1}{4} - \frac{l(l+1)}{\rho^2}\right]L = 0$$

$$\rho^2 L'' + \rho[2(s+1) - \rho]L' + \left[-\rho + 2s - \frac{s}{2}\rho - \frac{s}{2}\rho + s^2 - s + \bar{\lambda}\rho - l(l+1)\right]L = 0$$

$$\rho^2 L'' + \rho[2(s+1) - \rho]L' + \left[\rho(\bar{\lambda} - s - 1) + s(s+1) - l(l+1)\right]L = 0 \tag{2.8.8}$$

- This choice of solution $R(\rho)$ provides an acceptable form that allows us to rewrite the radial part of Schrödinger equation so that it takes the form of the Laguerre equation with well-known solutions.
- If we examine Eq. (2.8.8), at $\rho = 0$, where the solution must remain finite, a quadratic equation in s is obtained:

$$s^2 + s - l - l^2 = 0$$

$$(s - l)(s + l + 1) = 0$$

with $s = l$ or $s = -(l+1)$
Since $s \geq 0$
Only (Eq. 2.8.6)

$$s = l \tag{2.8.9}$$

can be acceptable, since the solution R has to be finite at $\rho = 0$ (Eq. (2.8.8)), reduced to

$$\rho L'' + [2(l+1) - \rho]L' + [\bar{\lambda} - l - 1]L = 0 \tag{2.8.10}$$

If the radial wave function:

$$\boldsymbol{\rho L'' + [2(l+1) - \rho]L' + [\bar{\lambda} - l - 1]L = 0} \tag{2.8.10}$$

then show that

a. $R(\rho) = \rho^l L(\rho) e^{-\rho/2}$,

b. $\dfrac{a_{\upsilon+1}}{a_\upsilon} = \dfrac{\upsilon + l + 1 - \bar{\lambda}}{(\upsilon + 1)(\upsilon + 2l + 2)}$

c. $l = n-1, n-2, n-3, \ldots\ldots\ldots, 0$

d. $E_n = \dfrac{\mu Z^2 e^4}{2(4\pi\epsilon_o)^2\hbar^2 n^2}$, consider $\bar{\lambda} = \dfrac{Ze^2}{(4\pi\epsilon_o)\hbar}\left(\dfrac{\mu}{2|E|}\right)^{1/2}$

e. $\rho = 2\dfrac{1}{a_o}\dfrac{zr}{n}$

- Answers:

a. $R(\rho) = \rho^l L(\rho) e^{-\rho/2}$,

$$\text{If}: R(\rho) = \rho^s L(\rho) e^{-\rho/2}, \tag{2.8.6}$$

$s \geq 0$, and $s = l$, therefore

$$\therefore R(\rho) = \rho^l L(\rho) e^{-\rho/2}, \tag{2.8.11}$$

where

$$L(\rho)=a_0 + a_1\rho + a_2\rho^2+,\dots\dots,a_0 \neq 0 \tag{2.8.7}$$

b. $\dfrac{a_{\upsilon+1}}{a_\upsilon}=\dfrac{\upsilon+l+1-\bar{\lambda}}{(\upsilon+1)(\upsilon+2l+2)}$

- If

$$L(\rho)=a_0+a_1\rho+a_2\rho^2+ \tag{2.8.7}$$

$$\therefore L(\rho)=\sum_{\upsilon=0}^{\infty}a_\upsilon\rho^\upsilon$$

$$L'(\rho)=a_1+2a_2\rho+3a_3\rho^2+\dots\dots$$

$$L'(\rho)=\sum_{\upsilon=0}^{\infty}(\upsilon+1)\,a_{\upsilon+1}\rho^\upsilon$$

$$L''(\rho)=2a_2 + 6a_3\rho + 12a_4\rho^2+\dots\dots$$

$$L''(\rho)=\sum_{\upsilon=0}^{\infty}\upsilon(\upsilon+1)\,a_{\upsilon+1}\rho^{\upsilon-1}$$

$$\rho L'(\rho)=\sum_{\upsilon=0}^{\infty}(\upsilon+1)\,a_{\upsilon+1}\rho^{\upsilon+1}=\sum_{\upsilon=0}^{\infty}\upsilon a_\upsilon\rho^\upsilon$$

- Then by substituting in Eq. (2.8.10):

$$\rho\sum_{\upsilon=0}^{\infty}\upsilon(\upsilon+1)\,a_{\upsilon+1}\rho^{\upsilon-1}+2(l+1)\sum_{\upsilon=0}^{\infty}(\upsilon+1)\,a_{\upsilon+1}\rho^\upsilon-\rho\sum_{\upsilon=0}^{\infty}(\upsilon+1)\,a_{\upsilon+1}\rho^\upsilon+\left[\bar{\lambda}-l-1\right]\sum_{\upsilon=0}^{\infty}a_\upsilon\rho^\upsilon=0$$

$$\sum_{\upsilon=0}^{\infty}\upsilon(\upsilon+1)\,a_{\upsilon+1}\rho^\upsilon+2(l+1)\sum_{\upsilon=0}^{\infty}(\upsilon+1)\,a_{\upsilon+1}\rho^\upsilon-\sum_{\upsilon=0}^{\infty}(\upsilon+1)\,a_{\upsilon+1}\rho^{\upsilon+1}+\left[\bar{\lambda}-l-1\right]\sum_{\upsilon=0}^{\infty}a_\upsilon\rho^\upsilon=0$$

$$\sum_{\nu=0}^{\infty}\left[\upsilon(\upsilon+1)\,a_{\upsilon+1}+2(l+1)(\upsilon+1)\,a_{\upsilon+1}-\upsilon a_\upsilon+\left(\bar{\lambda}-l-1\right)a_\upsilon\right]\rho^\upsilon=0 \tag{2.8.12}$$

- In the radial equation, we see that the recursion relation between the coefficient of the terms is

$$\frac{a_{\upsilon+1}}{a_\upsilon}=\frac{\upsilon+l+1-\bar{\lambda}}{\upsilon(\upsilon+1)+2(l+1)(\upsilon+1)}$$

$$\frac{a_{\upsilon+1}}{a_\upsilon}=\frac{\upsilon+l+1-\bar{\lambda}}{(\upsilon+1)(\upsilon+2l+2)} \tag{2.8.13}$$

c. $l=n-1,n-2,n-3,\dots\dots\dots,0$

as $\upsilon\longrightarrow\infty$, $\dfrac{a_{\upsilon+1}}{a_\upsilon}\longrightarrow\dfrac{1}{\upsilon}$ approaches zero.

Thus, the wave function goes to infinity at large υ, therefore the series L will terminate.

- The series $L(\rho)$ will terminate when the highest power of L at $\upsilon=n'$ is as follows:

$$n'+l+1-\bar{\lambda}=0$$

$$\bar{\lambda}-l-1=n'$$

or

$$\bar{\lambda}=n'+l+1=n \tag{2.8.14}$$

then at $n'=0,1,2,\dots\dots\dots,l$

$$\therefore l=n-1,n-2,n-3,\dots\dots\dots,0 \tag{2.8.15}$$

- The integer n' is called the radial quantum number, and the integer n is called the total quantum number. Note that the angular quantum number l ranges from 0 to $n-1$.

d. $E_n = \dfrac{\mu Z^2 e^4}{2(4\pi\epsilon_o)^2\hbar^2 n^2}$, consider $\bar{\lambda} = \dfrac{Ze^2}{(4\pi\epsilon_o)\hbar}\left(\dfrac{\mu}{2|E|}\right)^{1/2}$

The energy from

$$\bar{\lambda} = \frac{Ze^2}{(4\pi\epsilon_o)\hbar}\left(\frac{\mu}{2|E|}\right)^{1/2} \quad (2.8.4)$$

$$\text{If}: \bar{\lambda} = n \quad (2.8.14)$$

$$E_n = \frac{\mu Z^2 e^4}{2(4\pi\epsilon_o)^2\hbar^2 n^2} \quad (2.8.16)$$

e. To show $\rho = \dfrac{2\,zr}{a_o\,n}$

if $\rho = \alpha r$, and

$$\alpha^2 = \frac{8\mu|E|}{\hbar^2} \quad (2.8.2)$$

$$E = -\frac{Z^2 e^4 \mu}{8\varepsilon_o^2 n^2 h^2} \text{ (Bohr)} \quad (1.7.5)$$

$$\alpha^2 = \frac{8\mu}{\hbar^2}\frac{Z^2 e^4 \mu}{8\varepsilon_o^2 n^2 h^2}$$

$$\alpha^2 = \frac{4\pi^2}{h^4}\frac{Z^2 e^4 \mu^2}{\varepsilon_o^2 n^2}$$

$$\alpha = \frac{2\pi}{h^2}\frac{\mu e^2 Z}{\varepsilon_o n}$$

$$r = \frac{n^2\,\varepsilon_o h^2}{Z\,\pi\,\mu\,e^2} = \frac{n^2}{Z} x a_o' \text{ (Bohr)} \quad (1.7.4)$$

$$\alpha = \frac{2\,Z}{a_o\,n}$$

$$\rho = \alpha r = \frac{2Zr}{n a_0'} \quad (2.8.17)$$

The associated Laguerre polynomial of the degree $(q-k)$:

$$xY'' + [k+1-x]\,Y' + [q-k]Y = 0$$

has a known solution:

$$Y = x^{\frac{(k-1)}{2}} e^{-\frac{x}{2}} L_q^k,$$

where

$$L_q^k = \frac{\partial^k}{\partial x^k} L_q$$

$$L_q = e^{\rho}\frac{\partial^q}{\partial \rho^q}\left(e^{-\rho}\rho^q\right)$$

Find the appropriate solution of the radial wave function:

$$\boldsymbol{R'' + \frac{2}{\rho}R' + \left[\frac{\lambda}{\rho} - \frac{1}{4} - \frac{l(l+1)}{\rho^2}\right]R = 0} \tag{2.8.5}$$

- The solutions of the radial equation:

$$R'' + \frac{2}{\rho}R' + \left[\frac{\lambda}{\rho} - \frac{1}{4} - \frac{l(l+1)}{\rho^2}\right]R = 0 \tag{2.8.5}$$

  This equation can be transformed into

$$\rho L'' + [2(l+1) - \rho]L' + [n - l - 1]L = 0 \tag{2.8.10}$$

  by letting

$$R = \rho^l L e^{-\rho}/_2, \quad l \geq 0 \tag{2.8.6}$$

- Eq. (2.8.10) is comparable to the associated Laguerre polynomial of the degree $(q-k)$:

$$\underbrace{xY''}_{\rho L''} + \left[\underbrace{\overbrace{k}^{(2l+1)} + 1}_{2(l+1)} - \underbrace{x}_{\rho}\right]\underbrace{Y'}_{L'} + \left[\underbrace{\overbrace{q}^{(n+l)} - k}_{n-l-1}\right]\underbrace{Y}_{L} = 0$$

  that has a known solution:

$$Y = x^{\frac{(k-1)}{2}} e^{-\frac{x}{2}} L_q^k, \text{where}$$

$$L_q^k = \frac{\partial^k}{\partial x^k} L_q$$

$$L_q = e^{\rho} \frac{\partial^q}{\partial \rho^q}(e^{-\rho}\rho^q)$$

- The corresponding values k and q of the associated Laguerre function can be obtained for the transformed equation:

$$k + 1 = 2(l+1) = 2l + 2$$

$$k = 2l + 1$$

  and

$$q - k = n - l - 1$$

$$q = k + n - l - 1 = 2l + 1 + n - l - 1$$

$$q = n + l$$

  These values are the same as the radial equation.

- Then the radial solution has exactly the same form as that of the associated Laguerre function ($q = n + l$, and $k = 2l + 1$), thus:

$$\therefore L = L_{n+l}^{2l+1}$$

$$\therefore R_{nl}(\rho) = \rho^l e^{-\frac{\rho}{2}} L_{n+l}^{2l+1} \tag{2.8.18}$$

$$\therefore L_{n+l}^{2l+1} = \frac{\partial^{2l+1}}{\partial x^{2l+1}} L_{n+l} \tag{2.8.19}$$

$$\therefore L_{n+l} \equiv L_{n+l}^0 = e^{\rho} \frac{\partial^{n+l}}{\partial \rho^{n+l}}\left(e^{-\rho}\rho^{n+l}\right) \tag{2.8.20}$$

Write the normalized solution of the radial equation, $R_{nl}(\rho)$. What are the essential value of μ and the definition of the parameter a'_o? Find $R_{10}(\rho)$, $R_{20}(\rho)$, $R_{21}(\rho)$, and $R_{30}(\rho)$.

- The normalized solutions:

$$R_{nl}(\rho) = -\left\{\left(\frac{2Z}{na'_0}\right)^3 \frac{(n-l-1)!}{2n[(n+l)!]^3}\right\}^{1/2} e^{-\rho/2}\rho^l L_{n+l}^{2l+1}(\rho) \quad (2.8.21)$$

$$a'_0 = \frac{\hbar^2(4\pi\epsilon_o)}{\mu r^2} \quad (2.8.22$$

and

$$\rho = \frac{2Zr}{na'_0} \quad (2.8.23)$$

$$L_{n+l}^{2l+1} = \frac{\partial^{2l+1}}{\partial\rho^{2l+1}} L_{n+l}^0, \quad n,l=0,1,2, \quad (2.8.24)$$

$$L_{n+l}^0 \equiv L_{n+l} = e^\rho \frac{\partial^{n+l}}{\partial\rho^{n+l}}(e^{-\rho}\rho^{n+l}), \quad n=0,1,2, \quad (2.8.25)$$

- Due to the nucleus of the larger mass compared to the mass of the electron, the reduced mass μ is essentially the same as the electron mass m_e. The parameter a'_o is donated by a_o, the Bohr radius, and it has the value of 0.53 Å.
- Therefore:

$$R_{10}(\rho) = -\left\{\left(\frac{2Z}{a'_0}\right)^3 \frac{(1-0-1)!}{2[(1+0)!]^3}\right\}^{1/2} e^{-zr/a'_o}\left(\frac{2zr}{a'_0}\right)^0 L_1^1(\rho)$$

$$R_{10}(\rho) = -2\left(\frac{Z}{a'_0}\right)^{3/2} e^{-zr/a'_o} L_1^1(\rho)$$

$$L_1^1(\rho) = \frac{\partial}{\partial\rho} L_1^0$$

$$L_1^0 \equiv L_{n+l} = e^\rho \frac{\partial}{\partial\rho}(e^{-\rho}\rho) = e^\rho(e^{-\rho} - \rho e^{-\rho}) = 1-\rho$$

$$L_1^1(\rho) = (-1)^0 \frac{\partial}{\partial\rho} L_1^0 = (1)(-1) = -1$$

$$\therefore R_{10}(\rho) = 2\left(\frac{Z}{a'_0}\right)^{3/2} e^{-zr/a'_o}$$

- $$R_{20}(\rho) = -\left\{\left(\frac{2Z}{2a'_0}\right)^3 \frac{(2-0-1)!}{4[(2+0)!]^3}\right\}^{1/2} e^{-zr/2a'_o}\rho^0 L_2^1(\rho)$$

$$R_{20}(\rho) = -\frac{1}{4\sqrt{2}}\left(\frac{Z}{a'_0}\right)^{3/2} e^{-zr/2a'_o} L_2^1(\rho)$$

$$L_2^1(\rho) = \frac{\partial}{\partial\rho} L_2^0$$

$$L_2^0 = e^\rho \frac{\partial^2}{\partial\rho^2}(e^{-\rho}\rho^2) = e^\rho \frac{\partial}{\partial\rho}\left(2e^{-\rho}\rho - e^{-\rho}\rho^2\right) = 2 - 4\rho + \rho^2$$

$$L_2^1(\rho) = \frac{\partial}{\partial \rho} L_2^0 = -4 + 2\rho$$

$$\therefore R_{20}(\rho) = \frac{1}{2\sqrt{2}} \left(\frac{Z}{a_0'}\right)^{3/2} e^{-zr/2a_o'} (2 - \rho)$$

- $$R_{21}(\rho) = -\left\{ \left(\frac{2Z}{2a_0'}\right)^3 \frac{(2-1-1)!}{4[(2+1)!]^3} \right\}^{1/2} e^{-zr/2a_o'} \rho L_3^3(\rho)$$

$$R_{21}(\rho) = -\frac{1}{12\sqrt{6}} \left(\frac{Z}{a_0'}\right)^{3/2} e^{-zr/2a_o'} \frac{Zr}{a_0'} L_3^3(\rho)$$

$$L_3^3 = \frac{\partial^3}{\partial \rho^3} L_3^0$$

$$L_3^0 = e^{\rho} \frac{\partial^3}{\partial \rho^3} (e^{-\rho} \rho^3)$$

$$\frac{\partial}{\partial \rho} (e^{-\rho} \rho^3) = 3\rho^2 e^{-\rho} - e^{-\rho} \rho^3$$

$$\frac{\partial^2}{\partial \rho^2} (e^{-\rho} \rho^3) = 6\rho e^{-\rho} - 6e^{-\rho} \rho^2 + e^{-\rho} \rho^3$$

$$\frac{\partial^3}{\partial \rho^3} (e^{-\rho} \rho^3) = 6e^{-\rho} - 18\rho e^{-\rho} + 9e^{-\rho} \rho^2 - e^{-\rho} \rho^3$$

$$L_3^0 = e^{-\rho} \frac{\partial^3}{\partial \rho^3} (e^{-\rho} \rho^3) = 6 - 18\rho + 9\rho^2 - \rho^3$$

$$L_3^3 = \frac{\partial^3}{\partial \rho^3} L_3^0 = (-1)^2 \frac{\partial^2}{\partial \rho^2} (-18 + 18\rho - 3\rho^2)$$
$$= \frac{\partial}{\partial \rho} (18 - 6\rho) = -6$$

$$\therefore R_{21}(r) = \frac{1}{2\sqrt{6}} \left(\frac{Z}{a_0'}\right)^{3/2} e^{-zr/2a_o'} \frac{Zr}{a_0'}$$

- $$R_{30}(\rho) = -\left\{ \left(\frac{2Z}{3a_0'}\right)^3 \frac{(3-0-1)!}{2 \times 3[(3 \times 2)!]^3} \right\}^{1/2} e^{-zr/3a_o'} \rho^0 L_3^1(\rho)$$

$$R_{30}(\rho) = -\frac{1}{27\sqrt{3}} \left(\frac{Z}{a_0'}\right)^{3/2} e^{-zr/3a_o'} L_3^1(\rho)$$

$$L_3^1 = \frac{\partial}{\partial \rho} L_3^0$$

$$L_3^0 = e^{-\rho} \frac{\partial^3}{\partial \rho^3} (e^{-\rho} \rho^3) = 6 - 18\rho + 9\rho^2 - \rho^3$$

$$L_3^1 = \frac{\partial}{\partial \rho} (6 - 18\rho + 9\rho^2 - \rho^3) = -18 + 18\rho - 3\rho^2$$

$$\therefore R_{30}(\rho) = \frac{1}{9\sqrt{3}} \left(\frac{Z}{a_0'}\right)^{3/2} e^{-zr/3a_o'} (6 - 6\rho + \rho^2)$$

Summary:

$$R_{10}(\rho) = 2\left(\frac{Z}{a'_0}\right)^{3/2} e^{-zr/a'_o}$$

$$R_{20}(\rho) = \frac{1}{2\sqrt{2}}\left(\frac{Z}{a'_0}\right)^{3/2} e^{-zr/2a'_o}(2-\rho)$$

$$R_{21}(r) = \frac{1}{2\sqrt{6}}\left(\frac{Z}{a'_0}\right)^{3/2} e^{-zr/2a'_o}\frac{Zr}{a'_0}$$

$$R_{30}(\rho) = \frac{1}{9\sqrt{3}}\left(\frac{Z}{a'_0}\right)^{3/2} e^{-zr/3a'_o}(6-6\rho+\rho^2)$$

2.9 THE FINAL SOLUTION FOR THE FULL WAVE FUNCTION, $\psi_{NLM}(R,\theta,\phi)$

Write the final solution for the full wave function, $\psi_{nlm}(r, \theta, \phi)$. How does it relate to the atomic orbital?

- All hydrogenic atomic orbitals have the following form:

$$\psi_{nlm}(r,\theta,\phi) = R_{nl}(r)Y_{lm}(\theta,\phi) \quad (2.9.1)$$

$$\text{or,} \quad \psi_{n,l,m_l} = R_{n,l}Y_{l,m_l},$$

- Atomic orbitals and their energies:
  - An orbital is a one-electron wave function.
  - An atomic orbital is a one-electron wave function that describes the distribution of electron in an atom.

Characterize the electron that is described by the wave function ψ_{nlm}.

- Each hydrogenic atomic orbital (hydrogen like a single electron atom) is defined by three quantum numbers: n, l, and m_l.
- Thus, an electron described by the wave function $\psi_{1,0,0}$ is said to occupy the orbital with $n=1$, $l=0$, and $m_l=0$.
- When an electron is described by one of these wave functions, we say that it occupies that orbital.
- For a given value of n, the values of l can range from $l=0$ to $l=n-1$, thus: $l=0, 1, 2, 3, 4, \ldots, n-1$ (Eq. 2.8.15)).
- The orbital with the same value of n but different values of l form a subshell of a given shell, and generally referred to by the letters s, p, d, f, g, ... state with $l=0, 1, 2, 3, \ldots$, respectively.
- The quantum number m_l specifies the orientation of the electron's orbital angular momentum.
- For a given value of l, suggest that the component of angular momentum about the z-axis may take only $2l+1$ values. There are $2l+1$ different wave functions.
- The z component of angular momentum is restricted to $m_l\hbar$. The values of m_l in integral steps from $-l$ to $+l$, thus:

$$m_l = 0, \pm 1, \pm 2, \pm 3, \ldots\ldots, \pm l \quad (2.7.1.2)$$

Exercise

Compare quantitatively the ionization energies and the radii of hydrogen and deuterium.

Answer

$$E_n = \frac{\mu Z^2 e^4}{2(4\pi\epsilon_o)^2\hbar^2 n^2} \quad (2.8.16)$$

$n=1$ in both hydrogen and deuterium,

$$E_{\text{ionization}} = -\frac{\mu Z^2 e^4}{2(4\pi\epsilon_o)^2\hbar^2}$$

$$\frac{E_{\text{ionization}}\left({}^1_1H\right)}{E_{\text{ionization}}\left({}^2_1H\right)} = \frac{-\dfrac{\mu_H Z^2 e^4}{2(4\pi\epsilon_o)^2\hbar^2}}{-\dfrac{\mu_D Z^2 e^4}{2(4\pi\epsilon_o)^2\hbar^2}} = \frac{\mu_H}{\mu_D}$$

$$E_{\text{ionization}}\left({}^2_1H\right) = E_{\text{ionization}}\left({}^1_1H\right) \cdot \frac{\mu_D}{\mu_H}$$

Similarly, since the electron does not rotate about the fixed mass point specified by the center of the proton, but about the center of mass of the proton-electron system, then:

$$r = \frac{n^2\ \varepsilon_o h^2}{Z\ \pi\ \mu\ e^2} \tag{1.7.4}$$

Bohr's model, therefore:

$$r_D = r_H \cdot \frac{\mu_H}{\mu_D}$$

where

$$\frac{\mu_H}{\mu_D} = \frac{\dfrac{m_e m_P}{m_e + m_P}}{\dfrac{m_e \cdot 2m_P}{m_e + 2m_P}} = \frac{m_e + 2m_P}{2m_e + 2m_P} \le 1$$

Thus the ionization energy of D is very slightly larger, and the radius very slightly smaller.

Exercise

The G polynomial of some state of hydrogen-like atom has $a_2\cos^2\theta$ as its last term. The L polynomial for the same state contains just one constant term. The Φ function of the same state is $\frac{1}{\sqrt{2\pi}}e^{i\phi}$. Which state are we talking about? (1s, 2s, $2p_1$, $2p_0$, ...?)

Answer:

$$\Phi(\phi) = \frac{1}{\sqrt{2\pi}}e^{im\phi} \tag{2.6.2}$$

$$\Phi(\phi) = \frac{1}{\sqrt{2\pi}}e^{i\phi} \rightarrow,\ \text{therefore, } m = 1$$

$$G(\omega) = \sum_{\substack{j=0,\,2\ldots\\ j=1,\,3\ldots}}^{l-|m|} a_j\omega^j \tag{2.7.10}$$

if: $a_2\cos^2\theta$ is the last term,

$\therefore j = 2$, since $l = |m| + j$, l must be 3

$$L(\rho) = a_0 + a_1\rho + a_2\rho^2 + \ldots\ldots,\ a_0 \neq 0 \tag{2.8.7}$$

$L = \sum_{\upsilon} a_\upsilon P^\upsilon = a_0$, this means $\upsilon = 0$

Since

$$n = l + 1 + \upsilon \tag{2.8.14}$$

$$n = 3 + 1 + 0 = 4$$

Thus, the state in question is $4f_{+1}$.

Exercise

In discussing the H-atom, we will encounter the associate Laguerre polynomials. An associate Laguerre polynomial of degree $n-l-1$ and order $2l+1$ is defined as

$$L(\rho) = \sum_{\upsilon=0}^{\text{until the series stops}} a_\upsilon \rho^\upsilon$$

where

$$a_{\upsilon+1} = -\frac{n-l-1-\upsilon}{2(\upsilon+1)(l+1)+\upsilon(\upsilon+1)} a_\upsilon$$

Obtain in terms of constant, a_0, the associated Laguerre polynomial with $n=3$ and $l=1$.

Answer

$$n = l + 1 + \nu \tag{2.8.14}$$

Degree, $\nu = n - l - 1$
Degree, $v = 3 - 1 - 1 = 1$, order, $n = 3$
This function is part of the 3p function.

$$a_{\upsilon+1} = -\frac{n-l-1-\upsilon}{2(\upsilon+1)(l+1)+\upsilon(\upsilon+1)} a_\upsilon \tag{2.8.13}$$

$$a_1 = -\frac{3-1-1-0}{2(0+1)(1+1)+0(0+1)} a_0$$

$$a_1 = -\frac{1}{4} a_0$$

$$a_2 = -\frac{3-1-1-1}{2(1+1)(1+1)+1(1+1)} a_1 = 0$$

Thus, the associated Laguerre polynomial with $n=3$, $l=1$

$$L(\rho) = \sum_{\upsilon=0} a_\upsilon \rho^\upsilon \tag{2.8.7}$$

$$L(\rho) = a_0 - \frac{1}{4} a_0 \rho = a_0\left(1 - \frac{1}{4}\rho\right)$$

The angular functions, Y_{l,m_l}, are complex (contain i) when $m \neq 0$ (Table 2.2).
How is it possible to avoid these complexities?
Derive the real functions that represent

$$d_{xy}, d_{xz}, d_{yz}, d_{z^2}, d_{x^2-y^2}, p_x, p_y, \text{ and } p_z$$

and write the wave function that describes each.

- The angular functions, Y_{l,m_l}, are complex when $m \neq 0$, therefore, these solutions cannot be represented graphically.
- However, the linear combination of members of a $Y_l^{\pm m}$ set of common l is also an acceptable (real) function. Thus, it cannot make any difference whether we use Schrödinger's original $Y_l^{\pm m}$ or linear combinations of theme when calculating the atomic properties.
- The linear combinations required are simply

$$Y_l^{real} = \frac{1}{\sqrt{2}}\left(Y_l^m + Y_l^{-m}\right) \tag{2.9.2}$$

$$Y_l^{real} = \frac{-i}{\sqrt{2}}\left(Y_l^m - Y_l^{-m}\right) \tag{2.9.3}$$

- e.g.,

$$Y_2^2 = \sqrt{(^{15}/_{16})}\left(^1/_{\sqrt{2\pi}}\right)\sin^2\theta\, e^{2i\phi}$$

$$\text{if } e^{\pm im\phi} = \cos(m\phi) \pm i\sin(m\phi)$$

$$\cos 2\phi = \cos^2\phi - \sin^2\phi$$

$$\sin 2\phi = 2\sin\phi\cos\phi$$

then

$$Y_2^2 = \sqrt{(^{15}/_{16})}\left(^1/_{\sqrt{2\pi}}\right)\sin^2\theta\,(\cos 2\phi + i\sin 2\phi)$$

$$Y_2^{-2} = \sqrt{(^{15}/_{16})}\left(^1/_{\sqrt{2\pi}}\right)\sin^2\theta\,(\cos 2\phi - i\sin 2\phi)$$

$$(Y_2^2 + Y_2^{-2}) = 2\sqrt{(^{15}/_{16})}\left(^1/_{\sqrt{2\pi}}\right)\sin^2\theta\cos 2\phi$$

$$(Y_2^2 + Y_2^{-2}) = 2\sqrt{(^{15}/_{16})}\left(^1/_{\sqrt{2\pi}}\right)\sin^2\theta\,(\cos^2\phi - \sin^2\phi)$$

$$(Y_2^2 + Y_2^{-2}) = 2\sqrt{(^{15}/_{16})}\left(^1/_{\sqrt{2\pi}}\right)\left(\frac{x^2 - y^2}{r^2}\right)$$

  - Normalization:

$$N^2\int (Y_2^2 + Y_2^{-2})^*(Y_2^2 + Y_2^{-2})\partial\tau = 1$$

$$N^2\int (Y_2^{2*}Y_2^2 + Y_2^{2*}Y_2^{-2} + Y_2^{-2*}Y_2^2 + Y_2^{-2*}Y_2^{-2})\partial\tau = N^2(1+0+0+1) = 2N^2 = 1$$

$$N = \frac{1}{\sqrt{2}}$$

  - The real function, which is

$$Y_{2a}^{real} = \frac{1}{\sqrt{2}}(Y_2^2 + Y_2^{-2}) \tag{2.9.2}$$

  and

$$(Y_2^2 - Y_2^{-2}) = 2i\sqrt{(^{15}/_{16})}\left(^1/_{\sqrt{2\pi}}\right)\sin^2\theta\sin 2\phi$$

$$(Y_2^2 - Y_2^{-2}) = 4i\sqrt{(^{15}/_{16})}\left(^1/_{\sqrt{2\pi}}\right)\sin^2\theta\sin\phi\cos\phi$$

$$(Y_2^2 - Y_2^{-2}) = 4i\sqrt{(^{15}/_{16})}\left(^1/_{\sqrt{2\pi}}\right)\frac{xy}{r^2}$$

$$-i(Y_2^2 - Y_2^{-2}) = 4\sqrt{(^{15}/_{16})}\left(^1/_{\sqrt{2\pi}}\right)\frac{xy}{r^2}$$

  - Normalization:

$$N^2\int i^2(Y_2^2 - Y_2^{-2})^*(Y_2^2 - Y_2^{-2})\partial\tau = 1$$

$$i^2N^2\int \left(Y_2^2 - Y_2^{-2}\right)^* \left(Y_2^2 - Y_2^{-2}\right)\partial\tau = 1$$

$$= i^2N^2\int \left(Y_2^{2*}Y_2^2 - Y_2^{2*}Y_2^{-2} - Y_2^{-2*}Y_2^2 + Y_2^{-2*}Y_2^{-2}\right)\partial\tau$$

$$= i^2N^2(1+0+0+1) = 2i^2N^2 = 1$$

$$N = \frac{1}{i\sqrt{2}} = \frac{-i}{\sqrt{2}}$$

$$Y_{2b}^{real} = \frac{-i}{\sqrt{2}}\left(Y_l^2 - Y_l^{-2}\right) \tag{2.9.3}$$

- Table 2.3 summarizes the real angular function. The results suggest that the angular functions may be generated from the following equation:

$$A = \text{constant}\cdot\frac{x^a y^b z^c}{r^l} \tag{2.9.4}$$

where:
  - l is the l quantum number and a, b, and c may be positive (or zero) integer values such that $a+b+c=l$, and r is the distance from the origin;
  - all orbitals of lower l value of the same parity (even or odd l) will be generated; and
  - linear combinations of the solutions made to achieve an orthogonal (linearly independent) set of solutions.
- The real functions of d orbitals, if

$$\psi_{n,2,\pm m} \equiv R_{n,2}Y_2^{\pm m} \equiv nd_{\pm m} \tag{2.9.5}$$

  - By taking suitable combinations, the five d wave functions in their real forms are obtained as shown in Table 2.4:
  - Exactly the same procedure may be adopted with p orbitals (Table 2.5).

TABLE 2.3 Harmonic and Real Angular Functions

Harmonic	Real
Y_0^0	$1/\sqrt{4\pi}$
Y_1^0	$\sqrt{(3/4\pi)}\frac{z}{r}$
$Y_1^{\pm 1}$	$\sqrt{(3/4\pi)}\frac{x}{r}$ $\sqrt{(3/4\pi)}\frac{y}{r}$
Y_2^0	$\sqrt{(5/16\pi)}\frac{(2z^2 - x^2 - y^2)}{r^2}$
$Y_2^{\pm 1}$	$\sqrt{(60/16\pi)}\frac{xz}{r^2}$ $\sqrt{(60/16\pi)}\frac{yz}{r^2}$
$Y_2^{\pm 2}$	$\sqrt{(60/16\pi)}\frac{x^2 - y^2}{r^2}$ $\sqrt{(60/16\pi)}\frac{xy}{r^2}$

TABLE 2.4 The Five d Wave Functions in Their Real Forms

Complex Forms $R_{n,2}Y_2^{\pm m}$	Real Linear Combinations $R_{n,2}Y_2^{real}$
$d_{+2} = R_{n,2}Y_2^2$	$d_{x^2-y^2} = \frac{1}{\sqrt{2}}(d_{+2} + d_{-2})$
$d_{+1} = R_{n,2}Y_2^1$	$d_{xz} = \frac{1}{\sqrt{2}}(d_{+1} + d_{-1})$
$d_0 = R_{n,2}Y_2^0$	$d_{z^2} = d_0$
$d_{-1} = R_{n,2}Y_2^{-1}$	$d_{yz} = \frac{1}{\sqrt{2}}(d_{+1} - d_{-1})$
$d_{-2} = R_{n,2}Y_2^{-2}$	$d_{xy} = \frac{1}{\sqrt{2}}(d_{+2} - d_{-2})$

TABLE 2.5 The Three p Wave Functions in Their Real Forms

Complex Forms $R_{n,1}Y_1^{\pm m}$	Real Linear Combinations $R_{n,1}Y_1^{real}$
$p_{+1} = R_{n,1}Y_1^1$	$p_x = \frac{1}{\sqrt{2}}(p_{+1} + p_{-1})$
$p_0 = R_{n,1}Y_1^0$	$p_z = p_0$
$p_{-1} = R_{n,1}Y_1^{-1}$	$p_y = \frac{1}{\sqrt{2}}(p_{+1} - p_{-1})$

2.10 THE ORTHONORMAL PROPERTIES OF THE REAL WAVE FUNCTIONS

Write the real wave functions of d_{xy}, d_{xz}, d_{yz}, d_{z^2}, and $d_{x^2-y^2}$.

Verify the orthonormal properties of the real d wave functions.

- $$\text{If}: A = \text{constant} \cdot \frac{x^a y^b z^c}{r^l} \tag{2.9.4}$$

For d orbitals, $l=2$, the possible solutions for A are as follows:

$$\frac{x^2}{r^2}, \frac{y^2}{r^2}, \frac{z^2}{r^2}, \frac{xy}{r^2}, \frac{xz}{r^2}, \text{ and } \frac{yz}{r^2}.$$

Note that the s-orbital is included in the above solutions ($l=0$, the angular part does not depend on x, y, or z), because

$$\frac{x^2}{r^2} + \frac{y^2}{r^2} + \frac{z^2}{r^2} = 1$$

The two liner combinations we can construct from these three terms, which are orthogonal to the s-orbitals and to each other, are as follows:

$$d_{x^2-y^2} = \frac{x^2}{r^2} - \frac{y^2}{r^2}$$

$$d_{z^2} = \frac{2z^2}{r^2} - \frac{x^2}{r^2} - \frac{y^2}{r^2}$$

Note from Table 2.4:

- $d_{z^2} = d_0$
- $d_{yz} = \frac{1}{\sqrt{2}}(d_{+1} - d_{-1})$
- $d_{xz} = \frac{1}{\sqrt{2}}(d_{+1} + d_{-1})$
- $d_{xy} = \frac{1}{\sqrt{2}}(d_{+2} - d_{-2})$
- $d_{x^2-y^2} = \frac{1}{\sqrt{2}}(d_{+2} + d_{-2})$

- The orthonormal properties of the real d wave functions are easily verified:

$$\int d_{yz} * d_{yz}\, \partial\tau = \frac{1}{2}\int [d_{+1} - d_{-1}]^* [d_{+1} - d_{-1}]\partial\tau = \frac{1}{2}\int [d_{+1}{}^* d_{+1} - d_{+1}{}^* d_{-1} - d_{-1}{}^* d_{+1} + d_{-1}{}^* d_{-1}]\partial\tau$$
$$= \frac{1}{2}[1 - 0 - 0 + 1] = 1$$

and

$$\int d_{xy}{}^* d_{(x^2-y^2)}\, \partial\tau = \frac{1}{2}\int [d_{+2} - d_{-2}]^* [d_{+2} + d_{-2}]\partial\tau$$
$$= \frac{1}{2}\int [d_{+2}{}^* d_{+2} + d_{+2}{}^* d_{-2} - d_{-2}{}^* d_{+2} - d_{-2}{}^* d_{-2}]\partial\tau$$
$$= \frac{1}{2}[1 - 0 - 0 - 1] = 0$$

Exercise

Suppose that you have a table in which the expressions are given for:

R_{10}	R_{20}	R_{21}	R_{30}	R_{31}	R_{32}	R_{40}	
Θ_{00}	Θ_{10}	Θ_{11}	Θ_{20}	Θ_{21}	Θ_{22}	Θ_{30}	Θ_{31}
Φ_0	$\Phi_{\pm1}$	$\Phi_{\pm2}$	$\Phi_{\pm3}$	$\Phi_{\pm4}$			

Using only the functions given in this table, for which states of the hydrogen-like atom can the wave functions be written? Write them out.

Answer

	$R_{nl}\Theta_{lm}\Phi_m$
1s	$R_{10}\Theta_{00}\Phi_0$
2s	$R_{20}\Theta_{00}\Phi_0$
$2p_0$	$R_{21}\Theta_{10}\Phi_0$
$2p_{\pm1}$	$R_{21}\Theta_{11}\Phi_{\pm1}$
3s	$R_{30}\Theta_{00}\Phi_0$
$3p_0$	$R_{31}\Theta_{10}\Phi_0$
$3p_{\pm1}$	$R_{31}\Theta_{11}\Phi_{\pm1}$
$3d_0$	$R_{32}\Theta_{21}\Phi_0$
$3d_{\pm1}$	$R_{32}\Theta_{21}\Phi_{\pm1}$
$3d_{\pm2}$	$R_{32}\Theta_{22}\Phi_{\pm2}$
4 s	$R_{40}\Theta_{00}\Phi_0$

Exercise

a. **Using the recurrence formulas, obtain the wave functions (without normalization constants) for the 3s, $3p_1$, $3p_0$, $3p_{-1}$, $3d_2$, $3d_1$, $3d_0$, $3d_{-1}$, and $3d_{-2}$ state of hydrogen-like atom.**

b. **Combining the above functions to obtain the real expressions for the 3p, and 3d, states namely, $3p_x$, $3p_z$, $3p_y$, $3d_{x^2-y^2}$, $3d_{xy}$, $3d_{xz}$, $3d_{yz}$, $3d_{z^2}$.**

Answer

- We need to get: $R_{nl}\Theta_{lm}\Phi_m$

3s	$R_{30}\Theta_{00}\Phi_0$
$3p_0$	$R_{31}\Theta_{10}\Phi_0$
$3p_{\pm1}$	$R_{31}\Theta_{11}\Phi_{\pm1}$
$3d_0$	$R_{32}\Theta_{21}\Phi_0$
$3d_{\pm1}$	$R_{32}\Theta_{21}\Phi_{\pm1}$
$3d_{\pm2}$	$R_{32}\Theta_{22}\Phi_{\pm2}$

$$\Theta_{lm} = \left(1 - \cos^2\theta\right)^{m/2} G(\cos\theta) \qquad (2.7.11)$$

where

$$G(\omega) = \sum_{\substack{j=0,\,2\ldots \\ j=1,\,3\ldots}}^{l-|m|} a_j\omega^j \qquad (2.7.10)$$

$$G(\cos\theta) = \sum_j a_j \cos^j\theta,$$

and

$$l = |m| + j$$

- Let us obtain Θs:
  - $\Theta_{00} = \left(1 - \cos^2\theta\right)^{0/2} G(\cos\theta)$

    $G(\cos\theta) = a_0\cos^0\theta = a_0$

    $\therefore \Theta_{00} = a_0$
  - $\Theta_{10} = \left(1 - \cos^2\theta\right)^{0/2} G(\cos\theta)$

    $l = 1,\ |m| = 0,\ j = l - |m|,\ j = 1$

    $G(\cos\theta) = a_1\cos\theta$

    $\therefore \Theta_{10} = a_1\cos\theta$
  - $\Theta_{11} = \left(1 - \cos^2\theta\right)^{1/2} G(\cos\theta)$

    $l = 1,\ |m| = 1,\ j = l - |m|,\ j = 0$

    $G(\cos\theta) = a_0\cos^0\theta$

    $\therefore \Theta_{11} = a_0\sin\theta$
  - $\Theta_{20} = \left(1 - \cos^2\theta\right)^{0/2} G(\cos\theta)$

    $l = 2,\ |m| = 0,\ j = l - |m|,\ j = 2$

    $G(\cos\theta) = a_0 + a_2\cos^2\theta$

    $$a_{j+2} = \frac{(j+|m|)(j+|m|+1) - (l(l+1))}{(j+1)(j+2)} a_j \qquad (2.7.8)$$

    for $a_0 : j = 0$, and $m = 0$, then

    $$a_2 = \frac{(0+0)(0+0+1) - (2(2+1))}{(0+1)(0+2)} a_0 = -\frac{6}{2} a_0$$

    $G(\cos\theta) = a_0 - 3a_0\cos^2\theta$

    $\therefore \Theta_{20} = a_0\left(1 - 3\cos^2\theta\right)$

- $\Theta_{21} = (1 - \cos^2\theta)^{1/2}\, G(\cos\theta)$

  $l = 2,\ |m| = 1,\ j = l - |m|,\ j = 1$

  $G(\cos\theta) = a_1 \cos\theta$

  $\therefore \Theta_{21} = a_1 \sin\theta \cos\theta$

- $\Theta_{22} = (1 - \cos^2\theta)^{2/2}\, G(\cos\theta)$

  $l = 2,\ |m| = 2,\ j = l - |m|,\ j = 0$

  $G(\cos\theta) = a_0$

  $\therefore \Theta_{22} = a_0 \sin^2\theta$

- Let us now evaluate R's:

$$R_{nl}(\rho) = \rho^l e^{-\frac{\rho}{2}} L(\rho) \tag{2.8.11}$$

$$L(\rho) = a_0 + a_1\rho + a_2\rho^2 + \tag{2.8.7}$$

$$\rho = \frac{2r}{na_0} z$$

$$\bar{\lambda} = n = l + 1 + n'$$

- $R_{30} = \rho^0 e^{-\frac{\rho}{2}} L(\rho):\ n = 3,\ l = 0$

  $3 = 0 + 1 + n',\ \rightarrow n' = 2$

  $L(\rho) = a_0 + a_1\rho + a_2\rho^2$

$$\frac{a_{\upsilon+1}}{a_\upsilon} = \frac{\upsilon + l + 1 - \bar{\lambda}}{\upsilon(\upsilon + 1) + 2(l + 1)(\upsilon + 1)} \tag{2.8.13}$$

  for $a_0 : \upsilon = 0, \bar{\lambda} = n = 3,$ and $l = 0$

$$a_1 = \frac{0 + 0 + 1 - 3}{0(0 + 1) + 2(1)1} a_0 = -\frac{2}{2} a_0 = -a_0$$

$$a_2 = \frac{1 + 0 + 1 - 3}{1(1 + 1) + 2(1)2} a_1 = -\frac{1}{6} a_1 = \frac{1}{6} a_0$$

$$L(\rho) = a_0 - a_0\rho + \frac{1}{6} a_0\rho^2 = a_0\left(1 - \rho + \frac{1}{6}\rho^2\right)$$

$$\therefore R_{30} = a_0\left(1 - \rho + \frac{1}{6}\rho^2\right) e^{-\frac{\rho}{2}}$$

  where

$$\rho = \frac{2r}{3a_0} z$$

- $R_{31} = \rho e^{-\frac{\rho}{2}} L(\rho)$

  $\bar{\lambda} = n = l + 1 + n'$

  $3 = 1 + 1 + n' \rightarrow n' = 1$

  $L(\rho) = a_0 + a_1\rho$

$$a_1 = \frac{0+1+1-3}{0(0+1)+2(2)1}a_0 = -\frac{1}{4}a_0$$

$$L(\rho) = a_0 - \frac{1}{4}a_0\rho = a_0\left(1-\frac{1}{4}\rho\right)$$

$$\therefore R_{31} = a_0\left(1-\frac{1}{4}\rho\right)\rho e^{-\frac{\rho}{2}}$$

where

$$\rho = \frac{2r}{3a_0}Z$$

- $R_{32} = \rho^2 e^{-\frac{\rho}{2}} L(\rho)$

$$\lambda = n = l + 1 + n'$$

$$3 = 2 + 1 + n' \rightarrow n' = 0$$

$$L(\rho) = a_0$$

$$\therefore R_{32} = a_0\rho^2 e^{-\frac{\rho}{2}}$$

where

$$\rho = \frac{2r}{3a_0}Z$$

- Now we can write the entire functions:

$$\therefore 3s: R_{30}\theta_{00}\phi_0 : N\left(1-\rho+\frac{1}{6}\rho^2\right)e^{-\frac{\rho}{2}} : \rho = \frac{2r}{3a_0}Z$$

$$\therefore 3p_0 : R_{31}\theta_{10}\phi_0 : N\left(1-\frac{1}{4}\rho\right)\rho e^{-\frac{\rho}{2}}\cos\theta$$

$$\therefore 3p_{\pm1} : R_{31}\theta_{11}\phi_{\pm1} : N\left(1-\frac{1}{4}\rho\right)\rho e^{-\frac{\rho}{2}}\sin\theta\, e^{\pm i\phi}$$

$$\therefore 3d_0 : R_{32}\theta_{20}\phi_0 : N\rho^2 e^{-\frac{\rho}{2}}\left(1-3\cos^2\theta\right)$$

$$\therefore 3d_{\pm1} : R_{32}\theta_{21}\phi_{\pm1} : N\rho^2 e^{-\frac{\rho}{2}}\sin\theta\cos\theta e^{\pm i\phi}$$

$$\therefore 3d_{\pm2} : R_{32}\theta_{22}\phi_{\pm2} : N\rho^2 e^{-\frac{\rho}{2}}\sin^2\theta e^{\pm 2i\phi}$$

where N is the normalization constant.

c. From Tables 2.4 and 2.5:

$3d_{z^2} = 3d_0 \quad 3p_z = 3p_0$

$3d_{xz} = 3d_{+1} + 3d_{-1} \quad 3p_x = 3p_{+1} + 3p_{-1}$

$3d_{yz} = -i(3d_{+1} - 3d_{-1}) \quad 3p_y = -i(3p_{+1} - 3p_{-1})$

$3d_{xy} = -i(3d_{+2} - 3d_{-2})$

$3d_{x^2-y^2} = 3d_{+2} + 3d_{-2}$

So, if

$e^{i\phi} + e^{-i\phi} = 2\cos\phi$ and $e^{i\phi} - e^{-i\phi} = -2i\sin\phi$

$$3p_x = N\left(1-\frac{1}{4}\rho\right)\rho e^{-\frac{\rho}{2}}\sin\theta\cos\phi$$

$$3p_y = N\left(1-\frac{1}{4}\rho\right)\rho e^{-\frac{\rho}{2}}\sin\theta\sin\phi$$

$$3d_{xz} = N\rho^2 e^{-\frac{\rho}{2}}\sin\theta\cos\theta\cos\phi$$

$$3d_{yz} = N\rho^2 e^{-\frac{\rho}{2}}\sin\theta\cos\theta\sin\phi$$

$$3d_{xy} = N\rho^2 e^{-\frac{\rho}{2}}\sin^2\theta\sin 2\phi$$

$$3d_{x^2-y^2} = N\rho^2 e^{-\frac{\rho}{2}}\sin^2\theta\cos 2\phi$$

2.11 THE QUANTUM NUMBERS: *n, l,* and m_l

The Principle Quantum Number, *n*

Identify the physical meaning, specifications, and possible values, of the principle quantum numbers *n*.

- The quantum number n is called the principle quantum number, comes from the radial solution, and takes the values: $n = 1, 2, 3, 4\ldots$ which known as $K, L, M, N, \ldots$.
- The principle quantum number, n, specifies the energy of the electron through the following:

$$E_n = \frac{\mu Z^2 e^4}{32\pi^2 \epsilon_o^2 \hbar^2 n^2} \tag{2.8.16}$$

- All orbitals of a given value of n form a single shell of the atom, and have the same energy.

The Quantum Numbers *l* and Angular Momentum

If the radial equation in the electron-proton system:

$$\frac{1}{r^2}\frac{\partial}{\partial r}\left(r^2\frac{\partial R}{\partial r}\right) + \left\{\frac{2m}{\hbar^2}[E - V(r)] - \frac{\lambda}{r^2}\right\}R = 0 \tag{2.4.27}$$

where $\lambda = l(l+1)$

a. **Find:**
   i. **the components of the effective potential energy;**
   ii. **the centrifugal force;**
   iii. **the relationship between the centrifugal force and the angular momentum; and**
   iv. **the relationship between the centrifugal potential and the angular momentum.**
b. **Define the angular momentum *L*, and show that $L = \hbar\sqrt{l(l+1)}$.**
c. **Identify the type of forces that control the effective potential when $l = 0$, $l \neq 0$, and *r* is very small or very large.**

- The radial function:

$$\frac{1}{r^2}\frac{\partial}{\partial r}\left(r^2\frac{\partial R}{\partial r}\right) + \left\{\frac{2m}{\hbar^2}[E - V(r)] - \frac{\lambda}{r^2}\right\}R = 0 \tag{2.4.27}$$

  If

$$R = \frac{x(r)}{r}, \quad \text{then}$$

$$\frac{\partial R}{\partial r} = \frac{\partial}{\partial r}\frac{x(r)}{r} = \frac{1}{r}x' - \frac{1}{r^2}x$$

$$r^2\frac{\partial R}{\partial r} = rx' - x$$

$$\frac{\partial}{\partial r}\left(r^2\frac{\partial R}{\partial r}\right) = x' + rx'' - x' = rx''$$

$$\frac{1}{r^2}\frac{\partial}{\partial r}\left(r^2\frac{\partial R}{\partial r}\right) = \frac{x''}{r}$$

$$\frac{x''}{r} + \left\{\frac{2m}{\hbar^2}[E - V(r)] - \frac{\lambda}{r^2}\right\}\frac{x}{r} = 0$$

- This equation looks like that of a particle moving in one-dimensional potential.

$$\because \lambda = l(l+1), \text{ then:}$$

$$-\frac{\hbar^2}{2mr}x'' + \underbrace{\left(\underbrace{V(r)}_{\text{Potential energy}} + \underbrace{\frac{\hbar^2 l(l+1)}{2mr^2}}_{\text{Centrifugal potential energy}}\right)}_{\text{Effective potential energy}} x = Ex \tag{2.11.1}$$

- The first term $V(r)$ is coulomb potential energy of the electron in the field of the nucleus:

$$V(r) = -\frac{Ze^2}{(4\pi\epsilon_o)r} \tag{2.11.3}$$

- The second term is from the centrifugal force that arises from of the electron around the nucleus.
- Let us consider a particle is rotating about an axis at position r with angular velocity ω. The angular momentum L of the particle is given by $L = mr^2\omega$.
- This angular momentum is associated by a centrifugal force, $F_{\text{centrifugal}}$:

$$\because F_{\text{centrifugal}} = m\omega^2 r = \frac{m^2\omega^2 r^4}{mr^3} = \frac{L^2}{mr^3}$$

$$\text{if}: \frac{\partial V_{\text{centrifugal}}}{\partial r} = F_{\text{centrifugal}}$$

where V is the associate centrifugal potential energy, then:

$$\int \partial V_{\text{centrifugal}} = \int F_{\text{centrifugal}} \partial r = \int m\omega^2 r \, \partial r$$

- The associate centrifugal potential energy with this force have a value:

$$\therefore V_{\text{centrifugal}} = \frac{m\omega^2 r^2}{2} = \frac{m^2\omega^2 r^4}{2mr^2} = \frac{L^2}{2mr^2}$$

- This term has the same value as the extra potential energy in

$$-\frac{\hbar^2}{2mr}x'' + \left(V(r) + \frac{\hbar^2 l(l+1)}{2mr^2}\right)x = Ex \tag{2.11.1}$$

This shows identification for the angular momentum, L:

$$L^2 = \hbar^2 l(l+1)$$

$$L = \hbar\sqrt{l(l+1)} \tag{2.11.2}$$

- When $l = 0$, the electron has no angular momentum, and the effective potential energy is purely coulombic, and attractive at all radii.
- When $l \neq 0$, the centrifugal term gives a positive contribution to the attractive potential energy.
- When the electron is close to the nucleus (r = small), the repulsive term dominates the attractive coulombic component and the net effect is essentially a repulsion of the electron from the nucleus.
- The two effective potential energies, $l = 0$ and $l \neq 0$, are qualitatively very different close to the nucleus. However, they are very similar at large distances.

The Angular Momentum Quantum Numbers, *l* and *m*

Identify the physical meaning, specifications, and possible values, of the angular quantum numbers *l* and *m*.

- The two quantum numbers; l, and m_l, come from the angular solution.
- The value of l specifies the magnitude of the orbital angular momentum of the electron around the nucleus. An electron in an orbital with quantum number l has an angular momentum: $L = \hbar\sqrt{l(l+1)}$. The values of l can range from $l=0$ to $l=n-1$, thus: $l=0, 1, 2, 3, 4\ldots, n-1$.
- The orbital with the same value of n but different values of l form a subshell of a given shell, and generally referred to by the letters: $s, p, d, f, g\ldots$.
- On the other hand, the z-component of angular momentum is restricted to $m_l\hbar$, thus; m_l specifies the orientation of the electron's orbital angular momentum, L.
- For a given value of l, m_l can range from $-l$ to $+l$, thus $m_l=0, \pm 1, \pm 2, \pm 3, \ldots\ldots, \pm l$.
- The value of l suggests that the component of angular momentum about the z-axis may take only $2l+1$ values, therefore, there are $2l+1$ wave functions of different orbital angular momentum orientations.

Picture and Represent Precisely m_l Vectors of p- and d-Orbitals

- The angular momentum is represented by a vector of length proportional to its magnitude (i.e., of length $\sqrt{l(l+1)}$).
- Then to represent correctly the value of the component of angular momentum, the vector must be oriented so that its projection on the z-axis is of length m_l unit.
- A p-electron has a non-zero angular momentum, it actual magnitude:

$$L^2 = \hbar^2 l(l+1) = 1(1+1)\cdot\hbar^2,$$

$$L = \sqrt{2}\hbar$$

Where $L^2 = l_x^2 + l_y^2 + l_z^2$

- There are $2l+1$ possible orientations of L, corresponding to the $2l+1$ values of m (Fig. 2.3).
- The allowed orientations of angular momentum in the case of d-electron, $l=2$, are:
  - $m_l=+2, +1, 0, -1, -2$ (Fig. 2.4).
- The value of m_l has referred to the z-component of angular momentum (l_z, the component about arbitrary axis, which normally labeled z), and no reference has made to the x- or y-components (l_x, and l_y the components about two axes perpendicular to l_z).

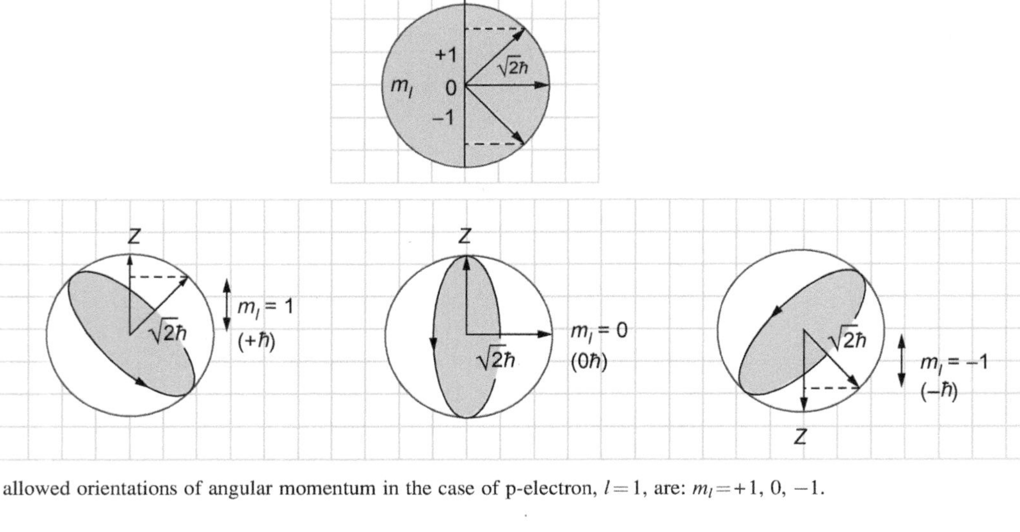

FIG. 2.3 The allowed orientations of angular momentum in the case of p-electron, $l=1$, are: $m_l=+1, 0, -1$.

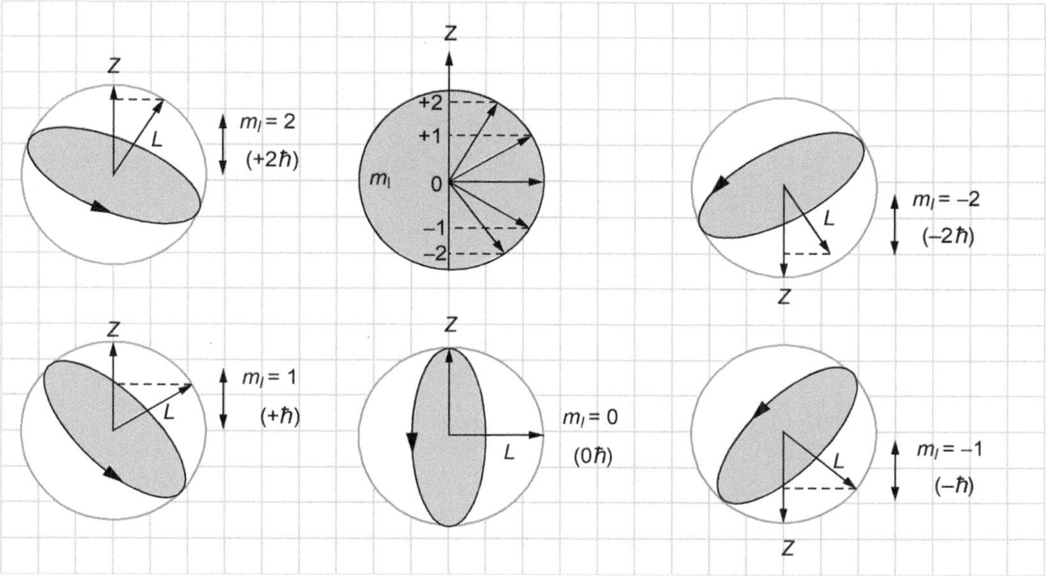

FIG. 2.4 The allowed orientations of angular momentum in the case of d-electron, $l = 2$, are: $m_l = +2, +1, 0, -1, -2$.

- The basis for this exclusion is that the uncertainty principle forbids the simultaneous and exact specification of more than one component.
- We cannot specify l_x and l_y; the vector L can lie anywhere on the surface of a cone whose axis is the z-axis, whose attitude is $m\hbar$, and whose slant height is $\sqrt{l(l+1)}$.
- Therefore in the vector model, cones are drawn with side $\sqrt{l(l+1)}$ units to represent the magnitude of angular momentum (Fig. 2.5).
- This means that the plane of rotation can take only a discrete range of orientations.
- This situation allows us to picture a precession of angular momentum vector L about the z-axis, given zero to x and y components on the average.
- Each cone has a definite projection (of m_l units) on the z-axis, representing the system's precise value of l_z (Fig. 2.5).

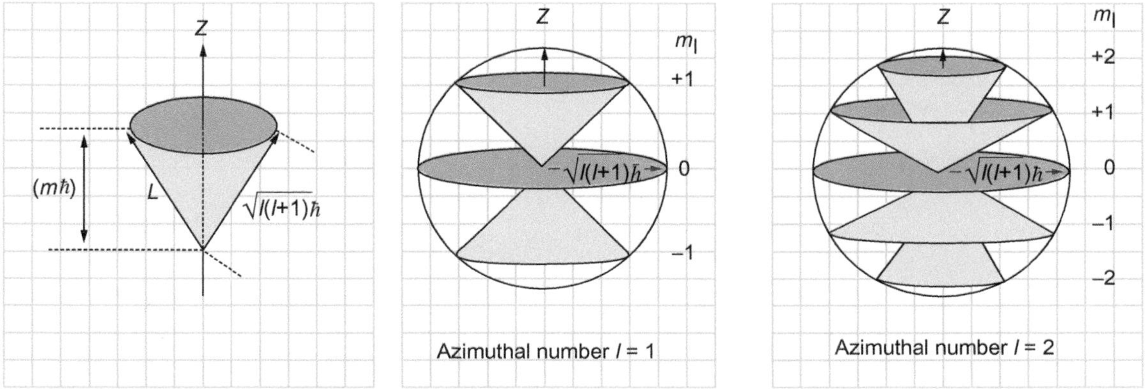

FIG. 2.5 Angular momentum cone models.

Define: l_x, l_y, l_z, L, and show that:

a. $i\hbar \dfrac{\partial}{\partial x}\psi = p_x\psi$

b. $l_z = -i\hbar \dfrac{\partial}{\partial \phi}$

c. $l_x = i\hbar\left(\cot\theta\cos\phi \dfrac{\partial}{\partial\phi} + \sin\phi\dfrac{\partial}{\partial\theta}\right)$

d. $l_y = i\hbar\left(\cot\theta\sin\phi \dfrac{\partial}{\partial\phi} - \cos\phi\dfrac{\partial}{\partial\theta}\right)$

e. $\hat{L}^2$ is identical with the angular part of the Laplacian operator in spherical coordinates:

$$\hat{L}^2 = -\hbar^2\left[\frac{1}{\sin\theta}\frac{\partial}{\partial\theta}\left(\sin\theta\frac{\partial}{\partial\theta}\right) + \frac{1}{\sin^2\theta}\frac{\partial^2}{\partial\phi^2}\right]$$

a. The z-component of orbital angular momentum, l_z, exhibited by d wave functions will be considered (l_x, and l_y are the components about two axes perpendicular to l_z).
- The quantum mechanical operator corresponding to the angular momentum l_z is constructed by writing down the classical expression and converting the expression to operator.
- Firstly, from the classical mechanics of one-particle angular momentum: consider a moving particle of mass m with velocity u, and let r be the vector from the origin to instantaneous position of the particle (Fig. 2.6).
  - In terms of unit vectors i, j, and k directed along the x, y, and z axes, respectively, the vector r can be written in terms of its Cartesian components as

$$r = ix + jy + kz \tag{2.11.4}$$

  - The velocity:

$$u = \frac{\partial r}{\partial t} = i\frac{\partial x}{\partial t} + j\frac{\partial y}{\partial t} + k\frac{\partial z}{\partial t}, \quad \text{where}$$

$$u_x = \frac{\partial x}{\partial t}, \quad u_y = \frac{\partial y}{\partial t}, \quad u = \frac{\partial z}{\partial t}$$

  - We define the particle's linear momentum vector p by

$$p_x = m\,u_x, \quad p_y = mu_y, \quad p_z = mu_z$$

  - The particle's angular momentum L with respect to the coordinate origin is defined as

$$L = r \cdot p \tag{2.11.5}$$

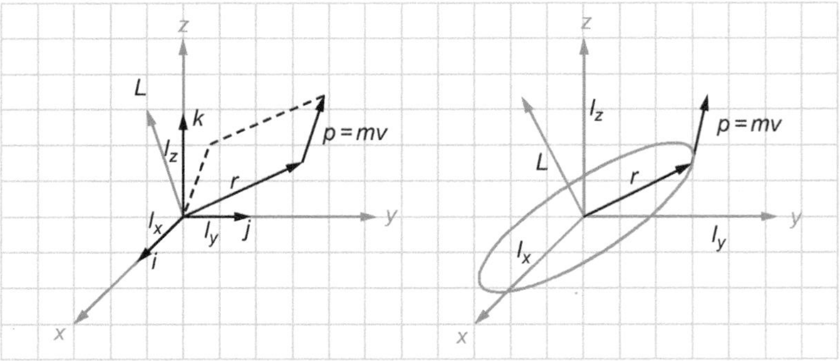

FIG. 2.6 The angular momentum L of a particle of mass m with velocity v, and r is the vector from the origin to the particle.

- Then, in terms of unit vectors i, j, and k,

$$L = (ix + jy + kz)(ip_x + jp_y + kp_z)$$

$$\because ixi = jxj = kxk = 0,$$

$$ixj = k,\ jxi = -k$$

$$jxk = i, kxj = -i$$

$$kxi = j,\ ixk = -j$$

$$L = il_x + jl_y + kl_z = i(yp_z - zp_y) - j(zp_x - xp_z) + k(xp_y - yp_x),\ \text{or}$$

$$L = \begin{vmatrix} i & j & k \\ x & y & z \\ p_x & p_y & p_z \end{vmatrix}$$

- Therefore,

$$l_x = yp_z - zp_y \tag{2.11.6}$$

$$l_y = zp_x - xp_z \tag{2.11.7}$$

$$l_z = xp_y - yp_x \tag{2.11.8}$$

$$L^2 = l_x^2 + l_y^2 + l_z^2 \tag{2.11.9}$$

where only the orbital angular momentum is considered.

- Secondly, from the quantum-mechanical: the operators for the components of the orbital angular momentum of a particle will be determined.
  - If the wave function of a moving particle in x-direction:

$$\psi(x, t) = a_0 e^{2\pi i\left(\frac{x}{\lambda} - \nu t\right)} \tag{2.1.1}$$

where a_0 is constant and λ and ν are wavelength and frequency of the wave which describe the behavior of the particle, and $i^2 = -1$.

  - Then the differential equation of the wave function ψ was derived, if:

$$y = ae^{bx},\ \text{then}$$

$$\frac{\partial \boldsymbol{y}}{\partial} = \boldsymbol{abe^{bx}},\ \textbf{consequently,}$$

$$\frac{\partial \boldsymbol{\psi(x,t)}}{\partial \boldsymbol{dx}} = \left(\frac{\boldsymbol{2i\pi}}{\boldsymbol{\lambda}}\right)\boldsymbol{\psi(x,t)}$$

  - If momentum: $p = mv$, and

$$\because \lambda = \frac{h}{p} \tag{1.12.1}$$

$$\therefore \frac{1}{\lambda} = \frac{p}{h},\ \hbar = \frac{h}{2\pi},\ \text{and}$$

$$\therefore \frac{\partial \psi(x, t)}{\partial x} = \left(\frac{ip_x}{\hbar}\right)\psi(x, t)$$

$$-i\hbar\frac{\partial}{\partial x}\psi(x, t) = p_x\psi(x, t) \tag{2.11.10}$$

b. Then

$$p_x \xrightarrow{\text{the operator}} -i\hbar\frac{\partial}{\partial x} \tag{2.11.11}$$

similarly,

$$p_y \xrightarrow{\text{the operator}} -i\hbar\frac{\partial}{\partial y} \quad (2.11.12)$$

and

$$p_z \xrightarrow{\text{the operator}} -i\hbar\frac{\partial}{\partial z} \quad (2.11.13)$$

- By replacing the coordinates and momenta in the classical equations by their corresponding operators:

$$\because l_z = xp_y - yp_x \quad (2.11.8)$$

$$\therefore l_z = -i\hbar\left(x\frac{\partial}{\partial y} - y\frac{\partial}{\partial x}\right)$$

$$x = r\sin\theta\cos\phi \qquad y = r\sin\theta\sin\phi$$

$$\frac{\partial}{\partial x} = \sin\theta\cos\phi\frac{\partial}{\partial r} + \frac{\cos\theta\cos\phi}{r}\frac{\partial}{\partial\theta} - \frac{\sin\phi}{r\sin\theta}\frac{\partial}{\partial\phi} \quad (2.4.17)$$

$$\frac{\partial}{\partial y} = \sin\theta\sin\phi\frac{\partial}{\partial r} + \frac{\cos\theta\sin\phi}{r}\frac{\partial}{\partial\theta} + \frac{\cos\phi}{r\sin\theta}\frac{\partial}{\partial\phi}$$

$$\frac{\partial}{\partial z} = \cos\theta\frac{\partial}{\partial r} - \frac{\sin\theta}{r}\frac{\partial}{\partial\theta} \quad (2.4.18\&19)$$

$$l_z = -i\hbar\left\{\begin{array}{l} r\sin\theta\cos\phi\left(\sin\theta\sin\phi\frac{\partial}{\partial r} + \frac{\cos\theta\sin\phi}{r}\frac{\partial}{\partial\theta} - \frac{\cos\phi}{r\sin\theta}\frac{\partial}{\partial\phi}\right) \\ -r\sin\theta\sin\phi\left(\sin\theta\cos\phi\frac{\partial}{\partial r} + \frac{\cos\theta\cos\phi}{r}\frac{\partial}{\partial\theta} - \frac{\sin\phi}{r\sin\theta}\frac{\partial}{\partial\phi}\right) \end{array}\right\}$$

$$l_z = -i\hbar\left(\cos^2\phi + \sin^2\phi\right)\frac{\partial}{\partial\phi}$$

$$l_z = -i\hbar\frac{\partial}{\partial\phi} \quad (2.11.14)$$

similarly:

c. $l_x = i\hbar\left(\cot\theta\cos\phi\frac{\partial}{\partial\phi} + \sin\phi\frac{\partial}{\partial\theta}\right)$, and

d. $l_y = i\hbar\left(\cot\theta\sin\phi\frac{\partial}{\partial\phi} - \cos\phi\frac{\partial}{\partial\theta}\right)$

e. $\hat{L}^2 = \hat{l}_x^2 + \hat{l}_y^2 + \hat{l}_z^2$ (2.11.9)

becomes

$$\hat{L}^2 = -\hbar^2\left[\frac{1}{\sin\theta}\frac{\partial}{\partial\theta}\left(\sin\theta\frac{\partial}{\partial\theta}\right) + \frac{1}{\sin^2\theta}\frac{\partial^2}{\partial\phi^2}\right] \quad (2.11.15)$$

Note that this expression is identical with the angular part of the Laplacian operator in spherical coordinates: a useful result.

Compute the orbital angular momentum that associated with the transform of:

a. d_{xy} **into** $d_{(x^2-y^2)}$ **b.** d_{xz} **into** d_{yz}, **c.** d_{xz} **into** d_{xz}
d. d_{yz} **into** d_{yz} **e.** d_{z^2} **into** d_{xy},

What are the expected requirements for these transformations?

- To find the observable value of l_z that corresponds to a wave function ψ, we have to evaluate the quantity:

$$\int \psi * l_z \psi \, \partial\tau$$

where,

$$\psi = \psi_{n,l,m} \equiv R_{n,l} Y_l^m,$$

$$\because p_\phi = m\hbar = \text{angular momentum in } z-\text{direction}$$

$$\therefore l_z \equiv m_l \hbar$$

Thus:

$$\underbrace{l_z}_{\text{operator}} \underbrace{\psi_{n,l,m'}}_{\text{wave function}} = \underbrace{m_l\hbar}_{\text{value}} \underbrace{\psi_{n,l,m'}}_{\text{wave function}}$$

and

$$\int \psi^*_{n,l,m} l_z \psi_{n,l,m'} \partial\tau = \int \psi^*_{n,l,m} m_l \hbar \psi_{n,l,m'} \partial\tau = m_l\hbar \int \psi^*_{n,l,m} \psi_{n,l,m'} \partial\tau$$

$$= m_l\hbar \rightarrow \text{when } \psi_{n,l,m} = \psi_{n,l,m'} \tag{2.11.21}$$

$$= 0 \rightarrow \text{when } \psi_{n,l,m} \neq \psi_{n,l,m'}$$

- If $\psi_{n,2,\pm m} \equiv R_{n,2} Y_2^{\pm m} \equiv nd_{\pm m}$, then from Table 2.4:

$$d_{z^2} = d_{m=0} = d_0$$

$$d_{yz} = \frac{1}{\sqrt{2}}(d_{+1} - d_{-1})$$

$$d_{xz} = \frac{1}{\sqrt{2}}(d_{+1} + d_{-1})$$

$$d_{xy} = \frac{1}{\sqrt{2}}(d_{+2} - d_{-2})$$

$$d_{x^2-y^2} = \frac{1}{\sqrt{2}}(d_{+2} + d_{-2})$$

- Thus:

$$\int d_{xy}{}^* l_z d_{xy} \, \partial\tau = \frac{1}{2} \int [(d_{+2} - d_{-2}]^* l_z [d_{+2} - d_{-2}] \partial\tau$$

$$= \frac{1}{2} \int [d_{+2}{}^* l_z d_{+2} - d_{+2}{}^* l_z d_{-2} - d_{-2}{}^* l_z d_{+2} + d_{-2}{}^* l_z d_{-2}] \partial\tau$$

$$= \frac{1}{2} \int [2d_{+2}{}^* d_{+2} + 2d_{+2}{}^* d_{-2} - 2d_{-2}{}^* d_{+2} - 2d_{-2}{}^* d_{-2}] \hbar \partial\tau = \frac{\hbar}{2} \int [2-0-0-2] = 0$$

while

$$\int d_{xy}{}^* l_z d_{(x^2-y^2)} \partial\tau = \frac{1}{2} \int [d_{+2} - d_{-2}]^* l_z [d_{+2} + d_{-2}] \partial\tau$$

$$= \frac{\hbar}{2} \int [2d_{+2}{}^* d_{+2} - 2d_{+2}{}^* d_{-2} - 2d_{-2}{}^* d_{+2} + 2d_{-2}{}^* d_{-2}] \partial\tau$$

$$= \frac{\hbar}{2}(2-0-0+2) = 2\hbar$$

Similarly:

$$\int d_{xz}{}^* l_z d_{xz} \partial\tau = \int d_{yz}{}^* l_z d_{yz} \partial\tau = 0$$

$$\int d_{xz}{}^* l_z d_{yz} \partial\tau = \hbar$$

while

$$\int d_{z^2}{}^* l_z d_{xy} \partial\tau, \text{ etc}\ldots = \int d_{z^2}{}^* l_z d_{z^2} \partial\tau = 0$$

- Transform $d_{(x^2-y^2)}$ into d_{xy} and d_{yz} into d_{xz}; the orbital angular momentum is to be associated with these two pairs of orbitals.
- The association of orbital angular momentum with such a rotation is conditional upon the fact that the two orbitals, which transform into each other under rotation, are degenerate.
- If the degeneracy is removed, the associated orbital angular momentum is lost.

2.12 THE SPIN QUANTUM NUMBER, *s*

Illustrate how Sten-Gerlach experiment led to the discovery of *s* quantum number; identify the physical meaning, specifications, and the possible values of *s* quantum number.

- When atoms are placed in a magnetic field, the energy levels of the electrons split into more than one component, as per the Sten-Gerlach experiment. The splitting can be seen in the line spectra of atoms; this is called the Zeeman Effect (Fig. 2.7).
- In the quantum theory of rotating systems, the direction of the angular momentum can only adopt one of $(2\,l+1)$ possible values.
- In the Sten-Gerlach experiment, a beam of silver split into two, and similar results are found for alkali atoms. Why are these lines not single as predicted by Bohr's model?
- Some things seem unacceptable with this:
  - Firstly, the outer electron in silver and alkali atoms is an s-orbital, with $l=0$, and so should not have any angular momentum at all.
  - Secondly, there is no way that $(2\,l+1=2)$ can yield two for any integral value of l.
- The explanation proposed is that the electron has its own intrinsic angular momentum.
- The value of this is specified by an additional quantum number s, spin quantum number, taking the value of

$$m_s = \pm\frac{1}{2} \tag{2.12.1}$$

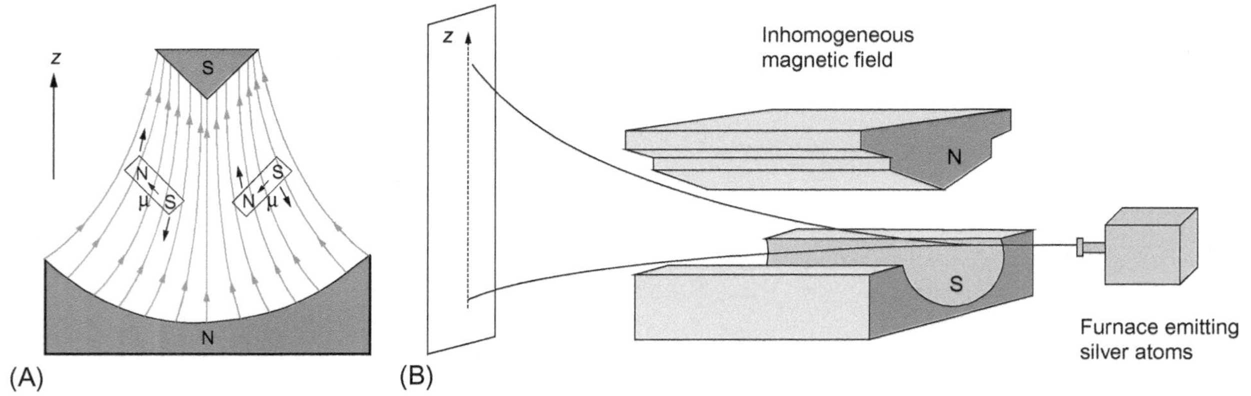

FIG. 2.7 (A) The deflection of magnetic dipoles in inhomogeneous magnetic field. (B) The Sten-Gerlach experiment, showing a beam of atoms splitting into two on passing through an inhomogeneous magnetic field.

- By analogy with orbital angular momentum:
  - the number of possible orientations of the spin in the magnetic field is:

$$2s+1=2(1/2)+1=2 \tag{2.12.2}$$

- Therefore, the operator, $\hat{s}$, that measure magnetic moment has only two eigenfunctions, α and β, with eigenvalues that have equal magnitude but opposite min sign.

  The total spin angular momentum:

$$\hat{s}\alpha=\hbar\sqrt{s(s+1)}\alpha=\hbar\sqrt{\frac{1}{2}\left(\frac{1}{2}+1\right)}\alpha=\hbar\sqrt{\frac{3}{4}}\alpha \tag{2.12.3}$$

$$\hat{s}\beta=\hbar\sqrt{s(s+1)}\beta=\hbar\sqrt{\frac{1}{2}\left(\frac{1}{2}+1\right)}\beta=\hbar\sqrt{\frac{3}{4}}\beta \tag{2.12.4}$$

The z-component of spin angular momentum:

$$\hat{s}_z\alpha=m_s\alpha=-\frac{\hbar}{2}\alpha \tag{2.12.5}$$

$$\hat{s}_z\beta=m_s\alpha=\frac{\hbar}{2}\beta \tag{2.12.6}$$

- The wave function that describes a single silver atom:

$$\psi_{\text{Ag}}=c_1\alpha+c_2\beta$$

with

$$c_1^2+c_2^2=1$$

- The relative number of silver atoms that was deflected upward and downward are equal, thus:

$$|c_1^2|_{\text{average}}=|c_2^2|_{\text{average}}=\frac{1}{2}$$

The average is over all the atoms that landed on the detector, the wave function of the silver atom:

$$\psi_{\text{Ag}}=\frac{1}{\sqrt{2}}(\alpha+\beta)$$

- The electron spin is essential for understanding the complex atom and the periodic table.
- To characterize the state of an electron in the hydrogenic atom, we need to specify not only the orbital it occupies but also its spin state.
- The pairing of electrons in many covalently bonded molecules relies upon spin.
- Molecules and solids wherein not all electrons are paired in bonds have magnetic properties.
- The elementary fermions such as quarks, leptons, neutrinos, and antiparticles can be in two spin states, indicated by $+\frac{1}{2}\hbar$ and $-\frac{1}{2}\hbar$, which represent the projection of its spin onto the z-coordinate axis.
- On the other hand, bosons such as photon, W^+, W^-, Z, and gluon have integral values of spin, $1\hbar$. These are called vector bosons, as opposed to a scalar boson with spin 0 (Higgs boson).

2.13 THE BOUNDARY SURFACE OF s-ORBITAL

Write the wave function of 1s orbital, and find the boundary surface of this orbital.

- The s-orbital is described by the wave function $\psi_{n,0,\ 0}$.
- When $n=1$, there is only one subshell, that with $l=0$, and that contains only one orbital, with $m_l=0$, and the wave function:

$$\psi_{1,0,\ 0}\equiv R_{10}Y_0^0\equiv R_{10}\Theta_{00}\Phi_0 \tag{2.9.5}$$

- If

$$R_{nl}(\rho) = -\left\{\left(\frac{2Z}{na_0'}\right)^3 \frac{(n-l-1)!}{2n[(n+l)!]^3}\right\}^{1/2} e^{-\rho/2}\rho^l L_{n+l}^{2l+1}(\rho) \tag{2.8.21}$$

where

$$\text{where, } a_0' = \frac{\hbar^2(4\pi\epsilon_o)}{\mu r^2} \tag{2.8.22}$$

$$\rho = \frac{2Zr}{na_0'} \tag{2.8.23}$$

$$L_{n+l}^{2l+1} = \frac{\partial^{2l+1}}{\partial\rho^{2l+1}} L_{n+l}^0, \quad n, l = 0, 1, 2, \tag{2.8.24}$$

and

$$L_{n+l}^0 \equiv L_{n+l} = e^\rho \frac{\partial^{n+l}}{\partial\rho^{n+l}}(e^{-\rho}\rho^{n+l}), \quad n = 0, 1, 2, \tag{2.8.25}$$

than

$$R_{10}(\rho) = -\left\{\left(\frac{2Z}{a_0'}\right)^3 \frac{(1-0-1)!}{2[(1+0)!]^3}\right\}^{1/2} e^{-zr/a_o'}\left(\frac{2zr}{a_0'}\right)^0 L_1^1(\rho)$$

$$R_{10}(\rho) = -2\left(\frac{Z}{a_0'}\right)^{3/2} e^{-zr/a_o'} L_1^1(\rho) \text{ and}$$

$$L_1^1(\rho) = \frac{\partial}{\partial\rho} L_1^0$$

$$L_1^0 \equiv L_{n+l} = e^\rho \frac{\partial}{\partial\rho}(e^{-\rho}\rho) = e^\rho(e^{-\rho} - \rho e^{-\rho}) = 1 - \rho$$

$$L_1^1(\rho) = (-1)^0 \frac{\partial}{\partial\rho} L_1^0 = (1)(-1) = -1$$

$$R_{10}(\rho) = 2\left(\frac{Z}{a_0'}\right)^{3/2} e^{-zr/a_o'}$$

$$Z = 1, \quad a_0' = a_0$$

$$R_{10}(\rho) = 2\left(\frac{1}{a_0}\right)^{3/2} e^{-r/a_o}$$

- $Y_l^m(\theta, \phi) = N_{ml} P_l^m \cos\theta \Phi_m(\phi)$

$$Y_l^m = (-1)^m \left[\frac{(2l+1)}{4\pi}\frac{(l-|m|)!}{(l+|m|)!}\right]^{1/2} P_l^{|m|}(\cos\theta)\, e^{im\phi} : m \geq 0 \tag{2.7.18}$$

$$P_l^{|m|}(\omega) = \frac{(1-\omega^2)^{|m|/2}}{2^l l!} \frac{\partial^{l+|m|}}{\partial\omega^{l+|m|}}(\omega^2 - 1)^l \tag{2.7.19}$$

$$P_0^0(\omega) = 1$$

$$Y_0^0 = (-1)^0 \left[\frac{(1)}{4\pi}\frac{(1)}{(1)}\right]^{1/2} (1)(1) = \left[\frac{1}{4\pi}\right]^{1/2}$$

- Then

$$\therefore \psi_{1,0,0} = R_{10}Y_0^0 = \left[\frac{1}{4\pi}\right]^{1/2}\left(2\left(\frac{1}{a_0}\right)^{3/2} e^{-r/a_o}\right) = \left(\frac{1}{\pi a_0^3}\right)^{1/2} e^{-r/a_0} \tag{2.13.1}$$

- This wave function is independent of angle; it has the same value at all points of constant radius. As a result, 1s is spherically symmetrical.
- Furthermore, the angular functions may be generated from the following equation:

$$A = \frac{x^a y^b z^c}{r^l}, \quad \text{for } l = 0,$$

$$A = \frac{x^0 y^0 z^0}{r^0} = 1$$

- The s-orbital has no angular dependence; it is spherically symmetric, and the radial functions completely describe the s-orbital (Fig. 2.8).

2.14 THE BOUNDARY SURFACE OF p-ORBITALS

Draw graphically the boundary surface of a p-orbital, and write the real wave functions of p_z, p_x, and p_y.

- The boundary surface of p-orbital is obtained by the angular function.
- The angular functions may be generated from the following equation:

$$A = \text{constant} \cdot \frac{x^a y^b z^c}{r^l}$$

  - For $l = 1$, the possibility for A are

$$A = \text{constant} \cdot \frac{x}{r},$$

$$A = \text{constant} \cdot \frac{y}{r}, \quad \text{and}$$

$$A = \text{constant} \cdot \frac{z}{r}$$

  - Making a plot of A versus $\frac{x}{r}$, $\frac{y}{r}$, and $\frac{z}{r}$ will establish their shape.

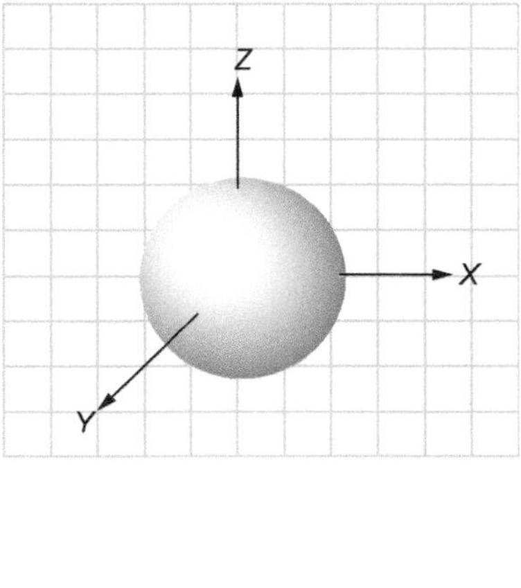

FIG. 2.8 The boundary surface of the s-orbital.

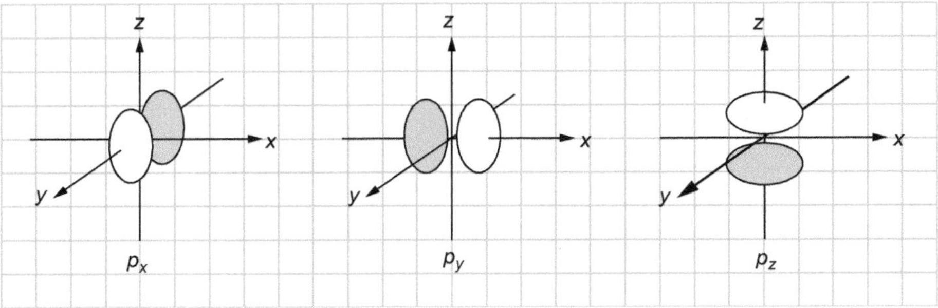

FIG. 2.9 The boundary surfaces of p-orbitals. The *light* and *dark lobes* have opposite sign of the wave function. The *nodal plane* goes through the nucleus, and separates the two lobes of each orbital.

- Drawing a plot of the function $A = \frac{x}{r}$.
  - $x = r \sin\theta \cos\phi$

    $\frac{x}{r} = \sin\theta \cos\phi$
  - Because the function, $\frac{x}{r}$, depends on two variables, θ and ϕ, we would need a three-dimensional plot to represent it.
  - Choosing a particular value of θ gives the projection of the function in a plane, $\theta = \frac{\pi}{2}$ gives xy plane (Fig. 2.9).
- Going up or down from the xy plane, the value of θ increases or decreases from its value of $\frac{\pi}{2}$ in the plane and $\sin\theta$ becomes <1. The value of the function is reduced by the factor $\sin\theta$ for each ϕ value until it is 0 for $\theta = 0$ and π. Thus, the function is the surface of a pair of tangent spheres.
- Note the real wave functions of p orbitals (Table 2.6):

$$P_z = P_0$$

$$P_y = \frac{1}{\sqrt{2}}(P_1 - P_{-1})$$

$$P_x = \frac{1}{\sqrt{2}}(P_1 + P_{-1})$$

TABLE 2.6 Data for the Function $A = \frac{x}{r} = \sin\theta \cos\phi$ in xy Plane

θ	ϕ	$A = \sin\theta \cos\phi = 1 \times \cos\phi$
$\frac{\pi}{2}$	0	1
$\frac{\pi}{2}$	$\frac{\pi}{4}$	0.707
$\frac{\pi}{2}$	$\frac{\pi}{2}$	0
$\frac{\pi}{2}$	$\frac{3\pi}{4}$	−0.707
$\frac{\pi}{2}$	π	−1
$\frac{\pi}{2}$	$\frac{5\pi}{4}$	−0.707
$\frac{\pi}{2}$	$\frac{3\pi}{2}$	0

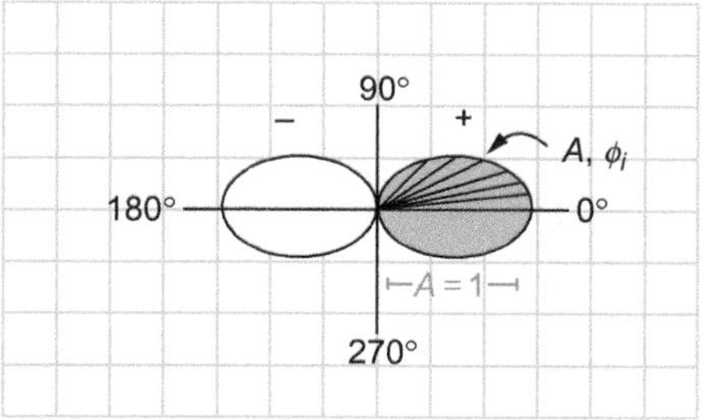

A plot of the function $A = \frac{x}{r} = \sin\theta\cos\phi$ and ϕ ° in the xy plane, where $\theta = \frac{\pi}{2}$.

Consider a hydrogen-like atom in ψ_{2p_z} state,

$$\psi_{2p_z} = \frac{1}{4\sqrt{2\pi}}\left(\frac{1}{a_0}\right)^{\frac{3}{2}}\frac{r}{a_0}e^{-\frac{r}{2a_0}}\cos\theta$$

Find the function that describes how the electron density varies along an arc of radius a_0 on the yz plane.

Answer:

- R is constant and equal to a_0.
  φ is constant and equal to $\pi/2$.
  As θ changes along from 0 to $\pi/2$, as we move along the arc, the density varies according to

$$\psi^2_{2p_z(a_0,\theta)} = \left[\frac{1}{4\sqrt{2\pi}}\left(\frac{1}{a_0}\right)^{\frac{3}{2}}\frac{a_0}{a_0}e^{-\frac{a_0}{2a_0}}\cos\theta\right]^2$$

$$= \frac{1}{16x2\pi}\cdot\frac{1}{a_0^3}\cdot e^{-\frac{1}{2}}\cdot\cos^2\theta = \text{constant}\cdot\cos^2\theta = A\cos^2\theta$$

Fig. 2.10 describes how the electron density varies along an arc of radius a_0 on yz plane.

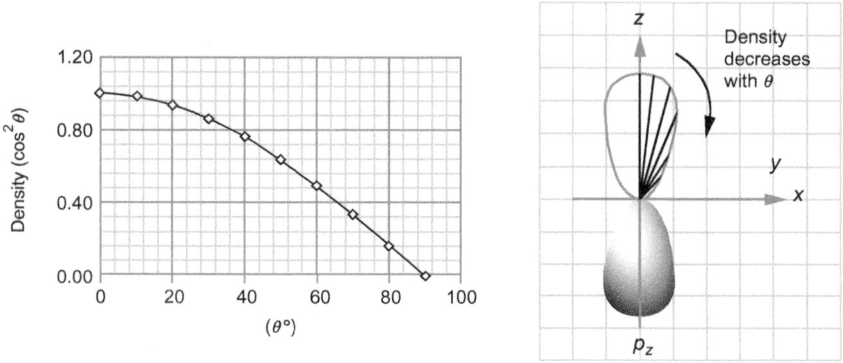

FIG. 2.10 The plot of $\cos^2\theta$ versus θ, which shows the decreases of the electron density as θ approaches 90 degrees for p_z orbital.

The wave function for the $2p_z$ state of H is

$$\psi_{2p_z}=\frac{1}{4\sqrt{2\pi}}\left(\frac{1}{a_0}\right)^{\frac{3}{2}}\frac{r}{a_0}e^{-\frac{r}{2a_0}}\cos\theta$$

a. **Evaluate the average of the inverse of the radius, $\left\langle\frac{1}{r}\right\rangle$, at this state.**

b. $H\psi_{2p_z}=-\frac{1}{4}\left(\frac{e^2}{2a_0}\right)\psi_{2p_z}$. **Given this fact, evaluate the average kinetic energy of the electron in a $2p_z$ state.**

Answer

a. If $\langle x\rangle=\int\psi^*x\psi\partial\tau$ (2.2.1)

Then

$$\left\langle\frac{1}{r}\right\rangle=\int\psi_{2p_z}{}^*\frac{1}{r}\psi_{2p_z}\partial\tau$$

$$\psi_{2p_z}=\frac{1}{4\sqrt{2\pi}}\left(\frac{1}{a_0}\right)^{\frac{3}{2}}\frac{r}{a_0}e^{-\frac{r}{2a_0}}\cos\theta,\ \text{and}$$

$\partial\tau=r^2\sin\theta\,\partial r\,\partial\theta\,\partial\phi$, then

$$\left\langle\frac{1}{r}\right\rangle=\left[\frac{1}{4\sqrt{2\pi}}\left(\frac{1}{a_0}\right)^{\frac{3}{2}}\frac{1}{a_0}\right]^2\int_{r=0}^{\infty}\int_{\theta=0}^{\pi}\int_{\phi=0}^{2\pi}\left(re^{-\frac{r}{2a_0}}\right)^2\cdot\frac{1}{r}\cos^2\theta.r^2\sin\theta\,\partial r\,\partial\theta\,\partial\phi$$

$$\left\langle\frac{1}{r}\right\rangle=\frac{1}{16\times2\pi}\frac{1}{a_0{}^5}\int_0^{\infty}r^3e^{-\frac{r}{a_0}}\partial r\int_0^{\pi}\cos^2\theta\sin\theta\,\partial\theta\int_0^{2\pi}\partial\phi$$

$$\because\int_0^{\infty}x^ne^{-ax}\partial r=\frac{n!}{a^{n+1}},\ \text{then}$$

$$\int_0^{\infty}r^3e^{-\frac{r}{a_0}}\partial r=\frac{3!}{\left(\frac{1}{a_0}\right)^4}$$

$$\because\int_0^{\pi}\sin^{2n+1}\theta\cdot\cos^m\theta\,\partial\theta=\int_0^{\pi}\cos^{2n+1}\theta\cdot\sin^m\theta\,\partial\theta=\frac{2^{n+1}\cdot n!}{(m+1)(m+3)\ldots\ldots(m+2n+1)},\ \text{then}$$

$$\int_0^{\pi}\cos^2\theta\sin\theta\,\partial\theta=\frac{2}{3},\ \text{and}$$

$$\int_0^{2\pi}\partial\phi=2\pi,\ \text{then}$$

$$\left\langle\frac{1}{r}\right\rangle=\frac{1}{16\times2\pi}\cdot\frac{1}{a_0{}^5}\cdot\frac{3!}{\left(\frac{1}{a_0}\right)^4}\cdot\frac{2}{3}\cdot2\pi=\frac{1}{16}\cdot\frac{1}{a_0}\cdot6\cdot\frac{2}{3}=\frac{1}{4a_0}$$

b. $\because H\psi=E\psi$ (2.1.7)

Then

$$\therefore H\psi_{2p_z}=-\frac{1}{4}\left(\frac{e^2}{2a_0}\right)\psi_{2p_z},\ \text{and}$$

$$\therefore \text{Total energy} = -\frac{1}{4}\left(\frac{e^2}{2a_0}\right)$$

$$\because \text{Potential energy} = -\frac{e^2}{4a_0}$$

$$\therefore KE = -\frac{1}{4}\left(\frac{e^2}{2a_0}\right) - \left(-\frac{e^2}{4a_0}\right) = \frac{e^2}{8a_0}$$

2.15 THE BOUNDARY SURFACE OF d-ORBITALS

How can you draw graphically the boundary surface of d-orbitals?

If $3d_{yz}$ function:

$$3d_{yz} = N\rho^2 e^{-\frac{\rho}{2}} \sin\theta \cos\theta \sin\phi$$

a. **At which points does the electron density become zero?**
b. **At which points does the electron density become maximized?**
c. **In the *yz* plane, between what values of θ is the electron density less than one half of the maximum density?**
d. **Show graphically how the electron density changes with the distance from the nucleus along the $\theta = 45$ degrees line in the *yz* plane.**
e. **Draw graphically the locus of the points in the *yz* plane where the electron density is equal to one half of the maximum density.**
f. **Select the region in the *yz* plane where the density is greater than one half of the maximum density.**

Answers:

a. $3d_{yz} = N\rho^2 e^{-\frac{\rho}{2}} \sin\theta \cos\theta \sin\phi$

- The probability density, $P(r)$:

  $P(r) = \psi^2$

  $\psi^2 = N^2\rho^4 e^{-\rho} \sin^2\theta \cos^2\theta \sin^2\phi$

- The electron density becomes zero when

  $[3d_{yz}]^2 = N^2\rho^4 e^{-\rho} \sin^2\theta \cos^2\theta \sin^2\phi = 0$

  - This happens when

    $\sin 0 = 0, \quad \cos\frac{\pi}{2} = 0, \quad \sin\pi = 0$

  then

    $[3d_{yz}]^2 = 0, \quad \text{when } \theta = 0, \frac{\pi}{2}, \pi, \text{ or when } \phi = 0, \pi, 2\pi$

  Physically, this means (Fig. 2.11):

$\phi = 0$ or $\phi = \pi$ or $\phi = 2\pi$	*xz*-plane
$\theta = 0$ or $\theta = \pi$	*z*-axis
$\theta = \frac{\pi}{2}$	*xy*-plane

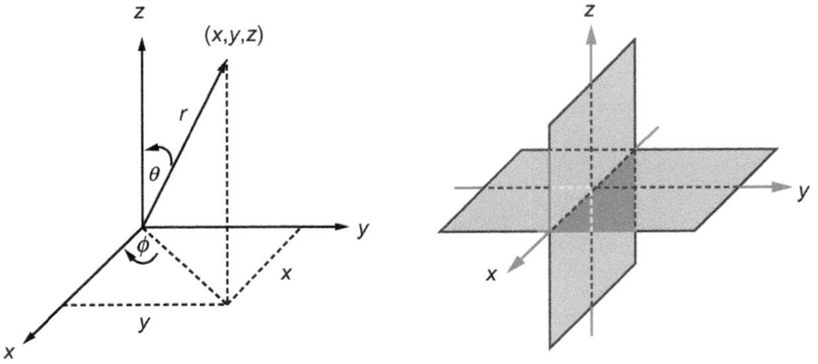

FIG. 2.11 The electron density becomes zero at the z-axis and the xz and xy planes.

b. $[3d_{yz}]^2 = N^2\rho^4 e^{-\rho}\sin^2\theta\cos^2\theta\sin^2\phi$

- becomes maximum when

  $\sin^2\phi$ becomes maximum,

  $\sin^2\theta\cos^2\theta$ becomes maximum, and

  $\rho^4 e^{-\rho}$ becomes maximum

  - $\sin^2\phi$: becomes maximum when, $\phi=\frac{\pi}{2}(90°)$ or $\phi=\frac{3\pi}{2}(270°)$
  - $\sin^2\theta\cos^2\theta = (\sin\theta\cos\theta)^2 = \left(\frac{1}{2}\sin 2\theta\right)^2$

    becomes maximum when, $\theta=\frac{\pi}{4}(45°)$ or $\theta=\frac{3\pi}{4}(135°)$
  - $\rho^4 e^{-\rho}$ becomes maximum when:

    $$\frac{\partial}{\partial\rho}(\rho^4 e^{-\rho})=0$$

    $$\frac{\partial}{\partial\rho}(\rho^4 e^{-\rho})=4\rho^3 e^{-\rho}-\rho^4 e^{-\rho}=\rho^3 e^{-\rho}(4-\rho)=0, \text{ then}$$

    $\rho=4$, where, $\rho=\frac{2}{a_0}\frac{z}{n}r$

    Take $z=1$, then for 3d state

    $$\rho=\frac{2r}{3a_0}, \text{ if } \rho=4$$

    $$\frac{2r}{3a_0}=4$$

    $$r=6a_0$$

- The electron density becomes maximum when (Fig. 2.12)

  $r=6a_0 \quad \theta=\frac{\pi}{4} \quad \phi=\frac{\pi}{2}$

  $r=6a_0 \quad \theta=\frac{\pi}{4} \quad \phi=\frac{3\pi}{2}$

  $r=6a_0 \quad \theta=\frac{3\pi}{4} \quad \phi=\frac{\pi}{2}$

  $r=6a_0 \quad \theta=\frac{3\pi}{4} \quad \phi=\frac{3\pi}{2}$

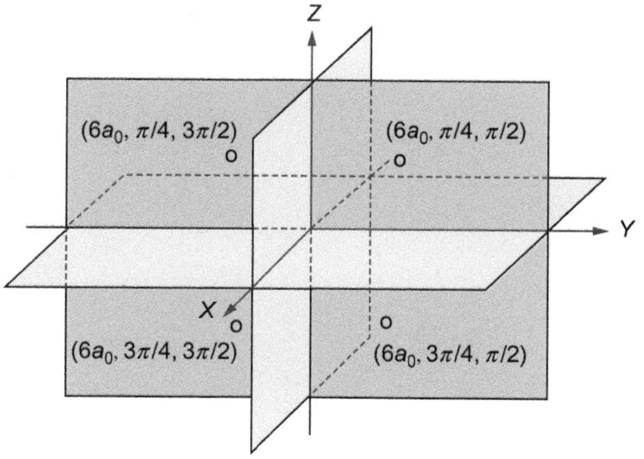

FIG. 2.12 Points at which the electron density becomes maximum.

c. Maximum density is

$$[3d_{yz}]^2 = N^2\rho^4 e^{-\rho}\sin^2\theta\cos^2\theta\sin^2\phi$$

$$= N^2\rho^4 e^{-\rho}\sin^2\frac{\pi}{4}\cos^2\frac{\pi}{4}\sin^2\frac{\pi}{2}$$

$$= N^2\rho^4 e^{-\rho}\left(\frac{\sqrt{2}}{2}\right)^2\left(\frac{\sqrt{2}}{2}\right)^2 1$$

$$= N^2\rho^4 e^{-\rho}\frac{1}{4}$$

- Therefore,

$$\frac{1}{2}\text{ maximum density} = \frac{1}{8}N^2\rho^4 e^{-\rho}\text{ or when}$$

$$\cancel{N^2\rho^4 e^{-\rho}}\sin^2\theta\cos^2\theta\sin^2\phi = \frac{1}{8}\cancel{N^2\rho^4 e^{-\rho}}$$

$$\cos^2\theta\sin^2\phi = \frac{1}{8}$$

- Let us see between what values of $\sin^2\theta\cos^2\theta$ is less than $\frac{1}{8}$.

First, $\sin^2\theta\cos^2\theta$ becomes equal to $\frac{1}{8}$ when

$$\sin^2\theta\cos^2\theta = \frac{1}{8}$$

$$\sin\theta\cos\theta = \pm\frac{1}{\sqrt{8}},\text{ then:}$$

$$\frac{1}{2}\sin 2\theta = \pm\frac{1}{\sqrt{8}},\quad \sin 2\theta = \pm\frac{2}{\sqrt{8}} = \pm\frac{2}{2.828} = \pm 0.707$$

$$2\theta = 45°,\ 135°,\ 225°,\text{ and } 315°$$

$$\text{then, } \theta = 22°30',\text{ or } \theta = 67°30',\text{ or } \theta = 112°30',\text{ or } \theta = 157°30'$$

The electron density never becomes equal to one half of the maximum density.

When

$$00° \; 00' < \theta < 22° \; 30'$$

$$67° \; 30' < \theta < 112° \; 30'$$

$$157° \; 30' < \theta < 180°$$

Let us make a plot of $\sin^2\theta \cos^2\theta$ versus θ (Table 2.7, Fig. 2.13).

d. Showing how the electron density changes with the distance from the nucleus along the $\theta = 45$ degrees line in the yz plane. We are required to plot (Table 2.8)

$$\psi^2_{3d_{yz}}(r, \theta = 45, \phi = 90) = N^2\rho^4 e^{-\rho}\sin^2 45 \cos^2 45 \sin^2 90$$
$$= N^2\rho^4 e^{-\rho}\frac{1}{4}$$
$$= \frac{1}{4}N^2\rho^4 e^{-\rho}$$

- Electron density varies as shown above. The units on the vertical scale is $N^2/4$ for the variation along the $\theta = 45$ degrees line (Fig. 2.14).

e. The points in the yz plane where the electron density is equal to 1/2 of the maximum density:
- when $\theta = 22°30'$, $\rho = 4$

$$\frac{1}{2}(\text{maximum density}) = N^2\rho^4 e^{-\rho}\sin^2\theta\cos^2\theta =$$
$$= \frac{1}{2}(4^4)(e^{-4})\left(\frac{1}{4}\right)N^2 = \frac{1}{2}(4.68)\left(\frac{1}{4}\right)N^2$$

TABLE 2.7 Data for θ ° and $\sin^2\theta \cos^2\theta$

θ	$\sin^2\theta \cos^2\theta$
0	0
5	0.007538
10	0.029244
15	0.0625
20	0.103294
25	0.146706
30	0.1875
35	0.220756
40	0.242462
45	0.25
50	0.242462
55	0.220756
60	0.1875
65	0.146706
70	0.103294
75	0.0625
80	0.029244
85	0.007538

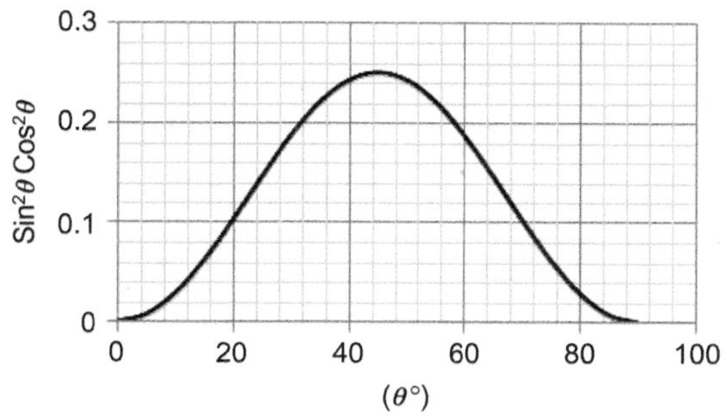

FIG. 2.13 The change in the electron density between $\theta = 0^\circ$–90° in the *yz*-plane.

TABLE 2.8 Values of ρ and $\rho^4 e^{-\rho}$

ρ	r	ρ^2	ρ^4	$\varepsilon^{-\rho}$	$\rho^4 e^{-\rho}$
0	0	0	0	1	0
1	1.5	1	1	0.368	0.368
2	3	4	16	0.135	2.165
3	4.5	9	81	0.05	4.033
4	6	16	256	0.018	4.689
5	7.5	25	625	0.007	4.211
6	9	36	1296	0.002	3.212
7	10.5	49	2401	9E-04	2.189
8	12	64	4096	3E-04	1.374
9	13.5	81	6561	1E-04	0.81
10	15	100	10,000	5E-05	0.454
	r (in units of a_o)				

Then

$$\cancel{N}^2 \rho^4 e^{-\rho} \sin^2\theta \cos^2\theta = (0.585)\cancel{N}^2$$

$$\rho^4 e^{-\rho} = \frac{0.585}{\sin^2\theta \cos^2\theta}$$

From the previous plot (Fig. 2.14), the values of ρ that correspond to $\rho^4 e^{-\rho}$, and by using $r = \frac{an}{2z}\rho$, $z = r\cos\theta$ and $y = r\sin\theta\sin\phi$, where $\phi = 90$, the values of *r*, *y*, and *z* can found in terms of a (Table 2.9). Fig. 2.15 exhibits the region in the *yz* plane where the density is greater than one half of the maximum density.

- There are three other loops symmetrically located in the three quadrants. The density is greater than half the maximum at all points inside the four loops.
- In the same way, the features for all d-orbitals can be presented (Fig. 2.16). The plots are simply surfaces in which each point on the surface defines a direction in space and distance of the surface from the origin.
- The d_{xy}-orbital has two nodal surfaces over the *xz* and *yz* planes which define surfaces over which d_{xy} has a zero value. In d_{xz}, d_{yz}, and $d_{x^2-y^2}$, there are two nodal planes meet at the nucleus, and separate the four lobes of each orbital.

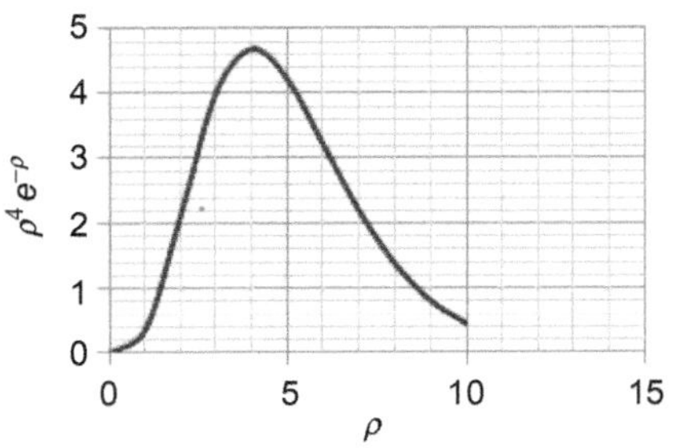

FIG. 2.14 The electron density changes with the distance from the nucleus along the $\theta = 45^\circ$ line in the *yz*-plane.

TABLE 2.9 The Locus of *y* and *z* Points in the *yz* Plane Where the Electron Density is Equal to One Half of the Maximum Density

θ	$\sin^2\theta\cos^2\theta$	$\rho^4 e^{-\rho}$	θ	r	y	z
θ	$\sin^2\theta\cos^2\theta$	$\rho^4 e^{-\rho}$	θ	r	y	z
25	0.146706	3.988	25	4.35	1.83839	3.942
25	0.146706	3.988	25	7.8	3.29642	7.069
30	0.1875	3.12	30	3.6	1.8	3.118
30	0.1875	3.12	30	9	4.5	7.794
35	0.220756	2.65	35	3.3	1.8928	2.703
35	0.220756	2.65	35	9.6	5.50633	7.864
40	0.242462	2.413	40	3.15	2.02478	2.413
40	0.242462	2.413	40	10.05	6.46002	7.699
45	0.25	2.34	45	3.15	2.22739	2.227
45	0.25	2.34	45	10.2	7.21249	7.212
50	0.242462	2.413	50	3.15	2.41304	2.025
50	0.242462	2.413	50	10.05	7.69875	6.46
55	0.220756	2.65	55	3.3	2.7032	1.893
55	0.220756	2.65	55	9.6	7.86386	5.506
60	0.1875	3.12	60	3.6	3.11769	1.8
60	0.1875	3.12	60	9	7.79423	4.5
65	0.146706	3.988	65	4.35	3.94244	1.838
65	0.146706	3.988	65	7.8	7.0692	3.296
67.5	0.125	4.68	67.5	6.15	5.68186	2.354
67.5	0.125	4.68	67.5	6.15	5.68186	2.354
22.5	0.125	4.68	22.5	6.15	2.3535	5.682
22.5	0.125	4.68	22.5	6.15	2.3535	5.682

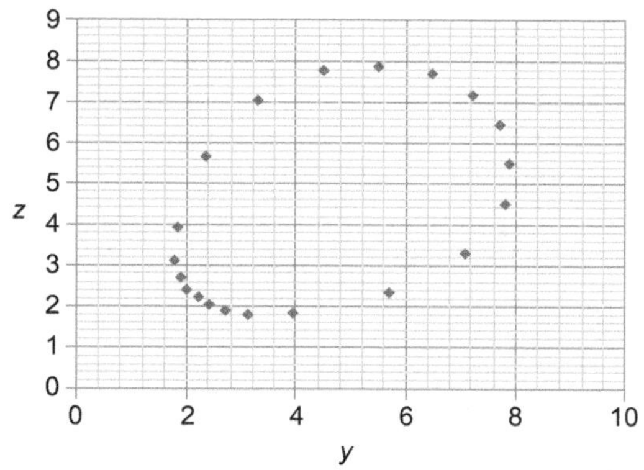

FIG. 2.15 The region in the *yz* plane where the density is greater than one half of the maximum density.

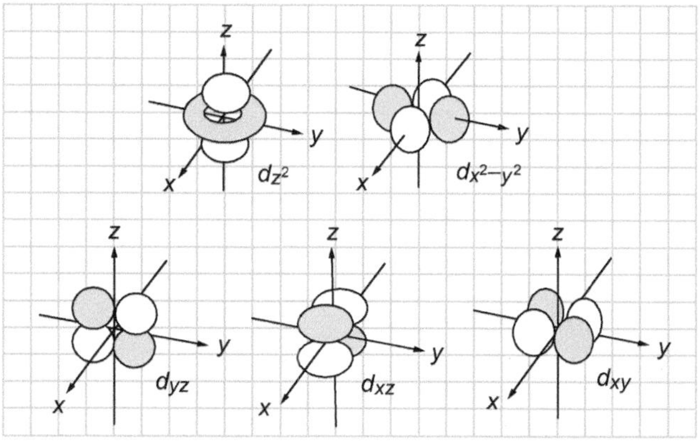

FIG. 2.16 The boundary surfaces of d-orbitals. The *light* and *dark lobes* have opposite sign of the wave function.

Exercise:

Evaluate $\langle r\rangle$, $\left\langle \frac{1}{r}\right\rangle$, $\langle KE\rangle$, $\langle PE\rangle$, $\langle$angular momentum$\rangle$, and $\langle \hat{L}\rangle$ for the $3d_{z^2}$, where

$$\psi_{3d_{z^2}} = \frac{1}{81\sqrt{6\pi}}\left(\frac{1}{a_0}\right)^{3/2}\left(\frac{r}{a_0}\right)^2 e^{-\frac{r}{3a_0}}(3\cos^2-1)$$

Is normalized wave function, $z = 1$

Answer:

$$\psi_{3d_{z^2}} = Nr^2 e^{-\frac{r}{3a_0}}(3\cos^2\theta - 1)$$

where

$$N = \frac{1}{81\sqrt{6\pi}}\left(\frac{1}{a_0}\right)^{7/2}$$

$$\because \langle r\rangle = \int \psi^*_{3d_{z^2}} r \psi_{3d_{z^2}} \partial\tau \quad (2.2.1)$$

$$\therefore \langle r\rangle = N^2 \iiint \left[r^2 e^{-\frac{r}{3a_0}}(3\cos^2\theta - 1)\right]^2 r\, r^2 \sin\theta\, \partial r\, \partial\theta\, \partial\phi,$$

$$\langle r\rangle = 2\pi N^2 \int_{r=0}^{\infty} r^7 e^{-\frac{2r}{3a_0}} \partial r \int_0^{\pi} [(3\cos^2\theta - 1)]^2 \sin\theta\, \partial\theta$$

$$\langle r\rangle = 2\pi N^2 \int_{r=0}^{\infty} r^7 e^{-\frac{2r}{3a_0}}\partial r \int_0^{\pi} \left[9\cos^4\theta \sin\theta - 6\cos^2\theta \sin\theta + \sin\theta\right]\partial\theta$$

$$\because \int_0^{\infty} x^n e^{-ax}\partial x = \frac{n!}{a^{n+1}}$$

$$\therefore \int_0^{\infty} r^7 e^{-\frac{2r}{3a_0}}\partial r = \frac{7!}{\left(\frac{2}{3a_0}\right)^8}, \text{ and}$$

$$\because \int_0^{\pi} \sin^{2n+1}\theta\cdot\cos^m\theta\,\partial\theta = \int_0^{\pi}\cos^{2n+1}\theta\cdot\sin^m\theta\,\partial\theta = \frac{2^{n+1}\cdot n!}{(m+1)(m+3)\ldots\ldots(m+2n+1)}$$

if $m=4$, $n=0$, then:

$$\int_0^{\pi} 9\cos^4\theta \sin\theta\,\partial\theta = \frac{(9)(2)(1)}{(5)} = \frac{18}{5}$$

$$\int_0^{\pi} 6\cos^2\theta \sin\theta\,\partial\theta = \frac{(6)(2)(1)}{(3)} = \frac{12}{3} = 4$$

$$\int_0^{\pi} \sin\theta\,\partial\theta = -\cos\theta|_0^{\pi} = 2$$

$$\langle r\rangle = 2\pi\left(\frac{1}{81\sqrt{6\pi}}\left(\frac{1}{a_0}\right)^{7/2}\right)^2\left(\frac{7!}{\left(\frac{2}{3a_0}\right)^8}\right)\left[\frac{18}{5} - 4 + 2\right]$$

$$\langle r\rangle = 2\pi\frac{7\cdot6\cdot5\cdot4\cdot3\cdot2\cdot1\cdot3^8\cdot a_0^8\cdot 8}{3^8\cdot 6\pi\cdot a_0^7\cdot 2^8\cdot 5} = \frac{21}{2}a_0$$

- The value of $\left\langle \frac{1}{r}\right\rangle$

$$\left\langle \frac{1}{r}\right\rangle = \int \psi_{3d_{z^2}}{}^* \frac{1}{r}\psi_{3d_{z^2}}\partial\tau$$

$$\left\langle \frac{1}{r}\right\rangle = N^2 \iiint \left[r^2 e^{-\frac{r}{3a_0}}(3\cos^2\theta - 1)\right]^2 \frac{1}{r} r^2 \sin\theta\,\partial r\,\partial\theta\,\partial\phi$$

$$\left\langle \frac{1}{r}\right\rangle = 2\pi N^2 \int_{r=0}^{\infty} r^5 e^{-\frac{2r}{3a_0}}\partial r \int_0^{\pi}\left[(3\cos^2\theta - 1)\right]^2 \sin\theta\,\partial\theta$$

$$\left\langle \frac{1}{r}\right\rangle = 2\pi\left(\frac{1}{81\sqrt{6\pi}}\left(\frac{1}{a_0}\right)^{7/2}\right)^2\left(\frac{5!}{\left(\frac{2}{3a_0}\right)^6}\right)\left[\frac{18}{5} - 4 + 2\right]$$

$$\left\langle \frac{1}{r}\right\rangle = 2\pi\frac{5\cdot4\cdot3\cdot2\cdot1\cdot3^6\cdot a_0^6\cdot 8}{3^8\cdot 6\pi\cdot a_0^7\cdot 2^6\cdot 5} = \frac{1}{9a_0}$$

You may note the following formulas for $\langle r\rangle$ and $\left\langle \frac{1}{r}\right\rangle$:

$$\langle r\rangle = n^2\left\{1 + \frac{1}{2}\left[1 - \frac{l(l+1)}{n^2}\right]\right\}\frac{a_0}{z}$$

$$\left\langle\frac{1}{r}\right\rangle = \frac{1}{a_0}\frac{1}{n^2}z$$

See that the above calculations agree with these formulas (we took $z=1$).

$$\langle r\rangle = 3^2\left\{1+\frac{1}{2}\left[1-\frac{2(2+1)}{3^2}\right]\right\}\frac{a_0}{1} = 3^2\left\{\frac{7}{6}\right\}\frac{a_0}{1} = \frac{21}{2}a_0$$

$$\left\langle\frac{1}{r}\right\rangle = \frac{1}{a_0}\cdot\frac{1}{9}\cdot 1 = \frac{1}{9a_0}$$

- $\langle PE\rangle$

$$\text{Total energy} = E_n = -\frac{z^2\,e^2}{2a_0\,n^2} = -\frac{1}{18a_0}, \text{ in atomic units}$$

$$\langle KE\rangle = \text{total energy} - \langle PE\rangle = -\frac{e^2}{18a_0} - \left(-\frac{e^2}{9a_0}\right) = \frac{e^2}{18a_0}$$

- $\langle$angular momentum$\rangle$ for the 3 d_{z^2}

$$\langle\text{angular momentum}\rangle = |L| = \sqrt{\langle L^2\rangle} = \frac{h}{2\pi}\sqrt{l(l+1)} \quad 2.11.2$$

$$\sqrt{\langle L^2\rangle} = \frac{h}{2\pi}\sqrt{2(2+1)} = \sqrt{6}\frac{h}{2\pi}$$

- $\langle\hat{L}\rangle$ for the 3 d_{z^2}

$$\hat{L}_Z\psi_{d_{z^2}} = L_Z\psi_{d_{z^2}} = 0\,\psi_{d_{z^2}}$$

$$\langle\hat{L}\rangle = 0$$

2.16 CALCULATING THE MOST PROBABLE RADIUS

Demonstrate how to calculate the most probable radius, r^*, and show that r^* of the hydrogenic 1s orbital is controlled by $r^* = \frac{a_0}{Z}$.

$$\psi_{1,0,0} = \psi = \left(\frac{Z^3}{\pi a_0^3}\right)^{1/2} e^{-zr/a_0}, \text{ and}$$

- The probability density, $P(r)$, at radius r in the ground state of the hydrogen atom is

$$P(r) = \psi^2$$

$$\text{if}: \psi_{1,0,0} = \psi = \left(\frac{Z^3}{\pi a_0^3}\right)^{1/2} e^{-zr/a_0}, \text{ and}$$

$$P(r) = \psi^2 = \frac{Z^3}{\pi a_0^3}e^{-2r/a_0}$$

- The probability, $P\partial r$, that the electron be found in a thin shell of volume ∂V, is the probability density at radius r multiplied by the volume between the inner and outer surface:

$$\because V = \frac{4\pi r^3}{3}$$

$$\therefore \partial V = 4\pi r^2\partial r$$

$$\therefore P\partial r = \psi^2 \times 4\pi r^2\partial r$$

Then

$$P = 4\pi r^2 \psi^2$$

- We can find the radius at which the radial distribution function of the hydrogenic 1s orbital has a maximum value by solving $\partial P/\partial r = 0$.

$$P = 4\pi r^2 \psi^2 = 4\pi r^2 \times \frac{Z^3}{\pi a_0^3} \times e^{-2Zr/a_0} = \frac{4Z^3}{a_0^3} \times r^2 e^{-2Zr/a_0}$$

$$\frac{\partial P}{\partial r} = \frac{4Z^3}{a_0^3}\left(2r^* - \frac{2Zr^{*2}}{a_0}\right)e^{--Zr^*/a_0} = 0 \ \ at\ r = r^*$$

$$\text{therefore,}\ \ 2r^* - \frac{2Zr^{*2}}{a_0} = 0,\ \text{and}$$

$$r^* = \frac{a_0}{Z}. \qquad (2.16.1)$$

Exercise

Calculate the most probable radius at which an electron will be found when it occupies a 1s orbital of hydrogenic atom of atomic number *Z*. Arrange the values for the one-electron species from H to Ne^{9+}.

Answer:

$$r^* = \frac{a_0}{Z}$$

With $a_0 = 52.9$ pm,

	H	He^+	Li^{2+}	Be^{3+}	B^{4+}	C^{5+}	N^{6+}	O^{7+}	F^{8+}	Ne^{9+}
r^*/pm	52.9	26.5	17.6	13.2	10.6	8.82	7.56	6.61	5.88	5.29

2.17 CALCULATING THE MEAN RADIUS OF AN ORBITAL

Demonstrate how to calculate the mean radius, $\langle r\rangle_{n,\,l}$, and compare your result with $\langle r\rangle_{n,\,l}$, which can be obtained from the general expression for the mean radius:

$$\langle r\rangle_{n,\,l} = n^2\left\{1 + \frac{1}{2}\left(1 - \frac{l(l+1)}{n^2}\right)\right\}\frac{a_0}{Z}.$$

- The mean value is the expectation value, and is given by

$$\langle r\rangle = \int_0^\infty \psi^* r\psi\, \partial\tau = \int_0^\infty r\psi^2 \partial\tau \qquad 2.2.1$$

where $\partial\tau = r^2\, \partial r\, \sin\theta\, \partial\theta\, \partial\phi$

$$\because \psi = RY$$

$$\langle r\rangle = \int rR^2 r^2 \partial r \int Y^2 \sin\theta\, \partial\theta\, \partial\phi$$

- The integration over the angular part gives 1, because the spherical harmonics are normalized.
  Consequently,

$$\langle r\rangle = \int_0^\infty rR^2 \times r^2 \partial r$$

For a 1s orbital,

$$R = \frac{2}{a_0^{3/2}} e^{-r/a_0}$$

Therefore,

$$\langle r \rangle = \frac{4}{a_0^3} \int_0^{\infty} r^3 e^{-2r/a_0} \partial r = \frac{3}{2} a_0 \tag{2.17.1}$$

$$\text{because,} \quad \int_0^{\infty} x^n e^{-ax} \partial x = \frac{n!}{a^{n+1}}$$

- The general expression for the mean radius of an orbital with quantum number l and n is

$$\langle r \rangle_{n,\,l} = n^2 \left\{ 1 + \frac{1}{2}\left(1 - \frac{l(l+1)}{n^2} \right) \right\} \frac{a_0}{Z} \tag{2.17.2}$$

If $n=1$, $l=0$, and $Z=1$, then

$$\langle r \rangle_{1,\,0} = 1^2 \left\{ 1 + \frac{1}{2}\left(1 - \frac{0(0+1)}{1^2} \right) \right\} \frac{a_0}{1} = \frac{3}{2} a_0$$

Note that for a given principle quantum number, the mean radius decreases as l increases.

How to find:

(a) **the probability density for an electron at some point;**
(b) **the occupation number for the infinitesimal small volume $\partial\tau$ about a point;**
(c) **the total occupation over all the space; and**
(d) **the surface density function for sphere of radius r.**

- The probability density for an electron at some point (r,θ,ϕ) is given by

$$\psi(r,\theta,\phi)^* \psi(r,\theta,\phi), \quad \text{unit is } 1/\text{volume}$$

$|\psi^2_{(r,\theta,\phi)}|$ is proportional to the probability of finding the electron in the region between r and $r + \partial r$.
- Multiplication by $\partial\tau$, where $\partial\tau$ represents infinitesimal small volume ($\partial x \partial y \partial z$ in the Cartesian coordinate):

$$\psi(r,\theta,\phi)^* \psi(r,\theta,\phi) \partial\tau, \quad \text{dimensionless}$$

gives the occupation number for the extremely small volume $\partial\tau$ about the point (r,θ,ϕ).
- The total occupation over all the space:

$$\int_0^{\infty} \psi(r,\theta,\phi)^* \psi(r,\theta,\phi) \partial\tau = 1$$

$$\because \psi(r,\theta,\phi) = R_{n,l}(r) Y_{l,m}(\theta,\phi)$$

Therefore, the total probability density function is given by

$$\therefore \int_0^{\infty} \psi * (r,\theta,\phi) \psi(r,\theta,\phi) \partial\tau = \int_0^{\infty} \underbrace{R^2{}_{n,l}(r)}_{\text{is normalized}} r^2 \partial r \int_0^{\pi} \int_0^{2\pi} \underbrace{Y^*_{l,m}(\theta,\phi) Y_{l,m}(\theta,\phi)\sin\theta\partial\theta\partial\phi}_{\text{is normalized}} = 1$$

or

$$\int_0^{\infty} R^2{}_{n,l}(r) \partial r^2 = 1$$

$$\int_0^{\pi}\int_0^{2\pi} Y^*_{l,m}(\theta,\phi)Y_{l,m}(\theta,\phi)\sin\theta\partial\theta\partial\phi = 1$$

- Thus, the probability of finding electron in the region between r and $r + \partial r$ from the nucleus, irrespective of direction (averaging over the angular variables), is

$$\psi*(r,\theta,\phi)\psi(r,\theta,\phi)\partial\tau = R^2_{n,l}(r)r^2\int_0^{\pi}\int_0^{2\pi} Y^*_{l,m}(\theta,\phi)Y_{l,m}(\theta,\phi)\sin\theta\partial\theta\partial\phi$$

But $Y_{l,m}*(\theta,\phi)Y_{l,m}(\theta,\phi)$ is normalized:

$$\int_0^{\pi}\int_0^{2\pi} Y^*_{l,m}(\theta,\phi)Y_{l,m}(\theta,\phi)\sin\theta\partial\theta\partial\phi = 1$$

$$\therefore\ \psi^*\psi(r,\theta,\phi)\partial\tau = R^2_{n,l}(r)r^2$$

$\therefore\ R^2_{n,l}(r)r^2$ is the electron occupation function for an infinitesimal thin spherical shell of radius r and thickness ∂r; it is 1/distance function.

$R^2_{n,l}(r)r^2$ is the surface density function for sphere of radius r.

Use the following radial functions (Ch. 2, p. 64) to draw $R_{n,l}$, $R^2_{n,l}$ and $r^2R^2_{n,l}$ *versus* r, and write your own conclusion, where $a_o = 0.5292$ Å, and $\rho = \dfrac{zr}{a_o}$:

1s:	$R_{1,0}$	$=N_{1,0}[e^{-\rho}]$	$N_{1,0}$	$=2\left(\frac{z}{a_o}\right)^{3/2}$
2s:	$R_{2,0}$	$=N_{2,0}\left[(2-\rho)e^{-\rho/2}\right]$	$N_{2,0}$	$=\frac{1}{2\sqrt{2}}\left(\frac{z}{a_o}\right)^{3/2}$
2p:	$R_{2,1}$	$=N_{2,1}\left[\rho e^{-\rho/2}\right]$	$N_{2,1}$	$=\frac{1}{2\sqrt{6}}\left(\frac{z}{a_o}\right)^{3/2}$
3s:	$R_{3,0}$	$=N_{3,0}\left[(27-18\rho+2\rho^2)e^{-\rho/3}\right]$	$N_{3,0}$	$=\frac{2}{81\sqrt{3}}\left(\frac{z}{a_o}\right)^{3/2}$
3p:	$R_{3,1}$	$=N_{3,1}\left[(6\rho-\rho^2)e^{-\rho/3}\right]$	$N_{3,1}$	$=\frac{4}{81\sqrt{6}}\left(\frac{z}{a_o}\right)^{3/2}$
3d:	$R_{3,2}$	$=N_{3,2}\left[\rho^2e^{-\rho/3}\right]$	$N_{3,2}$	$=\frac{4}{81\sqrt{30}}\left(\frac{z}{a_o}\right)^{3/2}$

Consider an electron in the ground state 1s of hydrogen atom. What is the probability that the electron is at a distance of 10 a_o? How does the Bohr model for the hydrogen atom differ from that of Schrödinger? What is the important electron behavior that exposed by the surface density plots? Explain the effect of this electron behavior on the stability of states in which it occurs.

- $R_{n,l}$ is the function of r and the distance is measured in Å (Figs. 2.17 and 2.18), these plots show:
  - all s-orbitals exhibit number of radial nodes, point at which the electron probability of being found almost vanish;
  - **number of nodes $= n - l - 1$,** where n and l are the principle and the azimuthal quantum numbers in each electronic state;
  - **number of nodes $= n - l - 1$,** where n and l are the principle and the azimuthal quantum numbers in each electronic state;

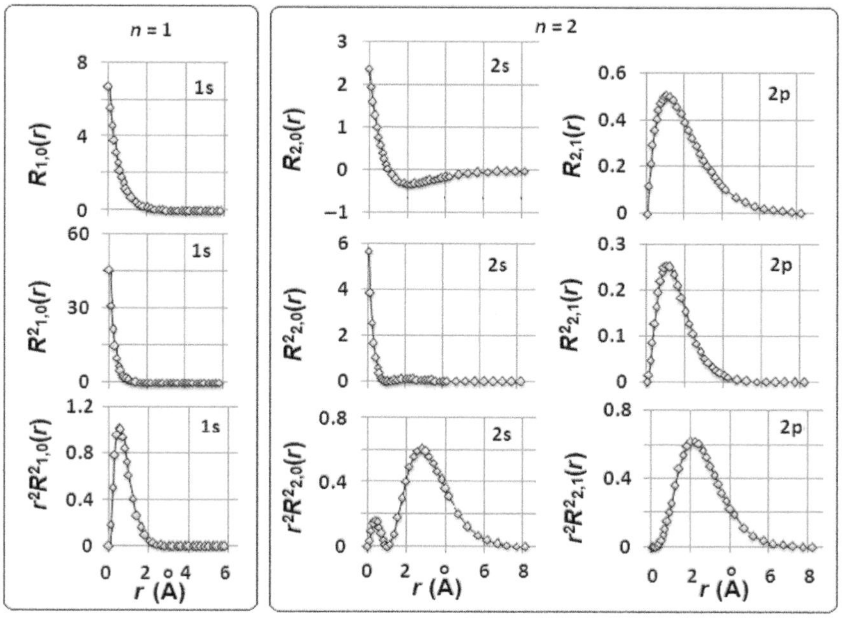

FIG. 2.17 The 1s, 2s, and 2p radial functions ($R_{n,l}$), radial density functions ($R^2_{n,l}$), and surface density functions ($r^2R^2_{n,l}$).

- the amplitude may be positive or negative; and
- the $R_{n,0}$ ($l=0$, s-orbital) function indicates maximum value at $r=0$. This confirms that s electrons have some probability of being found at distances close to the nucleus.

- $R_{n,l}$ contributes to the total probability, $R^2_{n,l}$, density function for the distance of the electron from the nucleus, irrespective of direction.
- Radial density distribution can be given either as $R^2(r)$ or as $R^2(r)r^2$.
  - $R^2(r)$ is the density function for the distance of the electron from the nucleus, and has the units of $\frac{1}{V}$ (density per volume unite).
  - $R^2(r)r^2$ is a $\frac{1}{\text{distance}}$ function (density per distance unite), is an electron occupation function for infinitesimally thin spherical shells of radius r and thickness dr (Fig. 2.18).
- The values of $R^2_{n,0}$ have none-zero amplitude at $r=0$:

$$R_{1,0} = 2\left(\frac{z}{a_o}\right)^{3/2}[e^{-\rho}], \quad \therefore R^2_{1,0}(r=0) = 4\left(\frac{1}{a_o}\right)^3, \quad \because Y^0_0 = \frac{1}{\sqrt{4\pi}}$$

$$\psi_{1,0} = R_{1,0}Y^0_0 = \frac{1}{\sqrt{\pi}}\left(\frac{1}{a_o}\right)^{3/2}$$

$$\therefore \psi^2_{1,0}(r=0) = \frac{1}{\pi a_o^3} = \frac{1}{\left(\frac{22}{7}\right)(0.5292)^3} = 2.147/\text{Å}^3$$

Therefore, the probability density at nucleus is 2.147/Å^3.

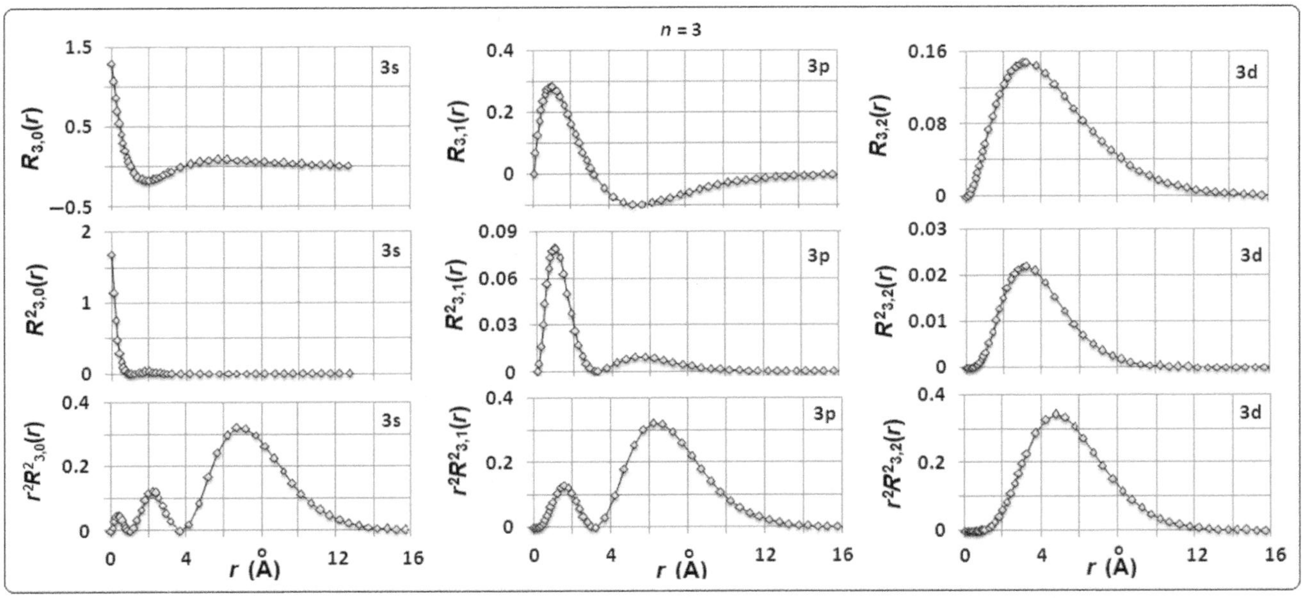

FIG. 2.18 The 3s, 3p, and 3d radial functions ($R_{n,l}$), radial density functions ($R^2_{n,l}$), and surface density functions ($r^2R^2_{n,l}$). Note: s-orbitals have a non-zero and finite value at the nucleus.

- The $R^2_{1,0}$ function (Figs. 2.17 and 2.18), shows a maximum at $r=0$, and decreases to zero at $r=\infty$, therefore, the electron with quantum number $n=1$, $l=0$, and $m=0$ is concentrated at the nucleus.
- The wave function of the electron in ground state of hydrogen have the quantum number: $n=1$, $l=0$, and $m=0$:

$$\psi_{1,0,0} = \frac{1}{\sqrt{\pi}}\left(\frac{z}{a_o}\right)^{3/2}[e^{-\rho}], \quad \text{or}$$

$$\psi_{1,0,0} = \frac{1}{\sqrt{\pi}}\left(\frac{1}{a_o}\right)^{3/2}\left[e^{-r/a_o}\right]$$

- The relative probability of finding the electron at 10 a_0 is given by

$$\frac{r^2\psi^2_{1,0,0}(r=10a_o)}{r^2\psi^2_{1,0,0}(r=a_o)} = \frac{100\,a_o^2\left(\frac{1}{\pi a_o^3}\right)\left[e^{-10a_o/a_o}\right]^2}{a_o^2\left(\frac{1}{\pi a_o^3}\right)\left[e^{-a_o/a_o}\right]^2} = 100\,e^{-18} = 1.52\times10^{-6}$$

This means that one out of 7×10^5 hydrogen atoms at any instant might have its electron at distance 10 a_o (it does not stay so big for a long time).

- Thus, the hydrogen atom in the quantum mechanics is not a static rigid structure of fixed size.
- The maxima in the $r^2R^2_{1,0}$ plot (Figs. 2.17 and 2.18) are the radius of the spherical surface over which the electron is most likely to be found.
- For 1s orbital, the position of the maximum probability for the electron does correspond to the value $r=a_0=0.5292$ Å. Notice the perfect agreement between Bohr and Schrödinger models on this value of distance for the hydrogen atom.

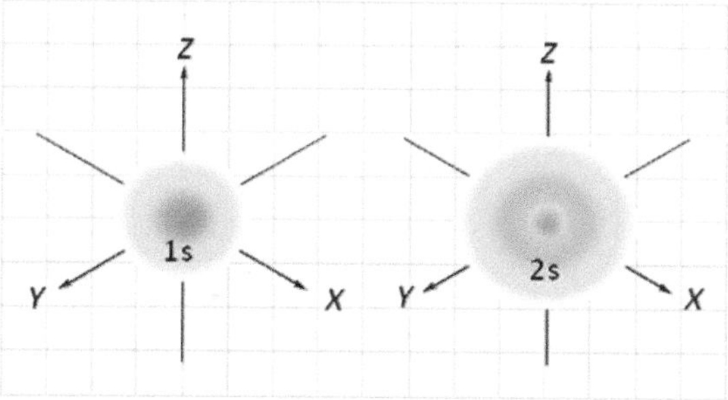

FIG. 2.19 Representations of the electron densities of 1s and 2s orbitals.

- In the Bohr model, the electron in this state revolves in a fixed circular orbital of radius $a_0=0.5292$ Å, while the Schrödinger model permits the full redial freedom.
- In the 2s state, $n=2$, $l=0$ (Figs. 2.17–2.19), in addition to the main peak in the surface density at $r=2.75$ Å, there is another little peak at much smaller value, $r=0.42$ Å.
- This electronic behavior is known as penetration, and shows that in certain states, the electron spends a small proportion of its time very close to the nucleus.
- When an electron penetrates to the nucleus, as a result:
  - the electron experiences high electrostatic attraction;
  - this attraction has a significant effect in lowering the energy of the electron; and
  - this increases the stability of states in which occur.
- For a given value of n, penetration is greater the smaller is the value of l.
- Penetration is greater for s electrons than for p electrons; p electrons, in turn, show greater penetration than d electrons, and so on (Fig. 2.17).

If: 2s-orbital has $L_{2,0}=2-\rho$, and 3s has $L_{3,0}=6-6\rho+\rho^2$, where, $\rho=\frac{2Zr}{na_0}$
Find out the number radial nodes of these orbitals.

- All s-orbitals are spherically symmetric, but differ in the number of radial nodes, the point at which the radial wave function passes through zero.
- The number of spherical nodal surfaces

$$= n - l - 1 \tag{2.17.3}$$

  l = number of nodal surfaces through the nucleus, for the real orbitals.
  $n-1$ = total number of nodal surfaces.
  (Each radial nodal point far from the nucleus forms a spherical surface in three dimensions.)
- For 2s with radial nodes where the polynomial $L_{2,0}$ is equal to zero:
  - $L_{2,0}=2-\rho=0$ at $\rho=2$,

    $\rho=\frac{2Zr}{na_0}$, which implies $r=2a_0$
    For 2s orbital with a radial node at $2a_0$ (1.06 Å Fig. 2.17).
    The number of spherical nodal surfaces $= n-l-1=2-0-1=1$.
- Similarly, 3s orbital with two nodes are found by solving the following:
  - $L_{3,0}=6-6\rho+\rho^2=0$ with $r=\frac{3}{2}\rho a_0$.

    One radial node at 1.90 a_0 and the other is at 7.10 a_0, (at 1.01 Å, and 3.76 Å Fig. 2.18).
    The number of spherical nodal surfaces $= n-l-1=3-0-1=2$.
- The energies of the s-orbitals increase as n increases, because the average distance from the nucleus increases and the electrons become less tightly bound.

Exercise

What is the radius of the sphere within which the probability of the finding of 1s electron of hydrogen atom is 90%?

Answer:

- If: $\psi_{1s} = Ne^{-\frac{r}{a_o}}$

The probability of finding the electron within a sphere of radius r_0 is

$$\int \psi_{1s}{}^{*} \psi_{1s} \partial \tau$$

$$\text{If} \int_0^{\pi} \sin\theta\,\partial\theta = -\cos\theta\Big|_0^{\pi} = 2, \quad \text{and} \int_0^{2\pi} \phi = 2\pi$$

- Then,

$$\int_0^{r_o}\int_0^{\pi}\int_0^{2\pi} (\psi_{1s})^2 r^2 \sin\theta\,\partial r\,\partial\theta\,\partial\phi = \int_0^{r_o}\int_0^{\pi}\int_0^{2\pi} \left(Ne^{-\frac{r}{a_o}}\right)^2 r^2 \sin\theta\,\partial r\,\partial\theta\,\partial\phi = 4\pi N^2 \int_0^{r_o} e^{-\frac{2r}{a_o}} r^2 \partial r$$

- Let $\frac{r}{a_o} = \varepsilon$, then $r = a_o\varepsilon$, $\partial r = a_o \partial \varepsilon$

$$4\pi N^2 \int_0^{r_o} e^{-\frac{2r}{a_o}} r^2 \partial r = 4\pi N^2 \int_0^{\varepsilon_o} e^{-2\varepsilon}(a_o\varepsilon)^2 a_o \partial\varepsilon = 4\pi N^2 a_o{}^3 \int_0^{\varepsilon_o} e^{-2\varepsilon}\varepsilon^2 \partial\varepsilon$$

where, $\varepsilon_o = \frac{r_o}{a_o}$

- We want to find ε_o such that

$$\frac{4\pi N^2 a_0{}^3 \int_0^{\varepsilon_o} e^{-2\varepsilon}\varepsilon^2 \partial\varepsilon}{4\pi N^2 a_0{}^3 \int_0^{\infty} e^{-2\varepsilon}\varepsilon^2 \partial\varepsilon} = 0.9$$

$4\pi N^2 a_0{}^3 \int_0^{\infty} e^{-2\varepsilon}\varepsilon^2 \partial\varepsilon \equiv 1$, probability of finding the electron in space

$$\frac{\int_0^{\varepsilon_o} e^{-2\varepsilon}\varepsilon^2 \partial\varepsilon}{\int_0^{\infty} e^{-2\varepsilon}\varepsilon^2 \partial\varepsilon} = 0.9$$

$$\int_0^{\varepsilon_o} e^{-2\varepsilon}\varepsilon^2 \partial\varepsilon = 0.9 \int_0^{\infty} \varepsilon e^{-2\varepsilon}\varepsilon^2 \partial\varepsilon$$

$$\because \int_0^{\infty} x^n e^{-ax} \partial r = \frac{n!}{a^{n+1}}$$

$$\therefore \int_0^{\varepsilon_o} \varepsilon^2 e^{-2\varepsilon}\,\partial\varepsilon = 0.225, \quad \text{and}$$

$$\int_0^{\varepsilon_o} e^{-2\varepsilon}\varepsilon^2 \partial\varepsilon = 0.9 \int_0^{\infty} \varepsilon e^{-2\varepsilon}\varepsilon^2 \partial\varepsilon = (0.9)\frac{2!}{2^3} = 0.225$$

$$\therefore \int_0^{\varepsilon_o} \varepsilon^2 e^{-2\varepsilon}\,\partial\varepsilon = 0.225$$

$$\because \int x^m e^{ax}\,\partial x = \frac{1}{a}x^m e^{ax} - \frac{m}{a}\int x^{m-1}e^{ax}\,\partial x,\quad m \geq 2,$$

$$\int x e^{ax}\,\partial x = \frac{1}{a^2}e^{ax}(ax-1)+C,\ \text{then}$$

$$\int_0^{\varepsilon_o} \varepsilon^2 e^{-2\varepsilon}\,\partial\varepsilon = 0.225,$$

$$-\frac{1}{2}\varepsilon^2 e^{-2\varepsilon} - \frac{2}{2}\int \varepsilon e^{-2\varepsilon}\,\partial\varepsilon \Big|_0^{\varepsilon_o} = -\frac{1}{2}\varepsilon^2 e^{-2\varepsilon} - \frac{1}{4}e^{-2\varepsilon}(-2\varepsilon-1)+C \Big|_0^{\varepsilon_o} = 0.225$$

$$-\frac{1}{2}e^{-2\varepsilon}\left[\varepsilon^2+\varepsilon+\frac{1}{2}\right]_0^{\varepsilon_0} = 0.225$$

$$-\frac{1}{2}e^{-2\varepsilon_0}\left[{\varepsilon_0}^2+\varepsilon_0+\frac{1}{2}\right]+\frac{1}{4} = 0.225$$

$$-\left[{\varepsilon_0}^2+\varepsilon_0+\frac{1}{2}\right]e^{-2\varepsilon_0} = -0.025$$

$$\left({\varepsilon_0}^2+\varepsilon_0+\frac{1}{2}\right)e^{-2\varepsilon_0} = 0.025$$

- In order to find the value of ε_0, we will plot $\left({\varepsilon_o}^2+\varepsilon_o+\frac{1}{2}\right)e^{-2\varepsilon_o}$ versus ε_o (Table 2.10, Fig. 2.20).

TABLE 2.10 Values of ε_0 and $\left({\varepsilon_o}^2+\varepsilon_o+\frac{1}{2}\right)e^{-2\varepsilon_0}$.

ε_o	ε_o^2	$\varepsilon_o^2+\varepsilon_o+0.5$	$e^{-2\varepsilon_0}$	$(\varepsilon_o^2+\varepsilon_o+0.5)\,e^{-2\varepsilon_0}$
1	1	2.5	0.135	0.338
2	4	6.5	0.018	0.119
3	9	12.5	0.002	0.031
3.1	9.61	13.21	0.002	0.027
3.12	9.73	13.3544	0.002	0.026
3.14	9.86	13.4996	0.002	0.025
3.16	9.99	13.6456	0.002	0.025
3.18	10.1	13.7924	0.002	0.024
3.2	10.2	13.94	0.002	0.023

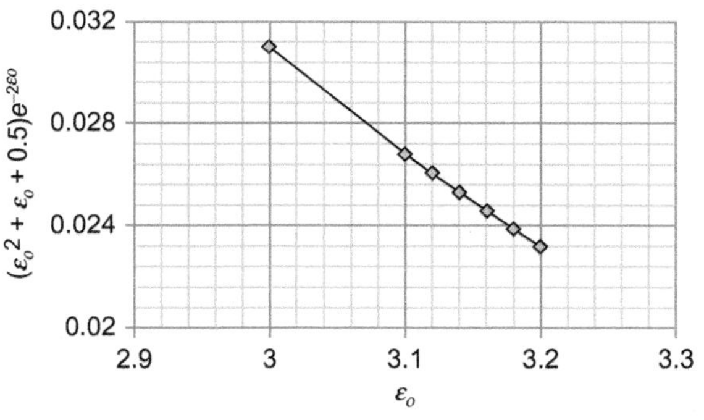

FIG. 2.20 Plot of $(\varepsilon_o^2+\varepsilon_o+0.5)e^{-2\varepsilon_o}$ versus ε_o.

When $(\varepsilon_o^2 + \varepsilon_o + 0.5)\, e^{-2\varepsilon_0} = 0.025$, the corresponding

$$\varepsilon_0 = 3.14a_0$$

At this value of ε_0, the probability of finding the electron e becomes 90%.

Consider an excited hydrogen atom with the electron in 2s orbital:

$$\psi_{2s} = \frac{1}{\sqrt{32\pi}}\left(\frac{1}{a_o}\right)^{3/2}\left(2 - \frac{r}{a_o}\right)e^{-r/2a_o}$$

a. **Calculate the probability of finding the electron in the volume that is defined by**

$$5.22\times10^{-10} \le r \le 5.26\times10^{-10}\text{m},\ \frac{\pi}{2} - 0.01 \le \phi \le \frac{\pi}{2} + 0.01,\ \text{and}\ \frac{\pi}{2} - 0.01 \le \theta \le \frac{\pi}{2} + 0.01$$

b. **Calculate the probability of finding the electron in the spherical shell that is defined by**

$$5.22\times10^{-10} \le r \le 5.26\times10^{-10}\text{m},\ a_o = 0.5292\times10^{-10}\text{m}$$

- The probability of finding the electron is

$$P = \int \psi_{2s}{}^{*}\psi_{2s}\partial\tau$$

a.

$$P = \frac{1}{\sqrt{32\pi}}\left(\frac{1}{a_o}\right)^3 \int_{5.22\times10^{-10}}^{5.26\times10^{-10}} r^2\left(2 - \frac{r}{a_o}\right)^2 e^{-r/2a_o}\partial r \int_{\frac{\pi}{2}-0.01}^{\frac{\pi}{2}+0.01} \sin\theta\partial\theta \int_{\frac{\pi}{2}-0.01}^{\frac{\pi}{2}+0.01} \phi\partial\phi$$

$$P = 0.00997 \times 6.75 \times 10^{-30}\text{m}^{-3} \times 3.43 \times 10^{-33}\text{m}^3 \times 0.02 \times 0.02 = 9.23 \times 10^{-8}$$

b.

$$P = \frac{1}{\sqrt{32\pi}}\left(\frac{1}{a_o}\right)^3 \int_{5.22\times10^{-10}}^{5.26\times10^{-10}} r^2\left(2 - \frac{r}{a_o}\right)^2 e^{-r/2a_o}\partial r \int_0^{\pi} \sin\theta\partial\theta \int_0^{2\pi} \phi\partial\phi$$

$$P = 0.00997 \times 6.75 \times 10^{-30}\text{m}^{-3} \times 3.43 \times 10^{-33}\text{m}^3 \times 2\pi \times 2\pi = 2.90 \times 10^{-3}$$

2.18 THE STRUCTURE OF MANY-ELECTRON ATOMS

What are the complications and the required procedures in order to apply the Schrödinger equations for many-electron atoms?

- The Schrödinger equations for many-electron atoms are extremely complicated, because all electrons interact with each other.
- Consequently, we are forced to make approximation, and the simple approached based on the structure of hydrogenic atoms was adopted.
- Central field approximation:
  - The electrons in many-electron atoms move in a spherically symmetric field of a shielded nucleus.
  - Consequently, the acceptable standing states of the electron in the atom can be regarded in terms of the four quantum numbers: n, l, m, and s.
- Orbital approximation:
  - The actual wave function of many-electron atoms, $\psi_{(e_1, e_2, \ldots..)}$, is a tangled function of all the electrons.
  - In the orbital approximation, the exact wave function is obtained by thinking of each electron as occupying its own ψ, and

$$\psi_{(e_1, e_2, \ldots..)} = \psi_{(e_1)}\psi_{(e_2)} \tag{2.18.1}$$

- The individual orbitals resemble the hydrogenic orbitals, wherein the nuclear charges are adjusted by the presence of all other electrons in the atom.
- We can then consider the number of orbitals and their angular dependence to be similar to that of hydrogen atom.

- Thus, the hydrogen orbitals are used to describe the electronic structure of an atom with more than one electron.
- The procedure is basically to select to each electron in the atom a set of the four quantum numbers n, l, m, and s. This requires that no more than two electrons may exist in any given orbital. If two electrons fill one orbital, then their spins must be paired, according to the Pauli Principle.
- The Pauli Exclusion Principle is the key to the structure of the complex atoms, to chemical periodicity, and to molecular structure.

2.19 THE PAULI EXCLUSION PRINCIPLE

What is the Pauli Exclusion Principle, and how does it apply?

Give examples of fermions and bosons, what happens when two identical fermions exchange or two identical bosons exchanges.

Suppose two electrons in an atom occupy an orbital ψ. Write the possible and the acceptable wave functions.

- The Pauli Exclusion Principle states that no two electrons in a given atom can exist in the same quantum state. This requests that no two electrons in the same atom can have the same set of all four quantum numbers n, l, m, and s (e.g., they cannot have the same wave function).
  - Only two electrons may occupy any given orbital, and if two do occupy one orbital, then their spins must be paired, which donated $\uparrow\downarrow$, and have zero net spin angular momentum.
  - If one electron has $m_s = +1/2$ then the other will have $m_s = -1/2$, and they are oriented on their appropriate processional cones so that the resultant spin is zero (Fig. 2.21).
- Since the electrons are indistinguishable particles, electrons might exchange to adopt an acceptable wave function.
  - When the labels of any two identical fermions ($s = 1/2$: electrons, protons, neutrons, and some atomic nuclei) are exchanged, the total wave function changes sign.
  - When the labels of any two identical bosons ($s = 1$, photons, and a number of atomic nuclei and atoms) are exchanged, the total wave function retains the same sign.
  - Consider the wave function for two electrons $\psi(1,2)$. The wave function must change sign if we interchange the labels 1 and 2: $\psi(1,2) = -\psi(2,1)$.
  - For a wave function to be acceptable (for electrons), it must change sign when the electrons are exchanged: $\psi \rightarrow -\psi$.
- Suppose the two electrons in an atom occupy an orbital ψ, then in orbital approximation the overall wave function is $\psi(1)\ \psi(2)$. Consider an α electron is an electron with $m_s = +1/2$; a β electron is an electron with $m_s = -1/2$.
- There are several possibilities for two spins:
  - both α, denoted $\alpha(1)\ \alpha(2)$;
  - both β, denoted $\beta(1)\ \beta(2)$; and

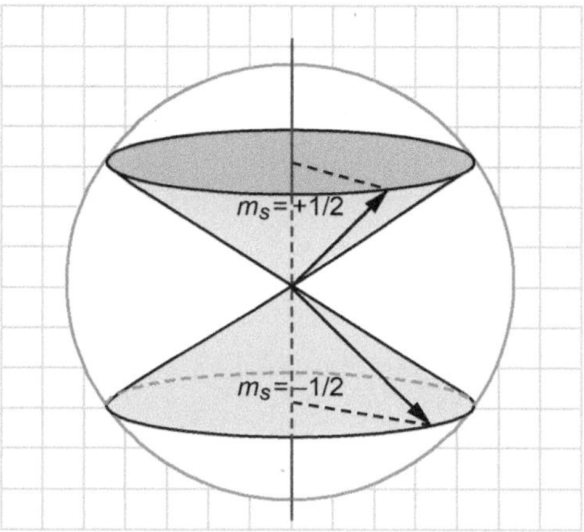

FIG. 2.21 An electron spin ($s = 1/2$) can take only two orientations with respect to a specified axis.

- ○ one α and the other β, denoted $\alpha(1)\,\beta(2)$ or $\beta(1)\,\alpha(2)$ (Fig. 2.22).

  Because we cannot tell which electron is α and which is β, it is therefore appropriate to express the spin states as normalized linear combinations:

$$\sigma_+(1,2) = \frac{1}{\sqrt{2}}[\alpha(1)\beta(2) + \beta(1)\alpha(2)] \quad (2.19.1)$$

$$\sigma_-(1,2) = \frac{1}{\sqrt{2}}[\alpha(1)\beta(2) - \beta(1)\alpha(2)] \quad (2.19.2)$$

  because these combinations allow one spin to be α and the other is β with equal probability.

- The total wave function of the system is the product of the orbital part and one of the spin states:

$$\psi(1)\psi(2)\alpha(1)\alpha(2)$$
$$\psi(1)\psi(2)\beta(1)\beta(2)$$
$$\psi(1)\psi(2)\sigma_+(1,2)$$
$$\psi(1)\psi(2)\sigma_-(1,2)$$

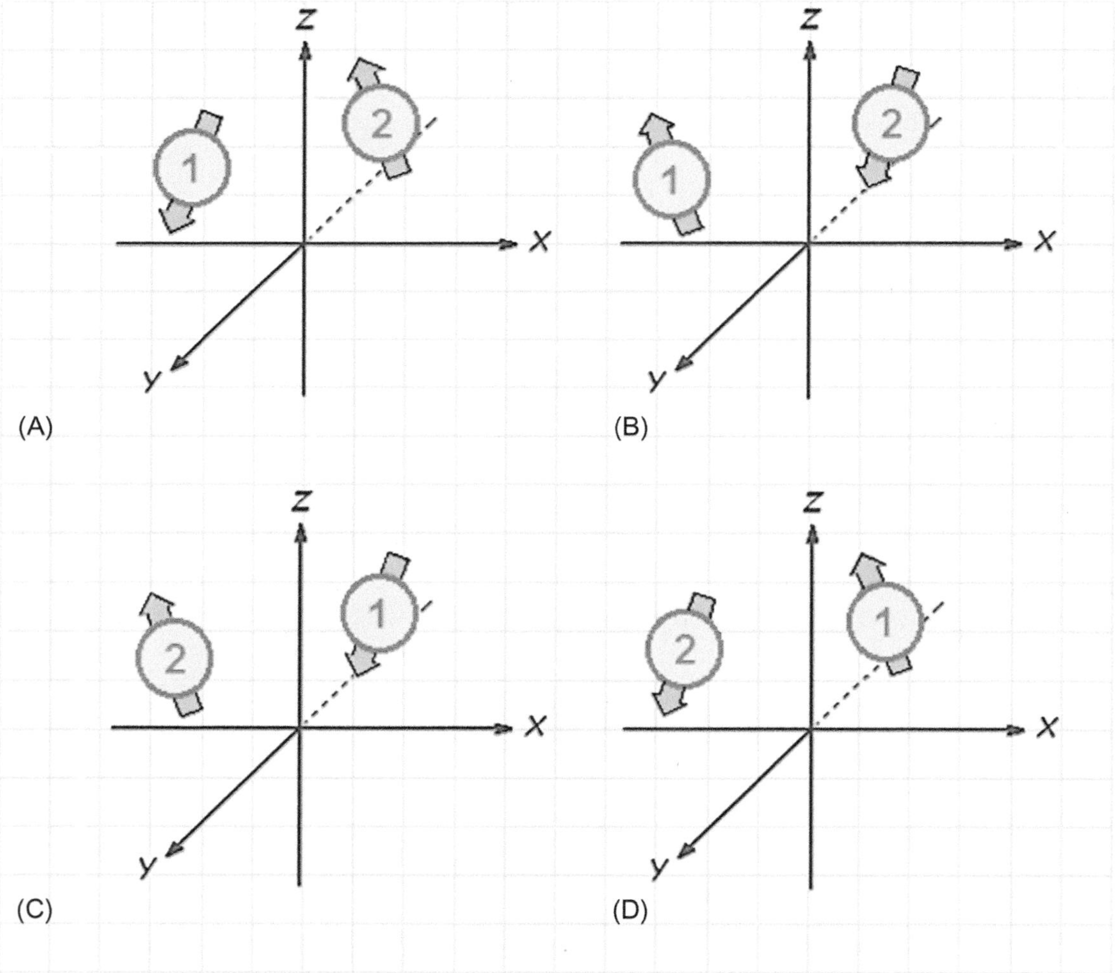

FIG. 2.22 Possible exchange of orbit and spin for two electrons: (A) original configuration; (B) spin exchanged; (C) orbital exchanged; (D) spin and orbitals exchanged.

- According to the Pauli Principle, for a wave function to be acceptable (for electrons), it must change sign when the electrons are exchanged
  - In each case, exchanging the labels 1 and 2 converts the factors:

$$\psi(1)\psi(2) \text{ into } \psi(2)\psi(1)$$
$$\alpha(1)\alpha(2) \text{ into } \alpha(2)\alpha(1), \text{ or}$$
$$\beta(1)\beta(2) \text{ into } \beta(2)\beta(1)$$

which are the same, because the order of multiplying the function does not change the value of the product.

Therefore, the first two overall products:

$$\psi(1)\psi(2)\alpha(1)\alpha(2)$$

and

$$\psi(1)\psi(2)\beta(1)\beta(2)$$

are not allowed.

  - The combination

$$\sigma_+(1,2) = \frac{1}{\sqrt{2}}[\alpha(1)\beta(2) + \beta(1)\alpha(2)]$$

changes to

$$\sigma_+(2,1) = \frac{1}{\sqrt{2}}[\alpha(2)\beta(1) + \beta(2)\alpha(1)] = \sigma_+(1,2)$$

because simply written in a different order, therefore, the third overall product is also disallowed.

  - Finally, consider the following:

$$\sigma_-(1,2) = \frac{1}{\sqrt{2}}[\alpha(1)\beta(2) - \beta(1)\alpha(2)]$$

When changed to

$$\sigma_-(2,1) = \frac{1}{\sqrt{2}}[\alpha(2)\beta(1) - \beta(2)\alpha(1)]$$
$$= -\frac{1}{\sqrt{2}}[\alpha(1)\beta(2) - \beta(1)\alpha(2)] = -\sigma_-(1,2)$$

the combination does change sign (it is antisymmetric).

The product $\psi(1)\psi(2)\sigma_-(1,2)$ also changes sign under electron exchange, and therefore it is acceptable.

- Only one of the four possible states is allowed by the Pauli Principle, and that one has paired α and β spins (↑↓).
- That acceptable product wave function $\psi(1)\psi(2)\sigma_-(1,2)$ can be expressed as a determinant:

$$\frac{1}{\sqrt{2}}\begin{vmatrix} \psi(1)\alpha(1) & \psi(2)\alpha(2) \\ \psi(1)\beta(1) & \psi(2)\beta(2) \end{vmatrix} = \frac{1}{\sqrt{2}}[\psi(1)\alpha(1)\psi(2)\beta(2) - \psi(2)\alpha(2)\psi(1)\beta(1)] = \psi(1)\psi(2)\sigma_-(1,2)$$

2.20 SLATER DETERMINANT

Suppose *N* electrons in an atom occupy $\psi_a, \psi_b \ldots$. What is the possible and the acceptable wave function?

Show how to use the Slater determinant to express acceptable wave function for *N* electrons occupy $\psi_a, \psi_b \ldots$

- Any acceptable wave function can be expressed as a Slater determinant, in general for N electrons in orbitals $\psi_a, \psi_b \ldots$:

$$\psi(1,2,\ldots\ldots,N)=\frac{1}{\sqrt{N}}\begin{vmatrix}\psi_a(1)\alpha(1) & \psi_a(2)\alpha(2) & \psi_a(3)\alpha(3) & \cdots & \psi_a(N)\alpha(N)\\ \psi_b(1)\beta(1) & \psi_b(2)\beta(2) & \psi_b(3)\beta(3) & \cdots & \psi_b(N)\beta(N)\\ \psi_c(1)\alpha(1) & \psi_c(2)\alpha(2) & \psi_c(3)\alpha(3) & \cdots & \psi_c(N)\alpha(N)\\ \vdots & \vdots & \vdots & \vdots & \vdots\\ \psi_z(1)\beta(1) & \psi_z(2)\beta(2) & \psi_z(3)\beta(3) & \cdots & \psi_z(N)\beta(N)\end{vmatrix} \quad (2.20.1)$$

- Writing a many-electrons wave function in this way ensures that it is antisymmetric under the interchange of any pair of electrons.

Exercise

Write all symmetric and antisymmetric wave functions for the 1s2s configurations. Normalize one of the antisymmetric-functions $\left(\int \alpha\beta = 0\right)$.

Answer:

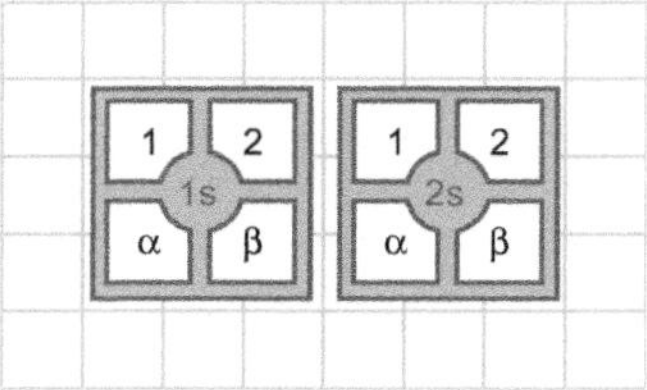

- Symmetric:

$S_1 = N_1[1s(1)\alpha(1)2s(2)\alpha(2) + 1s(2)\alpha(2)2s(1)\alpha(1)]$

$S_2 = N_2[1s(1)\beta(1)2s(2)\beta(2) + 1s(2)\beta(2)2s(1)\beta(1)]$

$S_3 = N_3[1s(1)\alpha(1)2s(2)\beta(2) + 1s(2)\alpha(2)2s(1)\beta(1)]$

$S_4 = N_4[1s(1)\beta(1)2s(2)\alpha(2) + 1s(2)\beta(2)2s(1)\alpha(1)]$

where S_1, S_2, S_3, and S_4: symbols just to designate the symmetric functions.
- Antisymmetric:

$D_1 = N_1[1s(1)\alpha(1)2s(2)\alpha(2) - 1s(2)\alpha(2)2s(1)\alpha(1)]$

$D_2 = N_2[1s(1)\beta(1)2s(2)\beta(2) - 1s(2)\beta(2)2s(1)\beta(1)]$

$D_3 = N_3[1s(1)\alpha(1)2s(2)\beta(2) - 1s(2)\alpha(2)2s(1)\beta(1)]$

$D_4 = N_4[1s(1)\beta(1)2s(2)\alpha(2) - 1s(2)\beta(2)2s(1)\alpha(1)]$
- Let us evaluate normalization constant for S_1:

$$\int S_1^* S_1\,\partial\tau = N_1^2\int [1s(1)\alpha(1)2s(2)\alpha(2) + 1s(2)\alpha(2)2s(1)\alpha(1)]^2\,\partial\tau$$

$$= N_1^2\int [1s(1)\alpha(1)2s(2)\alpha(2)]^2\,\partial\tau + N_1^2\int [1s(2)\alpha(2)2s(1)\alpha(1)]^2\,\partial\tau$$

$$+ 2N_1^2\int [1s(1)\alpha(1)2s(2)\alpha(2)][1s(2)\alpha(2)2s(1)\alpha(1)]\,\partial\tau$$

$$\int S_1^* S_1\,\partial\tau = N_1^2\left\{\begin{array}{c}\left[\int [1s(1)2s(2)]^2\,\partial\tau_{\text{spatial}}\right]\left[\int [\alpha(1)\alpha(2)]^2\,\partial\tau_{\text{spin}}\right]\\ +\left[\int [1s(2)2s(1)]^2\,\partial\tau_{\text{spatial}}\right]\left[\int [\alpha(2)\alpha(1)]^2\,\partial\tau_{\text{spin}}\right]\\ +2\int \left[1s(1)2s(2)1s(2)2s(1)\,\partial\tau_{\text{spatial}}\right][\alpha(1)\alpha(2)\alpha(2)\alpha(1)]\,\partial\tau_{\text{spin}}\end{array}\right\}$$

$$\int S_1^* S_1 \partial\tau = N_1^2 \left\{ \begin{array}{l} \left[\int [1s(1)]^2 \partial\tau_{\text{spatial1}} \int [2s(2)]^2 \partial\tau_{\text{spatial2}} \int [\alpha(1)]^2 \partial\tau_{\text{spin1}} \int [\alpha(2)]^2 \partial\tau_{\text{spin2}}\right] \\ + \left[\int [1s(2)]^2 \partial\tau_{\text{spatial2}} \int [2s(1)]^2 \partial\tau_{\text{spatial1}} \int [\alpha(2)]^2 \partial\tau_{\text{spin2}} \int [\alpha(1)]^2 \partial\tau_{\text{spin1}}\right] \\ +2\int 1s(1)2s(1)\partial\tau_{\text{spatial1}} \int 2s(2)1s(2)\partial\tau_{\text{spatial2}} \int \alpha(1)\alpha(1)\partial\tau_{\text{spin1}} \int \alpha(2)\alpha(2)\partial\tau_{\text{spin2}} \end{array} \right\}$$

$$\int S_1^* S_1 \partial\tau = N_1^2\{[1\times1\times1\times1] + [1\times1\times1\times1] + 2[0\times0\times1\times1]\}$$

$$\int S_1^* S_1 \partial\tau = 2N_1^2 = 1$$

$$N_1 = \frac{1}{\sqrt{2}}$$

Spatial 1, and spatial 2 are the electron coordinates.

All other normalizations will be $\frac{1}{\sqrt{2}}$ for the same reason.

Exercise

Indicate which of the following functions are symmetric, which are antisymmetric, and which are neither symmetric nor antisymmetric.

a. $\mathbf{1s(1)\alpha(1)2s(2)\alpha(2) - 1s(2)\alpha(2)2s(1)\alpha(1)}$

b. $\mathbf{1s(1)\beta(1)2s(2)\alpha(2)}$

c. $\mathbf{1s(1)2p_{+1}(2)\alpha(1)\beta(2) + 1s(2)2p_{+1}(1)\alpha(2)\beta(1)}$

d. $\mathbf{1s(1)\alpha(1)1s(2)\alpha(2)}$

Answer

a.	$1s(1)\alpha(1)2s(2)\alpha(2) - 1s(2)\alpha(2)2s(1)\alpha(1)$	Antisymmetric
b.	$1s(1)\beta(1)2s(2)\alpha(2)$	Neither
c.	$1s(1)2p_{+1}(2)\alpha(1)\beta(2) + 1s(2)2p_{+1}(1)\alpha(2)\beta(1)$	Symmetric
d.	$1s(1)\alpha(1)1s(2)\alpha(2)$	Symmetric

2.21 PENETRATION AND SHIELDING

Compute the expected total number of the degenerate orbitals in each shell of hydrogen atom when $n = 1, 2, \ldots, 5$. Why do you anticipate this orbital degeneracy and why has it been removed?

- All the hydrogen atom levels are expected to be degenerate. This arises from the fact that the energy depends on n, but not on l and m_l.
- The total degeneracy is summarized in Table 2.11.

$$g(n) = \sum_{n=1}^{n-1}(2l+1) \tag{2.21.1}$$

$$g(n) = \frac{2n(n-1)}{2} + n = n^2 \tag{2.21.2}$$

- When precise computing is performed, the degeneracy in l and n is lifted, as a result, of the spin-orbital interactions (p. 119). The resulting splitting is relatively small on the scale of the separation of the levels with different n values.

TABLE 2.11 Energy Levels and Orbital Degeneracy

		Number of Orbitals					
n		1	2	3	4	5	$(2l+1)$
$l=0$	s	1	1	1	1	1	1
$l=1$	p	0	3	3	3	3	3
$l=2$	d	0	0	5	5	5	5
$l=3$	f	0	0	0	7	7	7
$l=4$	g	0	0	0	0	9	9
Total		1	4	9	16	25	
$\sum_{n=1}^{n-1}(2l+1)$		1	4	9	16	25	

In many-electron atoms, why is the orbital degeneracy of the given shell lifted?

- Unlike the hydrogen atom, the 2s- and 2p-orbitals (all subshells of a given shell) are not degenerate in many-electron atoms. The reasons for the splitting are as follows:
  1. There is variation in penetration by electrons in orbitals of different l (p. 102).
     - Electrons are negatively charged and they feel attracted to the nucleus. The attraction that an electron feels is dependent on the penetration distance from the nucleus.
     - The radial distribution function, R (Figs. 2.17 and 2.18), reveals that s-electrons of outer shells have some possibility of existing at distances near to the nucleus that lie inside the most probable radii for electrons of lower principal quantum number, n.
     - Penetration is higher for s-electrons than for p-electrons; p-electrons, in turn, show higher penetration than d-electrons, and so on.
  2. The electron in a many-electron atom faces a coulombic repulsion from all the other surrounding electrons.
     - Negatively charged electrons are near to each other, and they can repel each other. The repulsion forces that an electron feels are dependent on the distance from the nearest negative charge.
     - If there is an electron at a distance r from the nucleus, it meets repulsion from a negative point charge located at the nucleus. The charge of this negative point is equal in magnitude to the total effect of the electrons within a sphere of radius r.
  3. The surrounding electrons do not block the full coulombic attraction of the nucleus.
     - The effect of the negative point charge, when averaged over all the location of the electrons, is to reduce the full charge of the nucleus from Ze to $Z_{eff}.e$, where Z_{eff} is the effective atomic number.

$$Z^*_{eff.} = Z - \sigma$$

σ = shielding constant.

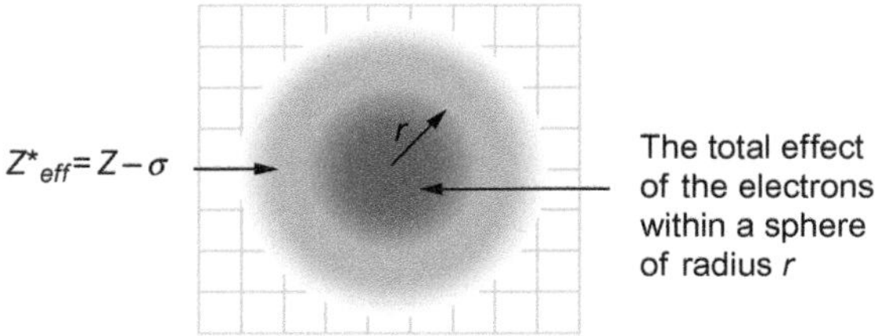

- The value of $Z_{eff.}$ is different for s- and p-electrons because they have different radial wave functions.

- An s-electron has a higher penetration through the inner shells than a p-electron, and an s-electron is more likely to be found close to the nucleus than a p-electron of the same shell (the p-orbital has a node at the nucleus, p. 87, 102).
- Therefore, an s-electron experiences less shielding (higher $Z_{eff.}$*) than a p-electron.
- The shielding effect describes the decrease in attraction between an electron and the nucleus in any atom with more than one electro shell.

- By combining these effects, the energy of the subshells in many-electron atom generally lies in the order: $s<p<d$.

What are Slater's empirical rules for calculating Z^*_{eff}, and σ?

a. **Sort the orbitals: 1s, 2s, 2p, 3s, 3p, 3d, 4s, 4p, 4d, 4f ... into groups.**
Complete:
b. **$\sigma=$ ...... for each electron in the same group except 1s; $\sigma=$ ......;**
c. **$\sigma=$ ...... for each d or f electrons in the underlying group; and**
d. **$\sigma=$ ...... for each s or p electrons in the underlying group.**

Answer:

- Slater suggested that only the highest-order term of each Schrödinger radial function is important:

$$R(r)=N_r r^{n^*-1}e^{-(Z^*/n^*)r} \quad \text{(Salter's radial function)} \quad (2.21.1)$$

N_r is the normalizing constant, and the value of n^* are not as n for larger n:

n	1	2	3	4	5	6
n^*	1	2	3	3.7	4.0	4.2

a. The electrons are sorted in the following order:

$$\underbrace{1s}_{1};\underbrace{2s2p}_{2};\underbrace{3s3p}_{3};\underbrace{3d}_{4};\underbrace{4s4p}_{5};\underbrace{4d}_{6};\underbrace{4f}_{7},......$$

ns and np electrons have always been considered as a single group.
Electrons in groups locating above that of a particular electron do not shield it at all.

b. $\sigma=0.35$ for each electron in the same group except 1s; $\sigma=0.30$;
c. $\sigma=1$ for each d or f electrons in the underlying group; and
d. $\sigma=0.85$ for each s or p electrons in the underlying group.

Calculate the ionization potential of Li, by using Slater approximations.

Answer:

Li: $1s^2 2s^1 \rightarrow Li^+$: $1s^2$

$$I=E^+_{Li}-E^0_{Li}$$

$$E_n=\frac{\mu Z^2 e^4}{2(4\pi\epsilon_o)^2\hbar^2 n^2} \quad (2.8.16)$$

$$E_n=-\frac{mZ^{*2}e^4}{2(4\pi\epsilon_o)^2\hbar^2 n^{*2}}=-(13.6)\left(\frac{Z^*}{n^*}\right)^2 \quad \text{electron volts}$$

$$I=-(13.6)\left\{2\left(\frac{Z^*}{n^*}\right)^2_{1s}\right\}_{ion}-\left[-(13.6)\left\{2\left(\frac{Z^*}{n^*}\right)^2_{1s}+\left(\frac{Z^*}{n^*}\right)^2_{2sp}\right\}_{atom}\right]$$

$$I=-(13.6)\left[\left\{2\left(\frac{Z^*}{n^*}\right)^2_{1s}\right\}_{ion}-\left\{2\left(\frac{Z^*}{n^*}\right)^2_{1s}+\left(\frac{Z^*}{n^*}\right)^2_{2sp}\right\}_{atom}\right]=+13.6\left(\frac{Z^*}{n^*}\right)^2_{2sp}$$

Each of the two 1s electrons of Li atom screens 0.85 charge unit from 2s electrons.

$$Z^* = Z - 2(0.85) = 3 - 1.7 = 1.3$$

$$I = +13.6\left(\frac{Z^*}{n^*}\right)^2_{2sp} = +13.6\left(\frac{1.3}{2}\right)^2_{2sp} = 5.7\,\text{eV}$$

This is comparable with the experimental value 5.4 eV.

Calculate the ionization potential of F, by using Slater approximations.

Answer:

F: $1s^2 2s^2 2p^5$ {or $(1s)^2(2sp)^7$} → F^+: $1s^2 2s^2 2p^4${or $(1s)^2(2sp)^6$}

$$I = E_F^+ - E_F^0$$

$$E_n = \frac{\mu Z^2 e^4}{2(4\pi\epsilon_o)^2\hbar^2 n^2} \quad (2.8.16)$$

$$E_n = -\frac{mZ^2e^4}{2(4\pi\epsilon_o)^2\hbar^2 n^2} = -(13.6)\left(\frac{Z^*}{n^*}\right)^2 \text{ electron volts}$$

$$I = -(13.6)\left[\left\{2\left(\frac{Z^*}{n^*}\right)^2_{1s} + 6\left(\frac{Z^*}{n^*}\right)^2_{2sp}\right\}_{ion} - \left\{2\left(\frac{Z^*}{n^*}\right)^2_{1s} + 7\left(\frac{Z^*}{n^*}\right)^2_{2sp}\right\}_{atom}\right]$$

$$I = -(13.6)\left[\left\{6\left(\frac{Z^*}{n^*}\right)^2_{2sp}\right\}_{ion} - \left\{7\left(\frac{Z^*}{n^*}\right)^2_{2sp}\right\}_{atom}\right]$$

Each of the two 1s groups and the five electrons of the 2sp group screen 2sp electrons.

$$Z^*_{ion} = Z - 5(0.35) - 2(0.85) = 9 - 1.75 - 1.7 = 5.55$$

$$Z^*_{atom} = Z - 6(0.35) - 2(0.85) = 9 - 1.75 - 1.7 = 5.20$$

$$I = -(13.6)\left[\left\{6\left(\frac{5.55}{2}\right)^2_{2sp}\right\}_{ion} - \left\{7\left(\frac{5.2}{2}\right)^2_{2sp}\right\}_{atom}\right] = 15.15\,\text{eV}$$

This is comparable with the experimental value 17 eV.

What is the energy of the 4 s electron in K?

Answer

K: $1s^2\ 2s^2\ 2p^6\ 3s^2\ 3p^6\ 4s^1$

Ten electrons of $1s^2\ 2s^2\ 2p^6$ with $n = 1$, and 2 contribute 1.00 each to the shielding.

Eight electrons of $3s^2\ 3p^6$ each contribute 0.85.

$\sigma = 10\ (1.00) + 8\ (0.85) = 16.8$

$$Z^* = 19 - 16.8 = 2.2$$

$$E_4 = -\frac{mZ^2e^4}{2(4\pi\epsilon_o)^2\hbar^2 n^2} = -\frac{me^4}{2(4\pi\epsilon_o)^2\hbar^2}\frac{Z^{*2}}{n^2} = -(13.6)\frac{(2.2)^2}{4^2} = -4.1\,\text{eV}$$

The experimental value is −4.3 eV.

2.22 THE BUILDING-UP PRINCIPLE

a. **Show that**

$$E_n = (-13.6\text{eV})\frac{Z^2}{n^2}$$

where Z is the core charge and E_n is the energy of outer electron.

b. **In Table 2.12, the observed energy of the outer electron of three electron ions draw**

$$E_{\text{Experimental}} - E_{\text{Theoretical}} \text{ versus } Z$$

c. **How can you explain the deviation of the experimental from calculated energy?**

- The relationship between the energy of the outer electron and the nuclear charge is theoretically given by

$$E_n = \frac{\mu Z^2 e^4}{2(4\pi\epsilon_o)^2\hbar^2 n^2} \quad (2.8.16)$$

$$\mu = m_e$$

$$E_n = -\frac{mZ^{*2}e^4}{2(4\pi\epsilon_o)^2\hbar^2 n^2} = -\frac{me^4}{2(4\pi\epsilon_o)^2\hbar^2}\frac{Z^{*2}}{n^2}$$

$$E_n = -\frac{me^4}{2(4\pi\epsilon_o)^2\hbar^2}\frac{Z^{*2}}{n^2} = -(2.179\times10^{-18}\text{J})\frac{Z^{*2}}{n^2} = -(13.6\text{eV})\frac{Z^{*2}}{n^2}$$

or

$$\sqrt{\frac{E}{-13.6}} = \frac{Z^*}{n}$$

With $Z^* = Z_{\text{core}}$

- If the positive nuclear charges do not screen and neutralize the outer electrons, the observed energy of the outer electron will be less than the theoretical one, where

$$\frac{Z^*}{n} = \sqrt{\frac{E_{\text{Theoretical}}}{-13.6}}$$

TABLE 2.12 The Observed Energy of the Outer Electron of Three Electron Ions

	Li^+	Be^{2+}	B^{3+}	C^{4+}	N^{5+}	O^{6+}
$1s^22s^1$	6.41	20.37	40.77	68.17	102.57	143.98
$1s^22p^1$	5.34	17.68	36.79	63.17	95.99	135.91
$1s^23s^1$	...	8	17.04	29.21	44.36	62.65
$1s^23p^1$	...	7.27	16.23	28.04	42.9	60.75
$1s^23d^1$	...	6.86	15.53	27	41.47	59.04
$1s^24s^1$, $1s^24p^1$	...	3.64	8.32	14.59	22.72	32.52
$1s^24d^1$, $1s^24f^1$	...	3.58	7.94	14.09	21.99	31.64

Thus, the difference between $(E_{\text{Theoretical}} - E_{\text{Experimental}})$ will measure the energy displacement of the outer electron.

- The deviation will indicate that the outer electron is displaced to a lower and more stable energy, and face a nuclear charge that is greater than Z*.

- Fig. 2.23 represents the difference $(E_{\text{Theoretical}} - E_{\text{Experimental}})$ in the energy of the outer electrons of three-electron ions (Li^+, Be^{2+}, B^{3+}, C^{4+}, N^{5+}, O^{6+}) as a function of the atomic number of these ions less two ($Z^* = Z_{\text{core}} = Z - 2$).
  - The two positive nuclear charges $(Z-2)$ screen and neutralize the two electrons of $1s^2$ of these three-electron ions ($1s^22s^1$, $1s^22p^1$, $1s^2\,3s^1$, $1s^2\,3p^1$, $1s^2\,3d^1$, $1s^2\,4\,s^1$, $1s^2\,4p^1$, and $1s^2\,4d^1$).
  - The value of

$$\frac{Z^*}{n} = \frac{Z-2}{n} = \sqrt{\frac{E_{\text{Theoretical}}}{-13.6}}$$

represents the theoretical ion energy, assuming that the $1s^2$ electrons completely neutralize two units of positive nuclear charge.

- The observed results indicate that the difference between $(E_{\text{Theoretical}} - E_{\text{Experimental}})$ (Fig. 2.23) depends on the principle quantum number of the unpaired electron.
- The shift is significantly large for the $1s^22s^1$ configuration, and decreases for higher ion energy; $1s^2\,nd^1$ and $1s^2\,nf^1$ configurations have almost full screening.
  - The energy of $1s^22s^1$configuration has moved to a lower or more stable energy. This indicates that the 2*s* electrons face a nuclear charge that is greater than $Z-2$.
  - This is in the same way as for the $1s^2p^1$ configuration, but there the 1s electrons have better screening.
  - The screening ability of 1s pair is least for 3s and greatest for 3d.
- The varying in the power of inner electrons to screen nuclear charge from the outer electrons is the basis to comprehend the electron configurations and their location in the periodic table.

What are the empirical rules for the order of the orbitals filling?

- The following are the empirical rules for the order of orbitals filling:
  1. Orbital energies increase as the sum of *n* and *l* increases.
  2. In the case where two orbitals have equal sum $(n+l)$, the one with the smaller *n* will be filled first.

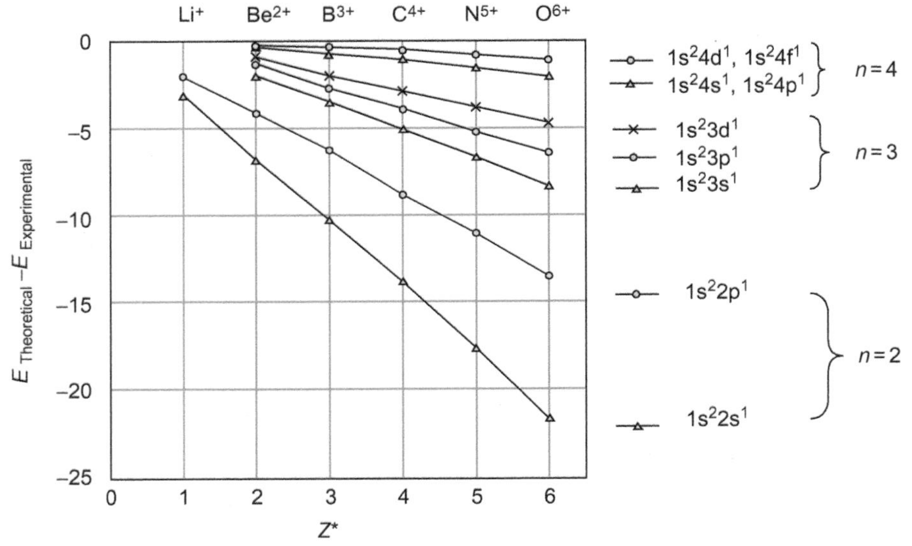

FIG. 2.23 The difference between the theoretical energy of the outer beyond $^1s^2$ and the observed one versus the atomic number of the ion less two.

The order of filling is as follows:

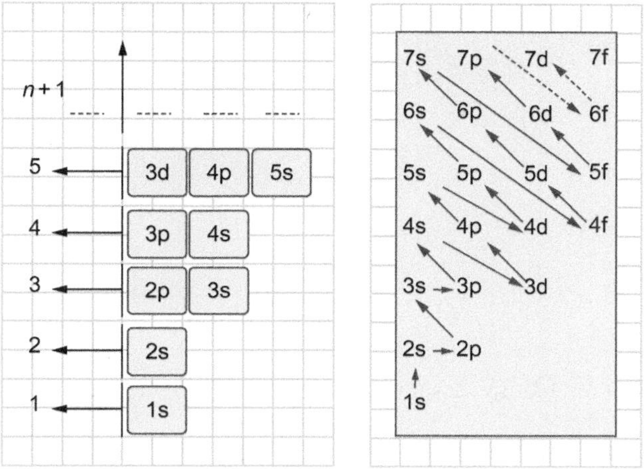

- Electrons singly occupy different orbitals of a given subshell before doubly occupying any one of them.
- Building up the periodic table starts with the hydrogen atom and successively adding a proton to the nucleus and an electron to the extranuclear shell, accounting for screening effects, and Hund's rule.

  Hund's rule:

An atom in its ground state adopts a configuration with the greatest number of unpaired electrons.

What are the ground state electronic configurations of the following?
N (Z = 7), Xe (Z = 54),Ge (Z = 32), U(Z = 92), Cu^{2+}

- N: $1s^2 2s^2 2p^3$
- Xe: $1s^2 2s^2 2p^6 3s^2 3p^6 4s^2 3d^{10} 4p^6 5s^2 4d^{10} 5p^6$
- Ge: $1s^2 2s^2 2p^6 3s^2 3p^6 4s^2 3d^{10} 4p^2$ or [Ar] $4s^2 3d^{10} 4p^2$
- U: $1s^2 2s^2 2p^6 3s^2 3p^6 4s^2 3d^{10} 4p^6 5s^2 4d^{10} 5p^6 6s^2 4f^{14} 5d^{10} 6p^6 7s^2 5f^4$ or [Rn] $7s^2 5f^4$
- Cu^{2+}: $1s^2 2s^2 2p^6 3s^2 3p^6 3d^9$

  In Cu^{2+} and all the first row of the transition metals ions, the 3d orbitals are drawn closer to the nucleus than they are in the neutral states, and the extra penetration of the 4 s orbital is insufficient to outweigh the increased stability of the 3d orbitals.

Define the:

a. **total orbital quantum number, $\vec{L}$;**
b. **total azimuthal quantum number, $\vec{M_l}$;**
c. **total spin quantum number, $\vec{S}$; and**
d. **total angular momentum, $\vec{J}$.**

- To identify the terms of a polyelectronic atom, we need to determine the appropriate quantum numbers (L, S, and J, where J is the spin-orbital interaction) for each term.

  a. L represents the total orbital quantum number of polyelectronic atoms,

$$\vec{L} = \sum \vec{l} = \vec{l}_1 + \vec{l}_2 + \cdots$$

where $\vec{l}$ is single electron orbital quantum number (Fig. 2.24).

Example: $2p^1\ 3p^1$

$$\vec{l}_1 = 1, \text{ and } \vec{l}_2 = 1, \quad \vec{L} = 2$$

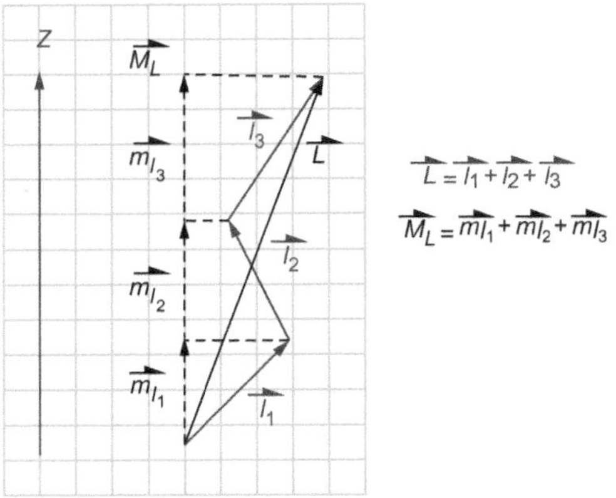

FIG. 2.24 Representation of orbital and the azimuthal quantum numbers of polyelectronic atoms.

b. M_L: represents the total azimuthal quantum number of polyelectronic atoms,

$$\vec{M}_L = \sum \vec{m}_l = \vec{m}_1 + \vec{m}_2 + \cdots = M_L$$

where $\vec{m}_l$ is the single electron azimuthal quantum number. The single electron m_l values correspond to the angular momentum vectors directed in the same orientation (the z-axis), thus the vector addition reduces to easy algebraic summation (Fig. 2.24).

- Summation of the angular momentum vectors of l and s is much more complicated because these vectors generally are not parallel.

c. The same applies to S or M_s:
$\vec{S}$ represents the total spin quantum number.

$$\vec{S} = \sum \vec{S} = \vec{s}_1 + \vec{s}_2 + \cdots, = \text{ and}$$

$$\vec{M}_s = \sum \vec{m}_s = \vec{m}_{s1} + \vec{m}_{s2} + \cdots = M_s$$

Example: two spins of +1/2:

$$\vec{m}_{s1} = \frac{1}{2}, \text{ and } \vec{m}_{s2} = \frac{1}{2}, \quad M_s = 1$$

- If we can estimate the value of M_S and M_L quantum numbers, we can write down the L and S quantum numbers.

d. The $\vec{L}$ and $\vec{S}$ vectors couple to produce a total angular momentum $\vec{J}$ (Fig. 2.25).
J: represents the total angular momenta, Russell–Saunders or L-S coupling, to assign all J values
$J = L+S, L+S-1, \ldots\ldots\ldots, L-S$
$2J+1 =$ number of microstates in each term, and $M_J = |M_L + M_S|$
It should be noted that J could never be less than zero.
For the cases where S is less than L, there are $(2S+1)$ values of J.
When L is less than S, there are $(2L+1)$ values of J.

- To identify the energy order of these terms, we need to consider Hund's rules.

Exercise:

To which configuration does each of the following antisymmetric wave functions (written as abbreviated Slater Determinants) belong?

$$\left|\left(100\tfrac{1}{2}\right)\left(100\tfrac{\bar{1}}{2}\right)\left(200\tfrac{1}{2}\right)\left(21\bar{1}\tfrac{1}{2}\right)\right|$$

$$\left|\left(100\tfrac{1}{2}\right)\left(100\tfrac{\bar{1}}{2}\right)\left(210\tfrac{1}{2}\right)\left(21\bar{1}\tfrac{1}{2}\right)\right|$$

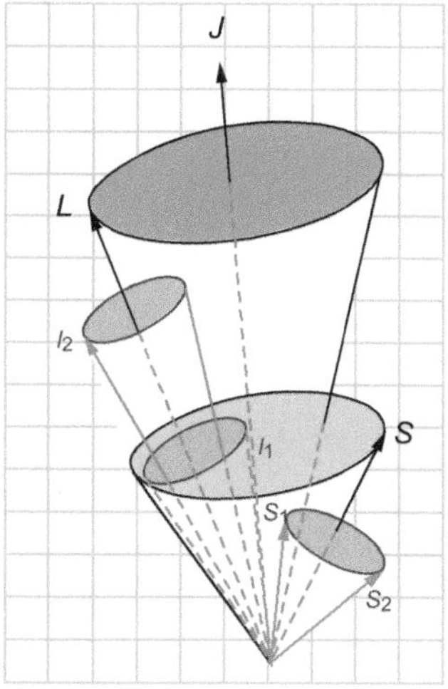

FIG. 2.25 An example of Russell-Saunders coupling.

$$\left|\left(100\frac{1}{2}\right)\left(200\frac{1}{2}\right)\left(200\frac{\bar{1}}{2}\right)\left(211\frac{1}{2}\right)\right|$$

Give the values of M_L and M_s for each of the above functions.

Answer:

- Slater determinant $= \left|(n/m_l s)(n'/m_l' s')\right|$
- $\left|\left(100\frac{1}{2}\right)\left(100\frac{\bar{1}}{2}\right)\left(200\frac{1}{2}\right)\left(21\bar{1}\frac{1}{2}\right)\right|$ $1s^2 2s2p$

  $M_L = -1 \quad M_S = 1$
- $\left|\left(100\frac{1}{2}\right)\left(100\frac{\bar{1}}{2}\right)\left(210\frac{1}{2}\right)\left(21\bar{1}\frac{1}{2}\right)\right|$ $1s^2 2p^2$

  $M_L = -1 \quad M_S = 1$
- $\left|\left(100\frac{1}{2}\right)\left(200\frac{1}{2}\right)\left(200\frac{\bar{1}}{2}\right)\left(211\frac{1}{2}\right)\right|$ $1s2s^2 2p$

  $M_L = 1 \quad M_S = 1$

Exercise

What is wrong with a function like this one?

$$\left|\left(100\frac{1}{2}\right)\left(100\frac{\bar{1}}{2}\right)\left(200\frac{1}{2}\right)\left(201\frac{\bar{1}}{2}\right)\right|$$

Answer

There cannot be an orbital like $\left(201\frac{\bar{1}}{2}\right)$, because m_l cannot exceed l.

For $l=0 \rightarrow m_l=0$

2.23 TERM STRUCTURE FOR POLYELECTRON ATOMS

Find out the total number of different ways for electron, *e*, to be placed in a given set of *n* orbital sites, and the number of terms of d^5.

- The electronic configurations of atoms or ions decide the number and kind of energy states.
- The energy states of atoms and ions are referred to as terms.

- As a number of electrons increases beyond two, the number of terms increases.
- The total number of terms for ion or atom in a magnetic field corresponds to the total number of different ways that e can be placed in a given set of n orbital sites ($n/2$ orbitals), and is expressed as follows:

$$\text{Number of microstates} = \frac{n!}{e!h!} \tag{2.23.1}$$

s-orbital: $n=2$, p-orbital: $n=6$, d-orbital: $n=10$, where h is the number of "holes" (sites available) and is equal to $(n-e)$. For d^5, the number of microstates $=252$.

$$\text{Number of microstates} = \frac{10!}{5! \times 5!} = \frac{10 \times 9 \times 8 \times 7 \times 6 \times 5!}{5 \times 4 \times 3 \times 2 \times 1 \times 5!} = 252$$

How can you identify the term symbols of the ground state and the different terms of the excited microstates of polyelectronic atom?

- The term symbols of the ground state and the different terms of the excited microstates can be identified according to Hund's rules: "An atom in its ground state adopts a configuration with the greatest number of unpaired electrons." Therefore:
  - Firstly, for an atom with partially occupied degenerate orbitals, the ground state is that with maximum spin momentum (S_{max}) and the maximum orbital momentum (L_{max}).
  - Secondly, if more than one term is obtained with the same largest value of S, that term with the largest L lies lowest. In other words, if there are two states with the highest spin multiplicity, the one of highest orbital multiplicity (L_{max}) will be of lower energy.
  - Finally, consider the spin-orbital interaction.
    - If the partially filled subshell is less than half-filled, the term with lowest J, $(L-S)$, value lies lowest in energy.
    - If the subshell is more than half-filled, the term with largest J, $(L+S)$, lies lowest in energy.
    - For exactly half-filled subshells, $L=0$ and only one value of J, $(=S)$, is possible.
- The term (state) is symbolized with spin multiplicity $(2S+1)$ and total orbital momentum L by: ${}^{2S+1}L$
- An analogous alphabetic code is used for L:

$$\begin{aligned} L = \; & 0 \rightarrow S \text{ term} \\ & 1 \rightarrow P \\ & 2 \rightarrow D \\ & 3 \rightarrow F \\ & 4 \rightarrow G \end{aligned}$$

Exercise

Identify the term symbols of the ground state and the different excited microstates of carbon atom, $2P^2$.

Answer

$$\text{Number of microstates} = \frac{n!}{e!h!} \quad (2.23.1)$$

$$\text{Number of microstates} = \frac{n!}{e!h!} = \frac{6 \times 5 \times 4!}{2 \times 1 \times 4!} = 15$$

- In order to identify the term symbols of the ground state and the different excited microstates of carbon atom, $2P^2$:
  - we firstly detect all probable microstates; then
  - categorize them by (M_L and M_S) values; and
  - trace the allowed L and S quantum numbers (Fig. 2.25).
- M_L and M_S are directly connected to the physical meaning of the quantum numbers, so these are used to describe terms rather than directly through L and S quantum numbers.

There are many distinct ways that two electrons can be assigned to three p orbitals (Fig. 2.26).

- Applying Hund's rules, the ordering of higher energy (Fig. 2.27):
  ${}^3P < {}^1D < {}^1S$
  The ground state is ${}^3P$, which splits by spin-orbital coupling in to: ${}^3P_2$, ${}^3P_1$, ${}^3P_0$.
- The states which arise by Russell-Saunders coupling from all p^n, d^n configurations are listed in Table 2.13.

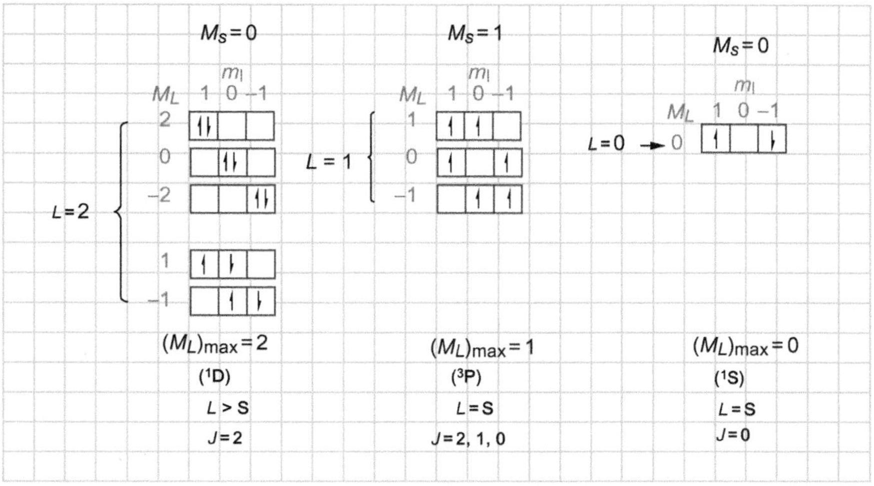

FIG. 2.26 The different microstates of carbon atom, $2P^2$

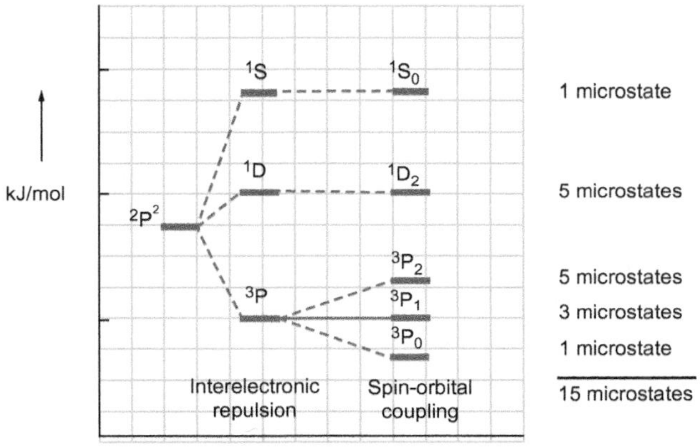

FIG. 2.27 The ordering of the microstates according to Hund's rules

TABLE 2.13 Russell-Saunders Coupling From All p^n, d^n Configurations

Configuration	Russell-Saunders States	Ground Term
p^1, p^5	^{2}P	^{2}P
p^2, p^4	(^{1}S, ^{1}D), ^{3}P	^{3}P
p^3	(^{2}P, ^{2}D), ^{4}S	^{4}S
p^0, p^6, d^0, d^{10}	^{1}S	^{1}S
d^1, d^9	^{2}D	^{2}D
d^2, d^8	(^{1}S, ^{1}D, ^{1}G), (^{3}P, ^{3}F)	^{3}F
d^3, d^7	^{2}D, (^{2}P, ^{2}D, ^{2}F, ^{2}G, ^{2}H), (^{4}P, ^{4}F)	^{4}F
d^4, d^6	(^{1}S, ^{1}D, ^{1}G), (^{1}S, ^{1}D, ^{1}F, ^{1}G, ^{1}I), (^{3}P, ^{3}F), (^{3}P, ^{3}D, ^{3}F, ^{3}G, ^{3}H), ^{3}D	^{3}D
d^5	^{2}D, (^{2}P, ^{2}D, ^{2}F, ^{2}G, ^{2}H), (^{2}S, ^{2}D, ^{2}F, ^{2}G, ^{2}I), (^{4}P, ^{4}D, ^{4}F, ^{4}G), ^{6}S	^{6}S

Exercise

a. **Give the spectroscopic symbols for all the terms of hydrogen atom for $n = 3$ state.**
b. **How do the energies of these terms compare when spin-orbit coupling is neglected?**
c. **How many subterms arise from each of the above terms when spin-orbit coupling is taken into consideration?**
d. **Indicate the degeneracy of each subterm.**

Answer:

a. Possible configurations are:

$3s^1$	$3p^1$	$3d^1$
From 3s		arises the term ^2S
From 3p		3p arises the term ^2P
From 3d		arises the term ^2D

b. They all have the same energy because in H atom exchange interaction is zero.
c. Two subterms arises from ^2D and ^2P:

$$^2D \rightarrow \ ^2D_{5/2} \ \text{and} \ ^2D_{3/2}$$

$$^2P \rightarrow ^2P_{3/2} \ \text{and} \ ^2P_{1/2}$$

One subterm arises from ^2S:

$$^2S \rightarrow ^2S_{1/2}$$

d. Degeneracies are:
$^2D_{5/2} \rightarrow$ 6 fold
$^2D_{3/2} \rightarrow$ 4 fold
$^2P_{3/2} \rightarrow$ 4 fold
$^2P_{1/2} \rightarrow$ 2 fold
$^2S_{1/2} \rightarrow$ 2 fold
Total = 18 fold
$n = 3$ level is 18 fold degenerate.

Exercise

To which of the following terms

$$\left|\left(100\tfrac{1}{2}\right)\left(100\tfrac{\overline{1}}{2}\right)\left(210\tfrac{1}{2}\right)\left(211\tfrac{1}{2}\right)\right|$$

can: ^2D, ^3P, ^4F, or ^1S not possibly belong?

Answer:

$$\left|\left(100\tfrac{1}{2}\right)\left(100\tfrac{\overline{1}}{2}\right)\left(210\tfrac{1}{2}\right)\left(211\tfrac{1}{2}\right)\right|$$

$$M_L = 1 \ \text{so} \ L \geq 1$$

$$M_S = 1 \ \text{so} \ S \geq 1$$

Therefore

$$2S + 1 \geq 3$$

^2D and ^1S do not satisfy these conditions.
Furthermore, it cannot belong to ^4F, because $\sum l = 2$, so L cannot be larger than 2.
Thus, it belongs to ^3P term

Exercise

A certain state of H atom is describe by the function

a. $\psi(r,\theta,\phi) = R(r)\Theta(\theta)\Phi(\phi)$

$$\Phi = \frac{1}{\sqrt{2\pi}}e^{2i\phi}$$

$$\Theta = 1-\cos^2\theta$$

$$R = \rho^l e^{-\frac{\rho}{2}}L(\rho)$$

The last term in $L(\rho)$ depends on the 2nd power of (ρ^2). What state of H does ψ describe?

b. **What is the spectroscopic symbol for this state?**

Answer:

$$\because \Phi = \frac{1}{\sqrt{2\pi}}e^{im\phi} \quad (2.6.2)$$

$$\therefore \Phi = \frac{1}{\sqrt{2\pi}}e^{im\phi} = \frac{1}{\sqrt{2\pi}}e^{2i\phi}$$

$$\therefore m=2,$$

$$\Theta_{l,\,m}(\theta) = (1-\cos^2\theta)^{\frac{|m|}{2}}\sum_{\substack{j=0,\,2\ldots\\ j=1,\,3\ldots}}^{l-|m|} a_j \cos^j\theta = 1-\cos^2\theta \quad (2.7.11)$$

$$\therefore a_0 = 1, \text{ and } j' = 0$$

$$j = l-|m| \quad 2.7.9$$

$$l = j' + |m| = 0+2 = 2$$

$$n = l + 1 + \upsilon' \quad (2.8.14)$$

$$L(\rho) = a_0 + a_1\rho + a_2\rho^2 \quad (2.8.7)$$

$$L(\rho) = a_0 + a_1\rho + a_2\rho^2$$

$$\upsilon' = 2$$

$$n = 2+1+2 = 5$$

The function describes $5d_{+2}$.
The spectroscopic symbol is ^2D.

Exercise

a. **Write all the wave functions arising from the $1s^2\,2s^2\,2p^6\,3s^1\,3p^1$ configuration of Mg as abbreviated Slater determinant, D_i.**
b. **Indicate the values of M_L and M_s for each wave function.**
c. **Give the spectroscopic symbols of all the terms arising from this [Ne] $3s^1\,3p^1$ configuration. Give the degeneracy of each term. If a function cannot uniquely be assigned to a particular L and S value, indicate this fact by a star.**

Answer:

$$3s \equiv 300$$

$$3p \equiv 311, 310, 31\bar{1}$$

TABLE 2.14 Arrangement of Terms According to M_L and M_S

	M_S		
M_L	1	0	−1
1	D_1	D_2D_7	D_8
0	D_3	D_4D_9	D_{10}
−1	D_5	D_6D_{11}	D_{12}

- All the wave functions arising from the electronic configuration of Mg as Slater determinant, D_i, and the M_L and M_s number of each wave function:

$$D_1 = \left|300\frac{1}{2}\ 311\frac{1}{2}\right| \quad M_L=1\ \ M_s=1$$

$$D_2 = \left|300\frac{1}{2}\ 311\frac{\bar{1}}{2}\right| \quad M_L=1\ \ M_s=0$$

$$D_3 = \left|300\frac{1}{2}\ 310\frac{1}{2}\right| \quad M_L=0\ \ M_s=1$$

$$D_4 = \left|300\frac{1}{2}\ 310\frac{\bar{1}}{2}\right| \quad M_L=0\ \ M_s=0$$

$$D_5 = \left|300\frac{1}{2}\ 31\bar{1}\frac{1}{2}\right| \quad M_L=-1\ \ M_s=1$$

$$D_6 = \left|300\frac{1}{2}\ 31\bar{1}\frac{\bar{1}}{2}\right| \quad M_L=-1\ \ M_s=0$$

$$D_7 = \left|300\frac{\bar{1}}{2}\ 311\frac{1}{2}\right| \quad M_L=1\ \ M_s=0$$

$$D_8 = \left|300\frac{\bar{1}}{2}\ 311\frac{\bar{1}}{2}\right| \quad M_L=1\ \ M_s=-1$$

$$D_9 = \left|300\frac{\bar{1}}{2}\ 310\frac{1}{2}\right| \quad M_L=0\ \ M_s=0$$

$$D_{10} = \left|300\frac{\bar{1}}{2}\ 310\frac{\bar{1}}{2}\right| \quad M_L=0\ \ M_s=-1$$

$$D_{11} = \left|300\frac{\bar{1}}{2}\ 31\bar{1}\frac{1}{2}\right| \quad M_L=-1\ \ M_s=0$$

$$D_{12} = \left|300\frac{\bar{1}}{2}\ 31\bar{1}\frac{\bar{1}}{2}\right| \quad M_L=-1\ \ M_s=1$$

- The wave functions terms are arranged according M_L and M_s (Table 2.14).
- The spectroscopic symbols of all the terms arising from this [Ne] $3s^1$ $3p^1$ configuration, and the degeneracy of each term is

$$^3P:\ \underbrace{D_1D_3D_5}_{M_S=1}\ \underbrace{D_2^*D_4^*D_6^*}_{M_S=0}\ \underbrace{D_8D_{10}D_{12}}_{M_S=-1}$$

$$^1P: D_7^*D_9^*D_{11}^*$$

Exercise

Antisymmetric wave function, D_i, arising from p^3 configuration can be arranged according to their M_L and M_s numbers (Table 2.15).

a. **Group these functions according to their L and S numbers. If a function cannot uniquely be assigned to particular L and S value, indicate this fact by a star.**

Answer

$L=2$	$S=1/2$	M_L range from 2 to -2	
		$D_1, D_3^*, D_8^*, D_{15}^*, D_{19}$	$M_S=1/2$
		$D_2, D_5^*, D_{11}^*, D_{17}^*, D_{20}$	$M_S=-1/2$
$L=1$	$S=1/2$	M_L range from 1 to -1	
		D_4^*, D_9^*, D_{16}^*	$M_S=1/2$
		$D_6^*, D_{12}^*, D_{18}^*$	$M_S=-1/2$
$L=0$	$S=3/2$	M_S values range from +3/2 to $-3/2$	
		$D_7, D_{10}^*, D_{13}^*, D_{14}$	$M_S=3/2$

b. **Write out the function D_7 in the abbreviated Slater determinant notation.**

$$M_S=1/2+1/2+1/2=3/2$$

$$M_L=\cdots+\cdots+\cdots=0$$

$$M_L=1+0+\overline{1}=0, L=1$$

$$D_7=\left|\left(n11\frac{1}{2}\right)\left(n1\overline{1}\frac{1}{2}\right)\left(n10\frac{1}{2}\right)\right|$$

c. **Write the spectroscopic symbols of the terms arising from the P^3 configuration and arrange them qualitatively on an energy level diagram**

$L=2 \quad S=1/2 \quad \rightarrow \quad {}^2D$

$L=1 \quad S=1/2 \quad \rightarrow \quad {}^2P$

$L=0 \quad S=3/2 \quad \rightarrow \quad {}^4S$

${}^4S \quad {}^2D \quad {}^2P$

$\Rightarrow$
energy

TABLE 2.15 Terms Arranged According to M_L and M_S

	M_S			
M_L	3/2	1/2	−1/2	−3/2
2		D_1	D_2	
1		D_3, D_4	D_5, D_6	
0	D_7	D_8, D_9, D_{10}	D_{11}, D_{12}, D_{13}	D_{14}
−1		D_{15}, D_{16}	D_{17}, D_{18}	
−2		D_{19}	D_{20}	

Exercise

A term is split into four subterms when the spin-orbital interaction is considered. Which of the following can be the spectroscopic symbol of this term: ^4P, ^3D, ^1F, ^3F, and ^4F?

Answer:

^4P	$S=3/2$	$L=1$	splits into 3
^3D	$S=1$	$L=2$	splits into 3
^1F	$S=0$	$L=3$	does not split
^3F	$S=1$	$L=3$	splits into 3
^4F	$S=3/2$	$L=3$	splits into 4

For ^4F, $J=|L+S|$ to $|L-S| = \frac{9}{2}, \frac{7}{2}, \frac{5}{2}, \frac{3}{2}$

Exercise

Give the spectroscopic symbol of a term which has a multiplicity of 3 (triplet), but which does not split under a spin-orbit interaction.

Answer:

$2S+1=3$

$S=1$, no spin-orbital interaction, $L=0$, the spectroscopic term: ^3S.

What is the Slater determinant wave functions formed from single electron orbital wave functions $(m_{l_1}, m_{l_2}, m_{l_3}, \ldots)$?

Write the single electron orbital wave functions (2, −1) as a Slater determinant.

What is the linear combination of the single electron orbital wave functions $(m_{l_1}, m_{l_2}, \ldots)$ and $\left(m'_{l_1}, m'_{l_2}, \ldots\right)$? Find the value of L_z.

What is the linear combination of the single electron spin wave functions $(m_{s_1}, m_{s_2}, \ldots)$ and $\left(m'_{s_1}, m'_{s_2}, \ldots\right)$? Find the value of S_z.

- The single electron wave function $(m_{l_1}, m_{l_2}, m_{l_3}, \ldots)$ has in Slater determinant:

$$\underbrace{(m_{l_1}, m_{l_2}, m_{l_3}, \ldots)}_{\substack{\text{Single electron wave functions}\\ \text{in which one electron has}\\ \text{a quantum number}\\ m_{l_1}\text{, the next } m_{l_2}\ldots}} = \frac{1}{\sqrt{n!}} \begin{vmatrix} m_{l_1}^1 & m_{l_1}^2 & m_{l_1}^3 & \cdots \\ m_{l_2}^1 & m_{l_2}^2 & m_{l_2}^3 & \cdots \\ m_{l_3}^1 & m_{l_3}^2 & m_{l_3}^3 & \cdots \\ \vdots & \vdots & \vdots & \cdots \end{vmatrix}$$

The superscript is the numbering of the electrons, for example, when $m_{l_1}^1 = 2$, and $m_{l_1}^2 = -1$:

$$(2, -1) = \frac{1}{\sqrt{2}} \begin{vmatrix} (2)^1 & (2)^2 \\ (-1)^1 & (-1)^2 \end{vmatrix} = \frac{1}{\sqrt{2}} \left[(2)^1(-1)^2 - (-1)^1(2)^2\right]$$

- $\langle M_L \rangle$ is a linear combination of the single electron orbital wave functions:

$$\langle M_L \rangle = c_1 (m_{l_1}, m_{l_2}, \ldots) + c_2 \left(m'_{l_1}, m'_{l_2}, \ldots\right)$$

$$c_1 c_1^* + c_2 c_2^* + \cdots = 1$$

$\langle M_L \rangle$ $\left(\text{in } \frac{h}{2\pi} \text{ unites}\right)$ is the observed value of L_z:

$$L_z = \int \langle M_L \rangle^* L_z \langle M_L \rangle = \langle \langle M_L \rangle^* | L_z | \langle M_L \rangle \rangle = M_L \frac{h}{2\pi}$$

- $\langle M_S \rangle$ is a linear combination of the single electron spin wave functions:

$$[M_S] = c_1(m_{s_1}, m_{s_2}, \ldots) + c_2\left(m'_{s_1}, m'_{s_2}, \ldots\right)$$

$$c_1c_1^* + c_2c_2^* + \cdots = 1$$

$$S_z = \int [M_S]^* S_z [M_S] = \langle M_S{}^* | S_z | M_S \rangle = M_S \frac{h}{2\pi}$$

2.24 TERM WAVE FUNCTIONS AND SINGLE ELECTRON WAVE FUNCTIONS

For d^2 configuration:

- **Set up a chart of macrostates. If a function cannot uniquely be assigned to particular *L* and *S* value; indicate this fact by a star.**
- **Resolve this group of microstates into the corresponding single electron wave function.**
- **Why is it necessary to know how to express the term wave functions of free ion into single electron wave functions?**
- **Compose the complete orbital wave function combinations for ^{3}F and ^{3}P terms and the spin wave functions of d^2.**

Answer:

- Number of microstates $= \dfrac{n!}{e!h!}$ (2.23.1)

$$N = \frac{n!}{e!\,h!} = \frac{10!}{(2!)(8!)} = 45$$

- For d-orbital:

  $m_l = 2, 1, 0, -1, -2$

- The maximum M_L for two electrons is 4.
- The allowed M_s is 1,0,−1.
- The Pauli allowed microstates are shown in Table 2.16.
- The d^2 configuration gives ^{1}G, ^{3}F, ^{1}D, ^{3}P, ^{1}S (45 microstates, ^{3}F: ground state term).
- It is essential to know how to express the term wave functions in order to study effect of the ligand field on the energy levels of the free ion (terms).
- The single electron wave functions and the associated values of M_L and M_S for the configuration d^2 are listed in Table 2.15.
- First: the wave function term is only one combination of the single electron wave functions.
  - The wave function for the component of the ^{1}G term with $L=4$, and $M_L=4$ is

$$\langle L, M_L \rangle [S, M_S] = \langle 4, 4 \rangle [0, 0] = \left(2\frac{1}{2}, 2 -\frac{1}{2}\right) \rightarrow$$

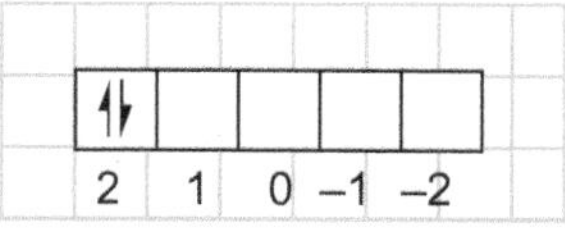

The ^{3}F term with $L=3$, $M_L=3$ and $M_L=2$, both $M_s=1$ are

$$\langle L, M_L \rangle [S, M_S] = \langle 3, 3 \rangle [1, 1] = \left(2\frac{1}{2}, 1\frac{1}{2}\right) \rightarrow$$

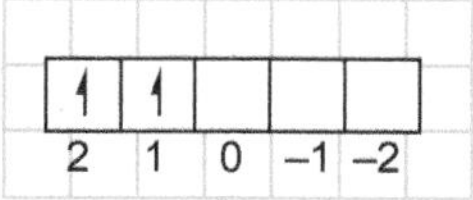

$$\langle L, M_L \rangle [S, M_S] = \langle 3, 2 \rangle [1, 1] = \left(2\frac{1}{2}, 0\frac{1}{2}\right) \rightarrow$$

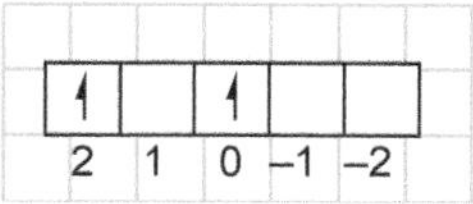

There is only one combination of the single electron wave functions.

TABLE 2.16 The Single Electron Wave Functions and the Associated Values of M_L and M_S for the Configuration d^2

	M_S		
M_L	1	0	−1
4	–	(2 ½, 2 − ½)[a]	–
3	(2½, 1 ½)	(2 ½, 1 − ½)*, (2 − ½, 1 ½)*	(2 − ½, 1 − ½)
2	(2 ½, 0 ½)	(2½,0 − ½)*,(2 − ½,0½)*,(1 ½,1 − ½)*	(2 − ½, 0 − ½)
1	(2 ½, −1 ½)*, (1 ½, 0 ½)*	(2½,−1 − ½)*,(2 − ½,−1 ½)*,(1 ½,0 − ½)*, (1 − ½, 0½)*	(2 − ½,−1 − ½)*, (1 − ½, 0 − ½)*
0	(2 ½, −2½)*, (1 ½, −1 ½)*	(2½,−2 − ½)*,(2 − ½,2½)*,(1 ½,−1 − ½)*, (1 − ½,−1 ½)*,(0½,0 − ½)*	(2 − ½,-2 − ½)*, (1 − ½,−1 − ½)*
−1	(−2 ½, 1 ½)*, (−1 ½, 0 ½)*	(−2½,1 − ½)*,(−2 − ½,1 ½)*,(−1 ½,0 − ½)*, (−1 − ½, 0½)*	(−2 − ½,1 − ½)*, (−1 − ½, 0 − ½)*
−2	(−2 ½, 0 ½)	(−2½,0 − ½)*,(−2-½,0½)*,(−1 ½,−1 − ½)*	(−2 − ½, 0 − ½)
−3	(−2½, −1 ½)	(−2 ½, −1 − ½)*, (−2 − ½, −1 ½)*	(−2−½, −1 − ½)
−4	–	(−2 ½, −2 − ½)	–

a: $M_L = 4$, $M_s = 0$ corresponds to (2½, 2 − ½) ⟶

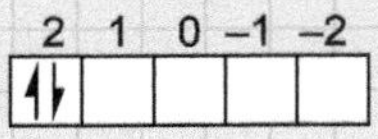

- Secondly: the wave function term is more than one combination of single electron wave function, examples:
  - The terms ^{1}G and ^{3}F with $M_L=3$, $M_S=0$:

$$\langle M_L\rangle[M_S] = \langle 3\rangle[0] = \left(2\frac{1}{2}, 1-\frac{1}{2}\right), \left(2-\frac{1}{2}, 1\frac{1}{2}\right)$$

The terms ^{3}P and ^{3}F with $M_L=3$, $M_S=0$:

$$\langle M_L\rangle[M_S] = \langle 1\rangle[1] = \left(2\frac{1}{2}, -1\frac{1}{2}\right), \left(1\frac{1}{2}, 0\frac{1}{2}\right)$$

- The complete orbital combinations for an F term from d^2:

$$\langle M_L\rangle = \langle \pm 3\rangle = \pm(\pm 2, \pm 1)$$

$$\langle \pm 2\rangle = \pm(\pm 2, 0)$$

$$\langle \pm 1\rangle = \underbrace{\pm\sqrt{\frac{2}{5}}\underbrace{(\pm 1, 0)}_{\substack{\text{between}\\ ^3\text{F and }^3\text{P}}} \pm \sqrt{\frac{3}{5}}\underbrace{(\pm 2, \mp 1)}_{\substack{\text{between}\\ ^3\text{F and }^3\text{P}}}}_{\text{linear combination}}$$

$$\langle 0\rangle = \underbrace{\sqrt{\frac{4}{5}}\underbrace{(1, -1)}_{\substack{\text{between}\\ ^3\text{F and }^3\text{P}}} + \sqrt{\frac{1}{5}}\underbrace{(2, -2)}_{\substack{\text{between}\\ ^3\text{F and }^3\text{P}}}}_{\text{linear combination}}$$

The sum of the squares of the coefficient that appear in $\langle \pm 1\rangle$ and $\langle 0\rangle$ in the wave functions of ^{3}F and ^{3}P must be unity.

- Similarly, the spin wave functions for the triplet term from d^2 are, for the ^{3}P term,

$$\langle \pm 1\rangle = \mp\sqrt{\frac{3}{5}}(\pm 1, 0) \pm \sqrt{\frac{2}{5}}(\pm 2, \mp 1)$$

$$\langle 0\rangle = -\sqrt{\frac{1}{5}}(1, -1) + \sqrt{\frac{4}{5}}(2, -2)$$

- In the same way, the spin wave function for a singlet term of d^2 is

$$[M_S]=[\pm 1]=\left(\pm\frac{1}{2},\pm\frac{1}{2}\right).$$

$$[0]=\sqrt{\frac{1}{2}}\left[\left(\frac{1}{2},-\frac{1}{2}\right)+\left(-\frac{1}{2},\frac{1}{2}\right)\right]$$

$$[0,0]=\sqrt{\frac{1}{2}}\left[\left(\frac{1}{2},-\frac{1}{2}\right)-\left(-\frac{1}{2},\frac{1}{2}\right)\right]$$

What are the single electron wave functions and the other terms that allow ^1S term wave function of d^2 to be a linear combination? (Use Table 2.16.)

Answer:

The single electron wave functions that ^1S term wave function of d^2 in a linear combination are

$(2½,-2-½)$, $(2-½,2½)$, $(1½,-1-½)$, $(1-½,-1½)$, and $(0½,0-½)$

The other terms that allow ^1S term wave function of d^2 in a linear combination are

$\langle 3,0\rangle$, $\langle 1,0\rangle[1,0]$, $\langle 4,0\rangle[0,0]$, and $\langle 2,0\rangle[0,0]$

2.25 SPIN-ORBITAL COUPLING

How does the spin-orbital coupling of heaver atoms ($Z\geq 30$) differ from that of light atoms?

a. For light atoms, splitting between terms with different values of J is typically small and occurs only in a magnetic field.
b. Russell-Saunders coupling is a good description of atoms having $Z\leq 30$.
c. Each term is split up into quantum number, J, in the Russell-Saunders coupling scheme. Each state is $(2J+1)$-fold degenerate.
d. In heaver atoms, $Z\geq 30$, a different coupling scheme is observed in which the orbital and the spin angular momenta of individual electrons first couple, giving $\vec{j}$ for each electron. The individual $\vec{j}$s are then coupled, producing an overall $\vec{j}$.

Show that:

a. $E_{j,l,s}=\frac{1}{2}\xi[j(j+1)-l(l+1)-s(s+1)]$ **measures the energy of the Russell-Saunders terms due to the interaction between the spin and orbital angular momenta of a single electron of a particular microstate ($\xi=$ constant);**
b. $E_{J,L,S}=\frac{1}{2}\lambda[J(J+1)-L(L+1)-S(S+1)]$ **is the corresponding property of the term ($\lambda=$ constant); and**
c. $\Delta E_{J,J+1}=\lambda(J+1)$ **is the energy difference between two adjacent spin-orbital states.**
d. **Define ξ and λ.**

a. The total angular momentum of an electron is described by the quantum numbers j and m_j, with $j=l+s$ (when two angular momenta are in the same direction) or $j=l-s$ (when they are opposed); see Ch. 2, p. 119, 122.
   - The energy of electron has magnetic moment μ in a magnetic field H is equal to their scalar product $(-\mu\cdot H)$.

$$E=-\underbrace{\mu}_{\propto s}\cdot\underbrace{H}_{\propto 1} \qquad (2.25.1)$$

   - If the magnetic field, H, arises from the orbital angular momentum of the electron, then it is proportional to l, and the magnetic moment μ is that of the electron spin, then it is proportional to s.
     - Subsequently, the energy is also proportional to $s\cdot l$

$$-\mu\cdot H\propto s\cdot l \qquad (2.25.2)$$

$$\therefore E \propto s \cdot l, \text{ and}$$
$$E = \text{constant} \cdot (s \cdot l)$$
$$\text{constant} = \xi$$
$$E = \xi \cdot (s \cdot l) \qquad (2.25.2)$$

If: $J = l + s$

Then $J \cdot J = (l+s) \cdot (l+s) = l \cdot l + s \cdot s + 2s \cdot l$

$$s \cdot l = \frac{1}{2}(J^2 - l^2 - s^2)$$

where we have used the fact that the scalar product of two vectors: $u \cdot \mu = u\mu \cos\theta$

- In quantum mechanics, we treat all the quantities as operators:

$$\hat{s} \cdot \hat{l} = \frac{1}{2}(\hat{j}^2 - \hat{l}^2 - \hat{s}^2) \qquad (2.25.3)$$

Then

$$\left\langle \vec{l}\ \vec{s} \right\rangle = \langle (j,l,s)|\hat{s}.\hat{l}|(j,l,s)\rangle$$

$$\begin{aligned}
\langle j,l,s|\hat{s}.\hat{l}|j,l,s\rangle &= \frac{1}{2}\left\langle j,l,s\left|\hat{j}^2 - \hat{l}^2 - \hat{s}^2\right|j,l,s\right\rangle \\
&= \frac{1}{2}\left[\left\langle j,l,s\left|\hat{j}^2\right|j,l,s\right\rangle - \left\langle j,l,s\left|\hat{l}^2\right|j,l,s\right\rangle - \langle j,l,s|\hat{s}^2|j,l,s\rangle\right] \\
&= \frac{1}{2}\left[\left\langle j\left|\hat{j}^2\right|j\right\rangle\langle l|l\rangle\langle s|s\rangle - \left\langle l\left|\hat{l}^2\right|l\right\rangle\langle j|li\rangle\langle s|s\rangle - \langle s|\hat{s}^2|s\rangle\langle j|j\rangle\langle l|l\rangle\right] \\
&= \frac{1}{2}[j(j+1) - l(l+1) - s(s+1)]\hbar^2
\end{aligned} \qquad 2.25.4$$

- By inserting this expression into the formula for the energy of interaction:
  The operator is $\xi\ \vec{l}\ \vec{s}$.
  Then

$$\therefore E_{j,l,s} = \frac{1}{2}\xi[j(j+1) - l(l+1) - s(s+1)] \qquad (2.25.5)$$

- The parameter ξ is used to describe the spin-orbital coupling generally for single electron.
- It measures the strength of the interaction between the spin and orbital angular momenta of a single electron of a particular microstate; it is a property of microstate and not of the term.
- The value of ξ is given by

$$\xi = \frac{Z_{eff} e^2}{2m^2c^2} r^{-3} \qquad (2.25.6)$$

where r^{-3} is the average value of r^{-3}, m is the mass of the electron, c is the speed of light, and Z_{eff} is the effective nuclear charge.

b. The parameter λ is used to describe the corresponding property of the term. The operator is $\lambda\ \vec{L}\ \vec{S}$.
- The values of λ and ξ are related by

$$\lambda = \pm\frac{\xi}{2S} \qquad (2.25.7)$$

  - The parameter ξ is a positive quantity.
  - The sign of λ is positive, if the shell is less than half-filled.
  - λ is negative, if the shell is more than half-filled.
  - Therefore, in the shell of less than half-filled, the lowest value of J corresponds to the lowest energy and λ is positive.

- A similar operator for $\hat{L}\cdot\hat{S}$:

$$\hat{L}\cdot\hat{S}=\frac{1}{2}(\hat{J}^2-\hat{L}^2-\hat{S}^2)$$

and the spin-orbital contribution to the energy:

$$E_{J,\,L,\,S}=\frac{1}{2}\lambda[J(J+1)-(L+1)-S(S+1)] \quad 2.25.8$$

The energy difference between two adjacent spin-orbital states in a term is given by

$$\Delta E_{J,J+1}=\frac{1}{2}\lambda[(J+1)(J+2)-L(L+1)-S(S+1)]-\frac{1}{2}\lambda[J(J+1)-L(L+1)-S(S+1)]$$
$$=\frac{1}{2}\lambda[(J+1)(J+2)-J(J+1)]=\frac{1}{2}\lambda[(J+1)(J+2-J)$$
$$\Delta E_{J,J+1}=\lambda(J+1) \quad (2.25.9)$$

2.26 SPIN-ORBITAL COUPLING IN EXTERNAL MAGNETIC FIELD

Show that:

a. **the coupling of the orbital angular momentum L with external magnetic field H has a potential energy:**

$$E=\beta LH,\quad \text{where } \beta=\frac{e\hbar}{2m_e}\ (\text{Bohr magneton})$$

b. **the coupling of the spin angular momenta S with external magnetic field H has a potential energy:**

$$E=\beta g_s SH$$

---

a. Answer:

- Any electron with orbital angular momentum ($l>0$) is actually a circulating current. This electron possesses:
  - a magnetic moment that arises from orbital momentum; and
  - a magnetic moment that arises from its spin.
- If an electron of charge-e is placed at the end of a vector r from the nucleus and is moving with velocity v:
  - the circulating current, I:

$$I=\frac{ev}{2\pi r} \quad (2.26.1)$$

  - the resulting magnetic moment, μ_m:

$$\mu_m=\frac{evr}{2} \quad (2.26.2)$$

  - the magnetic induction at the origin, $B(r)$, is given by:

$$B(r)=-\frac{\mu_o}{4\pi}\frac{e(r\cdot v)}{r^3} \quad (2.26.3)$$

If the orbital angular momentum, $\hbar\hat{L}=m_e r\cdot v$, and permeability of vacuum is μ_o, then

$$r\cdot v=\frac{\hat{L}\hbar}{m_e},\ \text{and}$$

$$B(r)=-\frac{\mu_o}{4\pi}\frac{e\hat{L}\hbar}{m_e r^3} \quad (2.26.4)$$

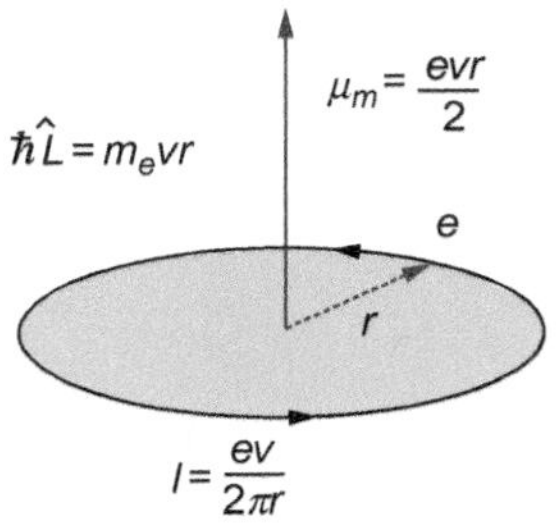

- The magnetic induction of a moving electron is equivalent to that of a tiny magnet with magnetic moment μ_L, the magnetic field of such magnet would be

$$B(r) = -\frac{\mu_o \mu_L}{2\pi r^3} \tag{2.26.5}$$

Then from Eqs. (2.26.4) and (2.26.5):

$$\therefore \frac{\mu_o \mu_L}{2\pi r^3} = \frac{\mu_o}{4\pi} \frac{e\hat{L}\hbar}{m_e r^3}$$

and

$$\mu_L = \hat{L}\frac{e\hbar}{2m_e} = \hat{L}\beta \tag{2.26.6}$$

$$\therefore \mu_L = \beta\sqrt{l(l+1)}$$

where $\beta = \frac{e\hbar}{2m_e}$ and is the natural unit of magnetic moment.
Or as a vector quantity:

$$\vec{\mu}_L = -\vec{L}\frac{e\hbar}{2m_e} = -\vec{L}\beta \tag{2.26.6}$$

The negative sign arising from the sign of the electron's charge shows that the orbital magnetic moment of the electron is antiparallel to its orbital angular momentum.

- The ratio of the magnetic moment to the angular momentum, called the magnetogyric ratio γ, is

$$\gamma_L = \frac{\mu_L}{\hat{L}} = \frac{e}{2m_e}$$

The vectors μ_L and L are parallel, directed along an axis normal to the plane of the current loop.

- Coupling of the orbital angular momentum with the external magnetic field H is the magnetic interaction of μ_L with H, which has a potential energy:

$$E = -\mu_L \cdot H\cos\theta,$$

where θ is the angle between the field direction z and the magnetic moment, when

$$\theta = 0°$$

$$E = -\mu_L \cdot H = \beta\hat{L}H \tag{2.26.7}$$

b. Answer

- The spin magnetic moment of an electron μ_S is also proportional to its angular momentum S. However, it is not given by $S\beta$, but by about twice this value:

$$\mu_S = (\text{about twice})\,\hat{S}\beta$$

$$\mu_S = g_s\beta\sqrt{s(s+1)} \tag{2.26.8}$$

$$\vec{\mu_S} = -g_s\,\vec{S}\,\beta$$

g_s is called the g-factor of the electron, $g_s = 2.0023$, and

$$E = -\mu_S \cdot H = \beta g_s SH \tag{2.26.9}$$

The magnetogyric ratio for electron spin is

$$g = \gamma_S = \frac{\mu_S}{S} = \frac{e}{m_e}$$

and is twice γ_L.

The effect of the external magnetic field H is to split each Russell-Saunders term into microstates. Prove that:

a. **the energy of the microstate E_J is**

$$E_J = gJH\beta, \quad \text{where } gJ = L + g_s S$$

b.
$$g = 1 + \frac{J(J+1) + S(S+1) - L(L+1)}{2J(J+1)}$$

---

a. Answer

- The interaction of the spin magnetic moment with the magnetic field arising from the orbital angular momentum is called spin-orbital coupling.
- Spin-orbital coupling is a magnetic interaction between spin and orbital magnetic moments, when angular momenta are parallel:

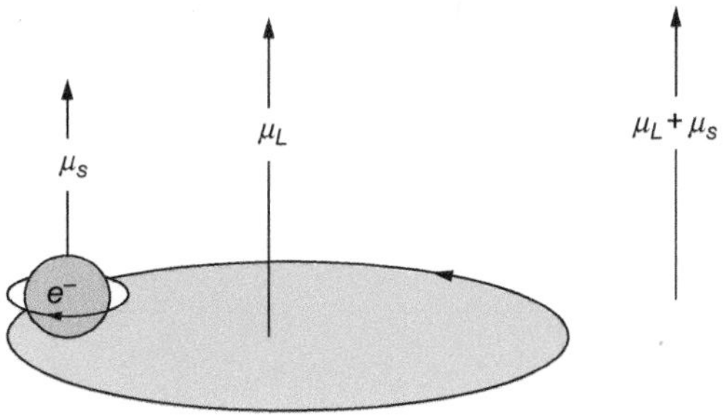

The magnetic moments are aligned (unfavorable).
When they are opposed:

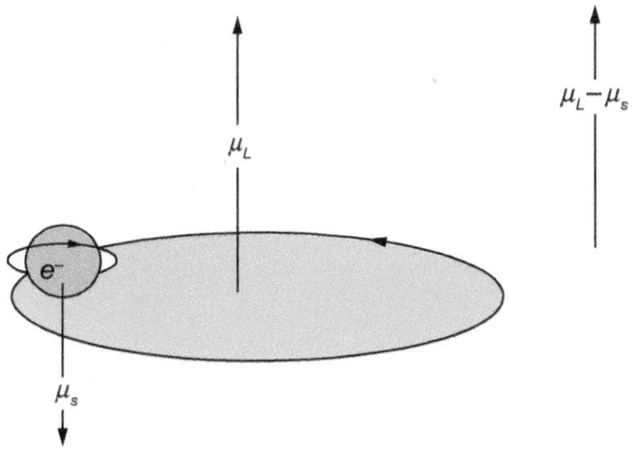

The interaction is favorable.

- The $\vec{L}$ and $\vec{S}$ vectors couple to produce a total angular momentum $\vec{J}$
  J: represents the total angular momenta, Russell-Saunders or L-S coupling.
  $J = |M_L + M_S|$, to assign all J values:

  $$J = L + S, L + S - 1, \ldots\ldots\ldots, L - S$$

  $2J+1$ = number of microstates in each term.
  It should be noted that J could never be less than 0; see Ch. 2, p. 119, 122.
- Applying a magnetic field, $H \neq 0$, the effect of H is to split each state of term, in turn.
  - Each component (J-level) is split into $2J+1$ equidistance levels.
  - Each level associated with magnetic quantum number m_j unit splits, $2J+1$. This leads to the Zeeman Effect (optical line split).
  - When $\Delta E = g\beta H \gg kT$, the magnetic coupling of

$$\mu_L = -L\beta \text{ of } E_L = LH\beta \quad (2.26.10)$$

and

$$\mu_S = -g_s S\beta \text{ of } E_S = g_s S H\beta \quad (2.26.11)$$

L-S coupling gives

$$E_J = E_L + E_S = LH\beta + g_s S H\beta$$

$$E_J = E_L + E_S = (L + g_s S)H\beta \quad (2.26.12)$$

$$\therefore E_J = gJH\beta$$

where

$$gJ = L + g_s S$$

b. Answer:

- In quantum mechanics, we treat all the quantities as operators:

$$\hat{L} + g_s\hat{S} = g\hat{J}, \quad \text{multiplies by } \hat{J}$$

$$\hat{L}\cdot\hat{J} + g_s\hat{S}\cdot\hat{J} = g\hat{J}^2, \quad \text{if } \hat{J}^2 = J(J+1)$$

$$\hat{L}\cdot\hat{J} + g_s\hat{S}\cdot\hat{J} = gJ(J+1) \quad (2.26.19)$$

By squaring $\hat{S} = \hat{J} - \hat{L}$ and $\hat{L} = \hat{J} - \hat{S}$, then

$$\hat{S}^2 = \hat{J}^2 - 2\,\hat{L}\hat{J} + \hat{L}^2$$

$$\hat{L}^2 = \hat{J}^2 - 2\,\hat{S}\hat{J} + \hat{S}^2$$

$$\therefore 2\hat{L}\hat{J} = \hat{J}^2 + \hat{L}^2 - \hat{S}^2$$

$$2\,\hat{L}\hat{J} = J(J+1) + L(L+1) - S(S+1)$$

$$\therefore \hat{L}\hat{J} = \frac{1}{2}(J(J+1) + L(L+1) - S(S+1)) \quad (2.26.13)$$

$$2\,\hat{S}\hat{J} = J(J+1) - L(L+1) + S(S+1)$$

$$\therefore \hat{S}\hat{J} = \frac{1}{2}(J(J+1) - L(L+1) + S(S+1)) \quad (2.26.14)$$

Substitute in

$$\hat{L}\cdot\hat{J} + g_s\hat{S}\cdot\hat{J} = gJ(J+1) \quad (2.26.19)$$

$$\therefore g = \frac{1}{2J(J+1)}\{(g_s+1)J(J+1) - (g_s-1)[L(L+1) - S(S+1)]\}$$

$$\because g_s \cong 2$$

then

$$g = \frac{3J(J+1) + S(S+1) - L(L+1)}{2J(J+1)}$$

$$\therefore g = 1 + \frac{J(J+1) + S(S+1) - L(L+1)}{2J(J+1)} \quad (2.26.15)$$

Example

Ground terms for d^n

$n=0$	1	2	3	4	5	6	7	8	9	10
^1S	^2D	^3F	^4F	^5D	^6D	^5D	^4F	^3F	^2D	^1S

What is the splitting of the ground state of d^3 in external magnetic field *H*?

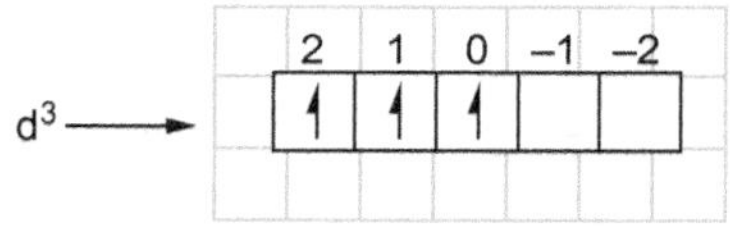

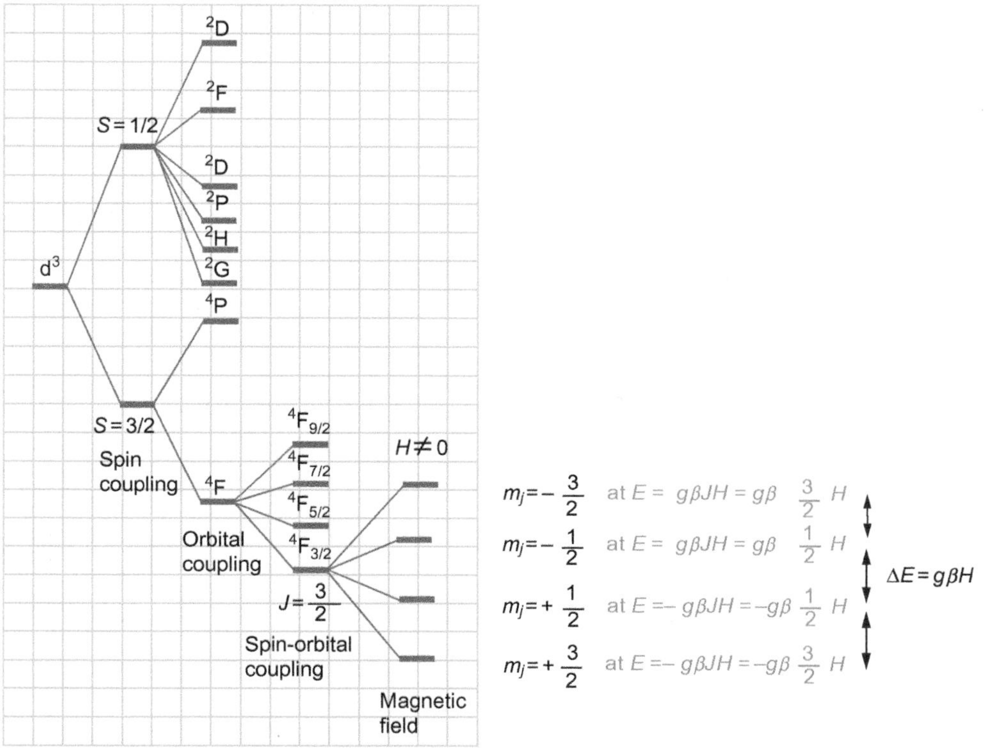

FIG. 2.28 The splitting of the ^{3}d configuration and $^4F_{3/2}$ term in the external magnetic field.

- $L = 3,\ S = \frac{3}{2}$
- $J = L + S, L + S - 1, \ldots\ldots, L - S = \frac{9}{2}, \frac{7}{2}, \ldots\ldots, \frac{3}{2}$
- In a shell that is less than half-filled, the lowest value of J corresponds to the lowest energy state, $J = \frac{3}{2}$. Applying the magnetic field, $H \neq 0$, the effect of H is to split each state of the term:

  $$^4F_{3/2} \overset{H}{\Rightarrow} -\frac{3}{2} < -\frac{1}{2} < +\frac{1}{2} < +\frac{3}{2}$$

  E—ranking (Fig. 2.28)

Example:

In d²: ^3F, ^1D, ^3P, ^1G, ^1S

What are the Russell–Saunders terms of d²-configuration and the splitting of ground state term by magnetic field H_o?

- The ground state of d^2: ^3F:
  $L = 3$, $S = 1$, then $J = 4, 3, 2$
  $J = 2$, is the lowest since the shell is less than half-filled, $2 < 5$.
- The complete ground state: $^3F_2$:
  Using

$$E_{J,L,S} = \frac{1}{2}\lambda[J(J+1) - L(L+1) - S(S+1)] \tag{2.25.8}$$

$$E_{2,3,1} = \frac{1}{2}\lambda[2(2+1) - 3(3+1) - 1(1+1)] = -4\lambda$$

(see Fig. 2.29)

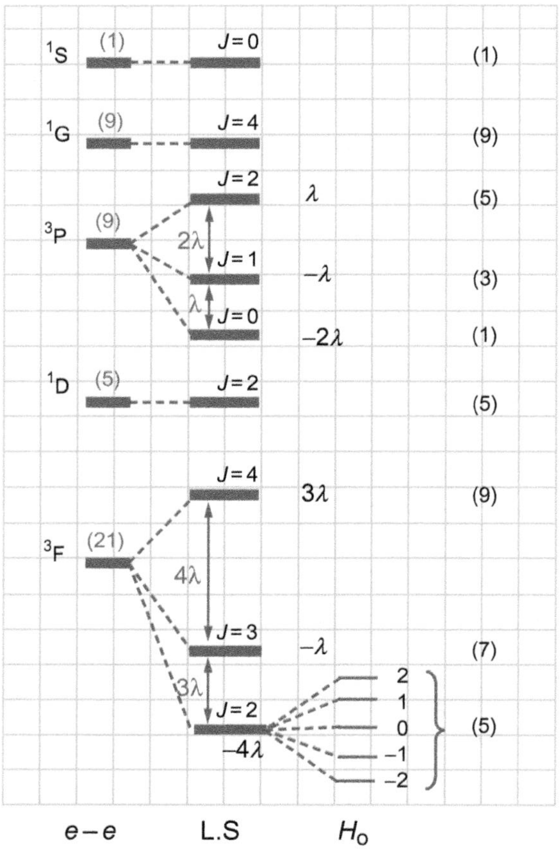

FIG. 2.29 Spin-orbital states from a d^2 configuration. The splitting of $^3F_3$ by magnetic field H_o is indicated on the far lower right.

- For $^3F_3$, $J=3$:

$$\Delta E_{J,J+1} = \lambda(J+1) \tag{2.25.9}$$

$$\Delta E_{2,3} = \lambda(J+1) = 3\lambda$$

$$E^3F_3 = -4\lambda + 3\lambda = -\lambda,$$

- For $^3F_4$, $J=4$:

$$\Delta E_{3,4} = \lambda(3+1) = 4\lambda$$

$$E^3F_4 = -\lambda + 4\lambda = 3\lambda$$

- ^1D: $L=2$, $S=0$, $J=2$, with one possible J value:

$$E_{2,2,0} = \frac{1}{2}\lambda[2(2+1) - 2(2+1) - 0(0+1)] = 0,$$

- ^3P: $L=1$, $S=1$, $J=0, 1, 2$:

$$E_{2,3,1} = \frac{1}{2}\lambda[0(0+1) - 1(1+1) - 1(1+1)] = -2\lambda,$$

- ^1G: $L=4$, $S=0$, $J=4$.
- ^1S: $L=0$, $S=0$, $J=0$.

Exercise

a. **Write all possible wave functions for the ground ($1s^2\ 2s^2\ 2p^6\ 3s^1$) and the first excited states of Na as Slater determinants. Use abbreviated notation and omit the closed shell orbitals.**

Answer:

- Na: $1s^2 2s^2 2p^6 3s^1$:

  $D_i = |(nlm_l m_s)|$

  $$D_1 = \left|\left(300\frac{1}{2}\right)\right|$$

  $$D_2 = \left|\left(300\frac{\overline{1}}{2}\right)\right|$$

- Na*: $1s^2 2s^2 2p^6 3p^1$:

  $$D_1 = \left|\left(311\frac{1}{2}\right)\right|,\ D_2 = \left|\left(311\frac{\overline{1}}{2}\right)\right|$$

  $$D_3 = \left|\left(310\frac{1}{2}\right)\right|,\ D_4 = \left|\left(310\frac{\overline{1}}{2}\right)\right|$$

  $$D_5 = \left|\left(31\overline{1}\frac{1}{2}\right)\right|,\ D_6 = \left|\left(31\overline{1}\frac{\overline{1}}{2}\right)\right|$$

b. **Derive the Russell-Saunders terms for these two configurations and give their spectroscopic symbols. Arrange them qualitatively on an energy level diagram.**

Answer

- For Na: $1s^2 2s^2 2p^6 3s^1$ (Table 2.17)
  Members of ^2S term:

  $$E_{J,\,L,\,S} = \frac{1}{2}\lambda[J(J+1) - L(L+1) - S(S+1)] \quad (2.25.8)$$

  ^2S: $L=0$, $S=½$, no spin coupling.
- Na*: $1s^2 2s^2 2p^6 3p^1$ (Table 2.18):

TABLE 2.17 Members of ^2S Term

	M_S	
M_L	1/2	−1/2
0	D_1	D_2

TABLE 2.18 Members of ^2P Term

	M_S	
M_L	1/2	−1/2
1	D_1	D_2
0	D_3	D_4
−1	D_5	D_6

FIG. 2.30 The Russell-Saunders terms for Na: $1s^22s^22p^63s^1$ and Na*: $1s^22s^22p^63p^1$configurations and their spectroscopic symbols.

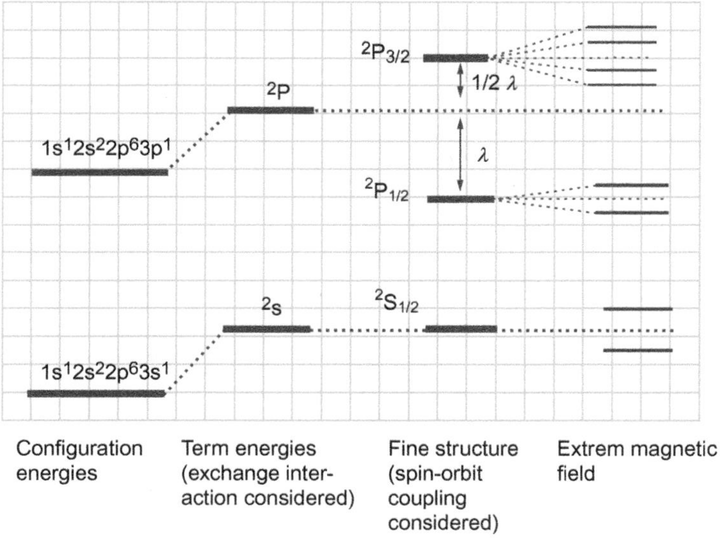

All are the members of ^{2}P term.
The effect of spin-orbital coupling on the term energies is presented in Fig. 2.30.

c. **Using Fig. 2.30, indicate quantitatively the effect of spin-orbital coupling on the term energies. Label the subterms arising from various terms. Express the energy of each subterm (as measured from the term energy in multiples of λ).**

Answer:

- ^{2}S does not split (*J* can only be 1/2).
- ^{2}P split into 2 (J can be 3/2 or 1/2).

$$E_{J,\,L,\,S} = \frac{1}{2}\lambda[J(J+1) - L(L+1) - S(S+1)] \tag{2.25.8}$$

$$E_{P_{1/2}} = \frac{1}{2}\lambda\left[\frac{1}{2}\left(\frac{1}{2}+1\right) - 1(1+1) - \frac{1}{2}\left(\frac{1}{2}+1\right)\right] = -\lambda$$

$$E_{P_{3/2}} = \frac{1}{2}\lambda\left[\frac{3}{2}\left(\frac{3}{2}+1\right) - 1(1+1) - \frac{1}{2}\left(\frac{1}{2}+1\right)\right] = \frac{1}{2}\lambda$$

$$E(^2p_{3/2}) - E(^2p) = 1/2\,\lambda$$

$$E(^2p_{1/2}) - E(^2p) = -\lambda$$

d. **Again using Fig. 2.30, show the effect of an external magnetic field on the subterm energies.**

Answer:

- See Fig. 2.30.

Exercise

As a result of two possible transitions from $1s^2\ 2s^2\ 2p^6\ 3p^1$ to $1s^2\ 2s^2\ 2p^6\ 3s^1$ configuration, sodium emits yellow light (the famous flame test) at 5895.9 and 5890.0 A°. Using this information, evaluate the magnitude of λ in the units of ergs and in units of cm^{-1}.

Answer: Fig. 2.31 shows the effect of the spin-orbital coupling on the term energies and the consequent electronic emissions.

$$E\left(^2P_{3/2}\right) - E\left(^2S_{1/2}\right) = h\nu = \frac{hc}{\lambda} = \frac{hc}{5890.0\times10^{-8}} = 3372\times10^{-12}\,\text{erg}$$

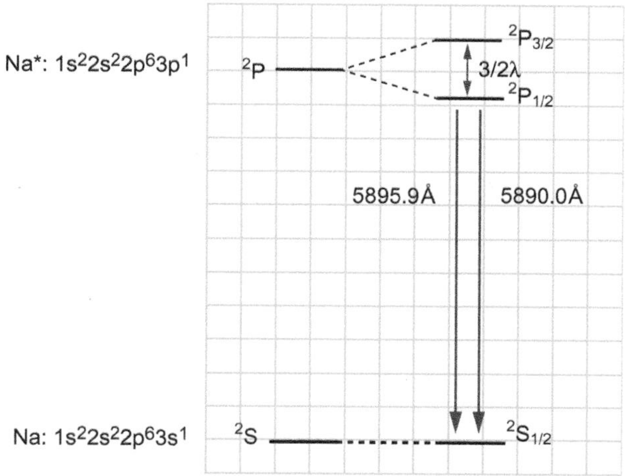

FIG. 2.31 The effect of spin-orbital coupling on the term energies of Na.

$$E\left({}^2P_{1/2}\right) - E\left({}^2S_{1/2}\right) = \frac{hc}{5895.9\times10^{-8}} = 3368\times10^{-12}\,\text{erg}$$

$$\Delta E_{J,J+1} = \lambda(J+1) \quad (2.25.9)$$

$$E\left({}^2P_{3/2}\right) - E\left({}^2P_{1/2}\right) = \left(\frac{3}{2}\right)\lambda$$

$$E\left({}^2P_{3/2}\right) - E\left({}^2P_{1/2}\right) = \frac{3}{2}\lambda = (3372-3368)\times10^{-12}\,\text{ergs} = 0.004\times10^{-12\,\text{ergs}}$$

$$\lambda = \frac{2}{3}\times0.004\times10^{-12} = \frac{8}{3}\times10^{-15}\,\text{ergs}$$

$$\lambda = \frac{8}{3}\times10^{-15}\times\frac{1}{hc} = 13.4\,\text{cm}^{-1}$$

Exercise

Consider a $^3P_2$ and a $^1D_2$ term. How does the spacing of the energy levels arising from $^3P_2$ in magnetic field of 1000 gauss compare with the spacing of the energy levels arising from $^1D_2$ in the same field? Be quantitative.

Answer:

- Separation between the levels arising in a magnetic field:

$$\because \Delta E_J = gH\beta,$$

$$\because g = 1 + \frac{J(J+1)+S(S+1)-L(L+1)}{2J(J+1)}$$

$$\therefore \Delta E = \left[1 + \frac{J(J+1)+S(S+1)-L(L+1)}{2J(J+1)}\right]\frac{eh}{4\pi mc}H \quad (2.26.15)$$

$$\because g = \left[1 + \frac{J(J+1)+S(S+1)-L(L+1)}{2J(J+1)}\right]$$

For $^3P_2$:

$$g = \left[1 + \frac{2(2+1)+1(1+1)-1(1+1)}{2\times2(2+1)}\right] = \frac{3}{2}$$

For $^1D_2$:

$$g = \left[1 + \frac{2(2+1)+0(0+1)-2(2+1)}{2\times2(2+1)}\right] = 1$$

Therefore, the spacing between the levels arising from $^3P_2$ is 1.5 times greater.

Exercise

a. **On an energy level diagram, show quantitatively how the spin-orbit coupling splits the ^{2}P and ^{2}D terms.**

Answer:

- $$E_J = \lambda \frac{J(J+1) - L(L+1) - S(S+1)}{2} \tag{2.25.8}$$
- For ^{2}D: $L=2$, $S=1/2$, $J=5/2, 3/2$:

$$E_{5/2} = \lambda \frac{\frac{5}{2}\left(\frac{5}{2}+1\right) - 2(2+1) - \frac{1}{2}\left(\frac{1}{2}+1\right)}{2} = \lambda \frac{\frac{35}{4} - 6 - \frac{3}{4}}{2} = 1\lambda$$

$$E_{3/2} = \lambda \frac{\frac{3}{2}\left(\frac{3}{2}+1\right) - 2(2+1) - \frac{1}{2}\left(\frac{1}{2}+1\right)}{2} = \lambda \frac{\frac{15}{4} - 6 - \frac{3}{4}}{2} = -\frac{3}{2}\lambda$$

- For ^{2}P: $L=1$, $S=1/2$, $J=3/2, 1/2$:

$$E_{3/2} = \lambda' \frac{\frac{3}{2}\left(\frac{3}{2}+1\right) - 1(1+1) - \frac{1}{2}\left(\frac{1}{2}+1\right)}{2} = \lambda' \frac{\frac{15}{4} - 2 - \frac{3}{4}}{2} = \frac{1}{2}\lambda'$$

$$E_{1/2} = \lambda' \frac{\frac{1}{2}\left(\frac{1}{2}+1\right) - 1(1+1) - \frac{1}{2}\left(\frac{1}{2}+1\right)}{2} = \lambda' \frac{\frac{3}{4} - 2 - \frac{3}{4}}{2} = -\lambda'$$

- Fig. 2.32 presents the effect of the spin-orbital coupling on the ^{2}P and ^{2}D terms.

b. **Suppose that λ for ^{2}P is 10 times larger than λ for ^{2}D. Evaluate the frequency difference between the largest and smallest frequency transitions from ^{2}D to ^{2}P (^{2}D → ^{2}P) in terms of λ (^{2}D).**

Answer:

- The frequency difference between the largest and smallest frequency transitions from ^{2}D to ^{2}P (Fig. 2.33) is given by

$$\Delta \upsilon = \frac{\Delta E}{h} = \frac{1}{h}(15\lambda)$$

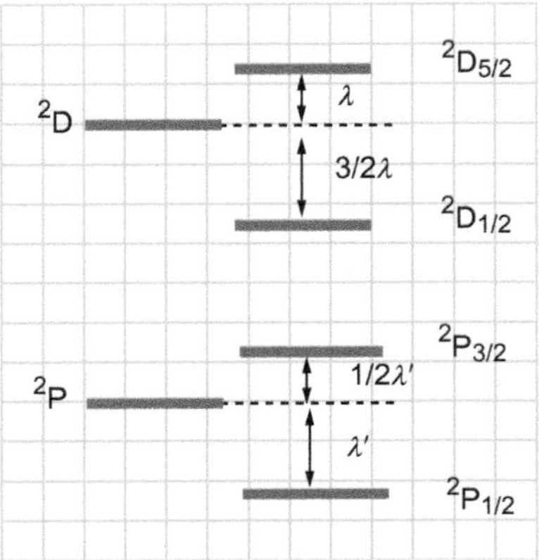

FIG. 2.32 The spin-orbital coupling and the splitting of $^2P_2$ and $^2D_2$ terms.

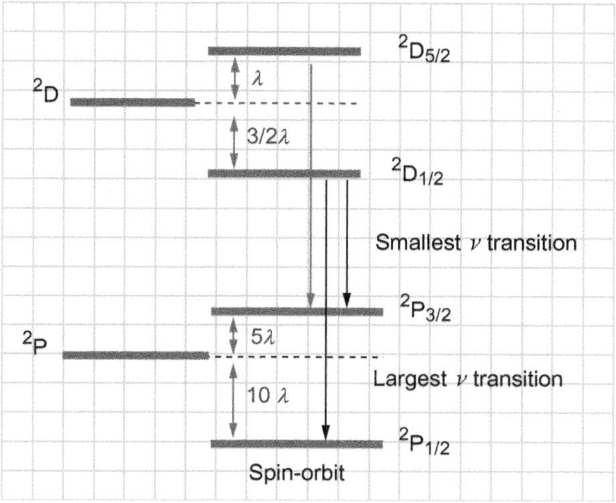

FIG. 2.33 The spin-orbital coupling splitting of $^2P_2$ and $^2D_2$ terms, when λ for ^2P is 10 times larger than λ for ^2D.

Exercise

On an energy level diagram, show quantitatively how the ^3F term splits when a spin-orbital interaction is considered. Give the spectroscopic symbols of the subterms, and indicate the spacing between them in terms of λ. Indicate the effect of an external magnetic field on each subterm.

Answer:

$$E_J = \lambda \frac{J(J+1) - L(L+1) - S(S+1)}{2} \tag{2.25.8}$$

- For ^3F $L=3$, $S=1$, $J=4, 3, 2$:

$$E\left({}^3F_4\right) = \lambda \frac{4(4+1) - 3(3+1) - 1(1+1)}{2} = 3\lambda$$

$$E\left({}^3F_3\right) = \lambda \frac{3(3+1) - 3(3+1) - 1(1+1)}{2} = -\lambda$$

$$E\left({}^3F_4\right) = \lambda \frac{2(2+1) - 3(3+1) - 1(1+1)}{2} = -4\lambda$$

- Let us see if the center of gravity rule is satisfied:

$$-4(5) - 1(7) + 3(9) = 0$$

- Fig. 2.34 shows the splitting of ^3F term in an external magnetic field.

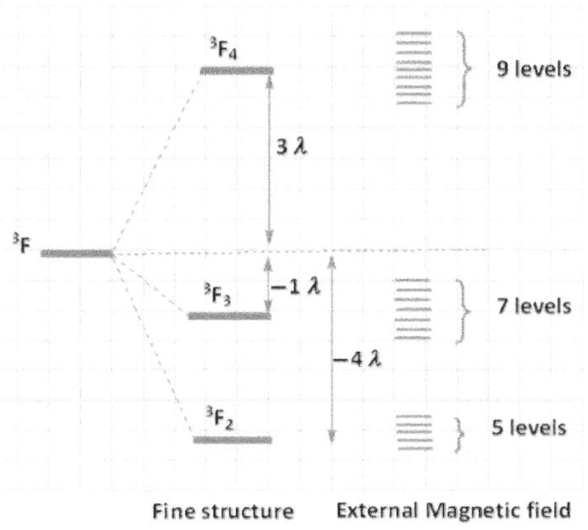

FIG. 2.34 Spin-orbital states from a ^3F term in an external magnetic field.

SUGGESTIONS FOR FURTHER READING

The wave function and Schrödinger equation of the hydrogen atom

E. Schrodinger, Quantisierung als Eigenwertproblem, Annalen der Phys. 384 (4) (1926) 361–376.
W. Heisenberg, Über quantentheoretische Umdeutung kinematischer und mechanischer Beziehungen, Z. Phys. 33 (1) (1925) 879–893.
P.A.M. Dirac, Foundations of quantum mechanics, Nature 203 (1964) 115–116.
D.J. Griffiths, Introduction to Quantum Mechanics, second ed., Benjamin Cummings, ISBN: 0-13-124405-1, 2004.
P.W. Atkins, Quanta: A Handbook of Concepts, Oxford University Press, 1991.
P.W. Atkins, Molecular Quantum Mechanics, Oxford University Press, 1983.
P.A.M. Dirac, The Principles of Quantum Mechanics, fourth ed., Oxford University Press, 1958.
B.H. Bransden, C.J. Joachain, Quantum Mechanics, second ed., Prentice Hall PTR, ISBN: 0-582-35691-1, 2000.
R. Liboff, Introductory Quantum Mechanics, fourth ed., Addison Wesley, ISBN: 0-8053-8714-5, 2002.
D. Halliday, Fundamentals of Physics, eighth ed., Wiley, ISBN: 0-471-15950-6, 2007.
E. Schrödinger, An undulatory theory of the mechanics of atoms and molecules, Phys. Rev. 28 (6) (1926) 1049–1070.
G. Teschl, Mathematical Methods in Quantum Mechanics; With Applications to Schrödinger Operators, American Mathematical Society, Providence, RI, ISBN: 978-0-8218-4660-5, 2009.
M. Jammer, The Conceptual Development of Quantum Mechanics, McGraw-Hill, New York, 1966.
E.H. Wichmann, Quantum Physics, vol. 4, McGraw-Hill, New York, 1971.
W.J. Moore, Schrodinger: Life and Thought, Cambridge University Press, 1989.
E. Schrödinger, Quantisierung als Eigenwertproblem (Dritte Mitteilung), Ann. Phys. 81 (1926) 109.
R.S. Barry, V - Atomic orbitals, J. Chem. Educ. 43 (1966) 283.
J. Singh, Quantum Mechanics: Fundamentals and Applications to Technology, John Wiley & Sons, Inc, New York, 1996. 171.
T. McDermott, G. Henderson, Spherical harmonic in Cartesian frame, J. Chem. Educ. 67 (1990) 915.
C.J. Ballhausen, Introduction to Crystal Field Theory, McGraw-Hill, New York, 1962.
H.L. Schlafer, G. Glimann, Basic Principles of Ligand Field Theory, John Wiley, New York, 1969.
B.N. Figgis, Introduction to Ligand Fields, Robert E. Krieger Publishing Company, Inc, Malabar, FL, 1996.
H.E. White, Introduction to Atomic Spectroscopy, McGraw-Hill, New York, 1934. p. 67.

The angular momentum

M.L. Boas, Mathematical Methods in Physical Sciences, John Wiley & Sons, Inc, New York, 1966. p. 203.
R.N. Zare, Angular Momentum, Wiley, New York, 1988.

The spin quantum number, *s*

R.E. Powell, Relativistic quantum chemistry: the electrons and the nodes, J. Chem. Educ. 45 (1968) 558.
J.L. Bills, The Pauli principle and electronic repulsion in helium, J. Chem. Educ. 51 (1974) 585.

The boundary surface of s-, p-, and d-orbitals

G.L. Breneman, Order out of chaos: shapes of hydrogen orbitals, J. Chem. Educ. 65 (1988) 31.
A. Szabo, Contour diagrams for relativistic orbitals, J. Chem. Educ. 46 (1969) 678.

Calculating the most probable and the mean radius of an orbital

J. Mason, Periodic contractions among the elements: or, on being the right size, J. Chem. Educ. 65 (1988) 17.

The structure of many-electron atoms

G. Herzberg, Atomic Spectra, Dover Publication, Inc., New York, 1944
D. Hartree, The Calculation of Atomic Structures, John Wiley & Sons, Inc., New York, 1957
R.S. Berry, V - Atomic orbitals, J. Chem. Educ. 43 (1966) 283.
R. Latter, Atomic energy levels for the Thomas-Fermi and Thomas-Fermi-Dirac potential, Phys. Rev. 99 (1955) 510.
F. Herman, Atomic Structure Calculation, Prentice-Hall, Inc., Englewood Cliffs, NJ, 1963
K.F. Purcell, J.C. Kotz, Inorganic Chemistry, W.B. Saunders Company, Philadelphia, PA, 1977. Ch. 1.

The Pauli exclusion principle

L. Pauling, E.B. Wilson, Introduction to Quantum Mechanics, McGraw-Hill Book Company, New York, 1935. pp. 184–185.

Penetration, shielding, and the building-up principle

H. Weinstein, P. Politzer, S. Srebrenik, A misconception concerning the electronic density distribution of an atom, Theoret. Chim. Acta 38 (159) (1975).
J.C. Slater, Atomic shielding constants, Phys. Rev. 36 (1930) 57.

Term structure for polyelectron atoms

C.J. Ballhausen, Introduction to Ligand Field Theory, McGraw-Hill, 1962. Ch. 2.
K.E. Hyde, Methods for obtaining Russell-Saunders term symbols from electronic configurations, J. Chem. Educ. 52 (1975) 87.
D.H. McDaniel, Spin factoring as an aid in the determination of spectroscopic terms, J. Chem. Educ. 54 (1977) 147.

B.E. Douglas, D.H. Hollingsworth, Symmetry in Bonding and Spectra, Academic Press, Orlando, FL, 1989. pp. 128–135.
N. Shenkuan, The physical basis of Hund's rule: orbital contraction effects, J. Chem. Educ. 69 (1992) 800.
P.G. Nelson, Relative energies of 3d and 4s orbitals, Educ. In. Chem. 29 (1992) 84.
S. Bashkin, J.O. Stonor Jr., Atomic Energy Levels and Grotrian Diagrams, North-Holland, Amsterdam, 1975.
E.M.R. Kiremire, A numerical algorithm technique for deriving Russell-Saunders (R-S) terms, J. Chem. Educ. 64 (1987) 951.

Spin-orbital coupling

C. J. Ballhausen, Introduction to Ligand Field Theory, McGraw-Hill, 1962 (Chapter 2).
J. Rubio, J.J. Perez, Energy levels in the jj coupling scheme, J. Chem. Educ. 86 (1986) 476.

Related Reading

J.C. Davis, Advanced Physical Chemistry, Ronald Press Co., New York, 1965
H. Eyring, J. Walter, G.E. Kimball, Quantum Chemistry, John Wiley & Sons, New York, 1944.
W.M. Hanna, Quantum Mechanics In Chemistry, W.A. Benjamin, Menlo Park, CA, 1970.
W. Kauzmann, Quantum Chemistry, Academic Press, New York, 1957.
E. Merzbacher, Quantum Mechanics, John Wiley & Sons, New York, 1961.
A. Messiah, Quantum Mechanics, North-Holland Publishing Co., Amsterdam, 1961

Chapter 3

Chemical Bonding

In the preceding chapter, we established the formal mathematical framework for the Schrödinger equation.

We applied this framework to set up the final real solution for the full orbital wave function and identify the four quantum numbers n, l, m, and s, also to compute the most probable radius, mean radius of an orbital, and the boundary surface of s, p, and d orbitals.

Then the orbital wave equations were used as the key feature in order to explain the orbital angular momentum, spin angular momentum, electronic configuration of many-electrons atoms, spin-orbital coupling, and magnetic moment.

In this chapter, we begin with a review of basic electron accounting procedures of bond formations and proceed to a model for predicting three-dimensional molecular structures.

This chapter also reports the basic concepts of metallic structure, describes the bonding, defines the role of free valence electrons, and relates the physical properties and theories of metallic bonding.

Emphasis is also placed on ionic bonding and the relationship among the lattice energy, thermodynamic parameters, and covalency. The text reviews the grounds of the ionic crystal structure, in which radii ratio reveals the geometrical arrangements, and the factors that influence the solubilities.

The foundation and basics of the coordination chemistry is laid out: first characterization, formulation, and then formation and stability including hard and soft interactions, chelate effect, macrocyclic effect, cavity size, solvation enthalpy, donor atom basicity, solvent competition, steric effect, metal oxidation state, and metal ionization potential.

This chapter concludes with a thorough discussion on intermolecular forces, in functions, consequences and significances. Consideration is given to the structural and chemical nature of the covalent networks in giant molecules such diamond, graphite, fullerenes, graphene, carbon nanotubes, and asbestos (Scheme 3.1).

The following are the main topic, which will be explored.

- 3.1: Electronegativity and electropositivity
- 3.2: Electronegativity trend
- 3.3: Molecular and nonmolecular compounds
- 3.4: Types of bonds
- 3.5: Metallic bonding and general properties of metals
  - Conductivity: mobility of electrons
  - Luster: free electron irradiation
  - Malleability, cohesive force: number of valence electrons
  - Theories of bonding in metals
  - Free electron theory
  - Bond lengths
  - Crystal structures of metals (metallic structures)
  - Alloy and metallic compounds
- 3.6: Ionic bonding
  - Lattice energy and cohesion of atomic lattice
  - Born-Haber cycle and heat of formation
  - Ionic crystal structures and the radius ratio
  - Stoichiometric and nonstoichiometric defects
  - Ionic character and covalency interference
  - Ionic character and melting point
  - Solubility of the ionic salts

Electrons, Atoms, and Molecules in Inorganic Chemistry. http://dx.doi.org/10.1016/B978-0-12-811048-5.00003-1

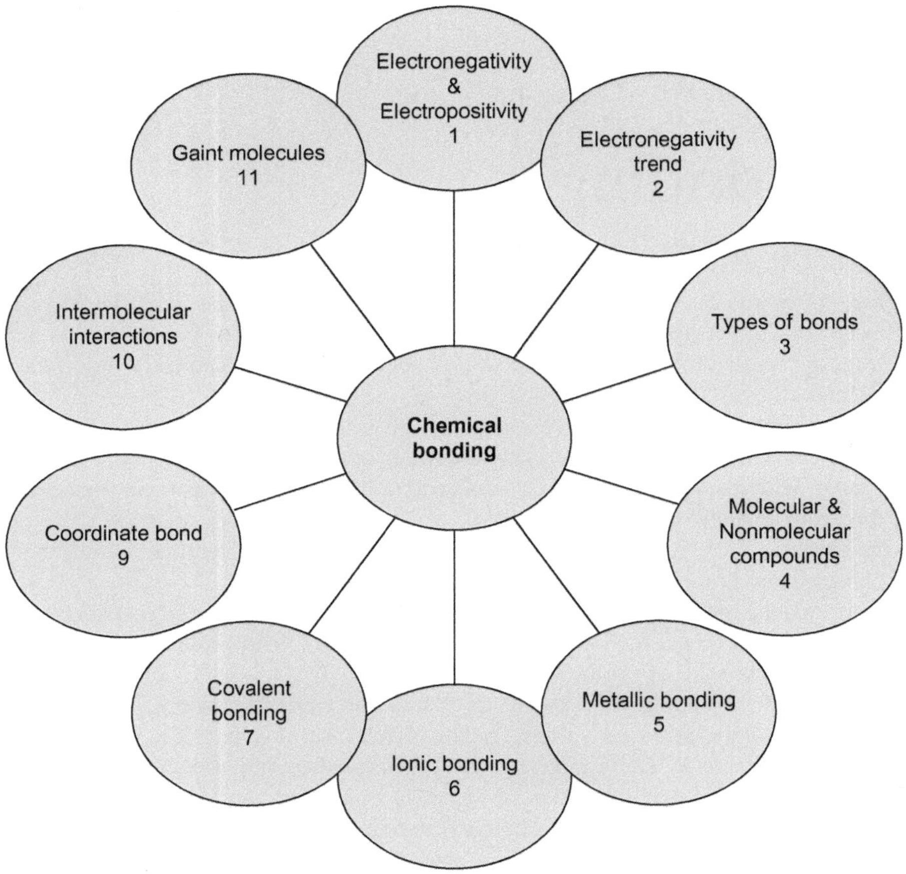

SCHEME 3.1 A schematic of the general approach used to introduce the chemical bonding.

- 3.7: Covalent bonding
  - The Lewis structures and octet rule
  - Exceptions to the octet rule
  - Bonding and polarity
- 3.8: Coordinate covalent bond (dative bonding)
  - Coordination number and "18-electron rule"
  - Ligand denticity
  - Nomenclature of complexes
- Complex formation
  - Coordinative comproportionation reaction
  - Complexation equilibrium
  - Multiligand complexation
  - Stepwise formation constants and the sequential analysis
- Stability of complexes
  - Hard and soft interactions, HSAB
    - Chemical features of hard and soft ions, and classification
    - Rule of interactions
    - Hard-Hard and soft-soft interactions
    - Hard-Soft interaction and anion polarizability
  - Chelate effect
    - Geometry of the chelate ring
    - Entropy and chelate formation

- Macrocyclic effect
- Cavity size
- Solvation enthalpy
- Donor atom basicity
- Solvent competition
- Steric effect
- Metal oxidation state
- Metal ionization potential
- 3.9: Intermolecular interactions
  - van der Waals forces
    - London dispersion force
    - Debye force
    - Keesom force
  - Ion-induced dipole forces
  - Ion-dipole forces
  - Hydrogen bonding
- 3.10: Covalent networks and Giant molecules
  - Graphite
  - Fullerenes
  - Graphene
  - Carbon nanotubes
  - Asbestos
- Suggestions for further reading

3.1 ELECTRONEGATIVITY AND ELECTROPOSITIVITY

What is meant by electronegativity and electropositivity?

- In order to be able to predict the way in which atoms, ions, and molecules will react with one another to proceed to a state of minimum energy, it is important to set up a model using the ionization energy, I_E, and electron affinity, E_A.
  - The electron affinity, E_A, of an atom or molecule is defined as the amount of energy released when an electron is added to a neutral state.

    Getting more ē, E_A or A:

$$F_{(g)} + \bar{e} \rightarrow F^- \qquad \Delta E = -E_A \; (\approx -20\,\text{kJ}\,\text{mol}^{-1})$$

    The more exothermic the reaction, the greater attraction for electrons become.
  - The ionization energy, I_E, is defined as the amount of energy needed to remove an electron from an atom or molecule in the gaseous state

    Keeping what atoms have, I_E:

$$K_{(g)} \rightarrow K^+ + \bar{e} \qquad \Delta E = I_E \; (\approx +100\text{kJ}\,\text{mol}^{-1})$$

- Absolute electronegativity is

$$\chi = \frac{1}{2}(I_E + E_A) \tag{3.1.1}$$

- Consequently, electronegativity is an important fundamental quantity.
  - Electronegativity, χ, is a measure of the tendency of an atom to attract ē charge (i.e., to get more, as well to keep what it already has). The opposite of this is electron positivity.

$$Y^+ + \bar{e} \rightarrow Y \quad \Delta H = -I_E$$

$$Y + \bar{e} \rightarrow Y^- \quad \Delta H = -E_A$$

$$Y^+ + 2\bar{e} \rightarrow Y^- \quad \Delta H = -(I_E + E_A)$$

- Thus, elements can be as follows:
  Electronegative elements: atoms can accept electrons.
  Electropositive elements: atoms tend to give up one or more electrons.
  Elements have little tendency to lose or gain electrons.

If $I_{E_H}=13.598\,eV$, $E_{A_H}=0.754\,eV$, $I_{E_{Cl}}=12.967\,eV$, and $E_{A_{Cl}}=1.462\,eV$, determine the sense of polarization in an HCl molecule.

- $$ZY \rightleftarrows Z^-(g)+Y^+(g)$$

  or

  $$Z^+(g)+Y^-(g) \rightleftarrows ZY$$

  then

  $$Z^+(g)+Y^-(g) \rightleftarrows ZY \rightleftarrows Z^-(g)+Y^+(g)$$

- The energy difference between the products on the right and left:

$$\left(I_{E_{Y^+}}-E_{A_{Z^-}}\right)-\left(I_{E_{Z^+}}-E_{A_{Y^-}}\right)=\left(I_{E_{Y^+}}+E_{A_{Y^-}}\right)-\left(I_{E_{Z^+}}+E_{A_{Z^-}}\right)=2\left(\chi_Y-\chi_Z\right)$$

$$2\left(\chi_H-\chi_{Cl}\right)=(13.598+0.754)-(12.967+1.462)=-0.077\,eV$$

  The negative sign indicates that the reaction is exothermic in

$$ZY \rightarrow Z^-(g)+Y^+(g),\ \text{in}$$

$$HCl \rightarrow H^+ + Cl^-$$

  The electron density in the molecule does not spread evenly along the molecular axis but is polarized in the sense of H^+Cl^-.
  H^+: electropositive element
  Cl^-: electronegative element

Assign the most electronegative element in SO_2 and SO_4^{2-} (the negative oxidation state):

- The sum of all atoms oxidation states equals the actual charge on the molecule:
  - SO_2: O is (−2) S is (+4)
  - SO_4^{2-}: O is (−2) S is (?)
    $x+4(-2)=-2$ S is (+6)
- In common chemistry, F and O are the most electronegative elements (4.0, ~3.5)
  - F is always (−1) unless joined to F.
  - O is (−2), except when there is a $O^{\delta+}-F^{\delta-}$ or $O^{\delta+}-O^{\delta-}$ bonds.

3.2 ELECTRONEGATIVITY AND ELECTROPOSITIVITY TRENDS

How can elements be chemically sorted in the periodic table?

- The periodic table is a sorting map of the chemical elements based on:
  - the atomic number (number of protons in the nucleus);
  - periodic occurrence of similar electron configurations;
  - chemical reactivity; and
  - the similarity in chemical properties.
- The columns of the table are called groups; the rows are called periods.
- The periodic table offers a useful basis for analyzing and understanding the logic behind the regularities in the chemical behaviors. The table can be used to derive relationships between the properties of the elements and predict the properties of new, yet to be discovered or synthesized elements.

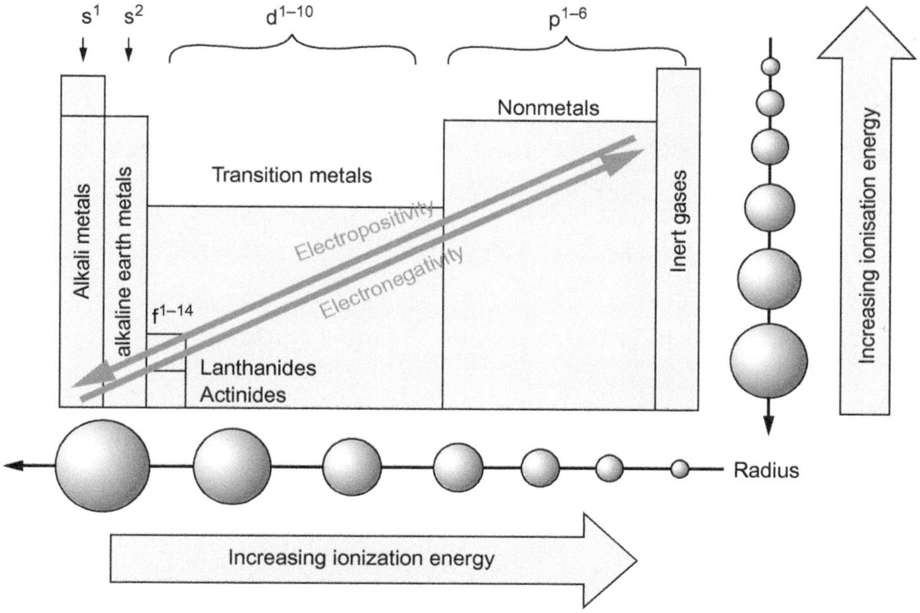

FIG. 3.1 Electronegativity, ionization energy, and radius trends in periodic table.

- Across raw: electrons are added to same shell or subshell, and the simultaneous increase in the nuclear charge causes contraction. Electrons become held more tightly, and attracted more strongly. As a result, there are increases in the ionization energy, electron affinity, and electronegativity (Fig. 3.1).
- Going down the group: jumping from row to row, the valence shell quantum jumps up by one to next "layer" of electron causes a bigger (sub)shell, and electrons become held more weakly. Therefore, I_E, E_A, and χ increase as we go up a group, and the net effect is that χ increases diagonally.
- Electronegative elements are at the top of groups on the right of periods that have small radii. Example: nonmetals (fluorine is the most electronegative element).

  Electropositive elements are at the bottom of groups on the left of periods that have large radii. Example: metals and metalloids, francium is the most electropositive element.

3.3 MOLECULAR AND NONMOLECULAR COMPOUNDS

Why can most atoms not exist by themselves and what are the possible consequences?

- Lower energy configurations happen when the atoms get an electronic configuration similar to the nearest inert gas.
- Atoms may achieve a lowest energy electronic configuration in three different ways:
  - by losing electrons, atoms may be turned into electropositive elements;
  - by gaining electrons, atoms may be turned into electronegative elements; or
  - by sharing electrons among elements that have little tendency to lose or gain electrons.
- As a consequence:
  - most atoms are more stable when they are combined than existing by themselves; and
  - atoms combine and form bonds when the electronic distribution of different atoms changes.
- A chemical bond is a mutual electrical attraction between the valence electrons and nuclei of different atoms that bind together.
- Atoms can form four types of components: free radicals, ions, molecules, and nonmolecular compounds.
  - Free radicals are highly reactive particles containing unpaired electrons.
  - The ion is a charged atom or group of chemically bound atoms that has an excess of electrons (anions) or a deficiency of electrons (cation). The positive cations and negative anions always exist together.
  - The molecule is the smallest neutral component of the matter that has its chemical properties. Molecules may be monoatomic, diatomic, triatomic, ... polyatomic. The noble gases are usually monoatomic molecules.
  - The nonmolecular solids are substances that exist as extended arrays rather than molecular units.

Give examples of nonmolecular solids.

- Nonmolecular solids could be:
  - metals in which delocalized electrons hold close-packed arrays of atoms together;
  - elements such as C, Si, Ge, red and black P, and B exist in infinite networks of localized bonds; or
  - compounds such as SiO_2 and SiC exist as extended array that held together by localized heteropolar bonds. Other examples include polymers, glasses, ceramics, superconductors, semiconductors, refractory compounds, alloys, etc.

3.4 TYPES OF BONDS

What is the type of chemical bond that formed in each the following reactions?

i. **Electropositive element reacts with electropositive element.**
ii. **Electropositive element reacts with electronegative element.**
iii. **Electronegative element reacts with electronegative element.**

Give examples. How are these types of bonding related to each other?

- There are three fundamental types of bonds:
  i. Metallic bonding:
     In metallic bonding, the valence electrons are free to move throughout the whole mineral.

$$\left.\begin{array}{c}\text{Electropositive element}\\+\\\text{Electropositive element}\end{array}\right\}\overset{\text{Metallic bond}}{\Rightarrow}\text{Metal}$$

  ii. Ionic bonding:
     Ionic bonding involves the complete transfer of one or more electrons from one atom to another.

$$\left.\begin{array}{c}\text{Electropositive element}\\+\\\text{Electronegative element}\end{array}\right\}\overset{\text{Ionic bond}}{\Rightarrow}\text{Ionic compound}$$

  iii. Covalent bonding and coordinate covalent bond (dative bonding):
     Covalent bonding involves the sharing of a pair of electrons between two atoms.

$$\left.\begin{array}{c}\text{Electronegative element}\\+\\\text{Electronegative element}\end{array}\right\}\xrightarrow{\text{Covalent bond}}\text{Covalent compound}$$

- These types of bonds are the extreme terminal limits. One type generally dominates, but in most substances, the bond type is somewhere between these extremes.
  - For example, lithium chloride is considered an ionic compound, but it is soluble in alcohol, implying that it also has some covalent character.
  - If the three extreme bond types are placed at the corners of a triangle, then compounds with bonds mainly of one type will be located as points near the corners. Compounds with bonds in-between two types will place along an edge of the triangle, while compounds with bonds that have all three types are presented as points inside the triangle (Fig. 3.2).

3.5 METALLIC BONDING AND GENERAL PROPERTIES OF METALS

What are the general physical properties of metals?

- All metals have the following physical properties:
  - They are good conductors of heat and electricity.
  - They have a metallic luster (bright, shiny and highly reflective).
  - They are ductile and malleable.
  - Their crystal structures are mostly cubic close-packed, hexagonal close-packed, or body-centered cubic.
  - They form alloys.

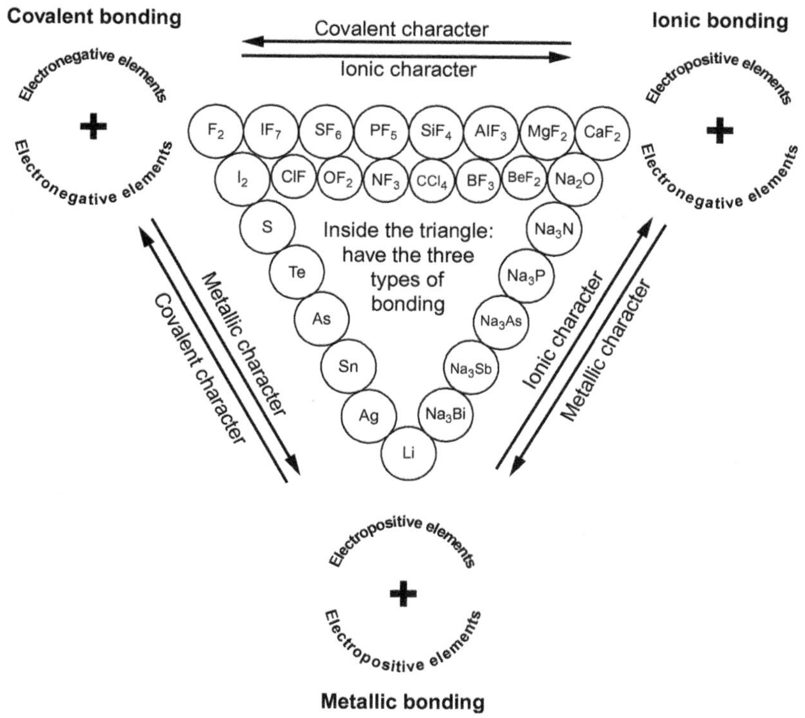

FIG. 3.2 The transitions between ionic, covalent, and metallic bonding.

Conductivity and Mobility of Electrons

There is an enormous change in the conductivity between metals and any other type of solid (Table 3.1).

How does the electrical conductivity of metals differ from that of liquid and other solids?

- The conductivity of a substance is defined as the ability or power to conduct or transmit heat, or electricity, and equal to the reciprocal of the resistance.
- Units of the electrical conductivity are Siemens per meter [S/m] and mho per meter [mho/m]. Its symbol is k or s, mho (reciprocal of ohm), 1 mho/m = 1 S/m. The dimensional of electrical conductivity is length^{-3} mass^{-1} time^{3} electrical current^{2}.
- Thermal conductivity is measured in watts per meter kelvin (W/(m K)). The dimension of thermal conductivity is $M^1L^1T^{-3}\Theta^{-1}$. These variables are mass (M), length (L), time (T), and temperature (Θ).
- All metals are good conductors of heat and electricity.

TABLE 3.1 Electrical Conductivity of Different Solids

Substance	Type of Bonding	Conductivity (mho/cm)
Zinc	Metallic	1.7×10^5
Sodium	Metallic	2.4×10^5
Copper	Metallic	6.0×10^5
Silver	Metallic	6.3×10^5
Sodium chloride	Ionic	10^{-7}
Diamond	Covalent giant molecule	10^{-14}
Quartz	Covalent giant molecule	10^{-14}

- The high electrical and thermal conductivities are the distinguishable feature of metals, which can be attributed to the mobility of electrons through the lattice.
- Metals show some degree of paramagnetism, which implies that they have unpaired electrons.
- Metallic elements have vacant orbitals. Both the number of electrons and the existence of vacant orbitals in the valence shell are important feature in explaining the conductivity and bonding of metals.
- The conductivity of metals increases with decreasing temperature.
- In the solid state, ionic compounds may conduct to a very small extent (semiconduction) if defects are present in the crystal.
- While in aqueous solution or fused melts of ionic compounds, the movement of ions is responsible for conduction.
  - Example: sodium chloride, sodium ions move toward the cathode, and chloride ions move toward the anode.

Luster and Free Electron Irradiation

Why are smooth surfaces of metals used as mirrors? Why do copper and gold have reddish and golden colors? Why do metals exhibit reflection, photoelectric effect, and thermionic emission?

- Smooth surfaces of metals normally have a shiny appearance.
- The shininess is observed at all viewing angles, while a few nonmetallic elements such as sulfur and iodine appear shiny only at low viewing angles.
- Because metals reflect light at all angles, they can be used as mirrors. Specifically, light of all wavelengths is absorbed, and directly re-emitted; almost all the light is reflected back. The free electrons in the metal absorb energy from light and re-emit it when the electrons drop back from the excited states to the original energy level.
- All metals except copper and gold have a silvery color. The reddish and golden colors of copper and gold occur because they absorb some colors more easily than others do.
- Copper, silver, and gold have similar electron configurations, but they have quite distinct colors.
- Electrons of outer levels absorb energy from incident light, and are excited from lower energy levels to higher vacant energy levels. The excited electrons then return to the ground energies and emit the difference of energy as photons.
- Copper has an intense absorption at the region (430–490 nm), so the reflected complementary light is orange.
- Gold absorbs blue light (490–530 nm) and reflects the rest of the visible spectrum. The complementary color we see is yellow.
- In silver and most all metals, the absorption peak lies in the ultraviolet region. Consequently, light is equally absorbed, re-emitted, and reflected evenly across the visible spectrum; we see it as a pure white.
- Some metals emit electrons when irradiated with wave-radiation (the photoelectric effect, Chapter 1.11a) and other emits electrons on heating (thermionic emission).

Malleability, Cohesive Force, Number of Valence Electrons

What is meant by malleable and ductile? Why are metals malleable and ductile? How can the strength of metallic bonding be measured and what factors affect this bonding?

- Metals are naturally soft and not rigid (malleable) and flexible (ductile). This means that:
  - there is not much resistance to deformation of the structure; and
  - a strong cohesive force holds the structure together (metallic bonds).

$$\left.\begin{array}{c}\text{Metallic bond}\\ \text{strength} =\\ \text{cohesive force}\end{array}\right\} \text{Metal}_{\text{crystal}} + \text{Heat of atomization} \Rightarrow \text{Metal}_{\text{gas}}$$

- The strength of the metallic bond (the cohesive force) may be evaluated as the heat of atomization.
- The factors that affect this bond are as follows:
  - The strength of the metallic bonding drops on descending a group in the periodic table, e.g., $Li \rightarrow Na \rightarrow K \rightarrow Rb \rightarrow Cs$, suggesting that they are inversely proportional to the internuclear distance.
  - The strength of the metallic bonding increases across the periodic table from Group I → Group II → Group III. This shows that the strength of metallic bonding depends on the number of valence electrons.

  - The strength of the metallic bonding increases at first across the transition series $Sc \rightarrow Ti \rightarrow V$ as the number of unpaired d electrons increases, $d^{1\rightarrow5}$. Continuing across the transition series, the number of electrons per atom involved in metallic bonding falls, until the d electrons are paired, reaching a minimum at Zn, $d^{5\rightarrow10}$.
  - The strength of the metallic bonding approaches the magnitude of the lattice energy that holds ionic crystals together (this chapter, page 159).
  - The strength of the metallic bonding is much larger than the weak van der Waals forces (this chapter, page 204) that hold covalent molecules together in the solid state.
  - The melting points and the boiling points of the metals follow the trends in the strength of the metallic bonding.
- There are two considerations influence the metallic bonding and structure of metals:
  - The metallic bond strength depends on the number of unpaired electrons accessible for bonding on each atom.
  - The crystal structure of metals depends on the number of s and p electrons on each atom that are involved with bonding.

How do metal atoms combine and form metals or alloys?

- In metals or alloys, the empty orbitals of the atoms' outer energy levels overlap. This overlapping allows the outer electrons of the atoms to form metal's "sea" that moves freely throughout the entire metal.
- The metal's sea of outer electrons binds the atoms together.
- The metallic bond varies with the number of electrons in the metal's electron sea as well with the nuclear charge of the metal atoms. The metal atoms are arranged in a three-dimensional structure.
- When a metal is vaporized, the bonded atoms in the normal (usually solid) state are converted to individual metal atoms in the gaseous state.
- The amount of heat required to vaporize the metal is a measure of the strength of the bonds that hold the metal together.

Theories of Bonding in Metals

What are the requirements and the main theories explaining the bonding in metals?

- The chemical bonding of metals and alloys is less understood than those with covalent and ionic compounds are.
- The acceptable theory of metallic bonding must describe both the bonding among a large number of similar atoms in a pure metal, as well as the bonding among various metal atoms in alloys.
- The theory should propose a none-directional bonds (valence electrons are attracted to the nuclei of neighboring atoms), because nearly all metallic properties remain the same even when the metal in the liquid state (such mercury), or when metal is dissolved in a suitable solvent (such the solutions of sodium in liquid ammonia).
- The theory should explain the great mobility of electrons.
- There are three theories are proposed to explain the metallic bonding:
  - Free electron theory.
  - Valence bond theory, which will be discussed in Chapter 5, page 288.
  - Molecule orbital or band theory, which will be discussed in Chapter 6, page 332.

Free Electron Theory

What is the free electron theory of metals?

- The first metal was regarded as a lattice with free electrons moving through it (Drude model).
- This concept was refined later (Lorentz model), suggesting that metals consist of a lattice of rigid spheres (positive ions), embedded in a gas of valence electrons which could freely move in the interstices.
- This model explains:
  - the free movement of electrons;
  - the cohesion that results from electrostatic attraction between the positive ions and the electron; and
  - the increase in the number of valence electrons is associated with an increase of the cohesive energy.
- Quantitative calculations are much less accurate than similar calculations for the lattice energies of ionic compounds.

Give examples of compounds that contain metallic bonds.

- Metallic elements frequently react with different metallic elements, yielding several types of alloys that have the feature of metals.

- Metallic bonding is found not only in metals and alloys, but also in other types of compounds:
  1. Interstitial borides, carbides, hydrides, nitrides synthesized by the transition elements (or by some of the lanthanides), and a number of low oxidation states of transition metal halides. These compounds demonstrate electrical conductivity, due to the presence of free electrons in conduction bands.
  2. Cluster compounds of boron, and metal cluster compounds of the transition metals, where the covalent bonding is delocalized over several atoms, and achieves a limited form of metallic bonding.
  3. A group of compounds such as the metal carbonyls contains a metal-metal bond.

  The cluster compounds, and compounds with metal-metal bonds, may help to explain the role of metals as catalysts.

Bond Lengths

Why are the interatomic distances in the metal crystal longer than that of the diatomic molecule of the alkali metals (Table 3.2)**? Why is the bonding energy greater in the metal crystal than that of diatomic molecules, in spite of the fact that the bonds in the metal are longer?**

- The alkali metals are diatomic molecules only in the vapor state, and the interatomic distances in the diatomic molecules are shorter than in the metal crystal (Table 3.2). In metal, the valence electrons are delocalized over a large number of bonds; every bond should be weaker and longer.
- Although the bonds in the metal are longer and weaker, there are many more of them than in the M_2 molecule, so the total bonding energy is greater in metal crystal. This can be understood by examining the enthalpy of dissociation of the M_2 molecules with the enthalpy of sublimation of the metal crystal (Table 3.2).

Crystal Structures of Metals (Metallic Structures)

What are the most common metal crystal structures?

- Metals are made up of positive ions packed together, usually in one of the three following arrangements:
  - hexagonal close-packed;
  - cubic close-packed (also called face-centered cubic); and
  - body-centered cubic.
- The two types of close packing depend on the arrangement of adjacent layers in the structure: hexagonal close packing ABAB and cubic close packing ABCABC (Fig. 3.3).
- Between two close-packed layers, ABAB…, there are octahedral and tetrahedral interstitial sites (Fig. 3.4). Tetrahedral sites are constructed by each sphere of a layer together with three spheres of the layer above or below. Staggered triangular groups of the two close-packed layers form the octahedral sites.
- The sequence ABCABC… describes a cubic close-packed arrangement (Fig. 3.5).
- Metallic elements with a close-packed structure usually have a coordination number of 12.
- Random packing such as ABABC or ACBACB is possible, but occurs rarely. Hexagonal ABABA and cubic ABCABC close packing are common.
- Some metals have a body-centered cubic type of structure, which fills the space slightly less efficiently, and have a coordination number of 14.

TABLE 3.2 Interatomic Distances in M_2 Molecules and Metal Crystals, and a Comparison of Enthalpies of Sublimation and Dissociation

	Distance in Metal Crystal (A°)	Distance in M_2 Molecule (A°)	Enthalpy of Sublimation of Metal (kJ/mol)	1/2 Enthalpy of Dissociation of M_2 Molecule (kJ/mol)
Li	3.04	2.67	161	54
Na	3.72	3.08	108	38
K	4.62	3.92	90	26
Rb	4.86	4.22	82	24
Cs	5.24	4.50	78	21

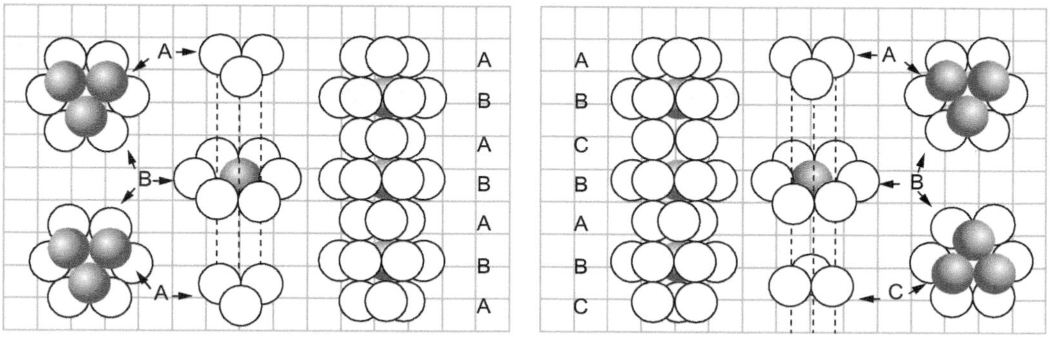

FIG. 3.3 ABA has hexagonal symmetry: every sphere in the third layer is exactly above a sphere in the first layer. ABC has cubic symmetry: all the spheres in the third layer are not exactly above spheres in the first layer.

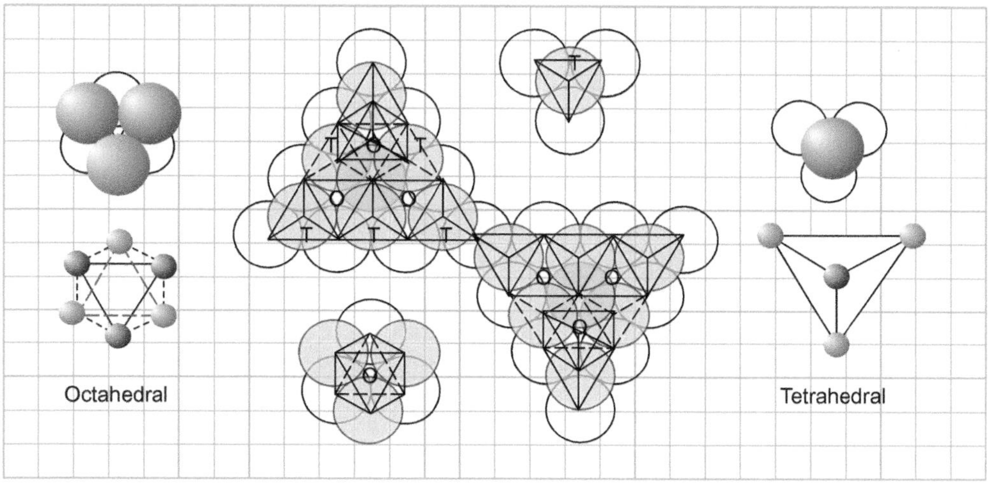

FIG. 3.4 Octahedral and tetrahedral sites in two closed-packed layers.

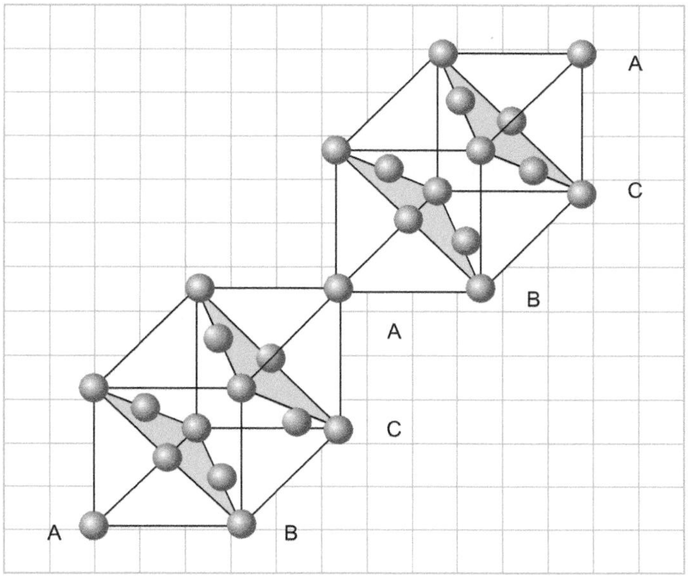

FIG. 3.5 Face-centered cubic, ABCABC.

Alloy and Metallic Compounds

What would happen if two metals heated together or a metal mixed with a nonmetallic element?

- When two metals are heated together, or a metal is mixed with a nonmetallic element, then one of the following may be formed:
  - an ionic compound;
  - a simple mixture may result; or
  - alloys.
- Alloys are materials that contain more than one element, and have the characteristic properties of metals.
  - It is one of the primary ways of modifying the properties of pure metallic elements.
  - Example: pure gold is too soft to be used in jewelry, whereas alloys of gold and copper are quite hard.
  - Alloys are used in the construction, transportation, and electrics industries.

What are the differences between the homogeneous and heterogeneous alloys?

- Alloys can be classified as homogeneous alloys, intermetallic compounds, or heterogeneous alloys.
- Homogeneous alloys: atoms of the different elements are distributed uniformly.
  - Brass: up to 40% zinc in copper.
  - Casting bronze: 10% Sn and 5% Pb in copper.
  - Cupronickel: 25% Ni in copper.
- Intermetallic compounds: are homogeneous alloys that have definite properties and compositions.
  - $CuAl_2$ plays many important roles in modern society.
  - Ni_3Al is a major component of jet aircraft engines because of its strength and low density.
  - Cr_3Pt is used in razor blade coatings, which adds hardness, allowing the blade to stay sharp longer.
  - Co_5Sm is used in the permanent magnets in lightweight headsets because of its high magnetic strength per unit weight.
- Heterogeneous alloys: the components are not dispersed uniformly
  - stainless steel: more than 12% chromium in iron; and
  - solder: tin and lead.
  - Mercury amalgam: 40%–70% Ag, 18%–30% Sn, 10%–30% Cu, and 2% Hg. The mercury amalgam sometimes used to fill teeth consists of a mixture of crystalline phases with different compositions.
  - Pewter: 6% antimony and 1.5% copper in tin solder tin.

Why is the ratio between atomic radii of the alloy components an important parameter in the formation and properties of alloys?

- The structures and the properties of alloys are more complicated than those of pure metals because they are made from elements with different atomic radii. There are two types:
  - a substitution alloy (atoms of the different components occupy equivalent sites); and
  - an interstitial alloy (atoms of the different components occupy nonequivalent sites).

    Each occurs depending on the chemical nature and on the relative sizes of the metal atoms and added atoms.

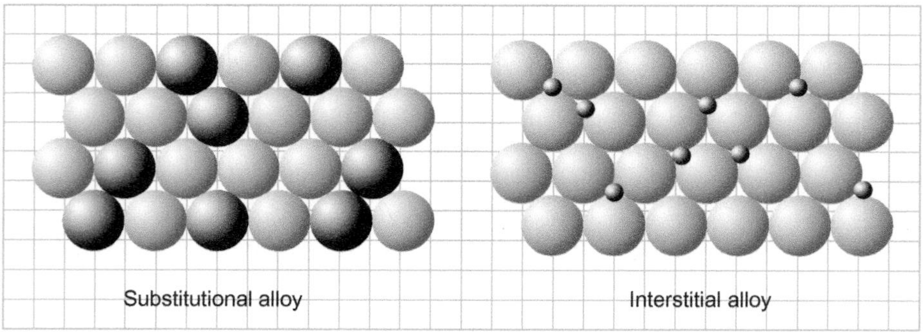

- Substitutional alloys:
  - Substitutional alloys are formed when the metallic components have similar atomic radii and chemical-bonding characteristics.
  - Elements that can form substitutional alloys have atoms with atomic radii that differ by no more than about 15%.

- When two metals differ in radii by more than about 15%, solubility is more limited.
- Because the metallic radii of the d-block elements are all similar, they can form an extensive range of alloys with one another with little distortion of the original crystal structure.
  - Examples:
    The copper-zinc alloy used for some "copper" coins. Because zinc atoms are nearly the same size as copper atoms and have similar electronic properties, zinc can take the place of some of the copper atoms in the alloy.
- Because there are slight differences in size and electronic structure:
  - the less abundant atoms in substitutional alloy distort the shape of the lattice
  - the more abundant atoms of the host metal hinder the flow of electrons.
  - Because the lattice is distorted, it is harder for one plane of atoms to slip past another.
  - Therefore, although a substitutional alloy has lower electrical and thermal conductivity than pure element, it is harder and stronger.

- Interstitial alloy
  - In interstitial alloys, atoms of the solute element fit into the interstices in a lattice formed by atoms of a metal with a larger atomic radius.
  - The atomic radius of the solute element must be less than 60% of the atomic radius of the host metal.
  - Normally, an interstitial element is a nonmetal that participates in bonding to neighboring atoms.
  - The presence of the extra bonds provided by the interstitial component causes the metal lattice to become harder, stronger, and less ductile.
  - The interstitial atoms interfere with the electrical conductivity and with the movement of the atoms forming the lattice. This restricted motion makes the alloy harder and stronger than the pure host metal would be.
  - Alloys of metals tend to be stronger and have lower electrical conductivity than pure metals.
  - Some alloy are softer than the component metal.
  - Examples
    - Steel is an alloy of iron that contains up to 3% carbon. Steel is much harder and stronger than pure iron.
    - Mild steels contain less than 0.2% carbon; mild steels are malleable and ductile and are used to make cables, nails, and chains.
    - Medium steels contain 0.2%–0.6% carbon; medium steels are tougher than mild steels and are used to make girders and rails.
    - High-carbon steel contains 0.6%–1.5% carbon, used in cutlery, tools, and springs.
    - The presence of big bismuth atoms helps to soften a metal and lower its melting point.
  - A low-melting point alloy of lead, tin, and bismuth is employed to control water sprinklers used in certain fire-extinguishing systems. The heat of the fire melts the alloys, which activates the sprinklers before the fire can spread.

3.6 IONIC BONDING

How does ionic bonding form?

- Ionic bonding is formed when electrons are transferred from one type of atom to another:
  - One of the reacting elements loss electrons and become positively charged ions.
  - Elements of the other reactant gain electrons and become negatively charged ions.
  - The electrostatic attractions between oppositely charged ions hold them in a crystal lattice.
  - An ionic crystal lattice is the periodic and systematic arrangement of ions that are found in crystal or the geometric pattern of a crystal.

Lattice Energy and Cohesion of Atomic Lattice

What is lattice energy? How can the stability of the ionic lattice be determined by using the Born treatment and crystal of NaCl as examples? Why does the lattice energy calculated using the Born treatment disagree with the experimental value from thermodynamic data?

- The lattice energy, U, is the necessary energy to separate ions of a mole of MX (crystal) as a gaseous ions from each other by infinite distances.
- The lattice energy is very valuable in relating properties of ionic substances since the formation and destruction of the crystal is the principal step in reactions involving ionic substances.
- The ionic model considers lattice to be held together by multidirectional electrostatic (coulombic) forces.

- The identity of a given M^+X^- (molecule) is lost:

$$Na^+Cl^-_{(g)} \rightarrow Na^+Cl^-_{(s)}$$

The lattice energy, U:

$$M^{q+}_{(g)} + X^{q-}_{(g)} \rightarrow MX_{(s)} \quad \Delta E \cong \Delta H = U \text{ (large negative value)}$$

- The ion pair $M^{q+}X^{q-}$ held together by attractive force:

$$\text{Attractive force} \propto \frac{q^+q^-}{\text{destance}^2} \quad \text{(coulomb law)} \tag{3.6.1}$$

The attractive energy is

$$E_c = \frac{q^+q^-}{4\pi\varepsilon_o r} \text{ (in joules)} \tag{3.6.2}$$

where q^+ and q^- in coulombs (C) initially at a distance r in meters to a distance infinity, $(4\pi\varepsilon_o)^{-1}$ is required in SI system of unites, ε_o is the permittivity (dielectric constant) of a vacuum that equals $8.854 \times 10^{-12}\ C^2\ m^{-1}\ J^{-1}$.

- For Na^+Cl^-, $q^+q^- = -1$

The ions move together from $r = \infty (E = 0)$, until they "touch." Closer approach is limited by the compressibility repulsive force, E_r (Fig. 3.6) is

$$E_r \propto \frac{1}{r^n} \quad (n > 2), \text{ then:}$$

$$E_r = \frac{Be^2}{r^n} \tag{3.6.3}$$

where B is the repulsion coefficient and n is the Born exponent.

The Born exponent is typically a number between 5 and 12, and varies with the nature of the ions. It is determined experimentally by measuring the compressibility of the solid, and expressing the steepness of the repulsive barrier.

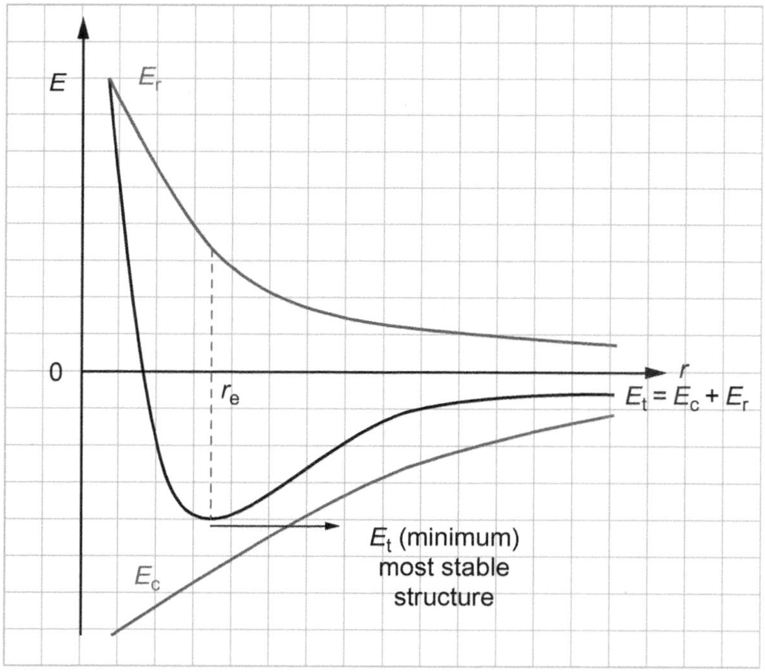

FIG. 3.6 Attraction, repulsion, and total energies.

- The total energy, E_t:

$$E_t = E_c + E_r \tag{3.6.4}$$

The most stable structure at $\frac{\delta E_t}{\delta r} = 0$, and $r = r_e$.
r_e is the equilibrium intermolecular distance.

- Consider E_c for Na^+Cl^- (Fig. 3.7) each Na^+ is surrounded by:
  - $6Cl^-$ at $r = r_e$ (negative);
  - $12Na^+$ at $r = \sqrt{2}r_e$ (positive); and
  - $8Cl^-$ at $r = \sqrt{3}r_e$ (negative).
- The total electrostatic terms depends on the geometrical properties of the lattice:

$$E_c = \frac{q^+q^-}{4\pi\varepsilon_o r} \text{ (in joules)} \tag{3.6.2}$$

$$E_c = 6\cdot\frac{q^+q^-e^2}{4\pi\varepsilon_o r_e} + 12\cdot\frac{(q^+)^2e^2}{4\pi\varepsilon_o\cdot\sqrt{2}r_e} + 8\cdot\frac{q^+q^-e^2}{4\pi\varepsilon_o\cdot\sqrt{3}r_e} + 6\cdot\frac{(q^+)^2e^2}{4\pi\varepsilon_o\cdot 2r_e} + \cdots$$

$$\text{For } \begin{array}{lcl} \text{NaCl}, q^+ & = & +1 \text{ and } q^- = -1 \\ q^+q^- & = & -|q^+q^-| \end{array}$$

$$E_c = \frac{e^2}{4\pi\varepsilon_o r_e}\left(-6 + \frac{12}{\sqrt{2}} - \frac{8}{\sqrt{3}} + \frac{6}{2} - \frac{24}{\sqrt{5}} + \frac{24}{\sqrt{6}}\cdots\right)$$

$$E_c = \frac{e^2}{4\pi\varepsilon_o r_e}(-6 + 8.485 - 4.618 + 3 - 10.733 + 9.798\cdots)$$

The term in parentheses is the sum of an infinite series converging toward 1.748. Therefore,

$$E_c = -\frac{e^2|q^+q^-|(1.748)}{4\pi\varepsilon_o r_e}$$

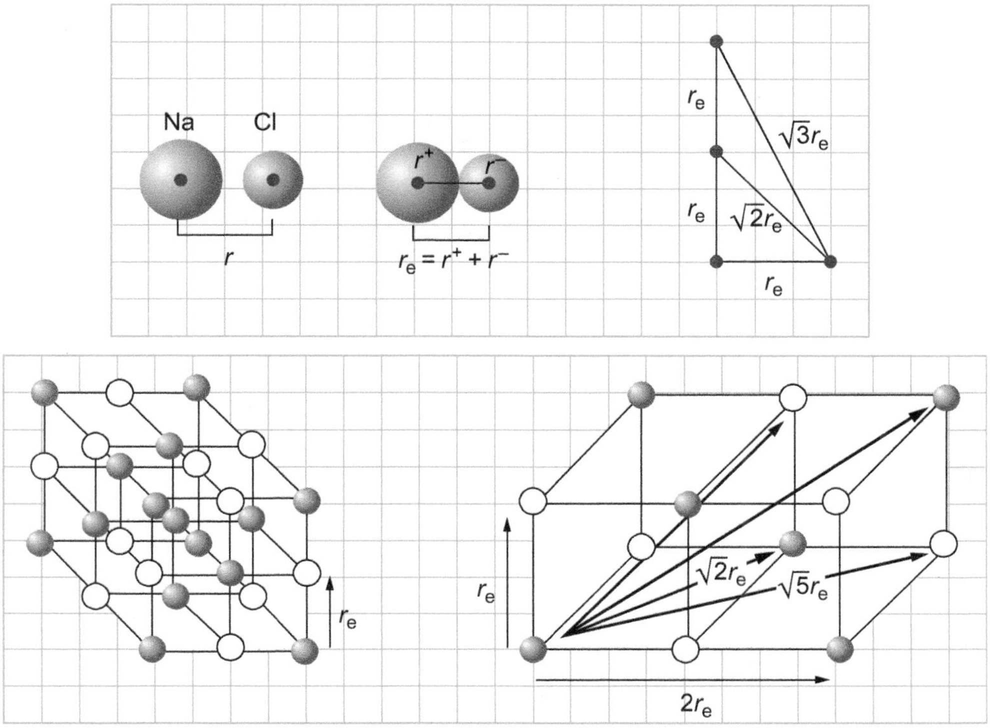

FIG. 3.7 The sodium chloride lattice and distances between a cation and some of its neighbors.

- In general,

$$E_c = -\frac{e^2|q^+q^-|}{4\pi\varepsilon_o r_e}\cdot A \quad \text{(Born equation)} \tag{3.6.5}$$

where A is Madelung constant, which depends on the geometrical arrangement of the ions. Examples:

	CsCl	NaCl	ZnS	CaF_2	TiO_2
$A=$	1.763	1.748	1.64	2.52	2.41

- The total energy, E_t (Fig. 3.6):

$$E_t = E_c + E_r \tag{3.6.4}$$

$$E_r = \frac{Be^2}{r^n} \tag{3.6.3}$$

$$E_t = -\frac{e^2|q^+q^-|}{4\pi\varepsilon_o r}\cdot A + \frac{Be^2}{r^n} \tag{3.6.6}$$

The most stable structure at $\frac{\delta E_t}{\delta r} = 0$, at $r = r_e$

$$\frac{\delta E_t}{\delta r} = \frac{e^2|q^+q^-|}{4\pi\varepsilon_o r_e^2}\cdot A - \frac{nBe^2}{r_e^{(n+1)}} = 0 \tag{3.6.7}$$

$$\therefore r_e^{(n-1)} = \frac{4\pi\varepsilon_o nB}{A|q^+q^-|}$$

r_e is the equilibrium intermolecular distance.

$$r_e = \left(\frac{4\pi\varepsilon_o nB}{A|q^+q^-|}\right)^{1/(n-1)} \tag{3.6.8}$$

$$B = \frac{r_e^{(n-1)}A\,|q^+q^-|}{4\pi\varepsilon_o n} \tag{3.6.9}$$

$$\because E_t = -\frac{e^2|q^+q^-|}{4\pi\varepsilon_o r}\cdot A + \frac{Be^2}{r^n} \tag{3.6.6}$$

$$\therefore (E_t)_o = -\frac{e^2|q^+q^-|}{4\pi\varepsilon_o r_e}\cdot A + \frac{r_e^{(n-1)}A\,|q^+q^-|}{4\pi\varepsilon_o n}\frac{e^2}{r_e^n}$$

$(E_t)_o$ is the total energy at $r = r_e$.

$$\therefore (E_t)_o = -\frac{e^2|q^+q^-|}{4\pi\varepsilon_o r_e}\cdot A\left(1 - \frac{1}{n}\right) \tag{3.6.10}$$

- The lattice energy, U, is the energy required to separate ions of a mole of MX (crystal) as gaseous ions from each other by infinite distances.

$$U = N_A (E_t)_o \tag{3.6.11}$$

where N_A is Avogadro's number.

$$U = -\frac{N_A e^2|q^+q^-|}{4\pi\varepsilon_o r_e}\cdot A\left(1 - \frac{1}{n}\right) \tag{3.6.12}$$

- For q^+ and q^- as simple signed integers, and r_e in Å:

$$U = -\frac{1460\,|q^+q^-|}{r_e} \cdot A \left(1 - \frac{1}{n}\right) \text{ in kJ/mol} \tag{3.6.13}$$

- This equation is derived from basic principles and demonstrates the significant dependences on charges, sizes, and configurations of ions.
- The lattice energy calculated using Born treatment disagrees with the experimental value from thermodynamic data due to:
  - ions being considered as hard spheres undistorting by neighboring ions; and
  - increasing the distortion corresponds to increasing covalent character.
- Better values of lattice energies are obtained using equations with additional terms, such as the Born-Mayer equation:

$$U = -\frac{Ne^2|q^+q^-|}{r_e} \cdot A \left(1 - \frac{34.5}{r_e}\right) \tag{3.6.14}$$

where r_e is in picometers.

Born-Haber Cycle and Heat of Formation

Why do sodium metal and chlorine gas combine and form NaCl? Calculate the enthalpy change ($\Delta H^o_{formation}$ = ?) for the formation of one mole of NaCl (s), using the Born-Haber cycle.

$$(\Delta H^o_{Na-sublimation} = 108\,kJ,\ \Delta H^o_{Na-ionization} = 468\,kJ,\ \Delta H^o_{Cl_2-dissociation} = 12\,kJ,$$
$$\Delta H^o_{Cl_2-electron\ affinity} = -348\,kJ,\ \Delta H^o_{NaCl-lattice\ energy} = -789\,kJ)$$

- The enthalpy change ($\Delta H^o_{formation}$) for the formation of one mole of NaCl (s), using the Born-Haber cycle (Scheme 3.2):

$$\left(\Delta H^o_{formation}\right)_{NaCl(s)} = \left(\Delta H^o_{sublimation}\right)_{Na(s)} + \frac{1}{2}\left(\Delta H^o_{dissociation}\right)_{Cl_2(g)} + \left(\Delta H^o_{ionization}\right)_{Na(g)}$$
$$+ \left(\Delta H^o_{electron\ affinity}\right)_{Cl(g)} + \left(\Delta H^o_{lattice\ energy}\right)_{NaCl(s)}$$
$$= +108\,kJ + 122\,kJ + 496\,kJ - 348\,kJ - 789\,kJ = -411\,kJ$$

- Therefore, the whole process is energetically favorable.

If

	$\Delta H_{sublimation}$ (kJ/mol)	$D_{dissociation}$F-F (kJ/mol)	E_I (kJ/mol)	$E_{affinaty}$ (F^-) (kJ/mol)	U (kJ/mol)
CaF	+176	+156	+589	−322	−795 (calc.)
CaF_2	+176	+156	+1733	−322	−2634

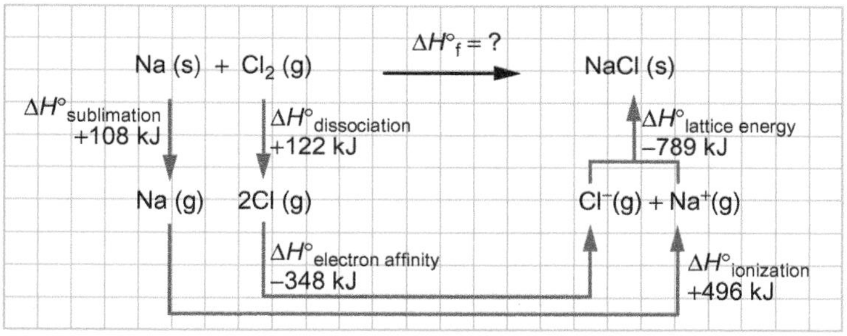

SCHEME 3.2 Born-Haber cycle of NaCl formation.

Explain why CaF is unstable but CaF_2 is stable.

- CaF:

$$\left(\Delta H^{o}_{\text{formation}}\right)_{\text{CaFl(s)}} = \left(\Delta H^{o}_{\text{sublimation}}\right)_{\text{Ca(s)}} + \frac{1}{2}\left(\Delta H^{o}_{\text{dissociation}}\right)_{\text{F}_2\text{(g)}} + \left(\Delta H^{o}_{\text{ionization}}\right)_{\text{Ca(g)}} + \left(\Delta H^{o}_{\text{electron affinity}}\right)_{\text{F(g)}} + \left(\Delta H^{o}_{\text{lattice energy}}\right)_{\text{CaF(s)}}$$

$$\Delta H^{o}_{\text{formation}} = -795 + 589 + 176 + 0.5(156) - 322 = -274\,\text{kJ/mol (stable)}$$

- CaF_2:

$$\Delta H^{o}_{\text{formation}} = -2634 + 1733 + 176 + 156 - 2(322) = -1213\,\text{kJ/mol (stable)}$$

- To explain the instability of CaF:

$$2\text{CaF(s)} \rightarrow \text{CaF}_2\text{(s)} + Ca\text{(s)} \quad \Delta H_{\text{reaction}} = -939\,\text{kJ/mol}$$

The formed CaF is spontaneously disproportionated to Ca and CaF_2.

Ionic Crystal Structures and the Radius Ratio

How could the radius ratio, $\frac{r^+}{r^-}$, be used to predict the stereochemistry of the ionic molecules?

- The structure of many ionic solids can be predicted by estimating the relative size of the positive and negative ions.
- Simple geometric calculations permit us to obtain how many ions of a given size can surround smaller ion. Thus, it is possible to predict the number of neighboring atoms and the geometrical structures from the relative sizes of the ions.

Calculate the limiting value of radius ratio, $\frac{r^+}{r^-}$, that appropriates to the following forms:

i. **planar triangle structure;**
ii. **tetrahedral structure;**
iii. **octahedral structure;**
iv. **square planer; and**
v. **centered cubic structure.**

i. Planar trigonal structure:
Three anions are surrounding one cation in planar trigonal structure (Fig. 3.8):

$$\cos 30 = \frac{r^-}{r^- + r^+} = 0.866$$

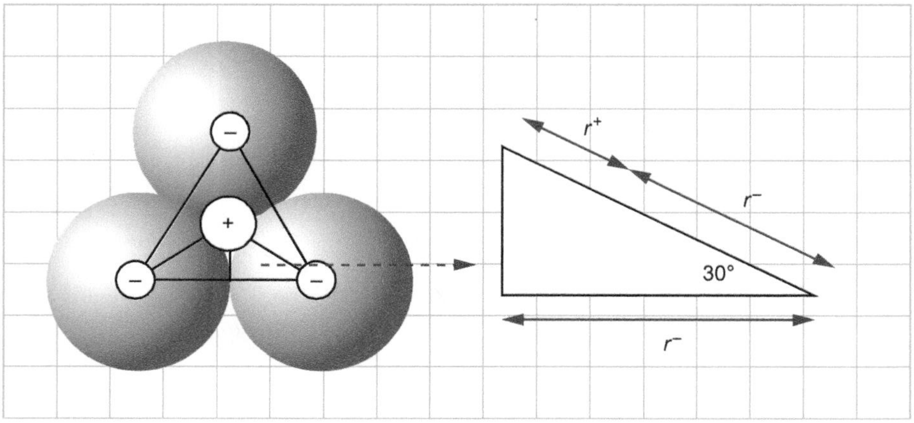

FIG. 3.8 Cation-anion and anion-anion contact for planar trigonal arrangement.

$$r^- + r^+ = \frac{r^-}{0.866} = 1.155\,r^-$$
$$r^+ = 0.155\,r^-$$
$$\frac{r^+}{r^-} = 0.155$$

- The cation will be stable in a trigonal planar hole only if it is at least large enough to keep the anions from touching, i.e., $r^+/r^- = 0.155$.

ii. Tetrahedral structure:
Four anions surround one cation in a tetrahedral structure (Fig. 3.9).

$$AC = CD = DB = x$$
$$CB = \sqrt{2}x = r^-$$
$$AB = \sqrt{3}x = r^- + r^+$$
$$\sin\theta = \frac{CB}{AB} = \frac{r^-}{r^- + r^+} = \frac{\sqrt{2}x}{\sqrt{3}x} = 0.816$$
$$r^- = 0.816\,r^- + 0.816\,r^+$$
$$0.816\,r^+ = 0.184\,r^-$$
$$\frac{r^+}{r^-} = 0.225$$

- The cation will be stable in a tetrahedral hole only if it is at least large enough to keep the anions from touching, i.e., $r^+/r^- = 0.225$.
- Smaller cations will preferentially fit into trigonal planar holes.

iii. Octahedral and square planer structures (Fig. 3.10):
Six anions surround one cation in octahedral structure or four anions surround one cation in square planer structure.
The diagonal of the square $= 2r^- + 2r^+$
The angle formed by the diagonal in the corner must be 45 degrees, so

$$\cos 45° = \frac{2r^-}{2r^- + 2r^+} = 0.707$$
$$r^- = 0.707\,r^- + 0.707\,r^+$$

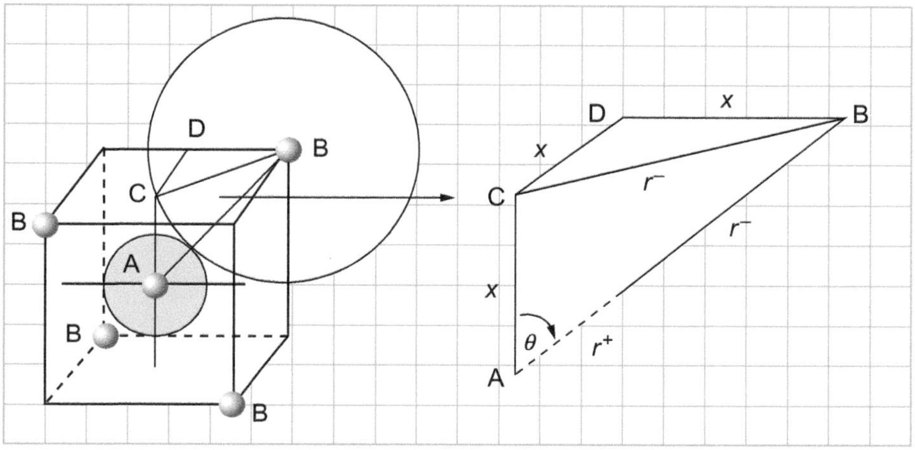

FIG. 3.9 Cation-anion and anion-anion contact for tetrahedral arrangement.

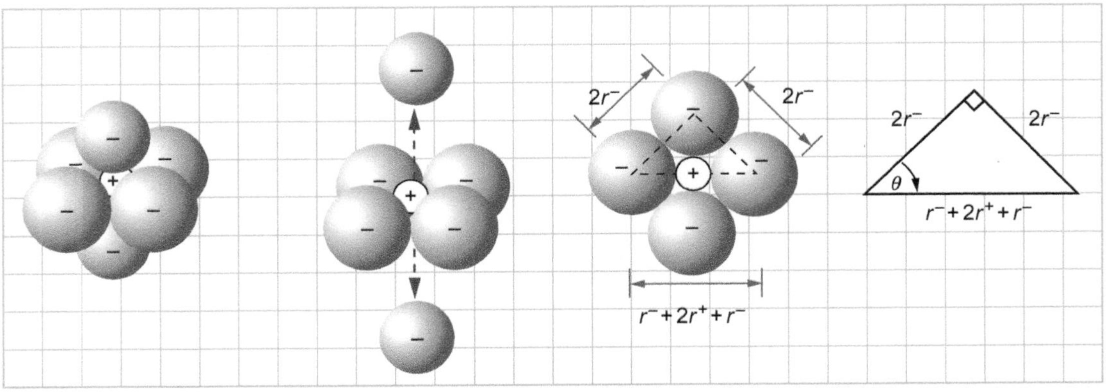

(A) (B) (C)

FIG. 3.10 Cation-anion and anion-anion contact for octahedral arrangement. (A) Octahedral structure, (B) square planar, (C) geometrical relationship between r^+ and r^-.

$$0.293r^- = 0.707r^+$$

$$\frac{r^+}{r^-} = 0.414$$

- The cation will be stable in an octahedral or square planar hole only if it is at least large enough to keep the anions from touching, i.e., $r^+/r^- = 0.414$.
- Smaller cations will preferentially fit into tetrahedral holes.

iv. Body-centered cubic structure:

Eight anions are surround one cation in a body-centered cubic structure (Fig. 3.11).

$$\mathrm{AC} = \mathrm{CD} = \mathrm{DB} = r^- = x$$

$$\mathrm{CB} = \sqrt{2}x$$

$$\mathrm{AB} = \sqrt{3}x = r^- + r^+$$

$$\cos\theta = \frac{\mathrm{AC}}{\mathrm{AB}} = \frac{r^-}{r^- + r^+} = \frac{x}{\sqrt{3}x} = 0.577$$

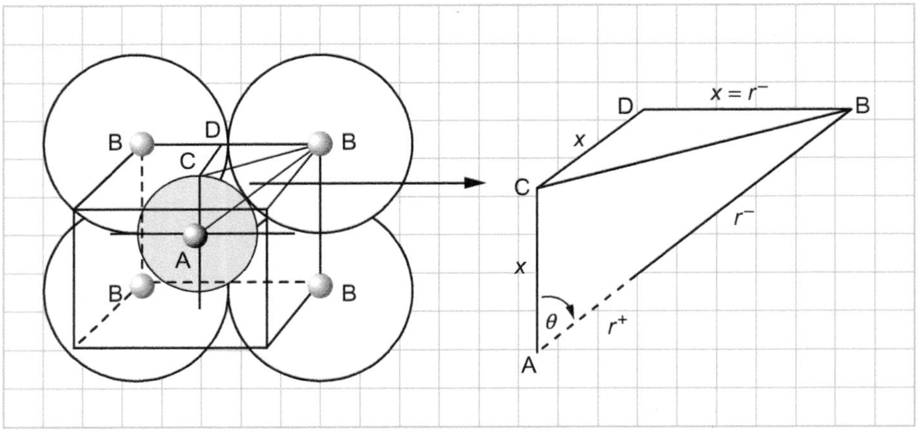

FIG. 3.11 Cation-anion and anion-anion contact for centered cubic arrangement.

$$0.577r^- + 0.577r^+ = r^-$$

$$0.577r^+ = 0.423r^-$$

$$\frac{r^+}{r^-} = 0.732$$

- The cation will be stable in a centered cubic hole only if it is at least large enough to keep the anions from touching, i.e., $r^+/r^- = 0.732$.
- Smaller cations will preferentially fit into octahedral or square planar holes.

Examples:

Use Table 3.3 to find the geometry of ZnS, NaCl, SrF_2, Rb_2O, and SnO_2.

Where $r_{Zn^{2+}} = 74$ pm, $r_{S^{2-}} = 184$ pm, $r_{Na^+} = 95$ pm, $r_{Cl^-} = 181$ pm, $r_{Sr^{2+}} = 113$ pm, $r_{F^-} = 136$ pm, $r_{Rb^+} = 148$ pm, $r_{O^{2-}} = 140$ pm, $r_{Sn^{4+}} = 69$ pm and $r_{O^{2-}} = 140$ pm.

- ZnS:
  $r^+/r^- = 74/184 = 0.4$,
  $0.225 < 0.4 < 0.414$
  Four anions surround one cation; the geometry is T_d:

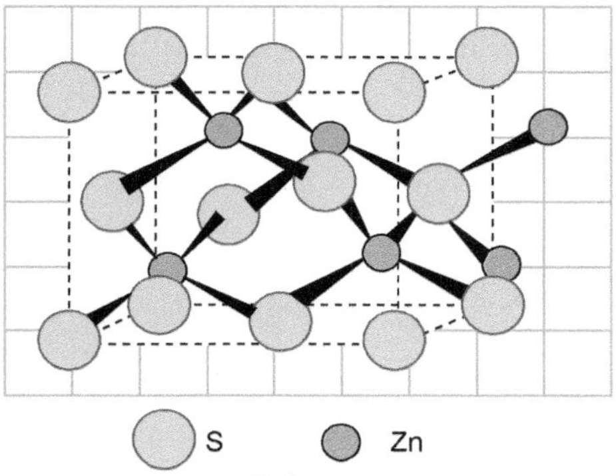

and similar to BeS.

TABLE 3.3 The Limiting Radius Ratios and Structures

		limiting Radius Ratio	
Maximum C.N.	Geometry	r^+/r^-	r^-/r^+
2	Linear	<0.155	>6.45
3	Planar triangle	0.155	6.45
4	Tetrahedral	0.225	4.44
4	Square planar	0.414	2.42
6	Octahedral	0.414	2.42
8	Cubic	0.732	1.37
12[a]	Dodecahedral	1	1

C.N.: number of adjacent atoms.
[a]*C.N. 12 is not found in simple ionic crystal, it occurs in metal oxides, and in closest-packed lattices of atoms.*

- NaCl:
  $r^+/r^- = 95/181 = 0.52$
  $0.414 < 0.52 < 0.732$
  Six anions surround one cation, the geometry O_h:

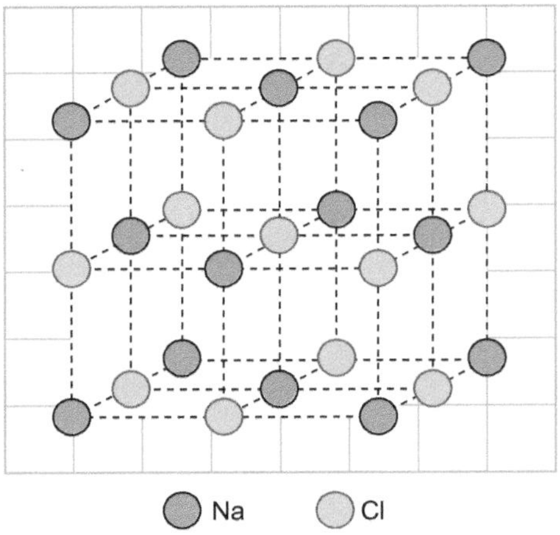

  Na^+Cl^-: C.N. for Na^+ is 6, so 1:1 stoichiometry gives 6:6 lattice (FCC).
  Cl^- atoms are octahedrally arrayed about Na^+.
- SrF_2:
  In 1:1 or 2:2 salts, the appropriate radius ratio is that of the smaller ion to the larger to determine how many of the latter will fit around the smaller ion.
  $r^+/r^- = 113/136 = 0.83$ $Sr^{2+}_{C.N.} = 8$
  $r^-/r^+ = 1.2$ $F^-_{C.N.} = 8$
  However, $Sr^{2+}:F^- = 1:2$
  Therefore, $Sr^{+2}_{C.N.} = 8$, and $F^-_{C.N.} = 4$, and similar to Cs^+Cl^- and CaF_2:

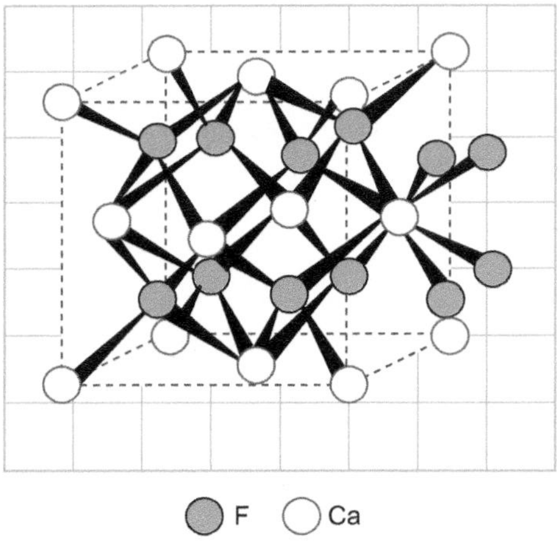

- Rb_2O:
  $r^+/r^- = 148/140 = 1.06$ maximum $Rb^+_{C.N.} = 8$
  $r^-/r^+ = 0.95$ maximum $O^{-2}_{C.N.} = 8$
  However, $Rb^+:O^{-2} = 2:1$.
  Therefore, $Rb^+_{C.N.} = 4$, and $O^{-2}_{C.N.} = 8$ and similar to CaF_2
- SnO_2:
  $r^+/r^- = 69/140 = 0.49$ maximum $Sn^{+4}_{C.N.} = 6$
  $r^-/r^+ = 2.03$ maximum $O^{-2}_{C.N.} = 6$
  However, $Sn^{+4}:O^{-2} = 1:2$.

Therefore, $Sn^{+4}_{C.N.} = 6$, and $O^{-2}_{C.N.} = 3$, and similar to TiO_2:

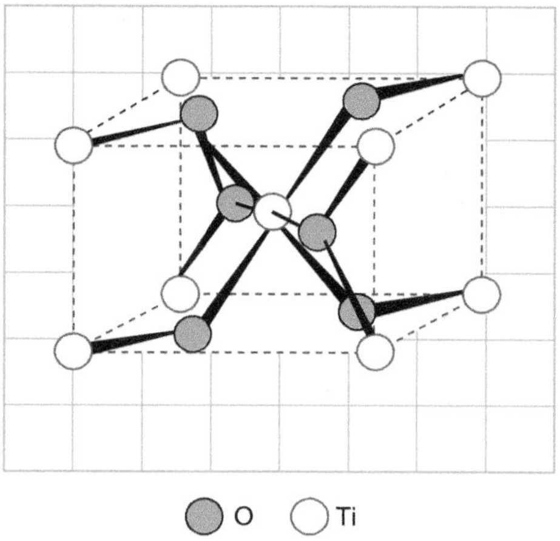

In M_2O, M = alkali metals, find the possible number of oxygen anions that can be fit around the central alkali ion.

	O^{-2}	Li^+	Na^+	K^+	Rb^+	Cs^+
r(Å):	1.20	0.6	0.95	1.33	1.48	1.74

- The radii ratio of the ions increases down the group:

	O^{-2}	Li^+	Na^+	K^+	Rb^+	Cs^+
r(Å)	1.20	0.6	0.95	1.33	1.48	1.74
r^+/r^-:		0.5	0.79	1.11	1.23	1.45
C.N.:		6	8	10	12	

Write a general formula to calculate the number of atoms surrounding a central atom.

- The maximum surrounding number (N) can be calculated using

$$N = \frac{2\pi}{\sqrt{3}}\left(\frac{d}{r}\right)^2\left(\frac{1}{1-\frac{r^2}{8d^2}}\right) \tag{3.6.15}$$

where d is the distance between two different atoms and r is the van der Waals radius of the surrounding atom.

Stoichiometric and Nonstoichiometric Defects

What are the types of possible defects in crystalline solids?

- Stoichiometric defects leave the stoichiometry unaffected.
  - In a Schottky defect, there is equal number of cation and anion vacancies (Fig. 3.12), e.g., NaCl and $CaCl_2$.
  - In a Frenkel defect, ion leaves its proper site vacant and occupies a vacant interstitial site (Fig. 3.12), e.g., AgBr.
- Nonstoichiometric defects:
  - These defects mostly occur in transition metal compounds, where the metal ion has the ability to exist in more than one oxidation state. Some sites of low oxidation become vacant, whereas others are sufficient to maintain electro-neutrality (Fig. 3.13), e.g., $Fe_{0.95}O$, $Ti_{0.74}O$, and $Ti_{1.67}O$.
  - Nonstoichiometric defects also occur when the metal has one oxidation state. In this defect, free electrons occupy a number of the anion sites (Fig. 3.13), e.g., $Na_{1+\gamma}Cl$, where γ is the number of Cl^- atoms that are replaced by electrons.

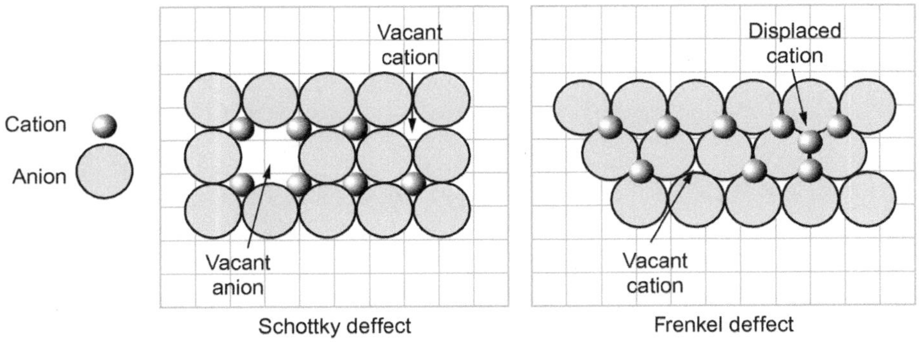

FIG. 3.12 Representation of Schottky and Frenkel defects. The defects are not restricted to one layer.

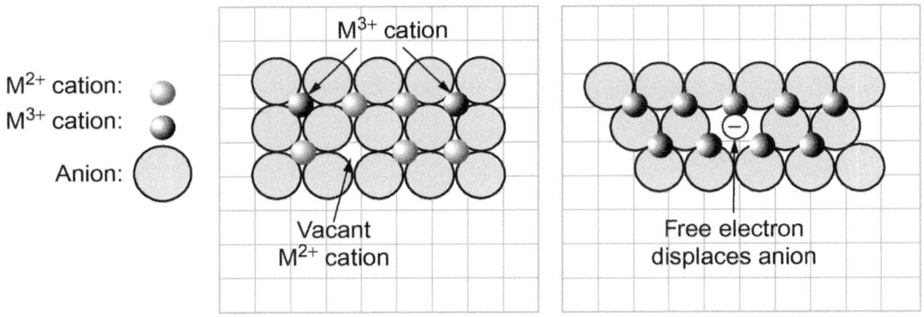

FIG. 3.13 Representation of nonstoichiometric defects.

Ionic Character and Covalency Interference

Explain the discrepancy between experimental and calculated lattice energies, $U_{exp.}$ and $U_{cal.}$, of NaBr and AgBr
Ag^+: $r^+ = 1.29$ Å, Na^+: $r^+ = 1.16$ Å, Br^-: $r^- = 1.82$ Å
NaBr: $r_e = 2.99$ Å, $U_{exp} = -728$ kJ/mol, $U_{cal} = -719$ kJ/mol
AgBr: $r_e = 3.11$ Å, $U_{exp.} = -907$ kJ/mol, $U_{cal.} = -723$ kJ/mol

- NaBr: $r_e = r^+ + r^- = 1.16 + 1.82 = 2.98$ Å:
  $U_{exp.} = -728$ kJ/mol, $U_{cal.} = -719$ kJ/mol
  The discrepancy between $U_{exp.}$ and $U_{cal.}$ is only about 1%
  Na^+: $r^+ = 1.16$ Å
  Br^-: $r^- = 1.82$ Å
  $\frac{r^-}{r^+} = 1.57$; octahedral (range: 1.37–2.42)
- AgBr: $r_e = 3.11$ Å:
  $r_e = r^+ + r^- = 1.29 + 1.82 = 3.11$ Å
  $U_{exp.} = -907$ kJ/mol, $U_{cal.} = -723$ kJ/mol
  Ag^+: $r^+ = 1.29$ Å
  Br^-: $r^- = 1.82$ Å
  $\frac{r^-}{r^+} = 1.41$; octahedral (range: 1.37–2.42)

The discrepancy between $U_{exp.}$ and $U_{cal.}$ is about 20%.

- This difference arises because of appreciably more covalency in AgBr, which increases U over ionic-only model.
- The lattice energy calculated using Born treatment disagrees with the experimental value from thermodynamic data due to:
  - ions being considered as hard spheres undistorting by neighboring ions; and
  - increasing the distortion corresponds to increasing covalent character.

- Note:
  - Covalency reduces the coordination number from 100% ionic model.
  - As the charge-to-radius ratio of a cation increases, its polarizing power increases.
  - As the charge and size of an anion increase, it becomes more polarizable (Fig. 3.14).

Ionic Character and Melting Point

Explain the difference in the melting points of:

a. **$MgBr_2$: mp = 700°C and $AlBr_3$: mp = 98°C; and**
b. **$CaCl_2$: mp = 772°C and $HgCl_2$: mp = 276°C.**

- As the charge-to-radius ratio of a cation increases, its polarizing power increases.
  a. The charge-to-radius ratio:

$$\left(\frac{q^+}{r^+}\right)_{Al^{3+}} > \left(\frac{q^+}{r^+}\right)_{Mg^{2+}}$$

  The polarizing power of Al^{3+} is more than Mg^{2+}, and Br^- is easy polarizable.
  $AlBr_3$ is expected to have less ionic character, and a lower melting point.
- b. The charge-to-radius ratio:

$$\left(\frac{q^+}{r^+}\right)_{Ca^{2+}} > \left(\frac{q^+}{r^+}\right)_{Hg^{2+}}$$

  $CaCl_2$ is expected to have a more ionic character, and a higher melting point.
  Both Ca^{2+} and Hg^{2+} are attached to a Cl^- anion that is hard to polarize.

Solubility of the Ionic Salts

What factors control the salt solubility?
Why are ionic substances slightly soluble in most common solvents, except for those that are extremely polar?
The dielectric constant of water is 78.5 (25°C). How does this affect the attractive forces between ions?
Explain why the solubility of AgCl, $PbCl_2$, Hg_2Cl_2, and $HgCl_2$ are low in water, but more soluble in ammonia.

- The factors that control the salt solubility are:
  1. The lattice energy:
     - The attractive forces between ions of the dissolved crystal must be dissociated.
     - This can be done only if the energy needed for the separation of the ions from the crystal is at least equivalent to the lattice energy.
  2. Dipole moment:
     - The energy required to separate ions comes from the solvation. The solvation energy increases when solvent molecules have a high dipole moment.
     - The forces between nonbonded neutral molecules (solvent) are weak unless the solvent molecule has a high dipole moment.
     - The larger the dipole moment, the stronger the attraction by ion, and the greater the solvation energy becomes.

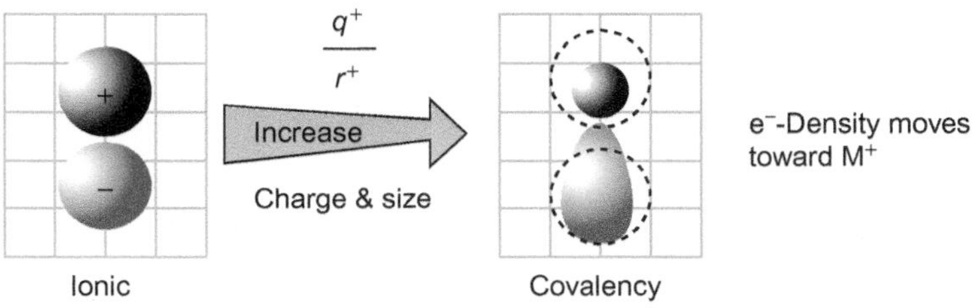

FIG. 3.14 Anion polarization.

3. Dielectric moment:
   The energy needed to separate ions is reduced by a decrease in the forces between ions, which are dependent on the dielectric constant of the medium.
   - Example: the dielectric constant of water is 78.5 (at 25°C); the attractive forces between ions in water is 1/78.5 of the force between the ions separated by the same distance in a vacuum.
   - Good solvents for ionic substances usually have high dipole moment and high dielectric constants, e.g., H_2O.
4. Heat of solution, L:
   - It is the energy change involved in dissolution of salt (Scheme 3.3):
     The energy change, L, from MX (s) to the solvated ions is state function and independent of the path.

$$L = \Delta H_{M^+} + \Delta H_{X^-} + U$$

   For soluble substances, the heat of solution is always negative (exothermic).
5. Entropy change that accompanies the dissolution:
   - Dissolution occurs when ΔG^o is negative:

$$\Delta G^o = \Delta H^o - T\Delta S^o$$

$$\Delta G^o = -RT \ln K_{sp.}$$

$$RT \ln K_{sp.} = -L + T\Delta S^o$$

where $K_{sp.}$ is the solubility product, ΔH^o is the heat of solution, $\Delta H^o = L$, and ΔS^o is the change in the entropy when one mole of solute dissolves to give one molar solution.
   - Example: the dissolution of NH_4Cl in H_2O is endothermic. The entropy change that associates the destruction of the ordered crystal lattice is account for the high solubility.
6. Charge density, z/r.
   - Size effect:
     - The entropy change associating the dissolving of a salt is usually less favorable for small ions, mainly for ions of high charge. Solvation energy and lattice energies are high for small ions.
     - The solubility of salts enhance as the size of the cations or anions increases, because of more favorable entropy changes on dissolution.
     - Salts are likely to be soluble if the size of the cation and anions differ substantially in size.
   - Charge effect:
     - Small ions of high charge density cause a great deal of ordering of solvent molecules, because of solvation.
     - Solubility of salts are high for low charges ions, e.g., salts of alkali metal and salts of NO_3^-, CH_3COO^-, ClO_4^-, and halides.
   - Size and charge effect:
     - Elements form soluble hydrated cations such as Cs^+(aq.) and Ca^{2+} have $Z/r < 0.03$ pm^{-1}.
     - Elements form soluble salts of oxoanions (e.g., SO_4^{2-} and PO_4^{3-}) have $Z/r > 0.12$ pm^{-1}.
     - Elements form insoluble oxides and hydroxides, but are soluble as halides and acetates have: $0.03 < Z/r < 0.12$ pm^{-1}

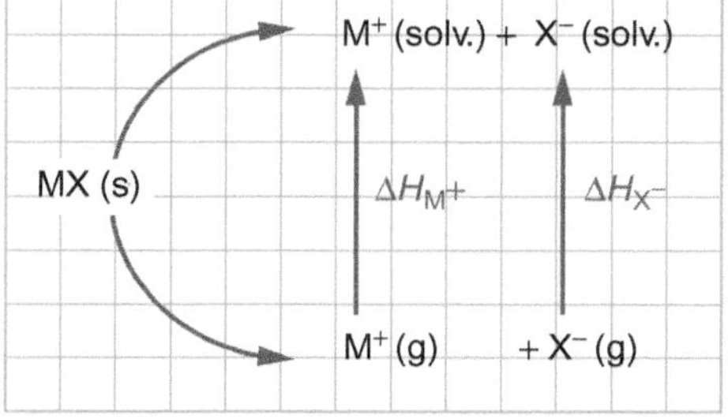

SCHEME 3.3 Energy change during dissolution, where U is the lattice energy of MX (s), ΔH_{M^+} and ΔH_{X^-} are the energies released as the result of the solvation of the gaseous positive and negative ions.

7. Polarization effect:
   - Solvation energies also enhance with the polarizability of the solvent molecules. Solubility depends on the relative polarizability of the salt anion and the solvent molecule.
   - If the cation is strong polarizing and the anion is easily polarizable, the solubility of the salt is low unless the solvent molecule are easy polarizable.
   - Example: the solubility of $AgCl$, $PbCl_2$, Hg_2Cl_2, and $HgCl_2$ are low in water, but more soluble in ammonia.
     - Solubility of salts containing 18- or (18+2)-electron cations is low in water because of the greater anion polarization than alkali and alkaline earth halides.
     - Water molecules have a higher dipole moment than ammonia, but are less polarizable, so water is a better solvent for ionic salts. However, ammonia is a better solvent than water for salts of strongly polarizing cations and salts of easily polarized anions.
8. Solvent interactions:
   - Solvent molecules may have donating electrons pairs (Lewis base) to the metal ions (Lewis acid).
     Example: a metal ion interacts strongly with a layer of water molecules to form the primary hydration sphere.
   - Hydrogen bonding examples:
     The primary hydration spheres form stronger hydrogen bonds to other H_2O molecules, and a secondary hydration sphere arises.
     The high hydration energy of F^- that results from the formation of strong hydrogen bonds contributes strongly to the great reactivity of F_2 to form solutions of metal fluoride.

3.7 COVALENT BONDING

How do covalent bonds form?

- Covalent bonding: electrons are shared and not transferred.
- A single covalent bond consists of a pair of electrons shared by two atoms.

The Lewis Structures and Octet Rule

Describe the octet rule and give examples. What are the exceptions of that rule?

- A stable arrangement is achieved when eight electrons surround the atom.

Examples: Fig. 3.15.

Exceptions to the Octet Rule

- Hydrogen is stable with only two electrons.
- Atoms such Be and B that have less than four outer electrons, BeF_2 and BF_3.

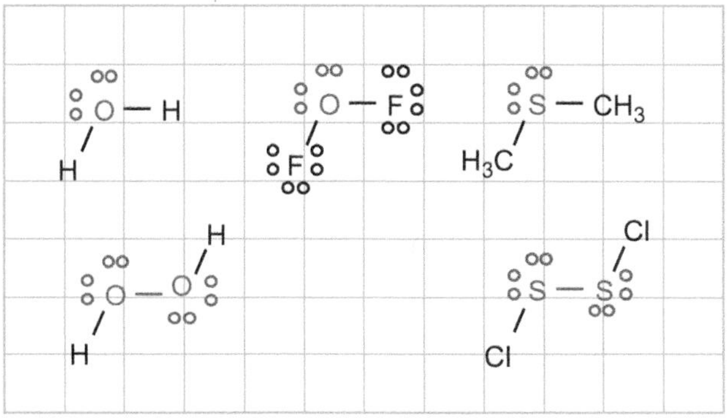

FIG. 3.15 Octet electron arrangements in some elected examples.

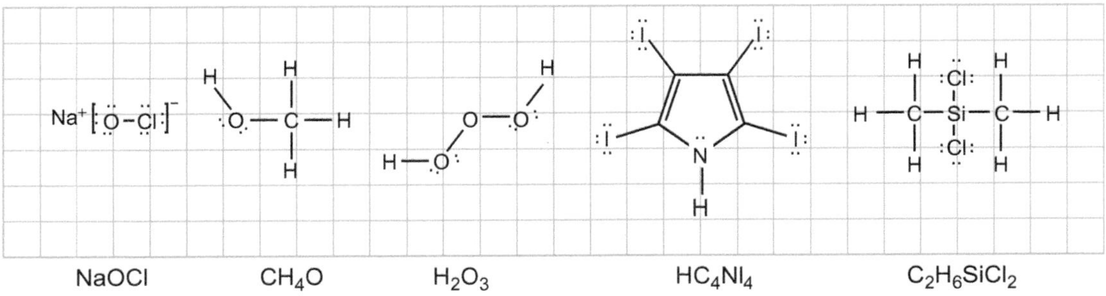

FIG. 3.16 Lewis electron dot structure of NaOCl, CH_4O, H_2O_3, HC_4NI, and $C_2H_6SiCl_2$.

- When atoms have an extra energy level, which is close in energy to the p level, may accept electrons and be used for bonding.
- In molecules, such as NO and ClO_2, which have an odd number of electrons.
- This does not explain why O_2 has two unpaired electrons and is paramagnetic.

What are the Lewis electron dot structures of the following?

i. **NaOCl**
ii. **CH_4O**
iii. **H_2O_3**
iv. **HC_4NI_4**
v. **$C_2H_6SiCl_2$**

Fig. 3.16 represents the above Lewis dot structures.

Bonding and Polarity

Define the nature of the chemical bond in LiNa, Li_4C, Li_2O, Na_2O, CO, CO_2, O_2, and Na, where $\chi_{Li}=0.97$, $\chi_{Na}=1.01$, $\chi_C=2.5$, $\chi_O=3.5$.

- This is essentially based on differences in electronegativities, $\Delta\chi$.
  - Ionic compound: when $\Delta\chi$ is large.
  - Metallic lattice: when $\Delta\chi$ is small and χs are low.
  - Covalent compound: when $\Delta\chi$ is small and χs are high.

LiNa	$\Delta\chi=0.04$	Alloy (metallic)
Li_4C	$\Delta\chi=1.53$	Ionic solid
Li_2O	$\Delta\chi=2.53$	Ionic solid
Na_2O	$\Delta\chi=2.49$	Ionic solid
CO	$\Delta\chi=1.00$	Covalent molecule
CO_2	$\Delta\chi=1.00$	Covalent molecule
O_2	$\Delta\chi=0.00$	Covalent molecule
Na	$\Delta\chi=0.00$	Metallic lattice

Predict the bonding type of the following diatomic molecules based on the bond polarity:

i. **F-F, H-F or Li-F;**
ii. **B-Cl or C-Cl; and**
iii. **P-F or P-Cl.**

$\chi_F=4, \chi_H=2.1, \chi_{Li}=1, \chi_{Cl}=3, \chi_B=2, \chi_C=2.5, \chi_P=2.1$

- The bonding types are listed in Table 3.4.
- The greater the difference in electronegativity between two atoms, the more polar their bond becomes.

TABLE 3.4 Bonding Types Based on the Electronegativities

i. Compound:	F-F	$H^{\delta+}$-$F^{\delta-}$	Li-F
Electronegativity difference:	4−4=0	4−2.1=1.9	4−1=3
Type of bond	Nonpolar covalent	Polar covalent	Ionic
ii Compound:	$B^{\delta+}$-$Cl^{\delta-}$	$C^{\delta+}$-$Cl^{\delta-}$	
Electronegativity difference:	3−2=1	3−2.5=0.5	
Type of bond	More polar covalent	Less polar covalent	
iii. Compound:	$P^{\delta+}$-$F^{\delta-}$	$P^{\delta+}$-$Cl^{\delta-}$	
Electronegativity difference:	4−2.1=1.9	4−3=1	
Type of bond	More polar covalent	Less polar covalent	

3.8 COORDINATE COVALENT BOND (DATIVE BONDING)

How do the coordination bonds form? Give examples.

A coordinate covalent bond (dative bonding) is formed by sharing an electron pair from atom A with atom B to form A → B.

Examples:

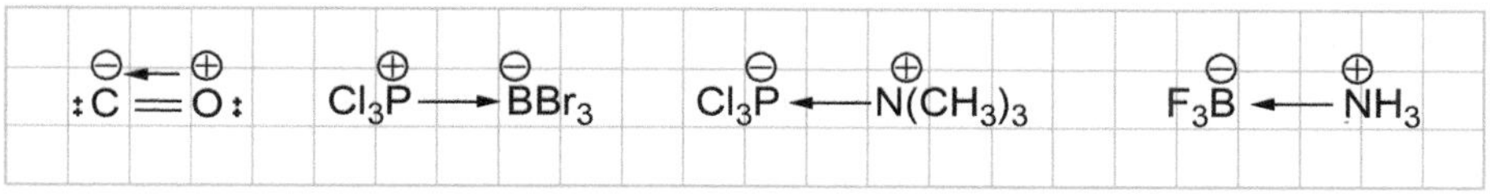

More examples of possible structures are given in Fig. 3.17:

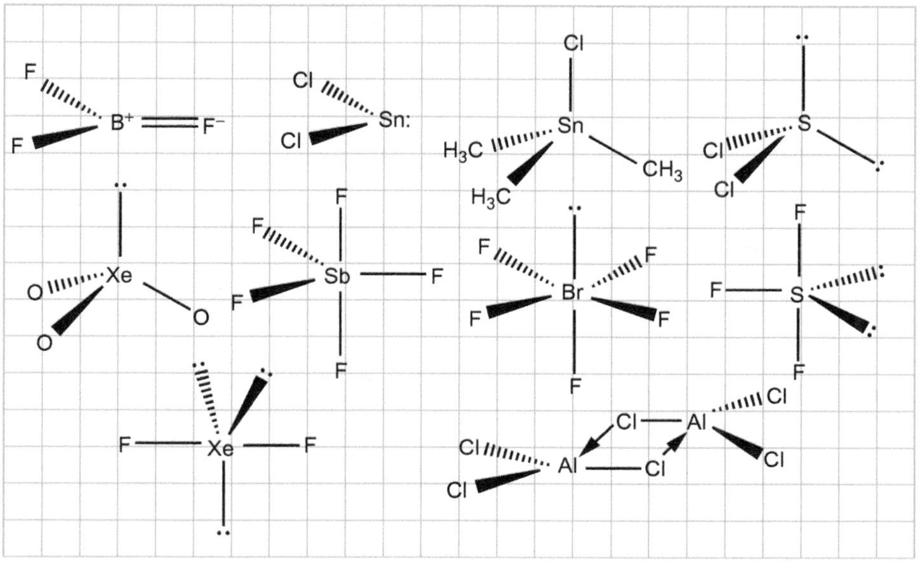

FIG. 3.17 Structures and bonding of some examples of coordination compounds.

Coordination Number and the "18-Electron Rule"

What formula would you consider for nickel carbonyl?

- The ground state electronic configuration of nickel is as follows:
  Ni: [Ar] $4s^2 3d^8$
  CO is neutral Lewis base that contributes two electrons, and nickel carbonyl is neutral.
  There are ten electrons in the outer 4s and 3d orbital of nickel atom.
  Eight more electrons are needed to satisfied the "18-electron rule."
  Nickel carbonyl: $Ni(CO)_4$
- The 18-electron rule is: "stable organometallic compounds of transition metals will have a total 18 valence electrons about the metal. They will have the 'effective atomic number, EAN: $(n-1)d^{10}$ ns^2 np^6' of the next higher inert gas."
- Exception: "molecules having only 16 valence electrons can often be just as stable as, or even more than, 18-electron molecules of the same metal."

Predict the formula expected for cobalt hydride carbonyl complex, $HCo(CO)_n$.

- Hydride: H^- Cobalt: Co^+
  The ground state electronic configuration of Co^+: $[Ar]3d^8$ contributes eight electrons to the required 18 electrons.
  CO: neutral Lewis base that contributes two electrons; there must be four CO to satisfy the 18-electron rule.
  Cobalt hydride carbonyl complex: $HCo(CO)_4$

Ligand Denticity

Give examples of unidentate, bidentate, and polydentate ligands.

- Uni- or monodentate ligands: occupy one coordination position in coordination sphere, and are "single-toothed."

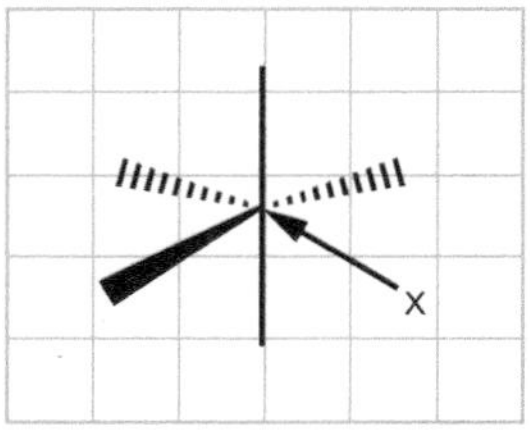

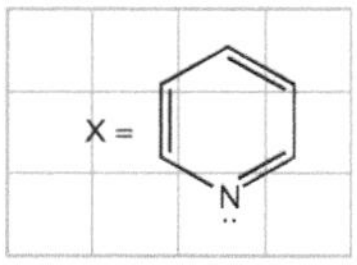

Pyridine

Examples:
H^-: hydrido, OH^-: hydroxo, O^{2-}: oxo, S^{2-}: thio, I^-: iodo, Br^-: bromo, Cl^-: chloro, F^-: fluoro, N_3^-: azido, NH_2^-: amido, $\ddot{N}O_2^-$: nitro, $\ddot{O}_2N^-$: nitrito, NO_3^-: nitrato, $\ddot{C}N^-$: cyano, $\ddot{N}CO^-$: cyanato-N, $\ddot{N}CS^-$: thiocyanato-N, $\ddot{S}CN^-$: thiocyanato-S, H_2O: aqua, NH_3: ammino, CO_3^{2-}: carbonato, and SO_4^{2-}: sulfato
- Bidentate (didentate) ligands: occupy two (adjacent) positions in coordination sphere; examples are given in Fig. 3.18.
- Ligands with more than two ligating atom are called tri, tetra-, etc., or polydentate ligands (Fig. 3.19).

Nomenclature of Complexes

Give named for the following binary compounds:

i. **NaCl, Li_3N, and K_2Te;**
ii. **N_2O, N_2O_4, $FeCl_2$, $FeBr_3$, SnO_2, Fe_3O_4;**
iii. **OH_2, NH_3, N_2H_4, B_2H_6, SiH_4, GeH_4, PH_3, AsH_3, SbH_3;**
iv. **MO^{n+}, VO^{2+}, NO^+, NO_2^+, PH_4^+; and**
v. **OH^-, O_2^-, O_3^-, N_3^-, CN^-, HS^-.**

- i. For binary compounds: name the more electropositive element first, then the more electronegative one with "—ide" as a suffix.
  NaCl: sodium chloride
  Li_3N: lithium nitride
  K_2Te: potassium telluride

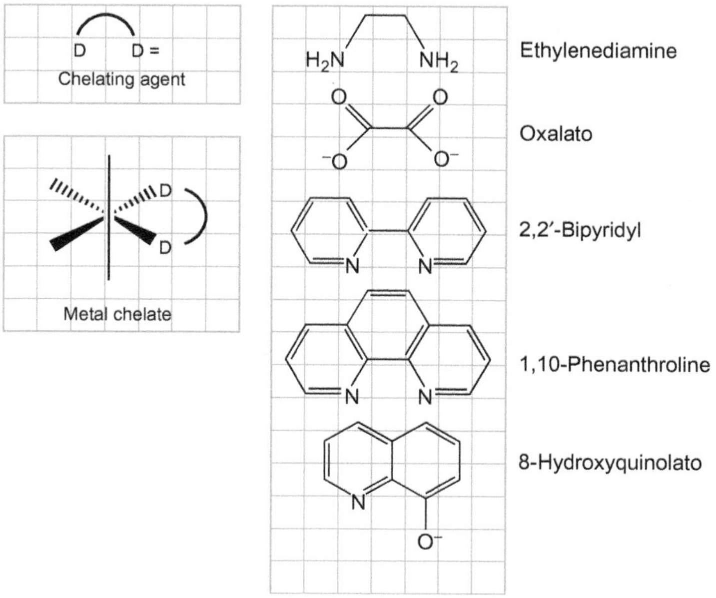

FIG. 3.18 Examples of bidentate ligands.

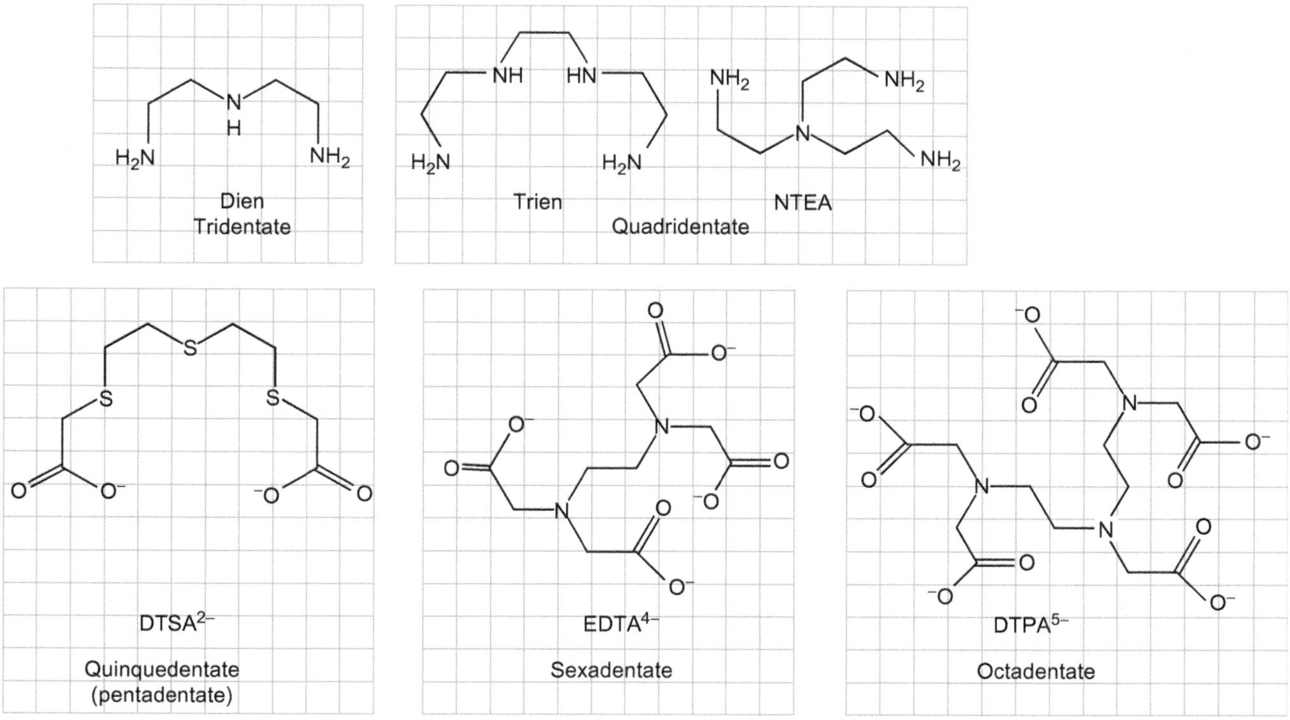

FIG. 3.19 Examples of polydentate ligands.

- ii. Ambiguity associated with different numbers of atoms is solved by using a number prefix or an oxidation state indicator: mono, bi (di), tri, tetra, penta, hexa, octa, ennea, deca, hendeca (undeca), dodeca, etc.

  N_2O: dinitrogen oxide

  N_2O_4: dinitrogentetraoxide

  $FeCl_2$: iron(II) chloride (ferrous)

$FeBr_3$: iron(III) bromide (ferric)
SnO_2: tin(IV) oxide
Fe_3O_4: triirontetraoxide

- iii. Hydrides are known by common names:
  OH_2: water
  NH_3: ammonia (or azane)
  N_2H_4: hydrazine(diazane or tetrahydridodinitrogen)
  B_2H_6: diborane (boron trihydride, boron(III) hydride, or boron hydride)
  SiH_4: silane
  GeH_4: germane (germanium tetrahydride)
  PH_3: phosphine (phosphane)
  AsH_3: arsine
  SbH_3: stibine
- iv. MO^{n+}: metalyl
  VO^{2+}: vanadyl (oxovanadium)
  NO^+: nitrosonium
  $NO_2{}^+$: nitryl (nitronium)
  $PH_4{}^+$: phosphonium
- Special anions have the suffix ....ide:
  OH^-: hydroxide
  $O_2{}^-$: superoxide
  $O_3{}^-$: ozonide
  $N_3{}^-$: azide
  CN^-: cyanide
  HS^-: hydrogen sulfide
  Others: named as complex anions or with an "... ate" suffix

Complex Formation

How are metal complexes named?

- Formula has central atom first: $[Co(NH_3)_6]^{3+}$.
- List the ligands alphabetically by ligand names.
- Neutral ligand name exceptions:
  H_2O: aqua NO: nitrosyl
  NH_3: ammine CO: carbonyl
  See that anionic ligand names end in "...o;" this is normally ...ido, ...ito, or ...ato.
  Elementide: elementido
  Nitride: nitrido
  Selenide: selenido
- Exceptions:
  H^-: hydrido- OH^-: hydroxo-
  Halide: halo- CN^-: cyano-
  $O_2{}^{2-}$: oxo- $C_6H_5{}^-$: phenyl-
  S^{2-}: thio- $C_5H_5{}^-$: cyclopentadienyl-
  Anions: most retain traditional names:
  $NO_2{}^-$: nitride $SO_2{}^-$: sulfite
  $S_2O_4{}^{2-}$: dithionite $S_2O_3{}^{2-}$: thiosulfate
- Use Greek prefixes for many of each ligand:

2 syllables ligand	3 or more syllables
bi...	bis...
tri...	tris...
tetra...	tetrakis...
penta...	pentakis...
hexa...	hexakis...

hepta... heptakis...
octa... octakis...

- Write as a continuous string, no spaces.
- The metal name and oxidation state are last:
  $[Co(NH_3)_6]^{3+}$: hexaamminecobalt(III)
  $[Ni(OH_2)_3(NH_3)_3](NO_3)_2{\cdot}4H_2O$: triamminetriaquanickel(II) nitratetetrahydrate
- You may need to indicate isomerism:
  $[Cr(NH_2\text{-}CH_2\text{-}CH_2\text{-}NH_2)_2Br_2](CH_3CO_2)$: *trans-* or *cis*-dibromobis(ethylenediamine) chromium(III) acetate
- Anions/Anionic complexes: replace "...um/...ium" of the central atom by "...ate," using Latin symbol name:
  $K_3[Rh(CN)_6]$: potassium hexacyanorhodate(III)
- Exceptions:

Antimonite	(for stibate)
Mercurate	(for hydrargyrate)
Tungstate	(for wolframate)
Potassiate	(for koliate)
Sodiate	(for natriate)

- Complex salts:
  $[Ag(NH_3)_2][W(phen)(SCH_3)_3(CN)_2]$: biamminesilver(I) dicyanotris(methane-thiolato) *o*-phenanthrolinetungustate(IV)
- Double salts:

$KMgF_3$:	magnesium potassium fluoride
$K_3Co(CN)_5$:	potassium cobalt(II) cyanide
$KAl(SO_4)_2{\cdot}12H_2O$:	potassium aluminum sulfate dodecahydrate

- Linkage isomers:

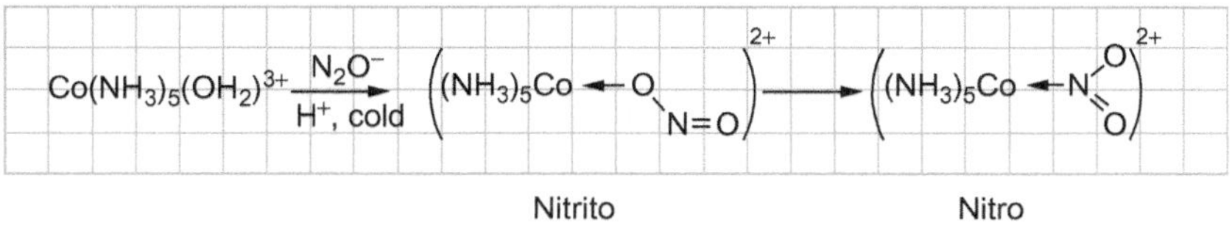

$M \leftarrow SCN$: thiocyanate – $M \leftarrow NCS$: isothiocyanate
Common ambidentate ligands:
NCO^-, NCS^-, $NCSe^-$, and N_2O^-

- Bridged centers: each bridge is denoted "μ"

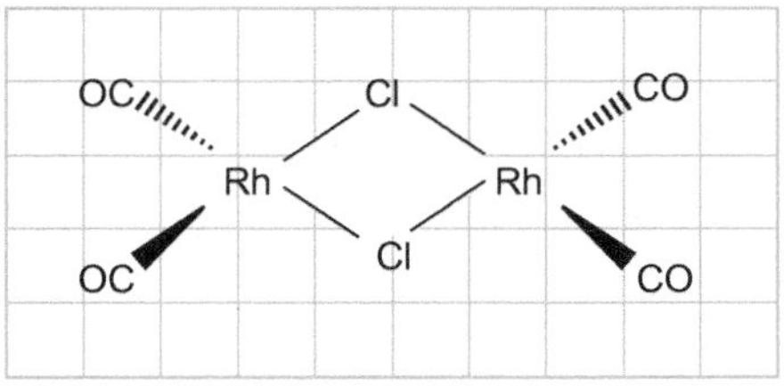

Bis(μ-chloride)bis(dicarbonyl-rhodium(I))

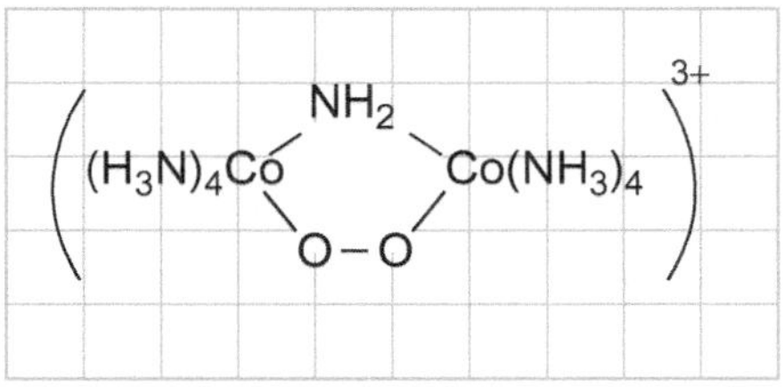

μ-Smido-μ-peroxobis(tetraammine)cobalt(II)

$[(H_3N)_5CO(NH_2)Co(NH_3)_5)]^{5+}$: μ – amidobis(pentaamminecobalt(III)

What are the proper names of the following?

i. $\mathbf{HCo(CO)_4}$
ii. $\mathbf{Ir(CO)(PPh_3)Cl}$
iii. $\mathbf{K_2[K(CN)_6]}$

- Answers:
  i. $HCo(CO)_4$: tetracarbonylhydridocobalt(I)
  ii. $Ir(CO)(PPh_3)_2Cl$: carbonylchloroditriphenylphosphineiridium(I)
  iii. $K_2[K(CN)_6]$: potassium hexacyanoferrate(II)

What is the chemical structure of 1-bromo-2,3,4-tricarbonyl-bis(triphenylphosphine)manganese(I)?

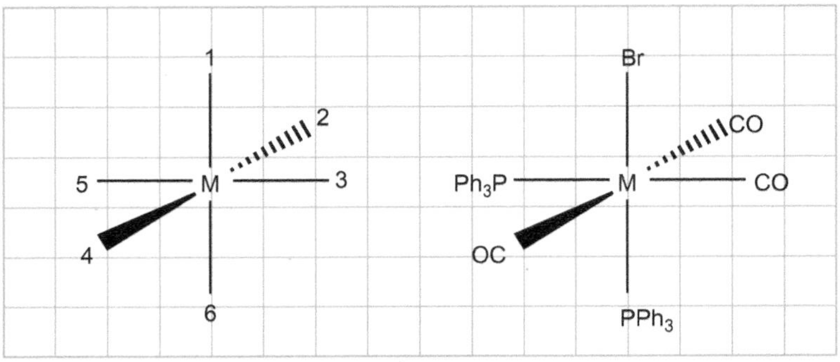

What are the chemical structures of the following?

i. **Hexasodium chloride fluoride bis(sulfate);**
ii. ***Cis*-bis(8-quinolinolato)silver(II);**
iii. **Potassium bis(dithiooxalato-*S,S′*)nickel(II);**
iv. **Potassium bis(dithiooxalato-*O,O′*)nickel(II);**
v. **[*N,N′*-bis(2-amino-κ*N*-ethyl)-1,2-ethanediamine-κ*N*]chloroplatinum(II);**
vi. **[*N*-(2-amino-κ*N*-ethyl)-*N′*-(2-aminoethyl)-1,2-ethanediamine-κ^2*N,N′*]chloroplatinum(II); and**
vii. **[2-(diphenylphosphino-κP)-phenyl-κC^1]hydrido-(triphenylphosphine-κP)nickel(II).**

- The answers:
  i. Hexasodium chloride fluoride bis(sulfate): $Na_6ClF(SO_4)_2$
  ii. *Cis*-bis(8-quinolinolato)silver(II)

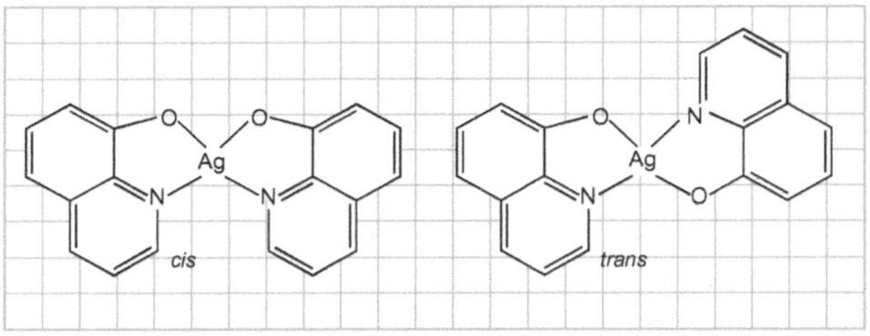

iii. Potassium bis(dithiooxalato-*S,S′*)nickel(II)

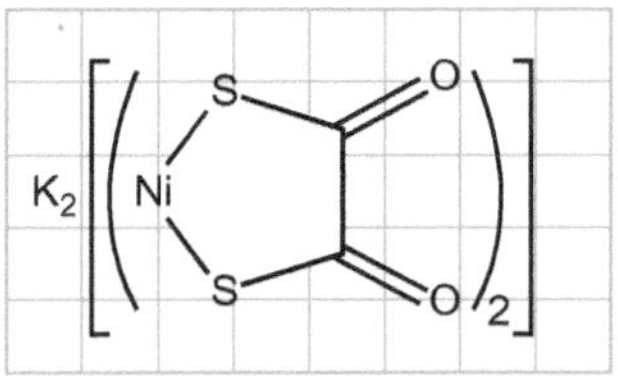

iv. Potassium bis(dithiooxalato-O,O′)nickel(II)

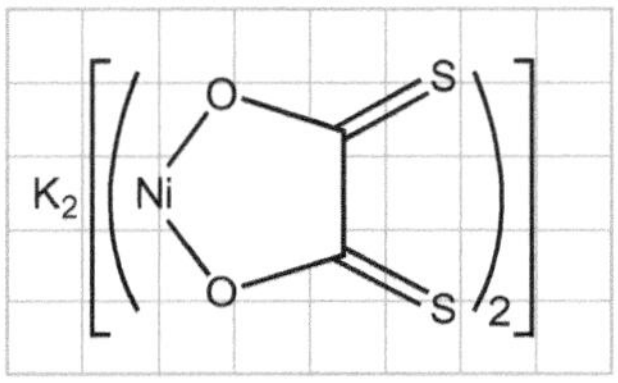

v. [*N,N*′-Bis(2-amino-κ*N*-ethyl)-1,2-ethanediamine-κ*N*]chloroplatinum(II)

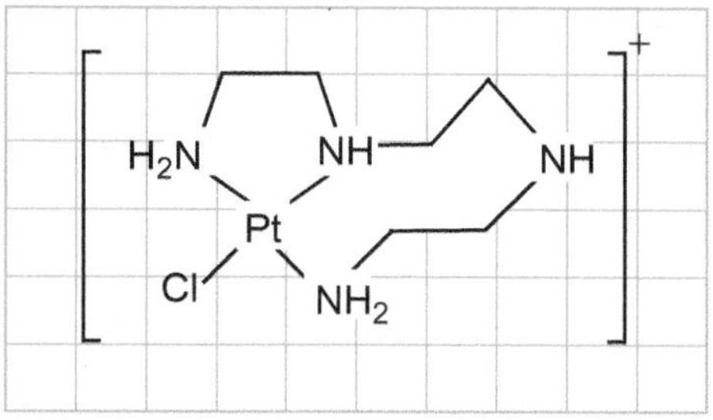

κ*N*: bonded through *N*.

vi. [*N*-(2-Amino-κ*N*-ethyl)-*N*′-(2-aminoethyl)-1,2-ethanediamine-κ^2N,N']chloroplatinum(II)

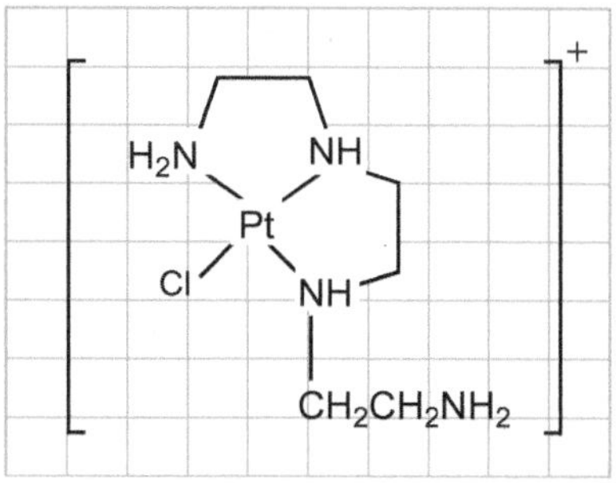

κ^2N,N': the right superscript to κ specifies the number of identically bound ligating atoms or ligands.

vii. [2-(Diphenylphosphino-κP)-phenyl-κC^1]hydrido-(triphenylphosphine-κP)nickel(II)

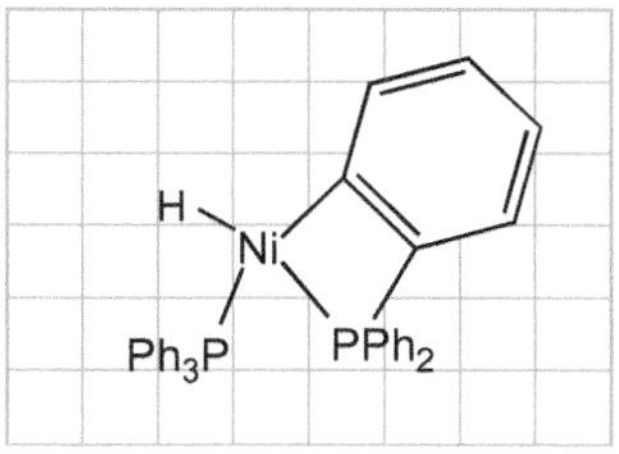

Coordinative Comproportionation Reaction

What are the expected products when Ni^+ interacts with NH_3 in H_2O?

- Complex formation in H_2O is a substitution process. H_2O is displaced, e.g.:

$$\text{"not}: Ni^{2+} + NH_3 \rightleftharpoons Ni(NH_3)^{2+}, \text{ but"}$$

$$Ni(OH_2)_6^{2+} + NH_3 \rightleftharpoons Ni(OH_2)_5(NH_3)^{2+}$$

Shorthand:

$$M + L \rightleftharpoons ML$$

$$K = K_f = \frac{[ML]}{[M][L]}$$

This is usually expressed as $pK_f = -\log_{10}K_f$

For the dissociation process,

$$K_d = K_i = \frac{1}{K_f}$$

- For Ni^{2+}/NH_3, we can get

$$Ni(OH_2)_5(NH_3)^{2+}, Ni(OH_2)_4(NH_3)_2^{2+}, Ni(OH_2)_3(NH_3)_3^{2+}, \ldots, Ni(NH_3)_6^{2+}$$

$$\text{Dynamic gains } NH_3 \leftarrow \underbrace{Ni(OH_2)_6^{2+}, \ldots, Ni(NH_3)_6^{2+}}_{\substack{Ni(OH_2)_3(NH_3)_3^{2+} \\ \text{Dominates}}} \rightarrow \text{Dynamic gains } H_2O$$

This is a comproportionation reaction, in fact:
$Ni(OH_2)_4(NH_3)_2^{2+}, Ni(OH_2)_2(NH_3)_4^{2+}$, and $Ni(OH_2)(NH_3)_5^{2+}$ are also present.
Speciation is the occurrence of a distribution of defined molecular species resulting from equilibriums involving instead of a single chemical species, e.g., Ni^{2+} in aqueous NH_3 solution.
Simple chemical analysis measure total nickel, regardless of chemical form: $Ni(OH_2)_6^{2+}, Ni(NH_3)_6^{2+}$ and all others in between.

- Some effective factors are:
  - i. charge effects;
  - ii. statistical effects; and
  - iii. chelate effects.
  - iv. d-electronic effects
  - v. nature of M^{n+}, donor atoms.

How can you explain the change in the complexation constant of the reaction

$$Cs^+ + Cl^- \rightleftharpoons Cs^+Cl^-_{(aq.)}$$

from $K = 0.3$ in H_2O to $K = 2000$ in dioxane at 25°C?

a. Metal cations generally have H_2O ligands in aqueous medium, $[Cs(H_2O)_n]^+$.
The association constant of $K_{Cs^+-H_2O} > K_{Cs^+-Dioxane}$; consequently, the replacement of H_2O by Cl^- is much more difficult than the replacement of dioxane.

b. Competition between H_2O and other Lewis bases, such dioxane and Cl^- for Cs^+, is pH dendence.

$$\text{Actually}: \quad \left.\begin{matrix} H^+, M^+\text{(acids)} \\ H_2O, OH^-, L, L^- \text{(bases)} \end{matrix}\right\} \begin{matrix} \text{pH} - \text{dependent} \\ \text{competition} \end{matrix}$$

The metal-ligand complexes formed at ultimate equilibrium are the result.

Complexation Equilibrium

If: $EDTA^{4-} + Cu^{2+} \rightleftharpoons [Cu(EDTA)]^{2-}$
$K_f = 7.25 \times 10^{18}$, $[Cu^{2+}_{Total}] = 5mM$, and $[EDTA^{4-}] = 0.1M$
What is the percentage of $[Cu(EDTA)]^{2-}$?

-

$$[Cu^{2+}_{Total}] = C_M = [M] + [ML]$$

$$[EDTA^{4-}] = C_L = [L] + [ML]$$

$$K = \frac{[ML]}{[M][L]}$$

$$[M] = C_M - [ML]$$

$$[L] = C_L - [ML]$$

$$K = \frac{[ML]}{(C_M - [ML])(C_L - [ML])} = 7.25 \times 10^{18}$$

$$[ML]^2 - (KC_M + KC_L + 1)[ML] + KC_MC_L = 0$$

Note: $M + nL \rightleftharpoons ML_n$, gives $(n+1)$th order equation.

If $C_L \gg C_M$, then $C_L \gg [ML]$, and $C_L \cong [L]$
(Analogously, if $C_M \gg C_L$)

$$K = \frac{[ML]}{[M](C_L)}, \text{ or}$$

$$K \cdot C_L = \frac{[ML]}{[M]}$$

$$\text{for } Cu^{2+}/EDTA^{4-}, \ \frac{[ML]}{[M]} = K \cdot C_L$$

$$\frac{[ML]}{[M]} = 7.25 \times 10^{18} \times 0.1 = 7.25 \times 10^{17}$$

$$[ML]\% = \frac{[ML]}{C_M} \times 100 = \frac{[ML]}{[M]+[ML]} \times 100 = \frac{7.25 \times 10^{17}}{1+7.25 \times 10^{17}} \times 100 \cong 99.9999\%$$

then

$$\left[Cu(EDTA)^{2-}\right] \cong 5\,mM$$

$$\left[Cu^{2+}\right] \cong 7 \times 10^{-21}\,mM$$

If: $2Gly^- + Zn^{2+} \rightleftharpoons Zn(Gly)_2$
$K_f = 9.1\times10^9$, $C_{Zn^{2+}} = 2\times10^{-6}$M, $C_{Gly} = 3\times10^{-6}$
Find $[ZnL_2]$, $[Zn^{2+}]$, and $[Gly^-]$.

- $2Gly^- + Zn^{2+} \rightleftharpoons Zn(Gly)_2$

	$2Gly^-$ +	Zn^{2+}	$\rightleftharpoons$	$Zn(Gly)_2$
Initial conc.	$2\,(3 \times 10^{-6})$	2×10^{-6}		0.0
Change	$-2x$	$-x$		$+x$
Equilibrium	$2\,(3 \times 10^{-6}) - 2x$	$(2 \times 10^{-6}) - x$		$+x$

-

$$[ZnL_2] = x$$

$$K = \frac{[ZnL_2]}{[Zn^{2+}][L]^2}$$

$$\frac{x}{\left(3 \times 10^{-6} - 2x\right)\left(2 \times 10^{-6} - x\right)} = 9.1 \times 10^9$$

$$x = 1.28 \times 10^{-7}$$

$$[ZnL_2] = 0.13 \times 10^{-6}\,M$$

$$\left[Zn^{2+}\right] = 1.87 \times 10^{-6}\,M$$

$$\left[Gly^-\right] = 2.74 \times 10^{-6}\,M$$

Multiligand Complexation

For the reaction:

$$\mathbf{Ni^{2+} + 6NH_3 \rightleftharpoons Ni(NH_3)_6^{\ 2+}}$$

Define the cumulative equilibrium constants and stepwise formation constants.

- $$Ni^{2+} + NH_3 \rightleftharpoons Ni(NH_3)^{2+}, \quad K = K_1$$
$$Ni(NH_3)^{2+} + NH_3 \rightleftharpoons Ni(NH_3)_2{}^{2+}, \quad K = K_2$$
$$Ni(NH_3)_2{}^{2+} + NH_3 \rightleftharpoons Ni(NH_3)_3{}^{2+}, \quad K = K_3$$
$$\ldots\ldots\ldots\ldots\ldots$$
$$Ni(NH_3)_5{}^{2+} + NH_3 \rightleftharpoons Ni(NH_3)_6{}^{2+}, \quad K = K_6$$
- The values of K_1, K_2, K_3, K_4, K_5, and K_6 are called stepwise formation constants or stepwise stability constants.
- Cumulative equilibrium constants: add the processes related to K_1 and K_2:

$$Ni^{2+} + 2NH_3 \rightleftharpoons Ni(NH_3)_2{}^{2+}, \quad K = K_1K_2 = \beta_2$$

K_1K_2 is symbolized as β_2

$$\beta_2 = \frac{\left[Ni(NH_3)_2{}^{2+}\right]}{\left[Ni^{2+}\right]\left[NH_3\right]^2}$$

The value of β_2 gives information about $Ni(NH_3)_2{}^{2+}$ non on $Ni(NH_3)^{2+}$.
- The values of β_2, β_3, β_4, β_5, and β_6 are called the overall formation constants or overall stability constants.

Stepwise Formation Constants and the Sequential Analysis

Why are there a steady decrease in the value of the stepwise formation constant K_i ($K_1 > K_2 > K_3 < \cdots < K_n$) as the number of ligands increase?

- There must be insignificant variations in metal-ligand bond energies as a function of i, which is usually the case.
- Reasons for the steady decrease in K_i values as the number of ligands increases are as follows:
  - Steric hindrance:
    If the ligands are bulkier than the replaced H_2O molecules, the steric hindrance increases as the number of ligands increase.
  - Statistical factors:
    For:

$$M(OH_2)_n^{2+} \overset{K_1}{\leftrightarrow} \ldots\ldots M(OH_2)_{n-i+1}^{2+}L_{i-1} \overset{K_i}{\leftrightarrow} M(OH_2)_{n-i}^{2+}L_i \overset{K_{i+1}}{\leftrightarrow} M(OH_2)_{n-i-1}^{2+}L_{i+1} \ldots\ldots \overset{K_n}{\leftrightarrow} M(L)_n^{2+}$$

where n is the coordination number.
The relative probability of passing from $M(OH_2)_{n-i+1}^{2+}L_{i-1}$ to $M(OH_2)_{n-i}^{2+}L_i$ is proportional to R_i:

$$R_i = \frac{n-i+1}{i}$$

The relative probability of passing from $M(OH_2)_{n-i}^{2+}L_i$ to $M(OH_2)_{n-i-1}^{2+}L_{i+1}$ is proportional to R_{i+1}:

$$R_{i+1} = \frac{n-i}{i+1}$$

On the basis of statistical considerations,

$$\frac{K_{i+1}}{K_i} = \frac{n-i}{i+1} \div \frac{n-i+1}{i} = \frac{i(n-i)}{(i+1)(n-i+1)}$$

For:

$$Ni(OH_2)_6^{2+} \rightleftharpoons \ldots\ldots\ldots Ni(OH_2)_{6-i}^{2+}L_i \ldots\ldots\ldots \rightleftharpoons Ni(L)_6^{2+}$$

$n = 6$

	K_2/K_1	K_3/K_2	K_4/K_3	K_5/K_4	K_6/K_5
Experimental	0.28	0.31	0.29	0.36	0.2
Statistical	0.417	0.533	0.562	0.533	0.417
	$K_2 < K_1$	$K_3 < K_2$	$K_4 < K_3$	$K_5 < K_4$	$K_6 < K_5$
	$K_1 >$	$K_2 >$	$K_3 >$	$K_4 >$	$K_5 > K_6$

Therefore, the dynamic nature of equilibrium means that successive Ks must be decreased in value for the L addition.

- Coulombic factors:
  Example:

$$Mn^{3+} + C_2O_4{}^{2-} \rightleftharpoons [Mn(C_2O_4)]^{+}, \quad \log K_1 = 9.98$$

$$[Mn(C_2O_4)]^{+} + C_2O_4{}^{2-} \rightleftharpoons [Mn(C_2O_4)_2]^{-}, \quad \log K_2 = 6.59$$

$$[Mn(C_2O_4)_2]^{-} + C_2O_4{}^{2-} \rightleftharpoons [Mn(C_2O_4)_2]^{3-}, \quad \log K_3 = 2.85$$

The steady decrease in the complexation constants, $\log K_1 > \log K_2 > \log K_3$, can be attribute to the following reasons:

i. the net charge of the complex becomes less positive; and
ii. inductive effect also present:

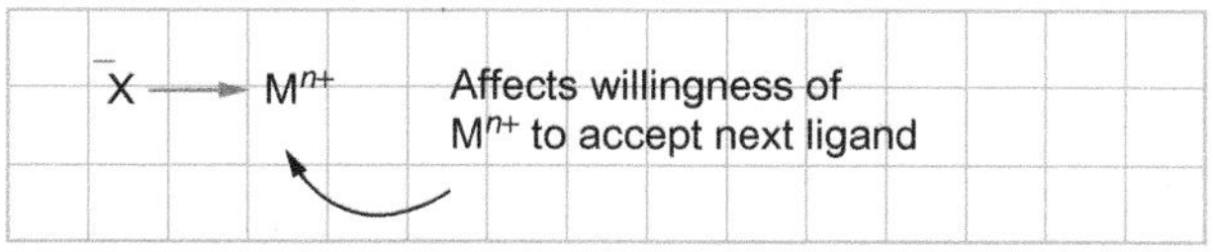

iii. anionic ligands generally bind better than neutral ones.

Complete and comment on the following:

$$\mathbf{M(OH_2)_6^{n+} + ...? \rightleftharpoons M(OH_2)_5(OH)^{(n-1)+} + H_2O, \quad K = K_{...?}}$$

$$\mathbf{M(OH_2)_6^{n+} \rightleftharpoons M(OH_2)_5(OH)^{(n-1)+} + ...?, \quad K = K_{...?}}$$

- $M(OH_2)_6^{n+} + OH^- \rightleftharpoons M(OH_2)_5(OH)^{(n-1)+} + H_2O$, $K = K_f$ (replacement of H_2O)
- $M(OH_2)_6^{n+} \rightleftharpoons M(OH_2)_5(OH)^{(n-1)+} + H^+$, $K = K_A$ (H^+-removal from H_2O-ligands), where $K_A = K_f \cdot K_W$
  - K_A is related strongly (but not solely) to ionic potential.

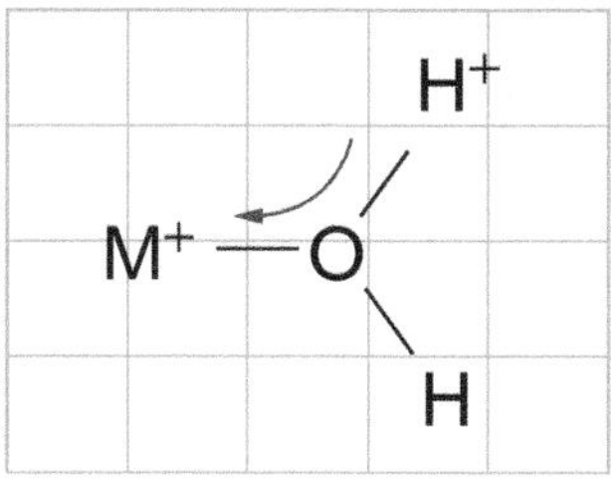

E.g.,

a. Reaction with pyrophosphoric acid:

	Mg^{2+}	Ca^{2+}	Sr^{2+}	Ba^{2+}
r (Å)	0.65<	0.94<	1.10<	1.29
$\log K_{P_2O_7{}^{4-}}$	7.2>	6.8>	5.4>	4.6

b. Solution containing Li^+, Mg^{2+}, Fe^{3+} and OH^-, get complexation in order $Fe^{3+} > Mg^{2+} > Li^+$, manifested by precipitation of $Fe(OH)_3$ first.
For the first step:

$$Fe(OH_2)_6{}^{3+} + OH^- \rightleftharpoons Fe(OH_2)_5(OH)^{2+} + H_2O$$

Replacements of H_2O or H^+-removal from H_2O-ligands for many cations are indistinguishable and they are certainly equivalent processes for all systems.

Complex Stability

What factors decide the complex stability?

- The stability of the complexes are controlled by:
  1. hard and soft interaction;
  2. chelate effect;
  3. macrocycle effect, cavity size;
  4. donor atom;
  5. steric effect;
  6. solvent competition; and
  7. metal oxidation state.

Hard and Soft Interactions, HSAB

Complete and comment on the following:

$$\ldots? + \underbrace{\mathbf{HF\,(aq.)}}_{pK_A=3.45} \rightleftharpoons \underbrace{\mathbf{NH_4^+\,(aq.)}}_{pK_A=9.25} + \ldots? \quad \log K_{eq.} = \ldots?$$

$$\ldots? + \underbrace{\mathbf{CaF^+\,(aq.)}}_{pK_A=0.51} \rightleftharpoons \underbrace{\mathbf{Ca(NH_3)^{2+}(aq.)}}_{pK_A=-0.2} + \ldots? \quad \log K_{displacement} = \ldots?$$

$$\ldots? + \underbrace{\mathbf{AgF\,(aq.)}}_{pK_A=0.36} \rightleftharpoons \underbrace{\mathbf{Ag(NH_3)^+(aq.)}}_{pK_A=3.32} + \ldots? \quad \log K_{displacement} = \ldots?$$

$$\mathbf{AgF + Ca(NH_3)^{2+} \rightleftharpoons \ldots? + \ldots? \quad K_{interchange} = \ldots?}$$

- A stronger base replaces a weaker base from its compounds:

$$\ddot{B}_{strong} + A\ddot{B}_{weak} \rightleftharpoons A\ddot{B}_{strong} + \ddot{B}_{weak}$$

where B is base and A is acid, then

$$NH_3(aq.) + \mathbf{HF(aq.)} \rightleftharpoons \mathbf{NH_4^+\,(aq.)} + F^-(aq.) \quad \log K_{eq.} = 9.25 - 3.45 = 5.80$$

$$NH_3(aq.) + \mathbf{CaF^+(aq.)} \rightleftharpoons \mathbf{Ca(NH_3)^{2+}(aq.)} + F^-(aq.) \quad \log K_{eq.} = -0.2 - 0.51 = -0.7$$

$$NH_3(aq.) + \mathbf{AgF\,(aq.)} \rightleftharpoons \mathbf{Ag(NH_3)^+(aq.)} + F^-(aq.) \quad \log K_{eq.} = 3.32 - 0.36 = 2.96$$

- Consider the following:
  Cations are electron acceptors.
  Ligands are electron donors.
  The calculated equilibrium constants indicate the following orders of acid strength in an aqueous solution:
  $Ca^{2+} < Ag^+ < H^+$ toward NH_3 (aqueous)
  $Ag^+ < Ca^{2+} < H^+$ toward F^- (aqueous)
- In order to explain the differences in relative acid strength toward various Lewis bases, the classification of hard and soft bases has been proposed.

- $K_{interchange}$ and the products for the reaction AgF and $Ca(NH_3)^{2+}$:

$$\begin{array}{lr} NH_3(aq.) + AgF(aq.) \rightleftharpoons Ag(NH_3)^+(aq.) + F^-(aq.) & K_{eq.} = 10^{2.96} \\ Ca(NH_3)^{2+}(aq.) + F^-(aq.) \rightleftharpoons NH_3(aq.) + CaF^+(aq.) & K_{eq.} = 10^{0.7} \\ \hline \mathbf{Ca(NH_3)^{2+} + AgF} \rightleftharpoons Ag(NH_3)^+ + CaF^+ & \mathbf{K_{interchange}} = 10^{2.96} \times 10^{0.7} = 10^{3.66} \end{array}$$

How can the Lewis concept of acid and base be used to explain complex formation?

- The coordination bond, dative bond, or donor-acceptor bond has a particular lone pair involved in the formation of a molecule from two others.
  - Lewis base: ē - donor.
  - Lewis acid: ē - acceptor.
  - Acceptor + Donor → Complex.

    E.g.,

    $NH_3 + H^+ \rightarrow NH_4^+$

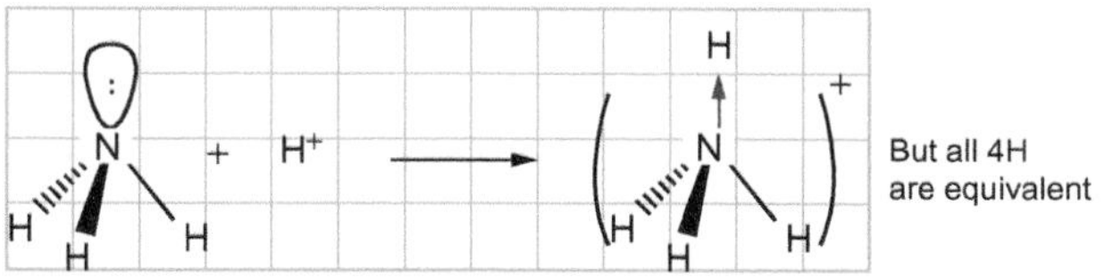

  - Coordination number = number of attached ligands.
  - Acceptor undergoes increase in the coordination number.
  - Each ligand has a donor atom (e.g., N, O, etc.).

Write down the electronic structure of the following, and assign which would be likely to act as Lewis acids and which are regarded as Lewis bases.

i. $\mathbf{PCl_3}$
ii. $\mathbf{AlBr_3}$
iii. $\mathbf{I^-}$
iv. $\mathbf{SO_3}$
v. $\mathbf{CN^-}$
vi. $\mathbf{I^+}$
vii. $\mathbf{SiCl_4}$
viii. $\mathbf{Ag^+}$
ix. $\mathbf{SCl_2}$

- Answer:

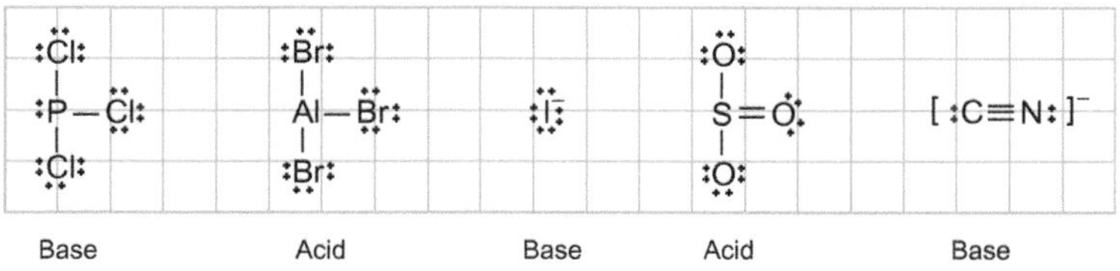

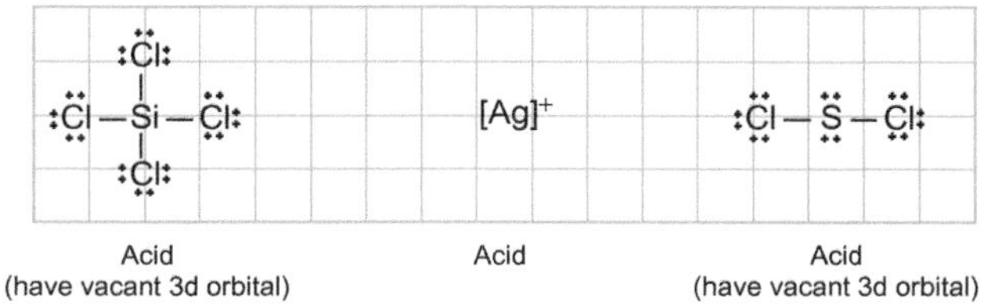

Arrange the following species according to their acidity: AsH_3, AsF_3, and AsF_5.

- Lewis acidity is strengthened by:
  - increasing the oxidation state of acceptor atom; and
  - increasing the electronegativity, χ, of acceptor atom by surrounding acceptor with other high χ-ligands, e.g.:

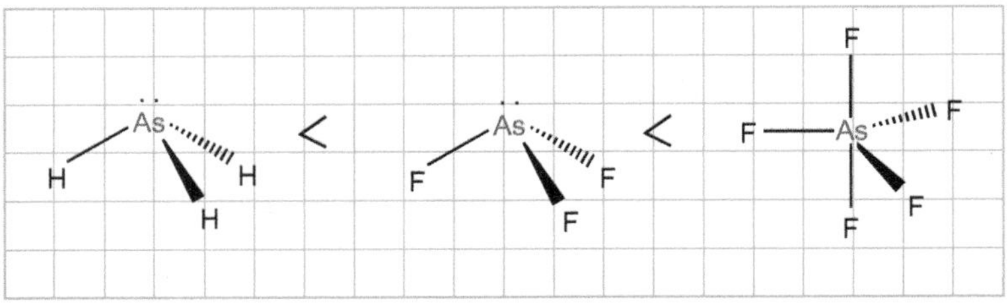

Lewis base (NH_3) 3d-acid Very strong 3d-acid (As is acceptor atom)

Give example to show that Lewis acid acts as electrophile

- Lewis acid = "electrophile"
  e.g.,

Nitronium cation

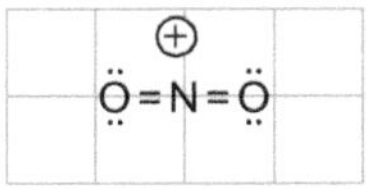

Base attaches
e.g., aryl carbanion

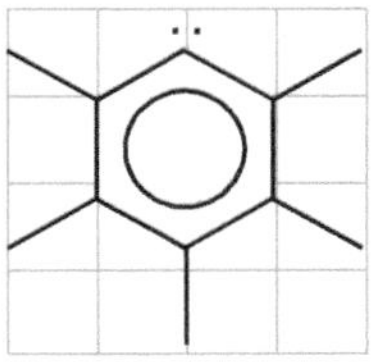

What are the main characters of Lewis bases and the factors that influence the basicity? Give examples to support your answers.

Examples:

- All elements and anions of groups IV–VI (14–16) are donor atoms, e.g., CH_3^-, CN^-, NH_2^-, NCS^-, N_3^-, OH^-, $OOCR^-$, SH^-, SR^-
- Halide anions, F^-, Cl^-, Br^-, I^-, and H^-
- Neutral molecules contain element of group V, VI (15, 16) as donor atoms, e.g., NH_3, NR_3, PR_3, AsR_3, OH_2, OHR, OR_2, SR_2.
  Group IV molecules usually do not have lone pair on group-IV atom.
  Group VII donors usually have too high χ to donate ē.
- Lewis basicity increases with:
  - increasing ē-density on donor atom;
  - decrease oxidation state of donor atom;
  - low-χ groups bound to donor atom; and
  - an increased negative charge on the molecule.

  E.g., C, N, and O make stronger σ-bases than their analogs from other rows:

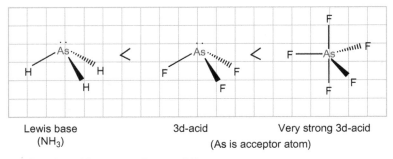

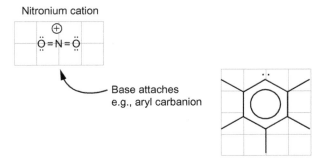

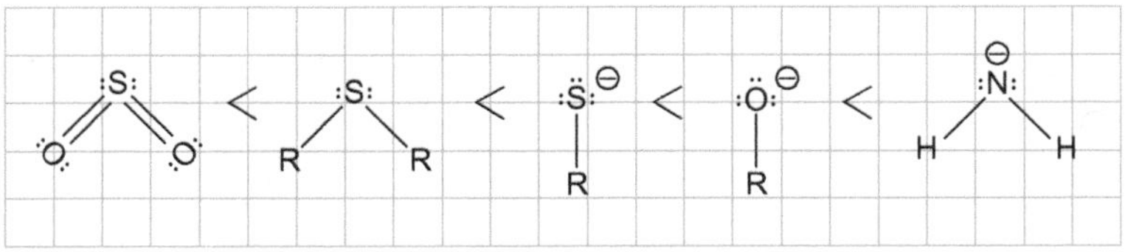

e. Lewis base = Lowrey-Brønsted base = nucleophile

TABLE 3.5 Conjugate Acids and pK_A

Base	Conj. Acid	pK_A
OH^-	H_2O	15.7
NH_3	NH_4^+	9.25
HS^-	H_2S	7
Pyridine	Pyr.H^+	5.4
F^-	HF	3.45
Br^-	HBr	−9

Which molecule of the following could form a coordinate bond with a proton?
OH^-, NH_3, HS^-, Pyridine, F^-, Br^-

- $\underbrace{\text{Base}}_{\text{Electron donor}} + \underbrace{H^+}_{\text{Electron acceptor}} \rightleftharpoons \underbrace{\text{Base} \rightarrow H^+}_{\text{Coordinate with proton}}$ (Table 3.5)

Chemical Features of Hard and Soft Ions, and Classification

What are the main characteristic of the hard and soft ions?

- *Hard ions*:
  - retain their valence electrons very strongly;
  - are not readily polarized; and
  - are a small size.
- *Soft ions*:
  - do not retain their valence electrons firmly.

According to hard-soft aspect, classify the cations, and give examples.

- Cations can be classified as follows:
  - *Hard acids*:
    - have higher oxidation states;
    - are on the left-hand side of periodic table; and
    - have low valence d-electron occupancy.

      Examples

      H^+, Li^+, Na^+, K^+, Rb^+, Cs^+
      Be^{2+}, Mg^{2+}, Ca^{2+}, Ba^{2+}, Sr^{2+}, Sc^{3+}, Ti^{4+}, Al^{3+}
      Mn^{3+}, Cr^{3+}, Fe^{3+}, Co^{3+}, Mo^{6+}
      Lanthanide, actinide M^{3+}, M^{4+-6+}
      P^{5+}, Si^{4+}, As^{5+}
  - *Soft acids*:
    - metals in lower oxidation states;
    - "heavy" metals;
    - are on the lower right center of the periodic table; and
    - have several d-electrons in valence shell.

      Examples:

      Co^+, Cu^+, Ti^+, Pd^{2+}, Ag^+, Au^+, Pt^{2+}, Cd^{2+}, Rh^+, Hg^+
      π-acids, such as dihalogens
      Metal atoms in zero oxidation state, e.g., pd^0.
  - *Borderline acids*: Fe^{2+}, Co^{2+}, Sn^{2+}, Zn^{2+}, Cu^{2+}, Ni^{2+}, Pb^{2+}.

According to hard-soft aspect, classify the anions, and give examples.

- Ligands can also be classified as follows:
  - *Hard bases*:
    - donor atoms of high electronegativity, χ; and
    - molecules without low-lying π^*-orbitals.

      Examples:

      O^{2-}, OH^-, ROH, OR^-, R_2O, $RCOO^-$, O_2^-, NH_3, NH_3^-, NR_3, NCS^-, Cl^-, PO_4^{3-}, SO_4^{2-}, F^-, NO_3^-, CO_3^{2-}.
  - *Soft bases*:
    - donor atoms of low electronegativity, χ
    - molecules have close-lying π^*-orbital
    - donor atom are readily polarizable

      Examples:

      RSH, RS^-, R_2S, SeR^-, R_3P, R_3As, CO, CN^-, NO, SCN^-, $S_2O_3^{2-}$, H^-, I^-, carbanions, R-C≡N.
  - Borderline bases:

    Aromatic amines (sp^2-N): imidazoles, pyridine, etc. RNH_2, N_2, N_3^-, NO_2^-, Br^-.

Rule of Interactions

What are stability criteria of complex formation in the hard-soft acid base model?

- General Rules:
  - A hard acid prefers to combine with hard base, and a soft acid prefers to combine with soft base, HSAB.

Hard-hard	Zr^{4+}, F^-	$K_f = 10^9$
Soft-hard	Cd^{2+}, F^-	$K_f = 10^o$
Hard-soft	UrO_2^{2+}, NCS^-	$K_f = 10^1$
Soft-soft	Hg^{2+}, NCS^-	$K_f = 10^9$

  - The formation of stable complexes results from interactions between hard acids and hard bases.
  - Hard acid-base interactions are mainly electrostatic.
  - Soft acid-base interactions involve significant polarization (covalency).

Hard-Hard and Soft-Soft Interactions

Identify the type of bonding in soft-soft and hard-hard interactions

- Thermodynamics in H_2O:

$$M - OH_2 + L \rightleftharpoons M - L + H_2O$$

  - Soft-soft interaction: the electron clouds are easily distorted or polarized.

    An M-L bond is more covalent, $D_{M-L} > D_{M-H_2O}$, where D = dissociation energy, and ΔH is negative (enthalpically driven).
    - Covalency in M-L depends (apparently) on polarizability of ligand and on polarizability of metal, not on the polarizing power of metal ion, e.g., Fe^{3+} harder than Fe^{2+}, Pt^{2+} softer than Ni^{2+}.
  - Hard-hard interactions: M-L bonds are rather ionic, $D_{M-L} \cong D_{M-H_2O}$, $\Delta H \cong 0$. Molecules of H_2O are released from the 2° sphere, and ΔS is positive (entropically driven).

    Note: ion-ion, M-L, replaces ion-dipole, M-OH_2.

Hard-Soft Interaction and Anion Polarizability

Ca^{2+} and Cd^{2+} are two metals of similar size and charge, but one is hard and the other is soft.

If:

$$\mathbf{CaO} \quad \Delta H_f^o = -635\,\text{kJ/mol, and } Ca(g) \rightarrow Ca^{2+}(g) + 2e \quad I_E = 1736\,\text{kJ}$$

$$\mathbf{CdO} \quad \Delta H_f^o = -255\,\text{kJ/mol, and } Cd(g) \rightarrow Cd^{2+}(g) + 2e \quad I_E = 2496\,\text{kJ}$$

What are the expected difference in the stability of CaO and CdO, and how can you explain the deviation?

Explain the following differences in the formation constants of CdX and CaX:

Anion	$\Delta H^o_{f(CdX)} - \Delta H^o_{f(CaX)}$ (kJ/mol)
None	760
F^-	543
O^{2-}	376
S^{2-}	314
Te^{2-}	167

Write your implications for the reaction:

$$CaS(g) + CdF_2 \rightarrow CaF_2(s) + CdS(s)$$

- CaO: $\Delta H^o_f = -635\,kJ/mol$
  CdO: $\Delta H^o_f = -255\,kJ/mol$
  Therefore, CaO (hard-hard) is more stable than CdO (soft-hard) by 380 kJ $(-635 - (-255) = -380kJ)$.
- From the Born-Haber cycle, we find that the main obvious difference is in the ionization energy:

$$Ca(g) \rightarrow Ca^{2+}(g) \quad I_E = 1736kJ$$

$$Cd(g) \rightarrow Cd^{2+}(g) \quad I_E = 2496kJ$$

  Accordingly, CaO should be more stable by $= 2496 - 1736 = 760$ kJ.
- However, why is CaO is only 380 kJ more stable?
  - CdO must have more lattice energy than expected for ionic CaO-type lattice.
  - Therefore, CdO has covalency.
  - CdO covalency arises from polarization anion; as a result, CdX covalency is more pronounced as anion *X* is more polarizable.
- Then:
  a. totally nonpolarizable anion, X:
     CaX more stable by 760 kJ (ΔE_I)
  b. very polarizable anion, X;
     CdX gains lattice energy (*U* becomes more negative)

Therefore, CaX stability advantage reduced from 760 kJ to a lower value

Anion	$\Delta H^o_{f(CdX)} - \Delta H^o_{f(CaX)}$ (kJ/mol)
None	760
F^-	543
O^{2-}	376
S^{2-}	314
Te^{2-}	167

i. Anion (base) softness increases as anion polarizability increases (probably synonymous).
ii. Softer cation is the one with higher I_E due to valence orbitals being at lower energy, so electron density is pulled down.
iii. The implications for the reaction:
    $CaS(g) + CdF_2 \rightarrow CaF_2(s) + CdS(s)$
    CaS (hard-soft) is more stable than CdS, by 314 kJ.
    The reaction is driven by: CaF_2 more stable than CdF_2 by 540 kJ.
    Therefore, the reaction is driven enthalpically by a hard-hard interaction.

Chelate Effect

$$M^{2+} + L \rightleftharpoons M(L)^{2+} \quad K_{M-L}$$

$$M = Ni^{2+},\ K_{Ni-NH_3} \cong 10^3,\ K_{Ni-En} \cong 10^{7.5},\ \text{where En} = NH_2 - CH_2 - CH_2 - NH_2$$

Specify the cause that affects the values of K_L, in spite of both ligands having similar N-donor atoms.

- Complexes of chelating ligands have greater thermodynamic stability. The chelate effect is a result of both enthalpic and entropic contributions.

$$\Delta G^o = \Delta H^o - T\Delta S^o$$

$$\Delta G^o = -RT \ln K$$

Therefore, K_f increases as ΔG^o becomes more negative. A more negative ΔG^o can result from making ΔH^o more negative or from making ΔS^o more positive.
- The stability constants increase as the basicity of the ligands increase.
- The complexes of NH_3 are generally less stable than those complexes of simple amines, and the stability increases with the basicity of the amine.
- Therefore, complex of chelate ligand, such as NH_2-CH_2-CH_2-NH_2, are much more stable than those of monodentate ligand like NH_3.
- This enhanced stability is known as the chelate effect (or chelate ring effect).

Rationalize the change in the stability of the following:

Complex	log K_f
$Cu(NH_3)_4^2$	11.9
$Cu(En)_2^{2+}$	20.0
$Cu(TriEn)^{2+}$	20.4

where En is ethylenediamine and TriEn = triethylenetetraamine.

- The stability increases with a greater number of chelate rings.
  $Cu^{2+} + 4NH_3 \rightleftharpoons Cu(NH_3)_4{}^{2+}$ $\log K = 11.9$, four coordinated nitrogen atoms with no chelate ring.
  $Cu^{2+} + 2En \rightleftharpoons Cu(En)_2{}^{2+}$ $\log K = 20.0$, four coordinated nitrogen atoms with two chelate ring.
  $Cu^{2+} + TriEn \rightleftharpoons Cu(TriEn)^{2+}$ $\log K = 20.4$, four coordinated nitrogen atoms with three chelate ring.

Entropy and Chelate Formation

What are the thermodynamic implications when you compare the complex formation of the following at 25°C?

Complex	Ligand	log K_f	ΔS^o (J/mol^{-1} deg^{-1})
$[Cd(CH_3NH_2)_4]^{2+}$ (aq)	$4CH_3NH_2$	6.52	−67.3
$[Cd(en)_2]^{2+}$ (aq)	2en	10.6	+14.1

- The chelate effect can be an entropy-driven process; additionally, there may be a substantial enthalpy influence for transition metal ions and this can be a guide to the electronic aspect of the reaction.
- The following two reactions illustrate a purely entropy-based chelate effect:

$$Cd^{2+}(aq) + 4CH_3NH_2 \rightleftharpoons [Cd(CH_3NH_2)_4]^{2+}(aq) \quad \log K_f = 6.52$$

$$Cd^{2+}(aq) + 2H_2NCH_2CH_2NH_2 \rightleftharpoons [Cd(en)_2]^{2+}(aq) \quad \log K_f = 10.6$$

By using: $\Delta G^o = \Delta H^o - T\Delta S^o$
$\Delta G^o = -2.3RT \log K$ then

Ligand	ΔH^o (kJ/mol)	ΔS^o (J/mol^{-1} deg^{-1})	$T\Delta S^o$ (J/mol^{-1})	ΔG^o (kJ/mol)
$4CH_3NH_2$	−57.3	−67.3	20.1	−37.2
2en	−56.5	+14.1	−4.2	−60.7

The enthalpy change are very close; the chelate effect can attributed to the entropy difference.
- After one NH_2 of En molecule has coordinated to the metal ion, the attachment probability of other NH_2 group is increased. If chain is too long, motional entropy works against chelation.

- Each En molecules replaces two H_2O molecules, increasing the total number of particles, and the randomness, and hence the entropy.

Stability and the Geometry of the Chelate Ring

How can you explain the change in the formation constants of the following zinc complexes?

L	$\log K_f$
Acetate	**1.0**
Oxalate	**5.0**
Malonate	**3.7**
Succinate	**2.3**

- Five-membered chelate rings have the greatest stability; however, six-membered chelate rings involving conjugation in the chelate ring are more stable. This is because the bond angles at the metal ion are 90 degrees in octahedral and planar complexes.
  Example: $\log K_f$ for Zn-chelates (Fig. 3.20):
- Complexes of chelating ligands have greater thermodynamic stability. The chelate effect is a result of both enthalpic and entropic contributions.

$$\Delta G^o = \Delta H^o - T\Delta S^o$$

$$\Delta G^o = -RT\ \ln K = -5.60 \log K \quad (\text{kJ/mol at } 25°\text{C})$$

Therefore, K_f increases as ΔG^o becomes more negative. A more negative ΔG^o can result from making ΔH^o more negative or from making ΔS^o more positive.

Macrocyclic Effect

Complete below table. Why are there differences in the formation constants of the following copper complexes? Discuss the possible stability sources.

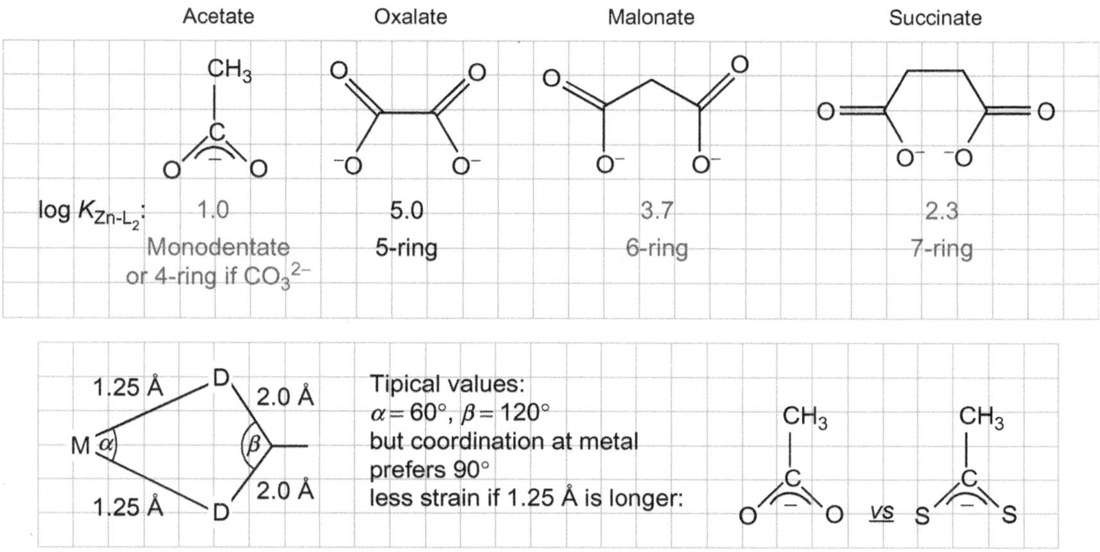

FIG. 3.20 Chelate ring geometry and complex stability; 4-ring has poor geometrical match to metal.

Formation Constants and Thermodynamic Parameters of Copper Chelate Complexes at 20°C

Complex	Log K_f	ΔH_f^o (kJ/mol)	ΔS_f^o (J/mol)
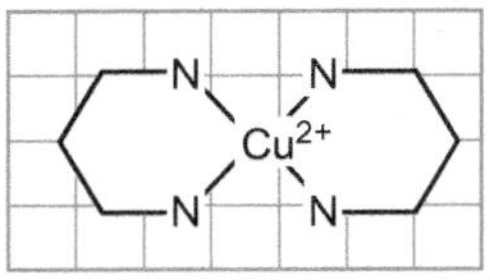	17.0	...?	−74
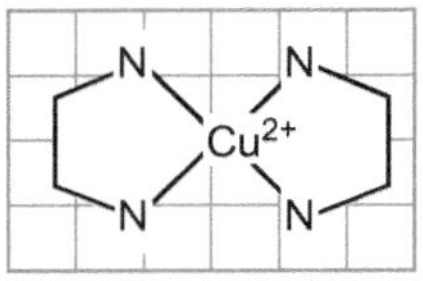	20.6	−114.3	...?
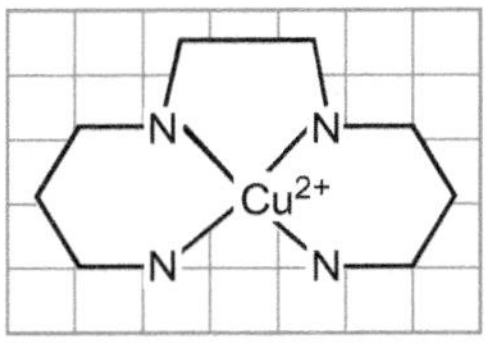	...?	−102	+67
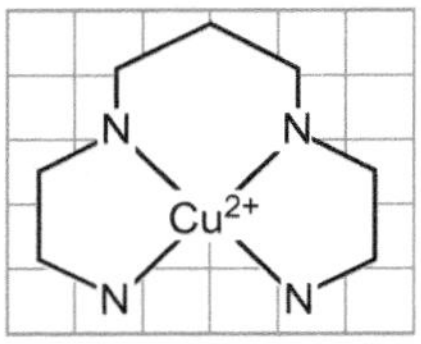	23.9	...?	+68
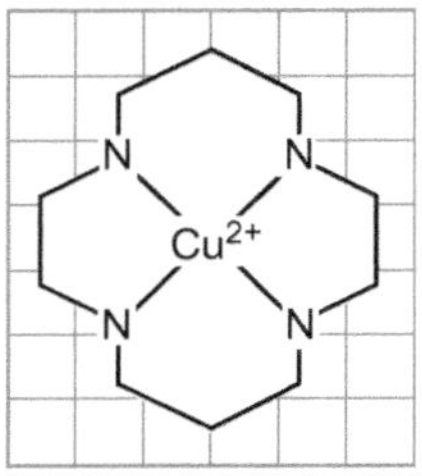	27.2	−133	...?
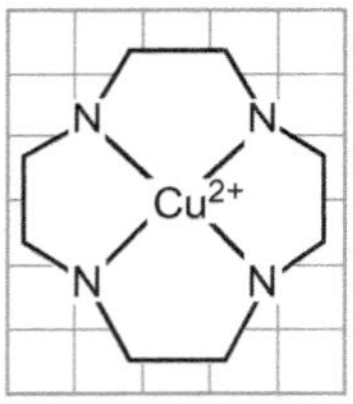	23	...?	+116
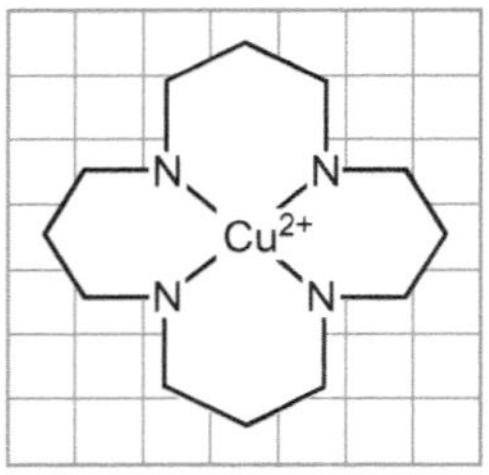	20.9	ΔG^o = ...? kJ/mol	
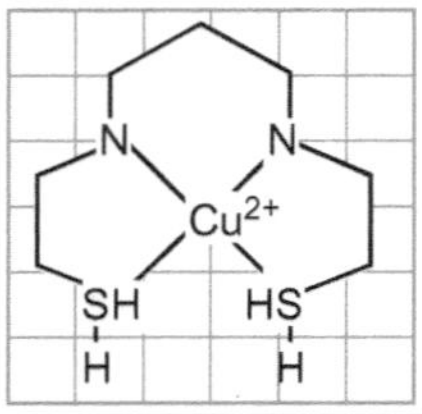	11.4	ΔG^o = ...? kJ/mol	
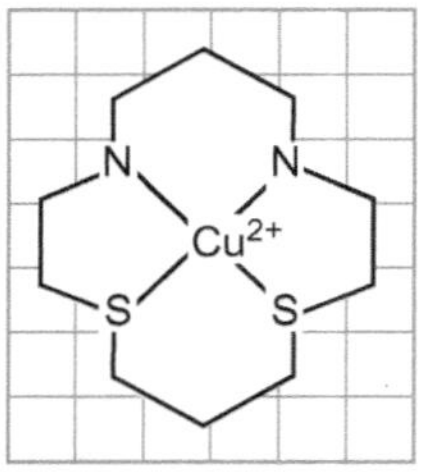	16	ΔG^o = ...? kJ/mol	

- Chelating n-dentate ligand forms more stable complex (more negative ΔG_f^o) than n-unidentate ligands of comparable type. Likewise, n-dentate macrocyclic ligand forms even more stable complexes than the most similar n-dentate open chain ligand (Table 3.6).
  Similar to the chelate effect, the macrocyclic effect results from favorable entropy change. However, the macrocyclic effect is usually associated with a favorable enthalpy change.
- Other possible stability source:

TABLE 3.6 Formation Constants and Thermodynamic Parameters of Copper Chelate Complexes at 20°C

Complex	Log K_f	ΔH_f^o (kJ/mol)	ΔS_f^o (J/mol)	Comment
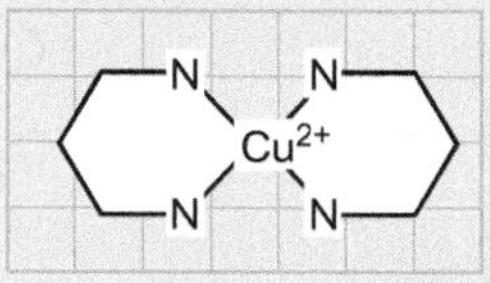	17.0	−117	−74	2(6-ring)
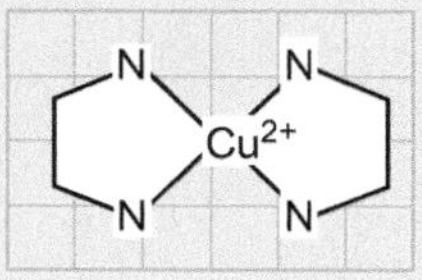	20.6	−114.3	+4	2(5-rings)
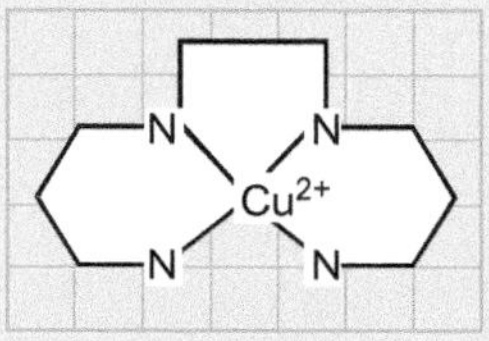	21.7	−102	+67	1(5-ring) +2(6-rings)
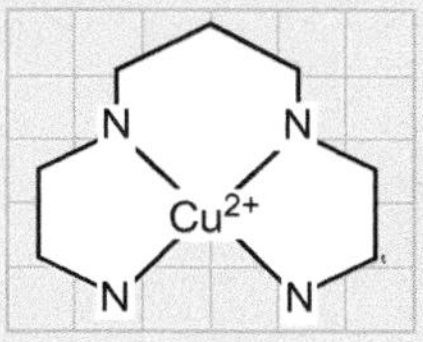	23.9	−114	+68	2(5-rings) +1(6-ring)
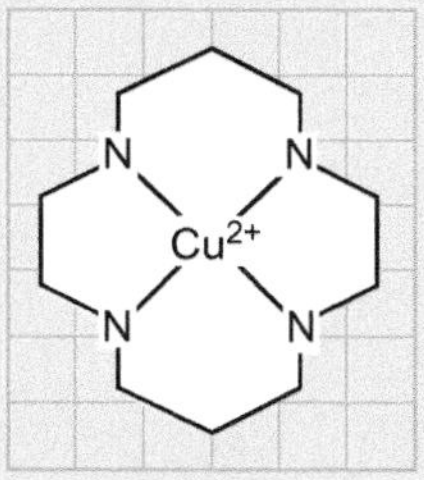	27.2	−133	+66	2(5-rings) +2(6-rings) Macrocyclic effect: when chelating agent itself a ring, stability is further enhanced.
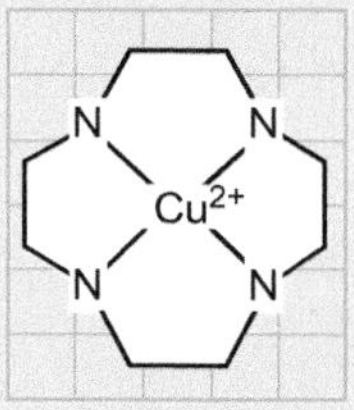	23	−95	+116	4(5-rings)
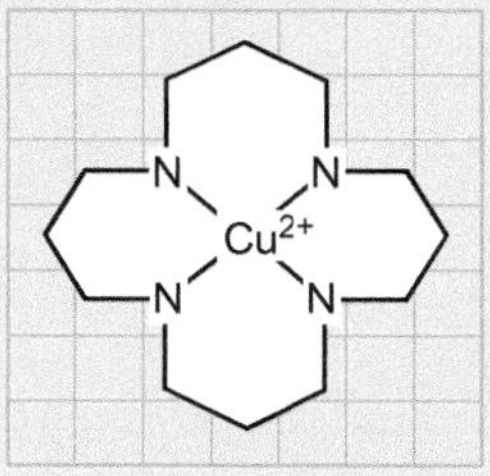	20.9	$\Delta G^o = -117$ kJ/mol		4(6-rings)

Continued

TABLE 3.6 Formation Constants and Thermodynamic Parameters of Copper Chelate Complexes at 20°C—cont'd

Complex	Log K_f	ΔH_f^o (kJ/mol)	ΔS_f^o (J/mol)	Comment
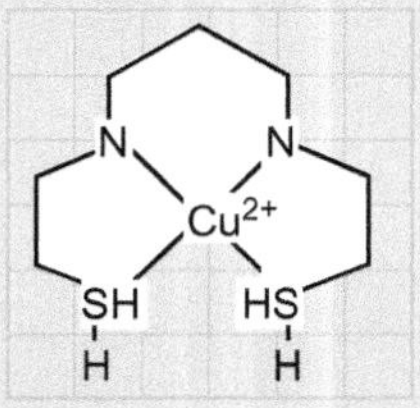	11.4	$\Delta G^o = -64$ kJ/mol		
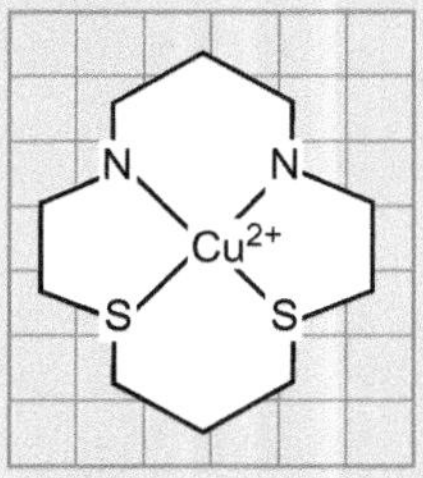	16	$\Delta G^o = -90$ kJ/mol		

Solvation Enthalpy Differences

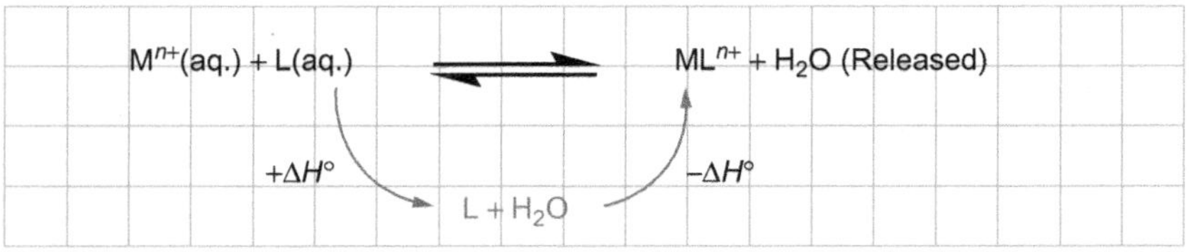

- Energy (+ΔH^o) input to remove H_2O that surround the ligand's donor atoms compensated by ΔS of extra released H_2O for a cyclic ligand.

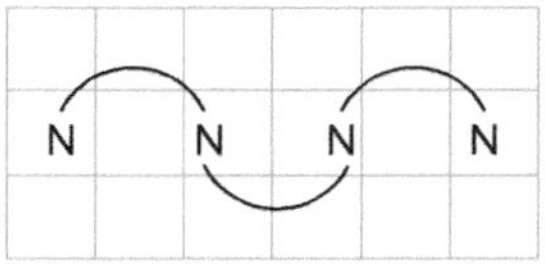

Open structure is well-solvated, more ΔH input to remove H_2O's

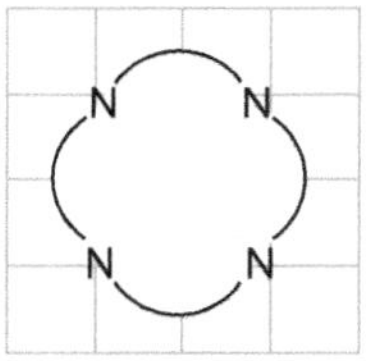

More compact, fewer H_2O's of solvation, less ΔH needed to remove H_2O's

Donor Atom Basicity

Also: 2° donor ligands better than 1° donors:

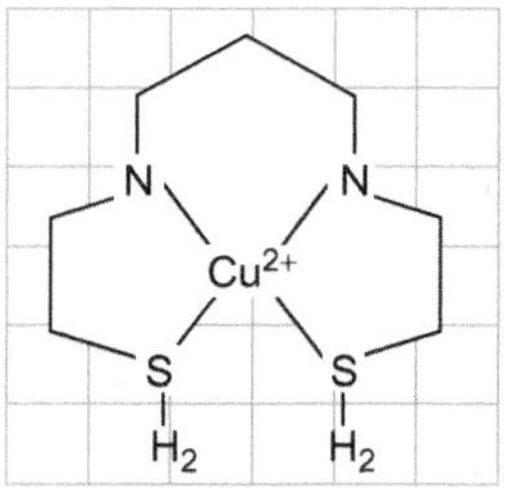 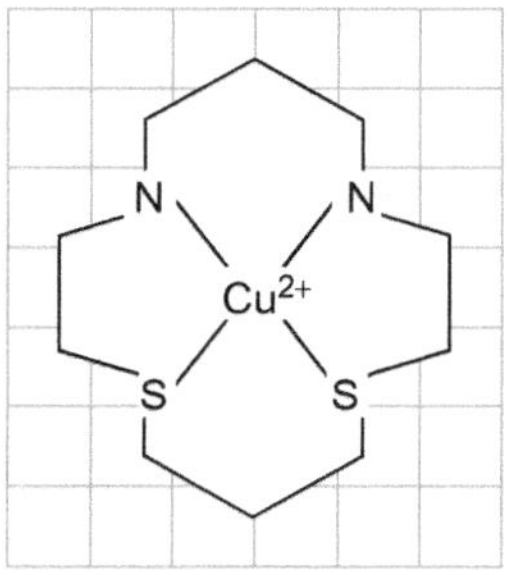

log K_f : 11.4 16.0

Cavity Size

What do you conclude from the following relative formation constants for the interactions of different cryptates with different cations?

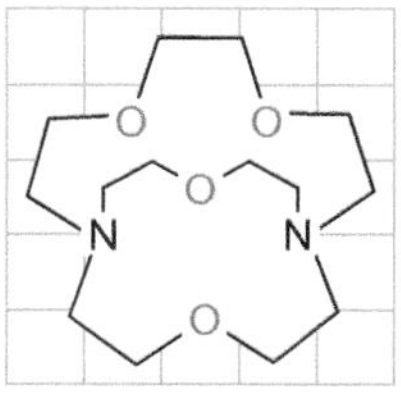

Cryptate: [2,1,1]

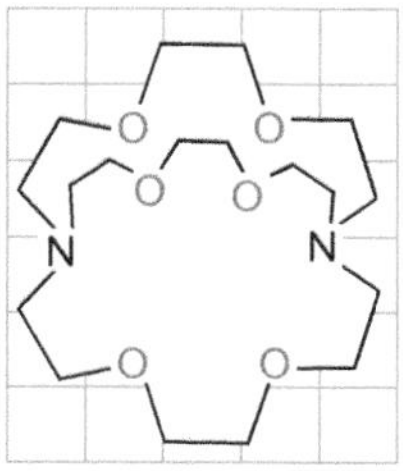

[2,2,2]

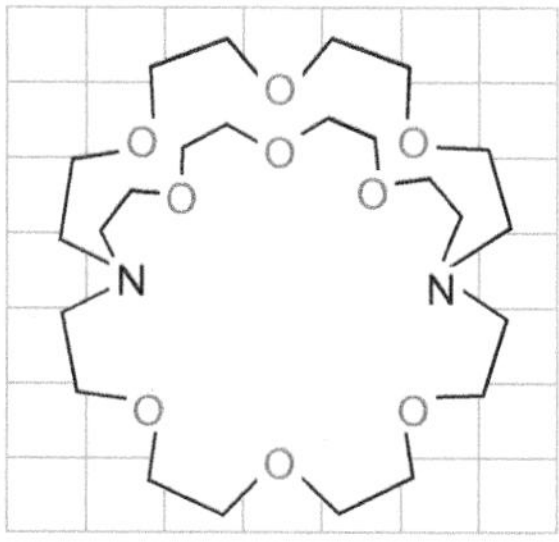

[3,3,3]

	Ionic radii (Å)	$\log K_f (CH_3OH)/\log K_f (H_2O)$
$[2,1,1]+Li^+$	0.60	7.7
[2,1,1] + Na	0.95	6.2
$[2,2,2]+Li^+$	0.60	1.8
$[2,2,2]+Na^+$	0.95	7.2
$[2,2,2]+K^+$	1.33	9.9
$[2,2,2]+Rb^+$	1.48	8.5
$[2,2,2]+Cs^+$	1.69	3.6
$[3,3,3]+K^+$	1.33	5.2
$[3,3,3]+Rb^+$	1.48	5.5
$[3,3,3]+Cs^+$	1.69	5.9

- The formation constant for interaction of cryptates with cation differs significantly from cation to cation, depending upon the relative sizes of cation and cavity.
- These cryptates differentiate among cations directly on a size basis by ligation cavity, which is a better match for one cation than others are.
  Note that the order of cavity-size increases: [2, 1, 1] < [2, 2, 2] < [3, 3, 3] matches the order of ionic-radii increase: $Li^+ < K^+ < Cs^+$.

Solvent Competition

Consider the reaction of dicyclohexano-18-crown-6 with Na^+, and K^+ in different solvents, e.g., H_2O, and CH_3OH.

	Ionic radii (Å)	$(\text{Log } K_f)_{H_2O}$	$(\text{Log } K_f)_{CH_3OH}$
$Na^+(\text{solvent})_n$ + dicyclohexane-18-crown-6	0.8	4.0	0.95
$K^+(\text{solvent})_n$ + dicyclohexane-18-crown-6	1.9	5.9	1.33

Explain how the solvent can influence the complex formation.

- In an aqueous solution, all ligands compete with H_2O (hard) for the cation (hard).
  - Hard acids gain nothing by binding soft base, so H_2O tends to stay hard-hard.
  - Soft acids gain a ΔH-advantage with a soft base, so H_2O is displaced at equilibrium.
  - Hard bases compete with H_2O for hard acids. So hard bases in H_2O may not form strong complexes, unless they are stronger bases than H_2O; e.g., F^-, NH_3 are more H^+-basic than H_2O.
- A complex formation requires stepwise substitution of the cation's solvation shell of Na^+, and K^+ by the donor atoms of dicyclohexano-18-crown-6. Such a substitution is sensitive to the nature of both solvent and cation.

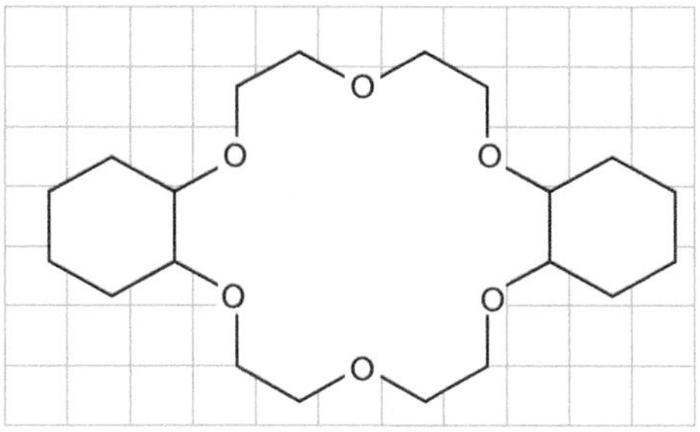

Dicyclohexane-18-crown-6

- Binding of the solvent molecules, H_2O or CH_3OH, by cations is determined by polarizing power of the cation, and decreases with an increase in the radius of the cation.
  As a result, cations of smaller size bind more strongly to H_2O molecules than to CH_3OH.
  Because: $r_{Na^+} < r_{K^+}$, then

$$K_{Na^+-H_2O} > K_{K^+-H_2O},\quad K_{Na^+-CH_3OH} > K_{K^+-CH_3OH}$$

$$K_{Na^+-H_2O} > K_{Na^+-CH_3OH},\quad K_{K^+-H_2O} > K_{K^+-CH_3OH}$$

- The oxygen donor atoms of dicyclohexano-18-crown-6 molecule substitute ligated oxygen-solvent molecules from the cation of larger size rather than the smaller one; therefore, H_2O competes very effectively. This explains the sequence of the observed formation constants:

$$K_{K^+-(18-crown-6)_{H_2O,\ CH_3OH}} > K_{Na^+-(18-crown-6)_{H_2O,\ CH_3OH}}$$

$$K_{Na^+-(18-crown-6)_{CH_3OH}} < K_{Na^+-(18-crown-6)_{H_2O}}$$

$$K_{K^+-(18-crown-6)_{CH_3OH}} < K_{K^+-(18-crown-6)_{H_2O}}$$

Note: dicyclohexano-18-crown-6, H_2O and CH_3OH are oxygen donors.

When the donor atoms of the solvent or the ligand are different, other factors that affect the displacement of the solvent molecules must be considered. These factors include the hard/soft acid-base concept (HSAB), and the entropy change associated with the randomization of released solute/solvent molecules.

Steric Effect

Consider the reaction of copper acetylacetonate with Lewis bases in nonaqueous solvent:

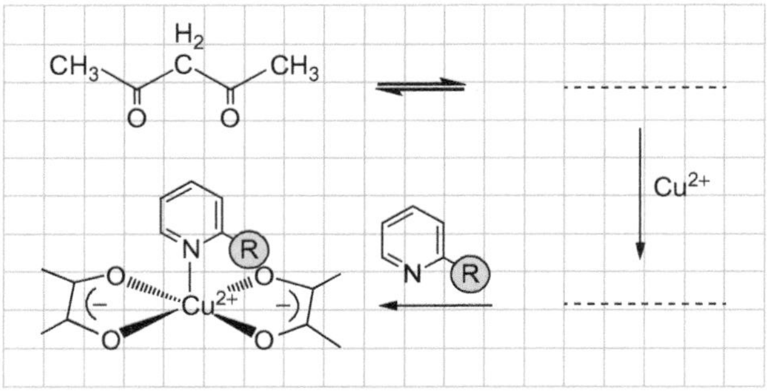

Complete and explain the change in the formation constants of copper acetyacetonate adducts:

R=	**2,6-Me$_2$**	**2-Me**	**Pyr**	**4-Me**
Log K_f=	**<1**	**3**	**7**	**11**

- In spite of the fact that 2,6-dimethylpyridine has higher pK_a than 2-methylpyridine, and 2-methylpyridine is stronger base than pyridine (expected from the inductive effect), we obtain the following (Scheme 3.4):

R=	2,6-Me$_2$<	2-Me <	Pyr<	4-Me
Log K_f=	<1	3	7	11 (inductive effect)

- The effect of the ortho-methyl groups in reducing the stability of the copper acetylacetonate adducts can be attributed to the strain between nonbounded groups attached to different atoms that conflict with the steric requirements.

Stability and Metal Oxidation State

Given:

$O_2 + 4H^+ + 4e^-$	**$\rightarrow 2H_2O$**	**$E^o = +1.229$ V**
$Fe^{3+} + e^-$	**$\rightarrow Fe^{2+}$**	**$E^o = 0.771$ V**

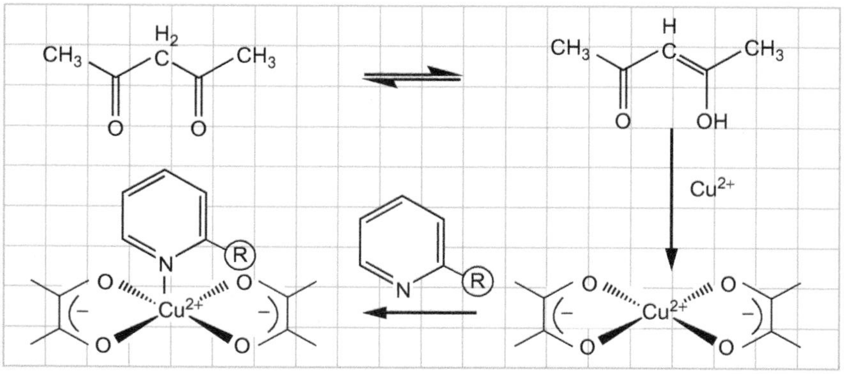

SCHEME 3.4 Reaction of copper acetylacetonate with pyridine derivatives.

What is the dominated oxidation state of iron in aqueous solution, Fe^{2+} or Fe^{3+}? What is the concentration of the dominate state at pH 7?

- In an aqueous environment:

$O_2+4H^++4e^-$	$\rightarrow$	$2H_2O$	$E^o=+1.229$ V
Fe^{2+}	$\rightarrow$	$Fe^{3+}+e^-$	$E^o=-0.771$ V
$4\ Fe^{2+}+O_2+4H^+$	$\rightarrow$	$4Fe^{3+}+2H_2O$	$E^o=+0.458$ V

$E^{o\prime}=E^o-0.059$ pH at 25°C
$E^{o\prime}=E^o-0.061$ pH at 37°C

- Therefore, at pH 7, and 25°C, $E^{o\prime}=0.045$ V,
  $\Delta G^o=-nFE^{o\prime}=-RT\ \text{Ln}\ K$
  $F=96{,}500$ C/mol, $R=8.3$ J/K mol
  $\Delta G^o=-4\times 96{,}500\times 0.045=-8.3\times 298\times \text{Ln}\ K$
  $K=1.1\times 10^3$
  and Fe^{+3} is the dominate oxidation state.
- $Fe^{3+}+OH^-\rightarrow Fe(OH)_3\downarrow$
  But, $Fe(OH)_3$ has $K_{sp}=10^{-38}$
  So for pH 7, $K_{sp}=[Fe^{3+}][OH^-]^3$
  $K_{sp}=[Fe^{3+}][10^{-7}]^3$
  $\therefore [Fe^{3+}]=10^{-17}$ m/L.

- Therefore, the concentration $[Fe^{3+}]$ is very low in aqueous solution.
- The hydrolysis of Fe^{3+} yields polymer formation and finally precipitates; Fe^{3+} is insoluble at a neutral pH in distal water. Only by suitable ligands is it possible to sustain soluble Fe^{3+} in neutral and basic solutions. Suitable ligands are EDTA, NTA, and citrate (Fig. 3.19).
- On the other hand, Fe^{2+} is soluble to the extent of 0.1 M at pH 7, but Fe^{2+} is spontaneously oxidized to Fe^{3+} in air.

If:	**Cu^+ (aq) + e**	$\rightarrow$	**Cu^0;**	**$E^o=0.52$ V**
	Cu^+ (aq)	$\rightarrow$	**Cu^{2+} (aq) + e;**	**$E^o=-0.153$ V**
For:	**Cu^+**	$\rightleftharpoons$	**$Cu^{2+}+e^-$**	

Find the concentration of Cu^+ ion in aqueous solution?

- The redox potentials of copper ion:

Cu^+ (aq) + e	$\rightarrow$	Cu^0	$E^o=0.52$ V
Cu^+ (aq)	$\rightarrow$	Cu^{+2} (aq) + e	$E^o=-0.153$ V
2 Cu^+ (aq)	$\rightleftharpoons$	Cu^0 (s) + Cu^{+2} (aq)	$E^o=0.37$ V

$\Delta G^o = -nFE^o$, and $\Delta G^o = -RT\ \ln K$, then

$$K = \frac{[Cu^{2+}][Cu]}{[Cu^+]^2} \cong 10^6$$

This shows that Cu^+ in aqueous solution disproportionate to Cu^{2+} and Cu^0.

Use E^o-diagram (Fig. 3.21) for manganese:

$Mn^{2+} + 2e \rightarrow Mn \quad E^o = -1.18$ V

$Mn^{3+} + e \rightarrow Mn^{2+} \quad E^o = 1.51$ V

Is Mn^{2+} stable in acidic aqueous solution?

- The sum for the half-reactions:

$$\begin{array}{llll} Mn^{2+} + 2e & \rightarrow & Mn & E^o = -1.18\,V \\ 2(Mn^{2+} & \rightarrow & Mn^{3+} + e) & E^o = -1.51\,V \\ \hline 3Mn^{2+} & \rightarrow & Mn + 2Mn^{3+} & E^o = -2.69\,V \end{array}$$

The reaction does not proceed and Mn^{2+} is not disproportionate because E° is negative.
A negative value of E^o means that the reaction will not be thermodynamically spontaneous, because $\Delta G^o = -nFE^o$, or

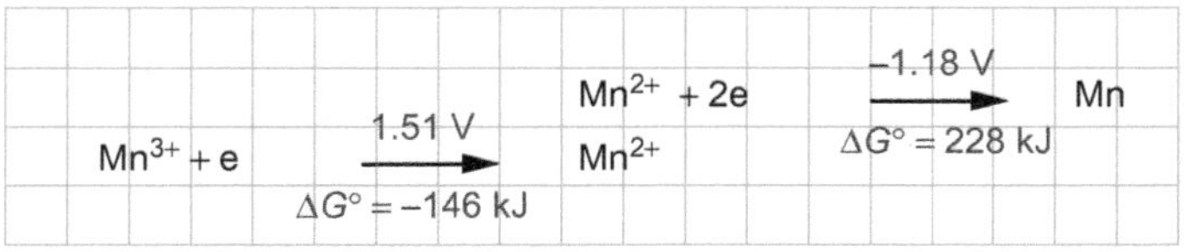

The above setup has more positive E^o value to its right ($\Delta G^o = -nFE^o = -146$ kJ, occurring spontaneously), than to its left ($\Delta G^o = -nFE^o = +228$ kJ, occurring in the reverse direction).

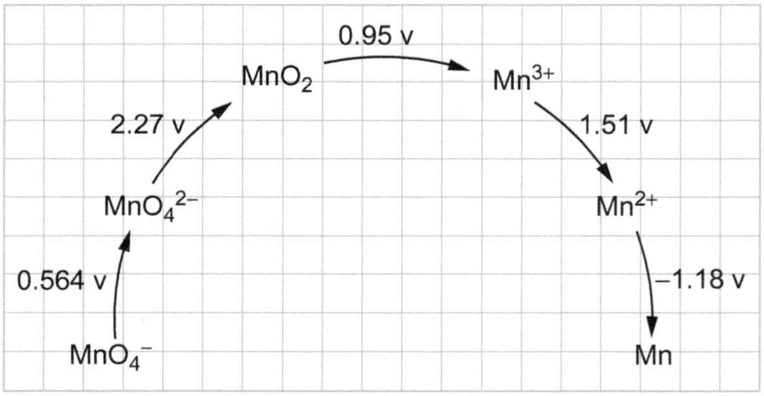

FIG. 3.21 E^o-diagram for manganese in acidic solution at 25°C, showing the half-reaction as reduction and only the substance that changes in oxidation.

E° for the half-reactions, e.g.,

$$MnO_4^{2-} + 4H^+ + 2e \xrightarrow{2.27\ v} MnO_2 + 2\,H_2O$$

$$MnO_2(s) + 4H^+ + e \xrightarrow{0.95\ v} Mn^{3+} + 2\,H_2O$$

$Hg^{2+} + 2e \rightleftarrows Hg^0$, $E^o = 0.792V$ find $[Hg^{2+}]$ at pH 7, and $T = 25°C$
If 0.1 M mercaptoacetate (MAH_2, at pH 7, pK_A of –SH is 9.8) is added:
$Hg^{2+} + 2MA^{2-} \rightleftarrows Hg(MA)_2{}^{2-}$ $\beta_2 = 2 \times 10^{47}$
Find $[Hg(MA)_2^{2-}]$, and comment.

- $$\begin{array}{llr} Hg^0 & \rightleftarrows Hg^{2+} + 2e & E^o = -0.792V \\ 2H^+ + 2e & \rightleftarrows H_2 & E^o = 0.0V \\ \hline Hg^0 + 2H^+ & \rightleftarrows Hg^{2+} + H_2 & E^o = -0.792V \end{array}$$

$$\Delta G^o = -nFE^o = -2 \times 96,500 \times -0.792 = 153kJ$$

$$K = e^{\frac{\Delta G^o}{RT}} = 1.47 \times 10^{-27} \text{ at } 298K$$

$$K = \frac{[Hg^{2+}]P_{H_2}}{[H^2]^2} = 1.47 \times 10^{-27}$$

At pH7, $P_{H_2} = 1\,atm$.

$$[Hg^{2+}] = 1.47 \times 10^{-27} \times 10^{-7} \times 10^{-7} = 1.47 \times 10^{-41}$$

Show that Hg^{2+} disproportionate to Hg^0.

- MAH_2 is added:

MAH_2 $\rightleftarrows$	$MA^{2-} + 2H^+$		$pK_A = 9.8$
Initial:	0.1 m	0	10^{-7}
Change:	$-x$	x	10^{-7}
Equilibrium:	$0.1 - x$	x	10^{-7}

$$K = \frac{[MA^{2-}][H^+]^2}{[MAH_2]} = \frac{(x)(10^{-7})}{(0.1 - x)}$$

$$[MA^{2-}] = 0.1 \times 10^{-9.8} \times 10^7 = 1.6 \times 10^{-4}M$$

- $$Hg^{2+} + 2MA^{2-} \rightleftarrows Hg(MA)_2{}^{2-} \quad \beta_2 = 2 \times 10^{47}$$

$$K = \frac{[Hg(MA)_2{}^{2-}]}{[Hg^{2+}][MA^{2-}]^2}$$

$$[Hg(MA)_2{}^{2-}] = K[Hg^{2+}][MA^{2-}]^2 = 2 \times 10^{47} \times 1.5 \times 10^{-41} \times (1.6 \times 10^{-4})^2$$

$$[Hg(MA)_2{}^{2-}] = 8 \times 10^{-2}M$$

Show that Hg^{2+} does not disproportionate to Hg^0, but trapped as $[Hg(MA)_2{}^{2-}]$

Stability and Metal Ionization Potential

How can the mode of ligation influence the $M^{n+}/M^{(n-1)+}$ redox cycle? Use a model compound.
For Fe/DTPA system, $E^o = 0.771$ V (Fe^{3+}/Fe^{2+}), log $K^{III} = 28.6$, log $K^{II} = 16.5$, find E_f for $Fe(DTPA)^{2-/3-}$ couple.

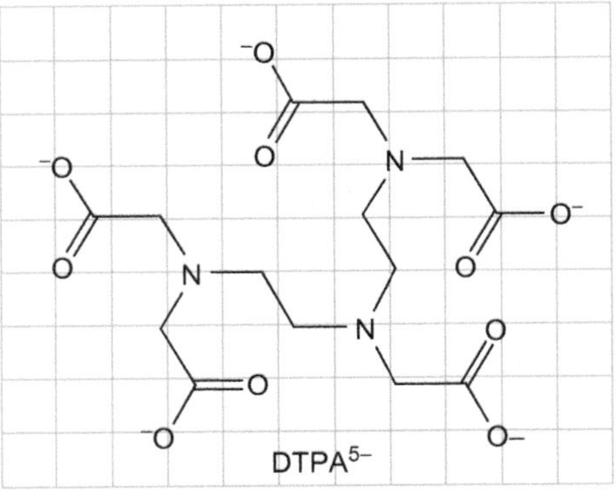

- The redox potential of $Fe(DTPA)^{2-/3-}$ is different from that of $Fe(OH_2)_6^{2-/3-}$.
- $DTAP^{5-}$ complexes Fe^{2+} and Fe^{3+}, but because it complex them differently ($K^{III} \neq K^{II}$), both of $\left[Fe(OH_2)_6{}^{2+}\right]$ and $\left[Fe(OH_2)_6{}^{3+}\right]$ are much less by different amount, $E_f \neq E^o$.
- Consider the redox cycle, $L = DTAP^{5-}$:

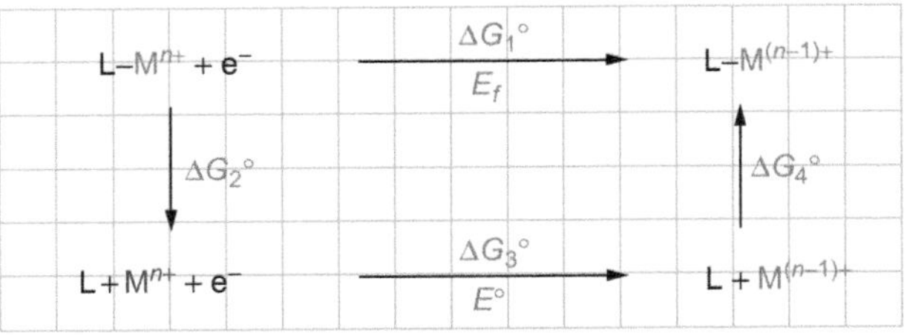

ΔG_1^o and ΔG_3^o are given by: $\Delta G^o = -nFE'^o$
$\Delta G_2^o = +RT\ \ln K_M^{n+}$ (decomplexation of $LM^{n+} = (-1)(-RT\ \ln\ K_M^{n+})$
$\Delta G_4^o\ \ln K_M^{(n-1)+}$ $\left(\text{complex formation of } LM^{(n-1)+}\right)$

$$\Delta G_1^o = \Delta G_2^o + \Delta G_3^o + \Delta G_4^o$$

$$-nFE_f = RT\ \ln K_M^{n+} - nFE^o - RT\ \ln K_M^{(n-1)+}$$

$$E_f = E^o - RT\ \ln \frac{K_M^{n+}}{K_M^{(n-1)+}}$$

At 25°C

$$E_f = E^o - \frac{0.059}{n} \log \frac{K_M^{n+}}{K_M^{(n-1)+}}$$

In case $K_M^{n+} > K_M^{(n-1)+}$, complexation here lowers E.
For

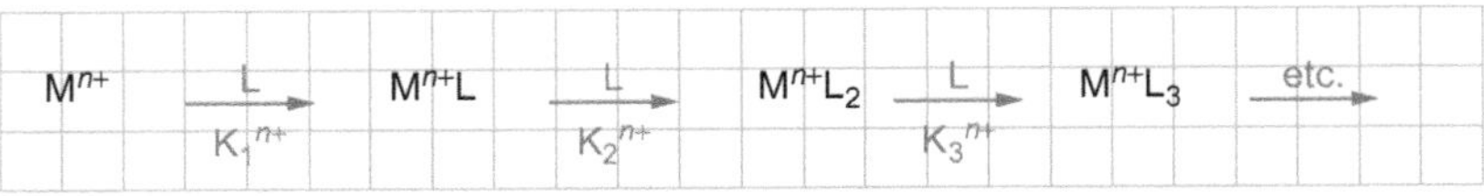

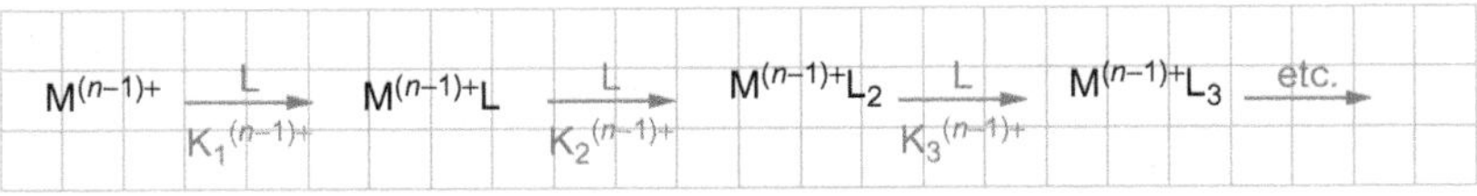

K values are the stepwise formation constants, while β is the overall formation constant:

$\beta_1 = K_1$, $\beta_2 = K_1K_2$, $\beta_3 = K_1K_2K_3$, $\beta_4 = K_1K_2K_3K_4$

$$E_{\text{Obs'd}} = E^{\text{o}} - \frac{RT}{nF} \text{Ln}\left(\frac{1 + \sum \beta_i^{n+}[\text{L}]^i}{1 + \sum \beta_j^{(n-1)+}[\text{L}]^j}\right)$$

If K_1 is only important, $n = 3$, and STD ($T = 273$ K, $P = 1$ atmosphere), [L] = 1 M.

$$E_{\text{f}} = E^{\text{o}} - \frac{RT}{nF} \ln\left(\frac{1 + K_1^{\text{III}}}{1 + K_1^{\text{II}}}\right)$$

For Fe/DTPA system, $E^{\text{o}} = 0.771$ V, $\log K^{\text{III}} = 28.6$, $\log K^{\text{II}} = 16.5$
$E_{\text{f}} = 0.057$ V
Complexation lower E by $\cong$700 mV, $Fe^{II}(DTPA)^{3-}$ is a considerably stronger reductant than $Fe(OH_2)_6{}^{2+}$

- When a ligand binds to a ferric ion stronger than it binds to a ferrous ion ($K_1^{\text{III}} > K_1^{\text{II}}$), the value of E_{f} drops (↓) versus that of E^{o}.

Use the E^{o}-diagram for manganese (Fig. 3.21):
Calculate E^{o} for $Mn^{3+} + 3e \rightarrow Mn$.

- The value of E^{o} can be calculated by:
  i. multiplying each of the E^{o} values of intervening couples by the number of the electrons involved:

$$Mn^{3+} + 1e \xrightarrow{1.51\,\text{V}} Mn^{2+}: \quad (1)(1.51)$$

$$Mn^{2+} + 2e \xrightarrow{-1.18\,\text{V}} Mn^{\text{o}}: \quad (2)(-1.18)$$

  ii. adding the products: $(1)(1.51) + (2)(-1.18) = -0.85$ V
  iii. dividing the sum by the total number of the electrons involved in the new-half-reaction:

$$E^{\text{o}} = \frac{-0.85}{3} = -0.283\,\text{V}$$

$[(1)(1.51) + (2)(-1.18)]/3 = -0.283$ V

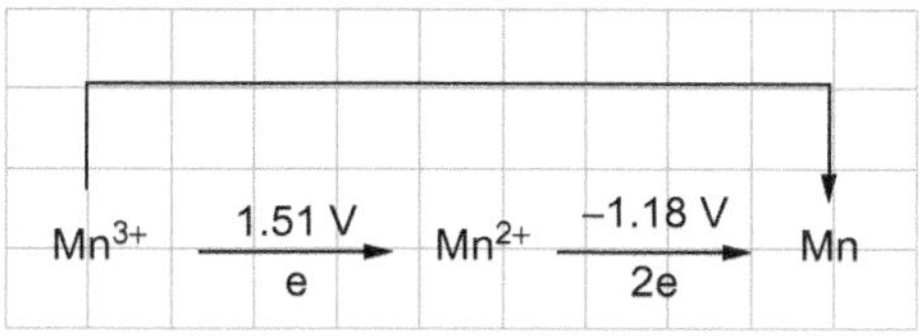

Given:

$$MnO_4{}^{2-} + 4H^+ + 2e \rightarrow MnO_2(s) + 2H_2O \quad E_A^{\text{o}} = 2.27\,\text{V}$$

What are the values of E as a function of pH, E_B^{o} at pH 14 and E_N^{o} at pH 7?

- Nernst equation:

$$E = E_A^{\text{o}} - \frac{RT}{nF} \ln Q$$

The value of E_A^{o} indicates that H^+ activity is unity, at 25°C, $F = 96{,}500$ C/mol, $R = 8.3$ J/K mol, 1 is used for the activity of solid MnO_2:

$$E = E_A^{\text{o}} - \frac{0.059}{2} \log \frac{1}{[H^+]^4 [MnO_4{}^{2-}]}$$

If the activity of MnO_4^{2-} is kept at 1:

$$E = E^{o}_{A} - \frac{(0.059)(4)}{2}\text{pH} = 2.27 - 0.118\text{pH}$$

The value of E^{o}_{B} is obtained by using the pH 14:

$$E^{o}_{B} = 2.27 - 0.118 \times 14 = 0.62\text{V}$$

For pH 7,

$$E^{o}_{N} = 2.27 - 0.118 \times 7 = 1.44\text{V}$$

3.9 INTERMOLECULAR INTERACTIONS

What is the difference between the intramolecular and intermolecular interactions?

- Intramolecular forces are the forces that hold together the atoms to form a molecule or compound. They include all types of chemical bond.
- Intermolecular forces are forces of attraction or repulsion that act between neighboring particles (atoms, molecules, or ions). They are weak compared to the intramolecular forces, or the forces that keep a molecule together.

Give examples of intermolecular interaction and arrange these forces from the weakest to the strongest.

- Intermolecular interactions are connected to bulk macroscopically observables such as melting point, boiling point, heat of vaporization, hardness, etc.
- Intermolecular forces have four major contributions, as follows:
    - van der Waals forces
        - London dispersion force
        - Debye force
        - Keesom force
    - Ion-induced dipole forces
    - Ion-dipole forces
    - A repulsive component resulting from the Pauli Exclusion Principle, which prevents the collapse of molecules.
- The arrangement of the forces:

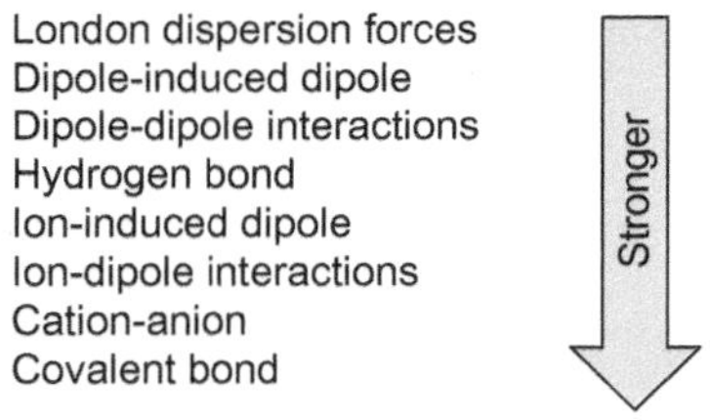

- Ion-dipole and ion-induced dipole forces are stronger than dipole-dipole interactions because the charge of any ion is much greater than the charge of a dipole moment. Ion-dipole bonding is stronger than hydrogen bonding.

van der Waals Forces

What are the van der Waals forces?

- van der Walls forces arise from:
    i. the weak attraction of nuclei from other molecules; and
    ii. the attraction exerted by a positive nucleus on electron beyond its own radius, e.g., F_2.
        - This is most noticeable for covalent molecules with low atomic weight.
        - The attractions are weaker than covalent and ionic bonds.

  - Molecules connected by van der Waals forces can be moved apart with energy input comparable to room temperature in range 10–300 K:
    i.e., Liquid ⇌ Gases
    H_2 (20 K), CH_4 (112 K), CF_4 (112 K)
  - van der Waals forces are additive, and the attraction increases with the number of electrons and opposite nuclear charge.
  - van der Waals forces are all short-range forces, have no directional characteristic, and only considered between the nearest particles.
- van der Waals forces play a fundamental role in:
  - supramolecular chemistry;
  - structural biology;
  - polymer science;
  - nanotechnology;
  - surface science;
  - condensed matter physics; and
  - defining many properties of organic compounds, including their solubility, boiling and melting points.
- van der Waals forces include (Fig. 3.22):
  - force between two instantaneously induced dipoles (London dispersion force);
  - force between a permanent dipole and a corresponding induced dipole (Debye force); and
  - force between two permanent dipoles (Keesom force).
- London dispersion forces: one molecule polarizes another (electron cloud temporarily distorted) given a brief dipole.
  - Examples:

$$H_2\,(\text{b.p. }20\text{K}) < CH_4\,(\text{b.p. }112\text{K}) \le CF_4\,(\text{b.p. }112\text{K}) \ll CCl_4\,(\text{b.p. }350\text{K}) < CBr_4\,(\text{b.p. }460\text{K})$$

  When constituent atoms are bigger (have a higher atomic number), then electrons are polarizable.
  - London dispersion forces are attractive interactions between any pair of molecules. These include nonpolar atoms that arise from the interactions of instantaneous multipoles.
  - These forces are fluctuating dipole-induced dipole, due to the nonzero instantaneous dipole moments of all atoms and molecules.

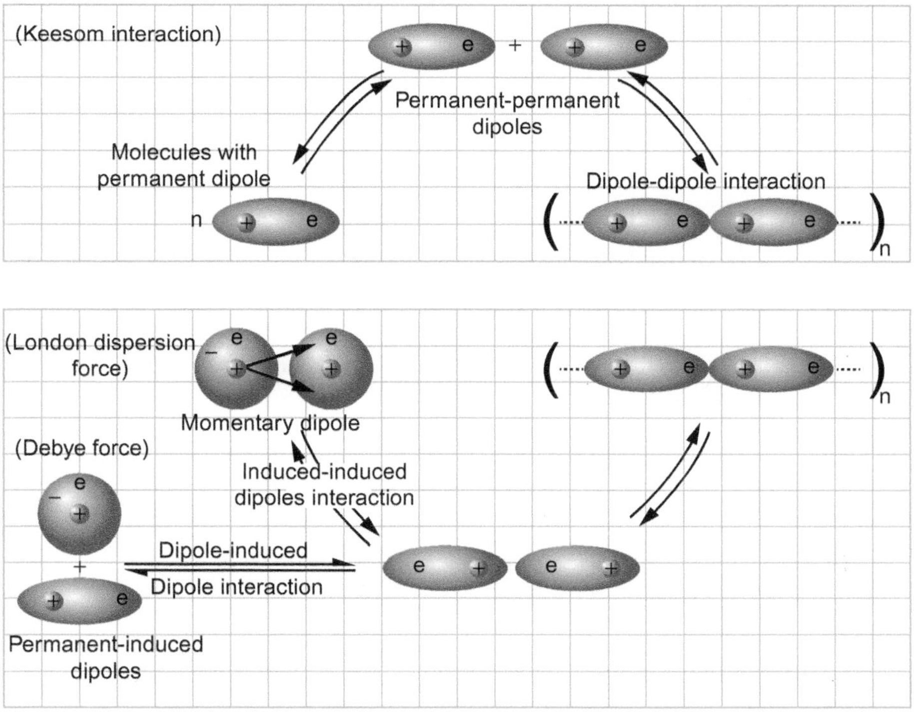

FIG. 3.22 The interactions between neutral molecules that contribute to van der Waals forces.

- Debye force (polarization):
  - This is an attractive interaction between a permanent dipole on one molecule with an induced dipole on another.
  - Examples: $Xe(H_2O)_x$, solvation of noble gases, between HCl and Ar.
  - Forces cannot occur between atoms, and are not temperature dependent.
- Keesom forces are attractive or repulsive electrostatic interactions between:
  - i. molecular ions: permanent charges;
  - ii. molecules without inversion center: dipoles;
  - iii. molecules with symmetry lower than cubic: quadrupoles; and
  - iv. dipolar molecules.

  - Examples:

			Boiling Point	Formula Weight
CO_2	Linear	$\mu=0$	194.7 K	44
CF_4	T_d	$\mu=0$	112 K	88
OF_2	Bent	$\mu=0.175$	128 K	54
NF_3	Triangular pyramid	$\mu=0.2$	144 K	71

  - These forces tend to align the molecules to increase attraction (reducing potential energy), and are temperature dependent.

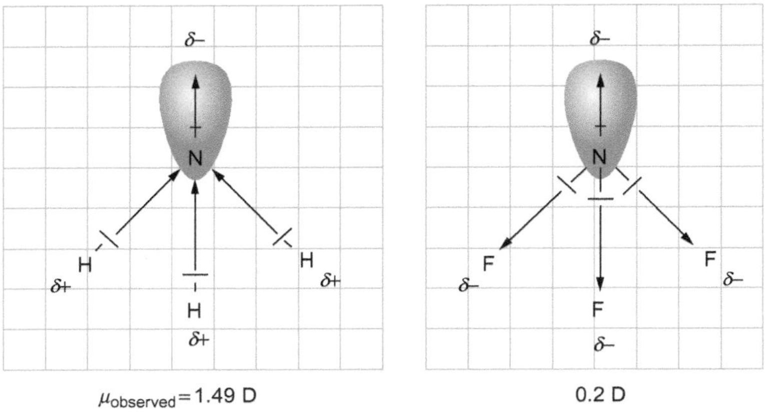

Ion-Induced Dipole Forces, Ion-Dipole Forces, and Hydrogen Bonding

Define and give example for ion-induced dipole forces, ion-dipole forces, and hydrogen bonding.

- Ion-induced dipole forces: electrostatic interactions between polar and nonpolar molecules.
  - Example: ion in molecular matrix.
  - An ion-induced dipole force consists of an ion and a nonpolar molecule interacting. As with a dipole-induced dipole force, the charge of the ion causes distortion of the electron cloud on the nonpolar molecule.
- Ion-dipole forces: electrostatic interactions between ion and polar molecules (stronger than hydrogen bonding), example: $K(OH_2)_6{}^+$.
  - They align so that the positive and negative groups are next to one another, allowing for maximum attraction.
  - Anion-dipole: solvation of anion by polar solvent such as H_2O, example: $[Cl(H_2O)_{\sim 4}]^-$.
  - Cation-dipole: solvation of cation by polar solvent, example: $[K(H_2O)_{6-7}]^+$.
    Note: mobility of aqueous Li^+, Na^+, K^+, and Cs^+ reverse of order expected from isolated ion's radii: $Li(OH_2)_4^+ < Na(OH_2)_6^+ < K(OH_2)_7^+ < Cs(OH_2)_8^+$, due to the graduate decrease in the stability of the secondary hydration sphere (Fig. 3.23).
- Hydrogen bonding: attraction between the lone pair of an electronegative atom and a hydrogen atom that is bonded to either nitrogen, oxygen, or fluorine,
  E.g., $(H_2O)_x$, $(HF)_x$, $(NH_3)_x$, alcohol, amines.

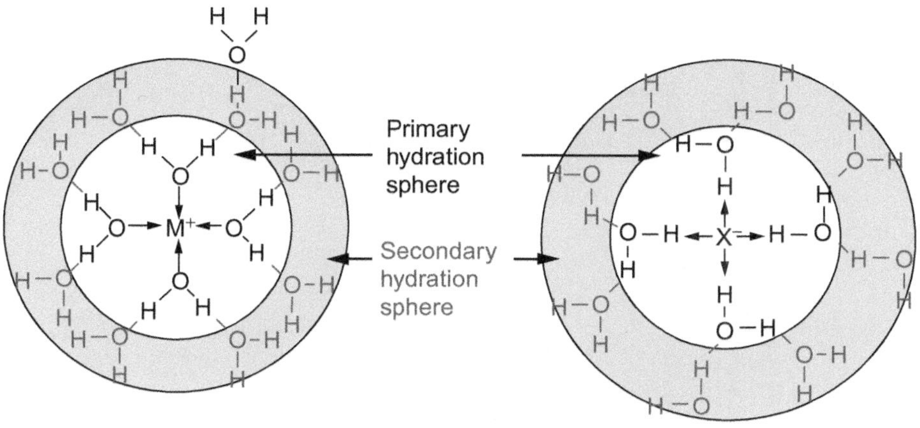

FIG. 3.23 Primary and secondary hydration sphere.

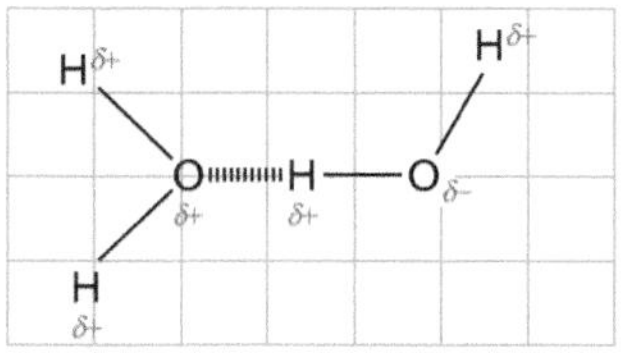

- Intermolecular hydrogen bonding is responsible for the high boiling point of water (100°C) compared to the other hydrides, which have no hydrogen bonds (Fig. 3.24).
- Intramolecular hydrogen bonding is responsible for the secondary, tertiary, and quaternary structures of proteins and nucleic acids. It also plays an important role in the structure of polymers, both synthetic and natural.
- The hydrogen bond has some features of covalent bonding.
  - it is directional;
  - stronger than a van der Waals interaction;
  - produces interatomic distances shorter than the sum of van der Waals radius; and
  - usually involves a limited number of interaction partners.

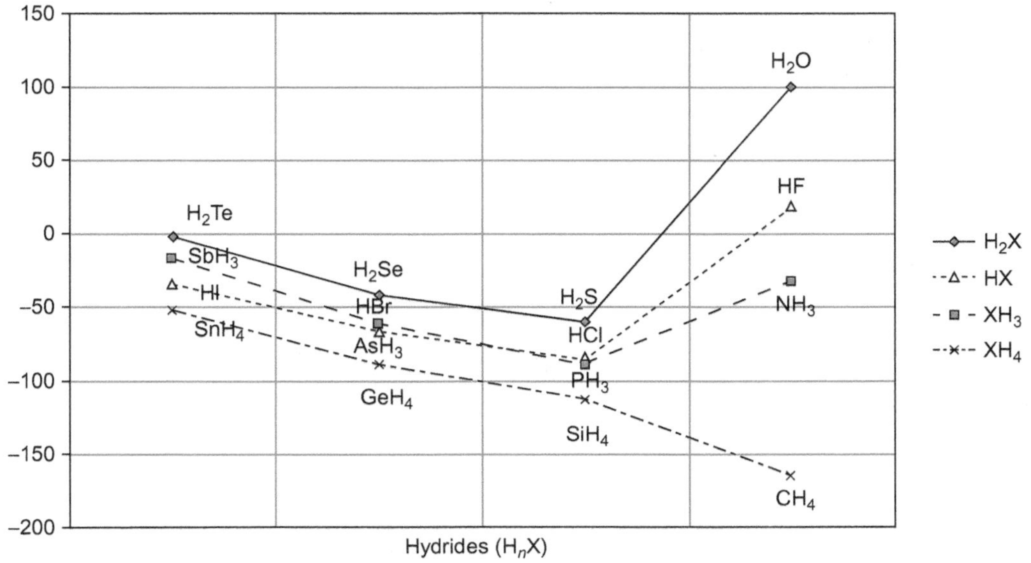

FIG. 3.24 Plot of the boiling points of some binary hydrogen compounds.

What is the maximum possible number of hydrogen bonds in which an H_2O molecule can form?

- The maximum possible number of hydrogen is four.

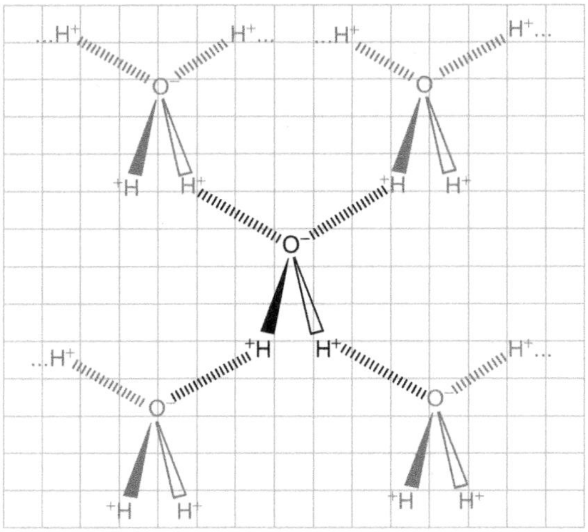

In the vapor state, methanol is associated to form a tetramer $(CH_3OH)_4$. What is the structure of the methanol tetramer?

- The expected structure of $(CH_3OH)_4$:

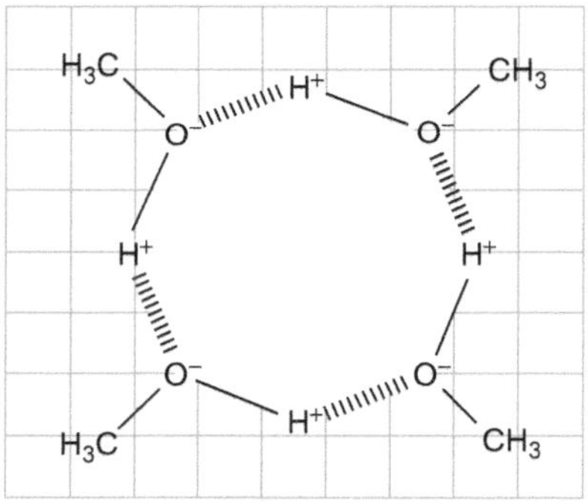

Why are $(CH_3)_2O$ quite soluble in water and fatty acids soluble in chloroform?

- Dimethyl ether can form two hydrogen bonds, and is completely miscible with water

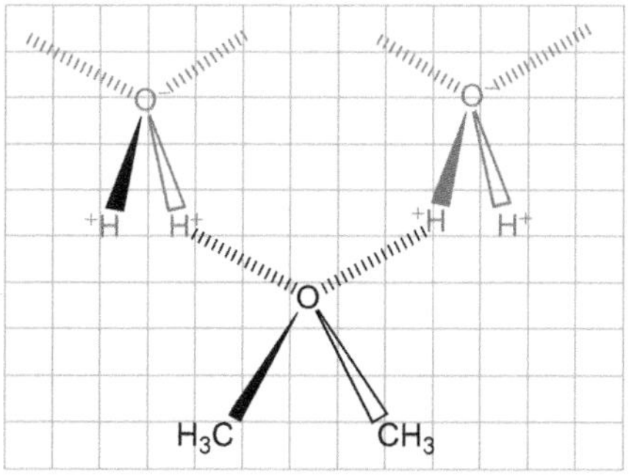

- Chloroform is a good solvent for fatty acids, because its polar C—H bond may engage in hydrogen bonding.

Graphically compare the boiling point of following binary hydrogen compounds:

(XH_4): $CH_4 = -164°C$, $SiH_4 = -111.5°C$, $GeH_4 = -88.5°C$, $SnH_4 = -51.8°C$
(XH_3): $NH_3 = -33.34°C$, $PH_3 = -87.7°C$, $AsH_3 = -62.5°C$, $SbH_3 = -17°C$
(H_2X): $H_2O = 100°C$, $H_2S = -60°C$, $H_2Se = -41.25°C$, $H_2Te = -2.2°C$
(HX): $HF = 19.5°C$, $HCl = -85.05°C$, $HBr = -66°C$, $HI = -35°C$.

a. **Which one of the represented elements would you expect to exhibit hydrogen bonding?**
b. **What is the chemical structure of hydrogen fluoride in pure state (vapor or liquid)?**
c. **Use the data given below to estimate the heat of sublimation of a hypothetical form of ice in which water molecules are not hydrogen bonded.**

If: Z = atomic number; H, Z = 1; C, Z = 6; O, Z = 8; Ge, Z = 32; Se, Z = 34; Sn, Z = 50; Te, Z = 52, and

	Heat of Sublimation (kcal/mol)
GeH_4	4.0
H_2Se	5.3
SnH_4	5.0
H_2Te	6.6
CH_4	2.0

a. Generally, boiling and melting points of the hydrides compounds regularly increase with increasing molecular weight within a group.
   - Fig. 3.24 indicates that boiling points of HF, H_2O, and NH_3 behave inconsistently, due to the occurrence of hydrogen bonds between molecules.
b. The chemical structure of hydrogen fluoride:

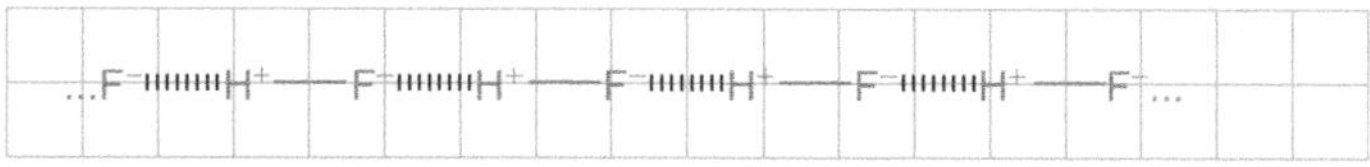

c. If: H, Z=1; C, Z=6; O, Z=8; Ge, Z=32; Se, Z=34; Sn, Z=50; Te, Z=52
   - The number of electrons in $CH_4 = 6 + 4(1) = 10$, CH_4 does not form hydrogen bonding.
     The number of electrons in $H_2O = 2(1) + 8 = 10$.
     Therefore, CH_4 and H_2O are isoelectronic.
   - Similarly, GeH_4 and H_2Se are isoelectronic (each has 36 electrons), and SnH_4 and H_2Te are isoelectronic (each has 54 electrons).
   - The ratio of the heat of sublimation of isoelectronic pair of molecules:

$$\frac{\Delta H_{\text{Sublimation }(GeH_4)}}{\Delta H_{\text{Sublimation }(H_2Se)}} = \frac{4.0}{5.3} = 0.755$$

$$\frac{\Delta H_{\text{Sublimation }(SnH_4)}}{\Delta H_{\text{Sublimation }(H_2Te)}} = \frac{5.0}{6.6} = 0.758$$

   - Fig. 3.24 indicates that CH_4, GeH_4, H_2Se, SnH_4, and H_2Te do not exhibit inconsistent boiling points, and are not hydrogen bonded.
   - Therefore, by assuming that a ratio of 0.76 would also apply to the isoelectronic pair, H_2O and CH_4:

$$\Delta H_{\text{Sublimation }(H_2O)} = \frac{\Delta H_{\text{Sublimation }(CH_4)}}{0.76} = \frac{2.0}{0.76} = 2.6\text{kca/mol}$$

What are the effects of temperature on the intermolecular interactions?

- All intermolecular forces depend on the relative orientation of the molecules, which means that they are anisotropic (except those between two noble gas atoms).

- The induction and dispersion interactions are always attractive, irrespective of orientation, but the electrostatic interaction changes sign upon rotation of the molecules. That is, the electrostatic force can be attractive or repulsive, depending on the mutual orientation of the molecules.
- When molecules are in thermal motion, as they are in the gas and liquid phase, the electrostatic force is averaged out, because the molecules thermally rotate and thus probe both repulsive and attractive parts of the electrostatic force.
- van der Waals forces are responsible for certain cases of pressure broadening (van der Waals broadening) of spectral lines and the formation of van der Waals molecules.

Design a schematic chart to present a systematic way of identifying the kinds of intermolecular forces in particular system.

- Fig. 3.25 represents a diagram for finding the type of intermolecular forces in particular system.

If the charges on the H and Cl atoms were 1+ and 1−, respectively, the bond length in the HCl molecule is 127 pm, and the experimentally measured dipole moment of $HCl_{(g)}$ is 1.08 D.
1 Debye = 3.34×10^{-30} C m, elementary charge = $\times 1.6\,10^{-19}$ C, pm = 10^{-12} m.

a. **Calculate the dipole moment, in Debye.**
b. **What magnitude of charge, in units of e, on the H and Cl atoms would lead to the experimentally measured dipole moment? What do you conclude?**
c. **If $\chi_{(H)} = 2.1$, and $\chi_{(Cl)} = 3$, find the partial charge on Cl and H atoms, and compare these results with the experimental magnitude of charges on each atoms.**

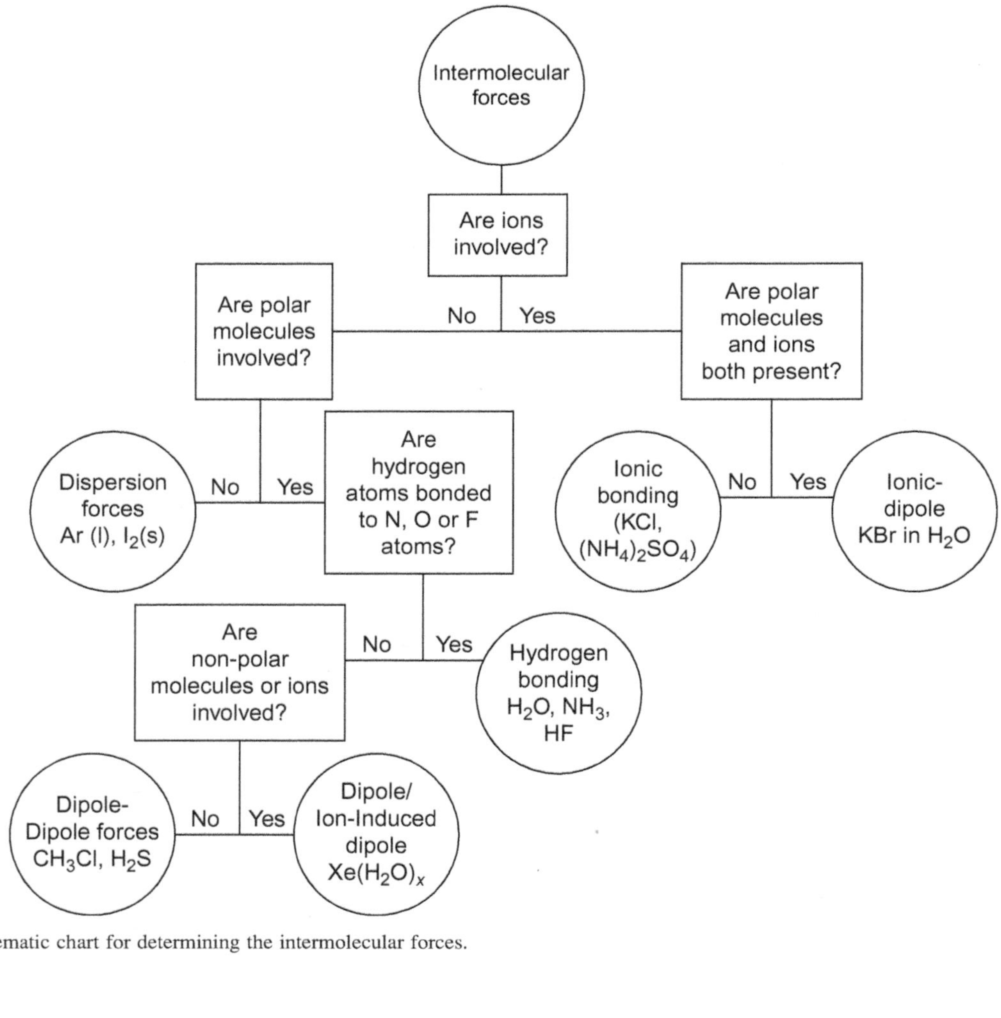

FIG. 3.25 Schematic chart for determining the intermolecular forces.

- a. Electronegativity causes inhomogeneous distribution of molecular charge that induces imbalances of electron distribution within the molecule.
  μ (dipole moment) = charge × separation = coulomb × meter
  1 Debye = 3.34×10^{-30} C m

$$\mu_{\text{Calculated}} = (1.6\times10^{-19}\text{C})(127\times10^{-12}\text{m})$$
$$= (1.6\times10^{-19}\text{C})(127\times10^{-12}\text{m})\left(\frac{1\text{D}}{3.34\times10^{-30}\text{Cm}}\right) = 6.08\text{D}$$

H − Cl : $\chi_H < \chi_{Cl}$, so: $H^{\delta+} \rightarrow Cl^{\delta-}$, the measured $\mu_{HCl} = 1.08\text{D}$

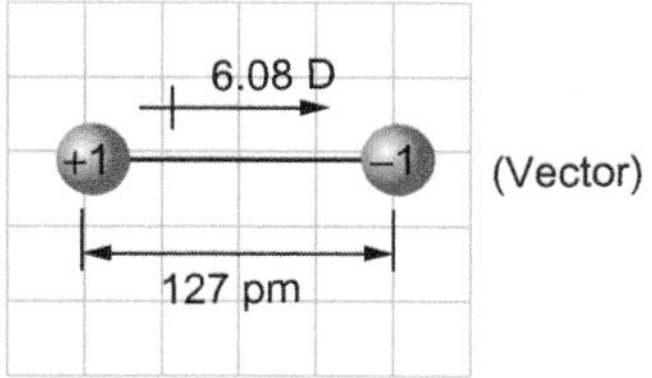

b. The charge in units of e:

$$\text{Charge in C} = \frac{\mu}{\text{bond length}} = \frac{1.08\text{D}}{127\times10^{-12}\text{m}} = 2.84\times10^{-20}\text{C}$$

$$\text{Charge in e} = \frac{2.84\times10^{-20}\text{C}}{1.6\times10^{-19}\text{Ce}^{-1}} = 0.178\text{e}$$

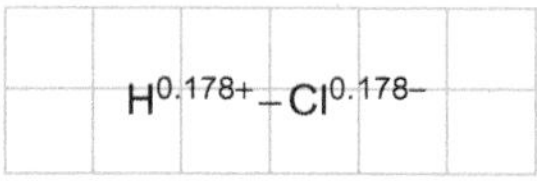

Since the H—Cl bond is polar covalent, the charges of the atoms are less than a full electronic charge.

- c. Electrons in covalent bonds are located according to the relative electronegativity of the bonded atoms:
  - For a Cl atom:
    The expected share of electric charge due to one electron of the bonding pair =
    $$= \frac{\chi_{Cl}}{\chi_{Cl} + \chi_H} = \frac{3}{3 + 2.1} = 0.59$$
    The Cl atom's share = 0.59 × 2 = 1.18 e (because 2 electrons are shared between H and Cl).
    Therefore, 0.18 e more than neutral Cl atom; this gives rise to a partial charge of −0.18 e on a Cl atom, which in a good agreement with the experimental measured of the partial charge −0.178 e on a Cl atom.
  - For H atom:
    The expected share of electric charge due to one electron of the bonding pair =
    $$= \frac{\chi_H}{\chi_{Cl} + \chi_H} = \frac{2.1}{3 + 2.1} = 0.41$$
    The H atom's share = 0.41 × 2 = 0.82 e ← (1 − 0.82 = 0.18)
    Therefore, 0.18 e less than a neutral H atom; this gives rise to a partial charge of +0.18 e on an H atom, which in a good agreement with the experimental measured of the partial charge +0.178 e on an H atom.

Arrange in order of decreasing boiling point: CO, H_2, $BaCl_2$, Ne, and HF.

- The boiling point depends on the intermolecular attractive forces in the liquid state.
- These substances need to be ordered according to the relative strengths of the different kinds of forces.
- Ionic substances have attractive forces stronger than for molecular ones; thus, $BaCl_2$ should have the highest boiling point.
- The intermolecular forces of CO, H_2, Ne, and HF depend on the molecular mass, polarity, and hydrogen bonding.

- The molecular masses are:
  $CO = 12\ u + 16\ u = 28\ u$, polar;
  $H_2 = 1\ u + 1\ u = 2\ u$, nonpolar;
  $Ne = 20\ u$, nonpolar; and
  $HF = 1\ u + 19\ u = 20\ u$, can form hydrogen bonds.
- The boiling point of H_2 should be the lowest because it is nonpolar and has the lowest molecular weight.
- Since HF can form hydrogen bonds, thus, it should have a higher boiling point than CO and Ne.
- Because CO is polar and has a higher molecular mass than those of Ne, therefore, CO should have a higher boiling point.
- The predicted order of boiling points is as follows:
  $H_2 < Ne < CO < HF < BaCl_2$

3.10 COVALENT NETWORKS AND GIANT MOLECULES

What are the structural differences among diamond, graphite, and fullerenes?

- The elements of giant crystals form are an extended covalent array (opposed to metallic).
- All group IV elements except lead, phosphorus, arsenic, selenium, and tellurium can form giant crystals, e.g., graphite, (Buckminster) fullerene.
  - Structure of diamond: each atom is surrounded by a regular tetrahedron of covalently bounded neighboring atoms. It needs to break C-C covalent bonds to melt or cleavage.

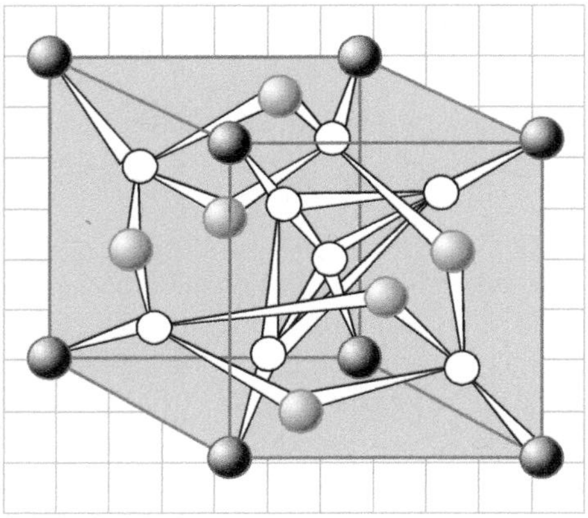

  - Graphite: infinite superbenzeno structure, the four valence electrons are used for conjugate double bonds through each layer, each carbon has one double and two single bonds.

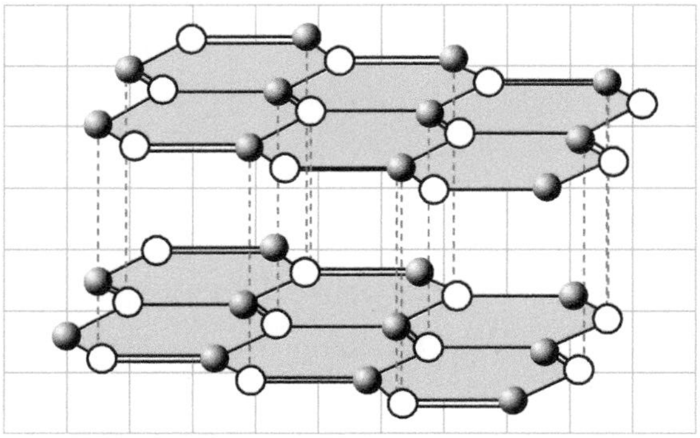

  - Layered structure "layer lattice" with weak interaction between layer planes.
  - It is used as a lubricant, 2D-conductor of electrons.
  - Fluorination to (C-F) gives a third dimension to the covalent structure of a C-atom.

- (Buckminster) Fullerene:
  - C_{60} and C_{70} condense from carbon vapor generated in electric arc discharge, soot from above is extracted with toluene to give red-purple solution (sooty flames, too). Chromatography on alumina separates 70% C_{60}, 25% C_{70}, and others. Most use is C_{60}.
  - C_{60} has soccer-ball type of structure: pairs of 6-ring alternate with 5-rings.
  - C_{60} and C_{70} can be oxidized and reduced, e.g., electrochemistry of C_{70} to yield $C_{70}^{\cdot-}$, C_{70}^{2-}, C_{70}^{3-} ($C_{60}^{\cdot-}$, C_{60}^{2-}, C_{60}^{3-}), and C_{70}^{4-} also observable.
  - All $C_n^{-/2-}$ show unpaired ē spins, so C_n^{2-} is diradical, not singlet.
  - Anions of all C_{60} and C_{70} have strong absorption bands in NIR and UV, tailing into visible region.

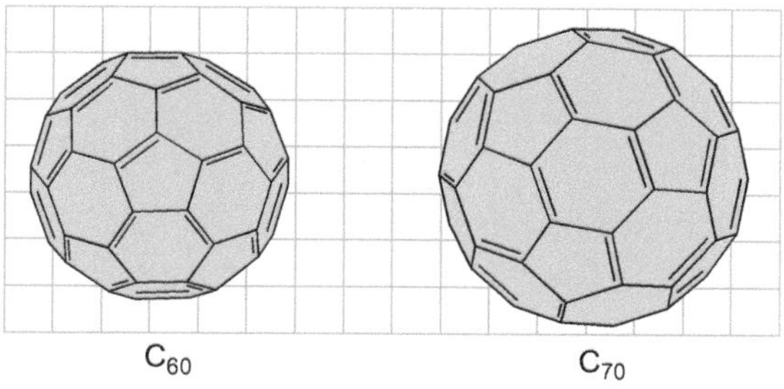

Are C_{60} and C_{70} aromatics or alkenes?

- X-ray crystallography of C_{60} and C_{70} were not successful, because in crystal the bucky balls are disordered by rotating among pseudorandom orientations.
- OsO_4 reacts with alkenes:

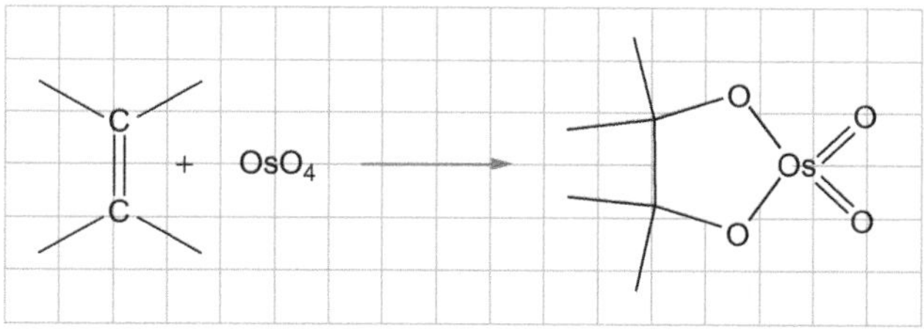

C_{60} reacts with OsO_4, and 1:1 can be isolated as 4-ᵗbutylpryidine. Adduct of Os improves solubility and crystallography.

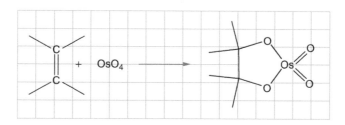

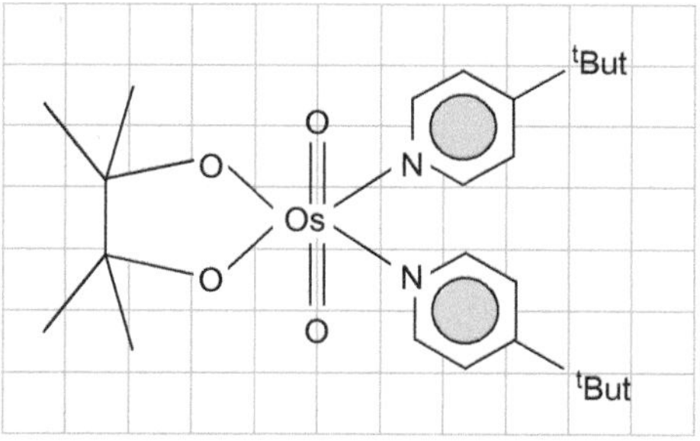

OsO_4 adds across C=C bond at 6/6 junction. This reduces the symmetry, I_h, to a lesser one coinciding with less disorder. Obtained structure: C_{60} shell is about 7.2 Å across (C-to-C), C-C distances are 1.39 Å (6/6) and 1.43 Å (6/5).

- C_{60} reacts with $(Ph_3P)_2Pt(C_2H_4)$ gives 1:1 adduct, typical of alkenes:

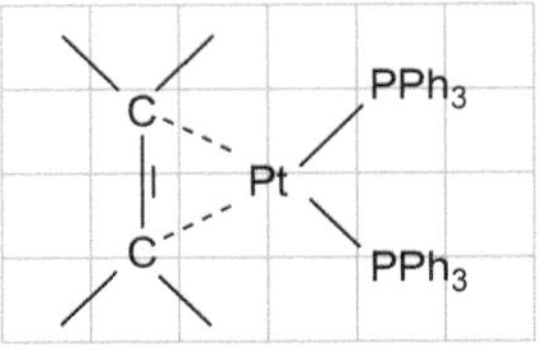

- $[(\eta^5 - C_5H_5)Ru(N{\equiv}C{-}CH_3)]^+$ reacts with arenes and gives us

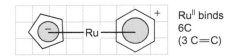

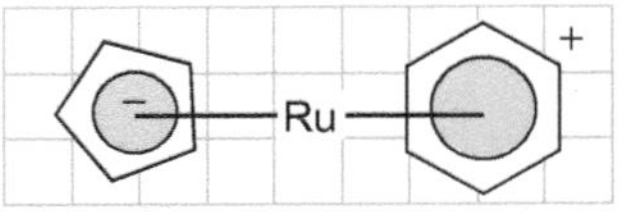

C_{60} yields $(\eta^5 - C_5H_5)Ru(N{\equiv}C{-}CH_3)C_{60}$. The CH_3CN shows that the Ru-C_{60} linkage probably involves only 2/3 of a C_6 unit, i.e., ties up only 2 C=C of C_{60}.

What happens when alkali metals are doped into Fullerite?

- Rb and K can be doped into fullerite, and yield C_{60}^-, and C_{60}^{2-} as M^+ salts.
- Ion formation quenches rotational disordering, and gets continuously varying composition phases, not just $Rb^+C_{60}^-$ but all possible $Rb_x^+C_{60}^-$ $(0 \geq x \geq 6)$.
- These solids become electrically superconducting at 5–30 K ($Rb_3^+C_{60}^-$ best).

What is meant by endohedral fullerenes?

- Fullerenes that have other atoms, ions, or clusters held within inner spheres are known as endohedral fullerenes, e.g., $C_{60}La$.

Graphite, Fullerenes, Graphene, Carbon Nanotubes, and Asbestos

What are the main chemical features of graphene?

- Carbon is able to form many allotropes due to its valency. Well-known forms of carbon allotropes include charcoal, diamond, graphite, buckminsterfullerene, graphene, nanotubes, nanobuds, nanoribbons, and other forms that exist at very high temperatures or extreme pressures.
- Graphene has a planar structure with edge hanging carbon atoms attached around, which are chemically reactive along with defects on the graphene plane.
- Graphene can be regarded as an infinitely flat large aromatic molecule.
- Graphene is a two-dimensional honeycomb lattice in which one atom forms each vertex.
  - Its carbon atoms are densely packed in a regular atomic-scale of hexagonal pattern.
  - Graphene's hexagonal lattice can be regarded as two interleaving triangular lattices.
  - Electron diffraction patterns showed the expected honeycomb lattice.
  - Graphene is a single freestanding atomic plane of graphite that is sufficiently isolated from its environment.
- Graphene has many surprising properties.
  - It is about 100 times stronger than the strongest steel.
  - It conducts heat and electricity efficiently:
  - It is nearly transparent.
  - It shows a large and nonlinear diamagnetism.
  - Electrons in graphene have a very high mobility due to delocalization of the pi bond electrons above and below the planes of the carbon atoms. It could possibly be used in electronic applications, such as: the bipolar transistor effect, ballistic transport of charges and large quantum oscillations in the material.
  - It can be used as material in the electrode, electrical lamps.
  - Graphene has a theoretical specific surface area of 2630 m^2/g. This is much larger than that reported for carbon black (>900 m^2/g) or for carbon nanotubes, from $\approx$100 to 1000 m^2/g and is similar to activated carbon.

- Graphene chemical structure:
  - Each atom has four bonds, one σ bond with each of its three neighbors and one π-bond that is oriented out of plane. The atoms are about 1.42 Å apart.
  - Graphene's stability is due to its tightly packed carbon atoms and a sp^2 orbital hybridization, Chapter 5, p. 290—a combination of orbitals s, p_x and p_y that constitute the σ-bond. The final p_z electron makes up the π-bond. The π-bonds hybridize together to form the π-band and π^*-bands, Chapter 6, p. 394. The gapless between these bands are responsible for most of graphene's notable electronic properties, via the half-filled band that permits free-moving electrons.
  - Graphene is the only form of carbon (or solid material) in which every atom is available for chemical reaction from two sides (due to the 2D structure).
  - Atoms at the edges of a graphene sheet have special chemical reactivity.
  - Graphene has the highest ratio of edge atoms of any allotrope.
  - Defects within a sheet increase its chemical reactivity.
  - The onset temperature of reaction between the basal plane of single-layer graphene and oxygen gas is below 260°C (530 K).
  - Graphene burns at very low temperature (e.g., 350°C (620 K)).
  - Graphene is commonly modified with oxygen- and nitrogen-containing functional groups, and is analyzed by infrared spectroscopy and X-ray photoelectron spectroscopy.
  - Graphene can self-repair holes in its sheets, when exposed to molecules containing carbon, such as hydrocarbons. Bombarded with pure carbon atoms, the atoms perfectly align into hexagons, completely filling the holes.
  - Physicists reported that single-layer graphene is a hundred times more chemically reactive than thicker sheets.
- Stability:
  - Graphene sheet is thermodynamically unstable if its size is less than 20 nm, about 6000 atoms, and becomes the most stable as within graphite only for molecules larger than 24,000 atoms.
- Thermal conductivity:
  - Graphene indicted a large thermal conductivity compared with the thermal conductivity of graphite at room temperature.
  - The isotopic composition of ^{12}C to ^{13}C has a considerable effect on the thermal conductivity. For instance, isotopically pure ^{12}C graphene exhibits higher thermal conductivity than either a 50:50 isotope ratio or the naturally occurring 99:1 ratio.
- Forms:
  - Monolayer sheets
  - Graphene oxide
  - Bilayer graphene
  - Graphene superlattices
  - Graphene nanoribbons
  - Graphene quantum dots

What are the resemblances and the differences between graphene and nanotubes?

- Carbon nanotubes has nearly the same chemical properties as graphene, and has similar extended conjugated polyaromatic system.
- Carbon nanotubes are a planar hexagonal assembly of carbon atoms arranged in a honeycomb lattice.
- Single-walled carbon nanotubes are elongated structure that can reach several micrometers of very small diameters. The end cape can be thought as an incomplete fullerene.

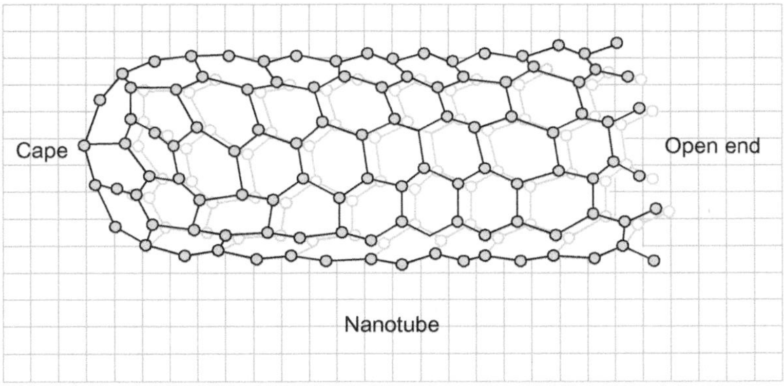

- A single-walled carbon nanotube can be pictured as a rolled graphene sheet (Fig. 3.26).
- Nanotubes are classified according to the rolling mode such as zigzag, chair, or armchair.

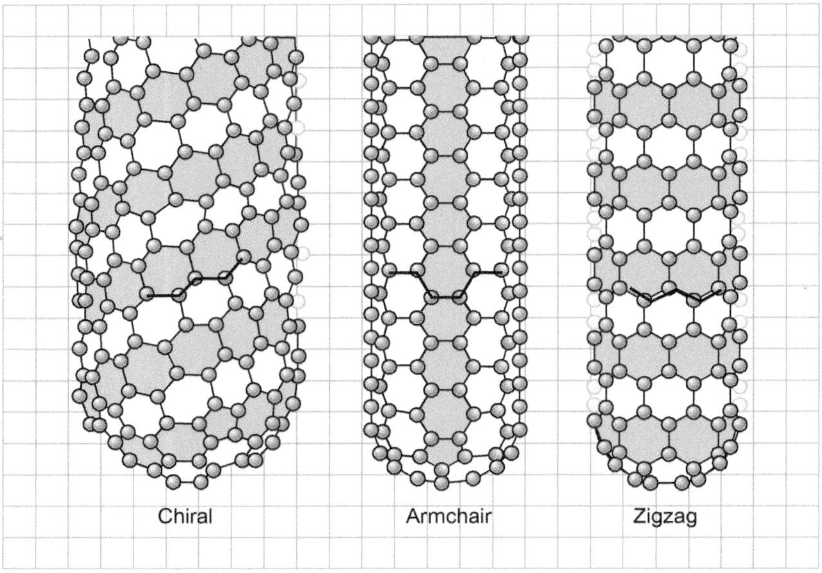

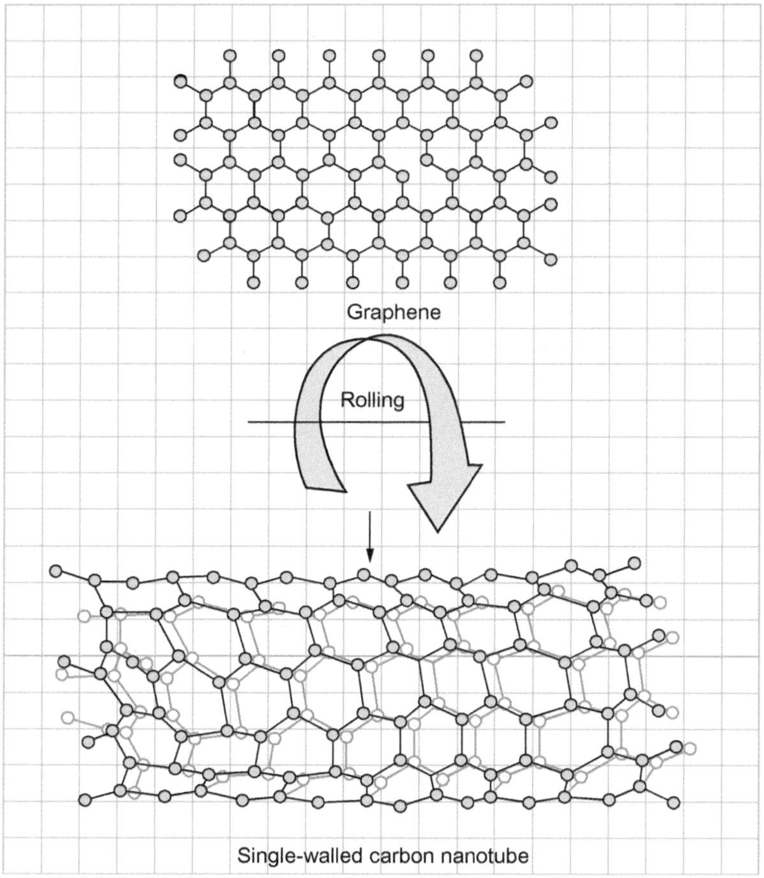

FIG. 3.26 Rolled-up sheet of graphene.

- The most noticeable features of single-walled carbon nanotube are:
  - semiconducting and metallic electronic properties;
  - mechanical properties; and
  - optical and chemical characteristics.
- Multiwalled nanotubes are assembly of concentric single-walled nanotube with different diameters.
- The observable remarks of multiwalled carbon nanotube are:
  - unique electrical properties;
  - extraordinary strength;
  - heat conduction efficiency; and
  - fastest known oscillators.

  Used in manufacturing:
  - high tensile strength fibers;
  - fire-resistant materials;
  - building blocks for molecular electronics;
  - artificial muscles; and
  - loudspeakers made from sheets of parallel carbon nanotubes.
- Carbon nanotubes can be synthesized by dipping hot graphite into water, graphene scrolling, SiC decomposition, flame synthesis, and ball milling,.
- The techniques for producing single- and multiwalled nanotubes include:
  - arch discharge;
  - laser ablation; and
  - chemical vapor deposition.
- Density gradient ultracentrifugation is a very important method for separating carbon nanotubes with different diameters, lengths, wall numbers, and chiralities, where surfactants used to well disperse carbon nanotubes through noncovalent interaction are very important.
- This procedure has been applied to the separation of graphene sheets with controlled thickness.

How is graphene prepared?

- The most common means to graphene involves the production of graphene oxide by extremely harsh oxidation chemistry (Fig. 3.27).

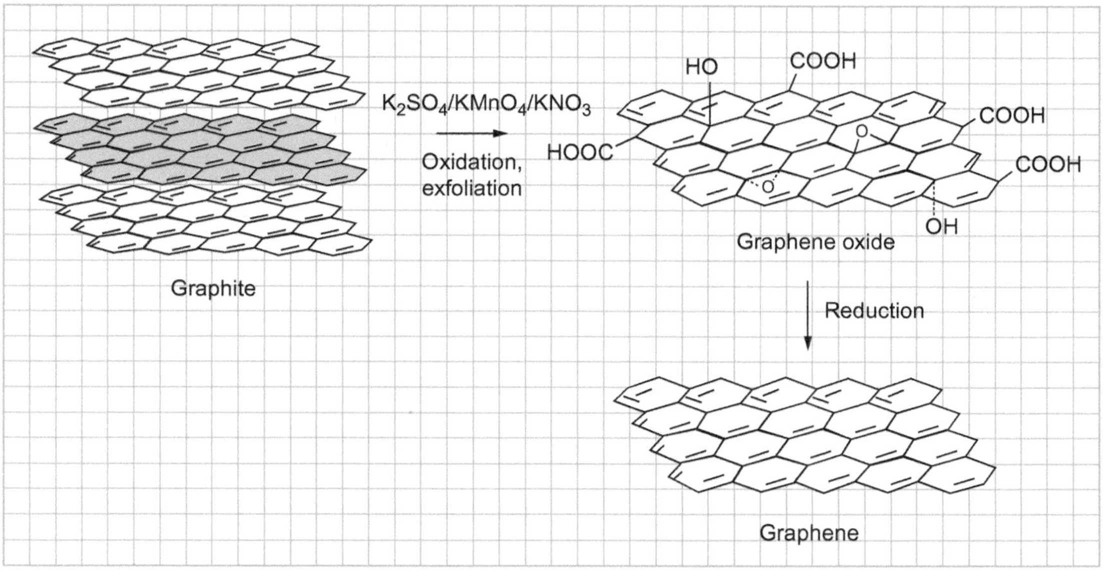

FIG. 3.27 Graphene can be produced from graphite oxidation and exfoliation to graphene oxide and subsequent reduction.

   - First, graphite is treated with an acidic mixture of sulfuric acid and nitric acid.
   - Graphene oxide can be simply prepared through exfoliation of graphite oxide.
- The chemical property of graphene oxide is related to the functional groups attached to graphene sheets.
- The structure of graphene oxide has been the subject of many studies; it contains epoxide functional groups along the basal plane of sheets as well as hydroxyl and carboxyl moieties along the edges.

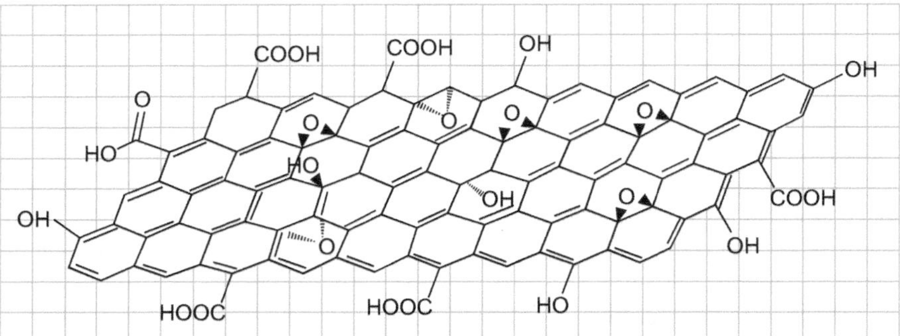

Graphene oxide

- Graphene oxide is water-soluble and also soluble in some organic solutions for further treatment or applications.
- Graphene oxide flakes show enhanced photo-conducting properties.
- As graphene oxide is electrically insulating, it must be converted by reduction to restore the electronic properties of graphene.
- Graphene oxide can be reduced to yield graphene sheets through:
  1. chemical reduction;
  2. thermal reduction; or
  3. electrochemical reduction.
- Chemically converted graphene is typically reduced by hydrazine or borohydride (Fig. 3.27).

$$\text{Graphene oxide} \xrightarrow[]{\substack{1.\ H_4N_2,\ \text{or} \\ 2.\ NaBH_4,\ \text{or} \\ 3.\ KOH/H_2O}} \text{Graphene}$$

- The properties of converted graphene can never fully match those of graphene for two reasons:
   - oxidation to graphene oxide introduces defects; and
   - chemical reduction does not fully restore the graphitic structure.
- Due to its low-cost, large-scale, and easy-processing, the reduced graphene oxide has been widely applied in many fields.

Graphene and carbon nanotubes have very wide potential applications in many areas, such as materials science and medicinal chemistry. What are the obstacles that limits these future applications and the alternative solutions?

- The perfect graphene and nanotubes are:
   - normally hydrophobic;
   - insoluble and chemically inert for many organic reagents; and
   - impermeable to all gases and liquids.
- The insolubility and difficult manipulation of graphene and carbon nanotubes in aqueous solutions and common organic solvents limits further applications.
- In order to solve this problem, functionalization of graphene and carbon nanotubes is an alternative key.
- In addition, the electrical properties can be tuned by edge-functionalization, and hetero-atom doping.

What are the possible means for functionalization of carbon nanotubes?

- There are three main means for functionalization of carbon nanotubes.
   - The first one is covalent interaction at the end groups or defects on carbon nanotubes.
     Both ends of carbon nanotubes, and the defects formed while growth, purification and dispersion, are more reactive compared with the side-wall of carbon nanotubes.

- The second one is noncovalent adsorption or wrapping around the side-wall of carbon nanotubes due to π-π interaction or van der Waals forces.
- The third possibility is filling in the inner cavities of carbon nanotubes with molecules have smaller diameters.
- Via the functionalization, water-soluble and organic-soluble carbon nanotube can be obtained for further manipulation and applications.

What are the possible approaches for functionalization of graphene? How are these different from those of carbon nanotube?

- Graphene can be functionalized similar to nanotubes by covalent interaction (Figs. 3.28 and 3.29), noncovalent adsorption, or wrapping through π–π interaction.
- The differences between their functionalization are based on their geometric structures.
- The covalent functionalization of graphene happens at the sites of edges and defects that are actively oxygenated groups caused in the process of synthesis or pretreatment.
- Both sides of graphene have similar reactivity, unlike the different reactivity of outer- and inner-wall for carbon nanotubes functionalization.
- Full hydrogenation from both sides of graphene sheet results in graphane, but partial hydrogenation leads to hydrogenated graphene.
- Similarly, both-side halogenation of graphene leads to graphene halides, while partial halogenation provides halogenated graphene.
- Graphene oxide can be further functionalized through covalent reactions of various reagents with oxygenated groups, as summarized in Fig. 3.28.
- Fig. 3.29 indicates some of the chemical similarity of the single-walled carbon nanotube and graphene.
  - Refluxing of the graphene oxide in solvents leads to reduction and folding of individual sheets as well as loss of carboxylic group functionality, indicating thermal instabilities of graphene oxide sheets.

$$\text{Graphene oxide} \xrightarrow{\text{Solvent reflux}} \text{Graphene}$$

  - Graphene oxide is converted to acid chloride groups by treatment with thionyl chloride, $SOCl_2$.

$$\begin{matrix} \text{Graphene} - \text{COOH} \\ \text{Nanotube} - \text{COOH} \end{matrix} \xrightarrow{SOCl_2} \begin{matrix} \text{Graphene} - \text{COCl} \\ \text{Nanotube} - \text{COCl} \end{matrix}$$

  - Graphene-COCl can be converted to the corresponding graphene amide via treatment with RNH_2.

$$\begin{matrix} \text{Graphene} - \text{COCl} \\ \text{Nanotube} - \text{COCl} \end{matrix} \xrightarrow{RNH_2} \begin{matrix} \text{Graphene} - \text{CONHR} \\ \text{Nanotube} - \text{CONHR} \end{matrix}$$

  The resulting material is soluble in tetrahydrofuran, tetrachloromethane, and dichloroethane.
- The covalent functionalization of graphene, graphene oxide, or carbon nanotubes can greatly improve their performance of solubility and process ability. However, it will also induce many defects, which may greatly decrease the intrinsic electrical and mechanical properties.
- The double-sided functional structure of graphene leads to 3D association structures for special applications (Fig. 3.30).
  - The cross-linkage of graphene oxide with polyallylamine through epoxy groups lead to increase stiffness and strength relative to unmodified graphene oxide paper.
- Noncovalent functionalization of graphene through π-π stacking or van der Waals interaction completely keeps the sp^2 network structure of graphene.
- The interaction of planar metallic complexes such Cu^{II}(bipyridine-(chlorestrol)$_2$)$_2$ complex can be used in the solubilization and debundling carbon nanotube (Fig. 3.31). The addition of the complex shows a reversible sol-gel phase transition on changing the redox state of Cu^I/Cu^{II} complexes.
- The noncovalent interaction is often reversible. Therefore, it has received extensive attention in many areas, especially as chemical sensors and biomedical materials.
- Several organic solvents, surfactants and biomolecules, e.g., DNA, sodium cholate, sodium deoxycholate, and sodium dodecyl sulfate, etc., have been encapsulated carbon nanotubes for dispersion and/or separation through noncovalent functionalization. These are also greatly utilized on the dispersion and separation of graphene for further applications.

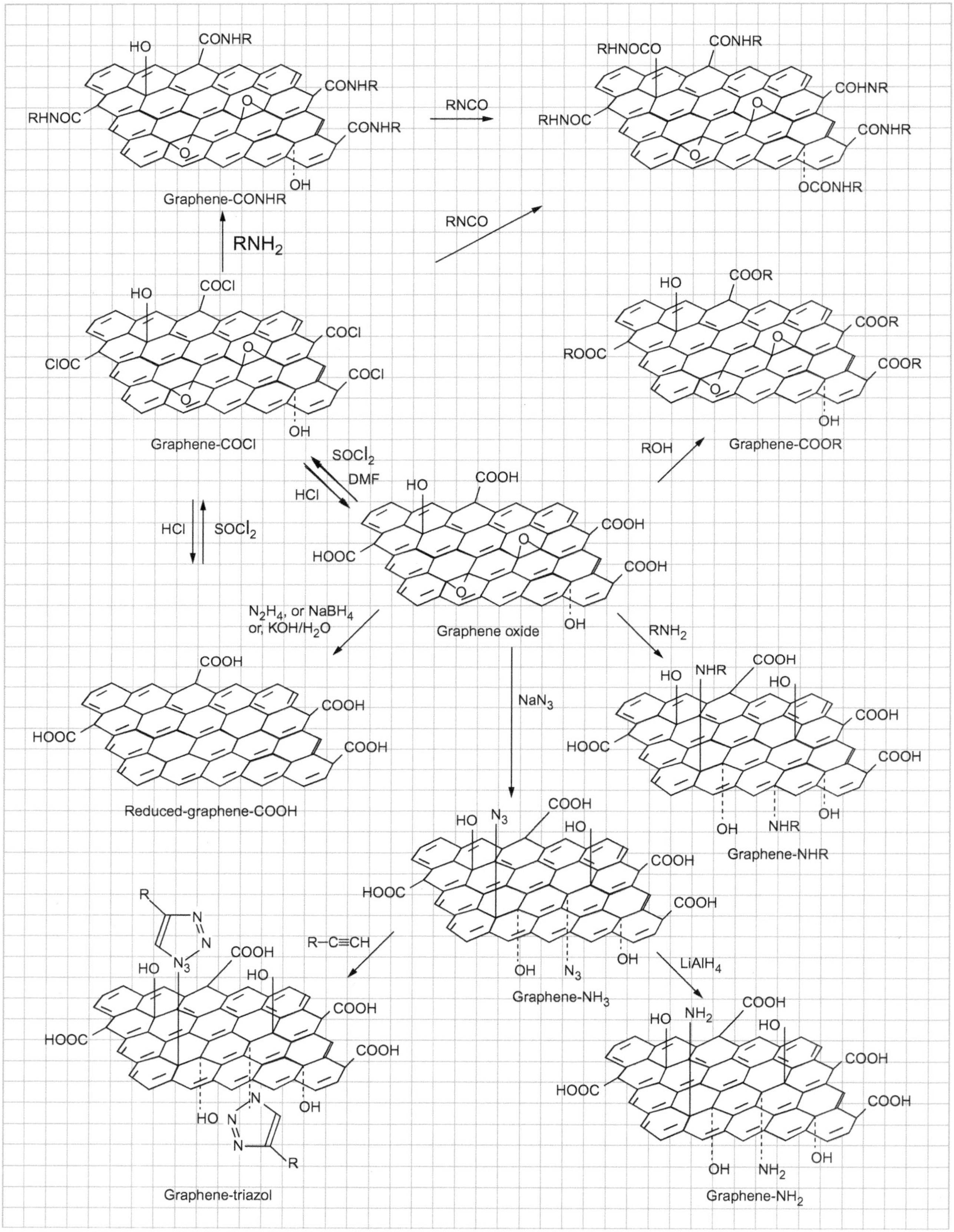

FIG. 3.28 A schematic diagram of several covalent functionalizations for graphene oxide.

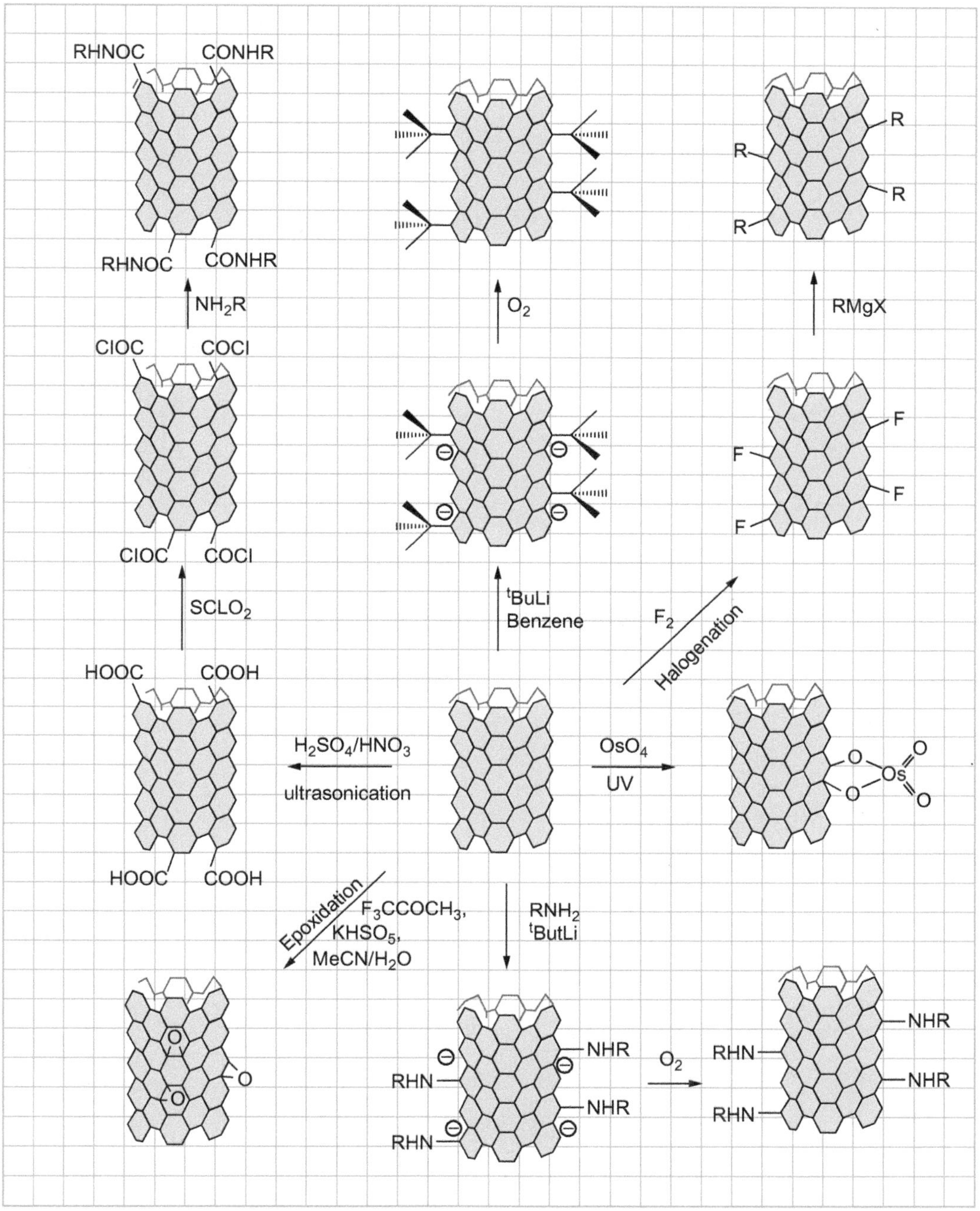

FIG. 3.29 A schematic diagram of several covalent functionalizations for carbon nanotube.

Why is CO_2 a gas at room temperature, while SiO_2 is a very hard, dense, high melting solid?

- In CO_2:
  i. CO_2 is O=C=O;
  ii. row-2/row-2 double bonding;
  iii. the valence forces are satisfies;
  iv. oxygen and carbon have close electronegativity; and
  v. low internal polarity.

  Therefore, only the weak van der Waals forces hold the molecules to one another.

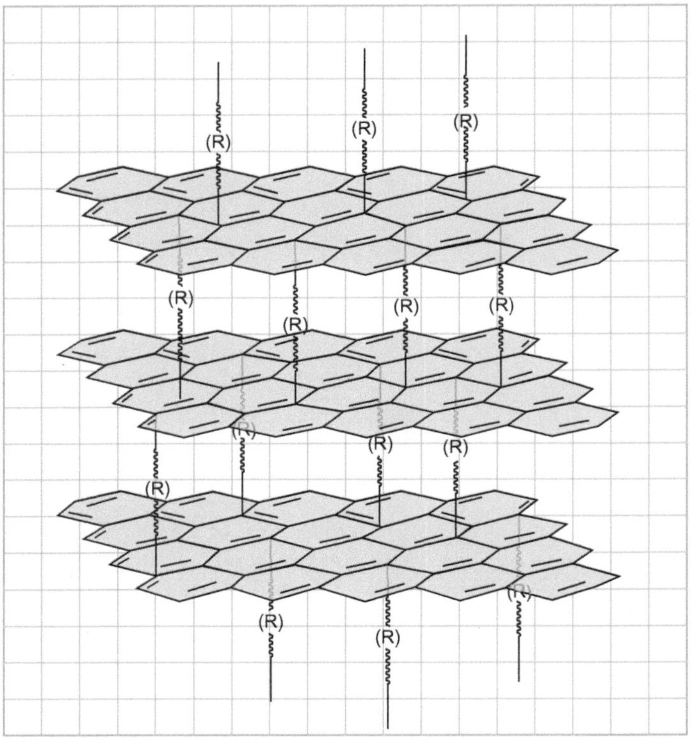

FIG. 3.30 Schematic structures of 3D supramolecular association of functionalized graphene by layer-by-layer cross linkage.

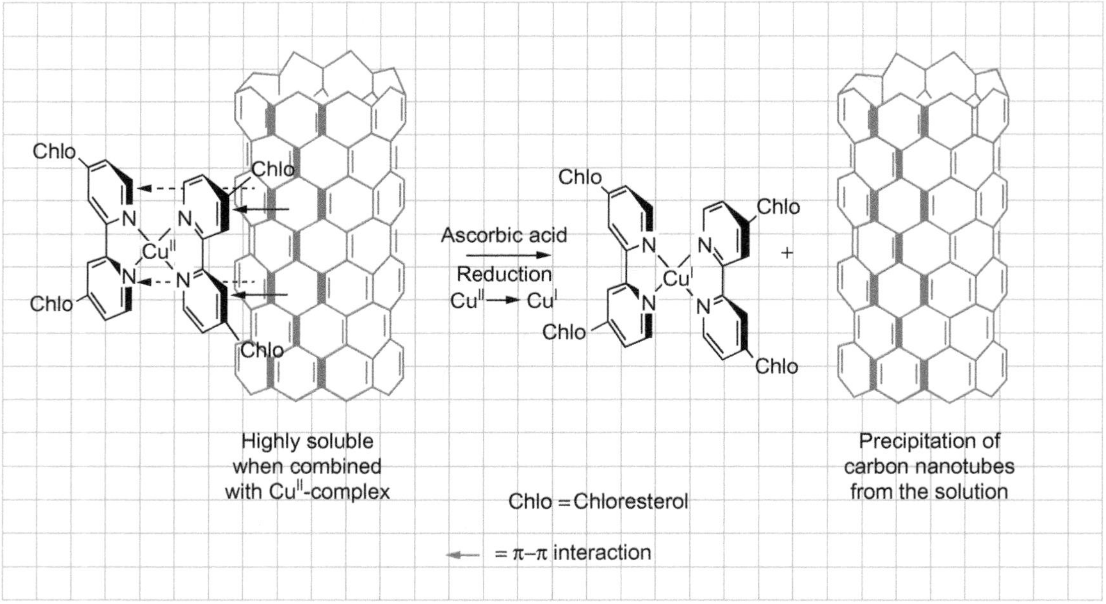

FIG. 3.31 Illustration of the presence and absence of the $\pi \rightarrow \pi$ interaction in (Cu^{2+}) complex-nanotube.

- In SiO_2:
  i. Row-2/row-3 less multiple bonding (double or triple bonds):
  ii. large difference in the electronegativity between O and Si;
  iii. more ionic character: O^{-}-Si^{++}-O^{-};
  iv. the large size of Si compare to C favors a higher coordination number;
  v. decrease the tendency of Si to form double bonds;
  vi. leads the formation of four Si—O single bonds;

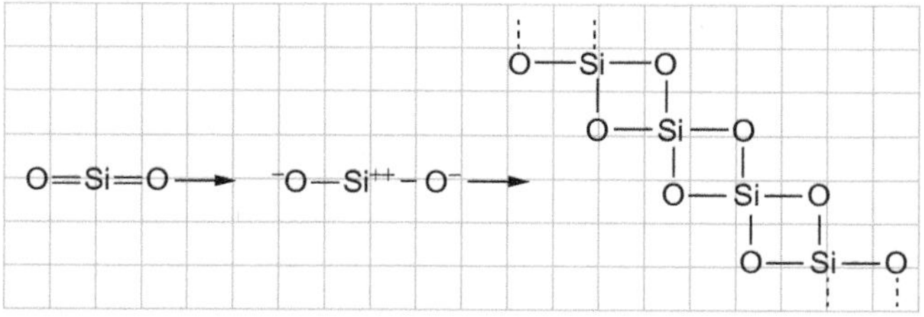

  vii. polymerized, no discrete SiO_2 molecules;
  viii. so each Si has tetrahedron of O, all covalently linked in three dimensions;
  ix. ions of Si(IV) are covalently unsaturated, and by adding the available O-donors become part of a covalent polymeric lattice; and
  x. can also bind other O-donors, such as:

$$SiO_2 + 2O^{-2} \xrightarrow{\Delta} SiO_4^{4-} \text{ (orthosilicate)}$$

Forsterite: Mg_2SiO_4
Phenacie: Be_2SiO_4.

The fundamental unit for all silicate structures is SiO_4^{4-} tetrahedron. Different structures are possible because the various ways in which tetrahedral can be joined together by sharing of oxygen atoms. What are the formulas of the following two anions:

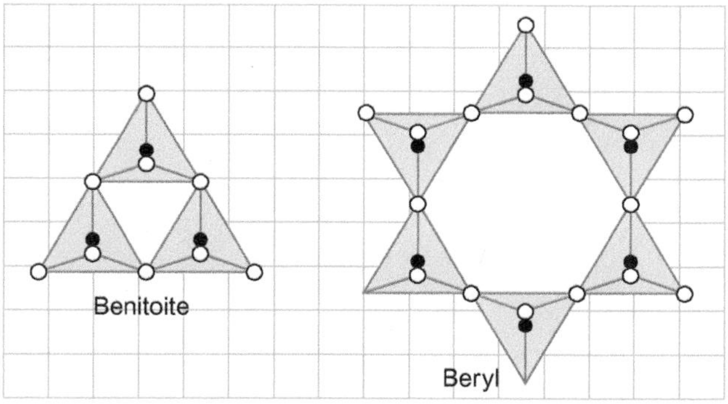

Where open circles are O atoms and closed circles are Si atoms?

- In Benitoite:
  Number of Si: 3 atoms
  Number of O: 9 atoms
  The fundamental unit is SiO_4^{-4} tetrahedron; each O has an oxidation number −2
  Then, Si has an oxidation number +4
  The calculated net charge:
  Si: 3 atoms at +4 = +12
  O: 9 atoms at −2 = −18
  The net charge = −6
  The anion formula of Benitoite anion: $Si_3O_9^{-6}$

- In Beryl:
  Number of Si: 6 atoms
  Number of O: 18 atoms
  Si: 6 atoms at +4 = +24
  O: 18 atoms at −2 = −36
  The net charge = −12
  The anion formula of Beryl anion: $Si_6O_{18}{}^{12-}$.

Asbestos is the general name for various fibrous silicate minerals. The main one is chrysotile. The silicate ions in chrysotile are double strands of SiO_4 as shown below:

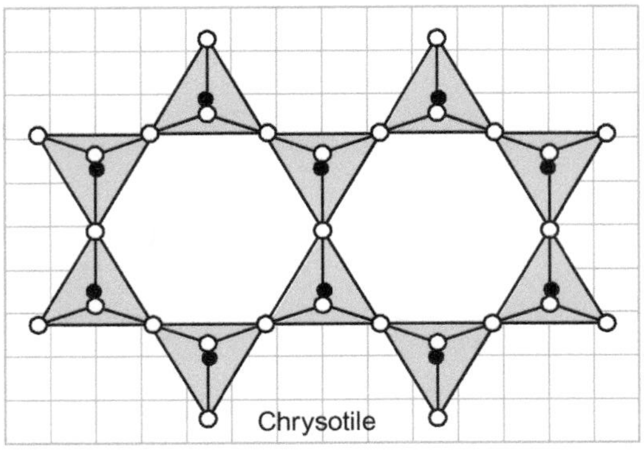

What is the general composition of the silicate ion in chrysotile?

- The simplest repeating unit of the double stranded silicate:

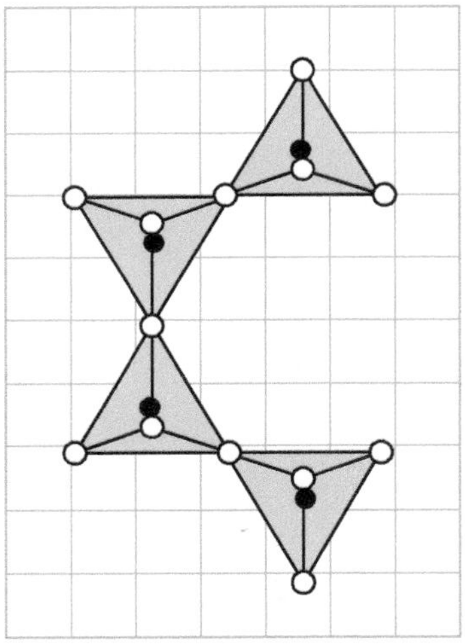

Number of Si: 4 atoms
Number of O: 11 atoms
Si: 4 atoms at +4 = +16
O: 11 atoms at −2 = −22
The net charge = −6
The composition of the silicate ions in chrysotile: $(Si_4O_{11})_n{}^{6n-}$
n: is the number of times the repeating unit $Si_4O_{11}{}^{6-}$ occurs

SUGGESTIONS FOR FURTHER READING

Periodic Trends in Atomic Size

J. Mason, J. Chem. Educ. 65 (1988) 17.
R.G. Pearson, J. Chem. Educ. 64 (1988) 561.
R.S. Mullikan, J. Chem. Phys. 2 (1934) 782.

Chemical Bonds

R.L. DeKock, H.B. Gary, Chemical Structure and Bonding, Benjamin/Cummings, Menlo Park, CA, 1980.
L. Pauling, The Nature of the Chemical Bond, third ed., Cornell University Press, Ithaca, NY, 1960.
W.E. Dasent, Inorganic Energetics, Penguin Books, London, 1970.
G.N. Lewis, J. Am. Chem. Soc. 38 (1916) 762.
N.N. Greenwood, Ionic Crystals Lattice Defects and Non-Stoichiometry, Butterworths, London, 1968.
A.B.P. Lever, J. Chem. Educ. 49 (12) (1972) 819.
N.B. Hannay, Solid-State Chemistry, Prentice-Hall, 1967.
W.E. Addison, Allotropy of the Elements, Oldbourne Press, 1966.
J. Donahue, The Structures of the Elements, Wiley, 1974.
F.S. Galasso, Structure and Properties of Inorganic Solids, Pergamon Press, 1970.
H. Krebs, Fundamentals of Inorganic Crystal Chemistry, McGraw-Hill, 1968.
M.F.A. Ladd, Structure and Bonding in Solid State Chemistry, Horwood-Wiley, 1979.
C.A. McDowell, in: H. Eyring, D. Henderson, W. Jost (Eds.), Physical Chemistry, vol. III, Academic Press, 1969, p. 496.
W.B. Pearson, The Crystal Chemistry and Physics of Metals and Alloys, Wiley, 1972.
V.I. Vedeneyev, et al., Bond Energies, Ionization Potentials, and Electron Affinities, Arnold, 1965.
A.W. Wells, Structural Inorganic Chemistry, fourth ed., Clarendon Press, 1975.
R.L. DeKock, H.B. Gray, Chemical Structure and Bonding, University Science Books, ISBN: 0-935702-61-X, 1989, p. 199.
H. PetrucciandH, Madura, General Chemistry Principles and Modern Applications, ninth ed., Macmillan Publishing Co., New Jersey, 1989.
J.T. Moore, Chemistry Made Simple, Random House Inc., New York, 2004.
N.A. Campbell, B. Williamson, R.J. Heyden, Biology: Exploring Life, Pearson Prentice Hall, Boston, Massachusetts, ISBN: 0-13-250882-6, 2006.
J.J. March, Advanced Organic Chemistry: Reactions, Mechanisms, and Structure, John Wiley & Sons, ISBN: 0-471-60180-2, 1992.
G.L. Miessler, D.A. Tarr, Inorganic Chemistry, Prentice Hall, ISBN: 0-13-035471-6, 2004.
L. Irving, The Arrangement of Electrons in Atoms and Molecules, J. Am. Chem. Soc. 41 (6) (1919) 868–934.
G.N. Lewis, The atom and the molecule, J. Am. Chem. Soc. 38 (4) (1916) 762–785.
H. Hettema, Quantum Chemistry: Classic Scientific Papers, World Scientific, ISBN: 978-981-02-2771-5, 2000 p. 140.
D.R. Stranks, M.L. Heffernan, K.C. Lee Dow, P.T. McTigue, G.R.A. Withers, Chemistry: A Structural View, Melbourne University Press, Carlton, Victoria, ISBN: 0-522-83988-6, 1970 p. 184.
N.G. Lewis, The Atom and the Molecule, J. Am. Chem. Soc 38 (4) (1916) 772.
K.J. Laidler, The World of Physical Chemistry, Oxford University Press, ISBN: 0-19-855919-4, 1993 p. 346.
H.H. James, A.S. Coolidge, The Ground State of the Hydrogen Molecule, J. Chem. Phys. 1 (12) (1933) 825–835.
P. Atkins, L. Jones, Chemistry: Molecules, Matter and Change, W. H. Freeman & Co, New York, ISBN: 0-7167-3107-X, 1997 pp. 294–295.
F. Weinhold, C. Landis, Valency and Bonding, Cambridge University Press, Cambridge, ISBN: 0-521-83128-8, 2005 pp. 96–100.
L. Pauling, The Nature of the Chemical Bond, Cornell University Press, 1960 p. 340–354.

Complex Formation

G.B. Kauffman, J. Chem. Educ. 36 (1956) 521.

Stability of Complexes

R.G. Pearson, Inorg. Chem. 27 (1988) 734.
R.G. Pearson, Coord. Chem. Rev. 100 (1988) 403.
R.J.P. Williams, J.D. Hale, C.J. Jørgensen (Ed.), Structure and Bonding, vol. 1, Springer-Verlag, Berlin, 1966, p. 255ff.
J. Huheey, R.S. Evans, J. Inorg. Nucl. Chem. 32 (1970) 383.
J. Huheey, R.S. Evans, Chem. Commun. 968 (1969).
D. Datta, Inorg. Chem. 31 (1992) 2797.
R.G. Pearson, Acc. Chem. Res. 26 (1993) 250.
R.G. Parr, Z. Zhou, Acc. Chem. Res. 26 (1993) 256.
J.J. Frausto da Silva, J. Chem. Educ. 60 (1983) 390.
J.I. Creaser, J.M. Harrowfield, A.J. Herlt, A.M. Sargeson, J. Springborg, R.J. Geue, M.R. Snow, J. Am. Chem. Soc. 99 (1977) 3181.
A. Hammershø, A.M. Sargeson, inorg. Chem. 22 (1983) 3554.
C.K. Jørgensen, Inorg. Chem. 3 (1964) 1201.
T.E. Sloan, A. Wojcicki, Inorg. Chem. 7 (1968) 1268.
T.L. Cottrell, The Strengths of Chemical Bonds, second ed., Butterworths, London, 1958.

Covalent Networks and Giant Molecules

R. Steudel, Chemistry of the Non-Metals, Walter de Gruyter, Berlin, 1977.
M.S. Dresselhaus, G. Dresselhaus, Intercalation Compounds of Graphite, Adv. Phys. 30 (1981) 139.
H.W. Kroto, J.R. Heath, S.C. O'Brien, R.F. Curl, R.E. Smalley, Nature 318 (1985) 162.
H.W. Kroto, Angew. Chem. Ed. Engl. 31 (1992) 111.
R.F. Curl, R.E. Smalley, Sci. Am 54 (1991).
R.D. Huffman, Phys. Today 22 (1992).
W.E. Billups, M.A. Ciufolini, Buckminsterfullerenes, VCH Publishers, New York, 1993.
M.S. Dresselhaus, G. Dresselhaus, P.C. Eklund, J. Mater. Res. 8 (1993) 2054.

Graphene and Carbon Nanotubes

D.M. Guldi, N. Martin, Carbon Nanotubes and Related Structures, Wiely-VCH Verlag GmbH & Co. KGaA, Weinheim, ISBN: 978-3-527-32406-4, 2010.
V.N. Popov, P. Lambin, Carbon Nanotubes, Springer, Dordrecht, 2006.
S. Reich, C. Thomsen, J. Maultzsch, Carbon Nanotube: Basic Concepts and Physical Properties, VCH, Weinheim, Germany, 2004.
Special issue on carbon nanotubes, Acc. Chem. Res. 35 (2002) 997.
P.J.F. Harris, Carbon Nanotubes and Related Structures: New Materials for the Twenty-First Century, Cambridge University Press, Cambridge, 2001.
L. Huang, B. Wu, G. Yu, Y. Liu, Roy. Soc. Chem. 21 (2011) 919–929.
M. Kanungo, H. Lu, G.G. Malliaras, G.B. Blanchet, Science 323 (2009) 234–237.
A.K. Geim, K.S. Novoselov, Nat. Mater. 6 (2007) 183.
K.S. Novoselov, A.K. Geim, S.V. Morozov, D. Jiang, Y. Zhag, S.V. Dubonos, I.V. Grigorieva, A.A. Firsov, Science 306 (2004) 666.
X. Wang, L. Zhi, K. Mullen, Nano Letter 8 (1) (2008) 323–327.

Chapter 4

Molecular Symmetry

In the previous chapter, we have discussed the formation of bonds and reviewed the process of predicting the molecule's structure. In order to deal with molecular structures, where many energy levels of atoms are involved, an extensive demand to symmetry must be made.

Consequently, it is suitable to establish a proper system for sorting molecules according to their structures. The approach in which symmetry arguments are framed is known as group theory.

The purpose of this chapter is to:

i. introduce some ideas and techniques that are essential for understanding the structure and properties of molecules and crystals;
ii. show how to identify the symmetries of the atomic orbitals; and
iii. show that the symmetry representation of electronic orbits form a basis for the wave function.

This chapter introduces the basis of molecular symmetry, and explores some immediate applications. Matrix representations that include symmetry operations, point group, translation, rotation motions, and atomic orbital representations are examined thoroughly. This leads on to the character tables, and finally the symmetry representation of the atomic orbital wave functions.

Succeeding chapters will explore the application of the wave model to these structures (Scheme 4.1). The following are the main topic that will be explored in this chapter.

- 4.1: Molecular symmetry
- 4.2: The symmetry elements
  - E: identity
  - C_n: proper rotation axis
  - σ: plane of symmetry
  - i: center of symmetry
  - S_n: improper rotation axis
- 4.3: The symmetry and point group
- 4.4: Some immediate applications
  - Dipole moments and polarity
  - Chirality
  - Equivalent atoms (or group of atoms)
  - Crystal symmetry
- 4.5: Group theory: properties of the groups and their elements
- 4.6: Similarity transforms, conjugation, and classes
- 4.7: Matrix representation
  - Matrices and vectors
  - Matrix representation of symmetry operation
  - Matrix representation of point group
  - Irreducible representations
  - Irreducible and degenerate representations
- 4.8: Motion representations of the groups
  - Translation motion
  - Rotational motion

Electrons, Atoms, and Molecules in Inorganic Chemistry. http://dx.doi.org/10.1016/B978-0-12-811048-5.00004-3

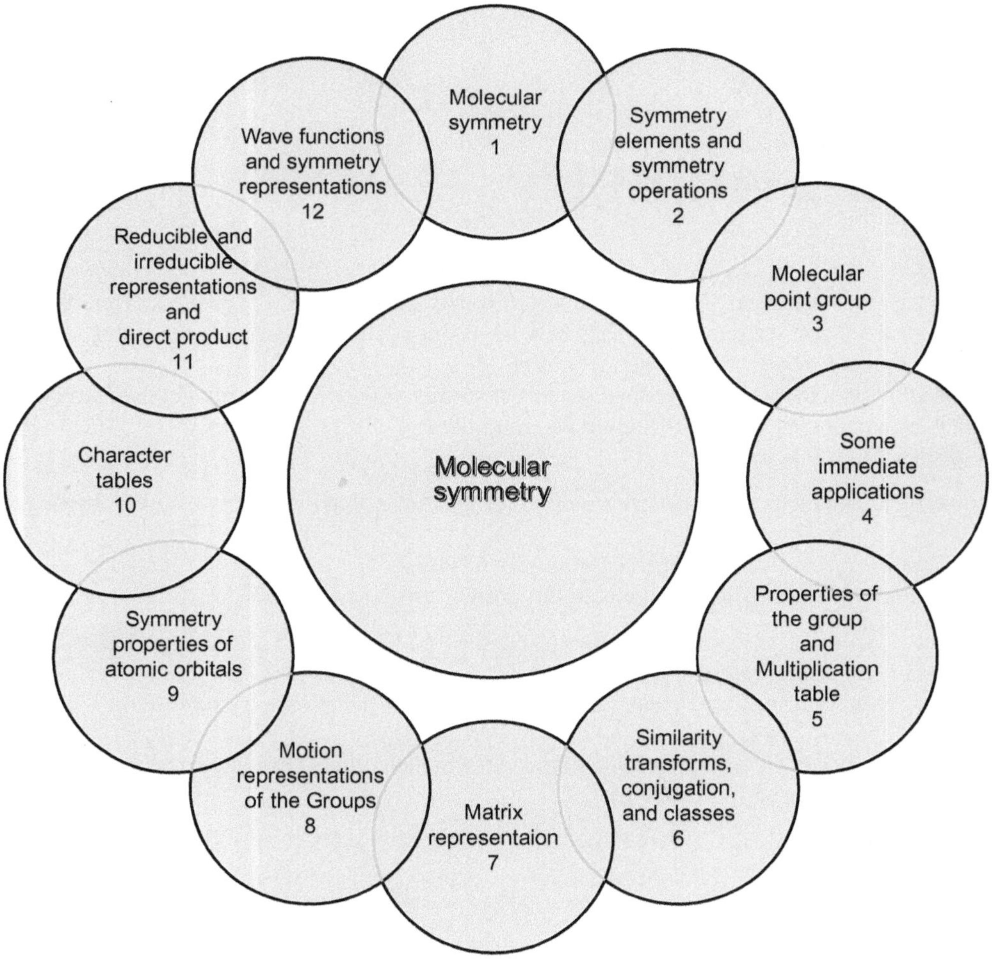

SCHEME 4.1 Approaches used to address the molecular symmetry.

- 4.9: Symmetry properties of atomic orbitals
  - Mullikan notation
  - Atomic orbital representation
- 4.10: Character tables
  - Properties of the characters of representations
- 4.11: Relation between any reducible and irreducible representations
  - Direct product
- 4.12: Group theory and quantum mechanics: irreducible representations and wave function
- Suggestions for further reading

The following questions will be answered in order to aid comprehension of the molecular symmetry and the relationship between the symmetry representations and the wave functions.

4.1 MOLECULAR SYMMETRY

Why is there an extensive demand made to molecular symmetry?

- Chemists use molecular symmetry to enable them in understanding:
  - chemical bonding;
  - molecular dynamics;
  - formulating hybrids orbitals;
  - forming molecular orbitals;

- selecting appropriate orbitals set under the action of a ligand field; and
- examining the vibrations of the molecules.

Which of the following molecules are more symmetrical?

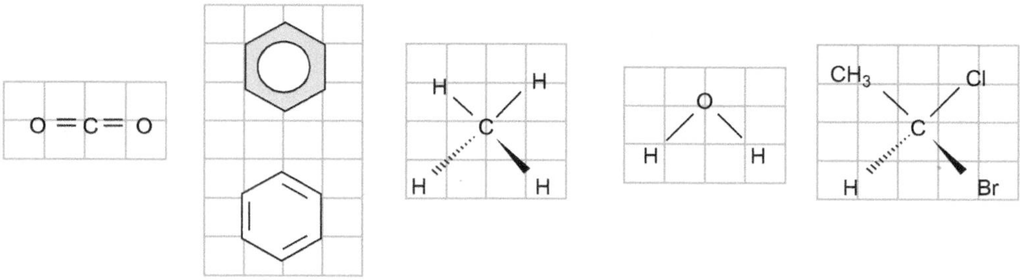

- Molecules like CO_2, C_6H_6, CH_4, and H_2O have high symmetry.

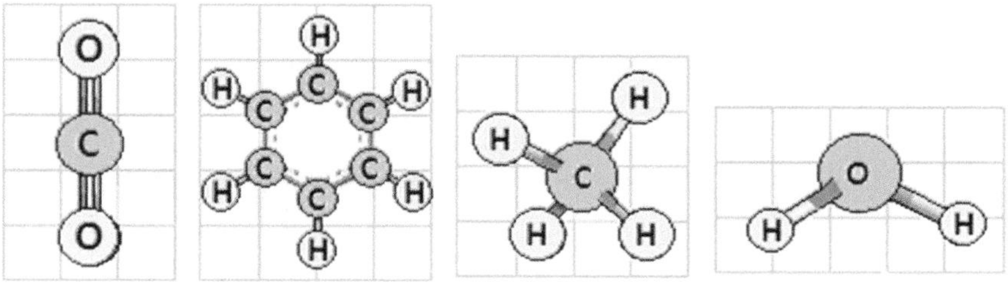

- Others, such as $CH_3CHBrCl$, have low symmetry.

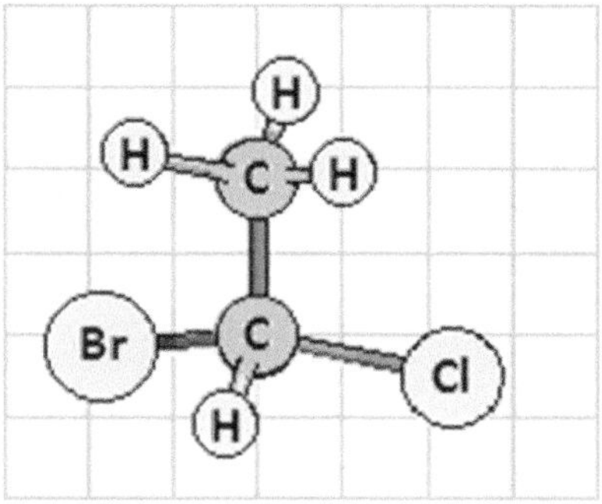

How can the molecular symmetry be probed?

- In order to understand better the idea of molecular symmetry, we must develop some rigid molecular criteria of symmetry.
- To do this, we must consider the kind of symmetry elements a molecule may have and symmetry operations generated by the symmetry elements.
- If an object can be carried from one orientation to another indistinguishable one by a spatial operation, then it possesses a symmetry element.
- The name of the element is closely related to (even the same as) the name of the operation.

What are the differences between the symmetry element and the symmetry operation?
What are the possible symmetry elements and what are the associated operations for each?

- The symmetry element has a geometrical entity such as a line, a plane, or a point with respect to which one or more symmetry operations may be carried out.

TABLE 4.1 Symmetry Elements and the Associate Symmetry Operations

Symmetry Symbol	Elements Description	Symmetry Operation
E	Identity	No change
C_n	Proper axis or Axis of symmetry	Rotation about the axis by $360/n$ degree
σ	Plane of symmetry	Reflection through the plane
i	Center of symmetry	Inversion of all atoms through the center
S_n	Rotation reflection axis or improper axis	Rotation about the axis by $360/n$ followed by reflection through the plane perpendicular to the rotation axis

- The symmetry operation is a movement of a body such that after the movement has been carried out, every point of the body is coincident with an equivalent point of the body in its original orientation.
  We can define a symmetry operation as an operation that effect is to take the body into an equivalent configuration identical to the original.
- Possible symmetry elements and the associate symmetry operations are summarized in Table 4.1.

4.2 THE SYMMETRY ELEMENTS

Identity, *E*

Name the symmetry operation that yields the original orientation as its result.
What are the symmetry elements possessed by OSClF?

- Identity is the operation that leaves a molecule unchanged in an orientation identical with the original.
- It is the operation of doing nothing.
- It produces not only an equivalent orientation, but also an identical one.
- This operation can be performed on any molecule, because every atom looks the same after doing nothing to it.
- It is required to be included in the description of the molecular symmetry in order to be able to apply the group theory and to satisfy certain mathematical requirements of groups.
- OSClF has only identity, *E*.

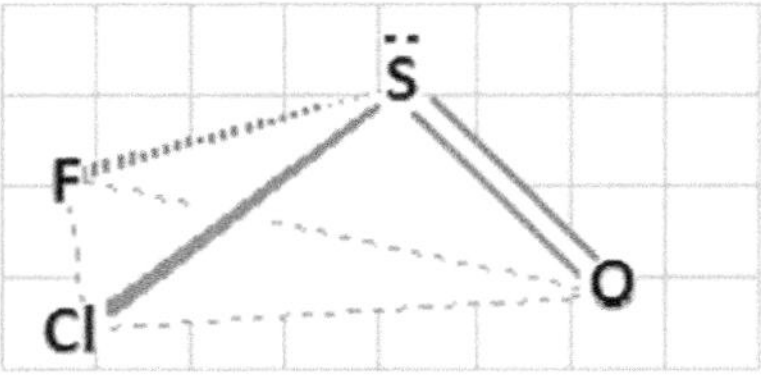

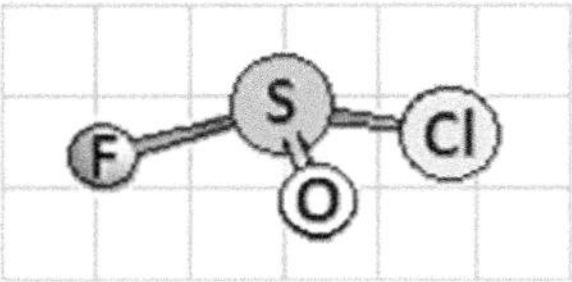

- All the molecules possess identity, *E*, which is analogous to "1" in multiplication of real numbers.

Proper Rotation Axis, C_n

What is the symbol for the following rotating symmetry operation?

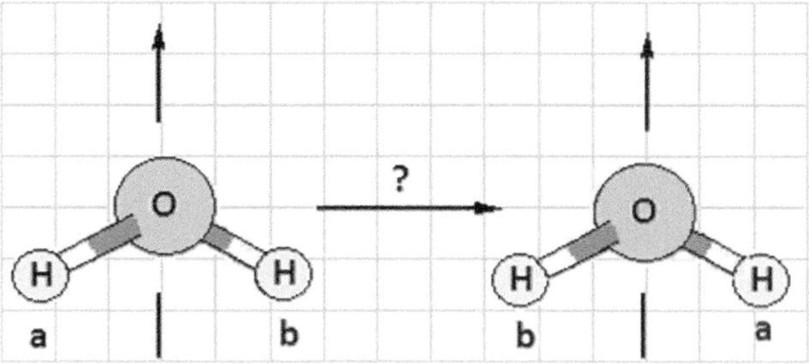

- It is rotated by 180°, $2\pi/2$, C_2.
  C_n: is the axis about which appropriate clockwise rotation gives an equivalent orientation indistinguishable from the original.
  n: order, the rotation through $2\pi/n$ give an equivalent configuration.

Detect the threefold C_3^1 and C_3^2 rotational operations in BF_3. What happens if we go one stage further to C_3^3?

- The BF_3 molecule has three orientations with respect to the rotation axis ($\perp$ to molecular plane) which are equivalent to a threefold axis $= C_3$ (Fig. 4.1).
- Each C_3 operation rotates the molecule by $360/3 = 120°$.
- The C_3 axis generates two operations: C_3 and C_3^2.
- All group theory rotations are clockwise, but in trigonometry, angles are measured anticlockwise: $C_3 = -\dfrac{2\pi}{3}$.
- The operation C_3^3 rotates the molecule back to being identical with the starting point ($C_3^3 = E$, identity operation, E).
- Note: $C_3 = C_3^1 = \text{clockwise rotation by } 120° = \dfrac{2\pi}{3}$

$$C_3^2 = C_3 \times C_3 = 240°$$

$$C_3^3 = C_3 \times C_3 \times C_3 = C_3 \times C_3^2 = 360°$$

How many twofold rotational operations are possible in the BF_3 molecule?

- BF_3 has three orientations with respect to rotation axis (along the B–F bond) which are equivalent to a twofold axis $= C_2$ (Fig. 4.2).

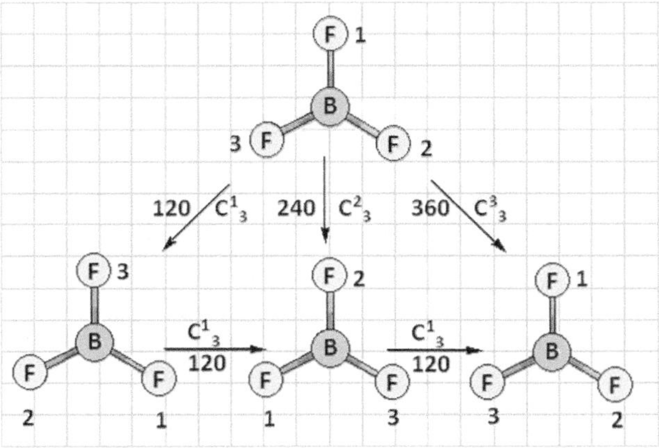

FIG. 4.1 Threefold rotational axis of symmetry in BF_3.

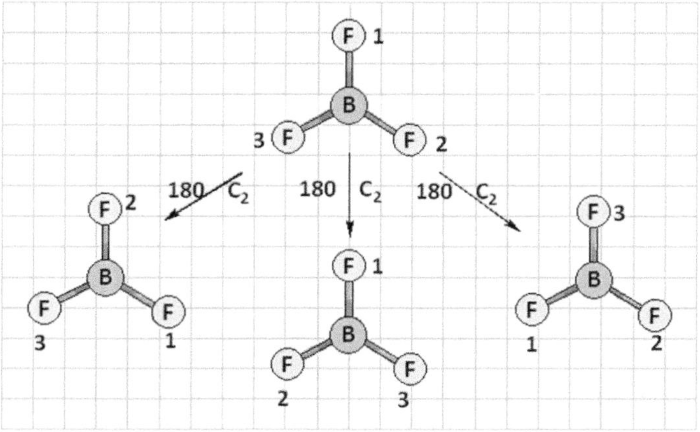

FIG. 4.2 Twofold rotational axis of symmetry in BF_3.

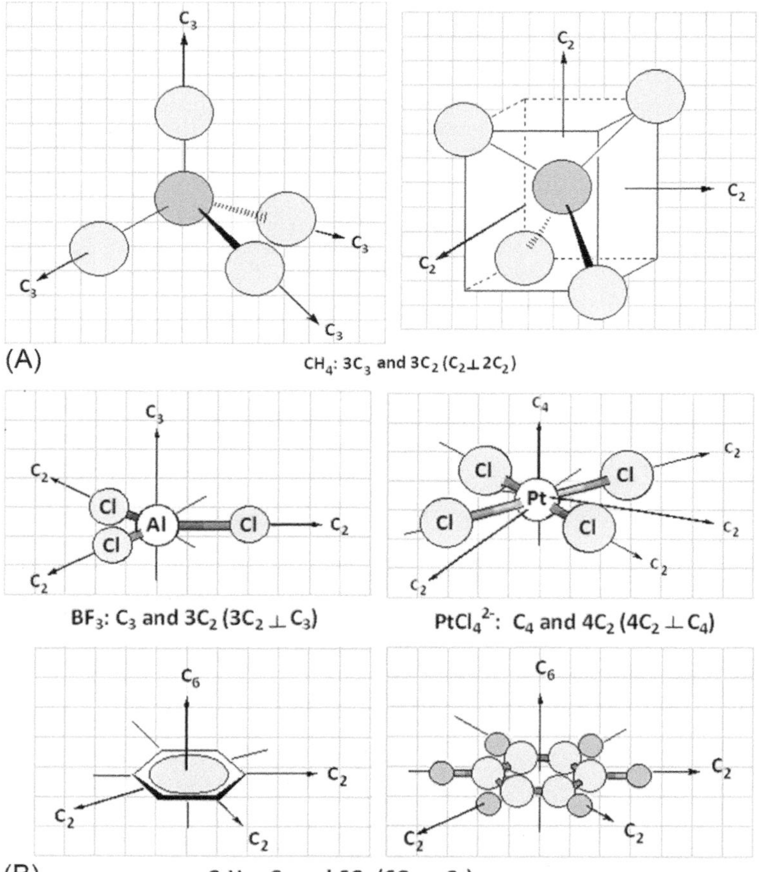

FIG. 4.3 (A) The rotation operations that possessed by CH_4: $3C_3$ and $3C_2$ ($C_2 \perp 2C_2$). (B) The rotation operations that possessed by $AlCl_3$, $PtCl_4^{2-}$, C_6H_6.

What are the rotational operations possessed by CH_4, $AlCl_3$, $PtCl_4^{2-}$, and C_6H_6, and what is the common feature among these molecules?

- The rotational operations possessed by CH_4, $AlCl_3$, $PtCl_4^{2-}$, C_6H_6 are represented in Fig. 4.3.
- All of these molecules contain $nC_2 \perp C_n$.

What is the principle axis of the molecules?

- The proper rotation axis of the highest order is defined as the z-direction of the Cartesian reference framework. E.g.,
  C_3 in BF_3 and
  C_∞ in a linear molecule.
- The ranking order is $z > x > y$
  e.g., in $PtCl_4^{2-}$: $C_4 = z$, $C_2 = z$, $2C_2'$ are x, y, and $2C_2''$ in between C_2'.

Plane of Symmetry, σ

What are the symmetry elements possessed by $SOCF_2$ and $SOCl_2$?

- $OSCF_2$ and $OSCl_2$ possess identity, E, and one plane of symmetry, σ (Fig. 4.4).
- The plane of symmetry, σ, must pass through the molecule and divides it in halves each being the other image (Fig. 4.4). Three possible types of planes:
  - σ_h: horizontal plane of symmetry is $\perp$ to the z-axis (principle plane);
  - σ_v: vertical to the x-y plane and pass through x and z or y and z; and
  - σ_d: vertical plane, which bisect the angles between adjacent pairs of axes-dihedral.

  This element generates only one operation, $\sigma^2 = \sigma\,\sigma = E$, because reflection in plane, followed by reflection again, places all points to identical location from which they stared.

What are the different planes of symmetry possessed by the following molecules: BF_3, CO_2, H_2O, CH_2Cl_2, $CHCl_3$, NH_3, CO_3^{2-}, NO_3^-, SO_3, $AuCl_4^{2-}$, and $PtCl_4^{2-}$?

- BF_3 has four mirror planes (Fig. 4.5): one in the molecular plane, σ_h and three vertical planes that are perpendicular ($\perp$) to the molecule plane, 3 σ_v.
- CO_2 has an infinite number of σ_v, and σ_h (Fig. 4.6):
- Molecules such: H_2O, $C_2H_2Cl_2$ have 2 σ_v and no σ_h (Fig. 4.7):
- NH_3 and $CHCl_3$ have 3 σ_v (Fig. 4.8):
- CO_3^{2-}, NO_3^-, SO_3 have σ_h and 3 σ_v (Fig. 4.9):
- $PtCl_4^{2-}$ and $AuCl_4^{2-}$ have one σ_h, two σ_v, and two σ_d (Fig. 4.10):

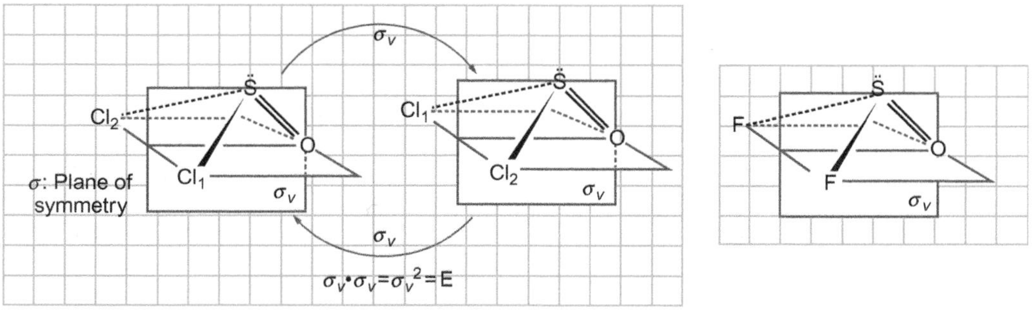

FIG. 4.4 Identity, E, and one plane of symmetry, σ, in $OSCF_2$ and $OSCl_2$.

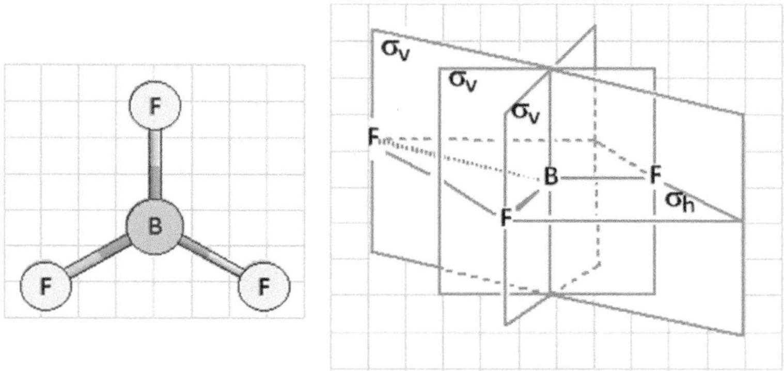

FIG. 4.5 One horizontal, σ_h, and three vertical, σ_v, planes of symmetry in a BF_3 molecule.

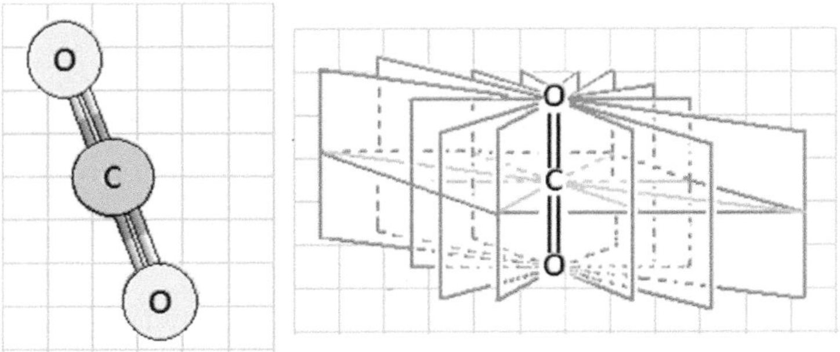

FIG. 4.6 Infinite number of σ_v in CO_2.

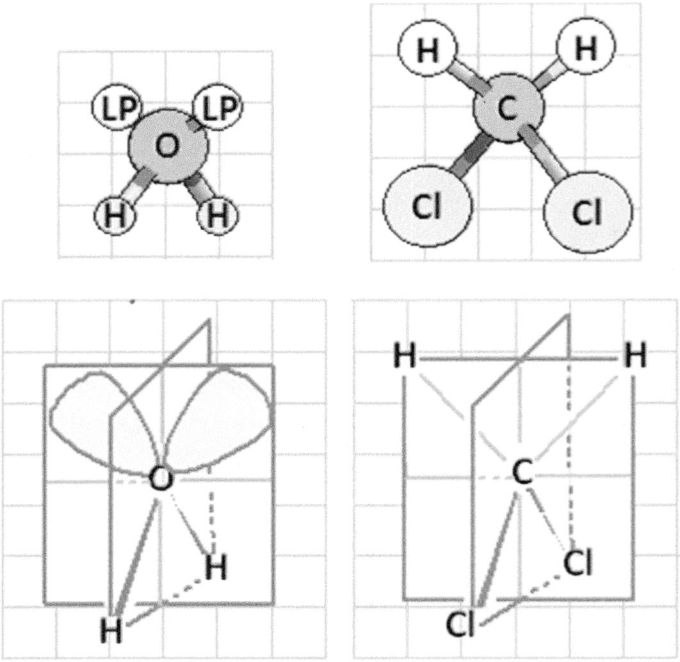

FIG. 4.7 Two vertical, $2\sigma_v$, and no horizontal, σ_h, planes of symmetry in H_2O and CH_2Cl_2 molecules.

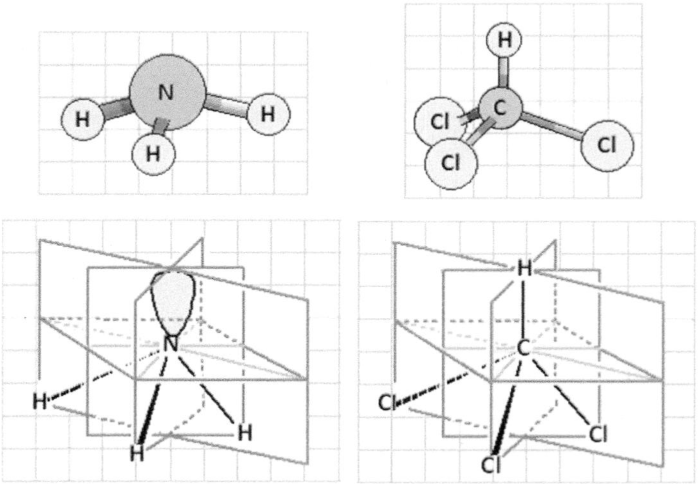

FIG. 4.8 Three vertical planes of symmetries, σ_v, in NH_3 and $CHCl_3$.

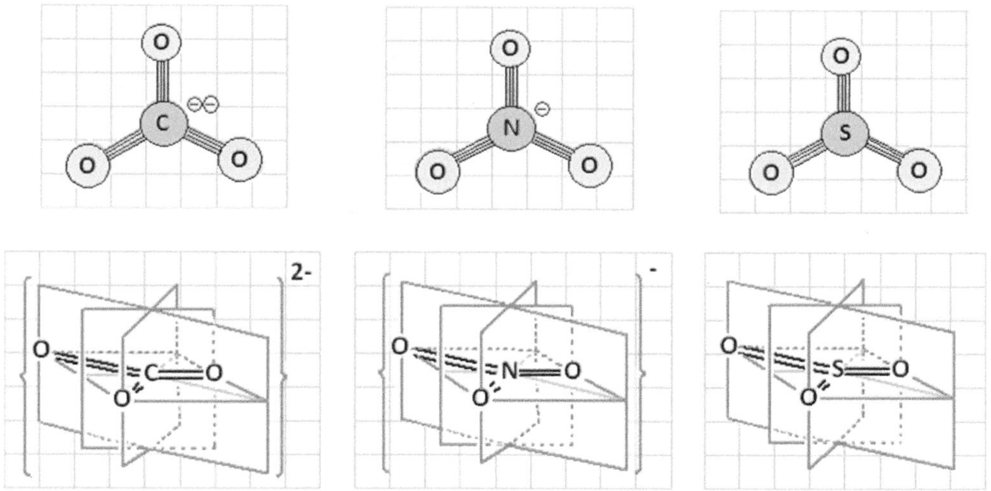

FIG. 4.9 One horizontal, σ_h, and three vertical, $3\sigma_v$, planes of symmetries in CO_3, NO_3^- and SO_3.

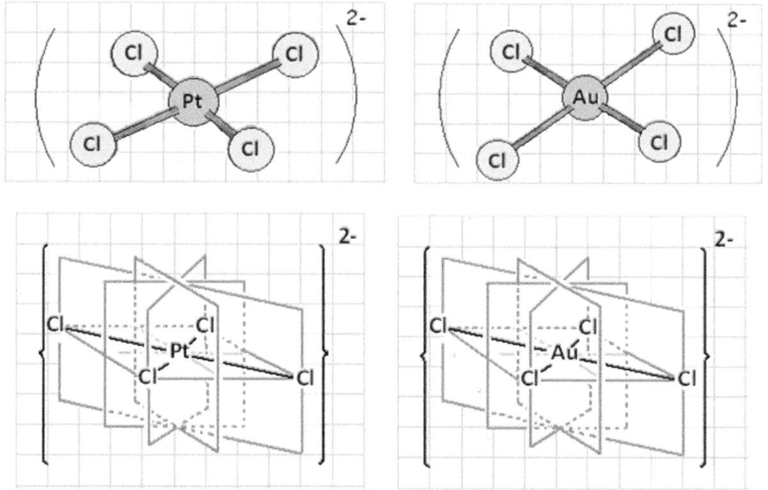

FIG. 4.10 One horizontal, σ_h, two vertical, σ_v, and two dihedral, σ_d, planes of symmetry in $PtCl_4^{2-}$ and $AuCl_4^{2-}$.

Center of Symmetry, *i*

Name the symmetry operation that inverts all atoms through the center of the molecule.

Which of the following have inversion centers: $PtCl^{2-}$, trans $PtCl_2(NH_3)_2$, AB_6, CO_2, C_2H_4, C_6H_6, $C_6H_2Cl_2Br_2$, C_5H_5, CCl_4, and cis $PtCl_2(NH_3)_2$?

- An inversion center, *i*, exists if each atom in the molecule can be moved toward and through this point then to an equivalent distance beyond where it meets another identical atom.
- Molecules that have an inversion center include: $PtCl_4^{2-}$, trans $PtCl_2(NH_3)_2$, AB_6, CO_2, C_2H_4, C_6H_6, and $C_6H_2Cl_2Br_2$ (Fig. 4.11).
- Molecules that do not have an inversion center include: C_5H_5, CCl_4, and cis $PtCl_2(NH_3)_2$ (Fig. 4.12).

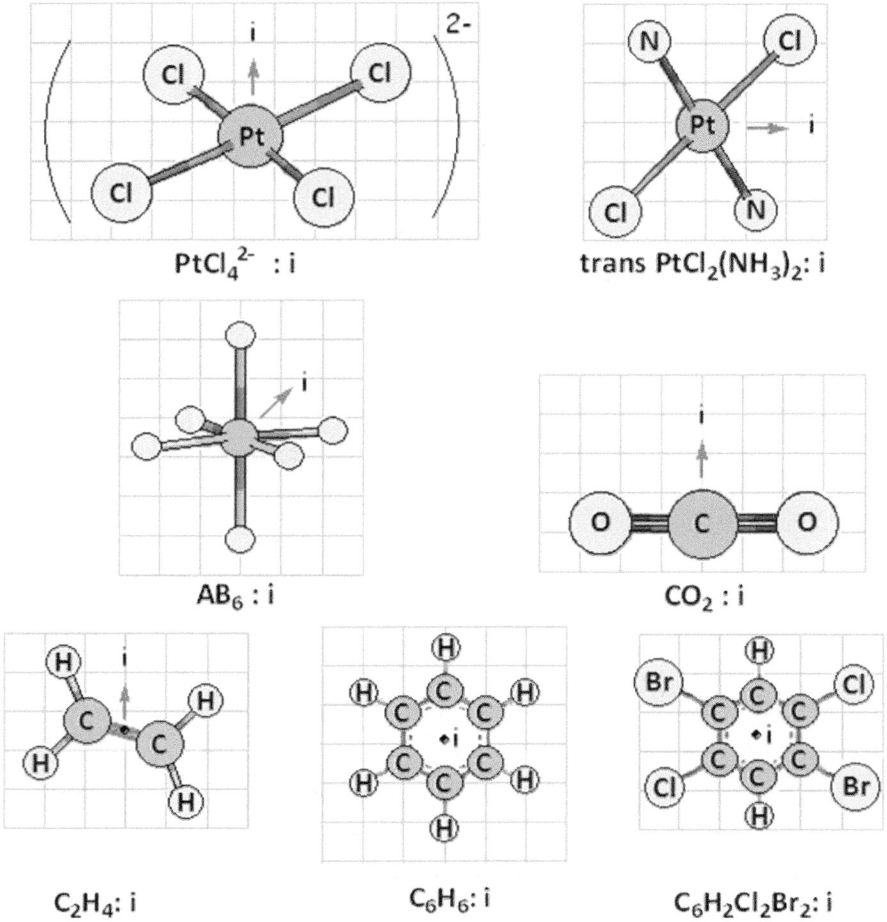

FIG. 4.11 Molecules that have a center of symmetry.

FIG. 4.12 Molecules do not possess center of symmetry.

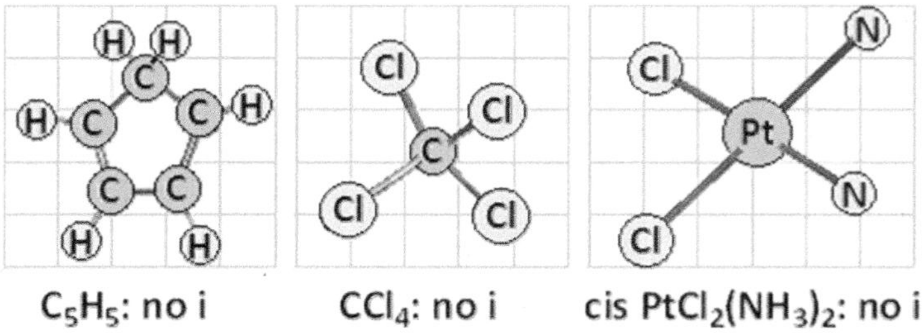

What is the product of the operation i^2?

- $i^2 = E$

S_n: Improper Rotation Axis

How can you define the following symmetry operations?

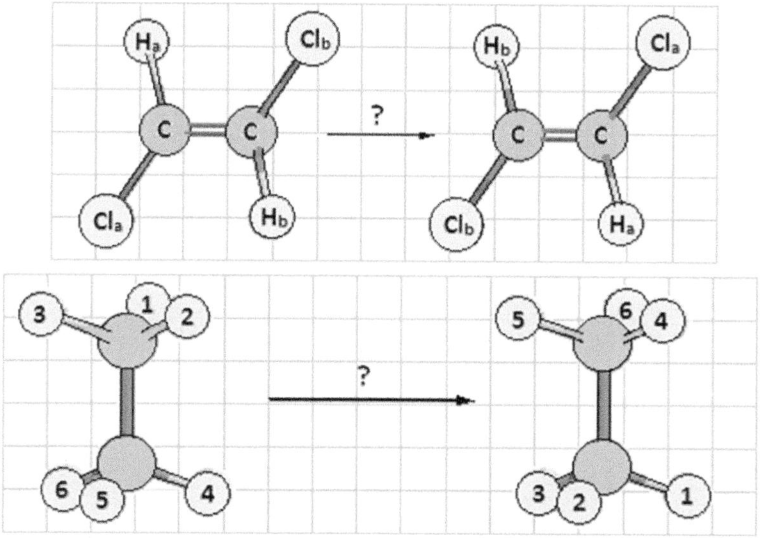

- Rotation reflection axis or improper axis, S_n: axis about which n fold rotation followed by reflection across a plane perpendicular to the rotation axis, producing a form indistinguishable from the original (Fig. 4.13).

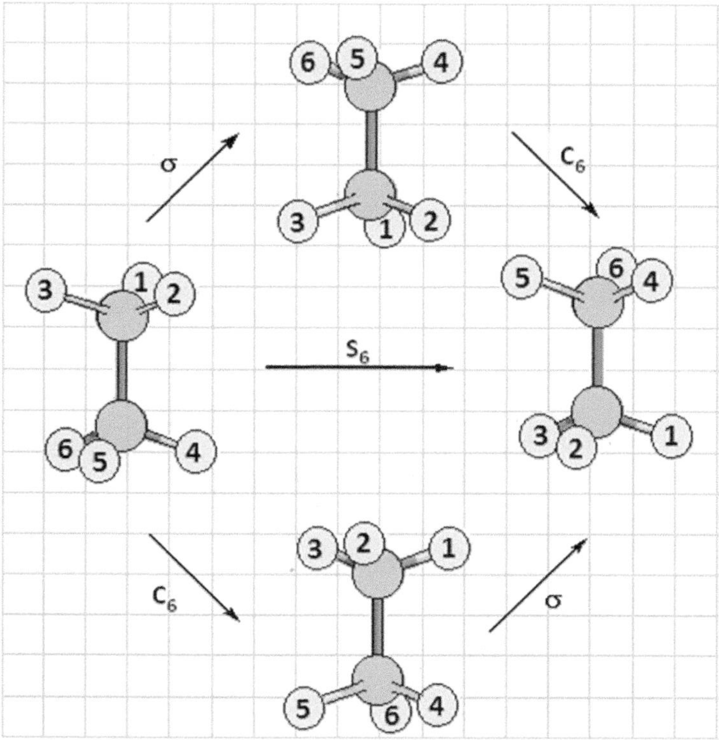

FIG. 4.13 The improper rotation axis, S_n, in staggered C_2H_6.

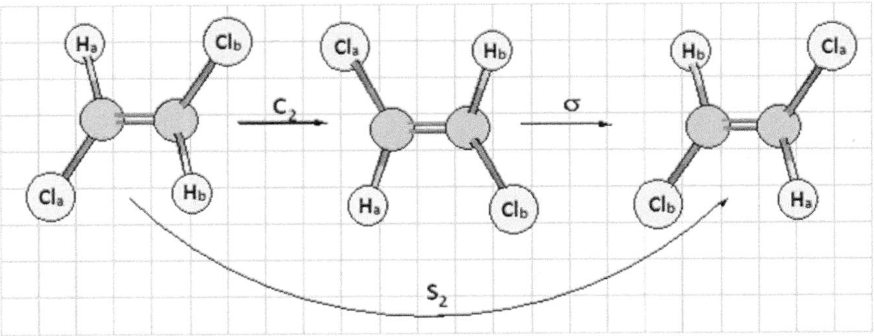

Write the equivalent operation for the following: C_n^n, $C_n^{n/2}$(even n), C_n^{-1}, σ^2, S_n, S_n^2, i^2, σ, i, S_1, S_2, and S_n(odd n).

- $C_n^n = E$ $\quad$ $C_n^{n/2} = C_2$ (even n) $\quad$ C_n^{-1} (anticlockwise) $= C_n^{n-1}$
- $\sigma^2 = E$
- $S_n = \sigma C_n$ (C_n then σ)
- ($S_1 = \sigma C_1 = \sigma$ $(C_1 = E)$)
- $S_n^2 = C_n^2$ $(\sigma^2 = E)$

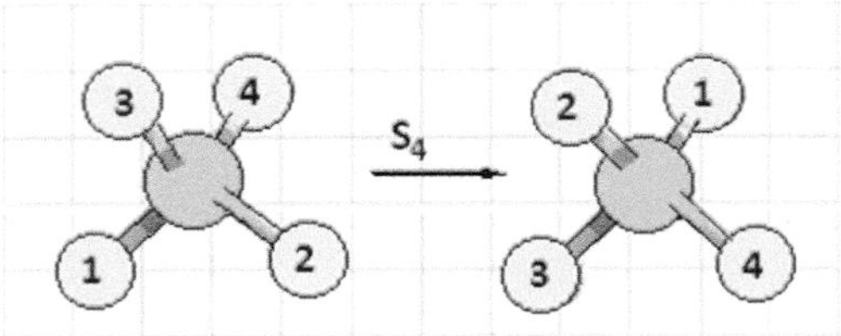

- $i^2 = E$
- $iC_2 = \sigma$
- $i = \sigma C_2 = S_2$
- $S_1 = \sigma C_1 = \sigma$
- $S_2 = i$

for odd n

- $S_n = \sigma_h C_n$

4.3 THE SYMMETRY AND POINT GROUP

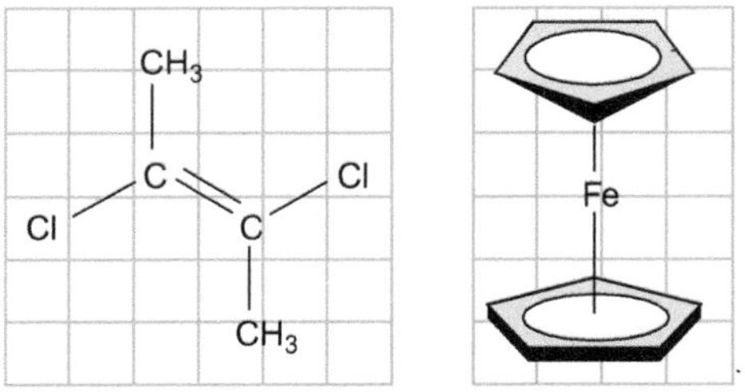

(A) (B)

For A and B, what is the geometrical identity (point, line, or plane) for the intersection of carried symmetry elements? What is the significance of this intersection?

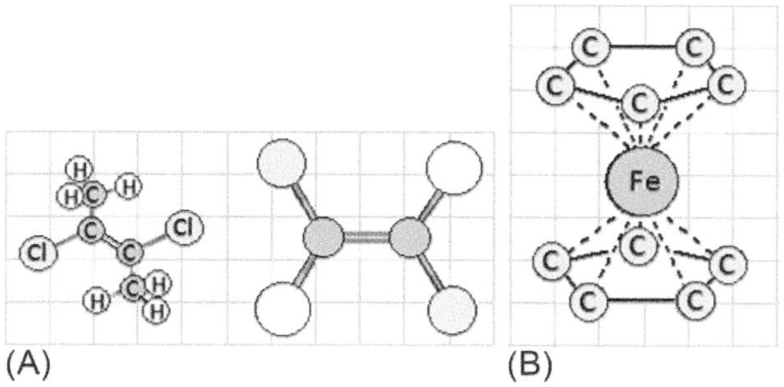

- All the symmetry elements of any molecule always pass through one common point (sometimes through a line or plane, but always through point).
  - A is the center of the C=C double bond.
  - B is the Fe atom.
  - That point carries all the symmetry elements of the molecule, which usually called a point-group, and represented by a symbol.
- Molecules can be classified by point groups.
  - It is probable to sort any given molecule into one of the point groups.
  - Each point group is a set of all the symmetry operations that can be carried out on a molecule belonging to this group.

What are the systematic procedures for classification of molecules by point groups?

- The following flow chart presents a systematic way to track the classification of molecules by point groups (Scheme 4.2):

  Notes:
  - The group I_h contains a regular dodecahedron and regular icosahedrons.
  - Only linear molecules belong to $C_{\infty v}$ or $D_{\infty h}$.

List the symmetry elements and classify the following molecules into their point group: $CH_3CHClBr$, H_2O_2, S_2Cl_2, CH_3—CHCl—CHCl—CH_3, NOCl, H_2O, CH_2Cl_2, NH_3, $CHCl_3$,HCN, OCN^-, SCN^-, CH_2=C=CH_2, trans CHCl=CHCl, C_6H_{12} (chair and boat forms), C_2H_4, BCl_3, C_4H_8, AB_4, C_6H_6, H_2, CO_2, C_2H_2, CH_4, CCl_4 and $PtCl_6^{2-}$.

- The symmetry elements, classifications, and suggested structures are summarized in Table 4.2.

4.4 SOME IMMEDIATE APPLICATIONS

Dipole Moments and Polarity

How can the symmetry operations be used to predict the polarity of molecules? What are the possible point groups of polar molecules?

- Polarity implies a separation of electric charge in a molecule or its chemical group causing a permanent electrical dipole moment.
- The permanent electrical dipole moment, μ, is a vector sum of bond dipole (and lone pair dipole) contributions.

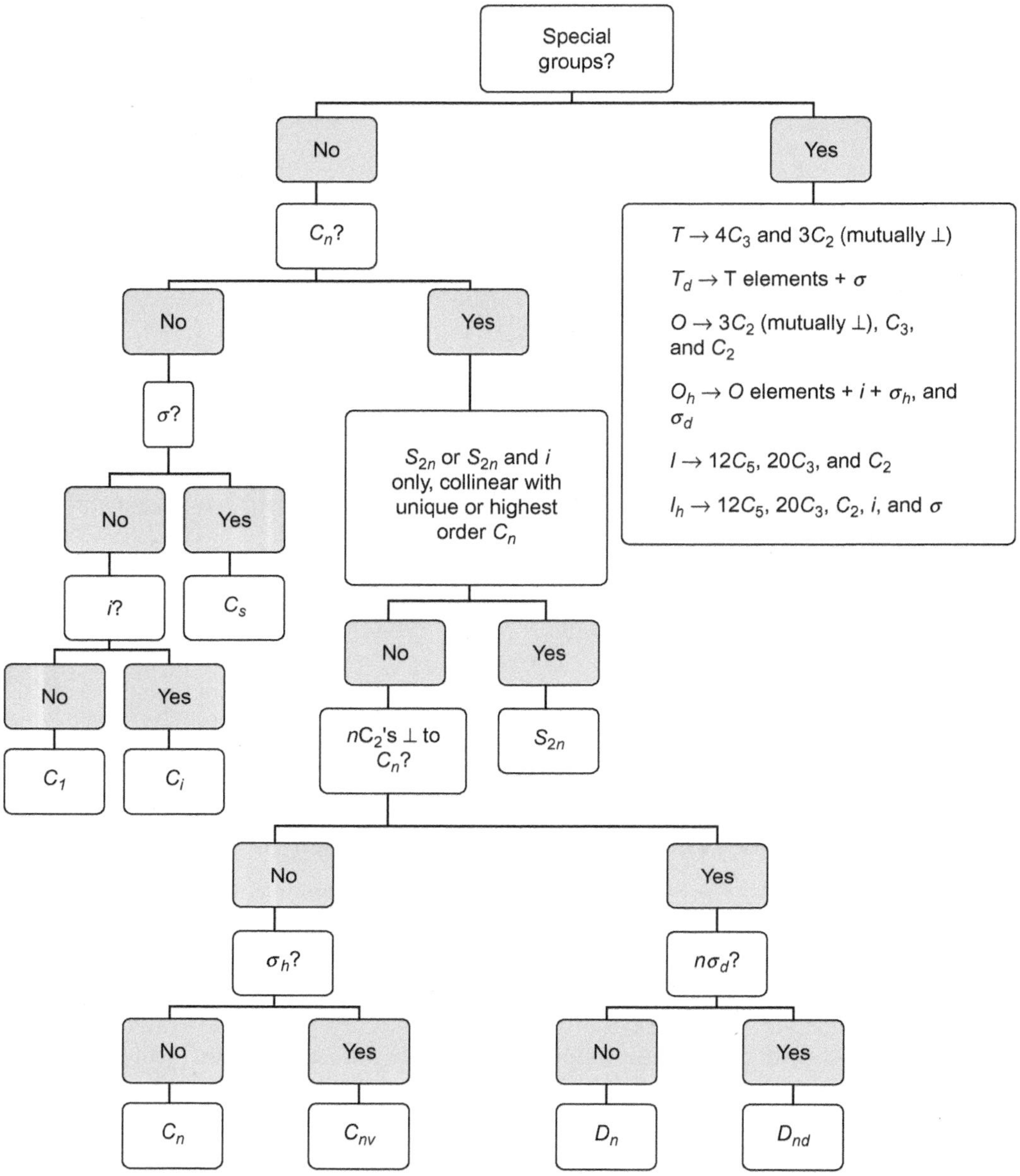

SCHEME 4.2 Classification of molecules by point group.

- Molecules with i have $\mu = 0$.

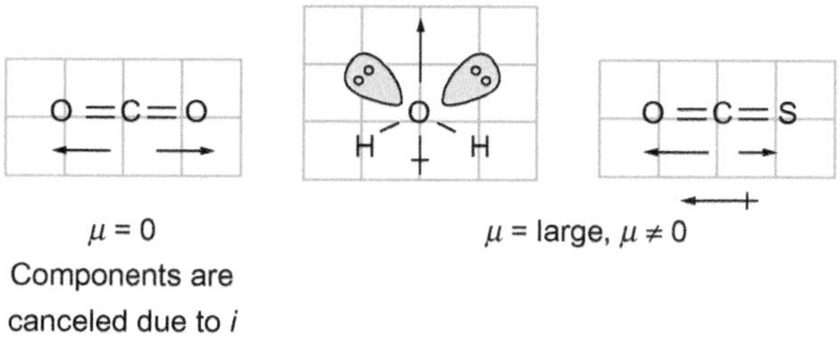

TABLE 4.2 List of Symmetry Elements and the Point Groups for Different Molecules

Example	Symmetry Elements	Point Group	Structure
$CH_3CHClBr$	E	C_1	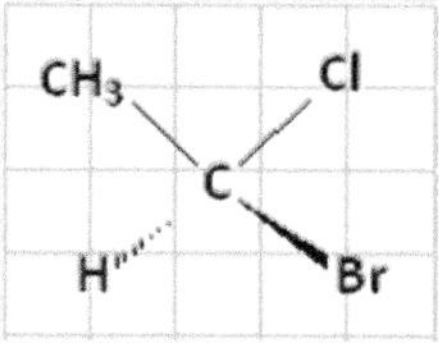 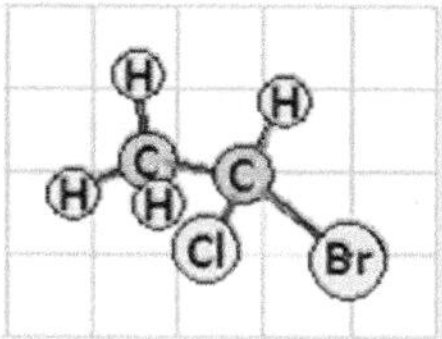
H_2O_2, S_2Cl_2	E, C_2	C_2	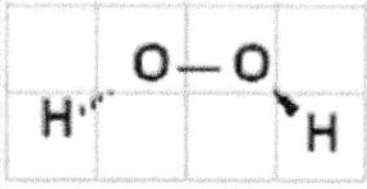 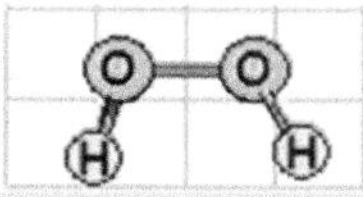
CH_3—$(CHCl)_2$—CH_3 Trans	E, i	C_i	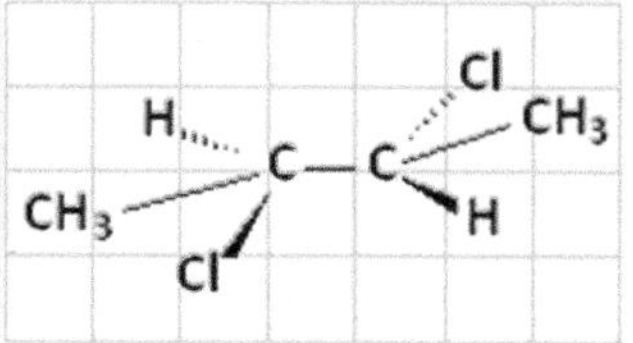 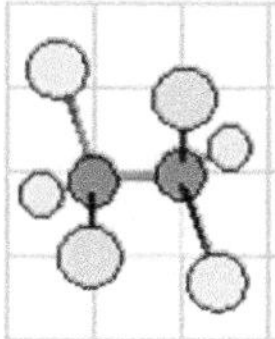
NOCl	E, σ	C_s	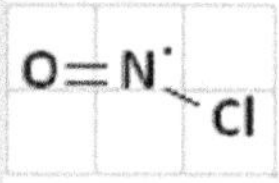 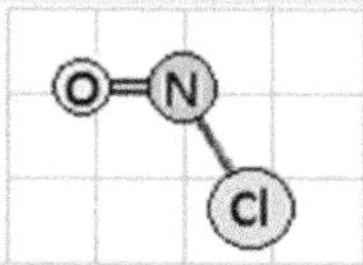
H_2O, CH_2Cl_2	E, C_2, $2\sigma_v$	C_{2v}	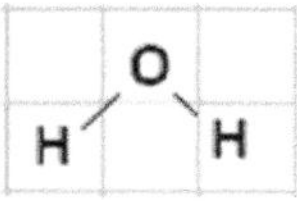 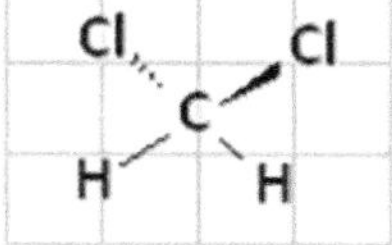 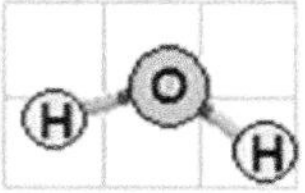 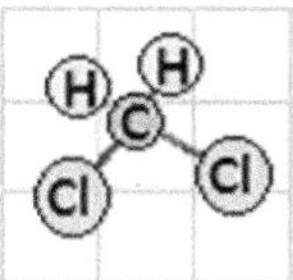
NH_3, $CHCl_3$	E, C_3, $3\sigma_v$	C_{3v}	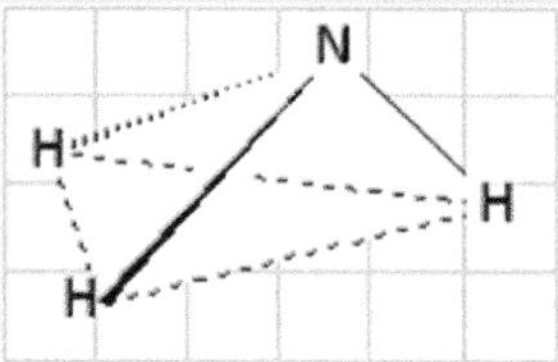 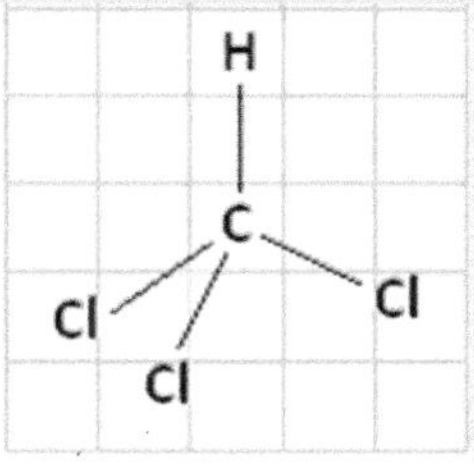 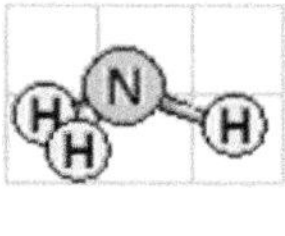 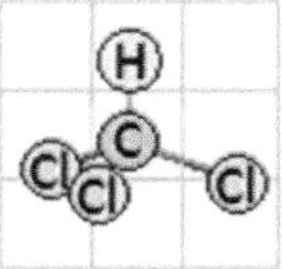
HCN, SCN, OCN	E, C_∞, $\infty\sigma_v$	$C_{\infty v}$	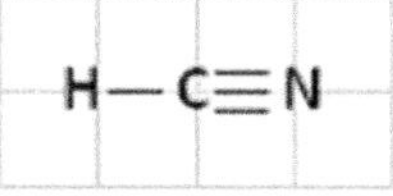 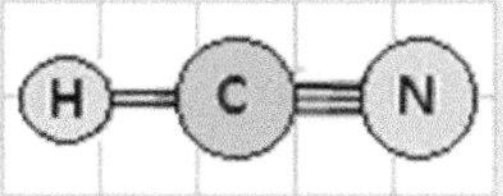
ClCH=CHCl Trans	E, C_2, σ_h, i	C_{2h}	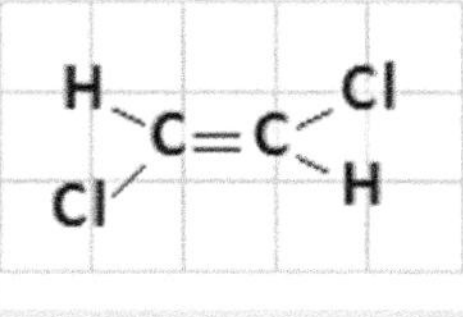 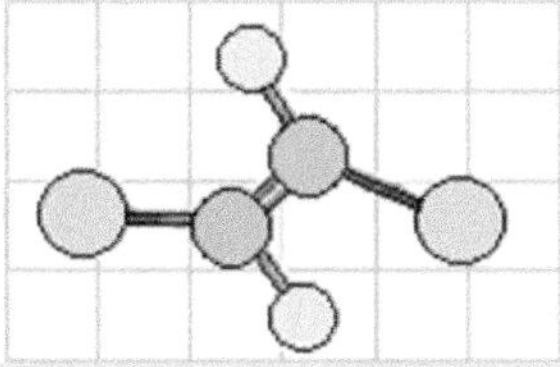

Continued

TABLE 4.2 List of Symmetry Elements and the Point Groups for Different Molecules—cont'd

Example	Symmetry Elements	Point Group	Structure
$H_2C{=}C{=}CH_2$	E, C_2 (⊥) S_4 coincident with one C_2, $2\sigma_d$	D_{2d}	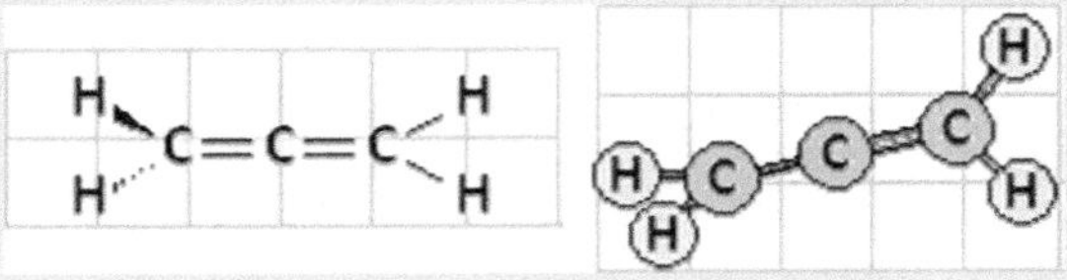
C_6H_{12}, cyclohexane Chair form	E, C_3, $3C_2$(⊥), $3\sigma_v$, S_6, i	D_{3d}	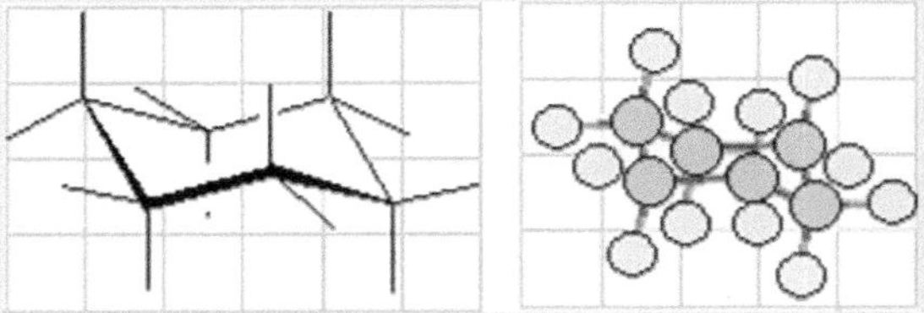
C_6H_{12}, cyclohexane Boat form	E, C_2, $2\sigma_v$	C_{2v}	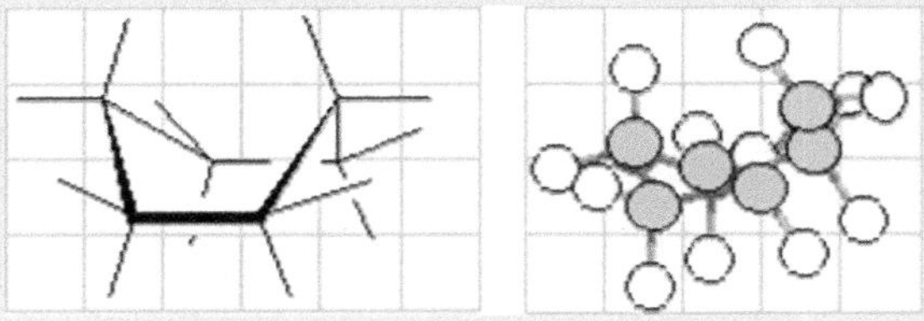
$H_2C{=}CH_2$	E, C_2, $2C_2$(⊥), σ_h, $2\sigma_d$, i, 2σ(⊥) σ_h	D_{2h}	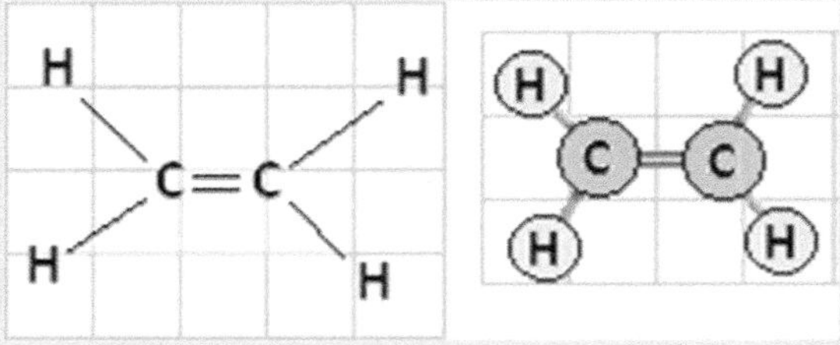
BCl_3	E, C_3, $3C_2$(⊥), σ_h, $3\sigma_v$	D_{3h}	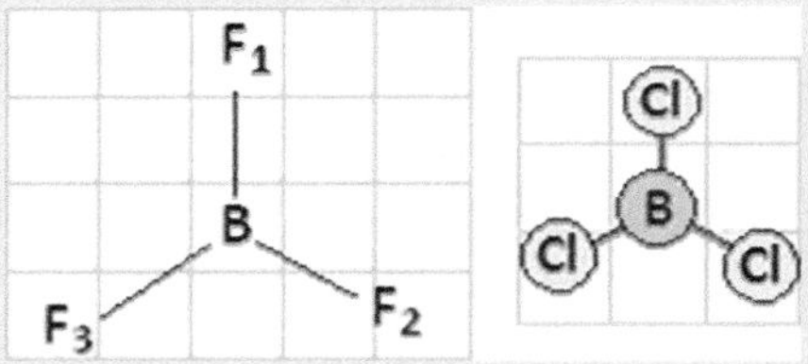
C_4H_8, AB_4	E, C_4, $4C_2$(⊥), σ_h, $4\sigma_v$, S_4	D_{4h}	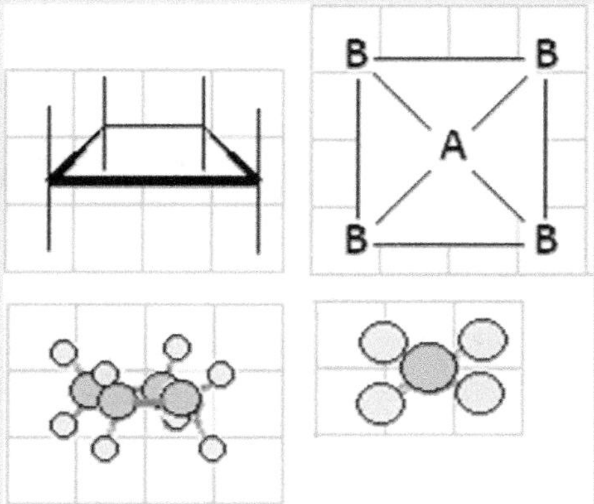
C_6H_6	E, C_6, $6C_2$(⊥), σ_h, $3\sigma_v$, $3\sigma_v$	D_{6h}	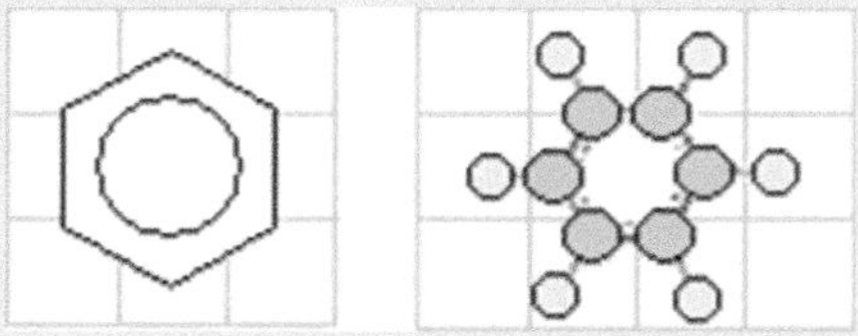

TABLE 4.2 List of Symmetry Elements and the Point Groups for Different Molecules—cont'd

Example	Symmetry Elements	Point Group	Structure
H_2, CO_2, HC≡CH	E, C_∞, $\infty C_2(\perp)$, $\infty\sigma_v$, σ_h	$D_{\infty h}$	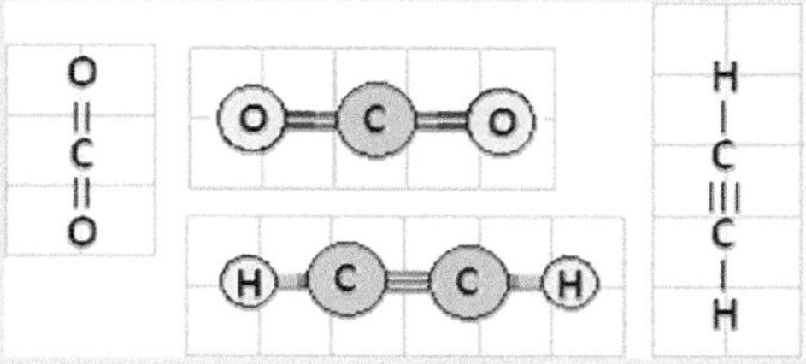
CH_4, CCl_4	E, $4C_3$, $3C_2(\perp)$, 6σ, $3S_4$ coincident with C_2	T_d	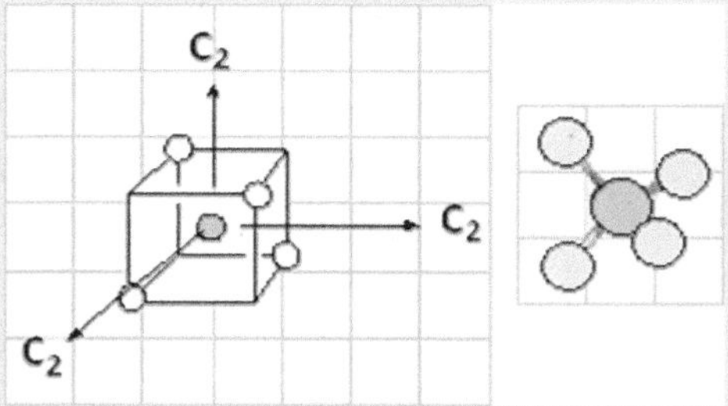
$PtCl_6^{2-}$	E, $3C_4(\perp)$, $4C_3$, $3S_4$, $3C_2$, $6C_2$, 9σ, $4S_6$, i	O_h	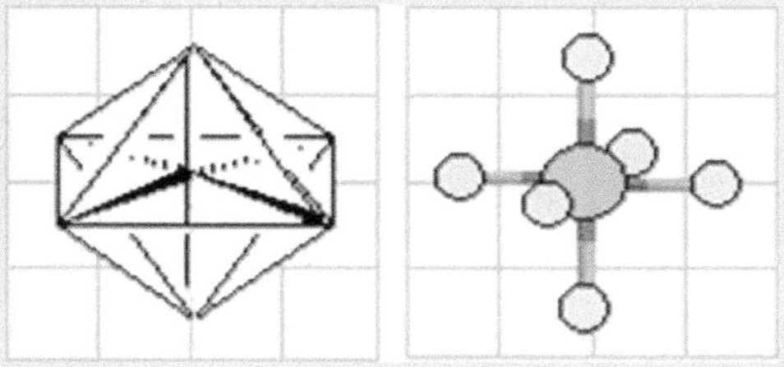

- If a molecule has two or more noncoincident proper rotation axes, $\mu=0$.

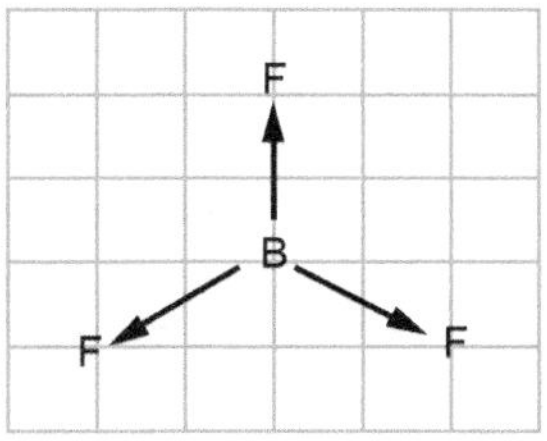

BF$_3$: C_3, $3C_2$, vectors sum $=0$, due to rotation symmetry $\mu=0$

- Predict $\mu \neq 0$ only for point groups: C_1, C_s, C_n, and C_{nv}.

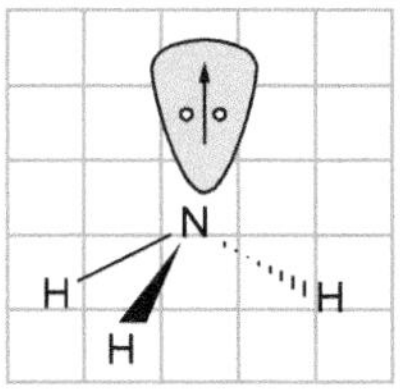

NH$_3$: C_{3v}

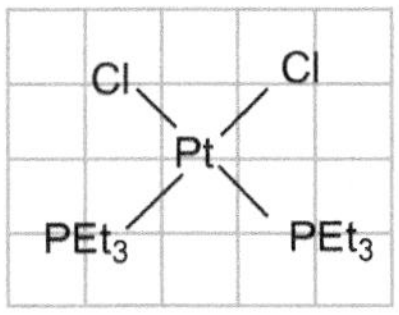

cis: C_{2v}, $\mu \neq 0$

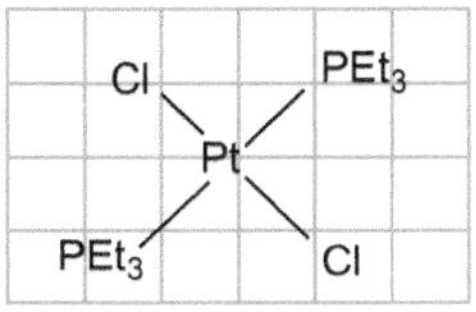

trans: D_{2h}, $\mu = 0$

PR$_3$ appears as cones of rotation

Polarity dictates a number of physical properties including solubility, surface tension, melting points, and boiling points.

Chirality

What are the possible point groups of chiral molecules?

- Handedness is detectable when molecules and its mirror image are not super-imposable. Chirality may be judged by internal symmetry:
  - An asymmetric molecule has only E, belongs to group C_1, and is chiral.

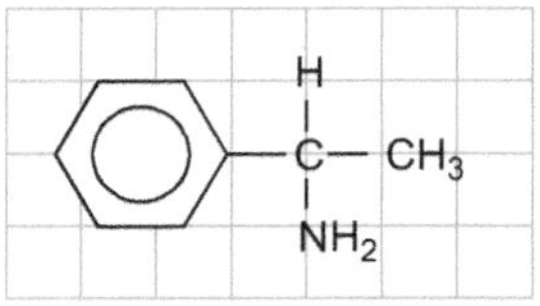

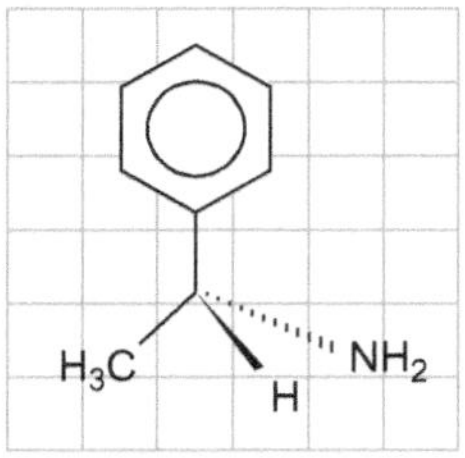

E.g., CHFClI, and most classical organics with "chiral center:" any carbon atom with four inequivalent ligands.

  - A dissymmetric molecule lacks S_n ($S_1 = \sigma$, $S_2 = i$, etc.) and belongs to C_n, D_h.

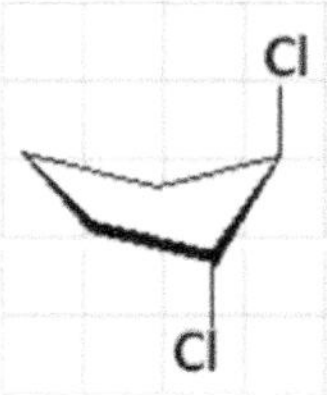

e.g. trans 1,2-dichlorocyclopentane has C_2, in C_2.

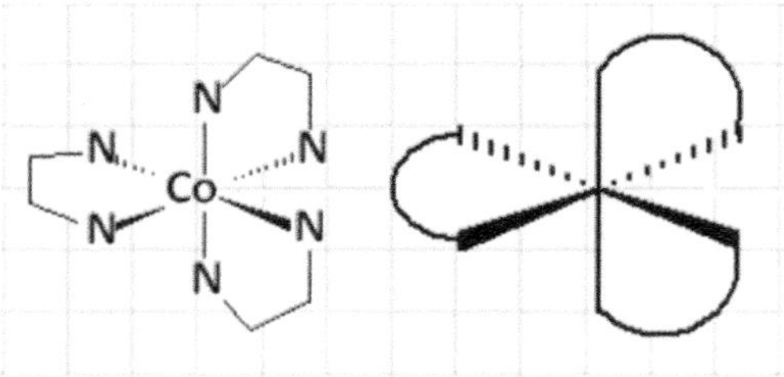

e.g. tris (ethylenediamine) cobalt (III), D_3: has D, L isomers (enantiomers)

Equivalent Atoms: (Or Group of Atoms)

How can you detect an equivalent atoms or group of atoms?

- Equivalent atoms (or group of atoms) are physically, chemically and spectroscopically indistinguishable and related by proper rotation operation of the molecule.

  In PF_5:

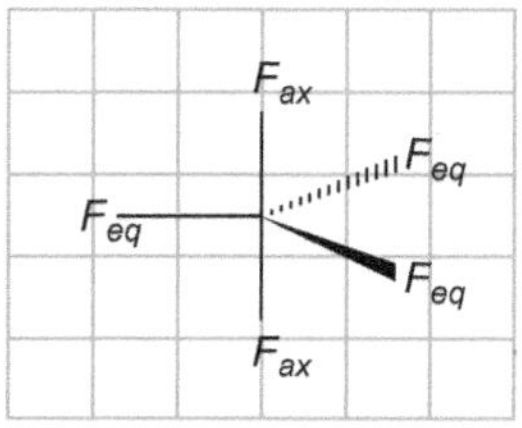

D_{3h} has $2F_{axial}$, $3F_{equatorial}$

$F_{axial} \overset{\text{exchangeable}}{\Longleftrightarrow} F_{axial}$ by C_2

$F_{equatorial} \overset{\text{exchangeable}}{\Longleftrightarrow} F_{equatorial}$ by C_3

$F_{axial} \neq F_{equatorial}$

Crystal Symmetry

What is the unit cell?

- Regular exterior morphology of crystal reflects order 3D arrangement of atom/molecules packed in it.
- A unit cell is the smallest possible fragment of lattice which, when stacked repeatedly in three dimensions, builds up the crystal lattice. Arrangement of atoms determines unit cell geometry.

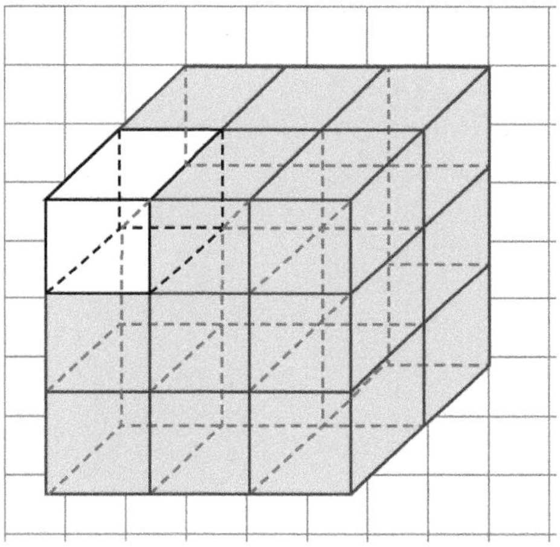

What are the structural elements of the unit cell?
How can these elements be employed as tools to classify the crystal structures in lattice systems?

- The molecule or atom is reproduces around the unit cell by symmetry operations of the crystal lattice (and of the unit cell).
  - Structures of CsCl, NaCl, and CO_2 unit cells:

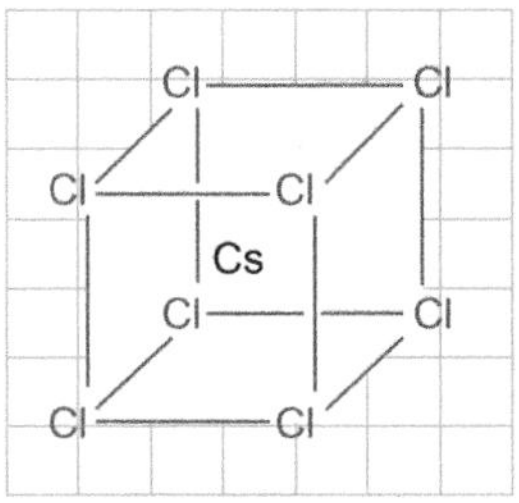

CsCl

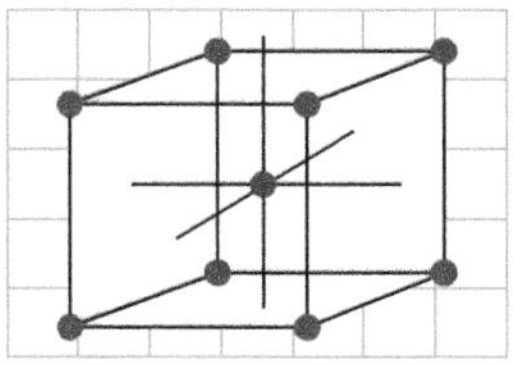

Body centered cubic

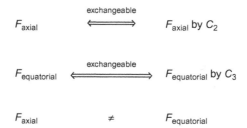

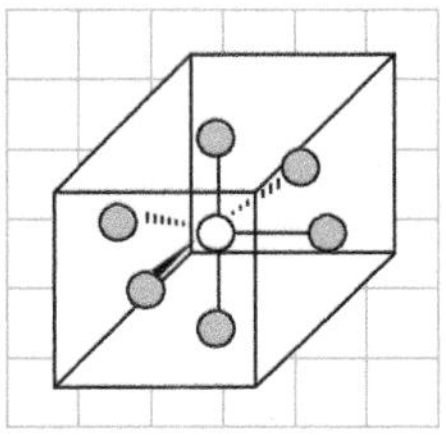

NaCl

CsCl unit cell is not $(1 \times Cs^+, 8 \times Cl^-)$
CsCl is $= 1 \times Cs^+, 8 \times (1/8)\ Cl^-$
NaCl unit cell is not $(1 \times Na^+, 6 \times Cl^-)$
NaCl is $= 1 \times Na^+, 6 \times (1/6)\ Cl^-$

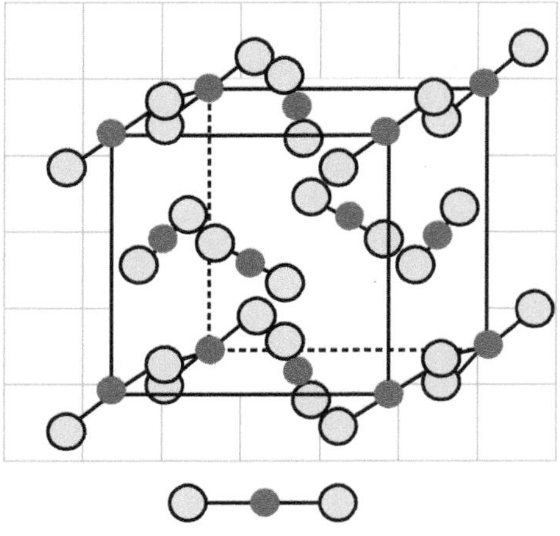

- The unit cell may or may not contain complete molecular units, e.g., CO_2 is a molecular crystal with a fraction of molecule at each site. Note the presence of symmetry relationships within the site, such as mirror plane.

- Unit cell dimensions: a, b, c, α, β, γ

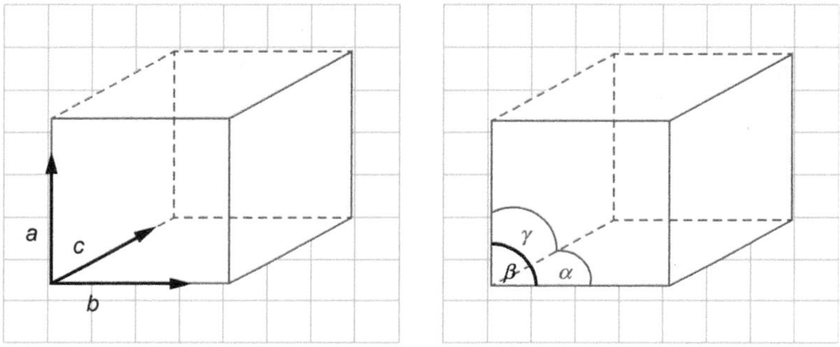

A cubic lattice is quite symmetric, and thus regular in shape.
Cubic: $a = b = c$, $\alpha = \beta = \gamma = 90°$

- Locations of atoms are given in terms of fractional values, with respect to unit cell dimensions:
  e.g., monoclinic: $\alpha = \gamma = 90°$, $\beta > 90°$; a, b, and c need not be equal.

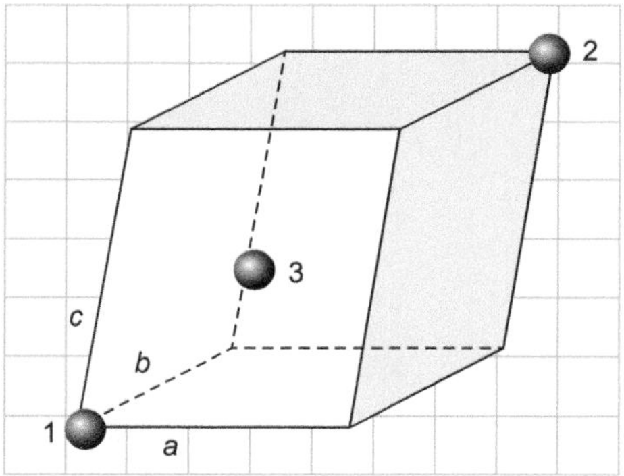

Point 1:	(0, 0, 0)		
Point 2:	(1, 1, 1)	→	(a/a, b/b, c/c)
Point 3:	(0.5, 0, 0.5)	→	($0.5a/a$, $0b/b$, $0.5c/c$), which is the center of front face.

- Coordinates of an atom at point 3 listed as (x, y, z), but these are not Cartesian coordinates.

	x/a	y/b	z/c
Then "A"	0.5	0	0.5
We also commonly see:	10^4x	10^4y	10^4z
A	5000	0	5000

- Crystal symmetry elements are listed in Table 4.3.
  E.g.,
  C_2 is 2 DIAD Axis
  C_3 is 3 TRIAD Axis
  σ is m
  i is I
  E is E
  $\overline{3}$ is inverse triad: C_3, i rather than C_3, σ
- Grouping of the crystal structures into lattice systems are achieved according to the inherent symmetries of the crystal unit.
- There are seven simple crystal systems or primitive lattices, labeled P (Table 4.4). In addition to these, there are seven other important lattice structures for crystals: body-centered, I, side-centered, A, B, or C, and face-centered, F, giving in total the fourteen so-called Bravias lattices.

How can the crystallographic point group be constructed?

- A characteristic assembly of the symmetry operation describes the crystallographic point group (Table 4.5). The symmetry designations of Hermann-Mauguin are constructed from corresponding symbols symmetry operations (Schöenflies notation):

  Draw the best guess structures in:

  i. **Ozone**
  ii. **$POCl_2Br$**
  iii. **Urea**
  iv. **N_2F_2**
  v. **Trans $Pt(NH_3)_2Cl_2$**

TABLE 4.3 Crystal Symmetry Elements, Associated Symbols, and Operations

Symmetry Element	Symbol	Operation
Monad	1	360° or 0° rotation about any line
Diad	2	180° rotation about the axis
Triad	3	120° rotation about the axis
tetrad	4	90° rotation about the axis
Hexad	6	60° rotation about the axis
Mirror plane	m	Reflection across the m plane
Inverse monad	$\overline{1}$	360° or 0° rotation about the axis + inversion through a point
Inverse diad	$\overline{2}$	180° rotation about the axis + inversion through a point
Inverse tiad	$\overline{3}$	120° rotation about the axis + inversion through a point
Inverse tetrad	$\overline{4}$	90° rotation about the axis + inversion through a point
Inverse hexad	$\overline{5}$	60° rotation about the axis + inversion through a point

TABLE 4.4 Symmetry Elements and Cell Dimensions of the Crystal Systems

Crystal System Name	Symmetry Elements	Parametral Plane Intercepts and Interaxial Angles
Triclinic	None (1)	$a \neq b \neq c$, $\alpha \neq \beta \neq \gamma$
Monoclinic	2 or $\overline{2}$ along y	$a \neq b \neq c$, $\alpha = \gamma = 90° \neq \beta$
Orthorhombic	Three ⊥ 2 or $\overline{2}$ along *x*, *y*, and *z*	$a \neq b \neq c$, $\alpha = \beta = \gamma = 90°$
Tetragonal	4 or $\overline{4}$ along *z*	$a = b \neq c$, $\alpha = \beta = \gamma = 90°$
Cubic	Four 3 along ⟨111⟩	$a = b = c$, $\alpha = \beta = \gamma = 90°$
Hexagonal	6 or $\overline{6}$ along *z*	$a = b \neq c$, $\alpha = \beta = 90°, \gamma = 120°$
Triagonal	3 along *z*	$a = b \neq c$, $\alpha = \beta = 90°, \gamma = 120°$
Triagonal	3 along ⟨111⟩	$a = b = c$, $\alpha = \beta = \gamma \neq 90°, < 120°$

TABLE 4.5 Three-Dimensional Crystallographic Point Groups

Hermann-Mauguinnotation	Schöenfliesnotation	Hermann-Mauguinnotation	Schöenfliesnotation
1	C_1	3*m*	C_{3v}
2	C_2	4*mm*	C_{4v}
3	C_3	6*mm*	C_{6v}
4	C_4	2/*m*	C_{2h}
6	C_6	$\overline{6}$	C_{3h}
$\overline{1}$	C_i, S_2	6/*m*	C_{6h}
$\overline{3}$	S_6	4/*m*	C_{4h}
$\overline{4}$	S_4	*mmm*	D_{2h}
222	D_2	$\overline{6}m2$	D_{3h}
32	D_3	$\frac{4}{m}mm$	D_{4h}
422	D_4	$\frac{6}{m}mm$	D_{6h}
622	D_6	$\overline{4}2m$	D_{2d}
23	*T*	$\overline{3}m$	D_{3d}
432	*O*	*m*3	T_h
m	C_s, S_1	$\overline{4}3m$	T_d
mm2	C_{2v}	*m*3*m*	O_h

and assign:

a. **the symmetry elements;**
b. **proper point group;**
c. **optical activity; and**
d. **solubility.**

- Answer:
  i. Ozone: O_3

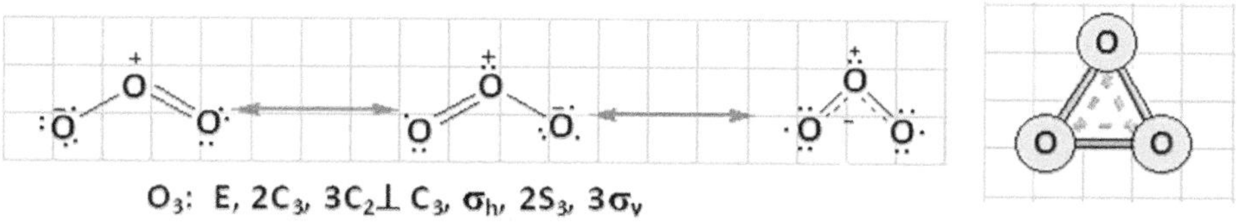

O_3: E, $2C_3$, $3C_2 \perp C_3$, σ_h, $2S_3$, $3\sigma_v$
Point group: D_{3h}
Optical activity: none
$\mu = 0$
Soluble in nonpolar solvent

  ii. $POCl_2Br$

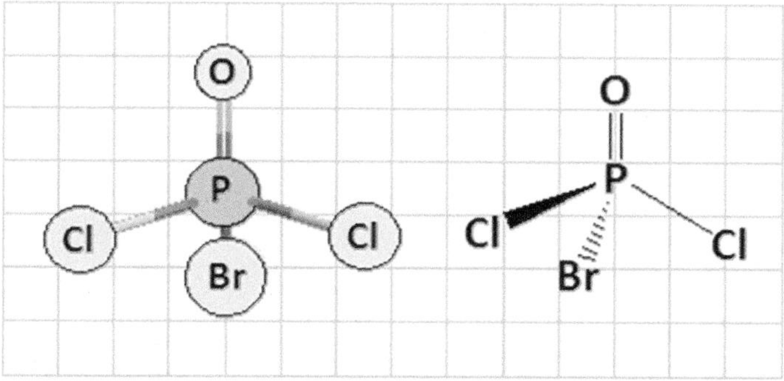

$POCl_2Br$: E, σ
Point group: C_s
Optical activity: none
$\mu \neq 0$
Soluble in polar solvent

  iii. Urea

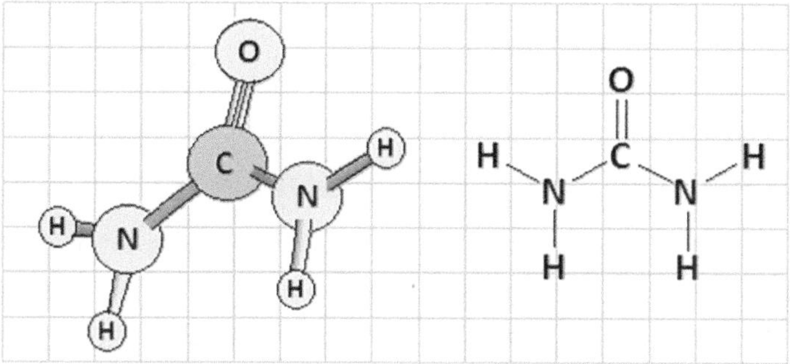

NH_2CONH_2: E, C_2, σ_v^{xz}, σ_v^{yz}
Point group: C_{2v}
Optical activity: none
$\mu \neq 0$
Soluble in polar solvent

iv. N_2F_2

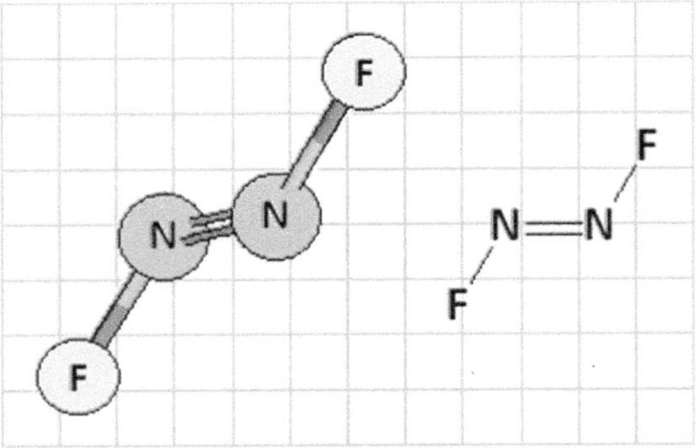

N_2F_2: E, C_2, i, σ_h
Point group: C_{2h}
Optical activity: none
$\mu = 0$
Soluble in nonpolar

v. Trans $Pt(NH_3)_2Cl_2$

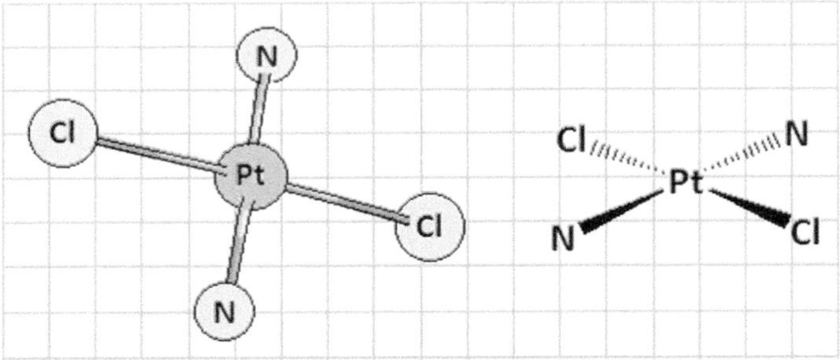

Trans Pt $(NH_3)_2Cl_2$: E, C_2, $2C_2 \perp C_2$, i, σ_h, $2\sigma_v$
Point group: D_{2h}
Optical activity: none
$\mu = 0$
Soluble in nonpolar

4.5 GROUP THEORY: PROPERTIES OF THE GROUPS AND THEIR ELEMENTS

What is the mathematic definition of the group, and the requirements for a set of elements to form a group?

- In mathematics, group theory studies is one of the fundamental algebraic structures known as groups.
- A group is a collection of elements that are related to each other according to certain rules.

- The group is equipped with an operation that combines any two elements to form a third element.
- The operation satisfies four conditions called the group axioms: closure, associativity, identity, and invertibility.
- An axiom or postulate is a statement that is taken to be true, to serve as a foundation or starting point for further reasoning and arguments.
- Specifying the element types or attributing any physical significance are not needed in order to discuss the group that they constitute.
- The group may be:
  - finite, containing a limited number of elements; or
  - infinite, containing an unlimited number of elements.
- The number of elements in a finite group is called its order, h.
- If we have a nonredundant and complete list of the h elements of a finite group, and all the possible products are known, then the group can be presented in the form of the group multiplication table.
- To explore groups, mathematicians have devised various notions to break groups into smaller, better-understandable pieces, such as subgroups, quotient groups and simple groups.
- In this work, we shall be concerned only with the group formed by the sets of symmetry operations that may be carried out on molecules. However, the theorem and basic definition and of group theory are more general.

What are the requirements to collect a set of symmetry elements into a group?

- A group is a collection of elements, which are interrelated according to certain rules.
- It satisfies the following conditions:
  a. The combination (product) of any element of the group with any other element, or with itself, yields an element in the group.

Example:	BF_3
Contains:	E, C_3, C_2, σ_h, σ_v, S_3
Point group:	D_{3h}
Example:	$C_2 . \sigma_v = E$ (Fig. 4.14)

Note: the first operation performed is listed on the right and subsequent operations to the lift.

  b. If A and B are two elements of a group, the combination is as follows:
     A. B = B. A (communicative)
  c. One element in the group must commute with all other elements and leave them unchanged.
     This element is the identity (E)
     $E. X = X. E = X$
  d. The associative law of multiplication must hold

$$\sigma_h \cdot C_2 \cdot \sigma_v = \sigma_h(C_2 \cdot \sigma_v) = (\sigma_h \cdot C_2) \cdot \sigma_v$$

  e. Each element must have an inverse (reciprocal) which is an element of the group.

$$BF_3 : C_3^2 \cdot C_3^1 = E$$

C_3^2 is the invers of C_3^1

$$C_3^2 \equiv C_3^{-1}$$

- The defining rules show that every group has what we may call a multiplication table.

Set up the group multiplication table of the symmetry operations for the water molecules.

- A multiplication table can be useful for understanding the properties of a group.
- The law of multiplication is to perform the operation from right to left or, in table form, to carry out the operation at the top of the table (labeling the columns) first and the operation at the left (labeling the rows) second.

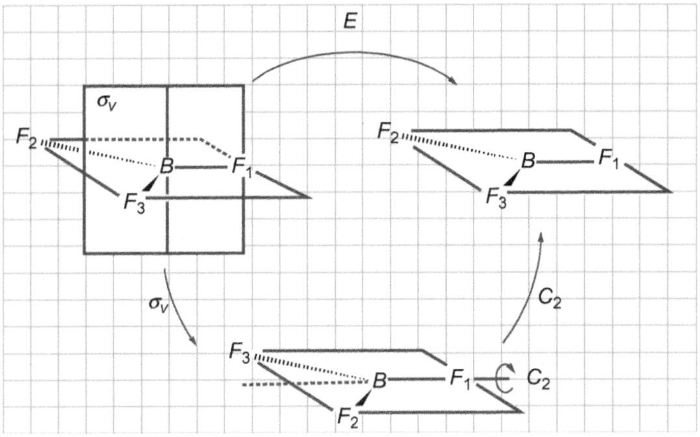

FIG. 4.14 Product of σ_v and C_2 yields E, $C_2 . \sigma_v = E$.

Example: the schematic pattern in the multiplication table of H_2O is shown in Table 4.6 and Figs. 4.15 and 4.16. The multiplication table is shown in Table 4.7.

i. The product $\sigma_v^{xz}\ C_2 = \sigma_v^{yz}$, and $\sigma_v^{xz}\sigma_v^{yz} = C_2$
ii. Note that each column and each row is a list of the element, each occurring once.
iii. Note that the product (E) is on the diagonal.
iv. The identity E occurs only once in each column or row.

In BF_3, which of the following pairs of operations commute?
C_3 and C_3^2, σ and C_3, σ and σ', E and C_3^2

- $C_3\ C_3^2 = E$; $C_3^2\ C_3 = E$; i.e., C_3 and C_3^2 commute
- $\sigma\ C_3 = \sigma'$; $C_3\ \sigma = \sigma''$ i.e., σ and C_3 do not commute
- $\sigma'\ \sigma = C_3$; $\sigma'\ \sigma = C_3^2$ i.e., σ and σ' do not commute
- $E\ C_3^2 = C_3^2$; $C_3^2\ E = C_3^2$ i.e., E and C_3^2 commute

4.6 SIMILARITY TRANSFORMS, CONJUGATION, AND CLASSES

Define the similarity transforms, the conjugated elements, and the class.
Find the similarity transform of C_3^2 by σ, the conjugate of C_3^2 and the constituents of C_3 class.

- If A and X are elements of group and also:

$$B = X^{-1}AX$$

B is called the **similarity transform** of ***A*** by ***X***, and ***A*** and ***B*** are said to be **conjugate** to each other (Fig. 4.17).

- Since $\sigma_v^{-1} = \sigma_v$ because: $\sigma_v . \sigma_v = E$

C_3^1 is the **similarity transform** of C_3^2 by σ_v, and C_3^1 and C_3^2 are **conjugated**.

  - Each element is conjugate with itself. There must be some element such that

TABLE 4.6 Schematic pattern in the Multiplication Table of H_2O

C_{2v}	E	C_2	σ_v^{xz}	σ_v^{yz}
E	$E.E = E$	$C_2.E = C_2$	$\sigma_v^{xz}.E = \sigma_v^{xz}$	$\sigma_v^{yz}.E = \sigma_v^{yz}$
C_2	$C_2.E = C_2$	$C_2.C_2 = E$	$C_2.\sigma_v^{xz} = \sigma_v^{yz}$	$\sigma_v^{yz}.\sigma_v^{yz} = \sigma_v^{xz}$
σ_v^{xz}	$\sigma_v^{xz}.E = \sigma_v^{xz}$	$\sigma_v^{xz}.C_2 = \sigma_v^{yz}$	$\sigma_v^{xz}.\sigma_v^{xz} = E$	$\sigma_v^{yz}.\sigma_v^{xz} = C_2$
σ_v^{yz}	$E.\sigma_v^{yz} = \sigma_v^{yz}$	$C_2.\sigma_v^{yz} = \sigma_v^{xz}$	$\sigma_v^{xz}.\sigma_v^{yz} = C_2$	$\sigma_v^{yz}.\sigma_v^{yz} = E$

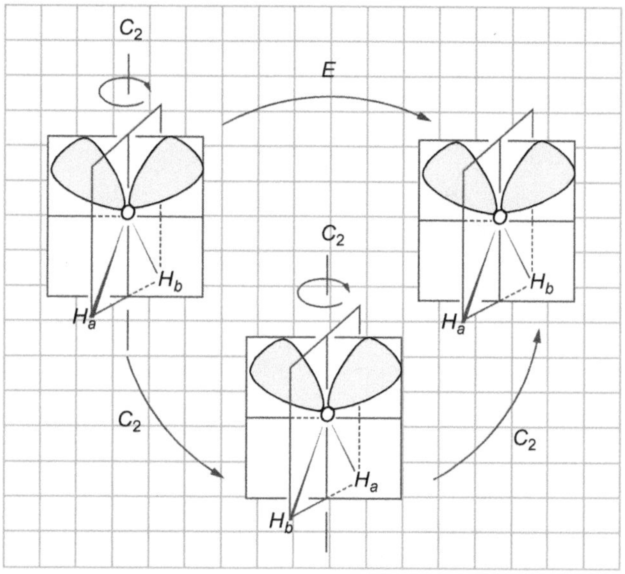

FIG. 4.15 The combination of C_2 with itself gives E, $C_2 . C_2 = E$.

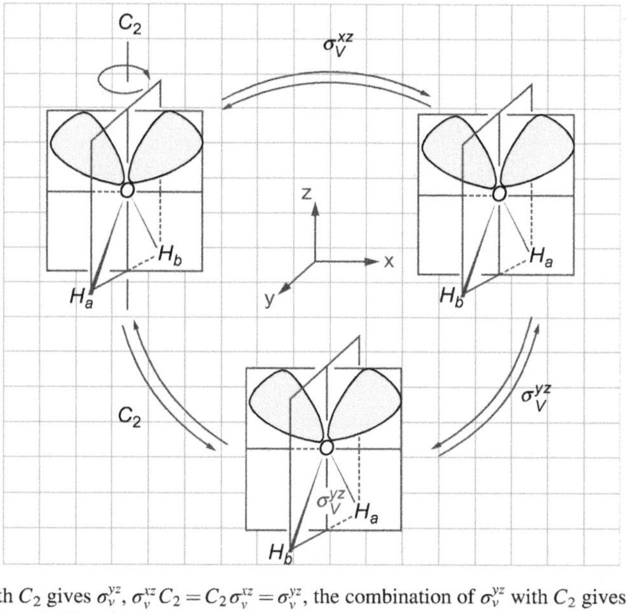

FIG. 4.16 The combination of σ_v^{xz} with C_2 gives σ_v^{yz}, $\sigma_v^{xz} C_2 = C_2 \sigma_v^{xz} = \sigma_v^{yz}$, the combination of σ_v^{yz} with C_2 gives σ_v^{xz}, $\sigma_v^{yz} C_2 = C_2 \sigma_v^{yz} = \sigma_v^{xz}$, and the combination of σ_v^{xz} with σ_v^{yz} gives C_2, $\sigma_v^{xz} \sigma_v^{yz} = \sigma_v^{yz} \sigma_v^{xz} = C_2$.

TABLE 4.7 Multiplication Table for C_{2v} Point Group

C_{2v}	E	C_2	σ_v^{xz}	σ_v^{yz}
E	E	C_2	σ_v^{xz}	σ_v^{yz}
C_2	C_2	E	σ_v^{yz}	σ_v^{xz}
σ_v^{xz}	σ_v^{xz}	σ_v^{yz}	E	C_2
σ_v^{yz}	σ_v^{yz}	σ_v^{xz}	C_2	E

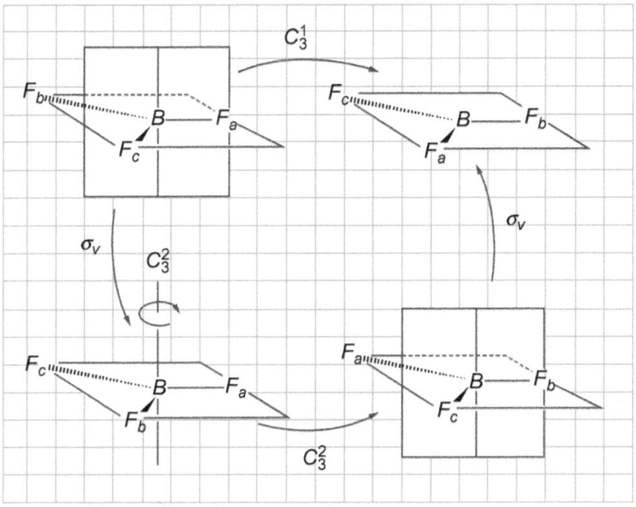

FIG. 4.17 The similarity transform of C_3^2 by σ_v is C_3^1, $\sigma_v C_3^2 \sigma_v = C_3^1 = \sigma_v^{-1} C_3^2 \sigma_v$.

$$X^{-1}AX = A(X \equiv E)$$

- If ***A*** is conjugate with ***B***, then ***B*** is **conjugate** with ***A***. If

$$A = X^{-1}BX$$

then there must be an element Y in the group such that

$$B = Y^{-1}AY$$

- If A is conjugate with B and C, then B and C are conjugate with each other.

- The elements of a group may be separated into smaller sets, and such sets called classes.
- A complete set of elements, which are conjugate to one another, is called a class of a group.
- In order to find the classes in any specific group, we start by one element and determine all the transforms, by means of all the elements of the group, as well as itself.
- We then take a next element that is not one of those found to be conjugated to the first, and define all its transforms, and so on until all elements in the group have been assigned to one class or another.
- The order of all classes must be integral factors of the order of the group.

Classify the symmetry elements of BF_3 into classes.

- The complete set of symmetry operations:
  $BF_3 : E, C_3^1, C_3^2, C_2, C_2', C_2'', \sigma, \sigma', \sigma'', \sigma_h, S_3$, and S_3^2
- Here the counter clockwise operation C_3^1 produce identical orientation obtained from C_3^2 (C_3^{-1}). C_3^1 and C_3^2 are similar symmetry operations differing only in direction of the rotation and consequently they belong to one class.
- The elements C_2, C_2', and C_2'' also σ_v, σ_v', and σ_v'' are transferred into one another by application of the operations C_3^1 and C_3^2.
- In listing all symmetry elements (or operations), it is sufficient to list one characteristic element of each class.
- Consequently: the list of symmetry operations for BF_3 (D_{3h}) is as follows:

$$E, 2C_3, 3C_2, \sigma_h, 3\sigma_V, 2S_3$$

4.7 MATRIX REPRESENTATION

Matrices and Vectors

What does the matrix have to do with a symmetry operation?

- A matrix is a regular array (see mathematics p. 721):

$$a = [a_{ij}] = \begin{bmatrix} a_{11} & a_{12} & \cdots & a_{1j} \\ a_{21} & a_{22} & \cdots & a_{2j} \\ \vdots & \vdots & \ddots & \vdots \\ a_{i1} & a_{i2} & \cdots & a_{ij} \end{bmatrix}$$

This is shorthand for representing sets of numbers, which can be combined (+, −, ×, ÷).

- A matrix can be used to define an initial position of a point or atom:

$$\begin{bmatrix} x \\ y \\ z \end{bmatrix} \text{ or } [x \ \ y \ \ z]$$

- We can use matrices as representations of symmetry operations.
- In addition, a matrix affords a way of representing a position vector.

  What does the vector have to do with a symmetry operation?
- A vector is defined by its Cartesian coordinates (x, y, z), which can be written in raw or column:

$$\vec{v_i} = [x_i \ \ y_i \ \ z_i]$$

or

$$\vec{v_i} = \begin{bmatrix} x_i \\ y_i \\ z_i \end{bmatrix}$$

- If an initial position vector $\vec{v}$ is subjected to a symmetry operation, R, it will be transformed into some resulting vector $\vec{v'}$ defined by the coordinates x', y' and z'. This transformation is expressed as follows:

$$\begin{bmatrix} \text{final position vector} \\ \text{matrix} \end{bmatrix} = \begin{bmatrix} \text{symmetry operation} \\ \text{matrix} \end{bmatrix} \begin{bmatrix} \text{initial position vector} \\ \text{matrix} \end{bmatrix}$$

- The vector $\vec{v_i}$ starts at the origin and terminates at the points (x_i, y_i, z_i).

  Note the following:
- Each of the coordinates of the outer terminus of the vector is numerically equal to the length of a projection of this vector on the axis concerned.
- The scalar is the product of two vectors. It is defined as the product of the lengths of the two vectors times the cosine of the angle between them.
- We use 3 × 3 matrixes to represent the effect of the symmetry operations on this vector.

Matrix Representation of Symmetry Operation

Write down the characters of the matrix representing the reflection of the point (x, y, z) through center in the space.

- For the inversion center, $(x, y) \xrightarrow{i} (x', y')$ (Fig. 4.18)

  $x' = -x$, and $y' = -y$

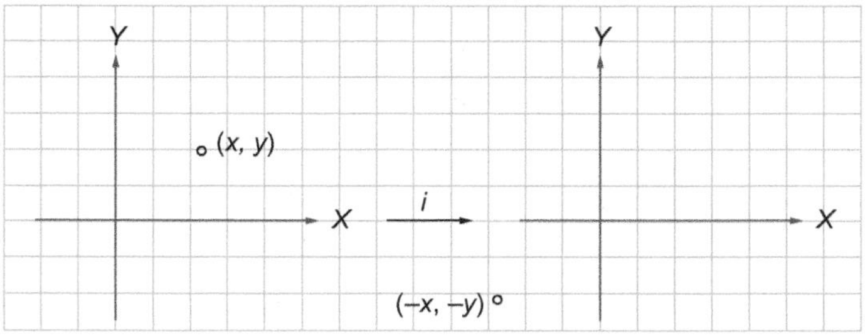

FIG. 4.18 The point (x,y) is inverted to $(-x,-y)$ by the operation i.

$$\begin{bmatrix} x' \\ y' \end{bmatrix} = \begin{bmatrix} -x \\ -y \end{bmatrix} = i \begin{bmatrix} x \\ y \end{bmatrix}$$

$$\vec{v}' = -\vec{v} = i\,\vec{v}$$

Where i is the inversion matrix, then

$$\underbrace{\begin{bmatrix} x' \\ y' \end{bmatrix}}_{\vec{v}'} = \underbrace{\begin{bmatrix} -1 & 0 \\ 0 & -1 \end{bmatrix}}_{i} \underbrace{\begin{bmatrix} x \\ y \end{bmatrix}}_{\vec{v}} = \underbrace{\begin{bmatrix} -x \\ -y \end{bmatrix}}_{-\vec{v}}$$

$$x' = -x = (-1)x + (0)y$$

$$y' = -y = (0)x + (-1)y$$

In three dimensions (Fig. 4.19):

$$\underbrace{\begin{bmatrix} x' \\ y' \\ z' \end{bmatrix}}_{\vec{v}'} = \underbrace{\begin{bmatrix} -1 & 0 & 0 \\ 0 & -1 & 0 \\ 0 & 0 & -1 \end{bmatrix}}_{i} \underbrace{\begin{bmatrix} x \\ y \\ z \end{bmatrix}}_{\vec{v}} = \underbrace{\begin{bmatrix} -x \\ -y \\ -z \end{bmatrix}}_{\vec{v}'}$$

$$\vec{v}' = i \cdot \vec{v}$$

- The inversion transformation matrix is $\begin{bmatrix} -1 & 0 & 0 \\ 0 & -1 & 0 \\ 0 & 0 & -1 \end{bmatrix}$.

Write down the characters of the matrices showing the rotation of the point (x, y, z) around z-axis by $\theta°$.

- For principle C_n:

$(x, y) \xrightarrow{C_n} (x', y')$ (Fig. 4.20):

$x' = l \cos(\theta - \alpha)$

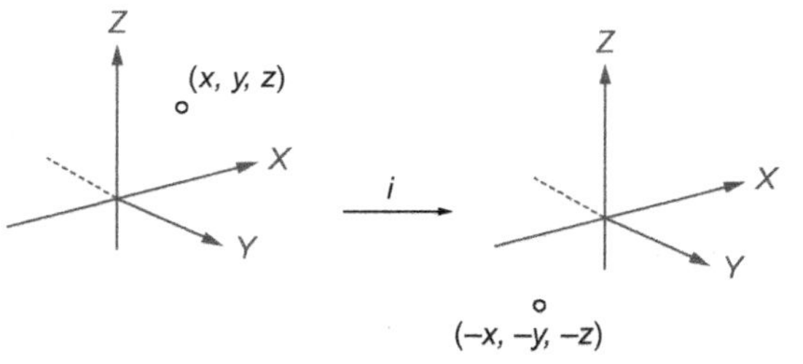

FIG. 4.19 The point (x,y,z) has inverted to $(-x,-y,-z)$ by the operation i.

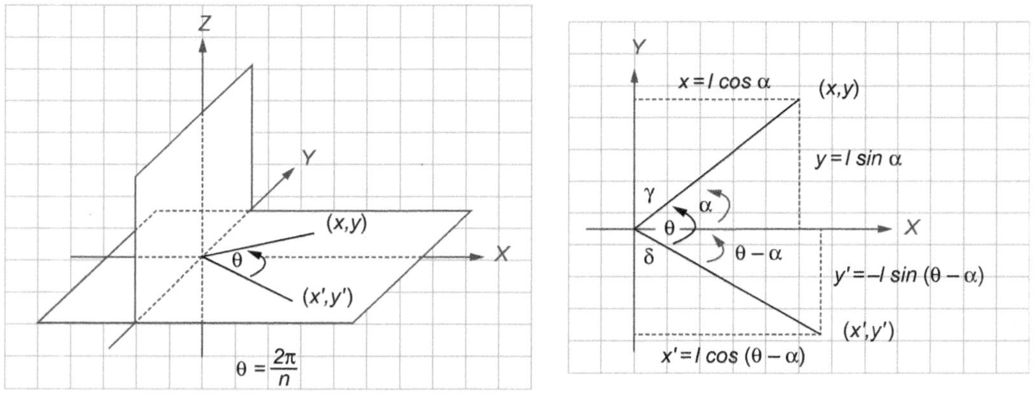

FIG. 4.20 The point (x,y) is rotated clockwise $\frac{2\pi}{n}$ by C_n to (x', y').

where l is the length of the vector $[x\ y]$.
Recall:

$$\sin(A \pm B) = \sin A \cos B \pm \cos A \sin B$$

$$\cos(A \pm B) = \cos A \cos B \mp \sin A \sin B$$

$x' = l \cos\theta \cos\alpha + l \sin\theta \sin\alpha$
$x' = x \cos\theta + y \sin\theta$
$y' = -l \sin(\theta - \alpha)$
$y' = -l \sin\theta \cos\alpha + l \cos\theta \sin\alpha$
$y' = -x \sin\theta + y \cos\theta$

for three dimensions,
$z' = z$, since z is C_n, so for C_n

$$\vec{v}' = C_n \vec{v}$$

$$\underbrace{\begin{bmatrix} x' \\ y' \\ z' \end{bmatrix}}_{\vec{v}'} = \underbrace{\begin{bmatrix} \cos\theta & \sin\theta & 0 \\ -\sin\theta & \cos\theta & 0 \\ 0 & 0 & 1 \end{bmatrix}}_{C_n} \underbrace{\begin{bmatrix} x \\ y \\ z \end{bmatrix}}_{\vec{v}}$$

- Different approach:

$$\theta = 180° - \gamma - \delta$$

$$x = l \sin\gamma$$

$$y = l \cos\gamma$$

$$x' = l \sin\delta$$

$$y' = -l \cos\gamma$$

$$\because x' = l \sin\delta$$

$$x' = l \sin(180° - (\gamma + \theta))$$

$$x' = l \sin 180° \cos(\gamma + \theta) - l \cos 180° \sin(\gamma + \theta)$$

$$\sin 180° = 0, \cos 180° = -1$$

$$x' = l \sin(\gamma + \theta)$$

$$x' = l \sin\gamma \cos\theta + l \cos\gamma \sin\theta$$

$$\therefore x' = x \cos\theta + y \sin\theta$$

$$\because y' = -l \cos\gamma$$

$$y = -l \cos(180° - (\delta + \theta))$$

$$y' = -l \cos 180° \cos(\delta + \theta) - l \sin 180° \sin(\delta + \theta)$$

$$y' = l \cos(\delta + \theta)$$

$$y' = -l \sin\gamma \sin\theta + l \cos\gamma \cos\theta$$

$$y' = -x \sin\theta + y \cos\theta$$

$$z' = (0)x + (0)y + (1)z$$

- For three dimensions

$$\begin{bmatrix} x' \\ y' \\ z' \end{bmatrix} = \begin{bmatrix} \cos\theta & \sin\theta & 0 \\ -\sin\theta & \cos\theta & 0 \\ 0 & 0 & 1 \end{bmatrix} \begin{bmatrix} x \\ y \\ z \end{bmatrix}$$

- Proper rotation matrix:
  e.g., for C_3

$$\theta = \frac{2\pi}{3}$$

It makes no difference which direction you use for the rotation, as long as successive rotations are preformed in the same direction. Choosing the x-axis determines the y-axis up to the direction.

$$C_3 = \begin{bmatrix} -\frac{1}{2} & \frac{\sqrt{3}}{2} & 0 \\ -\frac{\sqrt{3}}{2} & -\frac{1}{2} & 0 \\ 0 & 0 & 1 \end{bmatrix}$$

Matrix Representation of Point Group

What are the characters of the matrices representing each symmetry operations of C_{2v} point group?

- A set of matrices (one for each operation of the group) whose multiplication table is the same as that for the symmetry operations of the group is a representation of the group.
- H_2O: C_{2v} contains: E, C_2, σ_v, σ_v' (Fig. 4.21):
- Matrix describing effect of symmetry operations on x, y, and z:

$$\underbrace{\begin{bmatrix} 1 & 0 & 0 \\ 0 & 1 & 0 \\ 0 & 0 & 1 \end{bmatrix}}_{E} \underbrace{\begin{bmatrix} -1 & 0 & 0 \\ 0 & -1 & 0 \\ 0 & 0 & 1 \end{bmatrix}}_{C_2} \underbrace{\begin{bmatrix} 1 & 0 & 0 \\ 0 & -1 & 0 \\ 0 & 0 & 1 \end{bmatrix}}_{\sigma_v^{xz}} \underbrace{\begin{bmatrix} -1 & 0 & 0 \\ 0 & 1 & 0 \\ 0 & 0 & 1 \end{bmatrix}}_{\sigma_v^{yz}}$$

$$\begin{bmatrix} x \\ y \\ z \end{bmatrix} \rightarrow \begin{bmatrix} x \\ y \\ z \end{bmatrix} \quad \begin{bmatrix} x \\ y \\ z \end{bmatrix} \rightarrow \begin{bmatrix} -x \\ -y \\ z \end{bmatrix} \quad \begin{bmatrix} x \\ y \\ z \end{bmatrix} \rightarrow \begin{bmatrix} x \\ -y \\ z \end{bmatrix} \quad \begin{bmatrix} x \\ y \\ z \end{bmatrix} \rightarrow \begin{bmatrix} -x \\ y \\ z \end{bmatrix}$$

What are the characters of the matrices representing each symmetry operations of C_{3v} point group?

- The C_{3v} group contains: E, C_3, C_3^2, σ_v, σ_v', and σ_v''
- A matrix describing effect of symmetry operations on x, y, and z is as follows:

$$E = \begin{bmatrix} 1 & 0 & 0 \\ 0 & 1 & 0 \\ 0 & 0 & 1 \end{bmatrix}, \ \chi = 3$$

where the character χ is the trace of the matrix; $\chi = 1 + 1 + 1 = 3$

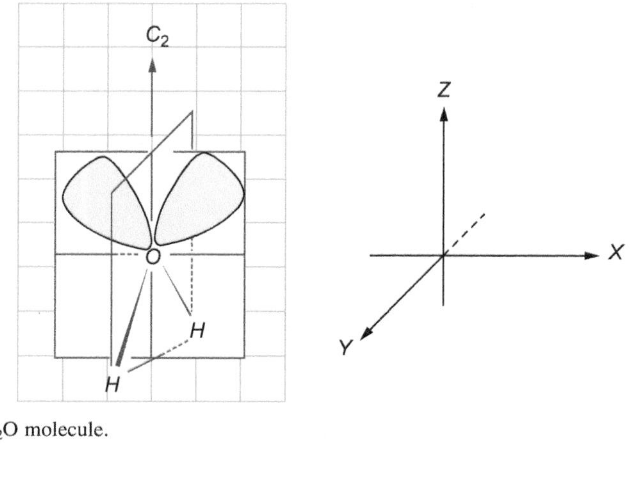

FIG. 4.21 Symmetry elements in H_2O molecule.

$$C_3 = \begin{bmatrix} \cos\frac{2\pi}{3} & \sin\frac{2\pi}{3} & 0 \\ -\sin\frac{2\pi}{3} & \cos\frac{2\pi}{3} & 0 \\ 0 & 0 & 1 \end{bmatrix} = \begin{bmatrix} -\frac{1}{2} & \frac{\sqrt{3}}{2} & 0 \\ -\frac{\sqrt{3}}{2} & -\frac{1}{2} & 0 \\ 0 & 0 & 1 \end{bmatrix}, \ \chi = 0$$

$$C_3^2 = \begin{bmatrix} \cos\frac{4\pi}{3} & \sin\frac{4\pi}{3} & 0 \\ -\sin\frac{4\pi}{3} & \cos\frac{4\pi}{3} & 0 \\ 0 & 0 & 1 \end{bmatrix} = \begin{bmatrix} -\frac{1}{2} & \frac{\sqrt{3}}{2} & 0 \\ -\frac{\sqrt{3}}{2} & -\frac{1}{2} & 0 \\ 0 & 0 & 1 \end{bmatrix}, \ \chi = 0$$

- Fig. 4.22 shows the transform of x- and y-vectors to x' and y' by reflection through a mirror plane σ:

$$x' = -x\cos 2\phi - y\sin 2\phi$$

$$y' = -x\sin 2\phi + y\cos 2\phi$$

Therefore,

$$\underbrace{\begin{bmatrix} x' \\ y' \\ z' \end{bmatrix}}_{\vec{v}'} = \underbrace{\begin{bmatrix} -\cos 2\phi & -\sin 2\phi & 0 \\ -\sin 2\phi & \cos 2\phi & 0 \\ 0 & 0 & 1 \end{bmatrix}}_{\sigma} \underbrace{\begin{bmatrix} x \\ y \\ z \end{bmatrix}}_{\vec{v}}$$

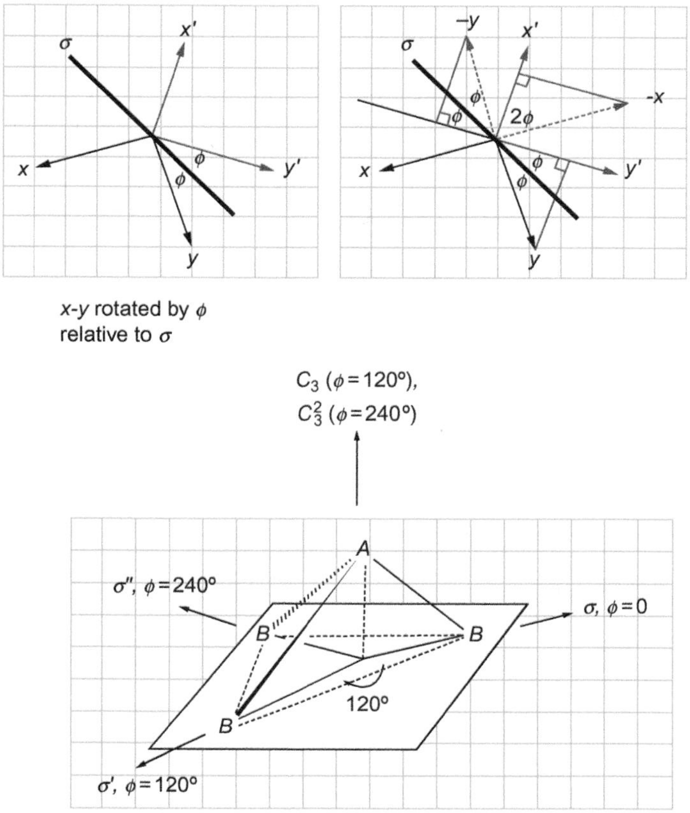

FIG. 4.22 Rotation by φ relative to the plane σ, and reflection through the plane σ.

$$\sigma = \begin{bmatrix} -\cos 0 & -\sin 0 & 0 \\ -\sin 0 & \cos 0 & 0 \\ 0 & 0 & 1 \end{bmatrix} = \begin{bmatrix} -1 & 0 & 0 \\ 0 & 1 & 0 \\ 0 & 0 & 1 \end{bmatrix}, \ \chi = 1$$

$$\sigma' = \begin{bmatrix} -\cos\left(2x\frac{2\pi}{3}\right) & -\sin\left(2x\frac{2\pi}{3}\right) & 0 \\ -\sin\left(2x\frac{2\pi}{3}\right) & \cos\left(2x\frac{2\pi}{3}\right) & 0 \\ 0 & 0 & 1 \end{bmatrix} = \begin{bmatrix} \frac{1}{2} & \frac{\sqrt{3}}{2} & 0 \\ \frac{\sqrt{3}}{2} & -\frac{1}{2} & 0 \\ 0 & 0 & 1 \end{bmatrix}, \ \chi = 0$$

$$\sigma'' = \begin{bmatrix} -\cos\left(2x\frac{4\pi}{3}\right) & -\sin\left(2x\frac{4\pi}{3}\right) & 0 \\ -\sin\left(2x\frac{4\pi}{3}\right) & \cos\left(2x\frac{4\pi}{3}\right) & 0 \\ 0 & 0 & 1 \end{bmatrix} = \begin{bmatrix} \frac{1}{2} & -\frac{\sqrt{3}}{2} & 0 \\ -\frac{\sqrt{3}}{2} & -\frac{1}{2} & 0 \\ 0 & 0 & 1 \end{bmatrix}, \ \chi = 0$$

Irreducible Representations

What are the characters describing the total representation of x, y, and z vectors in the C_{2v} point group?

- A matrix describing the effect of symmetry operations on x, y, and z is as follows:

$$\underbrace{\overset{E}{\begin{bmatrix} 1 & 0 & 0 \\ 0 & 1 & 0 \\ 0 & 0 & 1 \end{bmatrix}}}_{\substack{1+1+1 \\ \text{trace}=3}} \quad \underbrace{\overset{C_2}{\begin{bmatrix} -1 & 0 & 0 \\ 0 & -1 & 0 \\ 0 & 0 & 1 \end{bmatrix}}}_{\substack{-1-1+1 \\ \text{trace}=1}} \quad \underbrace{\overset{\sigma_v^{xz}}{\begin{bmatrix} 1 & 0 & 0 \\ 0 & -1 & 0 \\ 0 & 0 & 1 \end{bmatrix}}}_{\substack{1-1+1 \\ \text{trace}=1}} \quad \underbrace{\overset{\sigma_v^{yz}}{\begin{bmatrix} -1 & 0 & 0 \\ 0 & 1 & 0 \\ 0 & 0 & 1 \end{bmatrix}}}_{\substack{-1+1+1 \\ \text{trace}=1}}$$

$$\Gamma_{x,y,z} = [3, \ -1, \ +1, \ +1] \ \text{(reducible representation)}$$

- Each matrix is block diagonalized.
- For a vector describing symmetry operations in terms of one coordinate only, in terms of x-, y-, or z-coordinate, we could write:

$$\begin{array}{cccc} E & C_2 & \sigma_v & \sigma'_v \\ \underbrace{[1]} & \underbrace{[-1]} & \underbrace{[1]} & \underbrace{[-1]} \\ [x]\rightarrow[x] & [x]\rightarrow[-x] & [x]\rightarrow[x] & [x]\rightarrow[-x] \\ \chi(E)=1 & \chi(C_2)=-1 & \chi(\sigma_v)=1 & \chi(\sigma'_v)=-1 \end{array}$$

$$\begin{array}{cccc} E & C_2 & \sigma_v & \sigma'_v \\ \underbrace{[1]} & \underbrace{[-1]} & \underbrace{[-1]} & \underbrace{[1]} \\ [y]\rightarrow[y] & [y]\rightarrow[-y] & [y]\rightarrow[-y] & [y]\rightarrow[y] \\ \chi(E)=1 & \chi(C_2)=-1 & \chi(\sigma_v)=-1 & \chi(\sigma'_v)=1 \end{array}$$

$$\begin{array}{cccc} E & C_2 & \sigma_v & \sigma'_v \\ \underbrace{[1]} & \underbrace{[1]} & \underbrace{[1]} & \underbrace{[1]} \\ [z]\rightarrow[z] & [z]\rightarrow[z] & [z]\rightarrow[z] & [z]\rightarrow[zx] \\ \chi(E)=1 & \chi(C_2)=1 & \chi(\sigma_v)=1 & \chi(\sigma'_v)=1 \end{array}$$

or	E	C_2	σ_v	σ_v'
x	1	−1	1	−1
y	1	−1	−1	1
z	1	1	1	1

- These sets of 1D matrices describing the effects of symmetry operations is said to be a set of irreducible representations. E.g., (1, −1, 1, −1) as an irreducible representation of C_{2v}.

Irreducible and Degenerate Representations

Construct the matrices representing x, y, and z vectors in C_{4v} group. Which of these representations could be mixed, and what does that mean?

- Point group (Fig. 4.23): C_{4v}
  Contains: E, C_4, C_2, σ_{yz}, σ_{xz}, σ_d, σ_d'

$$\underset{\begin{bmatrix}x\\y\\z\end{bmatrix}\to\begin{bmatrix}x\\y\\z\end{bmatrix}}{\underbrace{\overset{E}{\begin{bmatrix}1&0&0\\0&1&0\\0&0&1\end{bmatrix}}}}\quad \underset{\begin{bmatrix}x\\y\\z\end{bmatrix}\to\begin{bmatrix}y\\-x\\z\end{bmatrix}}{\underbrace{\overset{C_4}{\begin{bmatrix}0&1&0\\-1&0&0\\0&0&1\end{bmatrix}}}}\quad \underset{\begin{bmatrix}x\\y\\z\end{bmatrix}\to\begin{bmatrix}-x\\-y\\z\end{bmatrix}}{\underbrace{\overset{C_2}{\begin{bmatrix}-1&0&0\\0&-1&0\\0&0&1\end{bmatrix}}}}\quad \underset{\begin{bmatrix}x\\y\\z\end{bmatrix}\to\begin{bmatrix}-x\\y\\z\end{bmatrix}}{\underbrace{\overset{\sigma_{yz}}{\begin{bmatrix}-1&0&0\\0&1&0\\0&0&1\end{bmatrix}}}}\quad \underset{\begin{bmatrix}x\\y\\z\end{bmatrix}\to\begin{bmatrix}x\\-y\\z\end{bmatrix}}{\underbrace{\overset{\sigma_{xz}}{\begin{bmatrix}1&0&0\\0&-1&0\\0&0&1\end{bmatrix}}}}\quad \underset{\begin{bmatrix}x\\y\\z\end{bmatrix}\to\begin{bmatrix}y\\x\\z\end{bmatrix}}{\underbrace{\overset{\sigma_d}{\begin{bmatrix}0&1&0\\1&0&0\\0&0&1\end{bmatrix}}}}\quad \underset{\begin{bmatrix}x\\y\\z\end{bmatrix}\to\begin{bmatrix}-y\\-x\\z\end{bmatrix}}{\underbrace{\overset{\sigma_d'}{\begin{bmatrix}0&-1&0\\-1&0&0\\0&0&1\end{bmatrix}}}}$$

The coordinate z never mixes with x or y by these operations, but x and y are mixed with each other. Both x and y belong to the same degenerate representation.

- The trace of each matrix is called its character:

$$\Gamma_{x,y}=\overset{E}{\begin{bmatrix}1&0\\0&1\end{bmatrix}}\quad\overset{C_4}{\begin{bmatrix}0&1\\-1&0\end{bmatrix}}\quad\overset{C_2}{\begin{bmatrix}-1&0\\0&-1\end{bmatrix}}\quad\overset{\sigma_{yz}}{\begin{bmatrix}-1&0\\0&1\end{bmatrix}}\quad\overset{\sigma_{xz}}{\begin{bmatrix}1&0\\0&-1\end{bmatrix}}\quad\overset{\sigma_d}{\begin{bmatrix}0&1\\1&0\end{bmatrix}}\quad\overset{\sigma_d'}{\begin{bmatrix}0&-1\\-1&0\end{bmatrix}}$$

$$\qquad\ \underset{1+1}{\underbrace{2}}\qquad\quad \underset{0+0}{\underbrace{0}}\qquad\quad \underset{-1-1}{\underbrace{-2}}\qquad\quad \underset{-1+1}{\underbrace{0}}\qquad\quad \underset{1-1}{\underbrace{0}}\qquad\quad \underset{0+0}{\underbrace{0}}\qquad\quad \underset{0+0}{\underbrace{0}}$$

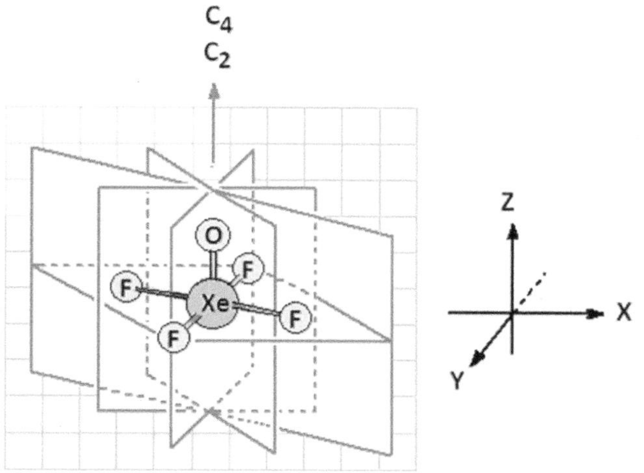

FIG. 4.23 Symmetry elements in an $OXeF_4$ molecule.

$$\begin{array}{lccccccc} & E & C_4 & C_2 & \sigma_{YZ} & \sigma_{XZ} & \sigma_d & \sigma_d' \\ \Gamma_Z = & [1] & [1] & [1] & [1] & [1] & [1] & [1] \\ & 1 & 1 & 1 & 1 & 1 & 1 & 1 \end{array}$$

$$\begin{array}{lccccc} & E & 2C_4 & 2C_2 & 2\sigma_V & 2\sigma_d \\ \Gamma_Z = & 1 & 1 & 1 & 1 & 1 \\ \Gamma_{X,Y} = & 2 & 0 & -2 & 0 & 0 \end{array}$$

Reducible representations furnish compound characters, which are sums of the simple characters of irreducible representations.

4.8 MOTION REPRESENTATIONS OF THE GROUPS

Translation Motion

Construct the matrices representing the translational motion in the *x*-, *y*-, and *z*-directions in the C_{2v} point group.

- H_2O: C_{2v} contains: E, C_2, σ_v^{xz}, σ_v^{yz}
- The transform matrix that describe the effect of E on translation in the x-, y-, and z-directions:

$$E: \begin{bmatrix} x \\ y \\ z \end{bmatrix} = \underbrace{\begin{bmatrix} 1 & 0 & 0 \\ 0 & 1 & 0 \\ 0 & 0 & 1 \end{bmatrix}}_{\text{Transformationmatrix}} \begin{bmatrix} x \\ y \\ z \end{bmatrix}, \quad \text{trace} = 3$$

E operation does not affect the translational motion in x, y, and z (Figs. 4.24 and 4.28):

- The transform matrix that describe the effect of C_2 on translation in the x-, y-, and z-directions:

$$C_2: \begin{bmatrix} -x \\ -y \\ z \end{bmatrix} = \underbrace{\begin{bmatrix} -1 & 0 & 0 \\ 0 & -1 & 0 \\ 0 & 0 & 1 \end{bmatrix}}_{\text{Transformation matrix}} \begin{bmatrix} x \\ y \\ z \end{bmatrix}, \quad \text{trace} = -1$$

- The directions of the translational motion in x and y will be reversed by the C_2 operation, but z has not (Figs. 4.25 and 4.28):
- The matrix that describe the effect of σ_v^{yz} on translation in the x-, y-, and z-directions:

$$\sigma_v^{yz}: \begin{bmatrix} -x \\ y \\ z \end{bmatrix} = \underbrace{\begin{bmatrix} -1 & 0 & 0 \\ 0 & 1 & 0 \\ 0 & 0 & 1 \end{bmatrix}}_{\text{Transform matrix}} \begin{bmatrix} x \\ y \\ z \end{bmatrix}$$

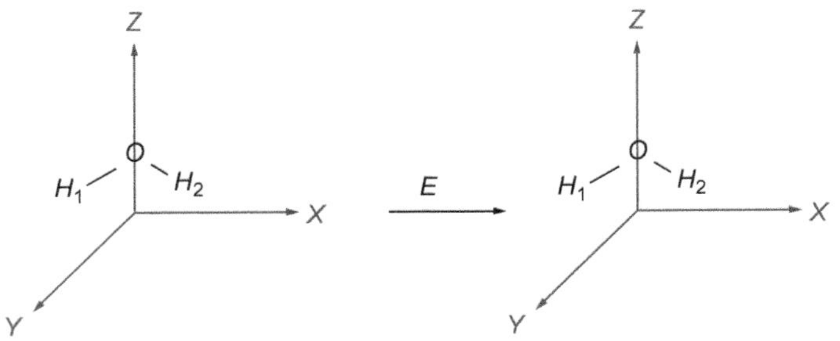

FIG. 4.24 The operation E has not changed the direction of the translational motion in x, y, and z.

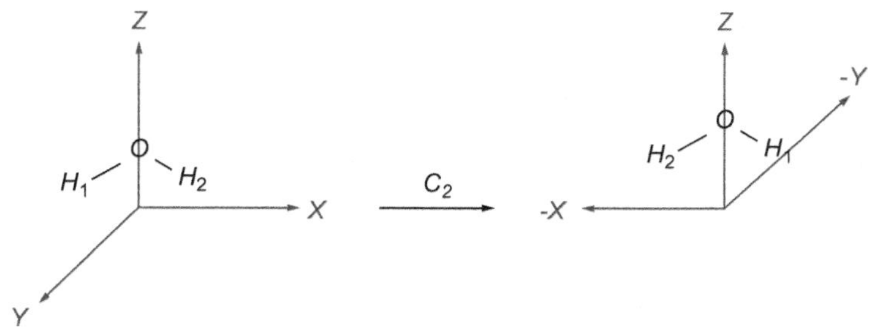

FIG. 4.25 The C_2 operation has reversed the direction in x and y.

- Translation motions along y, and z are not affected by the σ_v^{yz} operation, but will move in the opposite x-direction (Figs. 4.26 and 4.28).
- The resulting characters that describe the effect of σ_v^{xz} on a translation in the x-, y-, and z-directions:

$$\sigma_v^{xz}: \begin{bmatrix} x \\ -y \\ z \end{bmatrix} = \begin{bmatrix} x \\ y \\ z \end{bmatrix} \begin{bmatrix} 1 & 0 & 0 \\ 0 & -1 & 0 \\ 0 & 0 & 1 \end{bmatrix}$$

- The movement will be in the opposite y-direction by σ_v^{xz} (Fig. 4.27).
- Let us consider a translation in the x direction:
  - The identity E and σ_v^{xz} operations, the direction is unchanged.
  - For C_2 and σ_v^{yz} the direction reversed (Figs. 4.24–4.28).
  - The resulting characters for the four symmetry operations:

$$\begin{array}{ccccc} & E & C_2 & \sigma_v^{xz} & \sigma_v^{yz} \\ X & +1 & -1 & +1 & -1 \end{array}$$

  - These characters are obtained from the 1D diagonalized block of transformation matrices.

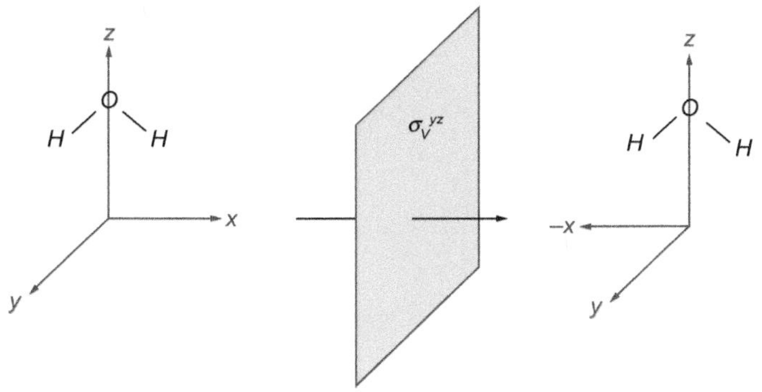

FIG. 4.26 The mirror image σ_v^{yz} will reverse the movement in the x-direction.

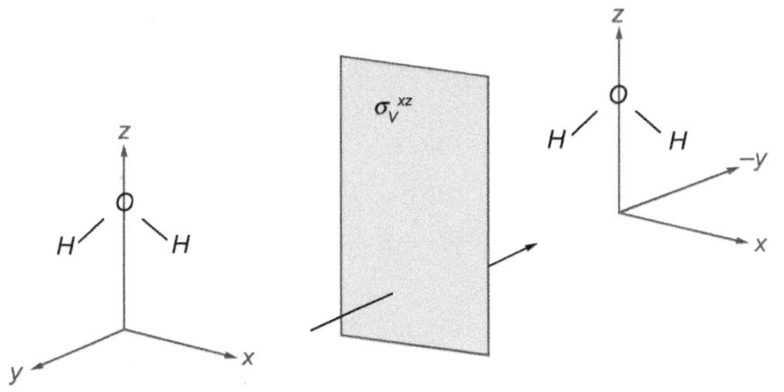

FIG. 4.27 The operation σ_v^{xz} will reverse the direction of movement in y.

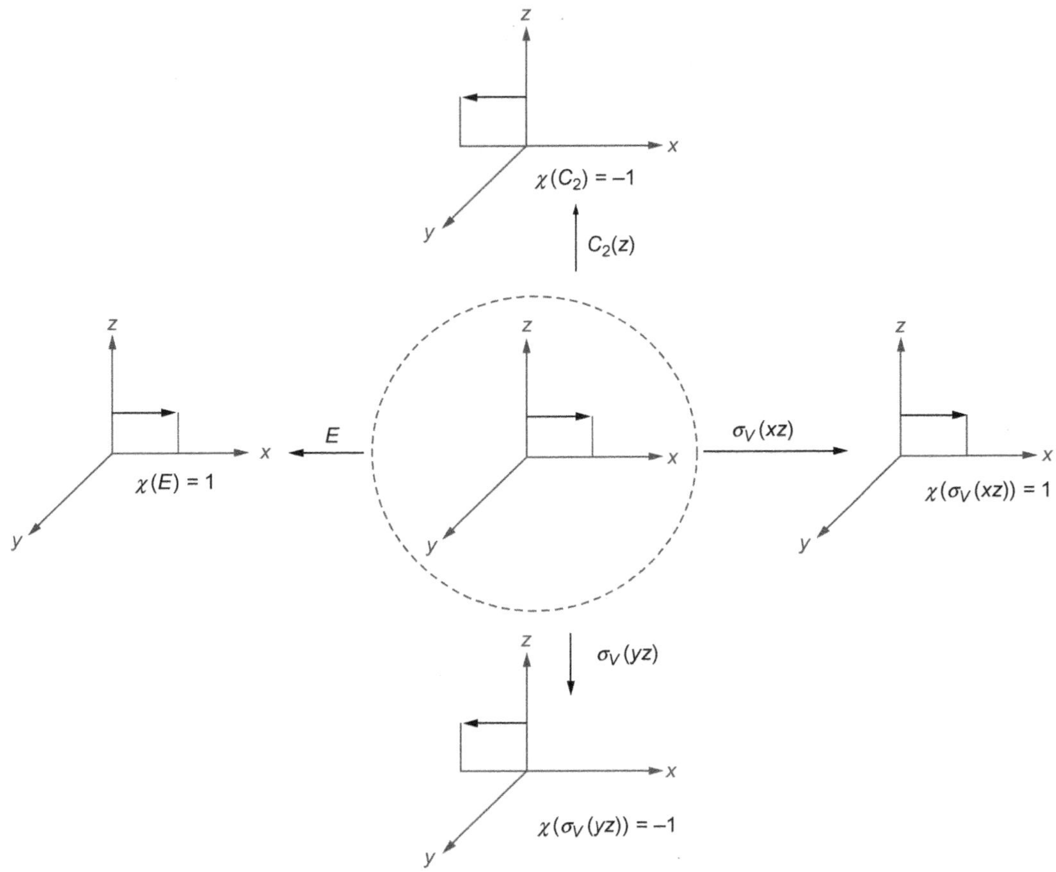

FIG. 4.28 Translational motion in x-direction of the C_{2v} point group.

- The full matrix representing the translational motion in the x-, y-, and z-directions can be constructed for the four symmetry operations of C_{2v} point group:

		E	C_2	σ_v^{XZ}	σ_v^{YZ}
Transition	Z	+1	+1	+1	+1
	X	+1	−1	+1	−1
	Y	+1	−1	−1	+1

- Note that the letters x, y, and z may be replaced by T_x, T_y, and T_z.

Rotational Motion

Construct the matrices representing the rotational motion around x, y, and z-axes in the C_{2v} group.

- Rotational around the z-axis:
  - The resulting characters for the four symmetry operations (Fig. 4.29):

	E	C_2	σ_v^{XZ}	σ_v^{YZ}
Z	+1	+1	−1	−1

- Rotational around y-axis:
  - The resulting characters for the four symmetry operations (Fig. 4.30):

	E	C_2	σ_v^{XZ}	σ_v^{YZ}
Y	+1	−1	+1	−1

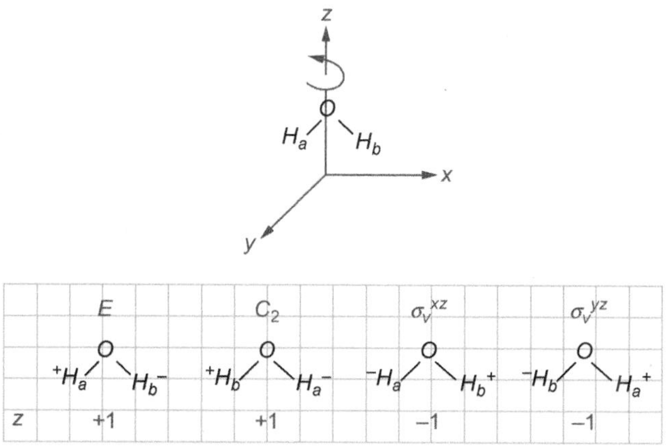

FIG. 4.29 The effect of the four symmetry operations of the C_{2v} point group on clockwise rotation about the z-axis.

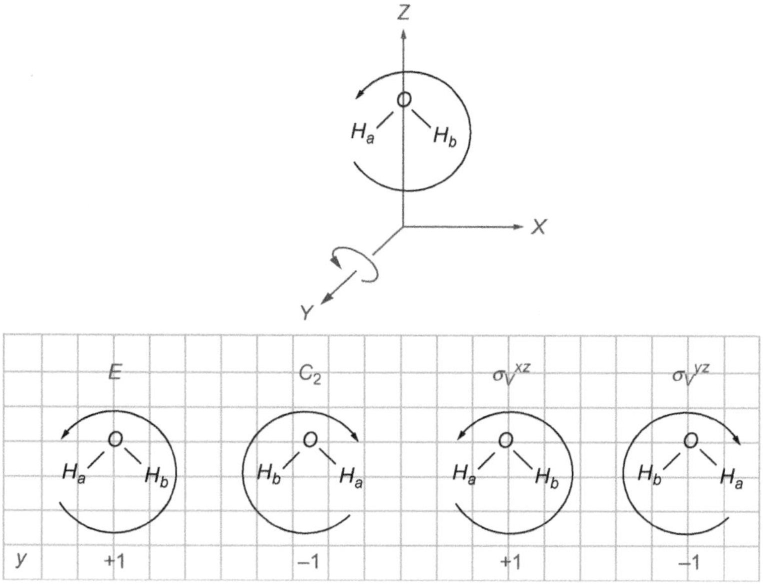

FIG. 4.30 The effect of the four symmetry operations of the C_{2v} point group on clockwise rotation about the y-axis.

- Rotational around the x-axis:
  - The resulting characters for the four symmetry operations (Fig. 4.31):

$$\begin{array}{ccccc} & E & C_2 & \sigma_v^{XZ} & \sigma_v^{YZ} \\ x & +1 & -1 & -1 & +1 \end{array}$$

- The resulting characters for the four symmetry operations are

$$\text{Rotational}\left|\begin{array}{ccccc} & E & C_2 & \sigma_v^{XZ} & \sigma_v^{YZ} \\ Z & +1 & +1 & -1 & -1 \\ X & +1 & -1 & -1 & +1 \\ Y & +1 & -1 & +1 & -1 \end{array}\right.$$

- Note that the letters x, y, and z may be replaced by R_x, R_y, and R_z.

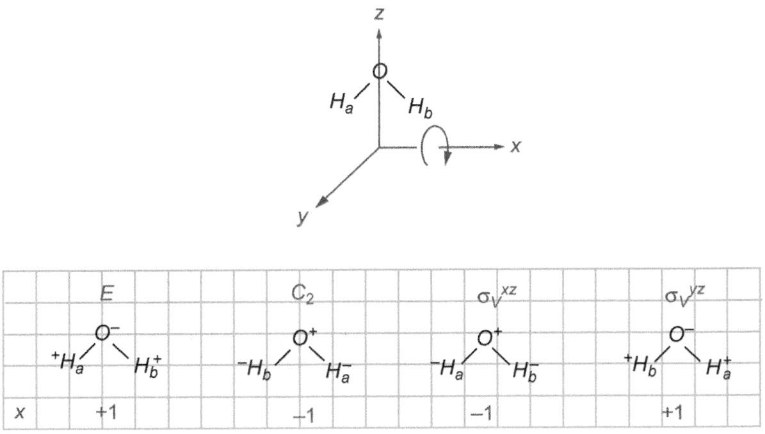

FIG. 4.31 The effect of the four symmetry operations of C_{2v} point group on clockwise rotation about the x-axis.

4.9 SYMMETRY PROPERTIES OF ATOMIC ORBITALS

Mullikan Notation

Summarize in a table the characters representing the C_{2v} point group, and the representations of the transitional and rotational motions:

C_{2v} point group	E	C_2	σ_v	σ_v'
X	+1	−1	+1	−1
Y	+1	−1	−1	+1
Z	+1	+1	+1	+1

Transition	E	C_2	σ_v^{XZ}	σ_v^{YZ}
Z	+1	+1	+1	+1
X	+1	−1	+1	−1
Y	+1	−1	−1	+1

Rotational	E	C_2	σ_v^{XZ}	σ_v^{YZ}
Z	+1	+1	−1	−1
X	+1	−1	−1	+1
Y	+1	−1	+1	−1

Employ Mullikan notation to symbolize each representation.

- Mullikan put a notations describing each distinct behavior patterns, or irreducible representation of the point group, +1 & −1 are the characters of the operation
- Mullikan notation:
  - A & B: denote 1-dimensional representation;
  - E: denote 2-dimensional irreducible representation;
  - T (F): denote 3-dimensional irreducible representation;
  - A: has $\chi = 1$, it is symmetric with respect C_n; and
  - B: has $\chi = -1$, it is antisymmetric with respect to C_n.

  Subscript:
  - 1: if representation is symmetric with respect to $C_2 \perp C_n$. Or if no $C_2 \perp C_n$, then with respect to σ_v.
  - 2: if representation is *antisymmetric* with respect to $C_2 \perp C_n$. Or if no $C_2 \perp C_n$, then with respect to σ_v.
  - g: if representation is *symmetric* with respect to i.
  - u: if representation is *antisymmetric* with respect to i.

TABLE 4-8 Mullikan notations and the characters of each representation for C_{2v} group.

C_{2v}	E	C_2	σ_v^x	σ_v^{yz}	
$\Gamma_1 \rightarrow A_1$	1	1	1	1	z
$\Gamma_2 \rightarrow A_2$	1	1	−1	−1	R_z
$\Gamma_3 \rightarrow B_1$	1	−1	1	−1	x, R_y
$\Gamma_4 \rightarrow B_2$	1	−1	−1	1	y, R_x
Mullikan Notation	The characters of irreducible representation of the group				x, y, z translation R_x, R_y, R_z Rotational

The symbols g and u stand for the German words *gerade*, *ungerade*, meaning even and odd (Chapter 10, p. 590).
Superscript:

- ′: if representation is *symmetric* with respect to σ_h.
- ″: if representation is antisymmetric with respect to σ_h.

- Mullikan notations, the characters representing of C_{2v} point group, and the representations of the transitional and rotational motions are summarized in Table 4.8.

Atomic orbital Representation

Write the character describing the representations of s, p_x, p_y, and p_z orbitals in C_{2v} point group.

- The s-orbital in C_{2v}

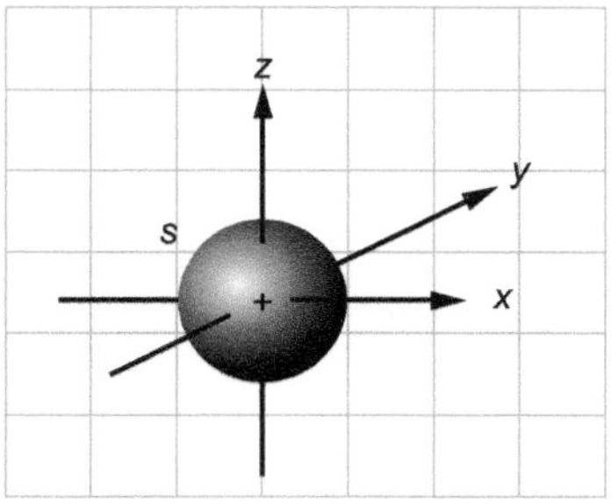

		E	C_2	σ_v^{XZ}	σ_v^{YZ}
s orbital	Z	+1	+1	+1	+1
	X	+1	+1	+1	+1
	Y	+1	+1	+1	+1

The s-orbital is symmetrical with respect to every operation.

- The p-orbitals:

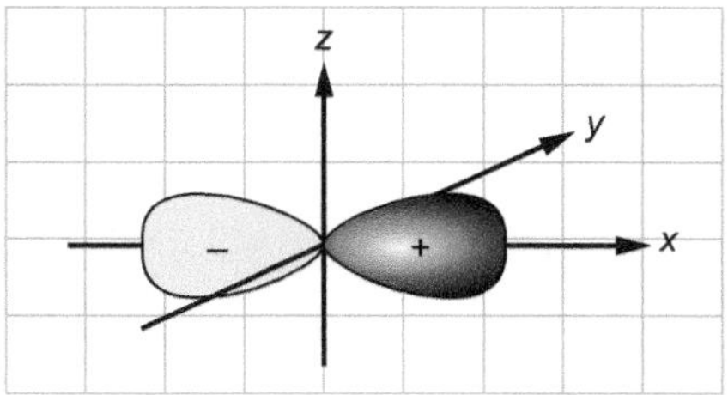

- p_x-orbital has a node in the yz-plane.
- The p_y-orbital has a node in the xz-plane.

- The p_z-orbital has a node in the xy-plane.
- The significant of the node is that there is a change in the sign in going from one lobe of the orbital into the other.
- The behaviors of p_x, p_y, and p_z orbitals with respect to the four C_{2v} symmetry operations are as follows:

C_{2v}	E	C_2	σ_v^{xz}	σ_v^{yz}	
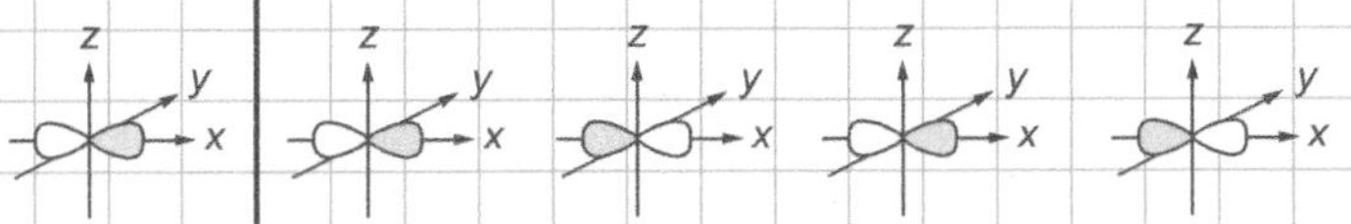					
p_x	+1	−1	+1	−1	B_1
p_y	+1	−1	−1	+1	B_2
p_z	+1	+1	+1	+1	A_1

- B_1, B_2, and A_1: we usually write b_1, b_2, and a_1 (lower-case Mullikan symbol) for the orbital representation of p_x, p_y, and p_z.
- Taking all three p-orbitals together:

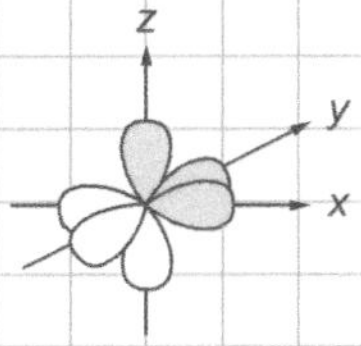

C_{2v}	E	C_2	σ_v^{xz}	σ_v^{yz}
Γ_p	3	−1	+1	+1

which reduces to $A_1 + B_1 + B_2$,or for orbital a_1, b_1, and b_2.

Write the character describing the representations of p_x and p_y orbitals in the C_{4v} point group.

C_{4v}	E	C_4	C_2	σ_v^{xz}	σ_v^{yz}	σ_d	$\sigma_{d'}$
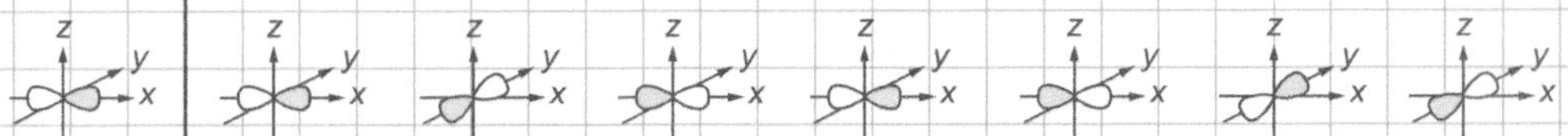							
p_x	+1	0	−1	+1	−1	0	0

C_{4v}	E	C_4	C_2	σ_v^{xz}	σ_v^{yz}	σ_d	$\sigma_{d'}$
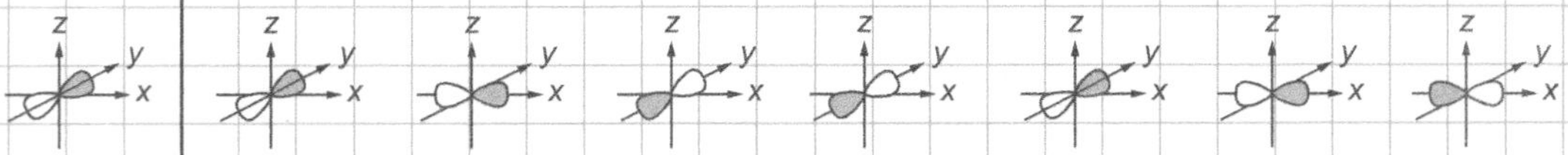							
p_y	+1	0	−1	−1	+1	0	0

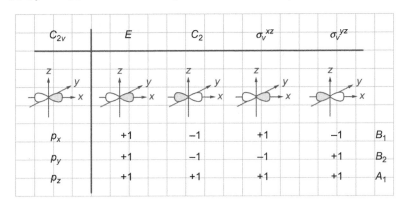

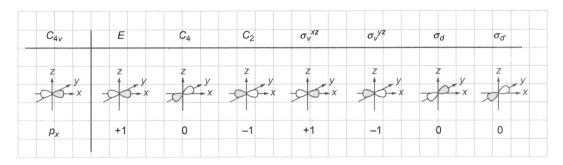

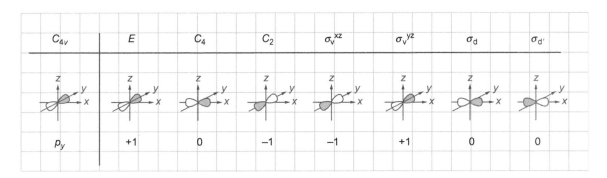

The orbitals p_x and p_y transforms as 2D irreducible representations; E

- They are doubly degenerate.
- C_{2v} and C_{4v} are examples of molecular frameworks, in which the threefold degeneracy of the p-orbitals is broken.
- Atomic orbital degeneracy may be broken by the imposition of symmetry lower than spherical, due to the atoms chemical environment.

Write the character describing the representations of d orbital in C_{2v} point group.

- Similarly, the *d*-orbitals can be presented.
- Using the Schrödinger equation (see Chapter 2), for $n=3$, $l=2$ gives a set of five wave functions.
- The fivefold degenerate set of 3d atomic orbitals has complex parts.
- Therefore, we usually find a set of five wave functions, which are linear combinations of the direct answers, taken to minimize imaginary parts of wave functions.

$$d_{xy},\ d_{xz},\ d_{yz},\ d_{x^2-y^2},\ \text{and}\ d_{z^2}$$

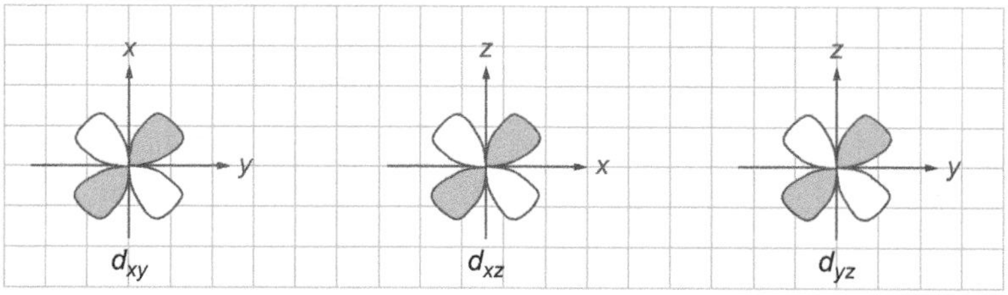

- It is obvious these three are symmetry related.

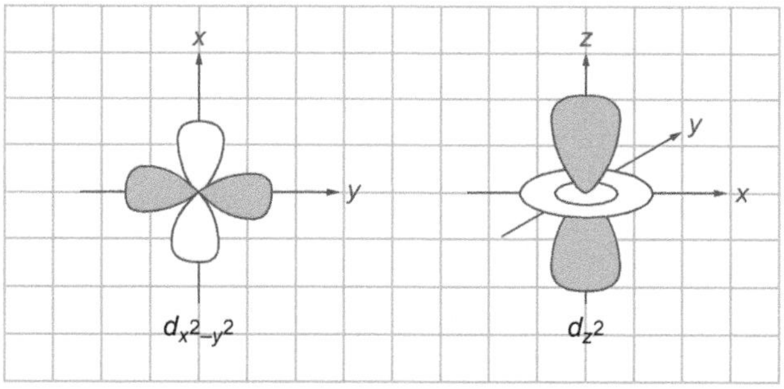

- These two are not obviously symmetry related.

C_2v		E	C_2	σ_v^{xz}	σ_v^{yz}
d_{z^2}	$\rightarrow A_1$	1	1	1	1
$d_{x^2-y^2}$	$\rightarrow A_1$	1	1	1	1
d_{xy}	$\rightarrow A_2$	1	1	−1	−1
d_{xz}	$\rightarrow B_1$	1	−1	1	−1
d_{yz}	$\rightarrow B_2$	1	−1	−1	1

4.10 CHARACTER TABLES

Tabulate the characters of all the representations of C_{2v} point group, translational, rotational motions, *s*, *p*, and *d* orbitals in one table. What are your observations?

Employ Mullikan notation to symbolize each representation, and identify that table.

- All the representations and the characters of C_{2v} point group are collected in Table 4.7.
- The full set of representations is included in that table, and called the character table of the group (Table 4.9).

TABLE 4.9 Character Table for the C_{2v} Point Group

C_{2v}	E	C_2	σ_v^{xz}	σ_v^{yz}		
$\Gamma_1 \rightarrow A_1$	1	1	1	1	z	x^2, y^2, z^2
$\Gamma_2 \rightarrow A_2$	1	1	−1	−1	R_z	xy
$\Gamma_3 \rightarrow B_1$	1	−1	1	−1	x, R_y	xz
$\Gamma_4 \rightarrow B_2$	1	−1	−1	1	y, R_x	yz
Mullikan notation	The characters of irreducible representation of the group				x, y, z translation R_x, R_y, R_z rotational	Atomic orbital representation

Note that:
$x^2 + y^2 \equiv s$ corresponds to the total symmetric irreducible representation;
$x, y, z \equiv p_x, p_y, p_z$ corresponds to translation in the x-, y-, and z-directions, respectively;
R_x, R_y, R_z corresponds to rotation about x, y, and z axes, respectively;
$x^2 - y^2 \equiv d_{x^2-y^2}$, $z^2 \equiv d_{z^2}$ are often written as $2z^2 - x^2 - y^2$; and
$xy \equiv d_{xy}$, $xz \equiv d_{xz}$, $yz \equiv d_{yz}$.

- The character table is a two-dimensional table whose rows are the irreducible Mullikan's representations of the entire group, and whose columns are the classes of these group elements. The entries consist of characters, the trace of the matrices representing group elements of the column's class in the given row's representation.
- We have found for distinct behavior patterns, or irreducible representation of the point group, that +1 and −1 are the characters of the operation.

Properties of the Characters of Representations

What are the correlations among the characters of the different representations of a group? Use the character table of the C_{2v} point group to confirm your answer.

- 1. The sum of the square of the dimensions of the irreducible representations of a group is equal to the order of the group.

$$\sum L_i^2 = l_1^2 + l_2^2 + l_3^2 + \cdots$$

Such $\chi_i\,(E)$ = the order of the representation:

$$\therefore \sum (\chi_i(E))^2 = h \tag{4.10.1}$$

Where h is the order of the group, the number of elements in finite group and i is the order of each of the motion.

- Example: H_2O

C_{2v}	E	C_2	σ_v^{xz}	σ_v^{yz}		
$\Gamma_1 \rightarrow A_1$	1	1	1	1	z	x^2, y^2, z^2
$\Gamma_2 \rightarrow A_2$	1	1	−1	−1	R_z	xy
$\Gamma_3 \rightarrow B_1$	1	−1	1	−1	x, R_y	xz
$\Gamma_4 \rightarrow B_2$	1	−1	−1	1	y, R_x	yz

$$\sum (\chi_i(E))^2 = h$$

C_{2v}	A_1	A_2	B_1	B_2	
$\Sigma(\chi_i(E))^2 =$	$1^2 +$	$1^2 +$	$1^2 +$	$1^2 =$	4

- 2. The sum of the square of the characters in any irreducible representation equal h:

$$\sum(\chi_i(R))^2 = h \quad (4.10.2)$$

C_{2v}	E	C_2	σ_v^{xz}	σ_v^{yz}	
$\Sigma(\chi(B_2))^2=$	1^2+	$(-1)^2+$	$(-1)^2+$	$1^2=$	4

- 3. The vectors whose components are characters of two different irreducible representation are orthogonal:

$$\sum\chi_i(R)\chi_j(R) = 0, \quad \text{if } i \neq j \quad (4.10.3)$$

Taking, for example, A_2 & B_1:

C_{2v}	E	C_2	σ_v^{xz}	σ_v^{yz}	
$A_2 \times B_1=$	$(1 \times 1)+$	$(1 \times -1)+$	$(-1 \times 1)+$	$(-1 \times -1)=$	0

- 4. In a given representation (reducible or irreducible), the characters of all matrices belonging to operations in the same class are identical.
- 5. The number of irreducible representations of a group is equal to the number of classes in the group.
  C_{2v} has four one-dimensional irreducible representations and four classes.

How can you deduce the characters of all the irreducible representations of the C_{3v} point group by using the correlations among the characters?

- Rule (5):
  We may take NH_3 as example of C_{3v} point group
  C_{3v} contains three classes: E, $2C_3$, $3\sigma_v$
  $\therefore$ Number of irreducible representations $=3$
  Let us call Γ_1, Γ_2, and Γ_3
- Rule (1):

$$\sum(\chi_i(E))^2 = h \quad (4.10.1)$$

If we denote their dimensions by l_1, l_2, and l_3, then:

$$l_1^2 + l_2^2 + l_3^2 = 6$$

The only values that will satisfy this requirement are 1, 1, 2, then:

C_{3v}	E	$2C_3$	$3\sigma_v$
Γ_1	1	?	?
Γ_2	1	?	?
Γ_3	2	?	?

- Rule (2):

$$\sum(\chi_i(R))^2 = h \quad (4.10.2)$$

$h=6$.

C_{3v}	E	$2C_3$	$3\sigma_v$	
Γ_1	1 +	$2(?)^2$ +	$3(?)^2$ =	6

Therefore,

C_{3v}	E	$2C_3$	$3\sigma_v$
Γ_1	1	1	1

- Appling rule (3) to find Γ_2,

$$\sum \chi_i(R)\chi_j(R) = 0, \text{ if } i \neq j \quad (4.10.3)$$

Γ_2 and Γ_1 are orthogonal

C_{3v}	E	$2C_3$	$3\sigma_v$
Γ_1	1	1	1
Γ_2	1	?	?
$\Gamma_2 \times \Gamma_1$:	$1 \times 1 + 2\ (1 \times ?) + 3\ (1 \times ?) = 0$		
	$1 \times 1 + 2\ (1 \times 1) + 3\ (1 \times -1) = 0$		

Then,

C_{3v}	E	$2C_3$	$3\sigma_v$
Γ_2	1	1	−1

The third representation will be of dimension 2

$$\chi_3(E) = 2$$

Rule (3):

$$\sum \chi_i(R)\chi_j(R) = 0, \text{ if } i \neq j \quad (4.10.3)$$

$$\sum \chi_1(R)\chi_3(R) = (1)(2) + 2(1)(\chi_3(C_3)) + 3\,(1)(\chi_3(\sigma_v)) = 0$$

$$\sum \chi_2(R)\chi_3(R) = (1)(2) + 2(1)(\chi_3(C_3)) + 3\,(-1)(\chi_3(\sigma_v)) = 0$$

Solving these two equations

$$\chi_3(C_3) = -1$$

$$\chi_3(\sigma_v) = 0$$

C_{3v}	E	$2C_3$	$3\sigma_v$
Γ_1 (A_1)	1	1	1
Γ_2 (A_2)	1	1	−1
Γ_3 (E)	2	−1	0

4.11 RELATION BETWEEN ANY REDUCIBLE AND IRREDUCIBLE REPRESENTATIONS

Generate the reducible representation, $\Gamma = A_1 + B_1 + B_2$, in the point group C_{2v}.

- Reducible representations furnish compound characters, which are sums of the simple characters of irreducible representations.

  Example: H_2O belongs to C_{2v}.

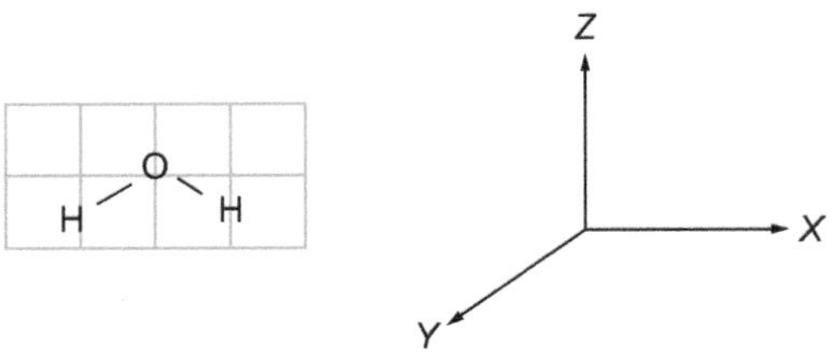

The total or reducible representation are as follows:

C_{2v}	E	C_2	σ_v^{xz}	σ_v^{yz}
Γ	3	−1	1	1

Γ: is a general symbol used when the symmetry is not known.

- A reducible representation can be decomposed into irreducible representations; it is the sum of irreducible representations.

C_{2v}	E	C_2	σ_v^{xz}	σ_v^{yz}
A_1	1	1	1	1
A_2	1	1	−1	−1
B_1	1	−1	1	−1
B_2	1	−1	−1	1
Γ	3	−1	1	1
$\Gamma=$	A_1 +	B_1 +	B_2	

- A_1, B_1, and B_2 are 1D matrices describing the effects of symmetry operations is a set of irreducible representations.

Use the reduction formula to reduce the representations Γ_a and Γ_b:

C_{3v}	E	$2C_3$	$3\sigma_v$
Γ_a	5	2	− 1
Γ_b	7	1	− 3

- The reduction formula is

$$\sigma_i = \frac{1}{h}\sum i \cdot \chi_R \cdot \chi_i \tag{4.11.1}$$

where

σ_i is the number of times the ith irreducible representation appears in the reducible representation;
χ_R is the character of reducible representation;
χ_i is the character of irreducible representation; and
i is the order of the class.

- Let us take group C_{3v}, the reducible representations Γ_1, Γ_2, and Γ_3 and two reducible representations Γ_a and Γ_b

C_{3v}	E	$2C_3$	$3\sigma_v$
Γ_1 (A_1)	1	1	1
Γ_2 (A_2)	1	1	−1
Γ_3 (E)	2	−1	0
Γ_a	5	2	−1
Γ_b	7	1	−3

- For Γ_a

$$\sigma_1 = 1/6(1\times 1\times 5 + 2\times 1\times 2 + 3\times 1\times -1) = 1$$

$$\sigma_2 = 1/6(1\times 1\times 5 + 2\times 1\times 2 + 3\times -1\times -1) = 2$$

$$\sigma_3 = 1/6(1\times 2\times 5 + 2\times -1\times 2 + 3\times 0\times -1) = 1$$

$$\Gamma_a = \Gamma_1 + 2\Gamma_2 + \Gamma_3 = A_1 + 2A_2 + E$$

- For Γ_b

$$\sigma_1 = 1/6(1\times 1\times 7 + 2\times 1\times 1 + 3\times 1\times -3) = 0$$

$$\sigma_2 = 1/6(1\times 1\times 7 + 2\times 1\times 1 + 3\times -1\times -3) = 3$$

$$\sigma_3 = 1/6(1\times 2\times 7 + 2\times -1\times 1 + 3\times 0\times -3) = 2$$

$$\Gamma_b = 3\Gamma_2 + 2\Gamma_3 = 3A_2 + 2E$$

The Direct Product

How is the direct product obtained? What is the significance of this product? In point group C_{4v}, find the direct product: B_1B_2 and EE.

- The direct products are the character representations that are obtained by taking the products of characters of individual sets of function.
- The direct product, $\Gamma_i\Gamma_j$, of the two irreducible representations Γ_i and Γ_j, is:

$$\Gamma_i\Gamma_j = \chi_{ki} \cdot \chi_{kj}$$

$$\begin{array}{lllll} \Gamma_i & \chi_{1i} & \chi_{2i} & \cdots & \chi_{ki} \\ \Gamma_j & \chi_{1j} & \chi_{2j} & \cdots & \chi_{kj} \\ \Gamma_i\Gamma_j & \chi_{1i}\cdot\chi_{1j} & \chi_{2i}\cdot\chi_{2j} & \cdots & \chi_{ki}\cdot\chi_{kj} \\ \Gamma_i\Gamma_j & \chi_1' & \chi_2' & \cdots & \chi_k' \end{array}$$

- The direct product of two irreducible representation may be reduced:

$$\Gamma_i\Gamma_j = \sum\nolimits_l n\Gamma_l$$

$$\sigma_i = \frac{1}{k}\sum i \cdot \chi_k' \cdot \chi_i \qquad (4.11.2)$$

- From Table 4.10, the direct product:

$$B_1B_2 = A_2$$

$$EE = A_1 + A_2 + B_1 + B_2$$

- Notice that the degenerated bases are transformed in pairs, with character corresponding to the trace of the transformation matrix for both.
  E in terms of x,y basis:

$$\begin{bmatrix} x \\ y \end{bmatrix} \underbrace{\begin{bmatrix} 1 & 0 \\ 0 & 1 \end{bmatrix}}_{\substack{E \\ \text{Transformation} \\ \text{matrix}}} = \begin{bmatrix} x \\ y \end{bmatrix}$$

TABLE 4.10 Character Table for C_{4v} Point Group

C_{4v}	E	$2C_4$	C_2	$2\sigma_v$	$2\sigma_d$	
A_1	1	1	1	1	1	
A_2	1	1	1	−1	−1	
B_1	1	−1	1	1	−1	
B_2	1	−1	1	−1	1	
E	2	0	−2	0	0	
B_1B_2	1	1	1	−1	−1	$= A_2$
$E.E$	4	0	4	0	0	$= A_1 + A_2 + B_1 + B_2$

In the C_{4v} point group, taking the cross product of $E \times E$ gives a total representation of four, because the cross product of two matrices increases the dimensionality of the matrix.

$$EE: \begin{bmatrix} x \\ y \\ x \\ y \end{bmatrix} \underbrace{\begin{bmatrix} 1 & 0 & 0 & 0 \\ 0 & 1 & 0 & 0 \\ 0 & 0 & 1 & 0 \\ 0 & 0 & 0 & 1 \end{bmatrix}}_{\substack{EE = \text{direct product} \\ \text{Transformation} \\ \text{matrix}}} = \begin{bmatrix} x \\ y \\ x \\ y \end{bmatrix}$$

What is the significance of the direct product?

- The direct product has a very important property. Consider:
  - the wave function ψ_i belongs to the irreducible representation Γ_i;
  - the wave function ψ_j belongs to the irreducible representation Γ_j; and
  - the operator α belongs to the irreducible representation Γ_α.

  The integral:

$$\int \psi_i \alpha \psi_j \partial\tau = 0$$

Unless the direct product,

$$\Gamma_i\Gamma_\alpha\Gamma_j = \Gamma_i(\Gamma_\alpha\Gamma_j) = (\Gamma_i\Gamma_\alpha)\Gamma_j = \text{a sum of irreducible representations}$$

contains after reduction the totally symmetric irreducible representation, e.g., invariant under all operations of the symmetry group to which the molecule belongs.

4.12 GROUP THEORY AND QUANTUM MECHANICS: IRREDUCIBLE REPRESENTATIONS AND WAVE FUNCTION

Show that:

i. **Any symmetry operation commutes with Hamiltonian operation.**
ii. **The wave functions form bases for irreducible representations in nondegenerate and degenerate eigenvalues, Γ.**

How to generate a representation of the group.
What do the irreducible representations have to do with wave function?

- The wave equation for any physical system (Chapter 2) is as follows:

$$\hat{H}\psi = E\psi$$

  - $\hat{H}$ is the Hamiltonian operator, which designates that certain operations are to be performed on a function written to its right.
  - If the function is an eigenfunction (ψ), the result of performing the operations indicated by $\hat{H}$ will yield the function itself multiplied by a constant, which is called an eigenvalue (E, the energy of the system).
- Special features must be considered:

1. If ψ is simultaneously an eigenfunction of two operators $\hat{R}$ and $\hat{S}$, that is, if

$$\hat{R}\psi = r\psi$$

$$\hat{S}\psi = s\psi$$

then a necessary condition for the existence of a complete set of simultaneous eigenfunctions of two operators is that the operators commute with each other:

$$\hat{R}\hat{S} = \hat{S}\hat{R}$$

We can then simultaneously assign definite values to the physical quantities r and s.

2. The Hamiltonian operator also commutes with any constant, c:

$$Hc\psi = cH\psi = cE\psi$$

3. In many incidents, several eigenfunctions yield the same eigenvalue:

$$H\psi_{i1} = E_i\psi_{i1}$$
$$H\psi_{i2} = E_i\psi_{i2}$$
$$\vdots \quad \vdots$$
$$H\psi_{ik} = E_i\psi_{ik}$$

In these instances, the eigenvalue is degenerate, and E_i is k-fold degenerate.

4. For a degenerate eigenvalue, not only does the initial set of eigenfunctions give the correct solution to the wave equation, but also any linear combination of these is also a solution:

$$H\sum_j a_{ij}\psi_{ij} = Ha_{i1}\psi_{i1} + Ha_{i2}\psi_{i2} + \cdots + Ha_{ik}\psi_{ik}$$
$$= E_i a_{i1}\psi_{i1} + E_i a_{i2}\psi_{i2} + \cdots + E_i a_{ik}\psi_{ik}$$
$$H\sum_j a_{ij}\psi_{ij} = E_i\sum_j a_{ij}\psi_{ij}$$

5. Similar to the irreducible representations, the eigenfunctions are orthogonal:

$$\int \psi_i^*\psi_j d\tau = \xi_{ij}$$

$\xi = 0$, when $i \neq j$ and
$\xi = 1$, when $i = j$
This shows that the wave functions form a basis for irreducible representations.

- In a nondegenerate eigenvalue, the following points must be considered:
  - A symmetry operation takes the system into an identical configuration, thus the energy of the system must be as before. Therefore, any symmetry operator commutes with Hamiltonian operator:

$$\hat{R}\hat{H} = \hat{H}\hat{R}$$

  - If we take the wave equation for a molecule and perform a symmetry operation, R, upon each side, then:

$$HR\psi_i = ER\psi_i$$

E commutes with $\hat{R}$ because E is constant

  - Therefore, eigen function $= R\psi_i$, and must be normalized, for we do not change the length of a vector by a symmetry operation:

$$\int_{-\infty}^{\infty} R\psi^* R\psi \, \delta\tau = R^2 \int_{-\infty}^{\infty} \psi^*\psi \, \delta\tau = 1$$

Since ψ is normalized, $R = \pm 1$, and $R\psi_i = \pm 1\psi_i$

By applying each of the operations in the group to an eigenfunction ψ_i belonging to a nondegenerate eigenvalue, we generate a representation of the group with all characters equal to ± 1, and the representations are one-dimensional.

  - Thus, the eigenfunctions (nondegenerate) for a molecule are bases for irreducible representations.
- In k-fold degenerate eigenvalues:
  - By carrying out a symmetry operation, R, upon each side of the wave equation:

$$HR\psi_{il} = E_i R\psi_{il}$$

$R\,\psi_{il}$ = linear combination of the ψ_{il}, then

$$R\psi_{ij} = \sum_{j=1}^{k} r_{jl}\psi_{jl}$$

- For some other operation, S:

$$S\psi_{ij} = \sum_{m=1}^{k} s_{mj}\psi_{im}$$

- Because R and S are members of a symmetry group, there must be an element as follows:
  $T = RS$, its effect on ψ_{il}:

$$SR\psi_{il} = T\psi_{il} = \sum_{m=1}^{k} t_{im}\psi_{im}$$

- By combining the preceding expressions for the separate effects of S and R:

$$SR\psi_{il} = S\sum_{j=1}^{k} r_{jl}\psi_{ij} = \sum_{j=1}^{k}\sum_{m=1}^{k} s_{mj}r_{jl}\psi_{im}$$

- By comparison:

$$t_{ml} = \sum_{j=1}^{k} s_{mj}r_{jl}$$

  This is the expression that gives the elements of a matrix, T, which is the product, SR, of two other matrices.
- Thus the matrixes, which describe the transformations of a set of k-eigenfunctions corresponding to the k-fold degenerate eigenvalue, are k-dimensional representations for the group.

If the wave functions of p_x and p_y orbitals of NH_3:

$$p_x = R\ \sin\theta\ \cos\phi$$

$$p_y = R\ \sin\theta\ \sin\phi$$

where R is constant, find the representations of these orbitals in the designated point group, e.g., C_{3v}, and compare your results with the reported presentation in the character table.

- Consider p_x and p_y orbitals of NH_3, which belong to C_{3v}.
  The eigenfunctions that describe these orbitals are as follows:

$$p_x = R\ \sin\theta\ \cos\phi$$

$$p_y = R\ \sin\theta\ \sin\phi$$

R is constant.
θ and φ are angles in a polar coordinate system.

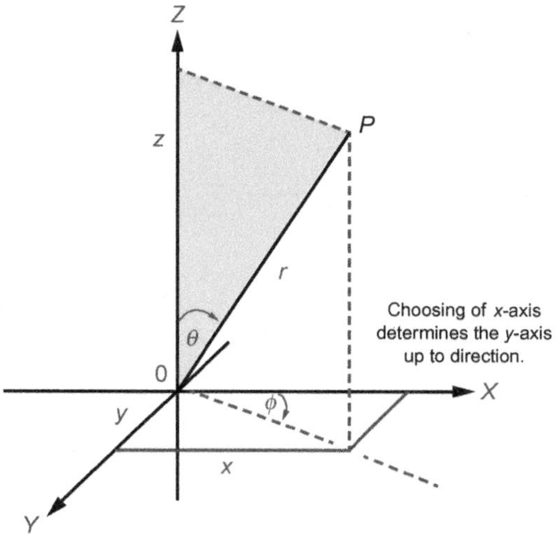

- Note that none of the operations in C_{3v} (E, C_3, σ_v) will affect θ, so that θ_2, the value of θ after application of a symmetry operation, will always equal θ_1:

$$\sin\theta_2 = \sin\theta_1$$

This will be used to work out the required matrices.

- E:

$$Ep_x = E(R\,\sin\theta\cos\phi)$$
$$Ep_y = E(R\,\sin\theta\sin\phi)$$
$$Ep_x = E(R\sin\theta_1\cos\phi_1) = R\sin\theta_2\cos\phi_2 = R\sin\theta_1\cos\phi_1 = p_x$$
$$Ep_y = E(R\sin\theta_1\sin\phi_1) = R\sin\theta_2\sin\phi_2 = R\sin\theta_1\sin\phi_1 = p_y$$

- C_3:
  - If we rotate by $2\pi/3$ about the z-axis:

$$\phi_2 = \phi_1 + \frac{2\pi}{3}, \text{ then:}$$

$$\cos\phi_2 = \cos\left(\phi_1 + \frac{2\pi}{3}\right) = \cos\phi_1\cos\frac{2\pi}{3} - \sin\phi_1\sin\frac{2\pi}{3}$$
$$= -\frac{1}{2}\cos\phi_1 - \left(\frac{\sqrt{3}}{2}\right)\sin\phi_1$$
$$\sin\phi_2 = \sin\left(\phi_1 + \frac{2\pi}{3}\right) = \sin\phi_1\cos\frac{2\pi}{3} + \cos\phi_1\sin\frac{2\pi}{3}$$
$$= -\frac{1}{2}\sin\phi_1 - \left(\frac{\sqrt{3}}{2}\right)\cos\phi_1$$

Therefore:

$$C_3p_x = C_3(R\,\sin\theta_1\,\cos\phi) = R\,\sin\theta_2\,\cos\phi_2$$
$$= R(\sin\theta_1)\underbrace{\left(-\frac{1}{2}\right)\left(\cos\phi_1 + \sqrt{3}\sin\phi_1\right)}_{\cos\phi_2}$$
$$= -\frac{1}{2}R\,\sin\theta_1\,\cos\phi_1 - \left(\frac{\sqrt{3}}{2}\right)R\,\sin\theta_1\,\sin\phi_1$$
$$= -\frac{1}{2}p_x - \left(\frac{\sqrt{3}}{2}\right)p_y$$
$$C_3p_y = C_3(R\,\sin\theta_1\,\sin\phi_1) = R\,\sin\theta_2\,\sin\phi_2$$
$$= R(\sin\theta_1)\underbrace{\left(-\frac{1}{2}\right)\left(\sin\phi_1 - \sqrt{3}\cos\phi_1\right)}_{\sin\phi_2}$$
$$= \left(\frac{\sqrt{3}}{2}\right)R\,\sin\theta_1\,\cos\phi_1 - \frac{1}{2}R\,\sin\theta_1\,\sin\phi_1$$
$$= \left(\frac{\sqrt{3}}{2}\right)p_x - \frac{1}{2}p_y$$

- σ_v:
  - If we reflect in xz plane:

$$\phi_2 = -\phi_1$$

and

$$\cos\phi_2 = \cos\phi_1$$
$$\sin\phi_2 = -\sin\phi_1$$

Therefore,

$$\sigma_v p_x = \sigma_v(R\sin\theta_1\cos\phi_1) = R\sin\theta_2\cos\phi_2 = R\sin\theta_1\cos\phi_1 = p_x$$
$$\sigma_v p_y = \sigma_v(R\sin\theta_1\sin\phi_1) = R\sin\theta_2\sin\phi_2 = R\sin\theta_1\sin\phi_1 = p_y$$

- Expressing these results in matrix notation:

$$Ep_x = p_x$$
$$Ep_y = p_y$$

Then,

$$\begin{bmatrix} 1 & 0 \\ 0 & 1 \end{bmatrix}\begin{bmatrix} p_x \\ p_y \end{bmatrix} = E\begin{bmatrix} p_x \\ p_y \end{bmatrix} \quad \text{and} \quad \chi(E) = 2$$

$$C_3 p_x = -\frac{1}{2}p_x - \left(\frac{\sqrt{3}}{2}\right)p_y$$

$$C_3 p_y = \left(\frac{\sqrt{3}}{2}\right)p_x - \frac{1}{2}p_y$$

Then,

$$\begin{bmatrix} -\frac{1}{2} & -\frac{\sqrt{3}}{2} \\ \frac{\sqrt{3}}{2} & -\frac{1}{2} \end{bmatrix}\begin{bmatrix} p_x \\ p_y \end{bmatrix} = C_3\begin{bmatrix} p_x \\ p_y \end{bmatrix} \quad \text{and} \quad \chi(C_3) = -1$$

$$\sigma_v p_x = p_x$$
$$\sigma_v p_y = -p_y$$

Therefore,

$$\begin{bmatrix} 1 & 0 \\ 0 & -1 \end{bmatrix}\begin{bmatrix} p_x \\ p_y \end{bmatrix} = \sigma_v\begin{bmatrix} p_x \\ p_y \end{bmatrix} \quad \text{and} \quad \chi(\sigma_v) = 0$$

- The obtained characters are

$$\chi(E) = 2,\ \chi(C_3) = -1, \ \text{and} \ \chi(\sigma_v) = 0$$

which are those of the E representation of C_{3v}.

C_{3v}	E	$2C_3$	$3\sigma_v$		
A_1	1	1	1	z	x^2+y^2, z^2
A_2	1	1	−1	R_z	
E	2	−1	0	$(x, y)(R_x, R_y)$	$(x^2-y^2, xy)(xz, yz)$

Therefore, p_x and p_y orbitals present a basis for the E representation.

- From the character table of group C_{3v}, the coordinate x, and y transform as E representation. For that reason, the functions $\sin\theta\cos\theta$ and $\sin\theta\sin\varphi$ transform in the same way as x and y. That is why; the p orbital, which has an eigenfunction $\sin\theta\cos\varphi$ is called p_x and the one which has an eigenfunction $\sin\theta\sin\varphi$ is called p_y.

SUGGESTIONS FOR FURTHER READING

J.P. Lowe, K. Peterson, Quantum Chemistry, third ed., Elsevier, AP, Amsterdam, ISBN: 0-12-457551-X, 2006.

D.A. McQuarrie, J.D. Simon, Physical Chemistry: A Molecular Approach, University Science Books, Sausalito, CA, ISBN: 0-935702-99-7, 1997.

J.N. Murrell, S.F.A. Kettle, J.M. Tedder, The Chemical Bond, second ed., John Wiley & Sons, Ltd, Chichester, ISBN: 0-471-90760-X, 1987.

P.W. Atkins, J. de Paula, Physical Chemistry, eighth ed., W.H. Freeman, New York, NY, ISBN: 0-7167-8759-8, 2006.

G.L. Miessler, D.A. Tarr, Inorganic Chemistry, second ed., Pearson/Prentice Hall College Div., Englewood Cliffs, New Jersey, Prentice-Hall, ISBN: 0-13-841891-8, 1998, (Chapter 4).

E.P. Wigner, Group Theory and its Application to the Quantum Mechanics of Atomic Spectra, Academic Press Inc., New York, ISBN: 978-0-1275-0550-3, 1959.

R.B. Shirts, J. Chem. Educ. 84 (2007).

H.C. Longuet-Higgins, Mol. Phys. 6 (5) (1963) 445–460.

P.R. Bunker, P. Jensen, Fundamentals of Molecular Symmetry, Institute of Physics Publishing, Bristol and Philadelphia, ISBN: 0-7503-0941-5, 2005.

S.L. Altmann, Induced Representations in Crystals and Molecules, Academic Press, London, New York, ISBN: 0120546507, 1977.

R.L. Flurry, Symmetry Groups, Prentice-Hall, Englewood Cliffs, NJ, ISBN: 0-13-880013-8, 1980, pp. 115–127.

D.A. Cotton, Chemical Applications of Group Theory, John Wiley & Sons, New York, 1971.

D.M. Bishop, Group Theory and Chemistry, Oxford University Press, London, 1973.

J.P. Fackler Jr., Symmetry in Coordination Chemistry, Academic Press, New York, 1971.

L.H. Hall, Group Theory and Symmetry in Chemistry, McGraw-Hill, New York, 1969.

R.M. Hochstrasser, Molecular Aspects of Symmetry, Benjamin, New York, 1966.

H.H. Jaffé, M. Orchin, Symmetry in Chemistry, John Wiley & Sons, New York, 1965.

A. Nussbaun, Applied Group Theory for Chemists, Physicists and Engineers, Prentice-Hall, Englewood Cliffs, NJ, 1971.

M. Tinkham, Group Theory and Quantum Mechanics, McGraw-Hill, New York, 1964.

C.J. Bradley, A.P. Cracknell, The Mathematical Theory of Symmetry in Solids: Representation Theory for Point Groups and Space Groups, Oxford University Press, Oxford, 1972.

I. Bernal, W.C. Hamilton, J.S. Ricci, Symmetry: A Stereoscopic Guide for Chemists, W.H. Freeman, San Francisco, 1972.

U. Fano, G. Racah, Irreducible Tensorial Sets, Academic Press, New York, 1959.

J.S. Griffith, The Theory of Transition-meal Ions, Cambridge University Press, Cambridge, 1961.

M. Hammermesh, Group Theory and Its Applications to Physical Problems, Addison-Wesley, Reading, MA, 1962.

V. Heine, Group Theory in Quantum Mechanics, Pergamon Press, Oxford, 1960.

H.C. Longuet-Higgins, Mol. Phys. 6 (1963) 445.

G.Y. Lyubarskii, The Application of Group Theory I Physics, Pergamon Press, Oxford, 1960, p. 71.

C.J. Slater, Phys. Rev. 34 (1929) 1293.

E.B. Wilson, C.C. Lin, D.R. Lide, J. Chem. Phys. 23 (1955) 136.

S.F.A. Kettle, Symmetry and Structure, Wiley, New York, 1985.

P.W. Atkins, M.S. Child, C.S.G. Phillips, Tables for Group Theory, Oxford University Press, London, ISBN: 13-9780198551317, 1970.

Crystal Structure Analysis

J.P. Glusker, K.N. Trueblood, Crystal Structure Analysis, third ed., Oxford University Press, London, ISBN: 0199576351, 2010.

D.E. Sands, Introduction to Crystallography, third ed., Benjamin, New York, ISBN: 0486678393, 1969.

L.S.D. Glasser, Crystallography and Applications, Van Nostrand Reinhold, London, 1977.

G.H. Stout, L.H. Jensen, X-ray Structure Determination, second ed., Macmillan, New York, ISBN: 0471607118, 1989.

Chapter 5

Valence Bond Theory and Orbital Hybridization

In order to deal with molecular structures, where many energy levels of atoms are involved, an extensive study was made into the relationship between Lewis structures and covalent bonding as well to the symmetry principles in the previous chapters. The molecular structures have been discussed from the point of view of symmetry elements and geometrical description, without considering how these features have their origin in the bonding.

This chapter will mainly be concerned with the electronic structures of molecules and how atoms bond together to form molecules. We will start with the concept of valence bond theory and then investigate how to predict the shapes and geometry of simple molecules using the valence shell electron-pair repulsion method (VSEPR). The process of predicting the molecule's structure will be reviewed. In addition, the relationships between the chemical bonds in molecules and its geometry using orbital hybridization theory will be addressed (Scheme 5.1). Special attention is devoted to:

i. the angles between the bonds formed by a given atom;
ii. dealing with certain aspects of multiple bonding, primary with bond lengths;
iii. using the symmetry to determine the σ/π hybridization of atomic orbitals; and
iv. using the symmetry adapted linear combination (SALC) to compute the contribution of each atomic orbit in the hybrid orbital and to find the SALC-MO (the symmetry adapted linear combination of molecular wave function).

The succeeding chapters will explore the application of the wave model to these structures. The following are the main topics that will be explored in this chapter.

Answering the following questions will be employed to comprehend the relationship between the symmetry representations and the electronic structures and wave functions.

Electrons, Atoms, and Molecules in Inorganic Chemistry. http://dx.doi.org/10.1016/B978-0-12-811048-5.00005-5

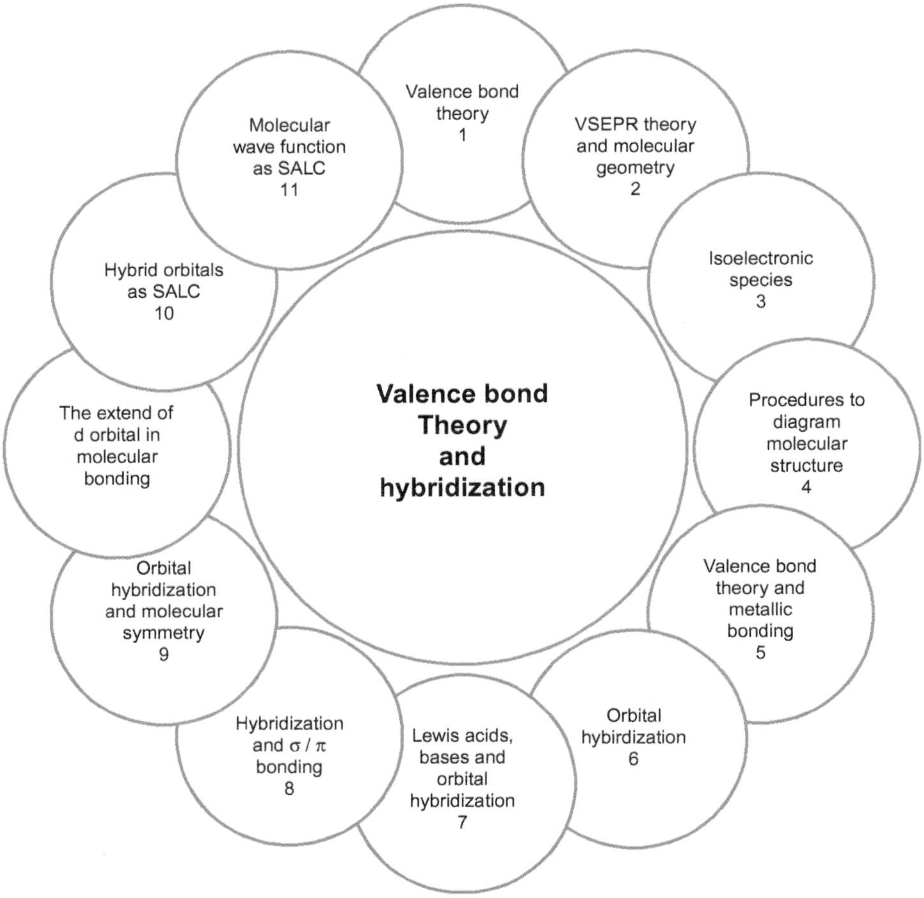

SCHEME 5.1 Approach used to address valence bond and hybridization theory.

5.1 VALENCE BOND THEORY

Use the valence bond theory to explain:

- **the bond formation;**
- **the possible number of bonds;**
- **the shape of the molecule; and**
- **the bond formation of HF, H_2O, and NH_3.**
- Atoms with unpaired electrons tend to combine with other atoms that also have unpaired electrons. Consequently:
  - The unpaired electrons are paired up, and reach a stable electronic arrangement, which is usually a full shell of electrons (nearest inert gas configuration).
  - A covalent bond results from the pairing of electrons (one from each atom). The spins of the two electrons must be opposite (antiparallel) because of the Pauli' exclusion principle that no two electrons in one atom can have all four quantum numbers the same.
- The number of bonds formed by an atom is generally equal to the number of unpaired electrons in the ground state (i.e., the lowest energy state).
- The directions in which the orbitals point determine the shape of the molecule.
- The formation of HF, H_2O, and NH_3:
  - In HF, the hydrogen atom has a singly filled s-orbital that overlaps with a singly occupied 2p-orbital of fluoride.
  - In H_2O, the oxygen atom has two singly filled 2p-orbitals, each of which overlaps with a singly occupied s-orbital of two hydrogen atoms.
  - In NH_3, there are three singly occupied p-orbitals on nitrogen atom, which overlap with s-orbitals of three hydrogen atoms.

5.2   VSEPR THEORY AND MOLECULAR GEOMETRY

What does "VSEPR" stand for and what is Gillespie-Nyholm theory?

- The three-dimensional arrangement of the molecule's atoms in space can be predicted through:
  - VSEPR, and
  - Orbitals hybridizations
- *VSEPR "Gillespie-Nyholm theory"*
  - VSEPR stands for "valence shell electron pair repulsion," referring to the repulsion between pair of valence electrons of the atoms in molecule.
  - VSEPR states that repulsion between the pairs of valence electrons surrounding an atom forces these pairs to be as far as possible.
  - As a result, repulsions between all the electron pairs present in the valence shell decide the structure of the molecule.
  - A lone pair of electrons takes up more space around the center atom than a bond pair, because the lone pair is connected to one nucleus while the bond pair is shared by two nuclei.
  - Consequently, the presence of lone pairs on the center atom causes slight distortion of the bond angles from the ideal shape.
  - The magnitude of repulsions between bonding pairs of electrons depends on the electronegativity difference between the central atom and the other atoms.
  - A double bond causes more repulsion than single bonds, and triple bonds cause more repulsion than a double bond.
  - Examples of predicted structures of some molecules based on the number of VSEPR are illustrated in Fig. 5.1.

Shape:	Linear	Bent	Trigonal-planar	Tetrahedral
Orbitals on central atom:	(2) $BeCl_2$	(3) CH_2	(3) BF_3	(4) CH_4

Shape:	Trigonal-pyramidal	Bent	Trigonal-bipyramidal	Octahedral
Orbitals on central atom:	(4) NH_3, NF_3	(4) H_2O, F_2O	(5) PCl_5	(6) SF_6

Shape:	Seesaw	T-shaped	Linear	Square pyramidal	Square planar
Orbitals on central atom:	(5) SF_4	(5) ClF_3	(5) XeF_2	(6) BrF_5	(6) XeF_4

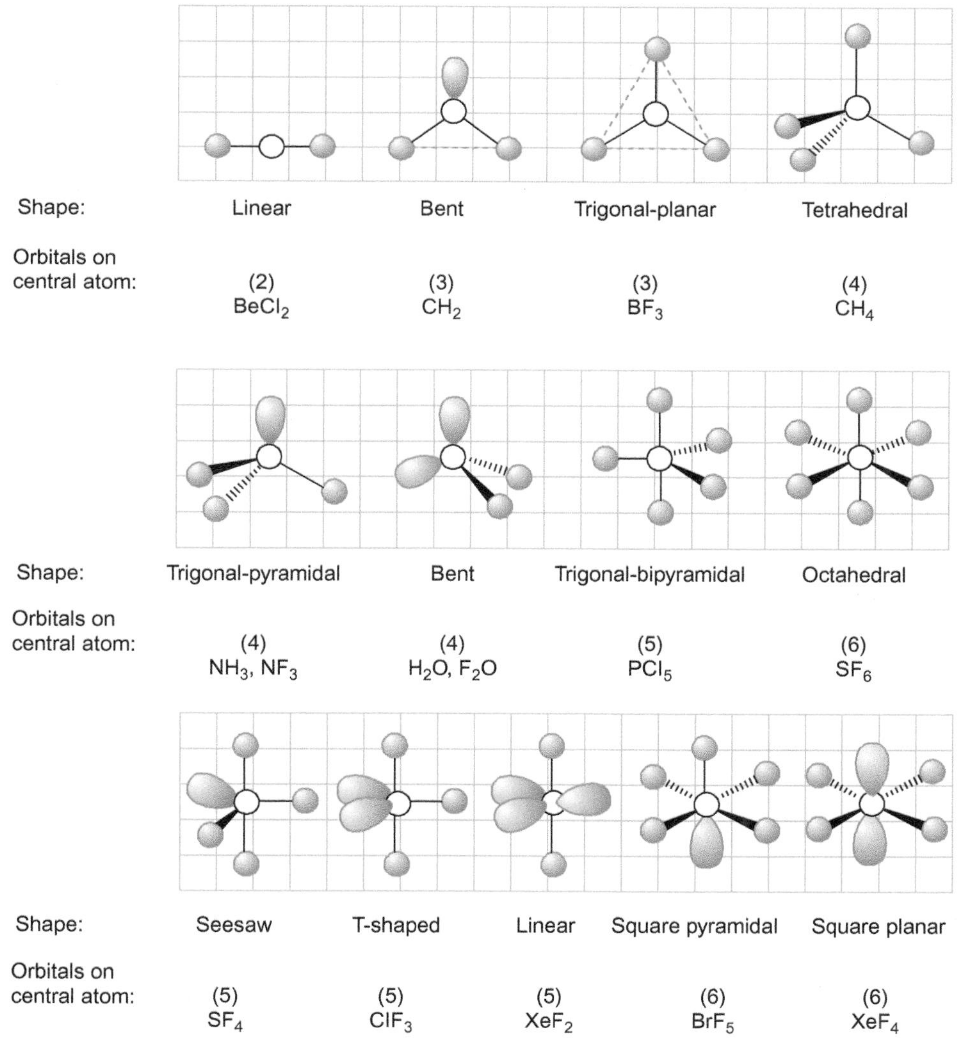

FIG. 5.1   Examples of some molecular structures based on VSEPR theory.

- BeF_2:
  - The beryllium atom forms a covalent bond with each fluorine atom. It is surrounded by only the two electron pairs that is shares with fluorine atoms.
  - According to VSEPR, the shared pairs are oriented as far away from each other as possible (Fig. 5.1).
  - The distance between electron pairs is maximized if the bonds to fluorine are on opposite sides of the beryllium atom, 180 degrees apart.
  - Thus, all three atoms lie on a straight line.
- NH_3:
  - The Lewis structure of ammonia shows that in addition to the three electron pairs it shares with the three hydrogen atoms, the central nitrogen atom has one unshared pair of electrons.
  - In NH_3, the electron pairs maximize their separation by filling the four corners of tetrahedron.
  - The nonbonding pair of electrons will occupy one of the four vertices of the tetrahedron.
  - The molecular geometry of NH_3 is trigonal pyramidal (Fig. 5.1).
  - Notice that the bond angles decrease as the number of nonbonding electron pairs increases.

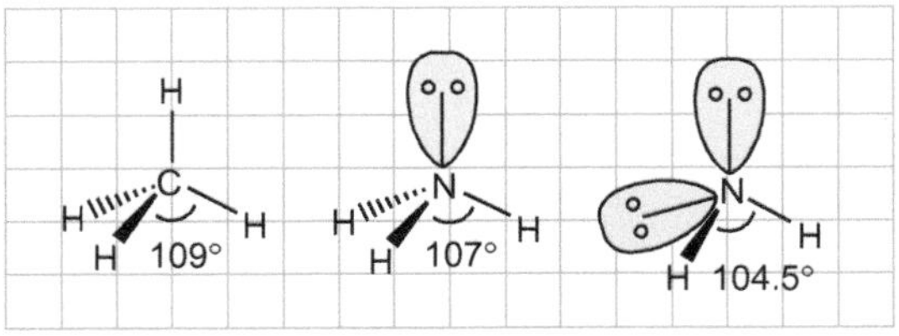

  - The bonding pair of electrons is attached by both nuclei of the bonded atoms.
  - By contrast, a nonbonding pair is attached primarily by only one nucleus.
  - Because a nonbonding pair experiences less nuclear attraction, its electron domain is spread out more in space than the electron domain for a bonding pair.

5.3 ISOELECTRONIC SPECIES

Which of the following compounds are isoelectronic?

ClI, Os^{6+}, Br_2, SF_6, GeF_4

F, Z = 9; S, Z = 16; Cl, Z = 17; Ge, Z = 32; Br, Z = 35; I, Z = 53; Os, Z = 76

- Isoelectronic compounds are those, which have the same total number of electrons, and same number of valence electrons.

Total no. of electrons in ClI	$= 53 + 17$	$= 70$
Total no. of electrons in Os^{6+}	$= 76 - 6$	$= 70$
Total no. of electrons in Br_2	$= 2 \times 35$	$= 70$
Total no. of electrons in SF_6	$= 16 + 6 \times 9$	$= 70$
Total no. of electrons in GeF_4	$= 32 + 4 \times 9$	$= 68$

ClI, Os^{6+}, Br_2, and SF_6 are isoelectronic, only GeF_4 is not because it does not have 70 electrons.

5.4 PROCEDURES TO DIAGRAM MOLECULAR STRUCTURE

What are the main guidelines to be considered to diagram the molecular structure?

1. Draw a draft based on analogies with known, similar, or isoelectronic compounds, which usually have the same structure.
2. The lower electronegativity atoms tend to be central, rather than terminal.
3. Heavier, larger elements tend to exhibit expanded valence shells (more than four pairs).
4. Every valence electron pair of each atom is stereochemically important. All valence shell electron pairs of the central atom must take into consideration the pairs that form bonds, or the pairs that are unshared (lone pair).
5. The repulsions between these pairs define the molecular shape. The electron pairs in the valence shell of the central atom take positions as far apart as possible.

6. The number of valence electron pairs (V) in molecule is equal to the number of σ pairs (n_σ), plus the number of π pairs (n_π), plus the number of lone pairs (n_l):

$$V = n_\sigma + n_\pi + n_l$$

7. The number of electrons that would be required to give two electrons to each H atom, and eight electrons to each of the other atoms is as follows:
   no. e^- for individual atoms $= 2$ (no. H atoms) $+ 8$ (no. other atoms)
   no. of bonding electrons $=$ (no. e^- for individual atoms) $-$ (total no. e^-)
   no. of bonds $= \frac{1}{2}$ (no. of bonding electrons)
8. The nonbonding pairs help to determine the positions of the atoms in the molecule or ion.
   no. of unshared electrons $= V -$ (no. of bonding electrons)
   Distribute the unshared electron among individual atoms of the molecule.
9. Check the atom formal charges:
   Formal charge $= +$(group no.) $-$ (no. bonds) $-$ (no. unshared e^-)
10. Notice that the formal charges added up to the charge of the ion.

What are the possible structures of ClO_3^-, SO_2, HNO_3, and N_2O?

- The possible structure of ClO_3^-:
  - Total no. of valence electrons:

$$V = 7_{\text{from Cl}} + 18_{\text{from O}} + 1_{\text{from ionic charge}} = 26$$

  - Number of electrons that would be required to give two electrons to each H atom, and eight electrons to each of the other atoms.
    no. e^- for individual atoms $= 2$(no. H atoms) $+ 8$(no. other atoms)
    $= 2(0) + 8(4) = 32$
  - no. of bonding electrons $=$ (no. e^- for individual atoms $-$ (total no. e^-))
    $= 32 - 26 = 6$
  - no. of bonds $= 6/2 = 3$
  - The possible structure:

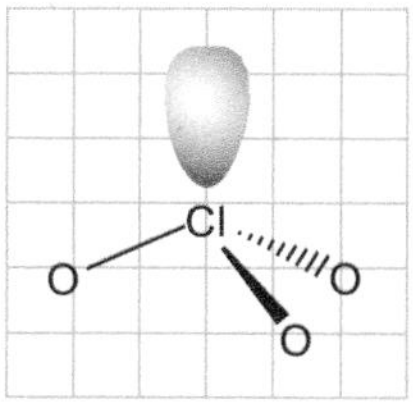

  - no. of unshared electrons $= 26 - 6 = 20$
    18 $e^- \rightarrow$ 3 O atoms, 2 $e^- \rightarrow$ chloride atom

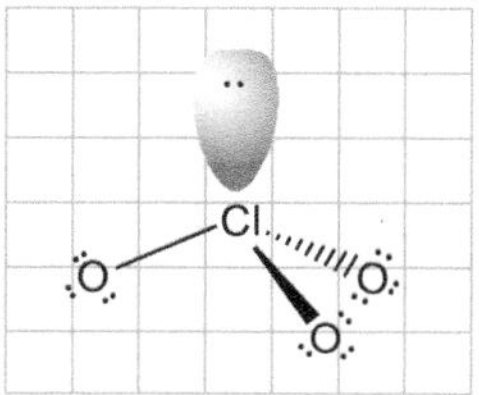

  - Formal charge $= +$(group no.) $-$ (no. bonds) $-$ (no. unshared e^-)
    The formal charge of Cl $= +7 - 3 - 2 = 2+$
    The formal charge of each O $= +6 - 1 - 6 = 1-$

- The structure is

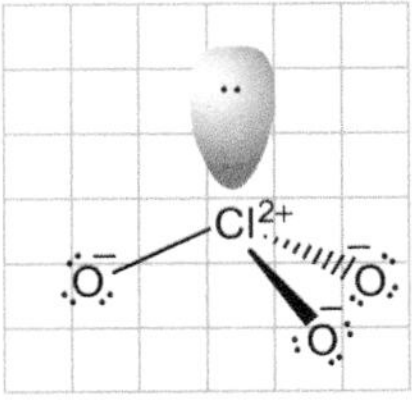

- The most possible structure of SO_2:
  - Total no. of valence electrons:

$$V = 6_{\text{from S}} + 12_{\text{from O}} = 18$$

  - $$\text{no. e}^- \text{for individual atoms} = 2(\text{no. H atoms}) + 8(\text{no. other atoms})$$
$$= 2(0) + 8(3) = 24$$
  - $$\text{no. of bonding electrons} = (\text{no. e}^- \text{for individual atoms} - (\text{total no. e}^-))$$
$$= 24 - 18 = 6$$
  - no. of bonds $= 6/2 = 3$
  - The structure:

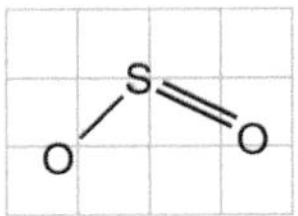

  - no. of unshared electrons $= 18 - 6 = 12$
    4 $e^- \rightarrow O$, 6 $e^- \rightarrow O$, and 2 $e^- \rightarrow S$

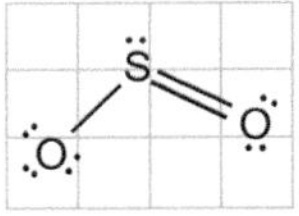

  - Formal charge = +(group no.) − (no. bonds) − (no. unshared e^-)
    The formal charge of S $= +6 - 3 - 2 = 1+$
    The formal charge of left-hand O $= +6 - 1 - 6 = 1-$
    The formal charge of right-hand O $= +6 - 2 - 4 = 0$
  - The structure is

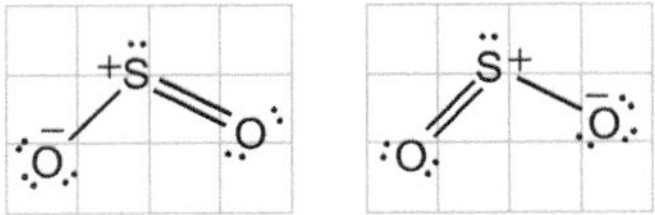

- The most possible structure of HNO_3:
  - Total no. of valence electrons:

$$V = 1_{\text{from H}} + 5_{\text{from N}} + 18_{\text{from O}} = 24$$

$$\text{no. e}^- \text{for individual atoms} = 2(\text{no. H atoms}) + 8(\text{no. other atoms})$$
$$= 2(1) + 8(4) = 34$$

$$\text{no. of bonding electrons} = (\text{no. e}^- \text{for individual atoms} - (\text{total no. e}^-))$$
$$= 34 - 24 = 10$$

  - no. of bonds $= 10/2 = 5$

- The structure:

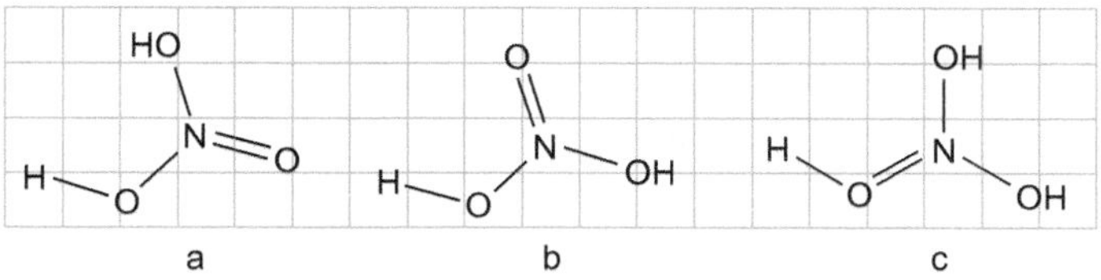

- no. of unshared electrons = 24 − 10 = 14

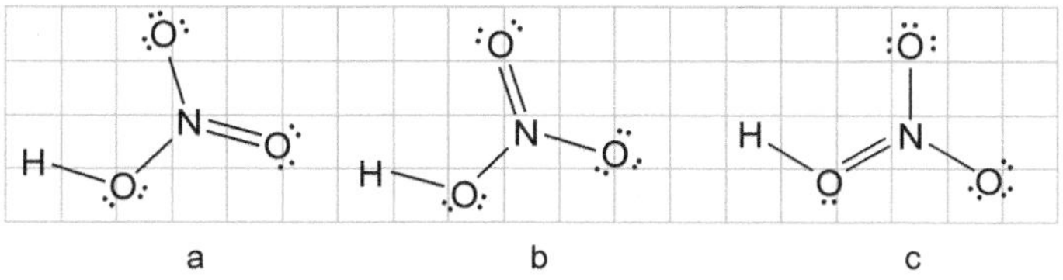

- Formal charge = +(group no.) − (no. bonds) − (no. unshared e^-)

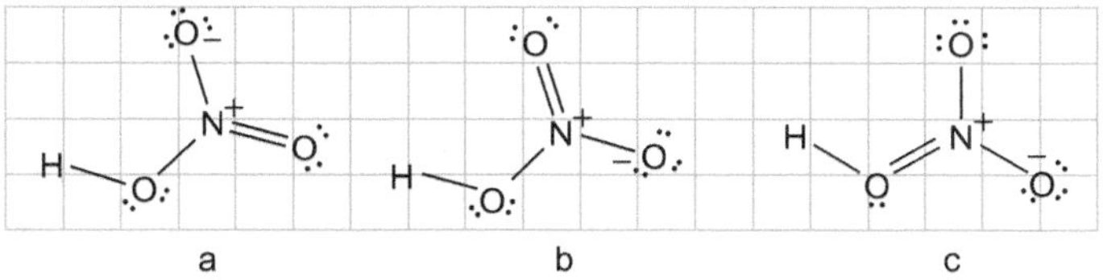

- Structure c must be discarded because adjacent atoms in the molecule carry a formal charge with the same sign.
- The structure is:

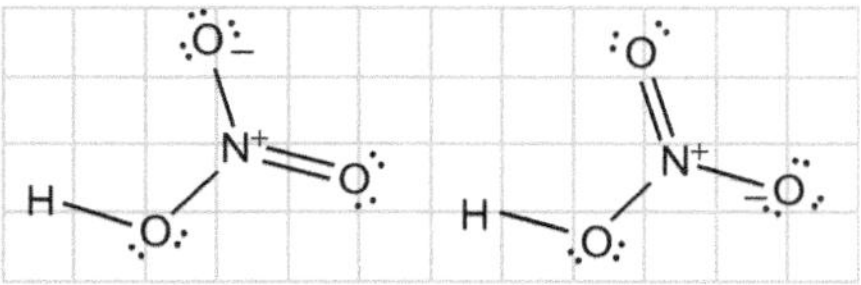

- The most possible structure of N_2O:
  - Total no. of valence electrons:

$$V = 10_{\text{from N}} + 6_{\text{from O}} = 16$$

- $\text{no. e}^-\text{for individual atoms} = 2(\text{no. H atoms}) + 8(\text{no. other atoms})$

$$= 2(0) + 8(3) = 24$$

- $\text{no. of bonding electrons} = (\text{no. e}^-\text{for individual atoms} - (\text{total no. e}^-))$

$$= 24 - 16 = 8$$

- no. of bonds = 8/2 = 4
- the structure:

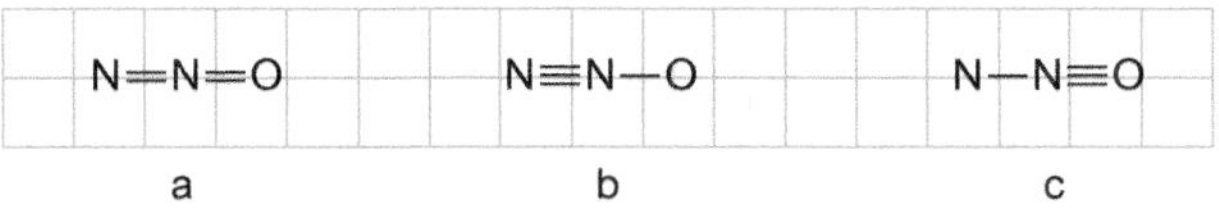

- no. of unshared electrons = 16 − 8 = 8

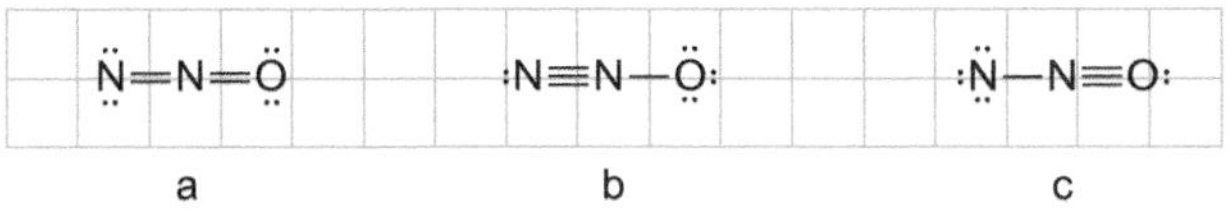

○ Formal charge = +(group no.) − (no. bonds) − (no. unshared e⁻)

      a   b   c

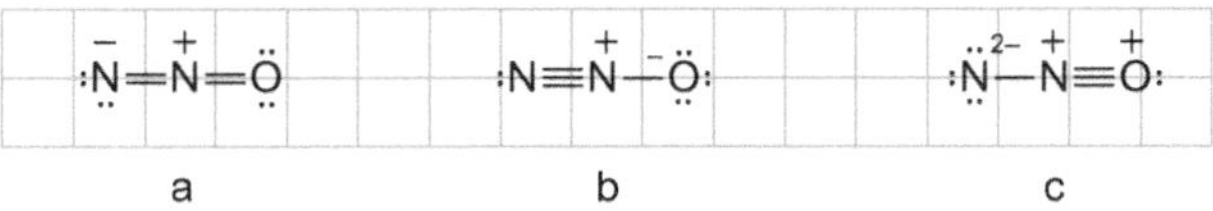

○ Structure c must be discarded because adjacent atoms in the molecule carry formal charge with the same sign.
○ The structure is

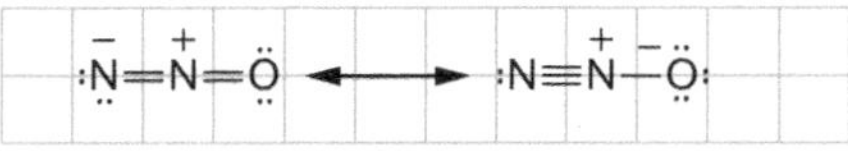

a   b

5.5 VALENCE BOND THEORY AND METALLIC BONDS

Use lithium as model to show how the valence bond theory could be used to explain the metallic bonding? How could the theory explain the metallic physical properties?

● Lithium has a body-centered cubic lattice structure, with eight nearest neighbors and six next-nearest neighbors at little greater distances.

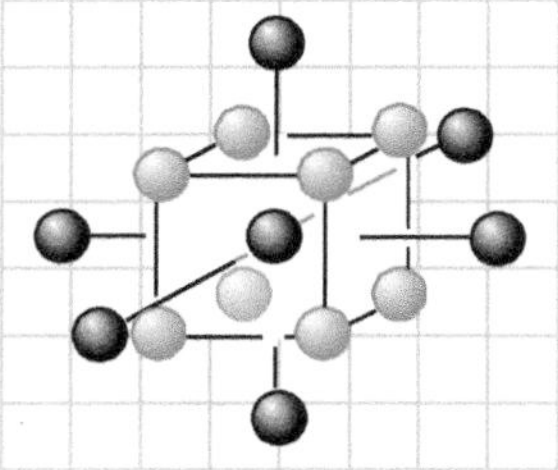

● A lithium atom has one electron in its valence shell that may be shared with one of its neighbors, producing a two-electron bond.

Li. + Li. ⟶ Li—Li

$$\left(Li\atop\cdot\right) + \left(Li\atop\cdot\right) \qquad \left(Li :\! Li\right)$$

    The lithium atom could be bonded to any of its neighbors; consequently, several different arrangements are possible (Fig. 5.2).

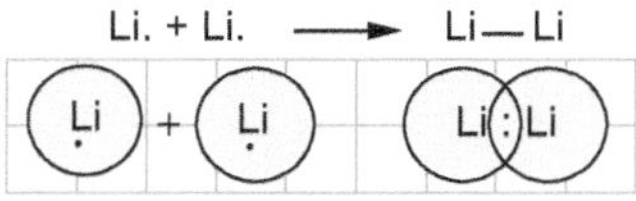

● A lithium atom also may form two bonds if it ionizes:

Li. + Li⁺ + Li. ⟶ Li—Li⁺—Li

$$\left(Li\atop\cdot\right) + \left(Li^+\right) + \left(Li\atop\cdot\right) \qquad \left(Li \cdot\! Li^+ \cdot\! Li\right)$$

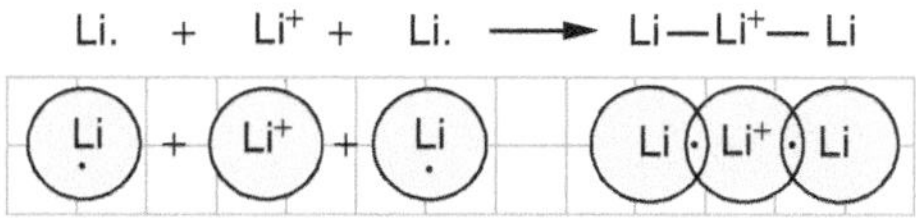

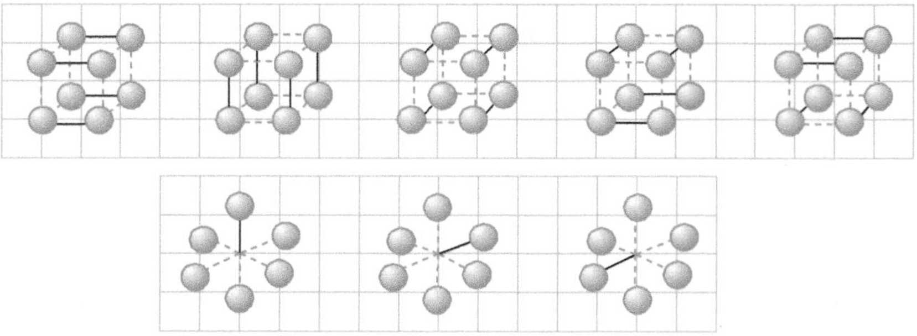

FIG. 5.2 Illustration of some two-electron bonding possibilities in lithium.

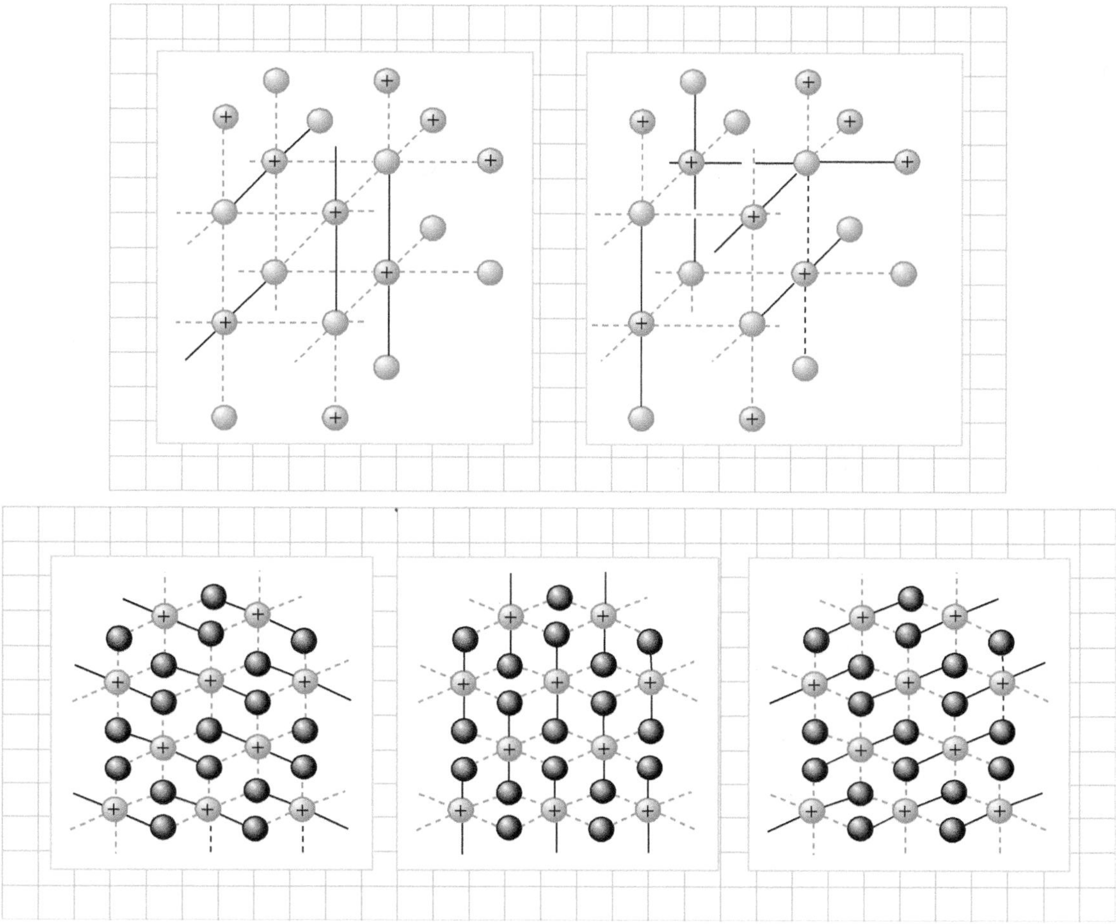

FIG. 5.3 Illustration of some bonding possibilities in ionized lithium.

It can then form many possible structures (Fig. 5.3).

- The theory suggests that the real structure is a mix of all possible bonding. The more possible structures are formed, the lower the potential energy of the system. As a result, the *cohesive force* that binds the structure together is large. In metallic lithium, the cohesive energy is three times greater than in Li molecules.
- The cohesive energy increases in intensity from Group I to II to III, because the atoms are able to form more bonds that give a larger number of possible structures.
- The presence of ions explains the electrical conduction, but the theory does not explain the thermal conductivity of solids, luster, or the preservation of metallic properties in the liquid state or in a solution.

What would valence bond theory not explain?

- The valence bond theory fails to explain the correct molecular structure. The s-orbital of hydrogen is spherically symmetrical, and the three p-orbitals p_x, p_y, and p_z are mutually perpendicular to each other. If s- and p-orbitals were used for bonding as in H_2O, NH_3 or any other organic compound, then the bond angle should be 90 degrees. however, the found bond angle are different:

$$H_2O : H-O-H = 104°27'$$

$$NH_3 : H-N-H = 107°48'$$

$$CH_4 : H-C-H = 109°28'$$

The chemical and physical data shows that in methane CH_4 have four comparable bonds point to the corners of a tetrahedron.

- The valence bond theory also fails to explain the formation of BeF_2, BF_3, CH_4, PCl_5, SF_6, and many other compounds.

5.6 ORBITAL HYBRIDIZATION

What is it meant by orbital hybridization?

- Hybridization is the mixing of two or more atomic orbitals of similar energies on the same atom to produce new hybridized orbitals of equal energies.
- The number of hybridized orbitals must equal to the number of pure orbitals taking place in the hybridization process.

	Number of outer orbitals
$s + p \Rightarrow sp \Rightarrow$ linear	2
$s + 2p \Rightarrow sp^2 \Rightarrow$ Trigonal planar	3
$s + 3p \Rightarrow sp^3 \Rightarrow$ Tetrahedral	4
$s + 3p + d_{z^2} \Rightarrow d_{z^2}sp^3 \Rightarrow$ Trigonal bipyramidal	5
$s + p + 3d \Rightarrow d^3sp \Rightarrow$ Trigonal bipyramidal	5
$s + 3p + d_{x^2-y^2} \Rightarrow d_{x^2-y^2}sp^3 \Rightarrow$ Square pyramidal	5
$s + 3p + 2d \Rightarrow d^2sp^3 \Rightarrow$ Octahedral	6
$s + 3p + 3d \Rightarrow d^3sp^3 \Rightarrow$ Pentagonal bipyramidal	7

- The shape of the hybridized orbital differs from those of the pure orbitals forming them, as they are shaped to have a better chance to overlap (react) with orbitals of another atoms.

Why and how do orbitals hybridize?

- Orbital hybridization was proposed to justify the molecular structure when the valence theory fails to explain the experimentally observed structure.
- Hybridization is the mixing atomic orbitals into new hybrid orbitals that appropriate for pairing of electrons to form electronic bonds as in valence bond theory. The new hybrid orbitals have different shape and energies than its constituent atomic orbitals.
  - By excitation of the atom, electrons that were paired in the ground state are unpaired and promoted into higher empty orbitals. This increases the number of unpaired electrons, and hence the number of bonds that can be formed.
  - Electron pairs repel each other, and the bonds and lone pairs around a central atom are generally separated by the largest possible angle.

The electronic configuration of Be is $1s^2\ 2s^2$, which does not have any unpaired electrons, how does BeF_2 is formed, and what is the expected molecular structure?

- The ground state electronic configuration of Be atom has no unpaired electrons ($1s^2\ 2s^2$). According to valence theory, no bonds can be formed. If energy is provided, an excited state will be produced by unpairing and promoting a 2s electron to an empty 2p level, giving $1s^2\ 2s^1\ 2p_x^1$. The two unpaired electrons can potentially form two bonds (Scheme 5.2).
- Electrons of sp-orbitals repel each other. The repulsion is minimized when these two hybrid orbitals are oriented at 180 degrees to each other. The new shape of the sp-orbitals are suitable to overlap more effectively. When sp-orbitals overlap with p-orbitals of fluoride atoms, a linear BeF_2 molecule is formed.

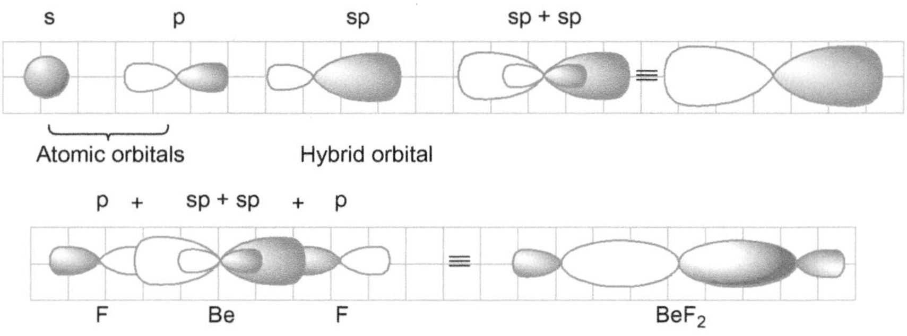

- The structure of a gaseous molecule of beryllium fluoride BeF_2 is linear F—Be—F.

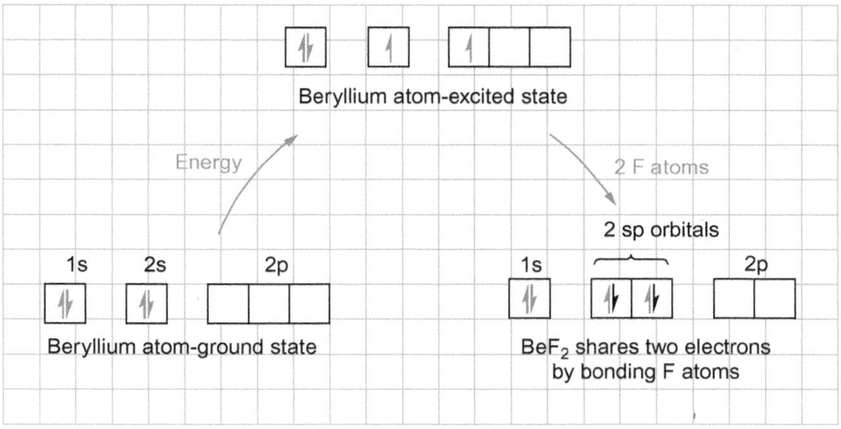

SCHEME 5.2 sp-orbital hybridization diagram for BeF_2 molecule, ⇂ = electron shared by F.

TABLE 5.1 Orbital Hybridization and Coordination Number

	IF_5	$[ZrF_7]^{3-}$	IF_7	$[Mo(CN)_8]^{4-}$	$[ReH_9]^{2-}$
Hybrid orbitals	d^2sp^2, d^4s, d^4p	d^5sp, d^3sp^3	d^5sp, d^3sp^3	d^4sp^3	d^5sp^3
Coordination number	5	7	7	8	9
Geometry	Triangular bipyramidal	Pentagonal bipyramidal	Pentagonal bipyramidal	Dodecahedral	Capped prism

What type of hybrid orbitals does the central atom adopt in IF_5, $[ZrF_7]^{3-}$, IF_7, $[Mo(CN)_8]^{4-}$, and $[ReH_9]^{2-}$? Rationalize the possible coordination number.

- Orbital hybridization based on s-, p-, and d-orbitals and the associated geometries are listed in Table 5.1. By using, a full set of s-, p-, and d-orbitals permits the prediction of the coordination number.

5.7 REHYBRIDIZATION AND COMPLEX FORMATION

How does BF_3 (g) bind to $N(CH_3)_3$ (g) to form BF_3-$N(CH_3)_3$ (s)?
Indicate the rehybridization in the formation of BF_3-$N(CH_3)_3$ (s).

- The coordination bond, or donor-acceptor bond: has a specific lone pair involved in formation of a molecule from two others.
  - Electron acceptor + Electron donor → Complex
  - Acceptor undergoes increase in the coordination number.
  - Each ligand has a donor atom (e.g., N, O, …)
  - Acceptor usually rehybridizes, e.g., from sp^2 to sp^3 for B in BF_3 (Scheme 5.3).

$$BF_3(g) + N(CH_3)_3(g) \rightarrow BF_3 - N(CH_3)_3(s)$$

  - The acceptor is originally coordinatively unsaturated, and can add ligands (donor) until a maximal coordination number is attained, becoming coordinatively saturated.
  - The acceptor must have vacant atomic orbits into which the donor can put electrons.

What is the coordination number in NaCl, $BeCl_2$, BF_3, $AlCl_3$, SiF_4, SiF_6^{2-}, and Si^{4+}?

- Using, a full set of s-, p-, and d-orbitals allows the prediction of the coordination number.

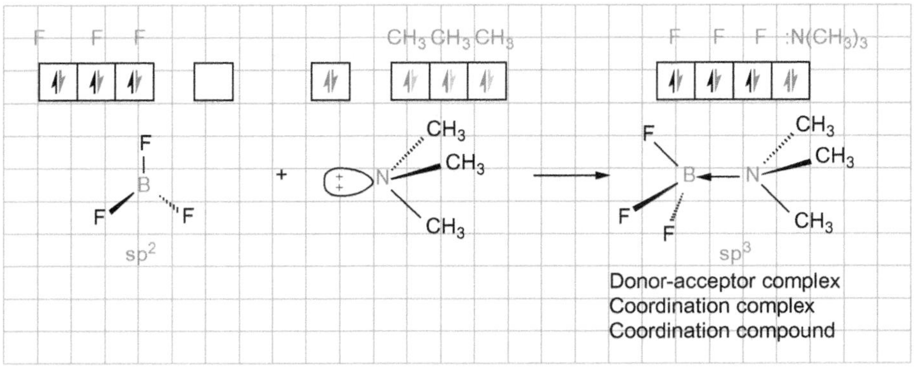

SCHEME 5.3 Rehybridization of sp^2 of B in BF_3 to sp^3 in BF_3-$N(CH_3)_3$ complex.

- Species with less than fully occupied valence shell p-orbitals have C.N. = 4 (maximum coordination, e.g., NaCl, $BeCl_2$, BF_3, and $AlCl_3$)

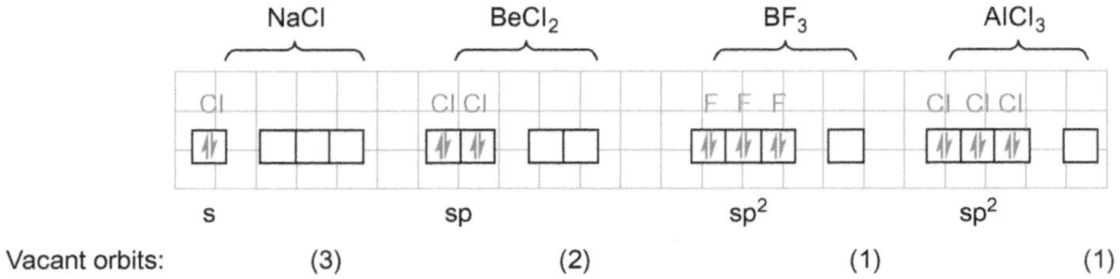

  - BF_3: sp^2, adds one ligand (such as $N(CH_3)_3$) to C.N. = 4 (Note: $\pi_{\perp}^2$ in BF_3)
    $AlCl_3$: sp^2, dimerized to form $(AlCl_3)_2$, C.N. = 4:

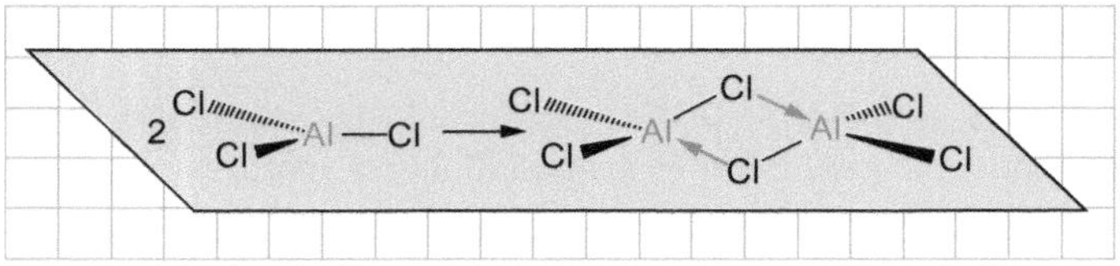

  - $BeCl_2$ as sp-hybridized accepts ē-pairs from donor, to increase C.N. to four
    $BeCl_2$: polymerized

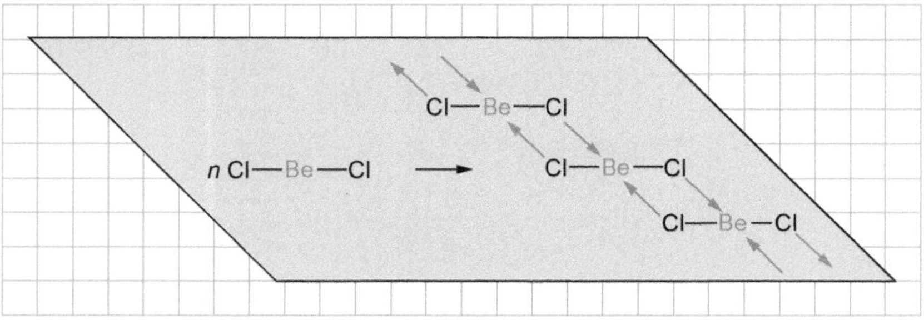

As does NaCl, which form 3-D polymer

- Species that have "available" d-orbital vacancies (close in energy) commonly achieve up to 6 ē-pairs.

	3s	3p	3d
SiF_4 sp^3	↿⇂	↿⇂ ↿⇂ ↿⇂	_ _ _ _ _

	2s	2p	
Add $2F^-$	↿⇂	↿⇂ ↿⇂ ↿⇂	

	3s	3p	3d
SiF_6^{2-} sp^3d^2	↿⇂	↿⇂ ↿⇂ ↿⇂	↿⇂ ↿⇂ _ _ _

Equivalently, $6F^-$ add to Si^{4+}

	3s	3p	3d
Si^{4+}	↿⇂	↿⇂ ↿⇂ ↿⇂	↿⇂ ↿⇂ _ _ _
	F	F F F	F F

- All "bare" cations (i.e., Na^+, Mg^{2+}, Fe^{3+}, Ca^{2+}, Cu^{2+}, etc.) are never found isolated in condensed phases, but always have surrounding ligands: Fe_2Cl_6, $Ca(OH_2)_6^{2+}$.

What is the coordination number of alkali cations?

- Metal cations act as Lewis acids, and water molecule as Lewis base, e.g.:

$$NaCl_{(s)} \xrightarrow{H_2O} Na^+_{(aq.)} + Cl^-_{(aq.)}$$

- Species with less than fully occupied valence shell p-orbitals usually have C.N. = 4:

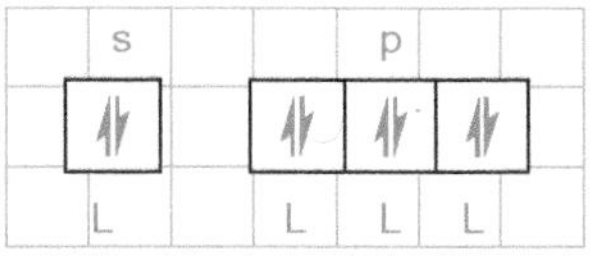

Coordination number = 4
Li^+, Na^+, K^+, Rb^+, Cs^+

Accordingly; formation of $Na(OH_2)_4^+$ is expected, but in solution Na^+ drags about 16 H_2O's around, due to development of a second coordination sphere of $r \cong 2.76$ Å.

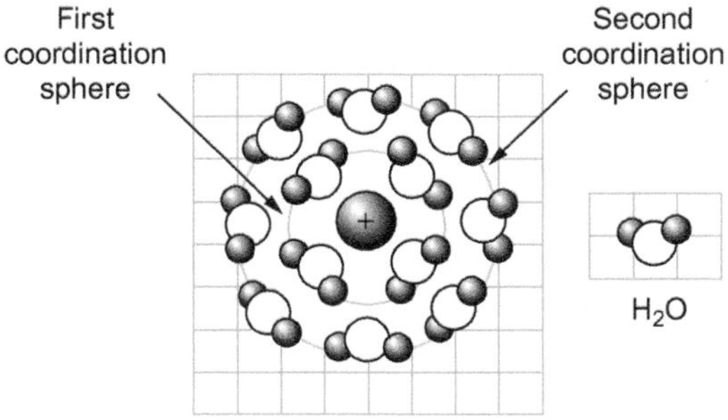

Similarly, $K(OH_2)_4^+$ with about 6–7 H_2O more, $r \cong 2.3$ Å.
$Li(OH_2)_4^+$ with 20–21 H_2O, $r \cong 2.4$ Å.
$Rb(OH_2)_6^+$ and $Cs(OH_2)_6^+$, $r \cong 2.28$ Å.

5.8 HYBRIDIZATION AND σ-/π-BONDING

How many σ and π bonds in tetracyanoethylene molecule, $C_2(CN)_4$? What are the hybridizations on the two carbons of C=C, and on the four carbons of C≡N groups?

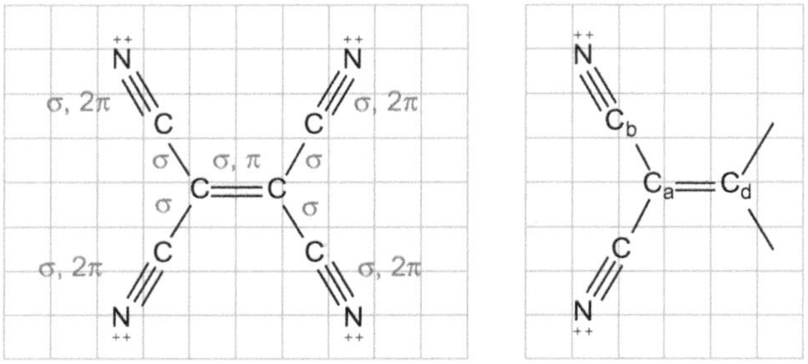

- σ-Bonds result from end to end overlap of orbitals, and the electron density is concentrated in between the two atoms, and on a line joining the two atoms.
- Double or triple bonds occur by the sideways overlap of orbitals, giving π bonds. In π-bonds, the electron density also concentrates between the atoms, but on either side of the line joining the atoms.
- The σ bonds (and lone pairs) determine the shape of the molecule but not by the π bonds, π bonds merely shorten the bond lengths.
- The Lewis structure of tetracyanoethylene molecule indicates nine σ and nine π bonds.
- The carbon atom, C_a, participates in 3σ and 1π bonds, C_a must be sp^2 hybridized and π bonds are not part of the hybridization scheme.
- However, C_b is involved in two σ and two π bonds, thus, C_b is sp hybridized.

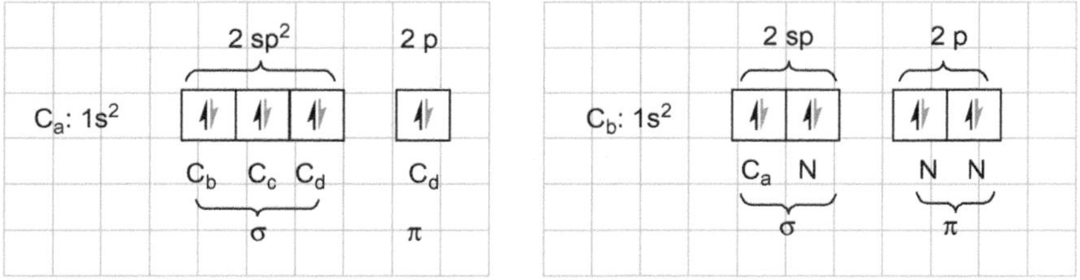

Explain the formation and distribution of the hybrid orbitals in molecular bonding of CO_2, SO_2, and SO_3.

- In CO_2, C is four-valent and O is two-valent, the bonding can be simply represented as O=C=O.
  - Triatomic molecules are either linear or bent. In CO_2, the C is excited and provided four unpaired electrons, which form two σ bonds and two π bonds in the molecule (Scheme 5.4).

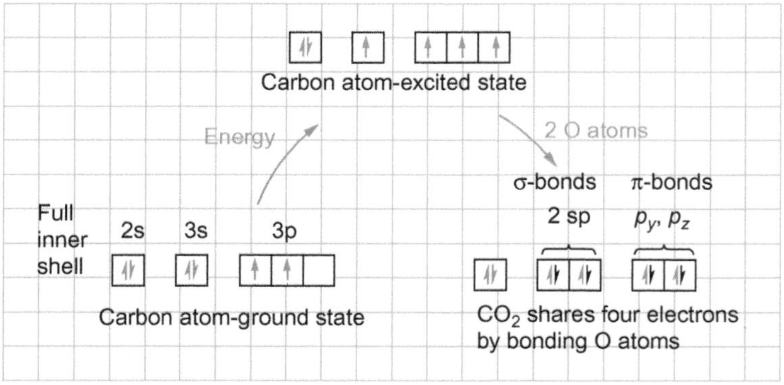

SCHEME 5.4 Orbital hybridization diagram for a CO_2 molecule.

- The s- and p-orbitals are used to form the σ bonds. These could be hybridized and the two sp-orbitals will point in opposite directions. These two orbitals overlap with p-orbitals from two O atoms, forming a linear molecule with a bond angle of 180 degrees.
- The remaining $2p_y$- and $2p_z$-orbitals on C used for π bonding are at right angles to the bond, and overlap sideways with p-orbitals on the O atoms at either side. This π overlap shortens the C—O distances, but does not affect the shape. The π orbitals are ignored in determining the shape of the molecule.

- In SO_2, S shows oxidation states of +2, +3, and +4, whilst O is −2. The structure may be represented as O=S=O.
  - Triatomic molecules are either linear or bent. The S atom is excited to provide four unpaired electrons (Scheme 5.5).
  - The two-electron pairs that form the π bonds do not influence the shape of the molecule, but just shorten the bond lengths. The remaining three orbitals point to the corners of a planar trigonal, with two corners occupied by O atoms and one corner occupied by a lone pair. The SO_2 is thus a bent molecule. The repulsion by the lone pair of electrons reduces the bond angle from the ideal value of 120° to 119°30′.

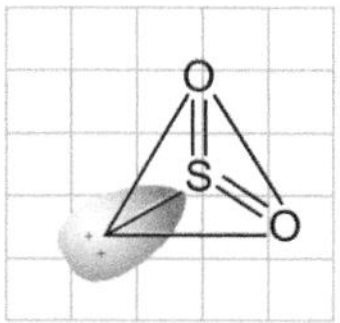

- In SO_3, valency requirements suggest the structure:

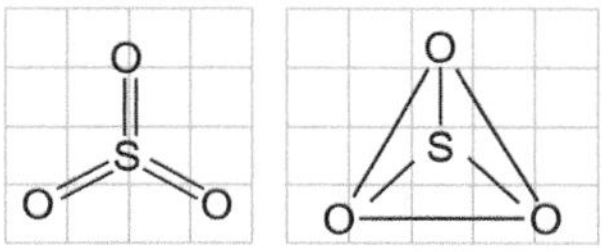

SO_3 molecule

  - The three σ orbitals are oriented towards the corners of an equilateral triangle. The 3π bonds shorten the bond lengths (Scheme 5.6), but do not influence the structure. It assumes that:
    (1) one 3p- and two 3d-orbitals on S are in the right direction to overlap with the $2p_y$, or $2p_z$-orbitals of three different O atoms, and
    (2) the π bonds formed are all of equal strength.
  - In molecules with more than one π bond, or molecules where the π bond could equally well exist in more than one position, it is better to treat the π bonding as being delocalized over several atoms rather than localized between two atoms.

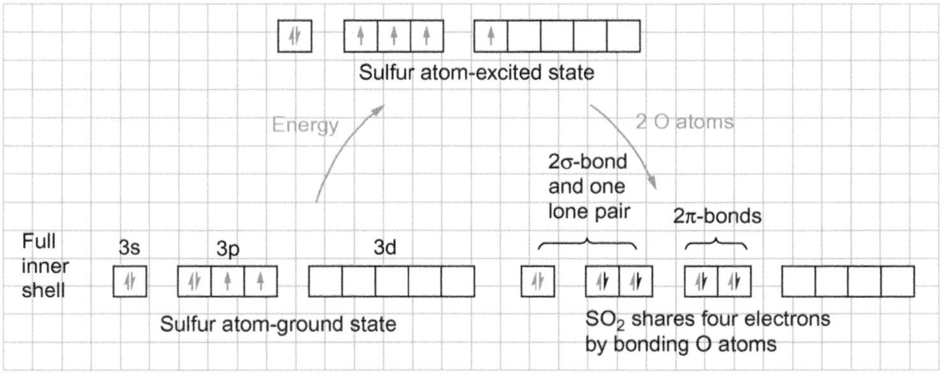

SCHEME 5.5 Orbital hybridization diagram for an SO_2 molecule.

SCHEME 5.6 Orbital hybridization diagram for an SO_3 molecule.

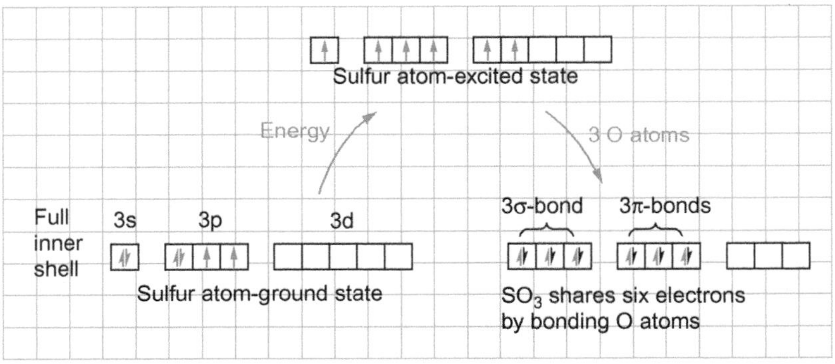

5.9 ORBITAL HYBRIDIZATION AND MOLECULAR SYMMETRY

Trigonal Planar Hybridization

A boron atom has only one unpaired electron in its outmost shell. How can you explain the formation and the shape of BF_3 molecule by using orbital hybridization and symmetry analysis?

- BCl_3: D_{3h} (trigonal planar)
  A set of three vectors (a_1, a_2, and a_3) could represent the three σ-bonds of BCl_3 (Fig. 5.4).
  - This set of vectors can be used as a basis to generate a reducible representation of D_{3h} point group. The operation of the D_{3h} are:

$$E \quad 2C_3 \quad 3C_2 \quad \sigma_h \ 2S_3 \quad 3\sigma_v$$

  - The character of a matrix representing each of these operations:

$$\vec{v}_i \xrightarrow{E} v_f$$

Where $\vec{v}_i$ and $\vec{v}_f$ represent the initial and final sets of the vectors.

$$\overset{E}{\begin{vmatrix} 1 & 0 & 0 \\ 0 & 1 & 0 \\ 0 & 0 & 1 \end{vmatrix}} \overset{\vec{v}_i}{\begin{vmatrix} a_1 \\ a_2 \\ a_3 \end{vmatrix}} = \overset{\vec{v}_f}{\begin{vmatrix} a_1 \\ a_2 \\ a_3 \end{vmatrix}}$$

$$\chi(E) = 3$$

$$\vec{v}_i \xrightarrow{C_3} \vec{v}_f$$

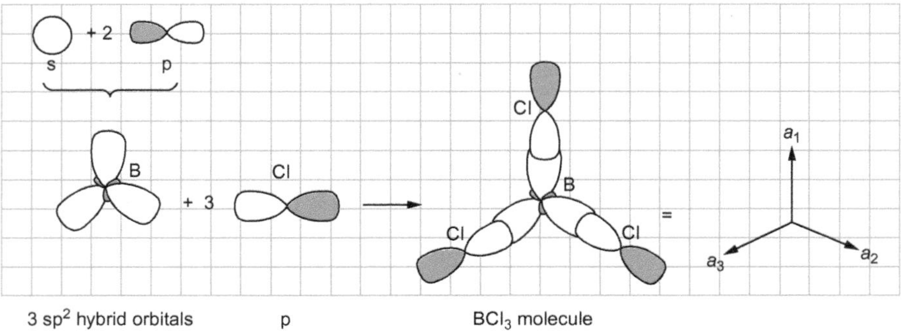

FIG. 5.4 Vector representation (→) of the σ-orbital for BCl_3.

$$\begin{matrix} C_3 & \vec{v}_i & & \vec{v}_f \\ \begin{vmatrix} 0 & 1 & 0 \\ 0 & 0 & 1 \\ 1 & 0 & 0 \end{vmatrix} & \begin{vmatrix} a_1 \\ a_2 \\ a_3 \end{vmatrix} & = & \begin{vmatrix} a_2 \\ a_3 \\ a_1 \end{vmatrix} \end{matrix}$$

$$\chi(C_3) = 0$$

$$\vec{v}_i \xrightarrow{C_2} \vec{v}_f$$

$$\begin{matrix} C_2 & \vec{v}_i & & \vec{v}_f \\ \begin{vmatrix} 1 & 0 & 0 \\ 0 & 0 & 1 \\ 0 & 1 & 0 \end{vmatrix} & \begin{vmatrix} a_1 \\ a_2 \\ a_3 \end{vmatrix} & = & \begin{vmatrix} a_1 \\ a_3 \\ a_2 \end{vmatrix} \end{matrix}$$

$$\chi(C_2) = 1$$

Likewise, $\chi\ (\sigma_h) = 3$, $\chi\ (S_3) = 0$, $\chi\ (\sigma_v) = 1$

- The character is equal to the number of vectors, which are unshifted by the operation.

$$\begin{matrix} E & 2C_3 & 3C_2 & \sigma_h & 2S_3 & 3\sigma_v \\ \begin{vmatrix} 1 & 0 & 0 \\ 0 & 1 & 0 \\ 0 & 0 & 1 \end{vmatrix} & \begin{vmatrix} 0 & 1 & 0 \\ 0 & 0 & 1 \\ 1 & 0 & 0 \end{vmatrix} & \begin{vmatrix} 1 & 0 & 0 \\ 0 & 0 & 1 \\ 0 & 1 & 0 \end{vmatrix} & \begin{vmatrix} 1 & 0 & 0 \\ 0 & 1 & 0 \\ 0 & 0 & 1 \end{vmatrix} & \begin{vmatrix} 0 & 1 & 0 \\ 0 & 0 & 1 \\ 1 & 0 & 0 \end{vmatrix} & \begin{vmatrix} 1 & 0 & 0 \\ 0 & 0 & 1 \\ 0 & 1 & 0 \end{vmatrix} \\ 3 & 0 & 1 & 3 & 0 & 1 \end{matrix}$$

- The complete set of characters of the reducible Γ_σ is as follows:

D_{3h}	E	$2C_3$	$3C_2$	σ_h	$2S_3$	$3\ \sigma_v$
Γ_σ-Reducible	3	0	1	3	0	1

This reducible representation can be reduced using the character table (Table 5.2):

$$\sigma_i = \frac{1}{h}\sum i \cdot \chi_R \cdot \chi_i$$

$$\text{Number of } A_1' = \frac{1}{12}(3+0+3+3+0+3) = 1$$

$$\text{Number of } A_2' = \frac{1}{12}(3+0-3+3+0-3) = 0$$

TABLE 5.2 Reducible and Irreducible Representations of σ-Bonding for the D_{3h} Point Group

D_{3h}	E	$2C_3$	$3C_2$	σ_h	$2S_3$	$3\sigma_v$		
A'_1	**1**	**1**	**1**	**1**	**1**	**1**		x^2+y^2, z^2
A'_2	1	1	−1	1	1	−1	R_z	–
E'	**2**	**−1**	**0**	**2**	**−1**	**0**	(x, y)	x^2-y^2, xy
A''_1	1	1	1	−1	−1	−1	–	–
A''_2	1	1	−1	−1	−1	1	z	–
E''	2	−1	0	−2	1	0	(R_x, R_y)	(xz, yz)
Γ_σ	**3**	**0**	**1**	**3**	**0**	**1**	$= A'_1 + E'$	

$$\text{Number of } E' = \frac{1}{12}(6+0+0+6+0+0) = 1 \text{ ... etc.}$$

$$\Gamma_\sigma = A_1' + E'$$

- The right side of the character table, D_{3h}, decides which orbitals belong to the symmetry species A'_1 and E'.

$$A_1' \xrightarrow{\text{represents}} s, d_{z^2}$$

$$E' \xrightarrow{\text{represents}} p_x,\ p_y - d_{xy},\ d_{x^2-y^2}$$

Note:

$$x^2 + y^2 \equiv s$$

$$x \equiv p_x, \quad y \equiv p_y, \quad z \equiv p_z$$

$$x^2 - y^2 \equiv d_{x^2-y^2}, \quad z^2 \equiv d_{z^2}$$

$$xy \equiv d_{xy}, \quad xz \equiv d_{xz}, \quad yz \equiv d_{xyz}$$

There are six hybrid orbitals, but we need three, $2A'_1$, E'.

- The criterion for choosing the appropriate contribution out of many available is that of relative energies of atomic orbitals.
- The orbitals of lowest energy, which are not filled, are the most likely.
- This because hybridization lowers energy of final bonded molecules more than if unhybrid orbitals are used.
- The atomic orbitals at a slightly elevated position in the energy will be used, if its energy is not too separated from that of the orbitals with which it is hybrid.

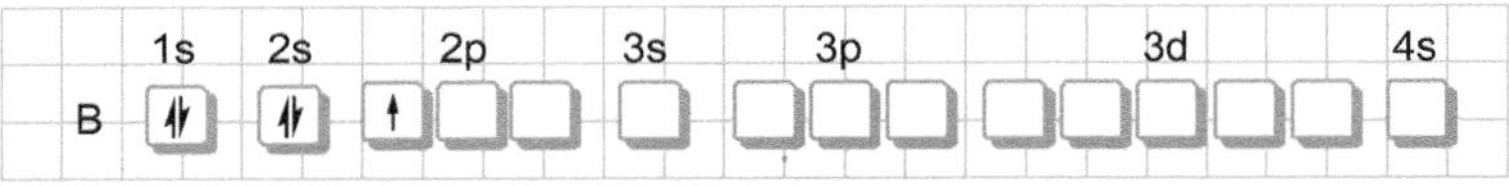

A'_1 includes either the d_{z^2} or the spherically symmetrical s-orbital.

E' includes p_x and p_y together or $d_{x^2-y^2}$ and d_{xy} because they are bracketed together in the two-degenerate E' representation.

- The most likely set of hybrids: $sp_xp_y = sp^2$

- In BF_3, B atom is the central atom in the molecule, when B is excited; three unpaired electrons are produced and available to form three covalent bonds (Scheme 5.7).
- Combining the wave functions of the 2s, $2p_x$, and $2p_y$ atomic orbitals give three hybrid sp^2 orbitals.

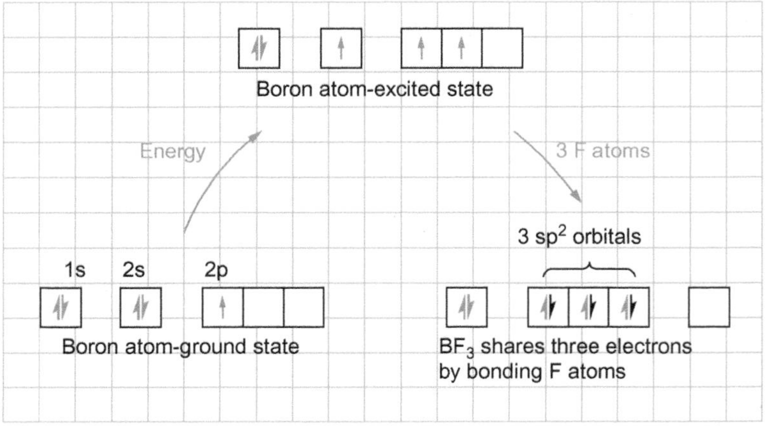

SCHEME 5.7 sp^2-orbital hybridization diagram for a BF_3 molecule.

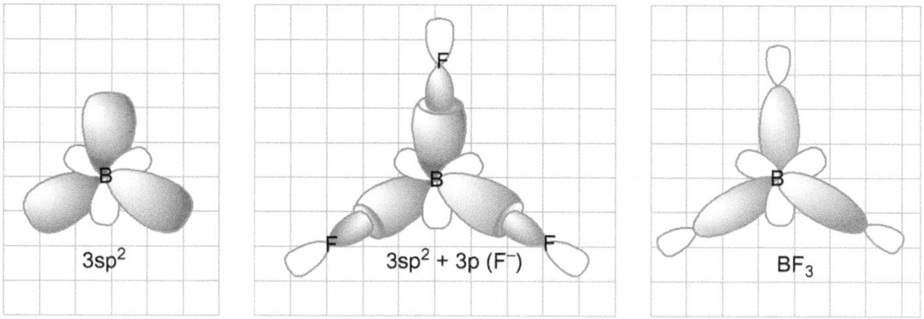

- These three orbitals are equivalent, and repulsion between them is minimized if they are distributed at 120 degrees to each other giving a planar triangle. Overlap of the sp² orbitals with p-orbitals from F atoms gives the planar triangular molecule BF_3 with bond angles of 120 degrees. In the hybrid orbitals, one lobe is bigger than the other is, so it can overlap more effectively.

How can you construct the appropriate hybrid orbital to describe the π-bonding in BF_3?

- The π-bond has a wave function whose sign differs in the two lobes.

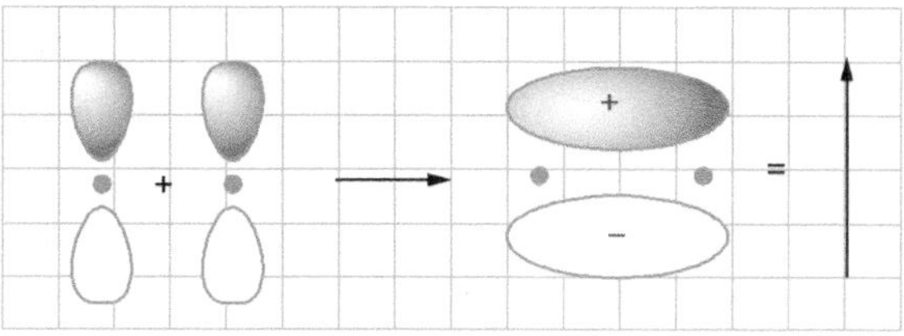

- The vector ↑ represents the symmetry of the π-orbital.
  In BF_3:
  ○ Now to get Γ_π (Fig. 5.5):
  Breaks down into:

$$\Gamma_\pi = \Gamma_{\pi_\perp} + \Gamma_{\pi_\parallel}$$

  ○ There are three hybrid $\Gamma_{\pi_\perp}$ orbitals and three hybrid $\Gamma_{\pi_\parallel}$; we must determine first the number of orbitals that are unshifted by each set of symmetry operations.

$$\text{Vector} \xrightarrow{\text{Symmetry operation}} \text{unmoved} \xrightarrow{\text{has}} \chi = 1$$

$$\text{Vector} \xrightarrow{\text{Symmetry operation}} (\text{moved}) \xrightarrow{\text{has}} \chi = 0$$

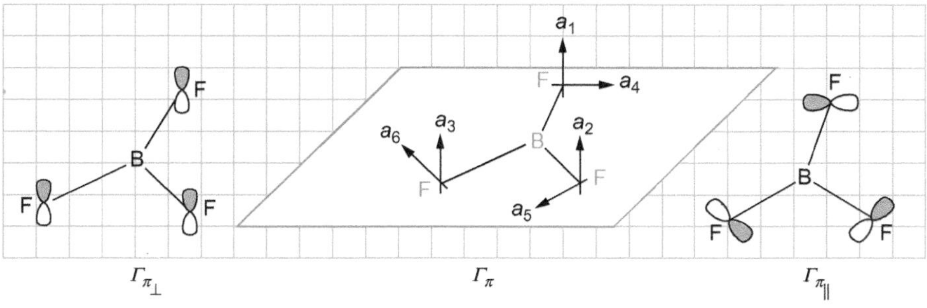

FIG. 5.5   Vector representation (→) of the π-orbital for BCl_3.

D_{3h}	E	$2C_3$	$3C_2$	σ_h	$2S_3$	$3\sigma_v$
$\Gamma_{\pi\perp}$	3	0	−1	−3	0	1
$\Gamma_{\pi\parallel}$	3	0	−1	3	0	−1

- Reduction using: $\sigma_i = \frac{1}{h}\sum i \cdot \chi_R \cdot \chi_i$, and Table 5.3.

$$\Gamma_{\pi\perp} = A_2'' + E''$$
$$\Gamma_{\pi\parallel} = A_2' + E'$$

- From character tables:

$$\Gamma_{\pi\perp} : A_2'' \xrightarrow{\text{represents}} p_z \text{ and } E'' \xrightarrow{\text{represents}} d_{xz},\, d_{yz}$$

$$\Gamma_{\pi\parallel} : A_2' \xrightarrow{\text{represents}} \text{none and } E' \xrightarrow{\text{represents}} \begin{cases} p_x,\, p_y \\ \text{and} \\ d_{xy},\, d_{x^2-y^2} \end{cases}$$

- The absence of A'_2 type atomic orbital means that the hybrid in the π-bonding system is contributed to only by the two E' orbitals. Therefore, if the $\pi_\parallel$ is fully occupied, then over the four atoms, there are only two bonding pairs.
- For both $\pi_\parallel$ and $\pi_\perp$, the nearest d-orbitals are high up in energy. Therefore, it is unlikely that d-orbitals would contribute to the π-bonding hybrids.
- For $\pi_\perp$, it will be dominated by $2p_z$.

$$\psi_\perp = c_1\Psi_{p_z} + c_2\left(\Psi_{d_{xz}} + \Psi_{d_{yz}}\right)$$

  Likely, only p_z orbital contributes to π-character, $c_1 \gg c_2$
- For $\Gamma_\sigma = A_1' + E'$

$$A_1' \xrightarrow{\text{represents}} s, \text{ and } E' \xrightarrow{\text{represents}} p_x,\, p_y$$

  However, $2p_x$, $2p_y$ become less important for $\pi_\parallel$, used up in σ-bonding orbitals, which are lower in energy.

TABLE 5.3 Reducible and Irreducible Representations of σ- and π-Bonding for the D_{3h} Point Group

D_{3h}	E	$2C_3$	$3C_2$	σ_h	$2S_3$	$3\sigma_v$		
A'_1	1	1	1	1	1	1		x^2+y^2, z^2
A'_2	**1**	**1**	**−1**	**1**	**1**	**−1**	R_z	–
E'	**2**	**−1**	**0**	**2**	**−1**	**0**	(x, y)	x^2-y^2, xy
A''_1	1	1	1	−1	−1	−1	–	–
A''_2	1	1	−1	−1	−1	1	z	–
E''	2	1	0	−2	1	0	(R_x, R_y)	(xz, yz)
$\Gamma_{\pi\perp}$	3	0	−1	−3	0	1	$=A_2''+E''$	
$\Gamma_{\pi\parallel}$	**3**	**0**	**−1**	**3**	**0**	**−1**	$=A_2'+E'$	
Γ_π	6	0	−2	0	0	0	$=\Gamma_{\pi\perp}+\Gamma_{\pi\parallel}$	
Γ_σ	5	2	1	3	0	3	$=A'_1+E'$	

The Extend of d-Orbital Participation in Molecular Bonding

Why is it unlikely for a d-orbital to take part in the orbital hybridization in PCl_5 and SF_6?
How can you account for involving the d-orbital?
Why does PH_5 not exist?

- The variation in size of phosphorus orbitals is obvious by the clear differences among the radial mean values: 3s = 0.47 Å, 3p = 0.55 Å, and 3d = 2.4 Å.
- Since the mean radial distance of an orbital is proportional to its energy, the 3d-orbital is not only much larger but also higher in energy than the 3s- and 3p-orbitals.
- Therefore, there seems to be some doubt as to whether d-orbitals can participate.
- Nevertheless, there are several factors that can cause a large contraction of the d-orbitals, and lower their energies, so mixing may be possible.
- First factor, when the central atom attaches to a highly electronegative element like F, O, or Cl, this element attracts more than shares their bonding electrons so it attains a δ^- charge. As a result, this leaves a δ^+ charge on the central atom that causes orbitals contraction. Since the 3d-orbitals contracts in size much more than the 3s- and 3p-orbitals, the energies of the 3s-, 3p-, and 3d-orbitals may become close enough to allow hybridization to occur.
    - The central P atom of PCl_5 is bonded to five highly electronegative Cl atoms that induce the orbitals contraction. The size and the energies of the 3s-, 3p-, and 3d-orbitals become close enough to allow hybridization to occur.
    - The presence of six highly electronegative F atoms of SF_6 causes a large contraction of the d-orbitals, and lowers their energy, so mixing may be possible.
    - In PH_5, hydrogen does not cause this large contraction, so PH_5 does not exist.
- A second factor affecting the size of d-orbitals is the number of d-orbitals occupied by electrons. If only one 3d-orbital is occupied on an S atom, the average radial distance is 2.46 Å, but when two 3d-orbitals are occupied, the distance drops to 1.60 Å.
- Thirdly, a further small contraction of d-orbitals may arise by coupling of the electrons occupying in different orbitals.

Trigonal Bipyramidal Hybridization

How does orbital hybridization work in the molecular bonding of PCl_5?
How can you construct the appropriate hybrid orbital to describe the σ-bonding in PCl_5?

- D_{3h}: AB_5 (trigonal bipyramidal)

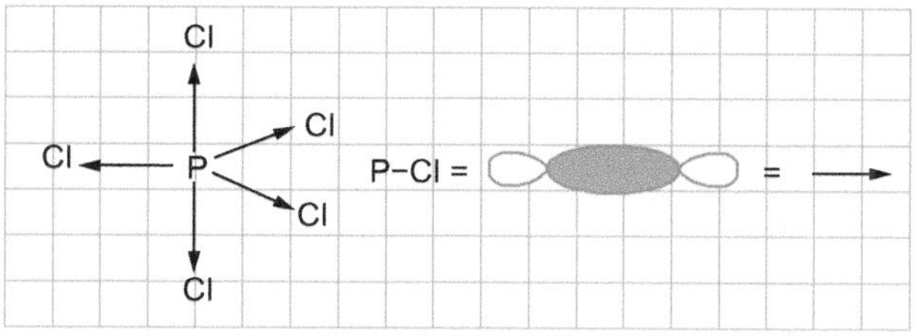

- There are five hybrid σ orbitals; we must determine first the number of orbitals that are unshifted by each set of symmetry operations.

$$\text{Vector} \xrightarrow{\text{Symmetry operation}} \text{unmoved} \xrightarrow{\text{has}} \chi = 1$$

$$\text{Vector} \xrightarrow{\text{Symmetry operation}} (\text{moved}) \xrightarrow{\text{has}} \chi = 0$$

D_{3h}	E	$2C_3$	$3C_2$	σ_h	$2S_3$	$3\sigma_v$
$\Gamma_{\sigma\text{-Reducible}}$	5	2	1	3	0	3

- Five hybrid orbitals decompose to irreducible using

$$\sigma_i = \frac{1}{h}\sum i \cdot \chi_R \cdot \chi_i$$

  and from the character Table 5.4:

$$\Gamma_\sigma = 2A_1' + A_2'' + E'$$

- This tells us which irreducible representations contribute to total representation and which atomic orbital could contribute to the hybrid orbital set on a P atom.

$$A_1' \xrightarrow{\text{represents}} s,\ d_{z^2}$$

$$A_2'' \xrightarrow{\text{represents}} p_z$$

$$E' \xrightarrow{\text{represents}} p_x, p_y - d_{xy},\ d_{x^2-y^2}$$

  There are seven hybrid orbitals, but we need only five, $2A'_1$, A''_2, E'.
- The relative energies of atomic orbitals will be considered for choosing the appropriate contribution out of many available

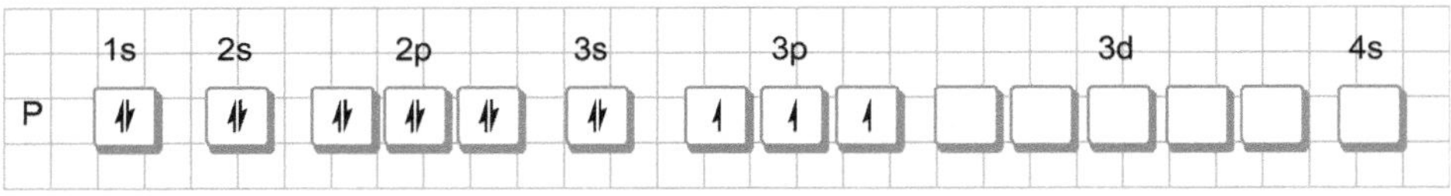

- For A_1', $2A_1' \xrightarrow{\text{represents}} s,\ d_{z^2}$
  It is unlikely to use 1s or 6s, nor a pair of s-type orbitals (3s+4s or 4s+5s).
  It is likely to use an s-type plus a d_{z^2} as $2A'_1$.
  For $A_2'', A_2'' \xrightarrow{\text{represents}} p_z$
  Use $3p_z$
  For $E', E' \xrightarrow{\text{represents}} p_x,\ p_y$ or $d_{xy},\ d_{x^2-y^2}$
  Use $(3p_x,\ 3p_y)$ or $(3d_{xy},\ 3d_{x^2-y^2})$
  So we have:

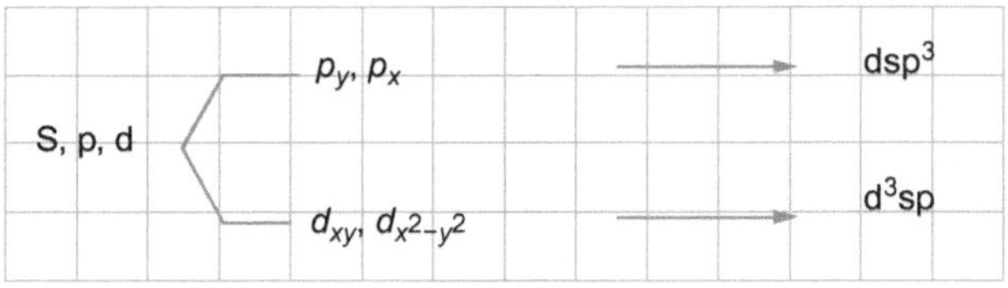

TABLE 5.4 Reducible and Irreducible Representations of σ-Bonding for the D_{3h} Point Group

D_{3h}	E	$2C_3$	$3C_2$	σ_h	$2S_3$	$3\sigma_v$		
A'_1	1	1	1	1	1	1		x^2+y^2, z^2
A'_2	1	1	−1	1	1	−1	R_z	–
E'	2	−1	0	2	−1	0	(x, y)	x^2-y^2, xy
A''_1	1	1	1	−1	−1	−1	–	–
A''_2	1	1	−1	−1	−1	1	z	–
E''	2	−1	0	−2	1	0	(R_x, R_y)	(xz, yz)
Γ_σ	5	2	1	3	0	3	$=2A'_1+A''_2+E'$	

- These two sets are close enough in energy that computation is needed to find the most stable.
  3p is singly occupied.
  3p rather than 3d.
  sp^3d is favored.
  However, we should remember that all D_{3h} molecules are not necessarily sp^3d hybridization.
- The bonding in PCl_5 may be explained by hybridization of the 3s, 3p, and 3d atomic orbitals of phosphorus (Scheme 5.8).

Tetragonal Pyramidal Hybridization

How can you construct the appropriate hybrid orbital to describe the σ-bonding in $XeOF_4$?

- $XeOF_4$: C_{4v} (tetragonal pyramidal)

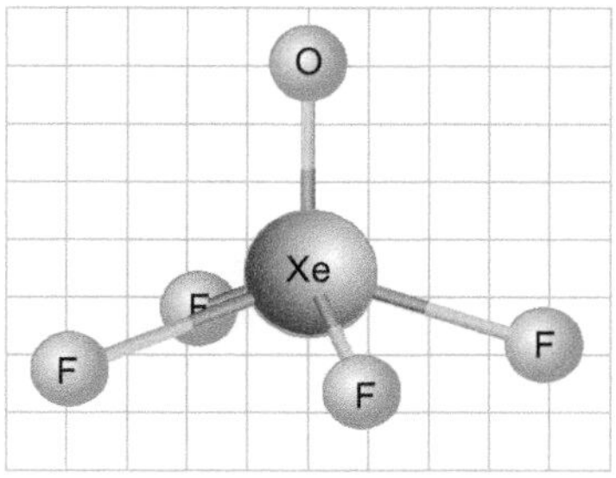

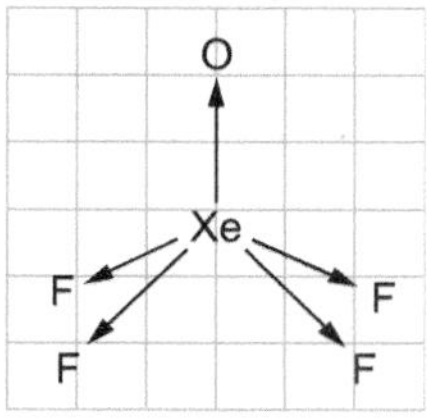

- There are five hybrid σ orbitals; we must determine first the number of orbitals that are unshifted by each set of symmetry operations.

$$\text{Vector} \xrightarrow{\text{Symmetry operation}} \text{unmoved} \xrightarrow{\text{has}} \chi = 1$$

$$\text{Vector} \xrightarrow{\text{Symmetry operation}} (\text{moved}) \xrightarrow{\text{has}} \chi = 0$$

- We obtain the following set of characters for the reducible representation Γ_σ:

C_{4v}	E	$2C_4$	C_2	$2\sigma_v$	$2\sigma_d$
Γ_σ-Reducible	5	1	1	3	1

This can be reduced to:

$$\Gamma_\sigma = 2A_1 + B_1 + E$$

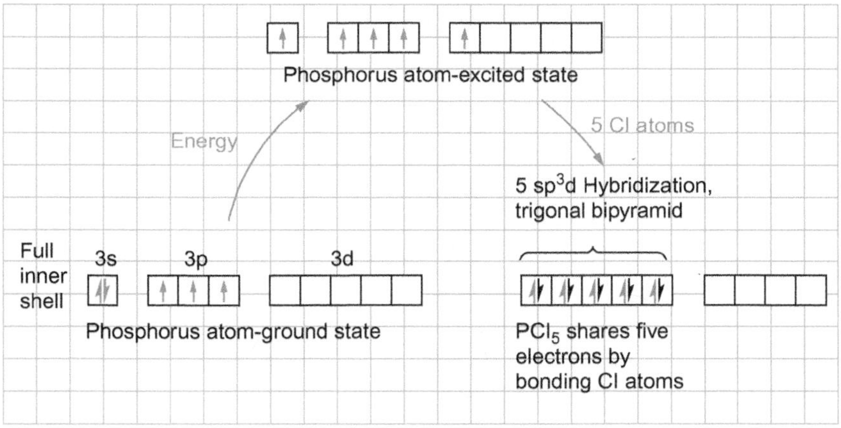

SCHEME 5.8 sp^3d-orbital hybridization diagram for a PCl_5 molecule.

- The character table (Table 5.5) shows

$$A_1 \xrightarrow{\text{represents}} s, p_z, d_{z^2}$$

$$B_1 \xrightarrow{\text{represents}} d_{x^2-y^2}$$

$$E \xrightarrow{\text{represents}} (p_x, p_y) \text{ or } (d_{xz}, d_{yz})$$

- The set of hybrid orbitals: dsp^3
  Other possible hybridization sd^4, spd^3, sd^2p^2, pd^4, and p^3d^2

Square Planar Hybridization

How can you construct the appropriate hybrid orbital to describe the σ and π bonding in $AuCl_4^-$?

- σ-Bonding in case of D_{4h}, square planar
  Examples: $AuCl_4^-$, XeF_4, $Ni(CN)_4^{-2}$, ML_4

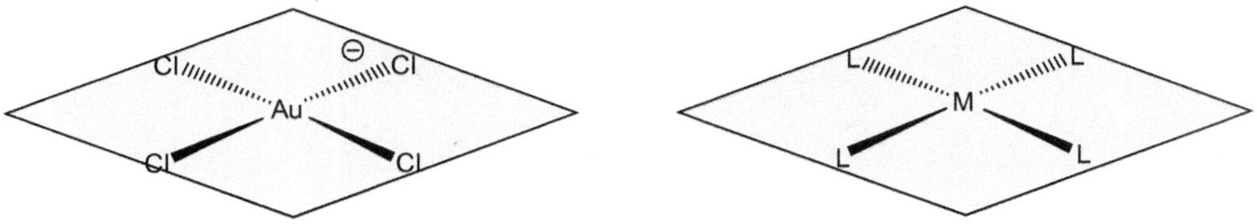

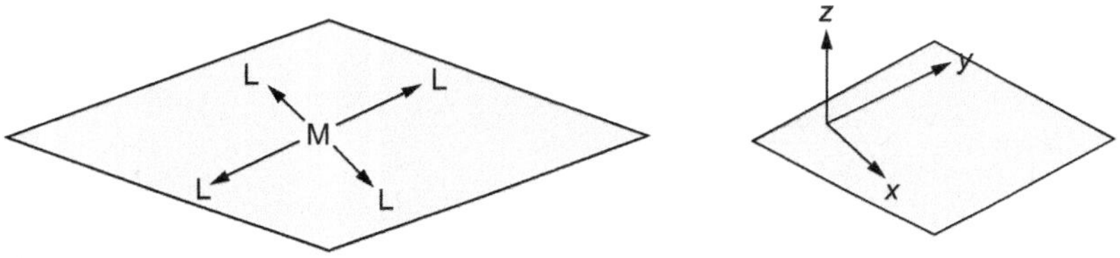

We obtain the following set of characters for the reducible representation Γ_σ:

D_{4h}	E	$2C_4$	C_2	$2C'_2$	$2C''_2$	i	$4S_4$	σ_h	$2\sigma_v$	$2\sigma_d$
Γ_σ-Reducible	4	0	0	2	0	0	0	4	2	0

This can be reduced to
$\Gamma_\sigma = A_{1g} + B_{1g} + E_u$ (Table 5.6).

TABLE 5.5 Reducible and Irreducible Representations of σ-Bonding for the C_{4v} Point Group

C_{4v}	E	$2C_4$	C_2	$2\sigma_v$	$2\sigma_d$		
A_1	**1**	**1**	**1**	**1**	**1**	z	x^2+y^2, z^2
A_2	1	1	1	−1	−1	R_z	—
B_1	**1**	**−1**	**1**	**1**	**−1**	—	x^2-y^2
B_2	1	−1	1	−1	1	—	xy
E	**2**	**0**	**−2**	**0**	**0**	$(x, y)(R_x, R_y)$	(xz, yz)
$\Gamma_{\sigma\text{-reducible}}$	5	1	1	3	1	$=2A_1+B_1+E$	

TABLE 5.6 Reducible and Irreducible Representations of σ-Bonding for the D_{4h} Point Group

D_{4h}	E	$2C_4$	C_2	$2C'_2$	$2C''_2$	i	$4S_4$	σ_h	$2\sigma_v$	$2\sigma_d$		
A_{1g}	**1**	**1**	**1**	**1**	**1**	**1**	**1**	**1**	**1**	**1**	–	x^2+y^2, z^2
A_{2g}	1	1	1	−1	−1	1	1	1	−1	−1	R_z	–
B_{1g}	**1**	**−1**	**1**	**1**	**−1**	**1**	**−1**	**1**	**1**	**−1**	–	x^2-y^2
B_{2g}	1	−1	1	−1	1	1	−1	1	−1	1	–	xy
E_g	2	0	−2	0	0	2	0	−2	0	0	(R_x, R_y)	(xz, yz)
A_{1u}	1	1	1	1	1	−1	−1	−1	−1	−1	–	–
A_{2u}	1	1	1	−1	−1	−1	−1	−1	1	1	z	–
B_{1u}	1	−1	1	1	−1	−1	1	−1	−1	1	–	–
B_{2u}	1	−1	1	−1	1	−1	1	−1	1	−1	–	–
E_u	**2**	**0**	**−2**	**0**	**0**	**−2**	**0**	**2**	**0**	**0**	(x, y)	–
Γ_σ	**4**	**0**	**0**	**2**	**0**	**0**	**0**	**4**	**2**	**0**	$=A_{1g}+B_{1g}+E_u$	

The possible hybridization: $(s, d_{x^2-y^2}, p_x, p_y)$ and $(d_{z^2}, d_{x^2-y^2}, p_x, p_y)$
Which are dsp^2 and d^2p^2

- π-Bonding in case of D_{4h}: square planar, ML_4
  - We obtain the following set of characters for the representations of Γ_σ, $\Gamma_{\pi\perp}$, and $\Gamma_{\pi\parallel}$ (Table 5.7).

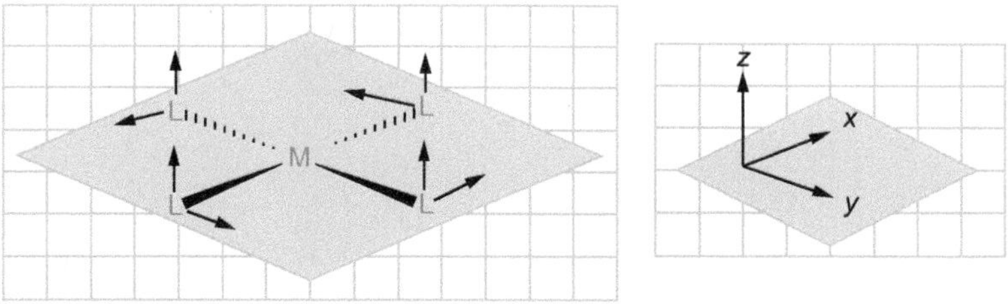

  - We obtain the following set of characters for the reducible representation Γ_σ, $\Gamma_{\pi\perp}$, and $\Gamma_{\pi\parallel}$.

D_{4h}	E	$2C_4$	C_2	$2C'_2$	$2C''_2$	i	$4S_4$	σ_h	$2\sigma_v$	$2\sigma_d$	
Γ_σ	4	0	0	2	0	0	0	4	2	0	$=A_{1g}+B_{1g}+E_u$
$\Gamma_{\pi\perp}$	4	0	0	−2	0	0	0	−4	2	0	$=A_{2u}+B_{2u}+E_g$
$\Gamma_{\pi\parallel}$	4	0	0	−2	0	0	0	4	−2	0	$=A_{1g}+B_{1g}+E_u$

This can be reduced to the following (see Table 5.7):

$$\Gamma_{\pi\perp} = A_{2u} + B_{2u} + E_g \xrightarrow{\text{represent}} p_z, (d_{xy}, d_{yz}) \xrightarrow{\text{from}} pd^2$$

$$\Gamma_{\pi\parallel} = A_{2g} + B_{2g} + E_u \xrightarrow{\text{represent}} d_{xy}, (p_x, p_y) \xrightarrow{\text{from}} dp^2$$

$$\Gamma_\sigma = A_{1g} + B_{1g} + E_u \xrightarrow{\text{represent}} (s, d_{x^2-y^2}, p_x, p_y) \text{ or } (d_{z^2}, d_{x^2-y^2}, p_x, p_y) \xrightarrow{\text{from}} sp^2d \text{ or } p^2d^2$$

There are three π-bonds shared among the four A–B sets.

TABLE 5.7 Reducible and Irreducible Representations of σ- and π-Bonding for the D_{4h} Point Group

D_{4h}	E	$2C_4$	C_2	$2C'_2$	$2C''_2$	i	$4S_4$	σ_h	$2\sigma_v$	$2\sigma_d$		
A_{1g}	1	1	1	1	1	1	1	1	1	1	–	x^2+y^2, z^2
A_{2g}	1	1	1	−1	−1	1	1	1	−1	−1	R_z	–
B_{1g}	1	−1	1	1	−1	1	−1	1	1	−1	–	x^2-y^2
B_{2g}	1	−1	1	−1	1	1	−1	1	−1	1	–	xy
E_g	2	0	−2	0	0	2	0	−2	0	0	(R_x, R_y)	(xz, yz)
A_{1u}	1	1	1	1	1	−1	−1	−1	−1	−1	–	–
A_{2u}	1	1	1	−1	−1	−1	−1	−1	1	1	z	–
B_{1u}	1	−1	1	1	−1	−1	1	−1	−1	1	–	–
B_{2u}	1	−1	1	−1	1	−1	1	−1	1	−1	–	–
E_u	2	0	−2	0	0	−2	0	2	0	0	(x, y)	–
Γ_σ	4	0	0	2	0	0	0	4	2	0	$=A_{1g}+B_{1g}+E_u$	
$\Gamma_{\pi\perp}$	4	0	0	−2	0	0	0	−4	2	0	$=A_{2u}+B_{2u}+E_g$	
$\Gamma_{\pi\parallel}$	4	0	0	−2	0	0	0	4	−2	0	$=A_{1g}+B_{1g}+E_u$	

Tetrahedral Hybridization

The carbon atom in ground state has only two unpaired electrons. How can you explain the formation of four bonds in CH_4?

How can you construct the appropriate hybrid orbital to describe how σ- and π-bonding in molecules have tetrahedral symmetry?

- σ-Bonding in T_d
  e.g., CH_4, MnO_4^-, MnO_4^{2-}, CrO_4^{2-}, or ML_4

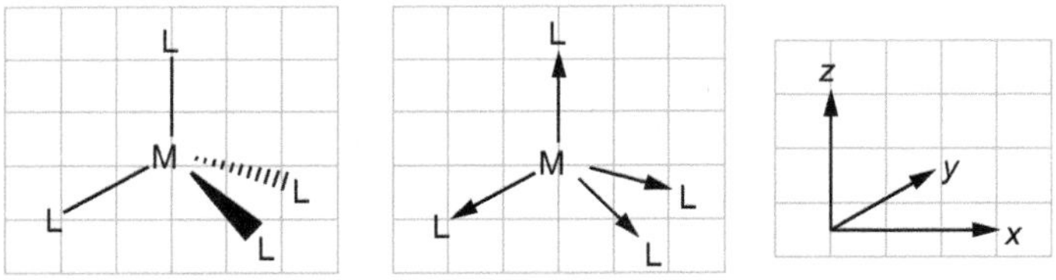

- We obtain the following set of characters for the representation for $\Gamma_{\sigma-\text{Reducible}}$:

T_d	E	$8C_3$	$3C_2$	$6S_4$	$6\sigma_d$
$\Gamma_{\sigma\text{-Reducible}}$	4	1	0	0	2

This can be reduced to $\Gamma_\sigma = A_1 + T_2$ (Table 5.8).

- The character table shows

$$A_1 \xrightarrow{\text{represents}} s$$

$$T_2 \xrightarrow{\text{represents}} (p_x, p_y, p_z) \text{ or } (d_{xy}, d_{xz}, d_{yz})$$

TABLE 5.8 Reducible and Irreducible Representations of σ-Bonding for the T_d Point Group

T_d	E	$8C_3$	$3C_2$	$6S_4$	$6\sigma_d$		
A_1	1	1	1	1	1	–	$x^2+y^2+z^2$
A_2	1	1	1	−1	−1	–	–
E	2	−1	2	0	0	–	$2z^2-x^2-y^2, x^2-y^2$
T_1	3	0	−1	1	−1	(R_x, R_y, R_z)	–
T_2	3	0	−1	−1	1	(x, y, z)	(xy, xz, yz)
Γ_σ	4	1	0	0	2	$=A_1+T_2$	

- The set of hybrid orbitals: sp^3 and sd^3
  sp^3: CH_4
  sd^3: MnO_4^-, MnO_4^{2-}, or CrO_4^{2-}
- In CH_4, the carbon atom in ground state has the electronic configuration: $1s^2$, $2s^2\,2p^1_x$, $2p^1_y$ that has only two unpaired electrons. Thus, carbon atoms possibly form only two bonds. When the carbon atom is excited, the 2s electrons may be unpaired to have $1s^2$, $2s^1$, $2p^1_x$, $2p^1_y$, $2p^1_z$. The four unpaired electrons may overlap with singly occupied s-orbitals of four hydrogen atoms (Scheme 5.9).
- The three p-orbitals p_x, p_y, and p_z are mutually at right angles to each other, and the s-orbital is spherically symmetrical. If the p-orbitals were used for bonding then the bond angle in water should be 90 degrees, and the bond angles in CH_4 should also be 90 degrees. The bond angles actually found 109 degrees 28′.
- The chemical and physical data shows that methane CH_4 has four comparable bonds. These bonds are equivalent, and the repulsion between its electron pairs will be the minimum when the bonds point to the corners of a tetrahedron.
- Mixing one s- and three p-orbitals gives four sp^3 hybrid orbitals.
- The shape of the sp^3-orbital shows that one lobe is expanded. This lobe can overlap more efficiently than either the s- or p-orbital. Thus, sp^3 hybrid orbitals have stronger bonds than the original atomic orbitals.

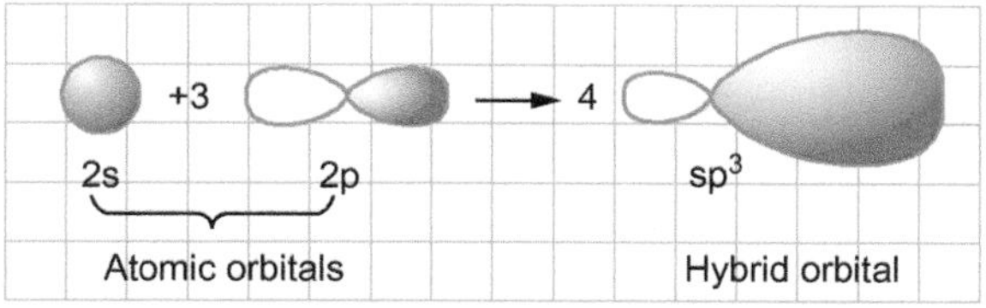

Combination of s and p atomic orbitals to give an sp^3 hybrid orbitals.

- π-Bonding in T_d
  In case: ML_4
  - We obtain the following set of characters for the representations for Γ_σ, and Γ_π (Table 5.9).

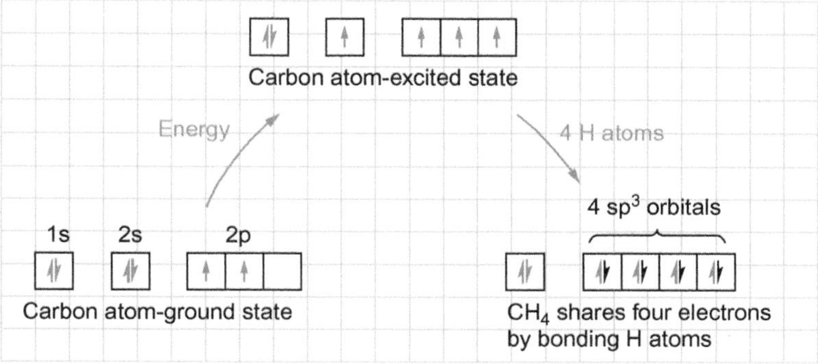

SCHEME 5.9 sp^3-orbital hybridization diagram for a CH_4 molecule.

TABLE 5.9 Reducible and Irreducible Representations of σ- and π-Bonding for the T_d Point Group

T_d	E	$8C_3$	$3C_2$	$6S_4$	$6\sigma_d$		
A_1	1	1	1	1	1	–	$x^2+y^2+z^2$
A_2	1	1	1	−1	−1	–	–
E	2	−1	2	0	0	–	$2z^2-x^2-y^2, x^2-y^2$
T_1	3	0	−1	1	−1	(R_x, R_y, R_z)	–
T_2	3	0	−1	−1	1	(x, y, z)	(xy, xz, yz)
Γ_σ	4	1	0	0	2	$=A_1+T_2$	
Γ_π	8	−1	0	0	0	$=E+T_1+T_2$	

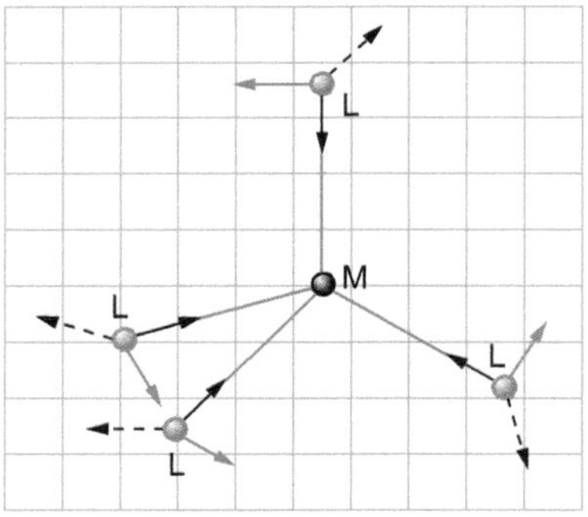

This can be reduced to:

$$\Gamma_\pi = E + T_1 + T_2$$

$$\Gamma_\sigma = A_1 + T_2$$

- The character table shows:

$$E \xrightarrow{\text{represents}} (d_{z^2}, d_{x^2-y^2})$$

$$T_1 \xrightarrow{\text{represents}} \text{none}$$

$$T_2 \xrightarrow{\text{represents}} (p_x, p_y, p_z), (d_{xy}, d_{xz}, d_{yz})$$

- There are two possible sets of hybrid orbitals: p^3d^2 or d^5.

- There are three situations:
  - A pure d^5 set for π bonding will form if the metal uses pure sp^3 hybrids for σ bonding.
  - A pure p^3d^2 set for π bonding will form if the metal uses pure sd^3 hybrids for σ bonding.
  - Intermediate cases in which σ orbitals are mixture of the sp^3 and sd^3, and the π orbitals are a complementary mixture of sp^3 and d^5.

Octahedral Hybridization

How can you construct the appropriate hybrid orbital to describe how the σ- and π-bonding in molecules belong to the O_h point group?

How do the orbitals hybridize in the molecular bonding of SF_6?

- σ-Bonding in O_h

  Examples: SF_6, PF_6^-, $Fe(CN)_6^{-3}$, or ML_4

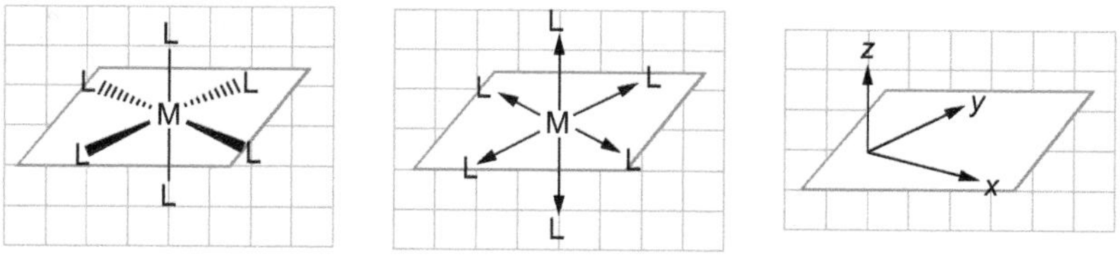

- ○ We obtain the following set of characters for the reducible representation Γ_σ:

O_h	E	$8C_3$	$6C_2$	$6C_4$	$3C_2$	i	$6S_4$	$8S_6$	$3\sigma_h$	$6\sigma_d$
Γ_σ-Reducible	6	0	0	2	2	0	0	0	4	2

Γ_σ can be reduced to:
$\Gamma_\sigma = A_{1g} + E_g + T_{1u}$ (Table 5.10)
The possible hybridization: sp^3d^2

- The structure of SF_6 can be described by mixing the 3s-, three 3p- and two 3d-orbitals, that is, sp^3d^2 hybridization (Scheme 5.10).
- π-Bonding in O_h: ML_6

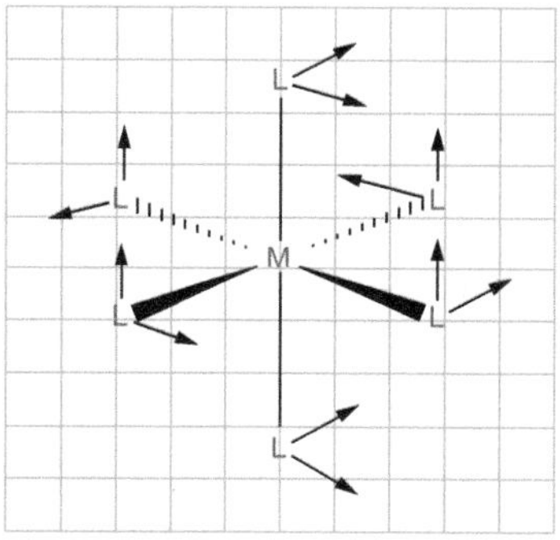

- We obtain the following set of characters for the reducible representations, Γ_σ and Γ_π (Table 5.11).

TABLE 5.10 Reducible and Irreducible Representations of σ-Bonding for the O_h Point Group

O_h	E	$8C_3$	$6C_2$	$6C_4$	$3C_2$	i	$6S_4$	$8S_6$	$3\sigma_h$	$6\sigma_d$		
A_{1g}	1	1	1	1	1	1	1	1	1	1	–	$x^2+y^2+z^2$
A_{2g}	1	1	−1	−1	1	1	−1	1	1	−1	–	–
E_g	2	−1	0	0	2	2	0	−1	2	0	–	$2z^2-x^2-y^2$, x^2-y^2
T_{1g}	3	0	−1	1	−1	3	1	0	−1	−1	(R_x, R_y, R_z)	–
T_{2g}	3	0	1	−1	−1	3	−1	0	−1	1		(xz, yz, xy)
A_{1u}	1	1	1	1	1	−1	−1	−1	−1	−1	–	–
A_{2u}	1	1	−1	−1	1	−1	1	−1	−1	1	–	–
E_u	2	−1	0	0	2	−2	0	1	−2	0	–	–
T_{1u}	3	0	−1	1	−1	−3	−1	0	1	1	(x, y, z)	–
T_{2u}	3	0	1	−1	−1	−3	1	0	1	−1	–	–
Γ_σ	6	0	0	2	2	0	0	0	4	2	$=A_{1g}+E_g+T_{1u}$	

SCHEME 5.10 sp^3d^2-orbital hybridization diagram for an SF_6 molecule.

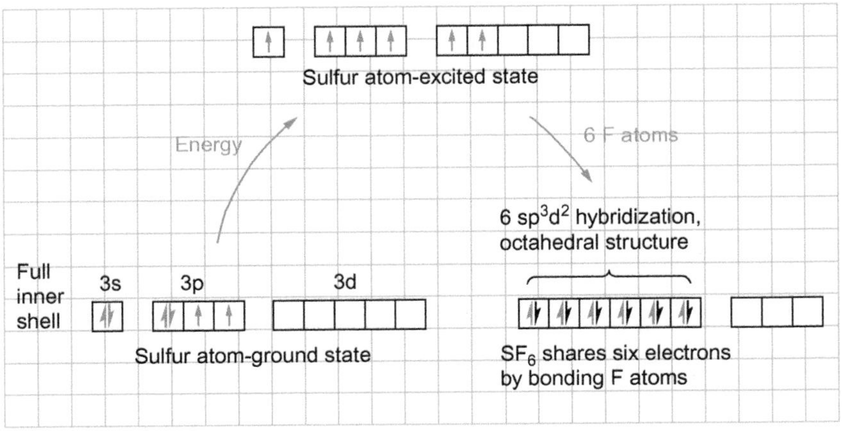

TABLE 5.11 Reducible and Irreducible Representations of σ- and π-Bonding for the O_h Point Group

O_h	E	$8C_3$	$6C_2$	$6C_4$	$3C_2$	i	$6S_4$	$8S_6$	$3\sigma_h$	$6\sigma_d$		
A_{1g}	1	1	1	1	1	1	1	1	1	1	–	$x^2+y^2+z^2$
A_{2g}	1	1	−1	−1	1	1	−1	1	1	−1	–	–
E_g	2	−1	0	0	2	2	0	−1	2	0	–	$2z^2-x^2-y^2$, x^2-y^2
T_{1g}	**3**	**0**	**−1**	**1**	**−1**	**3**	**1**	**0**	**−1**	**−1**	(R_x, R_y, R_z)	–
T_{2g}	**3**	**0**	**1**	**−1**	**−1**	**3**	**−1**	**0**	**−1**	**1**		(xz, yz, xy)
A_{1u}	1	1	1	1	1	−1	−1	−1	−1	−1	–	–
A_{2u}	1	1	−1	−1	1	−1	1	−1	−1	1	–	–
E_u	2	−1	0	0	2	−2	0	1	−2	0	–	–
T_{1u}	**3**	**0**	**−1**	**1**	**−1**	**−3**	**−1**	**0**	**1**	**1**	(x, y, z)	–
T_{2u}	**3**	**0**	**1**	**−1**	**−1**	**−3**	**1**	**0**	**1**	**−1**	–	–
Γ_π	**12**	**0**	**0**	**0**	**−4**	**0**	**0**	**0**	**0**	**0**	$=T_{1g}+T_{2g}+T_{1u}+T_{2u}$	
Γ_σ	6	0	0	2	2	0	0	0	4	2	$=A_{1g}+E_g+T_{1u}$	

Γ_π and Γ_σ can be reduced to:

$$\Gamma_\sigma = A_{1g} + E_{1g} + T_{1u} : sp^3d^2$$

$$\Gamma_\pi = T_{1g} + T_{2g} + T_{1u} + T_{2u}$$

$$T_{1g} : \text{none}$$

$$T_{2g} : (d_{xy}, d_{xz}, d_{yz})$$

$$T_{1u} : (p_x, p_y, p_z) \text{ involved } \sigma\text{-bonding}$$

$$T_{2u} : \text{none}$$

- Thus, only three d-orbitals on atom A of T_{2g} are available for π-bonding (Fig. 5.6).

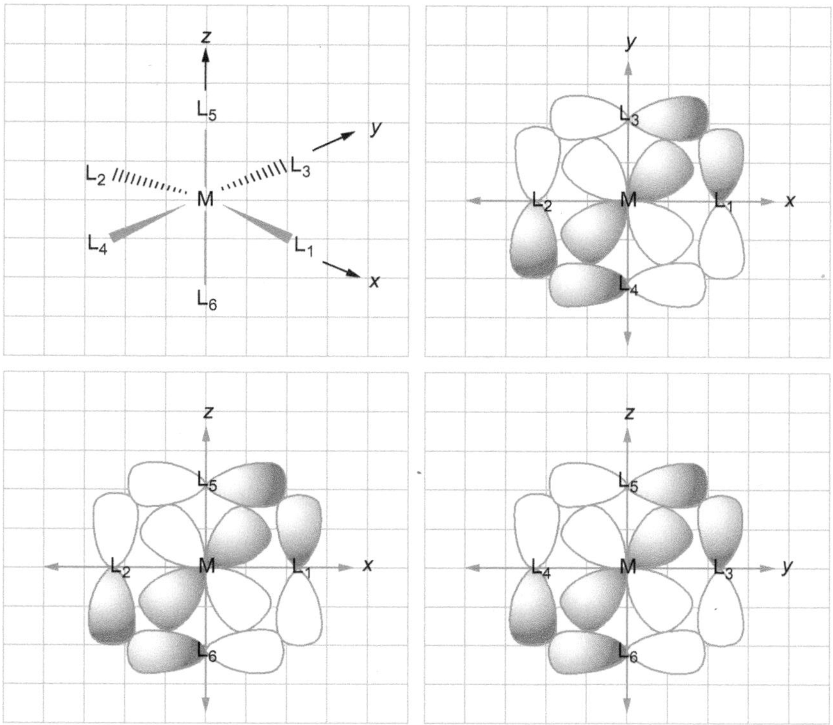

FIG. 5.6 The three d_π—p_π bonds in AB_6.

5.10 HYBRID ORBITALS AS SYMMETRY ADAPTED LINEAR COMBINATION OF ATOMIC ORBITALS (SALC)

What is the obstacle to employing molecular symmetry to understand chemical bonding and molecular dynamics? How can this problem be solved?

What is it meant by Symmetry Adapted Linear Combination (SALC)?

- The problem is to use one or more sets of orbital wave functions to obtain orthonormal linear combinations of them that form bases for irreducible representations of the symmetry group of a molecule.
- Group theory has a process that can be applied to assemble combinations of any basis set (of atomic orbitals in this case) that transform by specified symmetry.
- This process is called the symmetry-adapted linear combination (SALC).
- The formal way to find the functional form of the SALC is by using

$$\hat{P}_\Gamma(\psi_i) = \frac{l}{h}\sum \chi_R \hat{R}_\Gamma(\psi_i)$$

The operator $\hat{P}_\Gamma$ is called a projection operator. This operator shows the degree to which each element of the basis set $(\psi_1, \psi_2, \ldots, \psi_i)$ contributes to Γ.

$\Gamma_{\text{irred.}}$: the particular irreducible representation that are interested in.

l = dimension of irreducible representation Γ

χ_R = character of operation in point group

$\hat{R}_\Gamma$ represents the command to perform that symmetry operation.

The sum (Σ) is taken over all the operations of the point group.

- Such combinations are important because:
  - they represent acceptable solution for the orbital wave functions or the combined functions; and
  - the symmetry properties of the wave functions are precisely defined. This makes building molecular orbitals uncomplicated as it eliminates any combinations that would have zero overlap.

In BF_3, which types of atomic orbitals are suited by symmetry to contribute to the σ-bonding on B? Obtain the SALCs for the σ-bonding hybrid system on B, and express the hybrid wave functions as linear combinations of the unhybridized atomic orbitals.

- Previously, we had a method for finding (by symmetry considerations) which atomic orbitals could be candidate for hybridization to form σ and π bonding sets in various geometries.
  e.g., for BF_3

$$\psi_i = f\left(\psi_s, \psi_{2p_x}, \psi_{2p_y}\right), \quad i = 1,2, \text{ or } 3$$

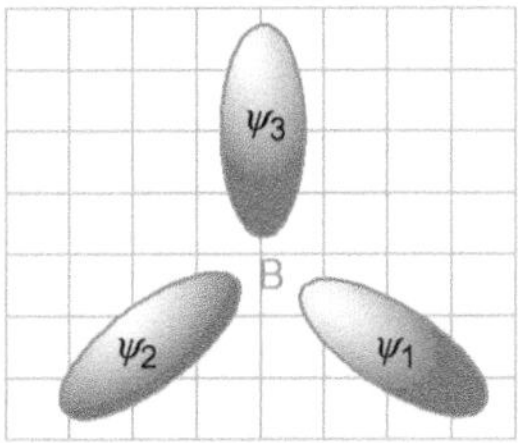

- These ψ_1, ψ_2, and $\psi_3$can be expressed as linear combinations of s, p_x, and p_y atomic orbitals:

$$\text{e.g.,}\ \psi_1 = C\psi_s + C'\psi_{2p_x} + C''\psi_{2p_y}$$

  Now we have to find values of constants C, C', and C''.
- Use symmetry considerations to find how each of s, p_x, and p_y is distributed among ψ_1, ψ_2, and ψ_3. Then take the inverse of the result to find how each ψ_1, ψ_2, and ψ_3 is made up.
- The general form ψ_i will be

$$\psi_i = a\psi_s + b\psi_{2p_x} + c\psi_{2p_y}$$

  where a, b, and c are usually integers, with one or more.
  With normalization, the criterion for correct absolute values of the coefficient is:

$$\int_0^\infty \psi^2 d\tau = 1$$

  having a value of unity (one), one needs to adjust all coefficients by a constant to make this true.

$$\psi_i = N\left(a\psi_s + b\psi_{2p_x} + c\psi_{2p_y}\right)$$

$$\psi_i = Na\psi_s + Nb\psi_{2p_x} + Nc\psi_{2p_y}$$

  Or

$$\psi_1 = C_{11}\psi_s + C_{12}\psi_{2p_x} + C_{13}\psi_{2p_y}$$

$$\psi_2 = C_{21}\psi_s + C_{22}\psi_{2p_x} + C_{23}\psi_{2p_y}$$

  Where ψ_1 and ψ_2 are orthogonal:

$$\int \psi_1 \cdot \psi_2 d\tau = 0$$

- Now to find values for Cs for BF_3, we have
  $\Gamma_\sigma = A_1' + E'$ (Ch. 5, p. 297)

- The projection operators $\hat{P}_{\Gamma}$: show the degree to which each element of the basis set (ψ_1, ψ_2, and ψ_3) contributes to A'_1 and E' (to s, p_x, p_y)

$$\hat{P}_{\Gamma}(\psi_i) = \frac{l}{h}\sum \chi_R \hat{R}_{\Gamma}(\psi_i)$$

i.e., carrying out $\hat{R}(\psi_1)$ and $\Gamma = A'_1$
and one is applying each operation to the orbital wave function (Fig. 5.7).
For A'_1; using $(\Gamma_\sigma = A'_1 + E')$
The x-axis is identical to C_2, A'_1 from character table

D_{3h}	E	$2C_3$	$3C_2$	σ_h	$2S_3$	$3\sigma_v$	
A'_1	1	1	1	1	1	1	x^2+y^2, z^2

$$\hat{P}_{A'_1}(\psi_1) = \frac{1}{12}\left\{\begin{array}{l}(1*\hat{E}*\psi_1) + (1*\hat{C}_3*\psi_1) + \left(1*\hat{C}_3^2*\psi_1\right) + (1*\hat{C}_2*\psi_1) \\ +\left(1*\hat{C}'_2*\psi_1\right) + \left(1*\hat{C}''_2*\psi_1\right) + (1*\hat{\sigma}_h*\psi_1) + (1*\hat{S}_3*\psi_1) \\ +\left(1*\hat{S}_3^2*\psi_1\right) + (1*\hat{\sigma}_v*\psi_1) + (1*\hat{\sigma}'_v*\psi_1) + (1*\hat{\sigma}''_v*\psi_1)\end{array}\right\}$$

$$\hat{P}_{A'_1}(\psi_1) = \frac{1}{12}[\psi_1 + \psi_2 + \psi_3 + \psi_1 + \psi_3 + \psi_2 + \psi_1 + \psi_2 + \psi_3 + \psi_1 + \psi_3 + \psi_2]$$

$$\hat{P}_{A'_1}(\psi_1) = \frac{1}{3}[\psi_1 + \psi_2 + \psi_3]$$

$$\frac{1}{3} \rightarrow \text{is often} = N^2$$

$$\psi_1 + \psi_2 + \psi_3 \rightarrow \text{the coefficients are characters } (\chi) \text{ of } E, C_3, C_3^2 \text{ for } A'_1$$

- Check for normalization:

$$\int \psi^2 d\tau = 1$$

$$\int \psi_1 \psi_2 d\tau = 0$$

$$N^2 \int (\psi_1 + \psi_2 + \psi_3)^2 d\tau = 1$$

$$\frac{1}{N^2} = \int \left(\psi_1^2 + \psi_2^2 + \psi_3^2 + 2\psi_1\psi_2 + 2\psi_1\psi_3 + 2\psi_2\psi_3\right) d\tau$$

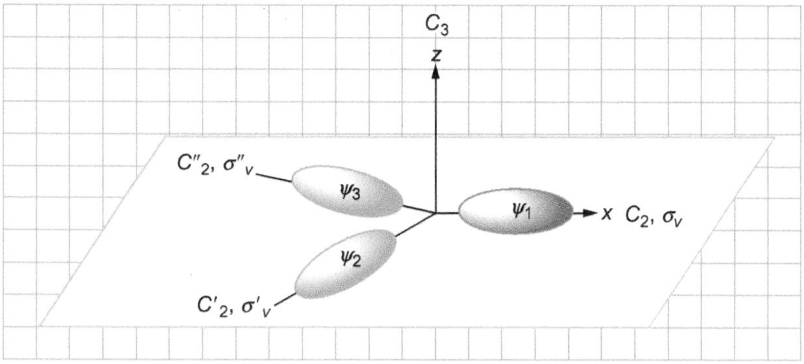

FIG. 5.7 The three molecular orbitals bases set for σ-molecular orbitals of BCl_3.

$$\frac{1}{N^2} = (1+1+1+0+0+0)$$

$$\frac{1}{N^2} = 3$$

$$N = \frac{1}{\sqrt{3}}$$

So the SALC for A'_1 is $\frac{1}{\sqrt{3}}(\psi_1 + \psi_2 + \psi_3)$

- The SALC for A'_2 we get: $0\psi_1 + 0\psi_2 + 0\psi_3$ (We find that $\hat{P}_{A'_2}(\psi_1) = 0$, because Γ_σ does not include A'_2)
  For E' (because $\Gamma_\sigma = A'_1 + E'$); using

$$\hat{P}_\Gamma(\psi_i) = \frac{l}{h}\sum \chi_R \hat{R}_\Gamma(\psi_i)$$

D_{3h}	E	$2C_3$	$3C_2$	σ_h	$2S_3$	$3\sigma_v$		
E'	2	−1	0	2	−1	0	(x, y)	$x^2 - y^2, xy$

$$\hat{P}_{E'}(\psi_1) = \frac{1}{\sqrt{6}}[2\psi_1 - \psi_2 - \psi_3]$$

E' is a 2D representation, E'_a, and E'_b
However, this is only one of the two functions which transform together as E'.

- How can we find the other? There are three methods:
1. Trial and errors need a lot of experience.
2. Use diagonal elements of the set of matrices, which make up E′ in 3D, rather the just working with their characters.

$$\begin{array}{ccccc} E & C_3 & & C_2 & C'_2 \\ \begin{bmatrix} 1 & 0 \\ 0 & 1 \end{bmatrix} & \begin{bmatrix} -1/2 & 0 \\ 0 & -1/2 \end{bmatrix} & but & \begin{bmatrix} 1 & 0 \\ 0 & -1 \end{bmatrix} & \begin{bmatrix} 0 & 1 \\ 1 & 0 \end{bmatrix} \\ 2 & -1 & & 0 & 0 \end{array}$$

The disadvantage is that we need to work out all the 2×2 matrices.

3. Pass to the pure rotational subgroups of D_{3h} (this is C_3, Table 5.12) where the characters of each operation in each representation are already split into two parts.
   The disadvantage we have to use complex numbers.
   - Note: for C_n:

$$\underset{[v']}{\begin{bmatrix} x' \\ y' \\ z' \end{bmatrix}} = \underset{[C_n]}{\begin{bmatrix} \cos\theta & \sin\theta & 0 \\ -\sin\theta & \cos\theta & 0 \\ 0 & 0 & 1 \end{bmatrix}} \underset{[v]}{\begin{bmatrix} x \\ y \\ z \end{bmatrix}}, \text{ and}$$

TABLE 5.12 Character Table for the C_3 Point Group

	ψ_1	ψ_2	ψ_3
C_3	E	C_3	C_3^2
A	1	1	1
E	$\left\{\begin{matrix}1\\1\end{matrix}\right.$	$\begin{matrix}\varepsilon\\\varepsilon^*\end{matrix}$	$\left.\begin{matrix}\varepsilon^*\\\varepsilon\end{matrix}\right\}$

$$C_3$$
$$\begin{bmatrix} \cos\frac{2\pi}{3} & \sin\frac{2\pi}{3} \\ -\sin\frac{2\pi}{3} & \cos\frac{2\pi}{3} \end{bmatrix}$$
$$\text{Trace} = \chi_{(C_3)} = 2\cos\theta$$

$\chi_{(C_3)} + \chi_{(C_3^2)} = \varepsilon + \varepsilon^*$ (character Table 5.12)

$$\text{If}\,\chi_{(C_3)} = \varepsilon \text{ then}\,\chi_{(C_3^2)} = \varepsilon^*$$

$$\text{If}\,\chi_{(C_3)} = \varepsilon^* \text{ then}\,\chi_{(C_3^2)} = \varepsilon$$

$$\varepsilon = e^{i\theta} = \cos\theta + i\sin\theta$$

$$\varepsilon^* = e^{-i\theta} = \cos\theta - i\sin\theta$$

$$e^{i\theta} + e^{-i\theta} = 2\cos\theta$$

$$e^{i\theta} - e^{-i\theta} = 2i\sin\theta$$

$$\varepsilon + \varepsilon^* = 2\cos\theta = 2\cos\frac{2\pi}{3} = -1$$

$$\varepsilon - \varepsilon^* = 2i\sin\theta = \sqrt{-3}$$

- For E' using Table 5.10, and

$$\hat{P}_\Gamma(\psi_i) = \frac{l}{h}\sum \chi_R \hat{R}_\Gamma(\psi_i)$$

D_{3h}	E	$2C_3$	$3C_2$	σ_h	$2S_3$	$3\sigma_v$		
E'	2	−1	0	2	−1	0	(x, y)	x^2-y^2, xy

$$\hat{P}_{E'_a}(\psi_1) = \left(1*\hat{E}*\psi_1\right) + \left(\varepsilon*\hat{C}_3*\psi_1\right) + \left(\varepsilon^**\hat{C}_3^2*\psi_1\right)$$

$$\hat{P}_{E'_a}(\psi_1) = \psi_1 + \varepsilon\psi_2 + \varepsilon^*\psi_3$$

$$\hat{P}_{E'_b}(\psi_1) = \left(1*\hat{E}*\psi_1\right) + \left(\varepsilon^**\hat{C}_3*\psi_1\right) + \left(\varepsilon*\hat{C}_3^2*\psi_1\right)$$

$$\hat{P}_{E'_b}(\psi_1) = \psi_1 + \varepsilon^*\psi_2 + \varepsilon\psi_3$$

- Now eliminate the imaginary coefficient by taking linear combinations of E'_a and E'_b:
  - Add the above wave functions:

$$2\psi_1 + (\varepsilon+\varepsilon^*)\,\psi_2 + (\varepsilon+\varepsilon^*)\,\psi_3$$

Thus $2\psi_1 - \psi_2 - \psi_3$ is the result of this linear combination.

and $N = \frac{1}{\sqrt{6}}$

  - Subtract the wave functions

$$0\psi_1 + (\varepsilon-\varepsilon^*)\psi_2 + (\varepsilon-\varepsilon^*)\psi_3$$

$$\varepsilon - \varepsilon^* = 2i\sin\theta = \sqrt{-3}$$

$$\sqrt{-3}\psi_2 - \sqrt{-3}\psi_3 = \sqrt{-3}(\psi_2 - \psi_3)$$

So the other linear combination is of the form $\psi_2 - \psi_3$; with normalization:

$$N^2\int(\psi_2-\psi_3)^2\delta\tau = 1$$

$$\int \left(\psi_2^2+\psi_3^2-2\psi_2\psi_3\right)\delta\tau=\frac{1}{N^2}$$

$$1+1-0=\frac{1}{N^2}$$

$$N=\frac{1}{\sqrt{2}}$$

So the other linear combination:

$$\frac{1}{\sqrt{2}}(\psi_2-\psi_3)$$

- Summary: determination of the SALCs' coefficient of $\psi_1,\psi_2,$ and ψ_3
  From previous work we have a set of SALCs:

1. Each is of the form:

$$a\psi_1+b\psi_2+c\,\psi_3$$

   where a, b, and c are the characters for E, C_3, and C_3^2 in the character table.
2. For E' representation we took linear combination to get SALCs with real coefficients.
3. The SALCs are mutually orthogonal
4. The SALCs are normalized, usually one can write (before normalization):

$$d\psi_1+e\psi_2+f\,\psi_3 \text{ so that } N=\sqrt{1/(d^2+e^2+f^2)}$$

$$A_1'\mapsto\psi_s \text{ or } s=\frac{1}{\sqrt{3}}\psi_1+\frac{1}{\sqrt{3}}\psi_2+\frac{1}{\sqrt{3}}\psi_3$$

$$E_a'\mapsto\psi_{Px} \text{ or } p_x=\frac{2}{\sqrt{6}}\psi_1-\frac{1}{\sqrt{6}}\psi_2-\frac{1}{\sqrt{6}}\psi_3$$

$$E_b'\mapsto\psi_{Py} \text{ or } p_y=\frac{1}{\sqrt{2}}\psi_2-\frac{1}{\sqrt{2}}\psi_3$$

   Note, ψ_1 and p_x are collinear and contribute to one another
   ψ_1 and p_y are perpendicular and not contribute to one another.
5. we want the inverse relationships:

$$\psi_1=C_{11}s+C_{12}p_x+C_{13}p_y$$

$$\psi_2=C_{21}s+C_{22}p_x+C_{23}p_y$$

$$\psi_3=C_{31}s+C_{32}p_x+C_{33}p_y$$

- Write the SALCs in matrix form:

$$\begin{bmatrix} s \\ p_x \\ p_y \end{bmatrix} = \begin{bmatrix} 1/\sqrt{3} & 1/\sqrt{3} & 1/\sqrt{3} \\ 2/\sqrt{6} & -1/\sqrt{6} & -1/\sqrt{6} \\ 0 & 1/\sqrt{2} & -1/\sqrt{2} \end{bmatrix} \begin{bmatrix} \psi_1 \\ \psi_2 \\ \psi_3 \end{bmatrix} \rightarrow \text{take the inverse}$$

(TRANSPOSE) of matrix coefficients $[a_{ij}\rightarrow a_{ji}]$

$$\begin{bmatrix} \psi_1 \\ \psi_2 \\ \psi_3 \end{bmatrix} = \begin{bmatrix} 1/\sqrt{3} & 2/\sqrt{6} & 0 \\ 1/\sqrt{3} & -1/\sqrt{6} & 1/\sqrt{2} \\ 1/\sqrt{3} & -1/\sqrt{6} & -1/\sqrt{2} \end{bmatrix} \begin{bmatrix} s \\ p_x \\ p_y \end{bmatrix} \rightarrow \text{write columns of}$$

- The SALCs' coefficient as raw of new matrix:

$$\psi_1 = \frac{1}{\sqrt{3}} s + \frac{2}{\sqrt{6}} p_X$$
$$\psi_2 = \frac{1}{\sqrt{3}} s - \frac{1}{\sqrt{6}} p_x + \frac{1}{\sqrt{2}} p_y$$
$$\psi_3 = \frac{1}{\sqrt{3}} s - \frac{1}{\sqrt{6}} p_X - \frac{1}{\sqrt{2}} p_Y$$

In tetrachloroplanate (II), which types of AOs are suited by symmetry to contribute to the σ-bonding, $\pi_{\perp}$-bonding, and $\pi_{\parallel}$-bonding hybrids on Pt.?

Obtain the SALCs for the σ-bonding hybrid system on Pt, and express the hybrid wave functions as linear combinations of the unhybridized AOs.

- A set of four vectors could represent the σ-bonds of $PtCl_4^{2-}$ (Fig. 5.8).
- We obtain the following set of characters for the reducible representation Γ_σ:

D_{4h}	E	$2C_4$	C_2	$2C'_2$	$2C''_2$	i	$4S_4$	σ_h	$2\sigma_v$	$2\sigma_d$
Γ_σ-Reducible	4	0	0	0	2	0	0	4	0	2

- Reduction using: $\sigma_i = \frac{1}{h}\sum i \cdot \chi_R \cdot \chi_i$, and Table 5.13.

$$\Gamma_\sigma = A_{1g} + B_{2g} + E_u$$

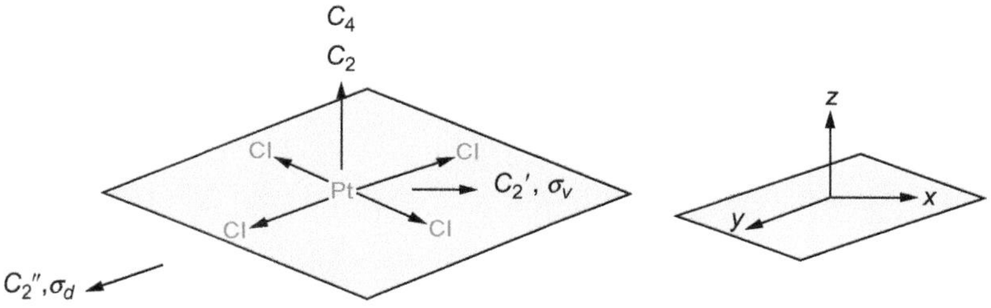

FIG. 5.8 The four vectors bases set for σ-molecular orbitals of $PtCl_4^{2-}$.

TABLE 5.13 Reducible and Irreducible Representations of σ Bonding for the D_{4h} Point Group

D_{4h}	E	$2C_4$	C_2	$2C'_2$	$2C''_2$	i	$4S_4$	σ_{hh}	$2\sigma_v$	$2\sigma_d$		
A_{1g}	1	1	1	1	1	1	1	1	1	1	–	x^2+y^2, z^2
A_{2g}	1	1	1	−1	−1	1	1	1	−1	−1	R_z	–
B_{1g}	1	−1	1	1	−1	1	−1	1	1	−1	–	x^2-y^2
B_{2g}	1	−1	1	−1	1	1	−1	1	−1	1	–	xy
E_g	2	0	−2	0	0	2	0	−2	0	0	(R_x, R_y)	(xz, yz)
A_{1u}	1	1	1	1	1	−1	−1	−1	−1	−1	–	–
A_{2u}	1	1	1	−1	−1	−1	−1	−1	1	1	z	–
B_{1u}	1	−1	1	1	−1	−1	1	−1	−1	1	–	–
B_{2u}	1	−1	1	−1	1	−1	1	−1	1	−1	–	–
E_u	2	0	−2	0	0	−2	0	2	0	0	(x, y)	–
Γ_σ	4	0	0	0	2	0	0	4	0	2	$=A_{1g}+B_{2g}+E_u$	

Suitable orbitals:	A_{1g}:	s or dz^2
	B_{2g}:	d_{xy}
	E_u:	P_x and P_y together
The set of hybrid orbitals:	dsp^2 or d^2p^2	

- $\pi_{\|}$-bonding (Fig. 5.9):
  We obtain the following set of characters for the reducible representation $\Gamma_{\pi_{\|}}$:

D_{4h}	E	$2C_4$	C_2	$2C'_2$	$2C''_2$	i	$4S_4$	σ_h	$2\sigma_v$	$2\sigma_d$
$\Gamma_{\pi_{\|}}$	4	0	0	0	−2	0	0	4	0	−2

Reduction using: $\sigma_i = \frac{1}{h}\sum i \cdot \chi_R \cdot \chi_i$, and Table 5.14.

$$\Gamma_{\pi_{\|}} = A_{2g} + B_{1g} + E_u$$

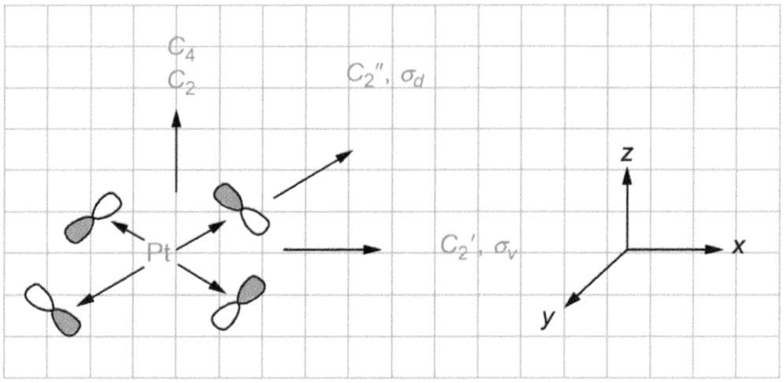

FIG. 5.9 The four atomic orbitals bases set for π-molecular orbitals of $PtCl^{2-}$.

TABLE 5.14 Reducible and Irreducible Representations of $\pi_{\|}$ Bonding for the D_{4h} Point Group

D_{4h}	E	$2C_4$	C_2	$2C'_2$	$2C''_2$	i	$4S_4$	σ_h	$2\sigma_v$	$2\sigma_d$		
A_{1g}	1	1	1	1	1	1	1	1	1	1	–	x^2+y^2, z^2
A_{2g}	1	1	1	−1	−1	1	1	1	−1	−1	R_z	–
B_{1g}	1	−1	1	1	−1	1	−1	1	1	−1	–	x^2-y^2
B_{2g}	1	−1	1	−1	1	1	−1	1	−1	1	–	xy
E_g	2	0	−2	0	0	2	0	−2	0	0	(R_x, R_y)	(xz, yz)
A_{1u}	1	1	1	1	1	−1	−1	−1	−1	−1	–	–
A_{2u}	1	1	1	−1	−1	−1	−1	−1	1	1	z	–
B_{1u}	1	−1	1	1	−1	−1	1	−1	−1	1	–	–
B_{2u}	1	−1	1	−1	1	−1	1	−1	1	−1	–	–
E_u	2	0	−2	0	0	−2	0	2	0	0	(x, y)	–
$\Gamma_{\pi_{\|}}$	4	0	0	0	−2	0	0	4	0	−2	$=A_{2g}+B_{1g}+E_u$	

Suitable orbitals: A_{2g}: None
B_{1g}: $d_{x^2-y^2}$
E_u: P_x, P_y
P_x, P_y are involved in σ-bonding

- $\pi_\perp$-bonding (Fig. 5.10):
We obtain the following set of characters for the reducible representation $\Gamma_{\pi\perp}$:

D_{4h}	E	$2C_4$	C_2	$2C'_2$	$2C''_2$	i	$4S_4$	σ_h	$2\sigma_v$	$2\sigma_d$
$\Gamma_{\pi\perp}$	4	0	0	0	−2	0	0	−4	0	2

Reduction using: $\sigma_i = \frac{1}{h}\sum i \cdot \chi_R \cdot \chi_i$, and Table 5.15.

$$\Gamma_{\pi\perp} = E_g + A_{2u} + B_{1u}$$

Suitable orbitals: E_g: d_{xz}, d_{yz}
A_{2u}: P_z
B_{1u}: None

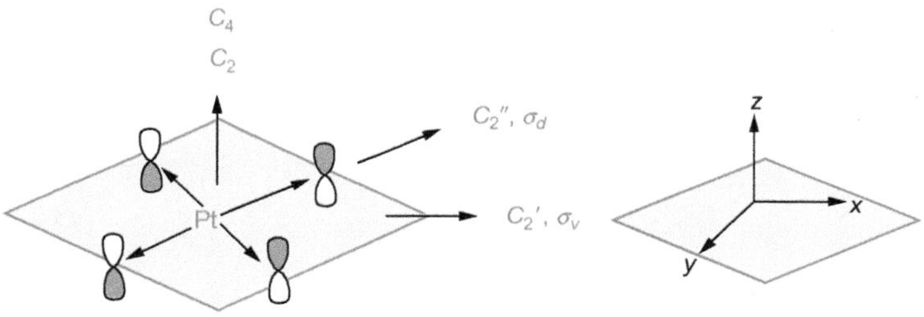

FIG. 5.10 The four atomic orbitals bases set for π-molecular orbitals of $PtCl_4^{2-}$.

TABLE 5.15 Reducible and Irreducible Representations of $\pi_\perp$ Bonding for the D_{4h} Point Group

D_{4h}	E	$2C_4$	C_2	$2C'_2$	$2C''_2$	i	$4S_4$	σ_h	$2\sigma_v$	$2\sigma_d$		
A_{1g}	1	1	1	1	1	1	1	1	1	1	–	x^2+y^2, z^2
A_{2g}	1	1	1	−1	−1	1	1	1	−1	−1	R_z	–
B_{1g}	1	−1	1	1	−1	1	−1	1	1	−1	–	x^2-y^2
B_{2g}	1	−1	1	−1	1	1	−1	1	−1	1	–	xy
E_g	2	0	−2	0	0	2	0	−2	0	0	(R_x, R_y)	(xz, yz)
A_{1u}	1	1	1	1	1	−1	−1	−1	−1	−1	–	–
A_{2u}	1	1	1	−1	−1	−1	−1	−1	1	1	z	–
B_{1u}	1	−1	1	1	−1	−1	1	−1	−1	1	–	–
B_{2u}	1	−1	1	−1	1	−1	1	−1	1	−1	–	–
E_u	2	0	−2	0	0	−2	0	2	0	0	(x, y)	–
$\Gamma_{\pi\perp}$	4	0	0	0	−2	0	0	−4	0	2	$=E_g+A_{2u}+B_{1u}$	

Summary:

$$\Gamma_{\pi\perp} = A_{2u} + B_{2u} + E_g \xrightarrow{\text{represent}} p_z, (d_{xz}, d_{yz}) \xrightarrow{\text{form}} \text{pd}^2$$

$$\Gamma_{\pi_{\|}} = A_{2g} + B_{2g} + E_u \xrightarrow{\text{represent}} d_{xy}, (p_x, p_y) \xrightarrow{\text{form}} \text{dp}^2$$

$$\Gamma_\sigma = A_{1g} + B_{1g} + E_u \xrightarrow{\text{represent}} (s, d_{x^2-y^2}, p_x, p_y) \text{ or } (d_{z^2}, d_{x^2-y^2}, p_x, p_y) \xrightarrow{\text{form}} \text{sp}^2\text{d or p}^2\text{d}^2$$

- The SALCs for the σ-bonding hybrid system on Pt (Fig. 5.11):

$$\Gamma_\sigma = A_{1g} + B_{2g} + E_u$$

- By using the projection operator and the representation of A_{1g}, B_{2g}, and E_u in D_{3h} (Table 5.16, Fig. 5.11), and:

$$\hat{P}_\Gamma(\psi_i) = \frac{l}{h}\sum \chi_R \hat{R}_\Gamma(\psi_i)$$

$$\hat{P}_{A1g}(\psi_1) = \frac{1}{16}(4\psi_1 + 4\psi_2 + 4\psi_3 + 4\psi_4)$$

- Normalization:

$$N^2 \int (\psi_1 + \psi_2 + \psi_3 + \psi_4)^2 = 1$$

$N = 1/2$

$$\hat{P}_{A_{1g}}(\psi_1) = \frac{1}{2}(\psi_1 + \psi_2 + \psi_3 + \psi_4)$$

Similarly

$$\hat{P}_{B_{2g}}(\psi_1) = \frac{1}{2}(\psi_1 - \psi_2 + \psi_3 - \psi_4)$$

$$\hat{P}_{E_u}(\psi_1) = \frac{1}{\sqrt{2}}(\psi_1 - \psi_3)$$

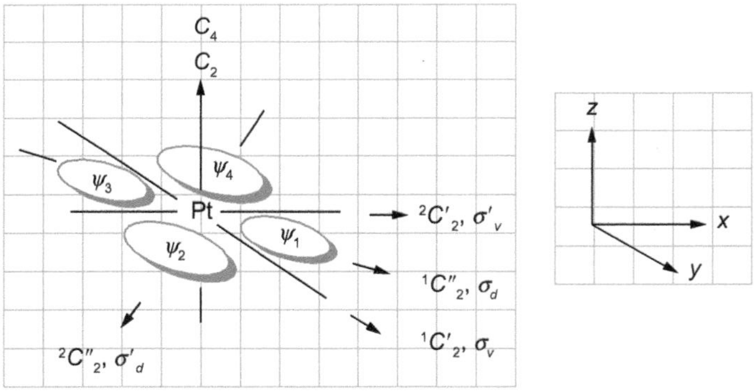

FIG. 5.11 The four molecular orbitals bases set for σ-molecular orbitals of $PtCl_4^{2-}$.

TABLE 5.16 The Product of the Operation $\hat{R}_\Gamma$, and the Representations of A_{1g}, B_{2g}, and E_u in D_{3h}.

D_{4h}	E	C_4	C'_4	C_2	$^1C'_2$	$^2C'_2$	$^1C''_2$	$^2C''_2$	i	$^1S_4$	$^2S_4$	σ_h	σ_v	σ'_v	σ_d	σ'_d
$\hat{R}_\Gamma(\psi_1)$	ψ_1	ψ_2	ψ_4	ψ_3	ψ_2	ψ_4	ψ_1	ψ_3	ψ_3	ψ_2	ψ_4	ψ_1	ψ_2	ψ_4	ψ_1	ψ_3
A_{1g}	1	1	1	1	1	1	1	1	1	1	1	1	1	1	1	1
B_{2g}	1	−1	−1	1	−1	−1	1	1	1	−1	−1	1	−1	−1	1	1
E_u	2	0	0	−2	0	0	0	0	−2	0	0	2	0	0	0	0

- To get the two components of E_u, pass to the pure rotational subgroups of D_{4h}, E, C_4, C'_4, and C_2, where the characters of each operation in each representation are already split into two parts.

  For C_n:

$$\begin{bmatrix} x' \\ y' \end{bmatrix} = \begin{bmatrix} \cos\theta & \sin\theta \\ -\sin\theta & \cos\theta \end{bmatrix} \begin{bmatrix} x \\ y \end{bmatrix}$$
$$2\cos\theta$$

Where

$$\varepsilon + \varepsilon^* = 2\cos\theta$$

Then

$$\chi(C_n) = 2\cos\theta = \varepsilon + \varepsilon^*$$

$$\varepsilon - \varepsilon^* = 2i\sin\theta$$

$$\theta = 90^\circ \text{ in } C_4,\ \ \theta = 270^\circ \text{ in } C'_4,\ \ \text{and } \theta = 180^\circ \text{ in } C_2$$

Then,

$$\begin{array}{ccccc} & E & C_4 & C'_2 & C_2 \\ \hat{R}_\Gamma(\psi_i) & \psi_1 & \psi_2 & \psi_3 & \psi_4 \end{array}$$

There are two choices: if one component is ε, then the other must be ε^*:

$$E_a = \psi_1 + \varepsilon\psi_2 + \varepsilon\psi_3 + \varepsilon\psi_4$$

$$E_b = \psi_1 + \varepsilon^*\psi_2 + \varepsilon^*\psi_3 + \varepsilon^*\psi_4$$

By addition:

$$\psi_1 + (\varepsilon + \varepsilon^*)\psi_2 + (\varepsilon + \varepsilon^*)\psi_3 + (\varepsilon + \varepsilon^*)\psi_4$$

$$\psi_1 + (2\cos 90^\circ)\psi_2 + (2\cos 270^\circ)\psi_3 + (2\cos 180^\circ)\psi_4$$

$$2\psi_1 - 2\psi_3$$

$$\psi_1 - \psi_3$$

After normalization

$$\hat{P}_{E_u}(\psi_1) = \frac{1}{\sqrt{2}}(\psi_1 - \psi_3)$$

By subtraction:

$$\psi_1 + (\varepsilon - \varepsilon^*)\psi_2 + (\varepsilon - \varepsilon^*)\psi_3 + (\varepsilon - \varepsilon^*)\psi_4$$

$$2i\psi_2 - 2i\psi_4$$

$$\psi_2 - \psi_4$$

After normalization

$$\hat{P}_{Eu}(\psi_1) = \frac{1}{\sqrt{2}}(\psi_2 - \psi_4)$$

- Therefore

$$s = \frac{1}{2}(\psi_1 + \psi_2 + \psi_3 + \psi_4)$$

$$d_{xy} = \frac{1}{2}(\psi_1 - \psi_2 + \psi_3 - \psi_4)$$

$$P_x = \frac{1}{\sqrt{2}}(\psi_1 - \psi_3)$$

$$P_y = \frac{1}{\sqrt{2}}(\psi_2 - \psi_4)$$

- Each is of these forms:

$$a\psi_1 + b\psi_2 + c\psi_3 + d\psi_4$$

where a, b, c and d are the characters for E, C_4, C_2 and C_4^3 in the character table of C_4 (Table 5.17).

$$\begin{bmatrix} s \\ d_{xy} \\ p_x \\ p_y \end{bmatrix} = \begin{bmatrix} \frac{1}{2} & \frac{1}{2} & \frac{1}{2} & \frac{1}{2} \\ \frac{1}{2} & -\frac{1}{2} & \frac{1}{2} & -\frac{1}{2} \\ \frac{1}{\sqrt{2}} & 0 & -\frac{1}{\sqrt{2}} & 0 \\ 0 & \frac{1}{\sqrt{2}} & 0 & -\frac{1}{\sqrt{2}} \end{bmatrix} \begin{bmatrix} \psi_1 \\ \psi_2 \\ \psi_3 \\ \psi_4 \end{bmatrix}$$

$$\begin{bmatrix} \psi_1 \\ \psi_2 \\ \psi_3 \\ \psi_4 \end{bmatrix} = \begin{bmatrix} \frac{1}{2} & \frac{1}{2} & \frac{1}{\sqrt{2}} & 0 \\ \frac{1}{2} & -\frac{1}{2} & 0 & \frac{1}{\sqrt{2}} \\ \frac{1}{2} & \frac{1}{2} & -\frac{1}{\sqrt{2}} & 0 \\ \frac{1}{2} & -\frac{1}{2} & 0 & -\frac{1}{\sqrt{2}} \end{bmatrix} \begin{bmatrix} s \\ d_{xy} \\ p_x \\ p_y \end{bmatrix}$$

$$\psi_1 = s + \frac{1}{2}d_{xy} + \frac{1}{\sqrt{2}}p_x$$

$$\psi_2 = s - \frac{1}{2}d_{xy} + \frac{1}{\sqrt{2}}p_y$$

$$\psi_3 = s + \frac{1}{2}d_{xy} - \frac{1}{\sqrt{2}}p_x$$

$$\psi_4 = s - \frac{1}{2}d_{xy} - \frac{1}{\sqrt{2}}p_x$$

5.11 MOLECULAR WAVE FUNCTION AS SYMMETRY ADAPTED LINEAR COMBINATION OF ATOMIC ORBITALS (SALC)

What are the molecular π-orbitals wave functions of the nitrite anion?

- The valence atomic orbitals of a given molecule are used as a basis set for reducible representation.
- Fig. 5.12 illustrates the three π-orbitals of the nitrite anion.
- NO_2^- belongs to C_{2v}

TABLE 5.17 Character Table for the C_4 Point Group

	ψ_1	ψ_2	ψ_3	ψ_4
C_4	E	C_4	C_2	C_4^3
A	1	1	1	1
B	1	−1	1	−1
E	$\left\{\begin{matrix}1\\1\end{matrix}\right.$	$\begin{matrix}i\\-i\end{matrix}$	$\begin{matrix}-1\\-1\end{matrix}$	$\left.\begin{matrix}-i\\i\end{matrix}\right\}$

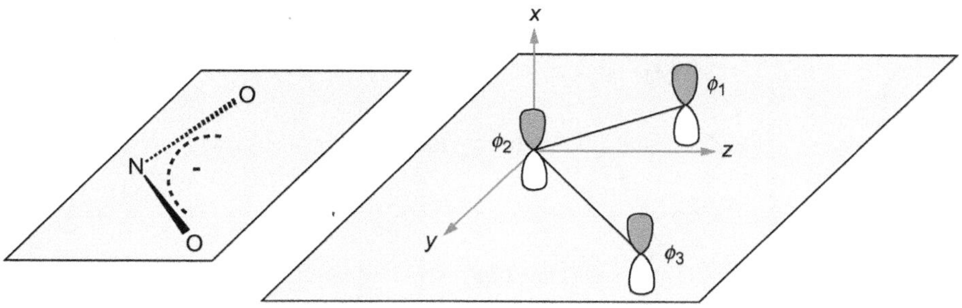

FIG. 5.12 The three atomic orbitals bases set for π-molecular orbitals of the nitrite anion.

The total representations of Γ_π:

C_{2v}	E	C_2	$\sigma_v(xz)$	$\sigma_v(yz)$
Γ_π	3	−1	1	−3

We factor the total representation using
$\sigma_i = \frac{1}{h}\sum i \cdot \chi_R \cdot \chi_i$ (Table 5.18).
Gives: $\Gamma_\pi = A_2 + 2B_1$
Therefore, A_2 and $2B_1$ from the three atomic orbitals of NO_2^-.

- The projection operator $(\hat{P})$ shows the degree to which each element of the basis set ($\phi_1, \phi_2,$ and ϕ_3) contributes to A_2, B_1, and B_1 (where $\phi_1$is the p-orbital of atom 1):

$$\hat{P}_\Gamma(\psi_i) = \frac{l}{h}\sum \chi_R \hat{R}_\Gamma(\psi_i)$$

i.e., carrying out $\hat{R}_\Gamma(\phi_1)$

- For A_2:

$$\psi(A_2) = \hat{P}(A_2)\phi_1 = \frac{1}{4}[(1)E\phi_1 + (1)C_2\phi_1 + (-1)\sigma_v\phi_1 + (-1)\sigma_v'\phi_1]$$

$$\hat{P}(A_2)\phi_1 = \frac{1}{4}[(1)(1)\phi_1 + (1)(-1)\phi_3 + (-1)(1)\phi_3 + (-1)(-1)\phi_1]$$

  - The factor ¼ shall be ignored; it is the relative values resulting from the operators, and shall be replaced by the absolute values for the normalized wave function.

$$\hat{P}(A_2)\phi_1 = N[\phi_1 - \phi_3]$$

TABLE 5.18 Reducible and Irreducible Representations of π Bonding for the C_{2v} Point Group

C_{2v}	E	C_2	$\sigma_v(xz)$	$\sigma_v(yz)$	–	–
A_1	1	1	1	1	z	x^2, y^2, z^2
A_2	1	1	−1	−1	R_z	xy
B_1	1	−1	1	−1	x, R_y	xz
B_2	1	−1	−1	1	y, R_y	yz
Γ_π	3	−1	1	−3	$=A_2+2B_1$	

- The normalized function:

$$\hat{P}(A_2)\phi_1 = \frac{1}{\sqrt{2}}[\phi_1 - \phi_3]$$

- Since there is no symmetry operation that interchanges the nitrogen orbital with oxygen, the nitrogen must form a separate irreducible representation.

- If we selected ϕ_2 to operate on:

$$\hat{P}(A_2)\phi_2 = \frac{1}{4}[(1)(1)\phi_2 + (1)(-1)\phi_2 + (-1)(1)\phi_2 + (-1)(-1)\phi_2] = 0$$

- Carrying out $\hat{R}_\Gamma(\phi_1)$, for B_1, we know that there are two molecular orbitals with B_1 symmetry (because $\Gamma_\pi = A_2 + 2B_1$):

$$\psi(B_1) = \hat{P}(B_1)\phi_1 = \frac{1}{4}[(1)(1)\phi_1 + (-1)(-1)\phi_3 + (1)(1)\phi_3 + (-1)(-1)\phi_1]$$

$$\psi(B_1) = N(\phi_1 + \phi_3)$$

When we use $\hat{P}(B_1)$ on ϕ_2 the following result is obtained:

$$\psi(B_1) = \hat{P}(B_1)\phi_2$$

$$\psi(B_1) = \frac{1}{4}[(1)(1)\phi_2 + (-1)(-1)\phi_2 + (1)(1)\phi_2 + (-1)(-1)\phi_2] = \phi_2$$

- Since the two B_1 molecular orbitals have similar symmetry, they can mix, so that any linear combination of these two molecular orbitals also has the appropriate symmetry to be one of the two B_1 molecular orbitals, the two molecular orbitals can be:

$$\psi_1 = a\phi_1 + b\phi_2 + a\phi_3$$

$$\psi_2 = a'\phi_1 - b'\phi_2 + a'\phi_3$$

- The magnitude of a and b will depend on the $\phi_1$and ϕ_2 atomic orbitals and their overlap integrals.
- We can now draw pictures of the molecular orbitals:

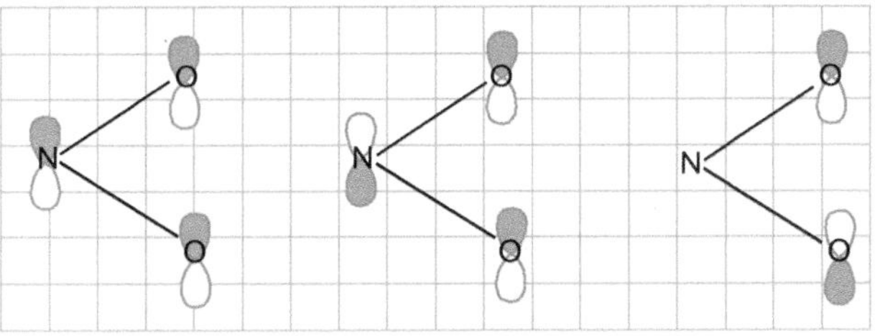

B_1 B_1 A_2

How do we find the total representations and the molecular orbital wave functions of NH_3 molecule?

- NH_3 belongs to the point group C_{3v}

First, we will find the total representation of the atomic orbital that involve in molecular formation (that includes: $2s$(N), $2p_z$(N), $2p_x$(N), $2p_y$(N), and $1s$(3H)) and factor it:

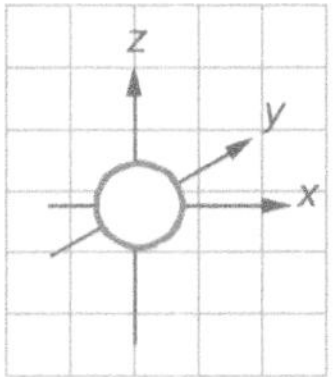

C_{3v}	E	$2C_3$	$3\sigma_v$
$\Gamma_{2s}(N)$	1	1	1

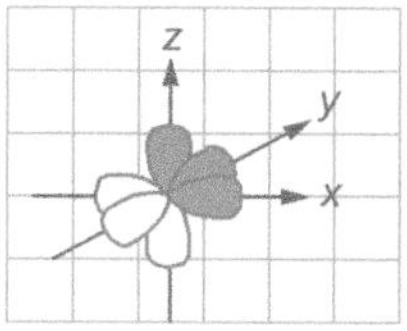

C_{3v}	E	$2C_3$	$3\sigma_v$
$\Gamma_{2p}(N)$	3	0	1

where C_3 for p_N-orbitals:

$$C_3 \begin{bmatrix} x \\ y \\ z \end{bmatrix} = \begin{bmatrix} -\frac{1}{2} & -\frac{\sqrt{3}}{2} & 0 \\ \frac{\sqrt{3}}{2} & -\frac{1}{2} & 0 \\ 0 & 0 & 1 \end{bmatrix} \begin{bmatrix} x \\ y \\ z \end{bmatrix}, \quad \text{trace} = 0$$

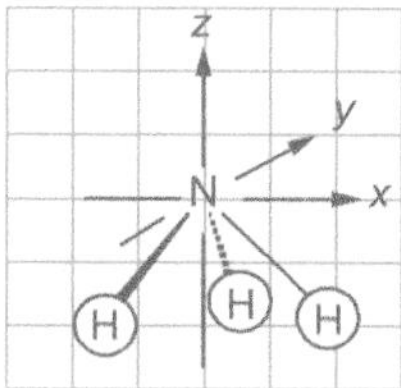

C_{3v}	E	$2C_3$	$3\sigma_v$
$\Gamma_{1s}(3H)$	3	0	1

The total representation:

Total	C_{3v}	E	$2C_3$	$3\sigma_v$
	Γ_{NH_3}	7	1	3

- Factoring the total representation using Table 5.19 and

$$\sigma_i = \frac{1}{h} \sum i \cdot \chi_R \cdot \chi_i$$

$$\sigma_{A_1} = \frac{1}{6}[1 \times 1 \times 7 + 2 \times 1 \times 1 + 3 \times 1 \times 3] = 3$$

$$\sigma_{A_2} = \frac{1}{6}[1 \times 1 \times 7 + 2 \times 1 \times 1 + 3 \times (-1) \times 3] = 0$$

$$\sigma_E = \frac{1}{6}[1 \times 2 \times 7 + 2 \times (-1) \times 1 + 3 \times (0) \times 3] = 2$$

$$\Gamma_{NH_3} = 3A_1 + 2E$$

TABLE 5.19 Reducible and Irreducible Representations of NH_3 for the C_{3v} Point Group

C_{3v}	E	$2C_3$	$3\sigma_v$		
A_1	1	1	1	z	x^2+y^2, z^2
A_2	1	1	−1	R_z	–
E	2	−1	0	(x, y) (R_x, R_y)	(x^2-y^2, xy) (xz, yz)
Γ_{NH_3}	7	1	3	$= 3A_1 + 2E$	

- We will define the atomic orbitals as following:

$$\phi_1 = N_{2s}$$
$$\phi_2 = N_{2p_x}$$
$$\phi_3 = N_{2p_y}$$
$$\phi_4 = N_{2p_z}$$
$$\phi_5 = H_{1s}(\sigma_v \text{ contains } H)$$
$$\phi_6 = H'_{1s}(\sigma'_v \text{ contains } H')$$
$$\phi_7 = H''_{1s}\left(\sigma''_v \text{ contains } H''\right)$$

- Then, we obtain the projection operators $\hat{P}(A_1)$ and $\hat{P}(E)$, there is no need to work with $\hat{P}(A_2)$.
  - In the C_{3v} point group, s and p_z orbitals belong to A_1 representation, while p_x and p_y orbitals belong to E representation (Fig. 5.13).

$$\hat{P}_\Gamma(\psi_i) = \frac{l}{h}\sum \chi_R \hat{R}_\Gamma(\psi_i)$$

- No symmetry operation can transform a hydrogen atom into a nitrogen atom, so we can factor the total representation formed form the hydrogen orbital (3, 0, 1), and find an A_1 and E for the hydrogen and nitrogen atoms separately.
- The projection operator for A_1:
  - Hydrogen atoms (ϕ_5, ϕ_6, and ϕ_7):

    $\hat{P}(A_1)$ on ϕ_5:

$$\psi_1 = \frac{1}{6}\left[1E\phi_5 + 1C_3\phi_5 + 1C_3^2\phi_5 + 1\sigma_v\phi_5 + 1\sigma'_v\phi_5 + 1\sigma''_v\phi_5\right]$$
$$\psi_1 = \frac{1}{6}[\phi_5 + \phi_7 + \phi_6 + \phi_5 + \phi_6 + \phi_7]$$
$$\psi_1 = \frac{1}{6}[2\phi_5 + 2\phi_6 + 2\phi_7]$$

    $\hat{P}(A_1)$ operating on ϕ_6 and ϕ_7 (equivalent atoms) will give the same wave function
  - Nitrogen atom ($\phi_4, \phi_1, \phi_2,$ and ϕ_3):

    $\hat{P}(A_1)$ on ϕ_4:

    ϕ_4 is not changed by any of the symmetry operations, it is a base set by itself and must belong to the irreducible representation A_1.

$$\psi_2 = \phi_4$$

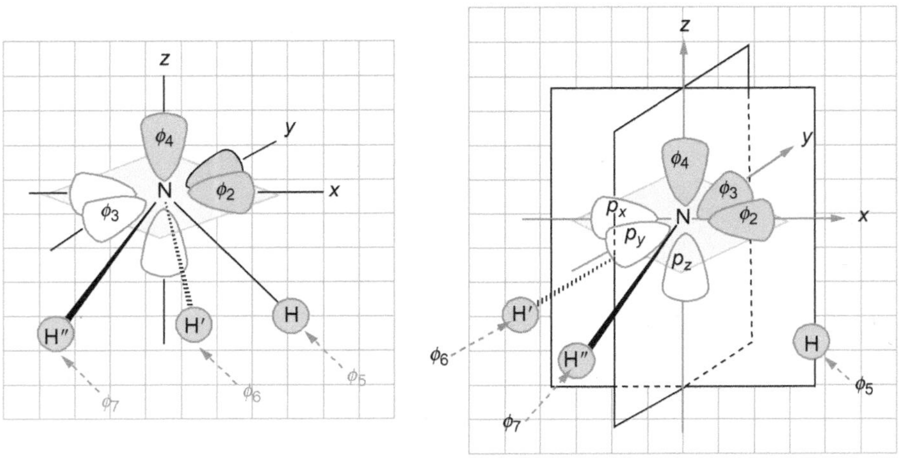

FIG. 5.13 The atomic orbitals locations of NH_3 in coordinate system, the zx plane passes through H and N.

When the projection operation is used:

$$\psi_2 = \frac{1}{6}\left[1E\phi_4 + 1C_3^*\phi_4 + 1C_3^2\phi_4 + 1\sigma_v\phi_4 + 1\sigma_v'\phi_4 + 1\sigma_v''\phi_4\right]$$

$$\psi_2 = \frac{1}{6}[\phi_4 + 1\times(0)\times\phi_4 + 1\times(0)\,\phi_4 + \phi_4 + \phi_4 + \phi_4]$$

$$\psi_2 = \frac{4}{6}\phi_4$$

$\hat{P}(A_1)$ on ϕ_1:

$$\psi_3 = \frac{1}{6}[\phi_1 + 1\times(0)\times\phi_1 + 1\times(0)\,\phi_1 + \phi_1 + \phi_1 + \phi_1]$$

$$\psi_3 = \frac{4}{6}\phi_1$$

$\hat{P}(A_1)$ on ϕ_2:
ϕ_2 and ϕ_3 form a doubly degenerate basis set and belong to E irreducible representation. Thus it is not necessary to carry out this operation. The projection operation confirms this conclusion:

$$\psi_{\phi_2} = \frac{1}{6}\left[1E\phi_2 + 1C_3\phi_2 + 1C_3^2\phi_2 + 1\sigma_v\phi_2 + 1\sigma_v'\phi_2 + 1\sigma_v''\phi_2\right]$$

$$= \frac{1}{6}\left[\phi_2 \underbrace{-\frac{1}{2}\phi_2 - \frac{\sqrt{3}}{2}\phi_3}_{C_3} \underbrace{-\frac{1}{2}\phi_2 + \frac{\sqrt{3}}{2}\phi_3}_{C_3^2} + \phi_2 + \underbrace{-\frac{1}{2}\phi_2 + \frac{\sqrt{3}}{2}\phi_3}_{\sigma_v'} + \underbrace{-\frac{1}{2}\phi_2 - \frac{\sqrt{3}}{2}\phi_3}_{\sigma_v''}\right] = 0$$

For C_3, and C_3^2 we use

$$\hat{C}_n\phi_2 = \phi_2 cos\theta + \phi_3 \sin\theta$$
$$\text{where } \hat{C}_n\phi_2 = \vec{x'}, \text{ and}$$

$$\begin{bmatrix} \vec{x'} \\ \vec{y'} \end{bmatrix} = \begin{bmatrix} \vec{x} \\ \vec{y} \end{bmatrix} \begin{bmatrix} \cos\theta & \sin\theta \\ -\sin\theta & \cos\theta \end{bmatrix}$$

A clockwise C_3 rotation, $\theta = -120°$; for $C_3{}^2$, we have $\theta = -240°$

$$\vec{x'} = \vec{x}\cos\theta + \vec{y}\sin\theta$$
$$\vec{y'} = -\vec{x}\sin\theta + \vec{y}\cos\theta$$

Using the following coordinate system:

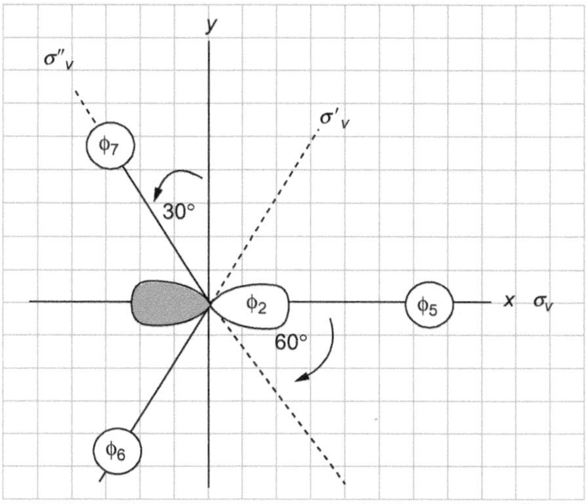

The matrix for σ_v is quite simple:

$$\sigma_v\phi_2 = \phi_2,$$

where

$$\begin{bmatrix} 1 & 0 & 0 \\ 1 & -1 & 0 \\ 0 & 0 & 1 \end{bmatrix} \begin{bmatrix} x \\ y \\ z \end{bmatrix} = \begin{bmatrix} x' \\ y' \\ z' \end{bmatrix}$$

The following matrices give the new x and y after reflection by σ_v' and σ_v'':

$$\sigma_v'\phi_2 = \phi_2 \cos 120 + \phi_3 \sin 120$$

$$\sigma_v''\phi_2 = \phi_2 \cos 240 + \phi_3 \sin 240, \text{ where}$$

$$\sigma_v' = \begin{bmatrix} -\frac{1}{2} & \frac{\sqrt{3}}{2} & 0 \\ \frac{\sqrt{3}}{2} & \frac{1}{2} & 0 \\ 0 & 0 & 1 \end{bmatrix} \text{ and } \sigma_v'' = \begin{bmatrix} -\frac{1}{2} & -\frac{\sqrt{3}}{2} & 0 \\ -\frac{\sqrt{3}}{2} & \frac{1}{2} & 0 \\ 0 & 0 & 1 \end{bmatrix}$$

Likewise, $\hat{P}(A_1)$ on ϕ_3 is 0.

- The projection operator for the degenerate irreducible representation E:

$$\hat{P}(E) \text{ on } \phi_2$$

$$\psi_4 = \frac{1}{3}\left[2E\phi_2 - 1C_3\phi_2 - 1C_3^2\phi_2\right]$$

$$\psi_4 = \frac{1}{3}\left[2\phi_2 - \left[\underbrace{-\frac{1}{2}\phi_2 - \frac{\sqrt{3}}{2}\phi_3}_{C_3}\right] - \left[\underbrace{-\frac{1}{2}\phi_2 + \frac{\sqrt{3}}{2}\phi_3}_{C_3^2}\right]\right] = \phi_2$$

$$\hat{P}(E) \text{ on } \phi_3$$

$$\psi_5 = \phi_3$$

$$\hat{P}(E) \text{ on } \phi_5 = \frac{1}{3}\left[2E\phi_5 - 1C_3\phi_5 - 1C_3^2\phi_5\right]$$

$$\hat{P}(E) = \frac{1}{3}[2\phi_5 - \phi_6 - \phi_7]$$

$$\hat{P}(E) \text{ on } \phi_6 = \frac{1}{3}[2\phi_6 - \phi_7 - \phi_5]$$

$$\hat{P}(E) \text{ on } \phi_7 = \frac{1}{3}[2\phi_7 - \phi_5 - \phi_6]$$

  - The results from ϕ_5, ϕ_6, and ϕ_7 are not orthogonal; we know there are only two wave functions in E. Linear combinations must be considered to get an orthonormal basis. There are three ways to do this:
  - Firstly, by using the matrix elements instead of with the trace.

    The diagonal elements are:

$$\underbrace{\overbrace{\begin{bmatrix} 1 & 0 \\ 0 & 1 \end{bmatrix}}^{E}}_{a_{11}=1,\ a_{22}=1} \underbrace{\overbrace{\begin{bmatrix} -\frac{1}{2} & \\ & -\frac{1}{2} \end{bmatrix}}^{C_3}}_{a_{11}=-\frac{1}{2},\ a_{22}=-\frac{1}{2}} \underbrace{\overbrace{\begin{bmatrix} -\frac{1}{2} & \\ & -\frac{1}{2} \end{bmatrix}}^{C^2{}_3}}_{a_{11}=-\frac{1}{2},\ a_{22}=-\frac{1}{2}} \underbrace{\overbrace{\begin{bmatrix} 1 & \\ & -1 \end{bmatrix}}^{\sigma_v}}_{a_{11}=1,\ a_{22}=-1} \underbrace{\overbrace{\begin{bmatrix} -\frac{1}{2} & \\ & \frac{1}{2} \end{bmatrix}}^{\sigma'_v}}_{a_{11}=-\frac{1}{2},\ a_{22}=\frac{1}{2}} \underbrace{\overbrace{\begin{bmatrix} -\frac{1}{2} & \\ & \frac{1}{2} \end{bmatrix}}^{\sigma''_v}}_{a_{11}=-\frac{1}{2},\ a_{22}=\frac{1}{2}}$$

where a_{11} corresponds to $\hat{P}(E_a)$ and a_{22} to $\hat{P}(E_b)$

$\hat{P}(E_a)$ on ϕ_5 gives:

$$\psi_a = N\left[E\phi_5 - \frac{1}{2}c_3\phi_5 - \frac{1}{2}C_3^2\phi_5 + 1\sigma_v\phi_5 - \frac{1}{2}\sigma_v'\phi_5 - \frac{1}{2}\sigma_v''\phi_5\right] = N[2\phi_5 - \phi_6 - \phi_7]$$

$\hat{P}(E_b)$ on ϕ_5 gives:

$$\psi_b = N\left[E\phi_5 - \frac{1}{2}c_3\phi_5 - \frac{1}{2}C_3^2\phi_5 - 1\sigma_v\phi_5 + \frac{1}{2}\sigma_v'\phi_5 + \frac{1}{2}\sigma_v''\phi_5\right] = 0$$

There is a node in ψ_b at ϕ_5.
The result of similar operations with $\hat{P}(E_b)$ on ϕ_6, or ϕ_7 gives the other wave function:

$$\psi_b = N[\phi_6 - \phi_7]$$

- The second way to obtain wave functions for degenerate irreducible representation involves working with a character table for a rotation group involving only the principal axis, i.e., C_n.
  For NH_3, we would use C_3. The E representations are given terms of the imaginaries:

$$\begin{array}{cccc} & E & C_3 & C_3^2 \\ E & \left\{\begin{array}{c} 1 \\ 1 \end{array}\right. & \begin{array}{c} \varepsilon \\ \varepsilon^* \end{array} & \left.\begin{array}{c} \varepsilon^* \\ \varepsilon \end{array}\right\} \end{array}$$

  where $\varepsilon = e^{i\varphi} = e^{\frac{2\pi i}{3}}$ and $\varepsilon^* = e^{-i\varphi} = e^{-\frac{2\pi i}{3}}$
  Thus:

$$\hat{P}_a\phi_6 = 1E\,\phi_6 + \varepsilon C_3\phi_6 + \varepsilon^* C_3^2\phi_6 = \phi_6 + \varepsilon\phi_7 + \varepsilon^*\phi_5 = \psi_a'$$

$$\hat{P}_b\phi_6 = 1E\,\phi_6 + \varepsilon^* C_3\phi_6 + \varepsilon C_3^2\phi_6 = \phi_6 + \varepsilon^*\phi_7 + \varepsilon\phi_5 = \psi_b'$$

  If we add the two together, we get one wave function.
  When we subtract the two equations and divide by i, we obtain the other wave function:
  Upon addition,

$$2\phi_6 + (\varepsilon + \varepsilon^*)\phi_7 + (\varepsilon + \varepsilon^*)\phi_5$$

$$\varepsilon = e^{i\phi} = \cos\phi + i\sin\phi$$

$$\varepsilon^* = e^{-i\phi} = \cos\phi - i\sin\phi$$

$$\varepsilon + \varepsilon^* = 2\cos\phi = 2\cos(-120) = 2\left(-\frac{1}{2}\right) = -1$$

$$\psi_a = 2\phi_6 - \phi_7 - \phi_5$$

  This is equivalent to the previous result, obtained by using the C_{3v} point group.
  When we subtract $\psi_a' - \psi_b'$ and divide by i

$$\frac{\psi_a' - \psi_b'}{i} = 0 + \frac{\varepsilon^* - \varepsilon}{i}\phi_7 + \frac{\varepsilon - \varepsilon^*}{i}\phi_5$$

$$\frac{\varepsilon^* - \varepsilon}{i} = \frac{2i}{i}\sin\phi = \sqrt{3}$$

$$\frac{\varepsilon - \varepsilon^*}{i} = -\frac{2i}{i}\sin\phi = -\sqrt{3}$$

$$\psi_b = \sqrt{3}\phi_7 - \sqrt{3}\phi_5$$

  The final wave function should be normalized:

$$\psi_b = \frac{1}{\sqrt{2}}(\phi_7 - \phi_5)$$

- The two E functions from ϕ_5, ϕ_6, and ϕ_7 can be produced by another method.
  i. Projects $\hat{P}(E)$ on ϕ_5 to get $2\phi_5 - \phi_6 - \phi_7$.
  ii. Carefully selecting for projection another function involving ϕ_6 and ϕ_7 (or sometimes ϕ_5, ϕ_6 and ϕ_7) that is orthogonal with the one just obtained, the desired function will result.

iii. If we try ϕ_6 and ϕ_7, we will see that $\phi_6 - \phi_7$is orthogonal to $2\phi_5 - \phi_6 - \phi_7$.
Note:
Projects $\hat{P}(E)$ on ϕ_6 to get $2\phi_6 - \phi_7 - \phi_5$ which is orthogonal to $\phi_7 - \phi_5$.
Projects $\hat{P}(E)$ on ϕ_7 to get $2\phi_7 - \phi_5 - \phi_6$ which is orthogonal to $\phi_5 - \phi_6$.

SUGGESTIONS FOR FURTHER READING

K.F. Purcell, J.C. Kotz, Inorganic Chemistry, W.B. Saunders Company, Philadelphia, PA/London/Toronto, 1977. ISBN: 0-7216-7407-0.

F.A. Cotton, Chemical Applications of Group Theory, third ed., John Wiley-Interscience, 1990. ISBN: 13:9780471510949.

I.S. Butler, J.F. Harrod, Inorganic Chemistry Principles and Applications, The Benjamin/Cummings Publishing Company, Inc, Redwood City, CA, 1989. ISBN: 0-8053-0247-6.

B. Douglas, D. McDaniel, J.J. Alexander, Concepts and Models of Inorganic Chemistry, third ed., John Wiley & Sons, New York, 1994. ISBN: 0-471-62978-2.

R.S. Drago, Physical Methods in Chemistry, W.B. Saunders Company, Philadelphia, PA, 1977. ISBN: 0-7216-3184-3.

H.B. Gray, Electrons and Chemical Bonding, W.A. Benjamin, Inc, New York, 1965.

D.K. Straub, J. Chem. Educ. 71 (1994) 750.

A.H. Cowley, N.C. Norman, Inorg. Chem. 34 (1986) 1.

C. Couret, J. Escudie, J. Satge, M. Lazraq, J. Amer. Chem. Soc. 109 (1987) 4411.

H. Meyer, G. Baum, W. Massa, S. Berger, A. Berndt, Angew. Chem. Int. Ed., Engl 26 (1987) 546.

G.H. Purser, J. Chem. Educ. 66 (1989) 710.

G.N. Lewis, J. Amer. Chem. Soc. 38 (1916) 762.

A.B.P. Lever, J. Chem. Educ. 49 (12) (1972) 819.

R.J. Gillespie, Molecular Structure, Van Nostrand Reinhold Co., London, 1972.

R.J. Gillespie, J. Chem. Educ. 51 (1974) 367.

R.J. Gillespie, J. Chem. Educ. 47 (1970) 18.

R.J. Gillespie, J. Amer. Chem. Soc. 82 (1960) 5973.

R.J. Gillespie, I. Hargittai, The VSEPR Model of Molecular Geometry, Allyn and Bacon, Boston, MA, 1991.

M.C. Favas, D.L. Kepert, Prog. Inorg. Chem. 27 (1980) 325.

J.L. Bills, R.L. Snow, J. Amer. Chem. Soc. 97 (1975) 6340.

R.L. Snow, J.L. Bills, J. Chem. Educ. 51 (1974).

L.S. Bartell, J. Chem. Educ. 45 (1968) 755.

R.D. Burbank, G.R. Jones, J. Amer. Chem. Soc. 96 (1974) 43.

L.S. Bartell, R.M. Gavin Jr., J. Chem. Phys. 48 (1968) 2460.

S.Y. Wang, L.L. Lohr, J. Chem. Phys. 61 (1974) 4110.

U.N. Nielsen, R. Haensel, W.H.E. Schwarz, J. Chem. Phys. 61 (1974) 3581.

K.S. Pitzer, L.S. Berstein, J. Chem. Phys. 63 (1975) 3849.

D.F. Shriver, P.W. Atkins, Inorganic Chemistry, Oxford University Press, 2010. ISBN: 978-0-19-923617-6. p. 74.

M.J. Winter, Chemical Bonding, Oxford Primer Series, Oxford University Press, 1993.

A. Zewail, The Chemical Bond: Structure and Dynamic, Academic Press, San Diego, CA, 1992.

B. Webster, Chemical Bonding Theory, Blackwell Scientific, Oxford, 1990.

J.N. Murrell, S.F. Kettle, J.M. Tedder, The Chemical Bond, Wiley, New York, 1985.

C.A. Coulson, The Shape and Structure of Molecules (revised by R. McWeeny), Oxford University Press, 1982.

R. McWeeny, Coulson's Valence, Oxford University Press, 1973.

L. Pauling, The Nature of the Chemical Bond, Cornell University Press, Ithaca, NY, 1960.

H. Krebs, Fundamentals of Inorganic Crystal Chemistry, McGraw-Hill, 1968.

Chapter 6

Molecular Orbital Theory

The molecular orbital or the orbital wave function represents the frame of the molecule that describes the distribution of spatial probability density for a particular electron bound to the group of nuclei.

The study of the electronic spectra, reactivity, and physical properties of substances will not be possible until we can interpret these phenomena in terms of the electron distribution in the molecule.

In the molecular orbital treatment, the electrons of the bonding shell of the central ion, as well as those of the ligand atoms, cannot be exclusively attached either with the central ion or with the ligand nuclei. The electrons are mutually linked, to some degree, with each of them.

The flexibility of the molecular orbital treatment is of great advantage in dealing with the magnetic and spectral properties of complexes.

When molecular orbital theory is understood, we can easily explain:

- the electronic structure of molecules;
- the symmetry of molecular orbitals;
- the relative energies of the molecular orbitals; and
- how the electron distribution changes upon going to some low-lying excited electronic state.

From this chapter, readers should gain an understanding of the distribution of electrons in some elected small molecules, and considerable thoughts regarding the relative energies of the molecular orbitals of those molecules. The chapter investigates how the electron distribution changes upon going to some low-lying excited electronic states. The theory is employed to obtain the energy change in chemical reactions, study stability and reactivity, find the delocalization energy, electron density, formal charge, bond order, ionization energy, equilibrium constant, and configuration interaction. It introduces the band theory concept, which makes it possible to rationalize conductivity, insulation, and semiconducting (Scheme 6.1).

In addition, a brief representation of molecular orbital theory is also developed, which should be a part of the background of any chemist.

Electrons, Atoms, and Molecules in Inorganic Chemistry. http://dx.doi.org/10.1016/B978-0-12-811048-5.00006-7

SCHEME 6.1 Outline of the approach used to present the molecular orbital theory.

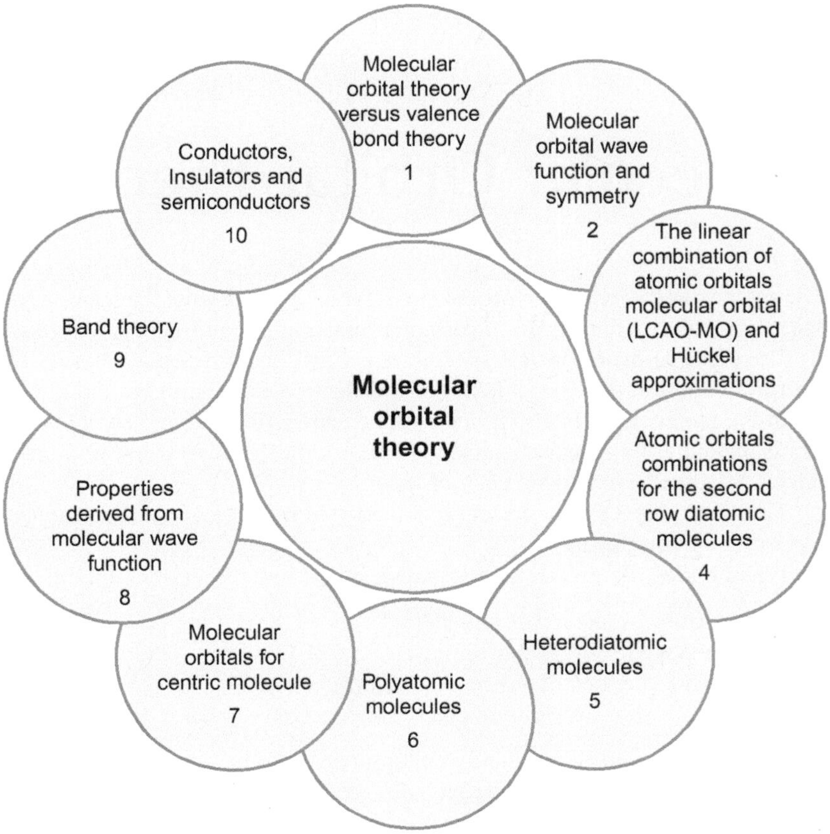

6.1 MOLECULAR ORBITAL THEORY VERSUS VALENCE BOND THEORY

What are the disadvantages of valence bond theory and the advantages of molecular orbital theory?

- The disadvantages of valence bond theory are as follows:
  1. All bonds are considered to be two-centered bonds.
  2. Less attention is taken for any interaction between such bonds.
  3. There is no a firm way to decide which pairs of atoms should be considered as bonded and which nonbonding, the selections are supported by chemical knowledge.
- The advantages of molecular orbital theory are as follows:
  1. All orbitals in a molecule spread over the entire molecule, e.g., the electrons occupying these orbitals may be delocalized over the whole molecule.
  2. Considerations of molecular symmetry properties are extremely informative in molecular orbital theory.
  3. With the symmetry properties of the molecular wave functions, it is often possible to reach many constructive conclusions about bonding without doing any actual quantum computations at all.
  4. The use symmetry properties in the molecular orbital theory can particularly reduce the effort involved by showing that many integrals must be identically equal to zero.

6.2 MOLECULAR ORBITAL WAVE FUNCTION AND SYMMETRY

What are the necessary procedures to find the linear combination of atomic orbitals—the molecular orbitals (LCAO-MO)—and to construct the molecular orbital energy diagrams?

Procedures:

- Draw a picture of the molecular shape.
- Note which types of atomic orbitals are available in valence shells.
- Divide these orbitals based on symmetry considerations, into σ, π and/or s, p, d sets: symmetry factoring.
- Find irreducible representations to which these sets of atomic orbitals give rise. (If it is a centric molecule, it usually helps to treat the central atom's orbitals in a separate step from that of the ligand's orbitals.)
- Form SALCs from the atomic orbitals by applying the projection operation $\hat{P}$'s, of the point group (Chapter 5, p 311). These SALCs are the LCAO-MOs.
- Compute energies of LCAO-MOs.
- Rank orbitals by energies.
- Method relies on LCAO approximation, starts with atomic orbitals, and adds them together.

6.3 THE LINEAR COMBINATION OF ATOMIC ORBITALS-MOLECULAR ORBITAL (LCAO-MO) AND HÜCKEL APPROXIMATIONS

a. **How can you construct the molecular wave function, ψ_j, resulting from a combination of i atomic orbitals?**
b. **Use a two-term LCAO-MO to write the secular equation and define: H_{ii}, H_{ij} and S_{ij} for a hydrogen molecule.**

- Each MO is written as a linear combination of atomic orbitals of various atoms:

$$\psi_j = \sum_i a_{ij}\phi_i \tag{6.3.1}$$

ϕ_is are a basis set, and it is convenient to be normalized, using

$$\int \phi_i\phi_i \partial\tau = 1$$

- For two term atomic orbitals:

$$\psi = a_1\phi_1 + a_2\phi_2$$

$$\because E = \frac{\langle \psi^* | \hat{H} | \psi \rangle}{\langle \psi^* | \psi \rangle}$$

$$E = \frac{\int (a_1\phi_1 + a_2\phi_2)\hat{H}(a_1\phi_1 + a_2\phi_2)\partial\tau}{\int (a_1\phi_1 + a_2\phi_2)(a_1\phi_1 + a_2\phi_2)\partial\tau}$$

$$E = \frac{\int \left[a_1\phi_1\hat{H}a_1\phi_1 + a_1\phi_1\hat{H}a_2\phi_2 + a_2\phi_2\hat{H}a_1\phi_1 + a_2\phi_2\hat{H}a_2\phi_2\right]\partial\tau}{\int [a_1\phi_1a_1\phi_1 + a_1\phi_1a_2\phi_2 + a_2\phi_2a_1\phi_1 + a_2\phi_2a_2\phi_2]\partial\tau}$$

$\because a_1$ and a_2 are constants:

$$E = \frac{a_1^2\int \phi_1\hat{H}\phi_1\partial\tau + 2a_1a_2\int \phi_1\hat{H}\phi_2\partial\tau + a_2^2\int \phi_2\hat{H}\phi_2\partial\tau}{a_1^2\int \phi_1\phi_1\partial\tau + 2a_1a_2\int \phi_1\phi_2\partial\tau + a_2^2\int \phi_2\phi_2\partial\tau} \tag{6.3.2}$$

- if:

$$H_{ii} = \int \phi_i H \phi_i \partial\tau \tag{6.3.3}$$

⇒ the energy of the atomic orbital ϕ_i.

$$H_{ij} = \int \phi_i H \phi_j \partial\tau \tag{6.3.4}$$

⇒ the energy of interaction between pairs of atomic orbitals.

$$S_{ij} = \int \phi_i \phi_j \partial\tau \tag{6.3.5}$$

⇒ the overlap integrals, and the value of $S_{11} = 1$ for normalized atomic orbitals.

$$\int \phi_i E \phi_i \partial\tau = E \int \phi_i \phi_i \partial\tau = ES_{ii}$$

$$\int \phi_i E \phi_j \partial\tau = E \int \phi_i \phi_j \partial\tau = ES_{ij}$$

$$\therefore E = \frac{a_1^2 H_{11} + 2a_1 a_2 H_{12} + a_2^2 H_{22}}{a_1^2 S_{11} + 2a_1 a_2 S_{12} + a_2^2 S_{22}}$$

- The variation principle can be used to estimate the coefficients a_1 and a_2 that minimized the energy and give the most stable system.
- Taking the partial derivatives $\frac{\partial E}{\partial a_1} = 0$, and $\frac{\partial E}{\partial a_2} = 0$ give the secular equations:

$$a_1(H_{11} - ES_{11}) + a_2(H_{12} - ES_{12}) = 0 \tag{6.3.6}$$

$$a_1(H_{21} - ES_{21}) + a_2(H_{22} - ES_{22}) = 0 \tag{6.3.7}$$

- The value of E can be found by solving the secular determinant:

$$\begin{vmatrix} H_{11} - ES_{11} & H_{12} - ES_{12} \\ H_{21} - ES_{21} & H_{22} - ES_{22} \end{vmatrix} = 0 \tag{6.3.8}$$

If $S_{11} = S_{22} = 1$ and the two atoms are the same:

$$(H_{11} - E)^2 - (H_{12} - ES_{12})^2 = 0$$

$$\therefore E_1 = \frac{H_{11} + H_{12}}{1 + S_{12}} \text{ and } E_2 = \frac{H_{11} - H_{12}}{1 - S_{12}} \tag{6.3.9}$$

- The secular equation can also be obtained from the wave equation:

$$\because H\psi = E\psi$$

$$\therefore \mathrm{H}\psi - E\psi = (H - E)\psi = 0, \Rightarrow \text{secular equation},$$

and if:

$$\psi_j = \sum_i a_{ij}\phi_i \tag{6.3.1}$$

then:

$$\sum_i a_i (H - E)\phi_i = 0, \Rightarrow \text{LCAO} - \text{MO}$$

- For two term atomic orbitals, e.g., an H_2-molecule, the LCAO-MO:

$$a_1(H - E)\phi_1 + a_2(H - E)\phi_2 = 0 \tag{6.3.10}$$

By multiplying by ϕ_1, and integrating over spatial coordinates of the wave functions:

$$a_1 \int \phi_1 (H - E)\phi_1 \partial\tau + a_2 \int \phi_1 (H - E)\phi_2 \partial\tau = 0,$$

$$a_1\underbrace{\int\phi_1 H\phi_1\partial\tau}_{H_{11}}-a_1\underbrace{\int\phi_1 E\phi_1\partial\tau}_{ES_{11}}+a_2\underbrace{\int\phi_1 H\phi_2\partial\tau}_{H_{12}}-a_2\underbrace{\int\phi_1 E\phi_2\partial\tau}_{ES_{12}}=0$$

and if

$$\therefore a_1(H_{11}-ES_{11})+a_2(H_{12}-ES_{12})=0 \tag{6.3.6}$$

- And again if:

$$a_1(H-E)\phi_1+a_2(H-E)\phi_2=0 \tag{6.3.10}$$

is multiplied by ϕ_2, and integrated over spatial coordinates of the wave functions, then

$$a_1\int\phi_2(H-E)\phi_1\partial\tau+a_2\int\phi_2(H-E)\phi_2\partial\tau=0,$$

$$a_1\int\phi_2 H\phi_1\partial\tau-a_1\int\phi_2 E\phi_1\partial\tau+a_2\int\phi_2 H\phi_2\partial\tau-a_2\int\phi_3 E\phi_2\partial\tau=0,$$

$$\therefore a_1(H_{21}-ES_{21})+a_2(H_{22}-ES_{22})=0 \tag{6.3.7}$$

- Eqs. (6.3.6) and (6.3.7) can be solved using the determinant:

$$\begin{vmatrix} H_{11}-ES_{11} & H_{12}-ES_{12} \\ H_{21}-ES_{21} & H_{22}-ES_{22} \end{vmatrix}=0 \tag{6.3.8}$$

$\because S_{11}=S_{22}=1$, therefore,

$$(H_{11}-E)(H_{22}-E)-(H_{12}-ES_{12})(H_{21}-ES_{21})=0$$

$$(1-S_{12}^2)E^2-(H_{11}+H_{22}-2H_{12}S_{12})E+H_{11}H_{22}-H_{12}^2=0$$

- The following Hückel approximations are then introduced:
  - $H_{11}=H_{22}=\alpha=0$
  - $H_{12}=\beta=1$, adjacent atoms
  - $S_{12}=0$
  - The linear transformation can be applied to calculate the E-values of the different levels of hydrogen molecule.

$$\begin{vmatrix} \alpha-E & 1 \\ 1 & \alpha-E \end{vmatrix}=0 \tag{6.3.11}$$

Use the symmetry considerations of H_2 to compute the molecular orbital wave functions, LCAO-MOs, the energies of these MOs, and the bond order.

- Dihydrogen molecule:

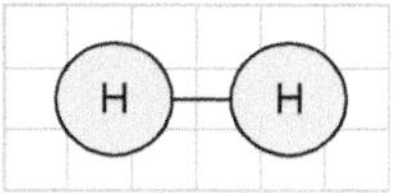

- Symmetry elements: E, $2C_\infty^\phi$, $\infty\sigma_v$, i, $2S_\infty^\phi$, and ∞C_2 (Fig. 6.1)
  Point group: $D_{\infty h}$
- The characters for the symmetry representation of the σ-bond between the two s-atomic orbitals, Γ_{σ_s}, in point group $D_{\infty h}$ is established in the same manner as in Chapter 5, p. 296?.

$D_{\infty h}$	E	$2C_\infty^\phi$	...	$\infty\sigma_v$	i	$2S_\infty^\phi$	...	C_2
Γ_{σ_s}	2	2	...	2	0	0	...	0

FIG. 6.1 Symmetry elements in the H_2-molecule.

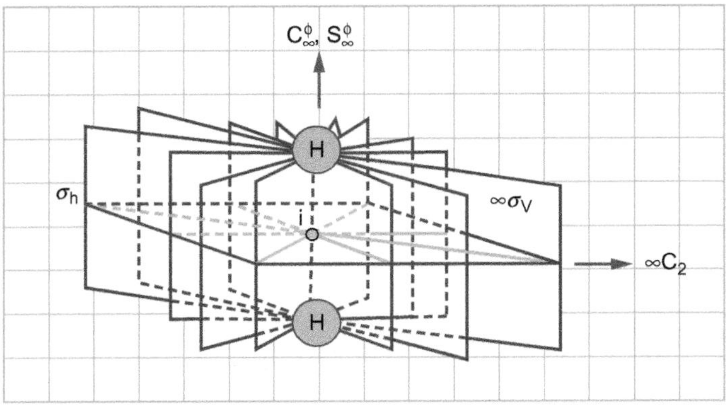

TABLE 6.1 Character Table for $D_{\infty h}$ and the Irreducible Representations of Γ_{σ_s}

$D_{\infty h}$	E	$2C^{\phi}_{\infty}$	...	$\infty\sigma_v$	i	$2S^{\phi}_{\infty}$	...	∞C_2		
Σ_g^+	**1**	**1**	...	**1**	**1**	**1**	...	**1**		x^2+y^2, z^2
Σ_g^-	1	1	...	−1	1	1	...	−1	R_z	
Π_g	2	$2\cos\phi$	...	0	2	$-2\cos\phi$	...	0	(R_x, R_y)	(xz, yz)
Δ_g	2	$2\cos 2\phi$	...	0	2	$2\cos 2\phi$	...	0		(x^2-y^2, xy)
...	...	...	...	...	...	...	...	...		
Σ_u^+	**1**	**1**	...	**1**	**−1**	**−1**	...	**−1**	z	
Σ_u^-	1	1	...	−1	−1	−1	...	1		
Π_u	2	$2\cos\phi$	...	0	−2	$2\cos\phi$	...	0	(x, y)	
Δ_u	2	$2\cos 2\phi$	...	0	−2	$-2\cos 2\phi$	...	0		
...	...	...	...	...	...	...	...	...		
Γ_{σ_s}	**2**	**2**		**2**	**0**	**0**		**0**	$=\sigma_g^+ + \sigma_u^+$	

Symmetric reduction is difficult, but inspection of the $D_{\infty h}$ table (Table 6.1) shows the following:

$$\Gamma_{\sigma_s} = \sigma_g^+ + \sigma_u^+$$

two wave equations: σ_g^+ and σ_u^+

- The molecular wave equation:

$$\psi_j = \sum_i a_{ij}\phi_i \qquad (6.3.1)$$

$$\psi = a_1\phi_{1sA} + a_2\phi_{1sB},$$

$$\because \sum_i a_i(H-E)\phi_i = 0, \text{ and } S_{11} = S_{22} = 1$$

$$\therefore a_1(H_{11}-E) + a_2(H_{12}-ES_{12}) = 0 \qquad (6.3.6)$$

$$a_1(H_{21}-ES_{21}) + a_2(H_{22}-E) = 0 \qquad (6.3.7)$$

- These two equations can be solved using

$$\begin{vmatrix} H_{11}-E & H_{12}-ES_{12} \\ H_{21}-ES_{21} & H_{22}-E \end{vmatrix} = 0 \tag{6.3.8}$$

Therefore,

$$(H_{11}-E)(H_{22}-E)-(H_{12}-ES_{12})(H_{21}-ES_{21})=0$$

$$(1-S_{12}^2)E^2-(H_{11}+H_{22}-2H_{12}S_{12})E + H_{11}H_{22}-H_{12}^2 = 0$$

- Using Hückel approximation:
$S_{12}=0$, $H_{11}=H_{22}=\alpha$, $H_{12}=H_{21}=\alpha$

$$\begin{vmatrix} \alpha-E & \beta \\ \beta & \alpha-E \end{vmatrix} = 0 \tag{6.3.11}$$

$$E^2-2\alpha E+\alpha^2-\beta^2=0$$

$$x=\frac{-b\pm\sqrt{b^2-4ac}}{2a}$$

$$E=\frac{2\alpha\pm\sqrt{4\alpha^2-4(\alpha^2-\beta^2)}}{2}$$

$$E=\alpha\pm\beta \text{ (two energy levels)} \tag{6.3.12}$$

- The two atomic orbitals mix and give two molecular orbitals, by in phase (bonding, σ_g^+) and out-phase (antibonding, σ_u^+) combinations (Fig. 6.2).
  - The molecular orbital, σ_g^+, is bonding via the electrons density that centered between the nuclei. Two $\dot{H}$'s are joined in σ_g^+, and go down in energy.
  - In the antibonding molecular orbital, σ_u^+, the electrons are more stable in the atomic orbitals, which decreases the electron density between nuclei. The antibonding molecular orbital, σ_u^+, is repulsive versus H—H interaction, because of ⊕. ..⊕ nuclear repulsion. To put two $\dot{H}$ together requires energy input.
- We plug the values of E for the bonding and antibonding orbitals, Eq. (6.3.12) to Eq. (6.3.8), to find a_1 and a_2:

$$a_1(H_{11}-E)+a_2(H_{12}-ES_{12})=0 \tag{6.3.6}$$

Therefore,

$$a_1(\alpha-(\alpha+\beta))+a_2\beta=0$$

or

$$a_1(\alpha-(\alpha-\beta))+a_2\beta=0$$

If $\alpha=0, \beta=1$ then

$$a_1\beta=-a_2\beta, \text{ and}$$

$$a_1\beta=a_2\beta,$$

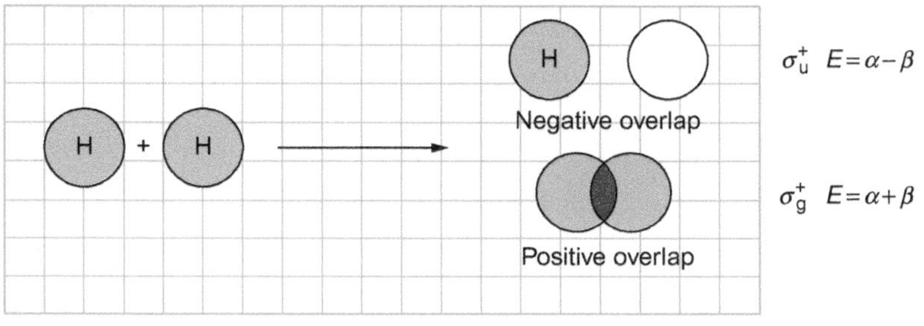

FIG. 6.2 Bonding and antibonding combinations of a hydrogen molecule.

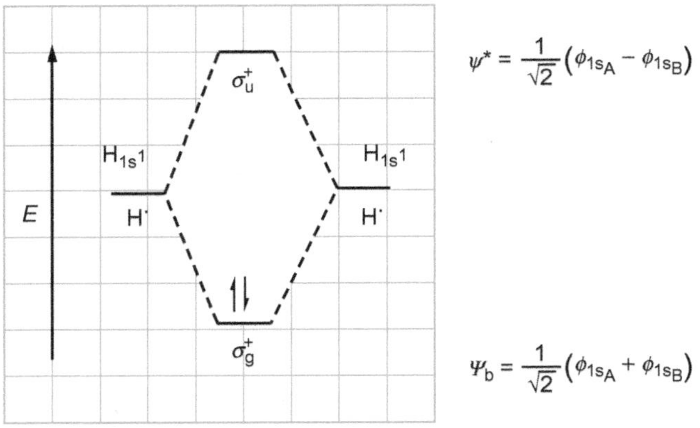

FIG. 6.3 Molecular energy diagram of a hydrogen molecule.

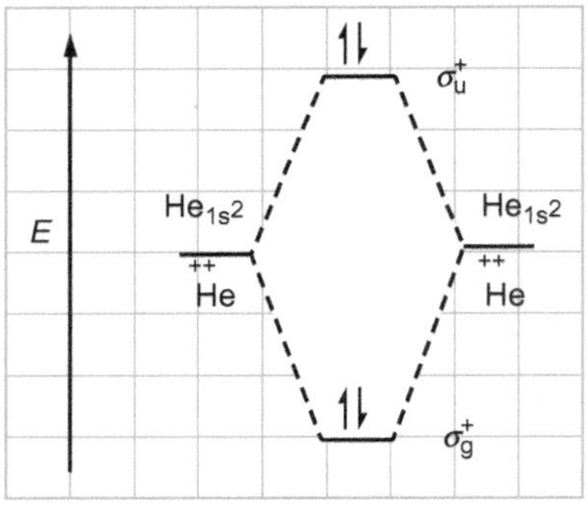

FIG. 6.4 Zero bond order prevents the formation of a helium molecule.

Then

$$a_1 = \pm a_2,$$

- Thus, after normalization, the molecular orbital wave equations are as follows:

$$\psi_b = \frac{1}{\sqrt{2}}(\phi_{1SA} + \phi_{1SB})$$

$$\psi^* = \frac{1}{\sqrt{2}}(\phi_{1SA} - \phi_{1SB})$$

- Molecular energy diagram (Fig. 6.3):
- H_2 is $(\sigma_g^+)^2$, has bond order = 1, H_2 has "σ-bond"

$$\text{Bond order} = \frac{(\text{number bonding e}^-) - (\text{number antibonding e}^-)}{2} \tag{6.3.13}$$

Why does He_2 not form, but He_2^+ does?

He_2 does not exist as stable species (bond order = 0, Eq. 6.3.13) (Fig. 6.4).

- He_2^+ is stable ($\sigma_{1s}^2\sigma_{1s}^1$), bond order = 1

6.4 ATOMIC ORBITALS COMBINATIONS FOR THE SECOND ROW DIATOMIC MOLECULES

For the second raw homopolar diatomic molecules, answer the following:

a. **What are the reducible representations of the orbital basis sets?**
b. **Find the molecular orbital wave functions, LCAO-MOs.**

c. Sketch the shapes of the MO.

d. Show the schematic changes in the shapes of diatomic molecular orbitals on going from the limit of separate atomic orbitals, to molecular orbitals, to the united atom.

e. Construct the correlation energy diagram for MOs of diatomic molecules. Discuss the implications of the energy levels crossing, and outline your conclusions.

- The atomic orbitals that one would logically start to combine are $1s_a$ and $1s_b$, $2s_a$ and $2s_b$, then $2p_a$ and $2p_b$ of the homopolar diatomic molecules ($X_2 = X_a - X_b$, $X_2 = Li_2$, Be_2, B_2, C_2, N_2, O_2, or F_2).
- It is not very useful to combine the 1s of the inner with the outer valence orbitals.
  - The operations of the point group, $D_{\infty h}$, first are applied to the two 1s atomic orbitals of the molecule (Fig. 6.5). As we solved before (Chapter 6, p. 335), the reducible presentation for which these two atomic orbitals form a basis is:

$D_{\infty h}$	E	$2C_{\infty}^{\phi}$	...	$\infty\sigma_v$	i	$2S_{\infty}^{\phi}$	...	C_2	
Γ_{σ_s}	2	2	...	2	0	0	...	0	$= \sigma_g^+ + \sigma_u^+$

  - It is useful to consider that "Starting with a given number of atomic orbitals, we must have the same number of molecular orbitals"
  - The molecular orbitals ψ_b (bonding) and ψ^* (antibonding) are derived by variation method.

$D_{\infty h}$	E	$2C_{\infty}^{\phi}$	...	$\infty\sigma_v$	I	$2S_{\infty}^{\phi}$	...	C_2	
ψ_b	1	1	...	1	1	1	...	1	$= \sigma_g^+$
ψ^*	1	1	...	1	1	−1	...	−1	$= \sigma_u^+$

  Each molecular orbital must be considered as a single entity (Fig. 6.6).
  - The character i for ψ^* is -1 because inversion changes the molecular orbital into minus itself.
- The eight valence orbitals of 2s- and 2p-orbitals are shown in Fig. 6.7.
  - The character under E is 8.
  - Upon rotation about the C_∞ axis, s and p_z are unchanged and contribute 4 to the trace.
  - On the other hand, p_x and p_y (Fig. 6.8) follow the rotation matrix:

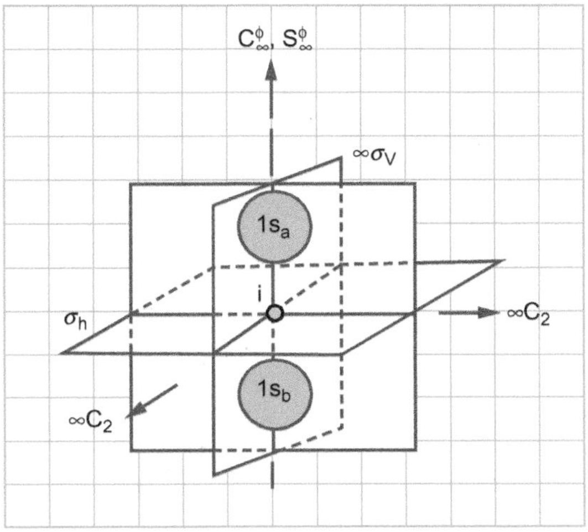

FIG. 6.5 The symmetry elements of point group $D_{\infty h}$ when two 1s orbitals are combined.

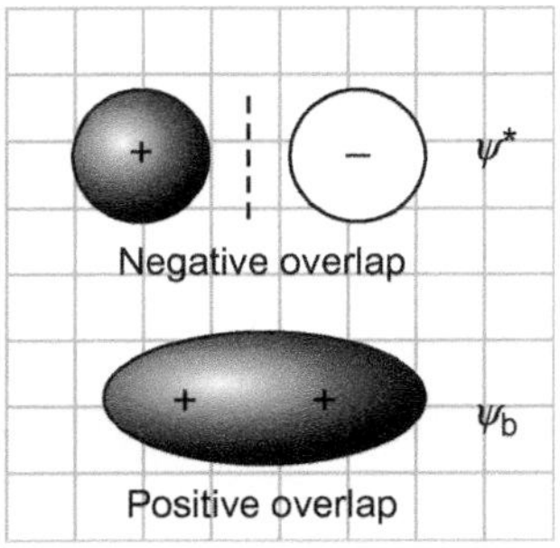

FIG. 6.6 Bonding and antibonding combinations of the two 1s atomic orbitals.

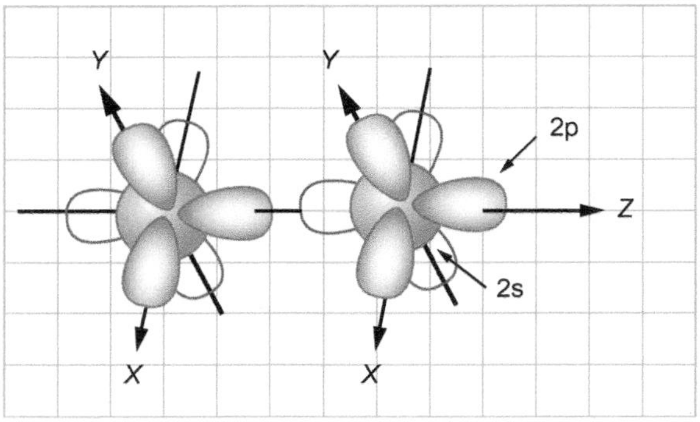

FIG. 6.7 The 2s and 2p basis atomic orbitals used to construct the molecular orbitals of diatomic molecules.

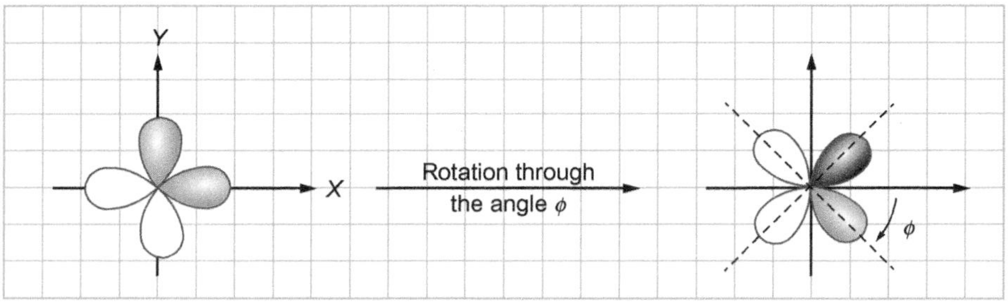

FIG. 6.8 Rotation of p_x and p_y orbitals about the C_∞ axis.

$$R_\phi \begin{bmatrix} x \\ y \end{bmatrix} = \begin{bmatrix} \cos\phi & \sin\phi \\ -\sin\phi & \cos\phi \end{bmatrix} \begin{bmatrix} x \\ y \end{bmatrix}$$

Trace = 2 cos ϕ

- Therefore the component of reducible representations is as shown in Table 6.2.
- The 2s-orbitals can be combined in the same way as the 1s-orbitals to give a σ_g^+ and σ_u^+ combination.

$D_{\infty h}$	E	$2C_\infty^\phi$	...	$\infty\sigma_v$	i	$2S_\infty^\phi$	...	∞C_2	
$\Gamma_{2s,2p_z}$	4	4	...	4	0	0	...	0	$= 2\sigma_g^+ + 2\sigma_u^+$
$\Gamma_{2p_x,2p_y}$	4	$4\cos\phi$	...	0	0	0	...	0	$= \pi_g + \pi_u$
$\Gamma_{2s,2p}$	8	$4+4\cos\phi$	...	4	0	0	...	0	

- The p_z orbitals can be combined in this manner as required.
- The combination 2s + 2s and $2p_z - 2p_z$ are transformed as σ_g^+ ($\chi(i) = 1$ (Fig. 6.9)).
- The combination 2s − 2s and $2p_z + 2p_z$ are transformed as σ_u^+ ($\chi(i) = -1$ (Fig. 6.10)).
- The combination $p_x + p_x$ and $p_y + p_y$ form a basis for the doubly degenerate representation π_u ($\chi(i) = -1$ (Fig. 6.11)).

TABLE 6.2 Character Table for $D_{\infty h}$ and the Irreducible Representations of $\Gamma_{2s,2p_z}$ and $\Gamma_{2p_x,2p_y}$

$D_{\infty h}$	E	$2C_{\infty}^{\phi}$	...	$\infty\sigma_v$	i	$2S_{\infty}^{\phi}$	...	∞C_2		
Σ_g^+	1	1	...	1	1	1	...	1		x^2+y^2, z^2
Σ_g^-	1	1	...	−1	1	1	...	−1	R_z	
Π_g	2	$2\cos\phi$	...	0	2	$-2\cos\phi$	...	0	(R_x, R_y)	(xz, yz)
Δ_g	2	$2\cos 2\phi$	...	0	2	$2\cos 2\phi$	...	0		(x^2-y^2, xy)
...	...	...	...	...	...	...	...	...		
Σ_u^+	1	1	...	1	−1	−1	...	−1	z	
Σ_u^-	1	1	...	−1	−1	−1	...	1		
Π_u	2	$2\cos\phi$	...	0	−2	$2\cos\phi$	...	0	(x,y)	
Δ_u	2	$2\cos 2\phi$	...	0	−2	$-2\cos 2\phi$	...	0		
...	...	...	...	...	...	...	...	...		
$\Gamma_{2s,2p_z}$	4	4		4	0	0		0	$=2\sigma_g^+ + 2\sigma_u^+$	
$\Gamma_{2p_x,2p_y}$	4	$4\cos\phi$		0	0	0		0	$=\pi_g + \pi_u$	

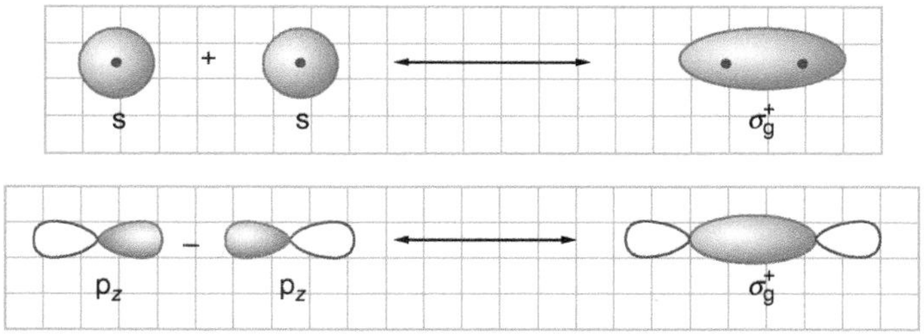

FIG. 6.9 Bonding combinations of 2s+2s and $2p_z - 2p_z$.

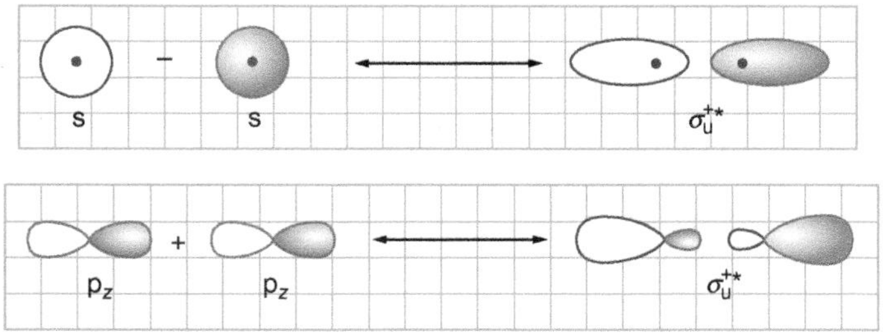

FIG. 6.10 Antibonding combinations of 2s − 2s and $2p_z + 2p_z$.

- ○ The combination $p_x - p_x$ and $p_y - p_y$ form a basis for the doubly degenerate representation π_g ($\chi(i) = 1$ (Fig. 6.12)).
- ○ σ_g^+ (2s) is lower than σ_u^+ (2s), and σ_g^+ ($2p_z$) is blow σ_u^+ ($2p_z$).
- ○ Further, π_u ($2p_x$, $2p_y$) are lower than π_g ($2p_x$, $2p_y$).
- In order to gain further insight into the possible relationships of the energy levels, one can consider *two extreme* possibilities.
- In one case, the *two atoms* are separated by an infinite distance. The 1s-, 2s-, and 2p-orbitals form just three energy levels.

- As the atoms are brought closer together, molecular orbitals may be formed by delocalization and spreading the atomic orbitals.
- When the two atoms come to zero separation, the united atom limit is achieved: a single atom of twice the nuclear charge of each of the original atoms.
- The path of transformation of each orbital is shown in Fig. 6.13.
- All these transformations can be collected in the orbitals correlation diagram (Fig. 6.14).
- By adding electrons into the orbitals correlation diagram, it is possible to set up the first row diatomic molecules.
  Note: 1 π_u and $3\sigma_g^+$ orbital energies cross between N_2 (diamagnetic) and O_2 (paramagnetic).
  Therefore, up to N_2 we use Scheme 6.2, and at O_2 we switch to Scheme 6.3.

FIG. 6.11 Bonding combination of $2p_x + 2p_x$.

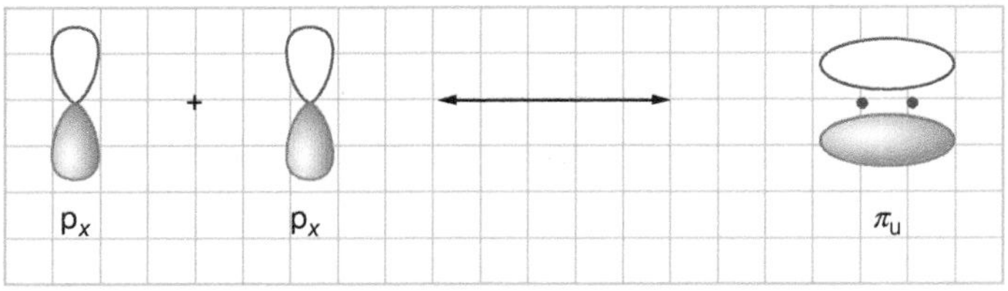

FIG. 6.12 Antibonding combinations of $2p_x - 2p_x$.

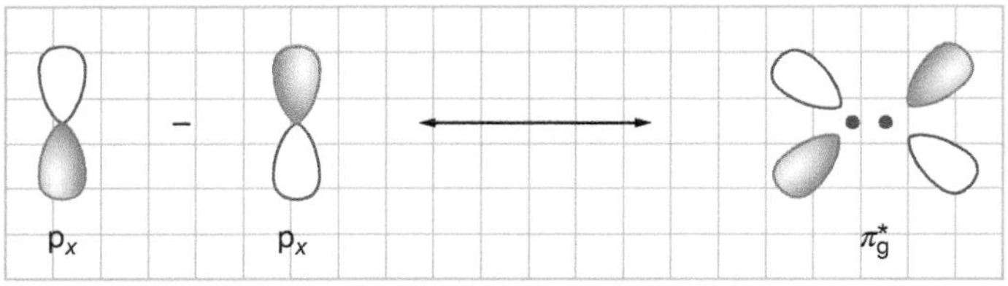

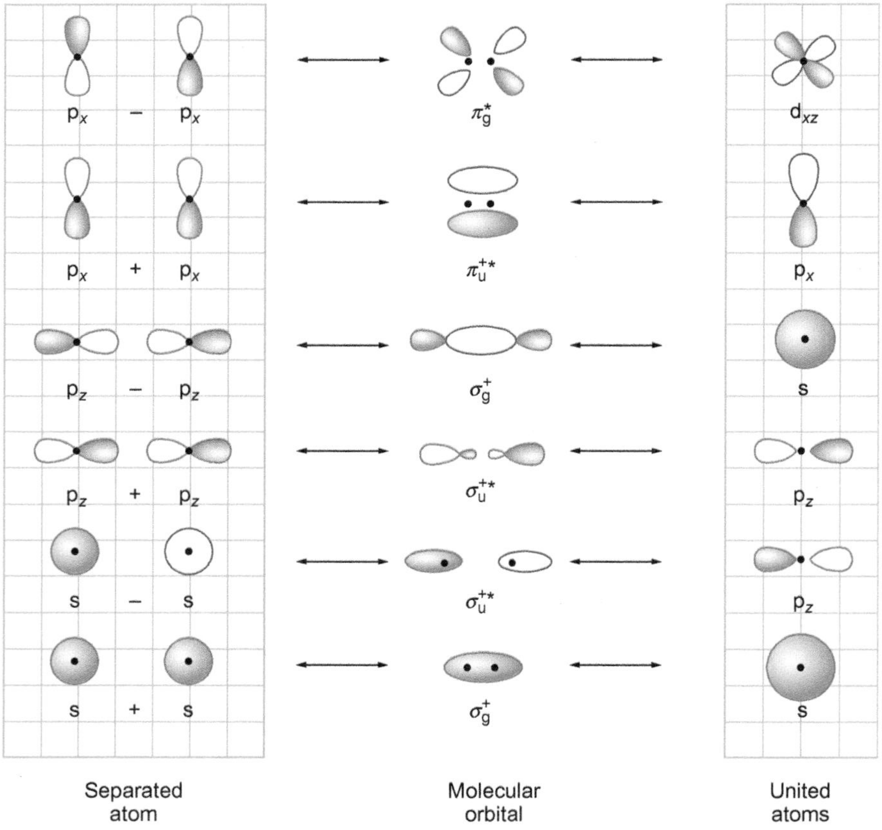

FIG. 6.13 The changes in the orbitals on going from the separate atomic orbitals, to molecular orbitals, to the united atom.

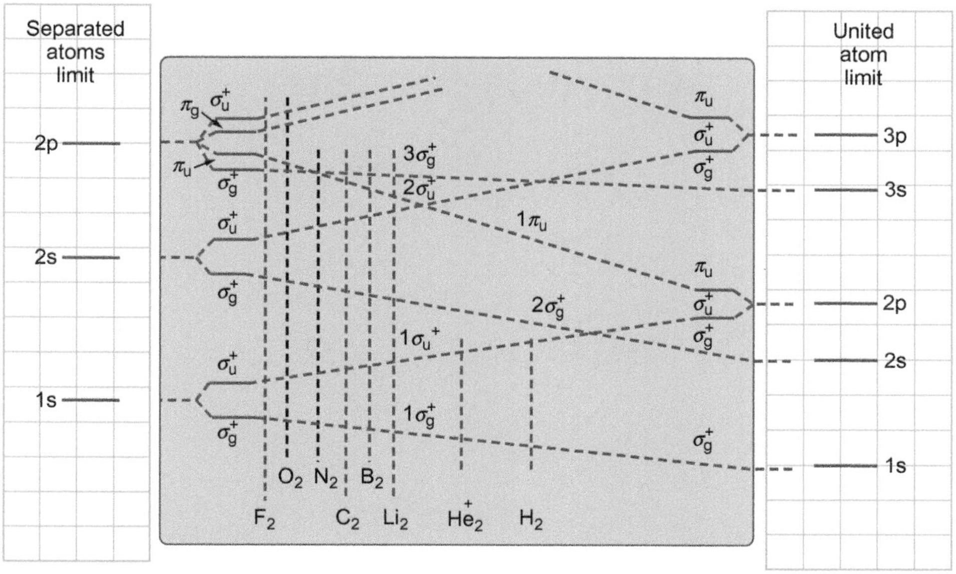

FIG. 6.14 Orbitals correlation diagram for diatomic molecules. *Dashed lines* specify the positions of the diatomic molecules. These lines only show the order of the orbitals for each molecule and have no quantitative implication.

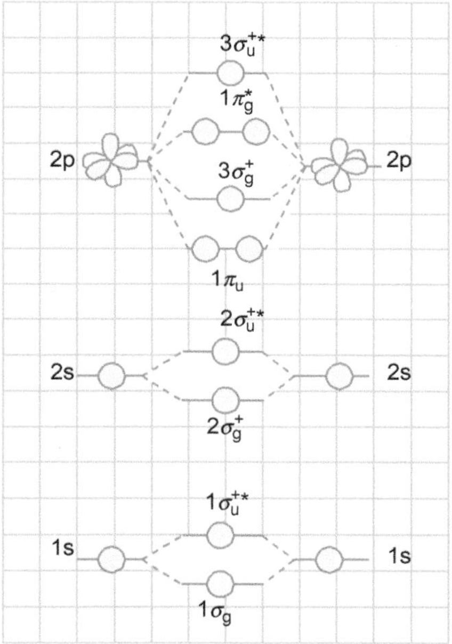

SCHEME 6.2 Energy diagram for Li_2, B_2, C_2, and N_2.

Explain the energy flip of $\pi_u(p_x, p_y)$ and $\sigma_g^+(p_z)$ levels in N_2 and O_2, and why does it occur?

How does this energy flip influence the electronic configuration for the second raw homopolar diatomic molecules.

- In Schemes 6.2 and 6.4A, we only considered the possibility of combination the atomic orbitals of the same symmetry.
- This treatment has totally disregarded the fact that molecular orbitals of the same symmetry and right energy can combine.

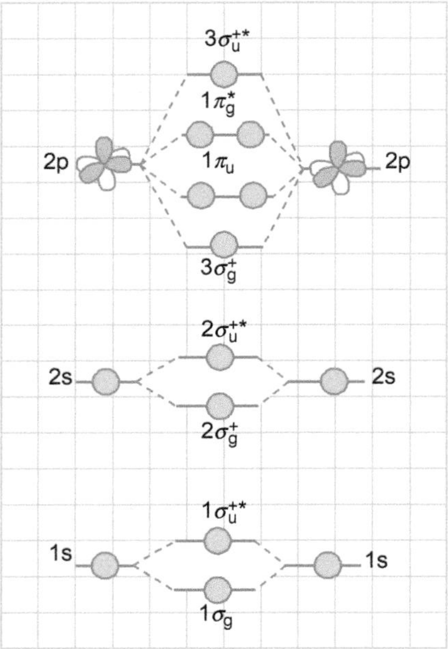

SCHEME 6.3 Energy diagram for O_2 and F_2.

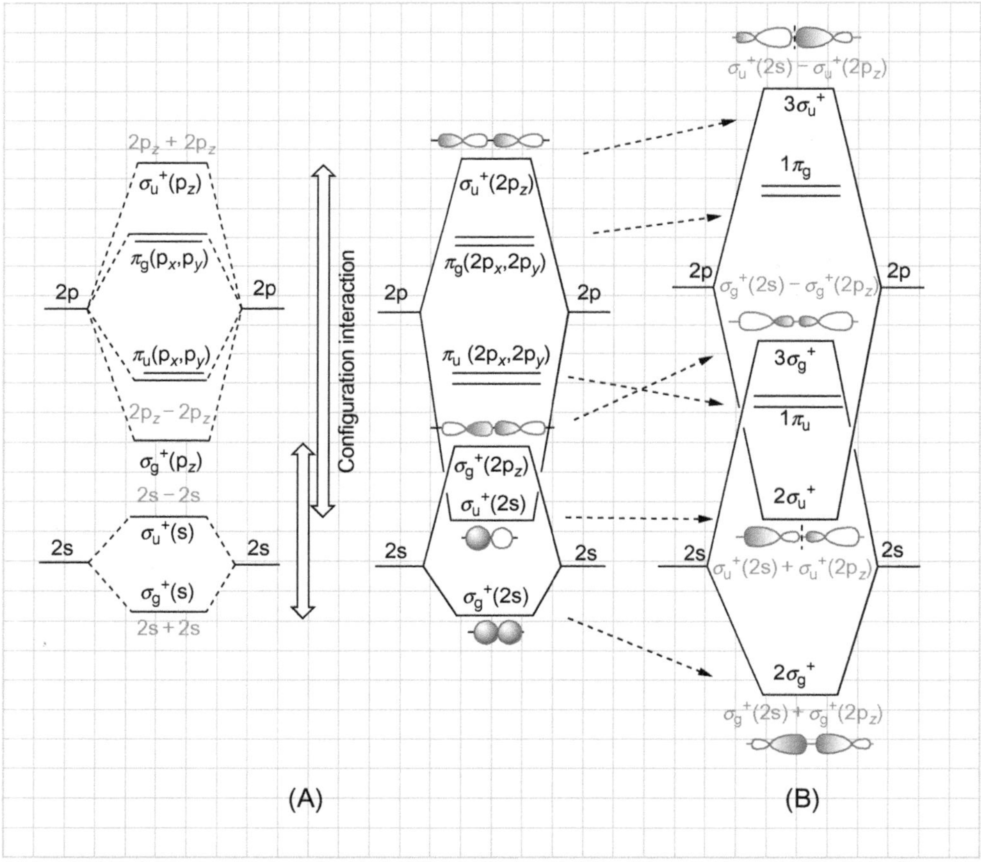

SCHEME 6.4 The molecular orbital scheme for homopolar diatomic: (A) without configuration interaction and (B) with configuration interaction.

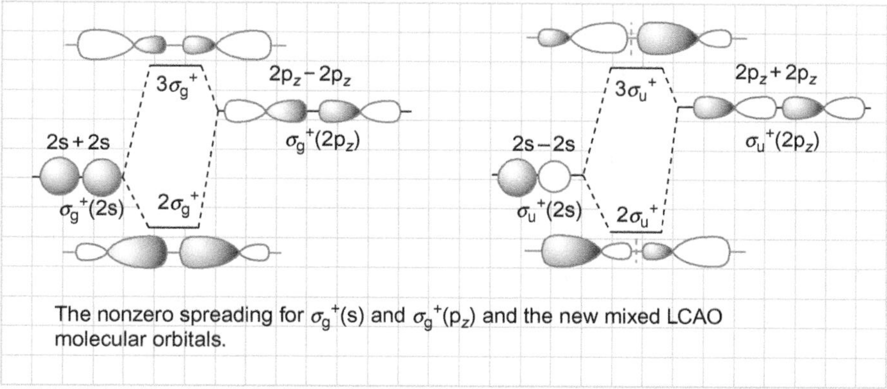

SCHEME 6.5 The nonzero spreading for $\sigma_g^+(2s)$ and $\sigma_g^+(2p_z)$ and the new mixed LCAO molecular orbitals.

- The molecular orbital diagram (Scheme 6.4B) can produced by further mixing of molecular orbitals have the same symmetry, and whose constituent atomic orbitals are different (2s and $2p_z$) (Scheme 6.5).
- The molecular orbital $\sigma_g^+(2s)$ can be mixed with $\sigma_g^+(2p_z)$ and $\sigma_u^+(2s)$ can be mixed with $\sigma_u^+(2p_z)$:

$$2\sigma_g^+ = \sigma_g^+(2s) + \sigma_g^+(2p_z), \text{ and } 3\sigma_g^+ = \sigma_g^+(2s) - \sigma_g^+(2p_z)$$

$$2\sigma_u^+ = \sigma_u^+(2s) + \sigma_u^+(2p_z), \text{ and } 3\sigma_u^+ = \sigma_u^+(2s) - \sigma_u^+(2p_z)$$

- The mixing of these two sets of molecular orbitals leads to lower energy of $2\sigma_g^+$ and $2\sigma_u^+$ and higher energy of $3\sigma_g^+$ and $3\sigma_u^+$ (Schemes 6.5 and 6.4B).
- The degree to which orbitals combined is strongly dependent on their difference in energy and falls rapidly as this energy difference increases.
- This kind of mixing or interaction is called configuration interaction.
- In the case of Li_2, B_2, C_2, and N_2, the overlap between 2s-orbitals is greater than that between 2p-orbitals, the configuration interaction in this case is to raise the energy of $3\sigma_g^+$ raises above $1\pi_u$, which is not subject to a configuration interaction effect.
- For Li_2, bond order = 1, Li_2: $\left(1\sigma_g^+\right)^2\left(1\sigma_u^+\right)^2\left(2\sigma_g^+\right)^2$ (Fig. 6.15)

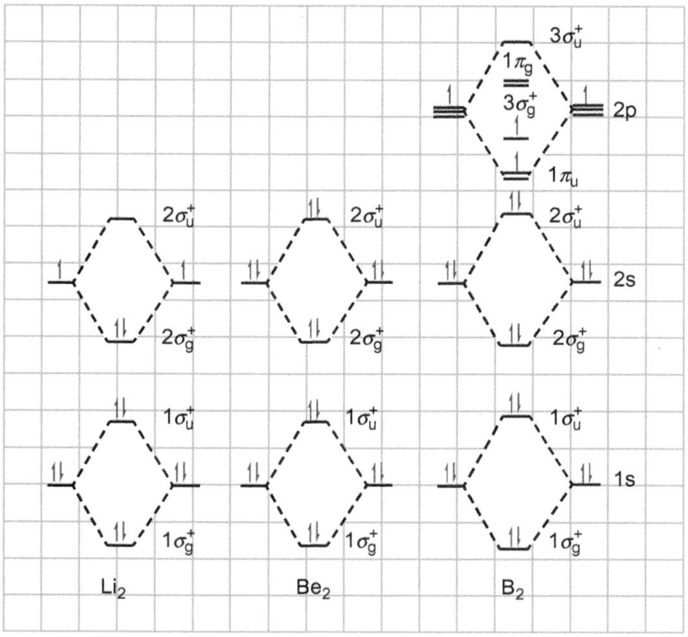

FIG. 6.15 Energy diagrams for Li_2, Be_2 (does not form), and B_2.

- Be_2: $\left(1\sigma_g^+\right)^2\left(1\sigma_u^+\right)^2\left(2\sigma_g^+\right)^2\left(2\sigma_u^+\right)^2$
  bond order = 0 (bond does not form, Eq. 6.3.13)
- B_2: $\left(1\sigma_g^+\right)^2\left(1\sigma_u^+\right)^2\left(2\sigma_g^+\right)^2\left(2\sigma_u^+\right)^1\left(1\pi_u\right)^2\left(3\sigma_g^+\right)^1$ instead of the expected electronic configuration of $\left(1\sigma_g^+\right)^2\left(1\sigma_u^+\right)^2\left(2\sigma_g^+\right)^2\left(2\sigma_u^+\right)^2\left(1\pi_u\right)^2$
  bond order = 1, paramagnetic (2 unpaired $\bar{e}$), so next M. O. must be doubly degenerate.
  The reason is the molecular orbitals $2\sigma_u^+$, $1\pi_u$, and $3\sigma_g^+$ are close together that the spin pairing energy is larger the energy gained by putting another electron in $2\sigma_u^+$ or $1\pi_u$.
  The spin pairing is the energy necessary to hold two electrons in the same orbitals.
- C_2: $\left(1\sigma_g^+\right)^2\left(1\sigma_u^+\right)^2\left(2\sigma_g^+\right)^2\left(2\sigma_u^+\right)^2\left(1\pi_u\right)^4$ (diamagnetic is known in gas phase at high temperature, bond order = 2)
- N_2: $\left(1\sigma_g^+\right)^2\left(1\sigma_u^+\right)^2\left(2\sigma_g^+\right)^2\left(2\sigma_u^+\right)^2\left(1\pi_u\right)^4\left(3\sigma_g^+\right)^2$ (bond order = 3)
- O_2:
  - In oxygen and fluorine atoms, the 2s and 2p levels are separated by considerable difference in energy, thus reducing the configuration interaction, and in this case, the energy of $1\pi_u$ raises above $3\sigma_g^+$ (Scheme 6.4A).
  - Oxygen is paramagnetic, and has two unpaired electrons. According to Hund's first rule, electrons are placed in each degenerate orbital before filling any one of them up.
  - The electron spins are aligned in the same direction.
  - The various oxygen species will have the following configuration:

$$\left(1\sigma_g^+\right)^2\left(1\sigma_u^+\right)^2\left(2\sigma_g^+\right)^2\left(2\sigma_u^+\right)^2\left(3\sigma_g^+\right)^2\left(1\pi_u\right)^4\left(1\pi_g\right)^2$$

$$O_2^+:\underbrace{\uparrow\ \_}_{1\pi_g},\quad O_2:\underbrace{\uparrow\ \uparrow}_{1\pi_g},\quad O_2^-:\underbrace{\uparrow\downarrow\ \uparrow}_{1\pi_g},\quad O_2^{2-}:\underbrace{\uparrow\downarrow\ \uparrow\downarrow}_{1\pi_g}$$

Write the ground state configurations of N_2^-, C_2^+, He_2^+, F_2^-, and O_2^-.

Give the spectroscopic symbols and the bond orders for the ground states of these species. Arrange them in the order of decreasing dissociation energy.

Answer:

- N_2^-: $\left(1\sigma_g^+\right)^2\left(1\sigma_u^+\right)^2\left(2\sigma_g^+\right)^2\left(2\sigma_u^+\right)^2\left(1\pi_u\right)^4\left(3\sigma_g^+\right)^2\left(1\pi_g\right)^1$
  Spectroscopic symbol: $^2\Pi_g$
  Bond order: 2.5 (Eq. 6.3.13)
- C_2^+: $\left(1\sigma_g^+\right)^2\left(1\sigma_u^+\right)^2\left(2\sigma_g^+\right)^2\left(2\sigma_u^+\right)^2\left(1\pi_u\right)^3$
  Spectroscopic symbol: $^2\Pi_u$
  Bond order: 1.5
- He_2^+: $\left(1\sigma_g^+\right)^2\left(1\sigma_u^+\right)^1$
  Spectroscopic symbol: $^2\Sigma_u$
  Bond order: 0.5
- F_2^-: $\left(1\sigma_g^+\right)^2\left(1\sigma_u^+\right)^2\left(2\sigma_g^+\right)^2\left(2\sigma_u^+\right)^2\left(3\sigma_g^+\right)^2\left(1\pi_u\right)^4\left(1\pi_g\right)^3\left(3\sigma_u^+\right)^1$
  Spectroscopic symbol: $^2\Sigma_u$
  Bond order: 0.5

- O_2^-: $\left(1\sigma_g^+\right)^2\left(1\sigma_u^+\right)^2\left(2\sigma_g^+\right)^2\left(2\sigma_u^+\right)^2\left(3\sigma_g^+\right)^2\left(1\pi_u\right)^4\left(1\pi_g\right)^3$
  Spectroscopic symbol: $^2\Pi_g$
  Bond order: 1.5
- Order of dissociation energy: $N_2^+ > C_2^+ \approx O_2^- > F_2^- \approx He_2^+$

6.5 HETERODIATOMIC MOLECULES

What are the differences between the molecular orbital treatment of heteropolar and that of homopolar molecules?
What are the consequent alternations in the molecular orbital correlation diagrams of the isoelectronic compounds: CN^-, CO, NO^+ and N_2?

- The molecular orbital treatment of heteropolar molecules follows exactly the same principles as that for homopolar molecules.
- However:
  - the symmetry labeled in the molecular orbital scheme are those appropriate to $C_{\infty v}$ instead of $D_{\infty v}$;
  - the valence atomic orbitals of the higher electronegative atom tend to be lower in energy than those orbitals of the lower electronegative atom;
  - the bonding molecular orbital has mostly the character of the atomic orbital of the higher electronegative atom; and
  - the antibonding molecular orbital has mostly the character of the atomic orbital of the lower electronegative atom.
- This means that:
  - the bonding electrons tend to be associated more with the more electronegative atom; and
  - the antibonding electrons are associated more with the less electronegative atom.
- Other heteronuclear molecular features are as follows:
  - There is a failing in the covalent bond energy of bonds formed from atomic orbitals of different energies (Scheme 6.6).
  - When the electronegativity difference between atoms of the heteronuclear is so great as to prevent covalent bonding, the two bonding electrons transfer to the bonding molecular orbital that is indistinguishable from the simple picture of an ionic bond.
- Bearing in mind these considerations, the energy level diagram can be represented in Scheme 6.4.
- The differences between the correlation diagrams of CN^-, CO, NO^+, and that of N_2.
  - These four molecules are isoelctronic, each with 14 electrons, and will have all levels up to $(5\sigma^+)^2$ and including $(1\pi)^4$ filled (Scheme 6.7).
    In an N_2 molecule, the electrons are evenly distributed over the two atoms and have the electronic configuration $(1\sigma^+)^2(2\sigma^+)^2(3\sigma^+)^2(4\sigma^+)^2(1\pi)^4(5\sigma^+)^2$.
  - In the case of hetero-atomic species, there is an unequal distribution of the electron between the two atoms as one passes from one energy level to the next.
  - The bonding orbitals, which are concentrated in the internuclear region, are biased toward the more electronegative atom.

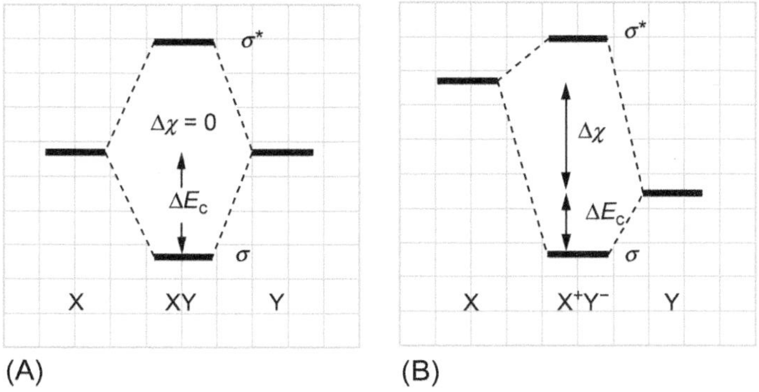

SCHEME 6.6 (A) Homonuclear diatomic molecule, XY, with insignificant electronegativity difference, $\Delta\chi \cong 0$, the covalent energy is maximized. (B) Homonuclear diatomic molecule, X^+Y^-, with large electronegativity difference, the covalent energy is maximized to limit extension.

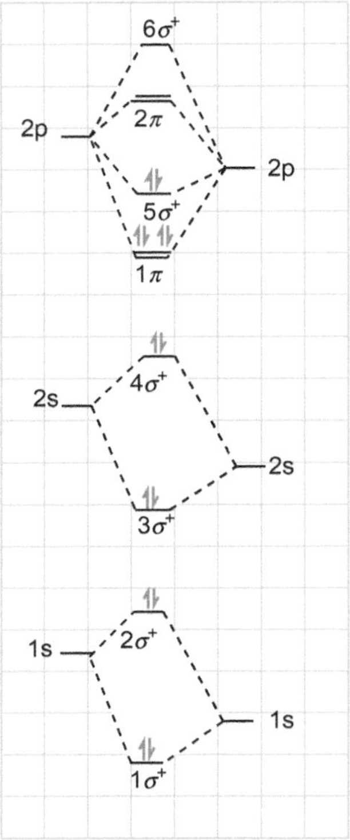

SCHEME 6.7 Energy diagram of hetero dipolar molecules.

- The antibonding orbitals, which are concentrated in the extra nuclear region, are biased toward the less electronegative atom.
- This extra nuclear electron density is associated with $4\sigma^+$, and the pair of electrons in this orbital explains the σ-donor properties of CN^-, CO, and NO^+ when they form metal complexes.
- In CO:
  i. Oxygen is more electronegative than carbon, thus the electron density on the oxygen atom is greater than that on the carbon atom.
  ii. The carbon atomic orbitals contribute less to the bonding molecular orbital, but contribute more to the antibonding molecular orbital.
  iii. The near-zero dipole moment of CO, although oxygen is more effective at attraction electron density than that of carbon. The electron density that projects beyond the end of the molecule has a larger effect on the molecular dipole.

- In BN:
  - the orbital 1π and $5\sigma^+$ are so close in energy that the energy needed to promote a 1πelectron to $5\sigma^+$ is less than the spin pairing energy; and
  - the BN configuration is $(1\sigma^+)^2(2\sigma^+)^2(3\sigma^+)^2(4\sigma^+)^2(1\pi)^3(5\sigma^+)^1$.
- HF:
  - The (1s) orbital of H is too high in energy to interact with any of 2p orbitals of F (Scheme 6.8).
  - However, only $2p_z$-orbital of F has the right (same) symmetry to interact with 1s orbital of H.
  - The $2p_x$- and $2p_y$-orbitals of F cannot interact with 1s of H; they stay nonbonding.
  - The 1s- and 2s-orbitals of F are also nonbonding because their energy is so low compared to 1s energy of H.
  - For less electronegative elements than F, the $2\sigma^+$ may be bonding since the 2s-orbital will be closer in energy to the 1s-orbital of H.

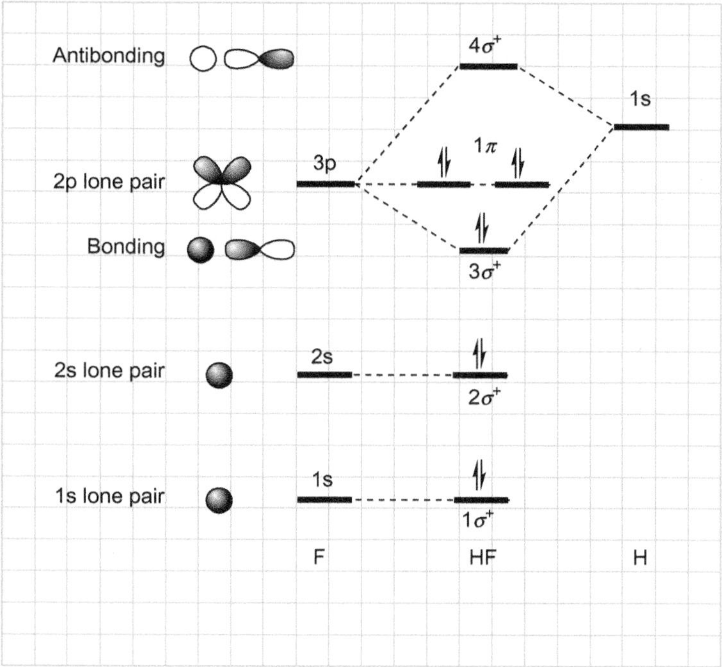

SCHEME 6.8 Molecular orbital diagram for HF.

6.6 POLYATOMIC MOLECULES

By using the Hückel molecular orbital theory, estimate the total energy of the following systems:

a. **H—H—H**

b. 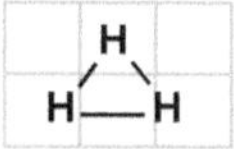

The answers:

a. Energy levels of linear H_3 according to Hückel molecular orbital theory can be determined using secular equations:

$$a_1(H_{11}-E)+a_2(H_{12}-ES_{12})+a_3(H_{13}-ES_{13})=0$$

$$a_1(H_{21}-ES_{21})+a_2(H_{22}-E)+a_3(H_{23}-ES_{23})=0$$

$$a_1(H_{31}-ES_{31})+a_2(H_{32}-ES_{32})+a_3(H_{33}-E)=0$$

The three equations can be solved using the secular determinant:

$$\begin{vmatrix} H_{11}-E & H_{12}-ES_{12} & H_{13}-ES_{13} \\ H_{21}-ES_{21} & H_{22}-E & H_{23}-ES_{23} \\ H_{31}-ES_{31} & H_{32}-ES_{32} & H_{33}-E \end{vmatrix}=0$$

If:

$$H_{11}=H_{22}=H_{33}=\alpha$$

$$H_{12}=H_{21}=H_{23}=H_{32}=\beta,\ \text{adjacent atoms}$$

$$H_{13}=H_{31}=0$$

$$S_{12}=S_{21}=S_{13}=S_{31}=S_{23}=S_{32}=0$$

The linear transformation can be applied to calculate the E-values of the different levels of a hydrogen molecule.

$$\begin{array}{ccc} 1 & 2 & 3 \\ \end{array}$$
$$\begin{vmatrix} \alpha - E & \beta & 0 \\ \beta & \alpha - E & \beta \\ 0 & \beta & \alpha - E \end{vmatrix} = 0$$

Let

$$\frac{\alpha - E}{\beta} = x$$

$$\begin{vmatrix} x & 1 & 0 \\ 1 & x & 1 \\ 0 & 1 & x \end{vmatrix} = 0$$

$$\because \begin{vmatrix} a_1 & b_1 & c_1 \\ a_2 & b_2 & c_2 \\ a_3 & b_3 & c_3 \end{vmatrix} = a_1 \begin{vmatrix} b_2 & c_2 \\ b_3 & c_3 \end{vmatrix} - b_1 \begin{vmatrix} a_2 & c_2 \\ a_3 & c_3 \end{vmatrix} + c_1 \begin{vmatrix} a_2 & b_2 \\ a_3 & b_3 \end{vmatrix} = 0$$

$$\therefore x \begin{vmatrix} x & 1 \\ 1 & x \end{vmatrix} - \begin{vmatrix} 1 & 1 \\ 0 & x \end{vmatrix} = 0$$

$$x\left(x^2 - 1\right) - x = 0$$

$$x\left(x^2 - 2\right) = 0$$

$$x = 0$$

$$x = \pm\sqrt{2}$$

$$\frac{\alpha - E}{\beta} = 0 \Rightarrow E = \alpha$$

$$\frac{\alpha - E}{\beta} = \sqrt{2} \Rightarrow E = \alpha - \sqrt{2}\beta$$

$$\frac{\alpha - E}{\beta} = -\sqrt{2} \Rightarrow E = \alpha + \sqrt{2}\beta$$

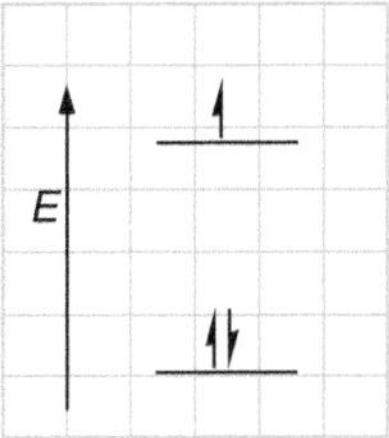

$$\text{Energy of H} - \text{H} - \text{H} = 2\left(\alpha + \sqrt{2}\beta\right) + \alpha - \sqrt{2}\beta = 3\alpha + \sqrt{2}\beta$$

b. 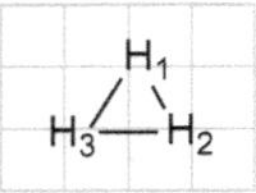

- Similar to above, the secular determinant is

$$\begin{vmatrix} H_{11} - E & H_{12} - ES_{12} & H_{13} - ES_{13} \\ H_{21} - ES_{21} & H_{22} - E & H_{23} - ES_{23} \\ H_{31} - ES_{31} & H_{32} - ES_{32} & H_{33} - E \end{vmatrix} = 0$$

If

$$H_{11} = H_{22} = H_{33} = \alpha$$

$$H_{12} = H_{13} = H_{21} = H_{23} = H_{31} = H_{32} = \beta, \text{ adjacent atoms}$$

$$S_{12} = S_{21} = S_{13} = S_{31} = S_{23} = S_{32} = 0$$

The linear transformation can be applied to calculate the E-values of the different levels of a hydrogen molecule.

- The secular determinant is reduced to

$$\begin{matrix} 1 & 2 & 3 \\ \end{matrix}$$
$$\begin{vmatrix} \alpha - E & \beta & \beta \\ \beta & \alpha - E & \beta \\ \beta & \beta & \alpha - E \end{vmatrix} = 0$$

Let $\dfrac{\alpha - E}{\beta} = x$

$$\begin{vmatrix} x & 1 & 1 \\ 1 & x & 1 \\ 1 & 1 & x \end{vmatrix} = 0$$

$$\therefore x\begin{vmatrix} x & 1 \\ 1 & x \end{vmatrix} - \begin{vmatrix} 1 & 1 \\ 1 & x \end{vmatrix} + \begin{vmatrix} 1 & x \\ 1 & 1 \end{vmatrix} = 0$$

$$x\left(x^2 - 1\right) - (x - 1) + (1 - x) = 0$$

$$(x - 1)[x(x + 1) - 1 - 1] = 0$$

$$(x - 1)\left(x^2 + x - 2\right) = 0$$

$$(x - 1)(x - 1)(x + 2) = 0$$

the roots are 1, 1, −2

$$\frac{\alpha - E}{\beta} = 1 \Rightarrow E = \alpha - \beta$$

$$\frac{\alpha - E}{\beta} = -2 \Rightarrow E = \alpha + 2\beta$$

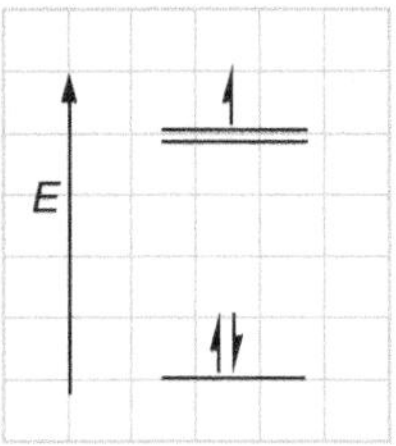

Energy of 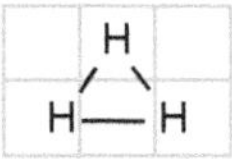 $= 2(\alpha + 2\beta) + \alpha - \beta = 3(\alpha + \beta)$

6.7 MOLECULAR ORBITALS FOR A CENTRIC MOLECULE

For BF_3:

a. **Select the atomic orbitals of B and F that can be combined.**
b. **Work out the linear combination of ligand group orbitals, LGOs, and sketch each of them.**

c. **Work out the linear combination of LGOs and the atomic orbitals of central atom.**
d. **Propose a molecular orbital diagram for the BF_3 molecule.**

Answers:

a. Orbitals that can be combined:

The valance orbitals of B and F are oriented by any choice of criterion, and combined to form σ, $\pi_{\|}$, and $\pi_{\perp}$ (Fig. 6.16).

- There are 16 atomic orbitals ready for combination (four for B and 12 for the three F atoms):

AO – type	B	F
2s	s_σ, A_1'	s_σ
$2p_x$, $2p_y$	$\left\{\begin{matrix} p_\sigma \\ p_{\pi\|} \end{matrix}\right\} E'$	$\left\{\begin{matrix} p_\sigma \\ p_{\pi\|} \end{matrix}\right\}$
$2p_z$	$p_{\pi\perp}, A_2''$	$p_{\pi\perp}$

The presentations A', E', and A_2'' are obtained from the character table for the point group D_{3h}.

b. The ligands have sets of equivalent atomic orbitals, called ligand group orbitals (LGOs), which are treated separately from the atomic orbitals of central atoms.

- The molecular orbitals are then constructed by linear combination of the LGOs with the atomic orbitals of central atom.
- LGOs:

  $2s^F$ and $2p^F$ are symmetry factorized into:

$$\sigma \rightarrow 2\,s_\sigma,\, 2p_\sigma,$$
$$\pi \rightarrow 2p_{\pi\|}, 2p_{\pi\perp}.$$

- LGOs: σ-bonding
  - σ_s: $2s^F$ form a set of LGOs:

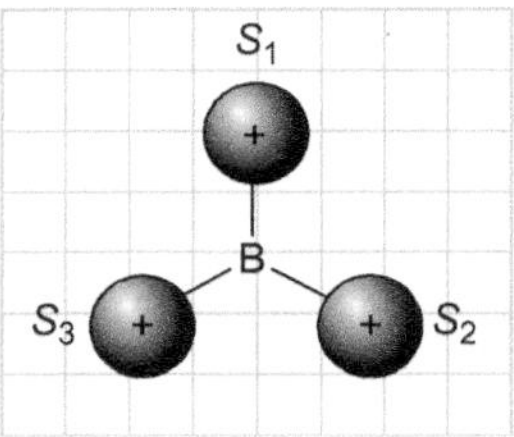

The basis set σ_s in point group D_{3h} (Chapter 5, p. 296):

D_{3h}	E	$2C_3$	$3C_2$	σ_h	$2S_3$	$3\sigma_v$	
Γ_s	3	0	1	3	0	1	$= A_1' + E'$

FIG. 6.16 Possible σ- and π-bonding of a p-orbital.

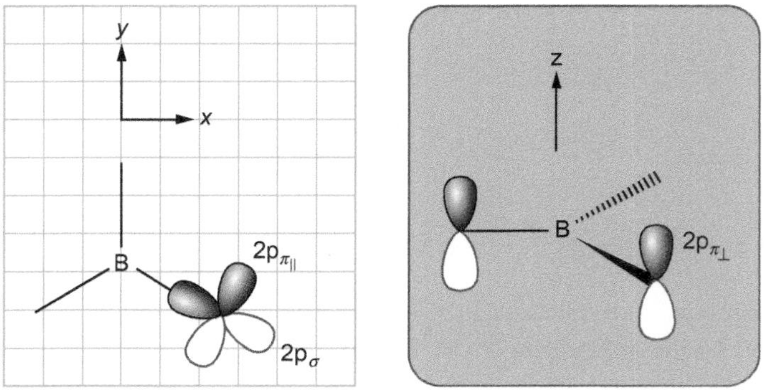

TABLE 6.3 The C_3 Character Table

C_3	E	C_3	C_3^2		
A	1	1	1	z, R_z	x^2+y^2, z^2
E	$\begin{Bmatrix}1\\1\end{Bmatrix}$	$\begin{matrix}\varepsilon\\ \varepsilon^*\end{matrix}$	$\begin{matrix}\varepsilon^*\\ \varepsilon\end{matrix}$	$(x, y)(R_x, R_y)$	$(x^2-y^2, xy)(yz, xz)$

- SALCs of $A_1{}'$ and E' are found by using the projection operators $\hat{P}_{A_1'}(s_1), \hat{P}_{E'}(s_1)$, which can be directly read from C_3-table (Table 6.3).

$$A_1' \rightarrow A = 1\,s_1 + 1\,s_2 + 1\,s_3$$

$$E' \rightarrow E = \begin{cases} 1s_1 + \varepsilon s_2 + \varepsilon^* s_3 \\ 1s_1 + \varepsilon^* s_2 + \varepsilon s_3 \end{cases}$$

The imaginary parts are removed by linear combinations, Chapter 5, p. 311.

$$E_a' = 2s_1 - s_2 - s_3$$

$$E_b' = s_2 - s_3$$

Now normalize, and draw the presentations of these LGOs:

$$a_1' : \sigma_1 = \frac{1}{\sqrt{3}}(s_1 + s_2 + s_3)$$

$$e' : \text{ i. } \sigma_2 = \frac{1}{\sqrt{6}}(2s_1 - s_2 - s_3)$$

$$\text{ii. } \sigma_3 = \frac{1}{\sqrt{2}}(s_2 - s_3) \text{ (Fig. 6.17)}$$

- σ: $2p^F(\sigma_4, \sigma_5, \sigma_6)$, again by using p_1, p_2, and p_3 as a basic set:

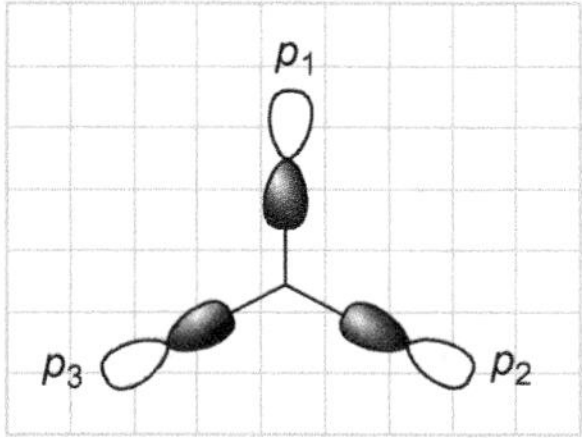

By analogy with previous, three LGOs are obtained:

$$a_1' : \sigma_4 = \frac{1}{\sqrt{3}}(p_1 + p_2 + p_3),$$

$$e' : \text{ i. } \sigma_5 = \frac{1}{\sqrt{6}}(2p_1 - p_2 - p_3), \text{ and,}$$

$$\text{ii. } \sigma_6 = \frac{1}{\sqrt{2}}(p_2 - p_3) \text{ (Fig. 6.18)}.$$

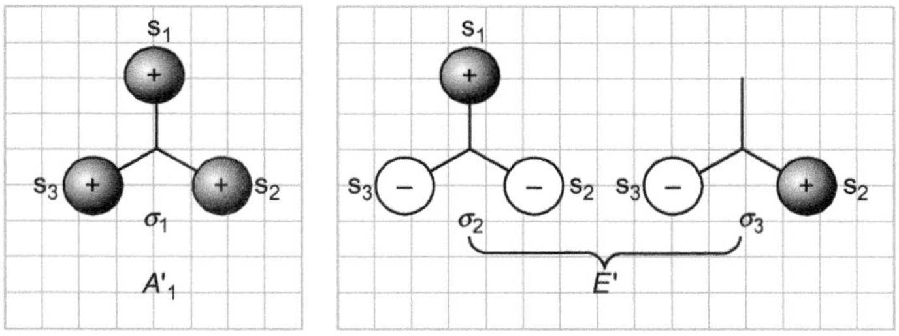

FIG. 6.17 F-$2s_\sigma$ forms a set of LGOs.

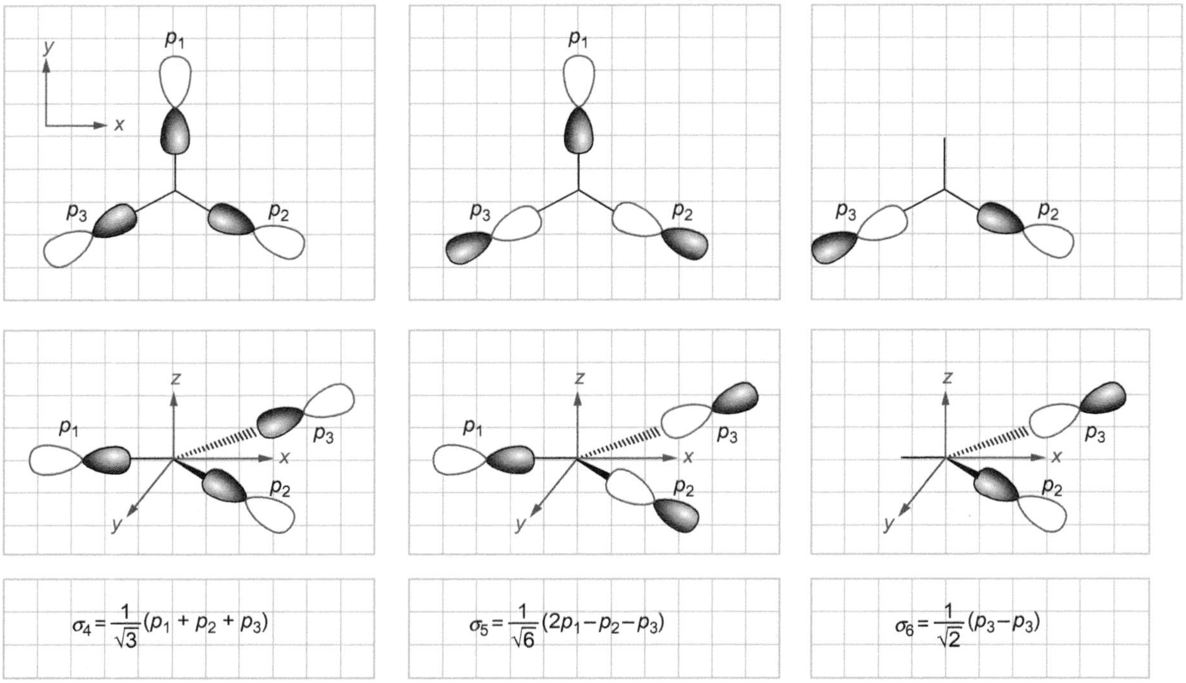

FIG. 6.18 F-2p$_\sigma$ form a set of LGOs.

- LGOs: π-bonding

  Similar treatments give F-2p$_{\pi\|}$ LGOs, and F-2p$_{\pi\perp}$ LGOs:

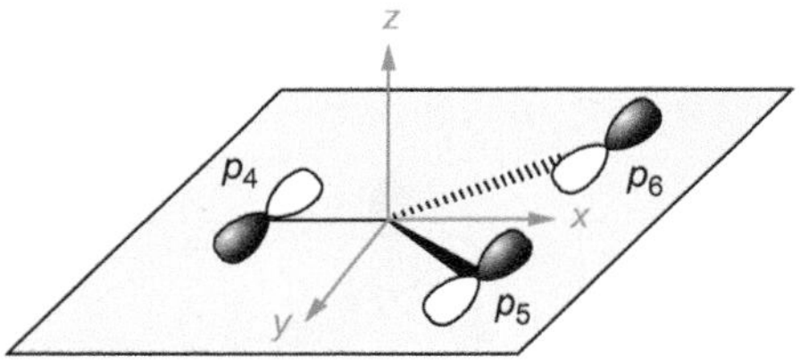

  - $\pi_{\|}$: 2p^F

D_{3h}	E	$2C_3$	$3C_2$	σ_h	$2S_3$	$3\sigma_v$	
$\Gamma_{p_{\pi\|}}$	3	0	−1	3	0	−1	$= A_2' + E'$

$$\Gamma_{p_{\pi\|}} = A_2' + E'$$

The SALCs of $A_2{}'$ and E' are found by using the projection operators $\hat{P}_{A_2'}(p_4), \hat{P}_{E'}(p_4)$ which can be directly read from C$_3$-Table (Table 6.2).

$$a_2': \quad \pi_1 = \frac{1}{\sqrt{3}}(p_4 + p_5 + p_6),$$

$$e': \quad \text{i.} \quad \pi_2 = \frac{1}{\sqrt{6}}(2p_4 - p_5 - p_6), \text{ and}$$

$$\text{ii.} \quad \pi_3 = \frac{1}{\sqrt{2}}(p_5 - p_6) \text{ (Fig. 6.19)}$$

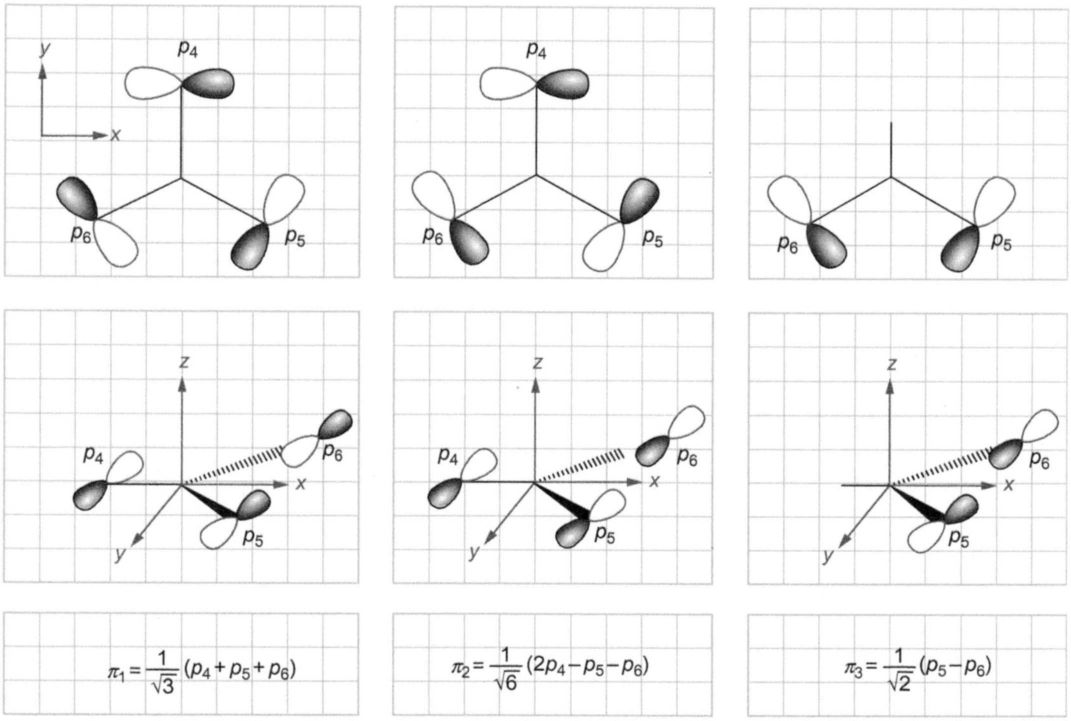

FIG. 6.19 F-$2p_{\pi\|}$ form a set of LGOs.

- $\pi_\perp$: $2p^F$
  The basis set in point group D_{3h}:

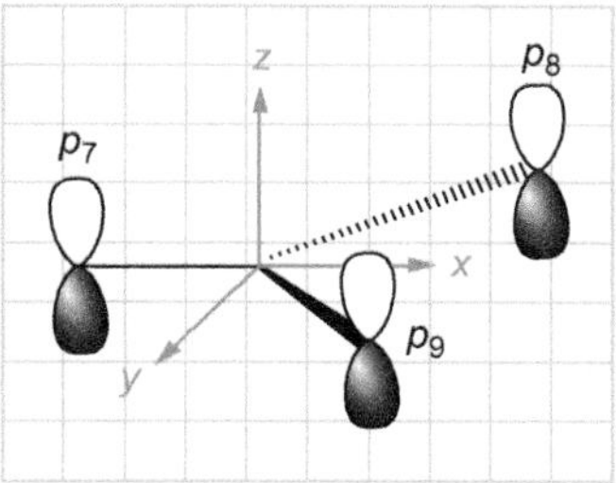

D_{3h}	E	$2C_3$	$3C_2$	σ_h	$2S_3$	$3\sigma_v$	
$\Gamma_{\pi\perp}$	3	0	−1	−3	0	1	$= A_2'' + E'$

$$\Gamma_{\pi\perp} = A_2'' + E'$$

SALCs of A_2'' and E' are found by using the projection operators:

$$a_2'' : \quad \pi_4 = \frac{1}{\sqrt{3}}(p_7 + p_8 + p_9):$$

$$e'' : \quad \text{i. } \pi_5 = \frac{1}{\sqrt{6}}(2p_7 - p_8 - p_9), \text{ and}$$

$$\text{ii. } \pi_6 = \frac{1}{\sqrt{2}}(p_8 - p_9) \text{ (Fig. 6.20)}$$

c. The linear combination of LGOs and the atomic orbitals of the central atom:
- B: $1s^2, 2s^2, 2p^1$:

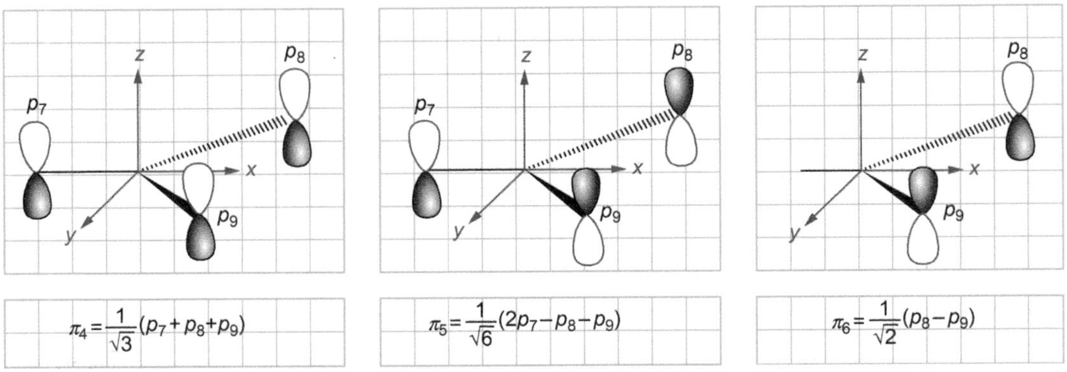

FIG. 6.20 F-2p$_{\pi\perp}$ form a set of LGOs.

There are four atomic orbitals ready for combination:

In D_{3h}

σ-bonding:	s^B	$\rightarrow$	a_1'	$\xrightarrow{\text{match}} \sigma_1^F, \sigma_4^F$
σ/π-bonding:	p_x^B, p_y^B	$\rightarrow$	e'	$\xrightarrow{\text{match}} \sigma_2^F, \sigma_3^F$
				$\xrightarrow{\text{match}} \sigma_5^F, \sigma_6^F$
				$\xrightarrow{\text{match}} \pi_2^F, \pi_3^F$
π-bonding:	p_z^B	$\rightarrow$	a_2''	$\xrightarrow{\text{match}} \pi_4^F$

Note that all boron atomic orbitals match symmetries with p_σ LGOs, and p_π LGOs:

s_σ:	σ_1^F	$\rightarrow$	a_1'
p_σ:	σ_4^F	$\rightarrow$	a_1'
$p_{\pi\|}$:	π_1^F	$\rightarrow$	a_2' (does not have matching, nonbonding)
s_σ:	σ_2^F, σ_3^F	$\rightarrow$	e'
p_σ:	σ_5^F, σ_6^F	$\rightarrow$	e'
$p_{\pi\|}$:	π_2^F, π_3^F	$\rightarrow$	e' (unfavorable)
$p_{\pi\perp}$:	π_4^F	$\rightarrow$	a_2''
$p_{\pi\perp}$:	π_5^F, π_6^F	$\rightarrow$	e'' (does not have matching, nonbonding)

- The ligands group orbitals a_2' (π_1^F) and e'' (π_5^F, π_6^F) have unique symmetries, and will be nonbonding.
- If any of p_π LGOs have the same symmetry patterns as the p_σ LGOs, the central atom atomic orbitals overlap more strongly with the p_σ sets, than p_π. That is involving of $p_{\pi\|}$-LGOs, e', are unfavorable (Fig. 6.21).
- B-F bonding is mostly formed through the terminal p_σ LGOs, a_1' and e' (Fig. 6.22), as a result, s_σ LGOs are regarded as stable lone orbitals.

  Note that $2p_x$- and $2p_y$-orbitals of boron atom are a degenerate pair in D_{3h} symmetry (e'). Consequently, all molecular orbitals involving them occur in degenerate pairs.
- Only one $p_{\pi\perp}$ LGOs (a_2'') matches symmetry (B, p_z) (Fig. 6.23).
- There are 16 SALCs:

$$\underbrace{s^B, \sigma_1^F, \sigma_4^F}_{A_1'},\ \underbrace{\pi_4^F}_{A_2'},\ \underbrace{p_x^B, \sigma_3^F, \sigma_5^F, \pi_3^F}_{E_a'},\ \underbrace{p_y^B, \sigma_2^F, \sigma_6^F, \pi_2^F}_{E_b'},\ \underbrace{p_z^B, \pi_4^F}_{A_2''},\ \underbrace{\pi_5^F}_{E_a''},\ \underbrace{\pi_6^F}_{E_b''}$$

  are used to construct the secular determinant. The entire determinant equals zero. The off-diagonal places contains zero for symmetry reason, thus each block must equal zero.
- The secular determinant has been factorized into the following blocks, each is to be solved for energies of the MOs:

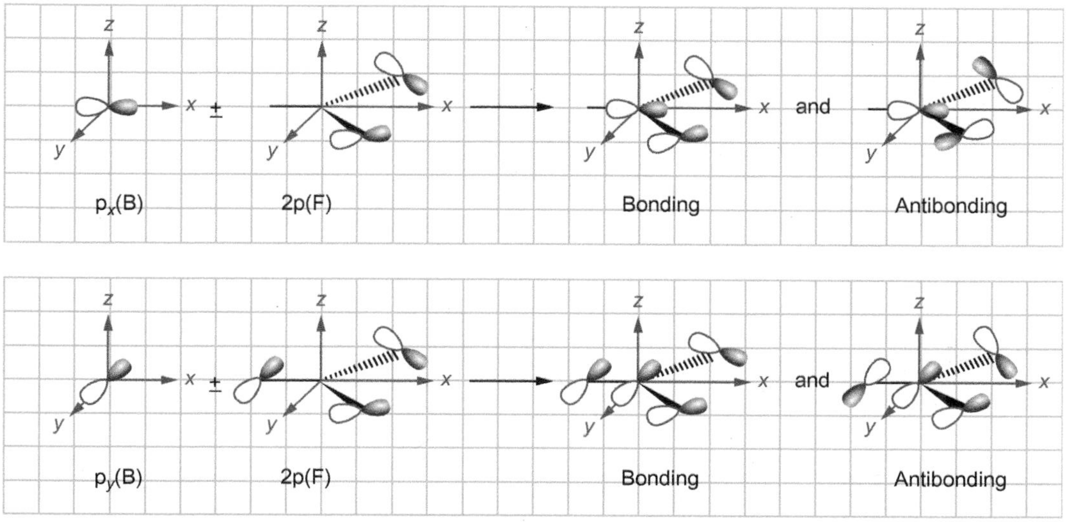

FIG. 6.21 Unfavorable overlapping of the central atom atomic orbitals with $p_{\pi\|}$-LGOs, nonbonding.

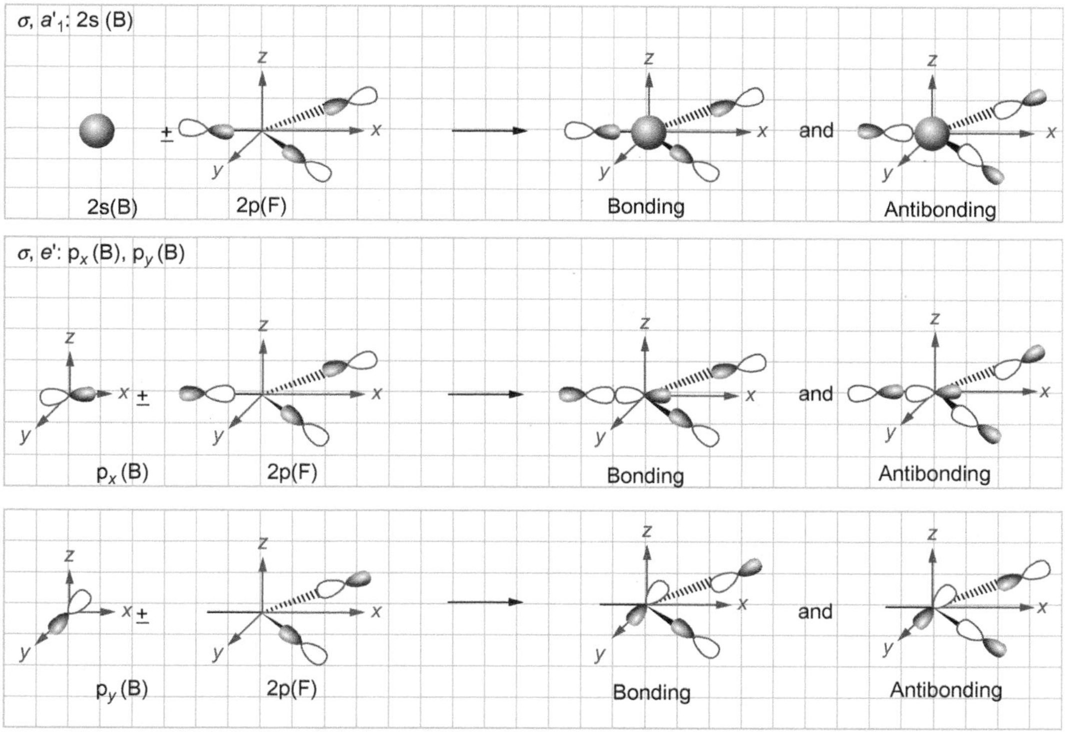

FIG. 6.22 Favorable σ-bond formation among the central atom atomic orbitals and terminal ligands.

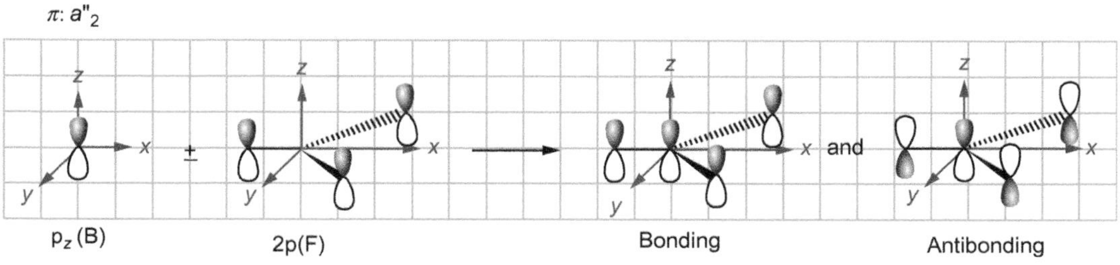

FIG. 6.23 $\pi_\perp$-bonding of the central atom atomic orbital, p_z, and$p_{\pi\perp}$-LGOs.

$$A_1' : \begin{bmatrix} H_{s^B,s^B}-E & H_{s^B,\sigma_1^F} & H_{s^B,\sigma_4^F} \\ H_{s^B,\sigma_1^F} & H_{\sigma_1^F,s\sigma_1^F}-\mathrm{E} & H_{\sigma_1^F,\sigma_4^F} \\ H_{s^B,\sigma_4^F} & H_{\sigma_1^F,\sigma_4^F} & H_{\sigma_4^F,\sigma_4^F}-E \end{bmatrix}$$

$$A_2' : E = H_{\pi_1^F,\,\pi_1^F}$$

$$E' : \begin{bmatrix} H_{p_x^B,p_x^B}-E & H_{p_x^B,\sigma_3^F} & H_{p_x^B,\sigma_5^F} & H_{p_x^B,\pi_3^F} \\ H_{p_x^B,\sigma_3^F} & H_{\sigma_3^F,\sigma_3^F}-E & H_{\sigma_3^F,\sigma_5^F} & H_{\sigma_3^F,\pi_3^F} \\ H_{p_x^B,\sigma_5^F} & H_{\sigma_5^F,\sigma_3^F} & H_{\sigma_5^F,\sigma_5^F}-E & H_{\sigma_5^F,\pi_3^F} \\ H_{p_x^B,\pi_3^F} & H_{\pi_3^F,\sigma_3^F} & H_{\pi_3^F,\sigma_5^F} & H_{\pi_3^F,\pi_3^F}-E \end{bmatrix}$$

$$E' : \begin{bmatrix} H_{p_y^B,p_y^B}-E & H_{p_y^B,\sigma_2^F} & H_{p_y^B,\sigma_6^F} & H_{p_y^B,\pi_2^F} \\ H_{p_y^B,\sigma_3^F} & H_{\sigma_2^F,\sigma_2^F}-E & H_{\sigma_2^F,\sigma_6^F} & H_{\sigma_2^F,\pi_2^F} \\ H_{p_y^B,\sigma_5^F} & H_{\sigma_5^F,\sigma_2^F} & H_{\sigma_5^F,\sigma_5^F}-E & H_{\sigma_6^F,\pi_2^F} \\ H_{p_y^B,\pi_2^F} & H_{\pi_2^F,\sigma_2^F} & H_{\pi_2^F,\sigma_5^F} & H_{\pi_2^F,\pi_2^F}-E \end{bmatrix}$$

$$A_2'' : \begin{bmatrix} H_{p_z^B,p_z^B}-E & H_{p_z^B,\,\pi_4^F} \\ H_{p_z^B,\,\pi_4^F} & H_{\pi_4^F,\,\pi_4^F}-E \end{bmatrix}$$

$$E'' : E = H_{\pi_5^F,\,\pi_5^F},\ E = H_{\pi_6^F,\,\pi_6^F}$$

d. The expected orbitals energy level diagram (Fig. 6.24):

Propose a molecular orbital scheme for the octahedral ML_6 transition metal complex. Draw a picture of each molecular orbital and suggest the energy level diagram.

- The central atom, M, has s, p, and d orbitals in its valence shell, while each L atom has s and p orbitals (Fig. 6.25).
- There are nine orbitals for M, and 24 for the set of six equivalent L atoms.

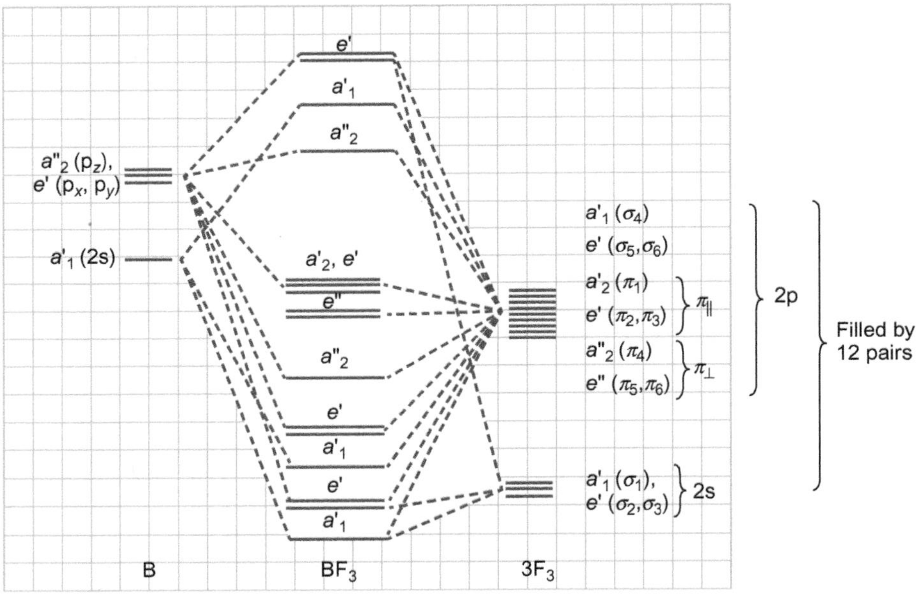

FIG. 6.24 Molecular energy diagram of BF_3.

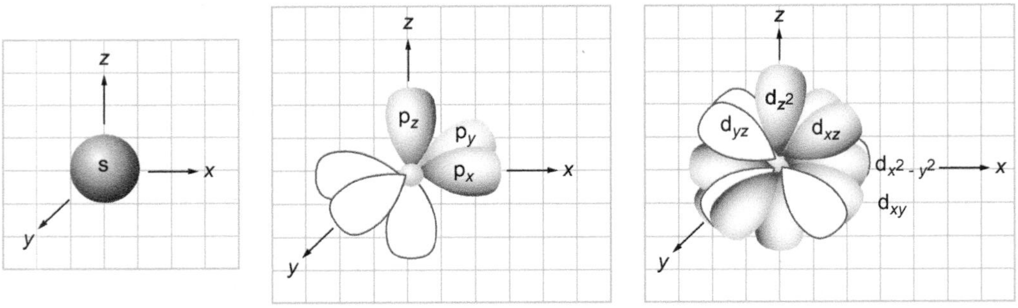

FIG. 6.25 The valence shell atomic orbitals s, p, and d of the central transition metal.

- The characters of the s, p, and d valence orbitals of M atom (Table 6.4) can be obtained directly from the O_h-character table (Table 6.5), or by using the matrix representations, e.g., the C_3 matrix:

$$C_3 \cdot \begin{bmatrix} d_{xz} \\ d_{yz} \\ d_{xy} \\ d_{z^2} \\ d_{x^2-y^2} \end{bmatrix} = \underbrace{\begin{bmatrix} 0 & 1 & 0 & 0 & 0 \\ 0 & 0 & 0 & 1 & 0 \\ 1 & 0 & 0 & 0 & 0 \\ 0 & 0 & 0 & -{}^1/_2 & {}^{\sqrt{3}}/_2 \\ 0 & 0 & 0 & -{}^{\sqrt{3}}/_2 & -{}^1/_2 \end{bmatrix}}_{\text{trac}=1} \begin{bmatrix} d_{xz} \\ d_{yz} \\ d_{xy} \\ d_{z^2} \\ d_{x^2-y^2} \end{bmatrix}$$

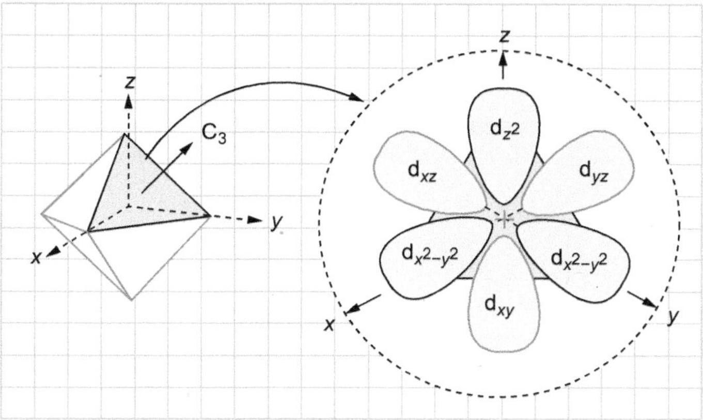

$$d_{xz} \xrightarrow{C_3} d_{yz},$$

$$d_{yz} \xrightarrow{C_3} d_{xy},$$

$$d_{xy} \xrightarrow{C_3} d_{xz},$$

TABLE 6.4 The Characters of the s, p, and d Valence Orbitals of the M Atom.

O_h	E	$8C_3$	$6C_2$	$6C_4$	$3C_2$	i	$6S_4$	$8S_6$	$3\sigma_h$	$3\sigma_d$	
Γ_s	1	1	1	1	1	1	1	1	1	1	$=A_{1g}$
Γ_p	3	0	−1	1	−1	−3	−1	0	1	1	$=T_{1u}$
Γ_d	5	−1	1	−1	1	5	−1	−1	1	1	$=E_g+T_{2g}$

TABLE 6.5 The Characters of the Ligand Sets, s_σ, p_σ, and p_π in O_h Symmetry

O_h	E	$8C_3$	$6C_2$	$6C_4$	$3C_2$	i	$6S_4$	$8S_6$	$3\sigma_h$	$3\sigma_d$	
Γ_{s_σ}	6	0	0	2	2	0	0	0	4	2	$=A_{1g}+E_g+T_{1u}$
Γ_{p_σ}	6	0	0	2	2	0	0	0	4	2	$=A_{1g}+E_g+T_{1u}$
Γ_π	12	0	0	0	−4	0	0	0	0	0	$=T_{1g}+T_{2g}+T_{1u}+T_{2u}$

$$\begin{bmatrix} d_{z^2} \\ d_{x^2-y^2} \end{bmatrix} = \begin{bmatrix} \cos 120 & -\sin 120 \\ \sin 120 & \cos 120 \end{bmatrix} \begin{bmatrix} d_{z^2} \\ d_{x^2-y^2} \end{bmatrix}$$

- Using Fig. 6.26, the ligand atom orbitals are sorted into three sets of ligand group orbitals (LGOs):

$$6s_\sigma, \left\{ \begin{matrix} 6p_x \\ 6p_y \end{matrix} \right\} p_\pi, \text{ and } 6\{p_z\}p_\sigma$$

The operations of the O_h point-group are employed to obtain the following set of characters for the representations, which reduced as indicated in Table 6.5.

- Examination of the O_h character table (Table 6.6) directly reveals that the s-, p-, and d-orbital span representations as follows:

$$s : A_{1g}$$

$$\left\{ \begin{matrix} d_{x^2-y^2} \\ d_{z^2} \end{matrix} \right\} : E_g$$

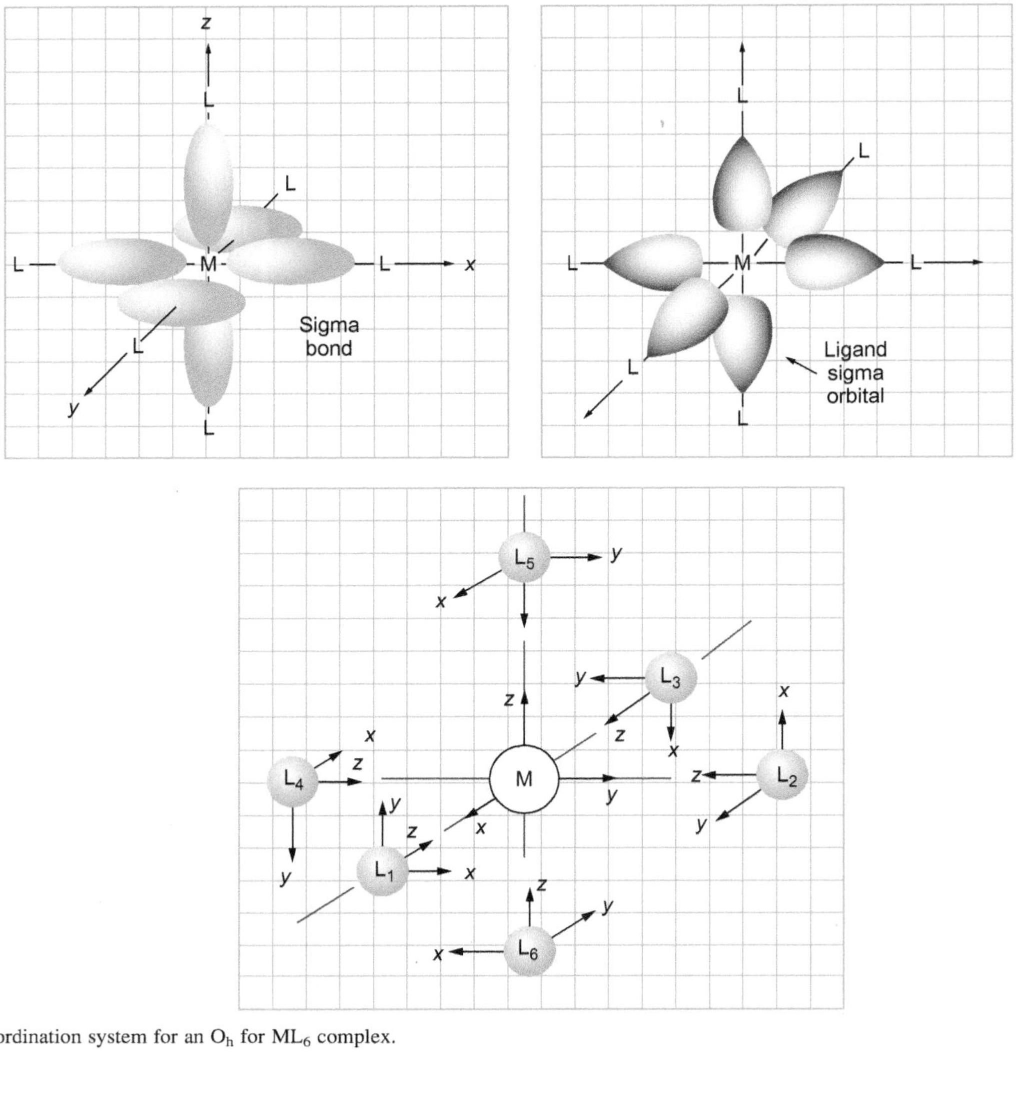

FIG. 6.26 Coordination system for an O_h for ML_6 complex.

TABLE 6.6 Character Table of the Octahedral Point Group

O_h	E	$8C_3$	$6C_2$	$6C_4$	$3C_2$	i	$6S_4$	$8S_6$	$3\sigma_h$	$6\sigma_d$		
A_{1g}	1	1	1	1	1	1	1	1	1	1	–	$x^2+y^2+z^2$
A_{2g}	1	1	−1	−1	1	1	−1	1	1	−1	–	–
E_g	2	−1	0	0	2	2	0	−1	2	0	–	$2z^2-x^2-y^2$, x^2-y^2
T_{1g}	3	0	−1	1	−1	3	1	0	−1	−1	(R_x, R_y, R_z)	–
T_{2g}	3	0	1	−1	−1	3	−1	0	−1	1	–	(xz, yz, xy)
A_{1u}	1	1	1	1	1	−1	−1	−1	−1	−1	–	–
A_{2u}	1	1	−1	−1	1	−1	1	−1	−1	1	–	–
E_u	2	−1	0	0	2	−2	0	1	−2	0	–	–
T_{1u}	3	0	−1	1	−1	−3	−1	0	1	1	(x,y,z)	–
T_{2u}	3	0	1	−1	−1	−3	1	0	1	−1	–	–

$$\left\{\begin{matrix} p_x \\ p_y \\ p_z \end{matrix}\right\} : T_{1u}$$

$$\left\{\begin{matrix} d_{xy} \\ d_{xz} \\ d_{yz} \end{matrix}\right\} : T_{2g}$$

$$\text{none} : T_{1g}$$

$$\text{none} : T_{2u}$$

- Only the usable SALCs are composed. The A_{1g} SALCs can be computed by projection operators:

$$\psi_{A_{1g}} = \frac{1}{\sqrt{6}}(\phi_1 + \phi_2 + \phi_3 + \phi_4 + \phi_5 + \phi_6)$$

where $\phi = $ s or p_z.

- The uncomplicated approach is to combine the ligand basis orbitals, treating those of each set separately, into the matched E_g, T_{1u}, and T_{2g} orbitals by tracking the central atom orbitals with which they are to interact on a one-to-one basis. For example:

$$s_\sigma : A_{1g}$$

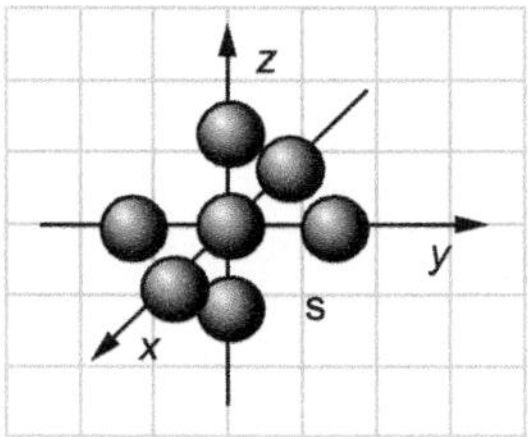

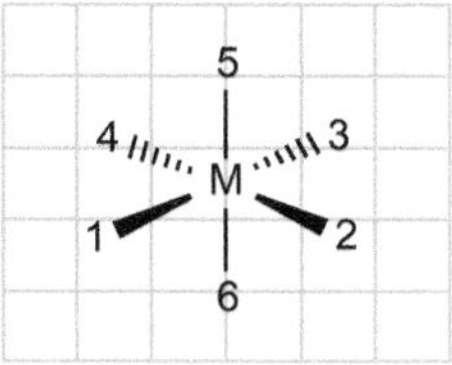

$$\psi_{A1g} = \frac{1}{\sqrt{6}}(\phi_1 + \phi_2 + \phi_3 + \phi_4 + \phi_5 + \phi_6) \Rightarrow s$$

$p_\sigma : T_{1u}$

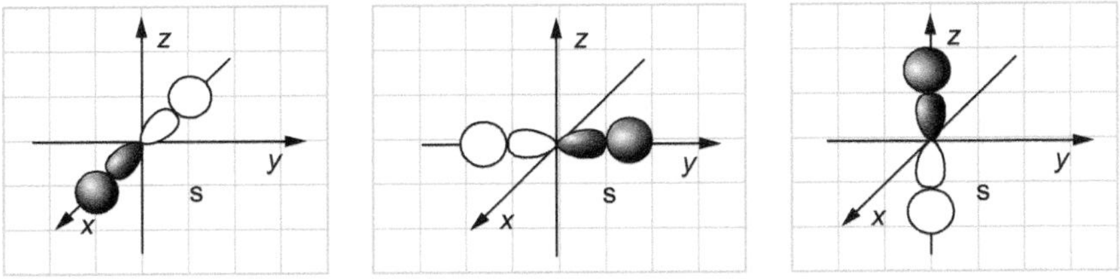

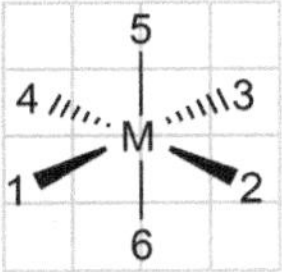

$$\psi_{T_{1u}} = \left\{ \begin{array}{l} \frac{1}{\sqrt{2}}(\phi_1 - \phi_3) \\ \frac{1}{\sqrt{2}}(\phi_2 - \phi_4) \\ \frac{1}{\sqrt{2}}(\phi_5 - \phi_6) \end{array} \right\} \Rightarrow \left\{ \begin{array}{l} p_x \\ p_y \\ p_z \end{array} \right\}$$

$d_\pi : T_{2g}$

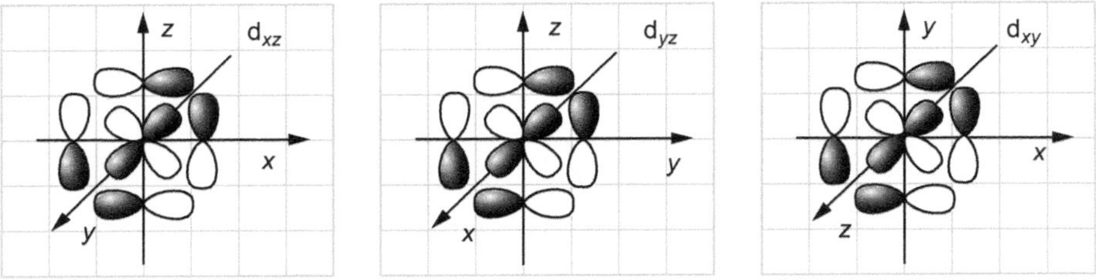

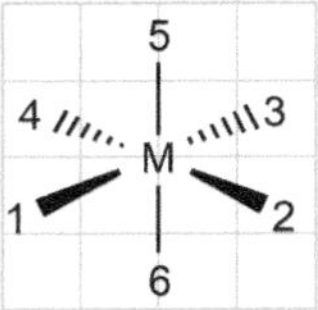

$$\psi_{T_{2g}} = \left\{ \begin{array}{l} \frac{1}{2}(p_y^1 + p_y^5 + p_y^3 + p_y^6) \\ \frac{1}{2}(p_y^2 + p_y^5 + p_y^4 + p_y^6) \\ \frac{1}{2}(p_y^1 + p_y^2 + p_y^3 + p_y^4) \end{array} \right\} \Rightarrow \left\{ \begin{array}{l} d_{xz} \\ d_{yz} \\ d_{xy} \end{array} \right\}$$

$d_\sigma : E_g$

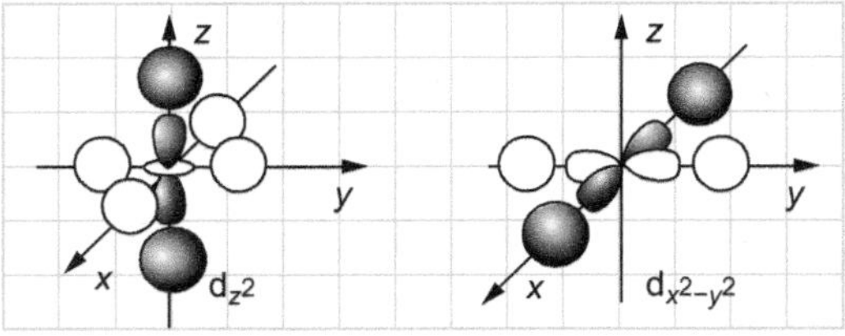

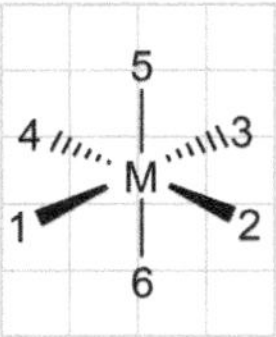

$$\psi_{E_g} = \left\{ \begin{array}{l} \frac{1}{\sqrt{12}}(2\phi_5 + 2\phi_6 - \phi_1 - \phi_2 - \phi_3 - \phi_4) \\ \frac{1}{2}(\phi_1 - \phi_2 + \phi_3 - \phi_4) \end{array} \right\} \Rightarrow \left\{ \begin{array}{l} d_{z^2} \\ d_{x^2-y^2} \end{array} \right\}$$

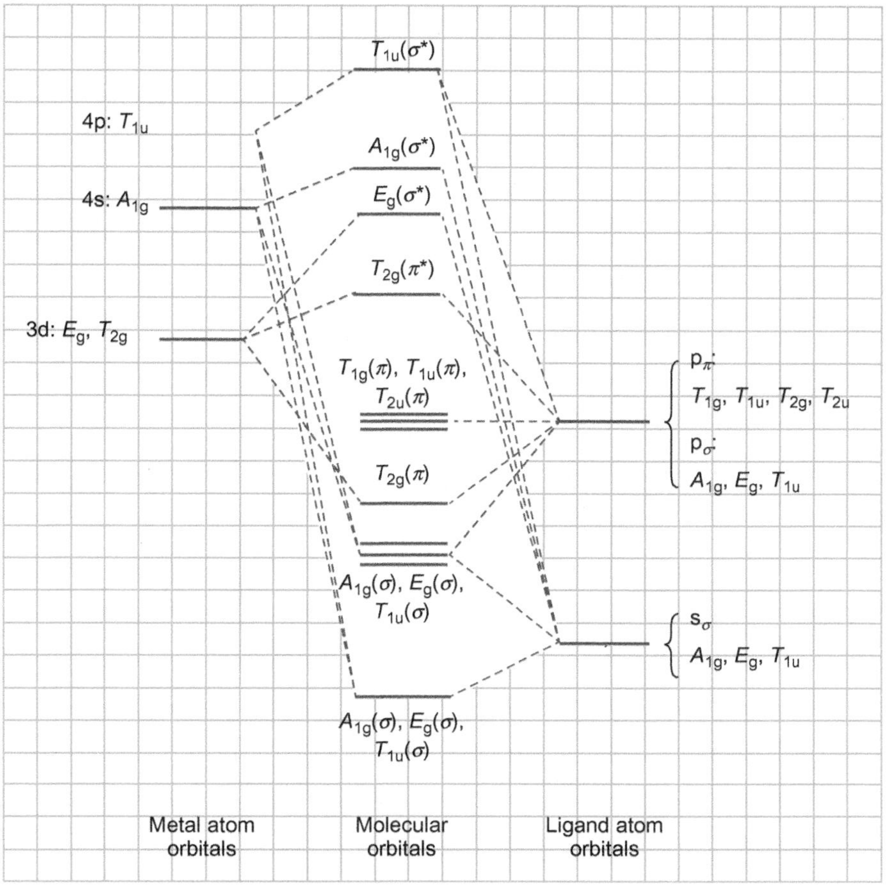

FIG. 6.27 Molecular orbital diagram for an octahedral ML_6.

These basis orbitals may be combined by LCAO method and an approximate MO diagram is constructed (Fig. 6.27).

- The forms of the ML_6 molecular orbitals are shown in Fig. 6.28.

Propose a molecular orbital scheme for a tetrahedral ML_4 molecule and construct the energy level diagram.

- For the ligand atoms, we shall use s_σ, p_σ and two p_π orbitals on each, and for the central atom, M, we will use s-, p-, and d-orbitals (Fig. 6.29).

  These provide basis of 25 orbitals, which are defined as follows:
  Set 1: four s_σ orbitals;
  Set 2: four p_σ orbitals; and
  Set 3: eight p_π orbitals.
- The symmetry operations of T_d group sort the ligand atom orbitals into sets of characters for the representations (Table 6.7).

  The SALCs can be computed by using the projection operators.
- The representation of the valence shell orbitals of the central atom are revealed from the T_d character table (Table 6.8).

$$s : A_1$$

$$\left\{\begin{matrix} d_{x^2-y^2} \\ d_{z^2} \end{matrix}\right\} : E$$

$$\left\{\begin{matrix} p_x \\ p_y \\ p_z \end{matrix}\right\} : T_2$$

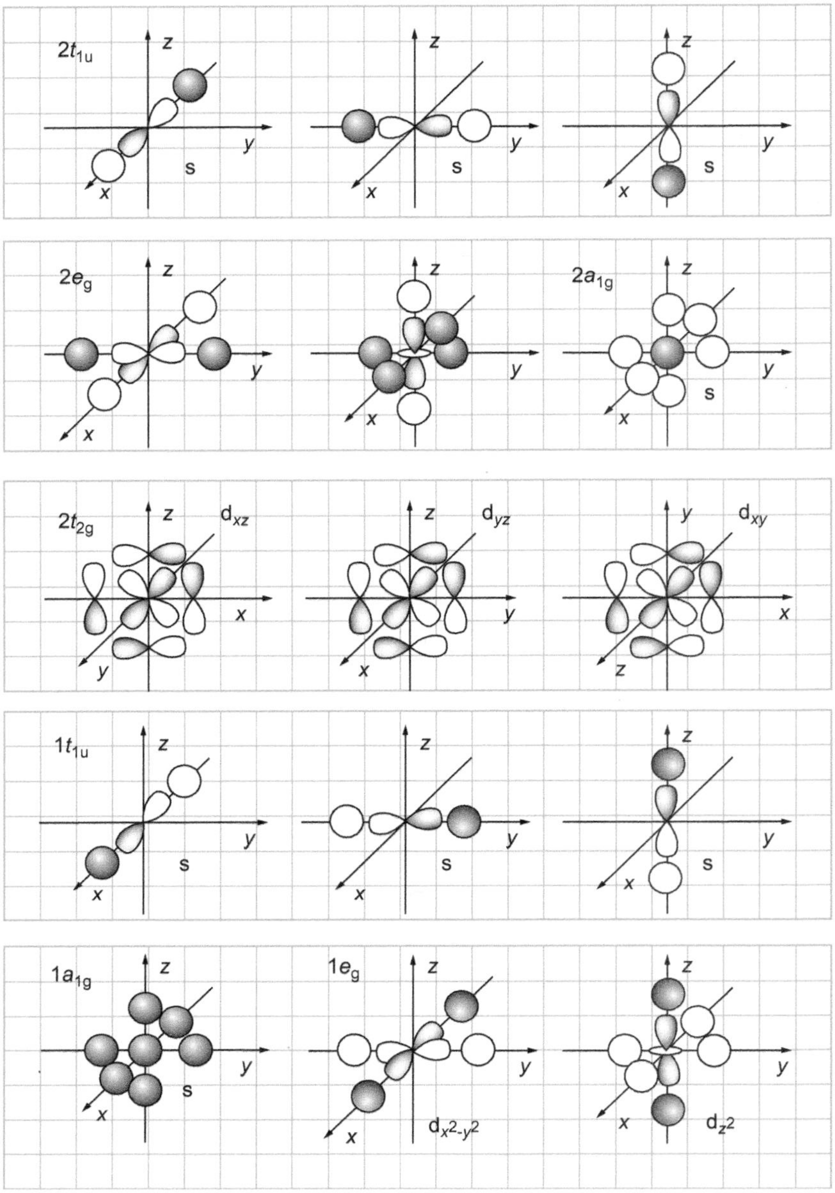

FIG. 6.28 Molecular orbitals of the ML_6 transition metal complex. Light and dark lobes are of different signs.

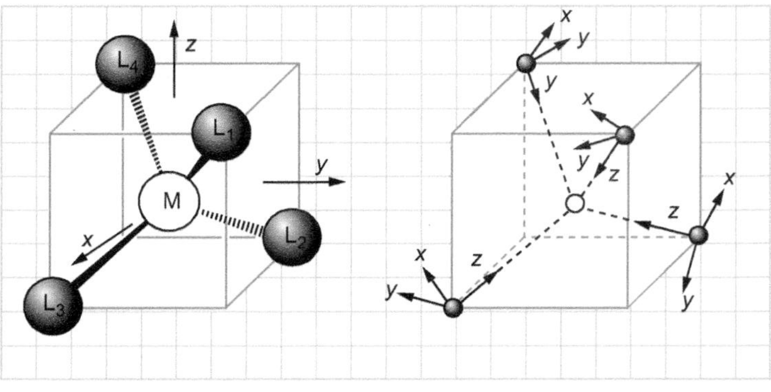

FIG. 6.29 Coordination system for tetrahedral ML_4 molecule. The x vector of each L atom and the z axis of atom M are coplanar.

TABLE 6.7 The Characters for Γ_{s_σ}, Γ_{p_σ}, and Γ_{p_π} Representations in the T_d Point Group

T_d	E	$8C_3$	$3C_2$	$6S_4$	$6\sigma_d$	
$\Gamma s_{\sigma\text{-reducible}}$	4	1	0	0	2	$=A_1+T_2$
$\Gamma p_{\sigma\text{-reducible}}$	4	1	0	0	2	$=A_1+T_2$
$\Gamma p_{\pi\text{-reducible}}$	8	−1	0	0	2	$=E+T_1+T_2$

TABLE 6.8 Character Table of the Tetrahedral Point Group

T_d	E	$8C_3$	$3C_2$	$6S_4$	$6\sigma_d$		
A_1	1	1	1	1	1	–	$x^2+y^2+z^2$
A_2	1	1	1	−1	−1	–	–
E	2	−1	2	0	0	–	$2z^2-x^2-y^2$, x^2-y^2
T_1	3	0	−1	1	−1	(R_x, R_y, R_z)	–
T_2	3	0	−1	−1	1	(x, y, z)	(xy, xz, yz)

$$\begin{Bmatrix} d_{xy} \\ d_{xz} \\ d_{yz} \end{Bmatrix} : T_2$$

- The SALCs of A_1, E, T_2 (p_σ), $T_2(p_\pi)$ can be computed by projection operators.
- The central atom orbitals combine with the ligand orbitals on a one-to-one basis:

s_σ: A_1

$$\psi_{A_1} = \frac{1}{2}(\phi_1+\phi_2+\phi_3+\phi_4) \text{ matchings}$$

p_σ: E

$$\psi_E = \begin{Bmatrix} \frac{1}{2}(p_x^1-p_x^2-p_x^3+p_x^4) \\ \frac{1}{2}(p_y^1-p_y^2-p_y^3+p_y^4) \end{Bmatrix} \text{ matching } \begin{Bmatrix} d_{z^2} \\ d_{x^2-y^2} \end{Bmatrix}$$

p_σ: T_2

$$\psi_{T_2} = \begin{Bmatrix} \frac{1}{2}(\phi^1-\phi^2+\phi^3-\phi^4) \\ \frac{1}{2}(\phi^1+\phi^2-\phi^3-\phi^4) \\ \frac{1}{2}(\phi^1-\phi^2-\phi^3+\phi^4) \end{Bmatrix} \text{ matching } \begin{Bmatrix} p_x \\ p_y \\ p_z \end{Bmatrix}$$

$$p_\pi : T_2 \rightarrow \psi_{T_2} = \begin{Bmatrix} \begin{Bmatrix} \frac{1}{4}(p_x^1+p_x^2-p_x^3-p_x^4) \\ +\frac{\sqrt{3}}{4}\left(-p_y^1-p_y^2+p_y^3+p_y^4\right) \end{Bmatrix} \\ \begin{Bmatrix} \frac{1}{4}(p_x^1-p_x^2+p_x^3-p_x^4) \\ +\frac{\sqrt{3}}{4}\left(p_y^1-p_y^2+p_y^3-p_y^4\right) \end{Bmatrix} \\ \left\{-\frac{1}{2}(p_x^1+p_x^2+p_x^3+p_x^4)\right\} \end{Bmatrix} \text{ matching } \begin{Bmatrix} d_{xz} \\ d_{yz} \\ d_{xy} \end{Bmatrix}$$

- An approximate MO diagram is constructed (Fig. 6.30).

FIG. 6.30 A molecular orbital energy diagram for tetrahedral, ML_4.

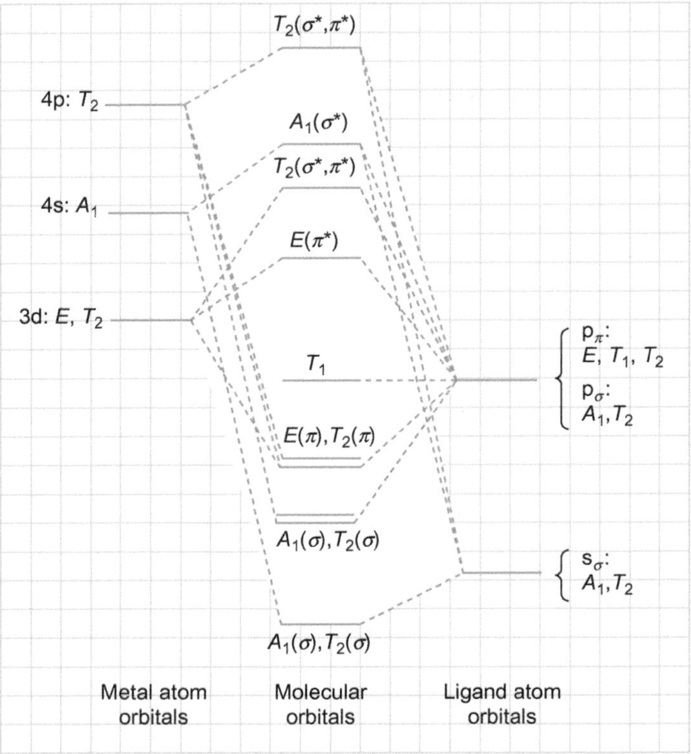

6.8 PROPERTIES DERIVED FROM MOLECULAR WAVE FUNCTION

- Beside the shape of the molecular orbital, the molecule will exist or not based on bond order, and the relative energies of the molecular orbitals, many other molecular properties are explored.

i. Energy Change in Chemical Reactions

Use the Hückel MO theory to estimate the change of the energies in the following reactions:

a. $\mathbf{H{-}H \rightarrow 2\dot{H}}$

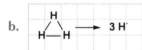

b. 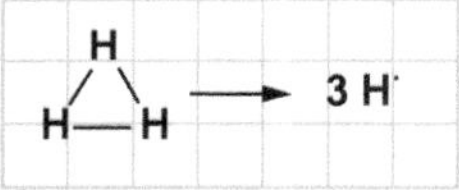

Answers:

a. $H-H \rightarrow 2\dot{H}$

- Energy levels of H_2 according to Hückel molecular orbital theory can be determined using

$$a_1(H_{11}-E)+a_2(H_{12}-ES_{12})=0$$

$$a_1(H_{21}-ES_{21})+a_2(H_{22}-E)=0$$

- These two equations can be solved using the secular determinant:

$$\begin{vmatrix} H_{11}-E & H_{12}-ES_{12} \\ H_{21}-ES_{21} & H_{22}-E \end{vmatrix} = 0 \qquad (6.3.10)$$

and

$H_{11}=H_{22}=\alpha$

$H_{12}=H_{21}=\beta$, adjacent atoms

$S_{12}=S_{21}=0$, then:

$$\begin{vmatrix} \alpha-E & \beta \\ \beta & \alpha-E \end{vmatrix}=0 \qquad (6.3.11)$$

Let $\dfrac{\alpha-E}{\beta}=x$

$$\begin{vmatrix} x & 1 \\ 1 & x \end{vmatrix}=0, \text{ then } x^2-1=0, \text{ and } x=\pm 1$$

$$\frac{\alpha-E}{\beta}=1 \Rightarrow E=\alpha-\beta$$

$$\frac{\alpha-E}{\beta}=-1 \Rightarrow E=\alpha+\beta$$

Energy of H—H $=2(\alpha+\beta)$

- $\text{H}-\text{H} \rightarrow 2\dot{\text{H}}$,

$$\Delta E=(\text{energy of } 2\dot{\text{H}})-(\text{energy of H}-\text{H}),$$

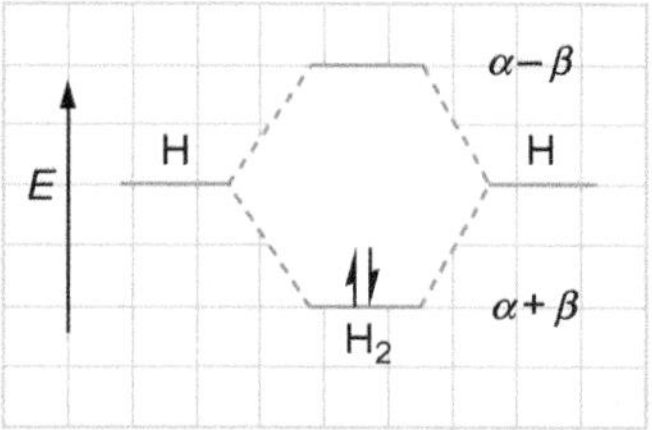

$$\Delta E=\left(2\alpha-\frac{e^2}{a_o D}\right)-(2\alpha+2\beta)$$

where:

$\alpha=H_{ii}$ is the energy of the electron in an atom (hydrogen atom) before interaction with others;

$\beta=H_{ij}$ is the energy of interaction between orbitals on adjacent atoms; and

$\dfrac{e^2}{a_o D}$ is nucleus-nucleus repulsive energy, where $a_o D$ is the distance between the nuclei.

$$\Delta E=-2\beta-\frac{e^2}{a_o D}$$

- 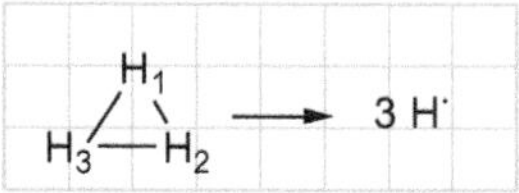

The secular determinant is reduced to the following:

$$\begin{vmatrix} x & 1 & 1 \\ 1 & x & 1 \\ 1 & 1 & x \end{vmatrix}=0$$

$$\therefore x\begin{vmatrix} x & 1 \\ 1 & x \end{vmatrix} - \begin{vmatrix} 1 & 1 \\ 1 & x \end{vmatrix} + \begin{vmatrix} 1 & x \\ 1 & 1 \end{vmatrix} = 0$$

$$x\left(x^2 - 1\right) - (x-1) + (1-x) = 0$$

$$(x-1)[x(x+1) - 1 - 1] = 0$$

$$(x-1)\left(x^2 + x - 2\right) = 0$$

$$(x-1)(x-1)(x+2) = 0$$

the roots are 1, 1, −2

$$\frac{\alpha - E}{\beta} = 1 \Rightarrow E = \alpha - \beta$$

$$\frac{\alpha - E}{\beta} = -2 \Rightarrow E = \alpha + 2\beta$$

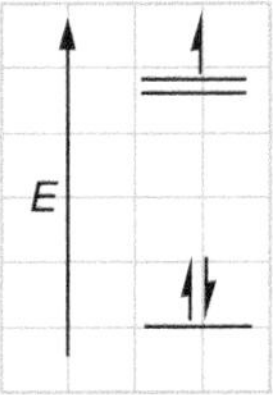

Energy of 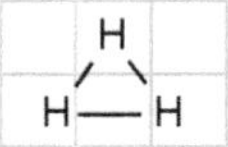 $= 2(\alpha + 2\beta) + \alpha - \beta = 3(\alpha + \beta)$

$\triangle E = (\text{energy of } 3\dot{H}) - \left(\text{energy of }\right.$ 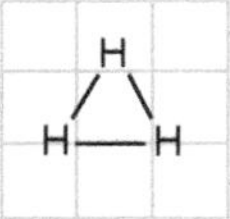 $\left.\right)$,

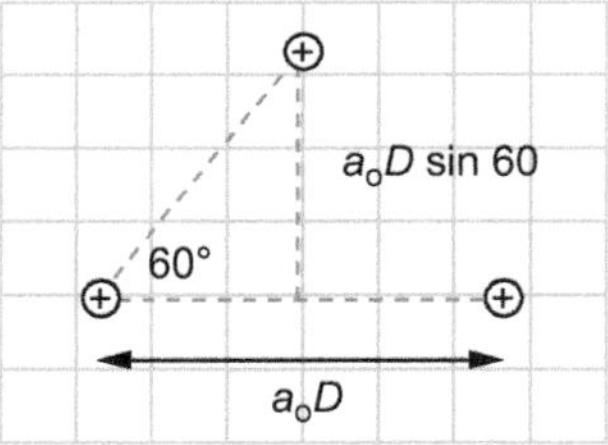

$$\Delta E = \left(3\alpha - \left(\frac{e^2}{a_oD} + \frac{2e^2}{a_oD\sin 60}\right)\right) - (2\alpha + 2\beta)$$

$\left(\frac{e^2}{a_oD} + \frac{2e^2}{a_oD\sin 60}\right)$: energy of three protons placed in the corners of equatorial triangle

$$\Delta E = -3\beta - \frac{e^2}{a_oD} - \frac{2e^2}{a_oD\sin 60}$$

ii. Chemical Stability

Calculate by the LCAO molecular orbital method whether the linear (H—H—H)$^+$ or the triangular (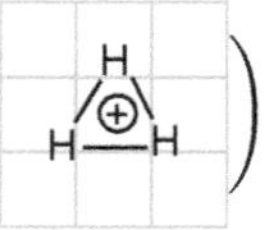) **state of H_3^+ is more stable.**

- The total energy of  $= 2(\alpha + 2\beta) + 1(\alpha - \beta) = 3(\alpha + \beta)$

- The energy diagram of the triangular state of H_3^+:

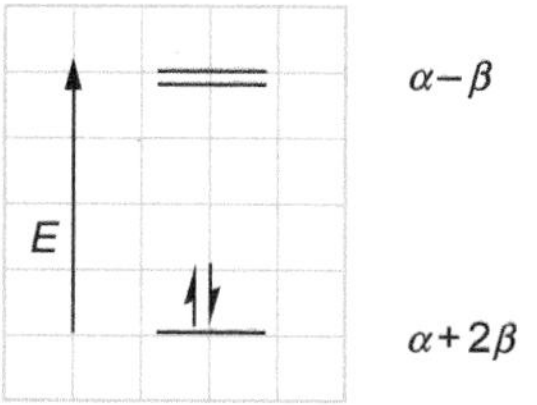

total energy of 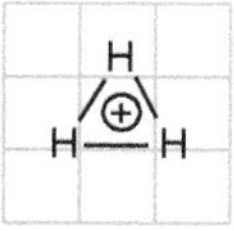 $= 2\alpha + 4\beta$

- The energy diagram of the linear state of H_3^+:

$H_{(1)}—H_{(2)}—H_{(3)}$, the secular determinant:

$$\begin{matrix} 1 & 2 & 3 \end{matrix}$$
$$\begin{vmatrix} \alpha - E & \beta & 0 \\ \beta & \alpha - E & \beta \\ 0 & \beta & \alpha - E \end{vmatrix} = 0$$

Let $\dfrac{\alpha - E}{\beta} = x$

$$\begin{vmatrix} x & 1 & 0 \\ 1 & x & 1 \\ 0 & 1 & x \end{vmatrix} = 0$$

$$\because \begin{vmatrix} a_1 & b_1 & c_1 \\ a_2 & b_2 & c_2 \\ a_3 & b_3 & c_3 \end{vmatrix} = a_1 \begin{vmatrix} b_2 & c_2 \\ b_3 & c_3 \end{vmatrix} - b_1 \begin{vmatrix} a_2 & c_2 \\ a_3 & c_3 \end{vmatrix} + c_1 \begin{vmatrix} a_2 & b_2 \\ a_3 & b_3 \end{vmatrix} = 0$$

$$\therefore x \begin{vmatrix} x & 1 \\ 1 & x \end{vmatrix} - \begin{vmatrix} 1 & 1 \\ 0 & x \end{vmatrix} = 0$$

$$x(x^2 - 1) - x = 0$$

$$x(x^2 - 2) = 0$$

$$x = 0$$

$$x = \pm\sqrt{2}$$

$$\frac{\alpha - E}{\beta} = 0 \Rightarrow E = \alpha$$

$$\frac{\alpha - E}{\beta} = \sqrt{2} \Rightarrow E = \alpha - \sqrt{2}\beta$$

$$\frac{\alpha - E}{\beta} = -\sqrt{2} \Rightarrow E = \alpha + \sqrt{2}\beta$$

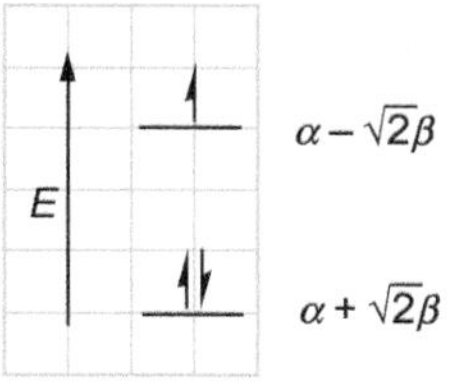

$$\text{Energy of H}-\text{H}-\text{H} = 2\left(\alpha + \sqrt{2}\beta\right) + \alpha = 3\alpha + 2\sqrt{2}\beta$$

- Energy of $\text{H}-\text{H}-\text{H}^+ = 2\left(\alpha + \sqrt{2}\beta\right) = 2\alpha + 2\sqrt{2}\beta$

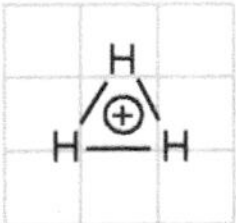
is more stable than H—H—H$^+$ by $\left(4 - 2\sqrt{2}\right)\beta$

iii. Chemical Reactivity
Consider the π-system of cyclobutadiene, then:

a. **calculate the total representation of π-orbitals for the four carbons and decompose these into irreducible representations;**
b. **find all the wave functions;**
c. **calculate the normalization factor for each wave functions, and indicate which wave functions can mix;**
d. **sketch the shapes of these π-molecular orbitals;**
e. **set the energy levels correlation diagram of the molecular wave functions; and**
f. **discuss the reactivity of C_4H_4 in compare to the trans-diphenyl derivative.**

The answers:

- The out of plane $\pi_\perp$-system of cyclobutadiene is formed by a set of four p_π orbitals; one on each carbon atom is taken as the basis for a representation of the group D_{4h}.

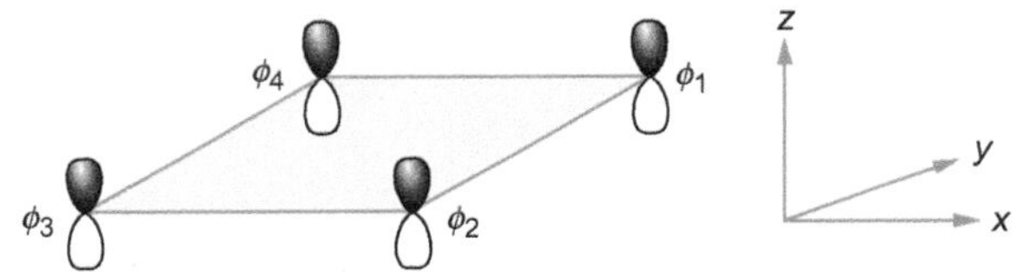

- The reducible and irreducible presentations for which these four atomic orbitals form a basis are obtained.

D_{4h}	E	$2C_4$	C_2	$2C_2'$	$2C_2''$	i	$4S_4$	σ_h	$2\sigma_v$	$2\sigma_d$
$\Gamma_{\pi\perp}$	4	0	0	−2	0	0	0	−4	2	0

$$\Gamma_{\pi\perp} = A_{2u} + B_{2u} + E_g$$

- SALCs from the atomic orbitals are formed by applying the projection operation $\hat{P}_{A_{2u}}(\phi_1)$ to get A_{2u} in terms of ϕ_1, ϕ_2, ϕ_3 and ϕ_4

Easier: by using the relative coefficient from the character table for C_4 (Chapter 5, p. 317 (Table 6.9)).

$$A_{2u} = \phi_1 + \phi_2 + \phi_3 + \phi_4$$

$$B_{2u} = \phi_1 - \phi_2 + \phi_3 - \phi_4$$

$$E_{ga} = \phi_1 + i\phi_2 - \phi_3 - i\phi_4$$

$$E_{gb} = \phi_1 - i\phi_2 - \phi_3 + i\phi_4$$

TABLE 6.9 Character Table for the C_4 Point Group

C_4	E	$2C_4$	C_2	$2C_4^3$
$A_{2u} \rightarrow A$	1	1	1	1
$B_{2u} \rightarrow B$	1	−1	1	−1
$E_g \rightarrow E$	$\left\{\begin{matrix}1 \\ 1\end{matrix}\right.$	$\begin{matrix}i \\ -i\end{matrix}$	$\begin{matrix}-1 \\ -1\end{matrix}$	$\left.\begin{matrix}-i \\ 1\end{matrix}\right\}$

- The linear combinations of E_{ga} and E_{gb} is employed to exclude the imaginary i:

$$E_{ga} + E_{gb} = E_{g1} = 2\phi_1 - 2\phi_2 = \phi_1 - \phi_2$$
$$E_{ga} - E_{gb} = E_{g2} = 2\phi_2 - 2\phi_4 = \phi_2 - \phi_4$$

- The sketch of relative atomic orbital contributions, viewed along C_4:

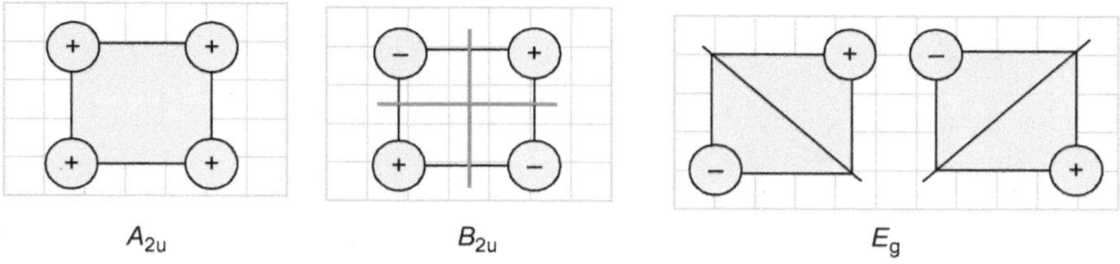

- After normalization:

$$a_{2u} = \frac{1}{2}(\phi_1 + \phi_2 + \phi_3 + \phi_4) \rightarrow \Psi_1$$

$$e_g = \begin{cases} 1/\sqrt{2}\,(\phi_1 - \phi_3) \rightarrow \Psi_2 \\ 1/\sqrt{2}\,(\phi_2 - \phi_4) \rightarrow \Psi_3 \end{cases}$$

$$b_{2u} = \frac{1}{2}(\phi_1 - \phi_2 + \phi_3 - \phi_4) \rightarrow \Psi_4$$

- If the entire determinant is to have the value of zero, each block factor separately must equal zero.

$$\begin{aligned} E_{a_{2u}} &= \int \Psi_1 H \Psi_1 d\tau \\ &= \frac{1}{4}\int (\phi_1 + \phi_2 + \phi_3 + \phi_4) H (\phi_1 + \phi_2 + \phi_3 + \phi_4)\, d\tau \\ &= \frac{1}{4}\int \phi_1 H \phi_1\, d\tau + \int \phi_1 H \phi_2\, d\tau + \int \phi_1 H \phi_4 d\tau \\ &+ \int \phi_2 H \phi_2\, d\tau + \int \phi_2 H \phi_1\, d\tau + \int \phi_2 H \phi_3\, d\tau \\ &+ \int \phi_3 H \phi_3\, d\tau + \int \phi_3 H \phi_2 d\tau + \int \phi_3 H \phi_4\, d\tau \\ &+ \int \phi_4 H \phi_4\, d\tau + \int \phi_4 H \phi_3\, d\tau + \int \phi_4 H \phi_1\, d\tau \end{aligned}$$

$$E_{a_{2u}} = \alpha + 2\beta$$

- Similarly: $E_{ga} = \alpha$

$$E_{gb} = \alpha$$
$$E_{b_{2u}} = \alpha - 2\beta$$

- Accordingly, the molecular electronic configuration of cyclobutadiene: $a_{2u}^2\ e_g^2$

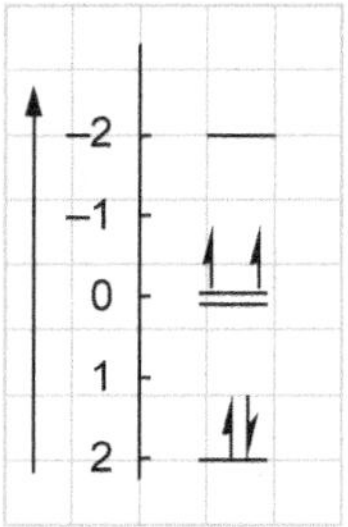

- Paramagnetic organics are usually reactive, so cyclobutadiene reacts with transition metals to form complexes, also forms stable dication dimerizes (by loss of e_g^2).
- However, trans-diphenyl cyclobutadiene belongs to the C_{2h} point group. The valence electrons are redistributed to have only paired electrons that form the more stable diamagnetic compound (Fig. 6.31).

iv. Delocalization Energy
By using the Hückel MO theory, evaluate:

a. the π-energies of butadiene and cyclobutadiene;
b. the delocalization energy; and
c. which of the two compounds is the more stable.
  - Butadiene:

$$\overset{1}{C}H_2{=}\overset{2}{C}H{-}\overset{3}{C}H{=}\overset{4}{C}H_2$$

FIG. 6.31 Change the symmetry involves redistribution of electrons.

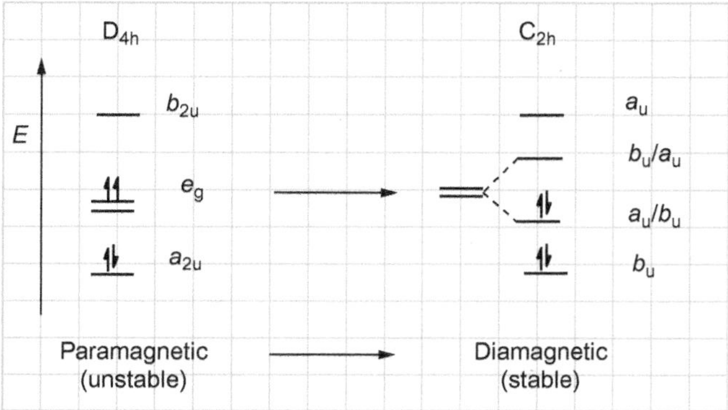

The secular determinant:

$$\begin{vmatrix} \alpha-E & \beta & 0 & 0 \\ \beta & \alpha-E & \beta & 0 \\ 0 & \beta & \alpha-E & \beta \\ 0 & 0 & \beta & \alpha-E \end{vmatrix} = 0$$

If $\dfrac{\alpha-E}{\beta} = x$, then the secular determinant:

$$\begin{vmatrix} x & 1 & 0 & 0 \\ 1 & x & 1 & 0 \\ 0 & 1 & x & 1 \\ 0 & 0 & 1 & x \end{vmatrix} = 0$$

$$x\begin{vmatrix} x & 1 & 0 \\ 1 & x & 1 \\ 0 & 1 & x \end{vmatrix} - 1\begin{vmatrix} 1 & 1 & 0 \\ 0 & x & 1 \\ 0 & 1 & x \end{vmatrix} = 0$$

$$x^2\begin{vmatrix} x & 1 \\ 1 & x \end{vmatrix} - x\begin{vmatrix} 1 & 1 \\ 0 & x \end{vmatrix} - \begin{vmatrix} x & 1 \\ 1 & x \end{vmatrix} + \begin{vmatrix} 0 & 1 \\ 0 & x \end{vmatrix} = 0$$

$$x^4 - x^2 - x^2 - x^2 + 1 = 0$$

$$x^4 - 3x^2 + 1 = 0, \quad \text{let } x^2 = y$$

$$y^2 - 3y + 1 = 0$$

$$y = \frac{-b \pm \sqrt{b^2 - 4ac}}{2a}$$

$$y = x^2 = \frac{3 \pm \sqrt{9-4}}{2} = \frac{3 \pm \sqrt{5}}{2} = \frac{3 \pm 2.24}{2} = 2.62 \text{ and } 0.38$$

$$x = \frac{\alpha - E}{\beta} = \pm 1.62 \text{ and } \pm 0.62$$

$E(\pi_1) = \alpha + 1.62\beta$, $E(\pi_2) = \alpha + 0.62\beta$, $E(\pi_3) = \alpha - 0.62\beta$
$E(\pi_4) = \alpha - 1.62\beta$

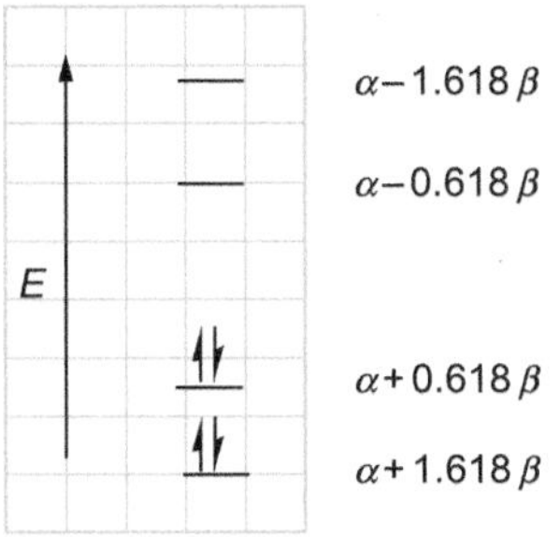

Energy of butadiene $= 2(\alpha + 1.62\beta) + 2(\alpha + 0.62\beta) = 4\alpha + 4.48\beta$
Remember that β is negative; we see that π-bonding has stabilized the molecule by $4\alpha + 4.48\beta$.

- The delocalization energy is the difference between the energies of two-ethylene double bond and the ground state energy of butadiene.

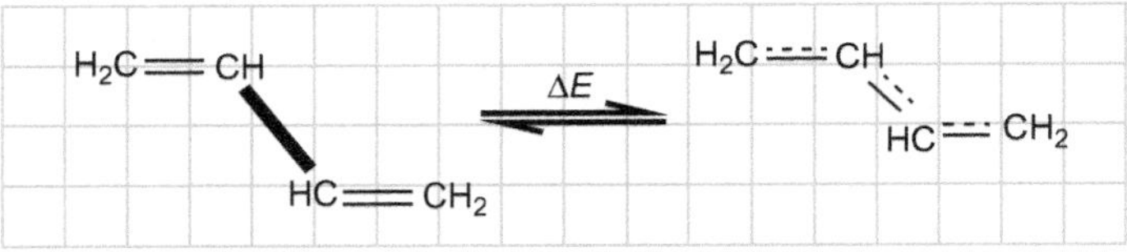

Two ethylene Butadiene

ΔE = delocalization energy

- Two ethylene π-bonds:

$$\overset{1}{\mathrm{CH}}=\overset{2}{\mathrm{CH}}$$

The secular determinant:

$$\begin{vmatrix} \alpha-E & \beta \\ \beta & \alpha-E \end{vmatrix}=0$$

If $\dfrac{\alpha-E}{\beta}=x$, then the secular determinant:

$$\begin{vmatrix} x & 1 \\ 1 & x \end{vmatrix}=0$$

$$x^2-1=0,\quad x=\pm 1$$

$$\frac{\alpha-E}{\beta}=1 \Rightarrow E=\alpha-\beta$$

$$\frac{\alpha-E}{\beta}=-1 \Rightarrow E=\alpha+\beta$$

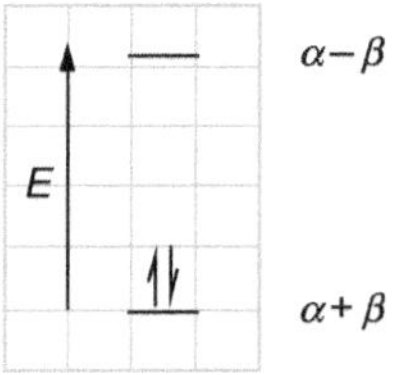

E_π of the ethylene π-bonds $=2\alpha+2\beta$
E_π of two ethylene π-bonds $=4\alpha+4\beta$

$$\text{The delocalization energy}=\begin{bmatrix}\text{the ground state}\\ \text{energy}\\ \text{of Butadiene}\end{bmatrix}-\begin{bmatrix}\text{the energies}\\ \text{of two}\\ \text{ethylene double bond}\end{bmatrix}$$

$$\text{Delocalization energy in butadiene}=4\alpha+4.48\beta-(4\alpha+4\beta)=0.48\beta$$

- Cyclobutadiene:

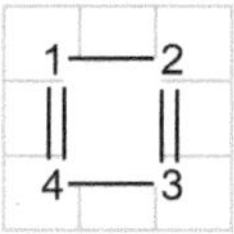

The secular determinant:

$$\begin{vmatrix} \alpha-E & \beta & 0 & \beta \\ \beta & \alpha-E & \beta & 0 \\ 0 & 0 & \alpha-E & \beta \\ \beta & 0 & \beta & \alpha-E \end{vmatrix}=0$$

If $\frac{\alpha-E}{\beta}=x$, then the secular determinant:

$$\begin{vmatrix} x & 1 & 0 & 1 \\ 1 & x & 1 & 0 \\ 0 & 1 & x & 1 \\ 1 & 0 & 1 & x \end{vmatrix}=0$$

$$x\begin{vmatrix} x & 1 & 0 \\ 1 & x & 1 \\ 0 & 1 & x \end{vmatrix}-\begin{vmatrix} 1 & 1 & 0 \\ 0 & x & 1 \\ 1 & 1 & x \end{vmatrix}-\begin{vmatrix} 1 & x & 1 \\ 0 & 1 & x \\ 1 & 0 & 1 \end{vmatrix}=0$$

$$x^2\begin{vmatrix} x & 1 \\ 1 & x \end{vmatrix}-x\begin{vmatrix} 1 & 1 \\ 0 & x \end{vmatrix}-\begin{vmatrix} x & 1 \\ 1 & x \end{vmatrix}+\begin{vmatrix} 0 & 1 \\ 1 & x \end{vmatrix}-\begin{vmatrix} 1 & x \\ 0 & 1 \end{vmatrix}+x\begin{vmatrix} 0 & x \\ 1 & 1 \end{vmatrix}-\begin{vmatrix} 0 & 1 \\ 1 & 0 \end{vmatrix}=0$$

$$x^4-x^2-x^2-x^2+1-1-1+x^2+1=0$$

$$x^4-4x^2=x^2(x^2-4)=0$$

$$x_1=0,\ x_2=0,\ x_3=2,\ x_4=-2$$

$$x_1=x_2=\frac{\alpha-E}{\beta}=0\Rightarrow E=\alpha$$

$$x_3=\frac{\alpha-E}{\beta}=2\Rightarrow E=\alpha-2\beta$$

$$x_4=\frac{\alpha-E}{\beta}=-2\Rightarrow E=\alpha+2\beta$$

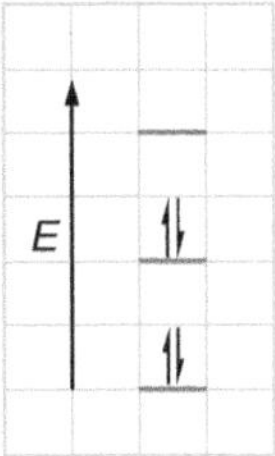

$$E_\pi=2(\alpha+2\beta)+2\alpha=4\alpha+4\beta$$

E_π of two ethylene π-bonds $=4\alpha+4\beta$
The π-bonding has stabilized the molecule by $4\alpha+4\beta$:

$$\text{Delocalization energy}=(4\alpha+4\beta)-(4\alpha+4\beta)=0$$

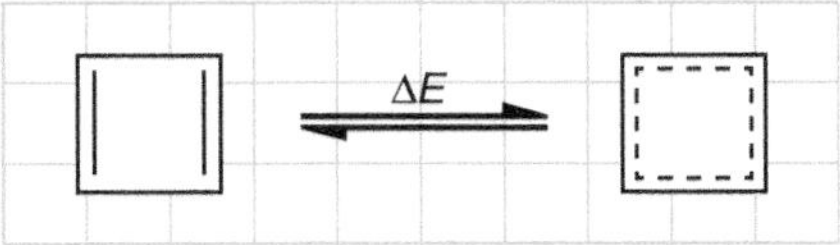

ΔE = delocalization energy

- Since the delocalization energy is 0:
  - the π system of cyclobutadiene is not more stable than that of two localized π-bonds;
  - the formation of a ring compound with 90 degree angles introduces a strain that should let cyclobutadiene be less stable than two ethylenes. This instability might explain why cyclobutadiene is difficult to synthesize.; and
  - butadiene is more stable than cyclobutadiene.

Estimate the delocalization energy in trimethylenemethane.

- Trimethylenemethane:

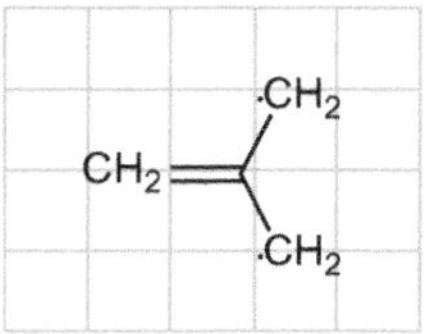

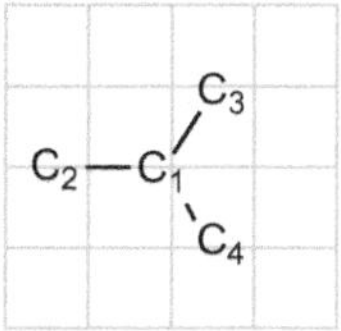

- The secular determinant:

$$\begin{vmatrix} \alpha-E & \beta & \beta & \beta \\ \beta & \alpha-E & 0 & 0 \\ \beta & 0 & \alpha-E & 0 \\ \beta & 0 & 0 & \alpha-E \end{vmatrix} = 0$$

- If $\dfrac{\alpha-E}{\beta}=x$, then the secular determinant:

$$\begin{vmatrix} x & 1 & 1 & 1 \\ 1 & x & 0 & 0 \\ 1 & 0 & x & 0 \\ 1 & 0 & 0 & x \end{vmatrix} = 0$$

The determinant expands to

$$x\begin{vmatrix} x & 0 & 0 \\ 0 & x & 0 \\ 0 & 0 & x \end{vmatrix} - \begin{vmatrix} 1 & 0 & 0 \\ 1 & x & 0 \\ 1 & 0 & x \end{vmatrix} + \begin{vmatrix} 1 & x & 0 \\ 1 & 0 & 0 \\ 1 & 0 & x \end{vmatrix} - \begin{vmatrix} 1 & x & 0 \\ 1 & 0 & x \\ 1 & 0 & 0 \end{vmatrix} = 0$$

$$x^4 - x^2 - x^2 - x^2 = 0$$

$$x^4 - 3x^2 = 0$$

$$x^2\left(x^2-3\right) = 0$$

The roots are $x=0$, $x=0$, $x=\sqrt{3}$, $x=-\sqrt{3}$
The energies: $\alpha+\sqrt{3}\beta, \alpha, \alpha, \alpha-\sqrt{3}\beta$

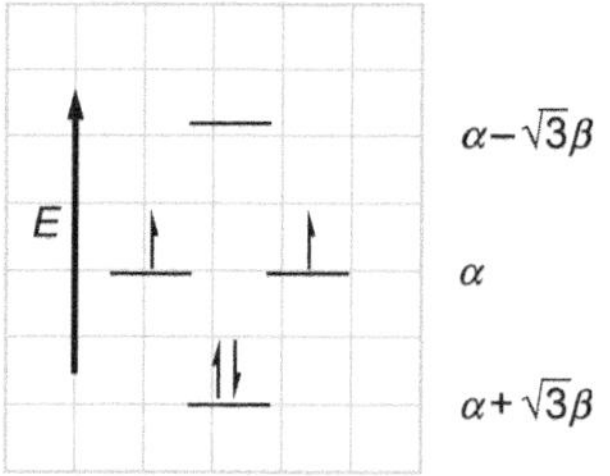

$$\text{Total } \pi - \text{energy} = 2\left(\alpha+\sqrt{3}\beta\right) + 2\alpha$$

- The delocalization energy is the difference between the energies of the 2p carbon electrons plus ethylene double bond and the ground state energy of trimethylenemethane.

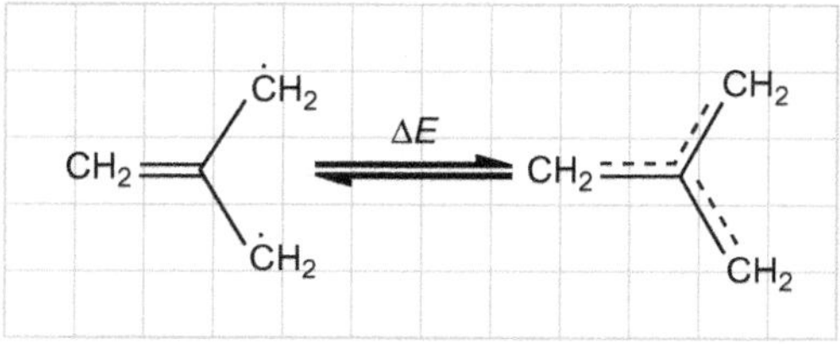

ΔE = delocalization energy

Energy of 2p-electrons of $\dot{C}H_2 = 2\alpha$
Energy of one double bond $= 2\alpha + 2\beta$

Ground state energy of 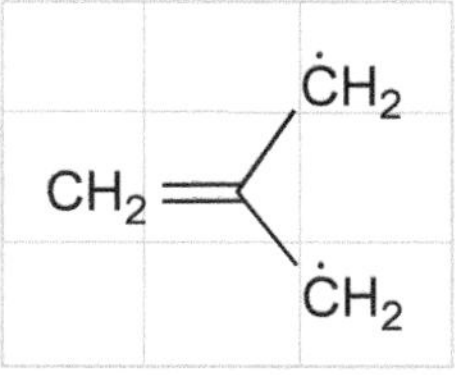 $= 4\alpha + 2\beta$

$$\text{The delocalization energy} = \left(4\alpha + 2\sqrt{3}\beta\right) - (4\alpha + 2\beta)$$
$$= 2\left(\sqrt{3} - 1\right)\beta$$

Consider the π-system of benzene and then:

a. **calculate the total representation of p_π-orbitals of the six carbons and decompose it into irreducible representations;**
b. **find all the wave functions;**
c. **calculate the normalization factor for each wave functions;**
d. **sketch the shapes of these π-molecular orbitals;**
e. **set the energy levels diagram of the molecular wave functions; and**
f. **find the delocalization energy of C_6H_6.**

- Benzene:

  Point group: D_{6h}

  Set of six p_π-orbitals, one on each carbon atom, is taken as the basis for a representation of the group D_{6h}. The reducible representation, Γ_π, is obtained, as in p. Chapter 5, p. 299:

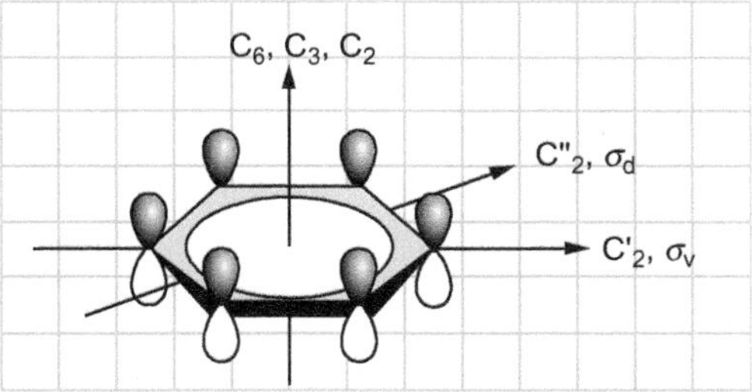

D_{6h}	E	$2C_6$	$2C_3$	C_2	$3C_2'$	$3C_2''$	i	$2S_3$	$2S_6$	σ_h	$3\sigma_v$	$3\sigma_d$
Γ_π	6	0	0	0	−2	0	0	0	0	-6	2	0

- This representation is reduced to

$$\Gamma_\pi = A_{2u} + B_{2g} + E_{1g} + E_{2u}$$

- All the essential symmetry properties of the LCAOs that we seek are determined by operations of the uniaxial rotational subgroup, C_6 (Table 6.10).

$$\Gamma_o = A + B + E_1 + E_2$$

where

$$\varepsilon + \varepsilon^* = 2\cos\theta = 2\cos\frac{2\pi}{6} = 1$$

$$\varepsilon - \varepsilon^* = 2i\sin\theta = i\sqrt{3}$$

- Using the projection operation: $\hat{P}_r(\phi) = \frac{l}{h}\sum \chi_R \hat{R}(\phi_i)$

TABLE 6.10 Character Table for the C_6 Point Group

C_6	E	C_6	C_3	C_2	C_3^2	C_6^5
A	1	1	1	1	1	1
B	1	−1	1	−1	1	−1
E_1	$\left\{\begin{matrix}1\\1\end{matrix}\right.$	$\begin{matrix}\varepsilon\\\varepsilon^*\end{matrix}$	$\begin{matrix}-\varepsilon^*\\-\varepsilon\end{matrix}$	$\begin{matrix}-1\\-1\end{matrix}$	$\begin{matrix}-\varepsilon\\-\varepsilon^*\end{matrix}$	$\left.\begin{matrix}\varepsilon^*\\\varepsilon\end{matrix}\right\}$
E_2	$\left\{\begin{matrix}1\\1\end{matrix}\right.$	$\begin{matrix}-\varepsilon^*\\-\varepsilon\end{matrix}$	$\begin{matrix}-\varepsilon\\-\varepsilon^*\end{matrix}$	$\begin{matrix}1\\1\end{matrix}$	$\begin{matrix}-\varepsilon^*\\-\varepsilon\end{matrix}$	$\left.\begin{matrix}-\varepsilon\\-\varepsilon^*\end{matrix}\right\}$
Γ_o	6	0	0	0	0	0

$$A:\ \psi_1 = \phi_1 + \phi_2 + \phi_3 + \phi_4 + \phi_5 + \phi_6$$

$$B:\ \psi_2 = \phi_1 - \phi_2 + \phi_3 - \phi_4 + \phi_5 - \phi_6$$

$$E_1:\ \begin{cases}\psi_3 = \phi_1 + \varepsilon\phi_2 - \varepsilon^*\phi_3 - \phi_4 - \varepsilon\phi_5 + \varepsilon^*\phi_6\\ \psi_4 = \phi_1 + \varepsilon^*\phi_2 - \varepsilon\phi_3 - \phi_4 - \varepsilon^*\phi_5 + \varepsilon\phi_6\end{cases}$$

$$E_2:\ \begin{cases}\psi_5 = \phi_1 - \varepsilon^*\phi_2 - \varepsilon\phi_3 + \phi_4 - \varepsilon^*\phi_5 - \varepsilon\phi_6\\ \psi_6 = \phi_1 - \varepsilon\phi_2 - \varepsilon^*\phi_3 + \phi_4 - \varepsilon\phi_5 - \varepsilon^*\phi_6\end{cases}$$

- The pairs of SALCs of E-type representations are changed into new linear combinations with real numbers as coefficients by adding them, and subtracting them, and dividing out i.

$$\psi(E_{1a}) = \psi_3 + \psi_4 = 2\phi_1 + (\varepsilon+\varepsilon^*)\phi_2 - (\varepsilon+\varepsilon^*)\phi_3 - 2\phi_4 - (\varepsilon+\varepsilon^*)\phi_5 + (\varepsilon+\varepsilon^*)\phi_6$$

if: $\varepsilon + \varepsilon^* = 1$

$$\psi(E_{1a}) = 2\phi_1 + \phi_2 - \phi_3 - 2\phi_4 - \phi_5 + \phi_6$$

$$\psi(E_{1b}) = \psi_3 - \psi_4 = [(\varepsilon-\varepsilon^*)\phi_2 - (\varepsilon^*-\varepsilon)\phi_3 - (\varepsilon-\varepsilon^*)\phi_5 + (\varepsilon^*-\varepsilon)\phi_6]$$

if: $\varepsilon - \varepsilon^* = i\sqrt{3}$

$$\psi(E_{1b}) = \sqrt{3}\phi_2 + \sqrt{3}\phi_3 - \sqrt{3}\phi_5 - \sqrt{3}\phi_6$$

Similarly:

$$\psi(E_{2a}) = \psi_5 + \psi_6 = 2\phi_1 - \phi_2 - \phi_3 + 2\phi_4 - \phi_5 - \phi_6$$

$$\psi(E_{2b}) = \psi_5 - \psi_6 = -\sqrt{3}\phi_2 + \sqrt{3}\phi_3 - \sqrt{3}\phi_5 + \sqrt{3}\phi_6$$

- Normalization these molecular orbitals wave function (Fig. 6.32):
- The energy of these molecular orbitals may now calculated, using

$$E_A = \int \Psi_1 H \Psi_1\, d\tau$$

$$E_A = \frac{1}{6}\int (\phi_1 + \phi_2 + \phi_3 + \phi_4 + \phi_5 + \phi_6) H (\phi_1 + \phi_2 + \phi_3 + \phi_4 + \phi_5 + \phi_6)\, d\tau$$

$$E_A = \frac{1}{6}(6\alpha + 12\beta)$$

$$E_A = \alpha + 2\beta$$

Similarly,

$$E_B = \alpha - 2\beta$$

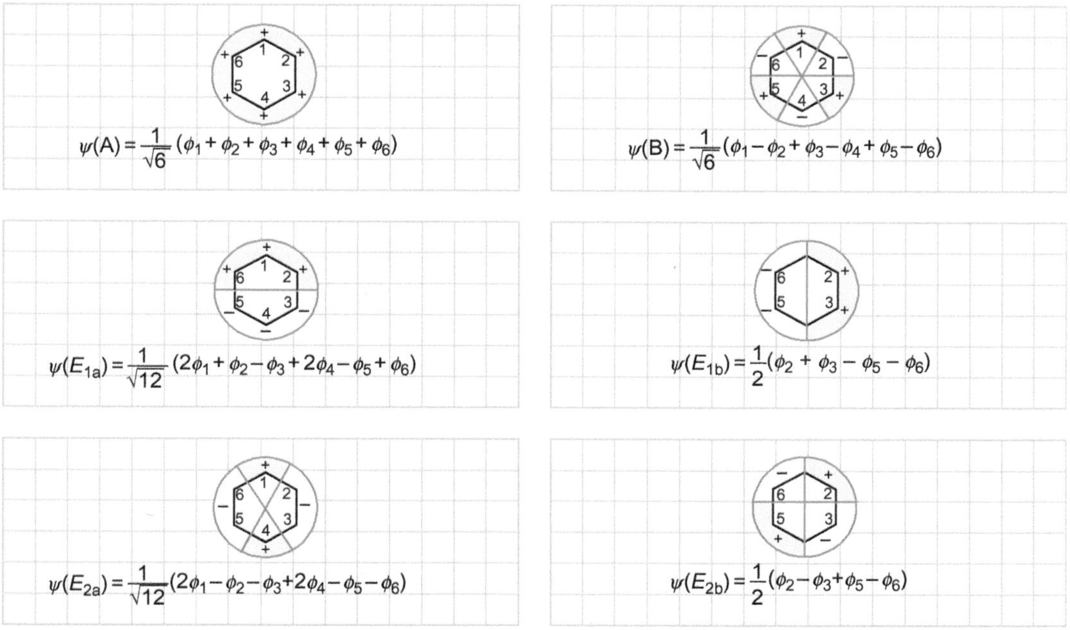
$\psi(A)=\frac{1}{\sqrt{6}}(\phi_1+\phi_2+\phi_3+\phi_4+\phi_5+\phi_6)$
$\psi(B)=\frac{1}{\sqrt{6}}(\phi_1-\phi_2+\phi_3-\phi_4+\phi_5-\phi_6)$
$\psi(E_{1a})=\frac{1}{\sqrt{12}}(2\phi_1+\phi_2-\phi_3+2\phi_4-\phi_5+\phi_6)$
$\psi(E_{1b})=\frac{1}{2}(\phi_2+\phi_3-\phi_5-\phi_6)$
$\psi(E_{2a})=\frac{1}{\sqrt{12}}(2\phi_1-\phi_2-\phi_3+2\phi_4-\phi_5-\phi_6)$
$\psi(E_{2b})=\frac{1}{2}(\phi_2-\phi_3+\phi_5-\phi_6)$

FIG. 6.32 Normalized molecular wave function of the benzene molecule.

$$E_{E1a}=E_{E1b}=\alpha+\beta$$

$$E_{E2a}=E_{E2b}=\alpha-\beta$$

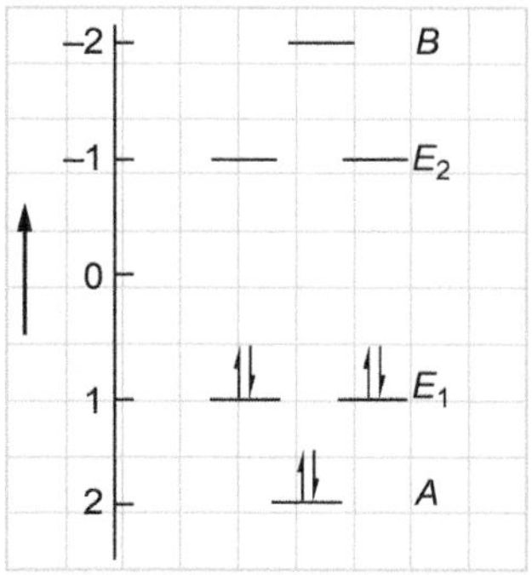
−2
−1
0
1
2
B
E_2
E_1
A

- When the atomic orbitals combined to form the molecular orbitals, the six electrons will occupy these molecular orbitals according to Hund's rule and Pauli's Exclusion Principle.
- The total energy of the system is then as follows:

$$E_T=2(2\beta)+4(\beta)=8\beta$$

β is negative, thus the π-bonding has stabilized the molecule by 8β.

- When two atomic π-orbitals, e.g., ϕ_1 and ϕ_2, interact to form a two-center bond, two molecular orbitals, ψ_1 and ψ_2, are formed.

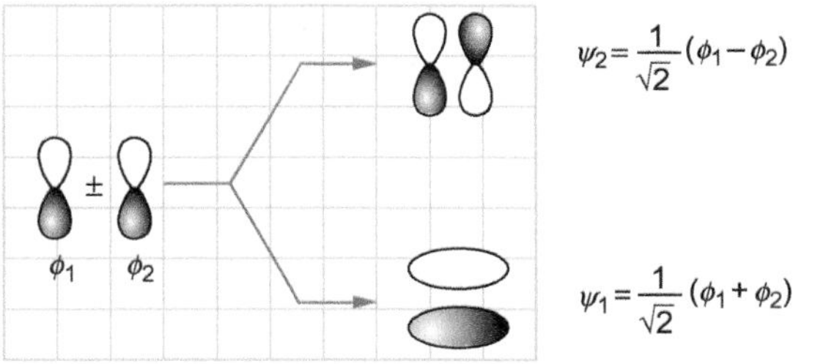
$\psi_2=\frac{1}{\sqrt{2}}(\phi_1-\phi_2)$
±
ϕ_1
ϕ_2
$\psi_1=\frac{1}{\sqrt{2}}(\phi_1+\phi_2)$

- In order that ψ_1 and ψ_2 be real, the following must be normalized:

$$\psi_1 = \frac{1}{\sqrt{2}}(\phi_1 + \phi_2)$$

$$\psi_2 = \frac{1}{\sqrt{2}}(\phi_1 - \phi_2)$$

- Their energies are as follows:

$$E_1 = \int \psi_1 H \psi_1 \partial\tau = \frac{1}{2}\left(\int \phi_1 H \phi_1 \partial\tau + \int \phi_1 H \phi_2 \partial\tau\right) + \int \phi_2 H \phi_1 \partial\tau + \int \phi_2 H \phi_2 \partial\tau$$

$$E_1 = \frac{1}{2}(2\alpha + 2\beta) \cong \beta$$

$$E_2 = -\beta$$

- The two π-electrons will occupy ψ_1 and their combined energy will be 2β.
- Thus, each pair of localized π-electrons in cyclohexatriene contributes $2\,\beta$ to the energy of the molecule, and the total energy of the localized cyclohexatriene structure equals 6β.

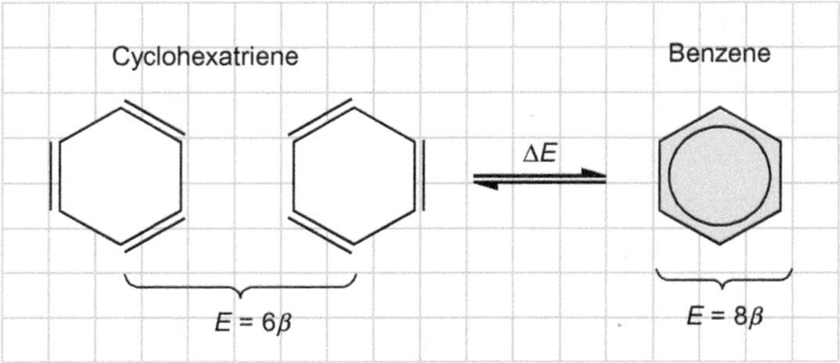

$\Delta E = 2\beta$ (delocalization energy)

- However, the actual energy of benzene is 8β, so the resonance or delocalization energy is 2β.
- Using the bond energies for C—C, C=C, and C—H to estimate the energy of cyclohexatriene. The value of $|\beta|$ obtained for benzene is 18–20 kcal/mol.

v. β-Evaluation

The peak of the lowest energy $\pi \rightarrow \pi^*$ transition for butadiene appears at 2170 Å. Evaluate β for this molecule.

- The energy level diagram of butadiene (see p. 372):

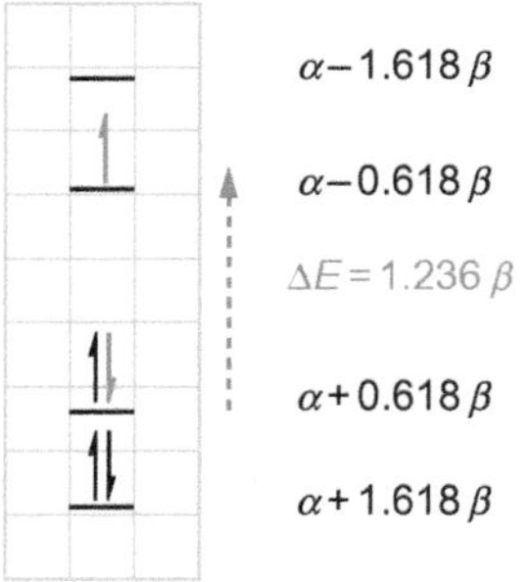

If: $E(\pi_1) = \alpha + 1.618\beta$, $E(\pi_2) = \alpha + 0.618\beta$, $E(\pi_3) = \alpha - 0.618\beta$, and $E(\pi_4) = \alpha - 1.618\beta$
$\pi \rightarrow \pi^*$ transition appears at wavelength 2170 Å

$$1.236\beta = h\upsilon = \frac{hc}{\lambda} = 6.62 \times 10^{-27} \frac{3 \times 10^{10}}{2.170 \times 10^{-5}}$$

$$\beta = 7.3 \times 10^{-12}\,\text{erg}$$

Evaluate the π-electron energy levels of benzene. Evaluate the total π-electron energy. Use the given information in Scheme 6.9 to evaluate β.

- Benzene (see page 377):

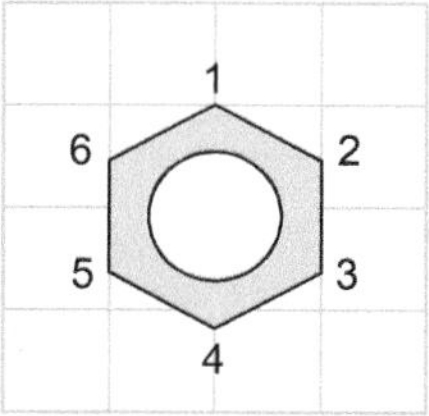

- The secular determinant:

$$\begin{vmatrix} \alpha-E & \beta & 0 & 0 & 0 & \beta \\ \beta & \alpha-E & \beta & 0 & 0 & 0 \\ 0 & \beta & \alpha-E & \beta & 0 & 0 \\ 0 & 0 & \beta & \alpha-E & \beta & 0 \\ 0 & 0 & 0 & \beta & \alpha-E & \beta \\ \beta & 0 & 0 & 0 & \beta & \alpha-E \end{vmatrix} = 0$$

- If $\dfrac{\alpha-E}{\beta} = x$, then:

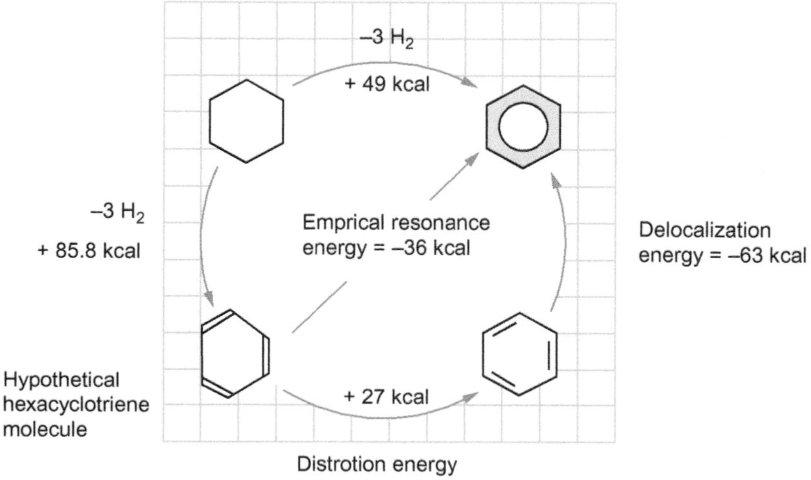

SCHEME 6.9 Conversion of cyclohexane to benzene.

$$\begin{vmatrix} x & 1 & 0 & 0 & 0 & 1 \\ 1 & x & 1 & 0 & 0 & 0 \\ 0 & 1 & x & 1 & 0 & 0 \\ 0 & 0 & 1 & x & 1 & 0 \\ 0 & 0 & 0 & 1 & x & 1 \\ 1 & 0 & 0 & 0 & 1 & x \end{vmatrix} = 0$$

The roots are: $x_{1,2} = \pm 2$, $x_{3,4} = 1$, $x_{5,6} = -1$

$$E_B = \alpha - 2\beta$$
$$E_{E2a} = E_{E2b} = \alpha - \beta$$
$$E_{E1a} = E_{E1b} = \alpha + \beta$$
$$E_A = \alpha + 2\beta$$

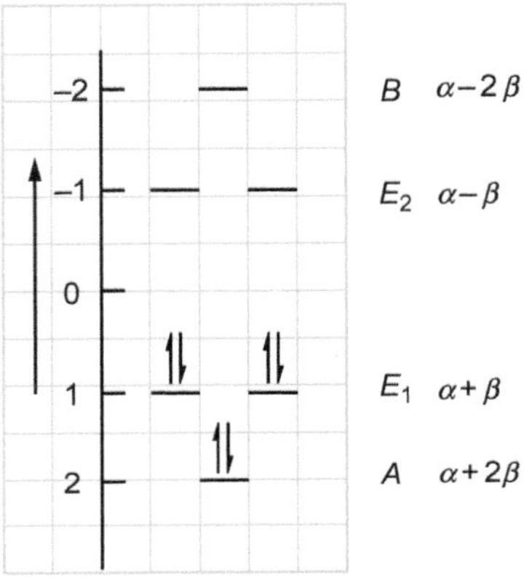

- The total energy of the system is then

$$E_T = 2(2\beta) + 4(\beta) = 8\beta$$

- Remember that β is negative, we see that β-bonding has stabilized the molecule by 8β.
-

$$\text{Delocalization energy} = \begin{bmatrix} \text{the energies of three} \\ \text{double bonds (p. 379)} \end{bmatrix} - \begin{bmatrix} \text{the total } \beta - \text{energy} \\ \text{of benzene} \end{bmatrix}$$
$$= 6\alpha + 6\beta - (6\alpha + 8\beta) = -2\beta$$

- According to the diagram this energy is +63 kcal/mol. Thus

$$-2\beta = 63\,\text{kcal/mol}$$

$$\beta = -31.5\,\text{kcal/mol}$$

vi. Electron Density
vii. Formal Charge
viii. Bond Order

By using the Hückel wave function, calculate the electron density, the formal charge, and the bond order for the allyl radical and allyl cationic.

- Firstly, we calculate molecular wave functions, and their energies for the allyl radical.
  The π-system of the allyl radical:

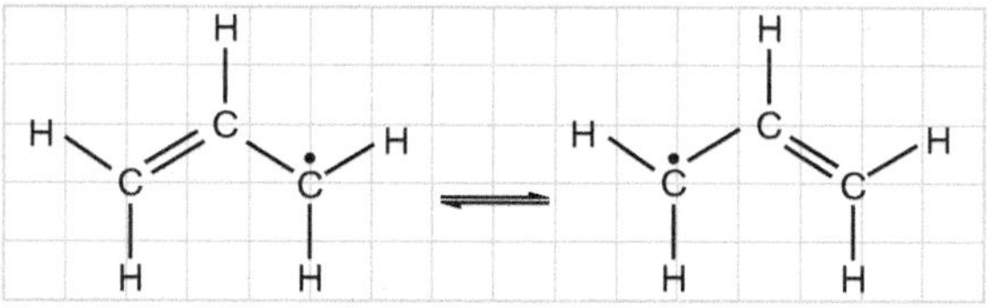

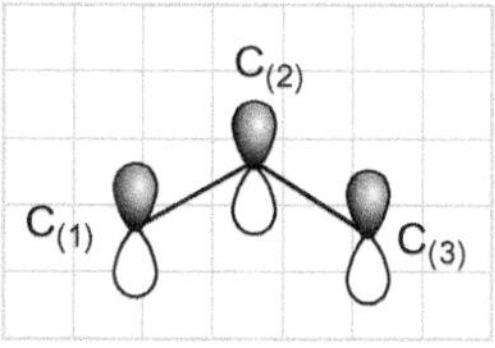

- The general secular determinant:

$$\begin{bmatrix} \alpha-E & \beta_{12}\cdots & \beta_{1n} \\ \vdots & \ddots & \vdots \\ \beta_{n1} & \cdots & \alpha-E \end{bmatrix}=0$$

The 3×3 secular determinant:

$$\begin{vmatrix} \alpha-E & \beta & 0 \\ \beta & \alpha-E & \beta \\ 0 & \beta & \alpha-E \end{vmatrix}=0$$

- Dividing by β and letting $\dfrac{\alpha-E}{\beta}=x$:

$$\begin{vmatrix} x & 1 & 0 \\ 1 & x & 1 \\ 0 & 1 & x \end{vmatrix}=0$$

$$x\begin{vmatrix} x & 1 \\ 1 & x \end{vmatrix}-1\begin{vmatrix} 1 & 1 \\ 0 & x \end{vmatrix}=0$$

$$x^3-2x=0$$

$$x=0,\ x=\pm\sqrt{2}$$

$$\frac{\alpha-E}{\beta}=0,\sqrt{2},-\sqrt{2}$$

- The three molecular energy levels:

$$E_1=\alpha+\sqrt{2}\beta$$

$$E_2=\alpha$$

$$E_3=\alpha-\sqrt{2}\beta$$

- By multiplying the column matrix (a_1, a_2, a_3) by the 3×3 secular determinant, the secular equations for the three atoms allyl π-system are:

$$a_1x + a_2 = 0 \quad 1$$

$$a_1 + a_2x + a_3 = 0 \quad 2$$

$$a_2 + a_3x = 0 \quad 3$$

  where $x = (\alpha - E)/\beta$
- For ψ_1, the bonding molecular orbital, $x = -\sqrt{2}$, so substitution into (1) and (3) gives:

$$-\sqrt{2}a_1 + a_2 = 0$$

  and

$$a_2 - \sqrt{2}a_3 = 0, \text{ then}$$

$$a_1 = \frac{a_2}{\sqrt{2}} \text{ and } a_3 = \frac{a_2}{\sqrt{2}}$$

  Using the normalization criterion:

$$a_1^2 + a_2^2 + a_3^2 = \left(\frac{a_2}{\sqrt{2}}\right)^2 + a_2^2 + \left(\frac{a_2}{\sqrt{2}}\right)^2 = 1$$

$$2a_2^2 = 1 \text{ and } a_2 = \frac{1}{\sqrt{2}}$$

$$a_1 = a_3 = \frac{a_2}{\sqrt{2}}$$

  Then: $a_1 = a_3 = \frac{1}{2}$
- With the coefficient defined, we can write:

$$\psi_1 = \frac{1}{2}\phi_1 + \frac{1}{\sqrt{2}}\phi_2 + \frac{1}{2}\phi_3 \quad \text{(bonding)}$$

- For ψ_2
  $x = 0$, and from equations 1, 2, and 3:
  $a_2 = 0$ and $a_1 = -a_3$
  $a_1^2 + a_3^2 = 1$ or $a_1 = \frac{1}{\sqrt{2}}$
  Accordingly, we can write:

$$\psi_2 = \frac{1}{\sqrt{2}}\phi_1 - \frac{1}{\sqrt{2}}\phi_3 \quad \text{(non-bonding)}$$

- Similarly,

$$\psi_3 = \frac{1}{2}\phi_1 - \frac{1}{\sqrt{2}}\phi_2 + \frac{1}{2}\phi_3 \quad \text{(antibonding)}$$

Electron density at atom r, q_r:

- The electron density q_i at an atom i in a molecule is given by:

$$q_i = -\sum_j n_j C_{ji}^2$$

where n_j is the number of the electrons in jth molecular orbital.
The summation is performed over all of the j occupied molecular orbitals and Cji is the coefficient of the atomic orbital i in the jth molecular orbital.

- By means of Hückel theory, wave functions for the molecular orbitals of allyl anion are filled by two electrons in each:

$$\psi_1 = \frac{1}{2}\phi_1 + \frac{1}{\sqrt{2}}\phi_2 + \frac{1}{2}\phi_3$$

$$\psi_2 = \frac{1}{\sqrt{2}}\phi_1 - \frac{1}{\sqrt{2}}\phi_3$$

- For the electrons density on carbon 1 (with $n=2$):

$$q_1 = -2\underbrace{\left(\frac{1}{2}\right)^2}_{\substack{\text{coefficient}\\ \text{of } \phi_1 \text{ in } \psi_1}} - 2\underbrace{\left(\frac{1}{\sqrt{2}}\right)^2}_{\substack{\text{coefficient}\\ \text{of } \phi_1 \text{ in } \psi_2}} = -1\frac{1}{2}$$

At carbon 2 we find:

$$q_1 = -2\underbrace{\left(\frac{1}{\sqrt{2}}\right)^2}_{\substack{\text{coefficient}\\ \text{of } \phi_2 \text{ in } \psi_1}} - 2\underbrace{(0)^2}_{\substack{\text{coefficient}\\ \text{of } \phi_2 \text{ in } \psi_2}} = -1$$

Formal charge, ∂

- The formal charge on the atom in the molecule is given by

$$\begin{bmatrix}\text{The formal charge}\\ \text{on the atom}\end{bmatrix} = \begin{bmatrix}\text{The electron density at}\\ \text{an atom, } q_i\end{bmatrix} - \begin{bmatrix}\text{The electron density}\\ \text{on a free atom}\end{bmatrix}$$

- Consider carbon: there is one electron in each of the orbitals of the atom to form four bonds.
- In Hückel theory, only p_z orbital is considered in the π-system.
- Since each carbon contains one electron, the charge on the "free atom" is -1.
- For allyl, the charge density calculated above at carbon 1:
  $q_1 = -1.5$

$$\partial_1 = -1.5 - (-1) = -0.5 \Rightarrow \text{charge of } -0.5 \text{ at carbon 1}$$

and

$$\partial_2 = -1 - (-1) = 0 \Rightarrow \text{charge of zero at carbon 2}$$

Bond order, r_{AB}

- The bond order between two adjacent atoms A and B is given by

$$r_{AB} = \sum_j n_j C_{Aj} C_{Bj}$$

$$\begin{bmatrix}\text{The bond order}\\ \text{between A and B}\end{bmatrix} = \sum \begin{bmatrix}\text{the number of}\\ \text{the electrons in}\\ \text{each the m.o.}\end{bmatrix}\begin{bmatrix}\text{the coefficients of}\\ \text{atom A}\\ \text{in each m.o.}\end{bmatrix}\begin{bmatrix}\text{the coefficients of}\\ \text{atom B}\\ \text{in each m.o.}\end{bmatrix}$$

- The π-bond order between carbon atom 1 and 2 of the allyl is

$$P_{AB} = 2\left(\frac{1}{2}\right)\left(\frac{1}{\sqrt{2}}\right) + 2\left(\frac{1}{\sqrt{2}}\right)(0) = \frac{\sqrt{2}}{2} = 0.7$$

ix. Ionization Energy

Hückel molecular orbital theory predicts that the energy of *j*th π-orbitals is: $E_j = \alpha + m_j\beta$. The m_j values for the highest occupied molecular orbital of some conjugated molecules are given below:

	m_j
Tetracene	0.293
1,2-Benzopyrene	0.363
Anthracene	0.412
Pyrene	0.443
1,2-Benzanthracene	0.450
1,2,5,6-Dibenzanthracene	0.475
Phenanthrene	0.603
Fluoranthrene	0.617
Naphthalene	0.616
Biphenyl	0.702
Benzene	1.000

Write the structural formulas for these hydrocarbons.
Estimate their ionization energies relative to that of benzene in terms of β.
Answer:

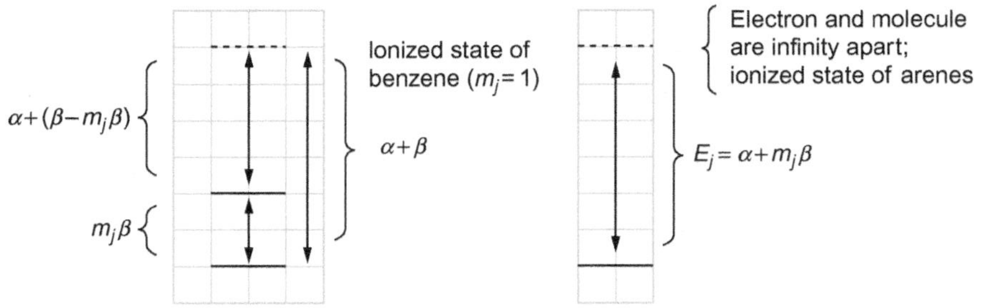

- Ionization energy = −[energy of the highest occupied molecular orbital]
  If: I_o = benzene ionization energy, I_j = others
  For benzene $E_{\text{highest}} = \alpha + \beta$, see pp. 379 and 382.
- The ionization energies relative to that of benzene:

$$I_j = -\alpha - m_j\beta = (-\alpha - \beta) + \beta - m_j\beta = I_o + (1 - m_j)\beta$$

β is negative (Table 6.11).

TABLE 6.11 The Change in the Ionization Energy of Polycyclic Aromatic Hydrocarbons

	Molecular Structures	$E_{highest}$	I_j
Naphthacene (Tetracene)	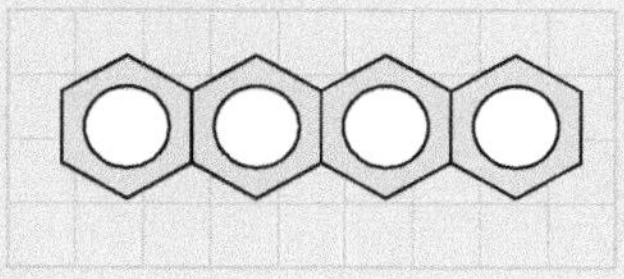	$=\alpha+0.293\beta$	$I_o+0.707\beta$
1,2-Benzopyrene	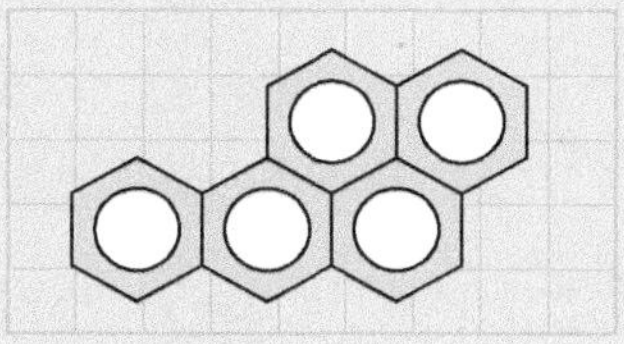	$=\alpha+0.363\beta$	$I_o+0.637\beta$
Anthracene	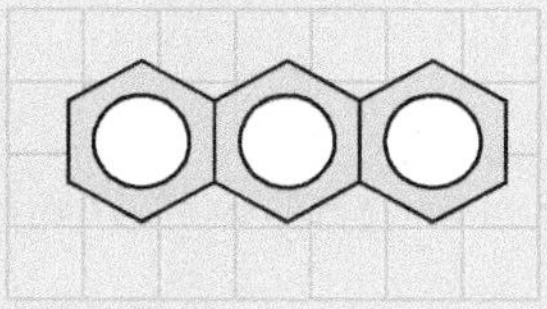	$=\alpha+0.412\beta$	$I_o+0.588\beta$
Pyrene	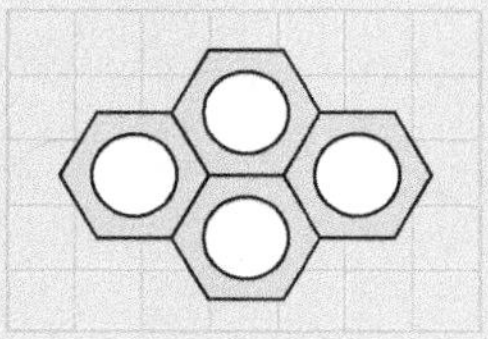	$=\alpha+0.443\beta$	$I_o+0.557\beta$
1,2-Benzanthracene	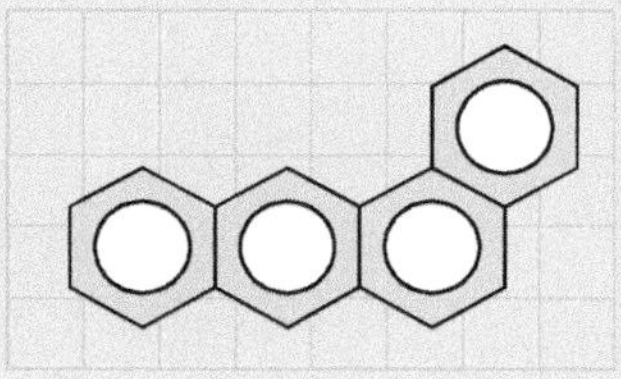	$=\alpha+0.450\beta$	$I_o+0.550\beta$
1,2,5,6-Dibenz-anthrecene	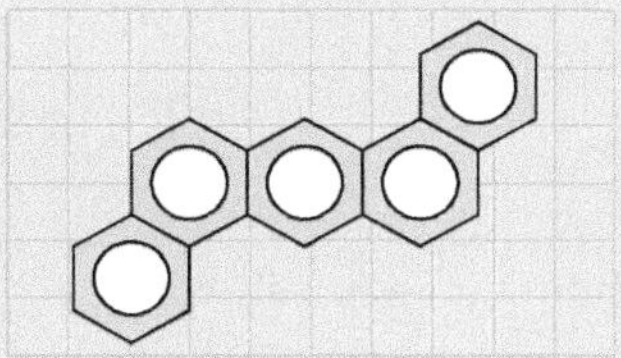	$=\alpha+0.475\beta$	$I_o+0.525\beta$
Phenathrene	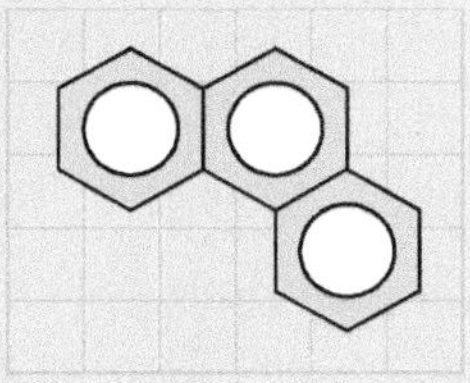	$=\alpha+0.603\beta$	$I_o+0.397\beta$
Fluranthene	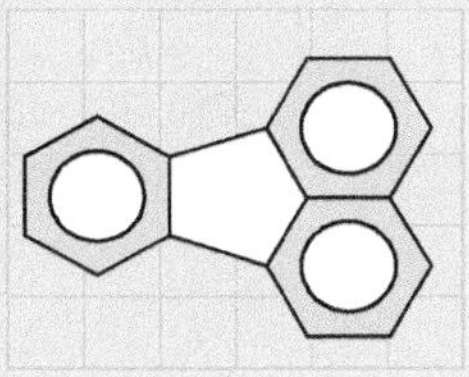	$=\alpha+0.617\beta$	$I_o+0.383\beta$

Continued

TABLE 6.11 The Change in the Ionization Energy of Polycyclic Aromatic Hydrocarbons—cont'd

	Molecular Structures	$E_{highest}$	I_j
Naphthalene	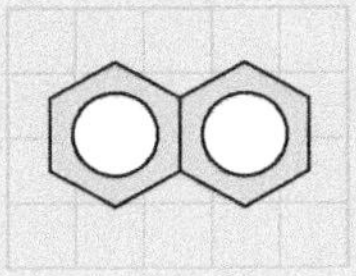	$=\alpha+0.616\beta$	$I_o+0.384\beta$
Biphenyl	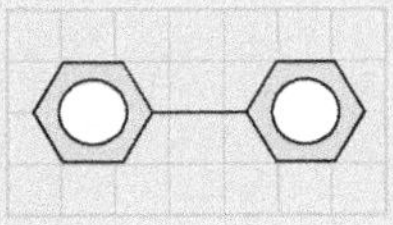	$=\alpha+0.702\beta$	$I_o+0.298\beta$

Note that since β is negative their ionization energies are all larger than benzene is.

TABLE 6.12 The Delocalization Energies of Different Polycyclic Hydrocarbons

Compound (Molecular Structures)	Delocalization Energy
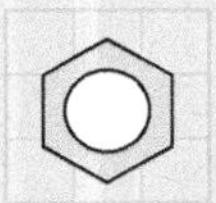	**2β**
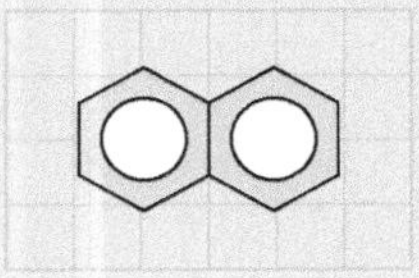	**3.683β**
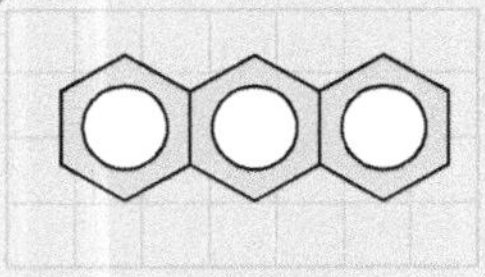	**5.314β**
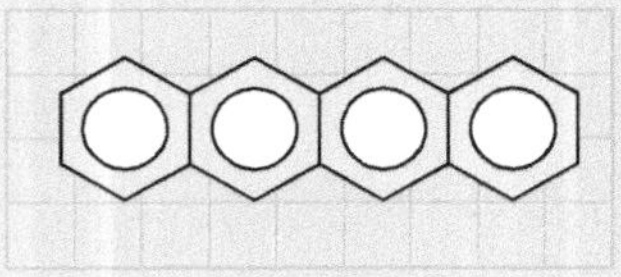	**6.932β**

x. Equilibrium Constant

Delocalization energies for naphthalene and anthracene are 3.683β and 5.314β, respectively. Using the value of β determined for benzene, evaluate the equilibrium constant of the following reaction at 250°C.

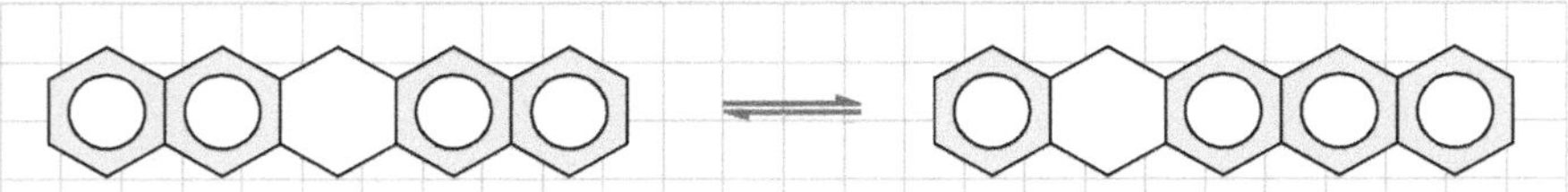

I II

The delocalization energies of benzene, naphthalene, anthracene, and tetracene are given in Table 6.12. What are the total π energies of these four compounds? (in terms of α and β)

What is the energy of the following reaction?

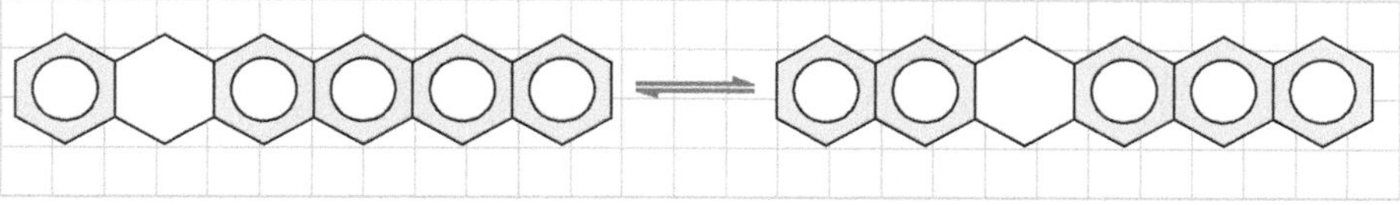

Answers:

- π-electron energy of the double bonds $= 10(\alpha+\beta)$ (see page 379)
- Delocalization energy of two naphthalenes $= 2 \times 3.683\beta$
- Delocalization energy of benzene $= 2\beta$
- Delocalization energy of anthracene $= 5.314\beta$

$$\pi - \text{electron energy of compound I} = 10(2\alpha + 2\beta) + 2 \times 3.683\beta$$

$$\pi - \text{electron energy of compound II} = 10(2\alpha + 2\beta) + 2\beta + 5.314\beta$$

- $\Delta E =$ energy difference between compound I & II $=$

$$= \text{total } \pi \text{ energy of compound II} - \text{ total } \pi \text{ energy of compound I}$$

$$= 10(2\alpha + 2\beta) + 7.314\beta - [10(2\alpha + 2\beta) + 7.366\beta] = -0.052\beta$$

$$\Delta E = \Delta G = -RT\text{Ln}K = -0.052\beta = 0.052 \times 31{,}500$$

$K = 0.2$ observed value close to this

xi. Configuration Interaction Among Orbitals of Similar Symmetry Representations in the Same Molecule (Interaction of molecular orbitals that have similar symmetry).

a. **Work out the symmetry adapted liner combination of π-orbitals for tetramethylenecyclobutane.**
b. **Use these linear combinations to obtain a set of orthonormal molecular orbitals.**
c. **Order the MOs by increasing energy as best as you can.**
d. **What are the atomic orbital coefficients?**
e. **Would any more information be obtained from a Hückel calculation? If so, what?**
   - Point group: D_{4h}

     Set of eight p_π orbitals, one on each carbon atom, is taken as the basis for a representation of the group D_{4h}; the results are in Table 6.13:

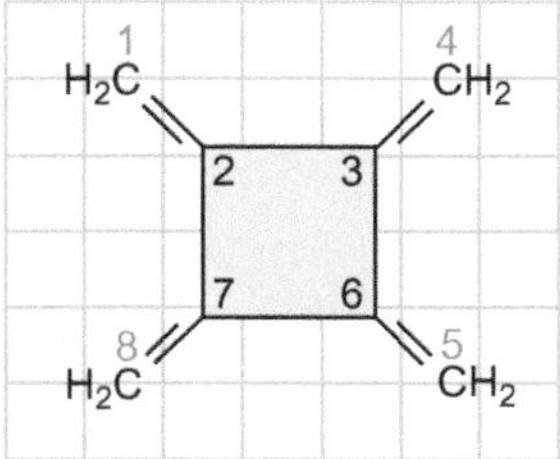

Tetramethylenecyclobutane

- Using: $\Gamma_i = \frac{1}{h}\sum i \cdot \chi_R \cdot \chi_i$, and the D_{4h} character table.

$$\Gamma_{\pi_{\text{inner}}} = A_{2u} + B_{1u} + E_g$$

$$\Gamma_{\pi_{\text{outer}}} = A_{2u} + B_{1u} + E_g$$

TABLE 6.13 The Characters for $\Gamma_{\pi inner}$, $\Gamma_{\pi outer}$, and $\Gamma_{\pi total}$ Representations in the D_{4h} Point Group

D_{4h}	E	$2C_4$	C_2	$2C'_2$	$2C''_2$	i	$4S_4$	σ_h	$2\sigma_v$	$2\sigma_d$
$\Gamma_{\pi inner}$	4	0	0	0	−2	0	0	−4	0	2
$\Gamma_{\pi outer}$	4	0	0	0	−2	0	0	−4	0	2
$\Gamma_{\pi total}$	8	0	0	0	−4	0	0	−8	0	4

TABLE 6.14 Character Table for the C_4 Point Group

C_4	E	$2C_4$	C_2	$2C_4^3$
$A_{2u} \to A$	1	1	1	1
$B_{2u} \to B$	1	−1	1	−1
$E_g \to E$	$\begin{cases}1\\1\end{cases}$	i $-i$	−1 −1	$\begin{matrix}-i\\1\end{matrix}\}$

- Using: $\hat{P}_r(\phi_i) = \frac{l}{h}\sum \chi_R \hat{R}(\phi_i)$

  or read the relative coefficient from Table 6.14.

  The molecular orbital wave functions, SALC (i = inner and o = outer molecular orbital):

$$A_i = \phi_2 + \phi_3 + \phi_6 + \phi_7$$

$$B_i = \phi_2 - \phi_3 + \phi_6 - \phi_7$$

$$E_{a/i} = \phi_2 + i\phi_3 - \phi_6 - i\phi_7$$

$$E_{b/i} = \phi_2 - i\phi_3 - \phi_6 + i\phi_7$$

- Take linear combinations of $E_{a/i}$ and $E_{b/i}$

$$E_{1/i} = E_{a/i} + E_{b/i} = 2\phi_2 - 2\phi_6 \Rightarrow \phi_2 - \phi_6$$

$$E_{2/i} = E_{a/i} - E_{b/i} = 2i\phi_3 - 2i\phi_7 \Rightarrow \phi_3 - \phi_7$$

- The normalized relative inner and outer atomic orbital contributions:

$$a_i = \frac{1}{2}(\phi_2 + \phi_3 + \phi_6 + \phi_7) \Rightarrow \Psi_{ai}$$

$$a_o = \frac{1}{2}(\phi_1 + \phi_4 + \phi_5 + \phi_8) \Rightarrow \Psi_{ao}$$

$$e_i = \begin{cases} {}^1/_{\sqrt{2}}(\phi_2 - \phi_6) \Rightarrow \Psi_{eai} \\ {}^1/_{\sqrt{2}}(\phi_3 - \phi_7) \Rightarrow \Psi_{ebi} \end{cases}$$

$$e_o = \begin{cases} {}^1/_{\sqrt{2}}(\phi_1 - \phi_5) \Rightarrow \Psi_{eao} \\ {}^1/_{\sqrt{2}}(\phi_4 - \phi_8) \Rightarrow \Psi_{ebo} \end{cases}$$

$$b_i = \frac{1}{2}(\phi_2 - \phi_3 + \phi_6 - \phi_7) \Rightarrow \Psi_{bi}$$

$$b_o = \frac{1}{2}(\phi_1 - \phi_4 + \phi_5 - \phi_8) \Rightarrow \Psi_{bo}$$

- For interaction of A orbitals, we have the equation:

$$\begin{bmatrix} H_{ai,ai}-E & H_{ai,ao} \\ H_{ao,ai} & H_{ao,ao}-E \end{bmatrix} = 0$$

$$H_{ai,ai} = \int \psi_{ai} H \psi_{ai} \partial\tau = \frac{1}{4}\int (\phi_2+\phi_3+\phi_6+\phi_7) H (\phi_2+\phi_3+\phi_6+\phi_7)\partial\tau$$

$$H_{ai,ai} = \frac{1}{4}\left(\int \phi_2 H\phi_2 \partial\tau + \int \phi_2 H\phi_3\partial\tau + \cdots + \int \phi_2 H\phi_6\partial\tau + \cdots + \int \phi_7 H\phi_7\partial\tau\right)$$

$$H_{ai,\,ai} = \frac{1}{4}(\alpha+\beta+0+\cdots+\alpha) = \frac{1}{4}(4\alpha+8\beta) = \alpha+2\beta$$

$$H_{ao,ao} = \int \psi_{ao} H\psi_{ao}\partial\tau = \frac{1}{4}\int (\phi_1+\phi_4+\phi_5+\phi_8) H (\phi_1+\phi_4+\phi_5+\phi_8)\partial\tau = \frac{1}{4}(4\alpha) = \alpha$$

$$H_{ai,ao} = \int \psi_{ao} H \psi_{ao}\partial\tau = \frac{1}{4}\int (\phi_2+\phi_3+\phi_6+\phi_7) H (\phi_1+\phi_4+\phi_5+\phi_8)\partial\tau = \frac{1}{4}(4\beta) = \beta$$

- As before, we chose $\alpha=0$, and $\beta=1$, the secular equations:

$$A: \quad \begin{bmatrix} 2-E & 1 \\ 1 & -E \end{bmatrix} = 0, \quad E^2-2E-1=0$$

$$E_A = \left(1+\sqrt{2}\right), \quad \left(1-\sqrt{2}\right)$$

In similar way:

$$B: \quad \begin{bmatrix} H_{bi,bi}-E & H_{bi,bo} \\ H_{bo,bi} & H_{bo,bo}-E \end{bmatrix} = \begin{bmatrix} -2-E & 1 \\ 1 & -E \end{bmatrix} = 0$$

$E_B = \left(\sqrt{2}-1\right), \left(-\sqrt{2}-1\right)$, and

$$E: \quad \begin{bmatrix} H_{ei,ei}-E & H_{ei,eo} \\ H_{eo,ei} & H_{eo,eo}-E \end{bmatrix} = \begin{bmatrix} -E & 1 \\ 1 & -E \end{bmatrix} = 0$$

$$E_E = \pm 1$$

- The molecular orbitals energy diagram of tetramethylenecyclobutane is represented in Fig 6.33.
- The total energy of the eight occupying the MOs:

$$\left[2\left(1+\sqrt{2}\right)+4(1)+2\left(\sqrt{2}-1\right)\right]\beta = 9.656\beta$$

- The resonance or delocalization energy $=9.656\beta - 8\times\beta = -1.66\beta$ (β for each π electron), taking $|\beta|=20$ kcal/mol. The delocalization energy $=20\times 1.66 \cong 33$ kcal/mol.
- In order to compute the actual expressions for the occupied MOs:

$$\psi_A = c_i\psi_{ai} + c_o\psi_{ao} \Rightarrow c_i = ?, \quad c_o = ?$$

- By returning to the equation:

$$\begin{bmatrix} H_{ai,ai}-E & H_{ai,ao} \\ H_{ao,ai} & H_{ao,ao}-E \end{bmatrix} = 0, \quad H_{ai,ai}=2, \quad H_{ai,ao}=1, \quad H_{ao,ao}=0$$

$$c_i(H_{ai,ai}-E) + c_o H_{ai,ao} = 0 = c_i(2-E) + c_o$$

and

$$c_i H_{ai,ao} + c_o(H_{ao,ao}-E) = 0 = c_i - c_o E$$

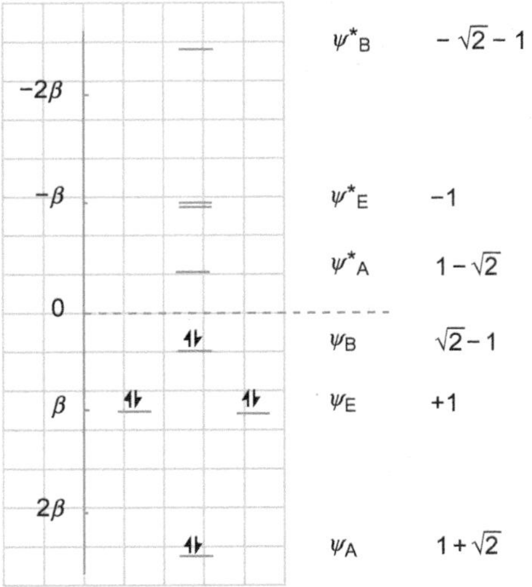

FIG. 6.33 Molecular orbital diagram for tetramethylenecyclobutane.

- The ratio c_i to c_o:

$$c_i/c_o = -1/(2-E)$$
$$c_i/c_o = E$$

- If we insert the energy $E_A = \left(1+\sqrt{2}\right)$, we obtain

$$c_i/c_o = -1/\left(2-\left(1+\sqrt{2}\right)-\right) = -1/\left(1-\sqrt{2}\right) = 1/0.414 = 2.414$$
$$c_i/c_o = 2.414$$

- The normalization condition:

$$c_i^2 + c_o^2 = 1$$

- Therefore, $c_i = 0.382$ and $c_o = 0.924$, and the final expression for the ψ_A:

$$\psi_A = c_i\psi_{ai} + c_o\psi_{ao} = (0.924)\left[\frac{1}{2}(\phi_2+\phi_3+\phi_6+\phi_7)\right] + (0.382)\left[\frac{1}{2}(\phi_1+\phi_4+\phi_5+\phi_8)\right]$$
$$\psi_A = c_i\psi_{ai} + c_o\psi_{ao} = 0.462(\phi_2+\phi_3+\phi_6+\phi_7) + 0.191(\phi_1+\phi_4+\phi_5+\phi_8)$$

- In the same way:

$$\psi_B = 0.462(\phi_1-\phi_4+\phi_5-\phi_8) + 0.191(\phi_2-\phi_3+\phi_6-\phi_7)$$
$$\psi_{Ea} = 0.500(\phi_1+\phi_2-\phi_5-\phi_6)$$
$$\psi_{Eb} = 0.500(\phi_3+\phi_4-\phi_7-\phi_8)$$

- Calculating the bond order:

$$\begin{bmatrix}\text{The bond order}\\ \text{between}\\ \text{C}_i \text{ and C}_{i+1}\end{bmatrix} = \sum \begin{bmatrix}\text{the number of}\\ \text{on each carbon}\\ \text{of the m.o.}\end{bmatrix}\begin{bmatrix}\text{the coefficients}\\ \text{of atom C}_i\\ \text{in each m.o.}\end{bmatrix}\begin{bmatrix}\text{the coefficients}\\ \text{of atom C}_{i+1}\\ \text{in each m.o.}\end{bmatrix}$$

- Thus, for one of the equivalent ring bond, e.g., between C_2 and C_3:

$$\begin{aligned}
\psi_A&:\ 2\times(0.462)\,(0.462) &&= 0.428\\
\psi_B&:\ 2\times(0.191)\,(-0.191) &&= -0.074\\
\psi_{Ea}&:\ 2\times(0.500)\,(0) &&= 0\\
\psi_{Eb}&:\ 2\times(0.500)\,(0) &&= 0\\
& &&\overline{\ \ 0.354\ \ }
\end{aligned}$$

For one of the exo-bonds, e.g., between C_1 and C_2:

ψ_A:	$2 \times (0.191)(0.462)$	$= 0.176$
ψ_B:	$2 \times (0.462)(0.191)$	$= 0.176$
ψ_{Ea}:	$2 \times (0.500)(0.500)$	$= 0.500$
ψ_{Eb}:	$2 \times (0)(0)$	$= 0$
		0.852

- From these numbers we can see that the p electrons are much more localized in the exo bonds than in the ring bonds.

Based on your results, for linear and cyclic compounds predict the systematic placement and number of molecular orbitals nodes in each of the energy levels of the correlation diagrams, and give examples.

- linear molecules:
  Each molecular orbital is distinguished by a specific nodal pattern of placement and number.
  - There is a precise link between the position of molecular orbital nodes in one orbital and that in the next higher orbital.
    e.g., a central molecular orbital node splits into two and moves outward by bond distance in the next higher orbital.
  - All nodes of the molecular orbital must be symmetrically sited.
  - The molecular orbital energy increases with increasing numbers of nodes because the net number of bonding interactions (between adjacent atoms) decreases by two as the number of molecular orbital nodes increases by one.
  - In all instances, successively higher energy orbitals have molecular orbital nodes symmetrically located and increasing in number by one in order from the most stable to least stable.
- Examples:
  - Triatomic:
    Consider three orbitals on three atoms: they yield three molecular orbitals, by linear combinations.

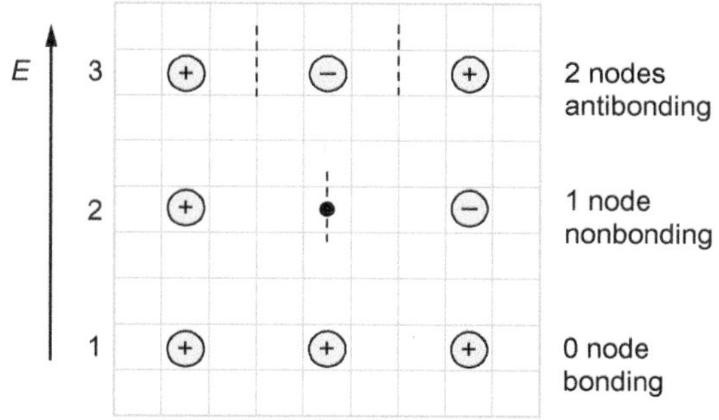

  - Tetraatomic:

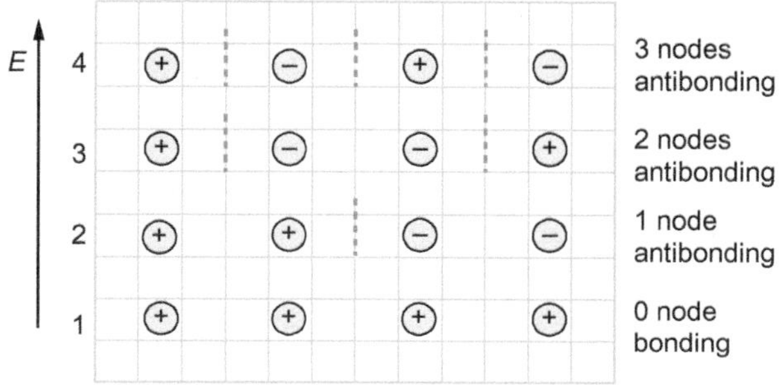

Molecular orbitals are in the same energy order as the number of nodes they possess.

  - Cyclic molecules:
    Nodal planes pass through center of polygon, but the pattern of contributions is the same.

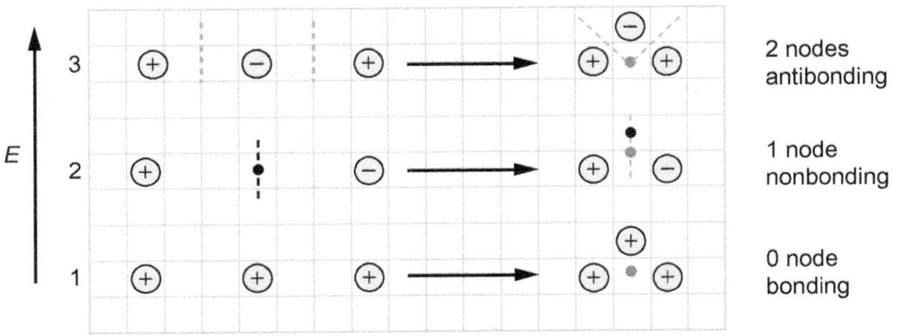

C_3H_3: D_{3h}

$$\Psi_{(A)} = \frac{1}{\sqrt{3}}(\phi_1 + \phi_2 + \phi_3) \quad E = \alpha + 2\beta$$

$$\Psi_{(Ea)} = \frac{1}{\sqrt{6}}(2\phi_1 - \phi_2 - \phi_3) \quad E = \alpha - \beta$$

$$\Psi_{(Eb)} = \frac{1}{\sqrt{2}}(\phi_2 - \phi_3) \quad E = \alpha - \beta$$

Count bonding (*B*) and antibonding (*A*) interactions between adjacent atoms.

$B > A$	Bonding molecular orbitals;
$B = A$	Nonbonding molecular orbitals; and
$B < A$	Antibonding molecular orbitals.

6.9 BAND THEORY: MOLECULE ORBITAL THEORY AND METALLIC BONDING ORBIT

Why are lithium metallic properties lost in the gas phase?

- A lithium atom contains one electron in its valence shell and two in its inner shell. The electronic configuration of a lithium atom is:

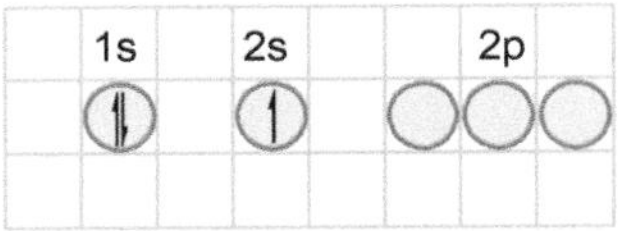

- A lithium molecule, Li_2, is formed in the vapor state. The bonding occurs through the 2s atomic orbital of the valence shell of the lithium atoms, while the three-2p orbitals are empty (Fig. 6.34). As a result, there are six electrons in the molecule that have the configuration: $\sigma_{1s}^2, \sigma^{*2}_{1s}, \sigma_{2s}^2, \sigma^{*0}_{2s}$.
- The molecular orbitals of the inner shell ($\sigma_{1s}^2, \sigma^{*2}_{1s}$) do not contribute to the bonding. The valence orbitals (2s) of two lithium atoms combine to give two molecular orbitals-one bonding (σ_{2s}) and one antibonding (σ^*_{2s}). Bonding occurs because the σ_{2s}^2 is full and σ^*_{2s} is empty.
- Only in the vapor state, bonding occurs to form Li_2, but in the solid, it is energetically more favorable for Li to have a metallic structure.

How could the band theory be used to explain the metallic bonding, thermal and electrical conductivity of lithium? If we start by N atoms being able to contribute one electron, define the Hückel secular determinant of the system, and the energy difference between the lowest and the highest molecular energy levels.

- Consider three Li atoms are connected together to form Li_3.
  - The three valence orbitals (2s) would combine to form three combined orbitals, one bonding, one nonbonding and one antibonding.
  - The energy of the nonbonding orbit intermediates that of the bonding and antibonding orbitals.
  - The three valence electrons from the three atoms would redistribute between the bonding (two electrons) and the nonbonding orbits (one electron) (Fig. 6.35).

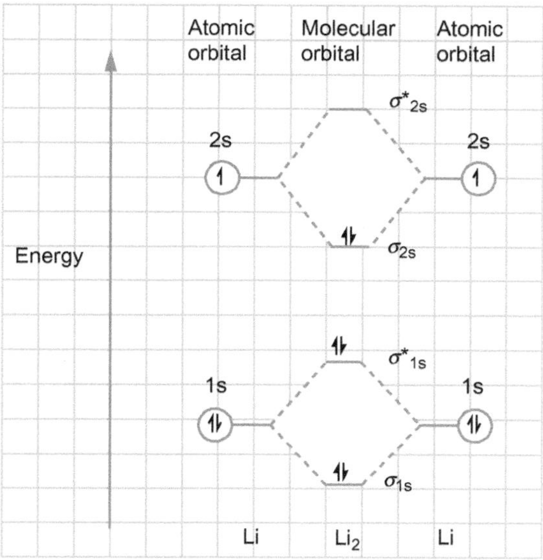

FIG. 6.34 Molecular orbitals for lithium.

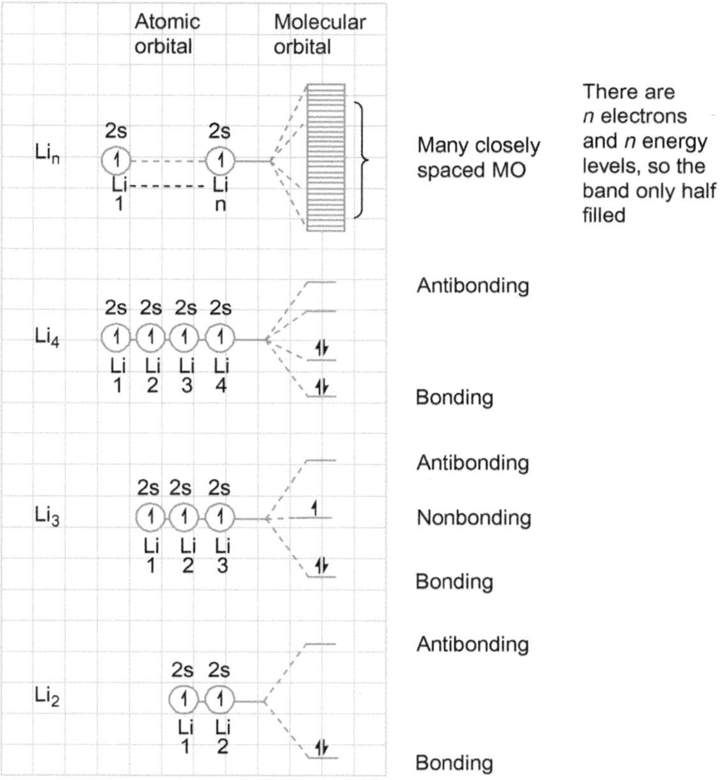

FIG. 6.35 Expansion of molecular orbitals into bands in metals.

- In Li_4:
  - the four valence orbitals would combine to give four combined orbits (two bonding and two antibonding orbits);
  - the existence of four orbits lowers the energy gap between the orbitals; and
  - the four valence electrons would occupy the two bonding orbitals (Fig. 6.35).

- In Li_n:
  - since the number of lithium atom in the cluster multiplies, the number of electrons increases;
  - the gaps among the energy levels of the different orbitals will further drop; and
  - because of the large number of atoms, the energy levels of the orbitals will be packed together, that they practically form a band (Fig. 6.35).
- The number of formed combined orbitals must be equal to the number of constituent atomic orbits.
- When N atoms are combined, there are N combined orbitals and the Hückel secular determinant:

$$\begin{vmatrix} E-\alpha & \beta & 0 & 0 & 0 & \cdots & 0 \\ \beta & E-\alpha & \beta & 0 & 0 & \cdots & 0 \\ 0 & \beta & E-\alpha & \beta & 0 & \cdots & 0 \\ 0 & 0 & \beta & E-\alpha & \beta & \cdots & 0 \\ \vdots & \vdots & \vdots & \vdots & \vdots & \cdots & \vdots \\ 0 & 0 & 0 & 0 & 0 & \cdots & E-\alpha \end{vmatrix} = 0 \quad (6.9.1)$$

Where β is the resonance integral, the roots of this symmetric tridiagonal determinant follow the expression:

$$E_k = \alpha + 2\beta \cos\left(\frac{k\pi}{N+1}\right), \quad k = 1, 2, \ldots, N \quad (6.9.2)$$

If N is infinity large, then, $\frac{\pi}{N+1} \cong 0$, and $\frac{N\pi}{N+1} \cong \pi$:

$$E_1 = \alpha + 2\beta \cos\left(\frac{\pi}{N+1}\right) = \alpha + 2\beta \quad (6.9.3)$$

$$E_N = \alpha + 2\beta \cos\left(\frac{N\pi}{N+1}\right) = \alpha - 2\beta \quad (6.9.4)$$

and

$$E_N - E_1 = 4\beta \quad (6.9.5)$$

- Because every orbit can be filled by two electrons, and there is only one valence electron in each lithium atom, then:
  - Only half of the 2s-valence band are occupied at $T=0$ (the bonding orbits (Fig. 6.36)).
  - A very small amount of energy will be required to excite an electron to an empty level.
- Consequently, electrons have a high degree of mobility because:
  - the combined orbits extend in three dimensions over all the atoms of the lithium lattice;
  - there is actually no energy gap between the occupied and unoccupied molecular orbitals; and
  - only half the orbitals in the valence band are occupied with electrons (Fig. 6.36).

  The mobile electrons explain the high thermal and electrical conduction of metals.

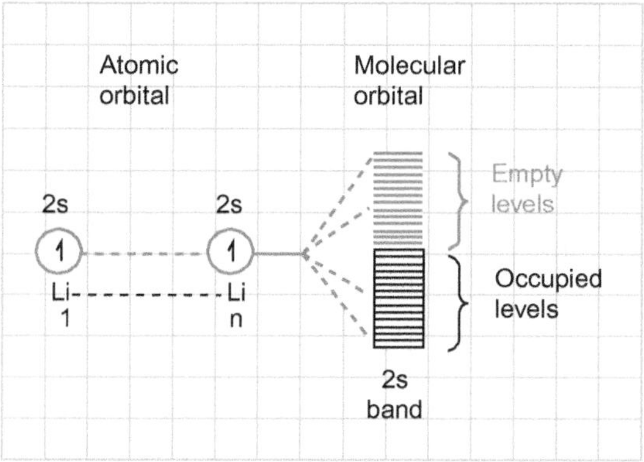

FIG. 6.36 Metallic orbitals for lithium showing a half-filled band.

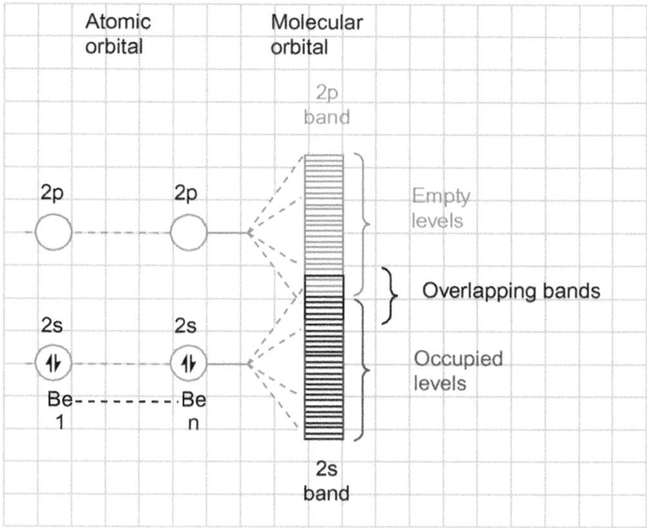

FIG. 6.37 Metallic molecular orbitals for beryllium showing overlapping bands.

In beryllium, there are two valence electrons; these electrons would just fill the 2s-valence band of combined orbitals. How could the band theory be used to explain the metallic bonding, thermal conductivity, and electrical conductivity of beryllium?

- The 2s and 2p atomic orbitals of beryllium form two bands of combined orbitals (Fig. 6.37).
- The upper levels of the 2s band overlaps with the lower levels of the 2p band (Fig. 6.37). Some of the 2p band will be occupied and some of the 2s band will be empty.
- It is likely and probable to promote electrons to empty levels of the conduction band.
- Then beryllium behaves as a metal only because the bands overlap, so perturbation from the occupied valence band to the empty conduction band can occur.

6.10 CONDUCTORS, INSULATORS, AND SEMICONDUCTORS

How does each of the following operate?

i. **conductors**
ii. **insulators**
iii. **semiconductors.**

- In electrical and thermal conductors (metals), either:
  - the valence band is partially full; or
  - the valence and conduction bands overlap.

    There is no significant gap between occupied and empty combined orbitals, so perturbation can easily occur (Fig. 6.38).
  - If an electrostatic potential is applied to metal, the potential energy of the states with the electron moving toward the positive charge is lower than those it moving toward the negative charge (Fig. 6.39).
  - The occupancy of the state in which the electrons moving toward the positive charge becomes more populated.
  - Thus, there is a net transfer of electrons into states moving toward the positive charge, and the metal is conductor.
- In electrical and thermal insulators (nonmetals), either:
  - the valence band is full, so perturbation within the band is impossible; or
  - there is a considerable energy gap (band gap) between the valence band and the next empty band (Fig. 6.38).

    Electrons cannot easily be promoted to empty levels.
  - There is no possibility of transfer of electrons even in the presence of a potential, and there is no net current and the material is an insulator.

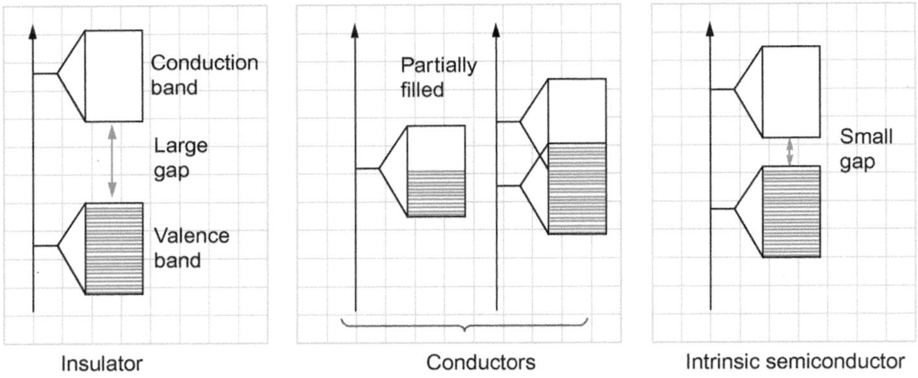

FIG. 6.38 Representations of the band theory for insulator, conductors, and intrinsic semiconductors.

FIG. 6.39 Effect of an electric field on the energy levels in a metal. When the field is applied, there is a net flow of electrons toward the positive charge.

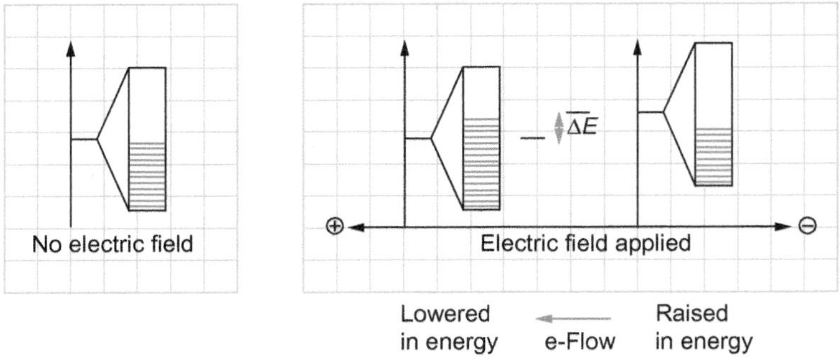

- Semiconductors are insulators (Fig. 6.38):
  - The energy gap between occupied and unoccupied bands is sufficient that thermal energy is capable to promote a small number of electrons from the valence band to the next empty band.
  - Both the unpaired electron left in the valence band as well the promoted electron in the conduction band can conduct electricity.

  Explain the paradox in the effect of temperature on the electrical conductivity of metallic solids and semiconductors.
- The population of the excited electron, P, in the conduction band will be determined by Boltzmann distribution as a function of temperature, the effect of Pauli principle and the band gap:

$$P = \frac{1}{e^{(E-E_F)/kT} + 1} \tag{6.9.6}$$

where E_F is the Fermi energy, the energy of the level for which $P = ½$.
Since $e^{(E-E_F)/kT} \gg 1$, then

$$P \cong e^{-(E-E_F)/kT} \tag{6.9.7}$$

- Because of the number of electrons moved to the conduction band increases as the temperature increases, the conductivity of semiconductors enhances with temperature (Fig. 6.40).
- However, the electrical conductivity of metals drop with increasing temperature even with more electrons is excited into empty orbitals. This inconsistency is resolved by the following:
  - The increasing in temperature escalates the thermal motions of the atoms and the electrons.
  - Collision between the moving electrons and atoms are more likely increases.
  - The electrons are deviated from their path through the solid.
  - Electrons become less efficient at transporting charge.

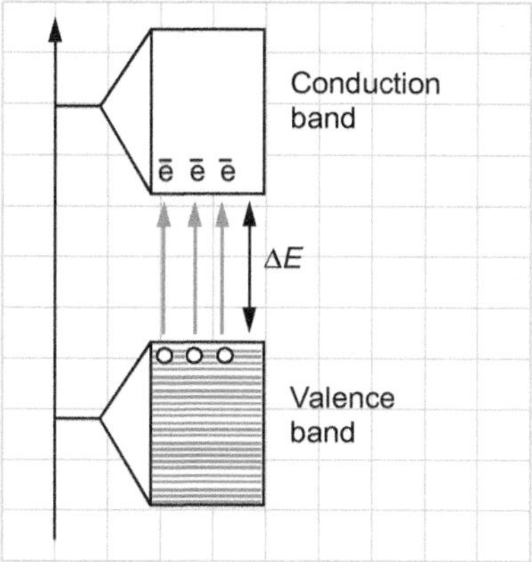

FIG. 6.40 Thermal excitation of electrons in an intrinsic semiconductor.

What are the differences between intrinsic and extrinsic semiconductors?

- Semiconductors that operate by thermal excitation across the energy gap are called intrinsic (pure) semiconductors, e.g., Si: [Ne] $3s^2$, $3p^2$.
  - The valence band composes of 3s- and 3p-orbitals.
  - The empty orbitals (conduction band) are results of the combination of 3d-, 4s- and 4p-orbitals.
  - Valence electrons are thermally excited crossing the energy gap to the lowest-lying levels of the conduction band.
  - For every electron excited, there will be left behind a hole in the valence band.
  - The electrons in both the valence band and conduction band will be free to move under potential (Fig. 6.39).
  - Because the number of electrons and holes in the conduction band is limited, the conductivity is not as high as in metal.
  - Like the thermal excitation, photons of light excite electrons from the valence band to the conduction band, and charge carriers are obtained. This is known as photoconductivity, and is used in photocells.
- Semiconductors that have their effective band gap reduced by impurities are known as extrinsic semiconductors.
  - Silicon can be exchanged by Al or Ga, or by P or As (Figs. 6.41 and 6.42).
  - The existence of Al, with one less electron than Si, generates a hole and the presence of P atom builds an excess electron.
  - Because the orbital overlap between Si atoms (host atoms) and either Al or P (foreign atoms) is different from the Si-Si (host-host) overlap, the energy levels of Si-Al or Si-P (host-foreign) do not localize in the conduction or valence band of Si (host), but in the band gap.
  - Atoms such as Al have an electron deficiency and generates a positive hole are called p-type, and those have an electron excess such P and injects negative charge carrier are called n-type (Figs. 6.41 and 6.42).

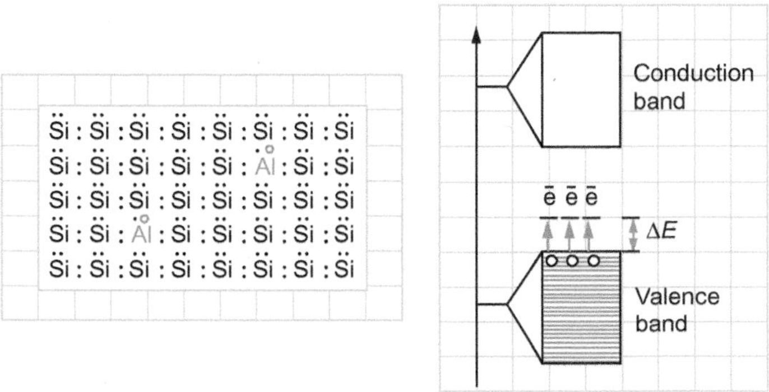

Conduction by holes
(p-type)

FIG. 6.41 Schematic representations of the band structure of p-type. Silicon is doped with traces of aluminum, and also gallium-doped germanium.

FIG. 6.42 Schematic representations of the band structure of n-type. Silicon is doped with traces of phosphorus, and also gallium-doped arsenic.

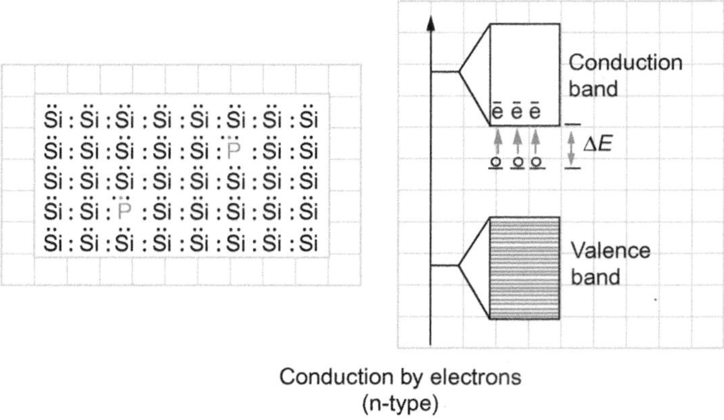

Cadmium sulfide photocells with a band gap of 236 kJ/mol are used for lighting up many cities though the world. What type of electromagnetic radiation emitted by the sun is most likely accountable for lighting up these cities at night? (1 eV = 96.49 kJ/mol = 8066 cm^{-1})

- For electrical conduction to occur, photo excited electrons must cross the forbidden energy gap.
- The energy of the forbidden gap in wave numbers, E:

$$E = 236\,\text{kJ}\,\text{mol}^{-1} \times \left(\frac{8066\,\text{cm}^{-1}}{96.49\,\text{kJ}\,\text{mol}^{-1}}\right) = 1.98 \times 10^4\,\text{cm}^{-1}$$

$$\lambda = \frac{10^7\,\text{nm}\,\text{cm}^{-1}}{1.98 \times 10^4\,\text{cm}^{-1}} = 505\,\text{nm}$$

This falls in the visible region (green region) of the electromagnetic spectrum.

TABLE 6.15 Semiconductors and Energy Gaps

Semiconductor	Energy Gap (kJ/mol)
PbS	36
Si	106
CdSe	168
GaP	217
ZnO	320

TABLE 6.16 Semiconductors and Energy Gaps

Semiconductor	Energy Gap (kJ/mol)	Energy Gap (cm^{-1})
PbS	36	3009
Si	106	8861
CdSe	168	14,044
GaP	217	18,140
ZnO	320	26,750

PbS is the suitable choice.

Semiconductors can be used to detect infrared radiation from the exhaust of missiles from several miles away. Which of the following semiconductors in Table 6.15 would you choose to design a detection system that operates in a 4000–3000 cm^{-1} region? (1 eV = 96.49 kJ/mol = 8066 cm^{-1})

- Answer:
  First the energy gap in the wave numbers must be calculated (Table 6.16)

SUGGESTIONS FOR FURTHER READING

C.J. Ballhausen, H.B. Gray, Molecular Orbital Theory, W. A. Benjamin, New York, 1965.
A.D. Baker, Betteridge, Photoelectron Spectroscopy, Pergamon Press, Oxford, 1972.
D.W. Turner, C. Baker, A.D. Baker, C.R. Brundle, Molecular Photoelectron Spectroscopy, John Wiley & Sons, New York, 1970.
R.L. Flurry Jr., Molecular Orbital Theory of Bonding in Organic Molecules, Marcel Dekker, New York, 1968.
H.B. Gray, Electrons and Chemical Bonding, W. A. Benjamin, New York, 1965.
W.L. Jorgenson, L. Salem, The Organic Chemist's Book of Orbitals, Academic Press, New York, 1973.
M. Ochin, H.H. Jaffé, Symmetry, Orbital and Spectra, John Wiley & Sons, New York, 1971.
A. Streitwieser Jr., Molecular Orbital Theory for Organic Chemists, John Wiley & Sons, New York, 1961.
A. Streitwieser Jr., P.H. Owens, Orbital and Electron Density Diagrams, Macmillan, New York, 1973.
G.G. Hall, J.E. Lennard-Jones, Foundations of Molecular Orbital Theory, Adv. Quantum Chem. 22 (1991) 1.
J. Daintith, Oxford Dictionary of Chemistry, Oxford University Press, New York, ISBN: 0-19-860918-3, 2004.
M.D. Licker, McGraw-Hill Concise Encyclopedia of Chemistry, McGraw-Hill, New York, ISBN: 0-07-143953-6, 2004.
G.L. Miessler, D.A. Tarr, Inorganic Chemistry, fifth ed., Pearson, Upper Saddle River, NJ, ISBN: 9780321811059, 2013, pp. 117–165, 475–534.
J.S. Griffith, L.E. Orgel, Ligand field theory, Q. Rev. Chem. Soc. 11 (1957) 381–383.
C.A. Coulson, Valence, Clarendon Press, Oxford, 1952.
E. Hückel, Trans. Faraday Soc. 30 (1934) 40–52.
C.A. Coulson, Math. Proc. Camb. Philos. Soc. 34 (2) (1938) 204–212.
G.G. Hall, Proc. R. Soc. Lond. A 202 (1070) (1950) 336–344.
F. Jensen, Introduction to Computational Chemistry, John Wiley and Sons, Chichester, West Sussex, England, ISBN: 978-0-471-98425-2, 1999.
R.H. Petrucci, W.S. Harwood, General Chemistry: Principles and Modern Applications, Pearson Prentice Hall, Upper Saddle River, NJ, 2007.
L. Dingrando, K. Tallman, N. Hainen, C. Wistrom, Chemistry, McGraw-Hill, School Pub Co, Glencoe, 2004.
J.C. Kotz, P.M. Treichel, G.C. Weaver, Bonding and Molecular Structure: Orbital Hybridization and Molecular Orbitals, Chemistry & Chemical Reactivity, Thomson Brooks/Cole, Belmont, CA, 2006, pp. 457–466.
R.S. Mulliken, Phys. Rev. 41 (1) (1932) 49–71.
F. Hund, Z. Phys. 36 (1926) 657–674.
F. Hund, Zur Deutung der Molekelspektren, Zeitschrift für Physik, Part I, vol. 40, pp. 742–764 (1927); Part II, vol. 42, pp. 93–120 (1927); Part III, vol. 43, pp. 805–826 (1927); Part IV, vol. 51, pp. 759–795 (1928); Part V, vol. 63, pp. 719–751 (1930).
R.S. Mulliken, Phys. Rev. 29 (1927) 637–649.
R.S. Mulliken, Phys. Rev. 32 (1928) 186–222.
F. Hund, W. Kutzelnigg, Angew. Chem. Int. Ed. 35 (1996) 573–586.
R.S. Mulliken, Science 157 (3785) (1967) 13–24.
J. Lennard-Jones, Trans. Faraday Soc. 25 (1929) 668–686.
G.L. Miessler, D.A. Tarr, Inorganic Chemistry, third ed., Pearson Prentice Hall, Upper Saddle River, NJ, ISBN: 978-1418001803, 2004.
C.E. Housecroft, A.G. Sharpe, Inorganic Chemistry, second ed., Pearson Prentice Hall, Upper Saddle River, NJ, ISBN: 9780130399137, 2005, pp. 29–33, 41–43.
P. Atkins, J. De Paula, Atkins Physical Chemistry, eighth ed., Oxford University Press, Oxford, 2006.
Y. Jean, F. Volatron, An Introduction to Molecular Orbitals, Oxford University Press, Oxford, 1993.
M. Munowitz, Principles of Chemistry, Norton & Company, New York, ISBN: 978-0-393-972887, 2000, pp. 229–233.
L. Gagliardi, B.O. Roos, Nature 433 (2005) 848–851.
P.W. Atkins, Inorganic Chemistry, fourth ed., W.H. Freeman, New York, ISBN: 978-0-7167-4878-6, 2006, p. 208.
R. Chang, Chemistry, seventh ed., McGraw Hill, New York, 2002.
R.L. deKock, H.B. Gray, Chemical Structure and Bonding, Benjamin/Cummings, Menlo Park, CA, 1980.
I.N. Levine, Quantum Chemistry, fourth ed., Prentice Hall, Englewood Cliffs, NJ, 1991.
T.W.G. Solomons, Fundamentals of Organic Chemistry, fifth ed., John Wiley, New York, 1997.

Chapter 7

Crystal Field Theory

A small number of bonds are exclusively ionic, covalent, or metallic. Almost all are partially mix of the three main types, and have some characteristics of at least of two, or sometimes of all three types (Scheme 7.1).

In the molecular orbital treatment, the valence electrons of the bonding shells of the ligand atoms and that of the central ion are not entirely associated either with the ligand nuclei or with the central ion. These shared electrons are delocalized to certain extent, which can vary continuously from zero to unity.

When the electrons of the free ion are located entirely on a central ion while the ligand atoms at the equilibrium distance in the coordination cluster, the electrons of the ligands also remain totally attached to those atoms during the assembly. This case is represented as the crystal field model.

Having identified the valence orbital's wave functions of the central ion in their real forms (Chapter 2), it is possible to explore the impact of various distributions of ligand atoms around the central ion upon its valence orbitals. We assert a quantitative basis of this effect only in the case of purely ionic model of coordination; other degrees of mixing will be addressed in the following chapter.

In this chapter, we start with why there is a need for another theory, and the bases of the crystal field theory, then look at how the theory differs from the ligand field theory. The effect of a cubic crystal field on d- and f-electrons are introduced, then compute the expressions of the Hamiltonian to find the crystal field potential that is experienced by electron in octahedral, square planar, tetragonally distorted octahedral, and tetrahedral ligand arrangements.

The perturbation theory for degenerate systems is used to explain how the crystal field potential of the surrounding ligands perturbs the degeneracy of d-orbitals of the central ion. The energies of the perturbed d-orbitals are calculated by solving the secular determinant. The obtained energies are fed back into secular equations that are derived from the secular determinant, to yield wave functions appropriate for the presence of the potential. The variation in the potential energy of each d-electron due to their crystal field is determined and the splitting of d-orbitals so deduced in octahedral, tetrahedral, tetragonally distorted D_{4h}, in terms of D_q, D_t, and D_s.

The crystal field treatment carefully examines the types of interactions that might affect the free ion. There are two crystal field type approaches in order to study the effects that the ligands of the metal complex have on the energies of the d-orbitals.

For free ions in weak crystal fields, problems and the required approximations are discussed. We then study the influence of a weak field on a polyelectronic configuration of free ion terms, and find the splitting in each term and the wave function for each state.

In the strong field approach, the first concern was how the strong field differs from the weak field approach; we define the determinantal, symmetry and the energy of each of basis states. We compute the appropriate Hamiltonian, and the diagonal and nondiagonal interelectronic repulsion in terms of A, B, and C parameters.

- 7.1: The advantages and disadvantages of valence bond theory
- 7.2: Bases of crystal field theory
  - d-Orbitals in cubic crystal field
  - f-Orbitals in cubic crystal field
- 7.3: The crystal field potential
  - Octahedral crystal field potential, V_{oct}
  - Square planar crystal field potential, $V_{sq.pl.}$
  - Tetragonal crystal field potential, $V_{Tetrag.}$
  - Tetrahedral crystal field potential, V_{T_d}
- 7.4: Zero-order perturbation theory (The effect of crystal field on the orbital wave functions of degenerate orbitals)
  - The linear combination of atomic orbitals, LCAO-MO, and energy calculation
  - The perturbation theory for degenerate systems
  - The splitting of d-orbitals in octahedral crystal field, V_{oct}

Electrons, Atoms, and Molecules in Inorganic Chemistry. http://dx.doi.org/10.1016/B978-0-12-811048-5.00007-9

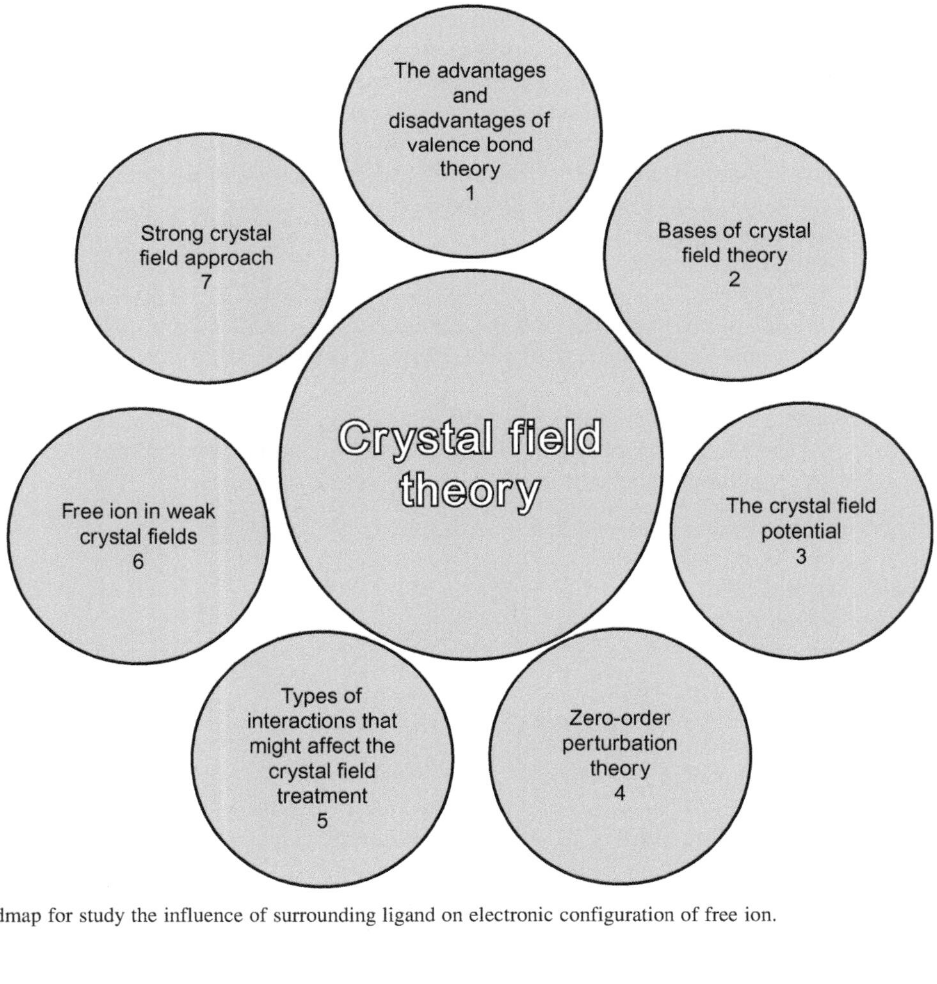

SCHEME 7.1 Roadmap for study the influence of surrounding ligand on electronic configuration of free ion.

7.1 THE ADVANTAGES AND DISADVANTAGES OF VALENCE BOND THEORY

What are the advantages of valence bond theory, and why was there a need for another theory?

- Valence bond theory explains chemical bonding using electron pairing and orbital hybridization (Chapter 5).
- Orbital hybridization is the concept of mixing atomic orbitals into new hybrid orbitals of different energies and shapes. These hybrid orbitals are suitable for pairing of electrons to form chemical bonds.
- Orbital hybridizations and valence shell electron-pair repulsion theory (VSEPR) are very useful in the explanation of molecular structure and atomic bonding properties.
- However, the valence bond theory does not provide an explanation for the spectral bands of metal complexes.
- The valence bond approach fails to explain the magnetic moment of most complexes.

7.2 BASES OF CRYSTAL FIELD THEORY

What are the bases of the crystal field theory? How does the crystal field theory differ from the ligand field theory?

- In crystal field theory, the coordination bond is regarded as a purely ionic model, wherein the anions and cations are rigid, nonpolarizable spheres of integral charge (point charges).
  - In such model, the electrons of the central metal ion are subject to a potential from the coordinated ligand atoms.
  - Each negative charge of the surrounding ligands induces an electrostatic potential on the electrons of the central ion, that potential can be calculated.
  - The potential formed nearby an ion is a characteristic feature of a lattice, and such a lattice could happen only in a crystal, and in coordination cluster of type AB: MO, NaCl or AB_2: MO_2, TiO_2, MX_2, and CaF_2.
- In contrast, the electrons of the ligand atoms and the central ion may be mutually shared during the molecular formation so that any individual electron exists with the equal probability on the ligands or the central ion. This sort of bonding has a completely covalent nature in the coordination cluster, e.g., $K_3Fe(CN)_6$.
- The term ligand field theory is employed for all nonzero degrees of electron mixing and the name crystal field theory is kept for the extreme in which there is no mixing of the ligand electrons and the central ion.
- In ligand field theory, there is an ion or atom under the influence of its nearest neighbors, or the central ion is influenced by attached array of ligand. Then the crystal field theory is a special case of the ligand field theory.
- Many of the results of ligand field theory depend only on the approximate symmetry of the ligands.

d-Orbitals in Cubic Crystal Field

Consider there are six negative points charges disposed at the apices of a regular octahedron d-metal complex. What are the induced effects of these charges upon the electronic structure of the central ion?

- Consider a d-metal atom or ion is concentric with *spherical shell* of uniformly distributed negative charge, and the total charge equals 6q units (Fig. 7.1).

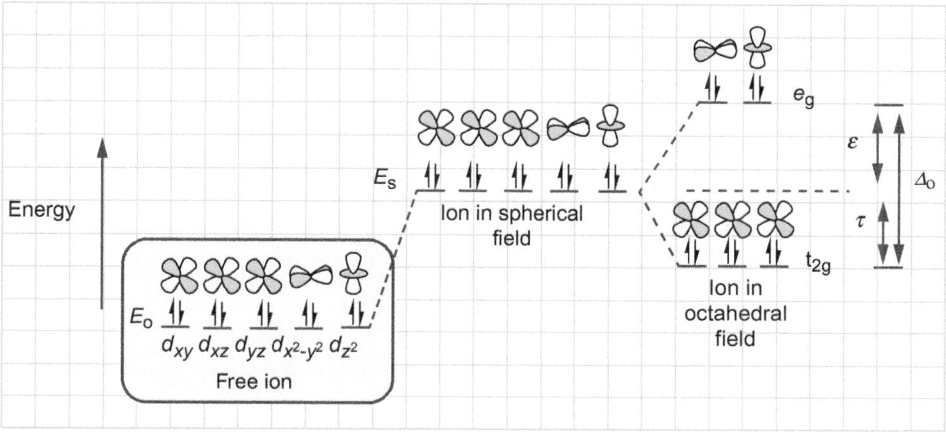

FIG. 7.1 The d-electrons in spherical field and the splitting due to an octahedral field.

- The 10 d-electrons of the central ion will now have an energy E_s, which higher than its energy E_o, where E_o is the energy of the 10 electrons of d-orbital of the free ion.
  Note:
  - $E_s > E_o$: as a result of the repulsive forces between the outer shell of negative charges and d-electrons.
  - All the five pairs of d-electrons have the same energy, since the charge distribution is spherical.
- If the total charge (6q) on the spherical shell is focused in six point charges, each q, at six apices of an octahedron:

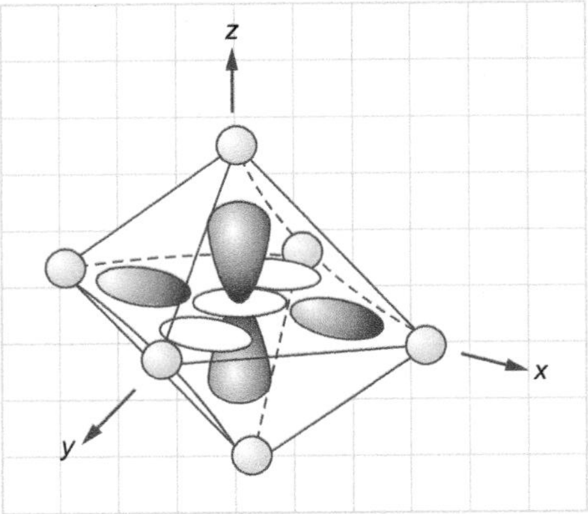

  - The d_{z^2} and $d_{x^2-y^2}$ orbitals, e_g, locate their electron density maxima toward the *Z*, *X* and *Y*-axes, respectively.
  - While, the d_{xy}, d_{xz}, and d_{yz} orbitals, t_{2g}, have their electron density maxima located in the regions between the Cartesian axes.
  - The d-electrons of the central ion tend to restrict themselves as far as possible in d_{xy}, d_{xz}, and d_{yz} orbitals avoiding d_{z^2} and $d_{x^2-y^2}$ orbitals.
- Thus in octahedral crystal field, the d-orbitals of the central ion are all influenced by the presence of the ligand atoms, but the d_{z^2} and $d_{x^2-y^2}$ orbitals are more destabilized than the d_{xy}, d_{xz}, and d_{yz} orbitals.
- Now three pairs of electrons are in t_{2g} orbitals and two pairs of electrons in e_g orbitals. These orbitals are differing in energy by Δ_o (Fig. 7.1).

$$6(E_s - \tau) + 4(E_s + \varepsilon) = 10E_s$$

$$\varepsilon + \tau = \Delta_o$$

Solving we obtain:

$$3\tau = 2\varepsilon$$

Therefore:

$$\varepsilon = \frac{3}{5}\Delta_o$$

$$\tau = \frac{2}{5}\Delta_o$$

- The electrons of s-orbital of the central ion are spherically distributed, so no orbital splitting is predicted.
- Since each of the three p-orbitals has its electron density focused toward two ligand atoms, the p-orbitals also do not split by an octahedral crystal field.

f-Orbitals in Cubic Crystal Field

How are the real forms of f-orbitals influenced by an octahedral array of point charge?

- Fig. 7.2 represents the real form of f-orbitals that will be influenced by an octahedral assembly of point charges.
- Direction of the electron clouds of f-orbitals toward the point charge will decide extent of destabilization:

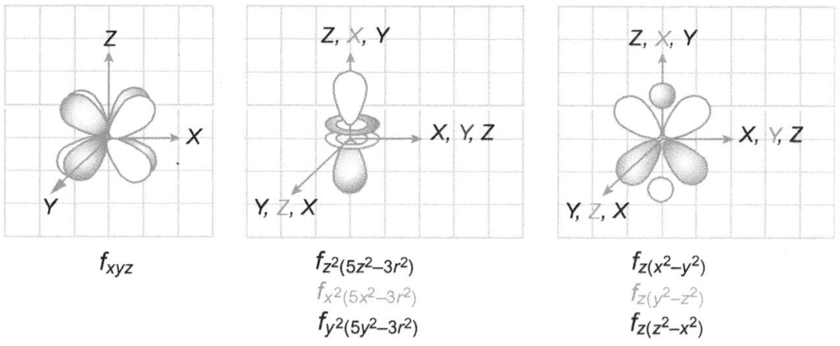

FIG. 7.2 f-Orbitals.

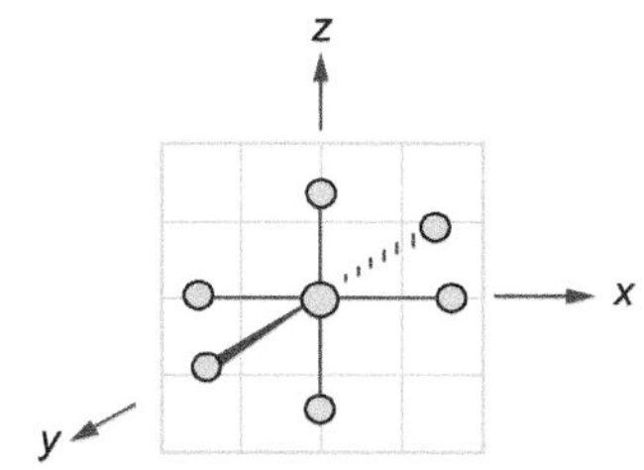

The octahedral array of point charges

- f_{xyz}-Orbital points to the octahedral faces and has the least destabilization by the charges on the axes (Fig. 7.3).
- The set, $f_{z(5z^2-3r^2)}$, $f_{x(5x^2-3r^2)}$, and $f_{y(5y^2-3r^2)}$, are directed along axes and the most destabilized by the ligand charges.
- The remaining set, $f_{z(x^2-y^2)}$, $f_{x(y^2-z^2)}$, and $f_{y(z^2-x^2)}$ symmetrically point toward the ligand charges to an extent, the amount of destabilization is in between the previous two sets.

7.3 THE CRYSTAL FIELD POTENTIAL

Octahedral Crystal field Potential, $V_{Oct.}$

Write an expression to find:

a. **the crystal field potentials for O_h arrangement;**
b. **the consequence of the crystal field potential; and**
c. **the distance between two points in the space in terms of the spherical harmonics.**

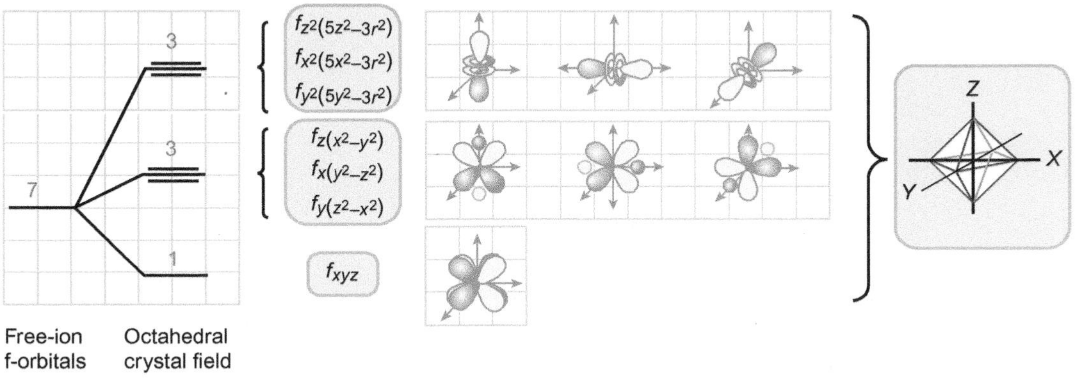

FIG. 7.3 The destabilization (splitting) of the f-orbitals by the octahedral crystal field.

What appropriate function, procedures, and simplifications are needed to find the crystal field potentials?

What is the general Hamiltonian operator that can be used to determine the energy of an electron within any crystal field environment?

Compute the octahedral Hamiltonian operator, $V_{Oct.}$, to find the crystal field potential that is experienced by electrons in an octahedron environment.

- Crystal field theory deals with the molecule as if it exists as a positive metal ion surrounded by negatively charged ions (ligands).
- By using crystal field theory, the *relative* energies of electronic transitions can be determined. The *absolute* energies cannot be generated by this simple theory.
- The model used for the calculation is a regular octahedral distribution of charges of magnitude *ze* about the central ion at a distance a (Fig. 7.4).
- In order to find the potential at any point in a space (x, y, z), which results from the octahedron of charges (i's), summing the potentials from the each charge must be done:

$$V_{(x,y,z)} = \sum_{i=1}^{6} V_{(i,x,y,z)} = \sum_{i=1}^{6} \frac{ez_i}{r_{ij}} \tag{7.3.1}$$

where r_{ij} is the distance from the ith charge to the point (x, y, z).

- Summing the potential over all the ligands permits the assessment of the potential experienced by an electron in any point of the molecular environment.
- Consequently, it is possible to obtain the effect of various distributions of ligand atoms around the central ion (crystal field potential) upon its d-orbitals as described in Scheme 7.2 by using the real forms of the d wave functions.
- The *additional potential* energy of each d-electron *due to their crystal field* can be determined and the splitting of d-orbitals thus deduced.
- Any orbital can be described in terms of spherical harmonics centered at the metal nucleus as origin, thus it is convenient to center the harmonics describing $\frac{1}{r_{ij}}$ at the metal nucleus.
- The potential energy, $\frac{ze}{r_{ij}}$, associated with an electron j (d-electron) at a distance r_{ij} from a negatively charged ligand i (ligand electron) can be specified in terms of spherical harmonics $Y_{lm}(\theta,\phi)$ centered at the origin.
- The distance, r_{ij}, between the two points, i and j, in space is given by (Fig. 7.5):

$$\frac{1}{r_{ij}} = \sum_{n=0}^{\infty}\sum_{m=-n}^{n} \frac{4\pi}{2n+1} \cdot \frac{r_<^n}{r_>^{n+1}} \cdot Y_{nj}^m \cdot Y_{ni}^{m*} \tag{7.3.2}$$

Where $r_<$ is the shorter of the distances from the origin to the point i and j, Y_{ni}^m refers to the ligand charge, and Y_{nj}^m refers to the electron (Table 7.1). The distance to point j from the central ion, $r_<$, can be written simply as r. $r_>$ is the distance a, $\frac{1}{r_{ij}}$ then takes the form:

$$\frac{1}{r_{ij}} = \sum_{n=0}^{\infty}\sum_{m=-n}^{n} \frac{4\pi}{2n+1} \cdot \frac{r^n}{a^{n+1}} \cdot Y_{nj}^m \cdot Y_{ni}^{m*} \tag{7.3.3}$$

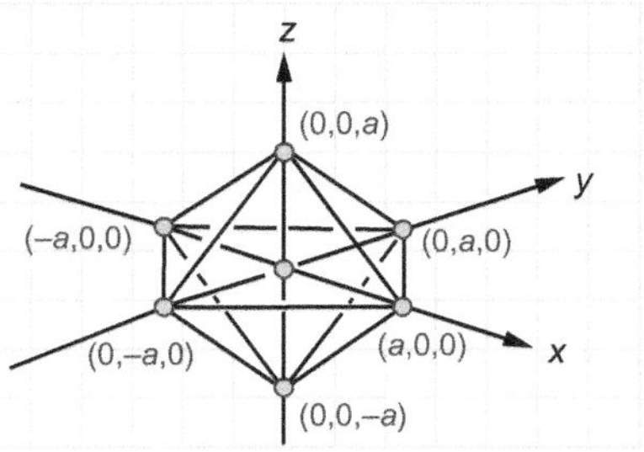

FIG. 7.4 Octahedral distribution of charges.

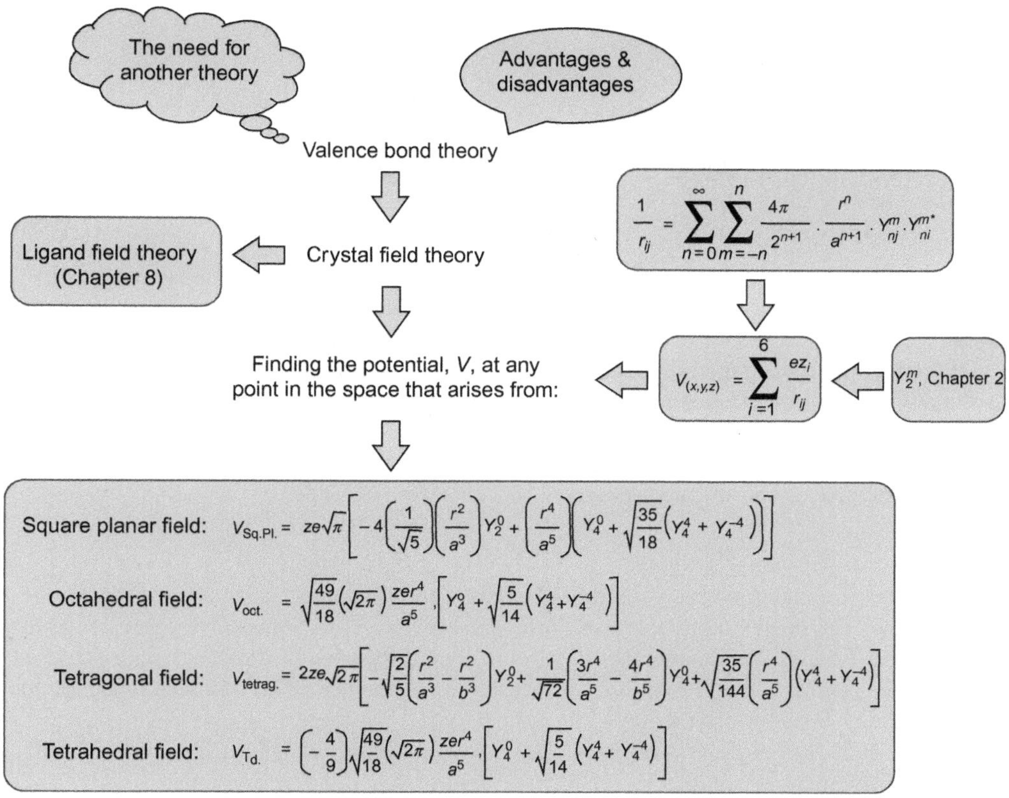

SCHEME 7.2 The approach used to find the crystal field potential that is experienced by electrons in a specific environment.

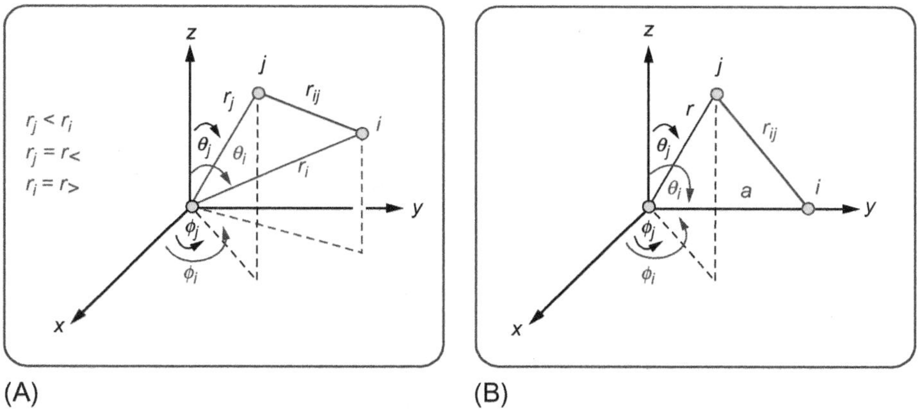

FIG. 7.5 (A) Distance r_{ij} from a negatively charged i to an electron j. (B) In the case of O_h, the six charges (i's) are located on the coordinate axes; θ_i and ϕ_i are 0, $\pm\frac{\pi}{2}$, or π.

TABLE 7.1 The Spherical Harmonic in Polar Coordinates

Harmonic	Polar Coordinates
$Y_0^0 = Y_0^{0*}$	$^1/_{\sqrt{2}}\,^1/_{\sqrt{2\pi}}$
$Y_2^0 = Y_2^{0*}$	$\sqrt{(^5/_8)}\left(^1/_{\sqrt{2\pi}}\right)(3\cos^2\theta - 1)$
$Y_2^{\pm 2} = Y_2^{\mp 2*}$	$\sqrt{(^{15}/_{16})}\left(^1/_{\sqrt{2\pi}}\right)\sin^2\theta\, e^{\pm 2i\phi}$
$Y_4^{\pm 2} = Y_4^{\mp 2*}$	$\sqrt{(^{45}/_{64})}\left(^1/_{\sqrt{2\pi}}\right)\sin^2\theta(7\cos^2\theta - 1)e^{\pm 2i\phi}$
$Y_4^{\pm 0} = Y_4^{\pm 0*}$	$\sqrt{(^9/_{128})}\left(^1/_{\sqrt{2\pi}}\right)(35\cos^4\theta - 30\cos^2\theta + 3)$
$Y_4^{\pm 4} = Y_4^{\mp 4*}$	$\sqrt{(^{315}/_{256})}\left(^1/_{\sqrt{2\pi}}\right)\sin^4\theta\, e^{\pm 4i\phi}$

- To carry out the summation from n = zero to n = infinity, will take some time. Therefore, consider first the terms that will vanish.
  - If the form of the Hamiltonian operator is identified, the total energy of any wave-function will be given by

$$E = \int \psi * H\psi \, \partial\tau$$

where ψ consists of radial component which is a scalar and angular components. H is the crystal field operator describing a specific geometry, and is constructed as a linear combination of spherical harmonics.
  - The crystal field operator must transform as totally symmetric representation in the group concerned. This is because the shape of a molecule is independent of whether it is reflected or rotated; thus all its characters under the operators of the group are unity.
  - The value of E of the function ψ perturbed by crystal field operator must not change if it is rotated or reflected.
  - Therefore, the direct product of the representation:

$$\Gamma_\psi \Gamma_H \Gamma_\psi \neq 0$$

Only if the direct product contains a totally symmetric representation.
  - Only even harmonics can contribute to the Hamiltonian, n = even (totally symmetric representation with respect to the inversion, e.g., d-orbital, see Chapter 10, p. 590). The contributions of the terms with n odd would no longer transform as totally symmetric representation of the group concerned. The odd spherical harmonics in V_{oct} do not interact with the d- or other orbitals.
  - Functions with $n > 2l$ do not couple and vanish. Consequently, integrals of spherical harmonics with $l > 4$ taken with the product of two spherical harmonics with $l = 2$ (the d wave functions) must vanish.
    The summation is now reduced to $n = 0, 2, 4$
  - The magnetic quantum number, m_l, is the z component of the angular quantum number l. When the principal axis (z) of the point group concerned is even, i.e., 2 or 4 for example, the odd values of the component m cannot contribute to the totally symmetric representation. These components only contribute to Hamiltonian involving groups whose principal axis is odd.
- The summation is now reduced to $n = 0$ ($m = 0$), $n = 2$ ($m = \pm 2$, 0) and, $n = 4$ ($m = \pm 4, \pm 2, 0$).
- Now consider:

$$\frac{1}{r_{ij}} = \sum_{n=0}^{\infty} \sum_{m=-n}^{n} \frac{4\pi}{2n+1} \cdot \frac{r^n}{a^{n+1}} \cdot Y_{nj}^m \cdot Y_{ni}^{m*} \tag{7.3.3}$$

and the summation of the potential of individual charges gives the general Hamiltonian:

$$\begin{aligned} V_{oct} = \sum \frac{ze}{r_{ij}} &= \underbrace{c_2^0 \cdot R_2 \cdot Y_2^0}_{n=2;\, m=0} + \underbrace{c_2^{\pm 2} \cdot R_2 \cdot \left(Y_2^2 + Y_2^{-2}\right)}_{n=2;\, m=\pm 2} \\ &+ \underbrace{c_4^0 \cdot R_4 \cdot Y_4^0}_{n=4;\, m=0} + \underbrace{c_4^2 \cdot R_4 \cdot \left(Y_4^2 + Y_4^{-2}\right)}_{n=4;\, m=\pm 2} + \underbrace{c_4^4 \cdot R_4 \cdot \left(Y_4^4 + Y_4^{-4}\right)}_{n=4;\, m=\pm 4} \end{aligned} \tag{7.3.4}$$

where

$$c_n^m = \sum \frac{4\pi\, ze}{2n+1} \sum Y_n^m \tag{7.3.5}$$

$$R_n = \frac{r^n}{a^{n+1}} \tag{7.3.6}$$

- For $n = 0$; $m = 0$ ($c_0^0 \cdot R_0 \cdot Y_0^{0*}$, need not to be considered?)
  For each charge there is the term containing Y_0^0 (Table 7.1):

$$c_0^0 \cdot R_0 \cdot Y_0^{0*} = \frac{4\pi ze}{a} Y_0^0 \cdot Y_0^{0*}$$

$$Y_0^0 = \frac{1}{\sqrt{2}}\frac{1}{\sqrt{2\pi}} = Y_0^{0*}$$

$$c_0^0 \cdot R_0 \cdot Y_0^{0*} = \frac{4\pi ze}{a}\frac{1}{\sqrt{2}}\frac{1}{\sqrt{2\pi}} \cdot \frac{1}{\sqrt{2}}\frac{1}{\sqrt{2\pi}} = \frac{ze}{a} \quad (7.3.7)$$

- So that the total contribution to $V_{(x,y,z)}$ from $n=0$ is $6\frac{ze}{a}$. This term is independent of the m wave function; consequently, it corresponds to an identical change in the energy of all the five d-orbitals (the first energy shift in Fig. 7.1).

- Now term-by-term ($n=2$ ($m=\pm2$, 0) and, $n=4$ ($m=\pm4$, ±2, 0)) and using Table 7.1:
- For $n=2$; $m=0$, ±2,

$$\underbrace{c_2^0 \cdot R_2 \cdot Y_2^0}_{n=2;\ m=0} = ?$$

Firstly, when $m=0$:

$$c_n^m = \sum \frac{4\pi\, ze}{2n+1}\sum Y_n^m$$

$$c_2^0 = \sum \frac{4\pi\, ze}{5}\sum Y_2^0$$

$$c_2^0 = \frac{4\pi\, ze}{5}\sqrt{(5/8)}\left(1/\sqrt{2\pi}\right)\sum(3\cos^2\theta - 1)$$

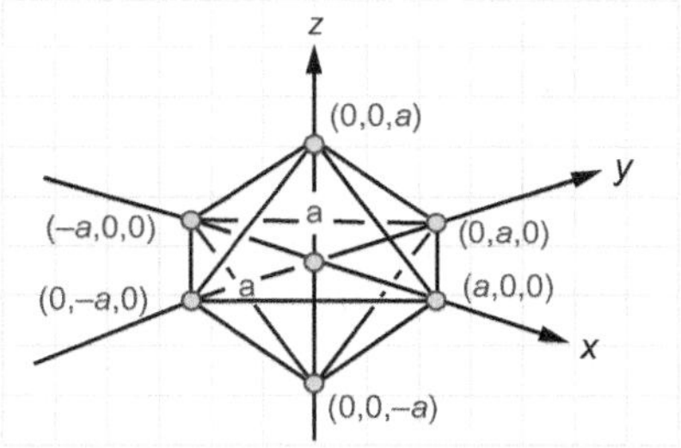

For the four points at ($\pm a$, 0, 0) and (0, $\pm a$, 0)

$$\theta = \frac{\pi}{2},\quad (3\cos^2\theta - 1) = -1$$

For the points at (0, 0, a)

$$\theta = 0,\quad (3\cos^2\theta - 1) = 2$$

For the points at (0, 0, $-a$)

$$\theta = \pi,\quad (3\cos^2\theta - 1) = 2$$

The summation of $(3\cos^2\theta - 1)$ over the six points yields zero.
Consequently, the contribution to $V_{(x,y,z)}$ from $n=0$, $m=0$ is zero.

- The c_n^m values are evaluated by summation over the ligands as follow:

$$c_2^{\pm2} = \frac{ze(4\pi)}{5}\sqrt{(15/16)}\left(1/\sqrt{2\pi}\right)\sum \sin^2\theta e^{\pm 2i\phi}$$

$$c_4^0 = \frac{ze(4\pi)}{9}\sqrt{(9/128)}\left(1/\sqrt{2\pi}\right)\sum(35\cos^4\theta - 30\cos^2\theta + 3)$$

$$c_4^{\pm 2} = \frac{ze(4\pi)}{9}\sqrt{(^{45}/_{64})}\left(^1/_{\sqrt{2\pi}}\right)\sum \sin^2\theta(7\cos^2\theta - 1)e^{\pm 2i\phi}$$

$$c_4^{\pm 4} = \frac{ze(4\pi)}{9}\sqrt{(^{315}/_{256})}\left(^1/_{\sqrt{2\pi}}\right)\sum \sin^4\theta e^{\pm 4i\phi}$$

$e^{\pm i\pi} = -1$, summing over the six ligands (Table 7.2).

- The total contribution to $V_{(x,y,z)}$:

$$V_{\text{oct}} = c_2^0 \cdot R_2 \cdot Y_2^0 + c_2^{\pm 2} \cdot R_2 \cdot \left(Y_2^2 + Y_2^{-2}\right) + c_4^0 \cdot R_4 \cdot Y_4^0 + c_4^2 \cdot R_4 \cdot \left(Y_4^2 + Y_4^{-2}\right) + c_4^4 \cdot R_4 \cdot \left(Y_4^4 + Y_4^{-4}\right) \tag{7.3.4}$$

  - $c_2^0 \cdot R_2 \cdot Y_2^0 = 0$
  - $c_2^{\pm 2} \cdot R_2 \cdot \left(Y_2^2 + Y_2^{-2}\right) = 0$
  - $c_4^0 \cdot R_4 \cdot Y_4^0 = \frac{ze(4\pi)}{9} \cdot \frac{r^4}{a^5}(28)\sqrt{(^9/_{128})} \cdot \left(^1/_{\sqrt{2\pi}}\right) \cdot Y_4^0$
  - $c_4^{\pm 2} \cdot R_4 \cdot \left(Y_4^2 + Y_4^{-2}\right) = 0$
  - $c_4^{\pm 4} \cdot R_4 \cdot \left(Y_4^4 + Y_4^{-4}\right) = \frac{ze(4\pi)}{9} \cdot \frac{r^4}{a^5}(4)\sqrt{(^{315}/_{256})}\left(^1/_{\sqrt{2\pi}}\right)\left(Y_4^4 + Y_4^{-4}\right)$
  - The sums for $c_2^{\pm 2}$ and $c_4^{\pm 2}$ are found to be zero.
- The octahedral Hamiltonian is then:

$$V_{\text{oct}} = \sqrt{(^{49}/_{18})} \cdot \left(\sqrt{2\pi}\right) \cdot \frac{zer^4}{a^5} \cdot \left[Y_4^0 + \sqrt{\frac{5}{14}}\left(Y_4^4 + Y_4^{-4}\right)\right] \tag{7.3.8}$$

Square Planar Crystal Field Potential, $V_{\text{Sq.Pl.}}$

Compute the square planar Hamiltonian operator, $V_{\text{Sq.Pl.}}$, to find the crystal field potential experienced by electrons in a square planar environment.

- The crystal field potential at any point in a space which arises from four charges arranged in square planar, $V_{\text{Sq.pl.}}$, from Eqs. (7.3.1) and (7.3.2):

TABLE 7.2 The Contributions of c_n^m Values and the Summation Over the Ligands

Ligand	θ_i	ϕ_i	$\sin^2\theta\ e^{\pm 2i\phi}$ or $\sin^2\theta\ e^{\pm 4i\phi}$	$35\cos^4\theta - 30\cos^2\theta + 3$	$\sin^4\theta\ e^{\pm 4i\phi}$
0, 0, *a*	0	–	0	8	0
0, 0, −*a*	π	–	0	8	0
0, *a*, 0	$\frac{\pi}{2}$	$\frac{\pi}{2}$	−1	3	1
0, −*a*, 0	$\frac{\pi}{2}$	$-\frac{\pi}{2}$	−1	3	1
a, 0, 0	$\frac{\pi}{2}$	0	1	3	1
−*a*, 0, 0	$\frac{\pi}{2}$	π	1	3	1
Total			0	28	4

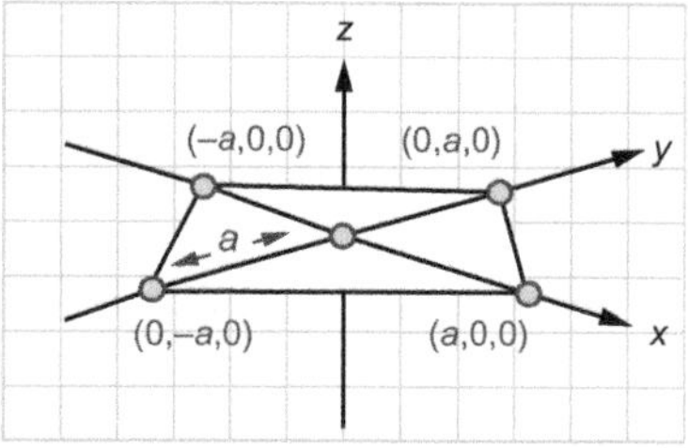

$$V_{\text{Sq.pl.}} = \sum \frac{ze}{r_{ij}} = \sum_{n=0}^{\infty}\sum_{m=-n}^{n} \frac{4\pi ze}{2n+1}\cdot\frac{r_<^n}{r_>^{n+1}}\cdot Y_{nj}^{m}\cdot Y_{ni}^{m*} \tag{7.3.9}$$

- Only even harmonics, $n=$even (totally symmetric representation), can contribute to the Hamiltonian.
- Functions with $n>2\,l$ vanish.

  The summation is now reduced to $n=0$, 2, 4

  When the principal axis (z) of the point group concerned is even, i.e., 2 or 4 for example, the odd values of the component m cannot contribute to the totally symmetric representation.

- The summation in Eq. (7.3.9) is now reduced to $n=0$ ($m=0$), $n=2$ ($m=\pm$ 2, 0) and $n=4$ (±4, ±2, 0).

$$V_{\text{Sq.pl.}} = \sum \frac{ze}{r_{ij}} = \underbrace{c_2^0\cdot R_2\cdot Y_2^0}_{n=2;\,m=0} + \underbrace{c_2^{\pm2}\cdot R_2\cdot\left(Y_2^2+Y_2^{-2}\right)}_{n=2;\,m=\pm2} + \underbrace{c_4^0\cdot R_4\cdot Y_4^0}_{n=4;\,m=0} + \underbrace{c_4^2\cdot R_4\cdot\left(Y_4^2+Y_4^{-2}\right)}_{n=4;\,m=\pm2} + \underbrace{c_4^4\cdot R_4\cdot\left(Y_4^4+Y_4^{-4}\right)}_{n=4;\,m=\pm4} \tag{7.3.10}$$

Where spherical harmonics Ys are given in Table 7.1:

$$c_n^m = \sum \frac{4\pi\, ze}{2n+1}\sum Y_n^m \tag{7.3.5}$$

$$R_n = \frac{r^n}{a^{n+1}} \tag{7.3.6}$$

For $n=0$; $m=0$ (not included?)
For each charge there is the term containing Y_0^0:

$$c_0^0\cdot R_0\cdot Y_0^{0*} = \frac{4\pi\, ze}{a}Y_0^0\cdot Y_0^{0*}$$

$$Y_0^0 = \frac{1}{\sqrt{2}}\,\frac{1}{\sqrt{2\pi}} = Y_0^{0*}$$

$$\frac{4\pi\, ze}{a}\,\frac{1}{\sqrt{2}}\,\frac{1}{\sqrt{2\pi}}\cdot Y_0^{0*} = \frac{ze}{a}$$

- The total contribution to $V_{D_{4h}}$ from $n=0$ is $4\frac{ze}{a}$. This term is independent of the m wave function, and consequently, it corresponds to an identical change in the energy of all the five d-orbitals.
- For $n=2$; $m=0$, ±2

$$c_2^0 = \frac{4\pi\, ze}{5}\sqrt{(^5/_8)}\left(^1/_{\sqrt{2\pi}}\right)\sum\left(3\cos^2\theta-1\right)$$

For the points at ($\pm a$, 0, 0) and (0, $\pm a$, 0)

$$\theta=\frac{\pi}{2},\quad \left(3\cos^2\theta-1\right)=-1$$

The summation of $\left(3\cos^2\theta-1\right)$ over the four points, yields: −4.

$$c_2^0 = -4ze\sqrt{\frac{\pi}{5}}$$

The contributions from the components +2, −2, +4, 0, −4, are computed in Table 7.3.

$$c_2^{\pm 2} = \frac{ze(4\pi)}{5}\sqrt{\left({}^{15}/_{16}\right)}\left({}^{1}/_{\sqrt{2\pi}}\right)\sum \sin^2\theta e^{\pm 2i\phi}$$

$$\therefore c_2^{\pm 2} = \frac{ze(4\pi)}{5}\sqrt{\left({}^{15}/_{16}\right)}\left({}^{1}/_{\sqrt{2\pi}}\right)(0) = 0$$

$$c_4^0 = \frac{ze(4\pi)}{9}\sqrt{\left({}^{9}/_{128}\right)}\left({}^{1}/_{\sqrt{2\pi}}\right)\sum\left(35\cos^4\theta - 30\cos^2\theta + 3\right)$$

$$\therefore c_4^0 = \frac{ze(4\pi)}{9}\sqrt{\left({}^{9}/_{128}\right)}\left({}^{1}/_{\sqrt{2\pi}}\right)(12) = ze\sqrt{\pi}$$

$$c_4^{\pm 2} = \frac{ze(4\pi)}{9}\sqrt{\left({}^{45}/_{64}\right)}\left({}^{1}/_{\sqrt{2\pi}}\right)\sum \sin^2\theta\left(7\cos^2\theta - 1\right)e^{\pm 2i\phi}$$

$$\therefore c_4^{\pm 2} = \frac{ze(4\pi)}{9}\sqrt{\left({}^{45}/_{64}\right)}\left({}^{1}/_{\sqrt{2\pi}}\right)(0) = 0$$

$$c_4^{\pm 4} = \frac{ze(4\pi)}{9}\sqrt{\left({}^{315}/_{256}\right)}\left({}^{1}/_{\sqrt{2\pi}}\right)\sum \sin^4\theta e^{\pm 4i\phi}$$

$$\therefore c_4^{\pm 4} = \frac{ze(4\pi)}{9}\sqrt{\left({}^{315}/_{256}\right)}\left({}^{1}/_{\sqrt{2\pi}}\right)(+4)$$

$$c_4^{\pm 4} = ze\sqrt{\pi}\sqrt{\frac{35}{18}}$$

- Summing over the four ligands yields the square planar Hamiltonian:

$$V_{\text{Sq.pl.}} = \sum \frac{ze}{r_{ij}} = c_2^0 \cdot R_2 \cdot Y_2^0 + \cancel{c_2^{\pm 2} \cdot R_2 \cdot \left(Y_2^2 + Y_2^{-2}\right)} + c_4^0 \cdot R_4 \cdot Y_4^0 + \cancel{c_4^{\pm 2} \cdot R_4 \cdot \left(Y_4^2 + Y_4^{-2}\right)} + c_4^4 \cdot R_4 \cdot \left(Y_4^4 + Y_4^{-4}\right)$$

and $R_n = \dfrac{r^n}{a^{n+1}}$: $R_2 = \dfrac{r^2}{a^3}$, and $R_4 = \dfrac{r^4}{a^5}$,

$$V_{\text{Sq.pl.}} = ze\sqrt{\pi}\left[-4\left(\frac{1}{\sqrt{5}}\right)\left(\frac{r^2}{a^3}\right)Y_2^0 + \left(\frac{r^4}{a^5}\right)\left(Y_4^0 + \sqrt{\frac{35}{18}}\left(Y_4^4 + Y_4^{-4}\right)\right)\right] \qquad (7.3.11)$$

TABLE 7.3 The Contributions c_n^m Values and the Summation Over the Ligands

Ligand	θ_i	ϕ_i	$\sin^2\theta\, e^{\pm 2i\phi}$ or $\sin^2\theta\, e^{\pm 4i\phi}$	$35\cos^4\theta - 30\cos^2\theta + 3$	$\sin^4\theta\, e^{\pm 4i\phi}$
0, a, 0	$\frac{\pi}{2}$	$\frac{\pi}{2}$	−1	3	1
0, −a, 0	$\frac{\pi}{2}$	$-\frac{\pi}{2}$	−1	3	1
a, 0, 0	$\frac{\pi}{2}$	0	1	3	1
−a, 0, 0	$\frac{\pi}{2}$	π	1	3	1
Total			0	12	4

Tetragonal Crystal Field Potential, V_{Tetrag}.

Compute the Hamiltonian operator, $V_{\text{Tetrag.}}$, to find the crystal field potential experienced by electrons in a tetragonally distorted octahedral environment.

- The two axial ligands are not the same as those in the plane, and from Eqs. (7.3.1) and (7.3.2):

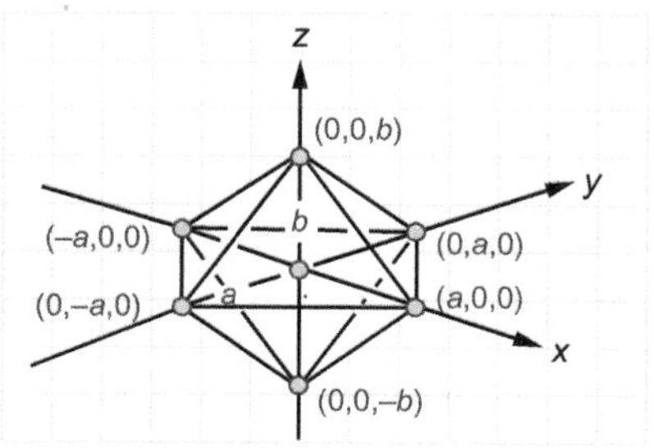

$$V_{\text{Tetrag}} = \sum \frac{ze}{r_{ij}} = \underbrace{c_2^0 \cdot R_2 \cdot Y_2^0}_{n=2;\, m=0} + \underbrace{c_2^{\pm 2} \cdot R_2 \cdot \left(Y_2^2 + Y_2^{-2}\right)}_{n=2;\, m=\pm 2} + \underbrace{c_4^0 \cdot R_4 \cdot Y_4^0}_{n=4;\, m=0} + \underbrace{c_4^2 \cdot R_4 \cdot \left(Y_4^2 + Y_4^{-2}\right)}_{n=4;\, m=\pm 2} + \underbrace{c_4^4 \cdot R_4 \cdot \left(Y_4^4 + Y_{\pm 4}^{-4}\right)}_{n=4;\, m=\pm 4} \quad (7.3.12)$$

where spherical harmonics Ys are given in Table 7.1:

$$c_n^m = \sum \frac{4\pi\, ze}{2n+1} \sum Y_n^m \quad (7.3.5)$$

$$R_n = \frac{r^n}{a^{n+1}}$$

- If the axial ligands have the coordinates $(b, \theta=0, \phi=0)$ and $(-b, \theta=\pi, \phi=0)$

$$c_2^0 = \frac{4\pi\, ze}{5} \sqrt{\left({}^5/_8\right)} \left({}^1/_{\sqrt{2\pi}}\right) \sum \left(3\cos^2\theta - 1\right)$$

$(0, 0, b)$: $a=b$, $\theta=0$, $\phi=0$, $\cos\theta=1$, then: $3\cos^2\theta - 1 = 2$ (Table 7.4).
$(0, 0, -b)$: $a=-b$, $\theta=\pi$, $\phi=0$, $\cos\theta=-1$, then: $3\cos^2\theta - 1 = 2$

TABLE 7.4 The Contributions c_n^m Values and the Summation Over the Ligands

Ligand	θ_i	ϕ_i	$\sin^2\theta\, e^{\pm 2i\phi}$ or $\sin^2\theta\, e^{\pm 4i\phi}$	$35\cos^4\theta - 30\cos^2\theta + 3$	$\sin^4\theta\, e^{\pm 4i\phi}$
0, 0, b	0	–	0	8	0
0, 0, $-b$	π	–	0	8	0
Total			0	16	0
0, a, 0	$\frac{\pi}{2}$	$\frac{\pi}{2}$	−1	3	1
0, −a, 0	$\frac{\pi}{2}$	$-\frac{\pi}{2}$	−1	3	1
a, 0, 0	$\frac{\pi}{2}$	0	1	3	1
−a, 0, 0	$\frac{\pi}{2}$	π	1	3	1
Total			0	12	4

$$c^0_{2,b} = \frac{4\pi ze}{5}\sqrt{(^5/_8)}\left(^1/_{\sqrt{2\pi}}\right)(4) = 4ze\sqrt{\frac{\pi}{5}}$$

$$R_{2,b} = \frac{r^2}{b^3}$$

If the equatorial ligands have the coordinates ($\pm a$, 0, 0) and (0, $\pm a$, 0)
($\pm a$, 0, 0): a, $\theta = 90$, $\phi = 0$, π, cos $\theta = 0$, then: $3\cos^2\theta - 1 = -1$
(0, $\pm a$, 0): a, $\theta = 90$, $\phi = \pm\pi/2$, cos $\theta = 0$, then: $3\cos^2\theta - 1 = -1$

$$c^0_{2,a} = \frac{4\pi ze}{5}\sqrt{(^5/_8)}\left(^1/_{\sqrt{2\pi}}\right)(4) = -4ze\sqrt{\frac{\pi}{5}}$$

$$R_{2,a} = \frac{r^2}{a^3}$$

$$\therefore c^0_2 \cdot R_2 \cdot Y^0_2 = 4ze\sqrt{\frac{\pi}{5}}\left(\frac{r^2}{b^3} - \frac{r^2}{a^3}\right)Y^0_2 = -2ze\sqrt{2\pi}\sqrt{\frac{2}{5}}\left(\frac{r^2}{a^3} - \frac{r^2}{b^3}\right)Y^0_2$$

$$c^{\pm 2}_2 = \frac{ze(4\pi)}{5}\sqrt{(^{15}/_{16})}\left(^1/_{\sqrt{2\pi}}\right)\sum \sin^2\theta e^{\pm 2i\phi}$$

$$\therefore c^{\pm 2}_{2:a,b} = \frac{ze(4\pi)}{5}\sqrt{(^{15}/_{16})}\left(^1/_{\sqrt{2\pi}}\right)(0) = 0$$

$$c^0_4 = \frac{ze(4\pi)}{9}\sqrt{(^9/_{128})}\left(^1/_{\sqrt{2\pi}}\right)\sum(35\cos^4\theta - 30\cos^2\theta + 3)$$

$$\therefore c^0_{4:b} = \frac{ze(4\pi)}{9}\sqrt{(^9/_{128})}\left(^1/_{\sqrt{2\pi}}\right)(16) = \frac{4}{3}ze\sqrt{\pi}$$

$$R_{4,b} = \frac{r^4}{b^5}$$

$$\therefore c^0_{4:a} = \frac{ze(4\pi)}{9}\sqrt{(^9/_{128})}\left(^1/_{\sqrt{2\pi}}\right)(12) = ze\sqrt{\pi}$$

$$R_{4,a} = \frac{r^4}{a^5}$$

$$\therefore c^0_4 \cdot R_4 . Y^0_4 = \frac{ze\sqrt{\pi}}{3}\left(\frac{4r^4}{b^5} + \frac{3r^4}{a^5}\right)\left(\frac{\sqrt{8}}{\sqrt{8}}\right) = \frac{ze2\sqrt{2\pi}}{\sqrt{72}}\left(\frac{4r^4}{b^5} + \frac{3r^4}{a^5}\right)Y^0_4$$

$$c^{\pm 2}_4 = \frac{ze(4\pi)}{9}\sqrt{(^{45}/_{64})}\left(^1/_{\sqrt{2\pi}}\right)\sum \sin^2\theta(7\cos^2\theta - 1)e^{\pm 2i\phi}$$

$$\therefore c^{\pm 2}_{4:a,b} = \frac{ze(4\pi)}{9}\sqrt{(^{45}/_{64})}\left(^1/_{\sqrt{2\pi}}\right)(0) = 0$$

$$c^{\pm 4}_4 = \frac{ze(4\pi)}{9}\sqrt{(^{315}/_{256})}\left(^1/_{\sqrt{2\pi}}\right)\sum \sin^4\theta e^{\pm 4i\phi}$$

$$\therefore c^{\pm 4}_{4:a} = \frac{ze(4\pi)}{9}\sqrt{(^{315}/_{256})}\left(^1/_{\sqrt{2\pi}}\right)(+4)$$

$$c_{4:a}^{\pm 4} = ze\sqrt{2\pi}\sqrt{\frac{9\times 35}{16\times 81}} = 2ze\sqrt{2\pi}\sqrt{\frac{35}{144}}$$

$$R_{4,a} = \frac{r^4}{a^5}$$

- The total, Eq. (7.3.12):

$$V_{\text{Tetrag.}} = c_2^0 \cdot R_2 \cdot Y_2^0 + \cancel{c_2^{\pm 2} \cdot R_2 \cdot (Y_2^2 + Y_2^{-2})} + c_4^0 \cdot R_4 . Y_4^0 + \cancel{c_4^2 \cdot R_4 \cdot (Y_4^2 + Y_4^{-2})} + c_4^4 \cdot R_4 \cdot (Y_4^4 + Y_4^{-4})$$

$$V_{\text{Tetrag.}} = -2ze\sqrt{2\pi}\sqrt{\frac{2}{5}}\left(\frac{r^2}{a^3} - \frac{r^2}{b^3}\right)Y_2^0 + \frac{ze2\sqrt{2\pi}}{\sqrt{72}}\left(\frac{4r^4}{b^5} + \frac{3r^4}{a^5}\right)Y_4^0 + 2ze\sqrt{2\pi}\sqrt{\frac{35}{144}}\frac{r^4}{a^5}(Y_4^4 + Y_4^{-4})$$

$$V_{\text{Tetrag}} = 2ze\sqrt{2\pi}\left[-\sqrt{\frac{2}{5}}\left(\frac{r^2}{a^3} - \frac{r^2}{b^3}\right)Y_2^0 + \frac{1}{\sqrt{72}}\left(\frac{3r^4}{a^5} + \frac{4r^4}{b^5}\right)Y_4^0 + \sqrt{\frac{35}{144}}\left(\frac{r^4}{a^5}\right)(Y_4^4 + Y_4^{-4})\right] \tag{7.3.13}$$

Tetrahedral Crystal Field Potential, V_{T_d}

What is the Hamiltonian operator, V_{T_d}, to determine the crystal field potential experienced by electrons in a tetrahedral environment?

$$\text{If}: \; V_{\text{oct}} = \sqrt{(^{49}/_{18})} \cdot \left(\sqrt{2\pi}\right) \cdot \frac{zer^4}{a^5} \cdot \left[Y_4^0 + \sqrt{\frac{5}{14}}(Y_4^4 + Y_4^{-4})\right] \tag{7.3.8}$$

Show that

$$V_{T_d} = \left(-\frac{4}{9}\right)V_{\text{oct.}} \text{ and } V_{\text{cube}} = \left(-\frac{8}{9}\right)V_{\text{oct.}}$$

- The quantitative treatment of tetrahedral problem is based upon Table 7.5 and Fig. 7.6:
- As described before, the total contribution to $V_{(x,\,y,\,z)}$:

$$V_{T_d} = c_2^0 \cdot R_2 \cdot Y_2^0 + c_2^{\pm 2} \cdot R_2 \cdot (Y_2^2 + Y_2^{-2}) + c_4^0 \cdot R_4 \cdot Y_4^0 + c_4^2 \cdot R_4 \cdot (Y_4^2 + Y_4^{-2}) + c_4^4 \cdot R_4 \cdot (Y_4^4 + Y_4^{-4}) \tag{7.3.14}$$

TABLE 7.5 The Contributions c_n^m Values and the Summation Over the Ligands

Ligand	$\cos\theta_i$	$\sin\theta_i$	ϕ	$\sin^2\theta\, e^{\pm 2i\phi}$	$e^{\pm 4i\phi}$	$35\cos^4\theta - 30\cos^2\theta + 3$	$\sin^4\theta\, e^{\pm 4i\phi}$
a	$\frac{1}{\sqrt{3}}$	$\frac{\sqrt{2}}{\sqrt{3}}$	$\frac{7\pi}{4}$	$\frac{2}{3}(-1)e^{\pm i\frac{\pi}{2}}$	$(-1)^7$	$-\frac{28}{9}$	$-\frac{4}{9}$
b	$-\frac{1}{\sqrt{3}}$	$\frac{\sqrt{2}}{\sqrt{3}}$	$\frac{\pi}{4}$	$\frac{2}{3}e^{\pm i\frac{\pi}{2}}$	-1	$-\frac{28}{9}$	$-\frac{4}{9}$
c	$-\frac{1}{\sqrt{3}}$	$\frac{\sqrt{2}}{\sqrt{3}}$	$\frac{3\pi}{4}$	$\frac{2}{3}(-1)e^{\pm i\frac{\pi}{2}}$	$(-1)^3$	$-\frac{28}{9}$	$-\frac{4}{9}$
d	$\frac{1}{\sqrt{3}}$	$\frac{\sqrt{2}}{\sqrt{3}}$	$\frac{5\pi}{4}$	$\frac{2}{3}(-1)^2e^{\pm i\frac{\pi}{2}}$	$(-1)^5$	$-\frac{28}{9}$	$-\frac{4}{9}$
Total				0	12	$28\left(-\frac{4}{9}\right)$	$4\left(-\frac{4}{9}\right)$

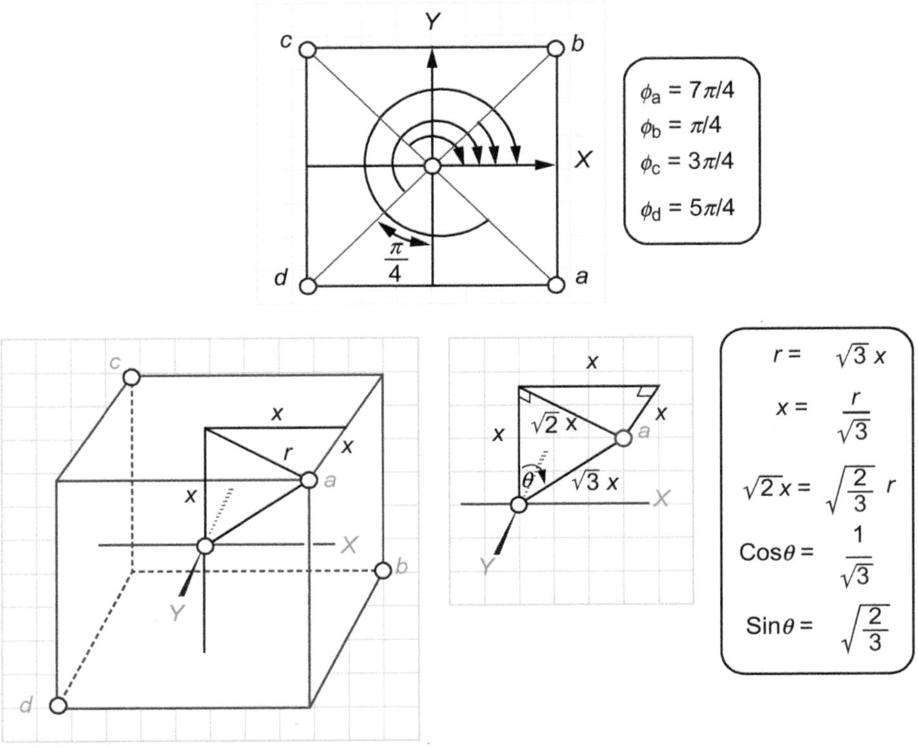

FIG. 7.6 The measure of cos θ, sin θ, and ϕ when the charged ligands a, b, c, and d form tetrahedron around the metal ion.

Ys are given in Table 7.1:

$$c_2^0 = \frac{4\pi ze}{5}\sqrt{(5/8)}\left(1/\sqrt{2\pi}\right)\sum(3\cos^2\theta - 1)$$

$$c_2^{\pm 2} = \frac{ze(4\pi)}{5}\sqrt{(15/16)}\left(1/\sqrt{2\pi}\right)\sum \sin^2\theta e^{\pm 2i\phi}$$

$$c_4^{\pm 2} = \frac{ze(4\pi)}{9}\sqrt{(45/64)}\left(1/\sqrt{2\pi}\right)\sum \sin^2\theta(7\cos^2\theta - 1)e^{\pm 2i\phi}$$

For the points a, b, c, and d

$$\cos\theta = \pm\frac{1}{\sqrt{3}}, \quad (3\cos^2\theta - 1) = 0$$

The summation of $(3\cos^2\theta - 1)$, $\sin^2\theta e^{\pm 2i\phi}$, and $\sin^2\theta(7\cos^2\theta - 1)$ over the four points yields zero (Table 7.5). Consequently, the contribution to $V_{(x,y,z)}$ from $n=0$, $m=0$ is zero.

- The c_n^m values are evaluated by summation over the ligands as follow:

$$c_4^0 = \frac{ze(4\pi)}{9}\sqrt{(9/128)}\left(1/\sqrt{2\pi}\right)\sum(35\cos^4\theta - 30\cos^2\theta + 3)$$

$$\therefore c_4^0 = \frac{ze(4\pi)}{9}\sqrt{(9/128)}\left(1/\sqrt{2\pi}\right)\left(28\left(-\frac{4}{9}\right)\right)$$

$$c_4^{\pm 4} = \frac{ze(4\pi)}{9}\sqrt{(315/256)}\left(1/\sqrt{2\pi}\right)\sum \sin^4\theta e^{\pm 4i\phi}$$

$$\therefore c_4^{\pm 4} = \frac{ze(4\pi)}{9}\sqrt{(315/256)}\left(1/\sqrt{2\pi}\right)\left[4\left(-\frac{4}{9}\right)\right]$$

- The total contribution to $V_{(x,y,z)}$:

$$V_{T_d} = c_2^0 \cdot R_2 \cdot Y_2^0 + c_2^{\pm 2} \cdot R_2 \cdot \left(Y_2^2 + Y_2^{-2}\right) + c_4^0 \cdot R_4 \cdot Y_4^0 + c_4^2 \cdot R_4 \cdot \left(Y_4^2 + Y_4^{-2}\right) + c_4^4 \cdot R_4 \cdot \left(Y_4^4 + Y_4^{-4}\right)$$

$$c_2^0 \cdot R_2 \cdot Y_2^0 = 0$$

$$c_2^{\pm 2} \cdot R_2 \cdot \left(Y_2^2 + Y_2^{-2}\right) = 0$$

$$c_4^0 \cdot R_4 \cdot Y_4^0 = -\left(\frac{4}{9}\right)\frac{ze(4\pi)}{9} \cdot \frac{r^4}{a^5}(28)\sqrt{\left({}^9/_{128}\right)} \cdot \left({}^1/_{\sqrt{2\pi}}\right) \cdot Y_4^0$$

$$c_4^2 \cdot R_4 \cdot \left(Y_4^2 + Y_4^{-2}\right) = 0$$

$$c_4^4 \cdot R_4 \cdot \left(Y_4^4 + Y_4^{-4}\right) = -\left(\frac{4}{9}\right)\frac{ze(4\pi)}{9} \cdot \frac{r^4}{a^5}(4)\sqrt{\left({}^{315}/_{256}\right)}\left({}^1/_{\sqrt{2\pi}}\right)\left(Y_4^4 + Y_4^{-4}\right)$$

The sums for $c_2^{\pm 2}$ and $c_4^{\pm 2}$ are found to be zero.

- The potential required is then:

$$V_{T_d} = \left(-\frac{4}{9}\right)\sqrt{\left({}^{49}/_{18}\right)} \cdot \left(\sqrt{2\pi}\right) \cdot \frac{zer^4}{a^5} \cdot \left[Y_4^0 + \sqrt{\frac{5}{14}}\left(Y_4^4 + Y_4^{-4}\right)\right] \quad (7.13.15)$$

$$\text{If}: \; V_{\text{oct}} = \sqrt{\left({}^{49}/_{18}\right)} \cdot \left(\sqrt{2\pi}\right) \cdot \frac{zer^4}{a^5} \cdot \left[Y_4^0 + \sqrt{\frac{5}{14}}\left(Y_4^4 + Y_4^{-4}\right)\right] \quad (7.3.8)$$

Then

$$V_{T_d} = \left(-\frac{4}{9}\right)V_{\text{oct.}} \quad (7.3.16)$$

and

$$V_{\text{cube}} = 2V_{T_d} = \left(-\frac{8}{9}\right)V_{\text{oct.}} \quad (7.3.17)$$

7.4 ZERO-ORDER PERTURBATION THEORY (THE EFFECT OF CRYSTAL FIELD ON THE ORBITAL WAVE FUNCTIONS OF DEGENERATE ORBITALS)

The Linear Combination of Atomic Orbitals, LCAO-MO, and Energy Calculation

a. **Use a two-term LCAO-MO to write the secular equation and define: H_{ii}, H_{ij}, and S_{ij}.**
b. **Write the extended secular determinant of LCAO-MO for n atoms, and the approximations made by Hückel.**

- The energy of the molecular orbitals, MO, can be obtained from the linear combination, LCAO.
- Each MO is written as a linear combination of atomic orbitals of molecular atoms:

$$\psi_j = \sum_i a_{ij}\phi_i$$

ϕ_is are a basis set, and it should be normalized, using

$$\int \phi_i \phi_i \partial\tau = 1$$

- **the secular equation:**

$$\because H\psi = E\psi$$

$$\therefore H\psi - E\psi = (H-E)\psi = 0$$

$$\because \psi_j = \sum_i a_{ij}\phi_i, \text{ then } \sum_i a_i(H-E)\phi_i = 0$$

- **For two-term LCAO-MO (diatomic molecule):**

$$a_1(H-E)\phi_1 + a_2(H-E)\phi_2 = 0$$

Multiplied by ϕ_1, and integrated over spatial coordinates of the wave functions:

$$a_1\int\phi_1(H-E)\phi_1\partial\tau + a_2\int\phi_1(H-E)\phi_2\partial\tau = 0,$$

$$a_1\int\phi_1 H\phi_1\partial\tau - a_1\int\phi_1 E\phi_1\partial\tau + a_2\int\phi_1 H\phi_2\partial\tau - a_2\int\phi_1 E\phi_2\partial\tau = 0, \text{ if}$$

$$H_{ii} = \int\phi_i H\phi_i\partial\tau$$

$$H_{ij} = \int\phi_i H\phi_j\partial\tau$$

$$S_{ij} = \int\phi_i\phi_j\partial\tau$$

$H_{ii} \rightarrow$ the energy of the atomic orbital ϕ_i
$H_{ij} \rightarrow$ the energy of interaction between pairs of atomic orbitals
$S_{ij} \rightarrow$ the overlap integrals, then:

$$\int\phi_i E\phi_i\partial\tau = E\int\phi_i\phi_i\,\partial\tau = E$$

$$\int\phi_i E\phi_j\partial\tau = E\int\phi_i\phi_j\partial\tau = ES_{ij}$$

$$\therefore a_1(H_{11}-E) + a_2(H_{12}-ES_{12}) = 0$$

- And again if:

$$a_1(H-E)\phi_1 + a_2(H-E)\phi_2 = 0$$

Multiplied by ϕ_2, and integrated over spatial coordinates of the wave functions, then:

$$a_1\int\phi_2(H-E)\phi_1\partial\tau + a_2\int\phi_2(H-E)\phi_2\partial\tau = 0,$$

$$a_1\int\phi_2 H\phi_1\partial\tau - a_1\int\phi_2 E\phi_1\partial\tau + a_2\int\phi_2 H\phi_2\partial\tau - a_2\int\phi_3 E\phi_2\partial\tau = 0,$$

$$\therefore a_1(H_{21}-ES_{21}) + a_2(H_{22}-E) = 0$$

- The two equations can be solved using 2×2 secular determinant:

$$\begin{vmatrix} H_{11}-E & H_{12}-ES_{12} \\ H_{21}-ES_{21} & H_{22}-E \end{vmatrix} = 0$$

Therefore,

$$(H_{11}-E)(H_{22}-E) - (H_{12}-ES_{12})(H_{21}-ES_{21}) = 0$$

$$(1-S_{12}^2)E^2 - (H_{11}+H_{22}-2H_{12}S_{12})E + H_{11}H_{22} - H_{12}^2 = 0$$

- The following scheme gives the correct expansions for determinants of order 2 and 3, respectively.

$$\begin{vmatrix} a_1 & b_1 \\ a_2 & b_2 \end{vmatrix} = a_1 b_2 - b_1 a_2 = 0$$

$$\begin{vmatrix} a_1 & b_1 & c_1 \\ a_2 & b_2 & c_2 \\ a_3 & b_3 & c_3 \end{vmatrix} = a_1 \begin{vmatrix} b_2 & c_2 \\ b_3 & c_3 \end{vmatrix} - b_1 \begin{vmatrix} a_2 & c_2 \\ a_3 & c_3 \end{vmatrix} + c_1 \begin{vmatrix} a_2 & b_2 \\ a_3 & b_3 \end{vmatrix}$$

- If we have n wave functions, the secular determinant:

$$\begin{vmatrix} H_{11}-E & H_{12}-ES_{12} & H_{13}-ES_{13} & \cdots & H_{1n}-ES_{1n} \\ H_{21}-ES_{21} & H_{22}-E & \cdots & \cdots & H_{2n}-E_{2n}S_{2n} \\ \vdots & \vdots & H_{33}-E & \ddots & \vdots \\ \cdots & \cdots & \cdots & \cdots & H_{nn}-E \end{vmatrix} = 0 \qquad (7.4.1)$$

- The following Hückel approximations are then introduced:
  - $H_{11}=H_{22}=H_{33}=\cdots=H_{nn}=\alpha=0$
  - $H_{ij}=0$, except for adjacent atoms $H_{ij}=\beta=1$
  - $S_{ij}=0$, then the secular determinant:

$$\begin{vmatrix} H_{11}-E & H_{12} & H_{13} & \ldots & H_{1n} \\ H_{21} & H_{22}-E & \ldots & \cdots & H_{2n} \\ \vdots & \vdots & H_{33}-E & \ddots & \vdots \\ H_{n1} & H_{n2} & \ldots & \ldots & H_{nn}-E \end{vmatrix} = 0 \qquad (7.4.2)$$

  - The linear transformation can be applied to calculate the E-values of the different levels of the molecule.

The Perturbation Theory for Degenerate Systems

How can the crystal field potential of the surrounding ligands perturb the degeneracy of d-orbitals of the central ion? What are the perturbed Hamiltonian, eigenvalue, and the corresponding secular determinant?

- Having now available the crystal field Hamiltonian ($V_{(x,y,z)}$), It is possible to investigate the effect of this potential upon the degeneracy of d-orbitals (Scheme 7.3).
- The crystal field induces a perturbation of degenerate system and takes the form known as zero-order perturbation theory.
- If E_0 is the energy before perturbation of n-fold degenerate eigenfunction ψ_i, (d-orbitals, $n=5$), and H_0 is the corresponding Hamiltonian:

$$H_0\psi_i = E_0\psi_i$$
$$i \rightarrow 1, 2, 3, \ldots, n$$

- When the Hamiltonian is slightly perturbed, the eigenfunctions ψ_i are no longer eigenfunctions of the new Hamiltonian (H_0+H') nor is E_0 an eigenvalue.

  The new eigenvalues of (H_0+H') are E_j' $(i\rightarrow 1,2,3, \ldots,n)$, and the corresponding eigenfunctions are ψ_j';

$$(H_0+H')\psi_j' = E_j'\psi_j'$$

- The wave functions ψ_j' can be described as a linear combinations of the wave functions ψ_i;

$$\psi_j' = c_{1j}\psi_1 + c_{2j}\psi_2 + \cdots + c_{nj}\psi_n = \sum_{i=1}^{n} c_{ij}\psi_i$$

  where for normalization,

$$\sum_{i=1}^{n} c_{ij} * c_{ij} = 1$$

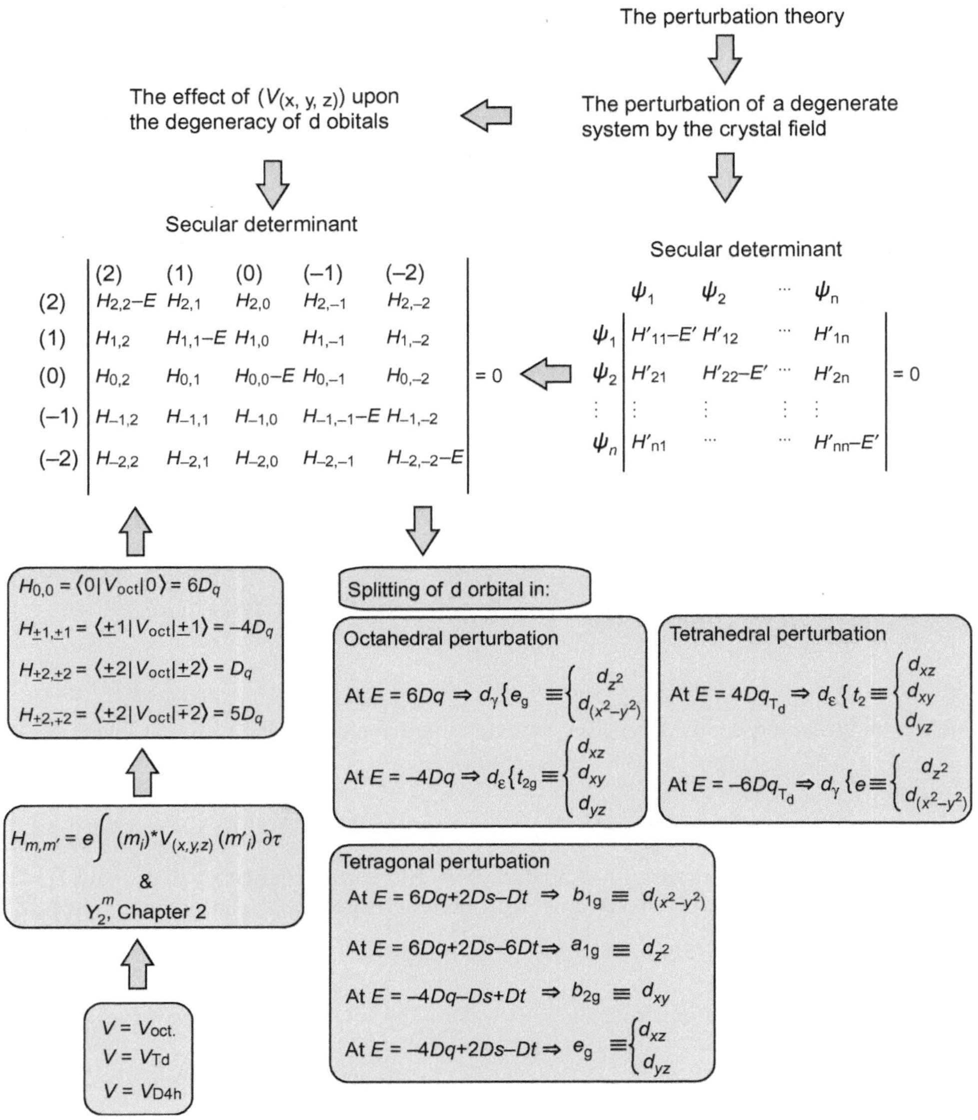

SCHEME 7.3 The approach used to find the change in the potential energy of each d-electron due to the crystal field of the surrounding.

- The secular determinant;

$$
\begin{array}{c|ccccc|}
 & \psi_1 & \psi_2 & \cdots & \psi_n \\
\psi_1 & H'_{11}-E' & H'_{12} & \cdots & H'_{1n} \\
\psi_2 & H'_{21} & H'_{22}-E' & \cdots & H'_{2n} \\
\vdots & \vdots & \vdots & \vdots & \vdots \\
\psi_n & H'_{n1} & \cdots & \cdots & H'_{nn}-E'
\end{array} = 0 \tag{7.4.3}
$$

Where

$$H'_{ij} = E' = \langle \psi'_i | H' | \psi_j \rangle = \int \psi'_i * H' \psi_j \partial\tau$$

- The set of eigenfunctions, ψ_i, is the five d wave functions (m_2, m_1, m_0, m_{-1}, m_{-2}), and the secular determinant is:

$$\begin{array}{c} \\ (2) \\ (1) \\ (0) \\ (-1) \\ (-2) \end{array} \begin{array}{c} \begin{array}{ccccc} (2) & (1) & (0) & (-1) & (-2) \end{array} \\ \left| \begin{array}{ccccc} H_{2,2}-E & H_{2,1} & H_{2,0} & H_{2,-1} & H_{2,-2} \\ H_{1,2} & H_{1,1}-E & H_{1,0} & H_{1,-1} & H_{1,-2} \\ H_{0,2} & H_{0,1} & H_{0,0}-E & H_{0,-1} & H_{0,-2} \\ H_{-1,2} & H_{-1,1} & H_{-1,0} & H_{-1,-1}-E & H_{-1,-2} \\ H_{-2,2} & H_{-2,1} & H_{-2,0} & H_{-2,-1} & H_{-2,-2}-E \end{array} \right| = 0 \end{array} \quad (7.4.4)$$

The matrix elements $H_{m,m'}$ are;

$$H_{m,m'} = e\int (m_i) * V_{(x,\, y,\, z)} \left(m_i'\right) \partial\tau \quad (7.4.5)$$

$\left(m_i'\right.$ may or may not be the same as $\left. m_i\right)$

- The energies of the perturbed d-orbitals are calculated by solving the secular determinant (Scheme 7.3).
- The obtained energies are fed back into secular equations that are derived from the secular determinant, to yield wave functions appropriate for the presence of the potential, $V_{(x,\, y,\, z)}$.
- The variation in the potential energy of each d-electron due to their crystal field is determined and the splitting of d-orbitals thus deduced.

The Splitting of d-Orbitals in Octahedral Crystal Field, $V_{oct.}$

$$\text{If:} \quad V_{oct.} = \sqrt{(^{49}/_{18})} \cdot \left(\sqrt{2\pi}\right) \cdot \frac{zer^4}{a^5} \cdot \left[Y_4^0 + \sqrt{\frac{5}{14}}\left(Y_4^4 + Y_4^{-4}\right)\right] \quad (7.3.8)$$

and

$$\theta_2^0 = \sqrt{\frac{5}{8}}\left(3\cos^2\theta - 1\right)$$

$$\theta_2^{\pm 2} = \sqrt{\frac{15}{16}}\sin^2\theta$$

$$\theta_4^{\pm 2} = \sqrt{\frac{45}{64}}\sin^2\theta\left(7\cos^2\theta - 1\right)$$

$$\theta_4^{\pm 0} = \sqrt{\frac{9}{128}}\left(35\cos^4\theta - 30\cos^2\theta + 3\right)$$

$$\theta_4^{\pm 4} = \sqrt{\frac{315}{256}}\sin^4\theta$$

Estimate the relative splitting in the energy of d-orbitals induced by octahedron of charges of potential V_{oct}.

- The appropriate secular determinant (Eq. 7.4.4):

$$\begin{array}{c} \\ (2) \\ (1) \\ (0) \\ (-1) \\ (-2) \end{array} \begin{array}{c} \begin{array}{ccccc} (2) & (1) & (0) & (-1) & (-2) \end{array} \\ \left| \begin{array}{ccccc} H_{2,2}-E & H_{2,1} & H_{2,0} & H_{2,-1} & H_{2,-2} \\ H_{1,2} & H_{1,1}-E & H_{1,0} & H_{1,-1} & H_{1,-2} \\ H_{0,2} & H_{0,1} & H_{0,0}-E & H_{0,-1} & H_{0,-2} \\ H_{-1,2} & H_{-1,1} & H_{-1,0} & H_{-1,-1}-E & H_{-1,-2} \\ H_{-2,2} & H_{-2,1} & H_{-2,0} & H_{-2,-1} & H_{-2,-2}-E \end{array} \right| = 0 \end{array} \quad (7.4.4)$$

where $H_{m,m'}$ are;

$$H_{mm} = \langle m|H|m'\rangle = e\int (m_i)^* V_{\text{oct.}} (m_i') \partial\tau \tag{7.4.5}$$

is solved for the energies of the five d wave functions (m_2, m_1, m_0, m_{-1}, m_{-2}) under the potential. Since

$$(m_i) = R_{n,2} Y_2^{m_l}$$

$$H_{m,m'} = e\int \underbrace{R_{n,2}^* Y_2^{m_l *}}_{m_i} V_{\text{oct.}} \underbrace{R_{n,2}' Y_2^{m_l'}}_{m_i} \partial\tau$$

$$\text{where}, \partial\tau = r^2 \sin\theta\, \partial\theta\, \partial\phi\, \partial r, \text{ and}$$

$$V_{\text{oct.}} = \sqrt{(49/18)} \cdot \left(\sqrt{2\pi}\right) \cdot \frac{zer^4}{a^5} \cdot \left[Y_4^0 + \sqrt{\frac{5}{14}}\left(Y_4^4 + Y_4^{-4}\right)\right] \tag{7.3.8}$$

r may be integrated out, if

$$\int_0^\infty R_{n,2}^* r^s R_{n,2} r^2 \partial r = \overline{r_2^s} \tag{7.4.6}$$

then

$$\int_0^\infty R_{n,2}^* r^4 R_{n,2} r^2 \partial r = \overline{r_2^4} \tag{7.4.7}$$

$\overline{r_2^4}$: the mean fourth power radius of d-electrons of central ion.

$$\int \underbrace{R_{n,2}^* Y_2^{m_l *}}_{m_i} V_{\text{oct.}} \underbrace{R_{n,2}' Y_2^{m_l'}}_{m_i} \partial\tau$$

$$= \sqrt{(49/18)}\left(\sqrt{2\pi}\right)\frac{ze}{a^5}\left[\int R_{n,2}^* r^4 R_{n,2} r^2 \partial r\right]\left[\iint Y_2^{m_l *}\left[Y_4^0 + \sqrt{\frac{5}{14}}\left(Y_4^4 + Y_4^{-4}\right)\right] Y_2^{m_l'} \sin\theta\, \partial\theta\, \partial\phi\right]$$

- The expression of the matrix elements, $H_{m,m'}$:

$$\int (m_i) * V_{(x,y,z)} (m_i') \partial\tau = \sqrt{(49/18)} \cdot \left(\sqrt{2\pi}\right) \cdot \overline{r_2^4} \cdot \frac{ze}{a^5} \cdot \int_0^n \int_0^{2\pi} [Y_2^{m_l *} Y_4^0 Y_2^{m_l'} \sin\theta\, \partial\theta\, \partial\phi$$
$$+ \sqrt{\frac{5}{14}}\left(Y_2^{m_l *} Y_4^4 Y_2^{m_l'} \sin\theta\, \partial\theta\, \partial\phi + Y_2^{m_l *} Y_4^{-4} Y_2^{m_l'} \sin\theta\, \partial\theta\, \partial\phi\right)] \tag{7.4.8}$$

- The Hamiltonian can only connect wave functions whose m_l values differ by 0, 4 or −4, because the octahedral Hamiltonian components (m_l) are 0, 4, and −4 (this chapter, p. 412).
- These integrations may be simplified by noting that:

$$\int_0^{2\pi} Y_{l1}^{m_{l1}} Y_{l2}^{m_{l2}} Y_{l3}^{m_{l3}} \partial\phi \neq 0, \quad \text{if}: m_{l_1} + m_{l_2} + m_{l_3} = 0 \tag{7.4.9}$$

Regardless of the value l_1, l_2, and l_3, as a result, the integrals ($Y_2^{m_l *} Y_4^0 Y_2^{m_l'}$) that include Y_4^0 are zero unless

$$(m_l)^* + 0 + m_l' = 0$$

$$\because m_l' = -m_l^*$$

$$\therefore m_l' = m_l$$

The matrix element is zero if not between a d wave function and itself (diagonal matrix elements), not one of other d wave functions (off-diagonal matrix elements).

Likewise, the integrals $Y_2^{m_l*}Y_4^4Y_2^{m_l'}$ that include Y_4^4 are zero except

$$m_l' + m_l{}^* = -4$$

$$m_l' - m_l = -4 \text{ (only even } m \text{ can contribute)}$$

or only if

$$m_l = -m_l' = -2$$

Also the integrals of Y_4^{-4} are zero if not:

$$m_l' = -m_l = 2$$

- The spherical harmonics $Y_l^{m_i}$ describe the directional properties of the orbital defined by the wave function, and are composed of two independent parts:

$$Y_l^{m_i} = \theta_2^{m_i} \cdot \phi_{m_i} = \theta_2^{m_i} \frac{1}{\sqrt{2\pi}} e^{im_i\phi}$$

Integration over ϕ_{m_i} yields 0 or $\frac{1}{\sqrt{2\pi}}$

$$\int_0^{2\pi} \phi_{m_{l_1}}^* \phi_{m_{l_2}} \phi_{m_{l_3}} \partial\phi = \left(\frac{1}{\sqrt{2\pi}}\right)^3 \int_0^{2\pi} e^{i(m_{l_1} + m_{l_2} + m_{l_3})\phi} d\phi$$

$$= \left(\frac{1}{\sqrt{2\pi}}\right)^3 \int_0^{2\pi} e^0 d\phi = \left(\frac{1}{\sqrt{2\pi}}\right)^3 \int_0^{2\pi} d\phi = \left(\frac{1}{\sqrt{2\pi}}\right)^3 2\pi = \frac{1}{\sqrt{2\pi}}$$

- The equation

$$\int (m_i) * V_{(x,y,z)} (m_i') \partial\tau = \sqrt{({}^{49}/_{18})} \cdot \left(\sqrt{2\pi}\right) \cdot \overline{r_2^4} \cdot \frac{ze}{a^5} \cdot \int_0^n \int_0^{2\pi} [Y_2^{m_l*} Y_4^0 Y_2^{m_l'} \sin\theta\, \partial\theta\, \partial\phi$$
$$+ \sqrt{\frac{5}{14}} \left(Y_2^{m_l*} Y_4^4 Y_2^{m_l'} \sin\theta\, \partial\theta\, \partial\phi + Y_2^{m_l*} Y_4^{-4} Y_2^{m_l'} \sin\theta\, \partial\theta\, \partial\phi \right)] \quad (7.4.8)$$

may now be reduced to:

$$\int (m_i) * V_{\text{oct}} (m_i) \partial\tau = ({}^{49}/_{18})^{1/2} \frac{ze\overline{r_2^4}}{a^5} \cdot \int_0^\pi \theta_2^{m_l*} \theta_4^0 \theta_2^{m_l} \sin\theta\, \partial\theta \quad (7.4.10)$$

Where $m_i = 0, \pm 1$ and ± 2

and

$$\int (\pm 2) * V_{\text{oct}} (\mp 2) \partial\tau = ({}^{35}/_{36})^{1/2} \frac{ze\overline{r_2^4}}{a^5} \cdot \int_0^\pi \theta_2^{2*} \theta_4^{\pm 4} \theta_2^2 \sin\theta\, \partial\theta \quad (7.4.11)$$

If: θ's functions (Chapter 2, p. 56):

$$\theta_2^0: \quad \sqrt{\frac{5}{8}}(3\cos^2\theta - 1)$$

$$\theta_2^{\pm 2}: \quad \sqrt{\frac{15}{16}}\sin^2\theta$$

$$\theta_4^{\pm 2}: \quad \sqrt{\frac{45}{64}}\sin^2\theta(7\cos^2\theta - 1)$$

$$\theta_4^{\pm 0}: \quad \sqrt{\frac{9}{128}}(35\cos^4\theta - 30\cos^2\theta + 3)$$

$$\theta_4^{\pm 4}: \quad \sqrt{\frac{315}{256}}\sin^4\theta$$

it follows that we need to evaluate the following integration:

$$\int_0^\pi \theta_2^{0*}\theta_4^0\theta_2^0 \sin\theta\,\partial\theta = \int_0^\pi (3\cos^2\theta - 1)^2(35\cos^4\theta - 30\cos^2\theta + 3)\sin\theta\,\partial\theta$$

$$= \int_0^\pi (9\cos^4\theta - 6\cos^2\theta + 1)(35\cos^4\theta - 30\cos^2\theta + 3)\sin\theta\,\partial\theta$$

$$= \int_0^\pi (315\cos^8\theta - 210\cos^6\theta + 35\cos^4\theta - 270\cos^6\theta$$
$$+180\cos^4\theta - 30\cos^2\theta + 27\cos^4\theta - 18\cos^2\theta + 3)\sin\theta\,\partial\theta$$

$$= \int_0^\pi (315\cos^8\theta - 480\cos^6\theta + 242\cos^4\theta - 48\cos^2\theta + 3)\sin\theta\,\partial\theta$$

Notes:

$$\int_0^\pi \sin^{2n}\theta\cdot\cos^{2m}\theta\,\partial\theta = \frac{(2m)!(2n)!\pi}{2^{2(m+n)}m!n!(m+n)!}$$

$$\int_0^\pi \sin^{2n+1}\theta\cdot\cos^m\theta\,\partial\theta = \int_0^\pi \cos^{2n+1}\theta\cdot\sin^m\theta\,\partial\theta = \frac{2^{n+1}\cdot n!}{(m+1)(m+3)\cdots(m+2n+1)}$$

$$\int_0^\pi \sin^{2n+1}\theta\cdot\cos^{2m+1}\theta\,\partial\theta = 0$$

Accordingly,

$$\int_0^\pi 315\cos^8\theta\sin\theta\,\partial\theta = (315)\frac{2}{9} = 70$$

$$-\int_0^\pi 480\cos^6\theta\sin\theta\,\partial\theta = -(480)\frac{2}{7} = -137.143$$

$$\int_0^\pi 242\cos^4\theta\sin\theta\,\partial\theta = (242)\frac{2}{5} = 96.8$$

$$-\int_0^\pi 48\cos^2\theta\sin\theta\,\partial\theta = -(48)\frac{2}{3} = -32$$

$$\int_0^\pi 3\sin\theta\,\partial\theta = 3\,[(-\cos\pi)-(\cos 0)]3\,[-(-1)-(-1)] = 6$$

- Therefore,

$$\int_0^\pi \theta_2^{0*}\theta_4^0\theta_2^0\sin\theta\,\partial\theta = 70 - 137.14 + 96.8 - 32 + 6 = \frac{5}{8}\times\sqrt{\frac{9}{128}}\times 3.657 = 0.6061 = \frac{(18)^{1/2}}{7}$$

- In the same way:

$$\int_0^\pi \theta_2^{1*}\theta_4^0\theta_2^1\sin\theta\,\partial\theta = -\frac{(8)^{1/2}}{7}$$

$$\int_0^\pi \theta_2^{2*}\theta_4^0\theta_2^2\sin\theta\,\partial\theta = \frac{(2)^{1/2}}{14}$$

$$\int_0^\pi \theta_2^{2*}\theta_4^4\theta_2^2\sin\theta\,\partial\theta = \frac{(35)^{1/2}}{7}$$

- Substitute in

$$\int (m_i)^* V_{oct}(m_i)\partial\tau = \left({}^{49}/{}_{18}\right)^{1/2}\frac{ze\overline{r_2^4}}{a^5}\cdot\int_0^\pi \theta_2^{m_i*}\theta_4^0\theta_2^{m_i'}\sin\theta\,\partial\theta, \tag{7.4.10}$$

using

$$\alpha_4 = \frac{ze^2\overline{r_2^4}}{a^5},$$

$$D_q = \frac{1}{6}\frac{ze^2\overline{r_2^4}}{a^5} = \frac{1}{6}\alpha_4 \text{ and}$$

The energy associated by the electron is defined by

$$H = eV$$

Then

$$\int (0)^* V_{oct}(0)\partial\tau = \frac{ze\overline{r_2^4}}{a^5},\quad H_{0,0} = \alpha_4 = 6D_q \tag{7.4.12}$$

$$\int (\pm1)^* V_{oct}(\pm1)\partial\tau = -\frac{2}{3}\frac{ze\overline{r_2^4}}{a^5},\quad H_{\pm1,\pm1} = -\frac{2}{3}\alpha_4 = -4D_q \tag{7.4.13}$$

$$\int (\pm2)^* V_{oct}(\pm2)\partial\tau = \frac{1}{6}\frac{ze\overline{r_2^4}}{a^5},\quad H_{\pm2,\pm2} = \frac{1}{6}\alpha_4 = D_q \tag{7.4.14}$$

And using:

$$\int (\pm 2) * V_{oct}(\mp 2)\partial\tau = \left({}^{35}/_{36}\right)^{1/2}\frac{ze\overline{r_2^4}}{a^5}\cdot\int_0^{\pi}\theta_2^{2*}\theta_4^{\pm 4}\theta_2^{2}\sin\theta\,\partial\theta \quad (7.4.11)$$

Then:

$$\int (\pm 2) * V_{oct}(\mp 2)\partial\tau = \frac{5}{6}\frac{ze\overline{r_2^4}}{a^5},\quad H_{\pm 2,\mp 2} = \frac{5}{6}\alpha_4 = 5D_q \quad (7.4.15)$$

- For simplicity the magnitudes of these elements will be presented as:

$$H_{0,0} = \langle 0|H|0\rangle = \langle 0|V|0\rangle = 6Dq \quad (7.4.12)$$

$$H_{\pm 1,\pm 1} = \langle \pm 1|H|\pm 1\rangle = \langle \pm 1|V|\pm 1\rangle = -4Dq \quad (7.4.13)$$

$$H_{\pm 2,\pm 2} = \langle \pm 2|H|\pm 2\rangle = \langle \pm 2|V|\pm 2\rangle = Dq \quad (7.4.14)$$

$$H_{\pm 2,\mp 2} = \langle \pm 2|H|\mp 2\rangle = \langle \pm 2|V|\mp 2\rangle = 5Dq \quad (7.4.15)$$

Which is not a strictly correct presentation since $m|H|m'$ is a unitless number. This is a short hand notation for example:

$$c\,\langle 0|V|0\rangle = 6Dq \times 1$$

- By substituting in matrix 7.4.4:

$$\begin{array}{c|ccccc|} & (2) & (1) & (0) & (-1) & (-2) \\ (2) & Dq-E & 0 & 0 & 0 & 5Dq-E \\ (1) & 0 & -4Dq-E & 0 & 0 & 0 \\ (0) & 0 & 0 & 6Dq-E & 0 & 0 \\ (-1) & 0 & 0 & 0 & -4Dq-E & 0 \\ (-2) & 5Dq & 0 & 0 & 0 & Dq-E \end{array} = 0 \quad (7.4.16)$$

- If the off-diagonal elements (H_{ij}) are zero, then the diagonal elements (H_{ii} to H_{jj} integrals) correspond directly with the energies of the real d-orbitals.
  - This determinant:

$$\begin{array}{|ccc|} (1) & (0) & (-1) \\ -4Dq-E & 0 & 0 \\ 0 & 6Dq-E & 0 \\ 0 & 0 & -4Dq-E \end{array} = 0 \quad (7.4.17)$$

    is solved to yield:

$$(1) \text{ and } (-1) \text{ at } E = -4Dq \quad (7.4.18)$$

$$(0) \text{ at } E = 6Dq \quad (7.4.19)$$

- If the off-diagonal elements (H_{ij}) are nonzero, then the secular determinant must be solved for its roots that correspond to the energy levels required.
  - The determinant:

$$\begin{array}{c|cc|} & (2) & (-2) \\ (2) & Dq-E & 5Dq \\ (-2) & 5Dq & Dq-E \end{array} = 0 \quad (7.4.20)$$

is solved:

$$Dq^2 - 2EDq + E^2 - 25\,Dq^2 = 0$$

$$E^2 - 2EDq - 24Dq^2 = 0$$

$$(E - 6Dq)(E + 4Dq) = 0$$

To give the energies,

$$E = 6\,Dq \text{ and } E = -4Dq \tag{7.4.21}$$

- The fitting secular equations are:

$$c_2(Dq - E) + c_{-2}(5Dq) = 0$$

$$c_2(5Dq) + c_{-2}(Dq - E) = 0$$

On substituting, and if: $E = -4Dq$
We have $c_2 = -c_{-2}$
- The wave function corresponding to the energy substituted is

$$\psi = c_2(2) + c_{-2}(-2)$$

and if:
$c_2{}^*c_2 + c_{-2}{}^*c_{-2} = 1$, for normalization
Then c_2 and c_{-2} are:

$$c_2 = -c_{-2} = \frac{1}{\sqrt{2}}$$

So the energy $-4Dq$ associates the wave function

$$\psi = \frac{1}{\sqrt{2}}((2) - (-2)) \tag{7.4.22}$$

or the wave function d_{xz}.
- Likewise, on substituting the energy $6Dq$ into:

$$c_2(Dq - E) + c_{-2}(5Dq) = 0$$

$$c_2(5Dq) + c_{-2}(Dq - E) = 0$$

the energy $6Dq$ is found to correspond:

$$\psi = \frac{1}{\sqrt{2}}((2) + (-2)) \tag{7.4.23}$$

or the wave function $d_{x^2-y^2}$.

- If the wave functions of d-orbitals:

$$d_{xz} = \frac{-1}{i\sqrt{2}}((1) + (-1))$$

$$d_{xy} = \frac{1}{i\sqrt{2}}((2) - (-2))$$

$$d_{yz} = \frac{-1}{\sqrt{2}}((1) - (-1))$$

$$d_{z^2} = 0$$

$$d_{(x^2-y^2)} = \frac{-1}{\sqrt{2}}((2) + (-2))$$

- Then, the energies of the five d-orbitals can be established for example:

$$\int d_{yz} V_{\text{oct}} d_{yz} d\tau = \frac{1}{2}\left[\int ((1) - (-1)) V_{\text{oct}}((1) - (-1)) d\tau\right]$$
$$= \frac{1}{2}\left[\int (1)V_{\text{oct}}(1)d\tau - \int (1)V_{\text{oct}}(-1)d\tau - \int (-1)V_{\text{oct}}(1)d\tau + \int (-1)V_{\text{oct}}(-1)\,d\tau\right] \tag{7.3.24}$$
$$H_{d_{yz},d_{yz}} = \frac{1}{2}[-4Dq - 0 - 0 - 4Dq] = -4Dq$$

- Collecting the results (Fig. 7.7):

$$\text{At } E = 6Dq \Rightarrow d_\gamma \{ e_g \equiv \begin{cases} d_{z^2} \\ d_{(x^2-y^2)} \end{cases}$$

$$\text{At } E = -4Dq \Rightarrow d_\varepsilon \{ t_{2g} \equiv \begin{cases} d_{xz} \\ d_{xy} \\ d_{yz} \end{cases}$$

Dq is defined so that the separation between e_g and t_{2g} orbitals is 10 Dq.

The Splitting of d-Orbitals in Tetrahedral Crystal Field, V_{T_d}

If

$$V_{T_d} = -\sqrt{\left({}^{392}/_{729}\right)} \cdot \left(\sqrt{2\pi}\right) \cdot \frac{zer^4}{a^5} \cdot \left[Y_4^0 + \sqrt{\frac{5}{14}}(Y_4^4 + Y_4^{-4})\right] \tag{7.3.15}$$

and

$$\theta_2^0 = \sqrt{\frac{5}{8}}(3\cos^2\theta - 1)$$

$$\theta_2^{\pm 2} = \sqrt{\frac{15}{16}}\sin^2\theta$$

$$\theta_4^{\pm 2} = \sqrt{\frac{45}{64}}\sin^2\theta(7\cos^2\theta - 1)$$

$$\theta_4^{\pm 0} = \sqrt{\frac{9}{128}}(35\cos^4\theta - 30\cos^2\theta + 3)$$

$$\theta_4^{\pm 4} = \sqrt{\frac{315}{256}}\sin^4\theta$$

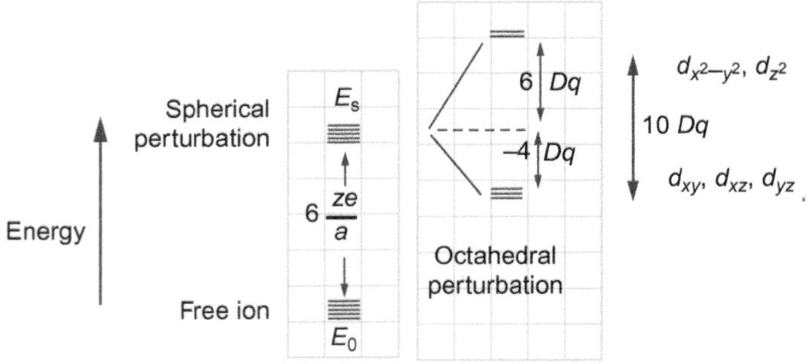

FIG. 7.7 The splitting of d-orbitals in an octahedral crystal field.

then evaluate the splitting in the energy of d-orbitals induced by a tetrahedron of charges (crystal field potential V_{Td}).

- The appropriate secular determinant:

$$\begin{array}{c} \\ (2) \\ (1) \\ (0) \\ (-1) \\ (-2) \end{array} \begin{array}{c} \begin{array}{ccccc} (2) & (1) & (0) & (-1) & (-2) \end{array} \\ \left| \begin{array}{ccccc} H_{2,2}-E & H_{2,1} & H_{2,0} & H_{2,-1} & H_{2,-2} \\ H_{1,2} & H_{1,1}-E & H_{1,0} & H_{1,-1} & H_{1,-2} \\ H_{0,2} & H_{0,1} & H_{0,0}-E & H_{0,-1} & H_{0,-2} \\ H_{-1,2} & H_{-1,1} & H_{-1,0} & H_{-1,-1}-E & H_{-1,-2} \\ H_{-2,2} & H_{-2,1} & H_{-2,0} & H_{-2,-1} & H_{-2,-2}-E \end{array} \right| \end{array} = 0 \tag{7.4.25}$$

where $H_{m,m'}$ are;

$$H_{m,m'} = e\int (m_i)^* V_{T_d}(m_i')\partial\tau \tag{7.4.26}$$

- To evaluate the matrix elements of:

$$(m_i) = R_{n,2}Y_2^{m_l}$$

$$H_{m,m'} = e\int R_{n,2}^* Y_2^{m_l *} V_{(x,\,y,\,z)} R_{n,2}' Y_2^{m_l'} \partial\tau \tag{7.4.27}$$

$$\partial\tau = r^2 \sin\theta\, \partial\theta\, \partial\phi\, \partial r$$

$$V_{T_d} = -\left(\frac{4}{9}\right) V_{\text{oct}}$$

$$V_{T_d} = -\left(\frac{4}{9}\right)\sqrt{\left({}^{49}/_{18}\right)} \cdot \left(\sqrt{2\pi}\right) \cdot \frac{zer^4}{a^5} \cdot \left[Y_4^0 + \sqrt{\frac{5}{14}}\left(Y_4^4 + Y_4^{-4}\right)\right] \tag{7.3.15}$$

r may be integrated out, if

$$\int_0^\infty R_{n,2}^* r^s R_{n,2} r^2 \partial r = \overline{r_2^s} \tag{7.4.6}$$

then

$$\int_0^\infty R_{n,2}^* r^4 R_{n,2} r^2 \partial r = \overline{r_2^4} \tag{7.4.7}$$

$\overline{r_2^4}$: the mean fourth power radius of d-electrons of central ion.

- The results expression of the matrix elements:

$$\begin{aligned} \int (m_i)^* V_{T_d}(m_i')\partial\tau = &-\left(\frac{4}{9}\right)\sqrt{\left({}^{49}/_{18}\right)} \cdot \left(\sqrt{2\pi}\right) \cdot \overline{r_2^4} \cdot \frac{ze}{a^5} \cdot \int_0^n \int_0^{2\pi} [Y_2^{m_l *} Y_4^0 Y_2^{m_l'} \sin\theta\, \partial\theta\, \partial\phi \\ &- \left(\frac{4}{9}\right)\sqrt{\frac{5}{14}}\left(Y_2^{m_l *} Y_4^4 Y_2^{m_l'} \sin\theta\, \partial\theta\, \partial\phi + Y_2^{m_l *} Y_4^{-4} Y_2^{m_l'} \sin\theta\, \partial\theta\, \partial\phi\right)\Big] \end{aligned} \tag{7.4.28}$$

- Following similar procedures of O_h, these integrations may be simplified by noting that, in order that

$$\int_0^{2\pi} Y_{l1}^{m_{l1}} Y_{l2}^{m_{l2}} Y_{l3}^{m_{l3}} \partial\phi \neq 0$$

is only if (p. 424)

$$m_{l_1} + m_{l_2} + m_{l_3} = 0 \tag{7.4.9}$$

and

$$Y_l^{m_i} = \theta_2^{m_i} \cdot \phi_{m_i} = \theta_2^{m_i} \frac{1}{\sqrt{2\pi}} e^{im_i\phi}$$

Integration over φ_{m_i} yields 0 or $\dfrac{1}{\sqrt{2\pi}}$

The equation:

$$\int (m_i) * V_{T_d}(m_i')\partial\tau = -\left(\frac{4}{9}\right)\sqrt{\left(^{49}/_{18}\right)} \cdot \left(\sqrt{2\pi}\right) \cdot r_2^4 \cdot \frac{ze}{a^5} \cdot \int_0^n \int_0^{2\pi} [Y_2^{m_i *} Y_4^0 Y_2^{m_i'} \sin\theta\,\partial\theta\,\partial\phi$$
$$- \left(\frac{4}{9}\right)\sqrt{\frac{5}{14}}\left(Y_2^{m_i *} Y_4^4 Y_2^{m_i'} \sin\theta\,\partial\theta\,\partial\phi + Y_2^{m_i *} Y_4^{-4} Y_2^{m_i'} \sin\theta\,\partial\theta\,\partial\phi\right)\Big] \tag{7.4.29}$$

is reduced to:

$$\int (m_i) * V_{T_d}(m_i)\partial\tau = -\left(\frac{4}{9}\right)\left(^{49}/_{18}\right)^{1/2} \frac{ze\overline{r_2^4}}{a^5.} \cdot \int_0^{\pi} \theta_2^{m_i *} \theta_4^0 \theta_2^{m_i} \sin\theta\,\partial\theta \tag{7.4.30}$$

Where $m_i = 0, \pm 1 \text{ and } \pm 2$
And

$$\int (\pm 2) * V_{T_d}(\mp 2)\partial\tau = -\left(\frac{4}{9}\right)\left(^{35}/_{36}\right)^{1/2} \frac{ze\overline{r_2^4}}{a^5} \cdot \int_0^{\pi} \theta_2^{2*} \theta_4^{\pm 4} \theta_2^2 \sin\theta\,\partial\theta \tag{7.4.31}$$

Following the same procedures of O_h (p. 426), if

$$\int_0^{\pi} \theta_2^{0*} \theta_4^0 \theta_2^0 \sin\theta\,\partial\theta = \frac{(18)^{1/2}}{7}$$

$$\int_0^{\pi} \theta_2^{1*} \theta_4^0 \theta_2^1 \sin\theta\,\partial\theta = -\frac{(8)^{1/2}}{7}$$

$$\int_0^{\pi} \theta_2^{2*} \theta_4^0 \theta_2^2 \sin\theta\,\partial\theta = \frac{(2)^{1/2}}{7}$$

$$\int_0^{\pi} \theta_2^{2*} \theta_4^4 \theta_2^2 \sin\theta\,\partial\theta = -\frac{(35)^{1/2}}{7}$$

- From the equation:

$$\int (m_i)^* V_{T_d}(m_i)\partial\tau = -\left(\frac{4}{9}\right)\left({}^{49}/_{18}\right)^{1/2} \frac{ze\overline{r_2^4}}{a^5} \cdot \int_0^{\pi} \theta_2^{m_l *}\theta_4^0\theta_2^{m_l'} \sin\theta\,\partial\theta, \tag{7.4.30}$$

- If: $H = eV$,

$$H_{m,m'} = \langle m|H|m'\rangle = \langle m|V|m'\rangle \text{ "shorthand representation", and}$$

$$Dq_{T_d} = \left(\frac{4}{9}\right)\left(\frac{1}{6}\right)\frac{ze^2\overline{r_2^4}}{a^5} \tag{7.4.32}$$

Then:

$$\int (0)^* V_{T_d}(0)\partial\tau = \frac{ze\overline{r_2^4}}{a^5}, \quad H_{0,0} = -6Dq_{T_d} \tag{7.4.33}$$

$$\int (\pm 1)^* V_{T_d}(\pm 1)\partial\tau = -\frac{2}{3}\frac{ze\overline{r_2^4}}{a^5}, \quad H_{\pm1,\pm1} = 4Dq_{T_d} \tag{7.4.34}$$

$$\int (\pm 2)^* V_{T_d}(\pm 2)\partial\tau = \frac{1}{6}\frac{ze\overline{r_2^4}}{a^5}, \quad H_{\pm2,\pm2} = -Dq_{T_d} \tag{7.4.35}$$

And from:

$$\int (\pm 2)^* V_{T_d}(\mp 2)\partial\tau = -\left(\frac{4}{9}\right)\left({}^{35}/_{36}\right)^{1/2} \frac{ze\overline{r_2^4}}{a^5} \cdot \int_0^{\pi} \theta_2^{2*}\theta_4^{\pm 4}\theta_2^{2} \sin\theta\,\partial\theta \tag{7.4.31}$$

Then:

$$\int (\pm 2)^* V_{T_d}(\mp 2)\partial\tau = \left(\frac{4}{9}\right)\left(\frac{5}{6}\right)\frac{ze\overline{r_2^4}}{a^5}, \quad H_{\pm2,\mp2} = -5Dq_{T_d} \tag{7.4.36}$$

- Then substitute in matrix (7.4.25):

$$\begin{array}{c} \\ (2) \\ (1) \\ (0) \\ (-1) \\ (-2) \end{array}\begin{array}{c} \begin{array}{ccccc} (2) & (1) & (0) & (-1) & (-2) \end{array} \\ \begin{vmatrix} -Dq_{T_d} - E & 0 & 0 & 0 & -5Dq_{T_d} \\ 0 & 4 - Dq_{T_d} - E & 0 & 0 & 0 \\ 0 & 0 & -6Dq_{T_d} - E & 0 & 0 \\ 0 & 0 & 0 & 4Dq_{T_d} - E & 0 \\ -5 - Dq_{T_d} & 0 & 0 & 0 & -Dq_{T_d} - E \end{vmatrix} \end{array} = 0 \tag{7.4.37}$$

  - This determinant may be reduced to:

$$(1) \text{ and } (-1) \text{ at } E = 4Dq_{T_d} \tag{7.4.38}$$

$$0 \text{ at } E = -6Dq_{T_d} \tag{7.4.39}$$

  - And the determinant

$$\begin{array}{c} \begin{array}{cc} (2) & (-2) \end{array} \\ \begin{vmatrix} -Dq_{T_d} - E & -5Dq_{T_d} \\ -5Dq_{T_d} & -Dq_{T_d} - E \end{vmatrix} \end{array} = 0 \tag{7.4.40}$$

$$Dq_{T_d}^2 + 2EDq_{T_d} + E^2 - 25Dq_{T_d}^2 = 0$$

$$E^2 + 2EDq_{T_d} - 24Dq_{T_d}^2 = 0$$

$$(E + 6Dq_{T_d})(E - 4Dq_{T_d}) = 0$$

the energy $4Dq$ associates the wave function:

$$\psi = \frac{1}{\sqrt{2}}((2) - (-2)) \text{ or the wave function } d_{xz} \tag{7.4.41}$$

- and energy $-6Dq$ is found to correspond:

$$\psi = \frac{1}{\sqrt{2}}((2) + (-2)) \text{ or the wave function } d_{(x^2-y^2)} \tag{7.4.42}$$

- Having the wave function of d-orbitals in the real form:

$$d_{xz} = \frac{-1}{i\sqrt{2}}((1) + (-1))$$

$$d_{yz} = \frac{-1}{\sqrt{2}}((1) - (-1))$$

The energies of the five d-orbitals can be established for example:

$$\int d_{xz} V_{T_h} d_{xz} \partial\tau = \frac{1}{2}\left[\int ((1) - (-1)) V_{T_h} ((1) - (-1)) \partial\tau\right]$$

$$= \frac{1}{2}\left[\int (1) V_{T_h} (1)\, \partial\tau - \int (1) V_{T_h} (-1)\, \partial\tau - \int (-1) V_{T_h} (1)\, \partial\tau + \int (-1) V_{T_h} (-1)\, \partial\tau\right]$$

$$H_{d_{xz}, d_{xz}} = \frac{1}{2}[4Dq - 0 - 0 - 4Dq] = -4Dq \tag{7.4.43}$$

- Collecting the results (Fig. 7.8):

$$\text{At } E = 4Dq_{T_d} \Rightarrow d_\varepsilon\{t_2 \equiv \begin{cases} d_{xz} \\ d_{xy} \\ d_{yz} \end{cases}$$

$$\text{At } E = -6Dq_{T_d} \Rightarrow d_\gamma\{e \equiv \begin{cases} d_{z^2} \\ d_{(x^2-y^2)} \end{cases}$$

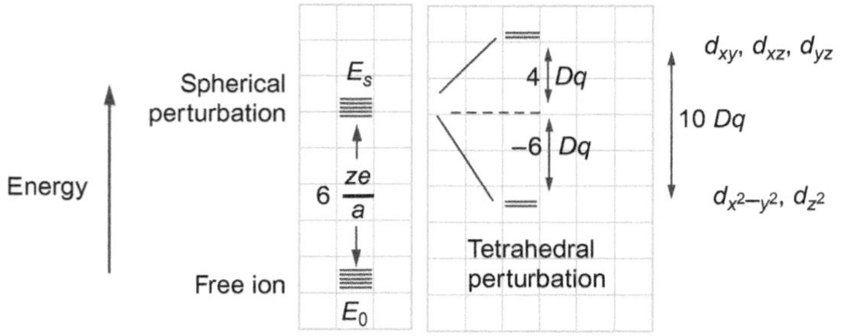

FIG. 7.8 The splitting of d-orbitals in a tetrahedral crystal field.

The Splitting of d-Orbitals in Tetragonal Crystal Field, $V_{D_{4h}}$

If

$$V_{D_{4h}} = 2ze\sqrt{2\pi}\left[-\sqrt{\frac{2}{5}}\left(\frac{r^2}{a^3}-\frac{r^2}{b^3}\right)Y_2^0+\frac{1}{\sqrt{72}}\left(\frac{3r^4}{a^5}+\frac{4r^4}{b^5}\right)Y_4^0+\sqrt{\frac{35}{144}}\left(\frac{r^4}{a^5}\right)\left(Y_4^4+Y_4^{-4}\right)\right] \quad (7.3.12)$$

show that

$$V_{D_{4h}} = V_{oct} - 2ze\left(\sqrt{2\pi}\right)\cdot\left[\sqrt{\frac{2}{5}}\left(\frac{r^2}{a^3}-\frac{r^2}{b^3}\right)Y_2^0\right] - 2ze\left(\sqrt{2\pi}\right)\cdot\left[\sqrt{\frac{2}{9}}\left(\frac{r^4}{a^5}-\frac{r^4}{b^5}\right)Y_4^0\right]$$

where

$$V_{oct} = \sqrt{\left({}^{49}/_{18}\right)}\cdot\left(\sqrt{2\pi}\right)\cdot\frac{zer^4}{a^5}\cdot\left[Y_4^0+\sqrt{\frac{5}{14}}\left(Y_4^4+Y_4^{-4}\right)\right] \quad (7.3.8)$$

and

$$\theta_2^0 = \sqrt{\frac{5}{8}}\left(3\cos^2\theta - 1\right)$$

$$\theta_2^{\pm1} = \sqrt{\frac{15}{4}}\sin\theta\cos\theta$$

$$\theta_2^{\pm2} = \sqrt{\frac{15}{16}}\sin^2\theta$$

$$\theta_4^{\pm2} = \sqrt{\frac{45}{64}}\sin^2\theta\left(7\cos^2\theta - 1\right)$$

$$\theta_4^{\pm0} = \sqrt{\frac{9}{128}}\left(35\cos^4\theta - 30\cos^2\theta + 3\right)$$

$$\theta_4^{\pm4} = \sqrt{\frac{315}{256}}\sin^4\theta$$

then find the splitting of d-orbital in tetragonal crystal field in terms of D_q, D_t, and D_s.

- The crystal field Hamiltonian of D_{4h} (this chapter, p. 415)

$$V_{D_{4h}} = 2ze\sqrt{2\pi}\left[-\sqrt{\frac{2}{5}}\left(\frac{r^2}{a^3}-\frac{r^2}{b^3}\right)Y_2^0+\frac{1}{\sqrt{72}}\left(\frac{3r^4}{a^5}+\frac{4r^4}{b^5}\right)Y_4^0+\sqrt{\frac{35}{144}}\left(\frac{r^4}{a^5}\right)\left(Y_4^4+Y_4^{-4}\right)\right] \quad (7.3.12)$$

$$V_{D_{4h}} = 2ze\sqrt{2\pi}\left[-\sqrt{\frac{2}{5}}\left(\frac{r^2}{a^3}-\frac{r^2}{b^3}\right)Y_2^0+\frac{1}{\sqrt{72}}\left(\frac{7r^4}{a^5}-\frac{4r^4}{a^5}+\frac{4r^4}{b^5}\right)Y_4^0+\sqrt{\frac{35}{144}}\left(\frac{r^4}{a^5}\right)\left(Y_4^4+Y_4^{-4}\right)\right]$$

$$V_{D_{4h}} = 2ze\sqrt{2\pi}\cdot\sqrt{\frac{35}{144}}\left(\frac{r^4}{a^5}\right)\left(Y_4^4+Y_4^{-4}\right) + 2ze\sqrt{2\pi}\frac{7}{\sqrt{72}}\frac{r^4}{a^5}Y_4^0$$

$$-2ze\sqrt{2\pi}\sqrt{\frac{2}{5}}\left(\frac{r^2}{a^3}-\frac{r^2}{b^3}\right)Y_2^0 - 2ze\sqrt{2\pi}\frac{1}{\sqrt{72}}\left(\frac{4r^4}{a^5}-\frac{4r^4}{b^5}\right)Y_4^0$$

$$V_{D_{4h}} = ze\sqrt{2\pi}\cdot\sqrt{\frac{5\times7\times7\times4}{144\times7}}\left(\frac{r^4}{a^5}\right)\left(Y_4^4+Y_4^{-4}\right)+ze\sqrt{2\pi}\frac{\sqrt{49}\,r^4}{\sqrt{18}a^5}Y_4^0$$
$$-2ze\sqrt{2\pi}\sqrt{\frac{2}{5}}\left(\frac{r^2}{a^3}-\frac{r^2}{b^3}\right)Y_2^0-2ze\sqrt{2\pi}\frac{\sqrt{2\times8}}{\sqrt{8\times9}}\left(\frac{r^4}{a^5}-\frac{r^4}{b^5}\right)Y_4^0$$

$$V_{D_{4h}} = ze\sqrt{2\pi}\cdot\sqrt{\frac{5\times49}{14\times18}}\left(\frac{r^4}{a^5}\right)\left(Y_4^4+Y_4^{-4}\right)+ze\sqrt{2\pi}\sqrt{\frac{49\,r^4}{18a^5}}Y_4^0$$
$$-2ze\sqrt{2\pi}\sqrt{\frac{2}{5}}\left(\frac{r^2}{a^3}-\frac{r^2}{b^3}\right)Y_2^0-2ze\sqrt{2\pi}\sqrt{\frac{2}{9}}\left(\frac{r^4}{a^5}-\frac{r^4}{b^5}\right)Y_4^0$$

$$V_{D_{4h}} = \sqrt{(^{49}/_{18})}\cdot\left(\sqrt{2\pi}\right)\cdot\frac{zer^4}{a^5}\cdot\left[Y_4^0+\sqrt{\frac{5}{14}}\left(Y_4^4+Y_4^{-4}\right)\right]$$
$$-2ze\left(\sqrt{2\pi}\right)\cdot\left[\sqrt{\frac{2}{5}}\left(\frac{r^2}{a^3}-\frac{r^2}{b^3}\right)Y_2^0\right]-2ze\left(\sqrt{2\pi}\right)\cdot\left[\sqrt{\frac{2}{9}}\left(\frac{r^4}{a^5}-\frac{r^4}{b^5}\right)Y_4^0\right]$$

Then:

$$V_{D_{4h}} = V_{oct}-2ze\left(\sqrt{2\pi}\right)\cdot\left[\sqrt{\frac{2}{5}}\left(\frac{r^2}{a^3}-\frac{r^2}{b^3}\right)Y_2^0\right]-2ze\left(\sqrt{2\pi}\right)\cdot\left[\sqrt{\frac{2}{9}}\left(\frac{r^4}{a^5}-\frac{r^4}{b^5}\right)Y_4^0\right] \quad (7.4.44)$$

- The appropriate secular determinant is:

$$\begin{array}{c} \\ (2)\\ (1)\\ (0)\\ (-1)\\ (-2)\end{array}\begin{array}{ccccc} (2) & (1) & (0) & (-1) & (-2)\\ \end{array}$$
$$\begin{vmatrix} H_{2,2}-E & H_{2,1} & H_{2,0} & H_{2,-1} & H_{2,-2}\\ H_{1,2} & H_{1,1}-E & H_{1,0} & H_{1,-1} & H_{1,-2}\\ H_{0,2} & H_{0,1} & H_{0,0}-E & H_{0,-1} & H_{0,-2}\\ H_{-1,2} & H_{-1,1} & H_{-1,0} & H_{-1,-1}-E & H_{-1,-2}\\ H_{-2,2} & H_{-2,1} & H_{-2,0} & H_{-2,-1} & H_{-2,-2}-E \end{vmatrix}=0 \quad (7.4.45)$$

(rows and columns labelled (2), (1), (0), (−1), (−2))

where $H_{m,m'}$ are;

$$H_{m,m'} = e\int(m_i)^*V_{D_{4h}}(m_i')\partial\tau$$

The Hamiltonian H includes the crystal field Hamiltonian, and the electric field experienced by the electrons is defined by: $H=eV$

- To evaluate the matrix elements of

$$(m_i) = R_{n,2}Y_2^{m_l}$$

$$H_{m,m'} = e\int R_{n,2}^*Y_2^{m_l*}V_{D_{4h}}R'_{n,2}Y_2^{m_l'}\partial\tau$$

$$\partial\tau = r^2\sin\theta\,\partial\theta\,\partial\phi\,\partial r,\ \text{and}$$

$$V_{D_{4h}} = \sqrt{(^{49}/_{18})}\cdot\left(\sqrt{2\pi}\right)\cdot\frac{zer^4}{a^5}\cdot\left[Y_4^0+\sqrt{\frac{5}{14}}\left(Y_4^4+Y_4^{-4}\right)\right]$$
$$-2ze\left(\sqrt{2\pi}\right)\cdot\left[\sqrt{\frac{2}{5}}\left(\frac{r^2}{a^3}-\frac{r^2}{b^3}\right)Y_2^0\right]-2ze\left(\sqrt{2\pi}\right)\cdot\left[\sqrt{\frac{2}{9}}\left(\frac{r^4}{a^5}-\frac{r^4}{b^5}\right)Y_4^0\right]$$

Then:

$$
\begin{aligned}
V_{\mathrm{D4h}} = & \sqrt{\left({}^{49}/_{18}\right)}\cdot\left(\sqrt{2\pi}\right)\cdot\frac{zer^4}{a^5}Y_4^0+\sqrt{\left({}^{49}/_{18}\right)}\cdot\sqrt{\frac{5}{14}}\cdot\left(\sqrt{2\pi}\right)\cdot\frac{zer^4}{a^5}\cdot Y_4^4 \\
& +\sqrt{\left({}^{49}/_{18}\right)}\cdot\sqrt{\frac{5}{14}}\cdot\left(\sqrt{2\pi}\right)\cdot\frac{zer^4}{a^5}\cdot Y_4^{-4}-2ze\left(\sqrt{2\pi}\right)\sqrt{\frac{2}{5}\frac{r^2}{a^3}}Y_2^0 \\
& +2ze\left(\sqrt{2\pi}\right)\sqrt{\frac{2}{5}\frac{r^2}{b^3}}Y_2^0-2ze\left(\sqrt{2\pi}\right)\sqrt{\frac{2}{9}\frac{r^4}{a^5}}Y_4^0+2ze\left(\sqrt{2\pi}\right)\sqrt{\frac{2}{9}\frac{r^4}{b^5}}Y_4^0
\end{aligned}
\tag{7.4.46}
$$

- If: $H=eV$,

$$H_{m,m'}=\langle m|H|m'\rangle=\langle m|V|m'\rangle$$

$$H_{m,m'}=e\int R_{n,2}^{*}Y_2^{m_l*}V_{\mathrm{D4h}}R'_{n,2}Y_2^{m'_l}r^2\sin\theta\,\partial\theta\,\partial\phi\,\partial r$$

$$
\begin{aligned}
H_{m,m'} = & \sqrt{\left({}^{49}/_{18}\right)}\cdot\left(\sqrt{2\pi}\right)\int_0^\infty R_{n,2}^{*}\frac{ze^2r^4}{a^5}R'_{n,2}r^2\,\partial r\int_0^\pi\int_0^{2\pi}Y_2^{m_l*}Y_4^0Y_2^{m'_l}\sin\theta\,\partial\theta\,\partial\phi \\
& +\sqrt{\left({}^{49}/_{18}\right)}\cdot\sqrt{\frac{5}{14}}\left(\sqrt{2\pi}\right)\int_0^\infty R_{n,2}^{*}\frac{ze^2r^4}{a^5}R'_{n,2}r^2\,\partial r\int_0^\pi\int_0^{2\pi}Y_2^{m_l*}Y_4^4Y_2^{m'_l}\sin\theta\,\partial\theta\,\partial\phi \\
& +\sqrt{\left({}^{49}/_{18}\right)}\cdot\sqrt{\frac{5}{14}}\left(\sqrt{2\pi}\right)\int_0^\infty R_{n,2}^{*}\frac{ze^2r^4}{a^5}R'_{n,2}r^2\,\partial r\int_0^\pi\int_0^{2\pi}Y_2^{m_l*}Y_4^{-4}Y_2^{m'_l}\sin\theta\,\partial\theta\,\partial\phi \\
& -2ze^2\left(\sqrt{2\pi}\right)\sqrt{\frac{2}{5}}\int_0^\infty R_{n,2}^{*}\frac{r^2}{a^3}R'_{n,2}r^2\partial r\int_0^\pi\int_0^{2\pi}Y_2^{m_l*}Y_2^0Y_2^{m'_l}\sin\theta\,\partial\theta\,\partial\phi \\
& +2ze^2\left(\sqrt{2\pi}\right)\sqrt{\frac{2}{5}}\int_0^\infty R_{n,2}^{*}\frac{r^2}{b^3}R'_{n,2}r^2\partial r\int_0^\pi\int_0^{2\pi}Y_2^{m_l*}Y_2^0Y_2^{m'_l}\sin\theta\,\partial\theta\,\partial\phi \\
& -2ze^2\left(\sqrt{2\pi}\right)\sqrt{\frac{2}{9}}\int_0^\infty R_{n,2}^{*}\frac{r^4}{a^5}R'_{n,2}2\,\partial r\int_0^\pi\int_0^{2\pi}Y_2^{m_l*}Y_4^0Y_2^{m'_l}\sin\theta\,\partial\theta\,\partial\phi \\
& +2ze^2\left(\sqrt{2\pi}\right)\sqrt{\frac{2}{9}}\int_0^\infty R_{n,2}^{*}\frac{r^4}{b^5}R'_{n,2}r^2\partial r\int_0^\pi\int_0^{2\pi}Y_2^{m_l*}Y_4^0Y_2^{m'_l}\sin\theta\,\partial\theta\,\partial\phi
\end{aligned}
\tag{7.4.47}
$$

- The Hamiltonian can only connect wave functions whose m_l values differ by 0, 2, 4 or −4, because the tetragonal Hamiltonian components (m_l) are 0, 2, 4, and −4.
- These integrations:

$$\int_0^{2\pi}Y_{l1}^{m_{l1}}Y_{l2}^{m_{l2}}Y_{l3}^{m_{l3}}\,\partial\phi\neq 0$$

only if

$$m_{l_1}+m_{l_2}+m_{l_3}=0 \tag{7.4.9}$$

The integral $H_{m,m'}$ may be reduced to (p. 421)

$$H_{m,m'} = \langle m|H|m'\rangle = \langle m|V|m'\rangle, \text{ “shorthand representation” and}$$

$$H_{m_i,m_i} = e\int (m_i)^* V_{D_{4h}}(m_i)\partial\tau$$

$$\begin{aligned}
&= \sqrt{(^{49}/_{18})}\cdot\left(\sqrt{2\pi}\right)\int_0^\infty R^*_{n,2}\frac{ze^2r^4}{a^5}R'_{n,2}r^2\partial r\int_0^\pi\int_0^{2\pi} Y_2^{m_i*}Y_4^0Y_2^{m'_i}\sin\theta\,\partial\theta\,\partial\phi\\
&-2ze^2\left(\sqrt{2\pi}\right)\sqrt{\frac{2}{5}}\int_0^\infty R^*_{n,2}\frac{r^2}{a^3}R'_{n,2}r^2\partial r\int_0^\pi\int_0^{2\pi} Y_2^{m_i*}Y_2^0Y_2^{m'_i}\sin\theta\,\partial\theta\,\partial\phi\\
&+2ze^2\left(\sqrt{2\pi}\right)\sqrt{\frac{2}{5}}\int_0^\infty R^*_{n,2}\frac{r^2}{b^3}R'_{n,2}r^2\,\partial r\int_0^\pi\int_0^{2\pi} Y_2^{m_i*}Y_2^0Y_2^{m'_i}\sin\theta\,\partial\theta\,\partial\phi\\
&-2ze^2\left(\sqrt{2\pi}\right)\sqrt{\frac{2}{9}}\int_0^\infty R^*_{n,2}\frac{r^4}{a^5}R'_{n,2}r^2\,\partial r\int_0^\pi\int_0^{2\pi} Y_2^{m_i*}Y_4^0Y_2^{m'_i}\sin\theta\,\partial\theta\,\partial\phi\\
&+2ze^2\left(\sqrt{2\pi}\right)\sqrt{\frac{2}{9}}\int_0^\infty R^*_{n,2}\frac{r^4}{b^5}R'_{n,2}r^2\,\partial r\int_0^\pi\int_0^{2\pi} Y_2^{m_i*}Y_4^0Y_2^{m'_i}\sin\theta\,\partial\theta\,\partial\phi
\end{aligned} \tag{7.4.48}$$

where $m_i = 0, \pm 1$ and ± 2
and

$$H_{\pm 2,\mp 2} = e\int (\pm 2)^* V_{D_{4h}}(\mp 2)\partial\tau$$

$$\begin{aligned}
&= \sqrt{(^{49}/_{18})}\cdot\sqrt{\frac{5}{14}}\left(\sqrt{2\pi}\right)\int_0^\infty R^*_{n,2}\frac{ze^2r^4}{a^5}R'_{n,2}r^2\,\partial r\int_0^\pi\int_0^{2\pi} Y_2^{m_i*}Y_4^4Y_2^{m'_i}\sin\theta\,\partial\theta\,\partial\phi\\
&+\sqrt{(^{49}/_{18})}\cdot\sqrt{\frac{5}{14}}\left(\sqrt{2\pi}\right)\int_0^\infty R^*_{n,2}\frac{ze^2r^4}{a^5}R'_{n,2}r^2\,\partial r\int_0^\pi\int_0^{2\pi} Y_2^{m_i*}Y_4^{-4}Y_2^{m'_i}\sin\theta\,\partial\theta\,\partial\phi
\end{aligned} \tag{7.4.49}$$

- Since

$$Y_l^{m_i} = \theta_2^{m_i}\cdot\phi_{m_i} = \theta_2^{m_i}\frac{1}{\sqrt{2\pi}}e^{im_i\phi}$$

Integration over ϕ_{m_i} yields 0 or $\frac{1}{\sqrt{2\pi}}$, and

r: may be integrated:

$$\int_0^\infty R^*_{n,2}r^2R_{n,2}r^2\partial r = \overline{r_2^2}$$

$$\int_0^\infty R^*_{n,2}r^4R_{n,2}r^2\partial r = \overline{r_2^4}$$

$\overline{r_2^4}$: the mean fourth power radius of d-electrons of central ion.

$$\alpha_2(\text{Eq.}) = ze^2\frac{\overline{r^2}}{a^3}\alpha_2(\text{Ax.}) = ze^2\frac{\overline{r^2}}{b^3},\quad \text{where, Eq.: equatorial, Ax.: axial}$$

$$\alpha_4(\text{Eq.}) = ze^2\frac{\overline{r^4}}{a^5}\alpha_4(\text{Ax.}) = ze^2\frac{\overline{r^4}}{b^5}$$

- Then

$$\begin{aligned}
H_{m_i,m_i} = e\int (m_i)^* V_{\text{D}_{4h}}(m_i)\partial\tau = \sqrt{(^{49}/_{18})}\cdot\alpha_4(\text{Eq.})\int_0^\pi \theta_2^{m_l*}\theta_4^0\theta_2^{m_l}\sin\theta\,\partial\theta \\
-2\sqrt{\frac{2}{5}}\alpha_2(\text{Eq.})\int_0^\pi \theta_2^{m_l*}\theta_2^0\theta_2^{m_l}\sin\theta\,\partial\theta \\
+2\sqrt{\frac{2}{5}}\alpha_2(\text{Ax.})\int_0^\pi \theta_2^{m_l*}\theta_2^0\theta_2^{m_l}\sin\theta\,\partial\theta \\
-2\sqrt{\frac{2}{9}}\alpha_4(\text{Eq.})\int_0^\pi \theta_2^{m_l*}\theta_4^0\theta_2^{m_l}\sin\theta\,\partial\theta \\
+2\sqrt{\frac{2}{9}}\alpha_4(\text{Ax.})\int_0^\pi \theta_2^{m_l*}\theta_4^0\theta_2^{m_l}\sin\theta\,\partial\theta
\end{aligned}\tag{7.4.50}$$

and

$$H_{\pm2,\mp2} = e\int(\pm2)^* V_{\text{D}_{4h}}(\mp2)\partial\tau = \sqrt{(^{49}/_{18})}\cdot\sqrt{\frac{5}{14}}\alpha_4(\text{Eq.})\int_0^\pi \theta_2^{2*}\theta_4^{\mp4}\theta_2^2\sin\theta\,\partial\theta \tag{7.4.51}$$

If:

$$\alpha_4(\text{Eq.}) = 6Dq$$

$$Ds = \frac{2}{7}(\alpha_2(\text{Eq.}) - (\alpha_2(\text{Ax.}))$$

$$Dt = \frac{2}{21}(\alpha_4(\text{Eq.}) - \alpha_4(\text{Ax.}))$$

$$\begin{aligned}
H_{m_i,m_i} = e\int (m_i)^* V_{\text{D}_{4h}}(m_i)\partial\tau = \sqrt{(^{49}/_{18})}\cdot 6Dq\int_0^\pi \theta_2^{m_l*}\theta_4^0\theta_2^{m_l}\sin\theta\,\partial\theta \\
-2\sqrt{\frac{27}{52}}Ds\int_0^\pi \theta_2^{m_l*}\theta_2^0\theta_2^{m_l}\sin\theta\,\partial\theta \\
-2\sqrt{\frac{221}{92}}Dt\int_0^\pi \theta_2^{m_l*}\theta_4^0\theta_2^{m_l}\sin\theta\,\partial\theta
\end{aligned}\tag{7.4.52}$$

and

$$\begin{aligned}
H_{\pm2,\mp2} = e\int (m_i)^* V_{\text{D}_{4h}}(m_i)\partial\tau = \left(\sqrt{(^{49}/_{18})}\cdot 6Dq - \frac{14}{\sqrt{2}}Dt\right)\int_0^\pi \theta_2^{m_l*}\theta_4^0\theta_2^{m_l}\sin\theta\,\partial\theta \\
-7\sqrt{\frac{2}{5}}Ds\int_0^\pi \theta_2^{m_l*}\theta_2^0\theta_2^{m_l}\sin\theta\,\partial\theta
\end{aligned}\tag{7.4.53}$$

If:

$$\int_0^\pi \theta_2^{0*}\theta_2^0\theta_2^0 \sin\theta\,\partial\theta = \left(\sqrt{\frac{5}{8}}\right)^3 \int_0^\pi (3\cos^2\theta - 1)^3 \sin\theta\,\partial\theta$$

$$= \left(\frac{5}{8}\right)^{3/2}\left[27\int_0^\pi \cos^6\theta\sin\theta\,\partial\theta - 27\int_0^\pi \cos^4\theta\sin\theta\,\partial\theta + 9\int_0^\pi \cos^2\theta\sin\theta\,\partial\theta - \int_0^\pi \sin\theta\,\partial\theta\right]$$

$$= \left(\frac{5}{8}\right)^{3/2}\left[27\times\left(\frac{2}{7}\right) - 27\times\left(\frac{2}{5}\right) + 9\times\left(\frac{2}{3}\right) - 2\right] = 0.452$$

$$\int_0^\pi \theta_2^{1*}\theta_2^0\theta_2^{1\prime} \sin\theta\,\partial\theta = \left(\frac{15}{4}\right)\left(\frac{5}{8}\right)^{1/2}\int_0^\pi \sin^2\theta\cos^2\theta(3\cos^2\theta - 1)\sin\theta\,\partial\theta$$

$$= \left(\frac{15}{4}\right)\left(\frac{5}{8}\right)^{1/2}\left[\int_0^\pi 3\sin^3\theta\cos^4\theta\,\partial\theta - \int_0^\pi \sin^3\theta\cos^2\theta\,\partial\theta\right]$$

$$= 2.965(12/35 - 4/15) = 0.226$$

$$\int_0^\pi \theta_2^{2*}\theta_2^0\theta_2^2 \sin\theta\,\partial\theta = \left(\frac{15}{16}\right)\left(\frac{5}{8}\right)^{1/2}\int_0^\pi \sin^2\theta(3\cos^2\theta - 1)\sin^2\theta\sin\theta\,d\theta$$

$$\int_0^\pi \theta_2^{2*}\theta_2^0\theta_2^2 \sin\theta\,\partial\theta = \left(\frac{15}{16}\right)\left(\frac{5}{8}\right)^{1/2}\left[\int_0^\pi 3\sin^5\theta\cos^2\theta\,d\theta - \int_0^\pi \sin^5\theta\,d\theta\right]$$

$$= 0.741(0.457 - 1.067) = -0.452$$

$$\int_0^\pi \theta_2^{0*}\theta_4^0\theta_2^0 \sin\theta\,\partial\theta = \frac{(18)^{1/2}}{7}, p.471??$$

$$\int_0^\pi \theta_2^{1*}\theta_4^0\theta_2^1 \sin\theta\,\partial\theta = -\frac{(8)^{1/2}}{7}$$

$$\int_0^\pi \theta_2^{2*}\theta_4^0\theta_2^2 \sin\theta\,\partial\theta = \frac{(2)^{1/2}}{14}$$

$$\int_0^\pi \theta_2^{2*}\theta_4^4\theta_2^2 \sin\theta\,\partial\theta = \frac{(35)^{1/2}}{7}$$

By substituting in Eqs. (7.4.52) and (7.4.53)

$$H_{0,0} = e\int (0)^* V_{D_{4h}}(0)\partial\tau = 6Dq - 2Ds - 6Dt \tag{7.4.54}$$

$$H_{\pm1,\pm1} = e\int (\pm1)^* V_{D_{4h}}(\pm1)\partial\tau = -4Dq + 4Dt - Ds \tag{7.4.55}$$

$$H_{\pm 2,\pm 2} = e\int(\pm 2)^* V_{D_{4h}}(\pm 2)\partial\tau = Dq - Dt + 2Ds \tag{7.4.56}$$

$$H_{\pm 2,\mp 2} = e\int(\pm 2)^* V_{D_{4h}}(\mp 2)\partial\tau = 5Dq \tag{7.4.57}$$

- For the real d-orbitals:

$$d_{xz} = \frac{-1}{i\sqrt{2}}((1) + (-1))$$

$$d_{xy} = \frac{1}{i\sqrt{2}}((2) - (-2))$$

$$d_{yz} = \frac{-1}{\sqrt{2}}((1) - (-1))$$

$$d_{z^2} = 0$$

$$d_{(x^2-y^2)} = \frac{-1}{\sqrt{2}}((2) + (-2))$$

$$\int (d_{x^2-y^2}) V_{D_{4h}} (d_{x^2-y^2})\, d\tau = \frac{1}{2}\int ((2) + (-2)) V_{D_{4h}} ((2) + (-2))\, d\tau$$

$$= \frac{1}{2}\left[\int (2)V_{D_{4h}}(2)\, d\tau + \int (2)V_{D_{4h}}(-2)\, d\tau + \int (-2)V_{D_{4h}}(2)\, d\tau + \int (-2)V_{D_{4h}}(-2)\, d\tau\right]$$

$$H_{d_{x^2-y^2}, d_{x^2-y^2}} = \frac{1}{2}[Dq - Dt + 2Ds + 5Dq + 5Dq + Dq - Dt + 2Ds]$$

$$H_{d_{x^2-y^2}, d_{x^2-y^2}} = 6Dq + 2Ds - Dt$$

Similarly (Fig. 7.9),

$$H_{d_{z^2}, d_{z^2}} = 6Dq - 2Ds - 6Dt$$

$$H_{d_{xy}, d_{xy}} = -4Dq + 2Ds - Dt$$

$$H_{d_{xz}, d_{xz}} = -4Dq - Ds + 4Dt$$

- *Note*:
  - Dq only depends upon a.
  - Dq possesses the same definition as in octahedral complexes.
  - In octahedral ML_6 and tetragonal trans ML_4Z_2, the $Dq(L)$ value remains roughly constant.

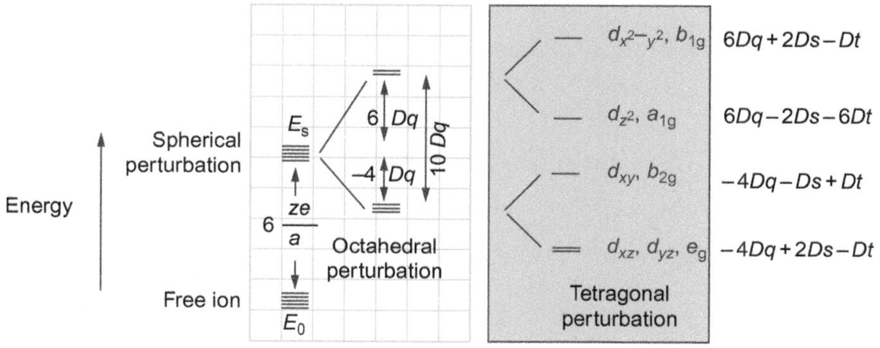

FIG. 7.9 The splitting of d-orbitals in tetragonal crystal field in comparison with that of octahedral.

7.5 TYPES OF INTERACTIONS THAT AFFECT THE CRYSTAL FIELD TREATMENT

What types of interactions might interfere with the crystal field treatment and the appropriate approach?

- The electrons of the metal ions in a complex undergo:
  - interelectronic repulsions; and
  - repelling by the electron density of the ligands (Lewis base).
- There are two crystal field type approaches in order to study the effects of the ligands of the metal complex have on the energies of the d-orbitals.
  - *Weak crystal field approach*; when the interelectronic repulsions are larger than the repulsion between the metal electrons and the electron density of the Lewis base (ligand).
  - *Strong crystal field approach*; when the interelectronic repulsions are small compare to the ligand electron-metal electron repulsions (strong Lewis base ligands).

7.6 FREE ION IN WEAK CRYSTAL FIELDS

Problems and the Required Approximations

What are the problems and the required approximations when we study the influence of weak crystal field potential on polyelectronic configuration of free ion terms?

- The constituents of free ion term wave functions are available, and expressed as functions of single electron wave functions (Chapter 2, p. 129).
- Thus, it is likely to resume the examination of the crystal field on polyelectron configuration.
- *The problem*: is to set up the secular determinant whose perturbing Hamiltonian is the sum of the crystal field, interelectronic repulsion, and spin-orbit coupling operator.
- *The approximation*: it considered that the crystal field causes a small perturbation compare to the electron repulsion effects within the configuration, but large compared to spin-orbit coupling:

$$\frac{e^2}{r_{ij}} > \text{crystal field} > \text{spin} - \text{orbital coupling}$$

- Consequently, in *the weak field approximation*, the ligand field *sorts out* the various wave functions of a free ion term, but *does not alter* them.
- Crystal field that cause splitting as large as, or larger than, the interelectronic repulsions are referred to as medium or strong.

$$\text{crystal field} \geq \frac{e^2}{r_{ij}} > \text{spin} - \text{orbital coupling}$$

The Effect of the Crystal Field on S Term

How the crystal field potential effect the S term?

- An S term is orbitally nondegenerate.
- The crystal field cannot split an S term, because it is concerned only with the orbital part of a wave function.

The Effect of the Cubic Crystal Field on P Term

How does the weak octahedron crystal field potential, $V_{oct.}$, effect the P term of the d^2 configuration?

- In the d^2 configuration, the orbital wave function for P term (Chapter 2: p. 130) are:

$$\langle 1, \pm 1\rangle = \mp\sqrt{\frac{3}{5}}(\pm 1,0) \pm \sqrt{\frac{2}{5}}(\pm 2, \mp 1)$$

$$\langle 1,0\rangle = -\sqrt{\frac{1}{5}}(1,-1)+\sqrt{\frac{4}{5}}(2,-2)$$

- In the weak field approximation, it is necessary to set up the secular determinant for the action of the ligand field potential ($V_{\text{oct.}}$) on the wave functions of the term being concerned as described in Scheme 7.4.
- The secular determinant:

$$
\begin{array}{c}
\quad\quad\quad \langle 1,1\rangle \quad\quad\quad \langle 1,0\rangle \quad\quad\quad \langle 1,-1\rangle \\
\begin{array}{c}\langle 1,1\rangle \\ \langle 1,0\rangle \\ \langle 1,-1\rangle\end{array}
\left|
\begin{array}{ccc}
H_{(1,1),(1,1)}-E & H_{(1,1)(1,0)} & H_{(1,1)(1,-1)} \\
H_{(1,0)(1,1)} & H_{(1,0),(1,0)}-E & H_{(1,0)(1,-1)} \\
H_{(1,-1)(1,1)} & H_{(1,-1)(1,0)} & H_{(1,-1),(1,-1)}-E
\end{array}
\right| = 0
\end{array}
\qquad (7.6.1)
$$

If: $H = eV$,

$$H_{m,m'} = \langle m|H|m'\rangle = \langle m|V|m'\rangle \quad \text{``shorthand representation''}$$

$$H_{(l,m),(l,m')} = \langle m|V|m'\rangle = e\int (l,m)^* V_{\text{oct}}(l,m')\,d\tau$$

- The first matrix element

$$\langle\langle 1,1\rangle|V_{\text{oct}}|\langle 1,1\rangle\rangle = \int\left[-\sqrt{\frac{3}{5}}(1,0)+\sqrt{\frac{2}{5}}(2,-1)\right]^* V_{\text{oct}}\left[-\sqrt{\frac{3}{5}}(1,0)+\sqrt{\frac{2}{5}}(2,-1)\right]\partial\tau$$

The induced perturbation of the weak octahedron crystal field potential, $V_{\text{oct.}}$, to the P term of d² configuration

$\Longleftarrow$ The perturbation theory

The secular determinant for the action of the ligand field potential ($V_{\text{oct.}}$) on the wave functions of the term being concerned.

$$
\begin{array}{c}
\quad\quad\quad \langle 1,1\rangle \quad\quad\quad \langle 1,0\rangle \quad\quad\quad \langle 1,-1\rangle \\
\begin{array}{c}\langle 1,1\rangle \\ \langle 1,0\rangle \\ \langle 1,-1\rangle\end{array}
\left|
\begin{array}{ccc}
H_{(1,1),(1,1)}-E & H_{(1,1),(1,0)} & H_{(1,1),(1,-1)} \\
H_{(1,0),(1,1)} & H_{(1,0),(1,0)}-E & H_{(1,0),(1,-1)} \\
H_{(1,-1),(1,1)} & H_{(1,-1),(1,0)} & H_{(1,-1),(1,-1)}-E
\end{array}
\right| = 0
\end{array}
$$

$H_{0,0} = \langle 0|V_{\text{oct}}|0\rangle = 6D_q$

$H_{\pm1,\pm1} = \langle\pm1|V_{\text{oct}}|\pm1\rangle = -4D_q$

$H_{\pm2,\pm2} = \langle\pm2|V_{\text{oct}}|\pm2\rangle = D_q$

$H_{\pm2,\mp2} = \langle\pm2|V_{\text{oct}}|\mp2\rangle = 5D_q$

$\Longrightarrow$

$$\int (m_l, m'_l)^* V_{\text{oct}}(m''_l, m'''_l)\,\partial\tau =$$

$$= \int (m'_l)^*(m''_l)\,\partial\tau_2 \int (m_l)^* V^1_{\text{oct}}(m''_l)\,\partial\tau_1$$

$$+ \int (m_l)^*(m''_l)\,\partial\tau_1 \int (m'_l)^* V^2_{\text{oct}}(m'''_l)\,\partial\tau_2$$

The orbital wave function for P term

$\langle 1,\pm1\rangle = \mp\sqrt{\frac{3}{5}}(\pm1,0)\pm\sqrt{\frac{2}{5}}(\pm2,\mp1)$

$\langle 1,0\rangle = -\sqrt{\frac{1}{5}}(1,-1)+\sqrt{\frac{4}{5}}(2,-2)$

$$H_{(l,m),(l,m')} = e\int (l,m)^* V_{\text{oct.}}(l,m')\,d\tau$$

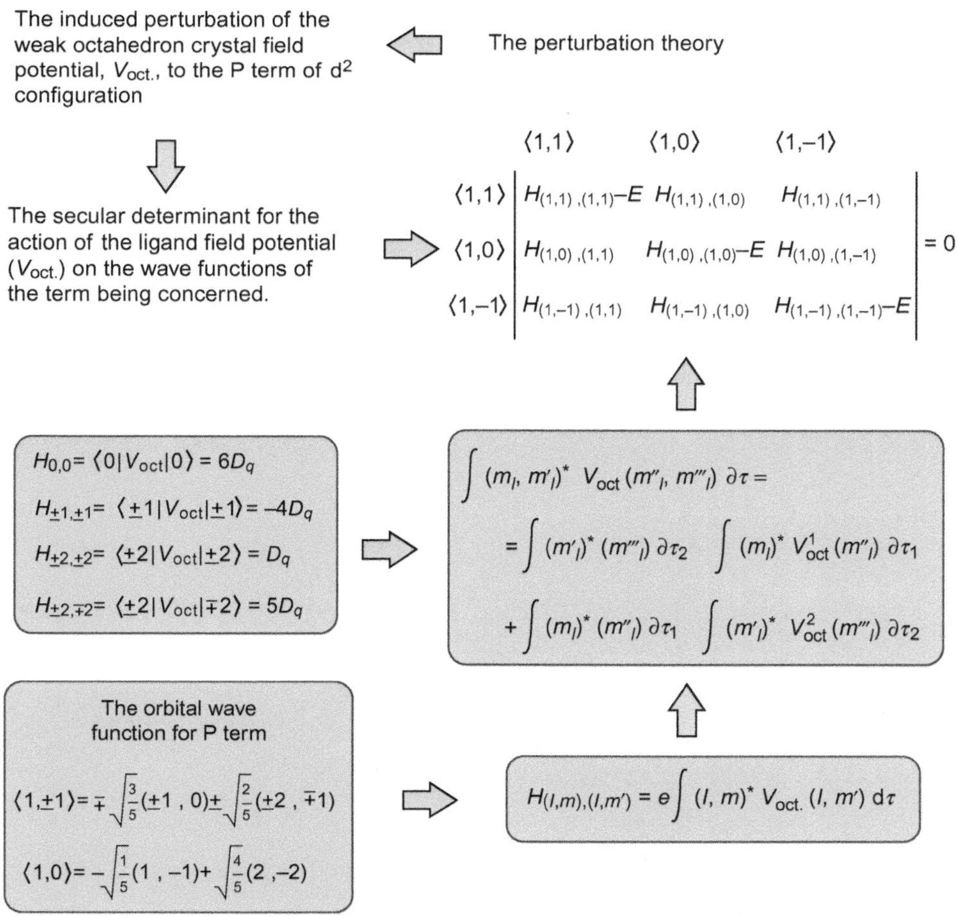

SCHEME 7.4 Approach used to estimate the effect of the weak octahedral crystal field potential, $V_{\text{Oct.}}$, on the P term of d² configuration.

$$\langle\langle 1,1\rangle|V_{\text{oct}}|\langle 1,1\rangle\rangle=\int\begin{bmatrix}\left(\frac{3}{5}\right)(1,0)^{*}V_{\text{oct}}(1,0)-\left(\frac{\sqrt{6}}{5}\right)(1,0)^{*}V_{\text{oct}}(2,-1)\\-\left(\frac{\sqrt{6}}{5}\right)(2,-1)^{*}V_{\text{oct}}(1,0)+\left(\frac{2}{5}\right)(2,-1)^{*}V_{\text{oct}}(2,-1)\end{bmatrix}\partial\tau \quad (7.6.2)$$

In the above equation, the coordinates of each electron are independently affected by V_{oct}. The value of V_{oct} is the sum of the components that acts on the coordinates of a particular electron.

$$V_{\text{oct}}=\sum V^{i}_{\text{oct}}=V^{1}_{\text{oct}}+V^{2}_{\text{oct}}+\cdots$$

The superscript specifies the numbering of the electrons.

The above integration (Eq. 7.6.2):

$$\int (m_l,m_l')^{*}V_{\text{oct}}(m_l'',m_l''')\partial\tau$$

can be expanded by insertion of the two-electron determinates. Each term is expanded and evaluated as follows:

$$\int (m_l,m_l')^{*}V_{\text{oct}}(m_l'',m_l''')\partial\tau=\int (m_l')^{*}(m_l''')\partial\tau_2\int (m_l)^{*}V^{1}_{\text{oct}}(m_l'')\partial\tau_1 + \int (m_l)^{*}(m''_l)\partial\tau_1\int (m'_l)^{*}V^{2}_{\text{oct}}(m'''_l)\partial\tau_2 \quad (7.6.3)$$

$$\int (m_l,m_l')^{*}V_{\text{oct}}(m_l'',m_l''')\partial\tau=0$$

Unless

$$m_l=m_l'' \quad \text{and} \quad \int (m_l')^{*}V^{2}_{\text{oct}}(m_l''')\partial\tau\neq 0$$

and/or

$$m_l'=m_l''' \quad \text{and} \quad \int (m_l)^{*}V^{1}_{\text{oct}}(m_l'')\partial\tau\neq 0$$

Also, when we examined the effect of V_{oct} on the d wave functions, we previously established that:

$$H_{0,0}=\langle 0|V_{\text{oct}}|0\rangle=6D_q \quad (7.4.12)$$

$$H_{\pm1,\pm1}=\langle \pm1|V_{\text{oct}}|\pm1\rangle=-4D_q \quad (7.4.13)$$

$$H_{\pm2,\pm2}=\langle \pm2|V_{\text{oct}}|\pm2\rangle=D_q \quad (7.4.14)$$

$$H_{\pm2,\mp2}=\langle \pm2|V_{\text{oct}}|\mp2\rangle=5D_q \quad (7.4.15)$$

Back to the first matrix element (Eq. 7.6.1):

$$\left\langle \langle 1,1|V_{\text{oct}}|\langle 1,1\rangle\rangle=\int\left\{\begin{matrix}\left(\frac{3}{5}\right)\underbrace{(1,0)^{*}V_{\text{oct}}(1,0)}_{\text{term1}}-\left(\frac{\sqrt{6}}{5}\right)\underbrace{(1,0)^{*}V_{\text{oct}}(2,-1)}_{\text{term2}}\\-\left(\frac{\sqrt{6}}{5}\right)\underbrace{(2,-1)^{*}V_{\text{oct}}(1,0)}_{\text{term3}}+\left(\frac{2}{5}\right)\underbrace{(2,-1)^{*}V_{\text{oct}}(2,-1)}_{\text{term4}}\end{matrix}\right\}\partial\tau\right.$$

From Eq. (7.6.3):

$$\int\left(\underbrace{(1,0)^{*}V_{\text{oct}}(1,0)}_{\text{term1}}\right)\partial\tau=\int((-1,0)^{*}V_{\text{oct}}(-1,0))\partial\tau$$

$$=\int((1)^{*}V^{1}_{\text{oct}}(1))\partial\tau_1+\int((0)^{*}V^{2}_{\text{oct}}(0))\partial\tau_2$$

From Eqs. (7.4.13) and (7.4.12)

$$H_{\langle 1,0\rangle,\langle 1,0\rangle} = H_{\langle -1,0\rangle,\langle -1,0\rangle} = -4D_q + 6D_q = 2D_q$$

$$\int\left(\underbrace{(\pm 1,0)^* V_{\text{oct}}(\pm 2,\mp 1)}_{\text{term2}}\right)\partial\tau = \int\left(\underbrace{(\pm 2,\mp 1)^* V_{\text{oct}}(\pm 1,0)}_{\text{term3}}\right)\partial\tau = 0$$

and

$$\int\left(\underbrace{(\pm 2,\mp 1)^* V_{\text{oct}}(\pm 2,\mp 1)}_{\text{term 4}}\right)\partial\tau = \int\left((\pm 2)^* V^1_{\text{oct}}(\pm 2)\right)\partial\tau_1 + \int\left((\mp 1)^* V^2_{\text{oct}}(\mp 1)\right)\partial\tau_2$$

From Eqs. (7.4.14) and (7.4.13):

$$H_{(\pm 2,\mp 1),(\pm 2,\mp 1)} = D_q - 4D_q = -3D_q$$

so that

$$H_{(1,1),(1,1)} = e\langle\langle 1,1\rangle|V_{oct}|\langle 1,1\rangle\rangle = -3\times\left(\frac{2}{5}\right)D_q + 2\times\left(\frac{3}{5}\right)D_q = 0 \tag{7.6.4}$$

The second matrix element (Eq. 7.6.2):

$$\langle\langle 1,0\rangle|V_{\text{oct}}|\langle 1,0\rangle\rangle = \int\left[-\sqrt{\frac{1}{5}}(1,-1) + \sqrt{\frac{4}{5}}(2,-2)\right]^* V_{\text{oct}}\left[-\sqrt{\frac{1}{5}}(1,-1) + \sqrt{\frac{4}{5}}(2,-2)\right]\partial\tau$$

$$= \int\left\{\begin{array}{l}\left(\frac{1}{5}\right)\underbrace{(1,-1)^* V_{\text{oct}}(1,-1)}_{\text{term 1}} - \left(\frac{2}{5}\right)\underbrace{(1,-1)^* V_{\text{oct}}(2,-2)}_{\text{term 2=0}} \\ -\left(\frac{2}{5}\right)\underbrace{(2,-2)^* V_{\text{oct}}(1,-1)}_{\text{term 3=0}} + \left(\frac{4}{5}\right)\underbrace{(2,-2)^* V_{\text{oct}}(2,-2)}_{\text{term 4}}\end{array}\right\}\partial\tau$$

From Eq. (7.6.3):

$$\int\left(\underbrace{(1,-1)^* V_{\text{oct}}(1,-1)}_{\text{term 1}}\right)\partial\tau = \int\left((1)^* V^1_{\text{oct}}(1)\right)\partial\tau_1 + \int\left((-1)^* V^2_{\text{oct}}(-1)\right)\partial\tau_2$$

From Eq. (7.4.13):

$$H_{(1,-1),(1,-1)} = -4D_q - 4D_q = -8D_q$$

$$\int\left(\underbrace{(2,-2)^* V_{\text{oct}}(2,-2)}_{\text{term 4}}\right)\partial\tau = \int\left((2)^* V^1_{\text{oct}}(2)\right)\partial\tau_1 + \int\left((-2)^* V^2_{\text{oct}}(-2)\right)\partial\tau_2$$

From Eq. (7.4.15):

$$H_{(2,-2),(2,-2)} = D_q + D_q = 2D_q$$

then

$$H_{(1,0),(1,0)} = e\langle\langle 1,0\rangle|V_{\text{oct}}|\langle 1,0\rangle\rangle = -\left(\frac{8}{5}\right)D_q - 0 - 0 + \left(\frac{2\times 4}{5}\right)D_q = 0 \tag{7.6.5}$$

- Similarly,

$$\langle\langle 1,\pm 1\rangle|V_{\text{oct}}|\langle 1,\mp 1\rangle\rangle = \langle\langle 1,\mp 1\rangle|V_{\text{oct}}|\langle 1,\pm 1\rangle\rangle = \langle\langle 1,0\rangle|V_{\text{oct}}|\langle 1,\pm 1\rangle\rangle = 1,\pm 1|V_{\text{oct}}|\langle 1,0\rangle\rangle = 0 \tag{7.6.6}$$

$$H_{(1,\pm 1),(1,\mp 1)} = H_{(1,\mp),(1,\pm 1)} = H_{(1,0),(1,\pm 1)} = H_{(1,\pm 1),(1,0)} = 0$$

- All matrix elements of P term are zero:

$$H_{(1,1),(1,1)} = 0$$

$$H_{(1,0),(1,0)} = 0$$

$$H_{(1,\pm 1),(1,\mp 1)} = 0$$

Accordingly, the crystal potential, $V_{oct.}$, does not split the P term.
- In the same way, it is possible to show that V_{oct} does not split the ^{4}P from d^3 configuration.
  The two electrons of a d^2 configuration are equivalent to the two holes in the half-filled d shell that results from d^3 configuration, with a reverse charge sign:

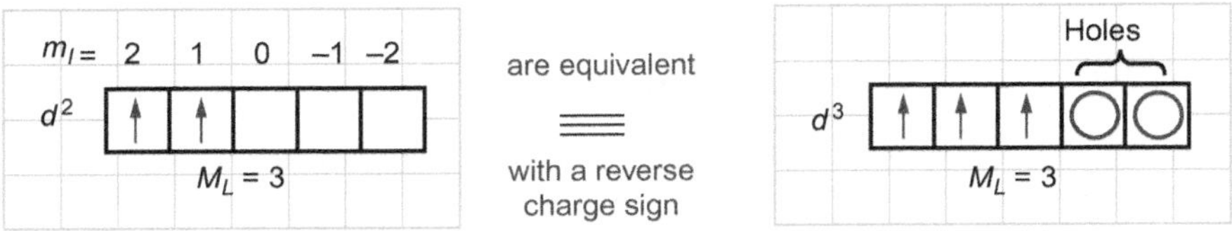

Thus the ^{4}P term of d^3, like the ^{3}P term of d^2, is unaffected by the weak crystal field.
- In general:
  - p- and s-orbitals set do not split by an octahedral ligand field; and
  - there is a one-to-one agreement in the influence of the ligand field on the term and on an orbital set that have the same litter (S term, s-orbital; P term, p-orbital).
- Since the matrix elements of the potentials for V_{oct} differ from those of tetrahedron or a cube of ligands, V_{Td} and V_{cube}, respectively, only by a numerical constant, likewise, V_{Td} and V_{cube} have no effect on an S or a P term in the weak field approximation.

The Effect of a Cubic Crystal Field on D Term

How can the weak octahedron crystal field potential, $V_{oct.}$, effect the ^{2}D term of d^1 configuration? How does this effect differ from ^{2}D of d^9 and ^{5}D of the d^4 and d^6 configurations?

- The orbital wave functions, $\langle l, m_l\rangle$, of the ^{2}D term of d^1: $\langle 2, \pm 2\rangle$, $\langle 2, \pm 1\rangle$, and $\langle 2, 0\rangle$ and the single electron orbital wave functions (± 2), (± 1), and (0), respectively, are the same, this chapter, p. 421.
- Thus, the ^{2}D term of d^1 must behave the same way as d-orbital, and split exactly in a similar way, p. 423:
  - The two wave functions: $\langle 2,0\rangle$ and $\frac{1}{\sqrt{2}}(\langle 2,2\rangle + \langle 2,-2\rangle) \Rightarrow$ at the energy $6Dq$.
  - The three wave functions: $\langle 2, \pm 1\rangle$ and $\frac{1}{\sqrt{2}}(\langle 2,2\rangle - \langle 2,-2\rangle) \Rightarrow$ at the energy $-4Dq$.
- The orbital splitting of ^{2}D term from the d^9 configuration can be obtained from d^1.
- Since the d^9 configuration corresponds to a single hole in the filled d^{10} shell, the effect of the ligand field can be derived from those of d^1 configuration by imagining the electron to have a positive charge.
  - The d^9 configuration is considered to be a hole in the filled d^{10} shell.
- When the charge reversed in sign, the matrix elements of V with the d-electron term wave functions would be reversed in sign.
- So, the ^{2}D term from d^9 configuration is split by a ligand field in the opposite sign to that for the ^{2}D term from d^1(D_q(d^9) = $-D_q$(d^1)):
  - The wave functions: $\langle 2,0\rangle$ and $\frac{1}{\sqrt{2}}(\langle 2,2\rangle + \langle 2,-2\rangle) \Rightarrow$ at the energy $-6Dq$
  - The wave functions: $\langle 2, \pm 1\rangle$ and $\frac{1}{\sqrt{2}}(\langle 2,2\rangle - \langle 2,-2\rangle) \Rightarrow$ at the energy $4Dq$
- Simply: the d^3 configuration is considered to be two holes in the d^5 half-filled shell.
- An additional ^{5}D term arises among others from d^4 and d^6 configurations.
  In the case of maximum multiplicity, it may be considered:
  - the d^4 configuration to be a hole in the d^5 half-filled shell; and
  - the d^6 configuration to be four holes in the filled d^{10} shell.

- As a result:
  - ^5D term of d^4 splits in the same manner as is the ^2D term of d^9, which reverses the splitting of the ^2D term of d^1 configuration.
  - ^5D term of d^6 splits in the opposite way to the ^5D term of d^4, which is similar to the splitting of the ^2D term of d^1 configuration.
- The splitting correlations of the D ground terms are concluded as follow:
  The splitting of the D term does not depend on its multiplicity.
  the splitting of d^n is the same as d^{n+5}, where $n<5$
  the splitting for d^{5-n} is the same as d^{10-n}, where $n<5$
  the splitting for d^{5-n} and d^{10-n} are reversed the for d^n and d^{n+5}, where $n<5$

The Effect of a Cubic Crystal Field on F Term

How does the octahedral crystal field Hamiltonian, $V_{(x,y,z)}$, in the weak field effect the F term from d^2? Find the energy and the wave function for each state.

- Scheme 7.5 describes the approach that used to see the response of the F term from d^2 to the weak octahedral crystal field, $V_{Oct.}$.
- The orbital wave functions, $\langle l, m_l\rangle$, of the ^3F term of d^2:

$$\langle 3, \pm 3\rangle, \langle 3, \pm 2\rangle, \langle 3, \pm 1\rangle, \text{ and } \langle 3, 0\rangle \tag{7.6.7}$$

- Since there is a parallel behavior between the splitting of an orbital set and the term having the same letter, it is expected that V_{oct} splits an F term similar to orbital splitting.
- The appropriate secular determinant:

$$\begin{array}{c} \\ \langle 3,3\rangle \\ \langle 3,2\rangle \\ \langle 3,1\rangle \\ \langle 3,0\rangle \\ \langle 3,-1\rangle \\ \langle 3,-2\rangle \\ \langle 3,-3\rangle \end{array}
\begin{array}{ccccccc}
\langle 3,3\rangle & \langle 3,2\rangle & \langle 3,1\rangle & \langle 3,0\rangle & \langle 3,-1\rangle & \langle 3,-2\rangle & \langle 3,-3\rangle \\
H_{\langle 3,3\rangle,\langle 3,3\rangle}-E & H_{\langle 3,3\rangle,\langle 3,2\rangle} & H_{\langle 3,3\rangle,\langle 3,1\rangle} & H_{\langle 3,3\rangle,\langle 3,0\rangle} & H_{\langle 3,3\rangle,\langle 3,-1\rangle} & H_{\langle 3,3\rangle,\langle 3,-2\rangle} & H_{\langle 3,3\rangle,\langle 3,-3\rangle} \\
H_{\langle 3,2\rangle,\langle 3,3\rangle} & H_{\langle 3,2\rangle,\langle 3,2\rangle}-E & H_{\langle 3,2\rangle,\langle 3,1\rangle} & H_{\langle 3,2\rangle,\langle 3,0\rangle} & H_{\langle 3,2\rangle,\langle 3,-1\rangle} & H_{\langle 3,2\rangle,\langle 3,-2\rangle} & H_{\langle 3,2\rangle,\langle 3,-3\rangle} \\
H_{\langle 3,1\rangle,\langle 3,3\rangle} & H_{\langle 3,1\rangle,\langle 3,2\rangle} & H_{\langle 3,1\rangle,\langle 3,1\rangle}-E & H_{\langle 3,1\rangle,\langle 3,0\rangle} & H_{\langle 3,1\rangle,\langle 3,-1\rangle} & H_{\langle 3,1\rangle,\langle 3,-2\rangle} & H_{\langle 3,1\rangle,\langle 3,-3\rangle} \\
H_{\langle 3,0\rangle,\langle 3,3\rangle} & H_{\langle 3,0\rangle,\langle 3,2\rangle} & H_{\langle 3,0\rangle,\langle 3,1\rangle} & H_{\langle 3,0\rangle,\langle 3,0\rangle}-E & H_{\langle 3,0\rangle,\langle 3,-1\rangle} & H_{\langle 3,0\rangle,\langle 3,-2\rangle} & H_{\langle 3,0\rangle,\langle 3,-3\rangle} \\
H_{\langle 3,-1\rangle,\langle 3,3\rangle} & H_{\langle 3,-1\rangle,\langle 3,2\rangle} & H_{\langle 3,-1\rangle,\langle 3,1\rangle} & H_{\langle 3,-1\rangle,\langle 3,0\rangle} & H_{\langle 3,-1\rangle,\langle 3,-1\rangle}-E & H_{\langle 3,-1\rangle,\langle 3,-2\rangle} & H_{\langle 3,-1\rangle,\langle 3,-3\rangle} \\
H_{\langle 3,-2\rangle,\langle 3,3\rangle} & H_{\langle 3,-2\rangle,\langle 3,2\rangle} & H_{\langle 3,-2\rangle,\langle 3,1\rangle} & H_{\langle 3,-2\rangle,\langle 3,0\rangle} & H_{\langle 3,-2\rangle,\langle 3,-1\rangle} & H_{\langle 3,-2\rangle,\langle 3,-2\rangle}-E & H_{\langle 3,-2\rangle,\langle 3,-3\rangle} \\
H_{\langle 3,-3\rangle,\langle 3,3\rangle} & H_{\langle 3,-3\rangle,\langle 3,2\rangle} & H_{\langle 3,-3\rangle,\langle 3,1\rangle} & H_{\langle 3,-3\rangle,\langle 3,0\rangle} & H_{\langle 3,-3\rangle,\langle 3,-1\rangle} & H_{\langle 3,-3\rangle,\langle 3,-2\rangle} & H_{\langle 3,-3\rangle,\langle 3,-3\rangle}-E
\end{array} = 0 \tag{7.6.8}$$

If: $H = eV$,

$$H_{(l,m),(l,m')} = \langle l, m|H|l, m'\rangle = \langle l, m|V|l, m'\rangle \text{ “shorthand representation, p. 421”, and}$$

where $H_{m,m'}$ are:

$$H_{(l,m),(l,m')} = \left\langle (l, m_i)|V_{oct}|(l, m_i')\right\rangle = e\int (l, m_i)^* V_{oct}(l, m_i')\partial\tau$$

is solved for the energies of d-orbitals under the potential.

Since

$$V_{oct} = \sqrt{(49/18)}\cdot\left(\sqrt{2\pi}\right)\cdot\frac{zer^4}{a^5}\cdot\left[Y_4^0 + \sqrt{\frac{5}{14}}\left(Y_4^4 + Y_4^{-4}\right)\right] \tag{7.3.8}$$

The expression of the matrix elements:

$$\int (m_i)^* V_{(x,y,z)}(m_i')\partial\tau$$
$$= \sqrt{(49/18)}\cdot\left(\sqrt{2\pi}\right)\cdot r_2^4\cdot\frac{ze}{a^5}\cdot\left[\langle (l,m)|Y_4^0|(l,m')\rangle + \sqrt{\frac{5}{14}}\langle (l,m)|\left(Y_4^4 + Y_4^{-4}\right)|(l,m')\rangle\right] \tag{7.6.9}$$

The perturbation theory ⇨ The induced perturbation of the weak octahedron crystal field potential, V_{Oct}., to the F term of d^2 configuration ⇨ The secular determinant for the action of the ligand field potential ($V_{Oct.}$) on the wave functions of the term being concerned.

⇩

$$\begin{array}{c|ccccccc|}
 & \langle 3,3\rangle & \langle 3,2\rangle & \langle 3,1\rangle & \langle 3,0\rangle & \langle 3,-1\rangle & \langle 3,-2\rangle & \langle 3,-3\rangle \\
\langle 3,3\rangle & H_{\langle 3,3\rangle,\langle 3,3\rangle}-E & H_{\langle 3,3\rangle,\langle 3,2\rangle} & H_{\langle 3,3\rangle,\langle 3,1\rangle} & H_{\langle 3,3\rangle,\langle 3,0\rangle} & H_{\langle 3,3\rangle,\langle 3,-1\rangle} & H_{\langle 3,3\rangle,\langle 3,-2\rangle} & H_{\langle 3,3\rangle,\langle 3,-3\rangle} \\
\langle 3,2\rangle & H_{\langle 3,2\rangle,\langle 3,3\rangle} & H_{\langle 3,2\rangle,\langle 3,2\rangle}-E & H_{\langle 3,2\rangle,\langle 3,1\rangle} & H_{\langle 3,2\rangle,\langle 3,0\rangle} & H_{\langle 3,2\rangle,\langle 3,-1\rangle} & H_{\langle 3,2\rangle,\langle 3,-2\rangle} & H_{\langle 3,2\rangle,\langle 3,-3\rangle} \\
\langle 3,1\rangle & H_{\langle 3,1\rangle,\langle 3,3\rangle} & H_{\langle 3,1\rangle,\langle 3,2\rangle} & H_{\langle 3,1\rangle,\langle 3,1\rangle}-E & H_{\langle 3,1\rangle,\langle 3,0\rangle} & H_{\langle 3,1\rangle,\langle 3,-1\rangle} & H_{\langle 3,1\rangle,\langle 3,-2\rangle} & H_{\langle 3,1\rangle,\langle 3,-3\rangle} \\
\langle 3,0\rangle & H_{\langle 3,0\rangle,\langle 3,3\rangle} & H_{\langle 3,0\rangle,\langle 3,2\rangle} & H_{\langle 3,0\rangle,\langle 3,1\rangle} & H_{\langle 3,0\rangle,\langle 3,0\rangle}-E & H_{\langle 3,0\rangle,\langle 3,-1\rangle} & H_{\langle 3,0\rangle,\langle 3,-2\rangle} & H_{\langle 3,0\rangle,\langle 3,-3\rangle} \\
\langle 3,-1\rangle & H_{\langle 3,-1\rangle,\langle 3,3\rangle} & H_{\langle 3,-1\rangle,\langle 3,2\rangle} & H_{\langle 3,-1\rangle,\langle 3,1\rangle} & H_{\langle 3,-1\rangle,\langle 3,0\rangle} & H_{\langle 3,-1\rangle,\langle 3,-1\rangle}-E & H_{\langle 3,-1\rangle,\langle 3,3-2\rangle} & H_{\langle 3,-1\rangle,\langle 3,-3\rangle} \\
\langle 3,-2\rangle & H_{\langle 3,-2\rangle,\langle 3,3\rangle} & H_{\langle 3,-2\rangle,\langle 3,2\rangle} & H_{\langle 3,-2\rangle,\langle 3,1\rangle} & H_{\langle 3,-2\rangle,\langle 3,0\rangle} & H_{\langle 3,-2\rangle,\langle 3,-1\rangle} & H_{\langle 3,-2\rangle,\langle 3,-2\rangle}-E & H_{\langle 3,-2\rangle,\langle 3,-3\rangle} \\
\langle 3,-3\rangle & H_{\langle 3,-3\rangle,\langle 3,3\rangle} & H_{\langle 3,-3\rangle,\langle 3,2\rangle} & H_{\langle 3,-3\rangle,\langle 3,1\rangle} & H_{\langle 3,-3\rangle,\langle 3,0\rangle} & H_{\langle 3,-3\rangle,\langle 3,-1\rangle} & H_{\langle 3,-3\rangle,\langle 3,-2\rangle} & H_{\langle 3,-3\rangle,\langle 3,-3\rangle}-E
\end{array} = 0$$

⇧

$$H_{(l,m),(l,m')} = e\int (l,m)^{*} V_{Oct.}\,(l,m')\,d\tau$$

&

$$V_{Oct.} = \sqrt{(49/18)}\cdot(\sqrt{2\pi})\cdot\frac{zer^4}{a^5}\cdot\left[Y_4^0 + \sqrt{\frac{5}{14}}\left(Y_4^4 + Y_4^{-4}\right)\right]$$

⇩

$$\begin{array}{c|ccccccc|}
 & \langle 3,3\rangle & \langle 3,2\rangle & \langle 3,1\rangle & \langle 3,0\rangle & \langle 3,-1\rangle & \langle 3,-2\rangle & \langle 3,-3\rangle \\
\langle 3,3\rangle & H_{\langle 3,3\rangle,\langle 3,3\rangle}-E & 0 & 0 & 0 & H_{\langle 3,3\rangle,\langle 3,-1\rangle} & 0 & 0 \\
\langle 3,2\rangle & 0 & H_{\langle 3,2\rangle,\langle 3,2\rangle}-E & 0 & 0 & 0 & H_{\langle 3,2\rangle,\langle 3,-2\rangle} & 0 \\
\langle 3,1\rangle & 0 & 0 & H_{\langle 3,1\rangle,\langle 3,1\rangle}-E & 0 & 0 & 0 & H_{\langle 3,1\rangle,\langle 3,-3\rangle} \\
\langle 3,0\rangle & 0 & 0 & 0 & H_{\langle 3,0\rangle,\langle 3,0\rangle}-E & 0 & 0 & 0 \\
\langle 3,-1\rangle & H_{\langle 3,-1\rangle,\langle 3,3\rangle} & 0 & 0 & 0 & H_{\langle 3,-1\rangle,\langle 3,-1\rangle}-E & 0 & 0 \\
\langle 3,-2\rangle & 0 & H_{\langle 3,-2\rangle,\langle 3,2\rangle} & 0 & 0 & 0 & H_{\langle 3,-2\rangle,\langle 3,-2\rangle}-E & 0 \\
\langle 3,-3\rangle & 0 & 0 & H_{\langle 3,-3\rangle,\langle 3,1\rangle} & 0 & 0 & 0 & H_{\langle 3,-3\rangle,\langle 3,-3\rangle}-E
\end{array} = 0$$

⇧

The orbital wave function for F term

$\langle 3,\pm 3\rangle = \pm(\pm 2, \pm 1)$

$\langle 3,\pm 2\rangle = \pm(\pm 2, 0)$

$\langle 3,\pm 1\rangle = \pm\sqrt{\frac{2}{5}}(\pm 1, 0) \pm \sqrt{\frac{3}{5}}(\pm 2, \mp 1)$

$\langle 3,0\rangle = \sqrt{\frac{4}{5}}(1,-1) + \sqrt{\frac{1}{5}}(2,-2)$

⇨

$$H_{(l,m),(l,m')} = e\int (l,m)^{*} V_{Oct.}\,(l,m')\,d\tau$$

&

$H_{0,0} = \langle 0|V_{Oct}|0\rangle = 6D_q$

$H_{\pm 1,\pm 1} = \langle \pm 1|V_{Oct}|\pm 1\rangle = -4D_q$

$H_{\pm 2,\pm 2} = \langle \pm 2|V_{Oct}|\pm 2\rangle = D_q$

$H_{\pm 2,\mp 2} = \langle \pm 2|V_{Oct}|\mp 2\rangle = 5D_q$

⇩

At $E = -6D_q$: $\frac{1}{\sqrt{24}}\left(\sqrt{15}\langle 3\rangle - 3\langle -1\rangle\right)$, $\langle 0\rangle$, and $\frac{1}{\sqrt{24}}\left(\sqrt{15}\langle -3\rangle - 3\langle 1\rangle\right)$

At $E = 2D_q$: $\frac{1}{\sqrt{24}}\left(3\langle 3\rangle + \sqrt{15}\langle -1\rangle\right)$, $\sqrt{\frac{1}{2}}\left(\langle 2\rangle - \langle -2\rangle\right)$, and $\frac{1}{\sqrt{24}}\left(3\langle -3\rangle + \sqrt{15}\langle 1\rangle\right)$

At $E = 12D_q$: $\sqrt{\frac{1}{2}}\left(\langle 2\rangle + \langle -2\rangle\right)$

SCHEME 7.5 Approach used to estimate the effect of the weak octahedron crystal field potential, $V_{Oct.}$, on the F term of d^2 configuration.

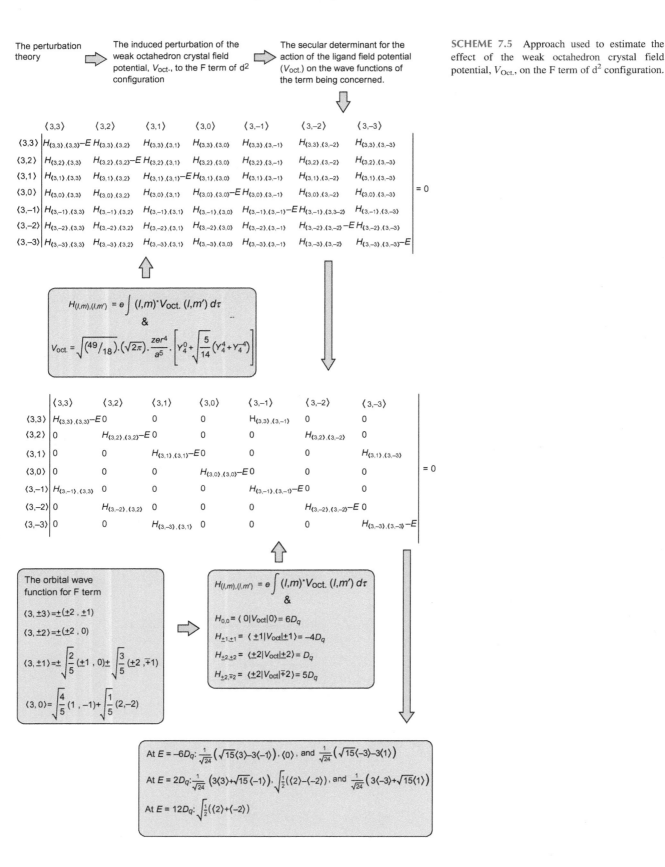

- These integrations may be simplified by noting that, in order that

$$\left\langle (l,m)|Y_{l'}^{m''}|(l,m')\right\rangle \neq 0$$

is only true if

$$m+m'+m''=0 \text{ regardless of the value } l \text{ and } l', \tag{7.4.9}$$

  - Consequently the integrals which involve $\langle (l,m)|Y_{l'}^{0}|(l,m')\rangle$ are zero unless

$$\begin{aligned} &m+m'=0\\ &m=-m'=m^{*}\\ &m+(-1\times m')=0,\quad m=m'=0 \end{aligned}$$

possible if m (from Eq. 7.6.7)

$$0+(-1\times 0)=0,\quad m_l=m'=0$$

$$1+(-1\times 1)=0,\quad m_l=m'=\pm 1$$

$$2+(-1\times 2)=0,\quad m_l=m'=\pm 2,$$

$$3+(-1\times 3)=0,\quad m_l=m'=\pm 3$$

That is the matrix element is zero unless it between an f-wave function and itself, not one of other f wave functions (a diagonal matrix).

Using the same argument, the integrals of $(l,m)|Y_{l'}^{4}|(l,m')$ are zero unless

$$m+m'+m''=0 \tag{7.4.9}$$

$$m+m'=-4$$

Which are only possible if

$$-2+(-1\times 2)=-4,\quad m=-2,\quad m'=2$$

$$-3+(-1\times 1)=-4,\quad m=-3,\quad m'=1$$

Those involving Y_4^{-4} are zero unless

$$2+(-1\times -2)=4,\quad m=2,\quad m'=-2$$

$$3+(-1\times -1)=4,\quad m=3,\quad m'=-1$$

- Now the secular determinant:

$$\begin{array}{c|ccccccc|c} & \langle 3,3\rangle & \langle 3,2\rangle & \langle 3,1\rangle & \langle 3,0\rangle & \langle 3,-1\rangle & \langle 3,-2\rangle & \langle 3,-3\rangle & \\ \langle 3,3\rangle & H_{\langle 3,3\rangle,\langle 3,3\rangle}-E & 0 & 0 & 0 & H_{\langle 3,3\rangle,\langle 3,-1\rangle} & 0 & 0 & \\ \langle 3,2\rangle & 0 & H_{\langle 3,2\rangle,\langle 3,2\rangle}-E & 0 & 0 & 0 & H_{\langle 3,2\rangle,3,-2\rangle} & 0 & \\ \langle 3,1\rangle & 0 & 0 & H_{\langle 3,1\rangle,\langle 3,1\rangle}-E & 0 & 0 & 0 & H_{\langle 3,1\rangle,\langle 3,-3\rangle} & \\ \langle 3,0\rangle & 0 & 0 & 0 & H_{\langle 3,0\rangle,\langle 3,0\rangle}-E & 0 & 0 & 0 & =0\\ \langle 3,-1\rangle & H_{\langle 3,-1\rangle,\langle 3,3\rangle} & 0 & 0 & 0 & H_{\langle 3,-1\rangle,\langle 3,-1\rangle}-E & 0 & 0 & \\ \langle 3,-2\rangle & 0 & H_{\langle 3,-2\rangle,\langle 3,2\rangle} & 0 & 0 & 0 & H_{\langle 3,-2\rangle,\langle 3,-2\rangle}-E & 0 & \\ \langle 3,-3\rangle & 0 & 0 & H_{\langle 3,-3\rangle,\langle 3,1\rangle} & 0 & 0 & 0 & H_{\langle 3,-3\rangle,\langle 3,-3\rangle}-E & \end{array}$$

- In the weak field approximation, it is necessary to set up the secular determinant for the action of the ligand field potential, $V_{oct.}$, on the wave functions of the term being concerned:

$$\langle 3,\pm 3\rangle V_{oct}\langle 3,\pm 3\rangle \tag{7.6.10}$$

$$\langle 3,\pm 2\rangle V_{oct}\langle 3,\pm 2\rangle \tag{7.6.11}$$

$$\langle 3,\pm 1\rangle V_{oct}\langle 3,\pm 1\rangle \tag{7.6.12}$$

$$\langle 3,0\rangle V_{oct}\langle 3,0\rangle \tag{7.6.13}$$

$$\langle 3,\pm 3\rangle V_{oct}\langle 3,\mp 1\rangle \tag{7.6.14}$$

$$\langle 3,\pm 2\rangle V_{oct}\langle 3,\mp 2\rangle \tag{7.6.15}$$

- The complete orbital combinations for an F term of single electron wave function of d^2, Chapter 2, p. 130:

$$\langle 3,\pm 3\rangle = \pm(\pm 2,\pm 1)$$

$$\langle 3,\pm 2\rangle = \pm(\pm 2,0)$$

$$\langle 3,\pm 1\rangle = \pm\sqrt{\frac{2}{5}}(\pm 1,0) \pm \sqrt{\frac{3}{5}}(\pm 2,\mp 1)$$

$$\langle 3,0\rangle = \sqrt{\frac{4}{5}}(1,-1) + \sqrt{\frac{1}{5}}(2,-2)$$

- Take the first matrix element (Eqs. 7.6.10 and 7.6.3):

$$\langle\langle 3,\pm 3\rangle|V_{oct}|\langle 3,\pm 3\rangle\rangle = \int(\pm 2,\pm 1)V_{oct}(\pm 2,\pm 1)\partial\tau$$

$$= \int(\pm 2)V_{oct}(\pm 2)\partial\tau + \int(\pm 1)V_{oct}(\pm 1)\partial\tau$$

From this chapter, p. 427:

$$H_{\pm 1,\pm 1} = \langle \pm 1|V_{oct}|\pm 1\rangle = -4D_q \tag{7.4.13}$$

$$H_{\pm 2,\pm 2} = \langle \pm 2|V_{oct}|\pm 2\rangle = D_q \tag{7.4.14}$$

$$H_{\langle 3,\pm 3\rangle,\langle 3,\pm 3\rangle} = D_q - 4D_q = -3D_q \tag{7.6.16}$$

- The second matrix element (Eqs. 7.6.11 and 7.6.3):

$$\langle\langle 3,\pm 2\rangle|V_{Oct.}|\langle 3,\pm 2\rangle\rangle = \int(\pm 2,0)V_{oct}(\pm 2,0)\partial\tau$$

$$= \int(\pm 2)V_{oct}(\pm 2)\partial\tau + \int(0)V_{oct}(0)\partial\tau$$

From:

$$H_{0,0} = \langle 0|V_{oct}|0\rangle = 6D_q \tag{7.4.12}$$

$$H_{\pm 2,\pm 2} = \langle \pm 2|V_{oct}|\pm 2\rangle = D_q \tag{7.4.14}$$

$$H_{\langle 3,\pm 2\rangle,\langle 3,\pm 2\rangle} = D_q + 6D_q = 7D_q \tag{7.6.17}$$

- The third matrix element (Eqs. 7.6.12 and 7.6.3):

$$\langle\langle 3,1\rangle|V_{oct}|\langle 3,1\rangle\rangle = \int\left\{\begin{array}{l} \left(\frac{2}{5}\right)\underbrace{(1,0)^*V_{oct}(1,0)}_{\text{term 1}} + \left(\frac{\sqrt{6}}{5}\right)\underbrace{(1,0)^*V_{oct}(2,-1)}_{\text{term 2}} + \\ \left(\frac{\sqrt{6}}{5}\right)\underbrace{(2,-1)^*V_{oct}(1,0)}_{\text{term 3}} + \left(\frac{3}{5}\right)\underbrace{(2,-1)^*V_{oct}(2,-1)}_{\text{term 4}} \end{array}\right\}\partial\tau$$

$$\int\left(\underbrace{(-1,0)^*V_{oct}(-1,0)}_{term\ 1}\right)\partial\tau = \int((1,0)^*V_{oct}(1,0))\partial\tau$$

$$= \int((1)^*V^1_{oct}(1))\partial\tau_1 + \int((0)^*V^2_{oct}(0))\partial\tau_2$$

From:

$$H_{0,0} = \langle 0|V_{\text{oct}}|0\rangle = 6D_q \quad (7.4.12)$$

$$H_{\pm1,\pm1} = \langle \pm1|V_{\text{oct}}|\pm1\rangle = -4D_q \quad (7.4.13)$$

$$H_{\langle 1,0\rangle,\langle 1,0\rangle} = H_{\langle -1.0\rangle,\langle -1,0\rangle} = -4D_q + 6D_q = 2D_q$$

$$\int\left(\underbrace{(\pm1,0)^* V_{\text{oct}}(\pm2,\mp1)}_{\text{term 2}}\right)\partial\tau = \int\left(\underbrace{(\pm2,\mp1)^* V_{\text{oct}}(\pm1,0)}_{\text{term 3}}\right)\partial\tau = 0$$

$$\int\left(\underbrace{(\pm2,\mp1)^* V_{\text{oct}}(\pm2,\mp1)}_{\text{term 4}}\right)\partial\tau = \int((\pm2)^* V^1_{\text{oct}}(\pm2))\partial\tau_1 + \int((\mp1)^* V^2_{\text{oct}}(\mp1))\partial\tau_2$$

$$H_{\pm1,\pm1} = \langle \pm1|V_{\text{oct}}|\pm1\rangle = -4D_q \quad (7.4.13)$$

$$H_{\pm2,\pm2} = \langle \pm2|V_{\text{oct}}|\pm2\rangle = D_q \quad (7.4.14)$$

$$H_{\langle \pm2,\mp1\rangle,\langle \pm2,\mp1\rangle} = D_q - 4D_q = -3D_q$$

So that

$$H_{\langle 3,1\rangle,\langle 3,1\rangle} = \langle\langle 3,1\rangle|V_{\text{oct}}|\langle 3,1\rangle\rangle = 2\left(\frac{2}{5}\right)D_q - 3\left(\frac{3}{5}\right)D_q = -D_q \quad (7.6.18)$$

- The fourth matrix element (Eqs. 7.6.13 and 7.6.3):

$$\langle\langle 3,0\rangle|V_{\text{oct}}|\langle 3,0\rangle\rangle = \int\left[\sqrt{\frac{4}{5}}(1,-1) + \sqrt{\frac{1}{5}}(2,-2)\right]^* V_{\text{oct}}\left[\sqrt{\frac{4}{5}}(1,-1) + \sqrt{\frac{1}{5}}(2,-2)\right]\partial\tau$$

$$= \int\left\{\begin{array}{l}\left(\frac{4}{5}\right)\underbrace{(1,-1)^* V_{\text{oct}}(1,-1)}_{\text{term 1}} + \left(\frac{2}{5}\right)\underbrace{(1,-1)^* V_{\text{oct}}(2,-2)}_{\text{term 2}=0} \\ +\left(\frac{2}{5}\right)\underbrace{(2,-2)^* V_{\text{oct}}(1,-1)}_{\text{term 3}=0} + \left(\frac{1}{5}\right)\underbrace{(2,-2)^* V_{\text{oct}}(2,-2)}_{\text{term 4}}\end{array}\right\}\partial\tau$$

$$\int\left(\underbrace{(1,-1)^* V_{\text{oct}}(1,-1)}_{\text{term 1}}\right)\partial\tau = \int((1)^* V^1_{\text{oct}}(1))\partial\tau_1 + \int((-1)^* V^2_{\text{oct}}(-1))\partial\tau_2$$

$$H_{\pm1,\pm1} = \langle \pm1|V_{\text{oct}}|\pm1\rangle = -4D_q \quad (7.4.13)$$

$$H_{\langle 1,-1\rangle,\langle 1,-1\rangle} = -4D_q - 4D_q = -8D_q$$

$$\int\left(\underbrace{(2,-2)^* V_{\text{oct}}(2,-2)}_{\text{term 4}}\right)\partial\tau = \int((2)^* V^1_{\text{oct}}(2))\partial\tau_1 + \int((-2)^* V^2_{\text{oct}}(-2))\partial\tau_2$$

$$H_{\pm2,\pm2} = \langle \pm2|V_{\text{oct}}|\pm2\rangle = D_q \quad (7.4.14)$$

$$H_{\langle 2,-2\rangle,\langle 2,-2\rangle} = D_q + D_q = 2D_q$$

then

$$H_{\langle 3,0\rangle,\langle 3,0\rangle} = \langle\langle 3,0\rangle|V_{\text{oct}}|\langle 3,0\rangle\rangle = -\left(\frac{8\times4}{5}\right)D_q - 0 - 0 + \left(\frac{2\times1}{5}\right)D_q = -6D_q \quad (7.6.19)$$

- The fifth matrix element (Eqs. 7.6.14 and 7.6.3):

$$\langle\langle 3, \pm 3\rangle|V_{oct}|\langle 3, \mp 1\rangle\rangle = \int (2,1)^* V_{oct}\left(\sqrt{\frac{2}{5}}(-1,0) + \sqrt{\frac{3}{5}}(-2,1)\right)\partial\tau$$

$$= \underbrace{\sqrt{\frac{2}{5}}\int (2,1)^* V_{oct}(-1,0)\partial\tau}_{\text{term 1}=0} + \underbrace{\sqrt{\frac{3}{5}}\int (2,1)^* V_{oct}(-2,1)\partial\tau}_{\text{term 2}}$$

$$\underbrace{\int (2,1)^* V_{oct}(-2,1)\partial\tau}_{\text{term 2}} = \int (2)^*(-2)\partial\tau \int (1)^* V_{oct}(1)\partial\tau + \int (1)^*(1)\partial\tau \int (2)^* V_{oct}(-2)\,\partial\tau$$

From:

$$H_{\pm1,\pm1} = \langle \pm 1|V_{oct}|\pm 1\rangle = -4D_q \quad (7.4.13)$$

$$H_{\pm2,\mp2} = \langle \pm 2|V_{oct}|\mp 2\rangle = 5D_q \quad (7.4.15)$$

$$H_{(2,1),(-2,1)} = 0 + 5D_q = 5D_q$$

$$\langle\langle 3, \pm 3\rangle|V_{oct}|\langle 3, \mp 1\rangle\rangle = \sqrt{\frac{2}{5}}\int (2,1)^* V_{oct}(-1,0)\partial\tau + \sqrt{\frac{3}{5}}\int (2,1)^* V_{oct}(-2,1)\partial\tau$$

$$H_{(3,\pm3),(3,\pm1)} = 0 + \sqrt{\frac{3}{5}} \times (5D_q) = \sqrt{15}D_q \quad (7.6.20)$$

- The sixth matrix element (Eqs. 7.6.15 and 7.6.3):

$$\langle\langle 3, \pm 2\rangle|V_{oct}|\langle 3, \mp 2\rangle\rangle = \int (2,0)^* V_{oct}(-2,0)\partial\tau$$

$$= \int (2)^*(-2)\partial\tau \int (0)^* V_{oct}(0)\,\partial\tau + \int (0)^*(0)\partial\tau \int (2)^* V_{oct}(-2)\,\partial\tau$$

$$H_{0,0} = \langle 0|V_{oct}|0\rangle = 6D_q \quad (7.4.12)$$

$$H_{\pm2,\mp2} = \langle \pm 2|V_{oct}|\mp 2\rangle = 5D_q \quad (7.4.15)$$

$$H_{(3,\pm2),(3,\mp2)} = 0 + 5D_q = 5D_q \quad (7.6.12)$$

- From the above, the expressions for the F term orbital wave functions are obtained:

$$H_{(3,\pm3),(3,\pm3)} = \langle\langle 3, \pm 3\rangle|V_{oct}|\langle 3, \pm 3\rangle\rangle = -3D_q \quad (7.6.16)$$

$$H_{(3,\pm2),(3,\pm2)} = \langle\langle 3, \pm 2\rangle|V_{oct}|\langle 3, \pm 2\rangle\rangle = 7D_q \quad (7.6.17)$$

$$H_{(3,\pm1),(3,\pm1)} = \langle\langle 3, \pm 1\rangle|V_{oct}|\langle 3, \pm 1\rangle\rangle = -D_q \quad (7.6.18)$$

$$H_{(3,0),(3,0)} = \langle\langle 3, 0\rangle|V_{oct}|\langle 3, 0\rangle\rangle = -6D_q \quad (7.6.19)$$

$$H_{(3,\pm3),(3,\mp1)} = \langle\langle 3, \pm 3\rangle|V_{oct}|\langle 3, \mp 1\rangle\rangle = \sqrt{15}D_q \quad (7.6.20)$$

$$H_{(3,\pm2),(3,\mp2)} = \langle\langle 3, \pm 2\rangle|V_{oct}|\langle 3, \mp 2\rangle\rangle = 5D_q \quad (7.6.21)$$

- All the other matrix elements, Eq. (7.6.8), of the series are zero and the secular determinant for the action of V_{oct} on the F-orbital wave functions:

$$\begin{array}{c|ccccccc|}
 & \langle 3,3\rangle & \langle 3,2\rangle & \langle 3,1\rangle & \langle 3,0\rangle & \langle 3,-1\rangle & \langle 3,-2\rangle & \langle 3,-3\rangle \\
\langle 3\rangle & -3Dq-E & & & & \sqrt{15Dq} & & \\
\langle 2\rangle & & 7Dq-E & & & & 5Dq & \\
\langle 1\rangle & & & -Dq-E & & & & \sqrt{15Dq} \\
\langle 0\rangle & & & & -6Dq-E & & & \\
\langle -1\rangle & \sqrt{15Dq} & & & & -Dq-E & & \\
\langle -2\rangle & & 5Dq & & & & 7Dq-E & \\
\langle -3\rangle & & & \sqrt{15Dq} & & & & -3Dq-E
\end{array} = 0 \quad (7.6.22)$$

- This reduced to subdeterminants:

$$\begin{matrix} \langle 3,\pm 3\rangle & \langle 3,\pm 1\rangle \end{matrix} \qquad \begin{matrix} \langle 3,2\rangle & \langle 3,-2\rangle \end{matrix} \qquad \langle 3,0\rangle$$

$$\begin{vmatrix} -3Dq-E & \sqrt{15}\,Dq \\ \sqrt{15}\,Dq & -Dq-E \end{vmatrix}=0, \quad \begin{vmatrix} 7Dq-E & 5\,Dq \\ 5\,Dq & 7Dq-E \end{vmatrix}=0, \text{ and } |-6Dq-E|=0$$

- Solving these determinants to obtain the energy of each level:

$$\langle 3,0\rangle$$

$$|-6Dq-E|=0$$

$$E=-6D_q \text{ occurring 1 times} \tag{7.6.23}$$

$$\begin{matrix} \langle 3,\pm 3\rangle & \langle 3,\mp 1\rangle \end{matrix}$$

$$\begin{vmatrix} -3Dq-E & \sqrt{15}\,Dq \\ \sqrt{15}\,Dq & -Dq-E \end{vmatrix}=0$$

$$(-3Dq-E)(-Dq-E)-15Dq^2=0$$

$$3Dq^2+4EDq+E^2-15Dq^2=0$$

$$E^2+4EDq-12Dq^2=0$$

$$(E+6Dq)(E-2Dq)=0$$

$$E=2D_q \text{ occurring 2 times} \tag{7.6.24}$$

$$E=-6D_q \text{ occurring 2 times} \tag{7.6.25}$$

$$\begin{matrix} \langle 3,2\rangle & \langle 3,-2\rangle \end{matrix}$$

$$\begin{vmatrix} 7Dq-E & 5Dq \\ 5Dq & 7Dq-E \end{vmatrix}=0$$

$$(7Dq-E)^2-25Dq^2=0$$

$$(7Dq-E+5Dq)(7Dq-E-5Dq)=0$$

$$E=12D_q \text{ occurring 1 time} \tag{7.6.26}$$

$$E=2D_q \text{ occurring 1 time} \tag{7.6.27}$$

- The corresponding wave functions from the sets of secular equations are obtained exactly as outlined for:
  At $E=-6D_q$

$$\begin{matrix} \langle 3,\pm 3\rangle & \langle 3\mp 1\rangle \end{matrix}$$

$$\begin{vmatrix} -3Dq-E & \sqrt{15}Dq \\ \sqrt{15}Dq & -Dq-E \end{vmatrix}=0$$

$$(-3Dq+6Dq)c_1+\sqrt{15}\,Dqc_2=0$$

$$3c_1+\sqrt{15}c_2=0$$

$$c_2=-\frac{3}{\sqrt{15}}c_1$$

$$c_1^2+c_2^2=1$$

$$c_1^2+\frac{9}{15}c_1^2=1$$

$$\frac{24}{15}c_1^2 = 1$$

$$c_1 = \frac{\sqrt{15}}{\sqrt{24}}$$

$$c_2 = -\frac{3}{\sqrt{24}}$$

The corresponding wave function:

$$c_1\langle \pm 3\rangle + c_2\langle \mp 1\rangle$$

$$\frac{\sqrt{15}}{\sqrt{24}}\langle \pm 3\rangle - \frac{3}{\sqrt{24}}\langle \mp 1\rangle$$

$$\frac{1}{\sqrt{24}}\left(\sqrt{15}\langle \pm 3\rangle - 3\langle \mp 1\rangle\right)$$

- Thus:
  - At $E = -6D_q$, the corresponding wave function:

$$\frac{1}{\sqrt{24}}\left(\sqrt{15}\langle 3\rangle - 3\langle -1\rangle\right) \quad (7.6.28)$$

$$\langle 0\rangle \quad (7.6.29)$$

$$\frac{1}{\sqrt{24}}\left(\sqrt{15}\langle -3\rangle - 3\langle 1\rangle\right) \quad (7.6.30)$$

  - At $E = 2D_q$, the corresponding wave function:

$$\frac{1}{\sqrt{24}}\left(3\langle 3\rangle + \sqrt{15}\langle -1\rangle\right) \quad (7.6.31)$$

$$\sqrt{\frac{1}{2}}(\langle 2\rangle - \langle -2\rangle) \quad (7.6.32)$$

$$\frac{1}{\sqrt{24}}\left(3\langle -3\rangle + \sqrt{15}\langle 1\rangle\right) \quad (7.6.33)$$

  - At $E = 12D_q$, the corresponding wave function:

$$\sqrt{\frac{1}{2}}(\langle 2\rangle + \langle -2\rangle) \quad (7.6.34)$$

If the corresponding wave-functions of ^{3}F term in octahedral crystal field:

At $E = -6D_q$: $\frac{1}{\sqrt{24}}\left(\sqrt{15}\langle 3\rangle - 3\langle -1\rangle\right)$, $\langle 0\rangle$, and $\frac{1}{\sqrt{24}}\left(\sqrt{15}\langle -3\rangle - 3\langle 1\rangle\right)$

At $E = 2D_q$: $\frac{1}{\sqrt{24}}\left(3\langle 3\rangle + \sqrt{15}\langle -1\rangle\right)$, $\sqrt{\frac{1}{2}}(\langle 2\rangle - \langle -2\rangle)$, and $\frac{1}{\sqrt{24}}\left(3\langle -3\rangle + \sqrt{15}\langle 1\rangle\right)$

At $E = 12D_q$: $\sqrt{\frac{1}{2}}(\langle 2\rangle + \langle -2\rangle)$

then draw the splitting diagram of ^{3}F term in $V_{oct.}$, and show:

- **the energy separations among these levels;**
- **the degeneracy of each level;**
- **the shift in the center of gravity;**
- **the effect of the multiplicity;**
- **the correlation in the splitting of F term among d^n, d^{n+5}, d^{5-n}, and d^{10-n} configurations; and**
- **the splitting of ^{4}F term of d^7, ^{4}F term of d^3, and ^{3}F term of d^8.**

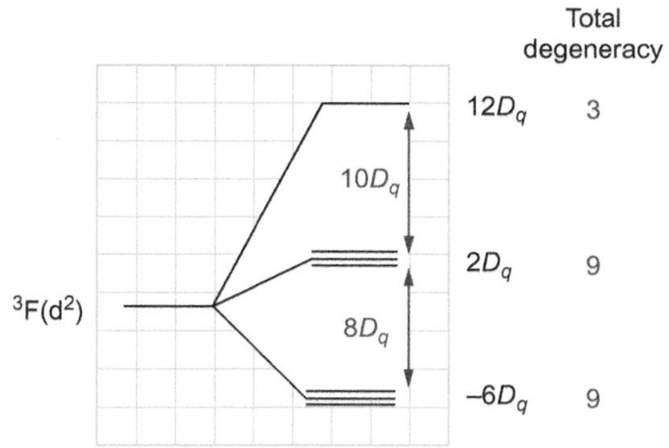

FIG. 7.10 The splitting of ^3F term of d^2 configuration in weak octahedral crystal field.

- Fig. 7.10 illustrates the splitting diagram of the ^3F term of d^2 by V_{oct}.
- The center of gravity of the F term $= 3 \times -6D_q + 3 \times 2D_q + 1 \times 12D_q = 0$
  There is no shift in the center of the gravity by the crystal field.
- There is a parallel behavior between the splitting of f-orbital set and F term, but reversed in sign.
- The splitting of F term is independent of the multiplicity.
- For d^{n+5}, the splitting of F term is identical to that of d^n ($n<5$)
- The splitting of F term is inverted relative to that of d^n for d^{5-n} and d^{10-n}.
- The splitting of ^4F term of d^7 (d^{2+5}) is similar to that of ^3F of d^2 configuration, but is inverted for ^4F term of $d^3(d^{5-2})$ and ^3F term of d^8 (d^{10-2}).

The Effect of a Cubic Crystal Field on G, H, and I

Fig. 7.11 **shows qualitatively the orbital splitting of ^1G, ^2H, and ^1I terms by weak $V_{Oct.}$. Write your comments on this.**

- G, H, and I terms may be split using the same approach as the terms previously handled, the splitting diagrams are:
  The numbers specify the total degeneracy of the level (spin-times-orbital).
- ^1G, ^2H, and ^1I are excited terms, and do not take place as ground terms of d-electron configurations, thus, the correlation $d^{5+n} \equiv d^n$, and $d^{5-n} \equiv d^{10-n}$, which are reversed relative to d^n, cannot be used.
- The presence of the crystal field does not alter the center of gravity of these terms.

If the splitting of the ground terms of d-orbitals in weak crystal field follows:
$d^{5+n} \equiv d^n$, $d^{5-n} \equiv d^{10-n}$ are inverted relative to d^n
Draw diagrams to show the splitting of the ground terms of d-orbitals, in an octahedral and in a tetrahedral weak crystal field.

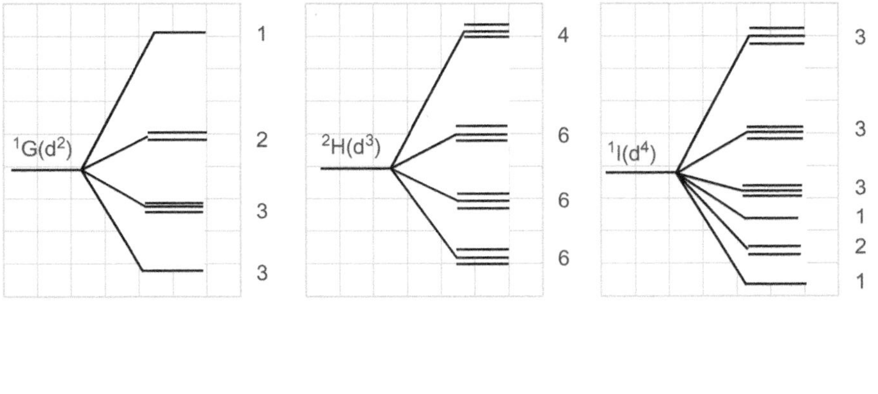

FIG. 7.11 The splitting of ^1G, ^2H, and ^1I terms by weak $V_{Oct.}$.

- For: d^0, d^5, and d^{10} configurations:

$$d^{5+0} \equiv d^0, \quad d^{5-0} \equiv d^{10-0} \equiv d^0 \quad \text{or} \quad d^0 \equiv d^5 \equiv d^{10}$$

Each has a nondegenerate ground state: ^6S for d^5, and ^1S for d^0 and d^{10}.

- For: d^1, d^4, d^6, and d^9 configurations, the splitting of the ground states are summarized in Fig. 7.12.
- For: d^2, d^3, d^7, and d^8 configurations, the splitting of the ground states are summarized in Fig. 7.13.
- The ligand field from tetrahedron or a cube the splitting diagrams are inverted relative to those for $V_{oct.}$, and the splitting is reduced by the appropriate factor, −4/9 and −8/9 time those of V_{oct}.

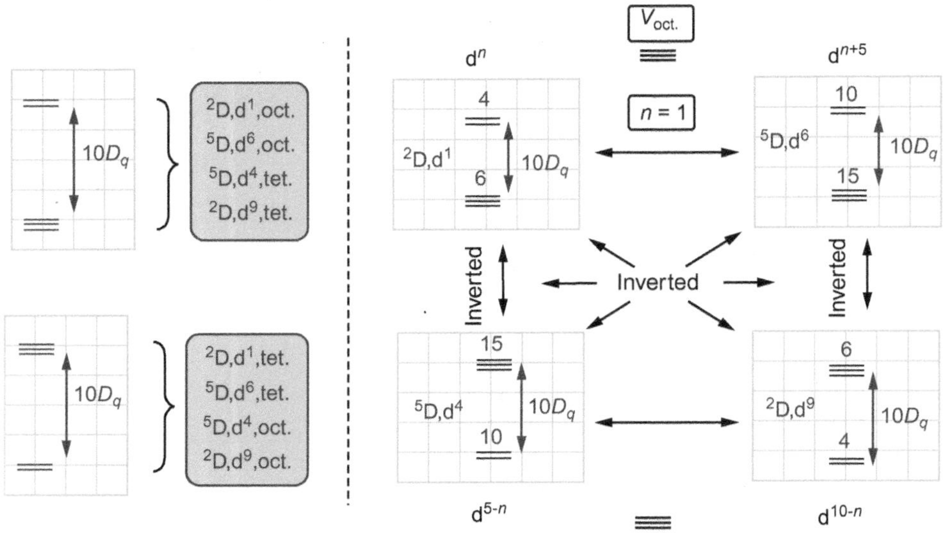

FIG. 7.12 The splitting pattern of various ground terms of d^1, d^4, d^6, and d^9 in octahedral field. The numbers specify the total degeneracy (total degeneracy = spin multiplicity-times-orbital degeneracy).

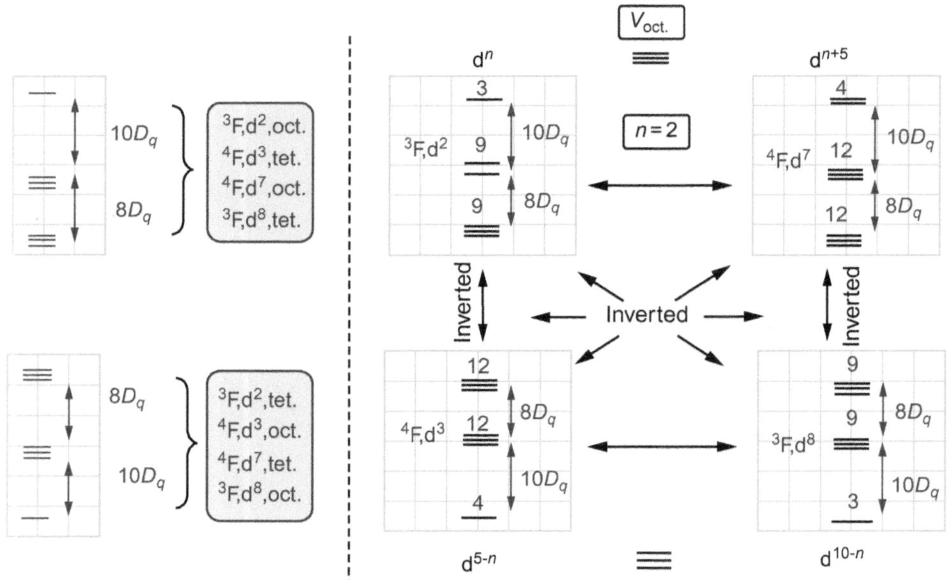

FIG. 7.13 The splitting pattern of various ground terms of d^2, d^3, d^7, and d^8 in octahedral field. The numbers specify the total degeneracy of the level (spin-times-orbital).

7.7 STRONG FIELD APPROACH

How does the strong field differ from the weak field approach?

- In the strong field approach:
  - real orbitals are used; and
  - interelectronic repulsions are included.

Determinantal Wave Functions

How should you write the determinantal form that describes many-electron functions? What is the d^2 wave function of spin +1/2 occupies each of d_{xz} and d_{yz} orbitals? What does "spinors" mean?

- Many-electron functions may written in a determinantal form, for n electrons:

$$\psi(A) = E\frac{1}{\sqrt{n!}}\begin{vmatrix} a_1(1) & a_2(1) & \dots & a_2(n) \\ a_1(2) & a_2(2) & \dots & a_2(n) \\ \vdots & \vdots & \ddots & \vdots \\ a_1(n) & a_2(3) & \dots & a_2(n) \end{vmatrix} \quad (7.7.1)$$

- For d_{xz} and $d_{yz} \rightarrow xz^+$ and yz^+
  The superscript + implies an electron of spin +1/2 occupies each of d_{xz} and d_{yz}.
  xz^+ and yz^+ are called spinors, spin-orbitals, because they describe both orbit and spin at the same time.

$$\psi_i = \frac{1}{\sqrt{2}}\begin{vmatrix} xz^+(1) & yz^+(1) \\ xz^+(2) & yz^+(2) \end{vmatrix} = \frac{1}{\sqrt{2}}(xz^+(1)yz^+(2) - yz^+(1)xz^+(2))$$

The Determinantal Wave Functions of d^2 in Strong Field of Tetragonal Structure, Trans-ML_4Z_2

What are the determinantal wave functions for the spin-triplet of d^2 configurations in a strong field of tetragonal structure, trans-ML_4Z_2?
What is the symmetry designation of each of these basis states?

- The energy level diagram (Fig. 7.14) will be used to describe the strong field orbital populations of various energy levels.

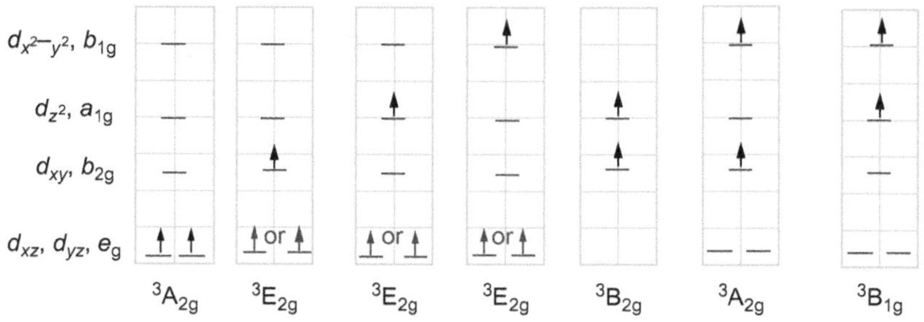

FIG. 7.14 The spin-triplet configurations of d^2 electrons, in a tetragonal structure.

- The determinantal wave functions and the symmetry designation of d^2 in spin-triplet configurations:

$B_1 = (xz^+, yz^+)$	$^3A_{2g}$
$B_2 = \frac{1}{\sqrt{2}}[(xz^+, xy^+) + (yz^+, xy^+)]$	^3E_g
$B_3 = \frac{1}{\sqrt{2}}[(xz^+, xy^+) - (yz^+, xy^+)]$	^3E_g
$B_4 = \frac{1}{\sqrt{2}}\left[\left(xz^+, z^{2+}\right) + \left(yz^+, z^{2+}\right)\right]$	^3E_g
$B_5 = \frac{1}{\sqrt{2}}\left[\left(xz^+, z^{2+}\right) - \left(yz^+, z^{2+}\right)\right]$	^3E_g
$B_6 = \frac{1}{\sqrt{2}}\left[\left(xz^+, x^2 - y^{2+}\right) + \left(yz^+, x^2 - y^{2+}\right)\right]$	^3E_g
$B_7 = \frac{1}{\sqrt{2}}\left[\left(xz^+, x^2 - y^{2+}\right) - \left(yz^+, x^2 - y^{2+}\right)\right]$	^3E_g
$B_8 = xy^+,\ z^{2+}$	$^3B_{2g}$
$B_9 = xy^+, x^2 - y^{2+}$	$^3A_{2g}$
$B_{10} = z^{2+},\ x^2 - y^{2+}$	$^3B_{1g}$

  - Only spin-triplet configurations are considered.
- Group theoretical direct products produce the symmetry of each energy level.
  - In B_1: there is no distinction between placing an electron in d_{xz} or in d_{yz}.
    Similarly in B_9: there is no distinction between placing an electron in d_{xy} or in $d_{x^2-y^2}$.
  - In B_2 to B_7: linear combinations are the appropriate symmetries.
  - In B_8 and B_{10}: direct product treatments

The Symmetry and the Energy of Determinant Wave Functions of D^2 in a Strong Field of Trans-ML_4Z_2

What are the energy of each of the basis states, $B_1, B_2, \ldots, B_{10}$ (of the previous problem), using the Kronecker delta function:

$$\langle abc|F|abc\rangle = \langle a|F|a\rangle\langle b|b\rangle\langle c|c\rangle + \langle b|F|b\rangle\langle a|a\rangle\langle c|c\rangle + \langle c|F|c\rangle\langle a|a\rangle\langle b|b\rangle, \quad (7.7.2)$$

and the crystal field data, p. 440:

$$\langle x^2 - y^2|V_{D_{4h}}|x^2 - y^2\rangle = 6Dq + 2Ds - Dt$$

$$\langle z^2|V_{D_{4h}}|z^2\rangle = 6Dq - 2Ds - 6Dt$$

$$\langle xy|V_{D_{4h}}|xy\rangle = -4Dq + 2Ds - Dt$$

$$\langle xz|V_{D_{4h}}|zy\rangle = \langle yz|V_{D_{4h}}|yz\rangle = -4Dq - Ds + 4Dt$$

- The energies of these basis states are obtained using:
  - Kronecker delta function:

$$\langle abc|F|abc\rangle = \langle a|F|a\rangle\langle b|b\rangle\langle c|c\rangle + \langle b|F|b\rangle\langle a|a\rangle\langle c|c\rangle + \langle c|F|c\rangle\langle a|a\rangle\langle b|b\rangle, \quad (7.7.2)$$

  - the crystal field data, p. 440:

$$\int \left(d_{x^2-y^2}\right) V_{V_{D_{4h}}} \left(d_{x^2-y^2}\right) d\tau = \langle x^2 - y^2|V_{D_{4h}}|x^2 - y^2\rangle = 6Dq + 2Ds - Dt$$

$$\int (d_{z^2}) V_{V_{D_{4h}}} (d_{z^2})\, d\tau = \langle z^2 | V_{D_{4h}} | z^2 \rangle = 6Dq - 2Ds - 6Dt$$

$$\int (d_{xy}) V_{V_{D_{4h}}} (d_{xy})\, d\tau = \langle xy | V_{D_{4h}} | xy \rangle = -4Dq + 2Ds - Dt$$

$$\int (d_{xz}) V_{V_{D_{4h}}} (d_{xz})\, d\tau = \int (d_{yz}) V_{D_{4h}} (d_{yz})\, d\tau = \langle xz | V_{D_{4h}} | zy \rangle = \langle yz | V_{D_{4h}} | yz \rangle = -4Dq - Ds + 4Dt$$

$$\begin{aligned} \langle B_1 | V_{V_{D_{4h}}} | B_1 \rangle &= \langle xz^+, yz^+ | V_{D_{4h}} | xz^+, yz^+ \rangle \\ &= \langle xz^+ | V_{V_{D_{4h}}} | xz^+ \rangle \langle yz^+ | yz^+ \rangle + \langle yz^+ | V_{V_{D_{4h}}} | yz^+ \rangle \langle xz^+ | xz^+ \rangle \\ &= -4Dq - Ds + 4Dt - 4Dq - Ds + 4Dt \\ &= -8Dq - 2Ds + 8Dt \end{aligned} \tag{7.7.3}$$

$$\begin{aligned} \langle B_2 | V_{V_{D_{4h}}} | B_2 B_2 \rangle &= \left\langle \frac{1}{\sqrt{2}} \frac{1}{\sqrt{2}} [(xz^+, xy^+) + (yz^+, xy^+)] | V_{D_{4h}} | \frac{1}{\sqrt{2}} [(xz^+, xy^+) + (yz^+, xy^+)] \right\rangle \\ &= \left\langle \frac{1}{\sqrt{2}} [(xz^+, xy^+)] | V_{D_{4h}} | \frac{1}{\sqrt{2}} [(xz^+, xy^+)] \right\rangle \\ &\quad + \left\langle \frac{1}{\sqrt{2}} [(yz^+, xy^+)] | V_{D_{4h}} | \frac{1}{\sqrt{2}} [+(yz^+, xy^+)] \right\rangle \\ &= \langle xz^+ | V_{D_{4h}} | xz^+ \rangle \langle xy^+ | xy^+ \rangle + \langle xy^+ | V_{D_{4h}} | xy^+ \rangle \langle yz^+ | yz^+ \rangle \\ &\quad + \langle yz^+ | V_{D_{4h}} | yz^+ \rangle \langle xy^+ | xy^+ \rangle + \langle xy^+ | V_{D_{4h}} | xy^+ \rangle \langle yz^+ | yz^+ \rangle \\ &= \frac{1}{2} [-4Dq - Ds + 4Dt - 4Dq + 2Ds - Dt - 4Dq - Ds + 4Dt - 4Dq + 2Ds - Dt] \\ &= -8Dq + Ds + 3Dt \end{aligned} \tag{7.7.4}$$

Thus,

$\langle B_1 \| V_{D_{4h}} \| B_1 \rangle = 8Dq - 2Ds + 8Dt$	$^3A_{2g}$
$\langle B_2 \| V_{D_{4h}} \| B_2 \rangle = -8Dq + Ds + 3Dt$	$^3E_{2g}$
$\langle B_3 \| V_{D_{4h}} \| B_3 \rangle = -8Dq + Ds + 3Dt$	$^3E_{2g}$
$\langle B_4 \| V_{D_{4h}} \| B_4 \rangle = 2Dq - 3Ds - 2Dt$	$^3E_{2g}$
$\langle B_5 \| V_{D_{4h}} \| B_5 \rangle = 2Dq - 3Ds - 2Dt$	$^3E_{2g}$
$\langle B_6 \| V_{D_{4h}} \| B_6 \rangle = 2Dq + Ds + 3Dt$	$^3E_{2g}$
$\langle B_7 \| V_{D_{4h}} \| B_7 \rangle = 2Dq + Ds + 3Dt$	$^3E_{2g}$
$\langle B_8 \| V_{D_{4h}} \| B_8 \rangle = 2Dq - 7Dt$	$^3B_{2g}$
$\langle B_9 \| V_{D_{4h}} \| B_9 \rangle = 2Dq + 4Ds - 2Dt$	$^3A_{2g}$
$\langle B_{10} \| V_{D_{4h}} \| B_{10} \rangle = 12Dq - 7Dt$	$^3B_{1g}$

The Appropriate Hamiltonian in Strong Field

Why do we care about the symmetry of the energy levels? Show that

$$H = \langle B_i | V_{D_{4h}} | B_j \rangle + \left\langle B_i \left| \frac{e^2}{r_{ij}} \right| B_j \right\rangle$$

where r_{ij} is distance between the ith and jth electrons, and B is an energy term, i and j take values from 1 to 10. Use basis states, B_1, B_2,..., B_{10} of the previous problem.

- When the symmetry designation takes place only once in a specified basis, as for $^3B_{1g}$ and $^3B_{2g}$, the energy are generated directly and no determinants are needed.
- States of identical symmetry may mixed by crystal field operator. This mixing is regulated by the magnitudes of the crystal field radial integrals (by off-diagonal matrix elements). The two $^3A_{2g}$ states are mixed in a 2 × 2 determinant, and the six $^3E_{2g}$ states are mixed in two equivalent 3 × 3 determinants.
- The expected Hamiltonian of free ion:

$$H = \underbrace{\left(-\frac{h^2}{8\pi^2 m}\right)\sum \nabla_i^2}_{\text{Sum of the kinetic energy over all electrons}} - \underbrace{\sum \frac{ze^2}{r}}_{\text{Sum the attraction of all the electrons by the nucleus}} + \underbrace{\sum \frac{e^2}{r_{ij}}}_{\text{Interelectronic repulsion}} + \underbrace{\sum \zeta_i(r_i) l_i s_i}_{\text{Spin – orbital coupling}} \quad (7.7.5)$$

where:
z is the effective nuclear charge;
r_i is the distance of the ith electron from the nucleus; and
r_{ij} is distance between the ith and jth electrons.

After ligation, the Hamiltonian is modified:

$$H = \underbrace{\left(-\frac{h^2}{8\pi^2 m}\right)\sum \nabla_i^2}_{\text{Sum of the kinetic energy over all electrons}} - \underbrace{\sum \frac{ze^2}{r}}_{\text{Sum the attraction of all the electrons by the nucleus}} + \underbrace{\sum \frac{e^2}{r_{ij}}}_{\text{Interelectronic repulsion}} + \underbrace{\sum \zeta_i(r_i) l_i s_i}_{\text{Spin – orbital coupling}} + \underbrace{V_i}_{\text{the energy of interaction between electrons and crystal field}} \quad (7.7.6)$$

Excluding kinetic, potential, and spin-orbital coupling terms that are common to all electrons, the appropriate Hamiltonian for two-electron function becomes

$$H = \sum \frac{e^2}{r_{ij}} + V_i$$

and the expanded Hamiltonian:

$$H = \langle B_i | V_{V_{D4h}} | B_j \rangle + \left\langle B_i \left| \frac{e^2}{r_{ij}} \right| B_j \right\rangle \quad (7.7.7)$$

The Diagonal Interelectronic Repulsion

Consider the determinantal functions (a^+,b^+) and (a^+,b^-), + implies an electron of spin +1/2 occupies either one of *a*- or *b*-orbital, implies an electron of spin −1/2.

How can the diagonal interelectronic repulsion for these two states, $\left\langle B_i \left| \frac{e^2}{r_{ij}} \right| B_j \right\rangle$, be evaluated?

- The diagonal interelectronic repulsion of (a^+,b^+):

$$E(a^+, b^+) = \left\langle (a^+, b^+) \left| \frac{e^2}{r_{ij}} \right| (a^+, b^+) \right\rangle$$

$$\text{if: } (a^+, b^+) = \frac{1}{\sqrt{2}} \begin{vmatrix} a^+(1) & b^+(1) \\ a^+(2) & b^+(2) \end{vmatrix} = \frac{1}{\sqrt{2}} (a^+(1)b^+(2) - b^+(1)a^+(2)) \quad (7.7.1)$$

then:

$$E(a^+, b^+) = \frac{1}{2} \left[\left\langle (a^+(1)b^+(2) - b^+(1)a^+(2)) \left| \frac{e^2}{r_{ij}} \right| (a^+(1)b^+(2) - b^+(1)a^+(2)) \right\rangle \right]$$

$$= \frac{1}{2} \left\{ \begin{array}{c} \left\langle (a^+(1)b^+(2)) \left| \frac{e^2}{r_{ij}} \right| (a^+(1)b^+(2)) \right\rangle + \\ \left\langle (b^+(1)a^+(2)) \left| \frac{e^2}{r_{ij}} \right| (b^+(1)a^+(2)) \right\rangle - 2 \left\langle (a^+(1)b^+(2)) \left| \frac{e^2}{r_{ij}} \right| (b^+(1)a^+(2)) \right\rangle \end{array} \right\}$$

Using Kronecker delta function:

$$\text{Then: } \left\langle (a^+(1)b^+(2))\left|\frac{e^2}{r_{ij}}\right|(a^+(1)b^+(2))\right\rangle = \left\langle (a(1)b(2))\left|\frac{e^2}{r_{ij}}\right|(a(1)b(2))\right\rangle \cdot \langle +(1)|+(1)\rangle \cdot \langle +(2)|+(2)\rangle$$
$$= \left\langle (a(1)b(2))\left|\frac{e^2}{r_{ij}}\right|(a(1)b(2))\right\rangle, \text{ and}$$

$$\left\langle (a^+(1)b^+(2))\left|\frac{e^2}{r_{ij}}\right|(b^+(1)a^+(2))\right\rangle = \left\langle (a(1)b(2))\left|\frac{e^2}{r_{ij}}\right|(b(1)a(2))\right\rangle \cdot \langle +(1)|+(1)\rangle \cdot \langle +(2)|+(2)\rangle$$
$$= \left\langle (a(1)b(2))\left|\frac{e^2}{r_{ij}}\right|(b(1)a(2))\right\rangle, \text{ and if:}$$

$$J(a,b) = \left\langle (a(1)b(2))\left|\frac{e^2}{r_{ij}}\right|(a(1)b(2))\right\rangle \text{ is the Coulomb integral} \tag{7.7.8}$$

$$K(a,b) = \left\langle (a(1)b(2))\left|\frac{e^2}{r_{ij}}\right|(b(1)a(2))\right\rangle \text{ is the exchange integral} \tag{7.7.9}$$

Then:

$$E(a^+,b^+) = J(a,b) - K(a,b) \tag{7.7.10}$$

- The diagonal interelectronic repulsion of (a^+,b^-):

$$E(a^+,b^-) = \left\langle (a^+,b^-)\left|\frac{e^2}{r_{ij}}\right|(a^+,b^-)\right\rangle$$

$$\text{if: } (a^+,b^-) = \frac{1}{\sqrt{2}}(a^+(1)b^-(2) - b^-(1)a^+(2), \text{ then:}$$

$$E(a^+,b^-) = \frac{1}{2}\left[\left\langle (a^+(1)b^-(2) - b^-(1)a^+(2))\left|\frac{e^2}{r_{ij}}\right|(a^+(1)b^-(2) - b^-(1)a^+(2))\right\rangle\right]$$
$$= \frac{1}{2}\left\{\begin{array}{c}\left\langle (a^+(1)b^-(2))\left|\frac{e^2}{r_{ij}}\right|(a^+(1)b^-(2))\right\rangle + \\ \left\langle (b^-(1)a^+(2))\left|\frac{e^2}{r_{ij}}\right|(b^-(1)a^+(2))\right\rangle - 2\left\langle (a^+(1)b^-(2))\left|\frac{e^2}{r_{ij}}\right|(b^-(1)a^+(2))\right\rangle\end{array}\right\}$$

$$\text{if: } \left\langle (a^+(1)b^-(2))\left|\frac{e^2}{r_{ij}}\right|(a^+(1)b^-(2))\right\rangle = \left\langle (a(1)b(2))\left|\frac{e^2}{r_{ij}}\right|(a(1)b(2))\right\rangle \cdot \langle +(1)|+(1)\rangle \cdot \langle -(2)|-(2)\rangle$$
$$= \left\langle (a(1)b(2))\left|\frac{e^2}{r_{ij}}\right|(a(1)b(2))\right\rangle, \text{ and}$$

$$\text{if: } \left\langle (a^+(1)b^-(2))\left|\frac{e^2}{r_{ij}}\right|(b^-(1)a^+(2))\right\rangle = \left\langle (a(1)b(2))\left|\frac{e^2}{r_{ij}}\right|(b(1)a(2))\right\rangle \cdot \langle +(1)|-(1)\rangle \cdot \langle -(2)|+(2)\rangle = 0$$

then

$$E(a^+,b^-) = J(a,b) \tag{7.7.11}$$

If

$J(a,b)$ is the Coulomb integral

$K(a,b)$ is the exchange integral

and the interelectronic repulsion is given by

$$E(a^+,b^+) = J(a,b) - K(a,b), \text{ and}$$
$$E(a^+,b^-) = J(a,b), \text{ and}$$

$$J(xy,xy) = J(xz,xz) = J(yz,yz) = J(x^2 - y^2) = J(z^2, z^2) = A + 4B + 3C$$

$$J(xy,yz) = J(xy,xz) = J(xz,yz) = J(x^2 - y^2,xz) = J(x^2 - y^2,yz) = A - 2B + C$$

$$J(z^2, xz) = J(z^2, yz) = A + 2B + C$$

$$J(z^2, xy) = J(z^2,x^2 - y^2) = A - 4B + C$$

$$J(x^2 - y^2,xy) = A + 4B + C$$

$$K(xy,xz) = K(xy,yz) = K(xz,yz) = K(x^2 - y^2,yz) = K(x^2 - y^2,xz) = 3B + C$$

$$K(z^2,x^2 - y^2) = K(z^2, xy) = 4B + C$$

$$K(z^2, yz) = K(z^2, xz) = B + C$$

$$K(x^2 - y^2,xy) = C$$

Evaluate the diagonal interelectronic repulsion for various bases terms of the d^2 configuration in a tetragonal structure.

To what do the parameters A, B, and C refer?

- A number of complicated integrals are involved in the calculation of the interelectronic repulsions. These integrals can be grouped into three specific types of combinations, which are called Racah parameters A, B, and C.
  A is roughly the same for all metals;
  B is approximately the bond strength between metal and ligand; and
  C is about $1/B$.
- The determinantal wave functions of d^2 in spin-triplet configurations:

$B_1 = (xz^+,yz^+)$ $\quad {}^3A_{2g}$

$B_2 = \frac{1}{\sqrt{2}}[(xz^+,xy^+) + (yz^+,xy^+)]$ $\quad {}^3E_g$

$B_3 = \frac{1}{\sqrt{2}}[(xz^+,xy^+) - (yz^+,xy^+)]$ $\quad {}^3E_g$

$B_4 = \frac{1}{\sqrt{2}}\left[\left(xz^+,z^{2+}\right) + \left(yz^+,z^{2+}\right)\right]$ $\quad {}^3E_g$

$B_5 = \frac{1}{\sqrt{2}}\left[\left(xz^+,z^{2+}\right) - \left(yz^+,z^{2+}\right)\right]$ $\quad {}^3E_g$

$B_6 = \frac{1}{\sqrt{2}}\left[\left(xz^+,x^2 - y^{2+}\right) + \left(yz^+,x^2 - y^{2+}\right)\right]$ $\quad {}^3E_g$

$B_7 = \frac{1}{\sqrt{2}}\left[\left(xz^+,x^2 - y^{2+}\right) - \left(yz^+,x^2 - y^{2+}\right)\right]$ $\quad {}^3E_g$

$B_8 = xy^+,z^{2+}$ $\quad {}^3B_{2g}$

$B_9 = xy^+,x^2 - y^{2+}$ $\quad {}^3A_{2g}$

$B_{10} = z^{2+},x^2 - y^{2+}$ $\quad {}^3B_{1g}$

And
$E(a^+, b^+) = J(a,b) - K(a,b)$, and
$E(a^+, b^-) = J(a,b)$
Then:

$$\left\langle B_1\left|\frac{e^2}{r_{ij}}\right|B_1\right\rangle = \left\langle (xz^+,yz^+)\left|\frac{e^2}{r_{ij}}\right|(xz^+,yz^+)\right\rangle = J(xz, yz) - K(xz,yz) = A - 2B + C - 3B - C = A - 5B$$

$$\left\langle B_2\left|\frac{e^2}{r_{ij}}\right|B_2\right\rangle=\frac{1}{2}\left\{\left\langle[(xz^+,xy^+)+(yz^+,xy^+)]\left|\frac{e^2}{r_{ij}}\right|[(xz^+,xy^+)+(yz^+,xy^+)]\right\rangle\right\}$$

$$=\frac{1}{2}\left\{\begin{array}{c}\left\langle(xz^+,xy^+)\left|\frac{e^2}{r_{ij}}\right|(xz^+,xy^+)\right\rangle+\left\langle(yz^+,xy^+)\left|\frac{e^2}{r_{ij}}\right|(yz^+,xy^+)\right\rangle\\+2\left\langle(xz^+,xy^+)\left|\frac{e^2}{r_{ij}}\right|(yz^+,xy^+)\right\rangle\end{array}\right\}$$

$$=\frac{1}{2}\left\{J(xz,xy)-K(xz,xy)+J(yz,xy)-K(yz,xy)+2\left\langle(xz^+,xy^+)\left|\frac{e^2}{r_{ij}}\right|(yz^+,xy^+)\right\rangle\right\}$$

$$=\frac{1}{2}\{A-2B+C-3B-C+A-2B+C-3B-C+0\}$$

$$=A-5B$$

- Thus:

$$\left\langle B_1\left|\frac{e^2}{r_{ij}}\right|B_1\right\rangle=A-5B$$

$$\left\langle B_2\left|\frac{e^2}{r_{ij}}\right|B_2\right\rangle=A-5B$$

$$\left\langle B_3\left|\frac{e^2}{r_{ij}}\right|B_3\right\rangle=A-5B$$

$$\left\langle B_4\left|\frac{e^2}{r_{ij}}\right|B_4\right\rangle=A+B$$

$$\left\langle B_5\left|\frac{e^2}{r_{ij}}\right|B_5\right\rangle=A+B$$

$$\left\langle B_6\left|\frac{e^2}{r_{ij}}\right|B_6\right\rangle=A-5B$$

$$\left\langle B_7\left|\frac{e^2}{r_{ij}}\right|B_7\right\rangle=A-5B$$

$$\left\langle B_8\left|\frac{e^2}{r_{ij}}\right|B_8\right\rangle=A-8B$$

$$\left\langle B_9\left|\frac{e^2}{r_{ij}}\right|B_9\right\rangle=A+4B$$

$$\left\langle B_{10}\left|\frac{e^2}{r_{ij}}\right|B_{10}\right\rangle=A-8B$$

The Nondiagonal Interelectronic Repulsion and the Energy of Each Level of the d^2 Configuration in Strong Field of Trans-ML_4Z_2

The matrix elements of $\left\langle ab\left|{}^{e^2}/{}_{r_{ij}}\right|cd\right\rangle$ **are given in** Table 7.6.

The interchange:

$$\left\langle ab\left|{}^{e^2}/{}_{r_{ij}}\right|cd\right\rangle=\left\langle ad\left|{}^{e^2}/{}_{r_{ij}}\right|cb\right\rangle=\left\langle cb\left|{}^{e^2}/{}_{r_{ij}}\right|ad\right\rangle=\left\langle bc\left|{}^{e^2}/{}_{r_{ij}}\right|da\right\rangle=\left\langle cd\left|{}^{e^2}/{}_{r_{ij}}\right|ab\right\rangle$$

If a match is not found, the element is zero.

Evaluate the nondiagonal interelectronic repulsion for various bases terms of the d^2 configuration in a tetragonal structure.

TABLE 7.6 The Matrix Elements of $\left\langle ab\left|{}^{e^2}/_{r_{ij}}\right|cd\right\rangle$

a	b	c	d	$\left\langle ab\left|{}^{e^2}/_{r_{ij}}\right|cd\right\rangle$
xz	z^2	xz	x^2-y^2	$-2\sqrt{3}B$
yz	z^2	yz	x^2-y^2	$2\sqrt{3}B$
yz	yz	z^2	x^2-y^2	$-\sqrt{3}B$
xz	xz	z^2	x^2-y^2	$\sqrt{3}B$
z^2	xy	xz	yz	$\sqrt{3}B$
z^2	xz	xy	yz	$-2\sqrt{3}B$
z^2	xy	yz	xz	$\sqrt{3}B$
x^2-y^2	xy	xz	yz	$3B$
x^2-y^2	xy	yz	xz	$-3B$

How are the energy levels of this configuration determined?

- The bases terms of d^2 configuration in tetragonal structure (pp. 457, 462):

$B_1 = (xz^+, yz^+)$ $\quad {}^3A_{2g} \quad \left\langle B_1\left|\frac{e^2}{r_{ij}}\right|B_1\right\rangle = A - 5B$

$B_2 = \frac{1}{\sqrt{2}}[(xz^+, xy^+) + (yz^+, xy^+)]$ $\quad {}^3E_g \quad \left\langle B_2\left|\frac{e^2}{r_{ij}}\right|B_2\right\rangle = A - 5B$

$B_3 = \frac{1}{\sqrt{2}}[(xz^+, xy^+) - (yz^+, xy^+)]$ $\quad {}^3E_g \quad \left\langle B_3\left|\frac{e^2}{r_{ij}}\right|B_3\right\rangle = A - 5B$

$B_4 = \frac{1}{\sqrt{2}}\left[\left(xz^+, z^{2+}\right) + \left(yz^+, z^{2+}\right)\right]$ $\quad {}^3E_g \quad \left\langle B_4\left|\frac{e^2}{r_{ij}}\right|B_4\right\rangle = A + B$

$B_5 = \frac{1}{\sqrt{2}}\left[\left(xz^+, z^{2+}\right) - \left(yz^+, z^{2+}\right)\right]$ $\quad {}^3E_g \quad \left\langle B_5\left|\frac{e^2}{r_{ij}}\right|B_5\right\rangle = A + B$

$B_6 = \frac{1}{\sqrt{2}}\left[\left(xz^+, x^2 - y^{2+}\right) + \left(yz^+, x^2 - y^{2+}\right)\right]$ $\quad {}^3E_g \quad \left\langle B_6\left|\frac{e^2}{r_{ij}}\right|B_6\right\rangle = A - 5B$

$B_7 = \frac{1}{\sqrt{2}}\left[\left(xz^+, x^2 - y^{2+}\right) - \left(yz^+, x^2 - y^{2+}\right)\right]$ $\quad {}^3E_g \quad \left\langle B_7\left|\frac{e^2}{r_{ij}}\right|B_7\right\rangle = A - 5B$

$B_8 = xy^+, z^{2+}$ $\quad {}^3B_{2g} \quad \left\langle B_8\left|\frac{e^2}{r_{ij}}\right|B_8\right\rangle = A - 8B$

$B_9 = xy^+, x^2 - y^{2+}$ $\quad {}^3A_{2g} \quad \left\langle B_9\left|\frac{e^2}{r_{ij}}\right|B_9\right\rangle = A + 4B$

$B_{10} = z^{2+}, x^2 - y^{2+}$ $\quad {}^3B_{1g} \quad \left\langle B_{10}\left|\frac{e^2}{r_{ij}}\right|B_{10}\right\rangle = A - 8B$

- The complete set of matrices of various energy levels of d^2 configuration:
  ${}^3A_{2g}$ states are mixed in a 2×2 determinant
  ${}^3E_{2g}$ states are mixed in two equivalent 3×3 determinants.
  ${}^3B_{2g}$ and ${}^3B_{1g}$

- The A_{2g} matrix (diagonal interelectronic, p. 495):

$$\begin{array}{c|cc|} & B_1 & B_9 \\ B_1 & -8Dq-2Ds+8Dt-5B-E & ? \\ B_2 & ? & 2Dq+4Ds-2Dt+4B-E \end{array} = 0 \quad (7.7.12)$$

The Racah A parameter is excluded since it equally moves all levels.

The off-diagonal for A_{2g} matrix: is $\left\langle B_1 \left| {e^2}/{r_{ij}} \right| B_9 \right\rangle$

From Eq. (7.7.1):

$$B_1 = \frac{1}{\sqrt{2}}(xz^+(1), yz^+(2) - yz^+(1), xz^+(2))$$

$$B_9 = xy^+, x^2-y^{2+} = \frac{1}{\sqrt{2}}\left(xy^+(1), x^2-y^{2+}(2) - x^2-y^{2+}(1), xy^+(2)\right)$$

The nondiagonal interelectronic repulsion:

$$\begin{aligned} \left\langle B_1 \left| {e^2}/{r_{ij}} \right| B_9 \right\rangle &= \frac{1}{2}\left\langle xz(1), yz(2) - yz(1), xz(2) \left| {e^2}/{r_{ij}} \right| xy(1), x^2-y^2(2) - x^2-y^2(1), xy(2) \right\rangle \\ &= \frac{1}{2}\left\langle xz, yz - yz, xz \left| {e^2}/{r_{ij}} \right| xy, x^2-y^2 - x^2-y^2, xy \right\rangle \\ &= \frac{1}{2}\left\{ \begin{array}{l} \left\langle xz, yz \left| {e^2}/{r_{ij}} \right| xy, x^2-y^2 \right\rangle - \left\langle xz, yz \left| {e^2}/{r_{ij}} \right| x^2-y^2, xy \right\rangle \\ - \left\langle yz, xz \left| {e^2}/{r_{ij}} \right| xy, x^2-y^2 \right\rangle + \left\langle yz, xz \left| {e^2}/{r_{ij}} \right| x^2-y^2, xy \right\rangle \end{array} \right\} \\ &= \frac{1}{2}\{-3B - 3B - 3B - 3B\} = -6B \end{aligned} \quad (7.7.13)$$

Note: the choice of writing the order of state (xz^+, yz^+) was arbitrary, if we chose (yz^+, xz^+), the off-diagonal would equal +6B, because $(xz^+, yz^+) = -(yz^+, xz^+)$.

The full A_{2g} matrix:

$$\begin{array}{c|cc|} & B_1 & B_9 \\ B_1 & -8Dq-2Ds+8Dt-5B-E & -6B \\ B_9 & -6B & 2Dq+4Ds-2Dt+4B-E \end{array} = 0 \quad (7.7.12)$$

The energies of B_1 and B_9 levels can be obtained from the root of the A_{2g} matrix.

- There are two 3×3 E_g determinants, each containing one component from each pair.

The E_g matrix:

$$\begin{array}{c|ccc|} & B_2 & B_4 \text{ or } B_5 & B_6 \text{ or } B_7 \\ B_2 & -8Dq+Ds+3Dt-5B-E & ? & ? \\ B_4 \text{ or } B_5 & ? & ? & ? \\ B_6 \text{ or } B_7 & ? & ? & ? \end{array} = 0 \quad (7.7.14)$$

  - The off-diagonal $\left\langle B_2 \left| {e^2}/{r_{ij}} \right| B_4 \right\rangle$:

$$B_2 = \frac{1}{\sqrt{2}}[(xz^+, xy^+) + (yz^+, xy^+)]$$

$$B_4 = \frac{1}{\sqrt{2}}\left[\left(xz^+, z^{2+}\right) + \left(yz^+, z^{2+}\right)\right]$$

$$\left\langle B_2\left|{}^{e^2}/_{r_{ij}}\right|B_4\right\rangle = \frac{1}{2}\left\langle (xz^+,xy^+) + (yz^+,xy^+)\left|{}^{e^2}/_{r_{ij}}\right|\left(xz^+,z^{2+}\right) + \left(yz^+,z^{2+}\right)\right\rangle$$
$$= \frac{1}{2}\left\langle (xz,xy) + (yz,xy)\left|{}^{e^2}/_{r_{ij}}\right|(xz,z^2) + (yz,z^2)\right\rangle$$
$$= \frac{1}{2}\left\{\begin{array}{l}\underbrace{\left\langle (xz,xy)\left|{}^{e^2}/_{r_{ij}}\right|(xz,z^2)\right\rangle}_{\text{term 1}} + \underbrace{\left\langle (xz,xy)\left|{}^{e^2}/_{r_{ij}}\right|(yz,z^2)\right\rangle}_{\text{term 2}} \\ + \underbrace{\left\langle (yz,xy)\left|{}^{e^2}/_{r_{ij}}\right|(xz,z^2)\right\rangle}_{\text{term 3}} + \underbrace{\left\langle (yz,xy)\left|{}^{e^2}/_{r_{ij}}\right|(yz,z^2)\right\rangle}_{\text{term 4}}\end{array}\right\}$$

$$\underbrace{\left\langle (xz,xy)\left|{}^{e^2}/_{r_{ij}}\right|(xz,z^2)\right\rangle}_{\text{term 1}} = 0, \text{ does not occur in single line}$$

$$\underbrace{\left\langle (xz,xy)\left|{}^{e^2}/_{r_{ij}}\right| + (xz,z^2)\right\rangle}_{\text{term 2}} = 0, \text{ does not occur in single line}$$

The second and the third may expand:

$$\underbrace{\left\langle (yz,xy)\left|{}^{e^2}/_{r_{ij}}\right|(xz,z^2)\right\rangle}_{\text{term 3}} = \frac{1}{2}\left\{\left\langle (xz(1),xy(2) - xy(1),xz(2))\left|{}^{e^2}/_{r_{ij}}\right|yz(1),z^2(2) - z^2(1)yz(2)\right\rangle\right\}$$
$$= \frac{1}{2}\left\{\left\langle xz,xy - xy,xz\left|{}^{e^2}/_{r_{ij}}\right|yz,z^2 - z^2,yz\right\rangle\right\}$$
$$= \frac{1}{2}\left\{\begin{array}{l}\left\langle xz,xy\left|{}^{e^2}/_{r_{ij}}\right|yz,z^2\right\rangle + \left\langle xz,xy\left|{}^{e^2}/_{r_{ij}}\right|z^2yz\right\rangle - \\ \left\langle xy,xz\left|{}^{e^2}/_{r_{ij}}\right|yz,z^2\right\rangle + \left\langle xy,xz\left|{}^{e^2}/_{r_{ij}}\right|z^2,yz\right\rangle\end{array}\right\}$$
$$= \frac{1}{2}\left\{-2\sqrt{3}B - \sqrt{3}B - \sqrt{3}B - 2\sqrt{3}B\right\} = -3\sqrt{3}B$$

$$\underbrace{\left\langle (yz,xy)\left|{}^{e^2}/_{r_{ij}}\right|(yz,z^2)\right\rangle}_{\text{term 4}} = \frac{1}{2}\left\{\left\langle yz(1),xy(2) - xy(1),yz(2)\left|{}^{e^2}/_{r_{ij}}\right|xz(1),z^2(2) - z^2(1),xz(2)\right\rangle\right\}$$
$$= \frac{1}{2}\left\{\left\langle yz,xy - xy,yz\left|{}^{e^2}/_{r_{ij}}\right|xz,z^2 - z^2,xz\right\rangle\right\}$$
$$= \frac{1}{2}\left\{\begin{array}{l}\left\langle yz,xy\left|{}^{e^2}/_{r_{ij}}\right|xz,z^2\right\rangle - \left\langle yz,xy\left|{}^{e^2}/_{r_{ij}}\right|z^2,xz\right\rangle - \\ \left\langle xy,yz\left|{}^{e^2}/_{r_{ij}}\right|xz,z^2\right\rangle + \left\langle xy,yz\left|{}^{e^2}/_{r_{ij}}\right|z^2,xz\right\rangle\end{array}\right\}$$
$$= \frac{1}{2}\left\{-2\sqrt{3}B - \sqrt{3}B - \sqrt{3}B - 2\sqrt{3}B\right\} = -3\sqrt{3}B$$

Then:

$$\left\langle B_2\left|{}^{e^2}/_{r_{ij}}\right|B_4\right\rangle = \frac{1}{2}\left\{0 - 3\sqrt{3}B - 3\sqrt{3}B + 0\right\} = -3\sqrt{3}B \quad (7.7.15)$$

- The off-diagonal $\left\langle B_2 \middle| {}^{e^2}/_{r_{ij}} \middle| B_5 \right\rangle$:

$$B_2 = \frac{1}{\sqrt{2}}[(xz^+, xy^+) + (yz^+, xy^+)]$$

$$B_5 = \frac{1}{\sqrt{2}}\left[\left(xz^+, z^{2+}\right) - \left(yz^+, z^{2+}\right)\right]$$

$$\left\langle B_2 \middle| {}^{e^2}/_{r_{ij}} \middle| B_5 \right\rangle = \frac{1}{2}\left\langle (xz^+, xy^+) + (yz^+, xy^+) \middle| {}^{e^2}/_{r_{ij}} \middle| \left(xz^+, z^{2+}\right) - \left(yz^+, z^{2+}\right) \right\rangle$$

$$= \frac{1}{2}\left\langle (xz, xy) + (yz, xy) \middle| {}^{e^2}/_{r_{ij}} \middle| (xz, z^2) - (yz, z^2) \right\rangle$$

$$= \frac{1}{2}\left\{ \begin{array}{l} \left\langle (xz, xy) \middle| {}^{e^2}/_{r_{ij}} \middle| (xz, z^2) \right\rangle - \left\langle (xz, xy) \middle| {}^{e^2}/_{r_{ij}} \middle| + (yz, z^2) \right\rangle \\ + \left\langle (yz, xy) \middle| {}^{e^2}/_{r_{ij}} \middle| (xz, z^2) \right\rangle - \left\langle (yz, xy) \middle| {}^{e^2}/_{r_{ij}} \middle| (yz, z^2) \right\rangle \end{array} \right\}$$

$$\left\langle B_2 \middle| {}^{e^2}/_{r_{ij}} \middle| B_5 \right\rangle = \frac{1}{2}\left\{0 - 3\sqrt{3}B + 3\sqrt{3}B + 0\right\} = 0 \qquad (7.7.16)$$

- The off-diagonal $\left\langle B_2 \middle| {}^{e^2}/_{r_{ij}} \middle| B_6 \right\rangle$:

$$B_2 = \frac{1}{\sqrt{2}}[(xz^+, xy^+) + (yz^+, xy^+)]$$

$$B_6 = \frac{1}{\sqrt{2}}\left[\left(xz^+, x^2 - y^{2+}\right) + \left(yz^+, x^2 - y^{2+}\right)\right]$$

$$\left\langle B_2 \middle| {}^{e^2}/_{r_{ij}} \middle| B_6 \right\rangle = \frac{1}{2}\left\langle (xz, xy) + (yz, xy) \middle| {}^{e^2}/_{r_{ij}} \middle| (xz, x^2 - y^2) + (yz, x^2 - y^2) \right\rangle$$

$$= \frac{1}{2}\left\{ \begin{array}{l} \underbrace{\left\langle (xz, xy) \middle| {}^{e^2}/_{r_{ij}} \middle| (xz, x^2 - y^2) \right\rangle}_{\text{term 1}} + \underbrace{\left\langle (xz, xy) \middle| {}^{e^2}/_{r_{ij}} \middle| (yz, x^2 - y^2) \right\rangle}_{\text{term 2}} \\ + \underbrace{\left\langle (xz, xy) \middle| {}^{e^2}/_{r_{ij}} \middle| (xz, x^2 - y^2) \right\rangle}_{\text{term 3}} + \underbrace{\left\langle (yz, xy) \middle| {}^{e^2}/_{r_{ij}} \middle| (yz, x^2 - y^2) \right\rangle}_{\text{term 4}} \end{array} \right\}$$

$$\underbrace{\left\langle (xz, xy) \middle| {}^{e^2}/_{r_{ij}} \middle| (xz, x^2 - y^2) \right\rangle}_{\text{term 1}} = 0, \text{ element does not occur in single line}$$

$$\underbrace{\left\langle (xz, xy) \middle| {}^{e^2}/_{r_{ij}} \middle| (yz, x^2 - y^2) \right\rangle}_{\text{term 2}} = 0, \text{ element does not occur in single line}$$

$$\underbrace{\left\langle (yz, xy) \middle| {}^{e^2}/_{r_{ij}} \middle| (xz, x^2 - y^2) \right\rangle}_{\text{term 3}} = \frac{1}{2}\left\langle xz(1), xy(2) - xy(1), xz(2) \middle| {}^{e^2}/_{r_{ij}} \middle| yz(1), x^2 - y^2(2) - x^2 - y^2(1), yz(2) \right\rangle$$

$$= \frac{1}{2}\left\langle xz, xy - xy, xz \middle| {}^{e^2}/_{r_{ij}} \middle| yz, x^2 - y^2 - x^2 - y^2, yz \right\rangle$$

$$= \frac{1}{2}\left\{ \begin{array}{l} \left\langle xz, xy \middle| {}^{e^2}/_{r_{ij}} \middle| yz, x^2 - y^2 \right\rangle - \left\langle xz, xy - xy \middle| {}^{e^2}/_{r_{ij}} \middle| x^2 - y^2, yz \right\rangle \\ \left\langle -xy, xz \middle| {}^{e^2}/_{r_{ij}} \middle| yz, x^2 - y^2 \right\rangle + \left\langle xy, xz \middle| {}^{e^2}/_{r_{ij}} \middle| x^2 - y^2, yz \right\rangle \end{array} \right\}$$

$$= \frac{1}{2}\{0 - 3B - 3B + 0\} = -3B$$

$$\underbrace{\left\langle (yz,xy)\left|{}^{e^2}/_{r_{ij}}\right|(yz,x^2-y^2)\right\rangle}_{\text{term 4}} = \frac{1}{2}\left\langle (yz(1),xy(2)-xy(1),yz(2))\left|{}^{e^2}/_{r_{ij}}\right|xz(1),x^2-y^2(2)-x^2-y^2(1),xz(2)\right\rangle$$

$$= \frac{1}{2}\left\langle yz,xy-xy,yz\left|{}^{e^2}/_{r_{ij}}\right|xz,x^2-y^2-x^2-y^2,xz\right\rangle$$

$$= \frac{1}{2}\left\{\begin{array}{l}\left\langle yz,xy\left|{}^{e^2}/_{r_{ij}}\right|xz,x^2-y^2\right\rangle - \left\langle yz,xy\left|{}^{e^2}/_{r_{ij}}\right|x^2-y^2,xz\right\rangle \\ \left\langle -xy,yz\left|{}^{e^2}/_{r_{ij}}\right|xz,x^2-y^2\right\rangle + \left\langle xy,yz\left|{}^{e^2}/_{r_{ij}}\right|x^2-y^2,xz\right\rangle\end{array}\right\}$$

$$= \frac{1}{2}\{0+3B+3B+0\} = 3B$$

$$\left\langle B_2\left|{}^{e^2}/_{r_{ij}}\right|B_6\right\rangle = \frac{1}{2}(0-3B+3B-0) = 0 \tag{7.7.17}$$

- The off-diagonal $\left\langle B_2\left|{}^{e^2}/_{r_{ij}}\right|B_7\right\rangle$:

$$B_2 = \frac{1}{\sqrt{2}}[(xz^+,xy^+)+(yz^+,xy^+)]$$

$$B_7 = \frac{1}{\sqrt{2}}\left[\left(xz^+,x^2-y^{2+}\right)-\left(yz^+,x^2-y^{2+}\right)\right]$$

$$\left\langle B_2\left|{}^{e^2}/_{r_{ij}}\right|B_7\right\rangle = \frac{1}{2}\left\langle (xz,xy)+(yz,xy)\left|{}^{e^2}/_{r_{ij}}\right|(xz,x^2-y^2)-(yz,x^2-y^2)\right\rangle$$

$$= \frac{1}{2}\left\{\begin{array}{l}\underbrace{\left\langle (xz,xy)\left|{}^{e^2}/_{r_{ij}}\right|(xz,x^2-y^2)\right\rangle}_{\text{term 1}} - \underbrace{\left\langle (xz,xy)\left|{}^{e^2}/_{r_{ij}}\right|(yz,x^2-y^2)\right\rangle}_{\text{term 2}} \\ +\underbrace{\left\langle (yz,xy)\left|{}^{e^2}/_{r_{ij}}\right|(xz,x^2-y^2)\right\rangle}_{\text{term 3}} - \underbrace{\left\langle (yz,xy)\left|{}^{e^2}/_{r_{ij}}\right|(yz,x^2-y^2)\right\rangle}_{\text{term}}\end{array}\right\}$$

$$\underbrace{\left\langle (xz,xy)\left|{}^{e^2}/_{r_{ij}}\right|(xz,x^2-y^2)\right\rangle}_{\text{term 1}} = 0, \text{ element does not occur in single line}$$

$$\underbrace{\left\langle (xz,xy)\left|{}^{e^2}/_{r_{ij}}\right|(yz,x^2-y^2)\right\rangle}_{\text{term 2}} = 0, \text{ element does not occur in single line}$$

$$\underbrace{\left\langle (yz,xy)\left|{}^{e^2}/_{r_{ij}}\right|(yz,x^2-y^2)\right\rangle}_{\text{term 3}} = \frac{1}{2}\left\langle xz(1),xy(2)-xy(1),xz(2\left|{}^{e^2}/_{r_{ij}}\right|yz(1),x^2-y^2(2)-x^2-y^2(1),yz(2)\right\rangle$$

$$= \frac{1}{2}\left\langle xz,xy-xy,xz\left|{}^{e^2}/_{r_{ij}}\right|yz,x^2-y^2-x^2-y^2,yz\right\rangle$$

$$= \frac{1}{2}\left\{\begin{array}{l}\left\langle xz,xy\left|{}^{e^2}/_{r_{ij}}\right|\left|z,x^2-y^2\right\rangle - \left\langle xz,xy-xy\left|{}^{e^2}/_{r_{ij}}\right|x^2-y^2,yz\right\rangle \\ -\left\langle xy,xz\left|{}^{e^2}/_{r_{ij}}\right|yz,x^2-y^2\right\rangle + \left\langle xy,xz\left|{}^{e^2}/_{r_{ij}}\right|x^2-y^2,yz\right\rangle\end{array}\right\}$$

$$= \frac{1}{2}\{0-3B-3B+0\} = -3B$$

$$\underbrace{\left\langle (yz, xy)\left|{}^{e^2}/_{r_{ij}}\right|(yz, x^2-y^2)\right\rangle}_{\text{term}} = \frac{1}{2}\left\langle (yz(1), xy(2) - xy(1), yz(2))\left|{}^{e^2}/_{r_{ij}}\right| xz(1), x^2-y^2(2) - x^2-y^2(1), xz(2)\right\rangle$$

$$= \frac{1}{2}\left\langle yz, xy - xy, yz\left|{}^{e^2}/_{r_{ij}}\right| xz, x^2-y^2 - x^2-y^2, xz\right\rangle$$

$$= \frac{1}{2}\left\{\begin{array}{l}\left\langle yz, xy\left|{}^{e^2}/_{r_{ij}}\right| xz, x^2-y^2\right\rangle - \left\langle yz, xy\left|{}^{e^2}/_{r_{ij}}\right| x^2-y^2, xz\right\rangle \\ -\left\langle xy, yz\left|{}^{e^2}/_{r_{ij}}\right| xz, x^2-y^2\right\rangle + \left\langle xy, yz\left|{}^{e^2}/_{r_{ij}}\right| x^2-y^2, xz\right\rangle\end{array}\right\}$$

$$= \frac{1}{2}\{0 + 3B + 3B + 0\} = 3B$$

$$\left\langle B_2\left|{}^{e^2}/_{r_{ij}}\right|B_7\right\rangle = \frac{1}{2}(0 + 3B + 3B - 0) = 3B \tag{7.7.18}$$

- in the same way:

$$\left\langle B_4\left|{}^{e^2}/_{r_{ij}}\right|B_7\right\rangle = -3\sqrt{3}B \tag{7.7.19}$$

- Summary:

$$\left\langle B_2\left|{}^{e^2}/_{r_{ij}}\right|B_4\right\rangle = -3\sqrt{3}B \tag{7.7.15}$$

$$\left\langle B_2\left|{}^{e^2}/_{r_{ij}}\right|B_5\right\rangle = 0 \tag{7.7.16}$$

$$\left\langle B_2\left|{}^{e^2}/_{r_{ij}}\right|B_6\right\rangle = 0 \tag{7.7.17}$$

$$\left\langle B_2\left|{}^{e^2}/_{r_{ij}}\right|B_7\right\rangle = 3B \tag{7.7.18}$$

$$\left\langle B_4\left|{}^{e^2}/_{r_{ij}}\right|B_7\right\rangle = -3\sqrt{3}B \tag{7.7.19}$$

- Thus, B_2, B_4, and B_7 belong to same 3×3 determinant, Eq. (7.7.20), but B_5 and B_6 belong to different determinants, Eq. (7.7.21) (diagonal interelectronic, p. 465):

$$\begin{array}{c|ccc|}
 & B_2 & B_4 & B_7 \\
B_2 & -8Dq + Ds + 3Dt - 5B - E & -3\sqrt{3}B & 3B \\
B_4 & 3\sqrt{3}B & -2Dq - 3Ds - 2Dt + B - E & -3\sqrt{3}B \\
B_7 & 3B & -3\sqrt{3}B & 2Dq + Ds + 3Dt - 5B - E
\end{array} = 0 \tag{7.7.20}$$

The energies of B_2, B_4 and B_7 levels can be estimated from the root of the matrix.

The off-diagonals of one matrix are the negative of those in the other, since the two components of each e-representation are orthogonal. The sign of an off-diagonal has no effect on the resultant energies.

$$\begin{array}{c|ccc|}
 & B_3 & B_5 & B_6 \\
B_3 & -8Dq + Ds + 3Dt - 5B - E & 3\sqrt{3}B & -3B \\
B_5 & 3\sqrt{3}B & -2Dq - 3Ds - 2Dt + B - E & -3\sqrt{3}B \\
B_6 & -3B & 3\sqrt{3}B & 2Dq + Ds + 3Dt - 5B - E
\end{array} = 0 \tag{7.7.21}$$

The energies of B_3, B_5 and B_6 levels can be evaluated from the root of the matrix.

- The remaining energy levels are simply obtained by summing the crystal field and the interelectronic repulsion energy:

$$E(^3B_{2g}) = E(B_8) = 2Dq - 7Dt - 8B$$

$$E(^3B_{1g}) = E(B_{10}) = 12Dq - 7Dt - 8B$$

- From above, all of the 10 energy levels of d^2 tetragonal ML_4Z_2 (D_{4h}) are computable, the most negative level will be the ground state.
- In O_h symmetry, $^3T_{1g}$ is the ground state. The axial distortion changes the symmetry of O_h to D_{4h}, and splits the degeneracy of $^3T_{1g}$ to E_g and A_{2g}. Therefore, the most negative root of either E_g or A_{2g} matrix will be the ground state.

SUGGESTIONS FOR FURTHER READING

H.A. Bethe, Ann. Physik. 3 (1929) 133.

J.H. Van Vleck, J. Chem. Phys. 3 (1935) 807.

C.J. Ballhausen, Moffitt, Ann. Rev. Phys. Chem. 7 (1959) 107.

H. Eyring, J. Walter, G. Kimball, Quantum Chemistry, John Wiley, New York, 1944.

L. Pauling, E.B. Wilson, Quantum Mechanics, McGraw Hill, New York, 1935.

J. Van Vleck, Phys. Rev. 41 (1932) 208.

S.S. Zumdahl, Chemical Principles, fifth ed, Houghton Mifflin Company, Boston, MA, ISBN: 0-669-39321-5, 2005.

M.S. Silberberg, Chemistry: The Molecular Nature of Matter and Change, fourth ed., McGraw Hill Company, New York, ISBN: 0-8151-8505-7, 2006.

D.F. Shriver, P.W. Atkins, Inorganic Chemistry, fourth ed., Oxford University Press, Oxford, ISBN: 0-8412-3849-9, 2001.

C.E. Housecroft, A.G. Sharpe, Inorganic Chemistry, second ed., Prentice Hall, Upper Saddle River, NJ, ISBN: 978-0130399137, 2004.

G.L. Miessler, D.A. Tarr, Inorganic Chemistry, third ed., Pearson Prentice Hall, Upper Saddle River, NJ, ISBN: 0-13-035471-6, 2003.

R.G. Shulman, S. Sugano, Phys. Rev. 130 (1963) 1517.

R.G. Shulman, S. Sugano, Phys. Rev. Lett. 7 (1961) 772.

A.B.P. Lever, Inorganic Electronic Spectroscopy, second ed., Elsevier, Amsterdam, ISBN: 0-444-42389-3, 1984.

C.J. Ballhausen, Introduction to Crystal Field Theory, McGraw Hill, New York, NY, 1962.

K.F. Purcell, J.C. Kotz, Inorganic Chemistry, W. B. Saunders Company, Philadelphia/London/Toronto, ISBN: 0-7216-7407-0, 1977.

F.A. Cotton, Chemical Applications of Group Theory, third ed., John Wiley-Interscience, Hoboken, NY, ISBN-13:9780471510949, 1990.

B.N. Figgis, Introduction to Ligand Field, Interscience, New York, 1966.

M. Gerloch, R.C. Slade, Ligand Field Parameters, Cambridge University Press, Cambridge, 1973.

F. Kauzmann, Quantum Chemistry, Academic Press, New York, 1957.

A.L. Companion, M.A. Komarynsky, J. Chem. Educ. 41 (1964) 257.

H. Hartmann, E. Konig, Z. Physik. Chem. 28 (1961) 425.

H. Hartmann, E. Konig, Theor. Chim. Acta 4 (1966) 148.

W.E. Hatfield, W.E. Parker, Symmetry in Chemical Bonding and Structure, Merrill Publ. Co., Columbus, OH, 1974.

T.M. Dunn, D.S. McClure, R.G. Pearson, Crystal Field Theory, Harper and Row, New York, 1965.

P. George, D.S. McClure, Prog. Inorg. Chem. 1 (1959) 381.

H.C. Rosenberg, C.A. Root, H.B. Gray, J. Am. Chem. Soc. 97 (1975) 21.

R.M. Hochstrasser, Molecular Aspects of Symmetry, Benjamin, New York, 1966.

R. McWeeny, Symmetry, An introduction to Group Theory and Its Applications, Macmillan, Pergamon, London, New York, 1963.

D.S. McClure, S. Kirschner, Advances in the Chemistry of the Coordination Compounds, Macmillan Co., New York, 1961.

H.L. Schlafer, G. Glisemann, Basic Principles of Ligand Field Theory, Wiley, New York, 1969.

J.S. Griffith, Theory of Transition Metal Ions, Cambridge University Press, Cambridge, 1961.

E.U. Condon, G.H. Shortley, The Theory of Atomic Spectra, Cambridge University Press, Cambridge, 1963.

C.J. Ballhausen, Molecular Electronic Structures of Transition Metal Complexes, McGraw Hill, New York, 1979.

I.S. Butler, J.F. Harrod, Inorganic Chemistry Principles and Applications, The Benjamin/Cummings Publishing Company, Inc., California, USA, ISBN: 0-8053-0247-6, 1989

B. Douglas, D. McDaniel, J.J. Alexander, Concepts and Models of Inorganic Chemistry, third ed., John Wiley & Sons, New York, ISBN: 0-471-62978-2, 1994.

R.S. Drago, Phys. Methods in Chemistry, W. B. Saunders Company, Philadelphia, PA, ISBN: 0-7216-3184-3, 1977.

H.B. Gray, Electrons and Chemical Bonding, W. A. Benjamin, Inc., New York, 1965.

Chapter 8

Ligand Field Theory

In the previous chapter, we showed that the states that arise from a specific electronic configuration of an ion are degenerate when the ion is free of disturbing influences. That degeneracy must split into two or more nondegenerate states when the ion is placed into a lattice.

The splitting of the free ion states are calculated assuming that all interactions between the ion and its surroundings are treated as purely electrostatic interactions between point charges. This assumption is an essential feature of the crystal field theory.

Crystal field theory has ignored any covalency in the metal-ligand bond. As a result, the quantitative derivation of Dq differs from that found experimentally. A ligand field takes into account the covalency in bond and treated Dq (and other parameters to be discussed shortly) as empirical parameters that are obtained from the electronic spectra.

If the covalence of the metal ligand bonds is comparatively small, energies will be given by equations identical in form to the derived in the crystal field theory. Most of the principles that will be developed are as much a part of crystal field theory as they are of ligand field theory.

The symmetry approach stays entirely applicable if we change the computational part from a purely electrostatic approach to one that admits the existence of some bond covalency between the metal ion and its neighbors.

In this chapter, we will show:

(1) how it is possible to use the symmetry and group theory to find what states will obtained when an ion of any electronic configuration is located into a crystalline environment of definite symmetry;
(2) the relative energies of these states will be investigated; and
(3) how the energies of the various states into which the free ion term are split depend on the strength of the interaction of the ion with its environment.

The relationship between the energy of the excited states and Dq are discussed using the correlation diagrams, Orgel diagrams, and Tanabe-Sugano diagrams.

- 8.1: The advantages and disadvantages of crystal field theory
- 8.2: Symmetry and orbital splitting by ligand field
- 8.3: Correlation tables
  - Orbital correlation table
  - Terms correlation table
- 8.4: Correlation diagrams of strong and weak fields
  - Octahedral diagrams
  - Method of descending symmetry
  - Tetrahedral diagrams
- 8.5: Orgel diagrams
  - Orgel diagram of *D* term
  - Orgel diagram of *F* term
  - Configuration and term interactions
- 8.6: Tanabe-Sugano diagrams (Scheme 8.1)
  - The advantages
  - d^n and d^{10-n} diagrams
  - d^5 diagram
- Suggestions for further reading

Electrons, Atoms, and Molecules in Inorganic Chemistry. http://dx.doi.org/10.1016/B978-0-12-811048-5.00008-0

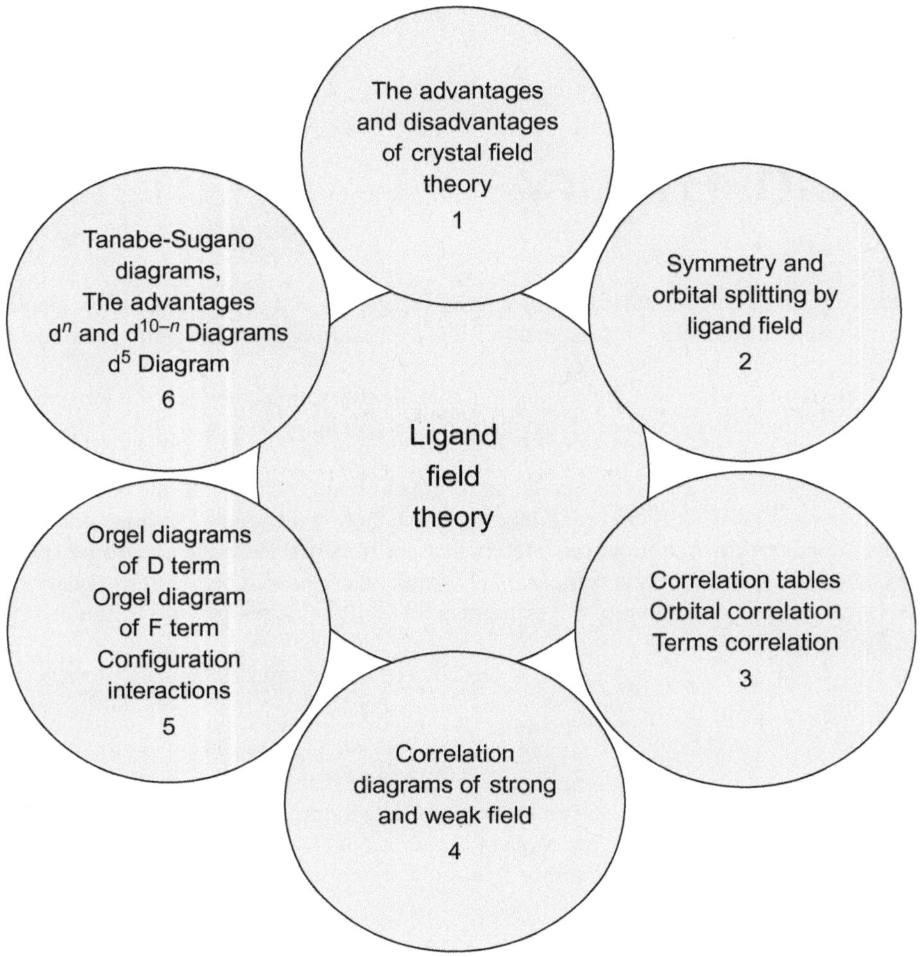

SCHEME 8.1 Approach used to present ligand field theory concept.

8.1 THE ADVANTAGES AND DISADVANTAGES OF CRYSTAL FIELD THEORY

What are the advantages of crystal field theory, and why was there a need for another theory?

- Crystal field theory was useful tool to describe the following in coordinated compounds:
  - optical spectra of transition metals ions;
  - the number of d-d transition and their relative energies;
  - some magnetic properties;
  - hydration enthalpies; and
  - spinel structures.

The success arose because crystal field theory is based firmly on molecular symmetry.

- However, the theory only considers d orbitals and falls to:
  - provide an adequate description of chemical bonding between neutral metal atoms and neutral or cationic ligands;
  - predict the absolute magnitude of transition energy;
  - explain the consequence of the ligand binding on the chemistry of the electrons of the central ion;
  - explain the quantitative calculation of Dq, which is differ considerably from that found experimentally; and
  - explain the charge transfer, which is between metal and ligand and more powerful than d-d transfer.

Why do the quantitative calculations of Dqs differ considerably from that found experimentally?

- In the point charge approximation, the radial parameter is defined as:

$$\mathrm{Dq} = ze^2 \frac{\overline{r^4}}{6a^5} = \frac{z\,e^2}{6a^5} \int_0^\infty R^*_{nl} r^4 R_{nl}\, r^2 \partial r$$

  where r is the distance from the origin at the center ion, a is the distance of the ligand from the origin, Chapter 7, p. 427.
- Therefore, the discrepancy between the calculated and found Dq reveals the variation in the metal-ligand distance, r, indicating that the metal-ligand bond does have some covalency (i.e., orbital overlapping).
- In the crystal field analysis, the covalency in the metal-ligand bond has been ignored.
- Ligand field theory realizes the covalency in the bond and treats Dq and other parameters as empirical parameters, which could be estimated from the electronic spectrum.
- When an orbital overlap between metal and ligand is considered, the metal orbitals can possibly form σ- or π-bonding.
- The term crystal field theory is kept for the extreme in which there is no mixing of the central ion and the ligand electrons, while the term ligand field theory for all non-zero degrees of mixing.
- The formulation in the entire details is the same.

What is the difficulty in determining the energies of the central ion terms and how can effectively approach this difficulty?

- Ligand field theory admits the covalency between the metal and its neighbors. Therefore, it is not simple to write down expressions for the Hamiltonian to compute the energies, since the orbitals involved are overlapped.
- On the other hand, the symmetry and group theory approach can straightforwardly be used to decide the states that result when an ion of any given electronic configuration is located into a surrounding environment of definite symmetry.
- Thus, in order to gain an understanding of this theory, symmetry considerations are essential approach.
- The ligand field theory cannot evaluate the absolute energies; however, experimental data can be fitted to the theory to compute semi-empirical radial parameters that manifest the interaction between ligands and metal.
- These interelectronic repulsion parameters (Racah B and C) are included in a well-defined manner to provide a precise fitting of the electronic spectrum.

8.2 SYMMETRY AND ORBITAL SPLITTING BY LIGAND FIELD

Find the splitting of d-orbital wave functions in an O_h environment using symmetry and group theory.

How can you obtain the symmetry species into which a set of degenerate orbitals split in symmetry lower than spherical?

Show that the effect of rotation by α:

$$\chi(\alpha) = \frac{\sin\left(l + \frac{1}{2}\right)\alpha}{\sin\left(\frac{\alpha}{2}\right)}$$

- The symmetry principles that already covered (Chapter 4) can be used to set up the procedures for determining all of the states taking place from different multiconfigurations in various geometries.
- First, we have to explore the effect of an O_h field on the total representation for which the set of d-orbital wave functions forms a basis.
- In order to find the total representation, we must obtain the elements of the matrices that describe the effect of each symmetry operations in the group upon the d-orbital basis set. The characters of these matrices will form the total representation.
- All the d-orbitals are gerade (i.e., symmetric to inversion); thus no changes will be happened as the result of the inversion symmetry operation. Therefore, the simpler pure rotational subgroup O can be used instead of O_h.
- The wave function for a single electron (Chapter 2):

$$\psi = R(r) \cdot \Theta(\theta) \cdot \Phi(\phi) \cdot \psi_s$$

$R(r)$: depends only on the radial distance from the nucleus, and is neglected. It is non-directional and unaffected by any symmetry operation;
$\Theta(\theta)$: depends only on the angle θ, and can be ignored. It is defined relative to the rotation axis, so it is unmoved by any rotation;
$\Phi(\phi)$: depends only on the angle ϕ, and change by rotation; and
ψ_s: is independent of r, θ, and ϕ.

- For d-orbital, $l=2$, m_l: 2, 1, 0, −1, −2. and $\Phi(\phi)=e^{im_l\phi}$

The effect of rotation by α on $e^{im_l\phi}$:

$$\begin{bmatrix} e^{2i\phi} \\ e^{i\phi} \\ e^{0} \\ e^{-i\phi} \\ e^{-2i\phi} \end{bmatrix} \xrightarrow{\text{rotation by}\,\alpha} \begin{bmatrix} e^{2i(\phi+\alpha)} \\ e^{i(\phi+\alpha)} \\ e^{0} \\ e^{-i(\phi+\alpha)} \\ e^{-2(\phi+\alpha)} \end{bmatrix}$$

$$\begin{bmatrix} e^{2i\phi} \\ e^{i\phi} \\ e^{0} \\ e^{-i\phi} \\ e^{-2i\phi} \end{bmatrix} \underbrace{\begin{bmatrix} e^{2i\alpha} & 0 & 0 & 0 & 0 \\ 0 & e^{i\alpha} & 0 & 0 & 0 \\ 0 & 0 & e^{0} & 0 & 0 \\ 0 & 0 & 0 & e^{-i\alpha} & 0 \\ 0 & 0 & 0 & 0 & e^{-2i\alpha} \end{bmatrix}}_{\text{the matrix that operate on d orbitals}} = \begin{bmatrix} e^{2i(\phi+\alpha)} \\ e^{i(\phi+\alpha)} \\ e^{0} \\ e^{-i(\phi+\alpha)} \\ e^{-2(\phi+\alpha)} \end{bmatrix}$$

Accordingly, the general matrix that operates on any set of orbits:

$$\begin{bmatrix} e^{li\alpha} & 0 & \cdots & 0 & 0 \\ 0 & e^{(l-1)i\alpha} & \cdots & 0 & 0 \\ \vdots & \vdots & \vdots & \vdots & \vdots \\ 0 & 0 & \cdots & e^{-(l-1)i\alpha} & 0 \\ 0 & 0 & \cdots & 0 & e^{-li\alpha} \end{bmatrix} \tag{8.2.1}$$

The sum of the diagonal elements:

$$\chi(\alpha) = \left[e^{i\alpha} + e^{2i\alpha} \cdots + e^{il\alpha}\right] + e^{0} + \left[e^{-i\alpha} + e^{-2i\alpha} \cdots + e^{-il\alpha}\right] \tag{8.2.2}$$

If

$$a\frac{1-r^n}{1-r} = a + ar + ar^2 + ar^3 + \cdots + ar^n$$

$$\chi(\alpha) = e^{i\alpha}\left(\frac{1-e^{il\alpha}}{1-e^{i\alpha}}\right) + 1 + e^{-i\alpha}\left(\frac{1-e^{-il\alpha}}{1-e^{-i\alpha}}\right)$$

$$\chi(\alpha) = \frac{e^{i\alpha} - e^{il\alpha}\cdot e^{i\alpha}}{1-e^{i\alpha}} + 1 + \frac{1}{e^{i\alpha}}\frac{\left(1-e^{-il\alpha}\right)}{\left(1-e^{-i\alpha}\right)}$$

$$\chi(\alpha) = -\frac{e^{i\alpha} - e^{il\alpha}\cdot e^{i\alpha}}{e^{i\alpha}-1} + \frac{e^{i\alpha}-1}{e^{i\alpha}-1} + \frac{1-e^{-il\alpha}}{e^{i\alpha}-1}$$

$$\chi(\alpha) = \frac{e^{il\alpha}\cdot e^{i\alpha} - e^{-il\alpha}}{e^{i\alpha}-1} = \frac{e^{(l+1)i\alpha} - e^{-il\alpha}}{e^{i\alpha}-1}$$

$$\text{if}: e^{i\alpha} = \cos\alpha + i\sin\alpha$$

and

$$e^{-i\alpha} = \cos\alpha - i\sin\alpha$$

then

$$\chi(\alpha)=\frac{\cos\,(l+1)\alpha+i\sin\,(l+1)\alpha-\cos\,l\alpha+i\sin\,l\alpha}{\cos\alpha+i\sin\alpha-1}$$

If:

$$\cos\alpha-\cos\beta=-2\sin\left(\frac{\alpha+\beta}{2}\right)\sin\left(\frac{\alpha-\beta}{2}\right)$$

$$\sin\alpha+\sin\beta=2\sin\left(\frac{\alpha+\beta}{2}\right)\cos\left(\frac{\alpha-\beta}{2}\right)$$

$$\cos\alpha=1-2\sin^2\left(\frac{\alpha}{2}\right)$$

$$\sin\alpha=2\sin\left(\frac{\alpha}{2}\right)\cos\left(\frac{\alpha}{2}\right)$$

$$\chi(\alpha)=\frac{-2i\sin\left(\frac{2l+1}{2}\right)\alpha\sin\left(\frac{\alpha}{2}\right)+2i\sin\left(\frac{2l+1}{2}\right)\alpha\cos\left(\frac{\alpha}{2}\right)}{\cos\alpha+i\sin\alpha-1}$$

$$\chi(\alpha)=\frac{2i\sin\left(\frac{2l+1}{2}\right)\alpha\left[\cos\left(\frac{\alpha}{2}\right)-\sin\left(\frac{\alpha}{2}\right)\right]}{2i\sin\left(\frac{\alpha}{2}\right)\cos\left(\frac{\alpha}{2}\right)+1-2\sin^2\left(\frac{\alpha}{2}\right)-1}$$

$$\chi(\alpha)=\frac{2i\sin\left(\frac{2l+1}{2}\right)\alpha\left[\cos\left(\frac{\alpha}{2}\right)-\sin\left(\frac{\alpha}{2}\right)\right]}{2i\sin\left(\frac{\alpha}{2}\right)\left[\cos\left(\frac{\alpha}{2}\right)-\sin\left(\frac{\alpha}{2}\right)\right]}$$

Then

$$\chi(\alpha)=\frac{\sin\left(l+\frac{1}{2}\right)\alpha}{\sin\left(\frac{\alpha}{2}\right)}\tag{8.2.3}$$

- To obtain the symmetry species into which a set of d-orbitals split in the rotational subgroup O, the appropriate characters are obtained by equations:

$$\chi[C(\phi)]=\frac{\sin\left(L+\frac{1}{2}\right)\phi}{\sin\left(\frac{\phi}{2}\right)}\tag{8.2.4}$$

For rotation by a angle ϕ radians

$$\chi(E)=2L+1\tag{8.2.5}$$

The character under the identity operation is the degeneracy of the orbital set. The resulting representation is reduced by means of the O character Table 8.1, and

$\Gamma_i=\frac{1}{h}\sum_R\chi(R)\chi_i(R)$ (Chapter 4, p. 273)

TABLE 8.1 O-Point Group Character Table

O	E	$6C_4$	$3C_2$	$8C_3$	$6C_2$	–	–
A_1	1	1	1	1	1	–	$x^2+y^2+z^2$
A_2	1	−1	1	1	−1	–	–
E	**2**	**0**	**2**	**−1**	**0**	–	$2z^2-x^2-y^2, x^2-y^2$
T_1	3	1	−1	0	−1	$(R_x, R_y, R_z);(x,y,z)$	–
$\mathbf{T_2}$	**3**	**−1**	**−1**	**0**	**1**	–	(xy, xz, yz)
Γ_d	5	−1	1	−1	1	$= E+T_2$	

The representations, s, p, d, ..., can be worked out in similar procedures, e.g.: In O, d-orbital ($L=2$):

$$\chi[E]=2L+1=2\times 2+1=5$$

$$\chi[C_3]=\frac{\sin\left(2+\frac{1}{2}\right)120}{\sin\left(\frac{120}{2}\right)}=\frac{\sin\,(300)}{\sin\,(60)}=\frac{-0.866}{0.866}=-1$$

- The following representation can be obtained:

O	E	$6C_4$	$3C_2$	$8C_3$	$6C_2$
Γ_d	5	−1	1	−1	1

We then reduce the outcome representation by means of the O character (Table **8.1**), and

$$\Gamma_i=\frac{1}{h}\sum_R \chi(R)\chi_i(R)$$

Since the d-orbitals are gerade, we can write:

$$\Gamma_d=\mathrm{E_g}+\mathrm{T_{2g}}$$

and

$$\mathrm{d}\xrightarrow{\text{splits in } \mathrm{O_h} \text{ symmetry to}}\mathrm{e_g}+\mathrm{t_{2g}}$$

where small letters are used to represent the states for a single electron in various symmetry environments, matching with the use of small letters, s, p, d, f,... to represent their states in the free atom.

- The total representation, Γ_d, can also be obtained by work directly with O_h instead of O, using

$$\chi[C(\phi)]=\frac{\sin\left(L+\frac{1}{2}\right)\phi}{\sin\left(\frac{\phi}{2}\right)} \tag{8.2.4}$$

For rotation by a angle ϕ radians

$$\chi[S(\phi)] = \pm \frac{\sin\left(L+\frac{1}{2}\right)(\phi+\pi)}{\sin\frac{1}{2}(\phi+\pi)} \tag{8.2.6}$$

For improper rotation by a angle ϕ radians, minus sign for ungerade states

$$\chi(E) = 2L+1 \tag{8.2.5}$$

The character under the identity operation in the degeneracy of the orbital set.

$$\chi(\sigma) = \pm\sin\left(L+\frac{1}{2}\right)\pi \tag{8.2.7}$$

Minus sign for ungerade states

$$\chi(i) = \pm(2L+1) \tag{8.2.8}$$

Minus sign for ungerade states

O_h	E	$8C_3$	$6C_2$	$6C_4$	$3C_2$	i	$6S_4$	$8S_6$	$3\sigma_h$	$2\sigma_d$
Γ_d	5	−1	1	−1	1	5	−1	−1	1	1

We then reduce the outcome representation by means of the O_h character (Table 8.2), and

$$\Gamma_i = \frac{1}{h}\sum_R \chi(R)\chi_i(R)$$

which splits in the O_h point group into $E_g + T_{2g}$

$$\Gamma_d = E_g + T_{2g}$$

and

$$d \xrightarrow{\text{splits in } O_h \text{ symmetry}} e_g + t_{2g}$$

TABLE 8.2 O_h-Point Group Character Table

O_h	E	$8C_3$	$6C_2$	$6C_4$	$3C_2$	i	$6S_4$	$8S_6$	$3\sigma_h$	$2\sigma_d$		
A_{1g}	1	1	1	1	1	1	1	1	1	1	–	$x^2+y^2+z^2$
A_{2g}	1	1	−1	−1	1	1	−1	1	1	−1	–	–
E_g	**2**	**−1**	**0**	**0**	**2**	**2**	**0**	**−1**	**2**	**0**	–	$2z^2-x^2-y^2$, x^2-y^2
T_{1g}	3	0	−1	1	−1	3	1	0	−1	−1	(R_x, R_y, R_z)	–
T_{2g}	**3**	**0**	**1**	**−1**	**−1**	**3**	**−1**	**0**	**−1**	**1**	–	(xz, yz,xy)
A_{1u}	1	1	1	1	1	−1	−1	−1	−1	−1	–	–
A_{2u}	1	1	−1	−1	1	−1	1	−1	−1	1	–	–
E_u	2	−1	0	0	2	−2	0	1	−2	0	–	–
T_{1u}	3	0	−1	1	−1	−3	−1	0	1	1	(x,y,z)	–
T_{2u}	3	0	1	−1	−1	−3	1	0	1	−1	–	–
Γ_d	**5**	**−1**	**1**	**−1**	**1**	**5**	**−1**	**−1**	**1**	**1**	$= E_g+T_{2g}$	

TABLE 8.3 Orbital Representations in O Symmetry

O	E	$6C_4$	$3C_2$	$8C_3$	$6C_2$		
A_1	1	1	1	1	1	–	$x^2+y^2+z^2$
A_2	1	−1	1	1	−1	–	–
E	2	0	2	−1	0	–	$2z^2-x^2-y^2$, x^2-y^2
T_1	3	1	−1	0	−1	(R_x, R_y, R_z);(x,y,z)	–
T_2	3	−1	−1	0	1	–	(xy, xz, yz)

Orbital						Irreducible representation	l	Orbital representation
s	1	1	1	1	1	$= A_1$	0	$= a_1$
p	3	1	−1	0	−1	$= T_1$	1	$= t_1$
d	5	−1	1	−1	1	$= E+T_2$	2	$= e+t_2$
f	7	−1	−1	1	−1	$= A_2+T_1+T_2$	3	$= a_2+t_1+t_2$
g	9	1	1	0	1	$= A_1+E+T_1+T_2$	4	$= a_1+e+t_1+t_2$
h	11	1	−1	−1	−1	$=2E+2T_1+T_2$	5	$=2e+2t_1+t_2$
i	13	−1	1	1	1	$= A_1+A_2+E+T_1+2T_2$	6	$= a_1+a_2+e+t_1+2t_2$

TABLE 8.4 Representations for the Various orbitals in O_h, T_d, D_{4h}, and D_3 Symmetries

Orbit	O_h	T_d	D_{4h}	D_3
s	a_{1g}	a_1	a_{1g}	a_1
p	t_{1u}	t_2	$a_{2u}+e_u$	a_2+e
d	e_g+t_{2g}	$e+t_2$	$a_{1g}+b_{1g}+b_{2g}+e_g$	a_1+2e
f	$a_{2u}+t_{1u}+t_{2u}$	$a_2+t_1+t_2$	$a_{2u}+b_{1u}+b_{2u}+2e_u$	a_1+2a_2+2e
g	$a_{1g}+e_g+t_{1g}+t_{2g}$	$a_1+e+t_1+t_2$	$2a_{1g}+a_{2g}+b_{1g}+b_{2g}+2e_g$	$2a_1+a_2+3e$
h	$e_u+2t_{1u}+t_{2u}$	$e+t_1+2t_2$	$a_{1u}+2a_{2u}+b_{1u}+b_{2u}+3e_u$	a_1+2a_2+4e
i	$a_{1g}+a_{2g}+e_g+t_{1g}+2t_{2g}$	$a_1+a_2+e+t_1+2t_2$	$2a_{1g}+a_{2g}+2b_{1g}+2b_{2g}+3e_g$	$3a_1+2a_2+4e$

Find the splitting of each of s, p, d, f, g, h, and i orbitals in O_h environment. Then compare the changes in the splitting of these orbits by varying the environment from O_h to T_d, D_{4h}, and D_3.

- Using a similar symmetry analysis, the splitting of each of these sets orbitals in octahedral environment are obtained by working directly with the simpler pure rotational subgroup O instead of O_h (Table 8.3).
- With similar procedures, we can work out the effect of other symmetry on electrons in other types of orbitals than d orbitals (Table 8.4).

8.3 CORRELATION TABLE

Orbital Correlation Table

Use the character tables to find the splitting of the sets of e_g and t_{2g} orbitals when the structure is changed from the O_h to C_{4v}.

What are the equivalents of the representations of group O_h in its subgroups when symmetry is changed to O_h, O, T_d, D_{4h}, D_{2d}, D_{3d}, D_3, C_{2v}, C_{4v}, and C_{2h}?

- The splitting of e_g and t_{2g} orbitals when the structure is changed from the O_h to C_{4v}:

	E	$8C_3$	$6C_2$	$6C_4$	$3C_2$	i	$6S_4$	$8S_6$	$3\sigma_h$	$6\sigma_d$	–
$(A_1)_{C_{4v}}$	1	0	0	1	1	0	0	0	0	1	–
$(A_2)_{C_{4v}}$	1	0	0	1	1	0	0	0	0	−1	–
$(B_1)_{C_{4v}}$	1	0	0	−1	1	0	0	0	0	−1	–
$(B_2)_{C_{4v}}$	1	0	0	−1	1	0	0	0	0	1	–
$E_{C_{4v}}$	2	0	0	0	−2	0	0	0	0	0	–
$(E_g)_{O_h}$	2	−1	0	0	2	2	0	−1	2	0	$(A_1+B_1)_{C_{4v}}$
$(T_{2g})_{O_h}$	3	0	1	−1	−1	3	−1	0	−1	1	$(B_2+E)_{C_{4v}}$

$$\text{If}: \Gamma_i = \frac{1}{h}\sum_R \chi(R)\chi_i(R), \text{ then}$$

$$\Gamma_{O_h} = \frac{1}{h_{O_h}}\sum_R \chi(R)_{O_h}\chi(R)_{C_{4v}}, \text{ and}$$

Therefore:

$$(E_g)_{O_h} = (A_1+B_1)_{C_{4v}}, \text{ and}$$

$$(T_{2g})_{O_h} = (B_2+E)_{C_{4v}}$$

or

$$(e_g)_{O_h} = (a_1+b_1)_{C_{4v}}$$

$$(t_{2g})_{O_h} = (b_2+e)_{C_{4v}}$$

- We can use similar procedures to set the correlation table (Table 8.5), which gives the equivalent of the representations of group O_h in its subgroups when symmetry is changed.

TABLE 8.5 Correlation Table

O_h	O	T_d	D_{4h}	D_{2d}	D_{3d}	D_3	C_{2v}	C_{4v}	C_{2h}
a_{1g}	a_1	a_1	a_{1g}	a_1	a_{1g}	a_1	a_1	a_1	a_g
a_{2g}	a_2	a_2	a_{1g}	a_1	a_{2g}	a_2	a_2	b_1	b_g
e_g	e	e	$a_{1g}+b_{1g}$	a_1+b_1	e_g	e	a_1+a_2	a_1+b_1	a_g+b_g
t_{1g}	t_1	t_1	$a_{2g}+e_g$	a_2+e	$a_{2g}+e_g$	a_2+e	$a_2+b_1+b_2$	a_2+e	a_g+2b_g
t_{2g}	t_2	t_2	$b_{2g}+e_g$	b_2+e	$a_{1g}+e_g$	a_1+e	$a_1+b_1+b_2$	b_2+e	$2a_g+b_g$
a_{1u}	a_1	a_2	a_{1u}	b_1	a_{1u}	a_1	a_2	a_2	a_u
a_{2u}	a_2	a_1	b_{1u}	a_1	a_{2u}	a_2	a_1	b_2	b_u
e_u	e	e	$a_{1u}+b_{1u}$	a_1+b_1	e_u	e	a_1+a_2	a_2+b_2	a_u+b_u
t_{1u}	t_1	t_2	$a_{2u}+e_u$	b_2+e	$a_{2u}+e_u$	a_2+e	$a_1+b_1+b_2$	a_1+e	a_u+2b_u
t_{2u}	t_2	t_1	$b_{2u}+e_u$	a_2+e	$a_{1u}+e_u$	a_1+e	$a_2+b_1+b_2$	b_1+e	$2a_u+b_u$

Term Correlation Tables

Find the splitting of each of S, P, D, F, G, H, and I terms in an O_h environment. Then compare the changes in the splitting of these terms by varying the environment from O_h to T_d, D_{4h}, and D_3.

- The protocol for single electrons in different sets of orbitals are also appropriated to the behavior of term arising from multielectron systems (Tables 8.6 and 8.7).

  Example:

  ^3F, ^3P, ^1G, ^1D, and ^1S-terms of d^2 configuration behave the same like f, p, g, d, and s orbitals.

  The F term is sevenfold degenerate similar to f orbital, and is defined by a precise wave function for each of the seven M_L values. Every one of these wave function has a ϕ angular function described by $e^{iM_L\phi}$, which is exactly similar to $e^{im_l\phi}$ in the wave function of single d-electron.
- Roles for the use of the subscripts g and u:
  - No subscripts are used in the case of the point group has no center of symmetry:

$$d \xrightarrow{\text{splits in } T_d \text{ symmetry}} e + t_2$$

for F-term of d^2 configuration

$$^3F \xrightarrow{\text{splits in } T_d \text{ symmetry}} {}^3A_2 + {}^3T_1 + {}^3T_2$$

  - If the surrounding does have a center of symmetry, type of the orbital (s, p, d, ...) determines the subscript.
  - All the representations of the atomic orbitals for which the quantum number l is even (s, d, g, ...) are centrosymmetric and have a g subscript:

TABLE 8.6 Term Splitting, When l is Even (s, d, g, ... Orbital)

Term	O_h	T_d	D_{4h}	D_3
S	A_{1g}	A_1	A_{1g}	A_1
P	T_{1g}	T_2	$A_{2g}+E_g$	A_2+E
D	E_g+T_{2g}	$E+T_2$	$A_{1g}+B_{1g}+B_{2g}+E_g$	A_1+2E
F	$A_{2g}+T_{1g}+T_{2g}$	$A_2+T_1+T_2$	$A_{2g}+B_{1g}+B_{2g}+2E_g$	A_1+2A_2+2E
G	$A_{1g}+E_g+T_{1g}+T_{2g}$	$A_1+E+T_1+T_2$	$2A_{1g}+A_{2g}+B_{1g}+B_{2g}+2E_g$	$2A_1+A_2+3E$
H	$E_g+2T_{1g}+T_{2gu}$	$E+T_1+2T_2$	$A_{1g}+2A_{2g}+B_{1g}+B_{2g}+3E_g$	A_1+2A_2+4E
I	$A_{1g}+A_{2g}+E_g+T_{1g}+2T_{2g}$	$A_1+A_2+E+T_1+2T_2$	$2A_{1g}+A_{2g}+2B_{1g}+2B_{2g}+3E_g$	$3A_1+2A_2+4E$

TABLE 8.7 Term splitting, When l is Odd (p, f, h, ... Orbital)

Term	O_h	T_d	D_{4h}	D_3
S	A_{1u}	A_1	A_{1u}	A_1
P	T_{1u}	T_2	$A_{2u}+E_u$	A_2+E
D	E_u+T_{2u}	$E+T_2$	$A_{1u}+B_{1u}+B_{2u}+E_u$	A_1+2E
F	$A_{2u}+T_{1u}+T_{2u}$	$A_2+T_1+T_2$	$A_{2u}+B_{1u}+B_{2u}+2E_u$	A_1+2A_2+2E
G	$A_{1u}+E_u+T_{1u}+T_{2u}$	$A_1+E+T_1+T_2$	$2A_{1u}+A_{2u}+B_{1u}+B_{2u}+2E_u$	$2A_1+A_2+3E$
H	$E_u+2T_{1u}+T_{2u}$	$E+T_1+2T_2$	$A_{1u}+2A_{2u}+B_{1u}+B_{2u}+3E_u$	A_1+2A_2+4E
I	$A_{1u}+A_{2u}+E_u+T_{1u}+2T_{2u}$	$A_1+A_2+E+T_1+2T_2$	$2A_{1u}+A_{2u}+2B_{1u}+2B_{2u}+3E_u$	$3A_1+2A_2+4E$

$$d \xrightarrow{\text{splits in } O_h \text{ symmetry}} e_g + t_{2g}$$

for D-term of d^2 configuration

$$D \xrightarrow{\text{splits in } O_h \text{ symmetry}} E_g + T_{2g}$$

- All the representations of the atomic orbitals for which the quantum number l is odd (p, f, h, ...) are antisymmetric to inversion and has u subscript:

$$f \xrightarrow{\text{splits in } O_h \text{ symmetry}} a_{2u} + t_{1u} + t_{2u}$$

- All the states that result from the splitting of a particular term have the same multiplicity as the parent term. For ^3F-terms of a d^2 configuration:

$$^3F \xrightarrow{\text{splits in } O_h \text{ symmetry}} {}^3A_{2g} + {}^3T_{1g} + {}^3T_{2g}$$

From Table 8.5, the splitting of different sets of one electron orbitals applies to the splitting of analogous term (Tables 8.6 and 8.7).

What are the free ion ground state terms for d^n ($n = 1, 2, \ldots, 9$) configurations in T_d and O_h fields, indicating the multiplicity and the superscripts u and g.

- The free ion ground state terms for d^n ($n=1, 2, \ldots, 9$) configurations in T_d and O_h fields, are collected in Table 8.8.
- The electronic configurations d^5 and d^{10} have S term as ground state, which are configured to $A_{1(g)}$ in octahedral and tetrahedral symmetries (Fig. 8.1).

8.4 CORRELATION DIAGRAMS OF STRONG AND WEAK FIELDS

Correlation Diagram of Strong and Weak Field States of O_h

Construct the energy level diagram for d^2 in O_h that shows how the energy of various states in which the free ion terms are split depending on the interaction of the ion with its environment.

- The interaction strength of the ion with its surrounding decides the energy of different states in which the free ion terms are split.
- The difference in energy between the two states into which the term of d-orbital has split in O_h or T_d symmetry (Δ_o or Δ_t) can be used as an appropriate measure of this interaction.

TABLE 8.8 Free Ion Ground States Terms for d^n Configurations in O_h and T_d Fields

		Configuration: Ground State			
		O_h		T_d	
Configuration	Free Ion State	High Spin	Low Spin	High Spin	Low Spin
d^1	^2D	t_{2g}^1: $^2T_{2g}$		e^1: ^2E	
d^2	^3F	t_{2g}^2: $^3T_{1g}$		e^2: $^3A_2$	
d^3	^4F	t_{2g}^3: $^4A_{2g}$		$e^2t_2^1$: $^4T_1$	e^3: ^2E
d^4	^5D	$t_{2g}^3e_g^1$: ^5E_g	t_{2g}^4: $^3T_{1g}$	$e^2t_2^2$: $^5T_2$	e^4: $^1A_1$
d^5	^6S	$t_{2g}^3e_g^2$: $^6A_{1g}$	t_{2g}^5: $^2T_{2g}$	$e^2t_2^3$: $^6A_1$	$e^4t_2^1$: $^2T_2$
d^6	^5D	$t_{2g}^4e_g^2$: $^5T_{2g}$	t_{2g}^6: $^1A_{1g}$	$e^3t_2^3$: ^5E	$e^4t_2^2$: $^3T_1$
d^7	^4F	$t_{2g}^5e_g^2$: $^4T_{1g}$	$t_{2g}^6e_{2g}^1$: ^2E_g	$e^4t_2^3$: $^4A_2$	
d^8	^3F	$t_{2g}^6e_g^2$: $^3A_{2g}$		$e^4t_2^4$: $^3T_1$	
d^9	^2D	$t_{2g}^6e_g^3$: ^2E_g		$e^4t_2^5$: $^2T_2$	

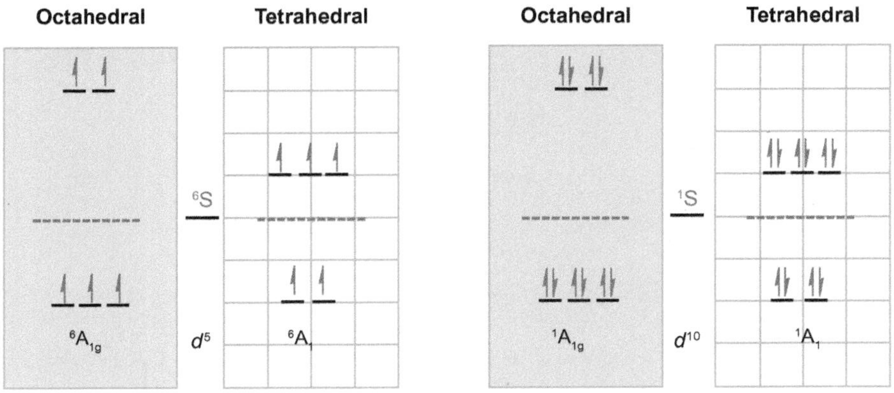

FIG. 8.1 Configurations and terms of d^5 and d^{10} in weak octahedral and tetrahedral fields.

- Constructing the energy level diagram will show how the energy of various states in which the free ion terms are split depending on the interaction of the ion with its environment.
- In the energy level diagram:
  - energy as ordinate and the magnitude of Δ_o or Δ_t as abscissa,
  - at the right side of the diagram, the free-ion term energy, where Δ_o or Δ_t is zero,
  - at the extreme left, the energy states where the interaction induces an energy difference much greater than that of the interelectronic interactions, π, which become negligible by compression, Δ_o or $\Delta_t >>> \pi$.
- The case of a d^2 ion in an octahedral environment will be considered to illustrate the way of constructing an energy level diagram (Fig. 8.2).
- The free ion terms of d^2 (and d^8) in order of increasing energy are: ^3F, ^1D, ^3P, ^1G, ^1S (Chapter 2, p. 123, 129).
- The splitting of these terms in an O_h environment (Table 8.6), in the weak field (left side):

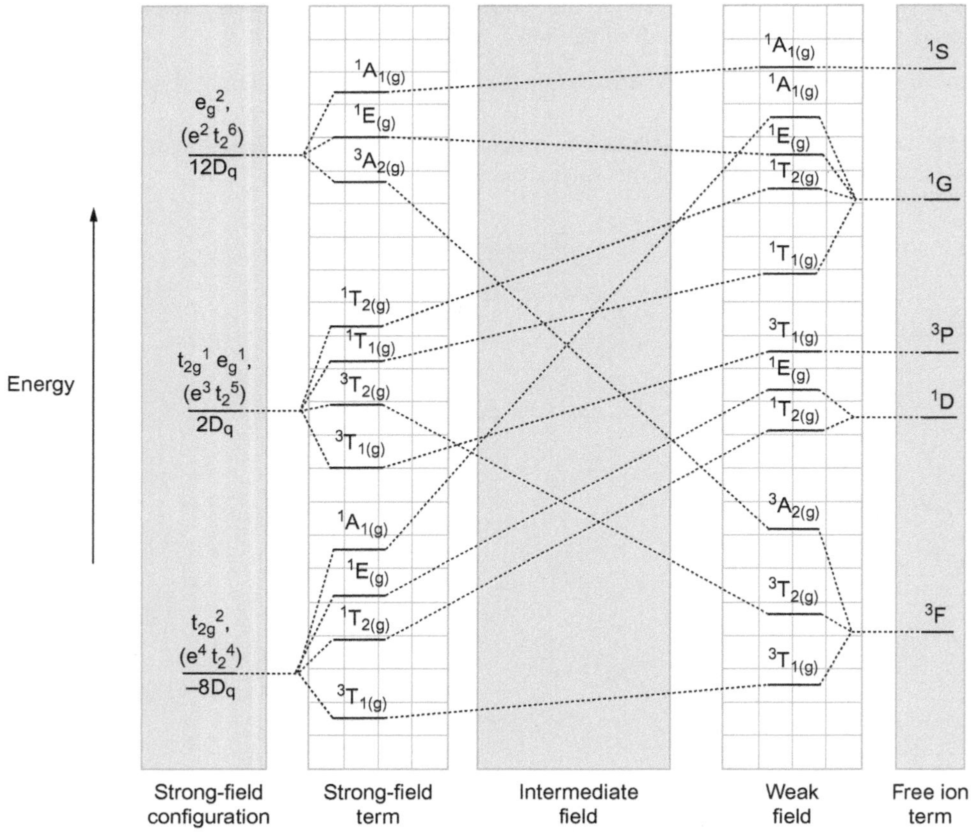

FIG. 8.2 Correlation diagram for a d^2 ion in O_h, and d^8 in T_d environments (the diagram is not to scale).

$$^{3}F \xrightarrow{\text{splits in } O_h \text{ symmetry to}} {}^{3}A_{2g} + {}^{3}T_{1g} + {}^{3}T_{2g}$$

$$^{1}D \xrightarrow{\text{splits in } O_h \text{ symmetry to}} {}^{1}E_{g} + {}^{1}T_{2g}$$

$$^{3}P \xrightarrow{O_h \text{ symmetry}} {}^{3}T_{1g}$$

$$^{1}G \xrightarrow{\text{splits in } O_h \text{ symmetry to}} {}^{1}A_{1g} + {}^{1}E_{g} + {}^{1}T_{1g} + {}^{1}T_{2g}$$

$$^{1}S \xrightarrow{O_h \text{ symmetry}} {}^{1}A_{1g}$$

- At the strong field limit, splitting of d orbitals is extremely large, and the following three configurations, in order of increasing energy, are possible: t_{2g}^2, $t_{2g}e_g$, and e_g^2.
- The symmetry properties of these states can be determined by taking the direct product representation (Chapter 4, p. 274) of the single electrons (Table 8.9).
- For the configuration t_{2g}^2, the direct product:

$$t_{2g} \times t_{2g} \xrightarrow{\text{decomposed into}} A_{1g} + E_g + T_{1g} + T_{2g}$$

  - Determining the multiplicities of these strong field states:
    The t_{2g}^2 may be regarded as a set of six boxes as shown below:

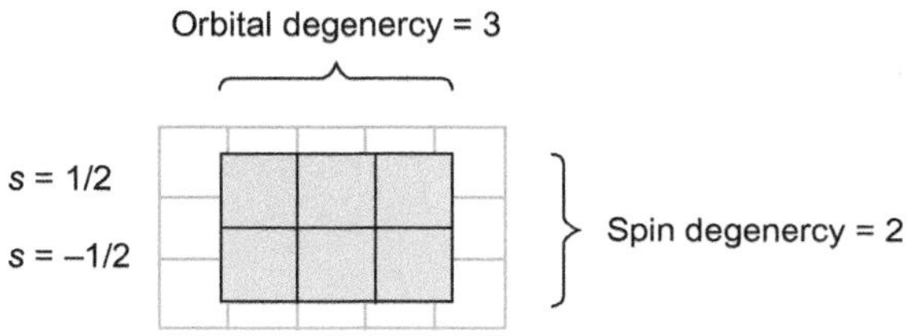

The number of ways in which two electrons may occupy these six boxes = 15

TABLE 8.9 The Direct Product Representations for t_{2g}^2, $t_{2g}e_g$, and e_g^2 Configurations

O_h	E	$8C_3$	$6C_2$	$6C_4$	$3C_2$	i	$6S_4$	$8S_6$	$3\sigma_h$	$6\sigma_d$		
A_{1g}	1	1	1	1	1	1	1	1	1	1	–	$x^2+y^2+z^2$
A_{2g}	1	1	−1	−1	1	1	−1	1	1	−1	–	–
E_g	2	−1	0	0	2	2	0	−1	2	0	–	$2z^2-x^2-y^2$, x^2-y^2
T_{1g}	3	0	−1	1	−1	3	1	0	−1	−1	(R_x, R_y, R_z)	–
T_{2g}	3	0	1	−1	−1	3	−1	0	−1	1		(xz, yz, xy)
A_{1u}	1	1	1	1	1	−1	−1	−1	−1	−1	–	–
A_{2u}	1	1	−1	−1	1	−1	1	−1	−1	1	–	–
E_u	2	−1	0	0	2	−2	0	1	−2	0	–	–
T_{1u}	3	0	−1	1	−1	−3	−1	0	1	1	(x, y, z)	–
T_{2u}	3	0	1	−1	−1	−3	1	0	1	−1	–	–
$t_{2g} \times t_{2g}$	9	0	1	1	1	9	1	0	1	1	$= A_{1g}+E_g+T_{1g}+T_{2g}$	
$t_{2g} \times e_g$	6	0	0	0	−2	6	0	0	−2	0	$= 2T_{1g}+2T_{2g}$	
$e_g \times e_g$	4	1	0	0	4	4	0	1	4	0	$= A_{1g}+A_{2g}+E_g$	

$$\text{Total degeneracy of the } t_{2g}^2 \text{ configuration} = \frac{n!}{e!h!} = \frac{6!}{2!4!} = \frac{(6\times5)}{2} = 15$$

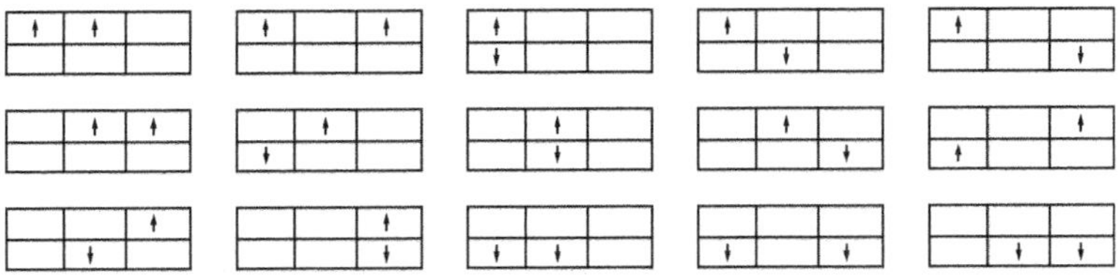

$$t_{2g} \times t_{2g} \Rightarrow {}^{a}A_{1g} + {}^{b}E_{g} + {}^{c}T_{1g} + {}^{d}T_{2g}$$

$$\text{The total degeneracy} = \underbrace{1\times a}_{^{a}A_{1g}} + \underbrace{2\times b}_{^{b}E_{g}} + \underbrace{3\times c}_{^{c}T_{1g}} + \underbrace{3\times d}_{^{d}T_{2g}} = 15$$

There are three sets of possible multiplicities:

	1 × a	+2 × b	+3 × c	+3 × d	=15	$^{a}A_{1g}+^{b}E_{g}+^{c}T_{1g}+^{d}T_{2g}$
I	1 × 1	+2 × 1	+3 × 1	+3 × 3	=15	$^{1}A_{1g}+^{1}E_{g}+^{1}T_{1g}+^{3}T_{2g}$
II	1 × 1	+2 × 1	+3 × 3	+3 × 1	=15	$^{1}A_{1g}+^{1}E_{g}+^{3}T_{1g}+^{1}T_{2g}$
III	1 × 3	+2 × 3	+3 × 1	=3 × 1	=15	$^{3}A_{1g}+^{3}E_{g}+^{1}T_{1g}+^{1}T_{2g}$

- For the configuration $t_{2g}\,e_g$, the direct product:

$$t_{2g} \times e_g \xrightarrow{\text{split into}} 2T_{1g} + 2T_{2g}$$

$$\text{The total degeneracy} = \frac{6!}{1!5!} \times \frac{4!}{1!3!} = \frac{(6\times4)}{1} = 24$$

$$t_{2g} \times e_g \Rightarrow {}^{1}T_{2g} + {}^{3}T_{1g} + {}^{1}T_{2g} + {}^{3}T_{2g}$$

- For the configuration e_g^2, the direct product:

$$e_g \times e_g \xrightarrow{\text{split into}} A_{1g} + A_{2g} + E_g$$

$$\text{The total degeneracy} = \frac{4!}{2!2!} = \frac{(4\times3\times2)}{2\times2} = 6$$

$$e_g \times e_g \Rightarrow {}^{a}A_{1g} + {}^{b}A_{2g} + {}^{c}E_g$$

$$1\times a + 1\times b + 2\times c = 6$$

There are two sets of possible multiplicity

	1 × a	+1 × b	+2 × c	=6	$^{a}A_{1g}+^{b}A_{2g}+^{c}E_{g}$
I	1 × 1	+1 × 3	+2 × 1	=6	$^{1}A_{1g}+^{3}A_{2g}+^{1}E_{g}$
II	1 × 3	+1 × 1	+2 × 1	=6	$^{3}A_{1g}+^{1}A_{2g}+^{1}E_{g}$

- In the case of more than one set of possible multiplicity, two principles must hold to identify the correct multiplicity:
  - There have to be one to one matching between the states at the two limits of the abscissa (strong and weak field).
  - As the strength of interaction changes, states of the identical symmetry and spin degeneracy never cross.
- In Fig. 8.2, on the extreme right-hand side are the states of free ion; on the left is the splitting of these states under the influence of an O_h environment.

Method of Descending Symmetry (Descending Multiplicities of the Orbital States)

How can you find the spin multiplicities of the orbital states as they arise from interelectronic interaction in the configuration e_g^2 and t_{2g}^2 by using the method of descending symmetry for d^2 in the strong field side of O_h environment?

- Lowering the symmetry does not change the spin multiplicities (degeneracies).
- For the configuration e_g^2, by lowering the symmetry to D_{4h} (correlation Table 8.5):

$$e_g \underset{\substack{\text{electron interactions} \\ \text{in } D_{4h} \text{ symmetry}}}{\overset{\text{split into}}{\Longrightarrow}} a_{1g} + b_{1g}$$

  - The possible ways to place two electrons: a_{1g}^2, $a_{1g}^1b_{1g}^1$, and b_{1g}^2
    The resulting states:

a_{1g}^2:	$a_{1g} \times a_{1g} = A_{1g}$	and must be singlet (exclusion principle): $^1A_{1g}$
$a_{1g}^1b_{1g}^1$:	$a_{1g} \times b_{1g} = B_{1g}$	singlet and triplet are permitted: $^1B_{1g}$ and $^3B_{1g}$
b_{1g}^2:	$b_{1g} \times b_{1g} = A_{1g}$	and must be singlet (exclusion principle): $^1A_{1g}$

    Therefore, the permitted multiplicities of D_{4h}: $^1A_{1g}$, $^1B_{1g}{}^3B_{1g}$, and $^1A_{1g}$.
  - For the configuration e_g^2, the direct product is as follows:

$$e_g \times e_g \underset{\substack{\text{electron interactions} \\ \text{in } O_h \text{ symmetry}}}{\overset{\text{split into}}{\Longrightarrow}} A_{1g} + A_{2g} + E_g$$

  - The states A_{1g}, A_{2g}, and E_g may change into states appropriate to D_{4h} symmetry correlation (Table 8.5):

$$A_{1g} \xrightarrow{\text{In } D_{4h}} A_{1g}$$

$$A_{2g} \xrightarrow{\text{In } D_{4h}} B_{1g}$$

$$E_g \xrightarrow{\text{In } D_{4h}} \begin{cases} A_{1g} \\ B_{1g} \end{cases}$$

  - Since the permitted multiplicities of D_{4h}: $^1A_{1g}$, $^1B_{1g}$, $^3B_{1g}$, and $^1A_{1g}$
    Lowering the symmetry cannot change the spin multiplicities,

$$A_{1g} \xrightarrow{\text{In } D_{4h}} {}^1A_{1g} \xrightarrow{\text{In } O_h} {}^1A_{1g}$$

$$A_{2g} \xrightarrow{\text{In } D_{4h}} {}^3B_{1g} \xrightarrow{\text{In } O_h} {}^3A_{2g}$$

$$E_g \xrightarrow{\text{In } D_{4h}} \left\{ \begin{matrix} {}^1A_{1g} \\ {}^1B_{1g} \end{matrix} \right\} \xrightarrow{\text{In } O_h} {}^1E_g$$

- For the configuration t_{2g}^2, by lowing the symmetry to C_{2h} (correlation Table 7.5):

$$t_{2g} \underset{\substack{\text{electron interactions} \\ \text{in } C_{2h} \text{ symmetry}}}{\overset{\text{split into}}{\Longrightarrow}} 2a_g + b_g = a_{g1} + a_{g2} + b_g$$

  - The possible ways to place 2 electrons: a_{g1}^2, $a_{g1}^1a_{g2}^1$, a_{g2}^2, $a_{g1}^1b_g^1$, $a_{g2}^1b_g^1$, and b_g^2

The resulting states:

a_{g1}^2:	$a_{g1} \times a_{g1} = A_g$	and must be singlet (exclusion principle): ^1A_g
$a_{g1}^1 a_{g2}^1$:	$a_{g1} \times a_{g2} = A_g$	singlet and triplet are permitted: ^1A_g and ^3A_g
a_{g2}^2:	$a_{g2} \times a_{g2} = A_g$	and must be singlet (exclusion principle): ^1A_g
b_g^2:	$b_g \times b_g = A_g$	and must be singlet (exclusion principle): ^1A_g
$a_{g1}^1 b_g^1$:	$a_{g1} \times b_g = B_g$	singlet and triplet are permitted: ^1B_g and ^3B_g
$a_{g2}^1 b_g^1$:	$a_{g2} \times b_g = B_g$	singlet and triplet are permitted: $^1B_{1g}$ and $^3B_{1g}$

- Therefore, the permitted multiplicities of C_{2h}: 4^1A_8, ^3A_g, 2^1B_g, and 2^3B_g.
- For the configuration t_{2g}^2, the direct product:

$$t_{2g} \times t_{2g} \underbrace{\xRightarrow{\text{split into}}}_{\substack{\text{electron interactions} \\ \text{in } O_h \text{ symmetry}}} A_{1g} + E_g + T_{1g} + T_{2g}$$

- The states A_{1g}, E_g, T_{1g}, and T_{2g} may change into states appropriate to C_{2h} symmetry (for correlation see Table 7.6):

$$A_{1g} \xrightarrow{\text{In } C_{2h}} A_g$$

$$E_g \xrightarrow{\text{In } C_{2h}} \begin{cases} A_g \\ B_g \end{cases}$$

$$T_{1g} \xrightarrow{\text{In } C_{2h}} \begin{cases} A_g \\ B_g \\ B_g \end{cases}$$

$$T_{2g} \xrightarrow{\text{In } C_{2h}} \begin{cases} A_g \\ A_g \\ B_g \end{cases}$$

- Since the permitted multiplicities of C_{2h}: 4^1A_8, ^3A_g, 2^1B_g, and 2^3B_g, and lowering the symmetry cannot change the spin multiplicities:

$$A_{1g} \xrightarrow{\text{In } C_{2h}} {}^1A_{1g} \xrightarrow{\text{In } O_h} {}^1A_{1g}$$

$$E_g \xrightarrow{\text{In } C_{2h}} \begin{Bmatrix} {}^1A_g \\ {}^1B_g \end{Bmatrix} \xrightarrow{\text{In } O_h} {}^1E_g$$

$$T_{1g} \xrightarrow{\text{In } C_{2h}} \begin{Bmatrix} {}^3A_g \\ {}^3B_g \\ {}^3B_g \end{Bmatrix} \xrightarrow{\text{In } O_h} {}^3T_{1g}$$

$$T_{2g} \xrightarrow{\text{In } C_{2h}} \begin{Bmatrix} {}^1A_g \\ {}^1A_g \\ {}^1B_g \end{Bmatrix} \xrightarrow{\text{In } O_h} {}^1T_{2g}$$

Correlation Diagram of Weak and Strong Field States of T_d

Construct the energy level diagram for d^2 in T_d symmetry that show how the energy of various states in which the free ion terms are split depending on the interaction of the ion with its environment.

- Similar procedures to those described for O_h symmetry.
- The free ion terms of d^2 in order of increasing energy are: $^3F, ^1D, ^3P, ^1G, ^1S$
- The splitting of these terms in a T_d environment (Table 8.6), in the weak field side (left side):

$$^3F \xrightarrow{\text{splits in } T_d \text{ symmetry}} {}^3A_2 + {}^3T_1 + {}^3T_2$$

$$^1D \xrightarrow{\text{splits in } T_d \text{ symmetry}} {}^1E + {}^1T_2$$

$$^3P \xrightarrow{\text{splits in } T_d \text{ symmetry}} {}^3T_1$$

$$^1G \xrightarrow{\text{splits in } T_d \text{ symmetry}} {}^1A_1 + {}^1E + {}^1T_1 + {}^1T_2$$

$$^1S \xrightarrow{\text{splits in } T_d \text{ symmetry}} {}^1A_1$$

- In strong field (right side), the following three configurations, in order of increasing energy, are: e^2, et_2, and t_2^2.

  The interelectronic coupling will cause the splitting of these states. The constituent of these can be obtained from the direct product representations. The same methods described for O_h case are employed to assign the spin multiplicities:

$$e \times e \Rightarrow {}^1A_1 + {}^3A_2 + {}^1E$$

$$t_2 \times e \Rightarrow {}^1T_1 + {}^3T_1 + {}^1T_2 + {}^3T_2$$

$$t_2 \times t_2 \Rightarrow {}^1A_1 + {}^1E + {}^3T_1 + {}^1T_2$$

- The states on the two sides of the diagram are correlated to obtain the complete correlation diagram (Fig. 8.3).

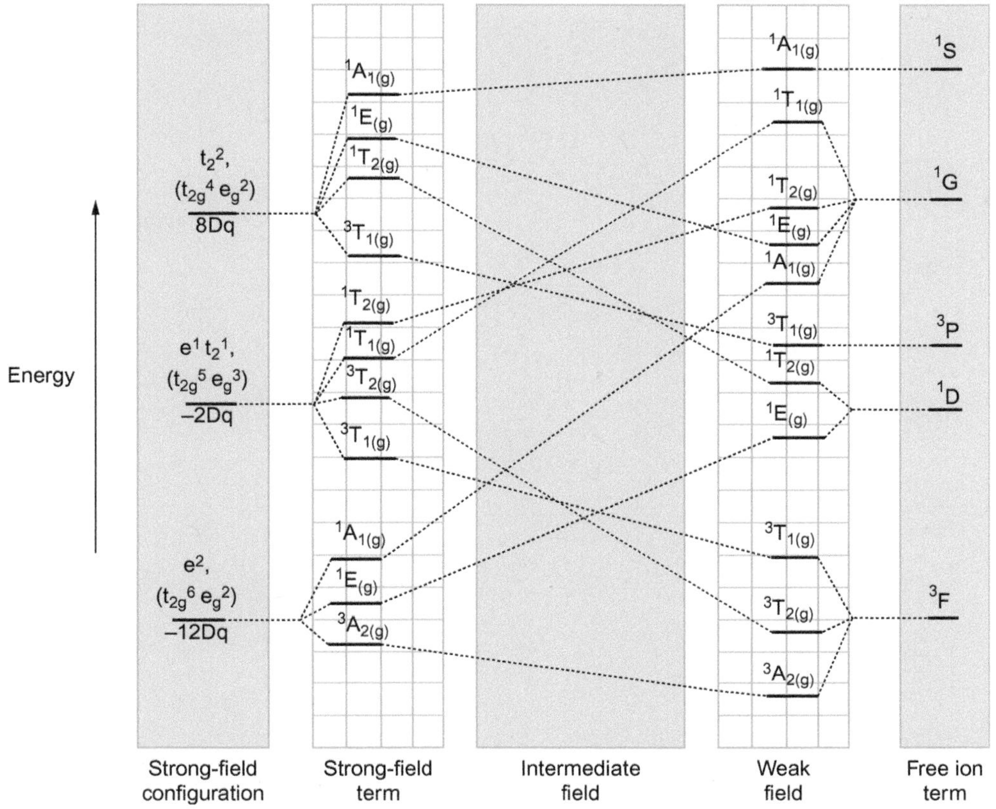

FIG. 8.3 Correlation diagram for a d^2 ion in T_d, and d^8 in O_h environments (the diagram is not to scale).

8.5 ORGEL DIAGRAM

Orgel Diagram of D Term Configuration

Determine the free ion ground state term symbol of d^1, d^9, and high-spin d^4 and d^6 configurations. Track the splitting of these terms in O_h and T_d ligand field environments. Compute the crystal field stabilization energy of these states. Summarize all the energy variations in one diagram. What is the common name of that diagram?

- The configurations of d^1, high-spin d^4, d^6, and d^9 have D term ground states (Table 8.10).
- The five folds of d^1, d^6, d^4 and d^9 in O_h as well as in T_d symmetry split into $T_{2(g)}$ and $E_{(g)}$:

$$D \xrightarrow{\text{in } O_h \text{ and } T_d} T_{2(g)} + E_{(g)}$$

- The electronic configurations d^1 and d^6 in O_h as well as d^4 and d^9 in T_d symmetry give rise to a $T_{2(g)}$ in the ground state. Promotion of an electron gives configurations corresponding to the exited $E_{(g)}$ state. As the value of Dq increases, so does the splitting between the states (Fig. 8.4).
- In contrast, d^4 and d^9 configurations in an O_h, also d^1 and d^6 in T_d field have $E_{(g)}$ terms in the ground state and $T_{2(g)}$ in the first excited state (Fig. 8.5).
- In a weak O_h field, the $T_{2(g)}$ term is the ground term for d^1 and d^6, with inverse splitting, $E_{(g)}$ ground term, for d^4 and d^9. In a weak T_d field, the $E_{(g)}$ term is the ground term for d^1 and d^6, with inverse splitting, $T_{(2g)}$ ground term, for d^4 and d^9. As a result, the splitting d^n is the same as d^{n+5} and the opposite with d^{10-n}.

TABLE 8.10 Configurations of d^1, $d^9$and High-Spin d^4 and d^6 Have D Term Ground States

m_l	2	1	0	−1	−2	*L*	*S*	Term
d^1	↑					2	1/2	^2D
d^4	↑	↑	↑	↑		2	2	^5D
d^6	↑↓	↑	↑	↑	↑	2	2	^5D
d^9	↑↓	↑↓	↑↓	↑↓	↑	2	1/2	^2D

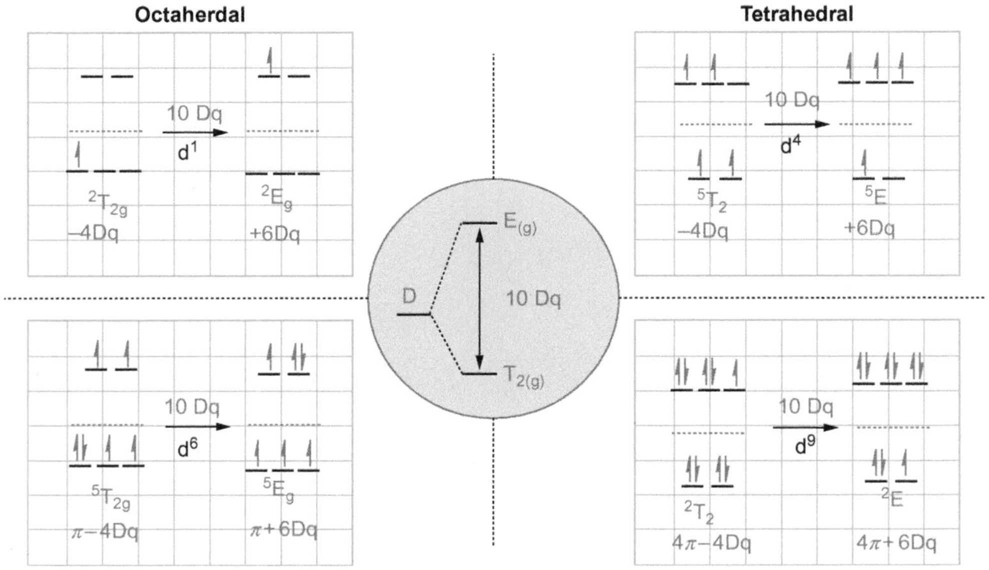

FIG. 8.4 Configurations and terms of d^1 and d^6 in weak octahedral and d^4, d^9 in tetrahedral fields.

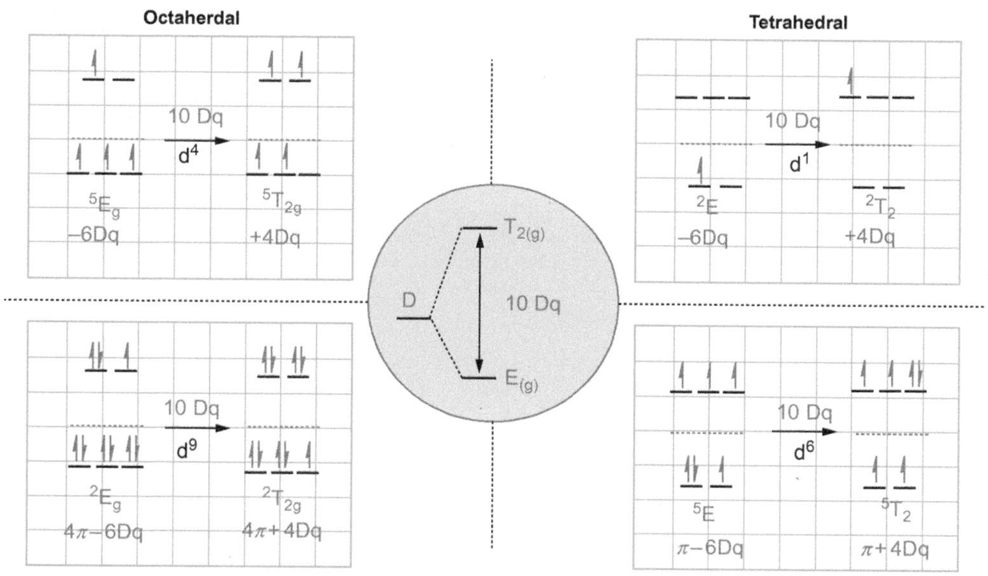

FIG. 8.5 Configurations and terms of d^4 and d^9 in weak octahedral and d^1, d^6 in tetrahedral fields.

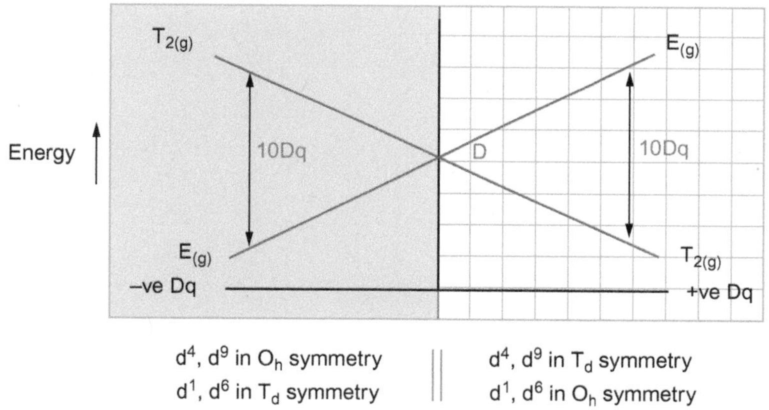

FIG. 8.6 Orgel diagram for ground and excited states of the D term configurations.

- Therefore, a single diagram, called an Orgel diagram, can summarize the change in energy with Dq for four different configurations: d^1, d^4, d^6, and d^9 (Fig. 8.6).
- The Orgel diagram is qualitative, because even for the same ligand $Dq_{Oh} > Dq_{Td}$, beside Dq depends on the particular metal ion.
- Orgel diagrams contain only those terms that have the same multiplicity as the ground state. Therefore, they are adequate for understanding the electronic spectra of multiplicity allowed transitions.

Orgel Diagram of F Term

Determine the free ion ground state term symbol of d^2, d^3, d^8, and high-spin d^7 configurations. Track the splitting of these terms in O_h and T_d ligand field environments in the presence of the nearby ^3P-term. Why can we not neglect this term? Compute the crystal field stabilization energy of these states, and summarize all the energy variations in one diagram. What is the common name of that diagram?

- All d^2, d^3, d^8, and high-spin d^7configuration have an F-ground term (Table 8.11):
- The five folds of d^2, d^7, d^3, and d^8 in O_h as well as in T_d symmetry split into $T_{1(g)}$, $T_{2(g)}$, and $A_{2(g)}$, (Table 8.6):

TABLE 8.11 Configurations of d^2, d^3, d^8, and High-Spin d^7 Have F Term Ground States

m_l	2	1	0	−1	−2	*L*	*S*	Term
d^2	↑	↑				3	1	^3F
d^3	↑	↑	↑			3	3/2	^4F
d^7	↑↓	↑↓	↑	↑	↑	3	3/2	^4F
d^8	↑↓	↑↓	↑↓	↑	↑	3	1	^3F

$$F \xrightarrow{\text{in } O_h \text{ and } T_d} T_{2(g)} + T_{1(g)} + A_{2(g)}$$

- The electronic configurations d^2 and d^7 in O_h as well as d^3 and d^8 in T_d symmetry give rise to a $T_{1(g)}$ in the ground state. Promotion of one electron gives configurations corresponding to the first exited $T_{2(g)}$ state. The promotions of two electrons to e_g generate the second excited state $A_{2(g)}$ (Fig. 8.7).
- On the other hand, d^3 and d^8 configurations in an O_h, as well, d^2 and d^7 in T_d field have $A_{2(g)}$ terms in the ground state and $T_{2(g)}$ in the first excited state. The promotions of two electrons to e_g produce the excited state $T_{1(g)}$ (Fig. 8.8).
- Close by the F ground term, there is also the P term of the same spin multiplicity at higher energy.

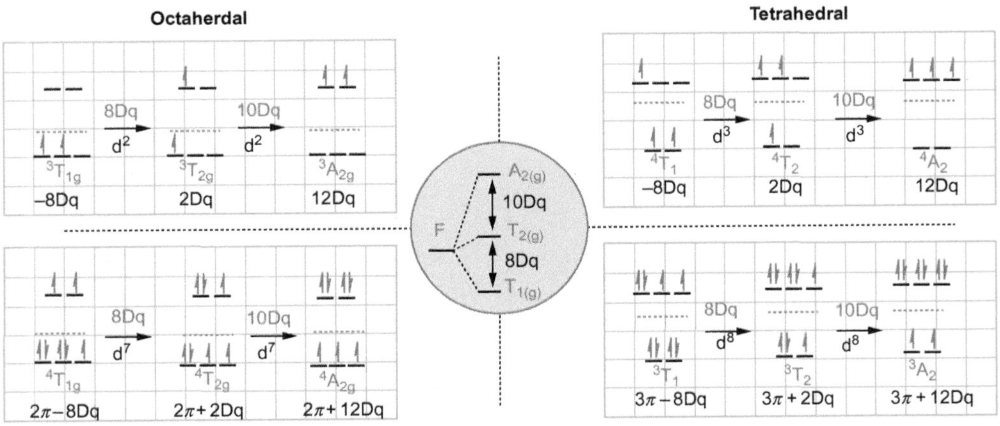

FIG. 8.7 Configurations and terms of d^2 and d^7 in weak octahedral and d^3, d^8 in tetrahedral fields.

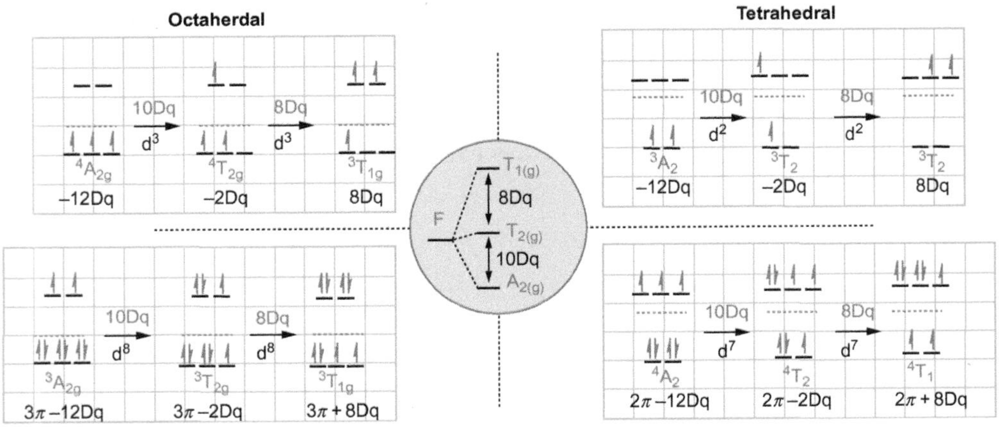

FIG. 8.8 Configurations and terms of d^3 and d^8 in weak octahedral and d^2, d^7 in tetrahedral fields.

- We cannot ignore this term for two reasons:
  i. The separation between the F and P term is the same order as the splitting field (crystal field).
  ii. Transitions between terms of the same spin multiplicity are allowed.
- When two electrons occupy a set of d-orbitals, there are two states of different energy, depending on whether or not the occupied orbitals are in the same plan.
  - If the two occupied orbitals are in the same plane:

$$\left.\begin{matrix} d_{xy}, d_{x^2-y^2} \\ d_{xz}, d_{z^2} \\ d_{yz}, d_{z^2} \end{matrix}\right\} \text{Three states, } \because\ 2l+1=3,\ \therefore\ l=1$$

These states belong to the ^3P term.
  - The coulombic repulsion between them is clearly greater than if they are in different planes:

$$\left.\begin{matrix} d_{xz}, d_{yz} & d_{xy}, d_{z^2} & \\ d_{xy}, d_{xz} & d_{xz}, d_{x^2-y^2} & d_{z^2}, d_{x^2-y^2} \\ d_{xy}, d_{yz} & d_{yz}, d_{x^2-y^2} & \end{matrix}\right\} \text{Seven states, } \because\ 2l+1=7,\ \therefore\ l=3$$

These states belong to the ^3F term.
  - The ^3P term is higher in energy than the ^3F term (Hund's rule).
- One more difficulty is that there are two excited states in F term ions, whereas there is one excited state in ^3P-term ions.

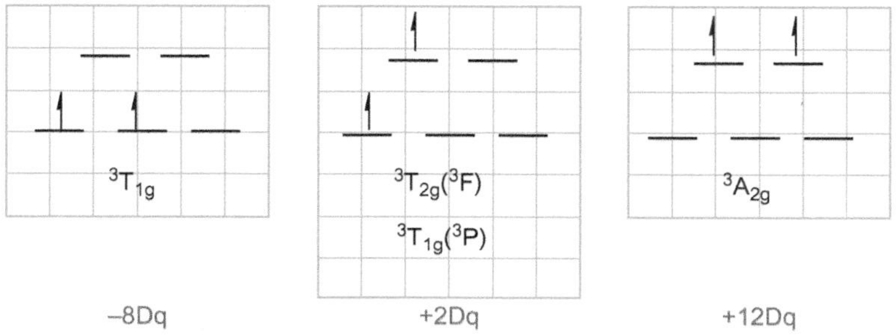

The energy terms of a d^2 ion in O_h field

The energy terms for a d^2 ion in O_h field.
- In the case of d^2 electronic configuration in O_h field:
  - The ground state is triply degenerate, T_{1g}, and crystal field energy $=-8$ Dq.
  - The first excited state is a sixfold degenerate, which arise from the two different terms, ^3F and ^3P, and crystal field energy $=2$ Dq.
  - The second excited state (A_{2g}), two electrons are promoted to e_g level, its energy $=18$ Dq.
- Energy levels of the same symmetry T_{1g} (F) and T_{1g} (P) cannot cross, and as crystal field increases as the repulsion between the two states becomes larger due to configuration interaction (Fig. 8.9).
- In an extremely weak O_h field, there are no curvatures of $^3T_{1g}$ states (Fig. 8.10).
- For stronger field (after curvature has set in), the energies of all the transitions will be increased by the amount that the ground-state energy has been lowered due to curvature.
- The d^3 case in O_h field, the ordering of the energy levels is reversed relative to d^2 case.
  - The ground state is nondegenerate ($^4A_{2g}$), the first excited sate contains six components [$^4T_{2g}(^3F)$ and $^4T_{1g}(^3P)$] and the second excited state is triply degenerate [$^4T_{1g}(^3F)$].
  - The Orgel diagram is constructed by back extrapolation of the d^2 diagram.
- Extending the analysis shows that d^7 case is produced by same diagram as d^2.
- On the other hand, d^8 behaves similarly to d^3.
- It is easy to show that ions in a T_d field have the same diagram but on the opposite side from their complements in O_h field (Fig. 8.11).

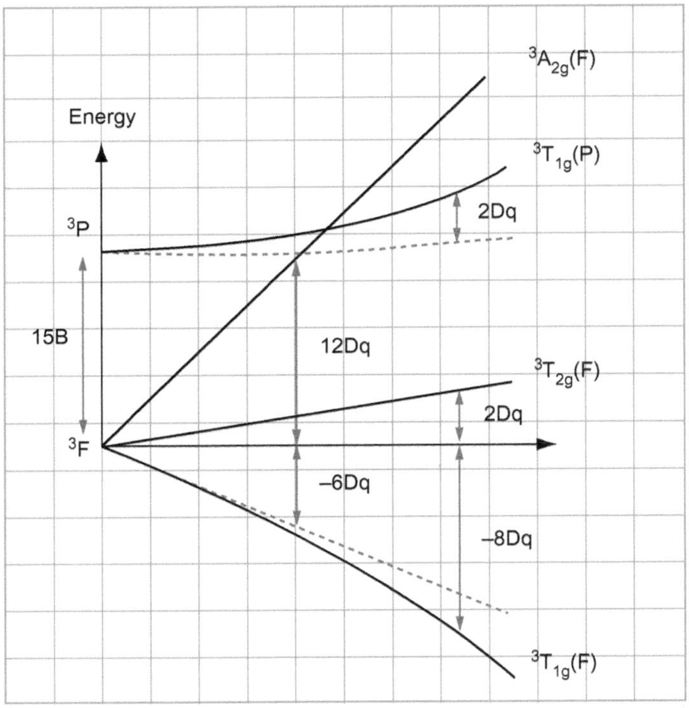

FIG. 8.9 Orgel diagram for d^2 ion in O_h symmetry.

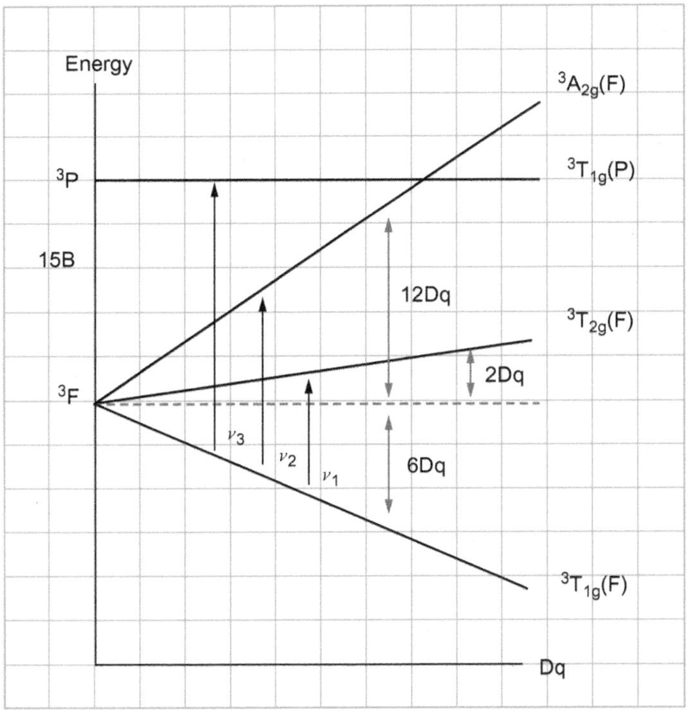

FIG. 8.10 d^2-Octahedral at weak field.

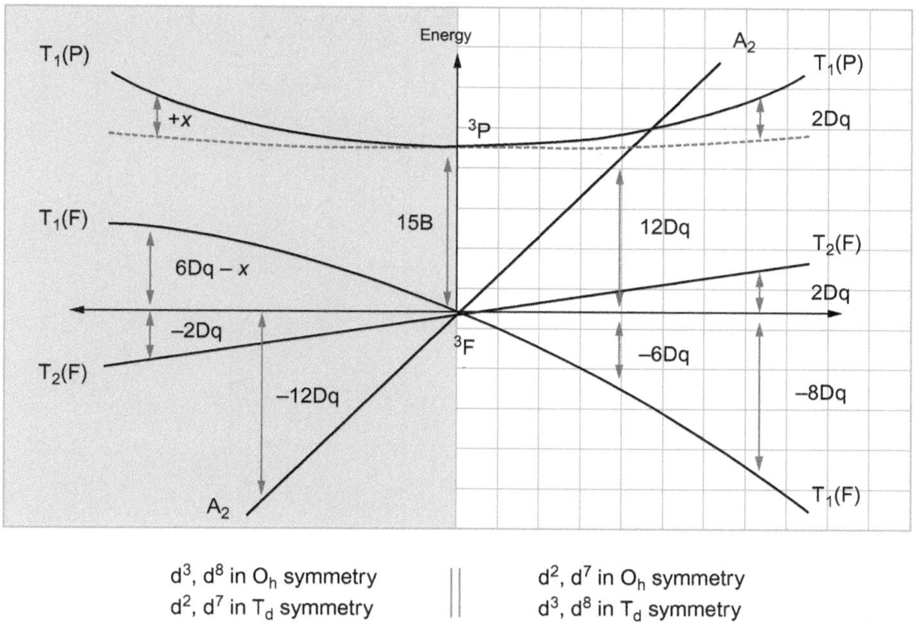

FIG. 8.11 Orgel diagram for ions with F-term ground states in O_h and T_d fields.

Configuration and Term Interactions

In F-state of d^2 configuration, if the linear combinations of spherical harmonic transforming in point group O_h (Chapter 7, p. 454):

$$\mathbf{T_2(F)}: \sqrt{\frac{5}{8}}|Y_3^1\rangle - \sqrt{\frac{3}{8}}|Y_3^{-3}\rangle,\ \sqrt{\frac{5}{8}}|Y_3^{-1}\rangle - \sqrt{\frac{3}{8}}|Y_3^{3}\rangle \text{ and } \sqrt{\frac{1}{2}}\left[|Y_3^2\rangle - |Y_3^{-2}\rangle\right]$$

$$\mathbf{T_1(F)}: \sqrt{\frac{3}{8}}|Y_3^1\rangle + \sqrt{\frac{5}{8}}|Y_3^{-3}\rangle,\ \sqrt{\frac{3}{8}}|Y_3^{-1}\rangle + \sqrt{\frac{5}{8}}|Y_3^{3}\rangle \text{ and } |Y_3^0\rangle$$

$$\mathbf{A_2(F)}: \sqrt{\frac{1}{2}}\left[|Y_3^2\rangle - |Y_3^{-2}\rangle\right]$$

and

$$\mathbf{T_1(P)}: |Y_1^{-1}\rangle,\ |Y_1^0\rangle \text{ and } -|Y_1^{-1}\rangle$$

Calculate the energies of the following:

$$E:\left({}^3\mathbf{T_{2g}(F)}\right), E:\left({}^3\mathbf{T_{1g}(F)}\right), E:\left({}^3\mathbf{A_{2g}(F)}, \text{ and } E:\left({}^3\mathbf{T_{1g}(P)}\right)\right.$$

How do the two terms ${}^3T_{1(g)}(F)$ and ${}^3T_{1(g)}(P)$ interact as a function of the field strength?

- The terms ${}^3T_{2g}(F)$, ${}^3T_{1g}(F)$, ${}^3A_{2g}(F)$ and ${}^3T_{1g}(P)$ originate from both the ground F term and the P term that ranks immediately above F of d^2 configuration.
- The energy

$$E\left[{}^3T_{2(g)}(F)\right] = \left\langle {}^3T_{2(g)}(F)|V|{}^3T_{2(g)}(F)\right\rangle \quad (8.5.1)$$

can be expanded, considering the free ion term:

$${}^3T_{2(g)}(F) = \sqrt{\frac{5}{8}}|Y_3^1\rangle - \sqrt{\frac{3}{8}}|Y_3^{-3}\rangle \quad (8.5.2)$$

The spherical harmonics, Y_l^m, define the total orbital angular momentum of a wave function that are represented by $|L, M_l, S, M_s\rangle$, therefore:

$$|Y_3^1\rangle = |3, 1, 1, 1\rangle$$

$$|Y_3^{-3}\rangle = |3, -3, 1, 1\rangle$$

these are two microstates of the orbital combinations for an ^{3}F term from d^2 (Chapter 2, p. 130):

$$|3,\pm 3,1,1\rangle = \pm(\pm 2,\ \pm 1)$$
$$|3,\pm 2,1,1\rangle = \pm(\pm 2,0)$$
$$|3,\pm 1,1,1\rangle = \pm\sqrt{\frac{2}{5}}(\pm 1,0) \pm \sqrt{\frac{3}{5}}(\pm 2,\mp 1)$$
$$|3,0,1,1\rangle = \sqrt{\frac{4}{5}}(1,-1) + \sqrt{\frac{1}{5}}(2,-2)$$

Then Eq. (8.5.1):

$$E\left[{}^3T_{2(g)}(F)\right] = \langle {}^3T_{2g}(F)|V|{}^3T_{2g}(F)\rangle = \left\langle \sqrt{\frac{5}{8}}|Y_3^1\rangle - \sqrt{\frac{3}{8}}|Y_3^{-3}\rangle \middle| V \middle| \sqrt{\frac{5}{8}}|Y_3^1\rangle - \sqrt{\frac{3}{8}}|Y_3^{-3}\rangle \right\rangle$$
$$= \left\langle \sqrt{\frac{5}{8}}|3,1,1,1\rangle - \sqrt{\frac{3}{8}}|3,-3,1,1\rangle \middle| V \middle| \sqrt{\frac{5}{8}}|3,1,1,1\rangle - \sqrt{\frac{3}{8}}|3,-3,1,1\rangle \right\rangle$$
$$= \frac{5}{8}\langle 3,1,1,1|V|3,1,1,1\rangle + \frac{3}{8}\langle 3,-3,1,1|V|3,-3,1,1\rangle - 2\frac{\sqrt{15}}{8}\langle 3,1,1,1|V|3,-3,1,1\rangle$$

Each term is expanded by inserting the two electrons determinants:

$$\frac{5}{8}\langle 3,1,1,1|V|3,1,1,1\rangle = \frac{5}{8}\left[\left\langle \sqrt{\frac{2}{5}}(1^+,0^+) + \sqrt{\frac{3}{5}}(2^+,-1^+)|V|\sqrt{\frac{2}{5}}(1^+,0^+) + \sqrt{\frac{3}{5}}(2^+,-1^+)\right\rangle\right]$$
$$= \frac{5}{8}\left[\frac{2}{5}\langle(1^+,0^+)|V|(1^+,0^+)\rangle + \frac{3}{5}\langle(2^+,-1^+)|V|(2^+,-1^+)\rangle + 2\frac{\sqrt{6}}{5}\langle(1^+,0^+)|V|(2^+,-1^+)\rangle\right]$$

Omitting the spin,

$$\frac{5}{8}\langle 3,1,1,1|V|3,1,1,1\rangle = \frac{5}{8}\left[\frac{2}{5}(\langle 1|V|1\rangle\langle 0|0\rangle + \langle 0|V|0\rangle\langle 1|1\rangle + 2\langle 1|V|0\rangle\langle 1|0\rangle)\right.$$
$$\left. + \frac{3}{5}(\langle 2|V|2\rangle\langle -1|-1\rangle + \langle -1|V|-1\rangle\langle 2|2\rangle + 2\langle 2|V|-1\rangle\langle 2|-1\rangle)\right]$$

we previously (Chapter 7, p. 427) established that:

$$\langle 0|V_{oct}|0\rangle = 6Dq$$
$$\langle \pm 1|V_{oct}|\pm 1\rangle = -4Dq$$
$$\langle \pm 2|V_{oct}|\pm 2\rangle = Dq$$
$$\langle \pm 2|V_{oct}|\mp 2\rangle = 5Dq$$

then:

$$\frac{5}{8}\langle 3,1,1,1|V|3,1,1,1\rangle = \frac{5}{8}\left[\frac{2}{5}(-4Dq + 6Dq + 0) + \frac{3}{5}(Dq - 4Dq + 0)\right]$$
$$= \frac{5}{8}\left[\frac{4}{5}Dq - \frac{9}{5}Dq\right] = -\frac{5}{8}Dq$$

$$\frac{3}{8}\langle 3,-3,1,1|V|3,-3,1,1\rangle = \frac{3}{8}[\langle(-1^+,-2^+)|V|(-1^+,-2^+)\rangle]$$
$$= \frac{3}{8}[\langle -1|V|-1\rangle\langle -2|-2\rangle + \langle -2|V|-2\rangle\langle -1|-1\rangle + 2\langle -1|V|-2\rangle\langle -1|-2\rangle]$$
$$= \frac{3}{8}[-4Dq + Dq + 0] = -\frac{9}{8}Dq$$

$$-2\frac{\sqrt{15}}{8}\langle 3,1,1,1|V|3,-3,1,1\rangle = -2\frac{\sqrt{15}}{8}\left[\left\langle\left(\sqrt{\frac{2}{5}}(1^+,0^+)+\sqrt{\frac{3}{5}}(2^+,-1^+)\right)|V|(-1^+,-2^+)\right\rangle\right]$$

$$= -2\frac{\sqrt{15}}{8}\left[\left\langle\left(\sqrt{\frac{2}{5}}(1^+,0^+)\right)|V|(-1^+,-2^+)\right\rangle+\left\langle\left(\sqrt{\frac{3}{5}}(2^+,-1^+)\right)|V|(-1^+,-2^+)\right\rangle\right]$$

$$= -2\frac{\sqrt{15}}{8}\left[0-\left\langle\left(\sqrt{\frac{3}{5}}(2^+,-1^+)\right)|V|(-2^+,-1^+)\right\rangle\right]$$

$$= 2\frac{\sqrt{15}}{8}\left[\sqrt{\frac{3}{5}}(\langle 2|V|-2\rangle\langle -1|-1\rangle+\langle -1|V|-1\rangle\langle 2|-2\rangle+0)\right]$$

$$= 2\frac{\sqrt{15}}{8}\left[\sqrt{\frac{3}{5}}(5\text{Dq}+0)\right] = \frac{30}{8}\text{Dq}$$

Therefore:

$$E\left[{}^3\text{T}_{2(g)}(\text{F})\right] = -\frac{5}{8}\text{Dq}-\frac{9}{8}\text{Dq}+\frac{30}{8}\text{Dq} = 2\text{Dq} \tag{8.5.3}$$

- Use the other components of ${}^3\text{T}_{2g}$: $\sqrt{\frac{5}{8}}|Y_3^{-1}\rangle-\sqrt{\frac{3}{8}}|Y_3^3\rangle$ and $\sqrt{\frac{1}{2}}[|Y_3^2\rangle-|Y_3^{-2}\rangle]$, would yield an energy of 2Dq.

$$E\left[{}^3\text{A}_{2(g)}(\text{F})\right] = \left\langle {}^3\text{A}_{2(g)}\text{F})|V|{}^3\text{A}_{2(g)}(\text{F})\right\rangle$$

$$= \left\langle\sqrt{\frac{1}{2}}[|Y_3^2\rangle-|Y_3^{-2}\rangle]|V|\sqrt{\frac{1}{2}}[|Y_3^2\rangle-|Y_3^{-2}\rangle]\right\rangle$$

$$= \frac{1}{2}\{\langle|3,2,1,1\rangle-|3,-2,1,1\rangle|V||3,2,1,1\rangle-|3,-2,1,1\rangle\rangle\}$$

$$= \frac{1}{2}\{\langle|3,2,1,1\rangle|V||3,2,1,1\rangle\rangle+\langle|3,-2,1,1\rangle|V||3,-2,1,1\rangle\rangle-2\langle|3,2,1,1\rangle|V||3,-2,1,1\rangle\rangle\}$$

$$= \frac{1}{2}\{\langle(2^+,0^+)|V|(2^+,0^+)\rangle+\langle(0^+,-2^+)|V|(0^+,-2^+)\rangle-2\langle(2^+,0^+)|V|(0^+,-2^+)\rangle\}$$

$$= \frac{1}{2}\left\{\begin{array}{c}\langle 2|V|2\rangle\langle 0|0\rangle+\langle 0|V|0\rangle\langle 2|2\rangle\\ +\langle 0|V|0\rangle\langle -2|-2\rangle+\langle -2|V|-2\rangle\langle 0|0\rangle\\ +2(\langle 2|V|-2\rangle\langle 0|0\rangle+\langle 0|V|0\rangle\langle 2|-2\rangle)\end{array}\right\}$$

$$E\left[{}^3\text{A}_{2(g)}(\text{F})\right] = \frac{1}{2}\{\text{Dq}+6\text{Dq}+6\text{Dq}+\text{Dq}+2(5\text{Dq})\} = 12\text{Dq} \tag{8.5.4}$$

- Since $\text{T}_{1(g)}\times\text{T}_{1(g)}$ in O_h or T_d contains $\text{A}_{1(g)}$:

O_h	E	$8C_3$	$6C_2$	$6C_4$	$3C_2$	i	$6S_4$	$8S_6$	$3\sigma_h$	$2\sigma_d$		
T_{1g}	3	0	−1	1	−1	3	1	0	−1	−1	(R_x, R_y, R_z)	–
$T_{1g}\times T_{1g}$	9	0	1	1	1	9	1	0	1	1	$= A_{1g}+E_g+T_{1g}+T_{2g}$	

T_d	E	$8C_3$	$3C_2$	$6S_4$	$6\sigma_d$		
T_1	3	0	−1	1	−1	(R_x, R_y, R_z)	–
$T_1\times T_1$	9	0	1	1	1	$= A_1+E+T_1+T_2$	

Therefore:

$$\left\langle {}^3\mathrm{T}_{1(g)}(\mathrm{F})\middle|\mathrm{H}\middle|{}^3\mathrm{T}_{1(g)}(\mathrm{P})\right\rangle \neq 0$$

and the two terms interact with each other.

- In order to find the energy of ${}^3\mathrm{T}_{1(g)}$ terms of d^2 during the transition from free ion ($V_{\text{field}}=0$, ${}^3\mathrm{T}^{\mathrm{o}}_{1(g)}$) to strong field, ($V_{\text{field}} \neq 0$) the secular determinant for the Hamiltonian operator has to be set up between the two free ion terms.

$$\begin{array}{c} \\ {}^3\mathrm{T}^{\mathrm{o}}_{1(g)}(\mathrm{F}) \\ {}^3\mathrm{T}^{\mathrm{o}}_{1(g)}(\mathrm{P}) \end{array}\begin{vmatrix} {}^3\mathrm{T}^{\mathrm{o}}_{1(g)}(\mathrm{F}) & {}^3\mathrm{T}^{\mathrm{o}}_{1(g)}(\mathrm{P}) \\ \langle \mathrm{H}_{11}\rangle - \mathrm{E} & \langle \mathrm{H}_{12}\rangle \\ \langle \mathrm{H}_{21}\rangle & \langle \mathrm{H}_{22}\rangle - \mathrm{E} \end{vmatrix} = 0$$

Let the label of the free ion term: ${}^3\mathrm{T}^{\mathrm{o}}_1$

- The Hamiltonian operator in the medium field strength of O_h symmetry:
  $H = H_o + V_{\text{oct}}$, and $V_{\text{oct}} = V$, then:

$$\langle H_{11}\rangle = \left\langle {}^3\mathrm{T}^{\mathrm{o}}_{1(g)}(\mathrm{F})|H_{11}|{}^3\mathrm{T}^{\mathrm{o}}_{1(g)}(\mathrm{F})\right\rangle = \left\langle {}^3\mathrm{T}^{\mathrm{o}}_{1(g)}(\mathrm{F})|H_o + V|{}^3\mathrm{T}^{\mathrm{o}}_{1(g)}(\mathrm{F})\right\rangle$$

$$= \left\langle {}^3\mathrm{T}^{\mathrm{o}}_{1(g)}(\mathrm{F})|H_o|{}^3\mathrm{T}^{\mathrm{o}}_{1(g)}(\mathrm{F})\right\rangle + \left\langle {}^3\mathrm{T}^{\mathrm{o}}_{1(g)}(\mathrm{F})|V|{}^3\mathrm{T}^{\mathrm{o}}_{1(g)}(\mathrm{F})\right\rangle$$

In the absence of a ligand field,

$$\left\langle {}^3\mathrm{T}^{\mathrm{o}}_{1(g)}(\mathrm{F})|H_o|{}^3\mathrm{T}^{\mathrm{o}}_{1(g)}(\mathrm{F})\right\rangle = 0$$

(Chapter 6), then:

$$\langle H_{11}\rangle = \left\langle {}^3\mathrm{T}^{\mathrm{o}}_{1(g)}(\mathrm{F})|H_{11}|{}^3\mathrm{T}^{\mathrm{o}}_{1(g)}(\mathrm{F})\right\rangle = \left\langle {}^3\mathrm{T}^{\mathrm{o}}_{1(g)}(\mathrm{F})|V|{}^3\mathrm{T}^{\mathrm{o}}_{1(g)}(\mathrm{F})\right\rangle \qquad (8.5.5)$$

- The energy of one of the three degenerate components of ${}^3\mathrm{T}^{\mathrm{o}}_{1(g)}(\mathrm{F})$:

$$E\left[{}^3\mathrm{T}_{1g}(\mathrm{F})\right] = \langle {}^3\mathrm{T}_{1g}(\mathrm{F})|V|{}^3\mathrm{T}_{1g}(\mathrm{F})\rangle$$

Consider:

$${}^3\mathrm{T}^{\mathrm{o}}_{1(g)}(\mathrm{F}) = |Y^0_3\rangle$$

$$\langle H_{11}\rangle = E\left[{}^3\mathrm{T}^{\mathrm{o}}_{1(g)}(\mathrm{F})\right] = \langle |Y^0_3\rangle|V||Y^0_3\rangle\rangle = \langle 3,0,1,1|V|3,0,1,1\rangle$$

$$= \left\langle\left(\sqrt{\frac{4}{5}}(1^+,-1^+) + \sqrt{\frac{1}{5}}(2^+,-2^+)\right)|V|\left(\sqrt{\frac{4}{5}}(1^+,-1^+) + \sqrt{\frac{1}{5}}(2^+,-2^+)\right)\right\rangle$$

$$= \frac{4}{5}\langle(1^+,-1^+)|V|(1^+,-1^+)\rangle + \frac{1}{5}\left\langle(2^+,-2^+)|V|(2^+,-2^+) + \frac{4}{5}\langle(1^+,-1^+)|V|(2^+,-2^+)\rangle\right\rangle$$

$$= \frac{4}{5}(\langle 1|V|1\rangle\langle -1|-1\rangle + \langle -1|V|-1\rangle\langle 1|1\rangle)$$

$$+ \frac{1}{5}(\langle 2|V|2\rangle\langle -2|-2\rangle + \langle -2|V|-2\rangle\langle 1|1\rangle) + \frac{4}{5}(0)$$

$$\langle \mathrm{H}_{11}\rangle = E\left[{}^3\mathrm{T}^{\mathrm{o}}_{1(g)}(\mathrm{F})\right] = \frac{4}{5}(-4\mathrm{Dq} - 4\mathrm{Dq}) + \frac{1}{5}(\mathrm{Dq} + \mathrm{Dq}) + \frac{4}{5}(0) = -6\mathrm{Dq} \qquad (8.5.5)$$

- On the other hand,

$$\langle H_{22}\rangle = \left\langle {}^3T^o_{1(g)}(P)|H_{22}|{}^3T^o_{1(g)}(P)\right\rangle = \left\langle {}^3T^o_{1(g)}(P)|H_o + V_{oct.}|{}^3T^o_{1(g)}(P)\right\rangle \tag{8.5.6}$$

$$= \left\langle {}^3T^o_{1(g)}(P)|H_o|{}^3T^o_{1(g)}(P)\right\rangle + \left\langle {}^3T^o_{1(g)}(P)|V|{}^3T^o_{1(g)}(P)\right\rangle$$

  - First, we will estimate: $\langle {}^3T^o_{1(g)}(P)|V|{}^3T^o_{1(g)}(P)\rangle$
    Consider:

$${}^3T^o_{1(g)}(P) = |Y_1^{-1}\rangle.$$

$$\left\langle {}^3T^o_{1(g)}(P)|V|{}^3T^o_{1(g)}(P)\right\rangle = \langle |Y_1^{-1}\rangle|V||Y_1^{-1}\rangle\rangle$$

$$= \langle |1,-1,1,1\rangle|V|\langle 1,-1,1,1\rangle$$

The orbital wave function for a P term from the d^2 configuration are (Chapter 2, p. 130):

$$|1,\pm 1,1,1\rangle = \mp\sqrt{\frac{3}{5}}(\pm 1,0) \pm \sqrt{\frac{2}{5}}(\pm 2,\mp 1)$$

$$|1,0,1,1\rangle = -\sqrt{\frac{1}{5}}(1,-1) + \sqrt{\frac{4}{5}}(2,-2)$$

then:

$$\left\langle {}^3T^o_{1(g)}(P)|V|{}^3T^o_{1(g)}(P)\right\rangle = \left\langle\left(\sqrt{\frac{3}{5}}(-1^+,0^+) + \sqrt{\frac{2}{5}}(1^+,-2^+)\right)|V|\sqrt{\frac{3}{5}}(-1^+,0^+) + \sqrt{\frac{2}{5}}(1^+,-2^+)\right\rangle$$

$$= \frac{3}{5}[\langle(-1^+,0^+)|V|(-1^+,0^+)\rangle] + \frac{2}{5}[\langle(1^+,-2^+)|V|(1^+,-2^+)\rangle] + 2\left(\frac{\sqrt{6}}{5}\right)\langle(-1^+,0^+)|V|(1^+,-2^+)\rangle$$

$$= \frac{3}{5}[\langle -1|V|-1\rangle\langle 0|0\rangle + \langle 0|V|0\rangle\langle -1|-1\rangle] + \frac{2}{5}[\langle 1|V|1\rangle\langle -2|-2\rangle + \langle -2|V|-2\rangle\langle 1|1\rangle] + 0$$

$$\left\langle {}^3T^o_{1(g)}(P)|V|{}^3T^o_{1(g)}(P)\right\rangle = \frac{3}{5}[-4Dq + 6Dq] + \frac{2}{5}[-4Dq + Dq] = 0$$

  - If:

$$\langle H_{22}\rangle = \left\langle {}^3T^o_{1(g)}(P)|H_{22}|{}^3T^o_{1(g)}(P)\right\rangle = \left\langle {}^3T^o_{1(g)}(P)|H_o + V_{oct.}|{}^3T^o_{1(g)}(P)\right\rangle$$

$$= \left\langle {}^3T^o_{1(g)}(P)|H_o|{}^3T^o_{1(g)}(P)\right\rangle + \left\langle {}^3T^o_{1(g)}(P)|V|{}^3T^o_{1(g)}(P)\right\rangle$$

$$\left\langle {}^3T^o_{1(g)}(P)|V|{}^3T^o_{1(g)}(P)\right\rangle = 0$$

while

$$\left\langle {}^3T^o_{1(g)}(P)|H_o|{}^3T^o_{1(g)}(P)\right\rangle = 15B$$

where 15B is the amount of interelectronic repulsion energy of $^3T^0_{1(g)}(P)$ relative to ^3F as zero.

$$\langle H_{22}\rangle = \left\langle {}^3T^o_{1(g)}(P)|H_{22}|{}^3T^o_{1(g)}(P)\right\rangle = 15B \tag{8.5.6}$$

- The of-diagonal element:

$$\langle H_{12}\rangle = \left\langle {}^3T^o_{1(g)}(F)|V|{}^3T^o_{1(g)}(P)\right\rangle = \langle |Y^0_3\rangle|V||Y^0_1\rangle\rangle$$

$$= \langle |3,0,1,1\rangle|V||1,0,1,1\rangle\rangle \tag{8.5.7}$$

$$= \left\langle \left(\sqrt{\frac{1}{5}}(2^+,-2^+)+\sqrt{\frac{4}{5}}(1^+,-1^+)\right)|V|\left(\sqrt{\frac{4}{5}}(2^+,-2^+)-\sqrt{\frac{1}{5}}(1^+,-1^+)\right)\right\rangle$$

$$=\frac{2}{5}[\langle(2^+,-2^+)|V|(2^+,-2^+)\rangle]-\frac{2}{5}[\langle(1^+,-1^+)|V|(1^+,-1^+)\rangle]$$

$$+\frac{3}{5}[\langle(2^+,-2^+)|V|(1^+,-1^+)\rangle]$$

$$=\frac{2}{5}[\langle 2|V|2\rangle\langle -2|-2\rangle+\langle -2|V|-2\rangle\langle 2|2\rangle]-\frac{2}{5}[\langle 1|V|1\rangle\langle -1|-1\rangle+\langle -1|V|-1\rangle\langle 1|1\rangle]+\frac{3}{5}[0]$$

$$\langle H_{12}\rangle = \left\langle {}^3T^o_{1(g)}(F)|V|{}^3T^o_{1(g)}(P)\right\rangle = \frac{2}{5}[Dq+Dq]-\frac{2}{5}[-4Dq-4Dq]+\frac{3}{5}[0]=4Dq$$

- The $^3T_{1g}$ matrix:

$$\begin{vmatrix} \langle H_{11}\rangle - E & \langle H_{12}\rangle \\ \langle H_{21}\rangle & \langle H_{22}\rangle - E \end{vmatrix} = \begin{vmatrix} -6Dq-E & 4Dq \\ 4Dq & 15B-E \end{vmatrix} = 0$$

$$(-6Dq-E)(15B-E)-16Dq^2=0$$

$$E^2+(6Dq-15B)E-16Dq^2-90DqB=0$$

$$E=\frac{-(6Dq-15B)\pm\sqrt{(6Dq-15B)^2-4(-16Dq^2-90DqB)}}{2}$$

- The Octahedral spin triplet d^2 energies:

$$E(^3T_{1g}(F)) = 7.5B - 3Dq - \frac{1}{2}(225B^2+100Dq^2+180DqB)^{1/2} \tag{8.5.8}$$

$$E(^3T_{2g}(F)) = 2Dq$$

$$E(^3A_{2g}(F)) = 12Dq \tag{8.5.4}$$

$$E(^3T_{1g}(P)) = 7.5B - 3Dq + \frac{1}{2}(225B^2+100Dq^2+180DqB)^{1/2} \tag{8.5.9}$$

8.6 TANABE-SUGANO DIAGRAMS

The Advantages

What are the disadvantages of Orgel and the correlation diagrams, and the advantages of Tanabe-Sugano diagrams, illustrating the bases and the simplifications that are made?

- Orgel diagrams are the simplest because there is one-to-one correspondence between electron configuration and states, without electron repulsion of low-spin configurations.
- The correlation diagrams of high d configurations are crowded and hard to follow. An alternative way of expressing the ground and the exciting state of d configurations in a cubic field has been developed by Tanabe and Sugano.

- For high-spin complexes, the Tanabe-Sugano diagrams are similar to Orgel diagrams except that they include states of lower-spin multiplicity.
- Unlike the correlation diagrams, Tanabe and Sugano used complete equations that consider the configurational interactions between states of same symmetry to construct their diagrams.
- The diagrams correlate the energy of states as a function of Dq. The energy of states require Racah parameters B and C in the analytical expression for their energies, where B is the unit of repulsion energy, and C is an additional repulsion factor. There is no theoretical relationship between the magnitudes of B and C, and assuming that $C = 4.5B$, which is not be true in every case (Chapter 7, p. 461).
- For convenience, E and Dq are expressed in relation to theoretically defined unit of electron/electron repulsion, B (as multiple of B). The ordinate is in units of E/B and the abscissa is in units of Dq/B.
- Configuration interaction is included as bending of lines for the excited states.
- The energies of states with spin multiplicity different from that of the ground state are also represented.
- Tanabe-Sugano diagrams differ from correlation diagrams in having the ground state as the constant reference (it becomes the horizontal axis), and are appropriate for low spin-states.
- The ground state is represented as horizontal baseline. A vertical line to an excited state is a direct measure of the transition energy (in terms of Dq and B).
- Electronic spectra can be mapped to these diagrams, and the values of Dq and B are figured.
- Tanabe-Sugano diagrams have qualitative significance, but they are valuable aids in interpreting spectra, and in finding the approximate value of Dq.
- Octahedral complexes with d^1 configurations are inverse that of d^9 with only two states (Fig. 8.12).
- Fig. 8.13 shows the energy level diagrams for $d^2$and d^3ions in an O_h field.
- The d orbital splitting diagrams for tetrahedral complexes are:
  - inverse that for octahedral complexes; and
  - omitting g and u subscripts.

d^n and d^{10-n} Diagrams

The Tanabe-Sugano diagram for d^4 is different from d^2 and d^3 diagrams, explain the differences. How do Tanabe and Sugano present both the high and low spin in one diagram? How does d^6diagram relate to that of d^4?

- The Tanabe-Sugano diagram for d^4 displays both high- and low-spin ground states; ^5E and $^3T_1$, respectively. When Dq/B is large enough, $^3T_1$ term takes place ^5E as ground state (Fig. 8.14).
- $^3T_1$ has crystal field energy, CF, of -16 Dq, while ^5E has only -6 Dq, as a result $^3T_1$ becomes more stable as Dq increases. When $^3T_1$ crosses ^5E, it becomes the new ground state, and is represented in the d^4 diagram as a horizontal line of zero slope instead of the steeply negative slope.

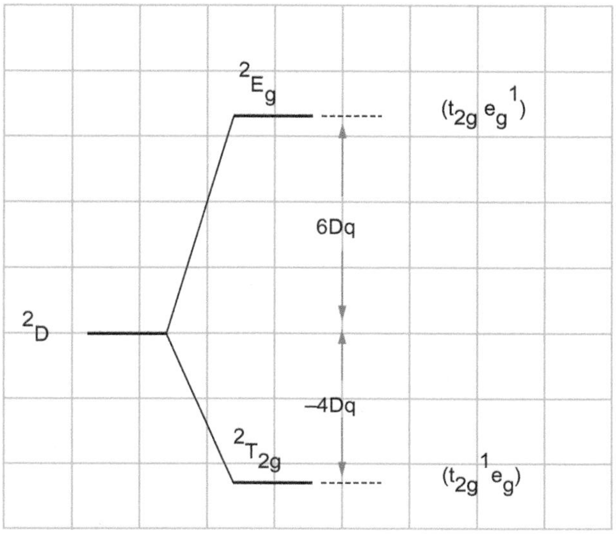

FIG. 8.12 Correlation diagram for a d^1 ion in O_h field.

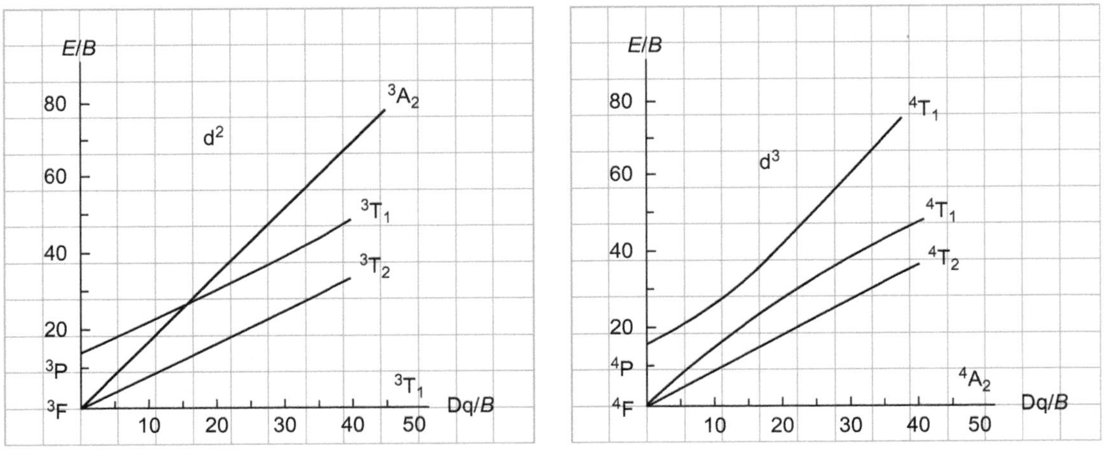

FIG. 8.13 Tanabe-Sugano diagrams for d^2 and d^3 ions in an O_h field.

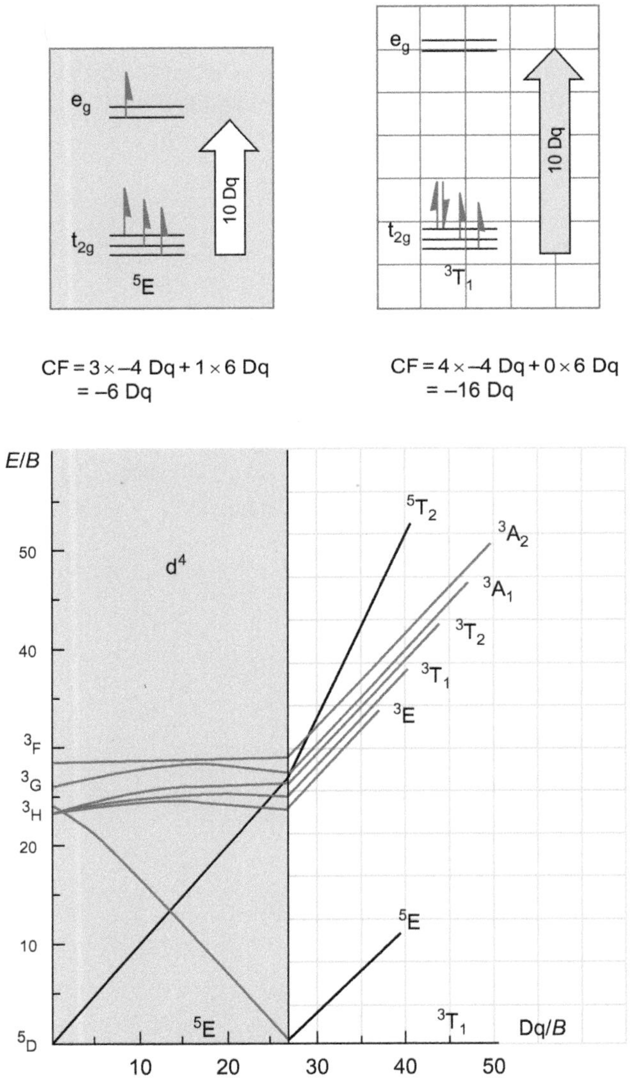

FIG. 8.14 Tanabe-Sugano diagrams for d^4 ion in O_h field.

- Consequently, there is an abrupt change in the excited state energies to compensate for that of the new ground state.
- The electronic spectra of high-spin d^4 complexes have only one spin-allowed transition; ${}^5T_2 \leftarrow {}^5E$. Unlike the low-spin d^4 complexes, many spin-allowed transitions occur.
- High-spin d^6 configuration is the hole equivalent of d^4, but the splitting is inverted. In the weak field, the term ${}^5D$ splits in O_h surrounding to the ground term ${}^5T_2$; $(t_{2g})^4(e_g)^2$, and the excited term ${}^5E_g$; $(t_{2g})^3(e_g)^3$ (Fig. 8.15).
- In the strong field, the term ${}^1I$ splits to completely filled ${}^1A_1$ term, $(t_{2g})^6$, and the excitations ${}^1T_1$ and ${}^1T_2$ of the configuration $(t_{2g})^5(e_g)^1$, in which the odd spins in the t_{2g} and e_g are spin opposed.

How do the Tanabe-Sugano diagrams for d^2 and d^3 ions relate to that of d^7 and d^8 configurations?

- The high-spin d^7 configurations:
  - are hole equivalents of d^3 (Fig. 8.16);
  - have similar correlation diagrams to that of d^3, except F term splitting is inverted; and
  - split similarly to d^2, since it differs from d^2 by 5 d electrons.

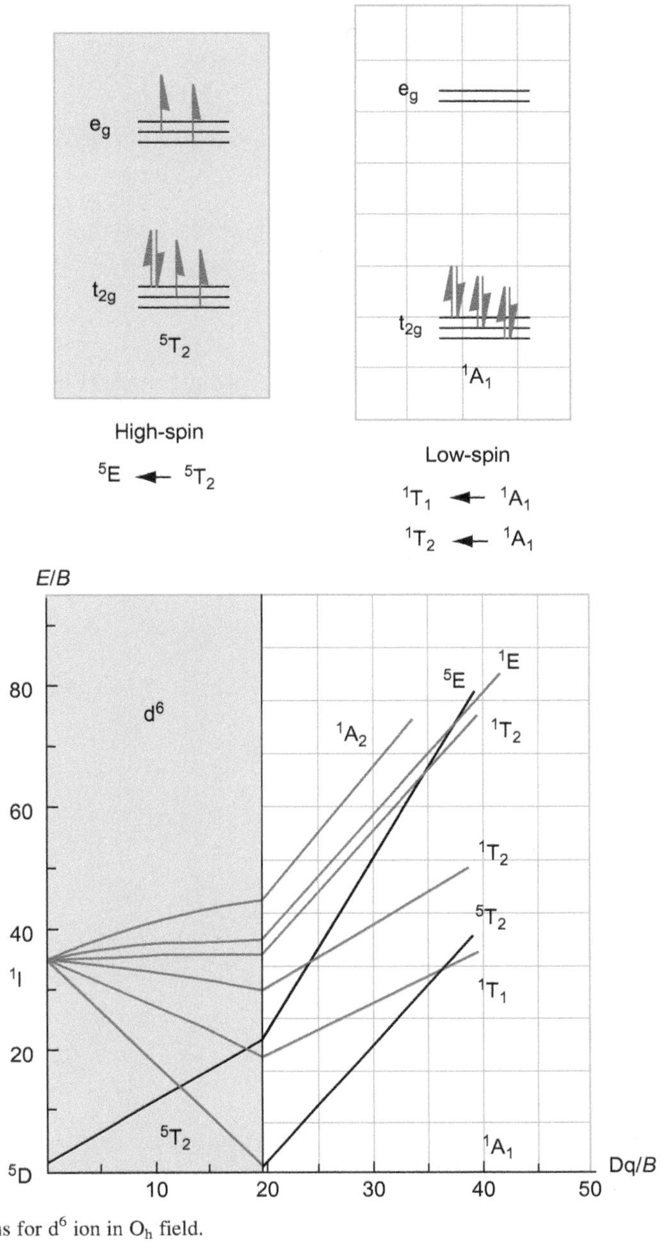

FIG. 8.15 Tanabe-Sugano diagrams for d^6 ion in O_h field.

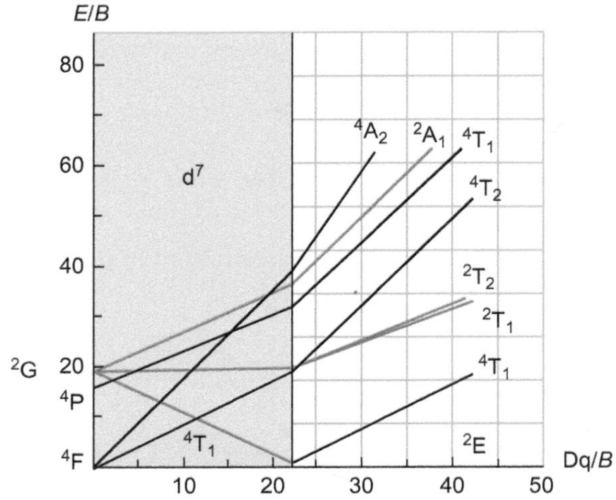

FIG. 8.16 Tanabe-Sugano diagrams for d^4 ion in O_h field.

- The low-spin state has the configuration $(t_{2g})^6(e_g)^1$ having an odd unpaired electron in the e_g subshell. Spin-allowed transitions are permitted from ^2E to other spin-doublet states (Fig. 8.16).
- Octahedral complexes with d^8 configuration:
  - can only exist with $S = 1$ arising from $(t_{2g})^6(e_g)^2$; and
  - are the same as for d^3 diagram (Fig. 8.17), except that the F term splitting is inverted.

d^5 Diagram

What do you conclude from the Tanabe-Sugano diagram for d^5 configuration (Fig. 8.18)? How can you explain the pale color of most manganese (II) complexes?

- In the weak field:
  - the state ^6S yields only one term, $^6A_1$, which has one unpaired electron in each of the five d-orbitals;
  - there are no possible spin-sextet excited states;
  - there are no spin-allowed transitions; and
  - all d-d excitations indicate the lower spin states.

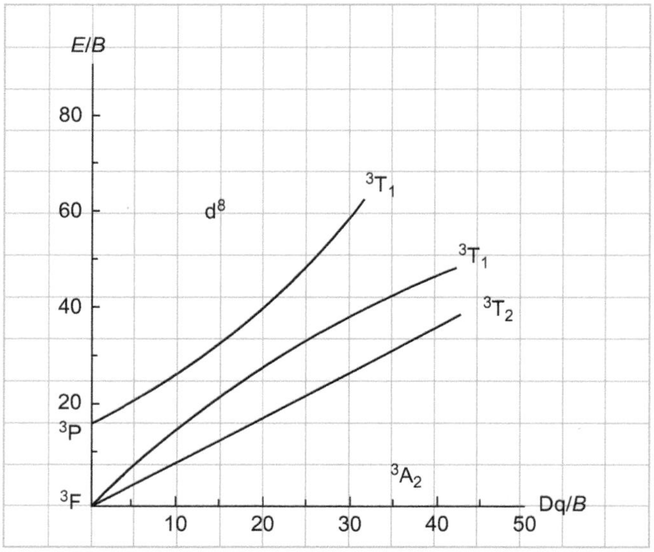

FIG. 8.17 Tanabe-Sugano diagrams for d^4ion in an O_h field.

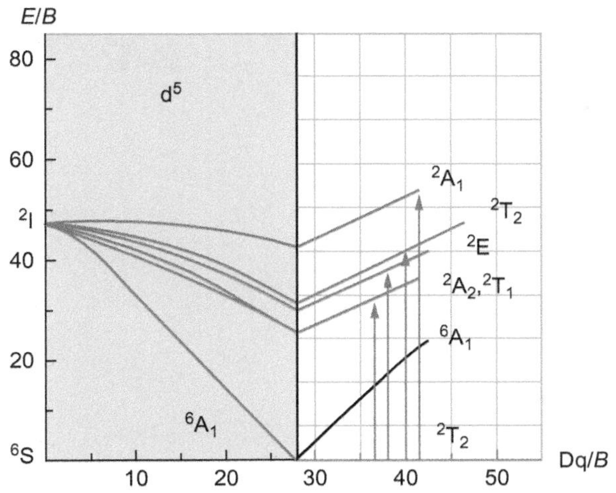

FIG. 8.18 Tanabe-Sugano diagrams for d^5 ion in O_h field.

- The pale color of manganese (II) complexes results from the weak d-d absorption due to spin-forbidden transition.
- In the strong field, the low-spin has the configuration $(t_{2g})^5$ in O_h, and yields the ground term $^2T_2$ with $S = 1/2$. Permitted transitions are spin-allowed from $^2T_2$ are illustrated in Fig. 8.18.

SUGGESTIONS FOR FURTHER READING

A.B.P. Lever, Inorganic Electronic Spectroscopy, second ed., Elsevier, Amsterdam, 1984. ISBN: 0-444-42389-3.
B.N. Figgis, Introduction to Ligand Field, Interscience, New York, 1966.
C.J. Ballhausen, J. Chem. Educ. 56 (1979) 194–197. 215–218, 357–361.
F.A. Cotton, G. Wilkinson, C.A. Murillo, Advanced Inorganic Chemistry, Wiley-Interscience, New York, NY, 1999. ISBN: 978-0471199571.
G.L. Miessler, P.J. Fischer, D.A. Tarr, Inorganic Chemistry, Prentice Hall, NJ, (2013). ISBN: 978-0321811059.
W.G. Jackson, J.A. McKeon, S. Cortez, Inorg. Chem. 43 (20) (2004) 6249–6254.
K. Bowman-James, Acc. Chem. Res. 38 (8) (2005) 671–678.
M.L.H. Green, J. Org. Chem. 500 (1–2) (1995) 127–148.
J.F. Hartwig, Organotransition Metal Chemistry, from Bonding to Catalysis, University Science Books, New York, 2010. ISBN: 1-891389-53-X.
A. von Zelewsky, Stereochemistry of Coordination Compounds, John Wiley, Chichester, 1995. ISBN: 047195599X.
J.S. Griffith, Theory of Transition Metal Ions, Cambridge University Press, London, New York, 1961. ISBN: 978-0-521-11599-5.
J.N. Murrell, S.F.A. Kettle, J.M. Tedder, Valence Theory, John Wiley, London, 1965.
E.U. Condon, G.H. Shortley, The Theory of Atomic Spectra, Cambridge University Press, London, New York, 1963. ISBN: 978-0-521-09209-8.
H.L. Schlafer, G. Glisemann, Basic Principles of Ligand Field Theory, Wiley, New York, 1969.
Y. Tanabe, S. Sugano, J. Phys. Soc. Jpn. 9 (753) (1954) 766.
L.E. Orgel, J. Chem. Phys. 23 (1955) 1004.
K.F. Purcell, J.C. Kotz, Inorganic Chemistry, W. B. Saunders Company, Philadelphia, London, Toronto, 1977. ISBN: 0-7216-7407-0.
F.A. Cotton, Chemical Applications of Group Theory, third ed., John Wiley-Interscience, New York, NY, 1990. ISBN: 9780471510949.
I.S. Butler, J.F. Harrod, Inorganic Chemistry Principles and Applications, The Benjamin/Cummings Publishing Company, Inc., Redwood, CA, 1989. ISBN: 0-8053-0247-6.
B. Douglas, D. McDaniel, J.J. Alexander, Concepts and Models of Inorganic Chemistry, third ed., John Wiley & Sons, New York, 1994. ISBN: 0-471-62978-2.
R.S. Drago, Phys. Methods in Chemistry, W.B. Saunders Company, Philadelphia, 1977. ISBN: 0-7216-3184-3.
C.R. Landis, F. Weinhold, Valence and extra-valence orbitals in main group and transition metal bonding, J. Comput. Chem. 28 (1) (2007) 198–203.

Chapter 9

Vibrational Rotational Spectroscopy

In this chapter, the concern is mainly with the vibrational transitions that are detected in infrared or Raman spectroscopy (Scheme 9.1). These transitions originate from vibrations of nuclei of the molecule, which appear in the 10^2–10^4 cm^{-1} region. Motion that repeats itself is called periodic, and the required time for each repetition is the period. The oscillations of atoms in molecules display periodic motions that occur frequently.

The energy levels of the electronic transition are far apart, and occur in the 10^4–10^6 cm^{-1} region (UV-visible region), whereas the levels of the rotational transitions are relatively close to each other, and are observed in the 1–10^2 cm^{-1} region (microwave region). However, pure electronic transitions may occur in the near-infrared region if electronic levels are closely spaced, while pure rotational spectra may appear in the far-infrared region if transitions to higher excited states are involved.

In this chapter, we review how to:

- obtain, monitor, and explain the vibrational-rotational excitations;
- find the quantum mechanical expression of the vibrational and rotational energy levels;
- predict the frequency of the bands in the vibrational-rotational spectrum; and
- compare between harmonic and anharmonic oscillation and between rigid and nonrigid rotor models for the possible excitations.

Visible light and infrared radiation are two sorts of electromagnetic radiation that consists of continuous electric and magnetic waves. Only the electric waves interact significantly with molecules and are important in explaining infrared absorption and Raman stretching. In this part, we focus on how the molecules interact with this radiation and the chemical information obtainable by the measurement of the infrared and Raman spectroscopy.

The vibrational spectra of diatomic molecules establish most of the essential principles that are used for complicated polyatomic molecules. Since infrared radiation will excite not only molecular vibration but also rotation, we need to comprehend both the rotation and vibration of diatomic molecules in order to analyze their spectra.

Molecular vibrations can be explained by classical mechanics using a simple ball and spring model, whereas vibrational energy levels and transitions between them are concepts taken from quantum mechanics. Both approaches are useful and will be treated here. The quantum mechanics of the translation, vibration, and rotation motions are explored in detail. As well, the expressions for the vibration-rotation energies of diatomic molecules for the harmonic and anharmonic oscillator models are introduced. Then, Schrödinger equation is identified for the vibration system of n-atom molecules. The symmetry of the vibrational wave functions of the harmonic oscillator is examined.

It was necessary to outline the general steps to determine the normal modes of vibration, and the symmetry representation of these modes. We examine the relationship and the differences among the Cartesian, internal, and normal coordinates that are used to characterize the stretching vibrations. We then show why the normal coordinates are used to calculate the vibrational energy of the polyatomic molecules. Lagrange's equation is used to show the change in the amplitude of displacement with time. We also examine how to calculate the relative amplitudes of motion, the kinetic, and potential energies for the vibrational motions of N-atom molecules. The significances of the force constants are discussed and how to calculate the force constants using GF-matrix method.

Theoretical analysis of the vibrational spectra of metal complexes is difficult, but recording of the spectra of a great number of different types of molecules shows a qualitative relationship between vibrational spectra and molecular structure. When ligands coordinate to metal ions, the newly formed coordinate bonds alter the symmetry of the ligand, the state of energy, and the electronic structure. These variations influence the vibration of the ligand and as a result change its vibrational spectrum. The formed complex will exhibit new vibrations that are not caused by the free ligand. The spectrum will also be affected by the interaction with the environment (solvent, ions bound in the outer sphere, and other molecules), as well as the symmetry and structure of the complex molecule. Recording and analysis of the vibrational spectra have become important tools in coordination chemistry.

Electrons, Atoms, and Molecules in Inorganic Chemistry. http://dx.doi.org/10.1016/B978-0-12-811048-5.00009-2

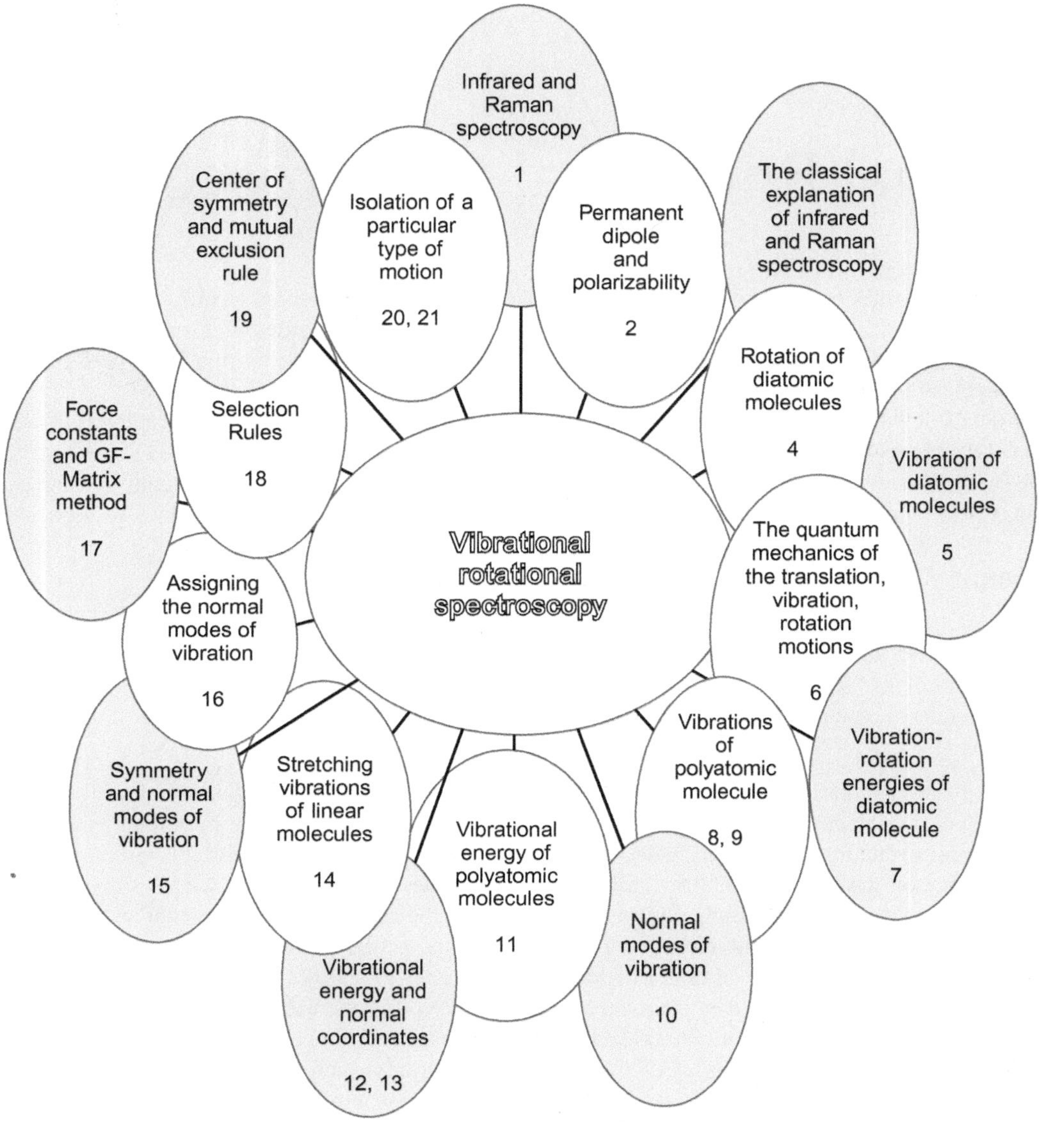

SCHEME 9.1 Outline of the principles of vibrational rotational spectroscopy.

The requirements and the selection rules for the allowed vibrational and rotational excitations are explored. We investigate the relationship between the center of symmetry and the mutual exclusion rule, how to distinguish among isomers, and binding modes, and define the forms of the normal modes of vibration and which of these modes are infrared and/or Raman active.

- 9.1: Infrared and Raman spectroscopy
- 9.2: Permanent dipole and polarizability
- 9.3: The classical explanation of infrared and Raman spectroscopy
- 9.4: Rotation of diatomic molecules
  - Rigid and nonrigid models

9.1 INFRARED AND RAMAN SPECTROSCOPY

How can the vibrational and rotational spectra be monitored?

- Vibrational spectra are normally monitored by two different techniques.
  - In infrared (IR) spectroscopy, light of different frequencies is passed through a sample and the intensity of the transmitted light is measured at each frequency (Fig. 9.1).
  - Raman spectroscopy is concerned with vibrational and rotational transitions. We observe not transmitted light but light scattered by the sample.
- In a Raman experiment, a monochromatic lasers beam of light illuminates the sample, and observations are made on the scattered light at right angles to the incident beam (Fig. 9.2).

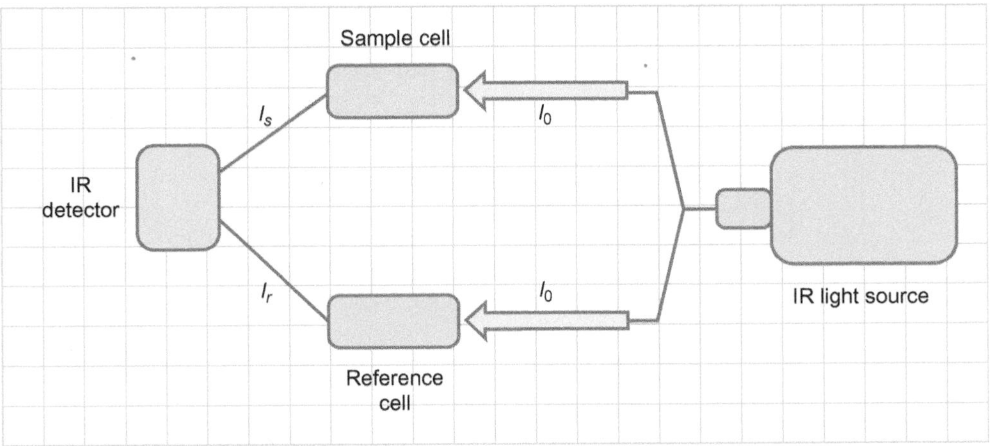

FIG. 9.1 Schematic representation of infrared spectrophotometer.

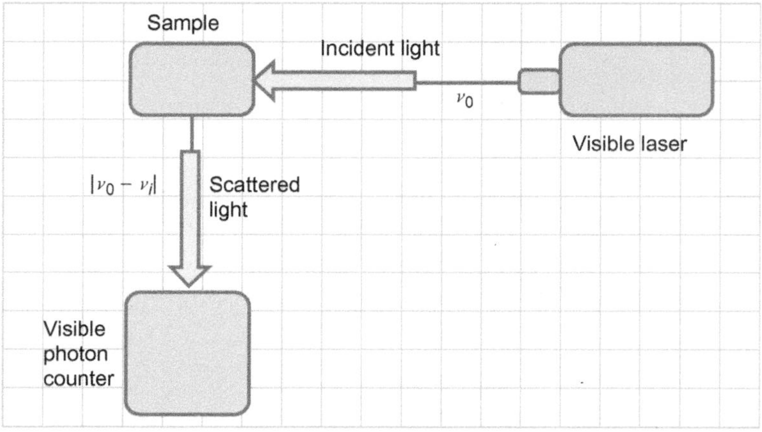

FIG. 9.2 Schematic representation of Raman set-up.

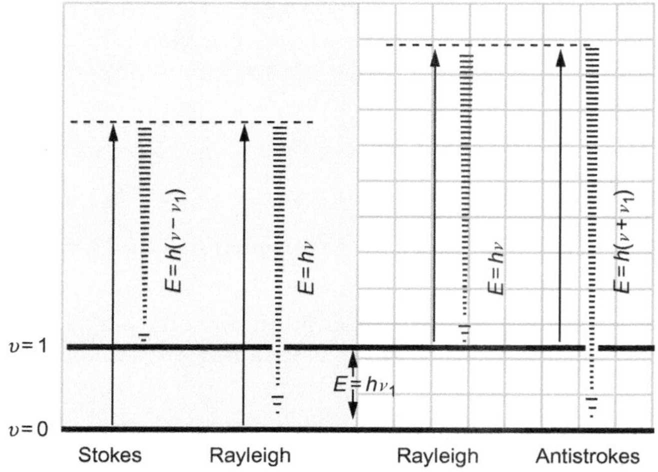

FIG. 9.3 Rayleigh, Stokes, and anti-Stokes radiations in the rotational-vibrational transitions.

What would happen if light of single frequency ν_0 and energy $h\nu_0$ were shone through a homogenous solution?

- If light of single frequency ν_0 and energy $h\nu_0$ is shone through homogenous solution, most of the light will pass directly through the sample.
- Some of the incident light collides with molecules of the sample and is scattered in all directions with unchanged frequency. This phenomenon is called Rayleigh scattering.
- A very small frication of scattered light (about 1/1000) has a frequencies ν_{vib} such that $\Delta E = h|\nu_0 - \nu_{vib}|$ corresponds to energies that are absorbed by the sample; this is *called Stokes radiation.*
- The frequency, ν_0, may be in any region of the spectrum (visible light), but the difference, $|\nu_0 - \nu_{vib}|$, is an infrared (vibrational) frequency. The process that emits light of a frequency other than ν_0 is called Raman scattering.
- The measured value of ν_{vib} is identical to the infrared frequency that would excite this vibrational mode if it were infrared-active.
- A molecule in the vibrational excited state $\upsilon = 1$ can collide with incident light quantum frequency ν_0. The molecule can return to the ground state by giving its additional energy $h\nu_{vib}$ to the photon. This photon, when scattered, will have a frequency $|\nu_0 + \nu_{vib}|$, and is referred to as anti-Stokes radiation.
- Because of the Boltzmann distribution, there are fewer molecules in $\upsilon = 1$ state than in $\upsilon = 0$, and the intensity of the anti-Stokes lines is much lower than that of the Stokes lines (Fig. 9.3).

9.2 PERMANENT DIPOLE AND POLARIZABILITY

How does the permanent dipole differ from the induced one? Give examples for each.

- Permanent dipole: if two particles of charges +e and −e are separated by distance r, the permanent dipole moment, μ, is defined by

$$\mu = er \tag{9.2.1}$$

- A molecule with a permanent dipole moment is called a polar molecule. Examples of molecules with a permanent dipole moment are as follows:

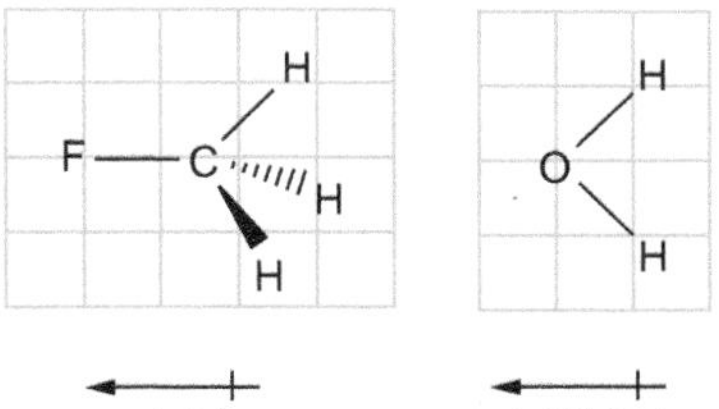

μ (Debye) 1.81 1.85 Debye

- Examples of molecules with no permanent dipole are as follows:

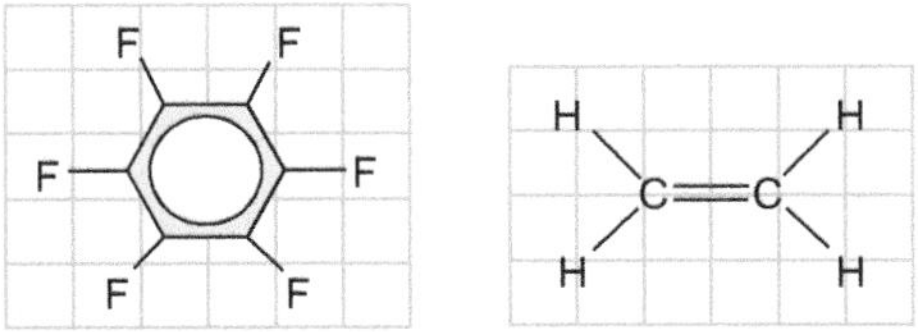

No permanent dipole moment

- If a molecule is placed in an electric field, the electrons are attracted toward the positive pole of the field, and the nuclei are attracted the opposite way. As a result, a dipole moment is induced in the molecule, μ_{ind}.
- The induced dipole is directly proportional to the field strength, ε, the proportionately constant, α, being named polarizability of the molecule:

$$\mu_{ind} = \alpha\varepsilon \tag{9.2.2}$$

What is the dimension of the polarizability constant, α?

- Dimension of

$$\alpha = \frac{\mu}{\varepsilon} = \frac{\text{charge} \cdot \text{length}}{\text{charge/length}^2} = \text{length}^3 = \text{volume}$$

- Examples:

	He	N_2	CFH_3	C_2H_4
α (Å^3)	0.203	1.72	3.84	4.5

9.3 THE CLASSICAL EXPLANATION OF INFRARED AND RAMAN SPECTROSCOPY

How could the infrared absorption be explained?

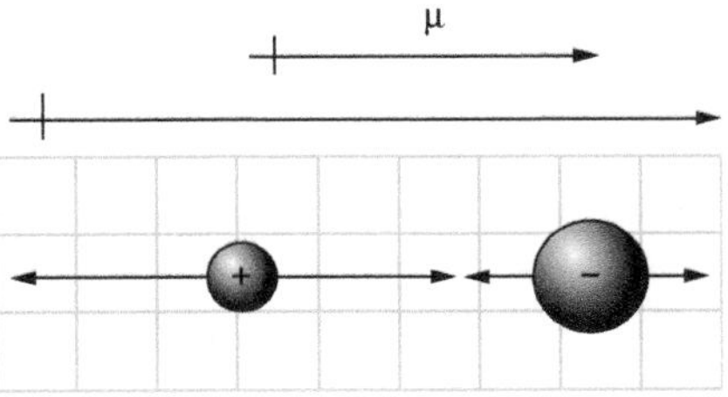

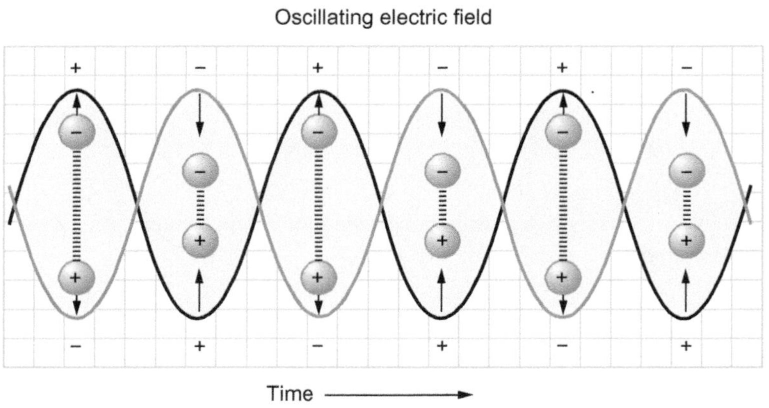

FIG. 9.4 Oscillating electric field and alternately increase and decrease the dipole spacing.

- Consider a heteronuclear diatomic molecule only vibrates at a certain frequency, where the two centers of the opposite charges (atoms) can move back and forth, and as a result, the molecular dipole oscillates about its equilibrium value.
- Consequently, the molecular dipole can absorb energy from vibrating electric field only if the field also vibrates at an identical frequency (Fig. 9.4).
- This resonance energy transfer does not happen if the frequencies of the light and the dipole vibrational frequency are not the same.
- The absorption of energy from the light wave by an oscillating permanent dipole is a molecular justification of infrared spectroscopy.
- In other words, the dipole moment of the molecule must change during the vibration, in order that a molecular vibration may have infrared absorption (Fig. 9.4).

Which of the following molecular vibrations can result in infrared absorption?

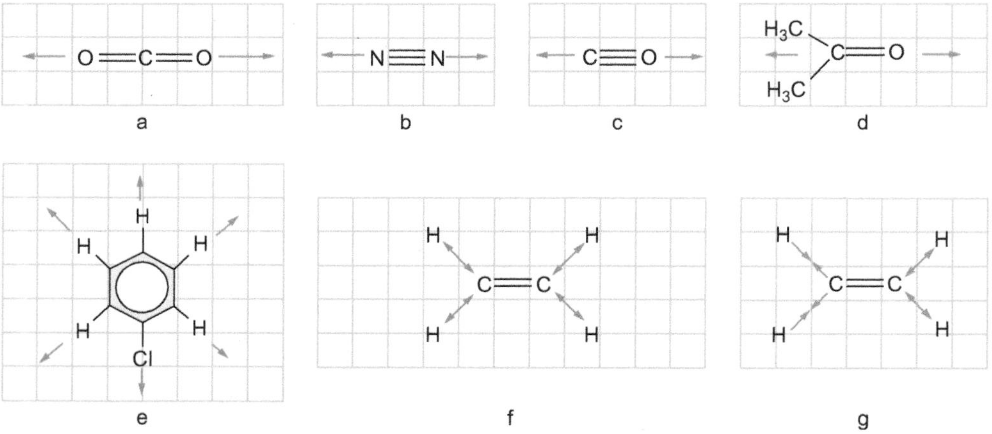

- c, d, e, and g are infrared active molecules.

How could the Raman excitations be explained?

- If a molecule is placed in the field of the electric vector of the radiation, a dipole moment is induced in the molecule, μ_{ind}.
- Consequently, the electrons in the molecule are forced into oscillations of the same frequency as the radiation. This oscillating dipole radiates energy in all directions and accounts for Rayleigh scattering.
- This scattering process corresponds to an elastic collision of the molecule with the photon. When the molecule in its ground vibrational state accepts energy from the photon being scattered, exciting the molecule into a higher vibration state.

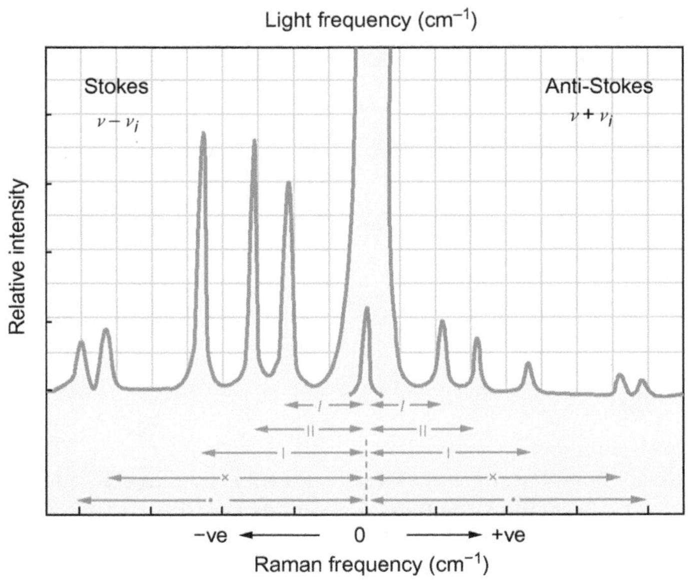

FIG. 9.5 Typical Raman spectrum of a small molecule.

- Consider a light wave whose electric field oscillates at a certain point in space according to the following:

$$\varepsilon = \varepsilon_0 \cos 2\pi \upsilon t \tag{9.3.1}$$

where ε_o is the maximum value of the field, ν is the frequency, and t is time.
- The induced dipole moment:

$$\mu_{\text{ind.}} = \alpha\varepsilon \tag{9.2.2}$$

$$\mu_{\text{ind.}} = \alpha\varepsilon_0 \cos 2\pi\upsilon t \tag{9.3.2}$$

- However, α is a molecular property whose value should vary as the molecule oscillates:

$$\alpha = \alpha_0 + (\Delta\alpha)\cos 2\pi\upsilon_0 t \tag{9.3.3}$$

where α_0 is the equilibrium polarizability, $\Delta\alpha$ is its maximum variation, and ν_0 is the natural vibration frequency.
- Then

$$\mu_{\text{ind.}} = [\alpha_0 + (\Delta\alpha)\cos 2\pi\upsilon_0 t][\varepsilon_0 \cos 2\pi\upsilon t]$$

which can be rearranged to

$$\mu_{\text{ind.}} = \alpha_0\varepsilon_0 \cos 2\pi\upsilon t + \frac{1}{2}\Delta\alpha\varepsilon_0[\cos 2\pi(\upsilon+\upsilon_0)t + \cos 2\pi(\upsilon-\upsilon_0)t] \tag{9.3.4}$$

because $\cos\theta\cos\phi = \frac{1}{2}[\cos(\theta+\phi) + \cos(\theta-\phi)]$
- Therefore, the induced dipole moment will oscillate with components of frequency ν (Rayleigh scattering), $\nu-\nu_0$ (Stokes radiation), and $\nu+\nu_0$ (anti-Stokes radiation) (Fig. 9.5).

9.4 ROTATION OF DIATOMIC MOLECULES

For diatomic molecules, find the quantum mechanical expression of the rotational energy levels, E_J, where Jth is the rotational quantum number.

- The rotational energy levels may be obtained by using the classical expression for the energy of a body rotating about axis x is

$$E = \frac{1}{2}I_{xx}\omega_x^2 \tag{9.4.1}$$

where I_{xx} is the moment of inertia and ω_x is the angular velocity.

A body allowed to rotate about all three axes has the energy

$$E=\frac{1}{2}I_{xx}\omega_x^2+\frac{1}{2}I_{yy}\omega_y^2+\frac{1}{2}I_{zz}\omega_z^2 \quad (9.4.2)$$

- The classical angular momentum of a body of moment of inertia I_{xx} and angular velocity ω_x is

$$J_x=I_{xx}\omega_x \quad (9.4.3)$$

$$\omega_x=\frac{J_x}{I_{xx}}$$

- Therefore, the energy of the body

$$E=\frac{J_x^2}{2I_{xx}}+\frac{J_y^2}{2I_{yy}}+\frac{J_z^2}{2I_{zz}} \quad (9.4.4)$$

- Suppose two nuclei of masses m_1 and m_2 are fixed at their equilibrium separation, r_e.
- The rotation of this diatomic molecule is equivalent to the motion of a single particle of reduced mass μ at a distance r_e from the center of rotation (Fig. 9.6). For such a rigid rotational motion, the three moment of inertia are identical:

$$I=I_{xx}=I_{yy}=I_{zz}$$

$$I=\mu r_e^2$$

$$\mu=\frac{m_1m_2}{m_1+m_2}$$

and the energy expression reduces to

$$E=\frac{1}{2I}\left(J_x^2+J_y^2+J_z^2\right)=\frac{J^2}{2I}=\frac{J^2}{2\mu r_e^2} \quad (9.4.5)$$

- The quantum magnitude of the angular momentum

$$J^2=J(J+1)\hbar^2$$

and the energy of the rotating molecule is limited to

$$E_J=J(J+1)\frac{\hbar^2}{2\mu r_e^2},\quad J=0,1,2,\ldots$$

$$E_J=J(J+1)\frac{\hbar^2}{2I}$$

$$E_J=BJ(J+1),\ \text{in joules} \quad (9.4.6)$$

where E_J is the energy (in joules) of the Jth rotational energy level, and J is the rotational quantum number.
- To convert the energy of rotation in joules to wave number in cm^{-1}, the energy will be divided by 100 hc.

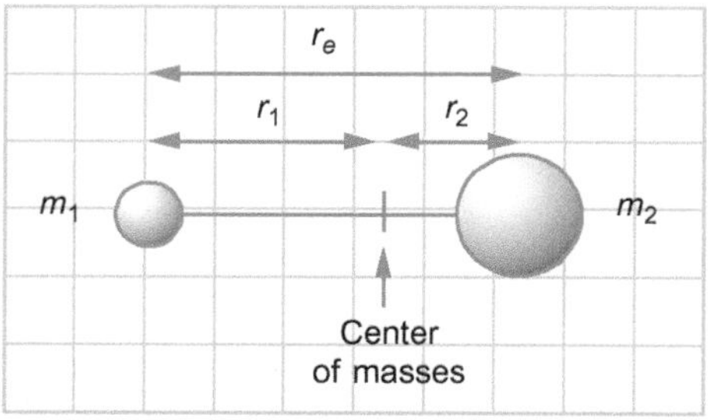

FIG. 9.6 The molecule will rotate about the center of mass define such that $m_1r_1=m_2r_2$.

$$\frac{\text{Energy (joules)}}{h(\text{joule s}) \cdot c(\text{m/s}) \cdot 100(\text{cm/m})} = \bar{E}\left(\text{cm}^{-1}\right)$$

The factor $\frac{\hbar^2}{2I}$ is usually written hcB, B is the rotational constant in joules, $\bar{B}$ is expressed in wave number, cm^{-1}.

$$\bar{E} = \bar{B}J(J+1), \text{ in cm}^{-1} \tag{9.4.7}$$

- The selection rules for the rigid rotor is

$$J \rightarrow J+1$$

where transition only is allowed.

- The separation in energy between neighboring rotational levels is

$$\Delta\bar{E} = \bar{E}_{J+1} - \bar{E}_J = 2\bar{B}(J+1)$$

$$\Delta\bar{E} = \bar{E}_J - \bar{E}_{J-1} = 2\bar{B}J$$

or

$$\bar{\nu}_{J+1} - \bar{\nu}_J = 2\overline{B}(J+1)$$

$$\bar{\nu}_J - \bar{\nu}_{J-1} = 2J\overline{B}$$

Predict the positions of the first four absorption bands in the rotational spectrum of CO ($r_e = 1.128$ A, $m_C = 12.000$ a.u., $m_O = 15.995$ a.u.)

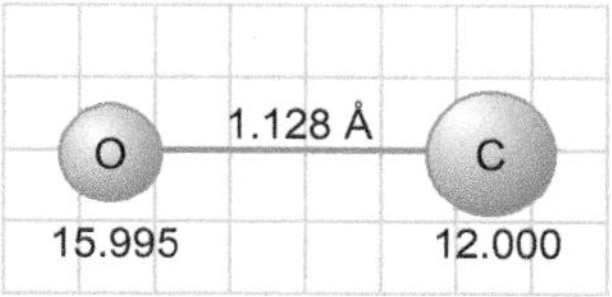

- The reduced mass of CO is

$$\mu = \frac{(12.000)(15.995)}{12.000 + 15.995} = 6.856\,\text{a.u.} = \frac{6.856}{6.022 \times 10^{26}} = 1.138 \times 10^{-26}\,\text{kg}$$

$$I = \mu r_e^2 = \left(1.138 \times 10^{-26}\right)\left(10128 \times 10^{-10}\right)^2 = 1.449 \times 10^{-46}\,\text{kg m}^2$$

The possible energy levels will be

$$E_J = J(J+1)\frac{\hbar^2}{2I}$$

$$E_J = \frac{\left(1.055 \times 10^{-34}\,\text{kg m}^2\text{s}^{-1}\right)^2}{2\left(1.449 \times 10^{-46}\,\text{kg m}^2\right)} J(J+1) = 3.84 \times 10^{-23} J(J+1)\,\text{joule}$$

$$E_J = 1.93J(J+1)\,\text{cm}^{-1}$$

$$\Delta\bar{E} = \bar{E}_{J+1} - \bar{E}_J = 2\bar{B}(J+1)$$

The energy levels of CO are shown in Table 9.1.

- The calculated values of the rotational energy levels of CO are kinetic because the potential energy in the rigid rotor model has not been considered.
- For the transition: $J = 1 \rightarrow J = 2$

  CO molecule absorbs a photon of wave number: 11.6–3.9 = 7.7 cm^{-1} (microwave range)

  It is expected to see rotational absorption spectrum for: $J \rightarrow J+1$ (allowed)

  Other transitions like: $J \rightarrow J+2$ are forbidden.

TABLE 9.1 The Calculated Rotational Energy Levels of CO

J	$J(J+1)$	$\bar{E}_J$ (cm^{-1})	$\Delta\bar{E}(J\rightarrow J+1)$
0	0	0	$2\bar{B}$
1	2	3.9	$4\bar{B}$
2	6	11.6	$6\bar{B}$
3	12	23.2	$8\bar{B}$
4	20	38.6	$10\bar{B}$

- The energy difference between the levels:

$$\Delta\bar{E} = \bar{B}(J+1)(J+2) - \bar{B}J(J+1) = 2\bar{B}(J+1)$$

That is, each absorption in the spectrum is separated from its neighbors by the energy $2\bar{B}$ for diatomic molecules.

Rigid and Nonrigid Models

Why does the rotation of real diatomic molecules not behave like rigid rotors?

Write the expression for the energy levels of nonrigid rotor.

Use the following data to compare between rigid and nonrigid rotor models for the transition from $J=3$ to $J=4$ rotational levels of HCl molecule: $\bar{\nu}_{obs.}=83.03\,\text{cm}^{-1}$; for rigid rotor $\bar{B}=10.34\,\text{cm}^{-1}$, and the centrifugal distortion constant $\bar{D}=0.0004\,\text{cm}^{-1}$.

- In the real diatomic molecules, the bond length increases as the rate of rotation increases, i.e., as J increases.
- The expression for a nonrigid rotor is

$$E_J = BJ(J+1) - DJ^2(J+1)^2 \tag{9.4.8}$$

where D is the centrifugal distortion constant that measures the increased bond length resulting from the rotation of the molecule.
- For a rigid rotor:

$$\bar{\nu}_{\text{Calc.}} = \bar{\nu}_{J+1} - \bar{\nu}_J = 2\bar{B}(J+1)$$

$$\bar{\nu}_{\text{Calc.}} = \bar{\nu}_4 - \bar{\nu}_3 = 2(10.34)(3+1) = 82.72\,\text{cm}^{-1}$$

where $\bar{\nu}_{\text{obs.}} = 83.03\,\text{cm}^{-1}$
- For a nonrigid rotor:

$$E_J = BJ(J+1) - DJ^2(J+1)^2$$

The change in the frequency due to the transition J to $J+1$ is given by

$$\bar{\nu}_{\text{Calc.}} = \bar{\nu}_{J+1} - \nu_J = \bar{B}(J+1)(J+2) - \bar{D}(J+1)^2(J+2)^2 - \bar{B}J(J+1) + \bar{D}J^2(J+1)^2$$

$$= \bar{B}(J+1)[(J+2)-J] - \bar{D}(J+1)^2\left[(J+2)^2 - J^2\right]$$

$$= 2\bar{B}(J+1) - \bar{D}(J+1)^2[(J+2)+J][(J+2)-J]$$

$$\bar{\nu}_{\text{Calc.}} = \bar{\nu}_{J+1} - \bar{\nu}_J = 2\bar{B}(J+1) - 4\bar{D}(J+1)^3 \tag{9.4.9}$$

$$\bar{\nu}_{\text{Calc.}} = \bar{\nu}_4 - \bar{\nu}_3 = 2(10.395)(4) - 4(0.0004)(4)^3 = 83.06\,\text{cm}^{-1}$$

where $\bar{\nu}_{\text{obs.}} = 83.03\,\text{cm}^{-1}$

9.5 VIBRATION OF DIATOMIC MOLECULES

Derive a formula to estimate the vibrational frequency for diatomic molecule by the use of the force constant and the reduced mass.

- Consider the molecule as two balls, with masses equal to the two atoms. These two balls are connected by a spring that represents the bond. During a vibration, each atom is displaced from its equilibrium position:

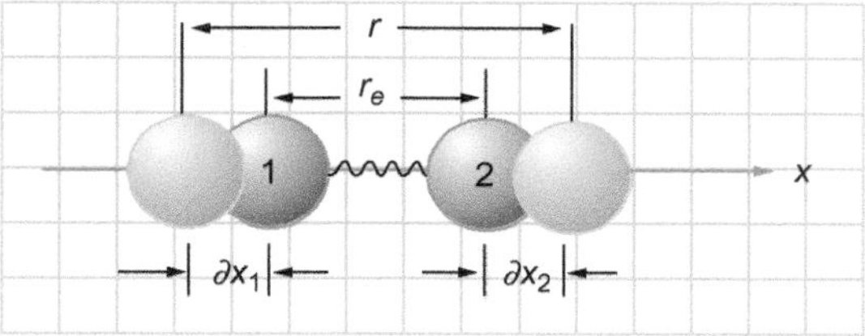

- According to Hooke's law equation, the force, F, needed to produce extension or compression is

$$F = -k(\partial x_2 - \partial x_1) = -k(r - r_e) = -kq \quad (9.5.1)$$

where k is the force constant.
- There will be two Newton's equations:

$$F = m_1\left(\frac{\partial(\partial x_1)}{\partial t^2}\right) \text{ for atom 1} \quad (9.5.2)$$

$$F = m_2\left(\frac{\partial^2(\partial x_2)}{\partial t^2}\right) \text{ for atom 2} \quad (9.5.3)$$

- The total stretching = vibration quantity $q = r - r_e = \partial x_2 - \partial x_1$
For the diatomic molecule

$$F = \left(\frac{m_1 m_2}{m_1 + m_2}\right)\left(\frac{\partial^2(r - r_e)}{\partial t^2}\right) = -k(r - r_e) \quad (9.5.4)$$

where $\mu = \frac{m_1 m_2}{m_1 + m_2}$
then

$$\mu\left(\frac{\partial^2 q}{\partial t^2}\right) = -kq \quad (9.5.5)$$

where q is the transverse displacement, varies in regular periodic time, and can be described by

$$q = A\cos 2\pi\nu t \quad (9.5.6)$$

where ν is the frequency:

$$\frac{\partial q}{\partial t} = A\frac{\partial \cos 2\pi\nu t}{\partial t} = -2\pi\nu A \sin 2\pi\nu t$$

$$\frac{\partial^2 q}{\partial t^2} = -2\pi\nu A\frac{\partial \sin 2\pi\nu t}{\partial t}$$

$$\frac{\partial^2 q}{\partial t^2} = -4\pi^2\nu^2 A\cos 2\pi\nu t \quad (9.5.7)$$

if

$$\mu\left(\frac{\partial^2 q}{\partial t^2}\right) = -kq \quad (9.5.5)$$

$$q = A\cos 2\pi\nu t \quad (9.5.6)$$

$\mu\left(\frac{\partial^2 q}{\partial t^2}\right) = -kA\cos 2\pi\nu t$, by substituting in (9.5.7):

$$\mu\left(-4\pi^2\nu^2 A\cos 2\pi\nu t\right) = -kA\cos 2\pi\nu t$$

$$k = 4\pi^2\nu^2\mu$$

$$\nu = \frac{1}{2\pi}\sqrt{\frac{k}{\mu}} \tag{9.5.8}$$

If $^{1}H^{35}Cl$ show a strong infrared ration at 2991 cm^{-1}:

Calculate the force constant *k* for $^{1}H^{35}Cl$ molecules. If the hydrogen is substituted by deuterium in this molecule, and the force constant is unaffected by this substitution, by what factor do you expect the frequency to shift?

- To calculate the force constant, *k*:

$$\because \nu = \frac{1}{2\pi}\sqrt{\frac{k}{\mu}} \tag{9.5.8}$$

$$\therefore k = 4\pi^2\mu\nu^2$$

$$\because \nu = \frac{c}{\lambda}$$

$$\therefore k = 4\pi^2\mu\left(\frac{c}{\lambda}\right)^2$$

$$\because \mu = \frac{m_A m_B}{m_A + m_B} = \frac{m_H m_{Cl}}{m_H + m_{Cl}} = -\frac{(1.008)(34.969)\,\text{a.u.}}{35.977}\left(\frac{1.661\times10^{-27}\,\text{kg}}{1\,\text{a.u.}}\right)$$

$$\therefore k = 4\pi^2(0.979976\,\text{a.u.})\left(2.991\times10^{8}\,\text{m/s}\right)^2\left(\frac{2991}{\text{cm}}\cdot\frac{100\,\text{cm}}{1\,\text{m}}\right)^2\left(\frac{1.661\times10^{-27}\,\text{kg}}{1\,\text{amu}}\right)$$

$$k = 514.7\,\text{N/m}$$

- The factor of the frequency to shift:

$$\frac{\nu_{DCl}}{\nu_{HCl}} = \sqrt{\frac{\mu_{HCl}}{\mu_{DCl}}} = \sqrt{\frac{m_H m_{Cl}(m_D + m_{Cl})}{m_D m_{Cl}(m_H + m_{Cl})}} = \sqrt{\left(\frac{1.0078}{2.0140}\right)\left(\frac{36.983}{35.977}\right)} = 0.717$$

The vibration frequency for DCl is: 2144.5 cm^{-1}, which is much lower than that of HCl.

Vibrational Energy Levels

Show that the quantum mechanical expression of the vibrational energy, E_υ, of a one-dimensional harmonic oscillator:

$$E_\upsilon = \left(\upsilon + \frac{1}{2}\right)h\nu_0$$

where υ is the vibrational quantum number.

- Consider the vibrational motion of a diatomic molecule AB; the distance r indicates the equilibrium position between A and B atoms, and *x* is the displacement (Fig. 9.7).
- There will be a force working to take back the system to the equilibrium position. This force, *F*, is proportional to the displacement, *x*.

$$F \propto x, \text{ then:}$$

$$F = -kx \tag{9.5.9}$$

where *k* is the force constant.

- The resulting motion, which proceeds when A and B are free to oscillate, is described as a simple harmonic motion.
- For harmonic oscillation of two atoms connected by a bond, the potential energy is given by

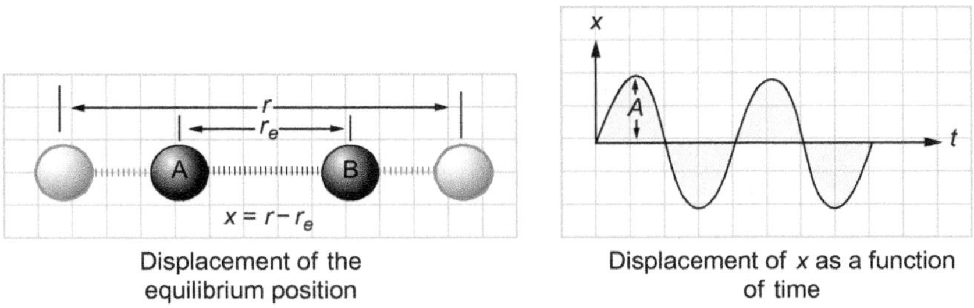

FIG. 9.7 Simple harmonic vibration of atoms A and B and the displacement of x as a function of time.

$$V(x) = \frac{1}{2}kx^2 \tag{9.5.10}$$

and from (9.5.8):

$$k = 4\pi^2\mu\nu_0^2$$

where μ is the reduced mass and ν_0 the fundamental vibration frequency (the frequency for transition from $\nu_0 \rightarrow \nu_1$.

- A molecular system can occupy only discrete quantized energy levels. Thus, a quantum mechanical treatment of molecular system leads to an exact solution.
- The Schrödinger equation for one-dimensional harmonic oscillator is as follows:

$$\frac{\partial^2\psi}{\partial x^2} + \frac{8\pi^2\mu}{h^2}(E - V)\psi = 0 \tag{9.5.11}$$

If

$$V(x) = \frac{1}{2}kx^2 \tag{9.5.10}$$

and

$$k = 4\pi^2\mu\nu_0^2$$

then

$$V(x) = \frac{4}{2}x^2\pi^2\mu\nu_0^2 = 2x^2\pi^2\mu\nu_0^2 \tag{9.5.12}$$

then

$$\frac{\partial^2\psi}{\partial x^2} + \frac{8\pi^2\mu}{h^2}(E - 2x^2\pi^2\mu\nu_0^2)\psi = 0$$

$$\frac{\partial^2\psi}{\partial x^2} + \left(\frac{8\pi^2\mu E}{h^2} - \frac{16\pi^4\mu^2\nu_0^2x^2}{h^2}\right)\psi = 0$$

$$\frac{\partial^2\psi}{\partial x^2} + \left(\frac{2\mu E}{\hbar^2} - \frac{4\pi^4\mu^2\nu_0^2x^2}{\hbar^2}\right)\psi = 0$$

- Multiply by $\frac{\hbar}{2\pi\mu\nu_0}$

$$\frac{\hbar}{2\pi\mu\nu_0}\frac{\partial^2\psi}{\partial x^2} + \left(\frac{E}{\hbar\pi\nu_0} - \frac{2\pi\mu\nu_0x^2}{\hbar}\right)\psi = 0$$

If

$$\alpha^2 = \frac{\hbar}{2\pi\mu\nu_0} \text{ and } \varepsilon = \frac{E}{\hbar\pi\nu_0}$$

then

$$\alpha^2\frac{\partial^2\psi}{\partial x^2}+\left(\varepsilon-\frac{x^2}{\alpha^2}\right)\psi=0 \tag{9.5.13}$$

- Next transform the independent variable x into a new variable y such that $x=\alpha y$, then

$$\frac{\partial^2}{\partial x^2}=\frac{\partial^2}{\alpha^2\partial y^2}$$

and (9.5.13)

$$\frac{\partial^2\psi}{\partial y^2}+(\varepsilon-y^2)\psi=0 \tag{9.5.14}$$

When y becomes very large, $\epsilon \ll y^2$:

$$\frac{\partial^2\psi}{\partial y^2}-y^2\psi=0$$

$$\therefore \psi=e^{\pm y^2/2}$$

The positive exponential does not behave properly:

$$\therefore \psi=e^{-y^2/2} \tag{9.5.15}$$

- We return to find a solution for Eq. (9.5.14):

$$\frac{\partial^2\psi}{\partial y^2}+(\varepsilon-y^2)\psi=0$$

Let the solutions be of the following form:

$$\psi=H(y)e^{-y^2/2}=He^{-y^2/2} \tag{9.5.16}$$

where

$$H=H(y)=\sum_u a_u y^u=a_0+a_1y+a_2y^2+a_3y^3 \tag{9.5.17}$$

$$\frac{\partial\psi}{\partial y}=H'e^{-y^2/2}-Hye^{-y^2/2}$$

$$\frac{\partial^2\psi}{\partial y^2}=H''e^{-y^2/2}-H'ye^{-y^2/2}-H'ye^{-y^2/2}-He^{-y^2/2}+Hy^2e^{-y^2/2}, \text{then}:$$

$$\frac{\partial^2\psi}{\partial y^2}+(\varepsilon-y^2)\psi=0$$

$$H''e^{-y^2/2}-H'ye^{-y^2/2}-H'ye^{-y^2/2}-He^{-y^2/2}+Hy^2e^{-y^2/2}+\varepsilon He^{-y^2/2}-Hy^2e^{-y^2/2}=0$$

$$H''e^{-y^2/2}-2H'ye^{-y^2/2}-He^{-y^2/2}+\varepsilon He^{-y^2/2}=0$$

$$H''-2H'y-H+\varepsilon H=0$$

$$H''-2yH'+(\varepsilon-1)H=0$$

- $H(y)$ can be express as power series in y:

$$H(y)=\sum_\upsilon a_\upsilon y^\upsilon=a_0+a_1y+a_2y^2+a_3y^3$$

$$H'(y)=\sum_\upsilon \upsilon a_\upsilon y^{\upsilon-1}=a_1+2a_2y+3a_3y^2$$

$$H''(y) = \sum_{v} v(v-1)a_v y^{v-2} = \sum_{v} (v+2)(v+1)a_{v+2}y^v = 2(1)a_2 + 3(2)a_3 y + 4(3)a_4 y^2$$

Then

$$H'' - 2yH' + (\varepsilon - 1)H = 0$$

change to

$$\sum_{v} v(v-1)a_v y^{v-2} - 2y\sum_{v} v a_v y^{v-1} + (\varepsilon - 1)\sum_{v} a_v y^v = 0$$

$$\sum_{v} (v+2)(v+1)a_{v+2}y^v - 2\sum_{v} v a_v y^v + (\varepsilon - 1)\sum_{v} a_v y^v = 0$$

$$(v+2)(v+1)a_{v+2} - 2va_v + (\varepsilon - 1)a_u = 0$$

$$(v+1)(v+2)a_{v+2} + (\varepsilon - 1 - 2v)a_v = 0$$

$$(v+1)(v+2)a_{v+2} = -(\varepsilon - 2v - 1)a_v$$

$$a_{v+2} = -\frac{(\varepsilon - 2v - 1)}{(v+1)(v+2)}a_u \tag{9.5.18}$$

- The termination of the series after the vth term can be reached by selection the energy parameter ϵ in such a way the numerator goes to zero for $\nu = v$:

$$\varepsilon - 2v - 1 = 0$$

$$\varepsilon = 2v + 1 \tag{9.5.19}$$

$$\varepsilon = \frac{E}{\hbar\pi\nu_0}, \quad \text{and} \quad \hbar\pi = \frac{1}{2}h$$

$$E_v = \hbar\pi\nu_0\varepsilon = \frac{1}{2}h\varepsilon\nu_0 = \frac{1}{2}h\nu_0(2v+1)$$

$$E_v = \left(v + \frac{1}{2}\right)h\nu_0 \tag{9.5.20}$$

where v is an integer 0, 1, 2, … representing the vibrational quantum number of different states, ν_0 is the fundamental frequency (cm^{-1}), and E_v is the quantum mechanical expression for a one dimensional harmonic oscillator.

If the fundamental absorption frequency ν_0=4159.5cm^{-1} for H_2 molecule, calculate the force constants assuming the harmonic oscillator model.

- The difference in the energy, ΔE, between two adjacent levels, E_v and E_{v+1} is given by

$$E_v = \left(v + \frac{1}{2}\right)h\nu_0 \tag{9.5.20}$$

$$\Delta E = h\nu_0\left(v + \frac{3}{2}\right) - h\nu_0\left(v + \frac{1}{2}\right) = h\nu_0 \tag{9.5.21}$$

$$\text{If}: \nu = \frac{1}{2\pi}\sqrt{\frac{k}{\mu}} \tag{9.5.8}$$

$2\pi\nu_0 = \sqrt{\frac{k}{\mu}}$, from (9.5.21):

$$\Delta E = \frac{h}{2\pi}\sqrt{\frac{k}{\mu}} \tag{9.5.22}$$

$$(\Delta E)^2 = \hbar^2 \frac{k}{\mu}$$

$$k = \frac{\mu(\Delta E)^2}{\hbar^2} \tag{9.5.23}$$

- H_2 molecule:

$$\Delta E = 4159.5\,\text{cm}^{-1} \times \frac{1.9864 \times 10^{-23}\,\text{joule}}{1\,\text{cm}^{-1}} = 8.2624 \times 10^{-20}\,\text{joule}$$

$$\mu = \frac{(1.008)(1.008)}{1.008 + 1.008} = 0.504\,\text{a.u.} = \frac{0.504}{6.02214 \times 10^{26}} = 8.3691 \times 10^{-28}\,\text{kg}$$

$$k = \frac{(8.3691 \times 10^{-28}\,\text{kg})\,(8.2624 \times 10^{-20})^2\,\text{j}^2}{(1.05457 \times 10^{-34})^2\,\text{j}^2\,\text{s}^2} = 513.7\frac{\text{kg}}{\text{s}^2} = 514\,\text{N}\,\text{m}^{-1}$$

$$1\ \text{N} = 1\ \text{kg}\,\text{m/s}^2,\ (\text{N: newton})$$

If the stretching frequency of C—C bond is about 1000 cm^{-1}, assuming the simple harmonic oscillator model, $m_H = 1$, $m_C = 12$, and $m_S = 32$ a.u.

(a) **Calculate the stretching frequency of C=C and C≡C bonds on the assumption that double and triple bonds have force constants which are two and three times those of the single bond, respectively.**

(b) **Calculate the frequencies of C—H and S—S bonds on assumption that all single bonds force constants are equal. Compare your results with the observed values:**

$$\nu_{C=C} = \mathbf{1650}, \nu_{CC} = \mathbf{2200}, \nu_{C-H} = \mathbf{3000}, \nu_{S=S} = \mathbf{500\ cm^{-1}}$$

- The vibration frequency of diatomic molecule, AB, is given by

$$\text{If}: \nu = \frac{1}{2\pi}\sqrt{\frac{k}{\mu}} \tag{9.5.8}$$

where $\mu = \frac{m_A m_B}{m_A + m_B}$, and $\frac{1}{\mu} = \frac{1}{m_A} + \frac{1}{m_B}$, then

$$\nu_{A-B} = \frac{1}{2\pi}\left[k\left(\frac{1}{m_A} + \frac{1}{m_B}\right)\right]^{1/2}$$

- For C—C

$$\nu_{C-C} = \frac{1}{2\pi}\left[k_{C-C}\left(\frac{1}{12} + \frac{1}{12}\right)\right]^{1/2} = \frac{1}{2\pi}\left[\frac{k_{C-C}}{6}\right]^{1/2}$$

If; $k_{C=C} = 2k_{C-C}$

$$\nu_{C=C} = \frac{1}{2\pi}\left[\frac{k_{C-C}}{3}\right]^{1/2}$$

$$\frac{\nu_{C=C}}{\nu_{C-C}} = [2]^{1/2}$$

$$\nu_{C=C} = [2]^{1/2}\nu_{C-C}$$

$\nu_{C=C} = 1.4142 \times 1000 = 1414\,\text{cm}^{-1}$: observed 1650 cm^{-1}
$\nu_{C\equiv C} = [3]^{1/2}\,\nu_{C-C} = 1.7320 \times 1000 = 1732\,\text{cm}^{-1}$: observed 2200 cm^{-1}

- If $k_{C-C} = k$

$$\nu_{X-Y} = \frac{1}{2\pi}\left[\frac{k}{\mu_{XY}}\right]^{1/2}$$

$$\mu_{XY} = \frac{m_X m_Y}{m_X + m_X}$$

$$\nu_{C-C} = \frac{1}{2\pi}\left[\frac{k}{\mu_{C-C}}\right]^{1/2}$$

$$\nu_{X-Y} = \left[\frac{\mu_{C-C}}{\mu_{XY}}\right]^{1/2} \nu_{C-C}$$

$$\mu_{C-C} = 6\ \text{a.u.}$$

$$\nu_{X-Y} = \left[\frac{6}{\mu_{XY}}\right]^{1/2} \times 1000$$

$$\mu_{C-H} = 0.92307$$

$$\nu_{C-H} = \left[\frac{6}{\mu_{C-H}}\right]^{1/2} \times 1000 = 2550\ \text{cm}^{-1}\ :\ \text{observed}\ 3000\ \text{cm}^{-1}$$

$$\mu_{S-S} = 16\ \text{a.u.}$$

$$\nu_{C-H} = \left[\frac{6}{\mu_{S-S}}\right]^{1/2} \times 1000 = 612\ \text{cm}^{-1}\ :\ \text{observed}\ 500\ \text{cm}^{-1}$$

Anharmonic Oscillation

Why do real molecules undergo anharmonic rather than harmonic oscillation?

- In the harmonic oscillations, the difference in the energy, ΔE, between two adjacent levels, E_υ and $E_{\upsilon+1}$ is given by:

$$\Delta E = h\nu_0\left(\upsilon + \frac{3}{2}\right) - h\nu_0\left(\upsilon + \frac{1}{2}\right) = h\nu_0, \text{ but}$$

$$2\pi\nu_0 = \sqrt{\frac{k}{\mu}}, \text{ then}$$

$$\Delta E = \frac{h}{2\pi}\sqrt{\frac{k}{\mu}} \qquad (9.5.22)$$

  Therefore, the vibrational energy levels are equally spaced.
- However, in real molecules the energy levels are converged and not equally spaced. This is because the molecules undergo anharmonic rather than harmonic oscillations (Fig. 9.8).
- At large displacement, the molecule approaches dissociation. The bond becomes easier to stretch than harmonic oscillator act because the restoring force is less than expected by

$$F = -kx \qquad (9.5.9)$$

- The anharmonic oscillation happens in all molecules and becomes greater as the vibrational quantum number increases.
- As $(r - r_e)$ decreases, the nuclei approach and repel each other. This repulsion becomes very strong at small values of $(r - r_e)$.

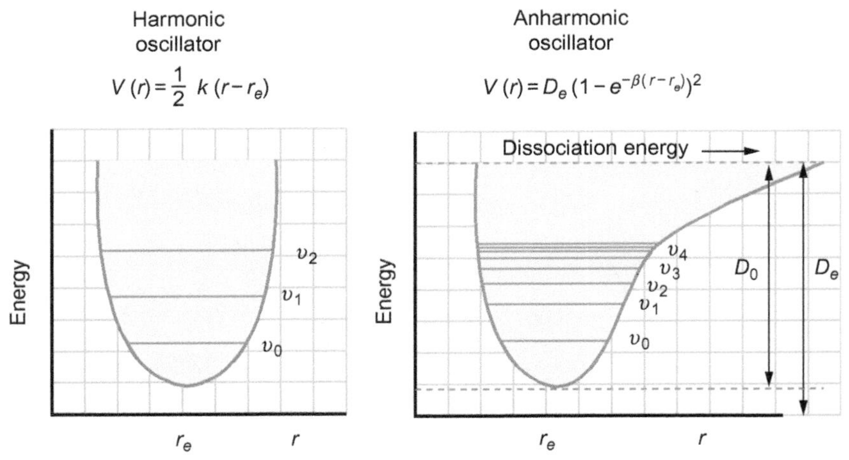

FIG. 9.8 Harmonic and anharmonic oscillators.

- Real molecules have asymmetric, anharmonic, and minimum potential (potential well, Fig. 9.8), which is controlled by

$$V(r) = D_e\left(1 - e^{-\beta(r-r_e)}\right)^2 \tag{9.5.24}$$

(the Morse potential)
where D_e is the depth of the potential well and β is a measure of the curvature at the bottom of the well.
- The calculated allowed energy levels:

$$\bar{E}_\nu = \bar{\nu}_e\left(\upsilon + \frac{1}{2}\right) - \bar{\nu}_e\rho_e\left(\upsilon + \frac{1}{2}\right)^2 \tag{9.5.25}$$

(in cm^{-1})
The value of $\bar{\nu}_e$ is almost the same as $\bar{\nu}_0$ of the harmonic oscillator, and $\bar{\nu}_e\rho_e$ is a small anharmonic correction.
- The transitions from the ground state $\upsilon = 0$ to higher energy levels are

$$\Delta\bar{E}_\nu = \bar{E}_\nu - \bar{E}_0 = \bar{\nu}_e\upsilon - \bar{\nu}_e\rho_e\upsilon(1+\upsilon) \tag{9.5.26}$$

where
$\upsilon = 0 \rightarrow \upsilon = 1$ is called the fundamental transition;
$\upsilon = 0 \rightarrow \upsilon = 2$ is the first overtone; and
$\upsilon = 0 \rightarrow \upsilon = 3$ is the second overtone.
- The dissociation energy $\bar{D}_0$

$$\bar{D}_0 = \bar{D}_e - \bar{E}_0 = \bar{D}_e - \frac{1}{2}\bar{\nu}_e + \frac{1}{4}\bar{\nu}_e\rho_e \tag{9.5.27}$$

If the vibrational frequencies of N_2 are

$$\bar{\nu}_0 = 2330.7\,cm^{-1},\ \bar{\nu}_e = 2359.61\,cm^{-1},\ \text{and}\ \bar{\nu}_e\rho_e = 14.456\,cm^{-1}$$

then estimate the energies of the fundamental and the first three overtones of N_2 using the anharmonic oscillator model, then compare these estimated values to those predicted by the harmonic oscillator model.

- Anharmonic oscillator model:

$$\Delta\bar{E}_\nu = \bar{E}_\nu - \bar{E}_0 = \bar{\nu}_e\upsilon - \bar{\nu}_e\rho_e\upsilon(1+\upsilon) \tag{9.5.26}$$

Fundamental transition, $\upsilon = 1$:

$$\Delta\bar{E}_\nu = \bar{E}_1 - \bar{E}_0 = \bar{\nu}_e(1) - \bar{\nu}_e\rho_e(1)(1+1)$$

$$\Delta\bar{E}_\nu = \bar{E}_1 - \bar{E}_0 = (2359.61)(1) - (14.456)(1)\,(1+1) = 2330.7\,cm^{-1}$$

The first overtone, $\upsilon = 2$:

$$\Delta\bar{E}_\nu = \bar{E}_2 - \bar{E}_0 = \bar{\nu}_e(2) - \bar{\nu}_e\rho_e(2)(2+1)$$

$$\Delta\bar{E}_\nu = \bar{E}_2 - \bar{E}_0 = (2359.61)(2) - (14.456)(2)(2+1) = 4632.5\,\text{cm}^{-1}$$

The second overtone, $\upsilon = 3$:

$$\Delta\bar{E}_\nu = \bar{E}_3 - \bar{E}_0 = \bar{\nu}_e(3) - \bar{\nu}_e\rho_e(3)\,(3+1)$$

$$\Delta\bar{E}_\nu = \bar{E}_3 - \bar{E}_0 = (2359.61)(3) - (14.456)(3)\,(3+1) = 6905.4\,\text{cm}^{-1}$$

The third overtone, $\upsilon = 4$:

$$\Delta\bar{E}_\nu = \bar{E}_4 - \bar{E}_0 = \bar{\nu}_e(4) - \bar{\nu}_e\rho_e(4)(4+1)$$

$$\Delta\bar{E}_\nu = \bar{E}_4 - \bar{E}_0 = (2359.61)(4) - (14.456)(4)\,(4+1) = 9149.3\,\text{cm}^{-1}$$

- Harmonic oscillator model:

$$E_\upsilon = \left(\upsilon + \frac{1}{2}\right)h\nu_0 \tag{9.5.20}$$

$$\bar{E}_\upsilon = \left(\upsilon + \frac{1}{2}\right)\bar{\nu}_0 \text{ in cm}^{-1}$$

Fundamental transition, $\upsilon = 1$:

$$\Delta\bar{E}_\nu = \bar{E}_1 - \bar{E}_o = \bar{\nu}_0\left(1+\frac{1}{2}\right) - \bar{\nu}_0\left(0+\frac{1}{2}\right)$$

$$\Delta\bar{E}_\nu = \bar{E}_1 - \bar{E}_o = (2330.7)\left(1+\frac{1}{2}\right) - (2330.7)\left(0+\frac{1}{2}\right) = 2330.7\,\text{cm}^{-1}$$

The first overtone, $\upsilon = 2$:

$$\Delta\bar{E}_\nu = \bar{E}_2 - \bar{E}_0 = \bar{\nu}_0\left(2+\frac{1}{2}\right) - \bar{\nu}_0\left(0+\frac{1}{2}\right)$$

$$\Delta\bar{E}_\nu = \bar{E}_2 - \bar{E}_o = (2330.7)\left(2+\frac{1}{2}\right) - (2330.7)\left(0+\frac{1}{2}\right) = 4661.4\,\text{cm}^{-1}$$

The second overtone, $\upsilon = 3$:

$$\Delta\bar{E}_\nu = \bar{E}_3 - \bar{E}_0 = \bar{\nu}_0\left(3+\frac{1}{2}\right) - \bar{\nu}_0\left(0+\frac{1}{2}\right)$$

$$\Delta\bar{E}_\nu = \bar{E}_3 - \bar{E}_0 = (2330.7)\left(1+\frac{1}{2}\right) - (2330.7)\left(0+\frac{1}{2}\right) = 6992.1\,\text{cm}^{-1}$$

The third overtone, $\upsilon = 4$:

$$\Delta\bar{E}_\nu = \bar{E}_4 - \bar{E}_0 = \bar{\nu}_0\left(4+\frac{1}{2}\right) - \bar{\nu}_0\left(0+\frac{1}{2}\right)$$

$$\Delta\bar{E}_\nu = \bar{E}_4 - \bar{E}_0 = (2330.7)\left(1+\frac{1}{2}\right) - (2330.7)\left(0+\frac{1}{2}\right) = 9322.8\,\text{cm}^{-1}$$

9.6 THE QUANTUM MECHANICS OF THE TRANSLATION, VIBRATION, AND ROTATION MOTIONS

Consider a diatomic molecule that consists of two atoms of masses m_1 and m_2 separated by a variable distance r (Fig. 9.9). What is the wave equation for each of the following?

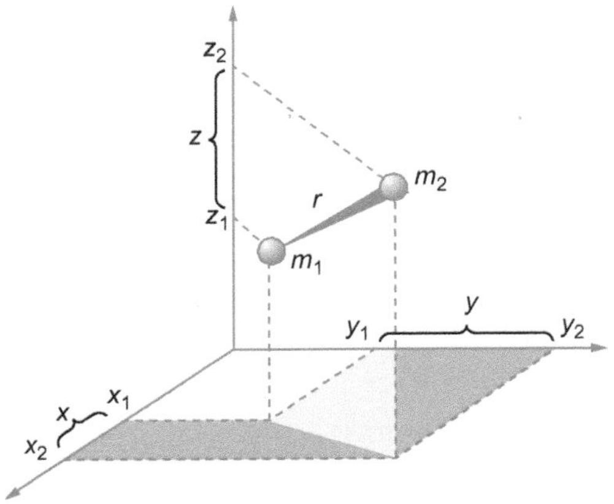

FIG. 9.9 A diatomic molecule and the Cartesian coordinates.

(a) **the total motions of the diatomic molecule,**
(b) **the translational motion,**
(c) **the rotational motion, and the rotational wave function for the first three rotational levels, and**
(d) **the vibrational motion, and the vibrational wave function for the first four vibrational wave functions.**

- In this diatomic molecule, the potential energy only depends on the internuclear separation, but not on place of the molecule in the space.
- The Schrödinger equation for this two-particle molecule will have a kinetic energy for each atom and a single potential energy:

$$\left[-\frac{\hbar^2}{2m_1}\nabla_1^2 - \frac{\hbar^2}{2m_2}\nabla_2^2 + V(r)\right]\psi_{\text{total}} = E_{\text{total}}\psi_{\text{total}} \tag{9.6.1}$$

where $-\dfrac{\hbar^2}{2m_i}\nabla_i^2$ is the kinetic energy for each atom.

$$\nabla_i^2 = \frac{\partial^2}{\partial x_i^2} + \frac{\partial^2}{\partial y_i^2} + \frac{\partial^2}{\partial z_i^2}$$

where ψ_{total} is the total wave function and E_{total} is the total energy.

- In order to separate the translation of all the molecule through space, the coordinates of the center of mass, R, are introduced:

$$R = \frac{1}{M}\sum_{i=1}^{N} m_i r_i$$

N is the total number of the particles, m_i is the mass, r_i is the position.

$M = \sum_{i=1}^{N} m_i$ is the total mass of the system, then

$$X = \frac{m_1 x_1 + m_2 x_2}{m_1 + m_2}$$

$$Y = \frac{m_1 y_1 + m_2 y_2}{m_1 + m_2}$$

$$Z = \frac{m_1 z_1 + m_2 z_2}{m_1 + m_2}$$

If

$$\psi_{\text{total}} = \psi_t \psi_{rv}$$

where ψ_t is the translational wave function and ψ_{rv} is the rotational and vibrational wave function.

- The Schrödinger equation can be rearranged to

$$\underbrace{-\frac{\hbar^2}{2M\psi_t}\left[\frac{\partial^2\psi_t}{\partial X^2}+\frac{\partial^2\psi_t}{\partial Y^2}+\frac{\partial^2\psi_t}{\partial Z^2}\right]}_{\text{depends only on the coordinates } X,\,Y,\,Z \text{ equal constant}} \underbrace{-\frac{\hbar^2}{2\mu\psi_{rv}}\left[\frac{\partial^2\psi_{rv}}{\partial X^2}+\frac{\partial^2\psi_{rv}}{\partial Y^2}+\frac{\partial^2\psi_{rv}}{\partial Z^2}\right]+V(x,y,z)}_{\text{depends only on } X,\,Y,\,Z \text{ equal constant}} = \underbrace{E_{\text{total}}}_{\text{sum of the two constants}} \tag{9.6.2}$$

$x = x_2 - x_1$, $y = y_2 - y_1$, and $z = z_2 - z_1$ (Fig. 9.9).
$M = m_1 + m_2$, μ is the reduced mass.

- The equation can be written as two independent equations. One describes the motion of the molecule in the space, where the translation energy is not quantized:

$$-\frac{\hbar^2}{2M}\left[\frac{\partial^2\psi_t}{\partial X^2}+\frac{\partial^2\psi_t}{\partial Y^2}+\frac{\partial^2\psi_t}{\partial Z^2}\right] = E_t\psi_t, \text{ translational motion equation} \tag{9.6.3}$$

The other equation describes the motion of the molecule in a coordinate system fixed at atom number 1:

$$-\frac{\hbar^2}{2\mu}\left[\frac{\partial^2\psi_{rv}}{\partial x^2}+\frac{\partial^2\psi_{rv}}{\partial y^2}+\frac{\partial^2\psi_{rv}}{\partial z^2}\right]+V(x,y,z) = E_{rv}\psi_{rv}, \text{ vibration} - \text{rotation equation} \tag{9.6.4}$$

This equation is solved by transforming into polar coordinates, as in Chapter 2, p. 41, in which

$$x = r\sin\theta\cos\phi$$

$$y = r\sin\theta\sin\phi$$

$$z = r\cos\theta$$

$$\partial\tau = \partial x\,\partial y\,\partial z = r^2\sin\theta\,\partial r\,\partial\theta\,\partial\phi$$

$$\nabla^2 = \frac{\partial^2\psi_{rv}}{\partial x^2}+\frac{\partial^2\psi_{rv}}{\partial y^2}+\frac{\partial^2\psi_{rv}}{\partial z^2} = \frac{1}{r^2}\frac{\partial}{\partial r}\left(r^2\frac{\partial}{\partial r}\right)+\frac{1}{r^2\sin\theta}\frac{\partial}{\partial\theta}\left(\sin\theta\frac{\partial}{\partial\theta}\right)+\frac{1}{r^2\sin^2\theta}\frac{\partial}{\partial\phi^2}$$

and ψ_{rv} in the form

$$\psi_{rv} = \underbrace{R(r)}_{\psi_{\text{vib.}}}\,\underbrace{\Theta(\theta)\Phi(\phi)}_{\psi_{\text{rot.}}}, \text{ then:}$$

$$-\frac{\hbar^2}{2\mu}\left[\frac{\partial^2\psi_{rv}}{\partial x^2}+\frac{\partial^2\psi_{rv}}{\partial y^2}+\frac{\partial^2\psi_{rv}}{\partial z^2}\right]+V(x,y,z) = E_{rv}\psi_{rv} \tag{9.6.4}$$

$$\frac{-\hbar^2}{2\mu r^2}\left[\Theta\Phi\frac{\partial}{\partial r}\left(r^2\frac{\partial R}{\partial r}\right)+\frac{R\Phi}{\sin\theta}\frac{\partial}{\partial\theta}\left(\sin\theta\frac{\partial\Theta}{\partial\theta}\right)+\frac{R\Theta}{\sin^2\theta}\frac{\partial^2\Phi}{\partial\phi^2}\right]+V(r)R\Theta\Phi = E_{rv}R\Theta\Phi$$

Multiplying by $\dfrac{-2\mu r^2\sin^2\theta}{R\Theta\Phi\,\hbar^2}$:

$$\frac{\sin^2\theta}{R}\frac{\partial}{\partial r}\left(r^2\frac{\partial R}{\partial r}\right)+\frac{\sin\theta}{\Theta}\frac{\partial}{\partial\theta}\left(\sin\theta\frac{\partial\Theta}{\partial\theta}\right)+\frac{1}{\Phi}\frac{\partial^2\Phi}{\partial\phi^2}+\frac{2\mu r^2\sin^2\theta}{\hbar^2}[E_{rv}-V(r)] = 0$$

$$\underbrace{\frac{\sin^2\theta}{R}\frac{\partial}{\partial r}\left(r^2\frac{\partial R}{\partial r}\right)+\frac{\sin\theta}{\Theta}\frac{\partial}{\partial\theta}\left(\sin\theta\frac{\partial\Theta}{\partial\theta}\right)+\frac{2\mu r^2\sin^2\theta}{\hbar^2}[E_{rv}-V(r)]}_{\text{depends on } \theta \text{ and } r} = \underbrace{-\frac{1}{\Phi}\frac{\partial^2\Phi}{\partial\phi^2}}_{\text{depends on } \phi} = m^2$$

where θ, ϕ, and r are independent variables.

- Then

$$\frac{1}{\Phi}\frac{\partial^2\Phi}{\partial\phi^2} = -m^2 \quad \Phi-\text{equation}$$

$$\frac{\sin^2\theta}{R}\frac{\partial}{\partial r}\left(r^2\frac{\partial R}{\partial r}\right) + \frac{\sin\theta}{\Theta}\frac{\partial}{\partial\theta}\left(\sin\theta\frac{\partial\Theta}{\partial\theta}\right) + \frac{2\mu r^2\sin^2\theta}{\hbar^2}[E_{rv} - V(r)] = m^2 \quad R\Theta\text{-equation}$$

- Φ - equation:Z

The solutions of Φ wave function are

$$\Phi_m = e^{im\phi}$$

and to normalize the Φ_m function:

$$N^2\int_0^{2\pi} \Phi_m{}^*\Phi_m\,\partial\phi = 1$$

$$N^2\int_0^{2\pi} \left(e^{-im\phi}\right)\left(e^{im\phi}\right)\partial\phi = 1$$

$$N^2\int_0^{2\pi} \partial\phi = 1$$

$$N^2(2\pi) = 1,\ \ N = \frac{1}{\sqrt{2\pi}}$$

The general normalized solution to Φ_m:

$$\Phi_m = \frac{1}{\sqrt{2\pi}}e^{im\phi} \tag{9.6.5}$$

- The wave function must satisfy the cyclic boundary condition:

$$\Phi_m(\phi) = \Phi_m(\phi + 2\pi)$$

The requirement that Φ and $\partial\Phi/\partial\phi$ continuous throughout the range of ϕ from 0 to 2π:

$$\Phi_m(\phi+2\pi) = \frac{1}{\sqrt{2\pi}}e^{im(\phi+2\pi)} = \frac{1}{\sqrt{2\pi}}e^{im\phi}e^{2\pi im} = e^{2\pi im}\,\Phi_m(\phi) = (-1)^{2m}\Phi_m(\phi) = \Phi_m(\phi)$$

where; $e^{\pi i} = -1$, and the value of $2m$ is an even number, therefore, in order to satisfy the cyclic boundary condition:

$$m = 0, \pm 1, \pm 2, \pm 3 \tag{9.6.6}$$

- $R\Theta$ - equationZ

$$\frac{\sin^2\theta}{R}\frac{\partial}{\partial r}\left(r^2\frac{\partial R}{\partial r}\right) + \frac{\sin\theta}{\Theta}\frac{\partial}{\partial\theta}\left(\sin\theta\frac{\partial\Theta}{\partial\theta}\right) + \frac{2\mu r^2\sin^2\theta}{\hbar^2}[E_{rv} - V(r)] = m^2$$

Divided by $\sin^2\theta$

$$\underbrace{\frac{1}{R}\frac{\partial}{\partial r}\left(r^2\frac{\partial R}{\partial r}\right) - \frac{2\mu r^2}{\hbar^2}[E_{rv} - V(r)]}_{\text{depends on } r} + \underbrace{\frac{-1}{\Theta\sin\theta}\frac{\partial}{\partial\theta}\left(\sin\theta\frac{\partial\Theta}{\partial\theta}\right) - \frac{m^2}{\sin^2\theta}}_{\text{depends on } \theta} = 0$$

or

$$\frac{1}{R}\frac{\partial}{\partial r}\left(r^2\frac{\partial R}{\partial r}\right)-\frac{2\mu r^2}{\hbar^2}\left[E_{rv}-V(r)\right]=\frac{-1}{\Theta\sin\theta}\frac{\partial}{\partial\theta}\left(\sin\theta\frac{\partial\Theta}{\partial\theta}\right)+\frac{m^2}{\sin^2\theta}=J(J+1)$$

Therefore,

$$\frac{1}{\Theta\sin\theta}\frac{\partial}{\partial\theta}\left(\sin\theta\frac{\partial\Theta}{\partial\theta}\right)-\frac{m^2}{\sin^2\theta}=-J(J+1)$$

$$\frac{1}{\sin\theta}\frac{\partial}{\partial\theta}\left(\sin\theta\frac{\partial\Theta}{\partial\theta}\right)-\frac{m^2}{\sin^2\theta}\Theta+J(J+1)\Theta=0\ \ \Theta\text{-equation} \tag{9.6.7}$$

$$\frac{1}{R}\frac{\partial}{\partial r}\left(r^2\frac{\partial R}{\partial r}\right)+\frac{2\mu r^2}{\hbar^2}\left[E_{rv}-V(r)\right]=J(J+1)$$

$$\frac{\partial}{\partial r}\left(r^2\frac{\partial R}{\partial r}\right)+\frac{2\mu r^2}{\hbar^2}\left[E_{rv}-V(r)\right]R-J(J+1)R=0\ \ R\text{-equation} \tag{9.6.8}$$

- The differential equation Θ is satisfied by functions of cos θ known as Legendre polynomials (Chapter 2, p. 56).

$$\Theta(\theta)=\left[\frac{(2J+1)}{2}\frac{(J-|m|)!}{(J+|m|)!}\right]^{1/2}P_J^{|m|}(\cos\theta) \tag{9.6.9}$$

$$P_J^{|m|}(\cos\theta)=\frac{(1-\cos^2\theta)^{|m|/2}}{2^J J!}\frac{\partial^{J+|m|}}{\partial\cos^{J+|m|}\theta}\left(\cos^2\theta-1\right)^J\ :-1\le\cos\theta\le 1 \tag{9.6.10}$$

J must have an integer value (0, 1, 2, 3,...) and for each value of J, $m=0, \pm1, \pm2, \pm3, \ldots, \pm J$ (9.6.6)

- The products $\Theta(\theta)\Phi(\phi)$ are the rotational wave function for diatomic molecule, which are known as spherical harmonics (Table 9.2).

$$\psi_{\text{rot.}}=\Theta(\theta)\Phi(\phi)=(-1)^m\left[\frac{(2J+1)}{4\pi}\frac{(J-|m|)!}{(J+|m|)!}\right]^{1/2}P_J^{|m|}(\cos\theta)e^{im\phi} \tag{9.6.11}$$

- Neither Θ nor Φ depends on the form of $V(r)$, therefore, the rotational wave function will be the same for the harmonic or anharmonic vibrational model of the molecule.
- In the harmonic oscillator model,

$$V(r)=\frac{1}{2}k(r-r_e)^2 \tag{9.5.10}$$

in R-equation

$$\frac{\partial}{\partial r}\left(r^2\frac{\partial R}{\partial r}\right)+\frac{2\mu r^2}{\hbar^2}\left[E_{rv}-V(r)\right]R-J(J+1)R=0\ \ R\text{-equation} \tag{9.6.8}$$

$$\frac{\partial}{\partial r}\left(r^2\frac{\partial R}{\partial r}\right)-J(J+1)R+\frac{2\mu r^2}{\hbar^2}\left[E_{rv}-\frac{1}{2}k(r-r_e)^2\right]R=0$$

This differential equation-R is satisfied by functions of q $(=r-r_e)$ called Hermite polynomials:

$$\frac{\partial}{\partial r}\left(r^2\frac{\partial\psi}{\partial r}\right)-J(J+1)\psi+\frac{2\mu r^2}{\hbar^2}\left[E_{rv}-\frac{1}{2}kq^2\right]\psi=0$$

TABLE 9.2 The Rotational Spherical Harmonics

J	*m*	Imaginary Wave Function	Real Wave Function
0	0	$\sqrt{\frac{1}{4\pi}}$	
1	0	$\sqrt{\frac{3}{4\pi}}\cos\theta$	
1	±1	$\mp\sqrt{\frac{3}{8\pi}}\sin\theta e^{\pm i\phi}$	$\sqrt{\frac{3}{4\pi}}\sin\theta\cos\phi$ $\sqrt{\frac{3}{4\pi}}\sin\theta\sin\phi$
2	0	$\sqrt{\frac{5}{16\pi}}(3\cos^2-1)$	
2	±1	$\mp\sqrt{\frac{15}{8\pi}}\sin\theta\cos\theta e^{\pm i\phi}$	$\sqrt{\frac{15}{4\pi}}\sin\theta\cos\theta\cos\phi$ $\sqrt{\frac{15}{4\pi}}\sin\theta\cos\theta\sin\phi$
2	±2	$\sqrt{\frac{15}{32\pi}}\sin^2\theta e^{\pm 2i\phi}$	$\sqrt{\frac{15}{16\pi}}\sin^2\theta\cos 2\phi$ $\sqrt{\frac{15}{16\pi}}\sin^2\theta\sin 2\phi$
3	0	$\sqrt{\frac{63}{16\pi}}\left(\frac{3}{5}\cos^3-\theta\cos\theta\right)$	
3	±1	$\mp\sqrt{\frac{21}{64\pi}}\sin\theta(5\cos^2-1)e^{\pm i\phi}$	$\sqrt{\frac{21}{32\pi}}\sin\theta(5\cos^2-1)\cos\phi$ $\sqrt{\frac{21}{32\pi}}\sin\theta(5\cos^2-1)\sin\phi$
3	±2	$\sqrt{\frac{105}{32\pi}}\sin^2\theta\cos\theta e^{\pm 2i\phi}$	$\sqrt{\frac{105}{16\pi}}\sin^2\theta\cos\theta\cos 2\phi$ $\sqrt{\frac{105}{16\pi}}\sin^2\theta\cos\theta\sin 2\phi$
3	±3	$\mp\sqrt{\frac{35}{64\pi}}\sin^3\theta e^{\pm 3i\phi}$	$\sqrt{\frac{35}{32\pi}}\sin^3\theta\cos 3\phi$ $\sqrt{\frac{35}{32\pi}}\sin^3\theta\sin 3\phi$

The functions have the form

$$\psi_v = N_v H_v e^{-\frac{\alpha q^2}{2}} \tag{9.6.12}$$

N_v is a normalization constant, H_v is the vth Hermite polynomial, $\alpha = \frac{\sqrt{\mu k}}{\hbar}$, v is the vibrational quantum number.

$$\psi_o = \left(\frac{\alpha}{\pi}\right)^{1/4} e^{-\frac{\alpha q^2}{2}}$$

$$\psi_1 = \left(\frac{4\alpha^3}{\pi}\right)^{1/4} q e^{-\frac{\alpha q^2}{2}}$$

$$\psi_2 = \left(\frac{\alpha}{4\pi}\right)^{1/4}(2\alpha q^2 - 1)e^{-\frac{\alpha q^2}{2}}$$

$$\psi_3 = \left(\frac{\alpha^3}{9\pi}\right)^{1/4}\left(2\alpha q^3 - 3q\right) e^{-\frac{\alpha q^2}{2}}$$

If the rotational wavefunction, $\psi_{J,m}(\theta,\phi)$:

J	m	Wavefunction
0	0	$\sqrt{\frac{1}{4\pi}}$
1	0	$\sqrt{\frac{3}{4\pi}}\cos\theta$
2	0	$\sqrt{\frac{5}{16\pi}}(3\cos^2-1)$

then show that the $\psi_{00}(\theta,\phi)\xrightarrow{J=0\to J=1}\psi_{10}(\theta,\phi)$ transition is allowed, and that $\psi_{00}(\theta,\phi)\xrightarrow{J=0\to J=2}\psi_{20}(\theta,\phi)$ transition is forbidden.

- If the electric field is along the z-axis, the dipole moment

$$\mu_z = \mu\cos\theta$$

and the transition dipole moment that associates $\psi_{00}\xrightarrow{0\to J}\psi_{J0}$ transition:

$$\langle\mu_z\rangle = \langle\psi_{00}(\theta,\phi)|\mu|\psi_{J0}(\theta,\phi)\rangle$$

$$\langle\mu_z\rangle = \mu\int_0^{\pi}\psi_{00}(\theta,\phi)\psi_{J0}(\theta,\phi)\cos\theta\ \sin\theta\ \partial\theta\int_0^{2\pi}\partial\phi$$

- The transition dipole moment that associates $\psi_{00}(\theta,\phi)\xrightarrow{J=0\to J=1}\psi_{10}(\theta,\phi)$ transition is

$$\langle\mu_z\rangle = \mu\sqrt{\frac{1}{4\pi}}\sqrt{\frac{3}{4\pi}}\int_0^{\pi}(\cos\theta)^2\sin\theta\ \partial\theta\int_0^{2\pi}\partial\phi = \frac{\mu\sqrt{3}}{4\pi}\left(\frac{2}{3}\right)(2\pi) = \frac{\mu}{\sqrt{3}} \neq 0$$

Therefore, $\psi_{00}(\theta,\phi)\xrightarrow{J=0\to J=1}\psi_{10}(\theta,\phi)$ transition is allowed.

- The transition dipole moment that associates $\psi_{00}(\theta,\phi)\xrightarrow{J=0\to J=2}\psi_{20}(\theta,\phi)$ transition is

$$\langle\mu_z\rangle = \mu\sqrt{\frac{1}{4\pi}}\sqrt{\frac{5}{16\pi}}\int_0^{\pi}\left[3(\cos\theta)^2 - 1\right]\cos\theta\ \sin\theta\ \partial\theta\int_0^{2\pi}\partial\phi$$

$$\langle\mu_z\rangle = \mu\sqrt{\frac{1}{4\pi}}\sqrt{\frac{5}{16\pi}}\left[\int_0^{\pi}3(\cos\theta)^3\sin\theta\ \partial\theta - \int_0^{\pi}\cos\theta\ \sin\theta\ \partial\theta\right]\int_0^{2\pi}\partial\phi$$

$$\langle\mu_z\rangle = \mu\frac{\sqrt{5}}{8}[0-0]\,[2\pi] = 0$$

Therefore, $\psi_{00}(\theta,\phi)\xrightarrow{J=0\to J=2}\psi_{20}(\theta,\phi)$ transition is forbidden.

9.7 VIBRATION-ROTATION ENERGIES OF DIATOMIC MOLECULES (VIBRATIONAL-ROTATIONAL STATE)

Find an expression for the vibration-rotation energies of diatomic molecule for the harmonic and anharmonic oscillator models

- The energy levels of the harmonic oscillator:

$$E_{\text{vib},J} = \underbrace{\left(v+\frac{1}{2}\right)h\nu_0}_{\substack{\text{vibrational energy of harmonic oscillator} \\ \text{Eq. (9.5.20)}}} \quad \underbrace{+BJ(J+1)-DJ^2(J+1)^2}_{\substack{\text{nonrigid rotational energy} \\ \text{Eq. (9.4.8)}}} \tag{9.7.1}$$

$$\nu_0 = \frac{1}{2\pi}\sqrt{\frac{k}{\mu}} \tag{9.5.8}$$

$B = \frac{\hbar^2}{2I} = \frac{\hbar^2}{2\mu r_e^2}$, (Chapter 9, p. 513)

The centrifugal distortion constant:

$$D = \frac{\hbar^4}{8\pi^2 v_o^2 I^3} \tag{9.7.2}$$

- In the anharmonic oscillator model, using Morse potential:

$$E_{\text{vib},J} = \underbrace{h\bar{v}_e\left(v+\frac{1}{2}\right) - h\bar{v}_e\rho_e\left(v+\frac{1}{2}\right)^2}_{\substack{\text{vibrational energy of anharmonic oscillator} \\ \text{Eq. (9.5.25)}}} \underbrace{+B_eJ(J+1)-D_eJ^2(J+1)^2}_{\substack{\text{nonrigid rotational energy} \\ \text{Eq. (9.4.10)}}} - \underbrace{\alpha_e\left(v+\frac{1}{2}\right)J(J+1)}_{\text{interaction between vobration and rotation}} \tag{9.7.3}$$

$$\bar{v}_e = \frac{\beta}{\pi}\sqrt{\frac{D_e}{2\mu}} \qquad \rho_e = \frac{h\bar{v}_e}{4D_e}$$

$$B_e = \frac{\hbar^2}{2I} = \frac{\hbar^2}{2\mu r_e^2} \qquad D = \frac{\hbar^4}{8\pi^2 \bar{v}_e{}^2 \mu^3 r_e^6} \tag{9.7.4}$$

$$\alpha_e = \frac{3\pi\hbar^3\bar{v}_e}{2\mu r_e^2 D_e}\left(\frac{1}{\beta r_e} - \frac{1}{\beta^2 r_e^2}\right)$$

- The selection rules for diatomic molecules, in transition from the state (v,J) to the state (v',J') (Fig. 9.10):

$$\Delta v = \pm 1 \tag{9.7.5}$$

$$\Delta J = \pm 1 \tag{9.7.6}$$

Calculate the centrifugal distortion constant, D, for $H^{35}Cl$ ($\bar{v}_e$=2988.9cm^{-1}, r_e = 1.2746 Å) and compare your answer to the experimental value 0.0004 cm^{-1}.

$$D = \frac{\hbar^4}{8\pi^2 \bar{v}_e{}^2 \mu^3 r_e^6} \tag{9.7.4}$$

$$\mu = \frac{1.008 \times 35.45}{1.008 + 35.45} = 0.9801\,\text{a.u.} = 1.627 \times 10^{-27}\,\text{kg}$$

$$\bar{v}_e = 2988.9\ \text{cm}^{-1} = \frac{2988.9}{\text{cm}} \times \frac{100\ \text{cm}}{1\,\text{m}} \times 2.998 \times 10^8 \times \frac{\text{m}}{\text{s}} = 8.960 \times 10^{13}\ \text{s}^{-1}$$

$$D = \frac{(1.05457 \times 10^{-34})^4\,\text{J}^4\text{s}^4}{8 \times (3.143)^2\,(8.960 \times 10^{13})^2\text{s}^{-2}(1.627 \times 10^{-27})^3\text{kg}^3(1.2746 \times 10^{-10})^6\text{m}^6}$$

$$D = \frac{(1.05457)^4 \times 10^{-21}}{8 \times (3.143)^2(8.960)^2(1.627)^3(1.2746)^6}\ \frac{\text{J}^4\text{s}^6}{\text{kg}^3\text{m}^6}$$

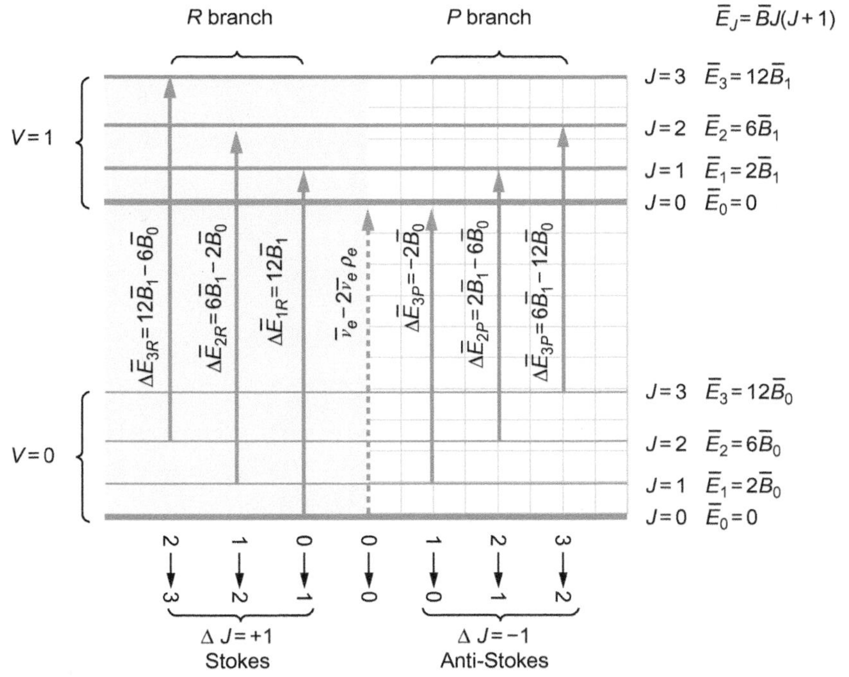

FIG. 9.10 Transitions of a rotating anharmonic oscillator.

$$D = 1.0556 \times 10^{-26}\frac{\text{J}^4}{\text{J}^3} = 0.00053\ \text{cm}^{-1}$$

where (1 J = 1 kg m^2/s^2, 1 cm^{-1} = 1.9864 × 10^{-23} J).

If the masses of ^{1}H and ^{35}Cl are 1.008 and 34.969 amu, respectively, and $\bar{B}$ = 10.593cm^{-1}, find the bond length r_e.

- The rotational constant is given by

$$B_e(\text{J}) = \frac{\hbar^2}{2I} = \frac{\hbar^2}{2\mu r_e^2} \tag{9.7.4}$$

$$hc\bar{B}\left(\text{cm}^{-1}\right) = \frac{h^2}{8\pi^2\mu r_e^2}$$

where $\bar{B}hc = B$, B is the rotational constant in joules, and $\bar{B}$ is expressed in wave number, cm^{-1}. Therefore, the bond length, r_e:

$$r_e = \sqrt{\frac{h}{8\pi^2\mu c}}$$

$$\because \mu = \frac{m_A m_B}{m_A + m_B} = \frac{m_H m_{Cl}}{m_H + m_{Cl}} = \frac{(1.008)(34.969)\,\text{amu}}{35.977}\left(\frac{1.661 \times 10^{-27}\,\text{kg}}{1\,\text{amu}}\right)$$

$$r_e = \sqrt{\frac{\left(6.626 \times 10^{-34}\,\text{Js}\right)(1\,\text{m})}{8\pi^2\left(2.998 \times 10^8\,\text{m/s}\right)\left(1.627 \times 10^{-27}\,\text{kg}\right)\left(10.593\,\text{cm}^{-1}\right)(100\,\text{cm})}} = 1.275 \times 10^{-10}\,\text{m}$$

9.8 VIBRATIONS OF POLYATOMIC MOLECULES

What are the differences between the vibration of diatomic and polyatomic molecules?

- In diatomic molecules, the nuclei only vibrate along the line connecting two nuclei.
- In polyatomic molecules:
  - all nuclei perform their own oscillations;
  - more complicated than diatomic molecules; and
  - may be represented as a superposition of a number of normal vibrations.

9.9 POLYATOMIC MOLECULAR MOTIONS AND DEGREES OF FREEDOM

What are the degrees of freedom for a polyatomic molecule and for each molecular motion?

- The number of degrees of freedom possessed by a molecule is the number of coordinates required to completely specify the position of the nuclei.
- For a single particle in three-dimensional space:
  - the three coordinates can describe the motion of the particle;
  - the particle has three degrees of freedom; and
  - each degree of freedom represents a translation of the particle motion through space (Fig. 9.11).
- Since each particle has three degree of freedom, diatomic molecules as a whole have six degrees of freedom:
  - Three of these are translation of the entire molecule in *x*-, *y*-, and *z*-directions; T_x, T_y, and T_z.
  - Only two degrees of freedom accounted for by rotation about the center of mass; R_x and R_y. Note that rotation about the molecular axis of a linear molecule is undefined because it does not represent any change of the nuclear coordinates. For nonlinear molecules, there will be a third axis of rotation, R_z.
  - The final one is vibrational degree of freedom (Fig. 9.12).

  The possible bulk and internal motions of molecules are translation, rotation, and vibration.
- The vector of motion on each atom in a molecule can be resolved into three components along the *x*-, *y*-, and *z*-axes.
- If there are *n* atoms in a molecule, there are 3*n* possible movements of its atoms.

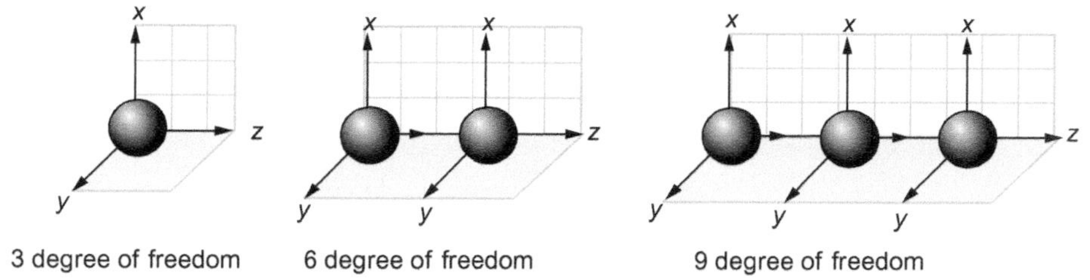

FIG. 9.11 The n-atom molecule have 3*n* degree of freedom.

FIG. 9.12 The six degrees of freedom of a diatomic molecule.

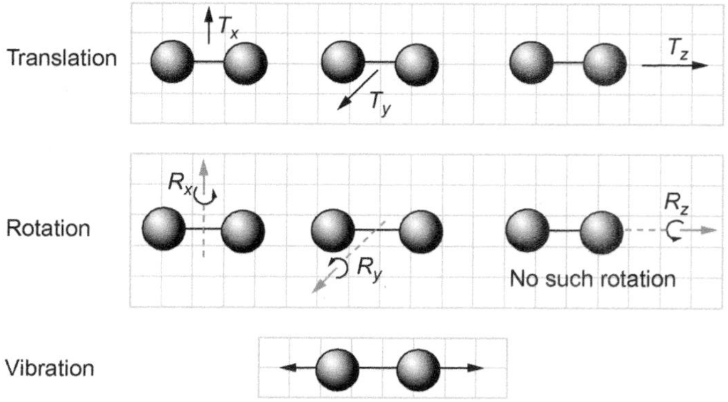

- Of the $3n$ movements, 3 will be concerted translation movements of the whole molecule along the three coordinate axes, and 3 (or 2 for linear molecule) will be concerted rotations about axes, the remaining $3n-6$ (or $3n-5$ for linear molecule) must be molecular vibrations.
- Translational and vibrational modes proceed without any change in interatomic dimensions.
- Consequently, there are $3n-6$ (or $3n-5$ for nonlinear molecule) normal modes of vibration that result in a change in bond lengths or angles in molecules.

9.10 NORMAL MODES OF VIBRATION, NORMAL COORDINATES, AND POLYATOMIC MOLECULES

What is the meaning of the normal mode of vibrations, and how can the normal coordinates be defined?

- Molecules are continually executing rotational and vibrational motions at all temperature, and it is possibly to obtain the molecular rotational and vibrational spectra. Consequently, the polyatomic molecules are present in different vibrational and rotational states of quantized energy.
- A diatomic molecule has just a single vibration, but polyatomic molecules undergo much more complex vibrations.
- These vibrational motions may be resolved into a superposition of a limited number of fundamental motions called normal modes of vibrations.
- Normal modes represent independent self-repeating motions in the molecule, and could take up energy independently of each other.
- In the normal modes of vibration (Fig. 9.13):
  - all atoms of the molecule oscillate at the same frequency;
  - all atoms reach the extremes of their amplitudes simultaneously;
  - all atoms reach their maximum velocity at the same instant;
  - all atoms of the molecule pass through their means positions at the same time, moving in phase with one another;
  - each mode can be assigned to an irreducible representation, or a symmetry species;
  - for nondegenerate modes, the normal coordinates and the vibrational wave functions are either symmetric or antisymmetric with respect to the symmetry operations of the point group of the molecule in its mean position; and
  - for degenerate modes, the symmetry operations will transform the degenerate set of vibrations into a linear combination of mutually degenerate normal coordinates or wave functions.
- In vibrational analysis, any arbitrary vibration of the molecule can be described by a superposition of the normal modes.
- In normal modes, the displacement of atoms from their equilibrium positions is defined in terms of the normal coordinates, Q_i, which are functions of distance and angles.
- Normal modes are determined by solving a secular determinant, and then the normal coordinates of the normal modes can be expressed as summation over the Cartesian of the atom positions.

What are the differences among the Cartesian *x*, internal, and normal coordinates used to characterize the stretching vibrations of a linear triatomic molecule?

- Fig. 9.14 illustrates three types of coordinates used to characterize the stretching vibrations of the linear triatomic molecules.

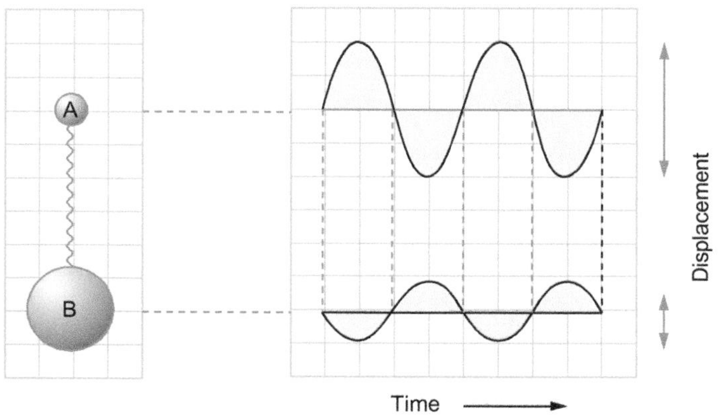

FIG. 9.13 Normal mode of vibration of a diatomic molecule A–B.

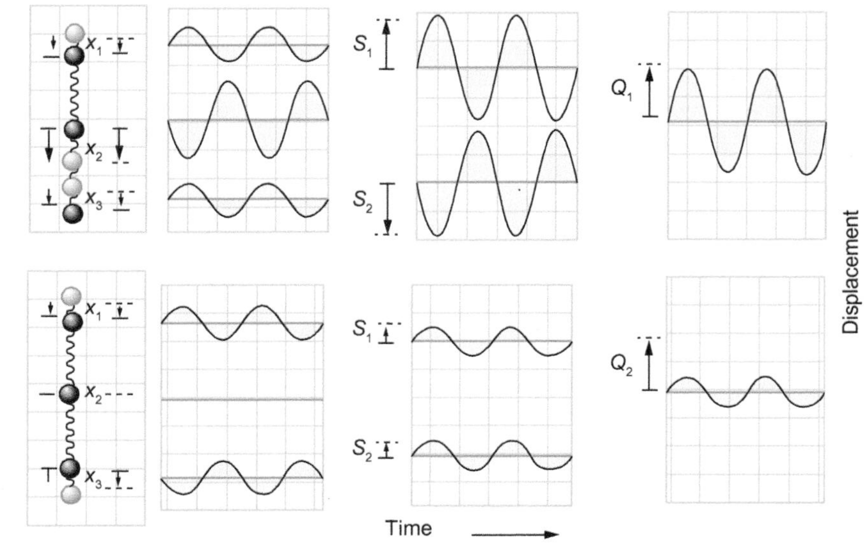

FIG. 9.14 The Cartesian x, the internal S, and normal Q coordinates.

- Cartesian displacement coordinates x measure displacement of each nuclear mass from equilibrium position by means of Cartesian coordinates. Every single atom has its Cartesian coordinate system, which has its origin at the equilibrium position of the atom. In Fig. 9.14, the Cartesian coordinates x are the mass displacements, x_1, x_2, and x_3.
- The internal coordinates, S, measure the variation in the shape of the molecule when compare to its equilibrium shape without pay attention for the molecules' position or orientation in space. Such variation can be change in a bond length or a bond angle from it equilibrium value. In Fig. 9.14, the internal coordinates S are bond length, $S_1 = x_1 - x_2$, and $S_2 = x_2 - x_3$.
- Every normal mode of vibration can be specified by a single normal coordinate, Q, which changes periodically. When one normal coordinate vibrates, each Cartesian displacement coordinate and each internal coordinate vibrates, each with an amplitude in specified proportion (+ or -) to the amplitude of the normal coordinate, the proportions being such that the resulting motion is a normal mode of vibration. In Fig. 9.14, $S_1/S_2 = -1$, $S_1 = 1Q_1$, $S_2 = -1Q_1$ for the, and $S_1/S_2 = 1$, $S_1 = 1Q_1$, $S_2 = 1Q_1$ for the symmetric mode of vibration.

Calculate the relative amplitudes of motion for the different types of atoms in the vibrational modes presented below:

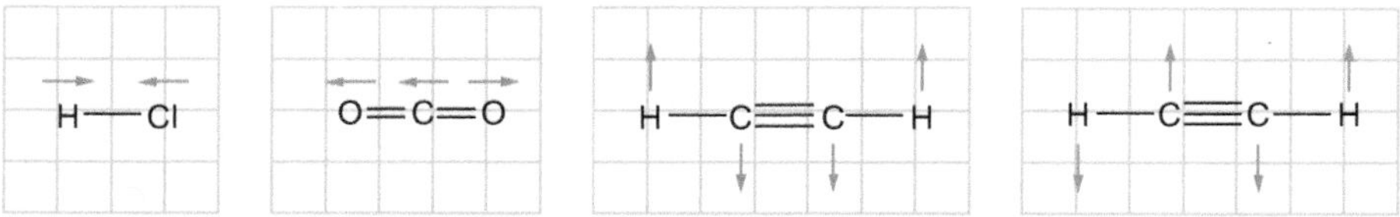

Considering such molecule vibration has neither linear nor angular momentum associated with it, $m_H = 1$ a.u., $m_C = 12$ a.u., $m_O = 16$ a.u., C≡C and C—H bond length are 0.120 and 0.106 nm, respectively.

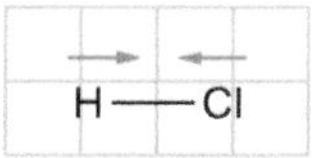

- Let the velocities of vibrating H and Cl atoms as they pass through the equilibrium position are represented by $\dot{x}_{\mathrm{H}}$ and $\dot{x}_{\mathrm{Cl}}$.
- Since the linear momentum is zero,

$$m_{\mathrm{H}}\dot{x}_{\mathrm{H}} + m_{\mathrm{Cl}}\dot{x}_{\mathrm{Cl}} = 0$$

$$\frac{\dot{x}_{\mathrm{H}}}{\dot{x}_{\mathrm{Cl}}} = -\frac{m_{\mathrm{Cl}}}{m_{\mathrm{H}}} = -\frac{35}{1}$$

- The velocities with which atom pass through their equilibrium positions are directly proportional to the amplitudes of the motion, then

$$\frac{\dot{x}_{\mathrm{H}}}{\dot{x}_{\mathrm{Cl}}}=\frac{x_{\mathrm{H}}}{x_{\mathrm{Cl}}}=-\frac{35}{1}$$

where x_{H} and x_{cl} are the extreme amplitudes of motions during the vibration.
Therefore, the hydrogen atom moves with amplitude 35 times that of the Cl atom and in opposite direction.

- Because the linear momentum is zero in

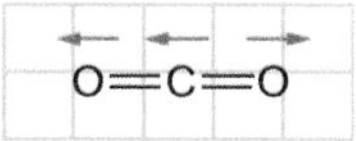

$$2m_{\mathrm{O}}\dot{x}_{\mathrm{O}}+m_{\mathrm{C}}\dot{x}_{\mathrm{C}}=0$$

$$\frac{\dot{x}_{\mathrm{O}}}{\dot{x}_{\mathrm{C}}}=\frac{x_{\mathrm{O}}}{x_{\mathrm{C}}}=-\frac{m_{\mathrm{C}}}{2m_{\mathrm{O}}}=-\frac{12}{32}=-\frac{3}{8}$$

- In the acetylene molecule:

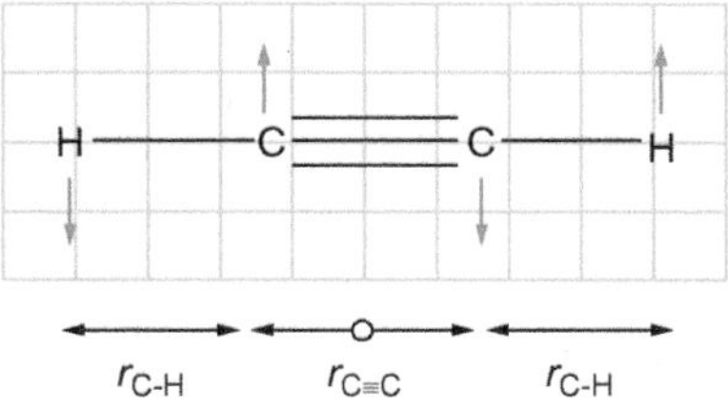

$$2m_H\dot{y}_H+2m_C\dot{y}_C=0$$

$\dot{y}_{\mathrm{H}}$ and $\dot{y}_{\mathrm{C}}$ are the velocities of H and O atoms perpendicular to the molecular axis:

$$\frac{\dot{y}_{\mathrm{H}}}{\dot{y}_{\mathrm{C}}}=\frac{y_{\mathrm{H}}}{y}=-\frac{m_{\mathrm{C}}}{m_{\mathrm{H}}}=-\frac{12}{1}$$

- If the angular momentum associated with the motion of an atom A at a distance r_{A} from the center of gravity is $m_{\mathrm{A}}r_{\mathrm{A}}\dot{y}_{\mathrm{A}}$

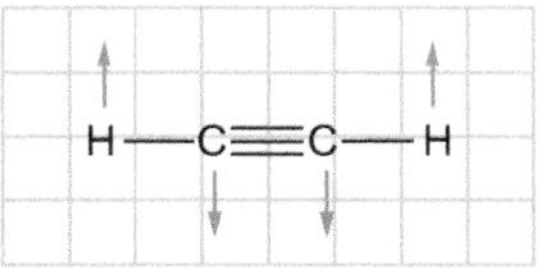

Then

$$2m_{\mathrm{C}}\left(\frac{1}{2}r_{\mathrm{C\equiv C}}\right)\dot{y}_C+2m_H\left(\frac{1}{2}r_{\mathrm{C\equiv C}}+r_{\mathrm{C-H}}\right)\dot{y}_{\mathrm{H}}=0$$

$$\frac{\dot{y}_{\mathrm{H}}}{\dot{y}_{\mathrm{C}}}=\frac{y_{\mathrm{H}}}{y_{\mathrm{C}}}=\frac{m_{\mathrm{C}}r_{\mathrm{C\equiv C}}}{mH(r_{\mathrm{C\equiv C}}+2r_{\mathrm{C-H}})}$$

$$\frac{\dot{y}_{\mathrm{H}}}{\dot{y}_{\mathrm{C}}}=\frac{y_{\mathrm{H}}}{y_{\mathrm{C}}}=\frac{12\times 0.120}{1\times(0.120+0.212)}=-4.3$$

9.11 VIBRATIONAL ENERGY OF POLYATOMIC MOLECULES

How can the kinetic and potential energies be estimated for the vibrational motions of an N-atom molecule?

- The molecule may be regarded as a superposition of a number of normal vibrations.
- Consider the displacement of each nucleus of the molecule be specified in terms of the Δx, Δy, and Δz with the origin of each system at the equilibrium position of each nucleus. Then the kinetic energy of am N-atom molecule is as follows:

$$T = \frac{1}{2}\sum_{N} m_N \left[\left(\frac{\partial \Delta x_N}{\partial t}\right)^2 + \left(\frac{\partial \Delta y_N}{\partial t}\right)^2 + \left(\frac{\partial \Delta z}{\partial \Delta t}\right)^2\right] \tag{9.11.1}$$

If generalized coordinates like

$$q_1 = \sqrt{m_1}\Delta x_1,\quad q_2 = \sqrt{m_1}\Delta y_1,\quad q_3 = \sqrt{m_1}\Delta z_1,\quad q_4 = \sqrt{m_2}\Delta x_2,\quad q_5 = \sqrt{m_2}\Delta y_2 \tag{9.11.2}$$

are used in terms of Cartesian coordinates, the kinetic energy is

$$T = \frac{1}{2}\sum_{i}^{3N} \dot{q}_i^2 \tag{9.11.3}$$

The generalized coordinates represent degrees of freedom of the system.

- The potential energy of a polyatomic molecule is a complex of all the nuclei displacement involved, and may be expanded in a Taylor series:

$$V(q_1, q_2, q_3, \ldots, q_{3N}) = V_0 + \sum_{i}^{3N}\left(\frac{\partial V}{\partial q_i}\right)_0 q_i + \frac{1}{2}\sum_{i,j}^{3N}\left(\frac{\partial^2 V}{\partial q_i \partial q_j}\right)_0 q_i q_j + \cdots \tag{9.11.4}$$

  - The subscript zero indicates the equilibrium state of the molecule.
  - The derivations are estimated at the equilibrium position, $q_i = 0$.
  - If the potential energy at $q_i = 0$ is regarded as a standard, the constant term V_0 is the potential in the equilibrium state, and can be taken as zero.
  - As well $\left(\frac{\partial V}{\partial q_i}\right)_0$ term corresponds to the minimum of the potential energy curve and also become zero, because V must be minimum at $q_i = 0$. The potential energy may be given by:

$$V = \frac{1}{2}\sum_{i,j}^{3N}\left(\frac{\partial^2 V}{\partial q_i \partial q_j}\right)_0 q_i q_j \tag{9.11.5}$$

$$V = \frac{1}{2}\sum_{i,j}^{3N} b_{ij} q_i q_j$$

Higher-order terms are neglected.

9.12 VIBRATIONAL DISPLACEMENTS

If:

$$T = \frac{1}{2}\sum_{i}^{3N}\dot{q}_i^2,\quad V = \frac{1}{2}\sum_{i,j}^{3N} b_{ij} q_i q_j,\quad \text{and}\quad b_{ij} = 0, \text{if}: i \neq j \tag{9.11.35}$$

Use Lagrange's equation to show that the change in the amplitude of displacement with time:

$$q_i = q_i^0 \sin\left(\sqrt{b_{ii}}t + \delta_i\right)$$

where q_i^0 is the amplitude and δ_i is the phase constant.

- If the potential energy does not include cross products such as q_iq_j, Lagrange's equation can be applied:

$$\because L = T - V \tag{9.12.1}$$

and

$$\because \frac{\partial}{\partial t}\left(\frac{\partial L}{\partial \dot{q}_i}\right) - \frac{\partial L}{\partial q_i} = 0 \tag{9.12.2}$$

V is independent of $\dot{q_i}$, T is independent of q_i, Eqs. (9.11.3), (9.11.5), and Lagrange equation becomes:

$$\frac{\partial}{\partial t}\left(\frac{\partial T}{\partial \dot{q}_i}\right) - \frac{\partial (T - V)}{\partial q_i} = 0$$

$$\frac{\partial}{\partial t}\left(\frac{\partial T}{\partial \dot{q}_i}\right) + \frac{\partial V}{\partial q_i} = 0, \quad i = 1, 2, \ldots, 3N \tag{9.12.3}$$

$$T = \frac{1}{2}\sum_i^{3N} \dot{q}_i^2, \rightarrow \frac{\partial T}{\partial q_i} = \ddot{q}$$

$$V = \frac{1}{2}\sum_{i,j}^{3N} b_{ij}q_iq_j, \rightarrow \frac{\partial V}{\partial q_i} = \sum_j b_{ij}q_j, \quad \text{then}$$

$$\frac{\partial}{\partial t}\left(\frac{\partial T}{\partial q_i}\right) + \frac{\partial V}{\partial q_i} = \ddot{q} + \sum_j b_{ij}q_j = 0, \quad j = 1, 2, \ldots, 3N \tag{9.12.4}$$

- if $b_{ij} = 0$, for $i \neq j$,

$$\ddot{q}_i + b_{ii}q_i = 0 \tag{9.12.5}$$

The solution is then given by

$$\begin{aligned} q_i &= q_i^0 \sin\left(\sqrt{b_{ii}}t + \partial_i\right) \\ \frac{\partial q_i}{\partial t} &= q_i^0 \cos\left(\sqrt{b_{ii}}t + \partial_i\right)\sqrt{b_{ii}} \\ \frac{\partial^2 q_i}{\partial t^2} &= -q_i^0 \sin\left(\sqrt{b_{ii}}t + \partial_i\right)\left(\sqrt{b_{ii}}\right)^2 = -b_{ii}q_i \end{aligned} \tag{9.12.6}$$

where q_i^0 is the amplitude and ∂_i is the phase constant.

9.13 VIBRATIONAL ENERGY AND NORMAL COORDINATES

Why are the normal coordinates, Q_i, used to calculate the vibrational energy of the polyatomic molecules, and what are the vibrational energies in terms of the normal coordinates?

- These kinetic and potential energies can be used to construct the Hamiltonian to be used in the Schrödinger equation. However, the cross-products of the type q_iq_j prevent the separation of the variables.
- This problem can be solved by using the normal coordinates.
- The advantage of working in normal modes is that:
  - each normal mode is an independent molecular vibration, associated with its own spectrum of quantum mechanical states; and
  - when the matrix govern the molecular vibrations is diagonalized, the cross products of the type q_iq_j are vanished.
- The coordinates q_i are set with the origin of each system at the equilibrium position of each molecule. This simplification is not applicable, and the coordinates q_i must be transformed into normal coordinates.

- Accordingly:

$$q_1 = \sum_i B_{1i}Q_i \tag{9.13.1}$$

$$q_2 = \sum_i B_{2i}Q_i \tag{9.13.2}$$

$$\vdots$$

$$q_k = \sum_i B_{ki}Q_i \tag{9.13.3}$$

- Using the proper coefficients B_{ij}, the kinetic and the potential energies can be written as:

$$T = \frac{1}{2}\sum_i^{3N} \dot{q}_i^2 \tag{9.11.3}$$

changes to:

$$T = \frac{1}{2}\sum_i \dot{Q}_i^2 \tag{9.13.4}$$

$$V = \frac{1}{2}\sum_{i,j}^{3N} b_{ii}q_iq_i$$

changes to:

$$V = \frac{1}{2}\sum_i \lambda_i Q_i^2 \tag{9.13.5}$$

λ_i is constant

How can the normal vibration equation be estimated? Find the general solution of the superposition of all normal vibrations.

- Lagrange's equation:

$$\ddot{q} + b_{ii}q_i = 0 \tag{9.12.5}$$

becomes

$$\ddot{Q}_i + \lambda_i Q_i = 0 \tag{9.13.6}$$

The solution for this equation:

$$Q_i = Q_i^0 \sin\left(\sqrt{\lambda_i}t + \partial_i\right) \tag{9.13.7}$$

and the frequency:

$$\nu_i = \frac{1}{2\pi}\sqrt{\lambda_i} \tag{9.13.8}$$

ν_i is known as normal vibration.

- In special case of one normal vibration:

$$Q_1^0 \neq 0, Q_2^0 = Q_3^0 = Q_4^0 = \cdots = 0$$

$$q_k = \sum_i B_{ki}Q_i \tag{9.13.9}$$

$$q_k = B_{k1}Q_1 = B_{k1}Q_1^0 \sin\left(\sqrt{\lambda_1}t + \partial_1\right)$$

$$q_k = A_{k1} \sin\left(\sqrt{\lambda_1}t + \partial_1\right) \quad (9.13.10)$$

As a result, the excitation of one normal vibration of the system causes:

- vibrations of all the nuclei in the system; and
- all the nuclei move with the same frequency and in phase.

- This is applicable for any other normal vibration, therefore, in general:

$$q_k = A_k \sin\left(\sqrt{\lambda_1}t + \partial\right) \quad (9.13.11)$$

If:

$$\ddot{q}_k + \sum_j b_{kj}q_j = 0 \quad (9.12.4)$$

$$\because \ddot{q}_k + \lambda_i q_k = 0$$

Then

$$-\lambda A_k + \sum_j b_{kj}A_j = 0 \quad (9.13.12)$$

- This can form a set of first-order simultaneous equations with respect to A. The order of this secular equation is 3 N.

$$-\lambda A_1 + b_{11}A_1 + b_{12}A_2 + b_{13}A_3 + \cdots + \cdots = 0$$

$$-\lambda A_2 + b_{21}A_1 + b_{22}A_2 + b_{23}A_3 + \cdots + \cdots = 0$$

$$-\lambda A_3 + b_{31}A_1 + b_{32}A_2 + b_{33}A_3 + \cdots + \cdots = 0$$

Then:

$$(b_{11} - \lambda)A_1 + b_{12}A_2 + b_{13}A_3 + \cdots + \cdots = 0$$

$$b_{21}A_1 + (b_{22} - \lambda)A_2 + b_{23}A_3 + \cdots + \cdots = 0$$

$$b_{31}A_1 + b_{32}A_2 + (b_{33} - \lambda)A_3 + \cdots + \cdots = 0$$

- The roots λ_i can be obtained by solving the secular determinant

$$\begin{matrix} A_1 & A_2 & A_3 & \cdots \end{matrix}$$
$$\begin{vmatrix} b_{11} - \lambda & b_{12} & b_{13} & \cdots \\ b_{21} & b_{22} - \lambda & b_{23} & \cdots \\ b_{31} & b_{32} & b_{33} - \lambda & \cdots \\ \vdots & \vdots & \vdots & \end{vmatrix} = 0 \quad (9.13.13)$$

For the root λ_1, the values of A_{k1}, A_{k2}, … for all the nuclei can be obtained; the same is true for the other roots. Therefore, the general solution of the superposition of all normal vibrations:

$$q_k = \sum_l B_{kl}Q_l^0 \sin\left(\sqrt{\lambda_l}t + \partial_l\right) \quad (9.13.14)$$

Example

In CO_2 molecules, which are constrained to move in only one direction, find:

- **the potential energy, and the kinetic energy; and**
- **the coefficients b_{ij}, λ_l, and the displacements q_i.**

Draw the normal coordinates Q_i, Q_2, Q_3, and find the ratio of displacements.

- If the mass, m_i, and the displacement, Δx_i, of each atom are identified as follows:

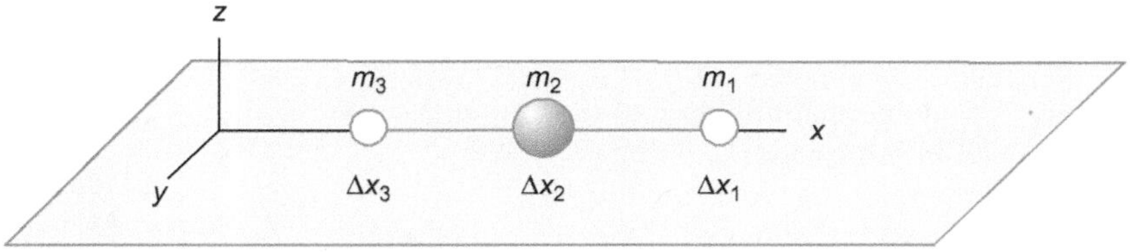

then the kinetic energy is given by

$$T = \frac{1}{2}\sum_N m_N\left[\left(\frac{\partial \Delta x_N}{\partial t}\right)^2 + \left(\frac{\partial \Delta y_N}{\partial t}\right)^2 + \left(\frac{\partial \Delta z}{\partial t}\right)^2\right] \tag{9.11.1}$$

Therefore,

$$T = \frac{1}{2}m_1\Delta\dot{x}_1^2 + \frac{1}{2}m_2\Delta\dot{x}_2^2 + \frac{1}{2}m_3\Delta\dot{x}_3^2$$

$$m_1 = m_3$$

$$T = \frac{1}{2}m_1\left(\Delta\dot{x}_1^2 + \Delta\dot{x}_3^2\right) + \frac{1}{2}m_3\Delta\dot{x}_2^2$$

The potential energy is given by:

$$V = \frac{1}{2}k\left[(\Delta x_1 - \Delta x_2)^2 + (\Delta x_2 - \Delta x_3)^2\right]$$

The potential energy is given by:

$$V = \frac{1}{2}k\left[(\Delta x_1 - \Delta x_2)^2 + (\Delta x_2 - \Delta x_3)^2\right]$$

$$q_1 = \sqrt{m_1}\,\Delta x_1, \quad q_2 = \sqrt{m_2}\Delta x_2, \quad \text{and} \quad q_3 = \sqrt{m_3}\Delta x_3 \tag{9.11.2}$$

$$V = \frac{1}{2}k\left[\left(\frac{q_1}{\sqrt{m_1}} - \frac{q_2}{\sqrt{m_2}}\right)^2 + \left(\frac{q_2}{\sqrt{m_2}} - \frac{q_3}{\sqrt{m_1}}\right)^2\right]$$

$$V = \frac{1}{2}k\left[\frac{q_1^2}{m_1} - \frac{2q_1q_2}{\sqrt{m_1m_2}} + \frac{q_2^2}{m_2} + \frac{q_2^2}{m_2} - \frac{2q_2q_3}{\sqrt{m_1m_2}} + \frac{q_3^2}{m_1}\right]$$

$$V = \frac{1}{2}\left[\frac{kq_1^2}{m_1} - \frac{kq_1q_2}{\sqrt{m_1m_2}} - \frac{kq_2q_1}{\sqrt{m_1m_2}} + \frac{2kq_2^2}{m_2} - \frac{kq_2q_3}{\sqrt{m_1m_2}} - \frac{kq_3q_2}{\sqrt{m_1m_2}} + \frac{kq_3^2}{m_1}\right]$$

If:

$$V = \frac{1}{2}\sum_{i,j}^{3N} b_{ij}q_iq_j \tag{9.11.5}$$

Then

$$b_{11} = \frac{k}{m_1}, \quad b_{22} = \frac{2k}{m_2}$$

$$b_{12} = b_{21} = -\frac{k}{\sqrt{m_1 m_2}}, \qquad b_{23} = b_{32} = -\frac{k}{\sqrt{m_1 m_2}}$$

$$b_{13} = b_{31} = 0, \qquad b_{33} = \frac{k}{m_1}$$

Substitute in

$$-\lambda A_k + \sum_j b_{kj} A_j = 0 \tag{9.13.12}$$

$$-\lambda A_1 + b_{11}A_1 + b_{12}A_2 + b_{13}A_3 = 0$$
$$-\lambda A_2 + b_{21}A_1 + b_{22}A_2 + b_{23}A_3 = 0$$
$$-\lambda A_3 + b_{31}A_1 + b_{32}A_2 + b_{33}A_3 = 0$$

then

$$(b_{11} - \lambda)A_1 + b_{12}A_2 + b_{13}A_3 = 0$$
$$b_{21}A_1 + (b_{22} - \lambda)A_2 + b_{23}A_3 = 0$$
$$b_{31}A_1 + b_{32}A_2 + (b_{33} - \lambda)A_3 = 0$$

If

$$q_k = A_k \sin\left(\sqrt{\lambda} t + \partial\right) \tag{9.13.11}$$

Therefore,

$$(b_{11} - \lambda)q_1 + b_{12}q_2 + b_{13}q_3 = 0$$
$$b_{21}q_1 + (b_{22} - \lambda)q_2 + b_{23}q_3 = 0$$
$$b_{31}q_1 + b_{32}q_2 + (b_{33} - \lambda)q_3 = 0$$

Substitute in

$$\begin{vmatrix} b_{11} - \lambda & b_{12} & b_{13} \\ b_{21} & b_{22} - \lambda & b_{23} \\ b_{31} & b_{32} & b_{33} - \lambda \end{vmatrix} = 0$$

- To find λ:

$$\begin{vmatrix} \frac{k}{m_1} - \lambda & -\frac{k}{\sqrt{m_1 m_2}} & 0 \\ -\frac{k}{\sqrt{m_1 m_2}} & \frac{2k}{m_2} - \lambda & -\frac{k}{\sqrt{m_1 m_2}} \\ 0 & -\frac{k}{\sqrt{m_1 m_2}} & \frac{k}{m_1} - \lambda \end{vmatrix} = 0$$

$$\left(\frac{k}{m_1} - \lambda\right)\left[\left(\frac{2k}{m_2} - \lambda\right)\left(\frac{k}{m_1} - \lambda\right) - \left(\frac{k}{\sqrt{m_1 m_2}}\right)\left(\frac{k}{\sqrt{m_1 m_2}}\right)\right] + \frac{k}{\sqrt{m_1 m_2}}\left[\left(-\frac{k}{\sqrt{m_1 m_2}}\right)\left(\frac{k}{m_1} - \lambda\right) - 0\right] = 0$$

$$\left(\frac{k}{m_1} - \lambda\right)\left[\frac{2k^2}{m_1 m_2} - \frac{2k\lambda}{m_2} - \frac{k\lambda}{m_1} + \lambda^2 - \frac{k^2}{m_1 m_2}\right] - \frac{k^2}{m_1 m_2}\left(\frac{k}{m_1} - \lambda\right) = 0$$

$$\left(\frac{k}{m_1} - \lambda\right)\left[\frac{2k^2}{m_1 m_2} - \frac{(2m_1 + m_2)k\lambda}{m_1 m_2} + \lambda^2 - \frac{k^2}{m_1 m_2} - \frac{k^2}{m_1 m_2}\right] = 0$$

If

$$\mu = \frac{2m_1 + m_2}{m_1 m_2}$$

$$\left(\frac{k}{m_1} - \lambda\right)\left[-\mu k\lambda + \lambda^2\right] = 0$$

$$\lambda\left(\frac{k}{m_1} - \lambda\right)(\lambda - \mu k) = 0$$

- Three roots are obtained:

$$\lambda_1 = \frac{k}{m_1}, \quad \lambda_2 = \mu k, \quad \lambda_3 = 0$$

- If

$$(b_{11} - \lambda)q_1 + b_{12}q_2 + b_{13}q_3 = 0$$
$$b_{21}q_1 + (b_{22} - \lambda)q_2 + b_{23}q_3 = 0$$
$$b_{31}q_1 + b_{32}q_2 + (b_{33} - \lambda)q_3 = 0$$

And

$$b_{11} = \frac{k}{m_1}, \qquad b_{22} = \frac{2k}{m_2}$$

$$b_{12} = b_{21} = -\frac{k}{\sqrt{m_1 m_2}}, \qquad b_{23} = b_{32} = -\frac{k}{\sqrt{m_1 m_2}}$$

$$b_{13} = b_{31} = 0, \qquad b_{33} = \frac{k}{m_1}$$

Then, when $\lambda_1 = \dfrac{k}{m_1}$

$$\left(\frac{k}{m_1} - \frac{k}{m_1}\right)q_1 - \frac{k}{\sqrt{m_1 m_2}}q_2 + (0)q_3 = 0, \qquad q_2 = 0$$

$$-\frac{k}{\sqrt{m_1 m_2}}q_1 + \left(\frac{2k}{m_2} - \frac{k}{m_1}\right)q_2 - \frac{k}{\sqrt{m_1 m_2}}q_3 = 0, \qquad q_1 = -q_3$$

$$(0)q_1 - \frac{k}{\sqrt{m_1 m_2}}q_2 + \left(\frac{k}{m_1} - \frac{k}{m_1}\right)q_3 = 0, \qquad q_2 = 0$$

Similarly

$$q_1 = q_3, \qquad q_2 = -2q_1\sqrt{\frac{m_1}{m_2}}, \qquad \text{for } \lambda_2 = \mu k$$

$$q_1 = q_3, \qquad q_2 = q_1\sqrt{\frac{m_1}{m_2}}, \qquad \text{for } \lambda_3 = 0$$

If

$$q_k = \sum_i B_{ki} Q_i$$

Then

$$q_1 = B_{11}Q_1 + B_{12}Q_2 + B_{13}Q_3$$
$$q_2 = B_{21}Q_1 + B_{22}Q_2 + B_{23}Q_3$$
$$q_3 = B_{31}Q_1 + B_{32}Q_2 + B_{33}Q_3$$

$$\begin{bmatrix} q_1 \\ q_2 \\ q_3 \end{bmatrix} = \begin{bmatrix} B_{11} & B_{12} & B_{13} \\ B_{21} & B_{22} & B_{23} \\ B_{31} & B_{32} & B_{33} \end{bmatrix} \begin{bmatrix} Q_1 \\ Q_2 \\ Q_3 \end{bmatrix}$$

$$\begin{bmatrix} Q_1 \\ Q_2 \\ Q_3 \end{bmatrix} = \begin{bmatrix} B_{11} & B_{21} & B_{31} \\ B_{12} & B_{22} & B_{32} \\ B_{13} & B_{23} & B_{33} \end{bmatrix} \begin{bmatrix} q_1 \\ q_2 \\ q_3 \end{bmatrix}$$

$$Q_1 = B_{11}q_1 + B_{21}q_2 + B_{31}q_3$$

$$Q_2 = B_{12}q_1 + B_{22}q_2 + B_{32}q_3$$

$$Q_3 = B_{13}q_1 + B_{23}q_2 + B_{33}q_3$$

- The relative displacements are as follows:

Normal Vibration, λ_i	Normal Coordinate, Q_i	Ratio of Displacements
$\lambda_1 = \frac{k}{m_1}$	Q_1	$B_{11}:B_{21}:B_{31} = 1:0:-1$
$\lambda_2 = \mu k$	Q_2	$B_{12}:B_{22}:B_{32} = 1:-2\sqrt{\frac{m_1}{m_2}}:1$

- Therefore, the mode of a normal vibration can be illustrated if the normal coordinate is converted into the corresponding rectangular coordinates.

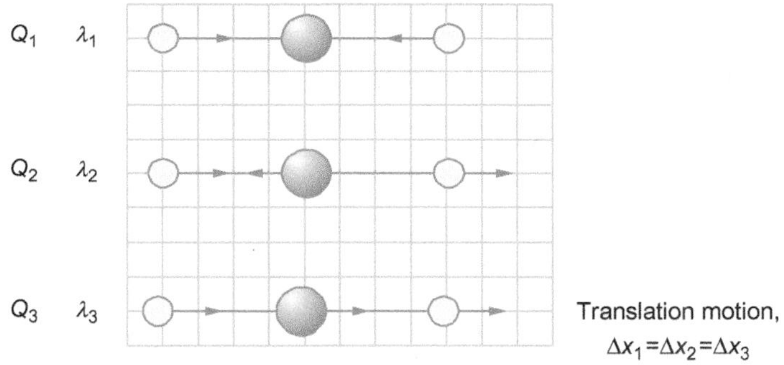

9.14 STRETCHING VIBRATIONS OF LINEAR MOLECULES

In the linear triatomic molecule A-B-A, show how to calculate the symmetric and the antisymmetric stretching frequencies.

- The two stretching vibrations in A-B-A will be mechanically coupled, and two new vibrations will result, which can be described as symmetric (in-phase) and antisymmetric (out-of-phase).

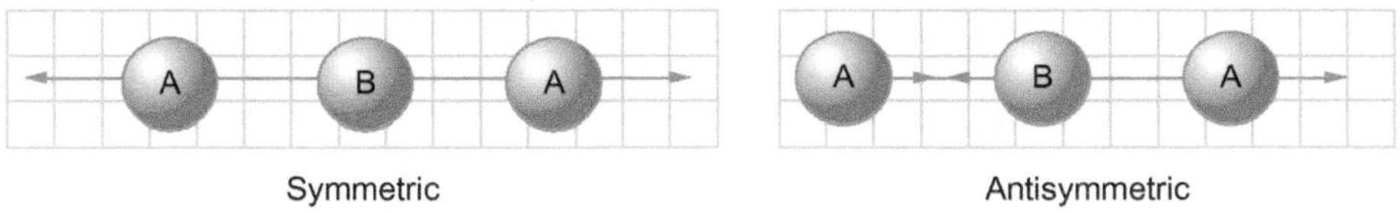

- Consider the following stretching:

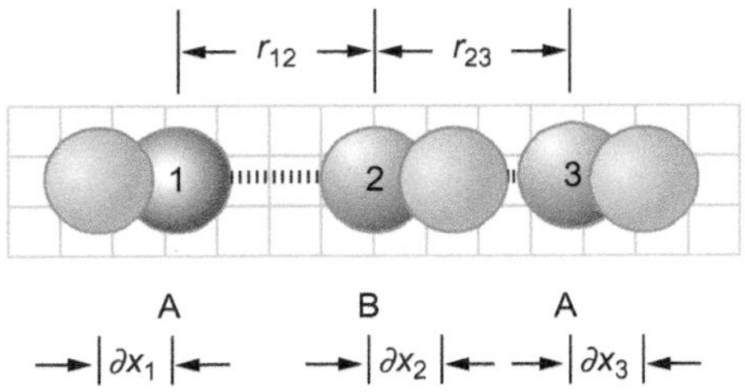

The three atoms A, B, and A are arranged on the x-axis
The equilibrium internuclear distance $= r_{12} = r_{23} = r$
The displacement of the atom i from the equilibrium position $= \partial x_i$
Assuming the force constant to be the same for both bonds,

$$\frac{d^2(\partial x_i)}{dt^2} = \partial \ddot{x}_i$$

$$F = m_i \partial \ddot{x}_i = -k \partial x_i, \text{ for atom } i \quad (9.5.9)$$

$$\begin{aligned} m_1 \partial \ddot{x}_1 &= -k(\partial x_1 - \partial x_2) && \text{for atom 1} \\ m_2 \partial \ddot{x}_2 &= -k(2\partial x_2 - \partial x_1 - \partial x_3) && \text{for atom 2} \\ m_1 \partial \ddot{x}_1 &= -k(\partial x_3 - \partial x_2) && \text{for atom 3} \end{aligned}$$

$$\partial x_i = \mathrm{A}_i \cos 2\pi \nu t = \mathrm{A}_i \cos \theta \quad (9.5.6)$$

$$\partial \ddot{x}_i = -4\pi^2 \nu^2 \mathrm{A}_i \cos 2\pi \nu t \quad (9.5.15)$$

$$\partial \ddot{x}_i = -\eta \mathrm{A}_i \cos \theta \quad (9.14.1)$$

where $\eta = 4\pi^2 \nu^2$, and $\theta = 2\pi\nu$, then

$$m_1 \eta \mathrm{A}_1 = k(\mathrm{A}_1 - \mathrm{A}_2)$$

$$m_2 \eta \mathrm{A}_2 = k(2\mathrm{A}_2 - \mathrm{A}_1 - \mathrm{A}_3)$$

$$m_1 \eta \mathrm{A}_3 = k(\mathrm{A}_3 - 2\mathrm{A}_2)$$

$$(k - m_1\eta)\mathrm{A}_1 - k\mathrm{A}_2 = 0$$

$$-k\mathrm{A}_1 + (2k - m_2\eta)\mathrm{A}_2 - k\mathrm{A}_3 = 0$$

$$-k\mathrm{A}_2 + (k - m_1\eta)\mathrm{A}_3 = 0$$

The determinant of the coefficient A_1, A_2, and A_3 is zero

$$\begin{vmatrix} (k - m_1\eta) & -k & 0 \\ -k & (2k - m_2\eta) & -k \\ 0 & -k & (k - m_1\eta) \end{vmatrix} = 0$$

$$(k - m_1\eta)^2(2k - m_2\eta) - k^2(k - m_1\eta) - k^2(k - m_1\eta) = 0$$

$$(k - m_1\eta)^2(2k - m_2\eta) - 2k^2(k - m_1\eta) = 0$$

$$(k - m_1\eta)\left[(k - m_1\eta)(2k - m_2\eta) - 2k^2\right] = 0$$

$$k - m_1\eta = 0, \because \eta = 4\pi^2\nu^2$$

$$\nu_1 = \frac{1}{2\pi}\sqrt{\frac{k}{m_1}} \quad (9.14.2)$$

This frequency corresponds to the symmetric stretching of the two A–B group because it involve no motion of atom B

$$(k - m_1\eta)(2k - m_2\eta) - 2k^2 = 0$$

$$m_1 m_2 \eta^2 - (2m_1 + m_2)k\eta = 0$$

$$\nu_2 = \frac{1}{2\pi}\sqrt{\frac{k(2m_1 + m_2)}{m_1 m_2}} \tag{9.14.3}$$

What is the Schrödinger equation for the vibration system of n-atom molecules?

- The Schrödinger equation:

$$\sum_i \frac{\partial^2 \psi_n}{\partial Q_i^2} + \frac{8\pi^2}{h^2}\left(E - \frac{1}{2}\sum_i \lambda_i Q_i^2\right)\psi_n = 0 \tag{9.14.4}$$

where
Q_i is the normal coordinates

$$\psi_n = \psi_1(Q_1)\psi_2(Q_2)\ldots\psi_v(Q_{3N-5,6})$$

$$E = E_1 + E_2 + E_3 + \cdots = \sum_i^{3N-5,6} E_i$$

$$E_i = h\nu_i(v_i + 1)$$

$$\nu_i = \frac{1}{2\pi}\sqrt{\lambda_i}$$

For each normal coordinate Q_i:

$$\frac{\partial^2 \psi_n}{\partial Q_i^2} + \frac{8\pi^2}{h^2}\left(E_i - \frac{1}{2}\lambda_i Q_i^2\right)\psi_i = 0 \tag{9.14.5}$$

9.15 SYMMETRY AND NORMAL MODES OF VIBRATION

How do the normal modes of vibration relate to the symmetry?

- In order to determine the symmetry of a normal vibration, the kinetic and potential energies must be considered:

$$T = \frac{1}{2}\sum_i \dot{Q}_i^2 \tag{9.13.4}$$

$$V = \frac{1}{2}\sum_i \lambda_i Q_i^2 \tag{9.13.5}$$

- If the molecule has only one normal vibration, Q_i:

$$T = \frac{1}{2}\dot{Q}_i^2$$

$$V = \frac{1}{2}\lambda_i Q_i^2$$

- These energies do not change when a symmetry operation, R, changes Q_i to RQ_i:

$$T = \frac{1}{2}\dot{Q}_i^2 = \frac{1}{2}\left(R\dot{Q}_i\right)^2 \tag{9.15.1}$$

$$V = \frac{1}{2}\lambda_i Q_i^2 = \frac{1}{2}\lambda_i (RQ_i)^2 \tag{9.15.2}$$

- Since a molecule is essentially not changed by applying a symmetry operation R, the normal mode RQ_i must have the same frequency as the normal mode Q_i.

- Thus if Q_i is nondegenerate, these require

$$(RQ_i)^2 = Q_i^2 \text{ or } RQ_i = \pm 1\, Q_i \tag{9.15.3}$$

for all Rs.
If $RQ_i = Q_i$, the vibration is symmetric.
If $RQ_i = -Q_i$, the vibration is antisymmetric
- If the vibration is doubly degenerate,

$$T = \frac{1}{2}\dot{Q}_{ia}^2 + \frac{1}{2}\dot{Q}_{ib}^2$$

$$V = \frac{1}{2}\lambda_i Q_{ia}^2 + \frac{1}{2}\lambda_i Q_{ib}^2 \tag{9.15.4}$$

These require:

$$(RQ_{ia})^2 + (RQ_{ib})^2 = Q_{ia}^2 + Q_{ib}^2$$

Or more conveniently by using the matrix:

$$R\begin{bmatrix} Q_{ia} \\ Q_{ib} \end{bmatrix} = \begin{bmatrix} a & b \\ c & d \end{bmatrix}\begin{bmatrix} Q_{ia} \\ Q_{ib} \end{bmatrix} \tag{9.15.5}$$

The values of a, b, c, and d are the elements of the matrix that represents R. In any situation, the normal vibration is either symmetric, antisymmetric, or degenerate for each symmetry operation.
- Therefore, $\hat{Q}$ forms the basis for irreducible representation in the molecular symmetry group, and the degenerate modes transform according to irreducible representations of dimensionality greater than one.
Or if a molecule belongs to a certain symmetry group, then the wave functions, which describe the molecule, must possess the same transformation properties under the symmetry operations of the group, as do the irreducible representations of the group.
- We can find the symmetries of all the possible molecular motions by using x, y, and z directions on each atom as a basis for a reducible representation of the group.
- For an n-atom molecule, this will produce a representation of order $3n$, i.e., the character of the identity representation will be $3n$, therefore, to have $3n$ degree of freedom.
- The character is equal to the extent to which the vectors in the basis are left unshifted by the operation.

What are the number and symmetry representations of the normal modes of vibration for an H_2O molecule?

- H_2O: belong to the point group C_{2v}:
C_{2v}: E, C_2, σ_v^{xz}, σ_v^{yz}
- We must design a coordinate system consistent with the character table.
- The only constraint is that the principal axis of rotation be the z-axis, but x and y can be oriented as we want.
- Calling the molecular plane xz, our challenge is to see how three displacement vectors on each atom, representing the three degrees of freedom of that atom, transform under the operations of the point group (Fig. 9.15).
Basis set: unit vectors of motion on each atom, nuclei as origins:
- For E: the E-matrix is a diagonal matrix, with all diagonal entries +1.

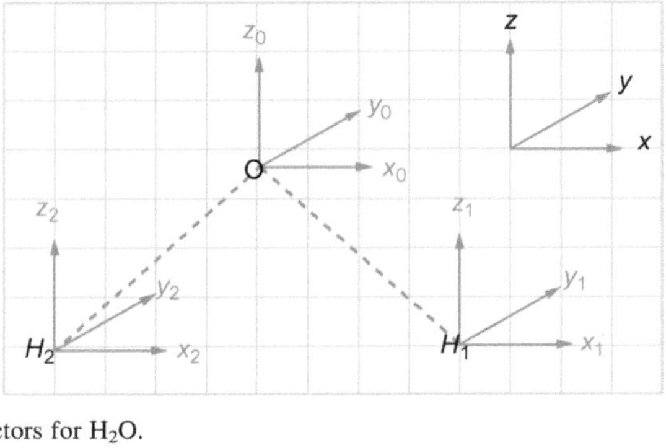

FIG. 9.15 The Cartesian displacement vectors for H_2O.

$$E=\begin{bmatrix}1&0&0\\0&1&0\\0&0&1\end{bmatrix}$$

The operation E accomplishes the following transformations:

$$x_0=1(x_0),\quad y_0=1(y_0),\quad z_0=1(z_0)$$
$$x_1=1(x_1),\quad y_1=1(y_1),\quad z_1=1(z_1)$$
$$x_2=1(x_2),\quad y_2=1(y_2),\quad z_2=1(z_2)$$

$$E\cdot\begin{bmatrix}x_0\\y_0\\z_0\\x_1\\y_1\\z_1\\x_2\\y_2\\z_2\end{bmatrix}=\begin{bmatrix}1&0&0&0&0&0&0&0&0\\0&1&0&0&0&0&0&0&0\\0&0&1&0&0&0&0&0&0\\0&0&0&1&0&0&0&0&0\\0&0&0&0&1&0&0&0&0\\0&0&0&0&0&1&0&0&0\\0&0&0&0&0&0&1&0&0\\0&0&0&0&0&0&0&1&0\\0&0&0&0&0&0&0&0&1\end{bmatrix}\begin{bmatrix}x_0\\y_0\\z_0\\x_1\\y_1\\z_1\\x_2\\y_2\\z_2\end{bmatrix}=\begin{bmatrix}x_0\\y_0\\z_0\\x_1\\y_1\\z_1\\x_2\\y_2\\z_2\end{bmatrix}$$

The character or the trace of E-matrix, the sum of the diagonal elements, is 9.

So, $\chi_E=3n$

where n = number of atoms in the molecule, for H_2O: $\chi_E=9$.

The basis of the representation is the set of nine arrows.

- For C_2: get $\chi_{C_2}=-1$.

The only atoms whose vectors contributed in nonzero way to χ_{C_2} was the O-atom, because the C_2-operation shifted the H-atoms' vectors.

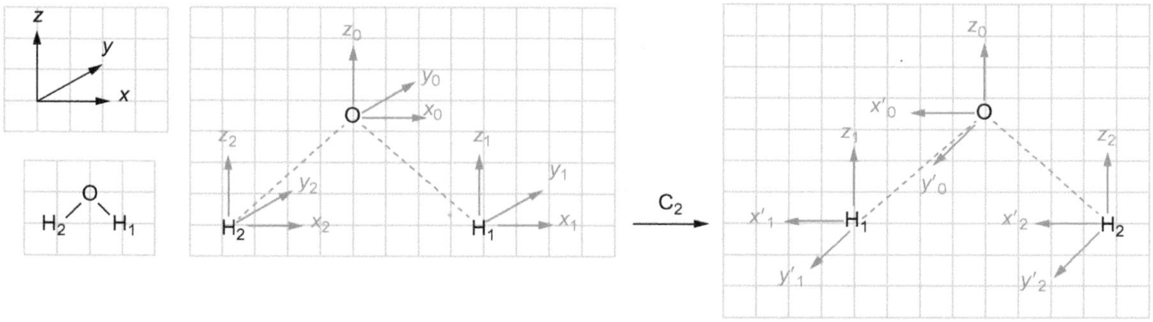

$$new\ \ x_0'=-1(x_0),\quad y_0'=-1(y_0),\quad z_0'=+1(z_0)$$
$$x_1'=-1(x_2),\quad y_1'=-1(y_2),\quad z_1'=+1(z_2)$$
$$x_2'=-1(x_1),\quad y_2'=-1(y_1),\quad z_2'=+1(z_1)$$

This gives the C_2 matrix:

$$\begin{bmatrix}x_0'\\y_0'\\z_0'\\x_1'\\y_1'\\z_1'\\x_2'\\y_2'\\z_2'\end{bmatrix}=\begin{bmatrix}-1&0&0&0&0&0&0&0&0\\0&-1&0&0&0&0&0&0&0\\0&0&1&0&0&0&0&0&0\\0&0&0&0&0&0&-1&0&0\\0&0&0&0&0&0&0&-1&0\\0&0&0&0&0&0&0&0&1\\0&0&0&-1&0&0&0&0&0\\0&0&0&0&-1&0&0&0&0\\0&0&0&0&0&1&0&0&0\end{bmatrix}\begin{bmatrix}x_0\\y_0\\z_0\\x_1\\y_1\\z_1\\x_2\\y_2\\z_2\end{bmatrix}$$

The character or the trace of C_2-matrix, the sum of the diagonal elements, is -1.

- So shifted atom contributes zero to X_R.
  Unshifted, unrotated atom contribute 3.
  Unshifted, rotated atom contributes according to trigonometry.
- e.g., for H_2O, σ_v^{xz}

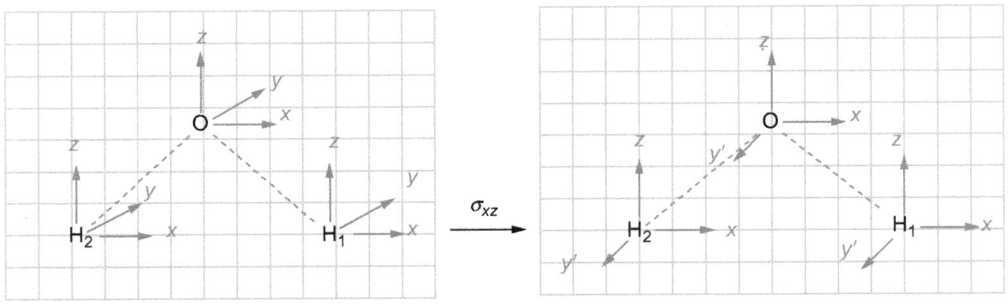

$$x's: +3 \quad y's: -3 \quad z's: +3$$

$$\chi_{\sigma_{xz}} = +3$$

- Using the same principal, $\chi_{\sigma_{yz}} = +1$, and

$$\Gamma_{\text{total}}(H_2O) = 9, -1, 3, 1$$

- Decomposed to (Table 9.3):

$$\Gamma_{\text{total}}(H_2O) = 3A_1 + A_2 + 3B_1 + 2B_2$$

Note:
$\Sigma l = 3+1+3+2 = \chi_E = 3n$, (9 of l-D representations, l is the order of each representation).
The molecule has 3n degrees of freedom of these ($\Gamma_{\text{tot.}}$), 3 are bulk/net translational: T_x, T_y, and T_z.

(a) For a linear molecule: two rotational modes, R_x and R_y, so there are $(3n-5)$ vibrational degrees of freedom.
(b) For a nonlinear molecules, such as H_2O, have R_x, R_y, and R_z, so there are $(3n-6)$ vibrational modes symmetries found by subtraction $\Gamma_{\text{tn.}}$, $\Gamma_{\text{rot.}}$ from $\Gamma_{\text{tot.}}$.

$\Gamma_{\text{tot.}} =$	$3A_1 +$	$A_2 +$	$3B_1 +$	$2B_2$
T_x			$-B_1$	
T_y				$-B_2$
T_z	$-A_1$			
R_x				$-B_2$
R_y			$-B_1$	
R_z		$-A_2$		
$\Gamma_{\text{vib.}} =$	$2A_1$		$+B_1$	

TABLE 9.3 The Character Table of the Point Group C_{2v}

C_2v	E	C_2	σ_v^{xz}	σ_v^{yz}	
A_1	1	1	1	1	z
A_2	1	1	−1	−1	R_z
B_1	1	−1	1	−1	x, R_y
B_2	1	−1	−1	1	y, R_x
Γ_{total} (H_2O)	9	−1	3	1	$=3A_1+A_2+3B_1+2B_2$

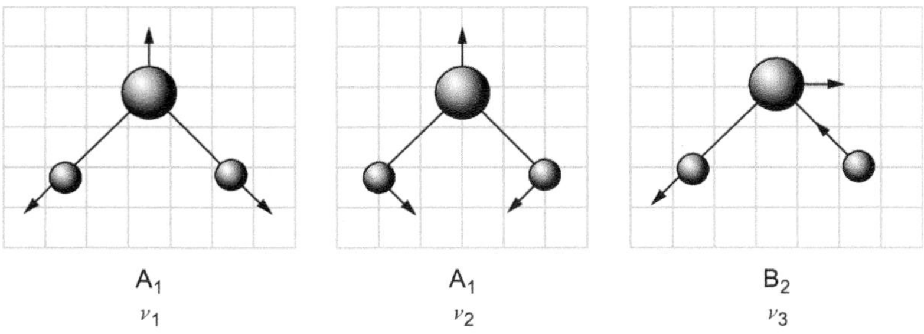

FIG. 9.16 The three modes of vibration of H_2O.

TABLE 9.4 Symmetry Representations of the Three Modes of Vibration of H_2O

C_2v	E	C_2	σ_v^{xz}	σ_v^{yz}	
A_1	1	1	1	1	v
A_1	1	1	1	1	v
B_2	1	−1	−1	1	v

- These represent the mechanically simple fundamental modes of molecular vibration, the normal modes.
- They may appear in the Raman or infrared spectrum, or both, or neither.
- Other modes are also observed; linear combinations of the normal modes.
- In general, to find the symmetry of a mode, we work out the effect of each operation of the group on the single unit picture, and see which operations of the group transform it into an equivalent picture.
- The symmetries of the three modes of vibration of H_2O can be illustrated (Fig. 9.16) by using the symmetry representation (Table 9.4).

9.16 ASSIGNING THE NORMAL MODES OF VIBRATION

What are the general steps to determine the normal modes of vibration?
How many normal modes of vibration does octahedral complex AB_6 have?

- Generation of Γ_{total} from motional vectors basis set:
  (1) An atom whose *nucleus* (vector origin) is *shifted* from its original position by an operation of the group contributes zero to the character of that operation's matrix in the total representation.
  (2) An atom which is *not shifted* by an operation, and of whose motional vectors, none is rotated about their local nuclear origin by the operation, contributes +3 to the character of that operation's matrix in the *total* representation.
  (3) If an atom is *unshifted* by an operation, but rotated about its own nucleus (vector origin), then it *contributes* to the character of that operation's matrix in accordance with:

Operation	Contribution to $\Gamma_{tot.}$	Operation	Contribution to $\Gamma_{tot.}$
C_n	$1+2\cos(2\pi/n)$	S_n	$-1+2\cos(2\pi/n)$ $=(C_n\text{'s value})-2$
C_2	−1	$S_1=\sigma$	1
C_3	0	$S_2=i$	−3
C_4	1	S_3	−2
C_6	2	S_4	−1
		S_6	0

- For O_h (AB_6): AB_6 belongs to the point group O_h:

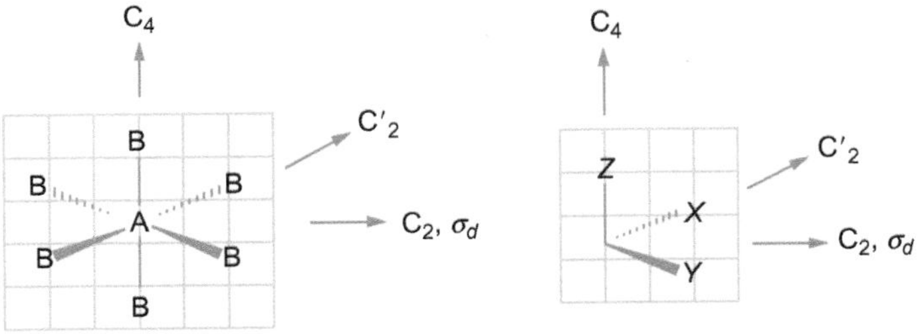

- The operation E accomplishes the following transformation:

$$x \rightarrow x \qquad y \rightarrow y \qquad z \rightarrow z$$

$$E \cdot \begin{bmatrix} x \\ y \\ z \end{bmatrix} = \begin{bmatrix} 1 & 0 & 0 \\ 0 & 1 & 0 \\ 0 & 0 & 1 \end{bmatrix} \cdot \begin{bmatrix} x \\ y \\ z \end{bmatrix}$$

The trace of this matrix is 3

- The C_3 operation:

$$x \rightarrow 0 \qquad y \rightarrow 0 \qquad z \rightarrow 0$$

The trace of this matrix is 0

- The C_2 operation:

$$x \rightarrow y \qquad y \rightarrow x \qquad z \rightarrow -z$$

The trace of this matrix is -1.

- Similarly, the traces of C_4, C_2', i, S_4, S_6, σ_h, σ_d are $\Gamma_{x,y,z}$: 1, -1, -3, -1, 0, 1, and 1, respectively.
- Determine the number of atoms, which do not change location during each symmetry operation.
- For each operation, multiply the number of unmoved atoms by the character of $\Gamma_{x,y,z}$ at the bottom of the character table.
- This gives the character of $\Gamma_{total.}$, the total representation of all degrees of freedom of the molecule, including transition, rotation, and vibration (Table 9.5).
- Using the formula

$$\Gamma_i = \frac{1}{h}\sum_R \chi_{(R)}\chi_i(R)$$

and O_h character (Table 9.6):

$$\Gamma_{A_{1g}} = (1/48)\,[1\times1\times21+8\times1\times0+6\times1\times(-1)+6\times1\times3+3\times1\times(-3)+1\times1\times(-3)$$
$$+6\times1\times(-1)+8\times1\times0+3\times1\times5+6\times1\times3]=1$$

- Similarly, we can find

$$\Gamma_{A_{2g}}=0, \Gamma_{E_g}=1, \Gamma_{T_{1g}}=1, \Gamma_{T_{2g}}=1, \Gamma_{A_{1u}}=0$$

TABLE 9.5 Determination of the Representations Spanned by the Twenty-One Degrees of Freedom of AB_6

O_h	E	$8C_3$	$6C_2$	$6C_4$	$3C_2$	i	$6S_4$	$8S_6$	$3\sigma_h$	$6\sigma_d$
$\Gamma_{x,y,z}$	3	0	−1	1	−1	−3	−1	0	1	1
$\Gamma_{unmoved}$	7	1	1	3	3	1	1	0	5	3
Γ_{total}	21	0	−1	3	−3	−3	−1	0	5	3

TABLE 9.6 The Character Table for the Octahedral Point Group

O_h	E	$8C_3$	$6C_2$	$6C_4$	C_2	i	$6S_4$	$8S_6$	$3\sigma_h$	$6\sigma_d$		
A_{1g}	1	1	1	1	1	1	1	1	1	1	–	$x^2+y^2+z^2$
A_{2g}	1	1	−1	−1	1	1	−1	1	1	−1	–	–
E_g	2	−1	0	0	2	2	0	−1	2	0	–	$2z^2-x^2-y^2,\ x^2-y^2$
T_{1g}	3	0	−1	1	−1	3	1	0	−1	−1	(R_x,R_y,R_z)	–
T_{2g}	3	0	1	−1	−1	3	−1	0	−1	1		$(xz,\ yz,xy)$
A_{1u}	1	1	1	1	1	−1	−1	−1	−1	−1	–	–
A_{2u}	1	1	−1	−1	1	−1	1	−1	−1	1	–	–
E_u	2	−1	0	0	2	−2	0	1	−2	0	–	–
T_{1u}	3	0	−1	1	−1	−3	−1	0	1	1	(x,y,z)	–
T_{2u}	3	0	1	−1	−1	−3	1	0	1	−1	–	–
Γ_{total}	21	0	−1	3	−3	−3	−1	0	5	3	$= A_{1g}+E_g+T_{1g}+T_{2g}+3T_{1u}+T_{2u}$	

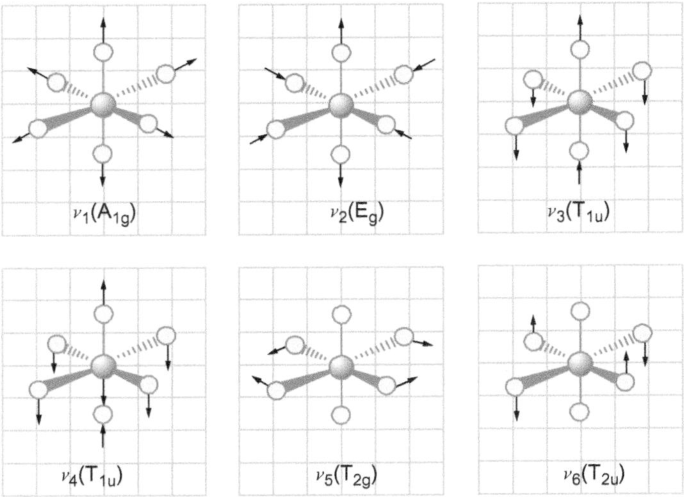

FIG. 9.17 AB_6 octahedral molecules.

$$\Gamma_{A_{2u}} = 0, \Gamma_{E_u} = 0, \Gamma_{T_{1u}} = 3, \Gamma_{T_{2u}} = 1$$

- Therefore,

$$\Gamma_{total} = A_{1g} + E_g + T_{1g} + T_{2g} + 3T_{1u} + T_{2u}$$

$$\Gamma_{trans} = T_{1u} \text{ contain electrical dipole moment in } x, y, \text{ and } z \text{ direction}$$

$$\Gamma_{rot.} = T_{1g} \text{ contain magnetic dipole moment } R_x, R_y, \text{ and } R_z$$

$$\Gamma_{vib.} = \Gamma_{total} - \Gamma_{rot.} - \Gamma_{vib.} = A_{1g} + E_g + T_{2g} + 2T_{1u} + T_{2u}$$

See Fig. 9.17.

Normal Modes of Vibration for Linear Triatomic Molecule

How many normal modes of vibration do CO_2 molecules have? What are the abbreviations used for the descriptions of these vibrations?

- CO_2: $D_{\infty h}$, Γ_{total} is derived (Table 9.7):
- Subtracting Γ_{trans} and $\Gamma_{rot.}$ from Γ_{total} to obtain $\Gamma_{vib.}$.
- We cannot use the formula

$$\Gamma_i = \frac{1}{h}\sum_R \chi_{(R)}\chi_i(R)$$

  to decompose Γ_{total}, because the order of the group is infinite.
- Looking at the $D_{\infty h}$ character table, you should suspect that π_u is involved.
- Subtracting from Γ_{total}, you will recognize the difference at the bottom of the table as the sum $\sigma_g^+ + \sigma_u^+$.
- The normal mode of vibration:

$$\Gamma_{vib.} = \sigma_g^+ + \sigma_u^+ + \pi_u$$

- Counting π_u as two representations, we find four representations for four degrees of freedom, $(3n-5)$.
- When two modes transform as a single degenerate irreducible representation (π or e), they must have the same energy and be degenerate.

TABLE 9.7 Vibrational Analysis of CO_2

$D_{\infty h}$	E	$2C_{\infty}^{\phi}$	$\infty\sigma_i$	i	$2S_{\infty}^{\phi}$	∞C_2	
Σ_g^+	1	1	1	1	1	1	
Σ_g^-	1	1	−1	1	1	−1	R_z
Π_g	2	$2c\varphi$	0	2	$-2c\varphi$	0	(R_x, R_y)
Δ_g	2	$2c2\varphi$	0	2	$2c2\varphi$	0	
Σ_u^+	1	1	1	−1	−1	−1	z
Σ_u^-	1	1	−1	−1	−1	1	
Π_u	2	$2c\varphi$	0	−2	$2c\varphi$	0	(x, y)
Δ_u	2	$2c2\varphi$	0	−2	$-2c2\varphi$	0	
$\Gamma_{x, y, z}$	3	$1+2c\varphi$	1	−3	$-1+2c\varphi$	−1	
Unmoved atoms	3	3	3	1	1	1	
Γ_{total}	9	$3+6c\varphi$	3	−3	$-1+2c\varphi$	−1	
Γ_{trans}	3	$1+2c\varphi$	1	−3	$-1+2c\varphi$	−1	
$\Gamma_{rot.}$	2	$2c\varphi$	0	2	$-2c\varphi$	0	
$\Gamma_{vib.} = \Gamma_{total} - \Gamma_{trans} - \Gamma_{rot.}$	4	$2+2c\varphi$	2	−2	$2c\varphi$	0	
π_u	2	$2c\varphi$	0	−2	$2c\varphi$	0	(x, y)
$\Gamma_{vib.} - \pi_u$	2	2	2	0	0	0	$=\sigma_g^+ + \sigma_u^+$

In which the symbol c is used for cosine when it appears in characters, φ is the arbitrary rotational angle.

- Since CO_2 has two bonds, we expect two stretching vibrations, and $4-2=2$ must be bending modes:

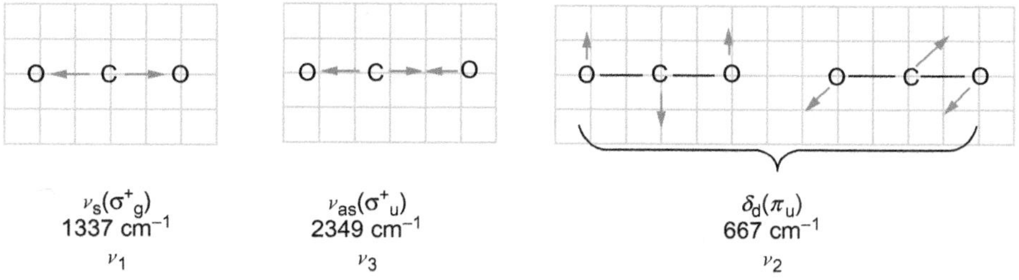

- The symbols used to label these vibrations are as follows:
  ν: stretching, δ: deformation (bending), ρ_w: wagging, ρ_r: rocking, ρ_t: twisting, s: symmetric, as: asymmetric, d: degenerate, π: out of plane
- To find the symmetry of a vibration, we apply the operations of the point group to its picture and write down the characters:

$D_{\infty h}$	E	$2C_{\infty}^{\phi}$	$\infty\sigma_i$	i	$2S_{\infty}^{\phi}$	∞C_2	
ν_s	1	1	1	1	1	1	$\rightarrow \sigma_g^+$
ν_{as}	1	1	1	−1	−1	−1	$\rightarrow \sigma_u^+$
∂_d	2	$2c\varphi$	0	−2	$2c\varphi$	0	$\rightarrow \pi_u$

- The vibration ν_{as} goes to minus itself upon inversion.

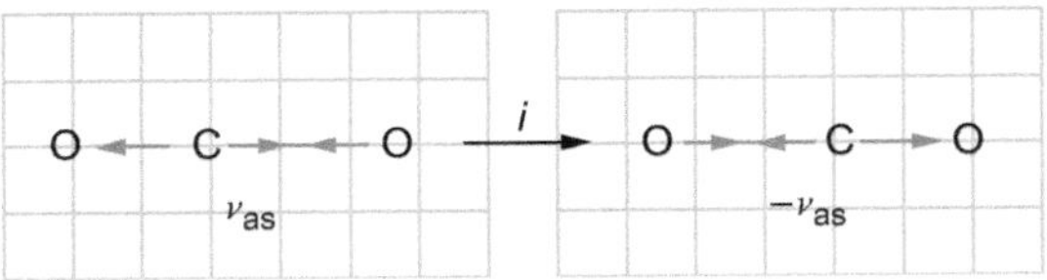

- The asymmetric stretching often (but not always) has a higher energy than the symmetric stretch.
- The energies of the bending mode generally fall below the energies of stretching modes.

BCl_3: D_{3h} point group:

How many modes of vibration does BCl_3 have?

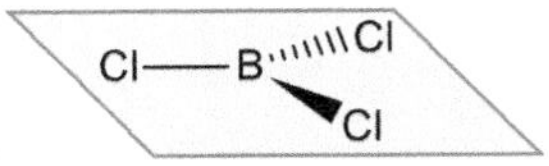

BCl_3 molecule

- $3n-6=6$ vibrational degrees of freedoms:
  Γ_{total} is driven in Table 9.8:
- The frequencies assigned to the a_1' modes of both the ^{10}B and ^{11}B species are identical because no B atom motion is involved in the vibration (Fig. 9.18).

TABLE 9.8 Determination of $\Gamma_{vib.}$ for BCl_3

D_{3h}	E	$2C_3$	$3C_2$	σ_h	$2S_3$	$3\sigma_v$
$\Gamma_{x,y,z}$	3	0	−1	1	−2	1
Unmoved atoms	4	1	2	4	1	2
Γ_{total}	12	0	−2	4	−2	2

$\Gamma_{total} = a_1' + a_2' + 3e' + 2a_2'' + e''$.
$\Gamma_{trans} = e' + a_2''$.
$\Gamma_{rot.} = a_2' + e''$.
$\Gamma_{vib.} = a_1' + 2e' + a_2''$.

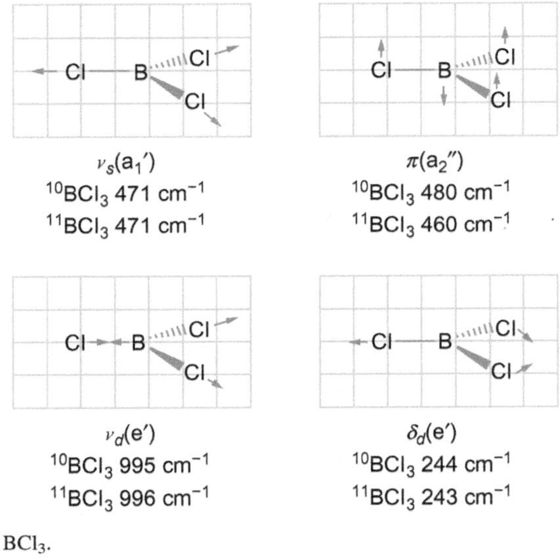

FIG. 9.18 Normal modes of vibration of BCl_3.

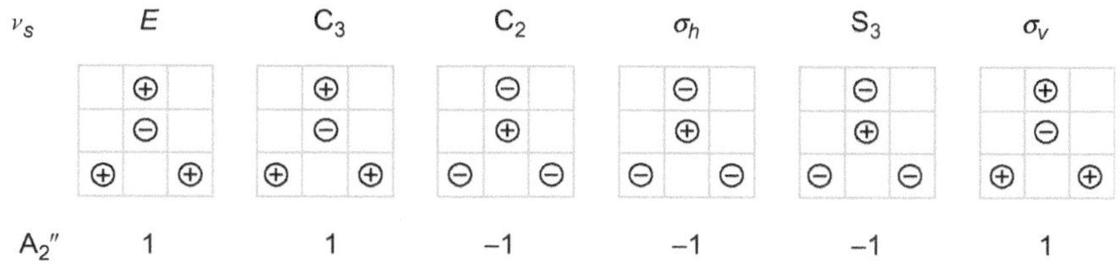

1, 1, −1, −1, −1, 1 is A_1' clearly, ν_1 is A_1' (most symmetric)

9.17 FORCE CONSTANTS AND THE GF-MATRIX METHOD

Why are force constants important?

- The kinetic and potential energies determine the frequency of the normal vibration.
- The masses of individual atoms and their geometrical displacements in the molecule decide the kinetic energy. The interaction between the individual atoms causes the potential energy that is described in terms of the force constants.
- Since the force constants provide valuable information about the nature of the interatomic forces, it is of great benefit to obtain the values of these constants from the observed frequencies.
- This typically proceeded by using an iterative computing approach. A starting set of assumed force constants is refined by successive approximations until a satisfactory agreement between calculated and observed frequencies. This specific set of the force constants is accepted as a representation of the potential energy of the system.
- For a diatomic molecule, the harmonic oscillator potential is

$$V = \frac{1}{2}F(r - r_e)^2 \tag{9.5.10}$$

where r is the internuclear distance, r_e is the equilibrium internuclear distance, and F is constant.

- From classical physics, the force

$$\text{Force} = -\frac{dV}{d(r - r_e)} = -F(r - r_e)$$

- Therefore, the force constant is equal to the second derivative of the potential energy:

$$F = \frac{d^2V}{d(r - r_e)^2} \tag{9.17.1}$$

- In general, the potential of polyatomic molecule as a function of the internal coordinates S is

$$V = \frac{1}{2}\sum_i \sum_j F_{ij} S_i S_j \tag{9.17.2}$$

and the force constant at equilibrium is

$$F_{ij} = \left(\frac{\partial^2 V}{\partial S_i \partial S_j}\right)_e \tag{9.17.3}$$

How can you calculate the force constants of a bent X_2Y molecule by using GF-matrix method?

Define:

(a) **the irreducible representations for the molecular vibrations;**
(b) **the complete set of normalized symmetry coordinates for vibrations;**
(c) **the potential energy in terms of internal coordinates and symmetry coordinates; and**
(d) **the kinetic energy in terms of internal coordinates and symmetry coordinates.**
(e) **combine both the potential and kinetic energies in Lagrange's equation.**

(a) *The irreducible representations for the molecular vibrations*

- In order to compute the vibrational frequencies, we need first to express both the potential and kinetic energies in terms of internal coordinates instead of the Cartesian coordinates.
- Force constants have clearer physical meaning than those expressed in terms of Cartesian coordinates.
- Bent X_2Y belongs to C_{2v}; the two X–Y distances (Δr_1, Δr_2), and the XYX angle ($\Delta\alpha$) are bases for the irreducible representations (Table 9.9):

$$\Gamma_{\Delta r_1,\Delta r_2}: A_1 + B_1$$

$$\Gamma_{\Delta\alpha}: A_1$$

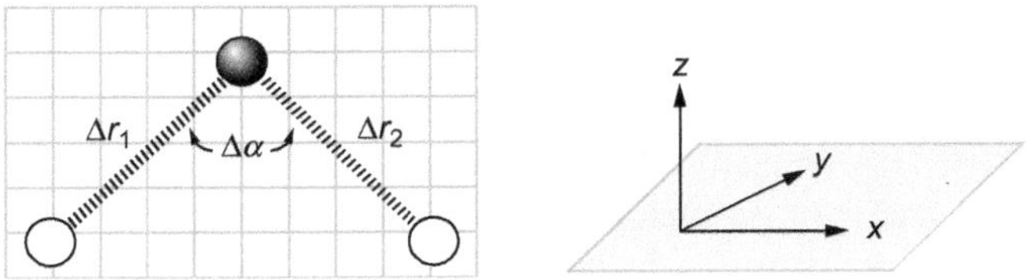

(b) *The complete set of normalized symmetry coordinates for vibrations*

- Using the symmetry coordinates as SALC's of the internal coordinates:
  $\Delta\alpha$: is A_1 symmetry coordinate
  Applying the projection operators to Δr_1, Δr_2:

$$\begin{aligned}\hat{P}_{A_1}(\Delta r_1) &\approx (1)\hat{E}\Delta r_1 + (1)\hat{C}_2\Delta r_1 + (1)\hat{\sigma}_v\Delta r_1 + (1)\hat{\sigma}_{v'}\Delta r_1\\ &\approx \Delta r_1 + \Delta r_2 + \Delta r_1 + \Delta r_2\\ &\approx 2(\Delta r_1 + \Delta r_2)\\ &\approx \Delta r_1 + \Delta r_2\end{aligned}$$

$$\begin{aligned}\hat{P}_{B_1}(\Delta r_1) &\approx (1)\hat{E}\Delta r_1 + (-1)\hat{C}_2\Delta r_1 + (1)\hat{\sigma}_v\Delta r_1 + (-1)\hat{\sigma}_{v'}\Delta r_1\\ &\approx \Delta r_1 - \Delta r_2 + \Delta r_1 - \Delta r_2\\ &\approx \Delta r_1 - \Delta r_2\end{aligned}$$

TABLE 9.9 Determination of $\Gamma_{\Delta r_1,\Delta r_2}$ and $\Gamma_{\Delta\alpha}$ for a Bent X_2Y Molecule

C_{2v}	E	C_2	σ_v^{xz}	σ_v^{yz}	
A_1	1	1	1	1	z
A_2	1	1	−1	−1	R_z
B_1	1	−1	1	−1	x, R_y
B_2	1	−1	−1	1	y, R_x
$\Gamma_{x,y,z}$	3	−1	1	1	
Unmoved atoms	3	1	3	1	
Γ_{Total}	9	−1	3	1	$= 3A_1 + A_2 + 3B_1 + 2B_2$
$\Gamma_{Trans.+Rotation}$	6	−2	0	0	$= A_1 + A_2 + 2B_1 + 2B_2$
$\Gamma_{vib.}$	3	1	3	1	$= 2A_1 + B_1$
$\Gamma_{\Delta r_1,\Delta r_2}$	2	0	2	0	$= A_1 + B_1$
$\Gamma_{\Delta\alpha}$	1	1	1	1	$= A_1$

- The complete set of normalized symmetry coordinates for vibrations is

$$\left.\begin{aligned} S_1 &= \Delta\alpha \\ S_2 &= \frac{1}{\sqrt{2}}(\Delta r_1 + \Delta r_2) \end{aligned}\right\} A_1$$

$$\left. S_3 = \frac{1}{\sqrt{2}}(\Delta r_1 - \Delta r_2) \right\} B_1$$

(c) *The potential energy in terms of internal coordinates*

- The potential energy, V_i, of stretching a given bond or bending a given angle is given by

$$2V_i = f_{ii}s_i$$

- The potential energy in terms of internal coordinates is

$$2V = \sum_{i,k} f_{ik}s_is_k \tag{9.17.2}$$

where f_{ik} is called the force constant, this cross term represents the energy of interaction between such motions, and $f_{ik} = f_{ki}$; s_i, and s_k are changes in internal coordinates.

- The sum extends over all values of i and k; there are nine force constant for the three internal coordinates:

	Δr_1	Δr_2	$\Delta\alpha$
Δr_1	f_{11}	f_{12}	f_{13}
Δr_2	f_{12}	f_{22}	f_{23}
$\Delta\alpha$	f_{13}	f_{23}	f_{33}

where the force constants for stretching an Y–X is $f_{11} = f_{22}$, the one for the bending XYX angle is f_{33}, and $f_{13} = f_{31} = f_{23} = f_{32}$.

$$2V = f_{11}(\Delta r_1)^2 + f_{22}(\Delta r_2)^2 + f_{33}(\Delta\alpha)^2 + 2f_{12}(\Delta r_1\Delta r_2) + 2f_{13}(\Delta r_1\Delta\alpha) + 2f_{23}(\Delta r_2\Delta\alpha)$$

- The potential energy in internal coordinates using the matrix notation

$$2V = \sum_{i,k} f_{ik}s_is_k \tag{9.17.2}$$

can be expressed in the following matrix notation:

$$2V = s'fs \tag{9.17.3}$$

where s_i can be written as a column matrix (vector) s, and s' as a row matrix

$$2V = (\Delta r_1 \;\; \Delta r_2 \;\; \Delta\alpha)\begin{bmatrix} f_{11} & f_{12} & f_{13} \\ f_{12} & f_{22} & f_{23} \\ f_{13} & f_{23} & f_{33} \end{bmatrix}\begin{bmatrix} \Delta r_1 \\ \Delta r_2 \\ \Delta\alpha \end{bmatrix} = (\Delta r_1 \;\; \Delta r_2 \;\; \Delta\alpha)\begin{bmatrix} f_{11}\Delta r_1 + f_{12}\Delta r_2 + f_{13}\Delta\alpha \\ f_{12}\Delta r_1 + f_{22}\Delta r_2 + f_{23}\Delta\alpha \\ f_{13}\Delta r_1 + f_{23}\Delta r_2 + f_{33}\Delta\alpha \end{bmatrix}$$

then

$$2V = f_{11}(\Delta r_1)^2 + f_{22}(\Delta r_2)^2 + f_{33}(\Delta\alpha)^2 + 2f_{12}(\Delta r_1\Delta r_2) + 2f_{13}(\Delta r_1\Delta\alpha) + 2f_{23}(\Delta r_2\Delta\alpha) \tag{9.17.4}$$

- *The potential energy in terms of symmetry coordinates:*

$$2V = \sum_{i,k} F_{jl}S_jS_l \tag{9.17.5}$$

where F_{jl} are the force constants that are described by symmetry coordinates S_j and S_l.
This potential can be expressed in matrix notation:

$$2V = S'FS \tag{9.17.5}$$

where S is written as a column matrix (vector), and S' is the corresponding row matrix (its transpose).

- The relation between the internal coordinates and the symmetry coordinates in matrix form:

$$S = Us \tag{9.17.6}$$

where U matrix is

$$S_1 = \Delta\alpha$$
$$S_2 = \frac{1}{\sqrt{2}}(\Delta r_1 + \Delta r_2)$$
$$S_3 = \frac{1}{\sqrt{2}}(\Delta r_1 - \Delta r_2)$$

$$\overset{S}{\begin{bmatrix} \Delta\alpha \\ 1/\sqrt{2}(\Delta r_1 + \Delta r_2) \\ 1/\sqrt{2}(\Delta r_1 - \Delta r_2) \end{bmatrix}} = \overset{U}{\begin{bmatrix} 0 & 0 & 1 \\ 1/\sqrt{2} & 1/\sqrt{2} & 0 \\ 1/\sqrt{2} & -1/\sqrt{2} & 0 \end{bmatrix}} \overset{s}{\begin{bmatrix} \Delta r_1 \\ \Delta r_2 \\ \Delta\alpha \end{bmatrix}}$$

- Because U describe a linear orthogonal transformation, the inverse of U is U':

$$S = Us \qquad 9.17.6$$

$$s = U^{-1}S = U'S$$

$$s' = (U'S)' = US' = S'U$$

- if the potential energy in matrix form:

$$2V = s'fs \tag{9.17.3}$$

then

$$2V = (S'U)f(U'S)$$

$2V = S'(UfU')S = S'FS$ (potential energy in symmetry coordinates)

$$UfU' = F$$

In X_2Y

$$f = \begin{bmatrix} f_{11} & f_{12} & f_{13} \\ f_{12} & f_{22} & f_{23} \\ f_{13} & f_{23} & f_{33} \end{bmatrix} = \begin{bmatrix} f_{11} & f_{12} & f_{13} \\ f_{12} & f_{11} & f_{13} \\ f_{13} & f_{13} & f_{33} \end{bmatrix} \tag{9.17.7}$$

where $f_{11}=f_{22}$, the one for the bending XYX angle is f_{33}, and the cross interactions: $f_{13}=f_{31}=f_{23}=f_{32}$.

$$F = UfU' = \begin{bmatrix} 0 & 0 & 1 \\ 1/\sqrt{2} & 1/\sqrt{2} & 0 \\ 1/\sqrt{2} & -1/\sqrt{2} & 0 \end{bmatrix} \begin{bmatrix} f_{11} & f_{12} & f_{13} \\ f_{12} & f_{11} & f_{13} \\ f_{13} & f_{13} & f_{33} \end{bmatrix} \begin{bmatrix} 0 & 1/\sqrt{2} & 1/\sqrt{2} \\ 0 & 1/\sqrt{2} & -1/\sqrt{2} \\ 1 & 0 & 0 \end{bmatrix}$$

$$= \begin{bmatrix} 0 & 0 & 1 \\ 1/\sqrt{2} & 1/\sqrt{2} & 0 \\ 1/\sqrt{2} & -1/\sqrt{2} & 0 \end{bmatrix} \begin{bmatrix} f_{13} & 1/\sqrt{2}(f_{11}+f_{12}) & 1/\sqrt{2}(f_{11}-f_{12}) \\ f_{13} & 1/\sqrt{2}(f_{12}+f_{11}) & 1/\sqrt{2}(f_{12}-f_{11}) \\ f_{33} & 2/\sqrt{2}(f_{13}) & 0 \end{bmatrix} \tag{9.17.8}$$

$$= \begin{bmatrix} f_{33} & \sqrt{2}f_{13} & 0 \\ \sqrt{2}f_{13} & f_{11}+f_{12} & 0 \\ 0 & 0 & f_{11}-f_{12} \end{bmatrix} = \begin{bmatrix} F_{11} & F_{12} & F_{13} \\ F_{21} & F_{22} & F_{23} \\ F_{31} & F_{32} & F_{33} \end{bmatrix}$$

- The F matrix is symmetry-factored into a 2×2 block for the two A_1 vibrations, and one-dimensional block for the single B_1 vibration.

(d) *The kinetic energy in terms of internal coordinates and in symmetry coordinates*

- The kinetic energy in terms of the internal coordinates (Wilson's equation) is

$$2T = \sum_{ij} \left(g^{-1}\right)_{ij} \dot{s}_i \dot{s}_j \tag{9.17.9}$$

or in matrix notation:

$$2T = \dot{s}' g^{-1} \dot{s} \tag{9.17.9}$$

$$2T = \begin{bmatrix} s_1 & s_2 & \ldots \end{bmatrix} \begin{bmatrix} (g^{-1})_{11} & (g^{-1})_{12} & \cdots \\ (g^{-1})_{21} & (g^{-1})_{22} & \cdots \\ (g^{-1})_{31} & \cdots & \cdots \end{bmatrix} \begin{bmatrix} s_1 \\ s_2 \\ \cdots \end{bmatrix} \tag{9.17.10}$$

where g^{-1} stands for the inverse of g matrix that is given by

$$g_{jl} = \sum_{i=1}^{3N} \frac{1}{M_i} B_{jl} B_{kl} \tag{9.17.11}$$

or in matrix:

$$g = BM^{-1}B' \tag{9.17.12}$$

- M^{-1} is a diagonal matrix whose components are μ_i, where μ_i is the reciprocal of the mass of the i atom.

$$M = \begin{bmatrix} \mu_1 & 0 & \cdots & 0 \\ 0 & \mu_2 & & 0 \\ \vdots & & \ddots & \vdots \\ 0 & 0 & \cdots & \mu_n \end{bmatrix} \tag{9.17.13}$$

- A unit change in the Cartesian coordinate x_j causes a B_{ji} change in the internal coordinate S_j.
- The relationship between the Cartesian displacement coordinate x and internal coordinate s is

$$s_j = \sum_{i=1}^{3N} B_{ji} x_j \tag{9.17.14}$$

or

$$S = BX$$

$$\begin{bmatrix} s_1 \\ s_2 \\ \cdots \end{bmatrix} = \begin{bmatrix} B_{ax1} & B_{ay1} & \cdots & \cdots \\ B_{bx1} & B_{by1} & \cdots & \cdots \\ \vdots & & \ddots & \vdots \\ \cdots & \cdots & \cdots & \cdots \end{bmatrix} \begin{bmatrix} x_1 \\ y_1 \\ \cdots \end{bmatrix} \tag{9.17.15}$$

- In X_2Y,

$$s_1 = \Delta r_1, s_2 = \Delta r_2, s_3 = \Delta\alpha$$

- The displacements in terms of x and y coordinates are illustrated in Figs. 9.19 and 9.20.

$$\Delta r_1 = -\Delta x_1 \sin\frac{\alpha}{2} - \Delta y_1 \cos\frac{\alpha}{2} + \Delta x_3 \sin\frac{\alpha}{2} + \Delta y_3 \cos\frac{\alpha}{2}$$

Notice $\sin\frac{\alpha}{2} = \cos\left(90° - \frac{\alpha}{2}\right)$ and $\cos\frac{\alpha}{2} = \sin\left(90° - \frac{\alpha}{2}\right)$.

$$\Delta r_2 = \Delta x_2 \sin\frac{\alpha}{2} - \Delta y_2 \cos\frac{\alpha}{2} - \Delta x_3 \sin\frac{\alpha}{2} + \Delta y_3 \cos\frac{\alpha}{2}$$

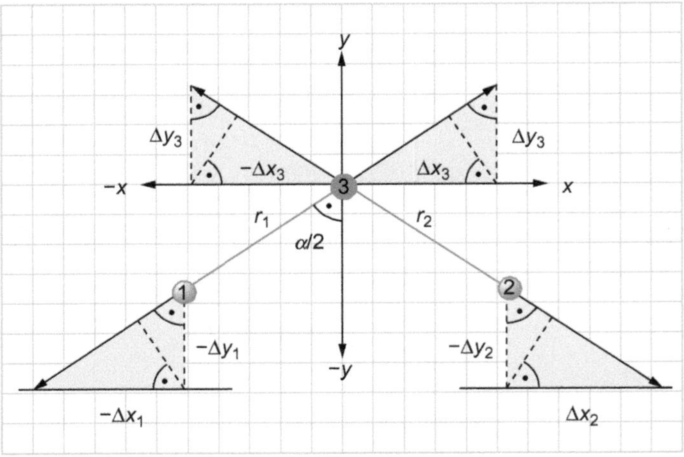

FIG. 9.19 Relationship between Cartesian and internal coordinates in the stretching vibration of a bent X_2Y molecule.

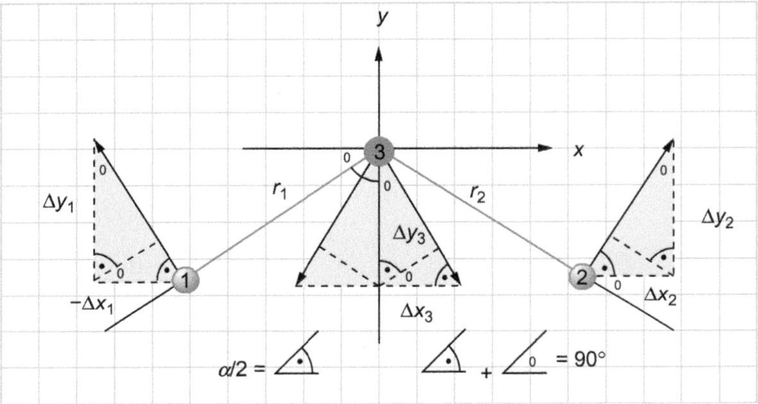

FIG. 9.20 Relationship between Cartesian and internal coordinates in the bending vibration of a bent X_2Y molecule.

$$r(\Delta\alpha) = -\Delta x_1 \cos\frac{\alpha}{2} + \Delta y_1 \sin\frac{\alpha}{2} + \Delta x_2 \cos\frac{\alpha}{2} + \Delta y_2 \sin\frac{\alpha}{2} - 2\Delta y_3 \sin\frac{\alpha}{2}$$

- When these equations are summarized in matrix form:

$$\begin{bmatrix} \Delta r_1 \\ \Delta r_2 \\ \Delta\alpha \end{bmatrix} = \begin{bmatrix} -\sin\frac{\alpha}{2} & -\cos\frac{\alpha}{2} & 0 & 0 & 0 & 0 & \sin\frac{\alpha}{2} & \cos\frac{\alpha}{2} & 0 \\ 0 & 0 & 0 & \sin\frac{\alpha}{2} & -\cos\frac{\alpha}{2} & 0 & -\sin\frac{\alpha}{2} & \cos\frac{\alpha}{2} & 0 \\ -\frac{1}{r}\cos\frac{\alpha}{0} & \frac{1}{r}\sin\frac{\alpha}{0} & 0 & \frac{1}{r}\cos\frac{\alpha}{2} & \frac{1}{r}\sin\frac{\alpha}{2} & 0 & 0 & -\frac{2}{r}\sin\frac{\alpha}{2} & 0 \end{bmatrix} \begin{bmatrix} \Delta x_1 \\ \Delta y_1 \\ \Delta z_1 \\ \Delta x_2 \\ \Delta y_2 \\ \Delta z_2 \\ \Delta x_3 \\ \Delta y_3 \\ \Delta z_3 \end{bmatrix} \qquad (9.17.16)$$

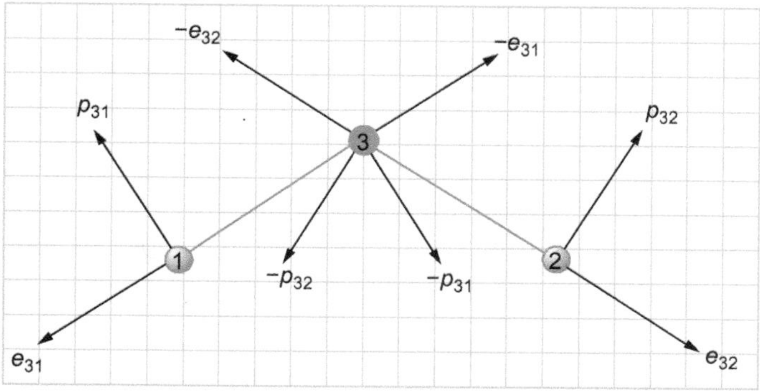

FIG. 9.21 Representation of the unit vectors in a bent YX_2 molecule.

- By using unit vector (Fig. 9.21) to compact this matrix, let:

$$e_{31} = \begin{bmatrix} \sin\frac{\alpha}{2} & \cos\frac{\alpha}{2} & 0 \end{bmatrix}$$

$$e_{32} = \begin{bmatrix} \sin\frac{\alpha}{2} & -\cos\frac{\alpha}{2} & 0 \end{bmatrix}$$

$$\frac{p_{31}}{r} = \begin{bmatrix} -\frac{1}{r}\cos\frac{\alpha}{2} & \frac{1}{r}\sin\frac{\alpha}{2} & 0 \end{bmatrix}$$

$$\frac{p_{32}}{r} = \begin{bmatrix} \frac{1}{r}\cos\frac{\alpha}{2} & \frac{1}{r}\sin\frac{\alpha}{2} & 0 \end{bmatrix}$$

$$\frac{-(p_{31}+p_{32})}{r} = \begin{bmatrix} 0 & -\frac{2}{r}\sin\frac{\alpha}{2} & 0 \end{bmatrix}$$

$$\rho_1 = \begin{bmatrix} \Delta x_1 \\ \Delta y_1 \\ \Delta z_1 \end{bmatrix}, \quad \rho_2 = \begin{bmatrix} \Delta x_2 \\ \Delta y_2 \\ \Delta z_2 \end{bmatrix}, \quad \text{and} \quad \rho_3 = \begin{bmatrix} \Delta x_3 \\ \Delta y_3 \\ \Delta z_3 \end{bmatrix}$$

where the displacement vectors of atoms 1, 2, and 3 are ρ_1, ρ_2 and ρ_3, respectively.
Then Eq. (9.17.16) changes to

$$\begin{bmatrix} \Delta r_1 \\ \Delta r_2 \\ \Delta \alpha \end{bmatrix} = \begin{bmatrix} e_{31} & 0 & -e_{31} \\ 0 & e_{32} & -e_{32} \\ \frac{p_{31}}{r} & \frac{p_{32}}{r} & \frac{-(p_{31}+p_{32})}{r} \end{bmatrix} \begin{bmatrix} \rho_1 \\ \rho_2 \\ \rho_3 \end{bmatrix} \tag{9.17.17}$$

- This equation is equivalent to

$$S = BX$$

in unit vectors, and in equation form,

$$\Delta r_1 = \Delta r_2 = e_{31} \cdot \rho_1 - e_{31} \cdot \rho_3$$

$$\Delta \alpha = \Delta \alpha_{132} = \frac{1}{r}\left[p_{31} \cdot \rho_1 + p_{32} \cdot \rho_2 - (p_{31}+p_{32}) \cdot \rho_3\right]$$

The dot represents the scalar product of the two vectors.

- If

$$\boldsymbol{g = BM^{-1}B'}$$

$$g = \begin{bmatrix} e_{31} & 0 & -e_{31} \\ 0 & e_{32} & -e_{32} \\ \frac{p_{31}}{r} & \frac{p_{32}}{r} & \frac{-(p_{31}+p_{32})}{r} \end{bmatrix} \begin{bmatrix} \mu_1 & 0 & 0 \\ 0 & \mu_1 & 0 \\ 0 & 0 & \mu_3 \end{bmatrix} \begin{bmatrix} e_{31} & 0 & \frac{p_{31}}{r} \\ 0 & e_{32} & \frac{p_{32}}{r} \\ -e_{31} & -e_{32} & \frac{-(p_{31}+p_{32})}{r} \end{bmatrix}$$

$$g = \begin{bmatrix} (\mu_3+\mu_1)e_{31}^2 & \mu_3 e_{31}.e_{32} & \frac{\mu_1}{r} e_{31}\cdot p_{31} + \frac{\mu_3}{r} e_{31} \quad (p_{31}+p_{32}) \\ 0 & (\mu_3+\mu_1)e_{31}^2 & \frac{\mu_1}{r} e_{32}\cdot p_{32} + \frac{\mu_3}{r} e_{32} \quad (p_{31}+p_{32}) \\ 0 & 0 & \frac{\mu_1}{r^2} p_{31}^2 + \frac{\mu_1}{r^2} p_{32}^2 + \frac{\mu_3}{r^2} \quad (p_{31}+p_{32})^2 \end{bmatrix} \quad (9.17.18)$$

- If

$$e_{31}\cdot e_{31} = \begin{bmatrix} \sin\frac{\alpha}{2} & \cos\frac{\alpha}{2} & 0 \end{bmatrix} \begin{bmatrix} \sin\frac{\alpha}{2} \\ \cos\frac{\alpha}{2} \\ 0 \end{bmatrix} = \sin^2\frac{\alpha}{2} + \cos^2\frac{\alpha}{2} = 1$$

Similarly:

$$e_{31}\cdot e_{31} = e_{32}\cdot e_{32} = p_{31}\cdot p_{31} = p_{32}\cdot p_{32} = 1$$

$$e_{31}\cdot p_{31} = \begin{bmatrix} \sin\frac{\alpha}{2} & \cos\frac{\alpha}{2} & 0 \end{bmatrix} \begin{bmatrix} -\frac{1}{r}\cos\frac{\alpha}{2} \\ \frac{1}{r}\sin\frac{\alpha}{2} \\ 0 \end{bmatrix} = 0$$

$$e_{31}\cdot p_{31} = e_{32}\cdot p_{32} = 0$$

$$e_{31}\cdot e_{32} = \cos\alpha$$

$$e_{31}\cdot p_{32} = e_{32}\cdot p_{31} = -\sin\alpha$$

$$(p_{31}+p_{32})^2 = 2(1-\cos\alpha)$$

- Therefore, g in terms of internal coordinates

$$g = \begin{bmatrix} \mu_3+\mu_1 & \mu_3\cos\alpha & -\frac{\mu_3}{r}\sin\alpha \\ \mu_3\cos\alpha & \mu_3+\mu_1 & -\frac{\mu_3}{r}\sin\alpha \\ -\frac{\mu_3}{r}\sin\alpha & -\frac{\mu_3}{r}\sin\alpha & \frac{2\mu_1}{r^2} + \frac{2\mu_3}{r^2}(1-\cos\alpha) \end{bmatrix} \quad (9.17.19)$$

- The g matrix in terms of symmetry coordinates set up is described by a procedure analogous to that used for the F matrix:

$$G = UgU'$$

$$G = UgU' = \begin{bmatrix} 0 & 0 & 1 \\ 1/\sqrt{2} & 1/\sqrt{2} & 0 \\ 1/\sqrt{2} & -1/\sqrt{2} & 0 \end{bmatrix} \begin{bmatrix} g_{11} & g_{12} & g_{13} \\ g_{12} & g_{11} & g_{13} \\ g_{13} & g_{13} & g_{33} \end{bmatrix} \begin{bmatrix} 0 & 1/\sqrt{2} & 1/\sqrt{2} \\ 0 & 1/\sqrt{2} & -1/\sqrt{2} \\ 1 & 0 & 0 \end{bmatrix}$$

$$= \begin{bmatrix} g_{33} & \sqrt{2}g_{13} & 0 \\ \sqrt{2}g_{13} & g_{11}+g_{12} & 0 \\ 0 & 0 & g_{11}-g_{12} \end{bmatrix} = \begin{bmatrix} G_{11} & G_{12} & G_{13} \\ G_{21} & G_{22} & G_{23} \\ G_{31} & G_{32} & G_{33} \end{bmatrix} \quad (9.17.20)$$

where

$$g_{11} = g_{22} = \mu_3 + \mu_1$$

$$g_{12} = \mu_3 \cos\alpha$$

$$g_{13} = -\frac{\mu_3}{r}\sin\alpha$$

$$g_{23} = -\frac{\mu_3}{r}\sin\alpha$$

$$g_{33} = \frac{2\mu_1}{r^2} + \frac{2\mu_3}{r^2}(1 - \cos\alpha)$$

- If the kinetic energy is expressed in matrix form:

$$2T = \dot{s}' g^{-1} \dot{s} \qquad (9.17.9)$$

then

$$2T = \left(\dot{S}' U\right) g^{-1} \left(U' \dot{S}\right)$$

$$2T = \dot{S}' \left(U g^{-1} U'\right) \dot{S} = \dot{S}' G^{-1} \dot{S} \text{ (kinetic energy in symmetric coordinates)}$$

$$U g^{-1} U' = G^{-1}$$

(e) *The potential and kinetic energies in Lagrange's equation*

- The kinetic and potential energies are described by internal coordinates:

$$2V = \sum_{j,l}^{3N} f_{jl} s_j s_l \qquad (9.17.2)$$

$$2T = \sum_{j,l}^{3N} g^{-1} \dot{s}_j \dot{s}_l \qquad (9.17.9)$$

- These equations are combined in Lagrange's equation:

$$\frac{\partial}{\partial t}\left(\frac{\partial T}{\partial \dot{S}_j}\right) + \frac{\partial V}{\partial S_j} = 0 \qquad (9.12.3)$$

$$\frac{\partial T}{\partial \dot{S}_j} = \sum_{j,l}^{3N} g^{-1} \dot{s}_j \qquad (9.17.21)$$

$$\frac{\partial}{\partial t}\left(\frac{\partial T}{\partial \dot{S}_j}\right) = \sum_{j,l}^{3N} g^{-1} \ddot{s}_j \qquad (9.17.22)$$

$$\frac{\partial V}{\partial S_j} = \sum_{j,l}^{3N} F_{jl} s_j \qquad (9.17.23)$$

- Substituting in Lagrange's equation:

$$\sum_{j,l}^{3N} g^{-1} \ddot{s}_j + \sum_{j,l}^{3N} f_{jl} s_j = 0$$

- These $3N$ equations have the general solution Eq. (9.13.7):

$$s_j = L_j \sin\left(\sqrt{\lambda}t + \alpha\right)$$
$$\dot{s}_j = \sqrt{\lambda}L_j \cos\left(\sqrt{\lambda}t + \alpha\right)$$
$$\ddot{s}_j = -\lambda L_j \sin\left(\sqrt{\lambda}t + \alpha\right)$$

where $\lambda = (2\pi\nu)^2$, and L_j is the maximum amplitude of the internal coordinate distortion.

- We substitute in

$$\sum_{j,l}^{3N} g^{-1}\ddot{s}_j + \sum_{j,l}^{3N} f_{jl}s_j = 0$$

$$\sum_{j,l}^{3N} f_{jl}L_j \sin\left(\sqrt{\lambda}t + \alpha\right) - \sum_{j,l}^{3N} g^{-1}\lambda L_j \sin\left(\sqrt{\lambda}t + \alpha\right) = 0$$

$$\sum_{j,l}^{3N} (f_{jl} - g^{-1}\lambda)L_j = 0 \tag{9.17.24}$$

- When this equation is written out,

$$(f_{11} - (g^{-1})_{11}\lambda)L_1 + (f_{12} - (g^{-1})_{12}\lambda)L_2 + \cdots + \left(f_{1,3N} - (g^{-1})_{1,3N}\lambda\right)L_{3N} = 0$$
$$(f_{21} - (g^{-1})_{21}\lambda)L_1 + (f_{22} - (g^{-1})_{22}\lambda)L_2 + \cdots + \left(f_{2,3N} - (g^{-1})_{2,3N}\lambda\right)L_{3N} = 0$$
$$\cdots + \cdots + \cdots + \cdots = 0$$
$$\left(f_{3N,1} - (g^{-1})_{3N,1}\lambda\right)L_1 + \left(f_{3N,2} - (g^{-1})_{3N,2}\lambda\right)L_2 + \cdots + \left(f_{3N,3N} - (g^{-1})_{3N,3N}\lambda\right)L_{3N} = 0$$

- the secular determinant is

$$\begin{vmatrix} (f_{11} - (g^{-1})_{11}\lambda) & (f_{12} - (g^{-1})_{12}\lambda) & \cdots \\ (f_{21} - (g^{-1})_{21}\lambda) & (f_{22} - (g^{-1})_{22}\lambda) & \cdots \\ \cdots & \cdots & \cdots \end{vmatrix} = 0 \tag{9.17.25}$$

- Matrix form of this determinant:

$$|f - g^{-1}\lambda| = 0$$

Multiplying by $|g|$

$$|g||f - g^{-1}\lambda| = |gf - gg^{-1}\lambda| = 0$$
$$|gf - I\lambda| = 0 \tag{9.17.26}$$

where the matrix times its inverse equals the identity matrix I (= 1).

$$\left|\begin{bmatrix} f_{11} & f_{12} & \cdots \\ f_{21} & f_{22} & \cdots \\ \cdots & \cdots & \cdots \end{bmatrix}\begin{bmatrix} g_{11} & g_{12} & \cdots \\ g_{21} & g_{22} & \cdots \\ \cdots & \cdots & \cdots \end{bmatrix} - \begin{bmatrix} \lambda & 0 & \cdots \\ 0 & \lambda & \cdots \\ \cdots & \cdots & \cdots \end{bmatrix}\right| = 0$$

$$\begin{vmatrix} \sum_i f_{i1}g_{1i} - \lambda & \sum_j f_{j2}g_{1j} & \cdots \\ \sum_i f_{i2}g_{2i} & \sum_j (f_{j2}\,g_{2j} - \lambda) & \cdots \\ \cdots & \cdots & \cdots \end{vmatrix} = 0 \tag{9.17.27}$$

Lagrange's Equation in Terms of Symmetry Coordinates

- If the molecule has some symmetry, the solution of the secular determinant can be simplified by the use of the symmetry coordinates.
- Symmetry coordinates are linear combination of the internal coordinates.
- The symmetry coordinates must be normal, orthogonal, and must transform by

$$S = Us$$

$$F = UfU', \text{ and}$$

$$G = UgU'$$

where

$$U = \begin{bmatrix} 0 & 0 & 1 \\ 1/\sqrt{2} & 1/\sqrt{2} & 0 \\ 1/\sqrt{2} & -1/\sqrt{2} & 0 \end{bmatrix}, \text{ and}$$

$$I = UU' = \begin{bmatrix} 0 & 0 & 1 \\ 1/\sqrt{2} & 1/\sqrt{2} & 0 \\ 1/\sqrt{2} & -1/\sqrt{2} & 0 \end{bmatrix} \begin{bmatrix} 0 & 1/\sqrt{2} & 1/\sqrt{2} \\ 0 & 1/\sqrt{2} & -1/\sqrt{2} \\ 1 & 0 & 0 \end{bmatrix} \begin{bmatrix} 1 & 0 & 0 \\ 0 & 1 & 0 \\ 0 & 0 & 1 \end{bmatrix}$$

- Both gf and GF give the same roots, since

$$\begin{aligned} |GF - I\lambda| &= |UgU'UfU' - U\lambda U'| \\ &= |UgfU' - U\lambda U'| \\ &= |U||gf - \lambda||U'| \end{aligned}$$

$$|gf - \lambda| = 0 \qquad (9.17.26)$$

$$|GF - I\lambda| = 0 \qquad (9.17.28)$$

Some of the *G* Matrix Elements Have the Following General Formulas:

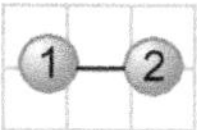

$$g^2_{rr} = \mu_1 + \mu_2$$

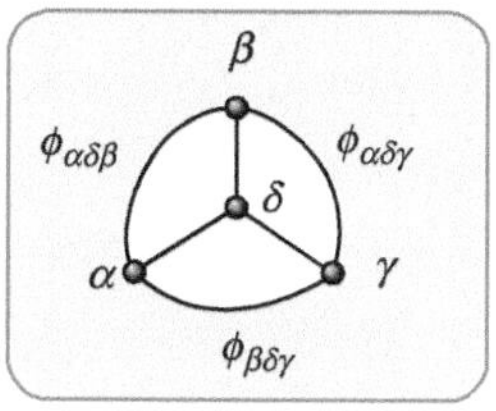

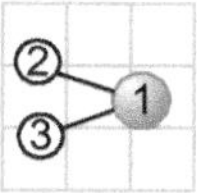

$$g^1_{rr} = \mu_1 \cos\phi$$

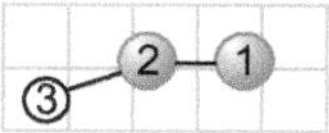

$$g^2_{r\phi} = -\frac{\mu_2}{r_{23}} \sin\phi$$

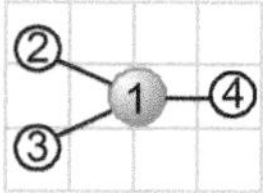

$$g^1_{r\phi}\begin{pmatrix}1\\1\end{pmatrix} = -\mu_1 \left(\frac{1}{r_{13}} \sin\phi_{213} \cos\psi_{234} + \frac{1}{r_{14}} \sin\phi_{214} \cos\psi_{243}\right)$$

$$g^3_{\phi\phi} = \frac{\mu_1}{r_{12}^2} + \frac{\mu_3}{r_{23}^2} + \left(\frac{1}{r_{12}^2} + \frac{1}{r_{23}^2} - \frac{2}{r_{12}r_{23}} \cos\phi\right)\mu_2$$

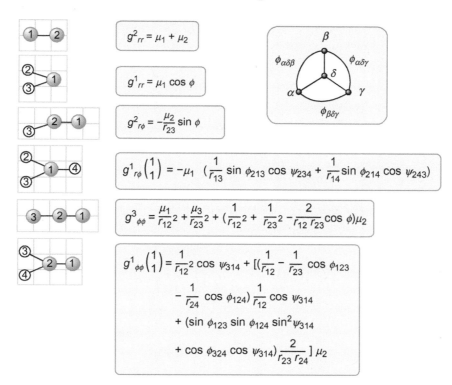

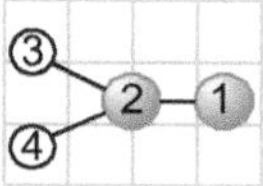

$$\begin{aligned} g^1_{\phi\phi}\begin{pmatrix}1\\1\end{pmatrix} = \frac{1}{r_{12}^2} \cos\psi_{314} + \Big[\Big(&\frac{1}{r_{12}} - \frac{1}{r_{23}} \cos\phi_{123} \\ &- \frac{1}{r_{24}} \cos\phi_{124}\Big) \frac{1}{r_{12}} \cos\psi_{314} \\ &+ (\sin\phi_{123} \sin\phi_{124} \sin^2\psi_{314} \\ &+ \cos\phi_{324} \cos\psi_{314}) \frac{2}{r_{23} r_{24}}\Big] \mu_2 \end{aligned}$$

Where the atoms in gray color are those common to both coordinates, the superscript indicates the number of common atoms, μ is the reciprocal of mass, and the spherical angle $\psi_{\alpha\beta\gamma}$ is defined as:

$$\cos\psi_{\alpha\beta\gamma}=\frac{\cos\phi_{\alpha\delta\gamma}-\cos\phi_{\alpha\delta\beta}\cos\phi_{\beta\delta\gamma}}{\sin\phi_{\alpha\delta\beta}\sin\phi_{\beta\delta\gamma}}$$

What are the formulas of the g matrix elements of the following?

(i) **a bent YX_2 molecule**
(ii) **a pyramidal YX_3 molecule**

- The g in terms of internal coordinates of a bent YX_2 molecule:

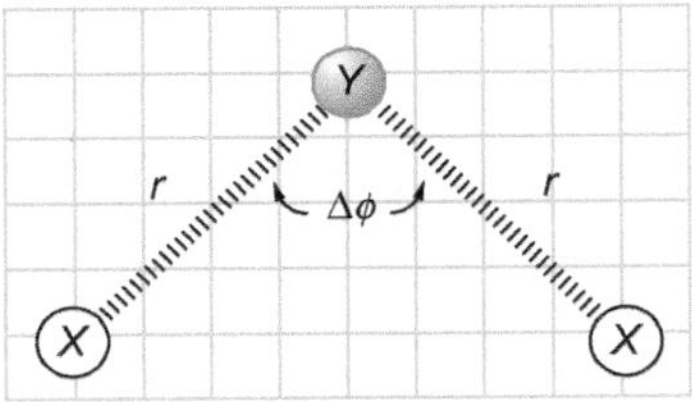

$$g=\begin{bmatrix} g_{11} & g_{12} & g_{13} \\ g_{12} & g_{11} & g_{13} \\ g_{13} & g_{13} & g_{33} \end{bmatrix}$$

$$g=\begin{bmatrix} g_{rr}^2 & g_{rr}^1 & g_{r\phi}^2 \\ g_{rr}^1 & g_{rr} & g_{r\phi}^2 \\ g_{r\phi}^2 & g_{r\phi}^2 & g_{\phi\phi}^2 \end{bmatrix}=\begin{bmatrix} \mu_Y+\mu_X & \mu_Y\cos\phi & -\frac{\mu_Y}{r}\sin\phi \\ \mu_Y\cos\phi & \mu_Y+\mu_X & -\frac{\mu_Y}{r}\sin\phi \\ -\frac{\mu_Y}{r}\sin\phi & -\frac{\mu_Y}{r}\sin\phi & \frac{2\mu_X}{r^2}+\frac{2\mu_Y}{r^2}(1-\cos\phi) \end{bmatrix}$$

- The g in terms of internal coordinates of a bent YX_2 molecule:

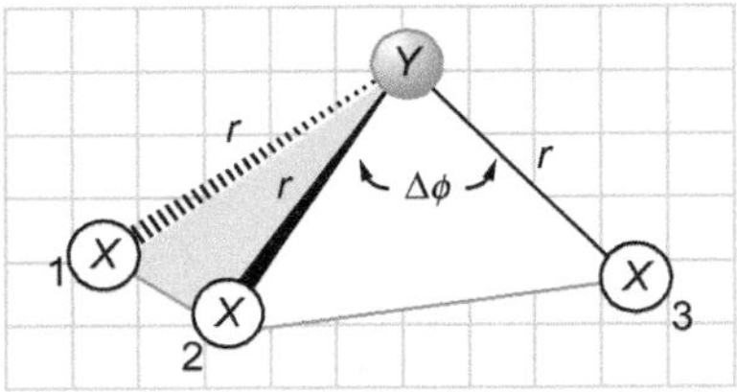

$$g=\begin{array}{c|cccccc|} & \Delta r_1 & \Delta r_2 & \Delta r_3 & \Delta\phi_{23} & \Delta\phi_{31} & \Delta\phi_{12} \\ \Delta r_1 & g_{rr}^2 & g_{rr}^1 & g_{rr}^1 & g_{r\phi}^1\begin{pmatrix}1\\1\end{pmatrix} & g_{r\phi}^2 & g_{r\phi}^2 \\ \Delta r_2 & & g_{rr}^2 & g_{rr}^1 & g_{r\phi}^2 & g_{r\phi}^1\begin{pmatrix}1\\1\end{pmatrix} & g_{r\phi}^2 \\ \Delta r_3 & & & g_{rr}^2 & g_{r\phi}^2 & g_{r\phi}^2 & g_{r\phi}^1\begin{pmatrix}1\\1\end{pmatrix} \\ \Delta\phi_{23} & & & & g_{\phi\phi}^3 & g_{\phi\phi}^2\begin{pmatrix}1\\1\end{pmatrix} & g_{\phi\phi}^2\begin{pmatrix}1\\1\end{pmatrix} \\ \Delta\phi_{31} & & & & & g_{\phi\phi}^3 & g_{\phi\phi}^2\begin{pmatrix}1\\1\end{pmatrix} \\ \Delta\phi_{12} & & & & & & g_{\phi\phi}^3 \end{array}$$

$$g_{rr}^{2} = \mu_X + \mu_Y$$

$$g_{rr}^{1} = \mu_X \cos\phi$$

$$g_{r\phi}^{1}\begin{pmatrix}1\\1\end{pmatrix} = -\frac{2}{r}\frac{\mu_X(1-\cos\phi)\cos\phi}{\sin\phi}$$

$$g_{r\phi}^{2} = -\frac{\mu_X}{r}\cos\alpha$$

$$g_{\phi\phi}^{3} = \frac{2}{r^2}\left[\mu_Y + \mu_X(1-\cos\phi)\right]$$

$$g_{\phi\phi}^{2}\begin{pmatrix}1\\1\end{pmatrix} = \frac{\mu_Y}{r^2}\frac{\cos\phi}{1+\cos\phi} + \frac{\mu_X}{r^2}\frac{(1+3\cos\phi)(1-\cos\phi)}{1+\cos\phi}$$

Estimate the vibrational frequencies for H_2O, assuming the generalized valence force are selected as: $f_{11} = 8.4280$, $f_{12} = -0.1050$, $f_{13} = 0.2625$, $f_{33} = 0.7680$.

- From above:

$$G = \begin{bmatrix} G_{11} & G_{12} & G_{13} \\ G_{21} & G_{22} & G_{23} \\ G_{31} & G_{32} & G_{33} \end{bmatrix} = \begin{bmatrix} g_{33} & \sqrt{2}g_{13} & 0 \\ \sqrt{2}g_{13} & g_{11}+g_{12} & 0 \\ 0 & 0 & g_{11}-g_{12} \end{bmatrix} \tag{9.17.20}$$

  where

$$g_{11} = \mu_3 + \mu_1$$

$$g_{12} = \mu_3 \cos\alpha$$

$$g_{13} = -\frac{\mu_3}{r}\sin\alpha$$

$$g_{23} = -\frac{\mu_3}{r}\sin\alpha$$

$$g_{33} = \frac{2\mu_1}{r^2} + \frac{2\mu_3}{r^2}(1-\cos\alpha)$$

- For H_2O molecule:

$$\mu_1 = \mu_H = \frac{1}{1.008} = 0.99206$$

$$\mu_3 = \mu_O = \frac{1}{15.995} = 0.06252$$

$$r = 0.96\text{Å}, \alpha = 105°$$

$$\cos\alpha = \cos 105° = -0.25882$$

$$G = \begin{bmatrix} 2.32370 & -0.08896 & 0 \\ -0.08896 & 1.03840 & 0 \\ 0 & 0 & 1.07062 \end{bmatrix}$$

- The generalized valence force are selected as

$$f_{11} = 8.4280, f_{12} = -0.1050, f_{13} = 0.2625, f_{33} = 0.7680$$

- Substituting in F matrix from:

$$F = \begin{bmatrix} F_{11} & F_{12} & F_{13} \\ F_{21} & F_{22} & F_{23} \\ F_{31} & F_{32} & F_{33} \end{bmatrix} = \begin{bmatrix} f_{33} & \sqrt{2}f_{13} & 0 \\ \sqrt{2}f_{13} & f_{11}+f_{12} & 0 \\ 0 & 0 & f_{11}-f_{12} \end{bmatrix} \quad (9.17.8)$$

$$F = \begin{bmatrix} f_{33} & \sqrt{2}f_{13} & 0 \\ \sqrt{2}f_{13} & f_{11}+f_{12} & 0 \\ 0 & 0 & f_{11}-f_{12} \end{bmatrix} = \begin{bmatrix} 0.70779 & 0.35638 & 0 \\ 0.35638 & 8.32300 & 0 \\ 0 & 0 & 8.5330 \end{bmatrix}$$

- From Lagrange's equation:

$$|GF - I\lambda| = 0 \quad (9.17.29)$$

$$\begin{bmatrix} F_{11} & F_{12} & F_{13} \\ F_{21} & F_{22} & F_{23} \\ F_{31} & F_{32} & F_{33} \end{bmatrix} \begin{bmatrix} G_{11} & G_{12} & G_{13} \\ G_{21} & G_{22} & G_{23} \\ G_{31} & G_{32} & G_{33} \end{bmatrix} - \begin{bmatrix} \lambda & 0 & 0 \\ 0 & \lambda & 0 \\ 0 & 0 & \lambda \end{bmatrix} = 0$$

$$\begin{bmatrix} 0.70779 & 0.35638 & 0 \\ 0.35638 & 8.32300 & 0 \\ 0 & 0 & 8.5330 \end{bmatrix} \begin{bmatrix} 2.32370 & -0.08896 & 0 \\ -0.08896 & 1.03840 & 0 \\ 0 & 0 & 1.07062 \end{bmatrix} - \begin{bmatrix} \lambda & 0 & 0 \\ 0 & \lambda & 0 \\ 0 & 0 & \lambda \end{bmatrix} = 0$$

- This secular equation can be factored into one of second order for the two A_1 vibrations and one for the B_1 vibration:

$$A_1: \begin{bmatrix} 0.70779 & 0.35638 \\ 0.35638 & 8.32300 \end{bmatrix} \begin{bmatrix} 2.32370 & -0.08896 \\ -0.08896 & 1.03840 \end{bmatrix} - \begin{bmatrix} \lambda & 0 \\ 0 & \lambda \end{bmatrix} = 0$$

$$B_1: [8.5330][1.07062] = \lambda$$

$$\lambda^2 - 10.22389\lambda + 13.86234 = 0$$

- If

$$\lambda = 4\pi^2 c^2 \bar{v}^2$$

$$\bar{v} = \frac{\sqrt{\lambda}}{2\pi c}$$

If the unit of mass is atomic weight, and unit of force constant is mdyn/Å (10^5 dyn/cm), then

$$\bar{v} = 1302.83\sqrt{\lambda}$$

$$A_1: \lambda_1 = 8.61475\left(\bar{v}_1 = 3824\,\text{cm}^{-1}\right), \lambda_2 = 1.60914\left(\bar{v}_2 = 1653\,\text{cm}^{-1}\right)$$

$$B_1: \lambda_3 = 9.13681\left(\bar{v}_3 = 3938\,\text{cm}^{-1}\right)$$

9.18 SELECTION RULES

IR-Selection Rules

What is meant by:

(a) **forbidden transition;**
(b) **allowed transitions;**
(c) **infrared active; and**
(d) **Raman active?**

- Forbidden transition: transition predicted to have zero intensity.
- Allowed transition: transition predicted to have nonzero intensity.
- Infrared active: transition predicted to be observed in the infrared spectrum.
- Raman active: transition predicted to be observed in the Raman spectrum.

What are the requirements for the allowed vibrating excitations?

- A vibration will be infrared active if the molecular dipole moment changes during vibrational oscillation.
- The infrared selection rule is predicted by following the transition moment integral for the one dimensional harmonic oscillator, $M_{\upsilon\upsilon'}$, for the transition $\upsilon \rightarrow \upsilon'$:

$$M_{\upsilon\upsilon'} = \int_0^\infty \psi^*(\upsilon')\mu\psi(\upsilon)dx$$

$$\mu = ex$$

where μ is the dipole moment, $\psi(\upsilon)$ is the wave function of the molecule at a certain mode of vibration, and υ is the vibrational quantum number.

- The probability of the transition occurring is proportional to the square of $M_{\upsilon\upsilon'}$

$$\text{Transition probability} \propto [M_{\upsilon\upsilon'}]^2$$

$$[M_{\upsilon\upsilon'}]^2 = \left[\int_0^\infty \psi^*(\upsilon')\mu\psi(\upsilon)dx\right]^2$$

- If μ were constant, independent of vibration, μ could be factored out of the integral:

$$M_{\upsilon\upsilon'} = \mu \int_0^\infty \psi^*(\upsilon')\psi(\upsilon)dx = 0$$

because $\psi(\upsilon') \neq \psi(\upsilon)$.
A transition is allowed if $M_{\upsilon\upsilon'} \neq 0$.
If the transition is to be allowed, μ must be a function of x and must change during the vibration.

- For a polyatomic molecule,

$$\mu_x = \sum_i e_i x_i$$

where e_i is the charge on the ith atom, and x_i is its x-coordinate.
There will be three Cartesian components of the dipole moment for any molecule.

$$\hat{\mu} = \sum_i (e_i x_i + e_i y_i + e_i z_i)$$

- Therefore, the transition will be allowed, if any one of these three components changes during the vibration, which transforms as x, y, and z.
- A vibration will be infrared active if it belongs to the same symmetry species as a component of dipole moment, i.e., to the same species as either x, y, or z.
Example:
A''_2 of BF_3 $\left(\Gamma_{\text{vib.}} = A_1' + 2E' + A_2''\right)$ (Chapter 9, p. 554)

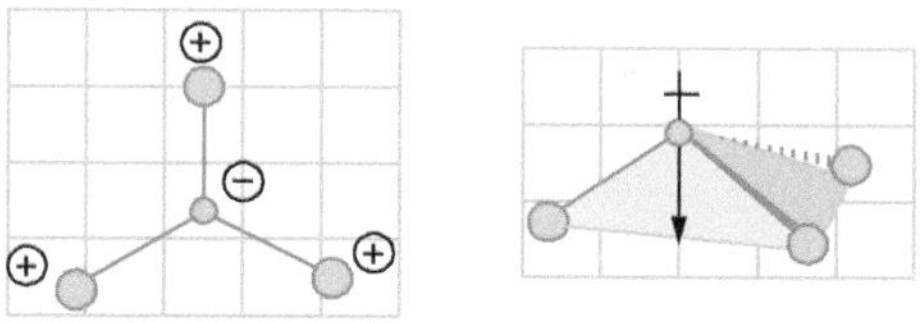

Distortion from D_{3h} ⟶ $\mu \neq 0$

  - A dipole moment is expressed as any one of, or any combination of dipole components in an x-, y-, or z-direction: must have an x-, y-, or z-component.
  - The mode is IR-active if representation transforms as any x, y, or z in D_{3h}: x, y: E', z: A_2'', and none: A_2'.
  - Therefore, for BF_3; ν_2, ν_3, and ν_4 are IR-active, see three bands (ν_{3a}, ν_{3b} at the same frequency).

TABLE 9.10 Character Table of D_{3h} Point Group and the Direct Products $e' \times a_2''$ and $e' \times e'$

D_{3h}	E	$2C_3$	$3C_2$	σ_h	$2S_3$	$3\sigma_v$		
A_1'	1	1	1	1	1	1		x^2+y^2, z^2
A_2'	1	1	−1	1	1	−1	Rz	–
E'	2	−1	0	2	−1	0	(x,y)	x^2-y^2, xy
A_1''	1	1	1	−1	−1	−1	–	–
A_2''	1	1	−1	−1	−1	1	z	–
E''	2	−1	0	−2	1	0	(R_x, R_y)	(xz, yz)
$e' \times a_2''$	2	−1	0	−2	1	0		
$e' \times e'$	4	1	0	4	1	0		

- Or if the direct product

$$\Gamma[\psi(v')] \times \Gamma[\hat{\mu}] \times \Gamma[\psi(v)]$$

contains the totally symmetric irreducible representation, A_1', of the point group, the transition $\upsilon \rightarrow \upsilon'$ is infrared active.

For a molecule belonging to a certain point group, the functions ϕ_a, ϕ_b, and ϕ_c will each form a basis for the irreducible representation of the group. When will the integral $\int_0^\infty \phi_a \phi_b d\tau$ be a nonzero or zero value?

- If ϕ_a forms a basis for irreducible representation Γ_a, and ϕ_b forms a basis for Γ_b,

$$\int_0^\infty \phi_a \phi_b d\tau \neq \text{zero}$$

only if the direct product $\Gamma_a \times \Gamma_b$ contains the totally symmetric irreducible representation of the point group.
Examples:
In the point group D_{3h}:

$$\int_0^\infty \phi_{e'} \phi_{a_2''} d\tau = \text{zero (odd function)}$$

because the direct product (Table 9.10)

$$e' \times a_2'' = e'' \neq a_1' \text{ (or contains } a_1'\text{)}$$

$$\int_0^\infty \phi_{e'} \phi_{e'} d\tau \neq \text{zero (even function)}$$

because the direct product

$$e' \times e' = a_1' + a_2' + e' \text{ (contains } a_1'\text{)}$$

- Unlike d-d, other electronic transitions, the ground, and the excited sates have the same symmetry.

What is meant by totally symmetric irreducible representation? Give an example.

- It is a particular case of the irreducible representation that is totally symmetric regardless the symmetry operation.

- Example:

D_{3h}	E	2C	$3C_2$	σ_h	$2S_3$	$3\sigma_v$	
A_1'	1	1	1	1	1	1	x^2+y^2, z^2

O_h	E	$8C_3$	$6C_2$	$6C_4$	$3C_2$	i	$6S_4$	$8S_6$	$3\sigma_h$	$6\sigma_d$		
A_{1g}	1	1	1	1	1	1	1	1	1	1	–	$x^2+y^2+z^2$

$D_{\infty h}$	E	$2C_\infty^\phi$	....	$\infty\sigma_i$	i	$2S_\infty^\phi$	.....	∞C_2
Σ_g^+	1	1	.....	1	1	1	......	1

$C_{\infty v}$	E	$2C_\infty^\phi$	$2C_\infty^{2\phi}$	$2C_\infty^{3\phi}$	.....	$\infty\sigma_v$
Σ^+	1	1	1	1	......	1

Raman Selection Rules

What are the selection rules for Raman spectroscopy?

- Excitation of vibration is associated with change in the polarizability of a molecule. Photons are not absorbed, but "stick" briefly to the molecule, and are scattered in all directions. They are typically observed at 90° to incident beam, and the induced dipole in one (x-) direction changes polarizability in x-, y-, or z-direction.
- A transition is Raman active if the polarizability, α, of the molecule changes during the vibration:

$$\mu_{\text{ind.}} = \alpha\varepsilon \tag{9.2.2}$$

$$\mu_{\text{ind.}} = \alpha_o\varepsilon_o\cos 2\pi\nu t + \frac{1}{2}\Delta\alpha\varepsilon_o[\cos 2\pi(\nu+\nu_o)t + \cos 2\pi(\nu-\nu_o)t] \neq 0L \tag{9.3.4}$$

- The electrical field in the x direction not only induces a dipole in x-direction, but also in y- and z-direction.

$$\mu_{\text{ind.}}(x) = \alpha_{xx}\varepsilon_x + \alpha_{xy}\varepsilon_y + \alpha_{xz}\varepsilon_z$$

$$\mu_{\text{ind.}}(y) = \alpha_{yx}\varepsilon_x + \alpha_{yy}\varepsilon_y + \alpha_{yz}\varepsilon_z$$

$$\mu_{\text{ind}}(z) = \alpha_{zx}\varepsilon_x + \alpha_{zy}\varepsilon_y + \alpha_{zz}\varepsilon_z$$

$$\begin{bmatrix}\mu_{\text{ind.}}(x)\\ \mu_{\text{ind.}}(y)\\ \mu_{\text{ind.}}(z)\end{bmatrix} = \underbrace{\begin{bmatrix}\alpha_{xx} & \alpha_{xy} & \alpha_{xz}\\ \alpha_{yx} & \alpha_{yy} & \alpha_{yz}\\ \alpha_{zx} & \alpha_{zx} & \alpha_{zx}\end{bmatrix}}_{\text{Polarizability tensor}} \begin{bmatrix}\varepsilon_x\\ \varepsilon_y\\ \varepsilon_z\end{bmatrix}$$

The tensor is symmetric, with $\alpha_{yx} = \alpha_{xy}$, $\alpha_{xz} = \alpha_{zx}$, and $\alpha_{yz} = \alpha_{zy}$.

$$\text{Transition probability} \propto [M_{vv'}]^2$$

$$[M_{vv'}]^2 = \left[\int_0^\infty \psi^*(v')\hat{\alpha}\psi(v)dx\right]^2$$

$$\hat{\alpha} = \begin{bmatrix}\alpha_{xx} & \alpha_{xy} & \alpha_{xz}\\ \alpha_{yx} & \alpha_{yy} & \alpha_{yz}\\ \alpha_{zx} & \alpha_{zx} & \alpha_{zx}\end{bmatrix}$$

- As a result, a transition will be Raman active if any of the six different components (α_{xx}, α_{yy}, α_{zz}, α_{xy}, α_{xz}, α_{yz}) of the polarizability tensor changes during the vibration.
- These six components transform as a binary product x^2, y^2, z^2, xy, xz, yz, or to combination of products such as $x^2 - y^2$.
- A vibration will be Raman active if it belongs to the same symmetry species as component of polarizability, i.e., to one of the binary products, x^2, y^2, z^2, xy, xz, yz, or to combination of products such as $x^2 - y^2$.
- E.g., in D_{3h}, $z = A_2''$

  Therefore, $z^2 = A_2''.\ A_2'' = A_1'$ (as in Table 9.10)

so A_1' modes are Raman-active.

BF_3:			R	IR
ν_1,	A_1'	z^2	+	−
ν_2,	A_2''	z	−	+
ν_3,	E'	x^2-y^2, xy, x, y	+	+
ν_4,	E'	x^2-E^2, xy, x, y	+	+

Two of the four bands are coincident: the E' bands occur in both spectra at the same ν.
Note: in molecules with a center of symmetry, there are no coincidences (mutual exclusion rule).

- The transition $\upsilon \rightarrow \upsilon'$ is Raman active, if the direct product

$$\Gamma[\psi(v')] \times \Gamma[\hat{\alpha}] \times \Gamma[\psi(v)]$$

contains the totally symmetric irreducible representation of the point group.

What is the symmetry of the following four vibrational wave functions of the harmonic oscillator (Chapter 9, p. 528)?

$$\psi_0 = \left(\frac{\alpha}{\pi}\right)^{1/4} e^{-\frac{\alpha q^2}{2}}$$

$$\psi_1 = \left(\frac{4\alpha^3}{\pi}\right)^{1/4} q e^{-\frac{\alpha q^2}{2}}$$

$$\psi_2 = \left(\frac{\alpha}{4\pi}\right)^{1/4} (2\alpha q^2 - 1) e^{-\frac{\alpha q^2}{2}}$$

$$\psi_3 = \left(\frac{\alpha^3}{9\pi}\right)^{1/4} (2\alpha q^3 - 3q) e^{-\frac{\alpha q^2}{2}}$$

- In a polyatomic molecule, each of the normal coordinates, q, forms a basis for a irreducible representation of the point group of the molecule.
- If q forms a basis of nondegenerate irreducible representation:
  - all the symmetry operations of the point group take q to positive or negative itself;
  - all q^2 will always remain plus itself;
  - the function ψ_0 will stay unaffected under any symmetry operation since it depends on q^2; and
  - the wave function ψ_0 transforms as the totally symmetric irreducible representation.
- If q forms a basis of nondegenerate irreducible representation, e.g., q_a and q_b are doubly degenerate irreducible representations.
  - Any symmetry operation, R, will take q_a into $\pm q_a'$ or some linear combination of $c_1 q_a + c_2 q_b$

$$Rq_a = q_a' = c_1 q_a + c_2 q_b$$

  c_1 and c_2 are constants, since q_a and q_b are normalized vibrational functions: $q_a^2 = 1$, and $q_b^2 = 1$ Then the coefficients c_1 and c_2 must be such that $q_a'2 = 1$.
  - Given that the wave function ψ_0 depends on the square of vibration coordinate, the operation has no net effect on ψ_0, even if q_a is degenerate.
- For the wave function ψ_1, we need to explore the symmetry product of $qe^{-\frac{\alpha q^2}{2}}$.
  - The exponential part $e^{-\frac{\alpha q^2}{2}}$ is totally symmetric regardless the symmetry of q.
  - The q has the symmetry of the vibration.
  - Therefore, the product $qe^{-\frac{\alpha q^2}{2}}$ has the symmetry of the vibration.
  - The wave function ψ_1 transforms as q, $\Gamma(\psi_1) = \Gamma(q)$
- The wave function ψ_2 has two components: $(2\alpha q^2 - 1)$ and $e^{-\frac{\alpha q^2}{2}}$
  - The exponential $e^{-\frac{\alpha q^2}{2}}$ transforms as q^2.
  - q^2 transforms as unity.
  - Therefore, ψ_2 is totally symmetric.

TABLE 9.11 Vibrational Analysis of N_2

$D_{\infty h}$	E	$2C_{\infty}^{\phi}$	$\infty\sigma_i$	i	$2S_{\infty}^{\phi}$	∞C_2		
Σ_g^+	1	1	1	1	1	1		x^2+y^2, z^2
Σ_g^-	1	1	−1	1	1	−1	Rz	
Π_g	2	$2c\varphi$	0	2	$-2c\varphi$	0	(R_x, R_y)	(xz, yz)
Δ_g	2	$2c2\varphi$	0	2	$2c2\varphi$	0		(x^2-y^2, xy)
Σ_u^+	1	1	1	−1	−1	−1	z	z^3
Σ_u^-	1	1	−1	−1	−1	1		
Π_u	2	$2c\varphi$	0	−2	$2c\varphi$	0	(x, y)	(xz^2, yz^2)
Δ_u	2	$2c2\varphi$	0	−2	$-2c2\varphi$	0		$(xyz, z(x^2-y^2))$
$\Gamma_{x, y, z}$	3	$1+2c\varphi$	1	−3	$-1+2c\varphi$	−1		
Unmoved atoms	2	2	2	0	0	0		
Γ_{total}	6	$2+4c\varphi$	2	0	0	0		
Γ_{trans}	3	$1+2c\varphi$	1	−3	$-1+2c\varphi$	−1		
$\Gamma_{rot.}$	2	$2c\varphi$	0	2	$-2c\varphi$	0		
$\Gamma_{vib.}=\Gamma_{total}-\Gamma_{trans}-\Gamma_{rot.}$	1	1	1	1	1	1	$=\sigma_g^+$	

- The wave function ψ_3, transforms as q, q^3, and $e^{-\frac{\alpha q^2}{2}}$. Both of q and q^3 have the symmetry of q, while $e^{-\frac{\alpha q^2}{2}}$ is totally symmetric. Then ψ_3 has the symmetry of q.
- In general, odd wave function has the symmetry of the vibration, but even wave functions are totally symmetric.

How many modes of vibration do N_2 and CO each have? Show whether the fundamental transitions are active or inactive in infrared or Raman spectroscopy.

- N_2: $D_{\infty h}$, $\Gamma_{vib.}$ is derived (Table 9.11):
- There is only stretching mode of N_2, ψ_1, has symmetry σ_g^+.
- ψ_0 is always totally symmetric has symmetry σ_g^+ (Table 9.11).
- To investigate if the fundamental transition $\upsilon_0 \rightarrow \upsilon_1$ is allowed in an infrared spectrum, the transition moment integral will be expended:

$$M_{01} = \int_{-\infty}^{\infty} \psi_1 \hat{\mu} \psi_0 d\tau$$

x and y transform as π_u, and z transforms as σ_u^+.

The symmetry of the integral is

$$\int_{-\infty}^{\infty} \psi_1 \hat{\mu} \psi_0 d\tau \xrightarrow{\text{transforms as}} \sigma_g^+ \begin{pmatrix} \sigma_u^+ \\ \pi_u \end{pmatrix} \sigma_g^+ = \begin{pmatrix} \sigma_u^+ \\ \pi_u \end{pmatrix}$$

where the direct products are

$$\sigma_g^+ \times \sigma_u^+ \times \sigma_g^+ = \sigma_u^+, \text{ and}$$

$$\sigma_g^+ \times \pi_u \times \sigma_g^+ = \pi_u$$

Neither of these triple direct products contains the totally symmetric representation σ_g^+ (Table 9.11).

Therefore, the integral is

$$\int_{-\infty}^{\infty} \psi_1 \hat{\mu} \psi_0 d\tau = 0$$

The fundamental transition, $\upsilon_0 \rightarrow \upsilon_1$, is forbidden or infrared inactive.

- For the Raman spectrum, the polarizability components (Table 9.11) transform as

$$z^2 \xrightarrow{\text{transforms as}} \sigma_g^+ \qquad x^2 + y^2 \xrightarrow{\text{transforms as}} \sigma_g^+$$

$$xz, yz \xrightarrow{\text{transforms as}} \pi_g \quad x^2 - y^2, xy \xrightarrow{\text{transforms as}} \delta_g$$

- The transition moment integral will be expended:

$$M_{01} = \int_{-\infty}^{\infty} \psi_1 \hat{\alpha} \psi_0 d\tau$$

$$\int_{-\infty}^{\infty} \psi_1 \hat{\alpha} \psi_0 d\tau \xrightarrow{\text{transforms as}} \sigma_g^+ \begin{pmatrix} \sigma_g^+ \\ \pi_g \\ \delta_g \end{pmatrix} \sigma_g^+ = \begin{pmatrix} \sigma_g^+ \\ \pi_g \\ \delta_g \end{pmatrix}$$

- The direct product

$$\sigma_g^+ \times \sigma_g^+ \times \sigma_g^+ = \sigma_g^+$$

contains σ_g^+, so the transition will be allowed in the Raman spectrum (2331 cm^{-1}).

- CO: $C_{\infty v}$, $\Gamma_{vib.}$ is derived (Table 9.12).
- The vibrational analysis for CO:
  - Infrared:

$$\int_{-\infty}^{\infty} \psi_1 \hat{\mu} \psi_0 d\tau \xrightarrow{\text{transforms as}} \sigma^+ \begin{pmatrix} \sigma^+ \\ \pi \end{pmatrix} \sigma^+ = \begin{pmatrix} \sigma^+ \\ \pi \end{pmatrix} : \text{allowed}$$

TABLE 9.12 The Vibrational Analysis of CO

$C_{\infty v}$	E	$2C_\infty^\phi$	$2C_\infty^{2\phi}$	$2C_\infty^{3\phi}$	...	$\infty\sigma_v$		
Σ^+	1	1	1	1	...	1	z	x^2+y^2, z^2
Σ^-	1	1	1	1	...	−1	R_z	
Π	2	$2c\varphi$	$2c2\varphi$	$2c3\varphi$	...	0	$(x, y), (R_x, R_y)$	(xz, yz)
Δ	2	$2c2\varphi$	$2c4\varphi$	$2c6\varphi$	...	0		$(x^2-y^2), xy$
Φ	2	$2c3\varphi$	$2c6\varphi$	$2c9\varphi$	...	0		
Γ	2	$2c4\varphi$	$2c8\varphi$	$2c12\varphi$	...	0		
⋮						⋮		
$\Gamma_{x,y,z}$	3	$1+2c\varphi$	$1+2c2\varphi$	$1+2c3\varphi$	...	1		
Unmoved atoms	2	2	2	2	...	2		
Γ_{total}	6	$2+4c\varphi$	$2+4c2\varphi$	$2+2c\varphi$	...	2		
Γ_{trans}	3	$1+2c\varphi$	$1+2c2\varphi$	$1+2c3\varphi$	...	1		
$\Gamma_{rot.}$	2	$2c\varphi$	$2c2\varphi$	$2c3\varphi$	...	0		
$\Gamma_{vib.}$	1	1	1	1	...	1	$=\sigma^+$	

- Raman:

$$\int_{-\infty}^{\infty} \psi_1 \hat{\alpha} \psi_0 d\tau \xrightarrow{\text{transforms as}} \sigma^+ \begin{pmatrix} \sigma^+ \\ \pi \\ \delta \end{pmatrix} \sigma^+ = \begin{pmatrix} \sigma^+ \\ \pi \\ \delta \end{pmatrix} : \text{allowed}$$

- Since the direct product:

$$\sigma^+ \times \sigma^+ \times \sigma^+ = \sigma^+$$

only z-component is allowed; in other words, CO molecule can absorb z-polarized light but not x- or y-polarized light.

9.19 CENTER OF SYMMETRY AND THE MUTUAL EXCLUSION RULE

How many modes of vibration does XeF_4 have? Show whether these modes are active or inactive in infrared or Raman spectroscopy.

XeF_4: D_{4h} point group:

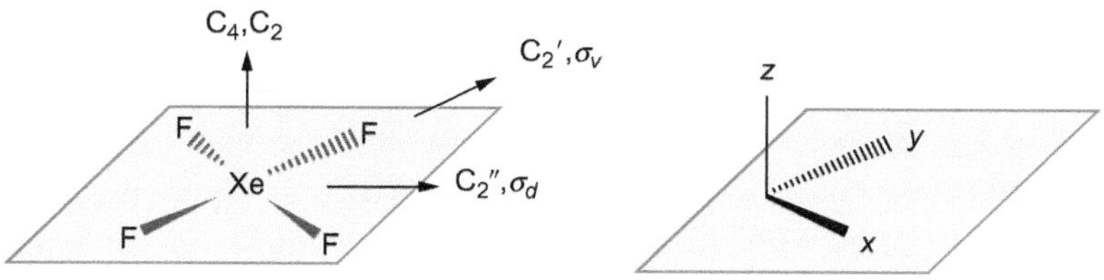

- $3n-6=9$ degrees of freedom are expected (Table 9.13).
- The representations $2e_u$ imply that there will be two sets of degenerate vibrations.

TABLE 9.13 Determination of $\Gamma_{vib.}$ for XeF_4

D_{4h}	E	$2C_4$	C_2	$2C_2'$	$2C_2''$	i	$4S_4$	σ_h	$2\sigma_v$	$2\sigma_d$		
A_{1g}	**1**	**1**	**1**	**1**	**1**	**1**	**1**	**1**	**1**	**1**	–	x^2+y^2, z^2
A_{2g}	1	1	1	−1	−1	1	1	1	−1	−1	R_z	–
B_{1g}	**1**	**−1**	**1**	**1**	**−1**	**1**	**−1**	**1**	**1**	**−1**	–	x^2-y^2
B_{2g}	1	−1	1	−1	1	1	−1	1	−1	1	–	xy
E_g	2	0	−2	0	0	2	0	−2	0	0	(R_x, R_y)	(xz, yz)
A_{1u}	1	1	1	1	1	−1	−1	−1	−1	−1	–	–
A_{2u}	1	1	1	−1	−1	−1	−1	−1	1	1	z	–
B_{1u}	1	−1	1	1	−1	−1	1	−1	−1	1	–	–
B_{2u}	1	−1	1	−1	1	−1	1	−1	1	−1	–	–
E_u	**2**	**0**	**−2**	**0**	**0**	**−2**	**0**	**2**	**0**	**0**	(x,y)	–
$\Gamma_{x,y,z}$	3	1	−1	−1	−1	−3	−1	1	1	1		
Unmoved atoms	5	1	1	3	1	1	1	5	3	1		
Γ_{total}	15	1	−1	−3	−1	−3	−1	5	3	1		

$\Gamma_{total} = a_{1g} + a_{2g} + b_{1g} + b_{2g} + e_g + 2a_{2u} + b_{2u} + 3e_u$.
$\Gamma_{trans.} = a_{2u} + e_u$.
$\Gamma_{rot.} = a_{2g} + e_g$.
$\Gamma_{vib.} = a_{1g} + b_{1g} + b_{2g} + a_{2u} + b_{2u} + 2e_u$.

- Therefore, there will be seven different vibrational energies for the nine normal modes of XeF_4.
- Since there are four chemical bonds, there are four stretching vibrations, and 9−4=5 bending vibrations.
- The vibrational analysis for XeF_4:
  - Infrared:

$$\int_{-\infty}^{\infty} \psi_1 \hat{\mu} \psi_0 d\tau \xrightarrow{\text{transforms as}} \underbrace{\forall}_{?} \begin{pmatrix} a_{2u} \\ e_u \end{pmatrix} a_{1g} = (= \text{or contains } a_{1g}) : \text{allowed for } \forall = a_{2u} \text{ or } e_u$$

    There are only three fundamental transitions in the infrared spectrum of XeF_4.
  - Raman:

$$\int_{-\infty}^{\infty} \psi_1 \hat{\alpha} \psi_0 d\tau \xrightarrow{\text{transforms as}} \underbrace{\forall}_{?} \begin{pmatrix} a_{1g} \\ b_{1g} \\ b_{2g} \\ e_g \end{pmatrix} a_{1g} = (= \text{or contains } a_{1g}) : \text{allowed for } \forall = a_{1g}, b_{1g} \text{ or } b_{2g}$$

    Only three fundamental transitions in the Raman spectrum of XeF_4.
- Note that for molecules with a center of symmetry:
  - components of $\hat{\mu}$ will always have u symmetry;
  - components of $\hat{\alpha}$ will always have g symmetry;
  - only u modes can be infrared active;
  - only g modes can be Raman active;
  - not all modes need to be active; and
  - no mode can be both infrared and Raman active.

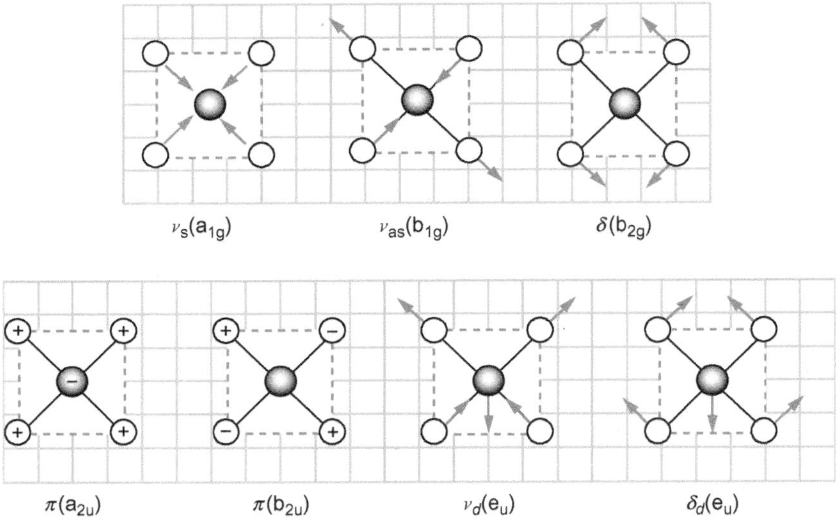

9.20 ISOLATION OF A PARTICULAR TYPE OF MOTION

How could you distinguish between the two isomers of $Pt(NH_3)_2Cl_2$?

- Formulate a limited basis set: define basis set by vectors describing the particular motion of interest.
- Trans-isomer: D_{2h}

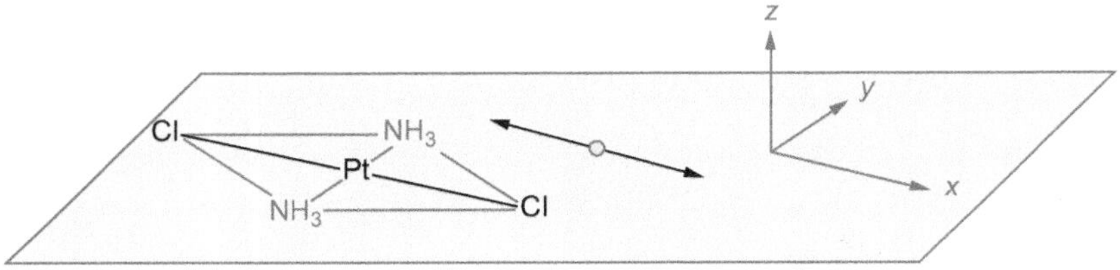

- ○ Draw an arrow between the Pt and Cl atoms in the molecule.
- ○ Determine how these arrows transform under each operation of the point group.

  Two-member basis set in D_{2h}

	E	$C_2(z)$	$C_2(y)$	$C_2(x)$	i	σ_{xy}	σ_{xz}	σ_{yz}	
$\Gamma_{Pt\text{-}Cl}$:	2	0	0	2	0	2	2	0	$= A_g + B_{3u}$

$\Gamma_{Pt\text{-}Cl} = a_g + b_{3u}$; noncoincident (center of symmetry and mutual exclusion rule)

a_g : x^2, y^2, z^2 ; is Raman active
b_{3u} : x ; is IR

Note: mutual exclusion via i.

- *cis* isomer: C_{2v}

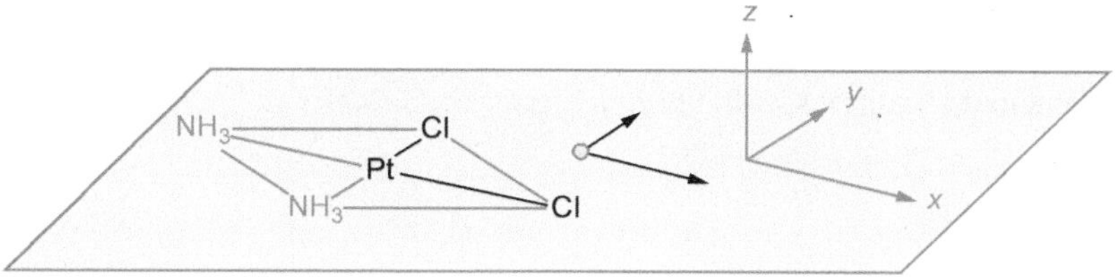

	E	C_2	σ_{xz}	σ_{yz}	
$\Gamma_{Pt\text{-}Cl}$:	2	0	2	0	$= A_1 + B_1$

$\Gamma_{Pt-Cl} = A_1 + B_1$, two coincident
$a_1 : z, x^2, y^2, z^2$ R, IR
$b_1 : x, xz$ R, IR

- We can easily tell the difference between two isomers in the 300–400 cm^{-1} region.

How many carbonyl stretching vibration bands are expected to be IR or Raman active for the complex $Mn(CO)_5I$?

- Carbonyl stretching in the transition metal complexes normally appears in the region 1700–2100 cm^{-1}.

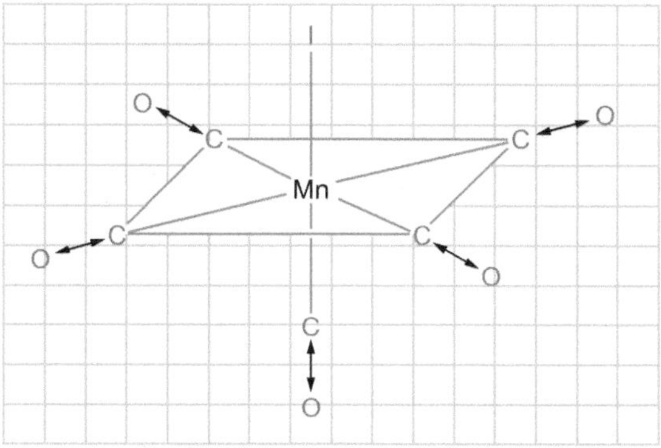

The five C—O stretching vectors

- $Mn(CO)_5I$: C_{4v}
- We shall only consider the stretching analysis of C≡O.

C_{4v}	E	$2C_4$	C_2	$2\sigma_v$	$2\sigma_d$		
A_1	1	1	1	1	1	z	x^2+y^2, z^2
A_2	1	1	1	−1	−1	R_z	
B_1	1	−1	1	1	−1		x^2-y^2
B_2	1	−1	1	−1	1		xy
E	2	0	−2	0	0	$(x, y), (R_x, R_y)$	(xz, yz)
$\Gamma_{stretch.}$	5	1	1	3	1	$= 2A_1 + B_1 + E$	

Therefore, there are:
three IR bands: $2a_1+e$; and
four Raman bands: $2a_1+B_1+e$.

- By using the metal-carbon stretching vibrations, we shall reach the same conclusion, since M-C stretching vectors have the same symmetries as C-O stretching vectors. However, the M-C stretching modes are usually mixed with the N-C-O bending modes.

How many carbonyl stretching vibration bands are expected to be IR or Raman active for the isomers of the octahedral complex $ML_3(CO)_3$?

If $Mo(CO)_3[P(OCH_3)_3]_3$ exhibits three IR bands at 1993, 1919, and 1890 cm^{-1}, and $Cr(CO)_3[CNCH_3]_3$ shows two IR bands at 1942, and 1860 cm^{-1}, assign the proper structure to these complexes.

- There are two possible geometrical isomers for the octahedral complex $ML_3(CO)_3$:
  - *mer* isomer: ligand are arranged on the meridian of the octahedron; and
  - *fac* isomer: ligand are arranged on the same face of the octahedron.

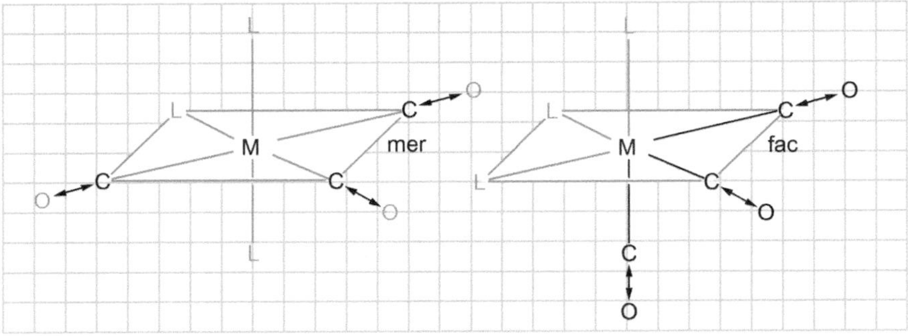

- The stretching analysis of *mer* isomer:

C_{2v}	E	C_2	$\sigma_{(xz)}$	$\sigma_{(yz)}$		
A_1	1	1	1	1	z	x^2, y^2, z^2
A_2	1	1	−1	−1	R_z	xy
B_1	1	−1	1	−1	x, R_y	xz
B_2	1	−1	−1	1	y, R_x	yz
$\Gamma_{stretch.}$	3	1	1	3	$= 2A_1+B_2$	

Thus, there are:
three IR stretching bands: $2a_1+b_2$; and
three Raman stretching bands: $2a_1+b_2$.

- The stretching analysis of *mer* isomer:

C_{3v}	E	$2C_3$	$3\sigma_v$		
A_1	1	1	1	z	x^2+y^2, z^2
A_2	1	1	−1	R_z	xy
B_1	1	−1	1	$(x,y), (R_x, R_x)$	$(x^2-y^2, xy), (xz, yz)$
$\Gamma_{stretch.}$	3	0	1	$= A_1+E$	

Thus, there are:
two IR stretching bands: a_1+e; and
two Raman stretching bands: a_1+e.

- $Mo(CO)_3[P(OCH_3)_3]_3$ exhibits three IR bands at 1993, 1919, and 1890 cm^{-1}, and is assigned as *mer* structure.
- $Cr(CO)_3[CNCH_3]_3$ with two IR bands at 1942 and 1860 cm^{-1}, and is assigned as *fac* structure.

The complexes $Fe(CO)_3[P(phenyl)_3]_2$, $Ru(CO)_3[P(phenyl)_3]_2$ and $Ru(CO)_3[As(phenyl)_3]_2$ each exhibit just a single IR active CO stretching mode. Assign the geometrical structure that consist with IR spectrum.

- There are six possible geometrical isomers for the complex $ML_2(CO)_3$:

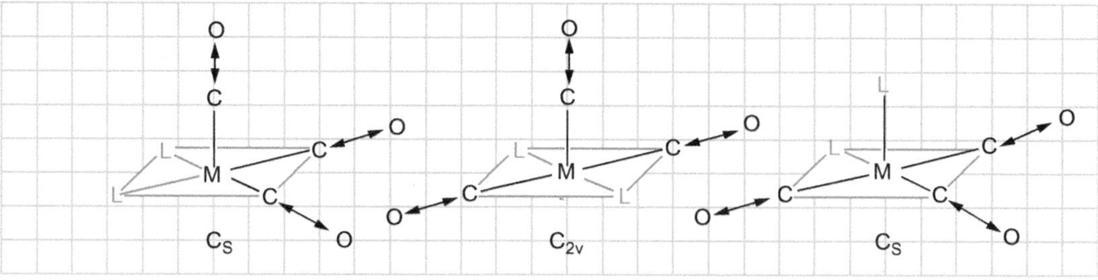

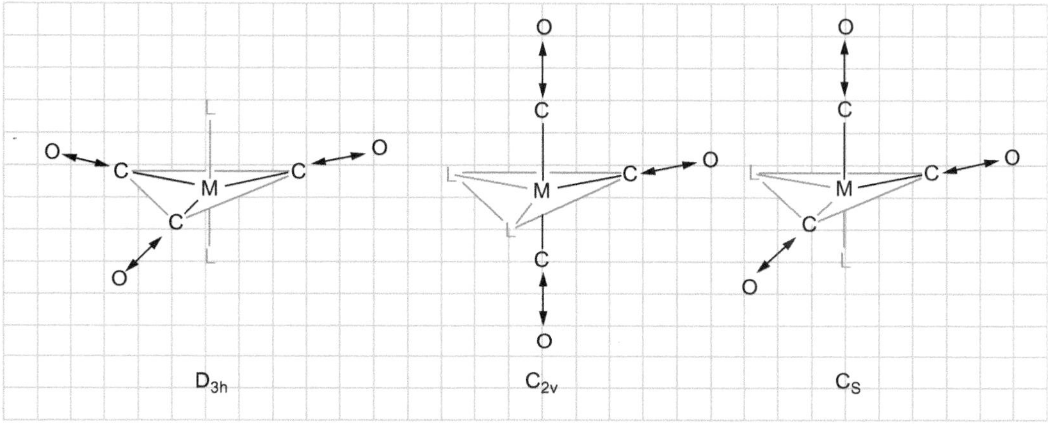

- Analysis of C_s structure:

C_s	E	σ_h		
A′	1	1	x, y, R_z	x^2, y^2, z^2, xy
A″	1	−1	z, R_x, R_y	yz, xz
$\Gamma_{stretch.}$	3	1	=2A′+A″	

Three IR stretching bands: 2a′+a″
Three Raman stretching bands: 2a′+a″

- Analysis of C_{2v} structure:

C_{2v}	E	C_2	$\sigma_{(xz)}$	$\sigma_{(yz)}$		
A_1	1	1	1	1	z	x^2, y^2, z^2
A_2	1	1	−1	−1	R_z	xy
B_1	1	−1	1	−1	x, R_y	xz
B_2	1	−1	−1	1	y, R_x	yz
$\Gamma_{Stretch.}$	3	1	3	1	$= 2A_1+B_1$	

Three IR stretching bands: $2A_1+B_1$
Three Raman stretching bands: $2A_1+B_1$

- Analysis of D_{3h} structure:

D_{3h}	E	$2C_3$	$3C_2$	σ_h	$2S_3$	$3\sigma_v$		
A_1'	1	1	1	1	1	1		x^2+y^2, z^2
A_2'	1	1	−1	1	1	−1	R_z	–
E'	2	−1	0	2	−1	0	(x,y)	x^2-y^2, xy
A_1''	1	1	1	−1	−1	−1	–	–
A_2''	1	1	−1	−1	−1	1	z	–
E''	2	1	0	−2	1	0	(R_x, R_y)	(xz, yz)
$\Gamma_{stretch.}$	3	0	1	3	0	1	$= A_1' + E'$	

One IR stretching bands: E'
Two Raman stretching bands: $A_1' + E'$

- Thus, the D_{4h} structure is assigned for $Fe(CO)_3[P(phenyl)_3]_2$, $Ru(CO)_3[P(phenyl)_3]_2$ and $Ru(CO)_3[As(phenyl)_3]_2$, since each exhibits just a single IR stretching bands.

For the linear symmetrical molecule *YX*$_2$ and unsymmetric of the type *YXX* or *YXZ*, define the forms of the normal modes of vibration and which of these modes are infrared and/or Raman active.

- Number of modes of vibration in the linear molecule $YX_2 = 3N-5$

$$N = 3$$

Number of modes of vibration $= 3 \times 3 - 5 = 4$ modes
Two modes are bond stretching and a doubly degenerate pair of angle-bending modes.
The forms are as follows:

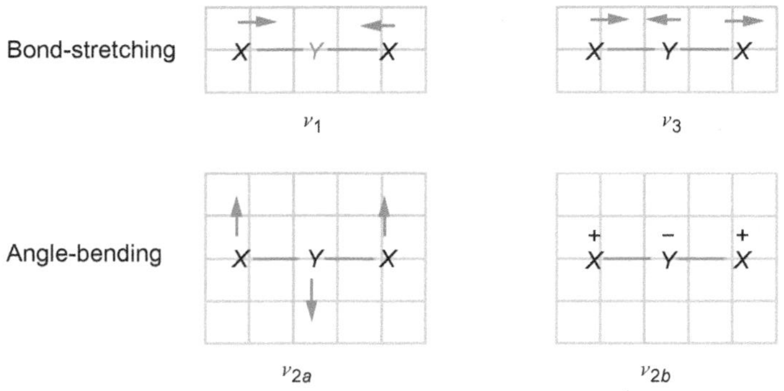

where + and − are vibrational amplitudes perpendicular to the plane of the paper.

- Infrared activity requires a change of the dipole moment between the two extremes of the vibration.
  The charge symmetry within the molecule are as follows:

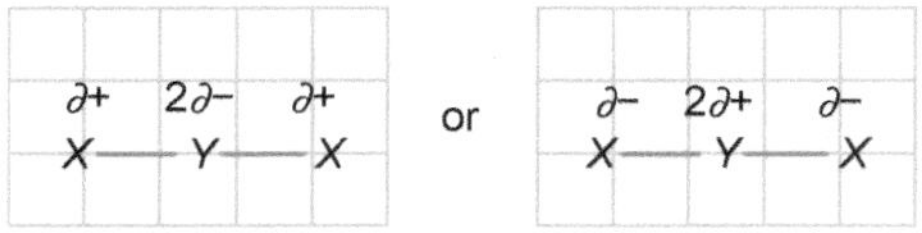

  - ν_1 does not cause a dipole change; infrared-inactive.
  - ν_{2a} and ν_{2b} give dipole changes perpendicular to the molecule axis: infrared-active.
  - ν_3 causes a dipole change parallel to the molecular axis: infrared-active.
- For the interaction with electromagnetic radiation, the electrical polarizability of molecules can be represented by ellipsoids. Note that the wavelength of electromagnetic radiation should be much greater than the molecular dimension.
- Activity in Raman spectra depends on a change of polarizability between the extremes of a vibration.
  - ν_1:

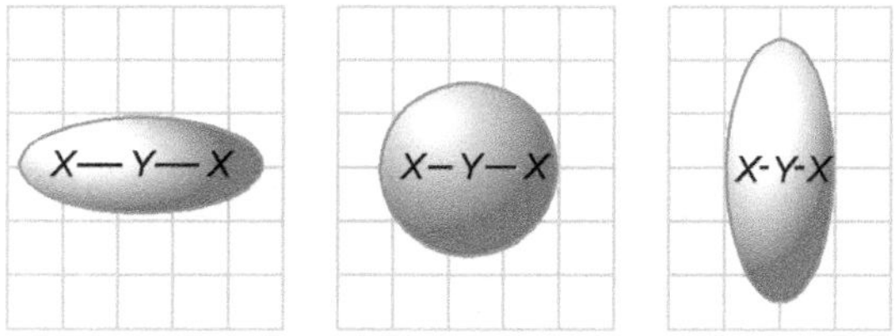

The polarizability ellipsoid is different at the two vibrational extremes, therefore, ν_1 is Raman-active.

- ν_{2a}:

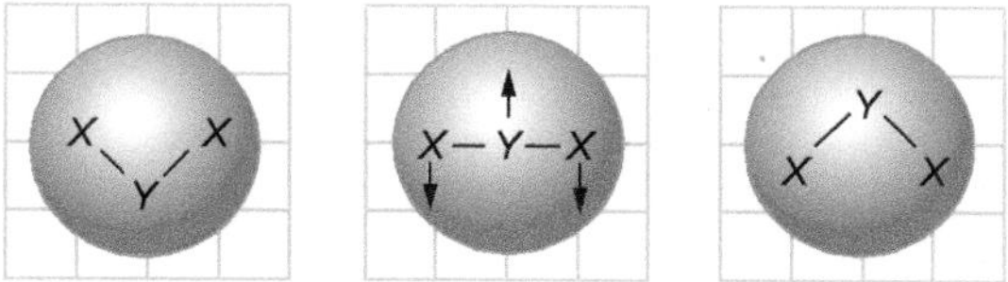

There is no variation in the polarizability ellipsoid at the two vibration extremes. The two extremes might not have similar polarizability to the molecule in the equilibrium position. However, individual bond-polarizability change are likely to cancel out, therefore, ν_{2a} (or ν_{2b}) is Raman-inactive.

- ν_3:

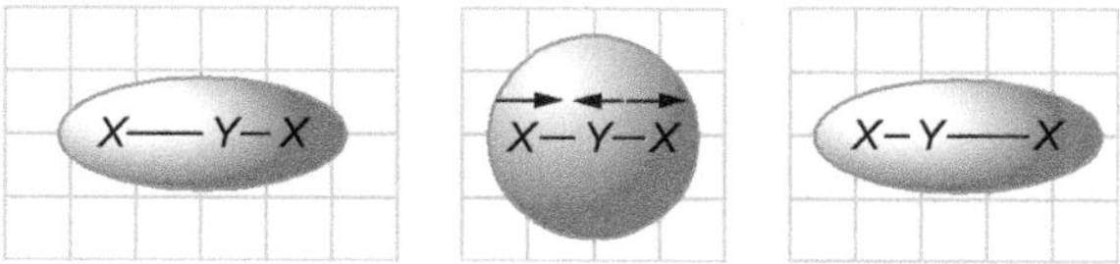

There is no change in the polarizability between the two vibration extremes, therefore, Raman-inactive.
Such behavior is predictable for the centrosymmetric molecules from the rule of mutual exclusion:
ν_1 is infrared-inactive and Raman-active;
$\nu_{2a/2b}$ is infrared-active ($\perp$) and Raman-inactive; and
ν_3 is infrared-active ($\|$) and Raman-inactive.

- Linear molecules *YXX* and *YXZ* will have different dipole moments and polarizabilities at the vibrational extremes, therefore, they should be infrared- and Raman-active.

9.21 DETECTING THE CHANGES OF SYMMETRY THROUGH REACTION

What are the types of binding modes in metallonitrate? How can you identify each mode of binding?

- NO_3^-Z is D_{3h}:
  $\Gamma_{vib.} = a_1' + 2e' + a_2''$ (Chapter 9: p. 554)
  can be ionic, monodentate, or bientate

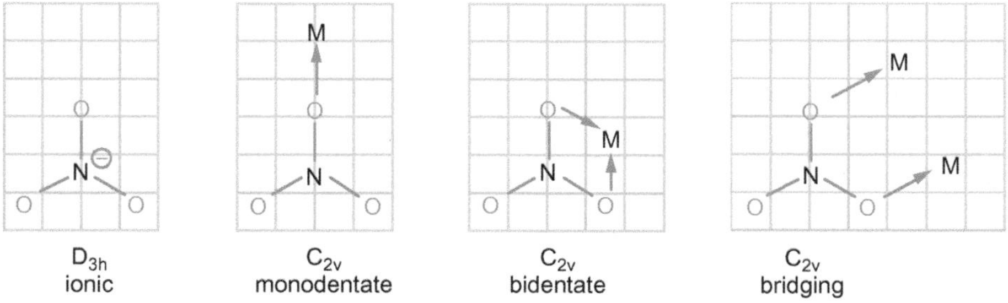

- To local symmetry about N, could work out for C_{2v}:

$$\Gamma_{vib.} = \Gamma_{total} - \Gamma_{trans.} - \Gamma_{rot.}$$

$$\Gamma_{vib.} = 3a_1 + b_1 + 2b_2$$

- Easier: the correlation table, Table 8.5, shows what happens to various presentations as point group symmetry of environment is reduced.

Table 8.5 Correlation Table for O_h Point Group

O_h	O	T_d	D_{4h}	D_{2d}	D_{3d}	D_3	C_{2v}	C_{4v}	C_{2h}
a_{1g}	a_1	a_1	a_{1g}	a_1	a_{1g}	a_1	a_1	a_1	a_g
a_{2g}	a_2	a_2	a_{1g}	a_1	a_{2g}	a_2	a_2	b_1	b_g
e_g	e	e	$a_{1g}+b_{1g}$	a_1+b_1	e_g	e	a_1+a_2	a_1+b_1	a_g+b_g
t_{1g}	t_1	t_1	$a_{2g}+e_g$	a_2+e	$a_{2g}+e_g$	a_2+e	$a_2+b_1+b_2$	a_2+e	a_g+2b_g
t_{2g}	t_2	t_2	$b_{2g}+e_g$	b_2+e	$a_{1g}+e_g$	a_1+e	$a_1+b_1+b_2$	b_2+e	$2a_g+b_g$
a_{1u}	a_1	a_2	a_{1u}	b_1	a_{1u}	a_1	a_2	a_2	a_u
a_{2u}	a_2	a_1	b_{1u}	a_1	a_{2u}	a_2	a_1	b_2	b_u
e_u	e	e	$a_{1u}+b_{1u}$	a_1+b_1	e_u	e	a_1+a_2	a_2+b_2	a_u+b_u
t_{1u}	t_1	t_2	$a_{2u}+e_u$	b_2+e	$a_{2u}+e_u$	a_2+e	$a_1+b_1+b_2$	a_1+e	a_u+2b_u
t_{2u}	t_2	t_1	$b_{2u}+e_u$	a_2+e	$a_{1u}+e_u$	a_1+e	$a_2+b_1+b_2$	b_1+e	$2a_u+b_u$

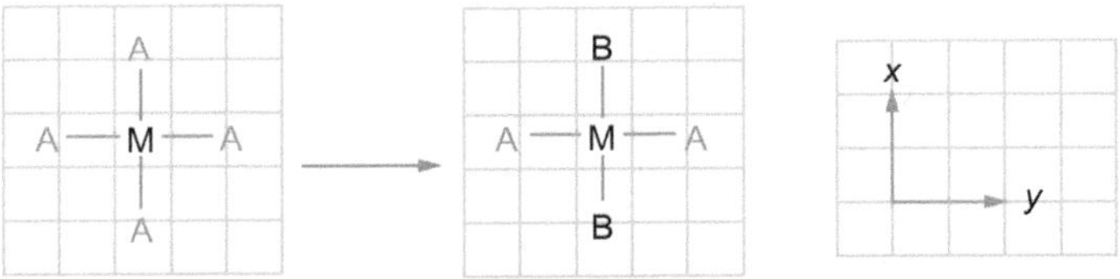

D_{4h}: x, y as e_u D_{2h}: $x=b_{3u}$, $y=b_{2u}$

- As $\frac{D_{4h} \rightarrow D_{2h}}{e_u \rightarrow b_{2u} + b_{3u}}$, decrease symmetry, which breaks degeneracies.

	D_{3h}		Change the symmetry	C_{2v}	
ν_1	a_1'	R 1068 cm^{-1}	→	a_1	R, IR becomes IR
ν_2	a_2''	IR 831 cm^{-1}	→	b_1	R, IR becomes R
ν_3	e'	R, IR 1400 cm^{-1}	→	a_1	R, IR 1420, 1305 cm^{-1}
		monodentate	→	b_2	R, IR 1500, 1270 cm^{-1}
ν_4	e'	R, IR 710 cm^{-1}	→	a_1	
				b_2	

- Note that shifts in ν_1, ν_2 also occur.

Ni $en_2NO_3^+$	bidentate	1290, 1476 cm^{-1}
Ni $en_2(NO_3)_2$	monodentate	1420, 1305 cm^{-1}

SUGGESTIONS FOR FURTHER READING

Theory of Molecular Vibrations

G.W. King, Spectroscopy and Molecular Structure, Holt, Rinehart, New York, 1964.

C.J.H. Schutte, The Theory of Molecular Spectroscopy, Vol. I, The Quantum Mechanics and Group Theory of Vibrating and Rotating Molecules, North Holland, Amsterdam, 1976.

P. Gans, Vibrating Molecules, Chapman and Hall, London, 1971.

N.B. Coithup, L.H. Daly, S.E. Wiberley, Introduction to Infrared and Raman Spectroscopy, Academic Press, New York, ISBN: 0-12-182552-3, 1975.

G.J. Bullen, D.J. Greenslade, Problems in Molecular Structure, Pion Limited, London, ISBN: 0 85086 083 0, 1983.

S. Akyüz, A.B. Dempster, R.L. Morehouse, An infrared and Raman spectroscopic study of some metal pyridine tetracyanonickelate complexes, J. Mol. Struct. 17 (1) (1973) 105–125.

E.T. Whittaker, Analytical Dynamics of Particles and Rigid Bodies, third ed., Cambridge Univ. Press, London/New York, 1927 (chapter 7).

S. Glasstone, Theoretical Chemistry, Van Nostrand-Reinhold, Princeton, NJ, 1944.

E.B. Wilson Jr., J.C. Decius, P.C. Cross, Molecular Vibrations, McGraw-Hill, New York, 1955.

G. Herzberg, Infrared and Raman Spectra of Polyatomic Molecules, Van Nostrand-Reinhold, Princeton, NJ, 1945.

G. Herzberg, Spectra of Diatomic Molecules, Van Nostrand-Reinhold, Princeton, NJ, 1945.

D. Steele, Theory of Vibrational Spectroscopy, Saunders, Philadelphia, PA, 1971.

J.C. Decius, Complete sets and redundancies among small vibrational coordinates, J. Chem. Phys. 17 (1949) 1315.

L.A. Woodward, Introduction to the Theory of Molecular Vibrations and Vibrational Spectroscopy, Oxford University Press, Oxford, 1972.

C.N. Banwell, Fundamentals of Molecular Spectroscopy, second ed., McGraw-Hill, London, 1972.

Infrared Spectroscopy

K. Nakamoto, Infrared and Raman of Inorganic and Coordination Compounds, fifth ed., Johns Wiley & Sons, Inc, New York, ISBN: 0-471-19406-9, 1997.

E.E. Bell, H.H. Nielsen, The infra-red spectrum of acetylene, J. Chem. Phys. 18 (1950) 1382.

T. Feldman, J. Romanko, H.L. Welsh, The Raman spectrum of ethylene, Can. J. Phys. 34 (1956) 737.

D.M. Gage, E.F. Barker, The infra-red absorption spectrum of boron trifluoride, J. Chem. Phys. 7 (1939) 455.

A. Langseth, J.R. Nielsen, Raman spectrum of nitrous oxide, Nature (London) 130 (1932) 92.

E.K. Plyler, E.F. Barker, The infrared spectrum and the molecular configuration of N_2O, Phys. Rev. 38 (1931) 1827.

E.K. Plyler, E.F. Barker, Phys. Rev. 41 (1932) 369.

D.M. Yost, D. DeVault, T.F. Anderson, E.F. Lassettre, The Raman spectrum of boron trifluoride gas, J. Chem. Phys. 6 (1938) 424.

E.B. Wilson Jr., A method of obtaining the expanded secular equation for the vibration frequencies of a molecule, J. Chem. Phys. 7 (1939) 1047.

E.B. Wilson Jr., Some mathematical methods for the study of molecular vibrations, J. Chem. Phys. 9 (1941) 76.

Raman Spectroscopy

J.G. Grasselli, B.J. Bulkin (Eds.), Analytical Raman Spectroscopy, Wiley, New York, 1991.

J.R. Ferraro, K. Nakamoto, Introductory Raman Spectroscopy, Academic Press, San Diego, CA, 1994.

H.C. Allen Jr., P.C. Cross, Molecular Vib-Rotors, John Wiley, New York, 1963. p. 235.

J.J. Barrett, N.I. Adams III, Laser-excited rotation–vibration Raman scattering in ultra-small gas samples, J. Opt. Soc. Am. 58 (1968) 311.

H.A. Szymanski, Raman Spectroscopy: Theory and Practice, Plenum Press, New York, Vol. I, 1967, Vol. 2, 1970.

J.A. Koningstein, Introduction to the Theory of the Raman Effect, D. Reidel, Dordrecht (Holland), 1973.

D.A. Long, Raman Spectroscopy, McGraw-Hill, New York, 1977.

J.G. Grasselli, M.K. Snavely, B.J. Bulkin, Chemical Applications of Raman Spectroscopy, Wiley, New York, 1981.

D.P. Strommen, K. Nakamoto, Laboratory Raman Spectroscopy, Wiley, New York, 1984.

R.J. Butcher, D.V. Willetts, W.J. Jones, On the use of a Fabry-Perot Etalon for the determination of rotational constants of simple molecules—the pure rotational Raman spectra of oxygen and nitrogen, Proc. R. Soc. Lond. Ser. A 324 (1971) 231.

J.M. Dowling, B.P. Stoicheff, High resolution Raman spectroscopy of gases: xii. Rotational spectra of C_2H_4 and C_2D_4 and the structure of ethylene molecule, Can. J. Phys. 37 (1959) 703.

S. Golden, An asymptotic expression for the energy levels of the rigid asymmetric rotor, J. Chem. Phys. 16 (1948) 78.

G. Herzberg, Infrared and Raman Spectra of Polyatomic Molecules, Van Nostrand, New York, 1945.

C.K. Ingold, et al., J. Chem. Soc. 912 (1936) (9 papers).

C.K. Ingold, et al., J. Chem. Soc. (1946) 222 (13 papers).

A.V. Jones, Infrared and Raman spectra of diacetylene, Proc. R. Soc. Lond. Ser. A 211 (1952) 285.

K.W.F. Kohlrausch, Phys. Z. 37 (1936) 58.

D.W. Lepard, D.E. Shaw, H.L. Welsh, The ν_{10} and ν_{11} Raman bands of gaseous ethane, Can. J. Phys. 44 (1966) 2353.

D.W. Lepard, D.M.C. Sweeney, H.L. Welsh, The Raman spectrum of ethane-d_6, Can. J. Phys. 40 (1962) 1567.

F.A. Miller, D.H. Lemmon, R.E. Witkowski, Observation of the lowest bending frequencies of carbon suboxide, dicyanoacetylene, diacetylene and dimethylacetylene, Spectrochim. Acta 21 (1965) 1709.

J. Romanko, T. Feldman, H.L. Welsh, The Raman spectrum of ethane, Can. J. Phys. 33 (1955) 588.

B.P. Stoicheff, High resolution Raman spectroscopy of gases: rotational spectra of C_6H_6 and C_6D_6 and internuclear distances in the benzene molecule, Can. J. Phys. 32 (1954) 339.

B.P. Stoicheff, High Resolution Raman spectroscopy, in: H.W. Thompson (Ed.), Advances in Spectroscopy, Volume I, Interscience, New York, 1959. p. 91.

C.H. Townes, A.L. Schawlow, Microwave Spectroscopy, McGraw-Hill, New York, 1955. p. 527.

A. Weber, E.A. McGinnis, The Raman spectrum of gaseous oxygen, J. Mol. Spectrosc. 4 (1960) 195.

Vibrational Spectroscopy

D.M. Adams, Metal-Ligand and Related Vibrations, St. Martin's Press, New York, 1968.

N.L. Alpert, W.E. Keiser, H.A. Szymanski, IR—Theory and Practice of Infrared Spectroscopy, Plenum Press, New York, 1970.

G.M. Barrow, Introduction to Molecular Spectroscopy, McGraw-Hill, New York, 1962.

L.J. Bellamy, Advanced in Infrared Group Frequencies, Methuen, London, 1968.

R.T. Conley, Infrared Spectroscopy, Allyn and Bacon, Boston, 1972.

F.R. Dollish, W.G. Fately, F.F. Bentley, Characteristic Raman Frequencies of Organic Compounds, John Wiley & Sons, New York, 1974.

J.R. Ferraro, Low Frequency Vibrations of Inorganic and Coordination Compounds, Plenum Press, New York, 1971.

S.K. Freeman, Applications of Laser Raman Spectroscopy, John Wiley & Sons, New York, 1974.

G. Herzberg, Spectra of Diatomic Molecules, Van Nostrand-Reinhold, New York, 1950.

K. Nakamoto, Infrared spectra of Inorganic and Coordination Compounds, John Wiley & Sons, New York, 1970.

F. Scheinmanm (Ed.), An Introduction to Spectroscopic Methods for the identification of Organic Compounds, Vol. I, Pergamon Press, Oxford, 1970.
R.M. Silverstein, G.C. Bassler, T.C. Morrill, Spectrometric Identification of Organic Compounds, John Wiley & Sons, New York, 1974.
D. Steele, Theory of Vibrational Spectroscopy, W. B. Saunders, Philadelphia, PA, 1971.
C.H. Townes, A.L. Schawlow, Microwave Spectroscopy, McGraw-Hill, New York, 1955.
H.A. Szymanski, Interpreted Infrared Spectra, Vols. I–III, Plenum Press, New York, 1964, 1966, 1967.
S. Waller, H. Straw, Spectroscopy, Vol. II, Chapman and Hall, Great Britain, 1967.
E.B. Wilson Jr., J.C. Decieus, P.C. Cross, Molecular Vibrations, McGraw-Hill, New York, 1955.
L.A. Woodward, Introduction to the Theory of Molecular Vibrations and Vibrational Spectroscopy, Oxford University Press, London, 1972.
D.J. Gardiner, Practical Raman Spectroscopy, Springer-Verlag, Berlin, ISBN: 978-0-387-50254-0, 1989.
P.R. Bunker, In: J.R. Durig (Ed.), Vibrational Spectra and Structure, vol. 3, Dekker, New York, 1975.

Chapter 10

Electronic Spectroscopy

Most of the molecules absorb light in the visible or ultraviolet region of the electromagnetic spectrum. This absorption will occur only when the energy is enough to cause state transition that involves electron transfers from the highest energy occupied orbitals to lowest unoccupied orbitals. Thus, the absorption of energy detects the electronic excitation of an electron from the ground state to a higher electronic excited state.

The absorption of visible or ultraviolet light induces a change in the energy of the electrons of the absorbing molecule. This change in the electronic structure also affects the vibrational, and possibly the rotational motions.

Molecules constantly vibrate and rotate about the axes between atoms. Such vibrational and rotational motions are quantized. Absorption of light also promotes transitions to higher vibrational and rotational energy levels. Every electronic excitation is certainly associated by precise transitions to higher vibrational and rotational energy levels in the molecule. Electronic transitions are significantly faster than the vibration of the molecule. Therefore, the distances between the atomic nuclei in the molecule are still unchanged in the instant of the electronic transition. In this stationary state of the excited molecule, the distances between the nuclei are different from those in the ground state. The excited molecule will be "compressed" or "stretched," that is, in a different vibrational mode than in the ground state. The readjustment of the nuclei distances corresponding to the excited state will give rise to further changes in the spectrum. Unfortunately, the absorption bands of most spectra are so broad that the vibrational and rotational information are obscured.

The symmetry of the chemical surroundings of an ion dictates the effect of this environment on the molecular energy levels. Molecular absorption in the visible and ultraviolet region of the spectrum is dependent on the electronic structure of the molecule. In the case of the transition metal complexes, the absorption spectra are generally used for investigation of bonding conditions in order to obtain information on the energy levels of the electron shells. The electronic structures of atoms or molecules and the energy levels involved are described in previous chapters.

The visible and ultraviolet spectrum is recorded directly from a device named spectrophotometer. The spectrum is simply a plot of wavelength (or frequency) of the absorption versus the absorption intensity. The data are regularly expressed as a graphical plot of wavelength versus the molar absorptivity. The use of the molar absorptivity as the unit of absorption intensity has advantage that all intensity values refer to the same number of absorbing molecules.

In this chapter, the focus on the chemical information obtained by recording the electronic transitions in the visible and ultraviolet regions. We examine the relationship between the solute concentration and light absorbance, as well the correlation among the molar extinction coefficient, integrated intensity, dipole strength, and the electronic transitions. The significance of the Born-Oppenheimer approximation is considered to illuminate the requirements and the symmetry considerations in order to allow electronic transition. The impact of the spin, orbital, and vibrational constraints is investigated to explore the **b**asis of the electronic absorption selection rules.

The electronic excitation of the function groups, donor-acceptor complexes, and porphyrins are spectroscopically characterized. The study outlines the roles of vibronic coupling, configuration interaction, and π-binding. Factors that affect the bandwidth, band intensity, and intensely colored metal complexes are also identified. The text addresses the impact of Jahn-Teller, temperature, low symmetry, and spectrochemical series. Comparative studies of octahedral versus tetrahedral, the low-spin versus high-spin, and d^n versus d^{10-n} configuration are conducted. We illustrate how to evaluate D_q, β, and the position of the absorption peaks, also advancing our understanding of the unexpected weak absorbance, the simultaneous pair excitations, and the absorption of the unpolarized light. The role of the magnetic dipole moment on the absorbance intensity is investigated, using circular dichroism spectroscopy and Kuhn anisotropy factor to examine the effects of lower symmetry, absolute configuration, and the energy levels within the molecule (Scheme 10.1).

Electrons, Atoms, and Molecules in Inorganic Chemistry. http://dx.doi.org/10.1016/B978-0-12-811048-5.00010-9

- 10.1: Beer-Lambert law
  - Molar extinction coefficient
  - Oscillator strength
  - Dipole strength
- 10.2: Allowed electronic transition
  - The transition moment and electronic transitions
  - The Born-Oppenheimer approximation
  - Even and odd functions and the symmetry considerations
  - Symmetry representations and the allowed transitions
- 10.3: Basis of the selection rules
- 10.4: Selection rules
  - Spin, orbital, and vibrational constraints
  - Spin selection rules
  - Orbital selection rules
  - Vibrational selection rules
- 10.5: Unexpected weak absorbance
  - Same total angular momentum of two spin-states
  - Vibronic coupling
  - Interference of magnetically allowed transitions
- 10.6: Spectroscopy of the electronic excitations
- 10.7: Electronic spectra of selected examples
  - Jahn-Teller theorem and vibronic coupling
  - The expected position of absorption peaks
  - Configuration interaction
  - Temperature and absorption spectra
  - Spectrochemical series
  - Octahedral versus tetrahedral
  - Simultaneous pair excitations: bridged dinuclear metal centers
  - Bandwidth in the electronic spectra
  - The low-spin versus high-spin
  - The effect of low symmetry
  - Band intensity and ligand-field
  - Tetrahedral complex
  - d^2 versus d^8 complexes
  - Calculation of D_q and β
  - π-binding
  - Intensely colored metal complexes
  - Donor-acceptor complexes
- 10.8: Spectroscopy of porphyrins
  - Configuration interaction
  - Absorption of the unpolarized light
  - The vibronic excitation
- 10.9: The magnetic dipole moment and the absorbance intensity
  - Circular dichroism spectroscopy
  - The effects of lower symmetry
  - Absolute configuration
  - Kuhn anisotropy factor
- Suggestions for further reading

10.1 BEER-LAMBERT LAW

Molar Extinction Coefficient, Oscillator Strength, and Dipole Strength

If a beam of monochromatic light of intensity I_o passes through a solution of path length L containing a solute of C molar concentration, and there is a reduction in the intensity of the transmitted light, ∂I.

What are the factors that decide the intensity of the transmitted light?

Derive a mathematical relationship between the solute concentration and light absorbance "Beer's law."

What is the relationship between the following?

i. **Molar extinction coefficient and integrated intensity (oscillator strength)**
ii. **Molar extinction coefficient and the dipole strength**
iii. **Integrated intensity and the dipole strength.**

- If ∂I: is the reduction in intensity that occurs when light passes through a layer of thickness ∂L containing an absorbing species J at molar concentration [C].

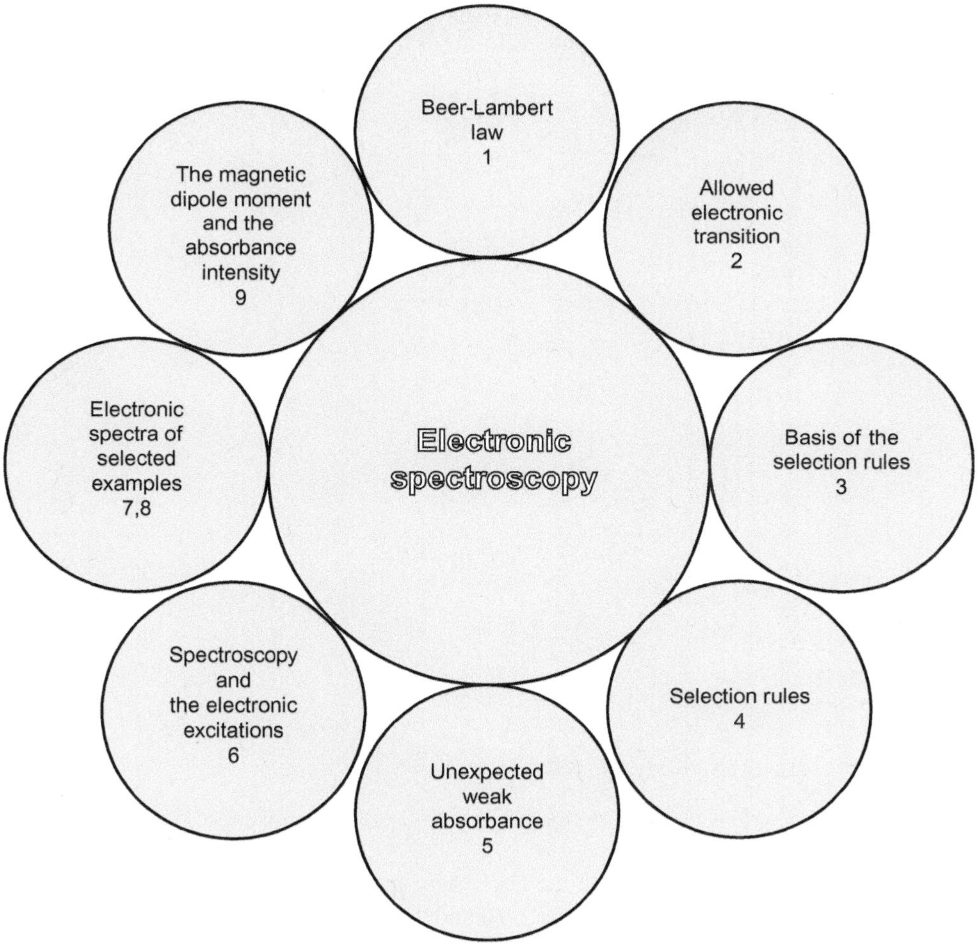

SCHEME 10.1 Approach used to introduce the electronic spectroscopic properties of the inorganic and organic compounds.

- The amount of light absorbed is proportional to the intensity of incident monochromatic light I, the solution path length L, and to the concentration of the solute C:

$$\partial I \propto -I$$

$$\partial I \propto -\partial L$$

$$\partial I \propto -[\mathrm{C}]$$

then

$$\partial I \propto -IC\,\partial L$$

$$\partial I = -\kappa IC\,\partial L \tag{10.1.1}$$

where κ is the proportionality coefficient.

$$\frac{\partial I}{I} = -\kappa\,C\,\partial L$$

This expression applies to each successive layer into which the sample can be regarded. Therefore, to obtain the intensity that emerges from a sample of thickness L:

$$\int_0^L \frac{\partial I}{I} = -\kappa C \int_0^L \partial L$$

$$\ln\frac{I}{I_o} = -\kappa CL$$

If $\kappa = \varepsilon \ln 10$, therefore,

$$\log\frac{I}{I_o} = -\varepsilon CL$$

$$\text{if}:\ T = \frac{I}{I_o}$$

where T is the transmittance, then

$$T = \frac{I}{I_o} = 10^{-\varepsilon CL}$$

The absorbance, A, is related to T by

$$A = -\log\frac{I}{I_o} = \log\frac{I_o}{I} = \varepsilon CL$$
$$A = \varepsilon CL \tag{10.1.2}$$

where ε is the molar extinction coefficient (in units of L cm^{-1} mol^{-1} or cm^2 mol^{-1}), commonly known as the molar absorptivity.

- The lambda max, λ_{max}, is the wavelength in the absorption spectrum where the absorbance is the maximum.
- The shape of an electronic absorption band is due to the vibrational sublevels of the electronic states. Therefore, the intensity of the interaction causing the electronic transition should be proportional to the area under the specific band rather than to any particular value of the absorbance.
- The molar extinction coefficient identifies the molar absorbance of a transition at the peak wavelength of an absorption band.
- Another unit that is used to measure the integrated intensity under the entire absorption band of the transition is the oscillator strength, f, which is related to ε by

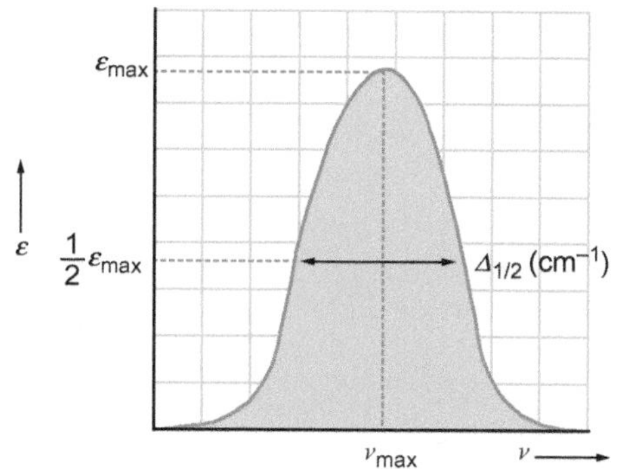

FIG. 10.1 Maximum molar absorptivity and bandwidth.

$$f = \frac{1000\,mc^2}{N\pi e^2}\ln(10)\int \varepsilon\partial\nu = \frac{2303\,mc^2}{N\pi e^2}\int \varepsilon d\nu = 4.315\times 10^{-9}\int \varepsilon\partial\nu \tag{10.1.3}$$

where m and e are the mass and charge of the electron, an v is the frequency of light in cm^{-1}. The oscillator strength is unitless; its value generally falls between zero and one, and is also known as integrated intensity.

- If the band stretches from ν_1 to ν_2 (in cm^{-1}) then:

$$f = 4.315\times 10^{-9}\int_{\upsilon_1}^{\upsilon_2} \varepsilon\,\partial\nu \approx 4.6\times 10^{-9}\varepsilon_{max}\Delta_{1/2} \tag{10.1.4}$$

$\Delta_{1/2}$ is the half bandwidth (in cm^{-1}), and ε has a value half its maximum (Fig. 10.1).

- The dipole strength D is another essential unit that has been given by Mulliken to characterize the intensity of the electronic transition.

$$D = \frac{3\,hc}{8\,\pi^3 N_i}\int \kappa(\lambda)\frac{\partial\lambda}{\lambda} = \frac{3000hc}{8\pi^3 N}\int \varepsilon(\lambda)\frac{\partial\lambda}{\lambda} \tag{10.1.5}$$

In which:
h is Planck's constant, c is the speed of light in vacuum;
N_i is the number of absorbing molecules per milliliter (cm^3) of solution;
N is Avogadro's number;
$\kappa(\lambda)$ is the partial absorption coefficient for the band in cm^{-1};
$\varepsilon(\lambda)$ is the molar absorption coefficient in units of cm^2 mol^{-1} for the band; and
λ is the wavelength of light.
The dipole strength is a convenient unit for the expression of theoretical calculations of the molar absorptivity.

- The integrated intensity, f, of absorption band due to the transition between the two states ψ_1 and ψ_2 is proportional to the dipole strength, D:

$$\begin{aligned} &f \propto D, \text{ and} \\ &f \propto \nu, \text{ then} \\ &f = \text{constant}\cdot\nu\cdot D \end{aligned} \tag{10.1.6}$$

10.2 ALLOWED ELECTRONIC TRANSITION

The Transition Moment and Electronic Transitions

If M is a dipole moment operator connecting states ψ_1 and ψ_2, what is the form of the transition moment integral?

What is the correlation between this integral and the measure of the integrated intensity under the absorption band?

- The transition moment, P, is usually given by:

$$P = \int \psi_1 \hat{M} \psi_2 \partial \nu = \langle \psi_1 | \hat{M} | \psi_2 \rangle = D^{1/2} \quad (10.2.1)$$

where $\hat{M}$ is dipole moment operator connecting states ψ_1 and ψ_2, and D is the dipole strength.

$$D = \left| \int \psi_1 \hat{M} \psi_2 \partial \nu \right|^2 = \langle \psi_1 | \hat{M} | \psi_2 \rangle^2,$$

Then the integrated intensity (oscillator strength, Eq. 10.1.6) is as follows:

$$f = \text{constant} \cdot \nu \langle \psi_1 | \hat{M} | \psi_2 \rangle^2 \quad (10.2.2)$$

The Born-Oppenheimer Approximation

What is the significance of the Born-Oppenheimer approximation?

- The nuclei are several thousand times more massive than the electrons, therefore, they will move much more slowly than the electrons. The nuclei may be assumed to remain stationary during the electronic transition. The separation of nuclear (vibrational) wave function, ψ_n, and total electronic, ψ_{el} wave function, is commonly known as the Born-Oppenheimer approximation.

$$\psi = \psi_n \psi_{el}, \text{ and}$$
$$f = \text{constant} \cdot \nu \langle \psi_{el} | \hat{M} | \psi_{el}^{ex} \rangle^2 \quad (10.2.3)$$

where ψ_{el} and ψ_{el}^{ex} are the electronic wave functions of the ground and excited states, respectively.

What are the requirements in order to allow electronic transitions?

- For electric dipole allowed transitions:

$$\vec{M}_r = q \sum_i \vec{r}_i \quad (10.2.4)$$

where q is the charge, and $\vec{r}_i$ is the displacement, with the summation carried out over all the electron in the molecule.
- The dipole moment operators M_x, M_y, and M_z, are the components of M, such that, for electric dipole allowed transitions:

$$\vec{M}_x = -e \sum_i \vec{x}_i$$

- Therefore, the intensity of an absorption band depends on the relative orientation of the electric vector of the incident light, and the symmetry axes of the molecule, then

$$f = \text{constant} \cdot \nu \left[\langle \psi_{el} | \hat{M}_x | \psi_{el}^{ex} \rangle^2 + \langle \psi_{el} | \hat{M}_y | \psi_{el}^{ex} \rangle^2 + \langle \psi_{el} | \hat{M}_z | \psi_{el}^{ex} \rangle^2 \right] \quad (10.2.5)$$

- In order to have allowed transition, at least one of the integrals must be nonzero. If all three of these integrals are zero, the transition is called forbidden. Note that the forbidden transitions show small values of the intensity integrals, in case of spin-orbit or vibronic coupling (this will be discussed separately).

Even and Odd Functions and the Symmetry Considerations

What are the differences between even and odd functions, and how could these functions be used to distinguish whether the electronic transition is allowed or forbidden?

- To evaluate the integrals:

$$\langle \psi_{el} | \hat{M}_x | \psi_{el}^{ex} \rangle, \quad \langle \psi_{el} | \hat{M}_y | \psi_{el}^{ex} \rangle, \quad \text{and} \quad \langle \psi_{el} | \hat{M}_z | \psi_{el}^{ex} \rangle$$

We need to be aware of even and odd functions. If we plot $y = f(x)$ in the xy plane:
  - An even function is one for which:

i. the yz plane is mirror plane; and
ii. the area under the curve of the even function over all the space has a finite value.
iii. symmetry consideration indicate A_1

- An odd function is one for which:
  i. the origin is a center of inversion;
  ii. the area under the curve of the odd function over all the space has a zero value; and
  iii. there is no A_1.

- This can be seen by looking at some simple mathematic functions (Fig. 10.2 and Table 10.1):
- The symmetry considerations can tell us if the integral; $\langle \psi_{el} | \hat{M}_x | \psi_{el}^{ex} \rangle$, $\langle \psi_{el} | \hat{M}_y | \psi_{el}^{ex} \rangle$, or $\langle \psi_{el} | \hat{M}_z | \psi_{el}^{ex} \rangle$ is different than zero. Such integrals are different from zero only when the integrand remains unchanged for any of the symmetry operations permitted by the symmetry of the molecule (totally symmetric irreducible representation, A_1).

Symmetry Representations and the Allowed Transitions

How could the symmetry representations be used to differentiate between the allowed and the forbidden transitions?

- The integral $\langle \psi_{el} | \hat{M}_r | \psi_{el}^{ex} \rangle$ can be nonzero only if the direct product (Chapter 4; p. 274), of the irreducible representations of the wave functions and operators belongs to symmetry species A_1. If the direct product is or contains A_1, the integral is nonzero and transition is allowed.

$$\langle \psi_{el} | \hat{M}_r | \psi_{el}^{ex} \rangle \neq \text{zero} \overset{\text{only if}}{\Rightarrow} \Gamma_{el} \Gamma_r \Gamma_{el}^{ex} = \text{or contains } A_1 \} \rightarrow \text{where,} \quad r = x,\ y,\ \text{or } z$$

For: O_h and T_d
$\Gamma_r \equiv x,\ y$, or z transforms as T_{1u} and T_2, respectively.

- Transitions in molecules without center of symmetry depend upon the symmetry of the initial and final states.
  If the *direct product* of the initial and final states and any one of x, y, or z is $\mathbf{A_1}$, the transition is *allowed*, where x, y, and z are the electric dipole moment operators.

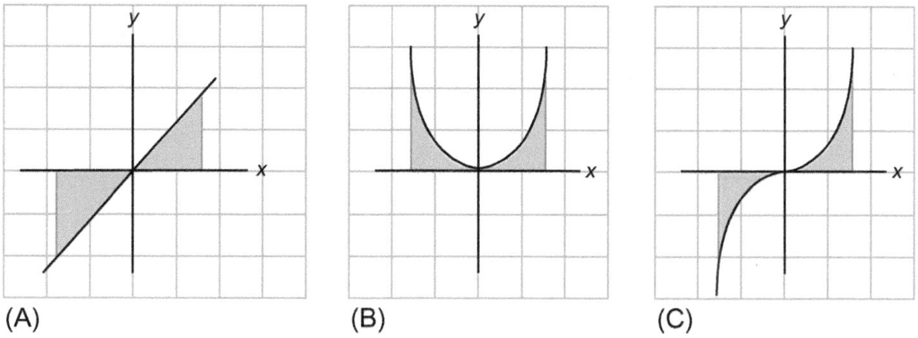

FIG. 10.2 The plots of (A) $y=x$ (odd), (B) $y=x^2$ (even), and (C) $y=x^3$ (odd).

TABLE 10.1 Odd and Even Functions and A_1 Representation

Plot	A	B	C
Function	$y=x$	$y=x^2$	$y=x^3$
Area under the curve $= \int_{-\infty}^{\infty} y(x)dx$	Area = zero	Area ≠ zero	Area = zero
Symmetric to σ^{yz} plane or to i	i	σ^{yz}	i
Even or odd function	Odd	Even	Odd
Symmetry consideration Dose contain A_1?	No A_1	A_1	No A_1

10.3 BASIS OF THE SELECTION RULES

What are the components that contribute to the intensity of the electronic absorption band?

- The electronic component of the wave function, ψ_{el}, may be broken down into:

$$\psi_{el} = \psi_{orbital/spin} \cdot \psi_{vibrational} \cdot \psi_{rotational} \cdot \psi_{transitional} \quad (10.3.1)$$

For the liquid and solid phases, $\psi_{rotational}$ and $\psi_{transitional}$ are not likely to change for a molecule in a time comparable to the lifetime of the excited electronic states, so that they integrate out to unity in the expression for ψ_{el}.

$$\psi_{el} = \psi_{orbital/spin} \cdot \psi_{vibrational} = \psi_{o/s} \cdot \psi_{v}$$

- The integrated intensity is proportional to the square of the transition moment integral (Eq. 10.2.3):

$$\sqrt{f} = \text{constant} \cdot \left\langle \psi_{o/s} \cdot \psi_{v} | \hat{M} | \psi_{s/o}^{ex} \cdot \psi_{v}^{ex} \right\rangle \quad (10.3.2)$$

The operator $\hat{M}$ can be divided into component, one depends on nuclear coordinate, $\hat{M}_n$, and the other depends on electron coordinates, $\hat{M}_e$:

$$\sqrt{f} = \text{constant} \cdot \left\langle \psi_{o/s} \cdot \psi_{v} | \hat{M}_n + \hat{M}_e | \psi_{o/s}^{ex} \cdot \psi_{v}^{ex} \right\rangle$$

$$\sqrt{f} = \text{constant} \cdot \left\{ \left\langle \psi_{o/s} \cdot \psi_{v} | \hat{M}_n | \psi_{o/s}^{ex} \cdot \psi_{v}^{ex} \right\rangle + \left\langle \psi_{o/s} \cdot \psi_{v} | \hat{M}_e | \psi_{o/s}^{ex} \cdot \psi_{v}^{ex} \right\rangle \right\}$$

By using Kronecker delta function:

$$\langle a \cdot b | F | a \cdot b \rangle = \langle a | F | a \rangle \langle b | b \rangle + \langle b | F | b \rangle \langle a | a \rangle$$

$$\sqrt{f} = \text{constant} \cdot \left\{ \begin{array}{l} \underbrace{\left\langle \psi_{o/s} | \hat{M}_n | \psi_{o/s}^{ex} \right\rangle}_{\substack{\text{Born-Oppenheimer} \\ \text{remain constant}}} \langle \psi_{v} | \psi_{v}^{ex} \rangle + \langle \psi_{v} | \hat{M}_n | \psi_{v}^{ex} \rangle \underbrace{\left\langle \psi_{o/s} | \psi_{o/s}^{ex} \right\rangle}_{\text{orthogonal=zero}} \\ + \left\langle \psi_{o/s} | \hat{M}_e | \psi_{o/s}^{ex} \right\rangle \langle \psi_{v} | \psi_{v}^{ex} \rangle + \langle \psi_{v} | \hat{M}_e | \psi_{v}^{ex} \rangle \underbrace{\left\langle \psi_{o/s} | \psi_{o/s}^{ex} \right\rangle}_{\text{orthogonal=zero}} \end{array} \right\}$$

According to the Born-Oppenheimer approximation, the electronic transition takes place in the very short time that the nuclei may be assumed to remain stationary during the Frank-Condon transition. If we consider the electronic component of the wave function:

$$\sqrt{f} = \text{constant} \cdot \left\langle \psi_{o/s} | \hat{M}_e | \psi_{o/s}^{ex} \right\rangle \underbrace{\langle \psi_{v} | \psi_{v}^{ex} \rangle}_{\text{Franck Condon factor}}$$

- The vibrational integral, $\langle \psi_{v} | \psi_{v}^{ex} \rangle$, is labeled by the Franck-Condon factor; it is not necessarily zero because the two vibrational wave functions do not belong to the same electronic state and therefore need not be orthogonal.
- The separation of electronic orbital and spin wave functions, $\psi_{o/s}$, is generally valid for the lighter elements of the periodic table, and therefore for most organic compounds. When the separation of spin and orbital wave functions is not a good approximation, there is said to be spin-orbital coupling.

$$\sqrt{f} = \text{constant} \cdot \langle \psi_{o} \cdot \psi_{s} | \hat{M}_e | \psi_{o}^{ex} \cdot \psi_{s}^{ex} \rangle \langle \psi_{v} | \psi_{v}^{ex} \rangle$$

$$\sqrt{f} = \text{constant} \cdot \left[\langle \psi_{o} | \hat{M}_e | \psi_{o}^{ex} \rangle \langle \psi_{s} | \psi_{s}^{ex} \rangle \langle \psi_{v} | \psi_{v}^{ex} \rangle + \underbrace{\langle \psi_{s} | \hat{M}_e | \psi_{s}^{ex} \rangle}_{\text{does not couple}=0} \langle \psi_{o} | \psi_{o}^{ex} \rangle \langle \psi_{v} | \psi_{v}^{ex} \rangle \right]$$

- Since the dipole moment operator does not couple electronic spins, they must be integrated out of the transition moment expression:

$$\sqrt{f} = \text{constant} \cdot \underbrace{\left\langle \psi_{\text{spin}} | \psi_{\text{spin}}^{\text{ex}} \right\rangle}_{\text{basis of spin selection rules}} \underbrace{\left\langle \psi_{\text{orbital}} | \hat{M} | \psi_{\text{orbital}}^{\text{ex}} \right\rangle}_{\text{basis of orbital selection rules}} \underbrace{\left\langle \psi_{\text{vibrational}} | \psi_{\text{vibrational}}^{\text{ex}} \right\rangle}_{\text{basis of vibrational selection rules}} \quad (10.3.3)$$

Therefore, there are orbital, spin, and vibrational selection rules for electronic transitions.

- In order for matter to absorb the electric field component of radiation:
  - the energy matching must be met; and
  - the energy transition in the molecule must be accompanied by a change in the electrical center of the molecule in order that an electrical work can be done by the electromagnetic radiation field.

10.4 SELECTION RULES

Spin, Orbital, and Vibrational Constraints

What are the spin, orbital, and vibrational selection rules?

Spin Selection Rules

- The integrated intensity, f, vanishes as a result of spin orthogonality, $\left\langle \psi_{\text{spin}} | \psi_{\text{spin}}^{\text{ex}} \right\rangle$, unless the spin functions, ψ_{spin}, of the ground and excited state are the same.
- Therefore, "Transitions between wave functions of different spin are forbidden," or transitions between states of different multiplicity are forbidden.

$$\Delta s = 0 \overset{\text{for}}{\longrightarrow} \text{spin allowed transitions} \quad (10.4.1)$$

Transition between *pure* singlet, $^1\psi$, and *pure* triplet, $^3\psi$, is forbidden, *spin forbidden*.

Orbital Selection Rules

- For molecules with center of symmetry:

  The dipole moment operator, $\hat{M}$, is odd in point group containing an inversion center, therefore, it can only couple an odd to an even wave function. The subscripts u and g refer to *ungerade* and *gerade*, which are German words for odd and even, respectively.

  For: O_h

  Γ_r transforms as T_{1u}, and if:

$$\text{u} \times \text{u} = \text{g} \times \text{g} = \text{u}$$

$$\text{g} \times \text{u} = \text{u} \times \text{g} = \text{g}, \quad \text{then in } \Gamma_{\text{u or g}} \times \Gamma_{\text{u}} \times \Gamma_{\text{u or g}}^{\text{ex}}$$

$$\text{u} \overset{\text{u}}{\longrightarrow} \text{u} \quad \text{u} \times \text{u} \times \text{u} = \text{u} \rightarrow \text{forbidden} \quad (10.4.2)$$

$$\text{u} \overset{\text{u}}{\longrightarrow} \text{g} \quad \text{u} \times \text{u} \times \text{g} = \text{g} \rightarrow \text{allowed, Laporte} - \text{allowed} \quad (10.4.3)$$

$$\text{g} \overset{\text{u}}{\longrightarrow} \text{g} \quad \text{g} \times \text{u} \times \text{g} = \text{u} \rightarrow \text{forbidden} \quad (10.4.4)$$

$$\text{g} \overset{\text{u}}{\longrightarrow} \text{u} \quad \text{g} \times \text{u} \times \text{u} = \text{g} \rightarrow \text{allowed} \quad (10.4.5)$$

Therefore, "electric dipole transitions between states of equal parity are forbidden."

In other words, p→p, d→d, and f→f transitions are forbidden, but p→d and d→f transitions would be allowed by this electric rule. This can be written in the form: $\Delta l = \pm 1$ for orbitally allowed transitions, and is known as Laporte's rule.

- Since the dipole moment operator is one electron operator, therefore, an electric dipole transition can only involve one electron excitation:

  "Transitions involving excitation of two or more electrons are forbidden."

Vibrational Selection Rules

- The shape of an electronic absorption band is mainly caused by the vibrational sublevels of the electronic states, ψ_v and ψ_v^{ex}.
- Then, the value of $\langle \psi_v | \psi_v^{ex} \rangle$ modulates the intensity of absorption bands whose general range of magnitude is determined primary by spin and orbital selection rule.
- The quantity $\langle \psi_v | \psi_v^{ex} \rangle$ was labeled as the Franck-Condon factor. This integral signifies the overlap of vibrational wave functions of the ground and excited electronic states. The overlap integral also is a measure of the degree of similarity of the vibrational wave functions.
- The expression of the Franck-Condon principle is "An electronic transition is so fast, compared to nuclear motion, that the nuclei still have nearly the same position and momentum immediately after the transition as before."
- The majority of ground state molecules are in the $\upsilon = 0$ vibrational state. The $\upsilon = 0$ wave function is considered as the ground state in the following discussion. In Fig. 10.3, a case is illustrated in which maximum overlap occurs between the $\upsilon = 0$ and $\upsilon' = 2$ vibrational wave functions.
  - The upper well has slightly greater internuclear separation than the lower well at the energy minimum (Fig. 10.3A).
  - There is a turning point in the vibration and the momentum must go through zero as it changes sign (when nuclei change their direction of motion).
  - A transition from the ground state to the vibronic level labeled A (which the internuclear separation is still nearly r_e^o) involves little change in the position or the momentum of nuclei, so this vibrational state is the most probable. However, this vibrational state is not the only final one because several near states have a significant probability of nuclei being at r_e^o.
  - A transition to the level B is less likely because at the internuclear separation $r = r_e^o$ there is a considerable momentum in the level B.
  - A transition to the level C is very unlikely because r cannot be equal to r_e^o according to the classic mechanics.

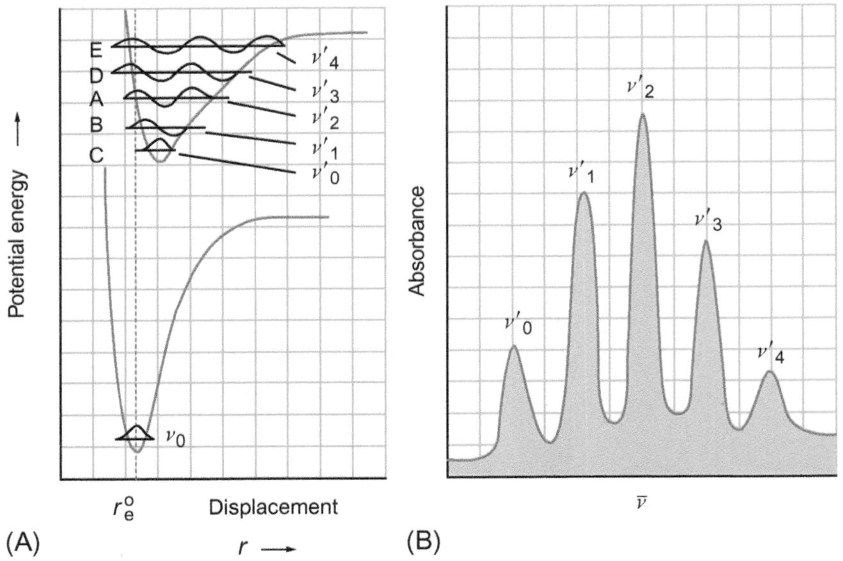

FIG. 10.3 (A) Illustration of maximal overlap between the $\upsilon = 0$ and $\upsilon' = 2$ vibrational wave functions, level A, where the internuclear separations are nearly equal. (B) Excitations to the vibrational sublevels impose the shape of the electronic absorption band.

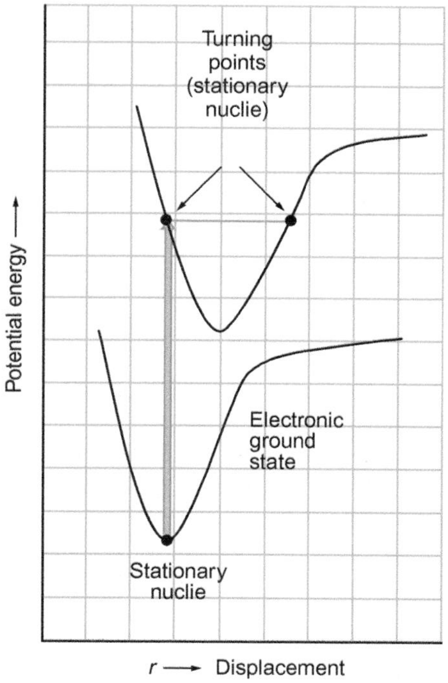

FIG. 10.4 The momentum that is near to zero in the ground state is also near to zero in the excited state at the separation r_e^o.

- The vibrational state whose turning point is vertically above the equilibrium internuclear separation of the ground state will give rise to the most intense absorption (Fig. 10.4).

10.5 UNEXPECTED WEAK ABSORBANCE

Why electrically forbidden transitions may exhibit unexpected weak absorbance?

- Electrically forbidden transitions exhibit unexpected weak absorbance due to the following:

i. *Same total angular momentum of two spin-states*

- If spin-orbital coupling is introduced, the singlet could have the same total angular momentum as the triplet, and the two states could interact:

$$\psi = a^1\psi + b^3\psi,$$

a and b are the relative contributions.

- If $a \gg b$, the ground state is mostly singlet, and the excited state will be mainly triplet.
- The small amount of singlet character in the essentially triplet excited state lead to intensity integral for the singlet-triplet transition that is not zero.

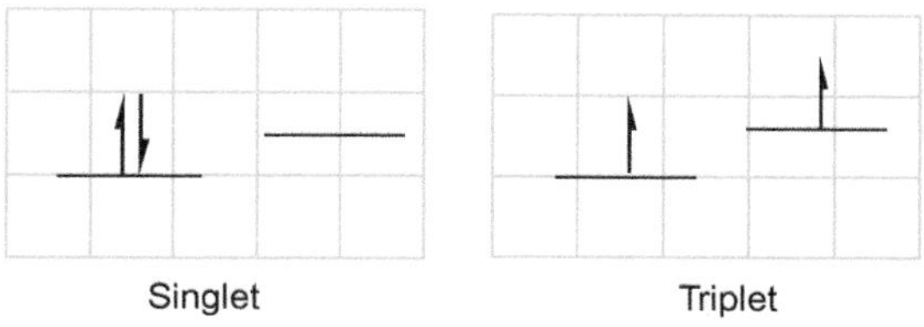

- Spin-orbital coupling is the interaction between the electron spin magnetic moment and the magnetic moment due to orbital motion of that electron.

- The orbital moment may either match or oppose the spin moment, inducing two different energy states. The doubly energy state of the electron (±½) is split, lowering the energy of one and raising the energy of the other.
- This is likely when an electron can occupy a set of degenerate orbitals that allow circulation about the molecule.
- For spin-orbital coupling, there must be:
  - no electron of similar spin in the other orbital, into which the electron should move; and
  - a rotating axis exists that permits to rotate one of the degenerate orbitals into another one.

  Examples:
  - If an electron can occupy either d_{xz} or d_{yz} orbitals of the metal ion, the electron can circle the nucleus around the z-axis (in O_h d^1).
  - On the other hand, high spin d^3 in O_h, there are electrons with the same spin quantum number in both d_{xz} and d_{yz}, so this ion does not have orbital angular momentum.

ii. *Vibronic coupling*
- Molecules are not rigid, and undergo vibrations.
- During certain vibrations, the molecular symmetry changes considerably, and the center of symmetry is removed.
- The states of d^n configurations no longer retain their g-character.
- The electronic transition can become slightly allowed by certain vibrational modes.
- This phenomenon is known as vibronic coupling.
- In the discussion of orbital and vibrational selection rules were based on:

$$\sqrt{f} = \text{constant} \cdot \underbrace{\left\langle \psi_{\text{spin}} | \psi^{\text{ex}}_{\text{spin}} \right\rangle}_{\text{basis of spin selection rules}} \underbrace{\left\langle \psi_{\text{orbital}} | M | \psi^{\text{ex}}_{\text{orbital}} \right\rangle}_{\text{basis of orbital selection rules}} \underbrace{\left\langle \psi_{\text{vibrational}} | \psi^{\text{ex}}_{\text{vibrational}} \right\rangle}_{\text{basis of vibrational selection rules}} \tag{10.3.3}$$

where only pure electronic transitions for which both $\psi_{\text{vibrational}}$ and $\psi^{\text{ex}}_{\text{vibrational}}$ are totally symmetric are considered.
- This equation can be arranged, if

$$\psi_{\text{el}} = \psi_{\text{orbital/spin}} \cdot \psi_{\text{vibrational}} = \psi_{\text{spin}} \cdot \psi_{\text{orbital}} \cdot \psi_{\text{vibrational}} \tag{10.3.1}$$

$$\psi_{\text{el}} = \psi_{\text{spin}} \cdot \psi_{\text{orbital, vibrational}}$$

$$\sqrt{f} = \text{constant} \cdot \left[\left\langle \psi_{\text{spin}} \cdot \psi_{\text{orbital, vibrational}} | M_{\text{el}} | \psi^{\text{ex}}_{\text{spin}} \psi^{\text{ex}}_{\text{orbital, vibrational}} \right\rangle \right]$$

$$\sqrt{f} = \text{constant} \cdot \left[\begin{array}{l} \left\langle \psi_{\text{spin}} | \psi^{\text{ex}}_{\text{spin}} \right\rangle \left\langle \psi_{\text{orbital, vibrational}} | M_{\text{el}} | \psi^{\text{ex}}_{\text{orbital, vibrational}} \right\rangle \\ + \left\langle \psi_{\text{orbital, vibrational}} | \psi^{\text{ex}}_{\text{orbital, vibrational}} \right\rangle \underbrace{\left\langle \psi_{\text{spin}} | M_{\text{el}} | \psi^{\text{ex}}_{\text{spin}} \right\rangle}_{=0} \end{array} \right]$$

$$\sqrt{f} = \text{constant} \cdot \underbrace{\left\langle \psi_{\text{spin}} | \psi^{\text{ex}}_{\text{spin}} \right\rangle}_{\text{basis of spin selection rules}} \underbrace{\left\langle \psi_{\text{orbital, vibrational}} | M | \psi^{\text{ex}}_{\text{orbital, vibrational}} \right\rangle}_{\text{basis of vibronic selection rules}}$$

The orbital and vibrational integrals are combined to allow for the case in which $\psi_{\text{vibrational}}$ and $\psi^{\text{ex}}_{\text{vibrational}}$ are not totally symmetric.
- The vibronic transition is allowed, if $\psi_{\text{orbital, vibrational}} |M| \psi^{\text{ex}}_{\text{orbital, vibrational}}$ be nonzero, or the direct product representation of:

$$\Gamma \left[\psi_{\text{orbital, vibrational}}(x, y, z)\, \psi^{\text{ex}}_{\text{orbital, vibrational}} \right]$$

has a symmetry component of normal vibrational modes (Chapter 9, p. 533, 546) of the concerned point group.

Example:

The coordinates x, y, and z in O_h point group, jointly form a basis for T_{1u} representation. Therefore, for $^1A_{1g} \rightarrow {^1T_{1g}}$ transition the direct product representation:

$$\Gamma\left[\psi_{\text{orbital, vibrational}}(x, y, z)\psi^{\text{ex}}_{\text{orbital, vibrational}}\right] = T_{1g} \times T_{1u} \times A_{1g}$$
$$= T_{1g} \times T_{1u}$$
$$= A_{1u} + E_u + T_{1u} + T_{2u}$$

The symmetries of the normal modes of octahedral molecules (Chapter 9) are:

$$A_{1g},\ E_g,\ 2T_{1u},\ T_{2g},\ T_{2u}$$

Although the pure electronic transition $^1A_{1g} \rightarrow {^1T_{1g}}$ is forbidden (g → g), there is simultaneous excitation of a vibration of T_{1u} or T_{1u} symmetry are allowed.

iii. *Interference of magnetically allowed transitions*
   - In molecules with center of symmetry, we can see that

$$g \rightarrow g$$
$$u \rightarrow u$$

only when electric dipole transitions are forbidden, these are very weak bands ($\varepsilon \cong 0.01\ M^{-1}\ cm^{-1}$), and complicate the spectra.

10.6 SPECTROSCOPY OF ELECTRONIC EXCITATIONS

Example: in the formaldehyde:

i. **What are the Lewis structure, the exact point group, and the possible types of bonding?**
ii. **What are the symmetry designations of the valence atomic orbitals and the suitable match of orbitals for bond formation in the carbonyl group?**
iii. **Define the oxygen orbitals that the two lone pairs most likely can occupy and the molecular orbital description of the valence electrons.**
iv. **Rank the molecular orbital of the molecule and draw qualitatively the boundary contours of the molecular orbitals.**
v. **What are the various excited states arising from the possible electronic excitations? Find out whether these transitions are allowed or forbidden.**
   - The functional groups in polyatomic molecules are commonly regarded as diatomic molecules, for instance, the carbonyl group of formaldehyde.
   - Using the coordinate system and the Lewis structure for the carbonyl of the formaldehyde (Fig. 10.5) directs us to anticipate one σ bonding, one σ*, one π bonding orbitals, one π* orbitals, and two oxygen lone pairs.
   - To designate these orbitals, it is necessary to find:
     - the symmetry of $H_2C{=}O$: C_{2v}; and
     - the symmetries of the valence atomic orbitals, where the plane yz contains the four atoms of the formaldehyde (Table 10.2).

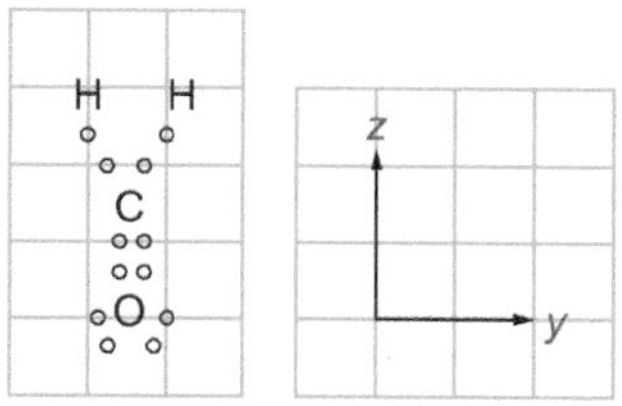

FIG. 10.5 Lewis structure of formaldehyde and coordinate system.

TABLE 10.2 Symmetry Designations of the Valence Atomic Orbitals of the Carbonyl Group in Formaldehyde

C_{2v}	E	C_2	σ_v^{xz}	σ_v^{yz}	Orbital
a_1	1	1	1	1	O($2p_z$)
b_2	1	−1	−1	1	O($2p_y$)
b_1	1	−1	1	−1	O($2p_x$)
a_1	1	1	1	1	O(2s)
a_1	1	1	1	1	C($2p_z$)
b_2	1	−1	−1	1	C($2p_y$)
b_1	1	−1	1	−1	C($2p_x$)
a_1	1	1	1	1	C(2s)

- The two lone pairs on oxygen are the nonbonding molecular orbitals, n_a^2 and n_b^2. Since there are no doubly degenerate irreducible representations in C_{2v}, the two lone pairs of oxygen of H_2CO do not have to be degenerate.
- The two n orbitals of the oxygen can be assigned as s and p_y orbitals; p_z is reserved to use in the σ bond and p_x in the π. The distinctive nature of s and p_y orbital promote the energies of the a_1 and b_2 lone pairs to be dissimilar.

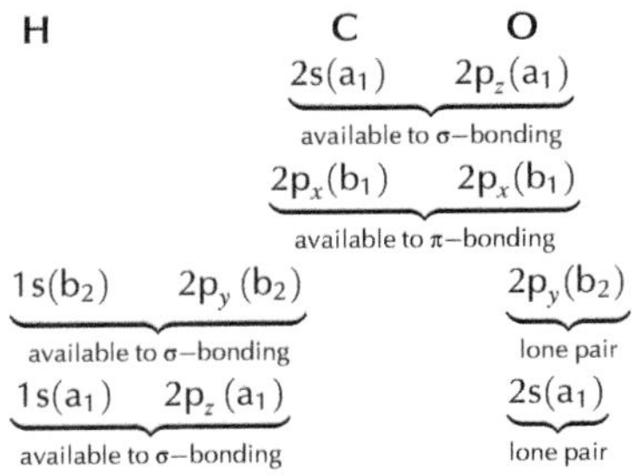

- A molecular orbital distribution of the valence electrons of the carbonyl group in formaldehyde molecule is: σ^2, π^2, n_a^2, n_b^2, π^*, σ^*.
- The ranking of these orbitals can be reached by chemical sense and by looking at the spectra of similar compounds. Fig. 10.6 shows the logical ordering of these orbitals, and represents qualitatively the boundary contours of the molecular orbitals.
- The symmetry operations E, C_2, σ_v^{xz}, and σ_v^{yz} (Table 10.3) are carried out on the molecular orbitals: σ, π, n_a, n_b, π^*, and σ^*, the designations are similar to that listed for the irreducible representations: $^1A_1$, $^1B_1$, $^1A_1$, $^1B_2$, $^1B_1$, $^1A_1$, respectively, superscripts are the multiplicities.
  - The multiplicities of these molecular orbitals are given by two times the sum of the individual spin, m_s, plus one: $2S + 1 = 2\sum m_s + 1$.
  - The lower case letters are used for an orbital and the upper case letters are reserved to describe the symmetry of the ground or excited state.
- Examples of possible excited states arising from electron excitation from the highest energy filled orbitals n_a, n_b, and π are given by the transition $n \rightarrow \pi^*$, $n \rightarrow \sigma^*$, $\pi \rightarrow \pi^*$, and $\sigma \rightarrow \sigma^*$ and are referred to as (1), (2), (3), and (4), respectively (Fig. 10.6). The transition $n \rightarrow \pi^*$ is the lowest in energy, which occur in formaldehyde and most carbonyl compounds.
- The symmetries of these excited states are the product of the symmetry species of each of the odd electron orbitals.
- The ground state is A_1, because each orbital contains a pair of electrons.

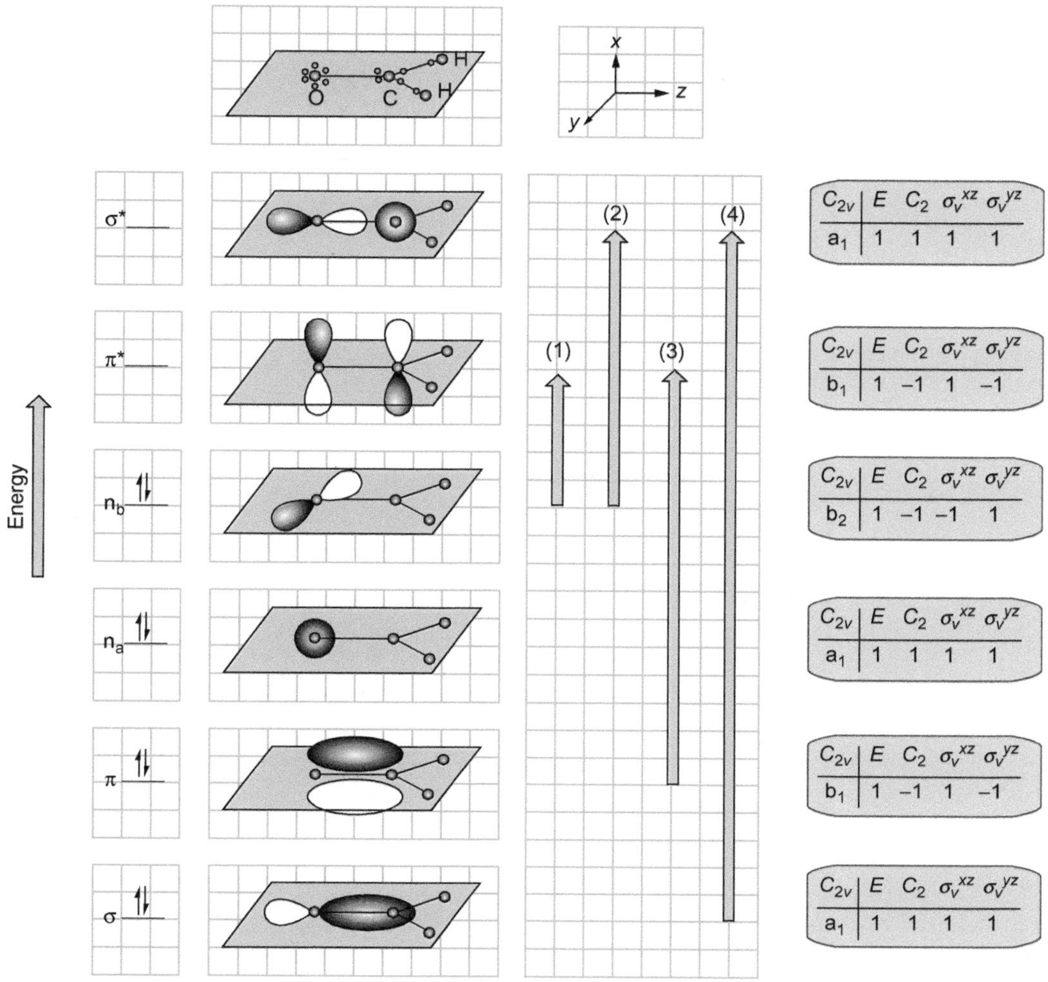

FIG. 10.6 Constructions and symmetries of the bonding molecular orbitals for the carbonyl group.

TABLE 10.3 Character Table of the C_{2v} Point Group

C_{2v}	E	C_2	σ_v^{xz}	σ_v^{yz}	
A_1	**1**	**1**	**1**	**1**	z
A_2	1	1	−1	−1	R_z
B_1	**1**	**−1**	**1**	**−1**	x, R_y
B_2	**1**	**−1**	**−1**	**1**	y, R_x
$\Gamma_{x,y,z}$	**3**	**−1**	**1**	**1**	$\Gamma_x(B_1)+\Gamma_y(B_2)+\Gamma_z(A_1)$

i. The electronic excitation: $(n_b \rightarrow \pi^*)$

$a_1^2b_1^2a_1^2b_2^1b_1^{1*}$	E	C_2	σ_v^{xz}	σ_v^{yz}	
$b_2 \times b_1$:	(1×1)	(-1×-1)	(1×-1)	(-1×1)	
Excited state	1	1	−1	−1	$={}^1A_2$

For $(n_b \rightarrow \pi^*)$	E	C_2	σ_v^{xz}	$\sigma_v^{yz}Z$	
$^1A_1 \rightarrow {}^1A_2$:	$(1\times3\times1)$	$(1\times-1\times1)$	$(1\times1\times-1)$	$(1\times1\times-1)$	
$({}^1A_1\Gamma^1_{x,y,z}A_2)$	3	-1	-1	-1	$={}^1A_2+{}^1B_1+{}^1B_2$
$^1A_1 \rightarrow {}^1A_2$ is	*Forbidden*	(No A_1)			

ii. The electronic excitation: $(n_b \rightarrow \sigma^*)$

$a_1^2b_1^2a_1^2b_2^1a_1^{1*}$	E	C_2	σ_v^{xz}	σ_v^{yz}	
$b_2 \times a_1$:	(1×1)	(1×-1)	(1×-1)	(1×1)	
Excited state	1	-1	-1	1	$={}^1B_2$

For $(n_b \rightarrow \sigma^*)$	E	C_2	σ_v^{xz}	σ_v^{yz}	
$^1A_1 \rightarrow {}^1B_2$:	$(1\times3\times1)$	$(1\times-1\times-1)$	$(1\times1\times-1)$	$(1\times1\times1)$	
$({}^1A_1\Gamma^1_{x,y,z}B_2)$	3	1	-1	1	$={}^1A_1+{}^1A_2+{}^1B_2$
$^1A_1 \rightarrow {}^1B_2$ is	*Allowed*				

iii. The electronic excitation: $(\pi \rightarrow \pi^*)$

$a_1^2b_1^1a_1^2b_2^2b_1^{1*}$	E	C_2	σ_v^{xz}	σ_v^{yz}	
$b_1 \times b_1$:	(1×1)	(-1×-1)	(1×1)	(-1×-1)	
Excited state	1	1	1	1	$={}^1A_1$

For $(\pi \rightarrow \pi^*)$	E	C_2	σ_v^{xz}	σ_v^{yz}	
$^1A_1 \rightarrow {}^1A_1$:	$(1\times3\times1)$	$(1\times-1\times1)$	$(1\times1\times1)$	$(1\times1\times1)$	
$({}^1A_1\Gamma^1_{x,y,z}A_1)$	3	-1	1	1	$={}^1A_1+{}^1B_1+{}^1B_2$
$^1A_1 \rightarrow {}^1A_1$ is	*Allowed*				

iv. The electronic excitation: $(\pi \rightarrow \sigma^*)$

$a_1^2b_1^1a_1^2b_2^2a_1^{1*}$	E	C_2	σ_v^{xz}	σ_v^{yz}	
$a_1 \times b_1$:	(1×1)	(1×-1)	(1×1)	(1×-1)	
Excited state	1	-1	1	-1	$={}^1B_1$

For $(\pi \rightarrow \sigma^*)$	E	C_2	σ_v^{xz}	σ_v^{yz}	
$^1A_1 \rightarrow {}^1B_1$:	$(1\times3\times1)$	$(1\times-1\times-1)$	$(1\times1\times1)$	$(1\times1\times-1)$	
$({}^1A_1\Gamma^1_{x,y,z}B_1)$	3	1	1	-1	$={}^1A_1+{}^1A_2+{}^1B_2$
$^1A_1 \rightarrow {}^1B_1$ is	*Allowed*				

v. The electronic excitation: $(n_a \rightarrow \pi^*)$

$a_1^1b_1^2a_1^2b_2^2b_1^{1*}$	E	C_2	σ_v^{xz}	σ_v^{yz}	
$a_1 \times b_1$:	(1×1)	(1×-1)	(1×1)	(1×-1)	
Excited state	1	-1	1	-1	$={}^1B_1$

For $(n_a \rightarrow \pi^*)$	E	C_2	σ_v^{xz}	σ_v^{yz}	
$^1A_1 \rightarrow {}^1B_1$:	$(1\times3\times1)$	$(1\times-1\times-1)$	$(1\times1\times1)$	$(1\times1\times-1)$	
$({}^1A_1\Gamma^1_{x,y,z}B_1)$	3	1	1	-1	$={}^1A_1+{}^1A_2+{}^1B_2$
$^1A_1 \rightarrow {}^1B_1$ is	*Allowed*				

vi. The electronic excitation: $(n_a \rightarrow \sigma^*)$

$a_1^1b_1^2a_1^2b_2^2a_1^{1*}$	E	C_2	σ_v^{xz}	σ_v^{yz}	
$a_1 \times a_1$:	(1×1)	(1×1)	(1×1)	(1×1)	
Excited state	1	1	1	1	$={}^1A_1$

For ($n_a \rightarrow \sigma^*$)	E	C_2	σ_v^{xz}	σ_v^{yz}	
$^1A_1 \rightarrow {}^1A_1$:	$(1\times3\times1)$	$(1\times-1\times1)$	$(1\times1\times1)$	$(1\times1\times1)$	
$(^1A_1\Gamma^1_{x,y,z}A_1)$	3	−1	1	1	$= {}^1A_1 + {}^1B_1 + {}^1B_2$
$^1A_1 \rightarrow {}^1A_1$ is	*Allowed*				

- The $^1A_1 \rightarrow {}^1A_2$ ($n_b \rightarrow \pi^*$) is forbidden and observed at 270 nm with $\varepsilon = 100$. Both $^1A_1 \rightarrow {}^1B_1$ ($n_a \rightarrow \pi^*$) and $^1A_1 \rightarrow {}^1A_1$ ($\pi \rightarrow \pi^*$) are allowed, and may share in the observed band at 185 nm.
- The integrands for $^1A_1 \rightarrow {}^1B_1$, $^1A_1 \rightarrow {}^1B_2$, and $^1A_1 \rightarrow {}^1A_1$ are all contain $^1A_1$, leading to allow transitions. These are expected to occur at very short wavelengths in the far ultraviolet region.

10.7 ELECTRONIC SPECTRA OF SELECTED EXAMPLES

Jahn-Teller Theorem and Vibronic Coupling: d^1 Configuration

Why do the aqueous solutions of titanium(III) salts have a purple color?
$Ti(H_2O)_6^{3+}$ is a d^1 configuration. How can you explain the asymmetric shape and the cause of this absorption?

- $Ti(H_2O)_6^{3+}$ is d^1 configuration, and the aqueous solutions of titanium(III) salts have a purple color.
- The visible spectrum of this violet solution shows an absorption band with maximum near 500 nm (20,000 cm^{-1}) and an extinction coefficient which suggests that transition is spin-allowed ($\Delta S = 0$) but orbitally forbidden ($g \rightarrow g$), $^2T_{2g}\left[(t_{2g})^1(e_g)^o\right] \rightarrow {}^2E_g\left[(t_{2g})^o(e_g)^1\right]$ transition (Fig. 10.7).
- This transition does not exhibit symmetric absorption band. An interpretation of this is provided by the Jahn-Teller theorem, according to which "Any nonlinear molecule, in an orbitally degenerate state (≥ 2, i. e. E and T states), will undergo a distortion that lowers the energy of the molecule and remove the degeneracy of that state."
- The ground state of $Ti(H_2O)_6^{3+}$, $^2T_{2g}$, is orbitally degenerate and a distortion in which two opposite ligands move closer to the Ti atom (Fig. 10.8).
- This reduces the symmetry of the complex to D_{4h} and gives the ground state $^2B_{2g}$.

Jahn-Teller:	O_h	→	D_{4h} (Table 10.4)
	$E_g(d_{x^2-y^2}, d_{z^2})$	→	$A_{1g}(d_{z^2}) + B_{1g}(d_{x^2-y^2})$
	T_{2g} (d_{xz}, d_{yz}, d_{xy})	→	B_{2g} (d_{xy}) + E_g (d_{xz}, d_{yz})

- The single $T_{2g} \rightarrow E_g$ transition of the O_h complex is replaced by two in the D_{4h} complex: $^2B_{2g} \rightarrow {}^2B_{1g}$ and $^2B_{2g} \rightarrow {}^2A_{1g}$.
- These two absorptions overlap to produce the asymmetric absorption band. In addition, a very low energy $^2B_{2g} \rightarrow {}^2E_g$ transition is expected in the infrared region of the spectrum.
- These d→d transitions are orbitally forbidden, but spin allowed. However, these transitions become allowed when the simultaneous electronic and vibrational transitions are examined.

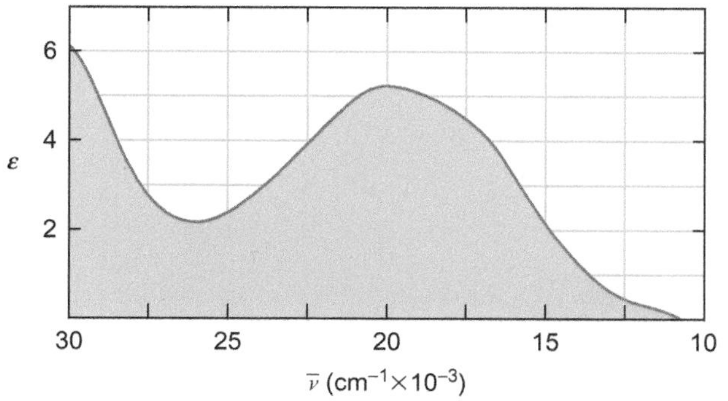

FIG. 10.7 The visible spectrum of $Ti(H_2O)_6^{3+}$. *(From A.B.P. Lever, Inorganic Electronic Spectroscopy, second ed., Elsevier Science Publishers B.V., Amsterdam, 1984, p. 381 (Chapter 4), reproduced with permission).*

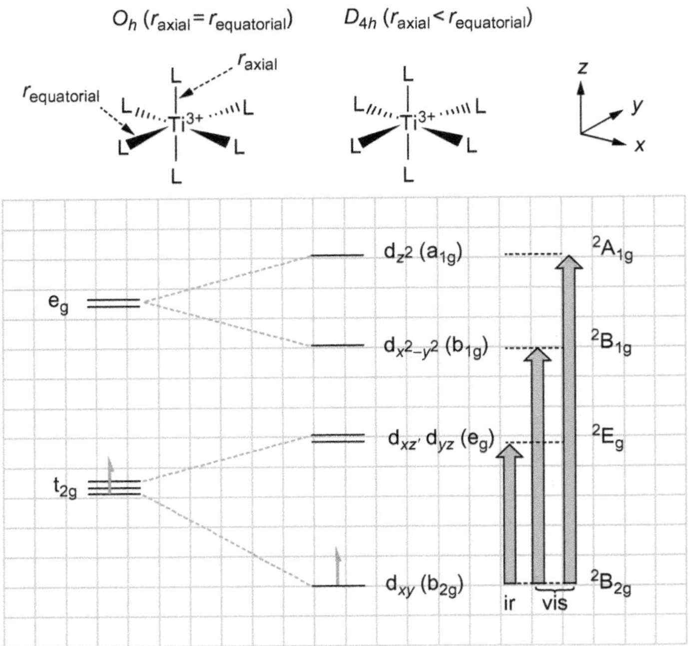

FIG. 10.8 Distortion of a d^1 octahedral complex by compression of axial bonds, the resulting energy level diagram and expected transition.

TABLE 10.4 Character Table of Point Group D_{4h}

D_{4h}	E	$2C_4$	C_2	$2C_2'$	$2C_2''$	i	$2S_4$	σ_h	$2\sigma_v$	$2\sigma_d$		
A_{1g}	1	1	1	1	1	1	1	1	1	1		x^2+y^2, z^2
A_{2g}	1	1	1	−1	−1	1	1	1	−1	−1	R_z	
B_{1g}	1	−1	1	1	−1	1	−1	1	1	−1		x^2-y^2
B_{2g}	**1**	**−1**	**1**	**−1**	**1**	**1**	**−1**	**1**	**−1**	**1**		xy
E_g	**2**	**0**	**−2**	**0**	**0**	**2**	**0**	**−2**	**0**	**0**	(R_x, R_y)	(xz, yz)
A_{1u}	1	1	1	1	1	−1	−1	−1	−1	−1		
A_{2u}	**1**	**1**	**1**	**−1**	**−1**	**−1**	**−1**	**−1**	**1**	**1**	z	
B_{1u}	1	−1	1	1	−1	−1	1	−1	−1	1		
B_{2u}	1	−1	1	−1	1	−1	1	−1	1	−1		
E_u	**2**	**0**	**−2**	**0**	**0**	**−2**	**0**	**2**	**0**	**0**	**(x,y)**	
$\langle ^2B_{2g}\|E_u\|^2E_g\rangle =$	**4**	**0**	**4**	**0**	**0**	**−4**	**0**	**−4**	**0**	**0**	$= A_{1u}+A_{2u}+B_{1u}+B_{2u}$	

- Since the electronic wave functions are both g, the direct product of irreducible representation of the ground state ($^2B_{2g}$), dipole moment operator [A_{2u}: z or E_u: x, y), and the excited representations (^2E_g, $^2B_{1g}$, and $^2A_{1g}$) contain representation of one of the normal mode vibrations (Chapter 9).

$$\text{e.g.}\quad \int \psi^*_{^2B_{2g}}(x,y)\,\psi_{^2E_g}\,\partial v \equiv \langle ^2B_g|E_u|^2E_g\rangle$$

D_{4h}	E	$2C_4$	C_2	$2C_2'$	$2C_2''$	i	$2S_4$	σ_h	$2\sigma_v$	$2\sigma_d$	
$\langle ^2B_g \vert E_u \vert ^2E_g \rangle =$	4	0	4	0	0	−4	0	−4	0	0	$=A_{1u}+A_{2u}+B_{1u}+B_{2u}$

- Using

$$\Gamma_i = \frac{1}{h}\sum_R \chi(R)\chi_i(R)$$

- This irreducible representation will be reduce to:
  $A_g = 1/16(1\times1\times4+0+1\times1\times4+0+0+1\times1\times-4+0+1\times1\times-4+0+0)=0$
  $A_{2g} = 1/16(1\times1\times4+0+1\times1\times4+0+0+1\times1\times-4+0+1\times1\times-4+0+0)=0$
  $B_{1g} = 1/16(1\times1\times4+0+1\times1\times4+0+0+1\times1\times-4+0+1\times1\times-4+0+0)=0$
  $B_{2g} = 1/16(1\times1\times4+0+1\times1\times4+0+0+1\times1\times-4+0+1\times1\times-4+0+0)=0$
  $E_g = 1/16(1\times2\times4+0+1\times-2\times4+0+0+1\times2\times-4+0+1\times-2\times-4+0+0)=0$
  $A_{1u} = 1/16(1\times1\times4+0+1\times1\times4+0+0+1\times-1\times-4+0+1\times-1\times-4+0+0)=1$
  $A_{2u} = 1/16(1\times1\times4+0+1\times1\times4+0+0+1\times-1\times-4+0+1\times-1\times-4+0+0)=1$
  $B_{1u} = 1/16(1\times1\times4+0+1\times1\times4+0+0+1\times-1\times-4+0+1\times-1\times-4+0+0)=1$
  $B_{2u} = 1/16(1\times1\times4+0+1\times1\times4+0+0+1\times-1\times-4+0+1\times-1\times-4+0+0)=1$
  $E_u = 1/16(1\times2\times4+0+1\times-2\times4+0+0+1\times-2\times-4+0+1\times2\times-4+0+0)=0$
  Therefore:

$$\int \psi^*_{^2B_{2g}}(x,y)\psi_{^2E_g}\partial\nu = \langle ^2B_g|E_u|^2E_g\rangle = A_{1u}+A_{2u}+B_{1u}+B_{2u}$$

- In D_{4h}, the normal modes of vibration are $2A_{1g}$, B_{1g}, B_{2g}, E_g, 2 A_{2u}, B_{1u}, and $3E_u$ (Chapter 9). Therefore, the transition $^2B_{2g} \rightarrow {}^2E_g$ become vibronically allowed.

The Expected Position of Absorption Peaks: d^2 Configuration

How many d-d transition may occur in aqueous solution of $V(H_2O)_6{}^{3+}$? Assuming $10D_q = 18{,}900\ cm^{-1}$, DQ = 51,389, and $B = 605\ cm^{-1}$, where DQ is the average global field strength, what will be the expected position of the absorption peak for these transitions?

- The d^2 configuration $(t_{2g})^2(e_g)^0$ has a $^3T_{1g}$ ground state.
- From the energy diagram (Fig. 10.9), the three triplet states $^3T_{1g}$, $^3T_{2g}$, and $^3A_{2g}$ are the higher energy configurations of the d^2 to which spin-allowed d-d transitions may occurs.
- The octahedral spin triplet d^2 energies (Chapter 8, p. 498):

$$E(^3T_{1g}(F)) = 7.5B - 3D_q - \frac{1}{2}\left(225B^2 + 100D_q{}^2 + 180D_qB\right)^{1/2}$$

$$E(^3T_{2g}(F)) = 2D_q$$

$$E(^3A_{2g}(F)) = 12D_q$$

$$E(^3T_{1g}(P)) = 7.5B - 3D_q + \frac{1}{2}\left(225B^2 + 100D_q{}^2 + 180D_qB\right)^{1/2}$$

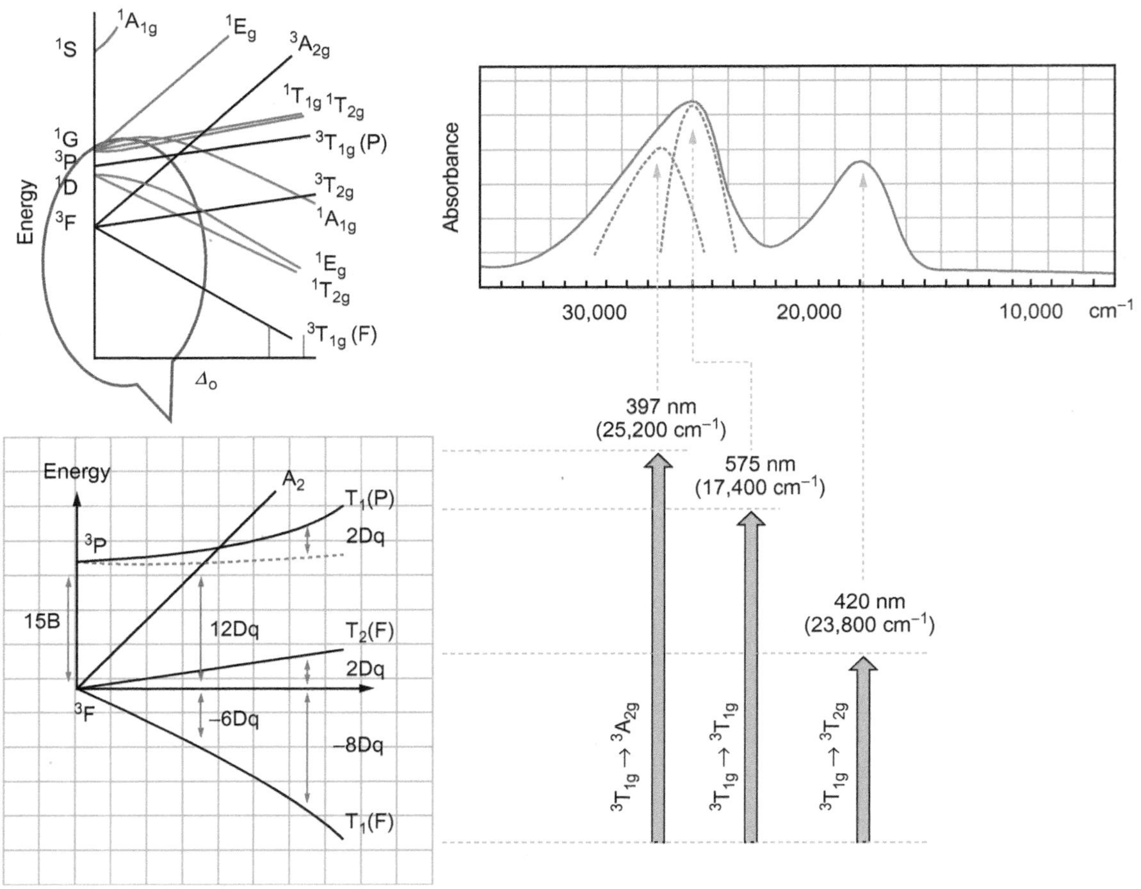

FIG. 10.9 The energies of the various states of a d^2 transition metal species as a function of Δ_O and the predicted transitions. Absorption spectrum of $V(H_2O)_6^{3+}$. *(From B.N. Figgis, Introduction to Ligand Fields, John Willey and Sons, Inc., 1986, p. 221 (Fig. 9.4 absorption spectra in solid, Chapter 9), ISBN 0-89874-819-4, reproduced with permission).*

By substitution for $D_q = 1890\ \text{cm}^{-1}$, $B = 605\ \text{cm}^{-1}$:

$$E\left({}^3T_{1g}(F)\right) = -13{,}835\ \text{cm}^{-1}$$

$$E\left({}^3T_{2g}(F)\right) = 3780\ \text{cm}^{-1}$$

$$E\left({}^3A_{2g}(F)\right) = 22{,}680\ \text{cm}^{-1}$$

$$E\left({}^3T_{1g}(P)\right) = 11{,}570\ \text{cm}^{-1}$$

- The three low energy d-d transitions for a d^2 species are observed for $V(H_2O)_6^{3+}$ at the following energies:

$$^3T_{1g} \rightarrow {}^3T_{2g}\ \ 568\ \text{nm}\ \left(17{,}615\ \text{cm}^{-1}\right)$$

$$^3T_{1g} \rightarrow {}^3T_{1g}\ \ 394\ \text{nm}\ \left(25{,}405\ \text{cm}^{-1}\right)$$

$$^3T_{1g} \rightarrow {}^3A_{2g}\ \ 273\ \text{nm}\ \left(36{,}515\ \text{cm}^{-1}\right)$$

The d^2 ions (V^{3+}) should have spectra with three peaks, and except in the region where the A_{2g} and T_{1g} (P) levels cross. These electrons transitions are expected to be weak and often obscured by charge transfer or ligand absorption.

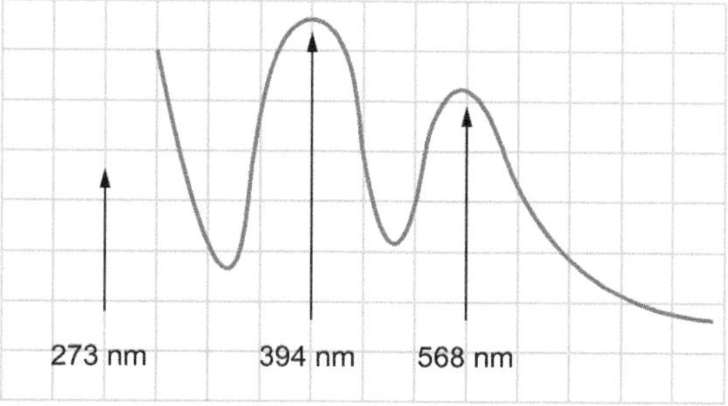

Configuration Interaction: d^3 Configuration

Name the ground and possible excited states for observed d-d transitions in the spectrum of VCL_6^{4-}, Fig. 10.10. Assuming $10D_q = 7330\ cm^{-1}$, calculate the configuration interaction between $T_{1g}(F)$ and $T_{1g}(P)$ states. What will be the value of Racah parameter B' that measures the interelectronic repulsion between ^4P and ^4F terms?

- V^{2+} in $[VCl_6]^{-4}$ is a d^3 ion.
- Fig. 10.10 represents the energy diagram and the expected transitions.
  - The ground state is $^4A_2$.
  - Three spin allowed transitions are predicted:

$$^4A_2\left(t_{2g}{}^3e_g{}^0\right) \rightarrow {}^4T_{2g}\left(t_{2g}{}^2e_g{}^1\right),\quad (\Delta_o)10D_q \approx 7200\ cm^{-1}$$

$^4A_2 \rightarrow {}^4T_{1g}$ (^4F), and
$^4A_2 \rightarrow {}^4T_{1g}$ (^4P), in order of increasing energy (Fig. 10.11).

- Energy of first excitation: $\nu_1 = 10D_q$
- The energy levels of the same symmetry $T_{1g}(F)$ and $T_{1g}(P)$ cannot cross, and as Δ_o increases as the separation between the two states becomes larger due to configuration interaction (Figs. 10.11 and 10.12).

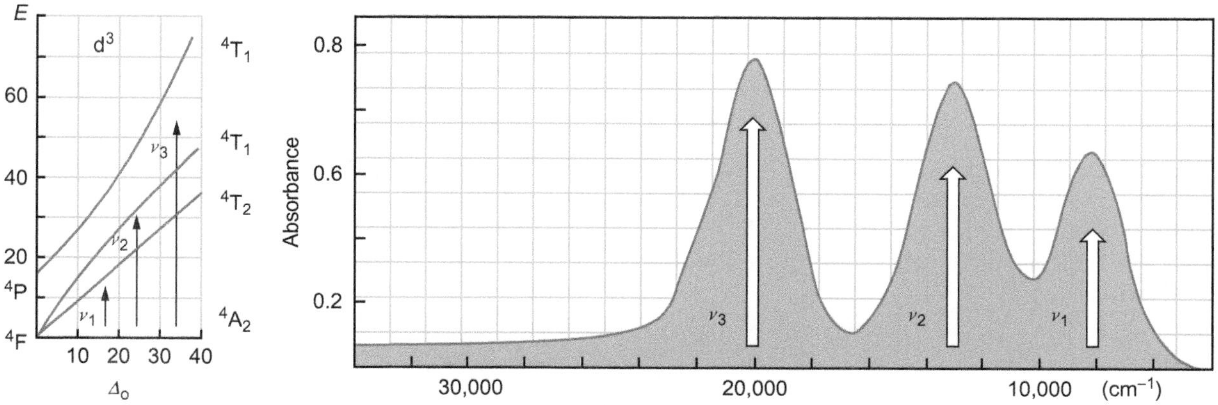

FIG. 10.10 Energy level diagram and expected transition for d^3 ions. Spectrum of $[VCl_6]^{4-}$. *(Sketched from A.B.P. Lever, Inorganic Electronic Spectroscopy, second ed., Elsevier Science Publishers B.V., Amsterdam, 1984, p. 416 (Chapter 6), reproduced with permission).*

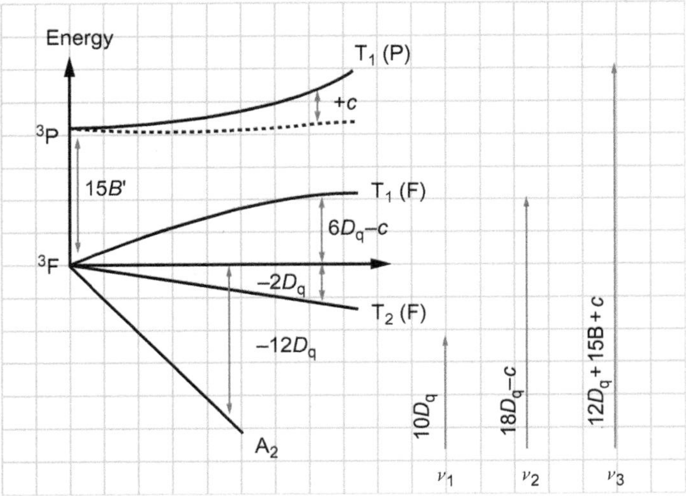

FIG. 10.11 Orgel diagram for d^3 ions in O_h field.

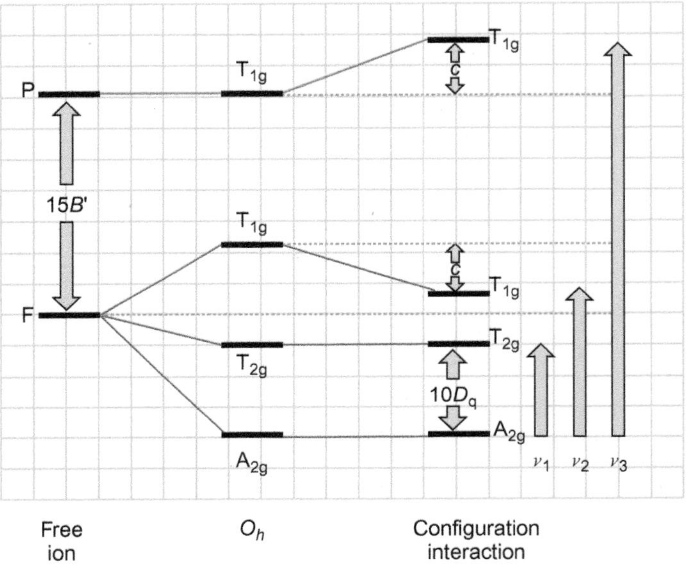

FIG. 10.12 Splitting and mixing of states for d^3 ions in an O_h field.

- Energy of second excitation: $\nu_2 = 18D_q - c$
  Where c is the deviation of $T_{1g}(F)$ and $T_{1g}(P)$ levels, the configuration interaction.
- Energy of third excitation: $\nu_3 = 12D_q + 15B' + c$
  where B' is the atomic interelectronic repulsion, Racah parameter of the complex.
  Then:

$$(\nu_2 - 18D_q + c) + (\nu_3 - 12D_q - 15B' - c) = 0$$

$$(\nu_2 - 18D_q + c) + (\nu_3 - 12D_q - 15B' - c) = 0$$

$15B' = \nu_2 + \nu_3 - 3\nu_1$ (Fig. 10.11)
Where $12D_q + 18D_q = 30D_q = 3\nu_1$

From the electronic spectrum, Fig. 10.10, B' can be calculated:

$$B' = \frac{12,080 + 20,180 - (3 \times 7,330)}{15} = 685\,\text{cm}^{-1}$$

- Comparable with the free ion value B (735 cm^{-1})
- This is a quantitative evidence for reduction of B in the complex ion.
- $c = 18D_q - \nu_2 = 1114\,\text{cm}^{-1}$

Temperature and Absorption Spectra

If the following figure is the electronic spectra of $[VCl_6]^{4-}$ at 296 and 77 K (Fig. 10.13), what are the observed impacts of temperature change on the band shapes? How can these variations be explained?

- The visible spectrum of $[VCl_6]^{4-}$ is strongly temperature dependent (Fig. 10.12). As temperature decreases, there are three aspects of the temperature dependence of the band shape that should be explained:
  (i) The band maxima move toward higher energy.
  (ii) The bands become sharper
  (iii) The bands become less intense.
- As the temperature decreases, ligands move toward the metal center, and D_q obviously increases.
- The slopes $\partial E/\partial D_q$ Fig. 10.10, are positive except for levels approaching crossover. Therefore, the bands maxima move toward higher energy.
- The variation of D_q values decreases as the vibrations decrease with decreasing temperature. Therefore, the bands become sharper.
- The band shape of a vibronically allowed transition will change with temperature because the change in the Boltzmann distribution of vibrational energy levels.
  - As the top vibrational levels of the ground electronic state become less populated, the lower energy side of the band will decrease in intensity, thus the observed band maximum will shift to higher energy.
  - Since the transition between ground vibrational levels in the two electronic states is forbidden ($g \rightarrow g$), the oscillator strength (band intensity) decreases as temperature decreases.

Spectrochemical Series

The following data indicates three electronic transitions in the visible spectra of $[CrF_6]^{3-}$, $[Cr(H_2O)_6]^{3+}$, $[Cr(NCS)_3]^{3+}$, $[Cr(NH_3)_6]^{3+}$, $[Cr(en)_3]^{3+}$, and $[Cr(CN)_6]^{3-}$ (Fig. 10.14):

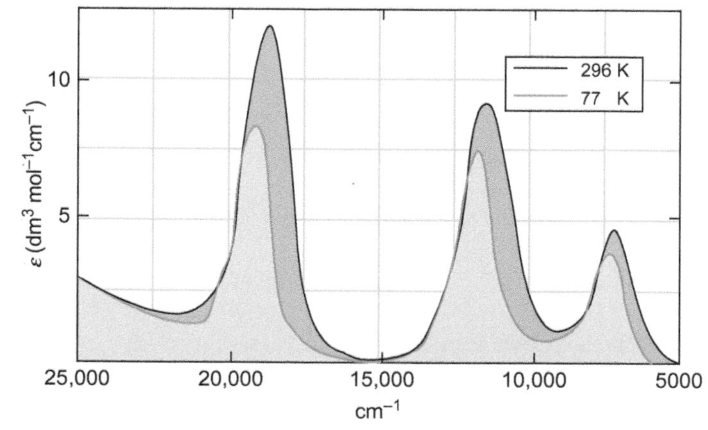

FIG. 10.13 The electronic spectra of $[VCl_6]^{4-}$ at 296 and 77 K, the peak positions at 19,100, 11,700, and 7200 cm^{-1}. *(From G.L. McPherson, M.R. Freedman, Inorg. Chem. 15 (2299) 1976, reproduced with permission).*

FIG. 10.14 Visible absorption spectra of some chromium(III) complexes in aqueous solution. $Cr(H_2O)_6^{3+}$ *(from B.N. Figgis, Introduction to Ligand Fields, John Willey and Sons, Inc., 1986, p. 224 (Fig. 9.7 absorption spectra in solid, Chapter 9), reproduced with permission. Spectrum of* $[CrF_6]^{3-}$ *sketched from A.B.P. Lever, Inorganic Electronic Spectroscopy, second ed., Elsevier Science Publishers B.V., Amsterdam, 1984, p. 192 (Chapter 4) reproduced with permission).*

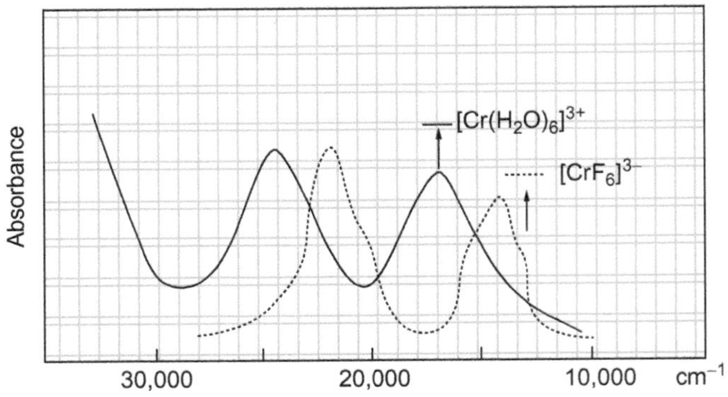

CrF_6^{3-}	**14,900 cm^{-1}**	**22,700 cm^{-1}**	**34,400 cm^{-1}**
$Cr(H_2O)_6^{3+}$	**17,400 cm^{-1}**	**24,600 cm^{-1}**	**37,800 cm^{-1}**
$Cr(NCS)_6^{3-}$	**17,700 cm^{-1}**	**23,800 cm^{-1}**	**32,400 cm^{-1}**
$Cr(NH_3)_6^{3+}$	**21,550 cm^{-1}**	**28,500 cm^{-1}**	
$Cr(en)_3^{3+}$	**22,300 cm^{-1}**	**28,670 cm^{-1}**	**en = NH_2—CH_2—CH_2—NH_2**
$Cr(CN)_6^{3-}$	**26,700 cm^{-1}**	**32,600 cm^{-1}**	

Name the ground and possible excited states for these transitions. How can you explain the correlation between band positions and ligand electronic properties? What are the induced spectral changes upon reduction of $[Cr(H_2O)_6]^{3+}$ to $[Cr(H_2O)_6]^{2+}$ (Fig. 10.16)?

- The spectra of $[CrF_6]^{3-}$, $[Cr(H_2O)_6]^{3+}$, $Cr(NCS)_6^{3-}$, $Cr(NH_3)_6^{3+}$, $[Cr(en)_3]^{3+}$, and $[Cr(CN)_6]^{3-}$ contain bands that are much sharper than those of other d^3 ions.
- The sharp bands in the spectra are transitions from the $^4A_2$ ground state to the two excited states $T_{1g}(F)$ and $T_{1g}(P)$ (Figs. 10.11 and 10.12). Three spin allowed transitions are predicted:

	$^4A_2 \rightarrow {}^4T_{2g}$, ($\Delta_o$) $10D_q$ (cm^{-1})	$^4A_2 \rightarrow {}^4T_{1g}(^4F)$ (cm^{-1})	$^4A_2 \rightarrow {}^4T_{1g}(^4P)$ (cm^{-1})
CrF_6^{3-}	14,900	22,700	34,400
$Cr(H_2O)_6^{3+}$	17,400	24,600	37,800
$Cr(NCS)_6^{3-}$	17,700	23,800	32,400
$Cr(NH_3)_6^{3+}$	21,550	28,500	
$Cr(en)_3^{3+}$	22,300	28,670	
$Cr(CN)_6^{3-}$	26,700	32,600	

- The positions of the d-d bands are dependent on the nature of the ligand.
- Each of F^-, H_2O, NCS^{1-}, NH_3, en, and CN^- have suitable p_π atomic orbitals or π molecular orbitals available for overlap with the $t_{2g}(d_\pi)$ orbitals of the metal ion, therefore, π-binding must be considered (Fig. 10.15).
- When ligands contain donor atoms like F^-, H_2O, and en, the only π orbitals available for coordination are t_{2g}, which are filled (saturated). The ligand 2p-orbitals (t_{2g}) will lie at lower energy than the unsaturated metal t_{2g} orbitals to an extent that matches the donor ability.
- Cyanide as a ligand has empty π-orbitals (unsaturated), the orbitals t_{2g} of the ligand will lie at higher energy than those of the corresponding metal orbital.
- It may be seen that the effect of F^-, H_2O, and en are to decrease Δ, while in CN^- it is increased.
- The ligands can be placed in order of their increasing crystal field, Δ, which is independent of the nature of metal.
- The sequence of the ligands placed in order of their Δ strength is:

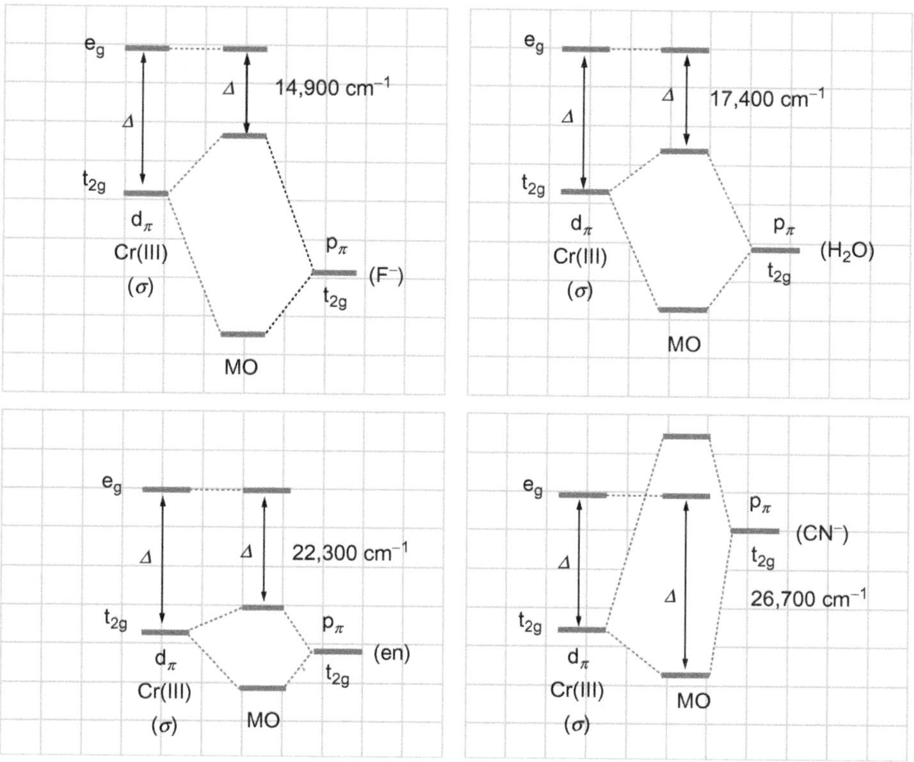

FIG. 10.15 Interaction of ligand orbitals and metal t_{2g} orbitals.

$$F^- < H_2O < en < CN^-$$

and are called spectrochemical series. Thus, π bonding is important in understanding the positions of ligands in the spectrochemical series. The order for some ligands is shown below:

$I^- < Br^- < Cl^- <$ —SCN$^-$ $< F^- < OH^- <$ Oxalate$^- \leq H_2O <$ —NCS$^- \leq NH_3 <$ Pyr$^- <$ Ethylenediamine $<$ Dipyr $<$ o-phenantholine $< NO_2^- < CN^- < CO$.

Configuration: $Cr(H_2O)_6^{2+}$ Versus $Cr(H_2O)_6^{3+}$

- Aqueous solutions of chromous ions are blue.
- There is a broad band at 14,100 cm^{-1} (709 nm, $\varepsilon = 4.2$) (Fig. 10.16).
- No interelectronic repulsions are involved in the splitting of the ^5D term.
- The band is attributed to the spin-allowed transition: $^5E_g \rightarrow {}^5T_{2g}$.

This corresponds to $10D_q$. Note that the value of D_q is larger for the trivalent ions, $[Cr(H_2O)_6]^{3+}$, than for the divalent ion, $[Cr(H_2O)_6]^{2+}$.

Octahedral Versus Tetrahedral: d^5 Configuration

What are the common and the different features in the electronic spectra of $[Fe(H_2O)_6]^{3+}$ and $[Fe(H_2O)_4]^{3+}$?

- $Fe(OH_2)_6^{3+}$, high-spin, the ground state of $3d^5$ is $^6A_{1g}(t_{2g})^3(e_g)^2$.
- Ligand field splitting, Δ_o, is larger for Fe^{3+} than Fe^{2+}. Therefore, the tendency toward a low-spin ground state structure is greater.

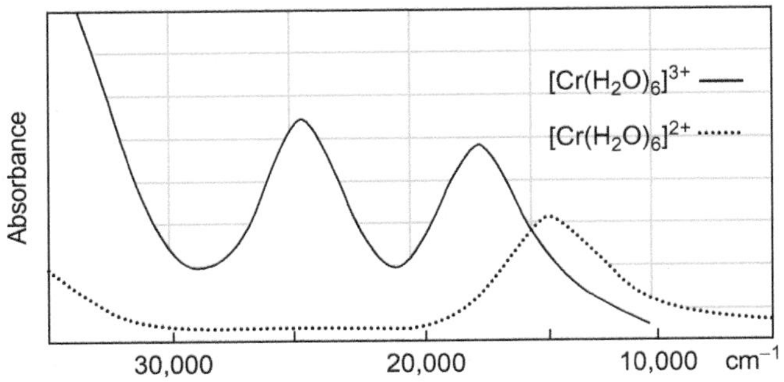

FIG. 10.16 Visible absorption spectra of $[Cr(H_2O)_6]^{3+}$ and $[Cr(H_2O)_6]^{2+}$. *(From B.N. Figgis, Introduction to Ligand Fields, John Willey and Sons, Inc., 1986, p. 224 (Fig. 9.7 absorption spectra in solid, Chapter 9), ISBN 0-89874-819-4, reproduced with permission).*

- The Tanabe-Sugano diagram for d^5 ions (Fig. 10.17) shows that $^4T_{1g}$, $^4T_{2g}$, the degenerate pair ^4E_g, $^4A_{1g}$, and $^4T_{2g}$ are the lowest electronic excited states.
- Consequently, all the transitions of $^6A_{1g}$ complexes are spin-forbidden, and weak absorption bands are expected. For example, the electronic spectra of $[Fe^{3+}O_6]$ in $Fe_2(SO_4)_3{\cdot}(NH_4)_2SO_4{\cdot}24H_2O$ (Fig. 10.18) is as follows:

$$^6A_{1g} \rightarrow {}^4T_{1g},\ 12{,}600\ cm^{-1}(794\ nm),\ \varepsilon = 0.05$$

$$^6A_{1g} \rightarrow {}^4T_{2g},\ 18{,}200\ cm^{-1}(540\ nm),\ \varepsilon = 0.01$$

$$^6A_{1g} \rightarrow {}^4E_g, {}^4A_{1g}\ 24{,}600\ cm^{-1}(406\ nm),\ \varepsilon = 1.3$$

$$^6A_1 \rightarrow {}^4A_2, {}^4A_{1g}\ 27{,}700\ cm^{-1}(361\ nm),\ \varepsilon = 1$$

- The molar extinction coefficients of these bands are very low. Note, d→d: $10^{-2} \leq \varepsilon \leq 5 \times 10^2$.
- It is important to point out the differences in the spectrum of octahedral $Fe^{3+}(H_2O)_6$ and tetrahedral $Fe^{3+}(H_2O)_4$.
- It is possible to use energy diagram for T_d as well as O_h, however, splitting of d orbital is inverse of that for T_d, and omitting g subscript, noncentrosymmetric (Fig. 10.19).
- From the calculated energy diagram, the spectrum of $Fe^{3+}(O)_4$ in orthoclase feldspar consists of three transitions (Fig. 10.20).

$$^6A_1 \rightarrow {}^4T_1,\ 22{,}500\ cm^{-1}(444\ nm),\ \varepsilon = 0.73$$

$$^6A_1 \rightarrow {}^4T_2,\ 23{,}900\ cm^{-1}(418\ nm),\ \varepsilon = 0.76$$

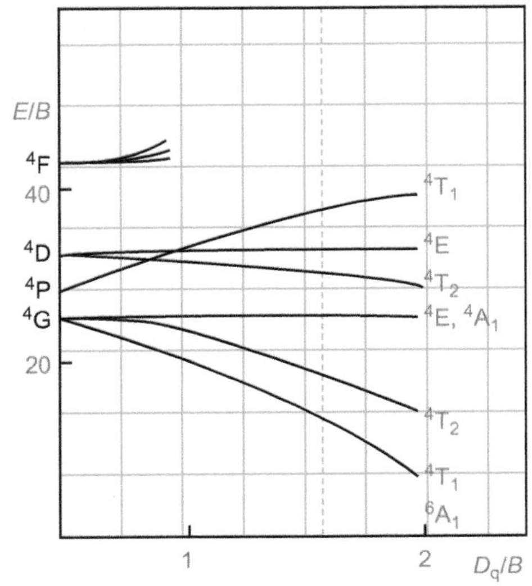

FIG. 10.17 Energy level diagram for high-spin d^5 ion in octahedral symmetry.

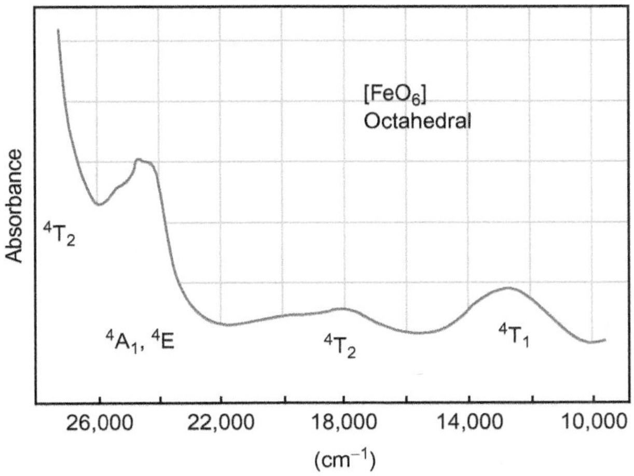

FIG. 10.18 Electronic spectra of $Fe_2(SO_4)_3{\cdot}(NH_4)_2SO_4{\cdot}24H_2O$, $[FeO_6]$ octahedral. *(From G.L. Eichhorn, Inorganic Biochemistry, vol. 1, Elsevier Scientific Publisher, Amsterdam, 1973, p. 16 (Chapter 3), reproduced with permission).*

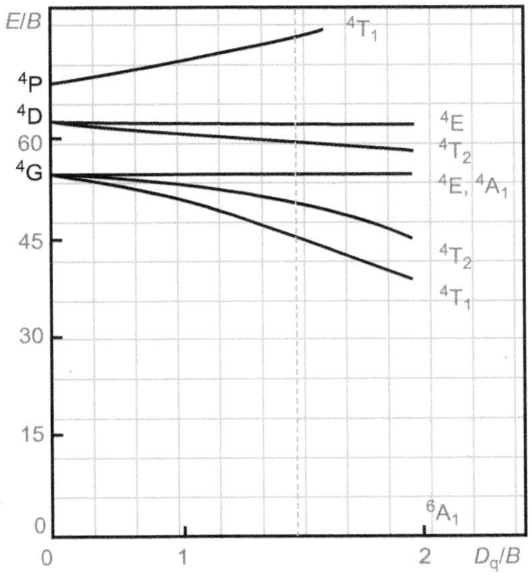

FIG. 10.19 Energy level diagram for high-spin d^5 ion in tetrahedral symmetry.

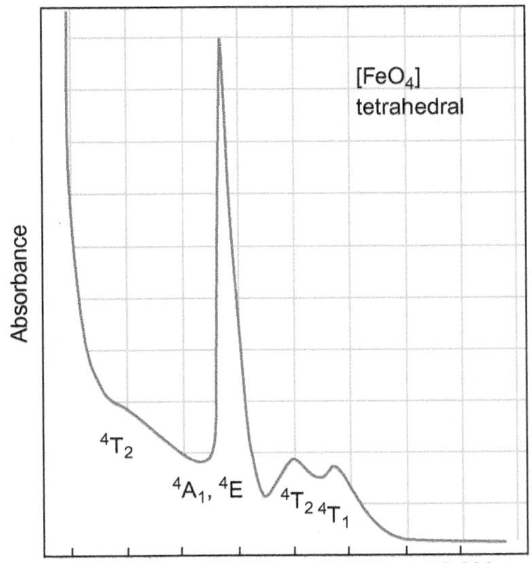

FIG. 10.20 Electronic spectra of orthoclase feldspar, $[FeO_4]$ tetrahedral. *(From G.L. Eichhorn, Inorganic Biochemistry, vol. 1, Elsevier Scientific Publisher, Amsterdam, 1973, p. 106 (Chapter 3), reproduced with permission).*

$$^6A_1 \rightarrow {}^4E, {}^4A_1 \quad 26{,}500\ cm^{-1}(377\ nm), \quad \varepsilon = 4.1$$

$$^6A_1 \rightarrow {}^4A_2, {}^4A_{1g} \quad 29{,}200\ cm^{-1}(342\ nm), \quad \varepsilon = 0.1$$

- The calculated values of $10D_q$ for octahedral is 13,700 cm^{-1}, while that of tetrahedral is 7350 cm^{-1}. The value of $10D_q$ is expected to be larger for octahedral complexes than for tetrahedral ones, as is found.
- The molar extinction coefficients of these bands are 10 times greater than those respective bands observed for $Fe^{3+}(H_2O)_6$.

Simultaneous Pair Excitations: Bridged Dinuclear Metal Centers

Use the Tanabe-Sugano diagram for d^5 ion to assign the spectrum of the six coordinate oxy-bridged dimer $[(HEDTA)Fe]_2O$ (Fig. 10.21).

- Ferric ion in the dimeric oxy-bridged $[(HEDTA)Fe]_2O$ is high-spin d^5 species:

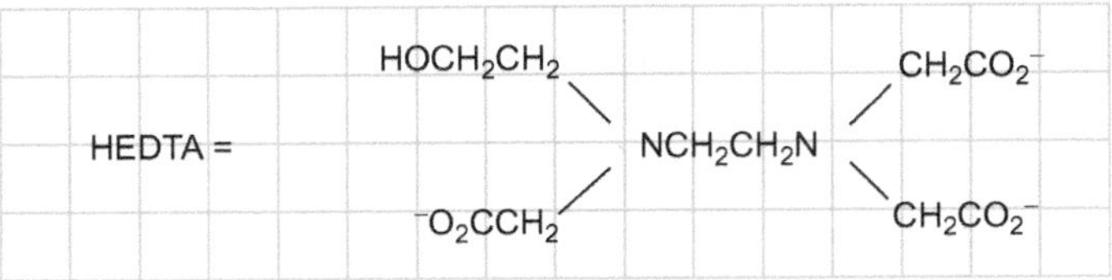

- The Tanabe-Sugano diagram for d^5 ions (Fig. 10.16) shows that $^4T_{1g}$, $^4T_{2g}$, the degenerate pair ^4E_g, $^4A_{1g}$, and $^4T_{2g}$ are the lowest electronic excited states.

  As a result, the first four bands labeled a through d can be assigned to the four d-d transitions:

Band a	$^6A_{1g}$	$\rightarrow$	$^4T_{1g}$,	11,200 cm^{-1}
Band b	$^6A_{1g}$	$\rightarrow$	$^4T_{2g}$,	18,200 cm^{-1}
Band c	$^6A_{1g}$	$\rightarrow$	$^4E_g, {}^6A_{1g}$	21,000 cm^{-1}
Band d	$^6A_1$	$\rightarrow$	$^4T_2$	24,400 cm^{-1}

weak sextet-quartet transitions are observed at low energy.

- These four bands (Fig. 10.21) are more intense than the typical bands of iron(III) complexes (Fig. 10.18). The intensity enhancement of these spin-forbidden bands is due to the spin coupling between the two iron centers.
- At higher energy, several other intense bands are observed, which cannot be interpreted in terms of isolated Fe(III) ions or charge transfer bands.

FIG. 10.21 The electronic spectrum of $[(HEDTA)Fe]_2O \cdot 6H_2O$ at 296 K. *(From G.L. Eichhorn, Inorganic Biochemistry, vol. 1, Elsevier Scientific Publisher, Amsterdam, 1973, p. 119 (Chapter 3), reproduced with permission).*

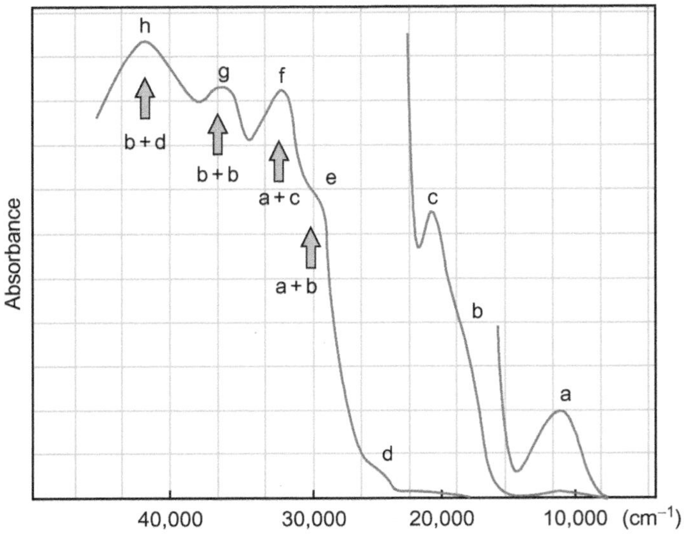

- The four intense bands (e-h) have been assigned as simultaneous d-d transitions on the two iron(III) centers. They are coupled so that the pair excitation becomes allowed.
- The energies of these bands are equal to sums of the energies of the spin-forbidden transitions.
- For example, band e is interpreted in terms of the simultaneous transition a+b.

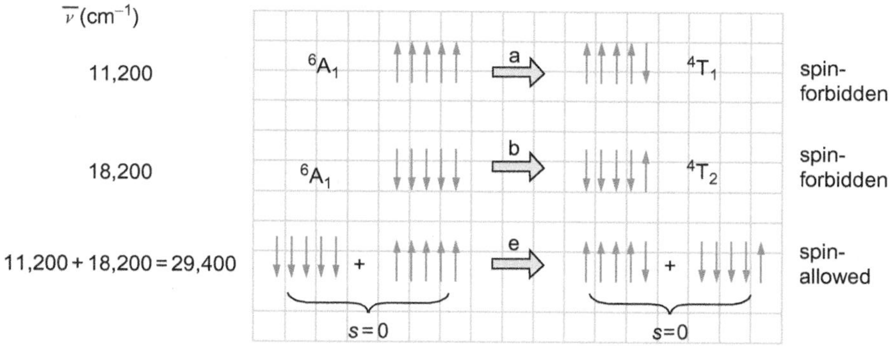

Both a and b are spin-forbidden and weak; however, e is spin-allowed and strong. Transitions such as *e* have been named *simultaneous pair excitations* (Table 10.5).

Give an example of low-spin iron(III) complexes. How many d-d spin-allowed electronic transitions do you expect?

- *Low-spin* ground states are obtained, with N-donor and CN^- ligands, e.g., $Fe(en)_3{}^{3+}$ and $Fe(CN)_6{}^{3-}$, which establish $^2T_{2g}(t_{2g})^5$ ground state. The energy level diagram is shown in Fig. 10.22:

TABLE 10.5 Assignment of Simultaneous Pair Excitations

Band	$\bar{\nu}$ (cm^{-1})	Assignment
e	29,200	a+b=11,200+18,200=29,400
f	32,500	a+c=11,200+21,000=32,200
g	36,800	b+b=18,200+18,200=36,400
h	42,600	B+d=18,200+24,400=42,600

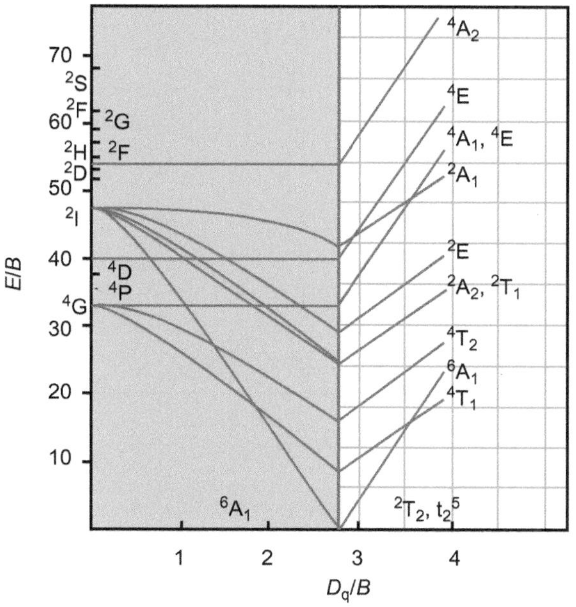

FIG. 10.22 Tanabe-Sugano diagram for d^5 ions in an octahedral field.

$$^2T_2 \rightarrow {}^2A_2, {}^2T_1, (t_{2g})^4(e_g)^1, \quad 326\,\text{nm}$$

Also, $Fe(CN)_6^{3}$ shows low energy charge transfer band $\pi(CN) \rightarrow Fe^{3+}$.

Bandwidth in the Electronic Spectra

If Fig. 10.23 represents the electronic spectra of $[Mn(H_2O)_6]^{2+}$ and MnF_2, how can you explain the unusual sharp bands in these spectra? What are the factors that contribute to the band width?

- $[Mn(H_2O)_6]^{2+}$ and MnF_2: d^5
- The ground state of d^5 configuration, (high-spin) has a unique spin multiplicity, ^6S.

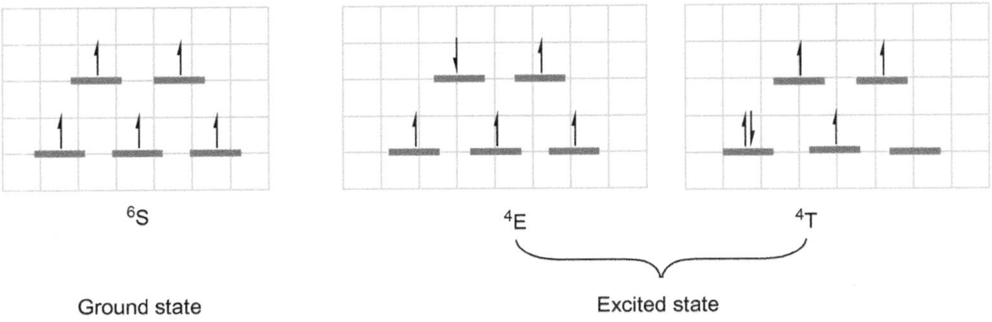

- Because any excitation of electrons within d-d orbital must lead to change in multiplicity, therefore, excitation of the ^6S state are spin-forbidden.
- The spectra of $[Mn(H_2O)_6]^{2+}$ and MnF_2 contain bands that are much sharper than those in other spectra shown in this section (Fig. 10.23).
- This is because the d^5configuration gives rise to an unusually large number of terms like $^4A_{1g}$, ^4E_g, and $^4T_{1g}$, which are paralleled to the ground term $^6A_{1g}$ (Fig. 10.24).
- Molecular vibrations modulate the crystal-field splitting Δ ($10D_q$). When the molecule is in the "compressed" phase of the vibration, the effective crystal-field of the ligand increased, and when the molecule is in the "stretched" phase, $10D_q$ is decreased.

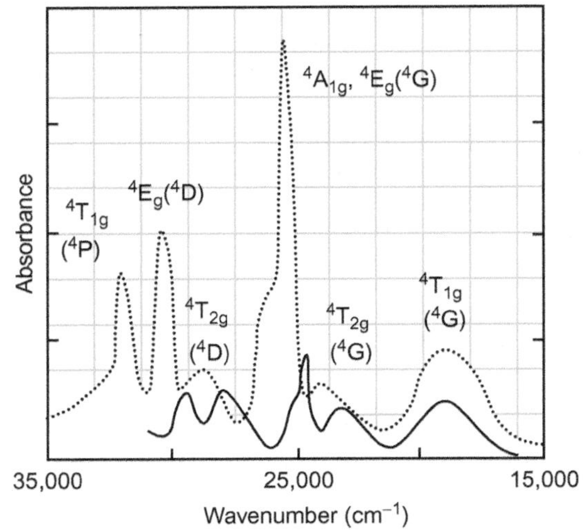

FIG. 10.23 The electronic spectra of $[Mn(H_2O)_6]^{2+}$ (___) and MnF_2 (......). *(From B.N. Figgis, Introduction to Ligand Fields, John Willey and Sons, Inc., 1986, p. 224 (Fig. 9.7 absorption spectra in solid, Chapter 9), ISBN 0-89874-819-4, reproduced with permission. From A.B.P. Lever, Inorganic Electronic Spectroscopy, second ed., Elsevier Science Publishers B.V., Amsterdam, 1984, p. 451 (Chapter 6), reproduced with permission).*

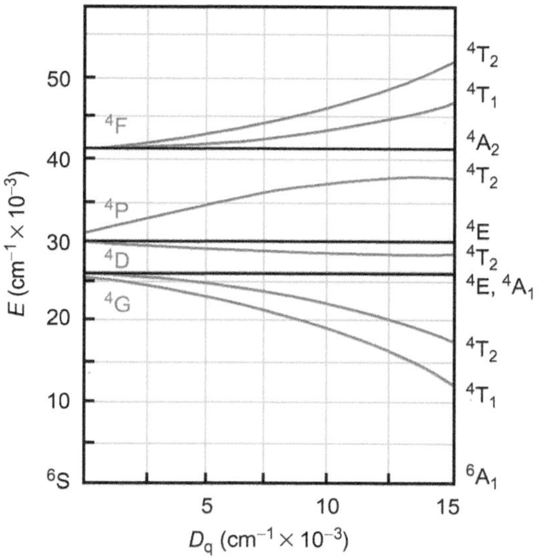

FIG. 10.24 Energy level diagram of high-spin d^5 ions.

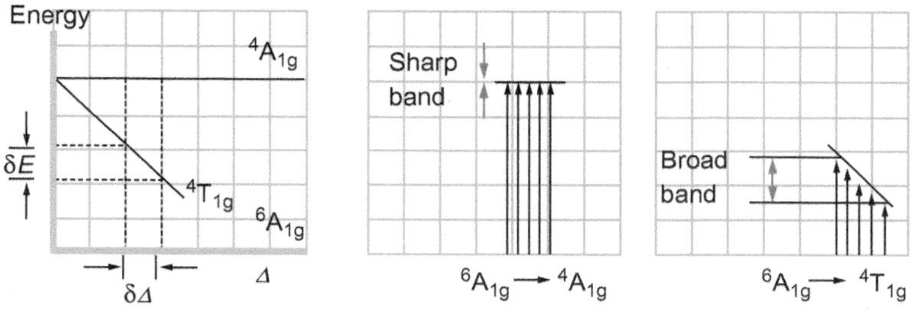

FIG. 10.25 The impact of the change in Δ on the energy of the electronic transitions.

- If the energy separation of the ground state and the excited state depends on the ligand field strength, the transition energy will have a range of values (∂E, $\partial\Delta$) dependent on vibration of the molecule.
- The width of the band will thus depend on the relative slopes of the ground state and excited state in a Tanabe-Sugano diagram (Fig. 10.25).
- For an octahedral Mn^{2+}, all transitions are spin-forbidden.
  - In the Tanabe-Sugano diagram for d^5, the $^4A_{1g}$ level runs parallel to $^6A_{1g}$ ground state, thus $^6A_{1g} \rightarrow {}^4A_{1g}$ is a sharp line.
  - However, the slope of $^4T_{1g}$ is very different from that of $^6A_{1g}$, and so $^6A_{1g} \rightarrow {}^4T_{1g}$ is a much broader band.
- The contributions to width of d-d absorption bands include:
  i. the Jahn-Teller effect;
  ii. variation of $10D_q$ with vibration;
  iii. vibrational structure due to simultaneous excitation of electrons and one or more vibrational modes;
  iv. spin-orbital coupling removing the degeneracy of a term (e.g., $^3T_{1g}$); and
  v. low-symmetry components due to the ligand field.

The Low-Spin Versus High-Spin: d^6 Configuration

How many d-d electronic transitions do you expect in absorption spectrum of $[Co(NH_3)_6]^{3+}$? Identify the ground and the possible excited states, and the type of these transitions.

- $[Co(NH_3)_6]^{3+}$: the ground state is $^1A_{1g}$.
- The excited states from d-d transitions are $^1T_{1g}$, and $^1T_{2g}$, which contain triply degenerate irreducible representations (Fig. 10.26).

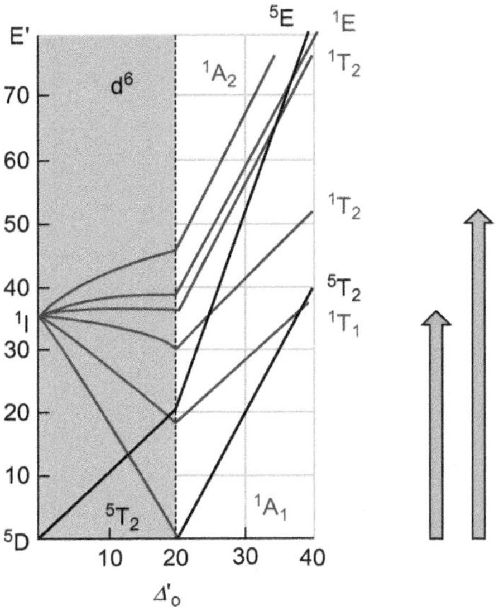

FIG. 10.26 Energy diagram for the configuration d^6.

- The pure electronic transitions:
  $^1A_{1g} \rightarrow {}^1T_{1g}$, and
  $^1A_{1g} \rightarrow {}^1T_{2g}$, are not allowed
- However, molecules are not rigid, and undergo vibrations. During certain vibrations, the molecular symmetry changes considerably, and the center of symmetry is removed.
- These states of d^n configurations no longer retain their g-character.
- Thus, the electronic transition can become slightly allowed by certain vibrational modes.
- The vibronic transition is allowed, if the direct product representation of

$$\Gamma\left(\psi_e'(x,y,z)\psi_e\right)$$

Tables 10.6 and 10.7 have a symmetry component of normal modes of the concerned point group, O_h for this situation (Chapter 9, p. 552)

TABLE 10.6 The Character Table for an Octahedral Point Group

O_h	E	$8C_3$	$6C_2$	$6C_4$	$3C_2$	i	$6S_4$	$8S_6$	$3\sigma_h$	$6\sigma_d$		
A_{1g}	**1**	**1**	**1**	**1**	**1**	**1**	**1**	**1**	**1**	**1**	–	$x^2+y^2+z^2$
A_{2g}	1	1	−1	−1	1	1	−1	1	1	−1	–	–
E_g	2	−1	0	0	2	2	0	−1	2	0	–	$2z^2-x^2-y^2$, x^2-y^2
T_{1g}	**3**	**0**	**−1**	**1**	**−1**	**3**	**1**	**0**	**−1**	**−1**	(R_x,R_y,R_z)	–
T_{2g}	3	0	1	−1	−1	3	−1	0	−1	1		(xz, yz, xy)
A_{1u}	1	1	1	1	1	−1	−1	−1	−1	−1	–	–
A_{2u}	1	1	−1	−1	1	−1	1	−1	−1	1	–	–
E_u	2	−1	0	0	2	−2	0	1	−2	0	–	–
T_{1u}	**3**	**0**	**−1**	**1**	**−1**	**−3**	**−1**	**0**	**1**	**1**	(x,y,z)	–
T_{2u}	3	0	1	−1	−1	−3	1	0	1	−1	–	–
$T_{1g}\times T_{1u}\times A_{1g}$	**9**	**0**	**1**	**1**	**1**	**−9**	**−1**	**0**	**−1**	**−1**	**$=A_{2u}+E_u+T_{1u}+T_{2u}$**	

TABLE 10.7 The Direct Product Representation of $\Gamma(T_{1g}(T_{1u})A_{1g})$

O_h	E	$8C_3$	$6C_2$	$6C_4$	$3C_2'$	i	$6S_4$	$8S_6$	$3\sigma_h$	$6\sigma_d$
T_{1g}	3	0	−1	1	−1	3	1	0	−1	−1
T_{1u}	3	0	−1	1	−1	−3	−1	0	1	1
A_{1g}	1	1	1	1	1	1	1	1	1	1
$T_{1g}\times T_{1u}\times A_{1g}$	9	0	1	1	1	−9	−1	0	−1	−1

- Thus for $^1A_{1g}\rightarrow{}^1T_{1g}$, in O_h, the coordinates x, y, and z are jointly form a basis for the T_{1u} representation (Table 10.6).

$$\Gamma\left(\psi_e'(x,y,z)\psi_e\right)=T_{1g}\times T_{1u}\times A_{1g}=T_{1g}\times T_{1u}$$

- Using

$$\Gamma_i=\frac{1}{h}\sum_R\chi(R)\chi_i(R)$$

and Table 10.6, this can be reduced to

$$\Gamma\left(\psi_e'(x,y,z)\psi_e\right)=A_{1u}+E_u+T_{1u}+T_{2u}$$

- Comparing these result with the symmetries of the normal modes of O_h, AB_6: A_{1g}, E_g, $2T_{1u}$, T_{2g}, T_{2u} (Chapter 9, p. 549).
- Thus, while the pure electronic transition $^1A_{1g}\rightarrow{}^1T_{2g}$ is not allowed, the vibronic excitations of symmetries T_{1u} and T_{2u} are allowed.
- Similarly, for $^1A_{1g}\rightarrow{}^1T_{2g}$ transition, we find

$$\begin{aligned}\Gamma\left(\psi_e'(x,y,z)\psi_e\right)&=T_{2g}\times T_{1u}\times A_{1g}\\&=T_{2g}\times T_{1u}\\&=A_{2u}+E_u+T_{1u}+T_{2u}\end{aligned}$$

- Thus, it can also occur because of the simultaneous excitations of T_{1u} or T_{2u} vibration.

Using the Tanabe-Sugano diagram for d^6 (Fig. 10.26), what are the electronic configuration of the ground and the spin allowed excited states in $[Fe(CN)_6]^{4-}$ and $[Fe(OH_2)_6]^{2+}$? How can you explain the splitting in the electronic spectrum of $Fe(ClO_4)_2{\cdot}6H_2O$ (Fig. 10.27)?

- $Fe(CN)_6^{\ 4-}$ is a low-spin state, the completely filled $(t_{2g})^6$ yields $^1A_1$, ground state.
- Two electronic spin-allowed excitations are expected (Fig. 10.26).
  - $^1A_{1g}\rightarrow{}^1T_1\ (t_{2g})^5(e_g)^1$, at 320 nm
  - $^1A_{1g}\rightarrow{}^1T_2\ (t_{2g})^5(e_g)^1$, at 270 nm

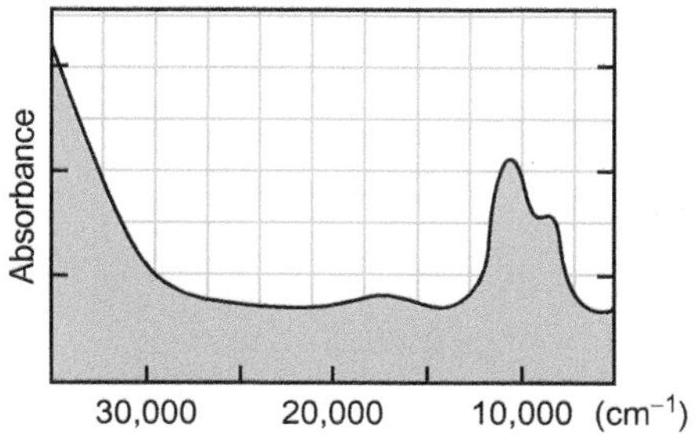

FIG. 10.27 Electronic spectra of $Fe(ClO_4)_26H_2O$. *(From B.N. Figgis, Introduction to Ligand Fields, John Willey and Sons, Inc., 1986, p. 224 (Fig. 9.7 absorption spectra in solid, Chapter 9), ISBN 0-89874-819-4, reproduced with permission).*

- Excitation to $^1T_1$, $^1T_2$ states have the configuration $(t_{2g})^5(e_g)^1$ in which the odd spin in t_{2g} and e_g configurations are spin opposed (Fig. 10.28).
- The electronic transitions of $[Fe(H_2O)_6]^{2+}$ (Fig. 10.26) are forbidden.
- Broad absorption in NIR, at 1000 nm is due to the transition $^5T_{2g} \rightarrow {}^5E_g$ (Fig. 10.26).
- This band usually shows some splitting, vibronically induced, due to Jahn-Teller distortion of ^5E_g (Fig. 10.29).

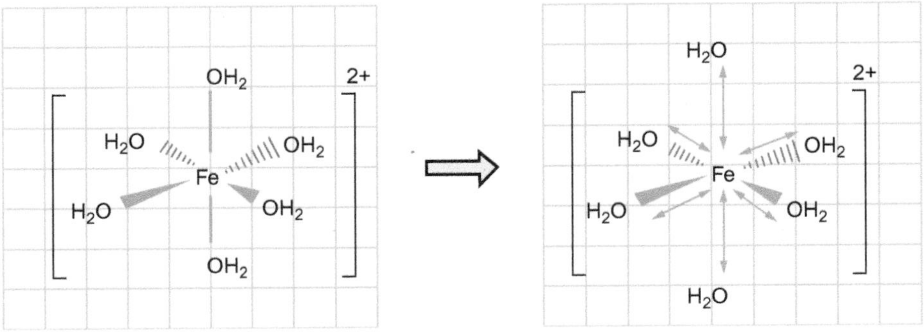

The Effect of Low Symmetry

If Fig. 10.30 represents the visible spectra of three (ethylenediamine)-cobalt(III) complexes, what are the spectral consequences of symmetry variation?

Use Fig. 10.31 to determine which transitions are allowed for each dipole operator in the electronic spectrum of *trans*-$[Co(en)_2F_2]^+$

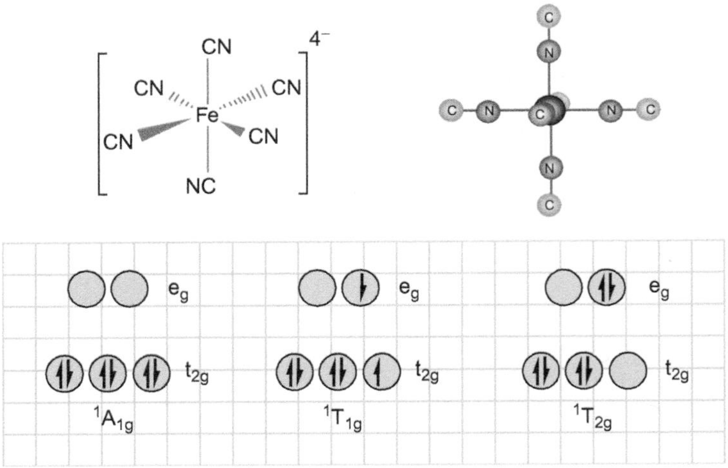

FIG. 10.28 The ground and spin allowed excitations in $[Fe(CN)_6]^{4-}$.

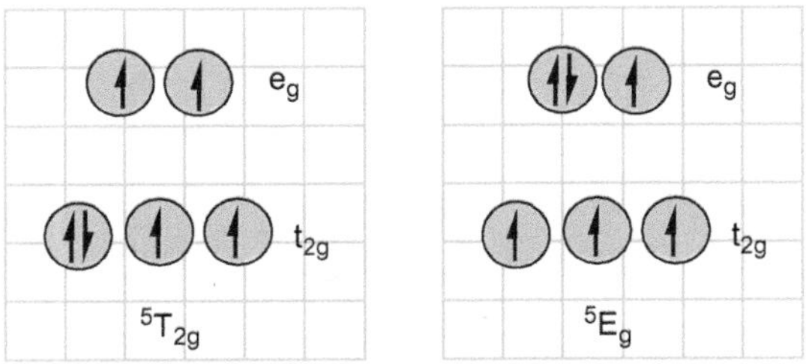

FIG. 10.29 The ground and spin allowed excitations in $[Fe(ClO_4)_2$.

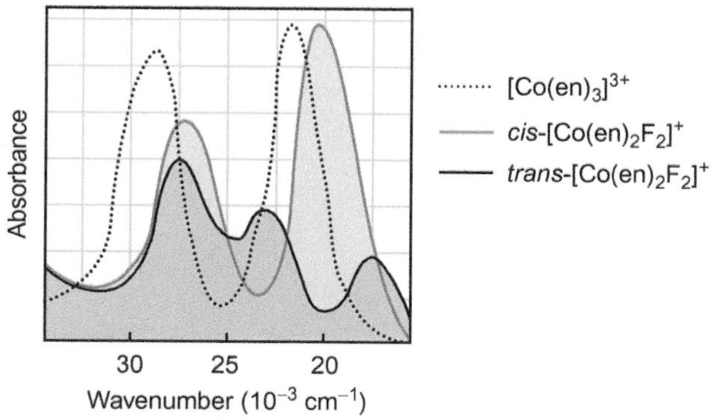

FIG. 10.30 The visible spectra of $[Co(en)_3]^{3+}$, *cis*-$[Co(en)_2F_2]^+$, and *trans*-$[Co(en)_2F_2]^+$. *(From B.N. Figgis, Introduction to Ligand Fields, John Willey and Sons, Inc., 1986, p. 234 (Fig. 9.7 absorption spectra in solid, Chapter 9), ISBN 0-89874-819-4, reproduced with permission).*

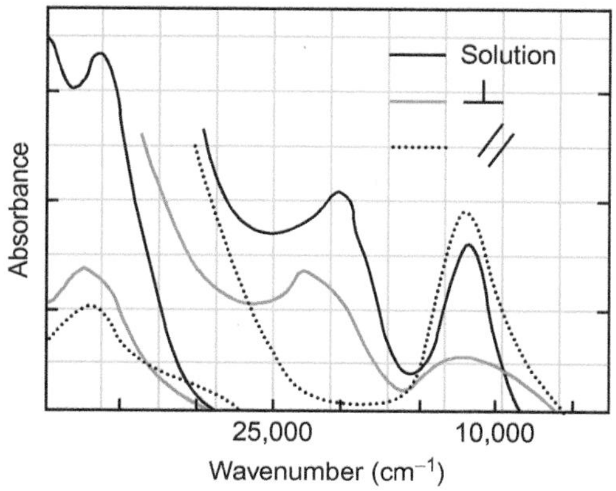

FIG. 10.31 The visible spectra of *trans*-$[Co(en)_2Cl_2]^+$ in solution and as a single crystal with perpendicular and paralleled orientation of Cl—Co—Cl axis with the plain of polarization. *(From B.N. Figgis, Introduction to Ligand Fields, John Willey and Sons, Inc., 1986, p. 231 (Fig. 9.6 absorption spectra in solid, Chapter 9), ISBN 0-89874-819-4, reproduced with permission).*

- The symmetries of these complexes:

$$[Co(en)_3]^{3+} \quad O_h$$
$$trans\text{-}[Co(en)_2F_2]^+ \quad D_{4h}$$
$$cis\text{-}[Co(en)_2F_2]^+ \quad C_{2v}$$

- The spectra of *cis*-$[Co(en)_2F_2]^+$ and *trans*-$[Co(en)_2F_2]^+$ with symmetry lower than octahedral, $[Co(en)_3]^{3+}$, can be analyzed by considering the lower symmetry as perturbation of the energy levels in an octahedral field.
- The representations of O_h field are decomposed into its subrepresentations when the symmetry is altered or lower (Table 10.8) (Chapter 8).
- The effect of the two symmetries on the energy levels of an octahedral field is shown in Fig. 10.32.
- The splitting of T_{1g} (O_h) level for complexes with ligand that differ considerably in the ligand field strength is lager in D_{4h} than C_{2v} symmetry.
- Thus, two separate bands are observed for ${}^1A_1 \rightarrow {}^1E_g$ and ${}^1A_1 \rightarrow {}^1A_{2g}$ in trans isomer, whereas a singly slightly asymmetric band is observed for ${}^1A_1 \rightarrow {}^1B_1$ and ${}^1A_2$, ${}^1A_1 \rightarrow {}^1B_2$ in the cis isomer.
- The splitting of ${}^1T_{2g}$ (O_h) level is small in both isomers, and the two transitions are not resolved.
- In D_{4h} symmetry, the electric dipole operators $-$x and $-$y transform as E_u, whereas the electric dipole operator z transforms as A_{2u}.

TABLE 10.8 Correlation Among the Representations in O_h and Those in the Lower Symmetries

O_h	O	T_d	D_{4h}	D_{2d}	D_{3d}	D_3	C_{2v}	C_{4v}	C_{2h}
A_{1g}	A_1	A_1	$\mathbf{A_{1g}}$	A_1	A_{1g}	A_1	$\mathbf{A_1}$	A_1	A_g
A_{2g}	A_2	A_2	$\mathbf{A_{1g}}$	A_1	A_{2g}	A_2	$\mathbf{A_2}$	B_1	B_g
E_g	E	E	$\mathbf{A_{1g}+B_{1g}}$	A_1+b_1	E_g	E	$\mathbf{A_1+A_2}$	A_1+B_1	A_g+B_g
T_{1g}	T_1	T_1	$\mathbf{A_{2g}+E_g}$	A_2+E	$A_{2g}+E_g$	A_2+E	$\mathbf{A_2+B_1+B_2}$	A_2+E	A_g+2B_g
T_{2g}	T_2	T_2	$\mathbf{B_{2g}+E_g}$	B_2+E	$A_{1g}+E_g$	A_1+E	$\mathbf{A_1+B_1+B_2}$	B_2+E	$2A_g+B_g$
A_{1u}	A_1	A_2	$\mathbf{A_{1u}}$	B_1	A_{1u}	A_1	$\mathbf{A_2}$	A_2	A_u
A_{2u}	A_2	A_1	$\mathbf{B_{1u}}$	A_1	A_{2u}	A_2	$\mathbf{A_1}$	B_2	B_u
E_u	E	E	$\mathbf{A_{1u}+B_{1u}}$	A_1+B_1	E_u	A	$\mathbf{A_1+A_2}$	A_2+B_2	A_u+B_u
T_{1u}	T_1	T_2	$\mathbf{A_{2u}+E_u}$	B_2+Ee	$A_{2u}+E_u$	A_2+E	$\mathbf{A_1+B_1+B_2}$	A_1+E	A_u+2B_u
T_{2u}	T_2	T_1	$\mathbf{B_{2u}+E_u}$	A_2+E	$A_{1u}+E_u$	A_1+E	$\mathbf{A_2+B_1+B_2}$	b_1+E	$2A_u+B_u$

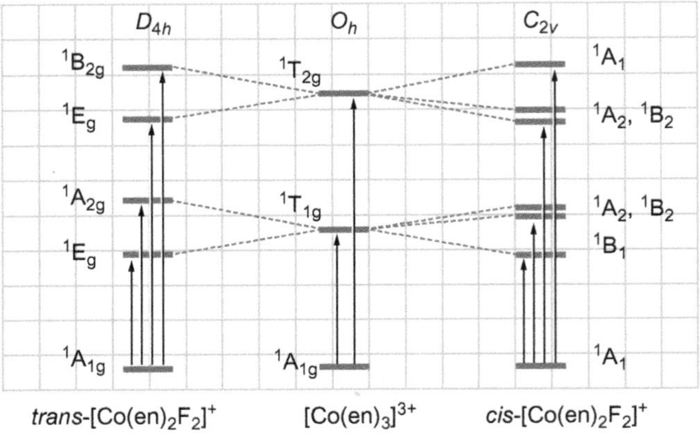

FIG. 10.32 Ground and excited states in $[Co(en)_3]^{3+}$, *cis*-$[Co(en)_2F_2]^+$, and *trans*-$[Co(en)_2F_2]^+$.

TABLE 10.9 Parallel and Perpendicular Transitions

Transition	Direct Product of Ground and Excited Representations	Direct Product With Dipole Moment Representation: A_{2u}	Direct Product With Dipole Moment Representation: E_u	Polaraization
$^1A_{1g} \rightarrow {}^1A_{2g}$	$A_{1g} \times A_{2g} = A_{2g}$	$A_{1g} \times A_{2u} \times A_{2g} = A_{1u}$	$A_{1g} \times E_u \times A_{2g} = E_u$	⊥
$^1A_{1g} \rightarrow {}^1B_{2g}$	$A_{1g} \times B_{2g} = B_{2g}$	$A_{1g} \times A_{2u} \times B_{2g} = B_{1u}$	$A_{1g} \times E_u \times B_{2g} = E_u$	⊥
$^1A_{1g} \rightarrow {}^1E_g$	$A_{1g} \times E_g = E_g$	$A_{1g} \times A_{2u} \times E_g = E_u$	$A_{1g} \times E_u \times E_g = A_{1u} + A_{2u}$ $A_{1g} \times E_u \times E_g = B_{1u} + B_{2u}$	‖, ⊥

‖ refers to *z* polarization, parallel to Cl—Co—Cl, ⊥ refers to *x,y* polarization, perpendicular to Cl—Co—Cl axis.

- To determine which transitions are allowed for which operator, we have to take the direct product of the ground and excited electronic wave functions with the dipole moment operators, and compare the results with the unsymmetrical vibrational modes of the molecules, $2A_{2u}$, B_{2u}, $3E_u$ for ML_4X_2 in D_{4h} (Table 10.9).
- The low-energy transition, $^1A_{1g} \rightarrow {}^1E_g$, occurs in both parallel (*x*, *y*) and perpendicular (*z*) spectra.
- The second transition, $^1A_{1g} \rightarrow {}^1A_{2g}$, occurs only in the perpendicular spectrum.
- The third transition, overlapping $^1A_{1g} \rightarrow {}^2E_g$ and $^1B_{2g} \rightarrow {}^1E_g$, occurs in both spectra.

Band Intensity and Ligand-Field: d^7 Configuration

Use the energy diagram in Fig. 10.33 to assign the possible electronic transitions in the visible spectra of $Co(BF_4)_2{\cdot}6H_2O$, and $[CoCl_4]^{2-}$ (Figs. 10.34–10.36, respectively).

- Pink $Co(H_2O)_6^{2+}$ is high-spin d^7 ion of configuration $(t_{2g})^5(e_g)^2$, yielding a $^4T_{1g}$ ground state.
- Examining the energy diagram leads to expect that the lowest energy transition is $^4T_{1g} \rightarrow {}^4T_{2g}$ (around 1200 nm) (Fig. 10.36).

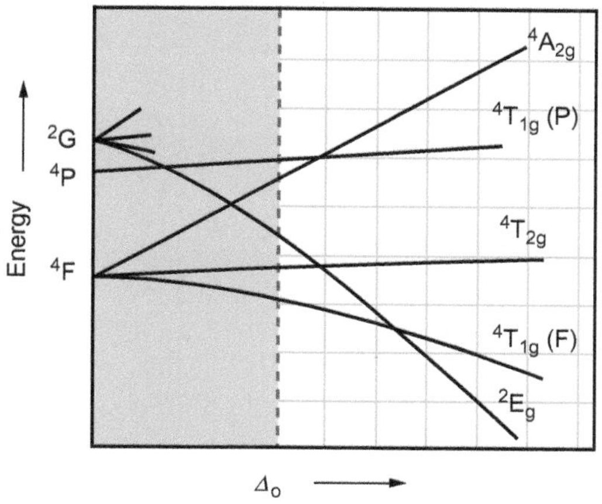

FIG. 10.33 Energy levels diagram for d^7 ion in an octahedral ligand field. The position of Δ_o for $Co(H_2O)_6^{2+}$ is the *dashed line*.

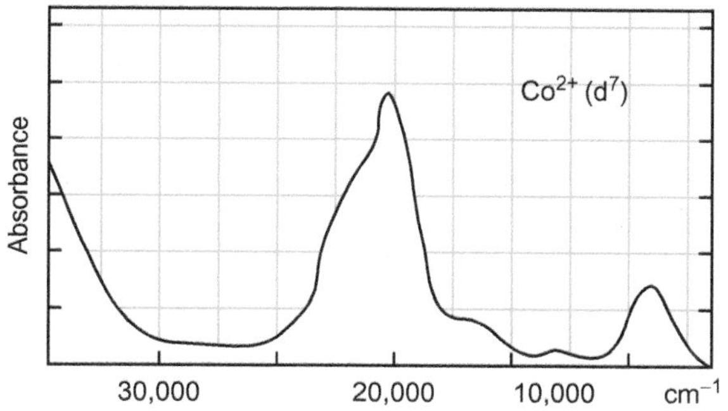

FIG. 10.34 The electronic spectrum of $Co(BF_4)_2{\cdot}6H_2O$. *(From B.P. Lever, Inorganic Electronic Spectroscopy, second ed., Elsevier Science Publishers B.V., Amsterdam, 1984, p. 483 (Chapter 6), ISBN 0-444-42389-3 (vol. 33), ISBN 0-444-41699-4 (Series), reproduced with permission).*

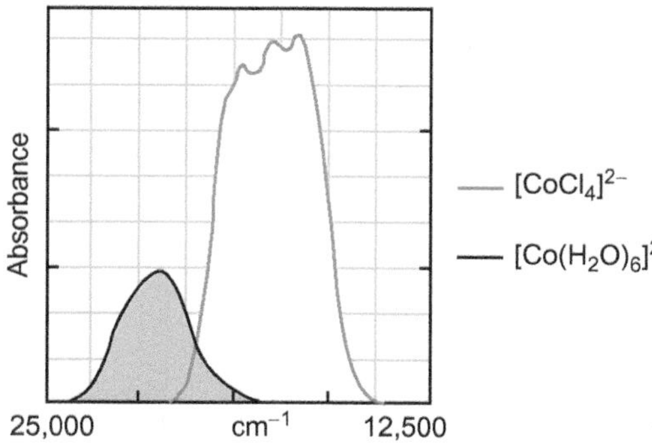

FIG. 10.35 The electronic spectra of $[CoCl_4]^{2-}$ and $[Co(H_2O)_6]^2[CoCl_4]^{2-}$, $[Co(H_2O)_6]^{2+}$. *$[CoCl_4]^{2-}$ and $[Co(H_2O)_6]^2[CoCl_4]^{2-}$: (From A.B.P. Lever, Inorganic Electronic Spectroscopy, second ed., Elsevier Science Publishers B.V., Amsterdam, 1984, p. 499 (Chapter 6), reproduced with permission. $[Co(H_2O)_6]^{2+}$: From B.P. Lever, Inorganic Electronic Spectroscopy, second ed., Elsevier Science Publishers B.V., Amsterdam, 1984, p. 483 (Chapter 6), ISBN 0-444-42389-3 (vol. 33), ISBN 0-444-41699-4 (Series), reproduced with permission).*

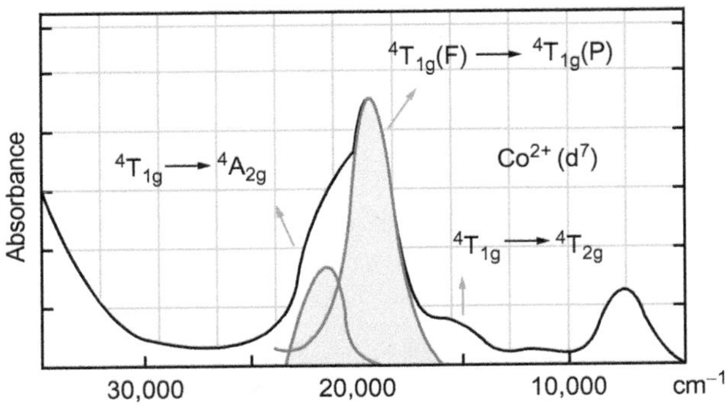

FIG. 10.36 Band assignments in the visible spectrum of $[Co(H_2O)_6]^{2+}$, $[Co(H_2O)_6]^{2+}$. *(From B.N. Figgis, Introduction to Ligand Fields, John Willey and Sons, Inc., 1986, p. 224 (Fig. 9.7 absorption spectra in solid, Chapter 9), ISBN 0-89874-819-4, reproduced with permission).*

- The $^4T_{1g}(F) \rightarrow {}^4T_{1g}(P)$ transition is the main feature of the visible spectrum.
- The $^4T_{1g} \rightarrow {}^4A_{2g}$ transition is expected at slightly lower energy ($\approx$560 nm) and is buried under stronger $^4T_{1g} \rightarrow {}^4T_{2g}$ absorption band.
- In tetrahedral complex of Co(II), the ground state belongs to A_2 representation, the excited states of T_1 and T_2 symmetries (Fig. 10.37).
- The coordinates *x*, *y*, and *z* form a basis for T_2 representation (Table 10.10):

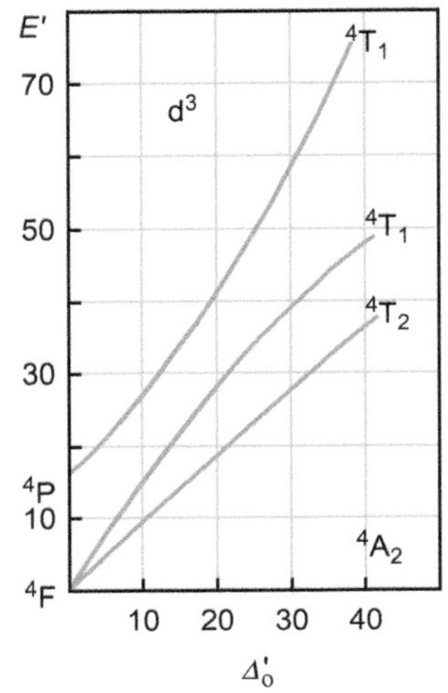

FIG. 10.37 Tanabe-Sugano diagram for d^3 ion in an octahedral ligand field and d^7 ion in a tetrahedral ligand.

TABLE 10.10 The Character Table of Tetrahedral Point Group

T_d	E	$8C_3$	$3C_2$	$6S_4$	$6\sigma_d$		
A_1	1	1	1	1	1		$x^2+y^2+z^2$
A_2	1	1	1	−1	−1		
E	2	−1	2	0	0		$2z^2-x^2-y^2, x^2-y^2$
T_1	3	0	−1	1	−1	R_x, R_y, R_z	
T_2	3	0	−1	−1	1	x,y,z	xy,xz,yz
$A_2 \times T_2 \times T_1$	9	0	1	1	1	$=A_1+E+T_1+T_2$	
$A_2 \times T_2 \times T_2$	9	0	1	−1	−1	$=A_2+E+T_1+T_2$	

- For the $A_2 \rightarrow T_1$

$$\Gamma[\psi_e'(x,y,z)\psi_e] = A_2 \times T_2 \times T_1 = A_1 + E + T_1 + T_2$$

- The direct product of $\Gamma[\psi_e'(x,y,z)\psi_e]$ contains A_1 representation, the transition $A_2 \rightarrow T_1$ is allowed by the symmetry of the purely electronic wave function.
- In the T_d point group, d_{xy}, d_{xz}, and d_{yz} orbitals, and p-orbitals transform as T_2 and therefore they can mix (Table 10.12).
- The two states A_2 and T_1 will have a differing amount of p-character.
- The intensity is gained by having some of the highly allowed $p \rightarrow d$ or $d \rightarrow p$ character associated with this transition.
- For the transition $A_2 \rightarrow T_2$, the direct product is

$$\Gamma[\psi_e'(x,y,z)\psi_e] = A_2 \times T_2 \times T_2 = A_2 + E + T_1 + T_2$$

- Therefore, the transition $A_2 \rightarrow T_2$ is not allowed by the symmetry of pure electronic wave function.
- Mixing of p-character into the wave function will not help.
- Comparing these result with the symmetries of the first excited states of the normal mode vibrations of T_d: A_1, E, and $2T_2$.
- The transitions $A_2 \rightarrow T_1$ and $A_2 \rightarrow T_2$ are vibronically allowed.
- However, the intensity of $A_2 \rightarrow T_2$ transition is 10–100 times smaller than the intensity of the $A_2 \rightarrow T_1$ transition.
- The greater intensity of the spectrum of $[CoCl_4]^{2-}$ than that of $[Co(H_2O)_6]^{2+}$ (Fig. 10.35) could be attributed to:
  i. intensity stealing
  ii. degree of covalency.
- Transitions between states of the same symmetry (g or u) are forbidden in centrosymmetric molecules.
- Tetrahedral molecules and *cis*-MX_4Y_2 molecules do not have a center of symmetry and thus d-d transitions in these compounds are not completely forbidden.

$$\varepsilon \approx 0.01 - 1.0\ \mathrm{dm^3\ mol^{-1}\ cm^{-1}} \rightarrow \text{spin} - \text{forbidden, Laporte forbidden}$$

$$\varepsilon \approx 10^2 - 10^4\ \mathrm{dm^3\ mol^{-1}\ cm^{-1}} \rightarrow \text{spin} - \text{allowed, Laporte forbidden}$$

d^8 Configuration: d^2 Versus d^8 Complexes

Assign the electronic transitions in the spectra of $Ni(H_2O)_6^{2+}$ and $Ni(H_2NCH_2CH_2NH_2)_3^{2+}$ (Fig. 10.38), and explain the splitting of the middle band in the spectrum of $Ni(H_2O)_6^{2+}$.

- The d^8 complexes, $Ni(H_2O)_6^{2+}$ and $Ni(H_2NCH_2CH_2NH_2)_3^{2+}$, have the same sets of states as the d^2 complexes, but in reverse order (Fig. 10.39).

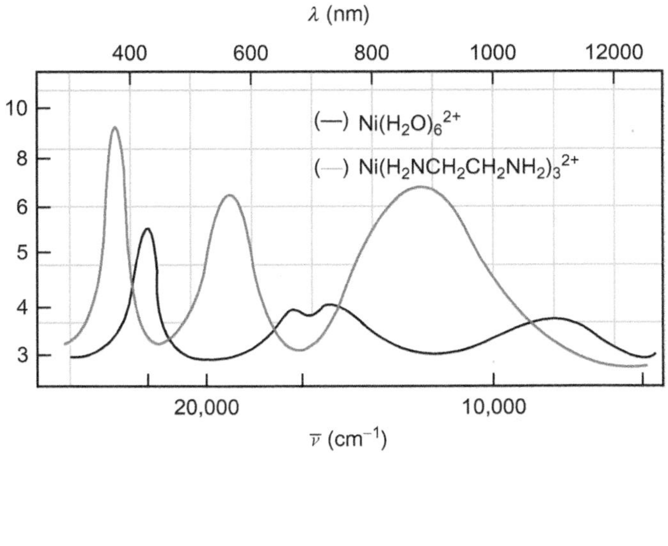

FIG. 10.38 Absorption spectra of $Ni(H_2O)_6^{2+}$ and $Ni(H_2NCH_2CH_2NH_2)_3^{2+}$. *(From F.A. Cotton, G. Wilkinson, Advanced Inorganic Chemistry: A Comprehensive Text, fourth ed., John Willey & Sons, New York, 1980, p. 787 (Chapter 21), reproduced with permission).*

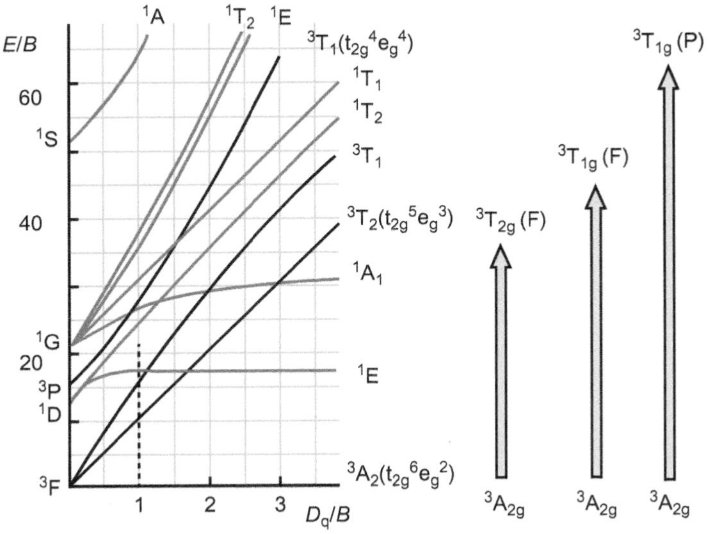

FIG. 10.39 Energy levels diagram for d^8 ion in an octahedral field.

- The spectra of two Ni(II) (d^8) complexes show three bands due to the spin-allowed d-d transitions:

$$^3A_{2g} \rightarrow {}^3T_{2g}(F),\ {}^3A_{2g} \rightarrow {}^3T_{1g}(F),\ \text{and}\ {}^3A_{2g} \rightarrow {}^3T_{2g}(P)$$

- Ligand-field strength, $10D_q$, can be estimated directly from the experimental spectra by using the peak maximum of the lowest-energy band ($^3A_{2g} \rightarrow {}^3T_{2g}$).
- For the aquo-complex, but not for other complexes, the ^1E_g state lies so close to $^3T_{1g}$ and mixing take place, leading to observation of a doublet band where the spin forbidden transition has stolen intensity from the spin allowed transition.
- According to the Tanabe-Sugano diagrams for O_h, at $\Delta_o/B \approx 1$, the ^1E_g and $^3T_{1g}$ levels cross (Fig. 10.39).

Calculation of D_q and β for Octahedral Ni(II) Complexes

The electronic spectrum of $Ni(Me_2SO)_6(ClO_4)_6$ indicates three bands at 7728, 1297, and 24,038 cm^{-1}.
Characterize the energy transitions for each of these bands.
Calculate the values of D_q, Racah parameter B', β, and β^o.
Why does the value of B' differ from that of B and what are the significances of β^o?

- The d^8 complexes have the same terms as the d^2 complexes, but in reverse order (Fig. 10.40).
- The energy of the various states contained in Orgel diagrams of d^8 complexes have the same terms as the d^2 complexes (Chapter 8, p. 488), but in reverse order (Fig. 10.40). These states are represented by:

$$E({}^3T_{2g}(F)) = -2D_q$$

$$E({}^3A_{2g}(F)) = -12D_q$$

and the root of

$$E^2 - (6D_q + 15B')E + 90D_qB' - 6D_q^2 = 0$$

are the energy of the two terms $^3T_{1g}$(F) and $^3T_{1g}$(P), then

$$E({}^3T_{1g}(F)) = 7.5B' + 3D_q - \frac{1}{2}\left(225B'^2 + 100D_q^2 - 180D_qB'\right)^{1/2}$$

$$E({}^3T_{1g}(P)) = 7.5B' + 3D_q + \frac{1}{2}\left(225B'^2 + 100D_q^2 - 180D_qB'\right)^{1/2}$$

B is the Racah parameter for the free gaseous ion, and B' is the same parameter for the complex. The ratio between B/B' is defined as β, and β^o as $(1-\beta) \times 100$:

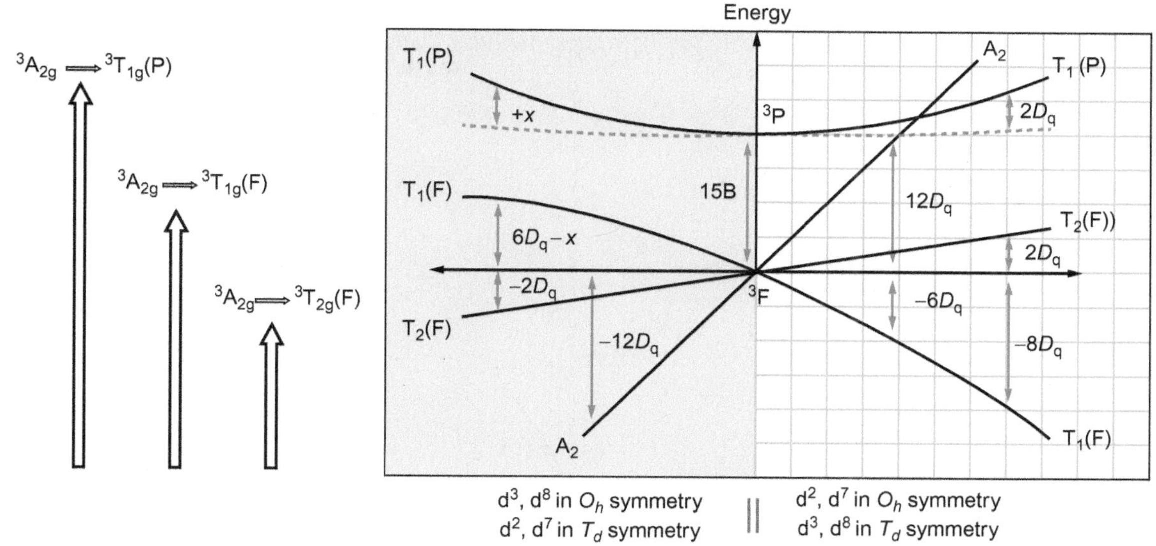

FIG. 10.40 Splitting and the interactions of the states for d^3 and d^8 ions in O_h and T_d environments, for a single value of D_q.

$$\beta = \frac{B'}{B}$$

$$\beta^o = (1-\beta) \times 100$$

- The value of D_q is estimated directly from the lowest energy transition, $^3A_{2g} \rightarrow {}^3T_{2g}(F)$, which occurs at 7728 cm^{-1}:

$$10D_q = 7728\ \text{cm}^{-1}$$

$$D_q = 773\ \text{cm}^{-1}$$

- For the energy transition, $^3A_{2g} \rightarrow {}^3T_{2g}(P)$, which occurs at 24,038 cm^{-1}, therefore:
  The observed energy for the transition, $^3A_{2g} \rightarrow {}^3T_{2g}(P) = 24{,}038\ \text{cm}^{-1}$

$$= \text{energy of } {}^3T_{2g}(P) - \text{energy of } {}^3A_{2g}$$

$$\text{Energy of } {}^3T_{2g}(P) = 24{,}038\ \text{cm}^{-1} + \text{energy of } {}^3A_{2g}$$

$$\text{Energy of } {}^3A_{2g} = -12D_q = -12 \times 773 = -9276\ \text{cm}^{-1}$$

$$\text{Energy of } {}^3T_{2g}(P) = 24{,}038\ \text{cm}^{-1} - 9276\ \text{cm}^{-1} = 14{,}762\ \text{cm}^{-1}$$

- By substituting for $D_q = 773$ in

$$E^2 - (6D_q + 15B')E + 90D_qB' - 16D_q^2 = 0$$

  the value B' is obtained:
  $B' = 921\ \text{cm}^{-1}$
- The value of B in complexes is always smaller than that of free ion. This phenomenon is identified as the nephelauxetic effect, which attributed to delocalization of the metal electron over molecular orbital that cover not only the metal but the ligand as well. As a result of this electron cloud spread or delocalization, the average interelectronic repulsion is reduced and B' is smaller.
  Consequently, the dropping of ^3P is a measure of the covalency, which could be expressed as β^o.
  The value of B for the nickel gaseous ion is 1056 cm^{-1}, then

$$\beta = \frac{B'}{B} = \frac{773}{921} = 0.872$$

$$\beta^o = (1 - \beta) \times 100 = 12.8\%$$

- The values of B' and D_q are substituted into

$$E^2 - (6D_q + 15B')E + 90D_qB' - 16D_q^2 = 0$$

and is solved for E. The obtained two roots are:
the energy of $^3T_{1g}(F) = 14{,}762\ cm^{-1}$; and
the energy of $^3T_{1g}(P) = 3694\ cm^{-1}$.

- The differences:

$$E(^3T_{1g}(F)) - E(^3A_{2g}(F)) \text{ and } E(^3T_{1g}(P)) - E(^3A_{2g}(F))$$

correspond to the transitions:

$$^3A_{2g}(F) \rightarrow {}^3T_{1g}(F) \text{ and} {}^3A_{2g}(F) \rightarrow {}^3T_{1g}(F)$$

$$^3A_{2g}(F) \rightarrow {}^3T_{1g}(F):\ 3694 - (-12 \times 773) = 12{,}970\ cm^{-1}$$

$$^3A_{2g}(F) \rightarrow {}^3T_{1g}(P):\ 14{,}762 - (-12 \times 773) = 24{,}038\ cm^{-1}$$

d^9 Configuration: π-Binding

Assign the following spectra of Cu(II) species (Figs. 10.41 and 10.42).
How could the electronic spectra of copper complexes characterize the structural variations?

- The five d-orbitals of the free metal ion are degenerate at the ground state ^2D.
- ^2D state splits into two states in O_h and in T_d ligand fields (Fig. 10.43).
- In O_h, the d^9 configuration is expected to give rise to a single d-d transition (Fig. 10.44):

$$^2E_g\left[(t_{2g})^6(e_g)^3\right] \rightarrow {}^2T_{2g}\left[(t_{2g})^5(e_g)^4\right]$$

in which the states are reversed from the d^1 case.

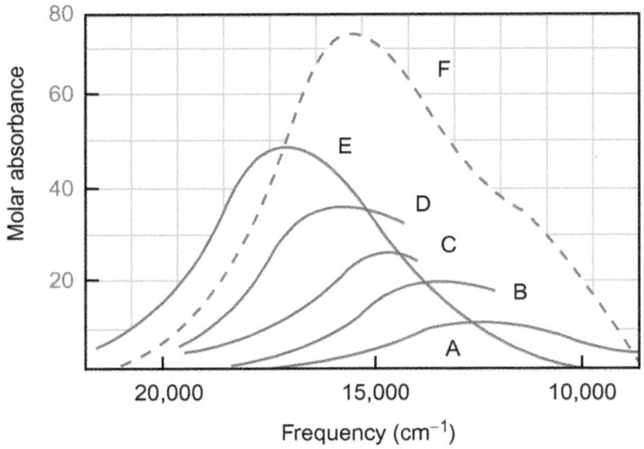

FIG. 10.41 (A) Absorption spectra of $[Cu(H_2O)_6]^{2+}$, and of the ammines in 2 M ammonium nitrate at 25°C: (B) $[Cu(NH_3)(H_2O)_5]^{2+}$, (C) $[Cu(NH_3)_2(H_2O)_4]^{2+}$, (D) $[Cu(NH_3)_3(H_2O)_3]^{2+}$, (E) $[Cu(NH_3)_4(H_2O)_2]^{2+}$, (F) $[Cu(NH_3)_5H_2O]^{2+}$. *(From F.A. Cotton, G. Wilkinson, Advanced Inorganic Chemistry: A Comprehensive Text, fourth ed., John Willey & Sons, New York, 1980, p. 816 (Chapter 21), reproduced with permission).*

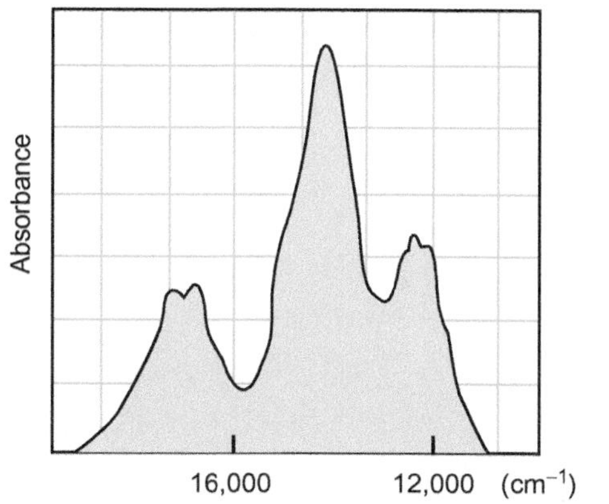

FIG. 10.42 The electronic absorption spectrum of $[CuCl_4]^{2-}$ at 77 K. *(From P. Casside, M.A. Hitchman, J. Chem. Soc. Chem. Commun. (1975) 837, reproduced with permission).*

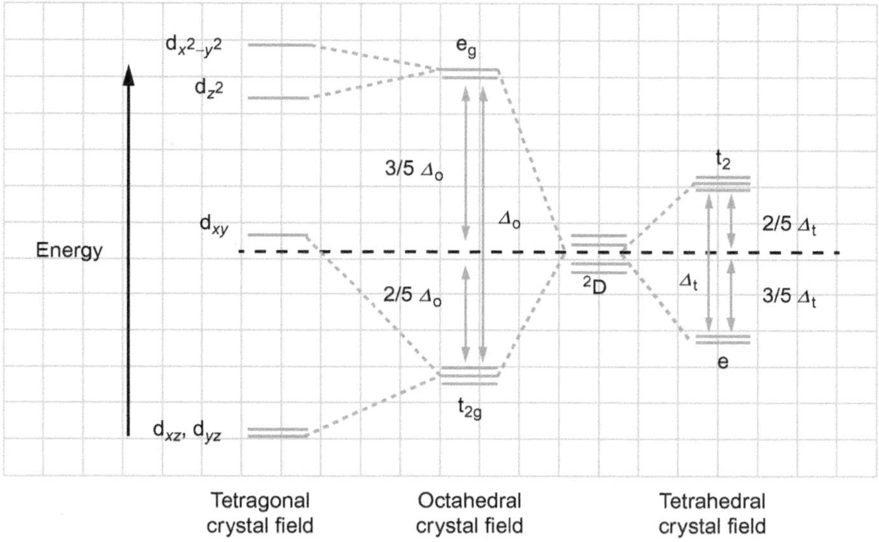

FIG. 10.43 Splitting of d orbitals by crystal field.

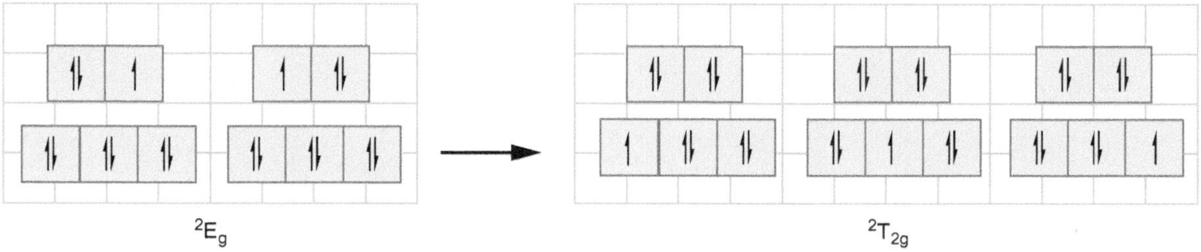

- Amine complexes of copper are much deeper blue than the aqua (Fig. 10.41).
- This is because the amines have a stronger ligand field that lets the absorption band to shift from the far red to the middle of the red region.
- In $[Cu(H_2O)_6]^{2+}$, the band maximum is at about 800 nm, while in $[Cu(NH_3)_4(H_2O)_2]^{2+}$ occurs at about 600 nm.
- π-Binding should be investigated, when the ligand has appropriate p_π or d_π atomic orbitals or π molecular orbitals accessible for overlap with the t_{2g} orbitals of the metal ion.
- When ligands contain empty π-orbitals (unsaturated), the orbitals t_{2g} of the ligand will be found at higher energy than of the corresponding metal π-orbital (Fig. 10.45A).

FIG. 10.44 Qualitative molecular orbital scheme for an octahedral complex.

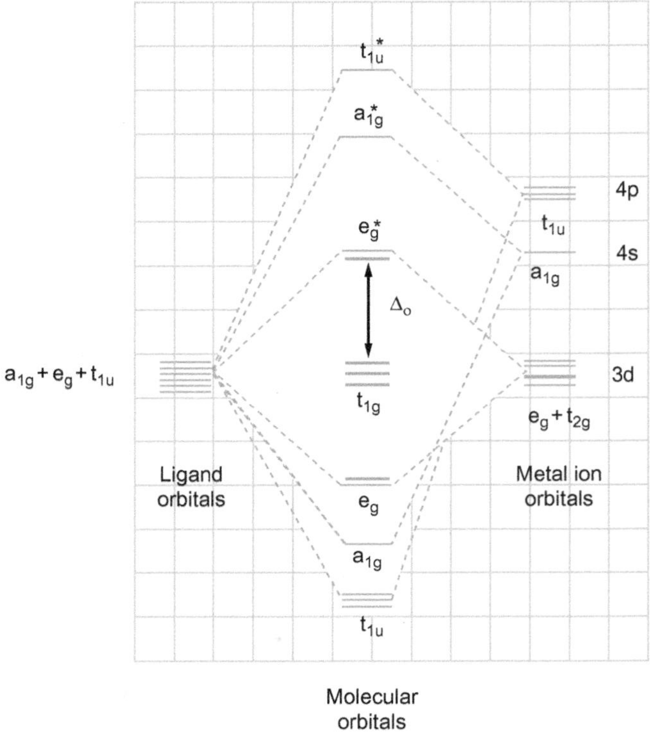

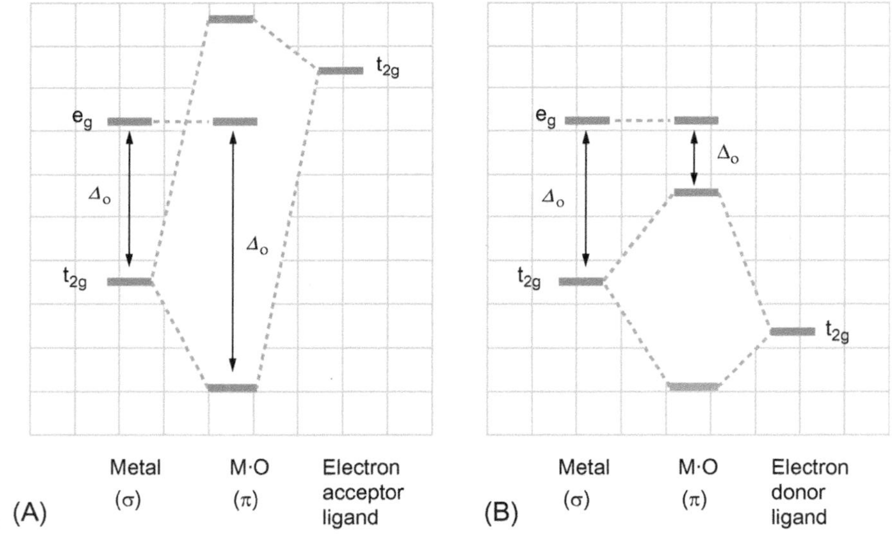

FIG. 10.45 Interaction of metal orbitals and metal orbitals. (A) The orbitals t_{2g} of the ligand at higher energy than of the corresponding metal t_{2g}-orbitals. (B) The orbitals t_{2g} of the ligand at lower energy than of the corresponding metal t_{2g}-orbitals.

- On other hand, ligands have donor atoms such as oxygen or fluorine, the available π-orbitals, t_{2g}, are occupied. The π-orbitals (t_{2g}) of the ligand locate at lower energy than the metal t_{2g} orbitals (Fig. 10.45B).
- In Fig. 10.45, It is possible to see that the effect of (a) is to increase Δ, while in (b) it is to decrease.
- Therefore, π bonding is the main key to figure out the value Δ, and the positions of ligands in the spectrochemical series.
- Spectrochemical series:

$$I^- < Br^- < Cl^- < -SCN^- < F^- < OH^- < \text{Oxalate}^- \leq H_2O < -NCS^- \leq NH_3 < \text{Pyr}^- < \text{ethylenediamine} < \text{Dipyr} < o-\text{phenantholine} < NO_2^- < CN^- < CO$$

- In Fig. 10.41, the binding impacts of fifth and sixth axial molecules of ammonia to $[Cu(NH_3)_4(H_2O)_2]^{2+}$ is complicated.
- The basis for this odd behavior is explained by the Jahn-Teller effect (Chapter 10, p. 601) (Figs. 10.46 and 10.47).
- This effect may be understood by considering the effect of unevenly filled e_g or t_{2g} shells on the symmetry of the Cu complexes.
- The symmetry of the Cu-complex is reduced to C_{4v} or D_{4h} and gives the ground state $^2B_2$ or $^2B_{2g}$.

Jahn-Teller:	O_h	→	C_{4v}	→	D_{4h} (Table 10.10)
	E_g	→	A_1+B_1	→	$A_{1g}+B_{1g}$
	T_{2g}	→	B_2+E	→	$B_{2g}+E_g$

- In D_{4h} (Fig. 10.46):
  Three electronic transitions are anticipated (Fig. 10.47):

D_{4h}	C_{4v}
$^2B_{1g}\rightarrow{}^2A_{1g}$	$^2B_1\rightarrow{}^2A_1$
$^2B_{1g}\rightarrow{}^2B_{2g}$	$^2B_1\rightarrow{}^2B_2$
$^2B_{1g}\rightarrow{}^2E_g$	$^2B_1\rightarrow{}^2E$

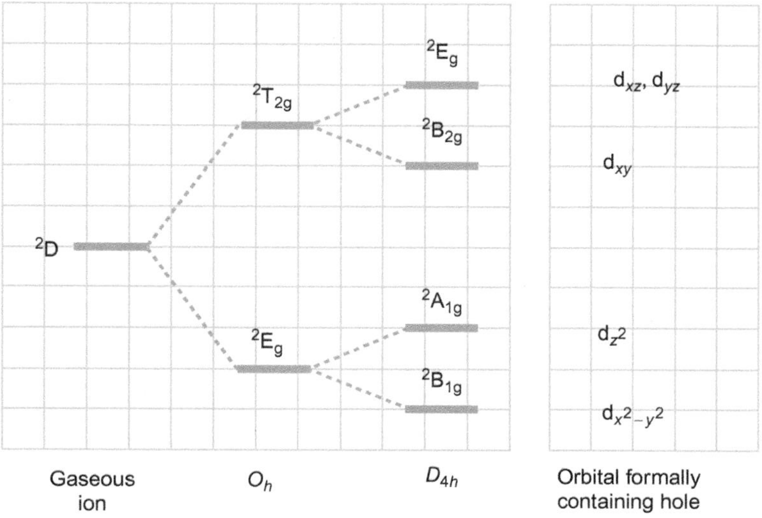

FIG. 10.46 The various energy states in O_h and D_{4h} fields.

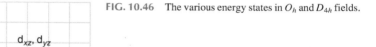

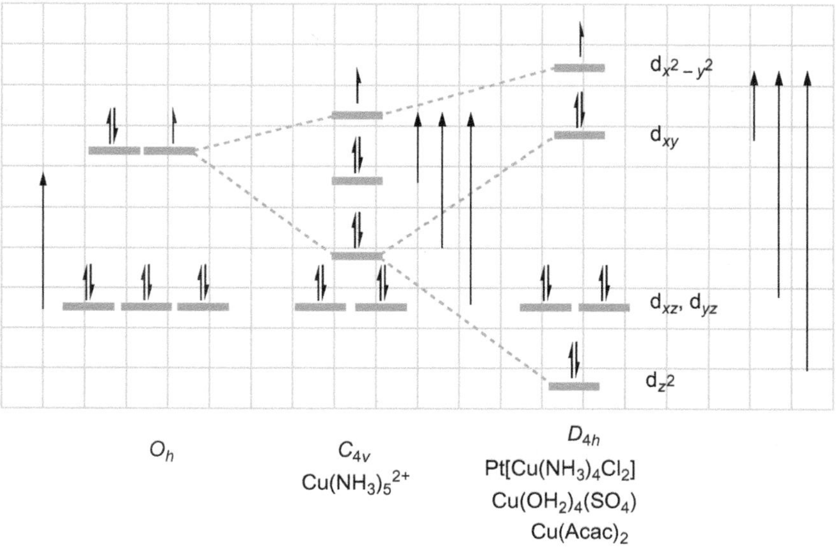

FIG. 10.47 The anticipated electronic transitions in O_h, C_{4v}, and D_{4h} environments.

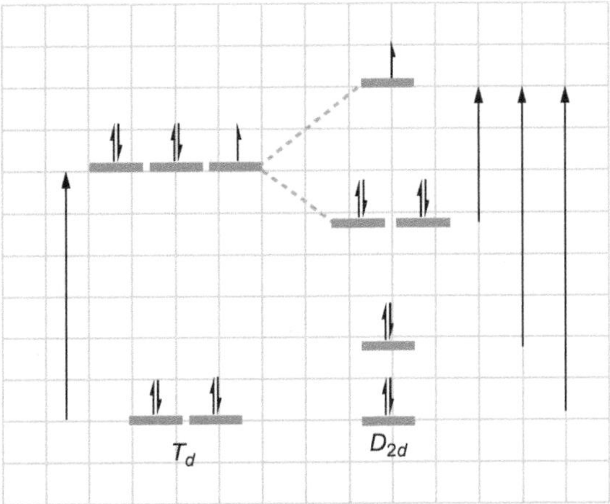

FIG. 10.48 The anticipated electronic transitions in T_d and D_{2d} environments.

- These transitions can be vibronically allowed.
- $[CuCl_4]^{2-}$ is a planar centrosymmetric molecule which gives three bands spectrum (Fig. 10.42).
- The unpaired electron will be in the $d_{x^2-y^2}$ orbital in the ground state.
- The vibrations required to make these transitions vibronically allowed are as follows:

Transition	Vibration Needed
$^2B_1 \rightarrow {}^2A_1$	$e_u + b_{2u}$
$^2B_1 \rightarrow {}^2B_2$	$e_u + a_{1u}$
$^2B_1 \rightarrow {}^2E$	$e_u + a_{1u} + a_{2u} + b_{1u} + b_{2u}$

- The observation of vibrational fine structure on the electronic transitions is conformation of the vibronic mechanism.
- On the other hand, in T_d, one transition is expected, while in D_{2d}, three bands will be observed (Fig. 10.48).

Intensely Colored Metal Complexes

What is the explanation for the intense color of HgI_2 and MnO_4 and for the changes in the intensity of color in the ferroin indicator reaction?

- All metal complexes exhibit charge-transfer spectra (CT; p → d, or d → p), but only if the CT absorption occurs into the visible region, it will affect the color of the compound. Examples:
  - HgI_2 (d^{10}, brick red)
  - MnO_4^- (d^0, intense purple)

  The bright colors of HgI_2 and MnO_4^- result from ligand to metal charge transfer in the visible region.
- Ferroin indicator:

$[Fe(phen)_3]^{2+}$	$\rightleftarrows$	$[Fe(phen)_3]^{3+} + e$
Blood red		Pale blue

  - $[Fe(phen)_3]^{2+}$ is low-spin d^6 and thus most likely to have a visible spectra similar to Co(III) complexes.
  - The possibility of oxidation of Fe(II) and the availability of low-lying empty orbitals on the phenanthroline give a very intense metal-to-ligand CT transition.
  - As a result, many Fe(II) complexes of unsaturated amines are intensely colored.
  - The CT bands in $[Fe(phen)_3]^{2+}$ have molar absorptivity larger than those for the allowed d-d transitions.

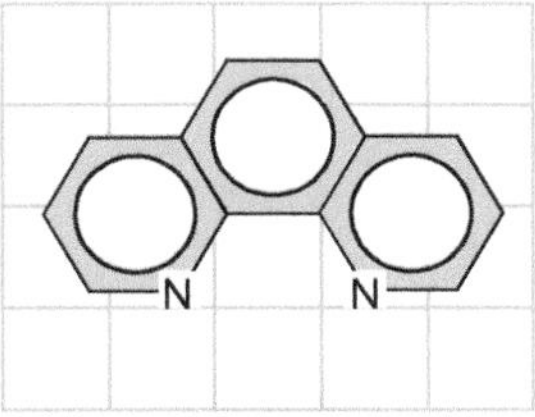

Phenanthroline

- ○ Fe^{3+}ion in $[Fe(phen)_3]^{3+}$ is isoelectronic with d^5 Mn(II), but the metal charge +3 and the strong ligand field stabilize the low-spin derivative.
- ○ Since Fe(III) is a strong oxidizing agent, electron acceptor, its CT bands would likely be ligand-to-metal than metal-to-ligand.
- ○ The unsaturated phenanthroline is more suitable for metal-to-ligand than ligand-to-metal charge transfer is, thus no CTs but only d-d transitions might be monitored.

- Consequently, the color change of the ferroin indicator is due to the difference in the intensity of d-d transitions and charge transfer band.

Donor-Acceptor Complexes

The reaction of tetracyanoethylene, TCNE, and thioanisole, PSM, gives the two rotational isomers X and Y (Scheme 10.2):

a. **Explain the formation of the TCNE-PSM complex.**
b. **The observed charge transfer bands at 27,100 cm^{-1} and 18,400 cm^{-1} are attributed to absorbance of the two rotational isomers. What do you think is the basis of this explanation?**
c. **Derive an equation connecting the logarithm of the absorptions of the two bands, A_X and A_Y, with temperature and energy difference ΔE, $\Delta E = \pm(E_x - E_y)$.**
d. **If $\varepsilon_X = \varepsilon_Y$ and $A_X/A_Y = 0.41$ at 296 K, what would be the predicted value of ΔE.**
e. **Is ΔE equal to the energy difference between the two charge transfer bands?**
   a. Two rotational configurations of donor-acceptor complexes for the interaction of tetracyanoethylene (TCNE, acceptor) with PSM (donor) are expected.
   The *Y* and *X* configurations are correspond to maximum overlap between the lowest vacant acceptor orbital of TCNE with highest and second highest donor orbitals of PSM, respectively.

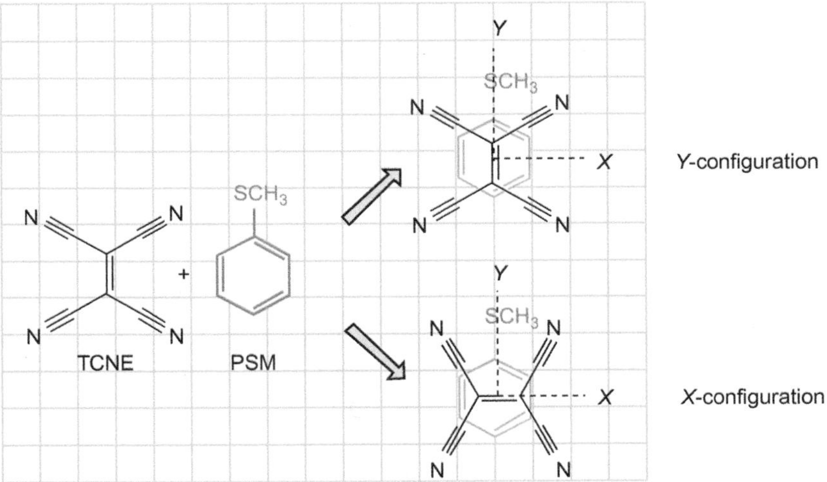

SCHEME 10.2 The two predicted conformations for TCNE-PSM complex.

b. The two charge transfer bands observed in these π-π^* electron donor-acceptor complexes are shown to be due to the existence of two rotational isomers in the ground electronic state.

The low-energy and high-energy bands are predicated to arise from the X and Y configurations, respectively.

c. Since the absorbance (A) is proportional to ε and the concentration c for each band (Beer's law) and the path length, l, is the same:

$$A = \varepsilon c l \tag{10.1.2}$$

$$\frac{A_X}{A_Y} = \frac{c_X}{c_Y} \cdot \frac{\varepsilon_X}{\varepsilon_Y} = \frac{N_X}{N_Y} \cdot \frac{\varepsilon_X}{\varepsilon_Y}$$

where N_X and N_Y are the numbers of complexes in each configuration.

If Boltzmann distribution:

$$\frac{N_X}{N_Y} = \exp\left(\frac{-\Delta E}{RT}\right)$$

$$\frac{A_X}{A_Y} = \frac{\varepsilon_X}{\varepsilon_Y} \cdot \exp\left(\frac{-\Delta E}{RT}\right)$$

$$\ln\frac{A_X}{A_Y} = \ln\frac{\varepsilon_X}{\varepsilon_Y} - \frac{\Delta E}{RT}$$

d. When $\varepsilon_X = \varepsilon_Y$, and $\frac{A_X}{A_Y} = 0.41$ at 296 K:

$$\ln\frac{\varepsilon_X}{\varepsilon_Y} = 0, \text{ and } \Delta E = -RT \ln\frac{A_X}{A_Y}$$

$$\Delta E = 2.2 \text{ kJ mol}^{-1}$$

e. If, 1 kJ mol^{-1} = 83.6 cm^{-1}, then the value of ΔE is not equal to the energy difference between the two charge transfer bands.

10.8 SPECTROSCOPY OF PORPHYRINS

Configuration Interaction, Absorption of the Unpolarized Light

- **In Fe(porphyrin) of hemoproteins, e.g., nitrosylmyoglobin in Fig. 10.49, the lowest empty orbitals are degenerate and have the symmetry label e_g in D_{4h} symmetry. The top-filled orbitals are almost degenerate, and represented by a_{2u}, a_{1u}, a'_{2u}, and b_{21u}. What are the symmetry of the ground state, the symmetries and the directions of the electronic dipole moments of the excited states ... $a_{1u}^2 a_{2u} e_g$ and ... $a_{2u}^2 a_{1u} e_g$, and the spectral consequences? Why can solutions of hemoproteins absorb the unpolarized light? Explain the bases of Q, Q_v, and B-bands in the electronic spectra of this protein.**

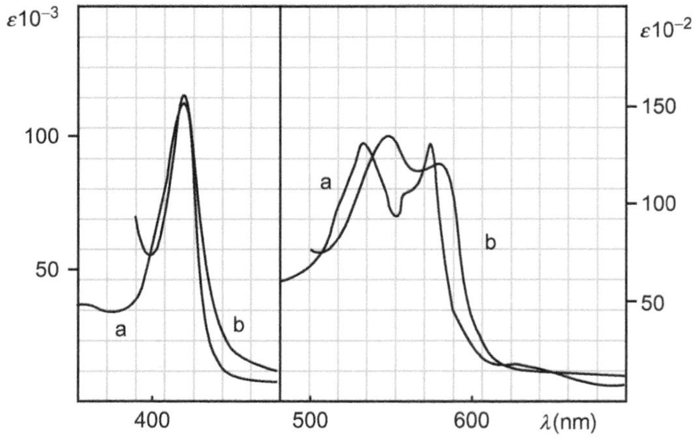

FIG. 10.49 Spectral properties of (a) Mb(III)NO, and (b) Mb(II)NO at pH 7 and 20°C. *(From, J.J. Stephanos, Hemoproteins: Reactivity and Spectroscopy (Ph.D. Thesis), Drexel University, Philadelphia, 1989, p. 45 (Chapter 2)).*

- In Fe(porphyrin) of hemoproteins, the lowest empty orbitals are degenerate and are of e_g symmetry in D_{4h}.

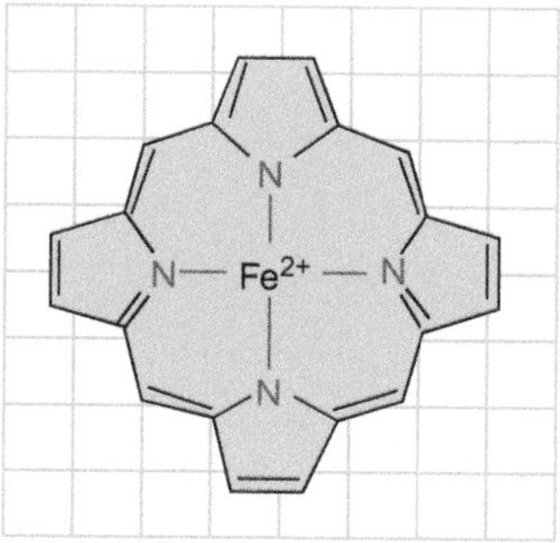

Fe(porphyrin)

- While the top-filled orbitals are almost degenerate, and are a_{2u}, a_{1u}, a'_{2u}, and b_{21u} in D_{4h} symmetry (Fig. 10.50).
- Since all orbitals of Fe(porphyrin) are occupied: … $a_{1u}^2 a_{2u}^2$, the ground state of porphyrin molecule has A_{1g} symmetry.

$$A_{1g} : \ldots a_{1u}^2 a_{2u}^2$$

- Take into consideration that only a single electron is excited from the orbital in the ground configuration to an orbital of higher energy.
  - The lowest-energy excited configurations are:

$$E_u : \ldots a_{1u}^2 a_{2u} e_g$$

$$E_u : \ldots a_{2u}^2 a_{1u} e_g$$

  - The symmetry of each excited state is E_u, which is established by taking the direct product of $a_{2u} \times e_g$ and $a_{1u} \times e_g$ representations (Table 10.11).
- The two excitations:
  $A_{1g} \rightarrow E_u$ (involves the electronic transition $a_{1u} \rightarrow e_g$); and
  $A_{1g} \rightarrow E_u$ (involves the electronic transition $a_{2u} \rightarrow e_g$)
  have almost equal energy change.
- In these two transitions, the excited states are irreducible representation, and have the same symmetry label E_u (Table 10.12).
- Consequently, these electronic promotions:
  - are dipole allowed, $g \rightarrow u$, with high transition moments;
  - have almost equal transition moments;
  - have excited configurations of the same symmetry, E_u; and
  - are strongly mixed by configuration interaction.

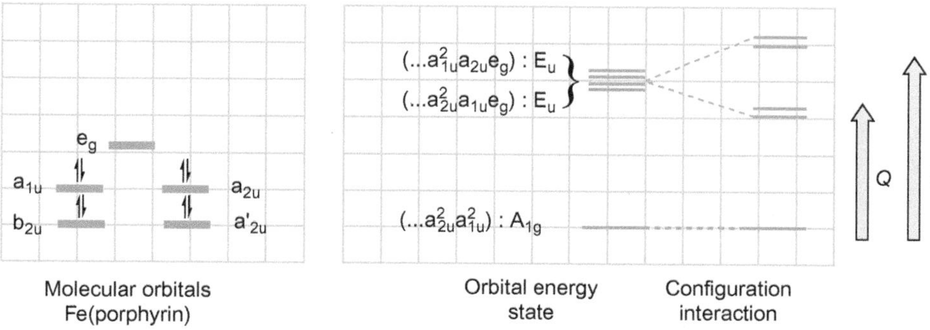

FIG. 10.50 The π-π* transitions of a metal porphyrin.

TABLE 10.11 The Character Table of the Point Group D_{4h}

D_{4h}	E	$2C_4$	C_2	$2C_2'$	$2C_2''$	i	$4S_4$	σ_h	$2\sigma_v$	$2\sigma_d$		
A_{1g}	1	1	1	1	1	1	1	1	1	1	–	x^2+y^2, z^2
A_{2g}	1	1	1	−1	−1	1	1	1	−1	−1	R_z	–
B_{1g}	1	−1	1	1	−1	1	−1	1	1	−1	–	x^2-y^2
B_{2g}	1	−1	1	−1	1	1	−1	1	−1	1	–	xy
E_g	2	0	−2	0	0	2	0	−2	0	0	(R_x, R_y)	(xz, yz)
A_{1u}	1	1	1	1	1	−1	−1	−1	−1	−1	–	–
A_{2u}	1	1	1	−1	−1	−1	−1	−1	1	1	z	–
B_{1u}	1	−1	1	1	−1	−1	1	−1	−1	1	–	–
B_{2u}	1	−1	1	−1	1	−1	1	−1	1	−1	–	–
E_u	2	0	−2	0	0	−2	0	2	0	0	(x, y)	–
$A_{1u} \times E_g$	2	0	−2	0	0	−2	0	2	0	0	$=E_u$	
$A_{2u} \times E_g$	2	0	−2	0	0	−2	0	2	0	0	$=E_u$	

TABLE 10.12 Vibrational Analysis of AB_4

D_{4h}	E	$2C_4$	C_2	$2C_2'$	$2C_2''$	i	$2S_4$	σ_h	$2\sigma_v$	$2\sigma_d$
$\Gamma_{x,y,z}$	3	1	−1	−1	−1	−3	−1	1	1	1
Unmoved atoms	5	1	1	3	1	1	1	5	3	1
Γ_{total}	15	1	−1	−3	−1	−3	−1	5	3	1

- The secular determinant for the mixing of these two states:

$$\begin{vmatrix} E_1^o - E & H_{12} \\ H_{21} & E_2^o - E \end{vmatrix} = 0$$

- Solution of the secular determinant gives us the two new energies after mixing

$$E_1 = \frac{1}{2}\left\{ E_1^o + E_2^o + \sqrt{(E_1^o)^2 + (E_2^o)^2 - 2E_1^oE_2^o + 4H_{12}^2} \right\}$$

$$E_2 = \frac{1}{2}\left\{ E_1^o + E_2^o - \sqrt{(E_1^o)^2 + (E_2^o)^2 - 2E_1^oE_2^o + 4H_{12}^2} \right\}$$

where E_1 and E_2 are the energy of the excited states.
- The E_u state has electronic dipole moment in the direction of the Cartesian coordinates, x and y (Table 10.12).
- Maximum absorption occurs when the electric dipole moment (transition moment) in the molecule is parallel to the direction of the electric vector of the light.
- As a result, for instance, in Fig. 10.49:
  - At E_1, the addition of the transition moments for these excitations gives a large net transition moment for the higher frequency (*B*-band).
  - However, at E_2, the subtraction of these moments is for the low frequency (*Q*-band).

- Solutions of hemoproteins absorb the unpolarized light because:
  - a beam of light is polarized in any direction;
  - in the solution, where molecules are randomly oriented absorb light polarized in any direction; and
  - polarized absorption and linear dichroism are techniques to study the optical properties of oriented systems.

The Vibronic Excitation of Q_v-Band (β) in Electronic-Spectra of Hemoproteins

In the previous question, what is the basis of the vibronic band Q_v (Fig. 10.49)?

- The normal modes of vibrations of Fe(porphyrin) will be calculated in the point group D_{4h}. as described in Chapter 9 (Table 10.12).

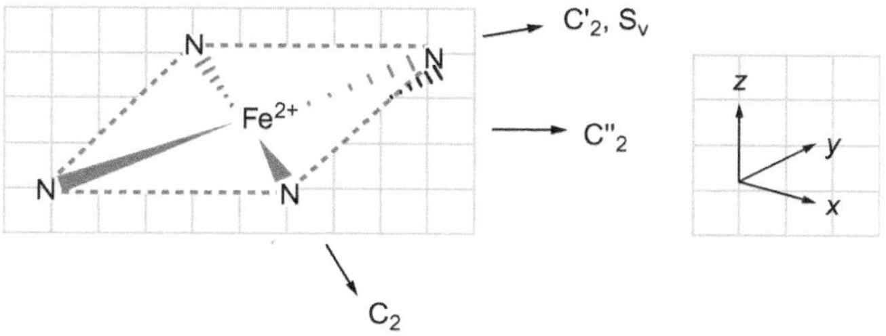

- Γ_{total} is a reducible representation, and is a sum of irreducible representation, Γ_i.
  By using the D_{4h} character table and

$$\Gamma_i = \frac{1}{h}\sum_R \chi(R)\chi_i(R), \quad \text{Chapter 4}$$

$$\Gamma_{\text{total}} = \mathrm{a_{1g} + a_{2g} + b_{1g} + b_{2g} + e_g + 2a_{2u} + b_{2u} + 3e_u}$$

$$\Gamma_{\text{transition}} = \mathrm{a_{2u} + e_u}$$

$$\Gamma_{\text{rotation}} = \mathrm{a_{2g} + e_g}$$

$$\Gamma_{\text{vibration}} = \mathrm{a_{1g} + b_{1g} + b_{2g} + a_{2u} + b_{2u} + 2e_u}$$

- The mode of vibration $\mathrm{e_u}$ has the symmetries of the excited configurations:

$$\mathrm{E_u} : \ldots \mathrm{a_{1u}^2 a_{2u} e_g}$$

$$\mathrm{E_u} : \ldots \mathrm{a_{2u}^2 a_{1u} e_g}$$

  These combine to yield Q- and B-bands.
  As a result, the vibronic excitations of symmetries $\mathrm{E_u}$ is also allowed.
- In a transition from ground state to the vibronic level takes place in a very short time (about 10^{-15} s):
  - The momentum, which is close to zero in the ground state, is close to zero in the excited state.
  - The electronic transition is fast, the molecule will retain the same vibrational kinetic energy and molecular configuration in the excited state that hold in the ground state at the time of the photon absorption.
  - The atoms in molecules do not change position significantly, Franck-Condon principle.

10.9 THE MAGNETIC DIPOLE MOMENT AND THE ABSORBANCE INTENSITY

What are the bases and requirements of the magnetically allowed transitions? How does it affect the electronic spectra? What is the right tool to detect these transitions?

- **Interference of magnetically allowed transitions; g → g and u → u complicates the absorption spectra (very weak bands, $\varepsilon \sim 0.01\ \mathrm{M^{-1}cm^{-1}}$).**
- The absorption of light by a chiral molecule causes a helical charge displacement in light by promotion of an electron from the ground state to an excited state with different charge distribution.

- A right-handed helical displacement of charge result in positive Cotton effect, whereas a left-handed displacement produces a negative effect.
- The displacement of an electron through a helical path involves the linear motion of the charge along the helix axis, generating a transient electric dipole moment; and a rotation of charge about that axis, producing a transient magnetic dipole moment.
- The magnetic dipole moment operators transform as R_x, R_y, and R_z.
- The rotation R_x, R_y, and R_z jointly form a basis for the T_{1g} representation in O_h.
- In the magnetically dipole allowed transitions, the direct product of the ground and the excited states must span a rotation in the point group concerned (contains T_{1g} representation in O_h).
- Therefore, the transitions $g \rightarrow g$, and $u \rightarrow u$ can be allowed magnetically, but not electrically.
- They complicate the assignment and have a very weak band in the electronic spectra; the value of ε is negligibly small ($\varepsilon \approx 0.01\ M^{-1}\ cm^{-1}$).
- Circular dichroism (CD) spectroscopy has been the principal method for identifying magnetic dipole allowed $d \rightarrow d$ transitions.

Circular Dichroism Spectroscopy

Define types of circularly polarized lights, specific rotation, and molar rotation.

What are the differences between the optically rotator dispersion (ORD) curve and circular dichroism spectroscopy (CD)? What are the bases of circular dichroism spectroscopy?

- Other kinds of polarized light exist beside the linearly polarized light. Fig. 10.51 shows the different types of light beams that are polarized.
- As a helpful example for picturing the rotation of linearly polarized light, consider the light as being formed of two oppositely rotating beams of circularly polarized light. Consequently, the linearly polarized light is the vector sum of the right and left circularly rotating constituents as represented in Fig. 10.52. The sums are illustrated at points A-E with resulting vectors having the properties of linearly polarized light (Fig. 10.52).
- An optical active substance rotates the plane of polarization of polarized light beam passing through it.

FIG. 10.51 Kinds of polarized light.

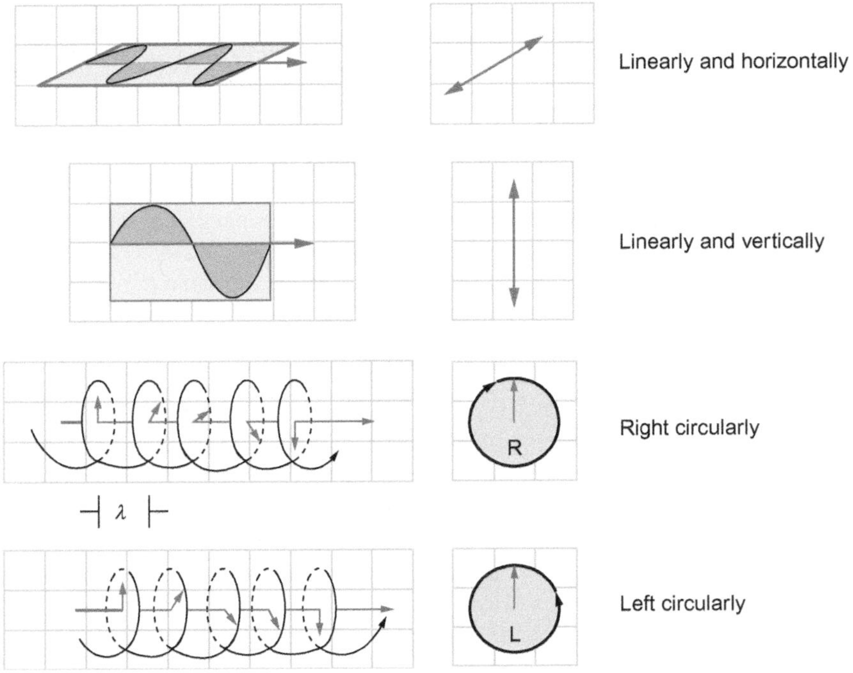

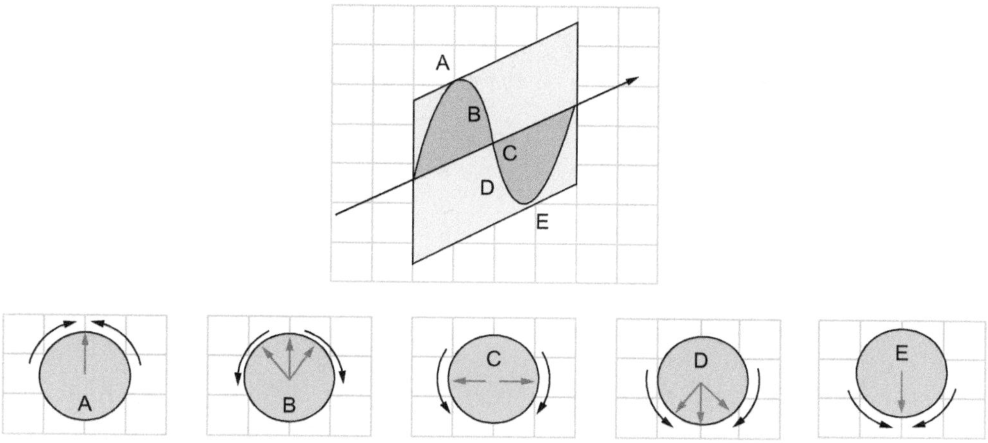

FIG. 10.52 Illustration of linearly polarized radiation as the vector sum of two oppositely rotating of circularly polarized radiation.

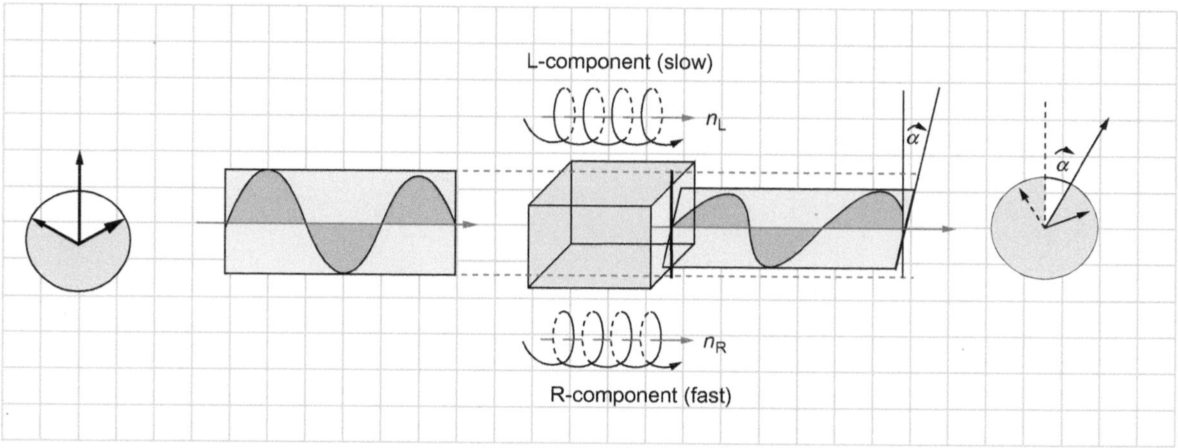

FIG. 10.53 Linearly polarized light is considered as the superposition of two oppositely rotating components which transmit with different velocities in an optical active medium. The polarization has been rotated through an angle α.

- The plane of polarization is rotated if the refractive index for left-handed circularly polarized light (n_L) is different from that for right-handed circularly polarized light (n_R). The plane of polarization rotates through an angle, $\boldsymbol{\alpha}$, and if λ is the wavelength (Fig. 10.53), then

$$\alpha = \frac{n_R - n_L}{\lambda} \tag{10.9.1}$$

- If the concentration of an optically active substance, c', is expressed in units of g cm^{-3}, the specific rotation $[\alpha]$ is defined as

$$[\alpha] = \frac{\alpha}{c'd'} \tag{10.9.2}$$

where d' is the thickness of the sample in decimeter.

- The molar rotation $[M]$ is defined as

$$[M] = M[\alpha] \times 10^{-2} = \frac{\alpha M}{c'd'} \times 10^{-2} \tag{10.9.3}$$

where M is the molecular weight of the optically active component.

- The optically rotator dispersion (ORD) curve is a plot of the molar rotation $[M]$ or α, against λ, Fig. 10.54.

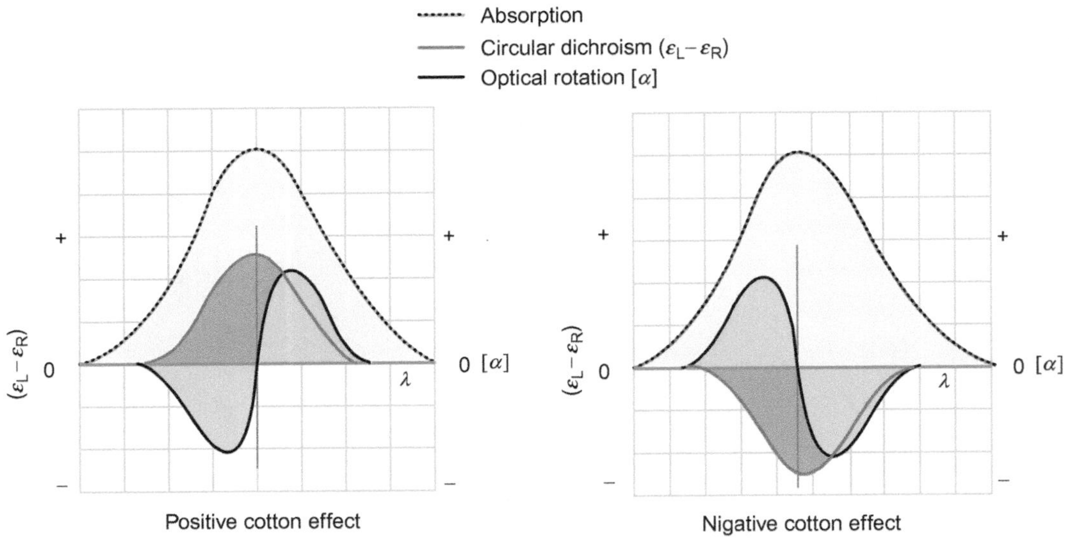

FIG. 10.54 Optical rotatory dispersion and circular dichroism are known as the Cotton effect.

- When optically active substance absorbs right and left circularly polarized light differently is called circular dichroism (CD) (Fig. 10.53).
- By plotting $\Delta\varepsilon$ versus λ, the CD curve results.
  where

$$\Delta\varepsilon = \varepsilon_R - \varepsilon_L = \alpha/(33 \cdot b \cdot c)$$

  Where:
  $\Delta\varepsilon$ molar in unit of cm^{-1} L mol^{-1};
  $\boldsymbol{\alpha}$ is the degrees of elasticity, and is read directly from the instrument;
  b is the bath length in cm; and
  c represents the molarity (moles per liter).
- The S-shaped curve is known as the Cotton effect. If the peak precedes the trough on measuring from longer to shorter wavelength, the Cotton effect is termed positive. Conversely, if a trough precedes a peak, it is a negative Cotton effect (Fig. 10.53).

What are the advantages of circular dichroism spectroscopy?

- The advantages of CD-spectroscopy are as follows:
  - Transitions that are electrically allowed and magnetically allowed will show a strong circular dichroism rotation.
  - Transitions that are electrically forbidden and magnetically allowed will show up quite clearly in CD spectrum.
  - The CD spectrum may contain clear features that are absent or obscured in the electronic spectrum.
- Circular dichroism spectroscopy offers valuable information for ordering the energy level within molecule.
  Octahedral Ni(II) complexes have three bands:

$$\mathbf{^3A_{2g} \rightarrow {}^3T_{2g}}$$

$$\mathbf{^3A_{2g} \rightarrow {}^3T_{1g}(F)}$$

$$\mathbf{^3A_{2g} \rightarrow {}^3T_{1g}(P)}$$

  in order of increasing energy, which of these transitions is magnetically dipole allowed transition?
- Since the magnetic dipole operator in O_h is T_{1g} (Table 10.13):

$$\Gamma\left[\psi'_{A_{2g}}\psi_{T_{2g}}\right] = A_{2g} \times T_{2g} = T_{1g}$$

$$\Gamma\left[\psi'_{A_{2g}}\psi_{T_{1g}}\right] = A_{2g} \times T_{1g} = T_{2g}$$

TABLE 10.13 The Character Table for an Octahedral Point Group

O_h	E	$8C_3$	$6C_2$	$6C_4$	$3C_2$	i	$6S_4$	$8S_6$	$3\sigma_h$	$6\sigma_d$		
A_{1g}	1	1	1	1	1	1	1	1	1	1	–	$x^2+y^2+z^2$
A_{2g}	1	1	−1	−1	1	1	−1	1	1	−1	–	–
E_g	2	−1	0	0	2	2	0	−1	2	0	–	$2z^2-x^2-y^2$, x^2-y^2
T_{1g}	3	0	−1	1	−1	3	1	0	−1	−1	(R_x, R_y, R_z)	–
T_{2g}	3	0	1	−1	−1	3	−1	0	−1	1		(xz, yz, xy)
A_{1u}	1	1	1	1	1	−1	−1	−1	−1	−1	–	–
A_{2u}	1	1	−1	−1	1	−1	1	−1	−1	1	–	–
E_u	2	−1	0	0	2	−2	0	1	−2	0	–	–
T_{1u}	3	0	−1	1	−1	−3	−1	0	1	1	(x, y, z)	–
T_{2u}	3	0	1	−1	−1	−3	1	0	1	−1	–	–
$A_{2g}\times T_{2g}$	3	0	−1	1	−1	3	1	0	−1	−1	$=T_{1g}$	
$A_{2g}\times T_{1g}$	3	0	1	−1	−1	3	−1	0	−1	1	$=T_{2g}$	

- Only for the $^3A_{2g}\rightarrow{}^3T_{2g}$ transition is a magnetic dipole allowed.
- It is found in the CD spectrum of $\mathbf{Ni(pn)_3}^{2+}$ (pn=propylene diamine) that the low-energy band has a maximum $(\varepsilon_R-\varepsilon_L)$ value of 0.8, while those for the other bands are less than 0.04. The CD bands are usually narrow and can be positive or negative.

The Effects of Lower Symmetry

Give examples of some optically active chelate structures, and the IUPAC nomenclature.

- Examples:

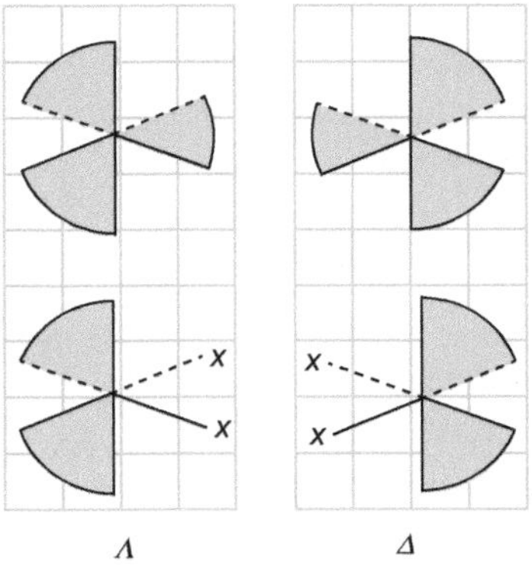

The main types of optically active chelate molecules

Fig. 10.55 represents the electronic and circular dichroism of some Co(III) complexes. What can you conclude from these spectra?

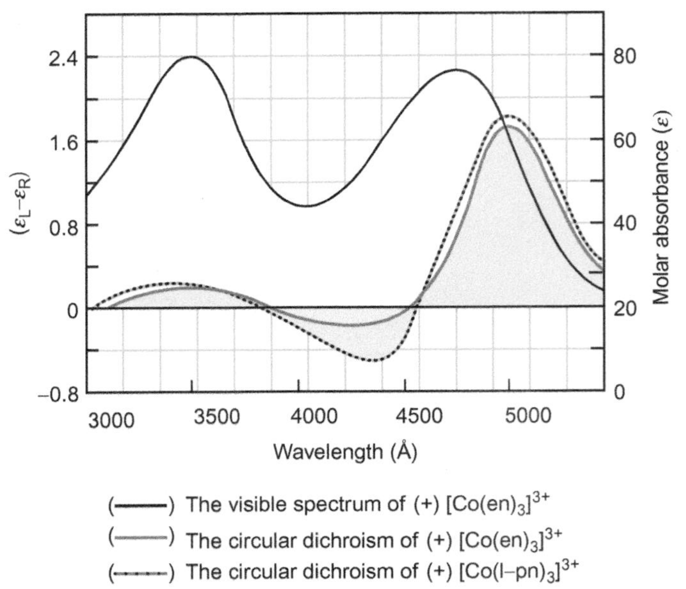

FIG. 10.55 The electronic absorption spectrum of (+)$[Co(en)_3]^{3+}$ and the circular dichroism of (+)$[Co(en)_3]^{3+}$ and (+)$[Co(l-pn)_3]^{3+}$, the electronic spectrum of (+)$[Co(l-pn)_3]^{3+}$ is almost identical with that of (+)$[Co(en)_3]^{3+}$, where en: $NH_2CH_2CH_2NH_2$, and pn: $NH_2CH_2CH_2CH_2NH_2$. *(From F.A. Cotton, G. Wilkinson, Advanced Inorganic Chemistry: A Comprehensive Text, fourth ed., John Willey & Sons, New York, 1980, p. 673 (Chapter 20), reproduced with permission).*

- Co(III) complexes with O_h symmetry have two absorptions assigned to

$$^1A_{1g} \rightarrow {}^1T_{1g}, \text{ and } {}^1A_{1g} \rightarrow {}^1T_{2g}.$$

- The symmetry of $Co(en)_3{}^{3+}$ or $Co(pn)_3{}^{3+}$ is D_3, so the transition to the T states are expected to show some splitting. This splitting is not detected in absorption spectra.
  Correlation among the representations in O_h and those in D_3

O_h	→	D_3
A_{1g}	→	A_1
T_{1g}	→	A_2+E
T_{2g}	→	A_1+E

- However, the effects of lower symmetry are observed in the CD spectrum.
  - The magnetic dipole allowed components, $^1A_1 \rightarrow {}^1A_2$, and $^1A_1 \rightarrow {}^1E_a$.
  - The $^1A_1 \rightarrow {}^1E_b$ transition of the high-energy band is magnetic dipole allowed, but the $^1A_1 \rightarrow {}^1A_1$ transition is not.
- In the CD, two components (+ and −) are seen in the low-energy band and one in the high-energy band, as predicted for a D_3 distortion.

Absolute Configuration

Tris(ethylenediamine) cobalt(III), tris(alaninato) cobalt(III), and bis(ethylenediamine) glutamate cobalt(III) have the Λ configuration. What can you conclude from their spectral properties in Figs. 10.54 and 10.56?

- Another application involves using the sign of the CD to obtain the absolute configuration of the molecule.
- Molecules that have the same absolute configuration:
  - must be similar molecules, e.g., enantiomers having the same sign of [α];
  - must contain similar electronic transitions; and
  - should have the same signs for ORD and CD effect.
- The configurations of enantiomers of tris(ethylenediamine) cobalt(III), tris(alaninato) cobalt(III), and bis(ethylenediamine) glutamate cobalt(III) are known from X-ray investigations. It has been found that the three D-enantiomers of these complexes have similar ORD spectra. Obviously, these three complex ions must have the same absolute configuration.
- Finally, optically active transitions are polarized, and polarization information can be used to support the assignment of the electronic spectrum.

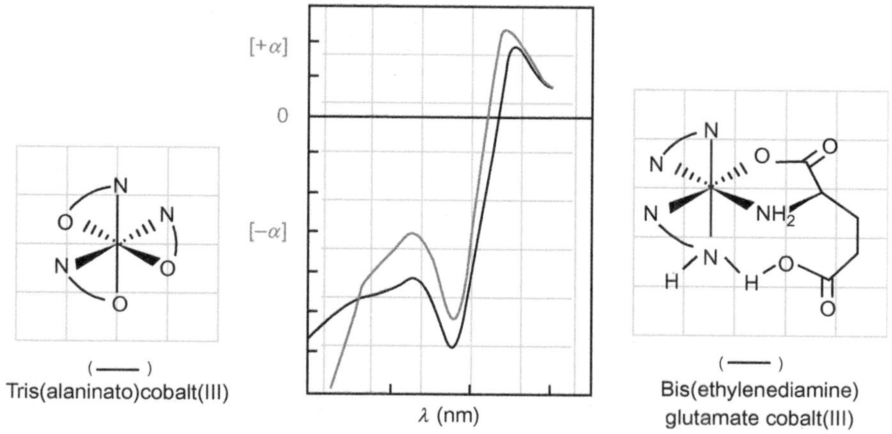

FIG. 10.56 Similar ORD spectra indicate the same optical configuration (Λ) for tris(alaninato)cobalt(III), and bis(ethylenediamine) glutamate cobalt(III).

Kuhn anisotropy Factor and Deducing the Energy Levels Within the Molecule

What are the advantages of using the ratio of the differential intensity ($\Delta\varepsilon$) in the CD spectrum to the intensity ε of the electronic transition?

- For a given transition, the ratio $\left|\frac{\Delta\varepsilon}{\varepsilon}\right|$ is known as Kuhn anisotropy (or dissymmetry) factor, and reflects the ratio of the rotational strength to the dipole strength, $\frac{4R}{D}$, where

$$4R = 91.6 \times 10^{-40} \int \left(\frac{\Delta\varepsilon}{\nu}\right) \partial\nu$$

$$D = 91.8 \times 10^{-40} \int \left(\frac{\varepsilon}{\nu}\right) \partial\nu,$$

where

$$\int \left(\frac{\varepsilon}{\nu}\right) \partial\nu = \ln 2 \cdot \sqrt{\pi}\, \varepsilon_{max} \frac{\Delta_{1/2}}{\nu^0}$$

ε_{max} is the maximum extinction coefficient, $\Delta_{1/2}$ is the bandwidth at half-height, and ν^0 is the frequency of the band intensity maximum.

The circular dichroism and Kuhn anisotropy spectra of ferrinitrosyl myoglobin are represented in Figs. 10.57 and 10.58. The electronic charge of Fe^{3+}—NO is redistribute toward Fe^{2+}—NO^+. Fe^{2+}—NO^+ and Fe^{2+}—CO are isoelectronic. Use the extended Hückel orbital energy scheme (Fig. 10.59) for carbonmonoxy hemoglobin in order to model the nitrosyliron(III)myoglobin. The main modification to the scheme in passing from CO to NO is that the

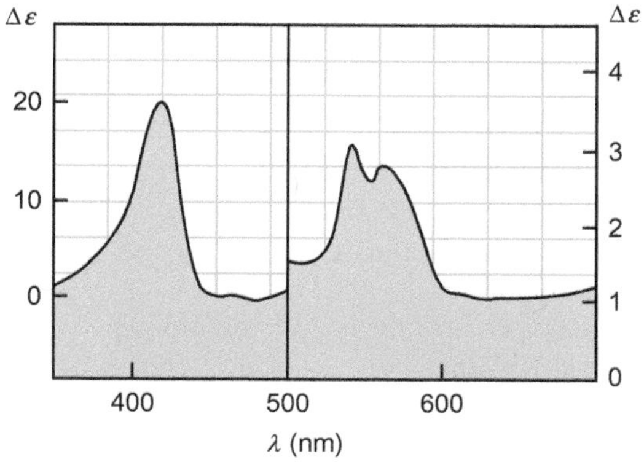

FIG. 10.57 Circular dichroism of ferrinitrosylmyoglobin. *(From A.W. Addison, J.J. Stephanos, Biochemistry 25 (1986) 4104, reproduced with permission).*

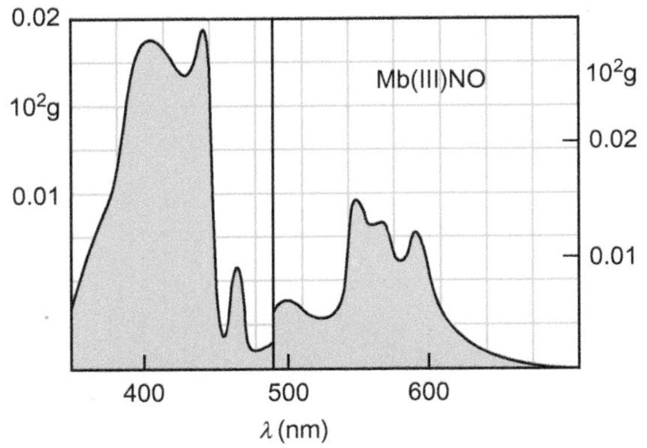

FIG. 10.58 Kuhn anisotropy spectrum for ferrimyoglobinnitrosyl. *(From A.W. Addison, J.J. Stephanos, Biochemistry 25 (1986) 4104, reproduced with permission).*

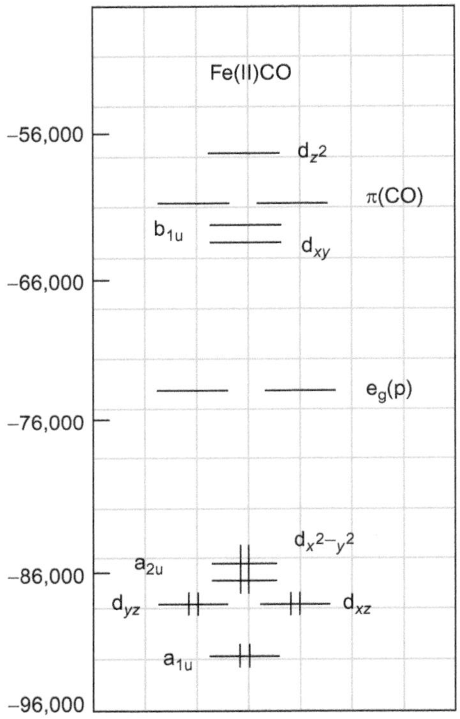

FIG. 10.59 Molecular orbital scheme for Fe(II)CO. *(From W. Addison, J.J. Stephanos, Biochemistry 25 (1986) 4104, reproduced with permission).*

doubly degenerate π* MO of the free NO molecule is at considerably lower energy ($-74{,}730\ cm^{-1}$) and thus close to the porphyrin $e_g(\pi^*)$ orbital at $-74{,}440\ cm^{-1}$. Corrections for exchange interactions energy of $4030\ cm^{-1}$ are considered to the orbital energy.

- The CD spectrum of Mb(III)NO (ferrimyoglobinnitrosyl) (Fig. 10.57) shows two positive dichroic bands at 563 and 547 nm in the visible region. While the Soret dichroic of Mb(III)NO exhibits only a positive sharp dichroic band.
- A number of bands and shoulders are observed in the CD spectrum of Mb(III)NO, which are not related to $\pi \rightarrow \pi^*$ transitions that dominate the simple absorption.
- The value of $\Delta\varepsilon$ is proportional to the product of the electric-dipole and magnetic-dipole, while the value of ε is proportional to the square of the electrical-dipole.
- Therefore, plotting the Kuhn anisotropy factor, g, $\left[g = \frac{\Delta\varepsilon}{\varepsilon}\right]$, versus wave length λ, will depict and emphasize the magnetically dipole transitions.
- The criteria of weak absorption and large Kuhn anisotropic factor are used to characterize the nature (electric dipole forbidden, magnetic dipole allowed) of these transitions.

TABLE 10.14 Orbital Origins of Transitions in Mb(III)NO

Orbital Promotion	Calculated Energy (10^{-3} ν cm^{-1})	Observed Energy (10^{-3} ν cm^{-1}, 10^4 g)
$d_{x^2-y^2} \rightarrow \pi_y^*(p)$	15.6	15.6
$a_{2u} \rightarrow \pi_y^*(NO)$	16.8	
$d_{xz} \rightarrow \pi_x^*(p)$	17.1	17.1 (1.2)
$a_{2u} \rightarrow \pi_x^*(p)$	17.2	17.3
$a_{1u} \rightarrow \pi_x^*(NO)$	18.6	17.8 (1.3)
$a_{2u} \rightarrow \pi_y^*(p)$	18.9	17.9
$d_{yz} \rightarrow \pi_y^*(NO)$	20.8	18.3
$d_{yz} \rightarrow \pi_y^*(p)$	21.1	20.0 (0.61)
$a_{1u} \rightarrow \pi_y^*(NO)$	22.6	21.7 (0.6)
$d_{x^2-y^2} \rightarrow b_{1u}$	22.8	22.9 (2.1)
$a_{1u} \rightarrow \pi_x^*(p)$	22.9	23.8
$a_{2u} \rightarrow d_{xy}$	23.9	24.3 (1.9)
$d_{x^2-y^2} \rightarrow d_{xy}$	25.2	24.9 (6.3)

- In Figs. 10.57 and 10.58, the observed weak absorption (ε) and large Kuhn anisotropy factor (g) are used to assign the origin (electric dipole forbidden, magnetic dipole allowed) of these transitions (Table 10.14).
- The vicinity chiral center contribution which dominates the heme protein CD shows generally low Kuhn anisotropies ($g = 10^{-3}$ for magnetically allowed, 10^{-4} for forbidden).
- When the selection rules for CD are taken into account, and corrections for exchange interactions are made, an exchange energy of 4030 cm^{-1} was added to the orbital energy difference for Fe(d) $\rightarrow$ π*(p), p(π) $\rightarrow$ NO (π*), and Fe (d) $\rightarrow$ NO (π*) transitions.
- The anticipated assignments fit the observed spectra adequately.
- For the d $\rightarrow$ d transitions of heme-iron, the exchange energy estimated by a crystal field calculation was applied, thus the transition energy is the difference between the energies of the acceptor and donor orbitals (Fig. 10.60).

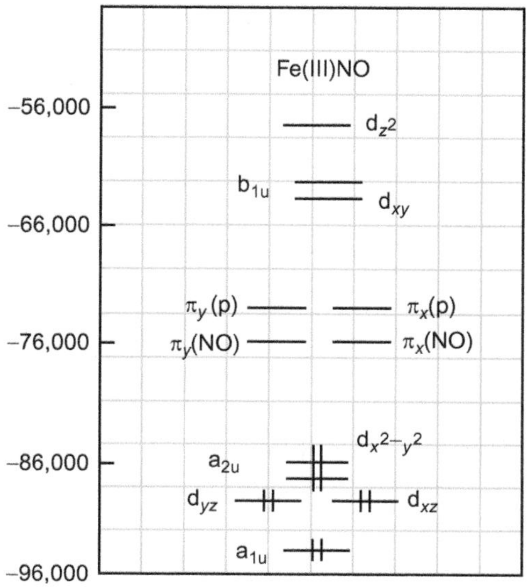

FIG. 10.60 Molecular orbital scheme for Fe(III)NO. *(From W. Addison, J.J. Stephanos, Biochemistry 25 (1986) 4104, reproduced with permission).*

SUGGESTIONS FOR FURTHER READING

G.M. Barrow, Introduction to Molecular Spectroscopy, McGraw-Hill, New York, NY, 1962.

F.A. Cotton, Chemical Applications to Ligand fields, John Wiley & Sons, New York, NY, 1971.

B.N. Figgis, An Introduction to Ligand Fields, John Wiley & Sons, New York, NY, 1966.

F. Grum, Visible and ultraviolet spectroscopy, in: A. Weissberger, B.W. Rossiter (Eds.), Techniques of Chemistry, John Wiley & Sons, New York, NY, 1972.

G. Herzberg, Electronic Spectra of Polyatomic Molecules, Van Nostrand Reinhold, New York, NY, 1966.

G. Herzberg, Spectra of Diatomic Molecules, Van Nostrand Reinhold, New York, NY, 1950.

G. Herzberg, The Spectra and Structure of Simple Free Radicals, Cornell University Press, Ithaca, NY, 1971.

H.H. Jaffé, M. Orchin, Theory and Applications of Ultraviolet Spectroscopy, John Wiley & Sons, New York, NY, 1962.

C.A. Parker, Photoluminescence of Solutions, Elsevier, Amsterdam, 1968.

M.B. Robin, Higher Excited States of Polyatomic Molecules, Academic Press, New York, NY, 1974.

J.I. Steinfeld, Molecules and Radiation, Harper and Row, New York, NY, 1974.

N.J. Turro, Molecular Photochemistry, W. A. Benjamin, New York, NY, 1967.

H.B. Abrahamson, C.C. Frazier, D.S. Ginley, H.B. Gray, J. Lilienthal, D.R. Tyler, M.S. Wrighton, Electronic spectra of dinuclear cobalt carbonyl complexes, Inorg. Chem. 16 (1977) 1554.

T.N. Bolotnikova, Opt. Spectrosc. (USSR) 8 (1953) 138.

E.A. Boudreaux, L.N. Mulay, Theory and Applications of Molecular Paramagnetic, John Wiley, New York, NY, 1976.

F.A. Cotton, Chemical Applications of Group Theory, John Wily, New York, NY, 1971 (Chapters 9 and 10).

P. Cassidy, M.A. Hitchman, Vibrational fine structure in the electronic spectrum of planar $CuCl_4^{2-}$, J. Chem. Soc. Chem. Commun. (1975) 837.

F.A. Cotton, G. Wilkinson, Advanced Inorganic Chemistry: A Comprehensive Text, third ed., John Wiley, New York, NY, 1972.

J.J. Dekkers, G.Ph. Hoornweg, G. Visser, C. MacLean, N.H. Velthorst, Chem. Phys. Lett. 47 (1977) 357.

P.B. Dorain, Symmetry in Inorganic Chemistry, Addison-Wesley, Reading, MA, 1965. p. 88.

R.C. Dunbar, R. Klein, Spectroscopy of radical cations. The McLafferty rearrangement product in fragmentation of n-butylbenzene and 2-phenylethanol ions, J. Am. Chem. Soc. 99 (1977) 3744.

P.P. Dymerski, E. Fu, R.C. Dunbar, Ion cyclotron resonance photodissociation spectroscopy spectra of substituted benzenes, J. Am. Chem. Soc. 96 (1974) 4109.

J. Ferguson, Spectroscopy of 3*d* complexes, Prog. Inorg. Chem. 12 (1970) 159.

B.N. Figgis, Introduction to Ligand Fields, John Wiley, Hoboken, NJ, 1966.

R. Gale, R.E. Godfrey, S.F. Mason, R.D. Peacock, B. Stewart, A complementation of the crystal field for tetrahedral intensities, J. Chem. Soc. Chem. Commun. (1975) 329.

L.H. Hall, Group Theory and Symmetry in Chemistry, McGraw-Hill, New York, NY, 1969 (Chapters 10 and 11).

B.J. Hathaway, D.E. Billing, The electronic properties and stereochemistry of mono-nuclear complexes of the copper(II) ion, Coord. Chem. Rev. 5 (1970) 143.

F.E. Ilse, H.Z. Hartmann, Z. Phys. Chem. (Leipzig) 197 (1951) 239.

A.B.P. Lever, Inorganic Electronic Spectroscopy, Elsevier, Amsterdam, 1968.

L.L. Lohr Jr., Spin-forbidden electronic excitations in transition metal complexes, Coord. Chem. Rev. 8 (1972) 241.

F.W. McLafferty, Interpretation of Mass Spectra, Benjamin, New York, NY, 1966. p. 123.

G.L. McPherson, M.R. Freedman, Electron paramagnetic resonance and ligand field spectra of vanadium(II) in crystals of cesium calcium trichloride. Spectroscopic properties of a perfectly octahedral hexachlorovanadate(4-) complex, Inorg. Chem. 15 (1976) 2299.

D.S. Martin, M.A. Tucker, A.J. Kassman, The absorption spectrum and dichroism of potassium tetrachloroplatinate(II) crystals. II, Inorg. Chem. 4 (1965) 1682.

M.J. Mobley, K.E. Rieckhoff, E.M. Voigt, The temperature dependence of multiple charge-transfer bands in .pi.-.pi. electron donor-acceptor complexes, J. Phys. Chem. 81 (1977) 809.

D. Nicholls, Comprehensive Inorganic Chemistry, Pergamon, New York, NY, 1973.

D. Nicholls, Complexes and First Row Transition Elements, Macmillan, London, 1975.

K. Ohbayashi, H. Akimoto, I. Tanaka, Emission spectra of methoxyl, ethoxyl, and isopropoxyl radicals, J. Phys. Chem. 81 (1977) 798.

U. Olsher, The electronic absorption spectrum of gaseous pyridine 1800 Å $\pi \rightarrow \pi^*$ transition, J. Chem. Phys. 66 (1977) 5242.

S. Onaka, D.F. Shriver, Application of Raman spectroscopy to bridged-nonbridged equilibria for polynuclear metal carbonyl derivatives. Metal-metal stretching frequencies of octacarbonyldicobalt, octacarbonyldiferrate(2-), dodecacarbonyltetracobalt, and bis[.eta.-cyclopentadienyl)dicarbonylruthenium], Inorg. Chem. 15 (1976) 915.

C.S.G. Phillips, R.J.P. Williams, Inorganic Chemistry, vol. 2, Oxford University Press, London, 1965.

D.A. Ramsay, J. Chem. Phys. 43 (1965) 18.

A.F. Schreiner, D.J. Hamm, Magnetic circular dichroism, crystal field, and related studies of nickel(II) amines, Inorg. Chem. 12 (1973) 2037.

A.F. Schreiner, D.J. Hamm, Cryogenic excited-state fine structure of $[Ni(NH_3)_6]A_2$ and magnetic circular dichroism temperature dependence of $^3T_{1g}(t_{2g}^5e_g^3)$, Inorg. Chem. 14 (1975) 519.

E.V. Shpolskii, Problems of the origin and structure of the quasilinear spectra of organic compounds at low temperature, Sov. Phys. Usp. 5 (1962) 522.

E.V. Shpolskii, New data on the nature of the quasilinear spectra of organic compounds, Sov. Phys. Usp. 6 (1963) 411.

J.T. Wrobleski, G.J. Long, The electronic spectra of the tetrahedral $[MnO_4]^{m-}$ chromophore, J. Chem. Educ. 54 (1977) 75.

D.R. Yarkony, H.F. Schaefer III, S. Rothenberg, Geometries of the methoxy radical (X ^2E and A $^2A_1$ states) and the methoxide ion, J. Am. Chem. Soc. 96 (1974) 656.

A. Cotton, Chemical Applications of Group Theory, John Wiley & Sons, New York, NY, 1990.

R. Drago, Physical Methods for Chemists, Surfside Scientific Publishers, Gainesville, FL, 1992.

D. Harris, M. Bertolucci, Symmetry and Spectroscopy, An Introduction to Vibrational and Electronic Spectroscopy, Dover Publications, Inc., New York, NY, 1989.

G. Miessler, D. Tarr, Inorganic Chemistry, Pearson Education Inc., New Jersey, 2004.

J.M. Chalmers, P. Griffiths (Eds.), In: Handbook of Vibrational Spectroscopy, 5 volume set, Wiley, New York, NY, 2002.

J. Workman, A. Springsteen (Eds.), Applied Spectroscopy: A Compact Reference for Practitioners, Academic Press, Boston, 1998.

Chapter 11

Magnetism

Atoms have magnetic moments due to the motion of their electrons. In addition each electron has an intrinsic magnetic moment associated with its spin. The net magnetic moment of an atom depends on the number the arrangement of electrons in the atom.

Rotation of an electron will generate a magnetic field that may be considered as small magnetic dipoles. When this rotated electron exists in a strong external magnetic field, parallel alignment will occur. The alignment of magnetic dipoles parallel to an external magnetic field tends to increase the field.

In a magnetically polarized material, the dipoles create a magnetic field parallel to the magnetic dipole vectors. Fig. 11.1 shows the observed magnetic behavior that relates to the number and orbital arrangement of the unpaired electron.

Magnetic measurement can provide information on the electronic configuration, oxidation state, and symmetry properties of center ion in the transition metal complexes. Theoretical principles and fundamentals of magnetochemistry are discussed and treated in this chapter to an extent that is essential in understanding its application in coordination chemistry.

This chapter begins by investigating the types of magnetic behaviors, key definitions, and the equations of magnetochemistry studies (Scheme 11.1). These provide a bridge to understand the spin and orbital magnetic moments. Then, thermal spreading using Boltzmann distribution is employed to investigate and estimate the magnetic moments and susceptibilities. The next section deals with Van Vleck treatment and the second-order Zeeman Effect to link the spin and orbital contributions to the magnetic susceptibility. Requirements and conditions for nonzero orbital contribution are discussed. The following section is devoted to study the effect of ligand field on spin-orbital coupling in A, E, and T ground terms. Curie law, deviations, and data presentations are presented. Next is a treatment and investigation of the magnetic behaviors of compounds that contain a unique magnetic center with spin crossover and thermal distortion. The interactions between magnetic centers in binuclear compounds, as well as the coupling mechanisms, are discussed. Finally, descriptions are given of Gouy's, Faraday's, Quincke's, and NMR methods that are most frequently used to measure the magnetic susceptibility.

- 11.1: Magnetic susceptibility
- 11.2: Types of magnetic behaviors
- 11.3: Diamagnetic behavior
- 11.4: Spin-only magnetic moment
- 11.5: Orbital magnetic moment
- 11.6: Second-order Zeeman Effect and Van Vleck equation
- 11.7: Spin-orbital coupling and magnetic susceptibility
  - Multiplet splitting $\gg kT$
  - Multiplet splitting $\ll kT$
  - Multiplet splitting comparable to kT
  - Requirements and conditions for zero and nonzero orbital contribution
- 11.8: Spin-orbital coupling: in A and E ground terms
- 11.9: Spin-orbital coupling: in T ground terms
- 11.10: Curie law, deviation, and data representations
- 11.11: The magnetic behaviors of compounds contain a unique magnetic center in
  - Spin crossover and spin equilibria
- 11.12: Structure-linked crossover, thermal isomerization
- 11.13: Interactions between magnetic centers in
  - Homobinuclear compounds
  - Heterobinuclear compounds
  - Coupling mechanisms

Electrons, Atoms, and Molecules in Inorganic Chemistry. http://dx.doi.org/10.1016/B978-0-12-811048-5.00011-0

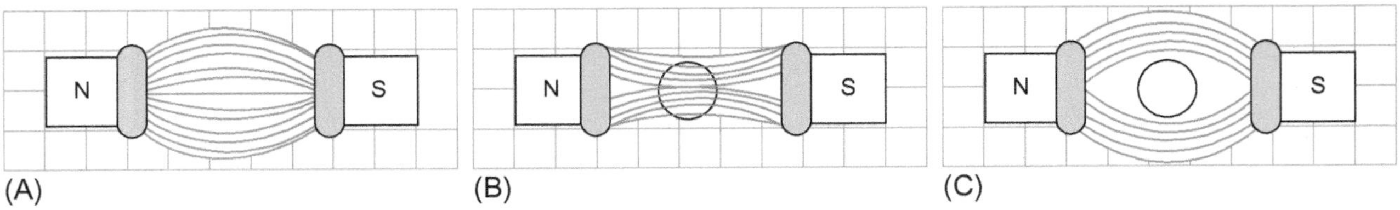

FIG. 11.1 (A) Magnetic field lines of flux; (B) paramagnetic substance and lines of flux; and (C) diamagnetic substance and lines of flux.

- 11.14: Measurement of the magnetic susceptibility
  - Gouy's methods
  - Faraday's method
  - Quincke's method
  - NMR method
- Suggestions for further reading

11.1 MAGNETIC SUSCEPTIBILITY

How can the magnetic behavior of a substance be examined and what does it mean by magnetic susceptibility?

- The magnetic behavior is examined by monitoring the induced magnetic polarization of the substance in a magnetic field, $\vec{H}_o$.

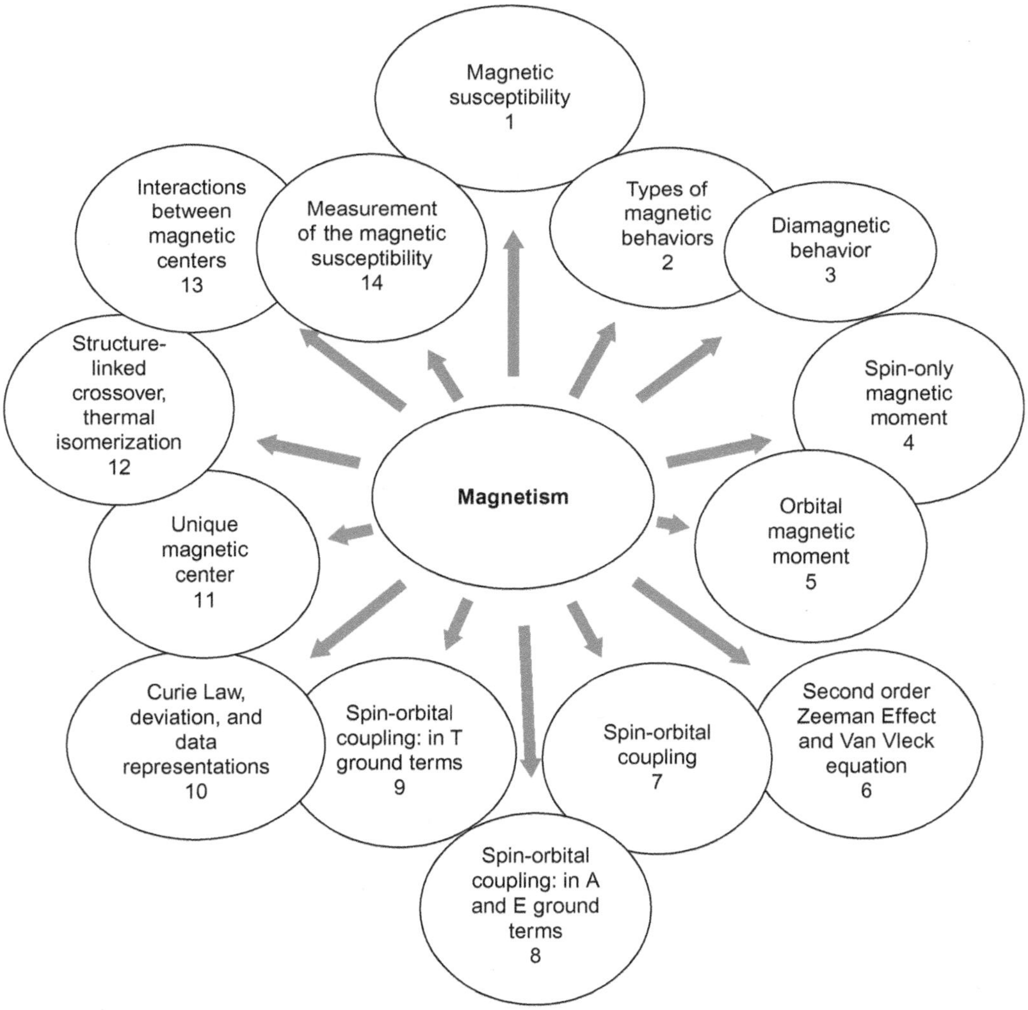

SCHEME 11.1 Elements used to introduce magnetic behaviors.

- Thus, the magnetic induction vector, $\vec{B}$, which usually called the magnetic flux density becomes

$$\vec{B}=\vec{H}_o+4\pi\vec{M} \tag{11.1.1}$$

where $\vec{H}_o$ is the applied magnetic field and $\vec{M}$ is magnetization, the intensity of magnetization per unit volume.

- If we assumed an isotropic system that the direction of $\vec{H}_o$ and $\vec{M}$ coincident, and factor out the directional property:

$$\frac{B}{H_o}=1+4\pi\frac{M}{H_o}=1+4\pi\chi_v \tag{11.1.2}$$

where

$\frac{B}{H_o}$ is the permeability of the medium, and

$$\frac{M}{H_o}=\chi_v, \text{ is the magnetic susceptibility per unit volume} \tag{11.1.3}$$

Divided by the density of the substance, *d*:

$$\frac{\chi_v}{d}=\chi_g\ (\text{cm}^3\,\text{g}^{-1}) \tag{11.1.4}$$

Then multiplying χ_g by the molecular weight, $M_{wt.}$, gives the molar susceptibility:

$$\chi_g\times M_{wt.}=\chi(\text{cm}^3\text{mol}^{-1}) \tag{11.1.5}$$

and

$$\chi=\frac{M}{H} \tag{11.1.6}$$

The induced magnetic moment, *M*, is proportional to the applied field, *H*, and χ is the constant of proportionality.

- Therefore, the magnetic susceptibility, χ, of a material is a measure of the accessibility to align electron spins with applied magnetic field.

11.2 TYPES OF MAGNETIC BEHAVIORS

What are the various types of magnetic behaviors?

- There are four main types of magnetic behaviors (Table 11.1).
- Only molecules containing unpaired electrons show paramagnetism.
  - It arises from the spin and orbital moments of the unpaired electrons.
  - These moments do not interact strongly with each other and are normally randomly oriented.
  - The external magnetic field tends to align this intrinsic magnetic moment of the molecule in the direction of the field.

TABLE 11.1 Main Magnetic Behaviors of Material

Type	Diamagnetism	Paramagnetism	Ferromagnetism	Antiferromagnetism
χ	Negative	Positive	Positive	Positive
Magnitude of χ (emu)	10^{-6}	0 to 10^{-4}	10^{-4} to 10^{-2}	0 to 10^{-4}
χ and the applied field	Independent	Independent	Dependent	Dependent
χ and temperature	None	Decreases	Decreases	Increases
Origin	Induced field by circulation of paired electrons	Spin- and orbitals-angular momentum of unpaired electrons	Spin alignment from dipole-dipole interaction of moments on adjacent atoms, ↑↑	Spin pairing, ↑↓, from dipole-dipole interactions

- At ordinary temperature and external field, only a very small fraction of the molecules are aligned because thermal motion tends to randomize there orientation. Therefore, the increase in the total magnetic field is very small.

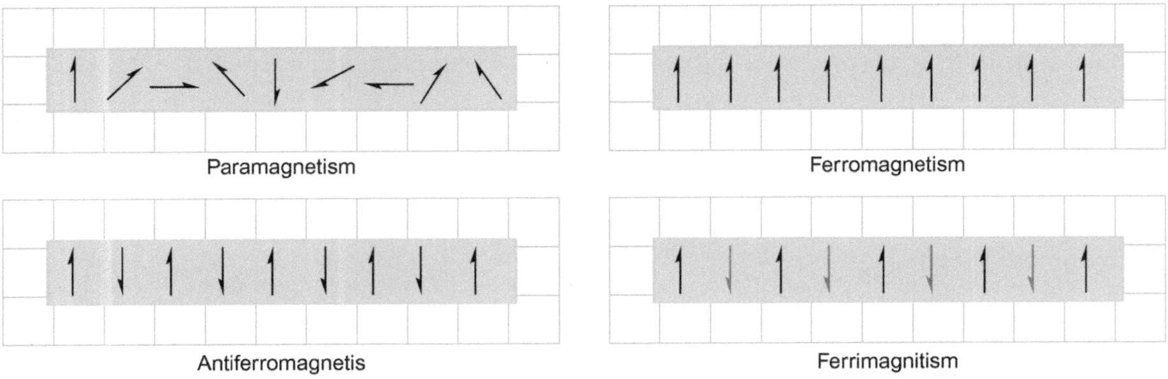

- In solid, the spins on neighboring centers may interact to produce magnetic behavior, such as ferromagnetism and antiferromagnetism.
  - Because of the strong interaction between neighboring magnetic centers, a high degree of magnetic alignment can be achieved even with weak external field.
  - This causes a very large increase in the total field.
  - Even when there is no external magnetic field, ferromagnetic compounds may have a magnetic dipole alignment as in permanent magnets.
- Diamagnetism is the result of an induced magnetic moment opposite in direction to the external field.
  - The induced dipoles thus weaken the resultant magnetic field.
- The temperature dependence of the susceptibility is distinguishing characteristic for these magnetic behaviors.
- In ferrimagnetism, ions are arranged with alternative spins as in antiferromagnetic, but because the individual spin moments are not the same, there is partial cancelation and a net overall moment is noticed.

What are the thermomagnetic behavior and distinctive features of paramagnetic, ferromagnetic, antiferromagnetic, and ferrimagnetic substances?

- Diamagnetic is nearly independent of temperature, because the large energy barrier between ground state and excited states of the paired electron.
- As a paramagnetic substance is cooled, the disordering of the thermal motion is reduced, more spins become aligned; as a result, the magnetic susceptibility increases (Fig. 11.2).
- In a ferromagnetic material, the spin on the various metal centers are coupled into parallel alignment, ↑↑↑↑ that is extended over neighbor atoms to form a magnetic domain.

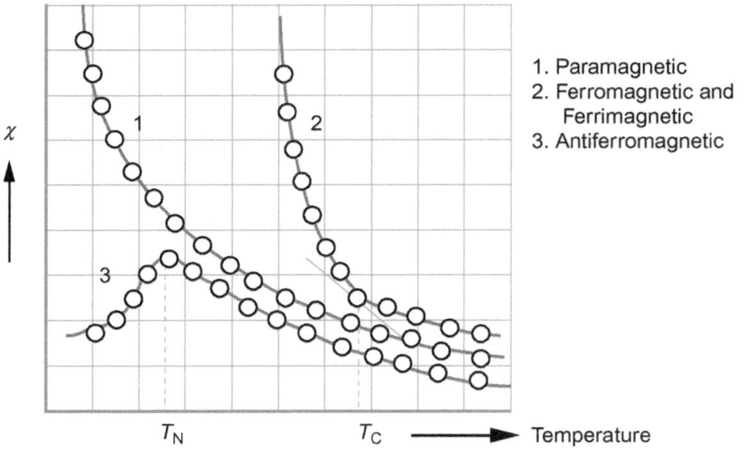

FIG. 11.2 The susceptibility as function of temperature. T_C: the temperature at which the break occurs in the plot of ferromagnetic behavior and is referred to as Curie temperature. T_N: the temperature at which the maximum takes place in the plot of antiferromagnetic behavior is known as the Neel temperature.

- The net magnetic moment and magnetic susceptibility become very large because the magnetic moments of individual spins added to each other.
- Below the Curie temperature, T_C, the magnetism is persevered after the applied magnetic field is removed because the spins are locked together.
- Ferromagnetism is exhibited by a substance holding unpaired electrons in d or f orbitals, which couple with unpaired electrons in similar orbitals on neighboring atoms.
- The coupling interaction is sufficient to align spins but not so strong to form covalent bond, where the electrons would be paired.
- At temperatures above T_C, the random thermal motion destroys the aligning of the interaction and material becomes paramagnetic (Fig. 11.2).
- The magnetization and the magnetic moment of ferromagnetic materials are not proportional to the field applied. Instead, a "hysteresis loop" is obtained (Fig. 11.3).
- A hysteresis loop is observed because the magnetization of the sample with an increasing field is not retraced as the field is decreased.
- Hard ferromagnetic has a broad loop; magnetization stays large even when the applied field reduced to zero, and is used for a permanent magnet when a field does not have to be inverted.
- Soft ferromagnetic has a narrower loop; it is much more sensitive to the magnetic field, and is used in transformers, where they must reply to a fast-oscillating field.

- Antiferromagnetic substance:
  - In an antiferromagnetic substance, the nearby spins are locked into an antiparallel alignment, as a result, the sum of the individual magnetic moments is canceled, and the material exhibits a low magnetic susceptibility and magnetic moments.
  - Antiferromagnetism is detected when a paramagnetic substance is cooled to a low temperature and is identified by sudden decrease in magnetic susceptibility at the T_C. Above T_N the magnetic susceptibility decreases as temperature is raised (Fig. 11.2).
- In ferrimagnetism, a net magnetic arrangement of the ions with different magnetic moments is detected below the T_C.
  - As with antiferromagnetism, these coupling are commonly transmitted through the ligands, for example: Fe_3O_4.

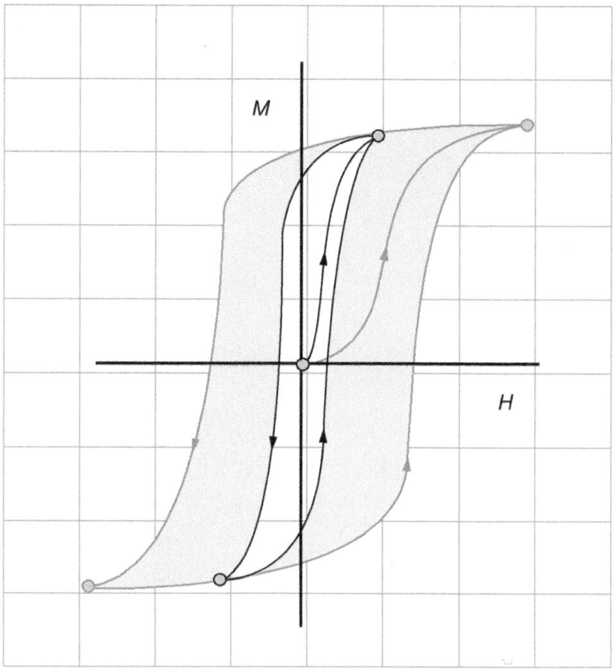

FIG. 11.3 Magnetization loops of ferromagnetic substance, *black curve* for soft ferromagnetic and *gray curve* for hard ferromagnetic.

11.3 DIAMAGNETIC BEHAVIOR

How can the diamagnetic susceptibility be calculated? What is the origin of this susceptibility? How does it influence the net susceptibility?

- This magnetic behavior is due to the action of an induced field caused by circulation of the paired electrons in an external field.
- When an electron rotates with angular velocity, ω, it will produce a current (I), and a magnetic field perpendicular to the plane of rotation. The current is given by

$$I = \frac{e\omega}{2\pi c} \quad (11.3.1)$$

where e is the fundamental electron charge, and c is the velocity of light.
- The generated magnetic moment is the product of the current and the area of the orbital:

$$\mu = \frac{e\omega}{2\pi c} \pi \overline{r_i^2} = \frac{e\omega \overline{r_i^2}}{2c} \quad (11.3.2)$$

$\overline{r_i^2}$ is the mean square radius of the projection of the orbit perpendicular to the direction of H.

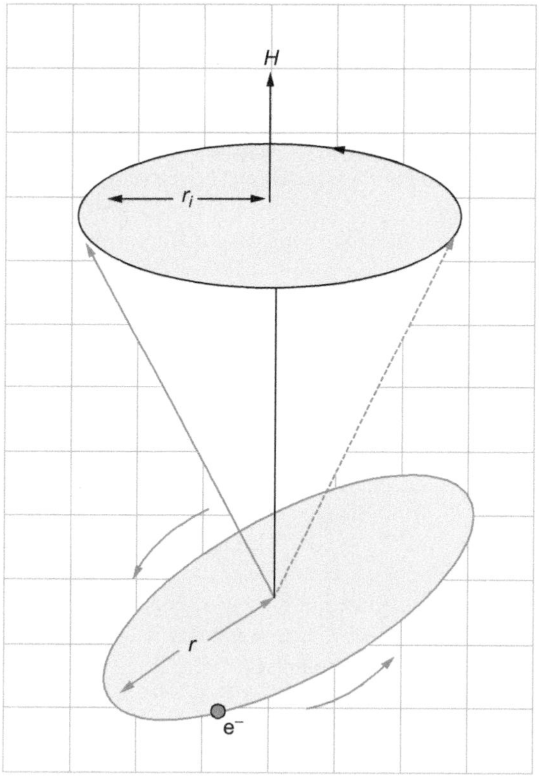

- Precession of an electron about the direction of H has angular velocity ω given by

$$\omega = \frac{eH}{2mc} \quad (11.3.3)$$

where m is the mass of the electron.

$$\mu = -\frac{e\omega \overline{r_i^2}}{2c} = -\frac{e\overline{r_i^2}}{2c}\frac{eH}{2mc} = -\frac{e^2 H}{4mc^2}\overline{r_i^2} \quad (11.3.4)$$

The negative sign pinpoints the opposite direction of μ.
- Consider, at a given instant, that the coordinates of the electron are x, y, and z, and the direction of H coincident with z-axis with the nucleus at the origin. Then the mean average radius is as follows:

$$\overline{r^2} = \overline{x^2} + \overline{y^2} + \overline{z^2}$$

Taking the projection of the orbit in xy plane (perpendicular to H), then

$$\overline{r_i^2} = \overline{x^2} + \overline{y^2} = \frac{2}{3}\overline{r^2}$$

Then

$$\mu = -\frac{e^2 H}{4mc^2}\frac{2}{3}\overline{r^2} = -\frac{e^2 H}{6mc^2}\overline{r^2} \tag{11.3.5}$$

- For multielectron atom

$$\mu(\text{atom}) = -\frac{e^2 H}{4mc^2}\frac{2}{3}\overline{r^2} = -\frac{e^2 H}{6mc^2}\sum\overline{r^2} \tag{11.3.6}$$

for μ in the unit of $cm^3\ mol^{-1}$

$$\mu_A(\text{gram atom}) = \frac{\mu \times M}{d} = -\frac{N_A e^2 H}{4mc^2}\frac{2}{3}\overline{r^2} = -\frac{N_A e^2 H}{6mc^2}\sum\overline{r^2} \tag{11.3.7}$$

where M and d are the molecular weight and the density.
- The diamagnetic molar susceptibility is given by

$$\chi_{Dia} = \chi_g \times M = \frac{\chi_V}{d} \times V = \chi_V \times V$$

$$\chi_{Dia} = \frac{I}{H} \times V = \frac{\mu_A}{V \times H} \times V = \frac{\mu_A}{H}$$

$$\chi_{Dia} = -\frac{N_A e^2}{6mc^2}\sum_{i=0}^{i=n}\overline{r_i^2} = -2.83\sum_{i=0}^{i=n}\overline{r_i^2} \tag{11.3.8}$$

where N_A is Avogadro's number and n is the number of the electron.
- The spinning electron and the spinning nucleus hold a similar diamagnetic behavior:
  - Essentially, this diamagnetic behavior is due to the induced change in angular momentum by an external magnetic field.
  - However, the electron spin hardly affects the diamagnetism of the orbiting electron because the spinning electron has much smaller radius $(\approx 10^{-13}\,\text{cm})$ as compared to the radius of the orbit $(\approx 10^{-8}\,\text{cm})$.
  - The nuclear spin barely disturbs the diamagnetism of the orbiting electron, since the nucleus has a mass much greater than that of the electron.
- Note the following facts about diamagnetic susceptibility (χ_{Dia}):
  - It is an intrinsic property of a compound containing paired or unpaired electrons. It can be evaluated in a substance having paired electrons. The unpaired electrons being paramagnetic overshadow diamagnetism.
  - It is independent of the strength of the applied magnetic field, H.
  - It is an induced effect that temporarily occurs only if H lasts.
  - It opposes the applied magnetic field, and has a negative value.
  - Its magnitude depends on the radius of the electron orbit, i.e., the larger the radius, the greater its magnitude.
  - It is independent of temperature. However, any small variation in the χ_{Dia} value with a change in temperature has to be considered in view of expansion or contraction of the radius of the electron orbit.
- Every molecule shows diamagnetism behavior. The molar susceptibility, χ_{Dia} can be calculated by summing the diamagnetic contributions from all atoms, χ_A, and from all bonds in functional group:

$$\chi_{Dia} = \sum_i \chi_A + \sum_i \chi_B \tag{11.3.9}$$

The values of χ_A and χ_B are known as Pascal's constants.
- For a transition metal complexes, the net susceptibility, χ:

$$\chi_{paramagnetic} = \chi_{measured} - \chi_{Diamagnetic} \tag{11.3.10}$$

if

$$\chi_{H} = -2.93 \times 10^{-6}\,cm^3\,mol^{-1}$$

$$\chi_{O} = -2.93 \times 10^{-6}\,cm^3\,mol^{-1}$$

$$\chi_{C}(\text{aromatic}) = -6.24 \times 10^{-6}\,cm^3\,mol^{-1}$$

$$\chi_{N}(\text{aromatic}) = -4.61 \times 10^{-6}\,cm^3\,mol^{-1}$$

$$\chi_{C=O} = +6.3 \times 10^{-6}\,cm^3\,mol^{-1}$$

Find χ for each of C_5H_5N, and $(CH_3)_2C{=}O$.

- C_5H_5N
  Sum of contributions to χ:

$$5 \times \chi_{H} = -14.6 \times 10^{-6}\,cm^3\,mol^{-1}$$

$$5 \times \chi_{C(\text{aromatic})} = -31.2 \times 10^{-6}\,cm^3\,mol^{-1}$$

$$1 \times \chi_{N(\text{aromatic})} = -4.6 \times 10^{-6}\,cm^3\,mol^{-1}$$

From Eq. (11.3.9):

$$\chi_{\text{Dia}} = \sum_i \chi_A + \sum_i \chi_B = -50.4 \times 10^{-6}\,cm^3\,mol^{-1}$$

- $(CH_3)_2C{=}O$
  Sum of contributions to χ:

$$6 \times \chi_{H} = -17.6 \times 10^{-6}\,cm^3\,mol^{-1}$$

$$3 \times \chi_{C} = -18.7 \times 10^{-6}\,cm^3\,mol^{-1}$$

$$1 \times \chi_{O} = -4.6 \times 10^{-6}\,cm^3\,mol^{-1}$$

$$1 \times \chi_{C=O} = +6.3 \times 10^{-6}\,cm^3\,mol^{-1}$$

$$\chi_{\text{Dia}} = \sum_i \chi_A + \sum_i \chi_B = -34.6 \times 10^{-6}\,cm^3\,mol^{-1}$$

$$\chi_{\text{Dia}} = \sum_i \chi_A + \sum_i \chi_B = -50.4 \times 10^{-6}\,cm^3\,mol^{-1}$$

11.4 SPIN-ONLY MAGNETIC SUSCEPTIBILITY, MAGNETIC MOMENT, AND THERMAL SPREADING

Consider a simple spherical system that contains only one electron and no orbital contribution to the momentum show that

$$\mu_n = -\frac{\partial E}{\partial H} = -m_s g \beta$$

where μ_n is the magnetic moment of an electron in state n, g is the electron g-factor, and β is the Bohr magneton. Use $H = 25$ kilogauss, and $g = 2.0023$ at 25°C, to show that $kT \gg \Delta E$, ΔE is energy spacing between spin states.

$$\chi = \frac{N_A g^2 \beta^2}{4kT}, \quad \text{for } m_s = \pm\frac{1}{2}$$

$$\chi = \frac{N_A g^2 \beta^2}{3kT} S(S+1), \quad \text{for } n \text{ unpaired electrons}$$

$$\mu_{\text{eff.}} = \left(\frac{3k}{N_A\beta^2}\right)^{\frac{1}{2}} (\chi T)^{\frac{1}{2}} = 2.828(\chi T)^{\frac{1}{2}} \text{ BM}$$

- The magnetic moment, $\vec{\mu}$, of this system:
  $\vec{\mu} = -g\beta\, \vec{S}$ (Chapter 2, p. 134)
  where g is the electron g-factor, $\vec{S}$ is the spin angular momentum, and β is the Bohr magneton of the electron $(\beta = 0.93 \times 10^{-20}$ erg gauss$^{-1})$.
- The Hamiltonian describes the interaction of moment, $\vec{\mu}$, with the applied magnetic field, $\vec{H}$, is given by

$$\hat{H} = -\vec{\mu} \cdot \vec{H} = g\beta \vec{S} \cdot \vec{H} \tag{11.4.1}$$

$\hat{H}$ is operating on two spin wave functions (Fig. 11.4), thus it has two eigenvalues:

$$E = m_s g\beta H \tag{11.4.2}$$

With $m_s = \pm\frac{1}{2}$, the energy difference is given by

$$\Delta E = g\beta H \tag{11.4.3}$$

- When the unpaired electron exists in a strong magnetic field, a parallel alignment will occur (ground state), which can be excited to antiparallel alignment (Fig. 11.4)
- In the case of $H = 25$ kilogauss, and $g = 2.0023$, the value of ΔE for free electrons is

$$\Delta E = g\beta H$$

$$\Delta E = 2.0023 \times 0.93 \times 10^{-20} \frac{\text{erg}}{\text{gauss}} \times 25{,}000 \text{ gauss}$$

$$\Delta E = 46.5535 \times 10^{-17} \text{erg.} = 46.5535 \times 10^{-24} \text{J} = 2.344 \text{ cm}^{-1}$$

where 1 cm^{-1} = 1.9864×10^{-23} J.

While the thermal energy, kT, at $T = 25°$C:

$$kT = 298 \text{K} \times 1.381 \times 10^{-23} \text{JK}^{-1} \times \frac{1 \text{ cm}^{-1}}{1.9864 \times 10^{-23} \text{J}} = 207 \text{ cm}^{-1}$$

where $k = 1.381 \times 10^{-23}$ JK^{-1}, $T = 25°\text{C} = 273 + 25 = 298$ K.

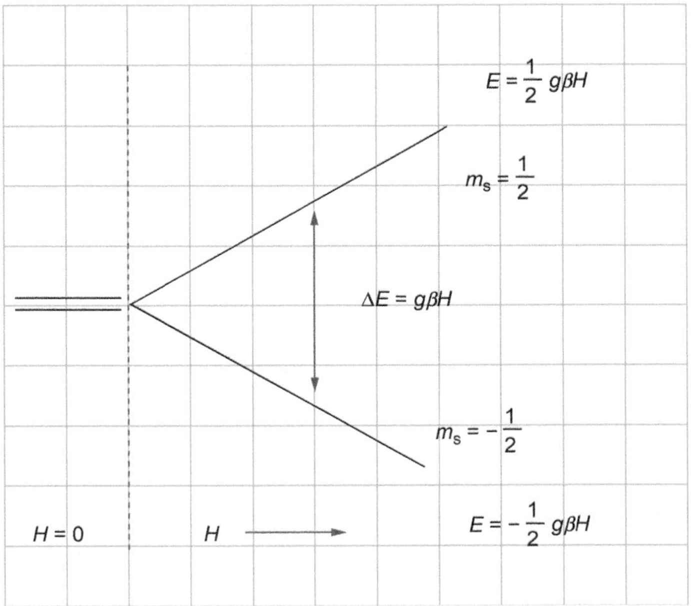

FIG. 11.4 Splitting of the spin states by external magnetic field.

$$\therefore kT \gg \Delta E$$

- Since ΔE is small enough compared to kT, the two states

$$E = +\frac{1}{2}g\beta H \text{ and } E = -\frac{1}{2}g\beta H$$

are populated at room temperature (Fig. 11.4).
- At a low temperature, the occupancy of the ground state is enhanced at the expense of the excited state.
- $\because E = m_s g\beta H$, Eq. (11.4.2), then

$$\frac{\partial E}{\partial H} = m_s g\beta \quad (11.4.4)$$

Therefore, the partial derivative of the energy state E_n with respect to the applied field H gives the magnitude of the projection of the magnetic moment, μ_n, along the field in quantum state n.

$$\mu_n = -\frac{\partial E}{\partial H} = -m_s g\beta \quad (11.4.5)$$

- The effective magnetic moment, μ_e, of a sample is the sum of individual electronic moments that are distributed over the various S levels.

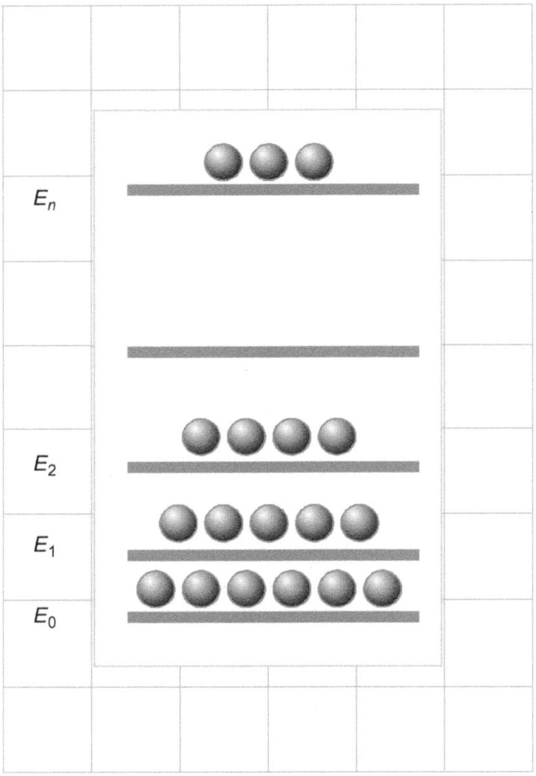

- Weighted by their Boltzmann populations, the probability, P_n, for populating specific states having energy levels E_n at thermal equilibrium is given by

$$P_n = \frac{N_n}{N} = \frac{e^{\frac{-E_n}{kT}}}{\sum_n e^{\frac{-E_n}{kT}}} \quad (11.4.6)$$

where N is the total population of all the states and N_n is the population of the state n.

- The magnetic moments over the individual states for one mole of material, M:

$$M = N_A \sum_n \mu_n P_n$$

$$M = N_A \frac{\sum_n \mu_n e^{\frac{-E_n}{kT}}}{\sum_n e^{\frac{-E_n}{kT}}} \tag{11.4.7}$$

N_A: Avogadro number.

- Then for: $m_s = \pm\frac{1}{2}$, $E = \pm\frac{1}{2}g\beta H$, and $\mu_n = -m_s g\beta$ (Eqs. 11.4.2 and 11.4.5)

$$M = N_A \frac{\sum_{m_s=-1/2}^{+1/2} \mu_n e^{\frac{-E_n}{kT}}}{\sum_{m_s=-1/2}^{+1/2} e^{\frac{-E_n}{kT}}} \tag{11.4.8}$$

$$M = N_A \left(\frac{\frac{g\beta}{2} e^{\frac{g\beta H}{2kT}} - \frac{g\beta}{2} e^{\frac{-g\beta H}{2kT}}}{e^{\frac{g\beta H}{2kT}} + e^{\frac{-g\beta H}{2kT}}} \right)$$

$$M = \frac{N_A g\beta}{2} \left(\frac{e^{\frac{g\beta H}{2kT}} - e^{\frac{-g\beta H}{2kT}}}{e^{\frac{g\beta H}{2kT}} + e^{\frac{-g\beta H}{2kT}}} \right)$$

at 25°C, $g\beta H = 2.344\,\text{cm}^{-1}$, $kT = 207\,\text{cm}^{-1}$, $\therefore \frac{g\beta H}{2kT} \ll 1$, then

$$e^{\frac{\pm g\beta H}{2kT}} \approx \left(1 \pm \frac{g\beta H}{2kT}\right) \tag{11.4.9}$$

$$M = \frac{N_A g\beta}{2} \left(\frac{\left(1 + \frac{g\beta H}{2kT}\right) - \left(1 - \frac{g\beta H}{2kT}\right)}{\left(1 + \frac{g\beta H}{2kT}\right) + \left(1 - \frac{g\beta H}{2kT}\right)} \right)$$

$$M = \frac{N_A g^2 \beta^2 H}{4kT} \tag{11.4.10}$$

If: $\chi = \frac{M}{H}$, then:

$$\chi = \frac{N_A g^2 \beta^2}{4k} \cdot \frac{1}{T} \tag{11.4.11}$$

- In case $m_s = S, S-1, \ldots, -S+1, -S$, $kT \gg \Delta E$, and Eq. (11.4.7):

$$M = N_A \frac{\sum_{m_s=-S}^{+S} \mu_n e^{\frac{-E_n}{kT}}}{\sum_{m_s=-S}^{+S} e^{\frac{-E_n}{kT}}} \tag{11.4.12}$$

$E=\pm m_s g\beta H$ and $\mu_n=-m_s g\beta$ (Eqs. 11.4.2 and 11.4.5)

$$M=N_A g\beta\frac{\sum_{m_s=-S}^{+S} m_s e^{\frac{g\beta H m_s}{kT}}}{\sum_{m_s=-S}^{+S} e^{\frac{g\beta H m_s}{kT}}}$$

$$M=N_A g\beta\left[\frac{Se^{\frac{g\beta HS}{kT}}+(S-1)e^{\frac{g\beta H(S-1)}{kT}}+\cdots+(-S+1)e^{\frac{g\beta H(-S+1)}{kT}}+(-S)e^{\frac{g\beta H(-S)}{kT}}}{e^{\frac{g\beta HS}{kT}}+e^{\frac{g\beta H(S-1)}{kT}}+\cdots+e^{\frac{g\beta H(-S+1)}{kT}}+e^{\frac{g\beta H(-S)}{kT}}}\right]$$

$$M=N_A g\beta\left[\frac{Se^{\frac{g\beta HS}{kT}}+(S-1)e^{\frac{g\beta H(S-1)}{kT}}+\cdots-(S-1)e^{\frac{-g\beta H(S-1)}{kT}}-Se^{\frac{-g\beta HS}{kT}}}{e^{\frac{g\beta HS}{kT}}+e^{\frac{g\beta H(S-1)}{kT}}+\cdots+e^{\frac{-g\beta H(S-1)}{kT}}+e^{\frac{-g\beta HS}{kT}}}\right]$$

at 25°C, $g\beta H=2.344\ \text{cm}^{-1}$, $kT=207\ \text{cm}^{-1}$, $\therefore \frac{g\beta H m_s}{kT}\ll 1$, and

$$e^{\frac{\pm g\beta H m_s}{kT}}\approx\left(1\pm\frac{g\beta H m_s}{kT}\right) \quad (11.4.9)$$

Then

$$M=N_A g\beta\left[\frac{S\left(1+\frac{g\beta HS}{kT}\right)+(S-1)\left(1+\frac{g\beta H(S-1)}{kT}\right)+\cdots-(S-1)\left(1-\frac{g\beta H(S-1)}{kT}\right)-S\left(1-\frac{g\beta HS}{kT}\right)}{\left(1+\frac{g\beta HS}{kT}\right)+\left(1+\frac{g\beta H(S-1)}{kT}\right)+\cdots+\left(1-\frac{g\beta H(S-1)}{kT}\right)+\left(1-\frac{g\beta HS}{kT}\right)}\right]$$

$$M=N_A g\beta\left[\frac{\left(\frac{g\beta HS^2}{kT}\right)+\left(\frac{g\beta H(S-1)^2}{kT}\right)+\cdots+\left(\frac{g\beta H(S-1)^2}{kT}\right)+\left(\frac{g\beta HS^2}{kT}\right)}{1+1+\cdots+1+1}\right]$$

$$1+1+1+\cdots=\sum_{-S}^{+S}1=(2S+1)\times 1 \quad (11.4.13)$$

$2S+1$ is the number of the energy levels, when degeneracy is removed by the magnetic field.

$$M=\frac{N_A g^2\beta^2 H}{kT}\left[\frac{2S^2+2(S-1)^2+\cdots+0}{2S+1}\right] \quad (11.4.14)$$

If: the summation formula of the quadratic series:

$$\sum_{k=0}^{S}k^2=0+2^2+3^2+\cdots+S^2=\frac{S(S+1)(2S+1)}{6} \quad (11.4.15)$$

Then

$$M=\frac{N_A g^2\beta^2 H}{3kT}\left[\frac{S(S+1)(2S+1)}{2S+1}\right]$$

$$M=\frac{N_A g^2\beta^2 H}{3kT}[S(S+1)] \quad (11.4.16)$$

$$\because \chi=\frac{M}{H},\ \text{then}$$

$$\chi=\frac{N_A g^2\beta^2}{3kT}[S(S+1)] \quad (11.4.17)$$

- Since the magnetic moment, μ_s, is related to the angular momentum by

$$\mu_s = g_s\sqrt{S(S+1)}\,\text{BM}, \qquad (11.4.18)$$
$$g_s = 2.0023, \text{ (Lande' factor)}$$

Then

$$\mu_e^2 = g^2[S(S+1)]$$

By substituting in Eq. (11.4.17), the paramagnetic molar susceptibility is given by

$$\chi_A = \frac{N\beta^2\mu_e^2}{3kT} \qquad (11.4.19)$$

$$\beta = \frac{eh}{4\pi mc} = 0.9273 \times 10^{-20}\ \text{erg}\,\text{G}^{-1} = \text{Bohr magneton (BM)}$$

$$k = 1.381 \times 10^{-16}\text{erg deg}^{-1} = \text{Boltzmann constant, therefore,}$$

$$\mu_e = \left(\frac{3k}{N_A\beta^2}\right)^{\frac{1}{2}}(\chi T)^{\frac{1}{2}}$$

$$\mu_e = 2.828\sqrt{\chi_A \cdot T}\,\text{BM} \qquad (11.4.20)$$

$$\mu_e^2 = 8\chi_A \cdot T$$

μ_e: effective magnetic moment of the paramagnetic center.

11.5 ORBITAL MAGNETIC MOMENT

Find the number of d electrons, number of unpaired electrons, ground states, and the expected μ_s. Compare and explain the deviation of the experimental $\mu_{exp.}$ from μ_s for the complexes listed in Table 11.2.

- In Table 11.3, the calculated spin-only magnetic moment, μ_s, is listed with the experimental values of $\mu_{exp.}$:

$$\mu_s = g_s\sqrt{S(S+1)} = \sqrt{n(n+1)}\,\text{BM} \qquad (11.4.18)$$
$$g_s = 2.0023 \approx 2$$

  where n is the number of the unpaired electrons.
- The data allow us to differentiate between high- and low-spin complexes, with the following examples (Table 11.3):
  d^4 complexes: $Cr(dipy)_3Br_2{\cdot}4H_2O$ is low-spin with $\mu = 3.3$, whereas $Cr(H_2O)_6SO_4$ is high-spin with $\mu = 4.8$.
  d^5 complexes: $K_4Mn(CN)_6{\cdot}3H_2O$ is low-spin with $\mu = 2.2$, whereas $K_2[Mn(H_2O)_6](SO_4)_2$ is high-spin with $\mu = 5.9$.
  d^7 complexes: $K_2PbCo(NO_2)_6$ is low-spin with $\mu = 1.8$, whereas $(NH_4)_2[Co(H_2O)_6](SO_4)_2$ is high-spin with $\mu = 5.1$.
- In Table 11.3, about half of $\mu_{exp.}$ values significantly deviate from μ_s values that are calculated from spin-only formula. These differences could attribute to the contribution of the orbital magnetic moment.
- On the other hand, complexes in which the values of $\mu_{exp.}$ and μ_s are close, the orbital magnetic moments are quenched

When can we expect to find orbital contribution to the magnetic moment?

- If an electron rotates about an axis, the magnetic moment will be given by

$$\mu_l = \frac{e\omega\overline{r_i^2}}{2c} \qquad (11.3.2)$$

  and the angular moment of the electron is given by
  $L = m\omega r^2$ Chapter 1 (Bohr), then

$$\mu_l = \frac{eL}{2mc}$$

TABLE 11.2 Experimental Values of Magnetic Moments, $\mu_{exp.}$

O_h Complexes		T_d Complexes	
Complex	$\mu_{exp.}$	**Complex**	$\mu_{exp.}$
$CsTi(SO_4)_2 \cdot 12\ H_2O$	1.8	$V(Nor)_4$[a]	1.82
$K_4Mn(CN)_6 \cdot 3H_2O$	2.2	$Co(Nor)_4$[a,b]	2.00
$K_2PbCo(NO_2)_6$[b]	1.8	$Co[P(O\text{-}i\text{-}C_3H_7)_3]_4$	2.0
$(NH_4)_2Cu(SO_4)_2 \cdot 6H_2O$	1.9	FeO_4^{2-}	2.8
$(NH_4)V(SO_4)_2 \cdot 12H_2O$	2.7	NiX_4^{2-}	3.2–4.1
$Cr(dipy)_2 \cdot 2H_2O$	3.3	$Mn(Nor)_4$[a]	3.78
$(NH_4)_2Ni(SO_4)_2 \cdot 6H_2O$	3.2	CoX_4^{-2}	4.4–4.8
$KCr(SO_4)_2 \cdot 12\ H_2O$	3.8	FeX_4^{-2}	5.0–5.2
$(NH_4)_2Co(SO_4)_2 \cdot 6H_2O$	5.1	$FeCl_4^{-2}$	5.9
$CrSO_4 \cdot 6H_2O$	4.9	VCl_4	1.62
$(NH_4)_2Fe(SO_4)_2 \cdot 6H_2O$	5.5	$[(C_2H_5)_4N]_2MnCl_4$	5.94
$K_2Mn(SO_4)_2 \cdot 6H_2O$	5.9	$[(C_2H_5)_4N]_2FeCl_4$	5.40
$K_3Mn(CN)_6$	3.5	$[(C_2H_5)_4N]FeCl_4$	5.92
$K_3Fe(CN)_6$	2.25	Cs_2CoCl_4	4.71
$[(C_2H_5)_4N]_2NiCl_4$	3.89		

[a]*Nor = 1-norbornyl group.*
[b]*Low-spin complex.*

TABLE 11.3 Calculated Spin Magnetic Moments and the Corresponding Experimental Values

Complex	No. of Unpaired Electrons	No. of d Electrons	Ground State	$'\mu_s = \sqrt{n(n+1)}$	$\mu_{exp.}$
O_h complexes					
$CsTi(SO_4)_2 \cdot 12H_2O$	1	1	$^2T_{2g}$	1.73	1.8
$K_4Mn(CN)_6 \cdot 3H_2O$[a]	1	1	$^2T_{2g}$	1.73	*2.2*
$K_2PbCo(NO_2)_6$	1	1	^2E_g	1.73	1.8
$(NH_4)_2Cu(SO_4)_2 \cdot 6H_2O$	1	1	^2E_g	1.73	*1.9*
$K_3Fe(CN)_6$	1	5	$^2T_{2g}$	1.73	2.3
$(NH_4)V(SO_4)_2 \cdot 12H_2O$	2	2	$^3T_{1g}$	2.83	2.8
$K_3Mn(CN)_6$	2	4	$^3T_{1g}$	2.83	*3.5*
$(NH_4)_2Ni(SO_4)_2 \cdot 6H_2O$	2	8	$^3T_{1g}$	2.83	*3.2*
$[(C_2H_5)_4N]_2NiCl_4$	2	8	$^3T_{1g}$	2.83	*3.9*
$KCr(SO_4)_2 \cdot 12H_2O$	3	3	$^4A_{2g}$	3.87	*3.8*
$CrSO_4 \cdot 6H_2O$	4	4	^5E_g	4.90	4.8
$Cr(dipy)_3Br_2 \cdot 4H_2O$	2	4	$^3T_{1g}$	2.83	*3.3*
$K_2Mn(SO_4)_2 \cdot 6H_2O$	5	5	$^6A_{1g}$	5.92	5.9

TABLE 11.3 Calculated Spin Magnetic Moments and the Corresponding Experimental Values—cont'd

Complex	No. of Unpaired Electrons	No. of d Electrons	Ground State	$\mu_s = \sqrt{n(n+1)}$	$\mu_{exp.}$
$(NH_4)_2Fe(SO_4)_2 \cdot 6H_2O$	4	6	$^5T_{2g}$	4.90	*5.5*
$(NH_4)_2Co(SO_4)_2 \cdot 6H_2O$	3	7	$^4T_{1g}$	3.87	*5.1*
T_d complexes					
$V(Nor)_4$[b]	1	1	^2E	1.73	1.82
VCl_4	1	1	^2E	1.73	*1.69*
$Co(Nor)_4$[a,b]	1	5	$^2T_2$	1.73	*2.00*
$Co[P(O\text{-}i\text{-}C_3H_7)_3]_4$	1	9	$^2T_2$	1.73	*2.0*
FeO_4^{2-}	2	2	$^3A_2$	2.83	2.8
NiX_4^{2-}	2	8	$^3T_1$	2.83	*3.2–4.1*
$Mn(Nor)_4$[b]	3	3	$^4T_1$	3.87	3.78
Cs_2CoCl_4	3	7	$^4A_2$	3.87	*4.71*
CoX_4^{-2}	3	7	$^4A_2$	3.87	*4.4–4.8*
$[(C_2H_5)_4N]_2FeCl_4$	4	6	^5E	4.9	*5.4*
FeX_4^{-2}	4	6	^5E	4.9	5.0–5.2
$[(C_2H_5)_4N]_2MnCl_4$	5	5	$^6A_1$	5.92	5.94
$[(C_2H_5)_4N]FeCl_4$	5	5	$^6A_1$	5.92	5.92
$FeCl_4^-$	5	5	$^6A_1$	5.92	5.9

Italizicied values indicate the disagreement between $\mu_{exp.}$ and μ_s.
[a]*Nor = 1-norbornyl group.*
[b]*low-spin complex.*

The actual magnitude of L in term of the quantum number, l, (Chapter 2, p. 76)

$$L = \frac{h}{2\pi}\sqrt{l(l+1)}$$

$$\therefore \mu_l = \frac{eh}{4\pi mc}\sqrt{l(l+1)}$$

- The orbital magnetic moment, μ_l is given by

$$\mu_l = g_l\sqrt{l(l+1)}\,\text{BM} \tag{11.5.1}$$

(Chapter 2, p. 134)
$g_l = 1$, BM $= 0.9273 \times 10^{-20}$ ergs gauss^{-1}, is known as the "Bohr magneton"

- In order to have orbital angular momentum (Fig. 11.5), an electron must be able to circulate about an axis.
- For that reason, an unfilled orbital must be available, in addition to the orbital containing the electron. This orbital has the following properties:
  - It must have the identical shape and same energy as the orbital containing the electron.
  - It has to be superimposable on the orbital containing the electron through rotation about an axis.
  - It cannot have an electron with the same spin as the first one.

FIG. 11.5 Orbital angular momentum of the p-orbital.

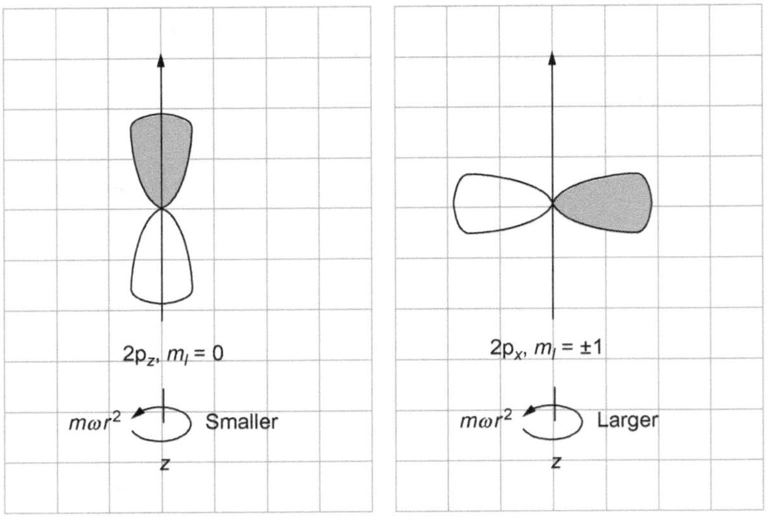

11.6 SECOND-ORDER ZEEMAN EFFECT AND VAN VLECK EQUATION

What is it meant by the second-order Zeeman Effect? How can the magnetic susceptibility be calculated when this effect is considered?

- The second-order Zeeman Effect is due to the distortion of electron distribution by applied field, raises χ, and is known as temperature independent paramagnetism (TIP), or Vleck paramagnetism. As a result, when this effect is considered, the modified value of χ_A will be given by

$$\chi = \frac{N_A \beta^2 \mu_e^2}{3k} \cdot \frac{1}{T} + N\alpha \tag{11.6.1}$$

and the exact value of $N\alpha$ may be found by standard numerical techniques.
- It is produced by higher ligand field levels in transition metals:

$$\text{For } T_d: \; N\alpha = \frac{8N_A\beta^2}{10Dq}, \; \text{and for } O_h: \; N\alpha = \frac{4N_A\beta^2}{10Dq}$$

It can be neglected in systems containing more than one unpaired electron.
- Zeeman effect is considerably significant for some complexes or ions, otherwise its dominant behavior is diamagnetic.
- In general, the contribution to the energy of a given state n:

$$E_n = E_n^{(0)} + \underbrace{H \cdot E_n^{(1)}}_{\text{First order Zeeman}} + \underbrace{H^2 \cdot E_n^{(2)}}_{\text{Second order Zeeman}} + \tag{11.6.2}$$

and if: $\mu = -\dfrac{\partial E}{\partial H}$ (Eq. 11.4.5), then:

$$\mu_n = -\frac{\partial E_n}{\partial H} = \underbrace{-E_n^{(1)}}_{\substack{\text{Contribute to} \\ \mu \text{ that is independent} \\ \text{on magnetic field}}} - \underbrace{2\, H \cdot E_n^{(2)}}_{\substack{\text{Contribute to} \\ \mu \text{ that is dependent} \\ \text{on magnetic field}}} - \tag{11.6.3}$$

Note that $E_n^{(0)}$ makes no contribution to the magnetic moment.
- The extent of the second-order contribution depends on multiplet separation, and can be significant, when the electronic excited and ground states are close in energy, and has the correct symmetry.
- Recalling the molar macroscopic magnetic moment:

$$M = N_A \frac{\sum_n \mu_n e^{\frac{-E_n}{kT}}}{\sum_n e^{\frac{-E_n}{kT}}} \tag{11.4.7}$$

Substituting for E_n (Eq. 11.6.2):

$$M = N_A \frac{\sum_n \mu_n e^{\frac{-E_n^{(0)} - HE_n^{(1)} - H^2 E_n^{(2)} \ldots}{kT}}}{\sum_n e^{\frac{-E_n^{(0)} - HE_n^{(1)} - H^2 E_n^{(2)} \ldots}{kT}}}, \text{ if}$$

$$e^{\frac{-a-bx-cx^2\ldots}{\alpha}} = \left(1 - \frac{bx}{\alpha}\right) e^{\frac{-a}{\alpha}} \tag{11.6.4}$$

Then

$$e^{\frac{-E_n^{(0)} - HE_n^{(1)} - H^2 E_n^{(2)} \ldots}{kT}} = \left(1 - \frac{HE_n^{(1)}}{kT}\right) e^{\frac{-E_n^{(0)}}{kT}}, \text{ and substituting for } \mu_n:$$

$$\mu_n = -E_n^{(1)} - 2H \cdot E_n^{(2)} \tag{11.6.3}$$

$$\therefore M = N_A \frac{\sum_n \left(-E_n^{(1)} - 2H \cdot E_n^{(2)}\right)\left(1 - \frac{HE_n^{(1)}}{kT}\right) e^{\frac{-E_n^{(0)}}{kT}}}{\sum_n \left(1 - \frac{HE_n^{(1)}}{kT}\right) e^{\frac{-E_n^{(0)}}{kT}}}$$

- For paramagnetic compounds, $M = 0$ at $H = 0$.

$$M = N_A \frac{\sum_n \left(-E_n^{(1)}\right) e^{\frac{-E_n^{(0)}}{kT}}}{\sum_n e^{\frac{-E_n^{(0)}}{kT}}} = 0, \text{ therefore:}$$

$$\sum_n \left(-E_n^{(1)}\right) e^{\frac{-E_n^{(0)}}{kT}} = 0 \tag{11.6.5}$$

and

$$M = N_A \frac{\sum_n \left(\frac{H\left(E_n^{(1)}\right)^2}{kT} - 2H \cdot E_n^{(2)}\right) e^{\frac{-E_n^{(0)}}{kT}}}{\sum_n e^{\frac{-E_n^{(0)}}{kT}}} \tag{11.6.6}$$

$$\text{and since}: \chi = \frac{M}{H}, \text{ therefore,}$$

$$\chi = N_A \frac{\sum_n \left(\frac{\left(E_n^{(1)}\right)^2}{kT} - 2E_n^{(2)}\right) e^{\frac{-E_n^{(0)}}{kT}}}{\sum_n e^{\frac{-E_n^{(0)}}{kT}}}, \text{ Van Vleck equation} \tag{11.6.7}$$

where $E_n^{(0)}$ is quantity of energy in absence of a field $E_n^{(0)}$. $E_n^{(1)}$ and $E_n^{(2)}$ are the multipliers of energy changes in H and H^2, respectively.

11.7 SPIN-ORBITAL COUPLING AND MAGNETIC SUSCEPTIBILITY

i. Multiplet splitting $\gg kT$

How do you calculate the magnetic moment and the susceptibility of ions have unquenched orbital angular contribution and large multiplet separation (multiplet splitting $\gg kT$)?

(a) Calculation of the magnetic moment:

- The coupling of L and S and is associated with the magnetic moment μ_{LS}, which is given by AB vector (Fig. 11.6). As a result, of the rapid precession of μ_S and μ_L about the direction of J, the component BC average out to zero and only component μ in the direction of J is considered.
- The components of $\hat{L}$ and $\hat{S}$ in the direction of $\hat{J}$ are $\hat{L}\cdot\cos\phi$ and $\hat{S}\cdot\cos\theta$, respectively, and

$$\mu=\beta\hat{L}\cos\phi+g\hat{S}\beta\cos\theta \tag{11.7.1}$$

$$g=2$$

$$\mu=\beta\hat{L}\cos\phi+2\hat{S}\beta\cos\theta$$

if

$$\hat{S}^2=\hat{J}^2+\hat{L}^2-2\hat{L}\hat{J}\cos\phi$$

$$\hat{L}^2=\hat{J}^2+\hat{S}^2-2\hat{S}\hat{J}\cos\theta$$

$$\hat{L}\cos\phi=\frac{\hat{J}^2+\hat{L}^2-\hat{S}^2}{2\hat{J}}$$

$$\hat{S}\cos\theta=\frac{\hat{J}^2+\hat{S}^2-\hat{L}^2}{2\hat{J}}$$

Then

$$\mu=\beta\hat{L}\cos\phi+2\hat{S}\beta\cos\theta$$

$$\mu=\left(\frac{\hat{J}^2+\hat{L}^2-\hat{S}^2}{2\hat{J}}+\frac{2\hat{J}^2+2\hat{S}^2-2\hat{L}^2}{2\hat{J}}\right)\beta \tag{11.7.2}$$

$$\mu=\left(\frac{3\hat{J}^2-\hat{L}^2+\hat{S}^2}{2\hat{J}}\right)=\left(\frac{3}{2}+\frac{\hat{S}^2-\hat{L}^2}{2\hat{J}^2}\right)\hat{J}\beta=g\beta\hat{J}$$

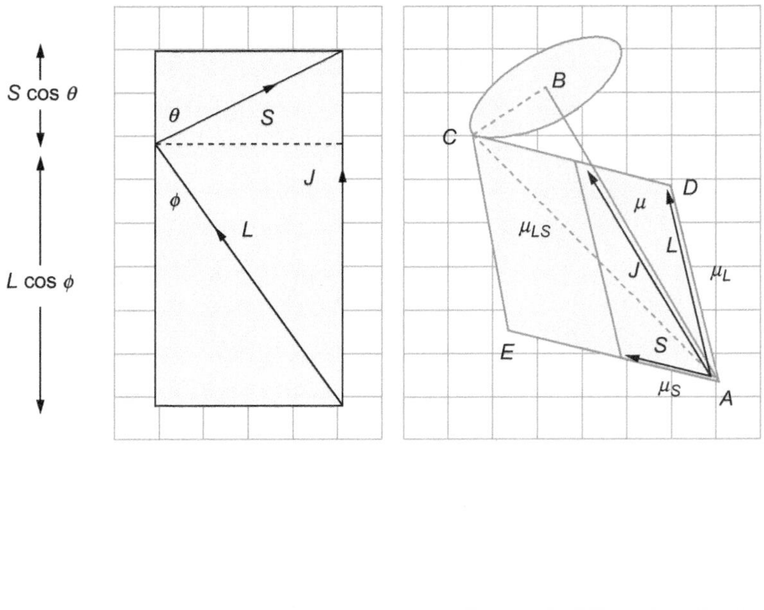

FIG. 11.6 The angular momentum vectors (*black*) and the magnetic moment vectors (*gray*).

$$\hat{J} = \hbar\sqrt{J(J+1)}$$

The term $\hbar$ is included in β

$$\therefore \mu = g\beta\sqrt{J(J+1)} \tag{11.7.3}$$

and

$$g = \left(\frac{3}{2} + \frac{S(S+1) - L(L+1)}{2J(J+1)}\right) \tag{11.7.4}$$

- The orbital magnetic moment only, $S=0$, so $J=\text{L}$, $g_l=1$:

$$\mu_L = \beta\sqrt{L(L+1)} \tag{11.7.5}$$

- The spin only magnetic moment, $L =$), so $J=S$, $g_S=2$:

$$\mu_S = 2\beta\sqrt{S(S+1)} \tag{11.4.18}$$

(b) Calculation the magnetic susceptibility:

- Consider as an example of multiplet splitting $\gg kT$, a metal ion with quantum number J (Russell-Saunders coupling, Chapter 2), the degeneracy of m_J levels ($m_J=J$, $(J-1)$, …, $-(J-1)$ and $-J$) is removed by applying a magnetic field, H (Fig. 11.7).
- In general, the number of levels $=2J+1$.
  Each is associated with m_J-value.
  As $m_J=J$ is the ground state, most atoms are therein.
- The $2J+1$ degeneracy is removed by the magnetic field, and $E_n^{(1)} = m_J g\beta$, $E_n^{(0)}$ and $E_n^{(2)}$ are taken as zero. Notice that $E_n^{(1)}$ is the multipliers of energy changes in H:

$$E_n^{(1)} = \frac{\partial H E_n^{(1)}}{\partial H}$$

- Recalling and substitute in Van Vleck equation:

$$\chi = N_A \frac{\sum_n \left(\frac{\left(E_n^{(1)}\right)^2}{kT} - 2E_n^{(2)}\right) e^{\frac{-E_n^{(0)}}{kT}}}{\sum_n e^{\frac{-E_n^{(0)}}{kT}}} \tag{11.6.7}$$

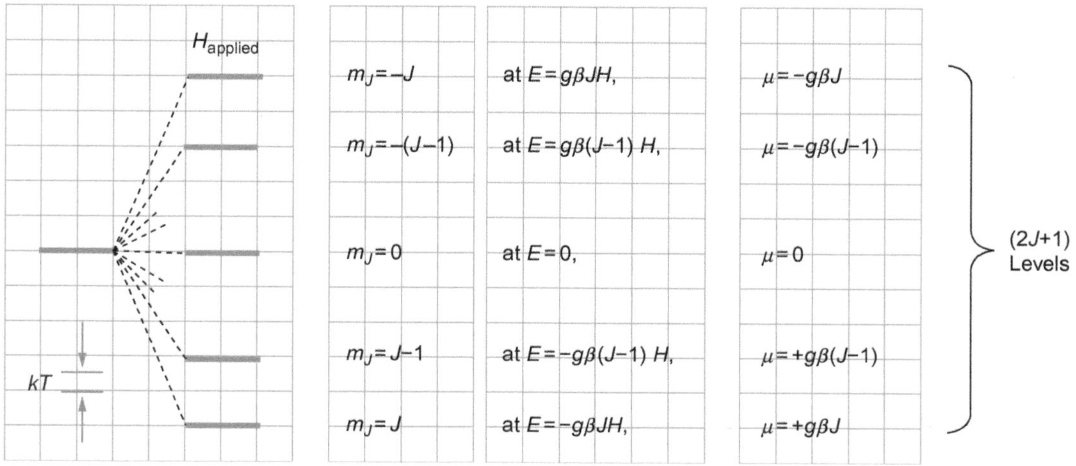

FIG. 11.7 Splitting of J-level into $2J+1$ levels.

$$\chi = N_A \frac{\sum_{m_J=-J,\, n=0}^{m_J=J} \left(\frac{(m_J g\beta)^2}{kT} - 2(0) \right) e^{\frac{0}{kT}}}{\sum_n e^{\frac{0}{kT}}}$$

$$\chi = \frac{N_A g^2 \beta^2}{kT} \frac{\sum_{m_J=-J,\, n=0}^{m_J=J} (m_J^2)}{(1+1+1\ldots)} \qquad (11.7.6)$$

$$1+1+1+\cdots = \sum_{-J}^{+J} 1 = (\text{number of levels}) \times 1) = (2J+1) \times 1 \qquad (11.4.13)$$

$$(1+1+1+\ldots) = 2J+1$$

$2J+1$ is the resulting number of the energy levels, when degeneracy is removed by the magnetic field.

$$\chi = \frac{N_A g^2 \beta^2}{kT} \left[\frac{J^2 + (J-1)^2 + \cdots + 0 + \cdots + (-J+1)^2 + (-J)^2}{2J+1} \right]$$

$$\chi = \frac{N_A g^2 \beta^2}{kT} \left[\frac{2J^2 + 2(J-1)^2 + \cdots + 0}{2J+1} \right]$$

- The summation of the quadratic series:

$$\sum_{k=0}^{J} k^2 = \frac{J(J+1)(2J+1)}{6} \qquad (11.4.15)$$

$$\chi = \frac{N_A g^2 \beta^2}{3kT} \left[\frac{J(J+1)(2J+1)}{2J+1} \right]$$

$$\chi = \frac{N_A g^2 \beta^2}{3kT} [J(J+1)] \qquad (11.7.7)$$

$$\text{but, } \chi = \frac{N_A \beta^2 \mu_e^2}{3kT}$$

$$\mu_e = g\sqrt{J(J+1)} \text{ BM} \qquad (11.7.3)$$

$$g = \frac{3}{2} + \frac{S(S+1) - L(L+1)}{2J(J+1)} \quad (\text{Chapter 2}) \qquad (11.7.4)$$

The electronic configuration of Nd^{3+}: $1s^2 2s^2\ 2p^6\ 3s^2\ 3p^6\ 3d^{10}\ 4s^2\ 4p^6\ 4d^{10}\ 5p^6\ 4f^3$. Find the ground state and the expected magnetic moment.

m_l:	3	2	1	0	−1	−2	−3	
	↑	↑	↑					Nd^{3+}: $4f^3$

$$S = \frac{3}{2} \quad L = 6 \quad J = 6 - \frac{3}{2} = \frac{9}{2}$$

so ${}^4I_{\frac{9}{2}}$ is the ground state

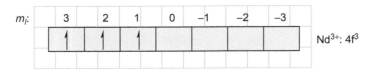

$$g=\frac{3}{2}+\frac{S(S+1)-L(L+1)}{2J(J+1)}=\frac{3}{2}+\frac{\frac{3}{2}\left(\frac{5}{2}\right)-6(7)}{2\left(\frac{9}{2}\right)\left(\frac{11}{2}\right)}=\frac{3}{2}+\frac{-\frac{153}{4}}{\frac{99}{2}}=\left(\frac{8}{11}\right)=0.727$$

$$\mu=g\sqrt{J(J+1)}=0.727\sqrt{\frac{9}{2}\cdot\frac{11}{2}}=3.62\text{ BM}$$

Experimental value is 3.5–3.6 BM at room temperature.

ii. Multiplet splitting $\ll kT$

How can you calculate the magnetic susceptibility of ions have unquenched orbital angular contribution and small multiplet separation (multiplet splitting $\ll kT$)?

- In ions with small multiplet separation:
  - Thermal energy is enough to fill all the lowest lying levels of the multiplet to equal extent.
  - The precession of S and L about J is so slow, and are not coupled to give J, and spin-orbital coupling does not take place.
  - On applying the magnetic field, S and L are oriented and quantized in the direction of the field.
- L and S will align independently with the applied magnetic field:

$$\chi_S=\frac{N\beta^2 g_s^2}{3kT}S(S+1)=\frac{N\beta^2}{3kT}4S(S+1), \tag{11.4.17}$$

where, $g_s=2$

$$\chi_L=\frac{N\beta^2 g_L^2}{3kT}L(L+1)=\frac{N\beta^2}{3kT}L(L+1),\quad \text{where,}\quad g_L=1,\quad \text{then}$$

$$\chi_{\text{Total}}=\chi_L+\chi_S$$

$$\chi=\frac{N\beta^2}{3kT}[L(L+1)+4S(S+1)] \tag{11.7.8}$$

and

$$\mu=\sqrt{L(L+1)+4S(S+1)} \tag{11.7.9}$$

If the experimental value for Cr^{3+} is 3.8 ± 0.1 BM, find out the contribution of the orbital angular momentum.

- For Cr^{3+}:

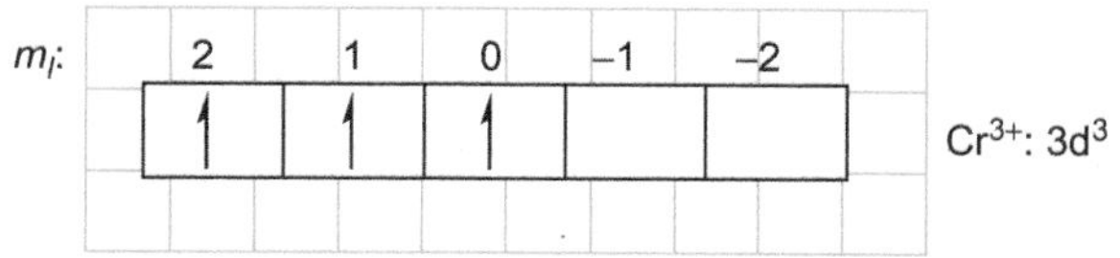

$$S=\frac{3}{2}\quad L=3\quad J=3-\frac{3}{2}=\frac{3}{2}$$

$$\text{so } {}^4F_{3/2} \text{ is the ground state}$$

$$\mu=\sqrt{L(L+1)+4S(S+1)}=\sqrt{3(3+1)+4\left(\frac{3}{2}\right)\left(\frac{3}{2}+1\right)}=\sqrt{27}=5.2\text{ BM}$$

The calculated and the experimental values of μ are far apart.

- If we use

$$\mu=g\sqrt{J(J+1)}\text{ BM}$$

$$g=\frac{3}{2}+\frac{S(S+1)-L(L+1)}{2J(J+1)}=\frac{3}{2}+\frac{\frac{3}{2}\left(\frac{5}{2}\right)-3(4)}{2\left(\frac{3}{2}\right)\left(\frac{5}{2}\right)}=0.4$$

$$\mu=g\sqrt{J(J+1)}=0.4\sqrt{\left(\frac{3}{2}\right)\left(\frac{5}{2}\right)}=0.775\,\text{BM}$$

Then the values of the calculated and the experimental values of μ are far apart.

- By setting $L=0$, quenching all orbital angular momentum:

$$\mu=\sqrt{L(L+1)+4S(S+1)}=\sqrt{4S(S+1)}=\sqrt{15}=3.87\,\text{BM}$$

Therefore, in this case the orbital angular momentum is quenched.

iii. Multiplet splitting $\cong kT$

What is the average magnetic susceptibility of a sample with multiplet widths comparable to kT?

- The levels of the lowest lying multiplet will be populated by different extents. The expression for susceptibility due to complete population is given by

$$\chi=\frac{N_A g^2\beta^2}{3kT}[J(J+1)] \tag{11.7.10}$$

- In this situation the contribution to the susceptibility, of each level, is given by N_J, where N_J is the number of molecules at the energy level J, then the susceptibility is sum a set of expressions:

$$\chi=\sum\frac{N_J g^2\beta^2}{3kT}[J(J+1)] \tag{11.7.11}$$

$$\because N_J\propto e^{\frac{-E}{kT}}$$

and each component is separate state and has $2J+1$ possible orientations, accordingly the contribution of each level:

$$N_J\propto(2J+1)e^{\frac{-E}{kT}}=A(2J+1)e^{\frac{-E}{kT}}$$

where A is the proportionality constant.

$$\chi=A\sum\frac{g^2\beta^2}{3kT}[J(J+1)(2J+1)]e^{\frac{-E}{kT}} \tag{11.7.12}$$

However, for 1 mol,

$$N_A=\sum N_J$$

$$N_A=A\sum(2J+1)e^{\frac{-E}{kT}}$$

$$A=\frac{N_A}{\sum(2J+1)e^{\frac{-E}{kT}}},\quad\text{then}$$

$$\chi_M=\frac{N_A}{3kT}\frac{\sum g^2\beta^2 J(J+1)(2J+1)e^{\frac{-E}{kT}}}{\sum(2J+1)e^{\frac{-E}{kT}}} \tag{11.7.13}$$

$$\chi_M=\frac{N_A\beta^2}{3kT}\frac{\sum g^2 J(J+1)(2J+1)e^{\frac{-E}{kT}}}{\sum(2J+1)e^{\frac{-E}{kT}}} \tag{11.7.14}$$

Requirements and conditions for zero and nonzero orbital contribution

What are the requirements of nonzero orbital angular momentum for d-electrons of free ion? Select the d-electron configurations that have nonzero orbital angular momentum in an octahedral and a tetrahedral field.

- The requirement for nonzero orbital angular momentum about a given axis is that the orbital occupied by electron should be transformable into equivalent degenerate orbital by rotation about that axis.
- This condition can be fulfilled in the case of d electrons of free transition metal ion (Fig. 11.8). Example, facile when:

$$d_{x^2-y^2} \xrightarrow{\text{rotation 45 degree around } z \text{ axis}} d_{xy} \text{ and } E_{d_{x^2-y^2}} \approx E_{d_{xy}}$$

$$d_{xz} \xrightarrow{\text{rotation 90 degree around z axis}} d_{yz} \text{ and } E_{d_{xz}} \approx E_{d_{yz}}$$

$$d_{xy} \xrightarrow{\text{rotation 90 degree around } x \text{ axis}} d_{yz} \text{ and } E_{d_{xy}} \approx E_{d_{yz}}$$

$$d_{xy} \xrightarrow{\text{rotation 90 degree around } y \text{ axis}} d_{xz} \text{ and } E_{d_{xy}} \approx E_{d_{xz}}$$

- Degeneracy maximizes spin-orbital coupling as $\bar{e}$ occupies indistinguishably processes about z.
- This coupling becomes limited or impossible as various configurations are examined in different field symmetry.
- In O_h and T_d, the interconvert of $d_{x^2-y^2} \leftrightarrow d_{xy}$ is lost, as these orbitals become nondegenerate.

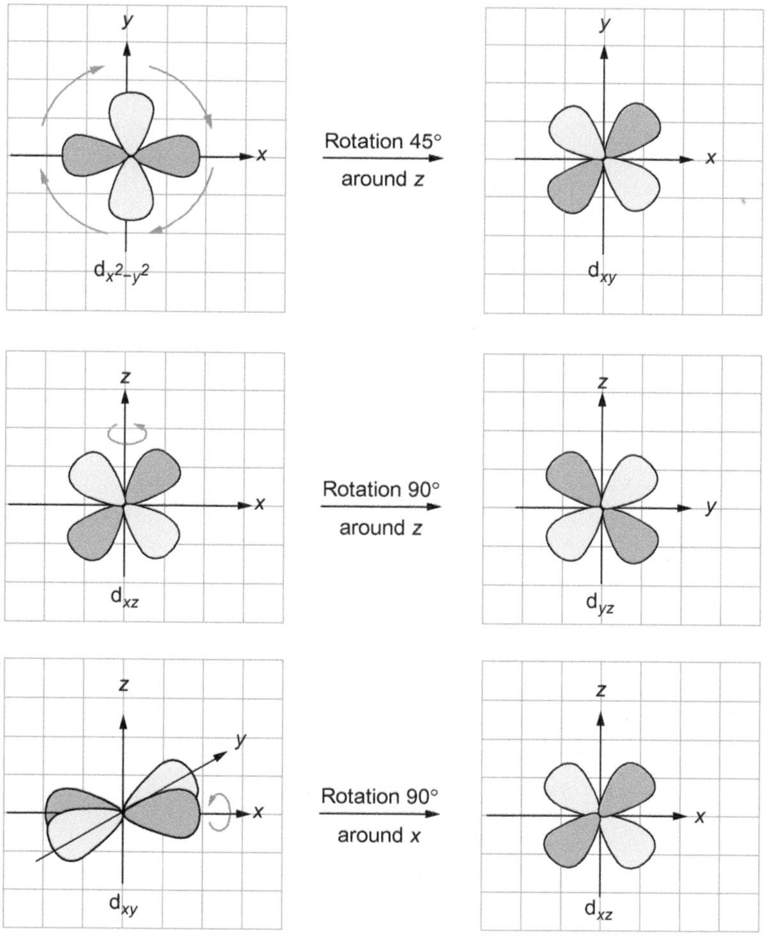

FIG. 11.8 Degenerate orbitals have superimposable shapes by rotation. This permits an electron to circulate and gain orbital angular momentum, $m\omega^2 r$.

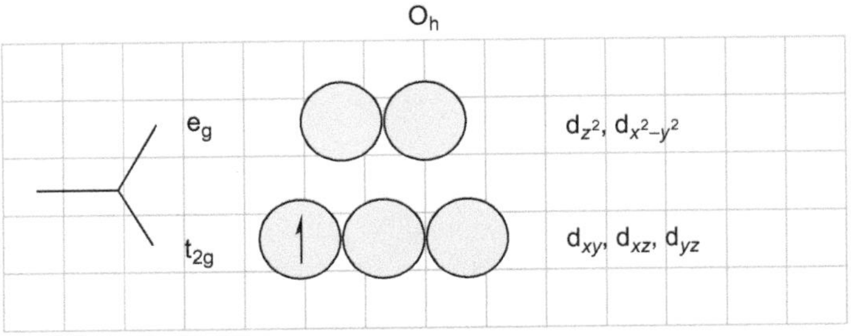

d$_{x^2-y^2}$ ⟷ d$_{xy}$ cannot interconvert
These orbitals are nondegenerate

- Only contribution to $L \neq 0$ is from t_{2g}-electrons (Fig. 11.9).
- Therefore, nonzero orbital angular momentum is expected in O_h for t_{2g}^1, t_{2g}^2, t_{2g}^4, and t_{2g}^5.
  e.g., high-spin O_h Co^{2+}: t_{2g}^5:
  $\mu_s = 3.87$; $\mu_{observed} \approx 5.0$ BM, angular momentum is observed.
  This implies that the effect of the orbital angular momentum is about 3 BM.
- Table 11.4 lists the electronic configuration of central atom that has nonzero angular momentum

What are the ground terms (states) of the d electrons in zero and octahedral fields? Which of these states would enhance the spin-orbital coupling for the complexes listed in Table 11.3?

- The ground terms of d electrons in zero and octahedral fields are listed in Table 11.5.
- The ground states of d electrons in octahedral field are $^1A_{1g}$, $^3A_{2g}$, $^4A_{2g}$, $^6A_{1g}$, ^2E_g, ^5E_g, $^3T_{1g}$, $^4T_{1g}$, $^5T_{1g}$, $^2T_{2g}$, $^3T_{2g}$, $^4T_{2g}$, and $^5T_{2g}$ (or d orbitals that have A, E, and T symmetries).
- In Table 11.3, all the complexes for which $\mu_{exp.}$ differs markedly from μ_s have T ground states.
- However, smaller but significant differences between $\mu_{exp.}$ and μ_s are listed in complexes with A_2, A_{2g}, E, and E_g.
- These differences are produced by angular momentum contributed to the ground state by mixing of higher T states. E.g., d^4 has T term of the same multiplicity at higher energy, and this can thus mix into the ground state.

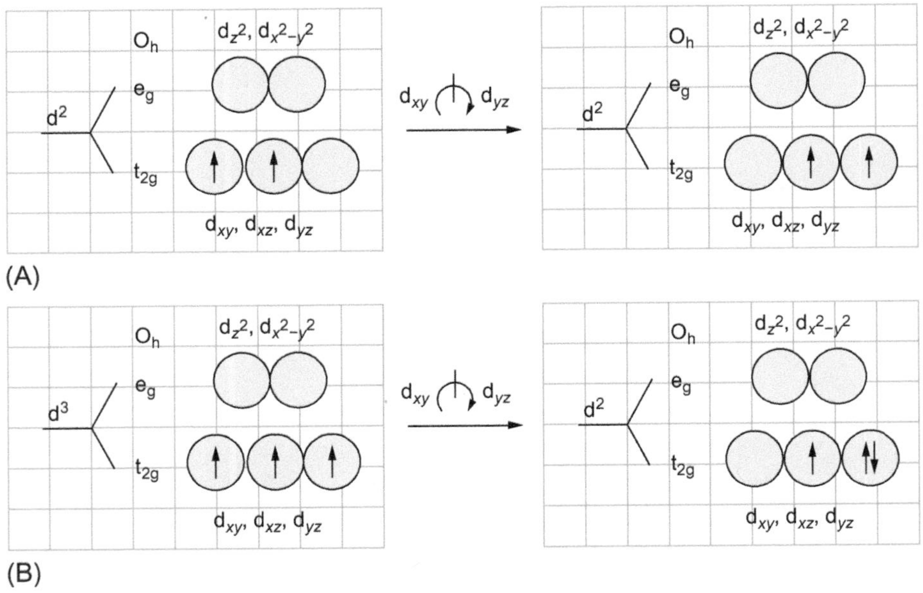

FIG. 11.9 (A) Nonzero orbital angular momentum and (B) zero orbital angular momentum.

TABLE 11.4 Nonzero Orbital Contribution

T_d	High-Spin O_h	Low-Spin O_h
–	d^1 (t_{2g}^1)	–
–	d^2 (t_{2g}^2)	–
d^3 ($e^2t_2^1$)	–	–
d^4 ($e^2t_2^2$)	–	d^4 (t_{2g}^4)
–	–	d^5 (t_{2g}^5)
–	d^6 ($t_{2g}^4e_g^2$)	–
–	d^7 ($t_{2g}^5e_g^2$)	–
d^8 ($e^4t_2^4$)	–	–
d^9 ($e^4t_2^5$)	–	–

TABLE 11.5 Ground States of d Electrons in an Octahedral Field

		Octahedral Field	
d Electron	Zero Ligand Field	Weak Ligand Field	Strong Ligand Field
d^1	^2D	$^2T_{2g}<^2E_g$	
d^2	^3F	$^3T_{1g}<^3T_{2g}<^3A_{2g}$	
d^3	^4F	$^4A_{2g}<^4T_{2g}<^4T_{1g}$	
d^4	^2D	$^5E_g<^5T_{2g}$	$^5T_{1g}$
d^5	^6S	$^6A_{1g}$	$^2T_{2g}$
d^6	^5D	$^5T_{2g}$	$^1A_{1g}$
d^7	^4F	$^4T_{1g}$	^2E_g
d^8	^3F	$^3A_{2g}$	
d^9	^2D	^2E_g	

11.8 SPIN-ORBITAL COUPLING: IN A AND E GROUND TERMS

The magnetic moment of complexes with A and E ground states can be calculated from the relations:

$$\mu_e = \mu_s\left(1 - \frac{\alpha\lambda}{10Dq}\right) \tag{11.8.1}$$

α=2 for E ground term
α=4 For A_2 ground term
λ is the spin-orbital coupling constant for particular ion terms (free ions values)

Find μ_e for $(NH_4)_2Ni(H_2O)_6(SO_4)_2$, $10Dq = 8900\ cm^{-1}$, $\lambda = -315 cm^{-1}$.

- For $Ni(H_2O)_6^{2+}$: $\alpha = 4$ for $^3A_{2g}$ (ground term):
  Number of unpaired electrons =2
  S=1

$$\mu_e = \mu_s\left(1 - \frac{\alpha\lambda}{10Dq}\right) = \sqrt{4S(S+1)}\left(1 - \frac{\alpha\lambda}{10Dq}\right) = \sqrt{4(1)(1+1)}\left(1 + \frac{(4)(315)}{8900}\right)$$

$$\mu_e = \sqrt{8}\left(1 + \frac{1260}{8900}\right) = 3.23 \text{ BM}$$

The observed $\mu_{exp.} = 3.2$ BM is comparable with the calculated μ_e, but depart from spin-only value, $\mu_s = 2.83$ BM by small extend.

- The amount of deviation depends on the relationship between spin-orbital coupling and the ligand field, $\frac{\alpha\lambda}{10Dq}$.
- Notice that λ increases μ_e for d^6-d^9, decreases it for d^1-d^4. The contribution is independent of temperature as $10Dq \gg kT$.

11.9 SPIN-ORBITAL COUPLING: IN T GROUND TERMS

In Ti^{3+} (d^1), the ^2D term undergoes splitting by crystal field, spin-orbital coupling, and magnetic field (Fig. 11.10). As a result of Boltzmann distribution over these states to the 2nd order in H, show that

$$\mu_e^2 = \frac{8 + \left(\frac{3\lambda}{kT} - 8\right)e^{\frac{-3\lambda}{2kT}}}{\frac{\lambda}{kT}\left(2 + e^{\frac{-3\lambda}{2kT}}\right)}$$

- Recalling the Van Vleck equation:

$$\chi = N_A \frac{\sum_n \left(\frac{\left(E_n^{(1)}\right)^2}{kT} - 2E_n^{(2)}\right) e^{\frac{-E_n^{(0)}}{kT}}}{\sum_n e^{\frac{-E_n^{(0)}}{kT}}} \qquad (11.6.7)$$

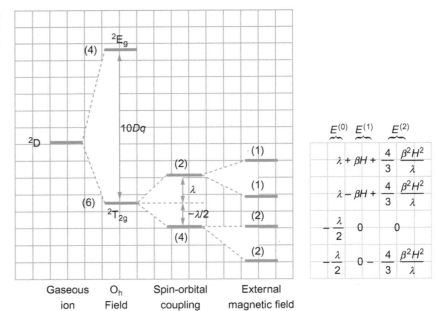

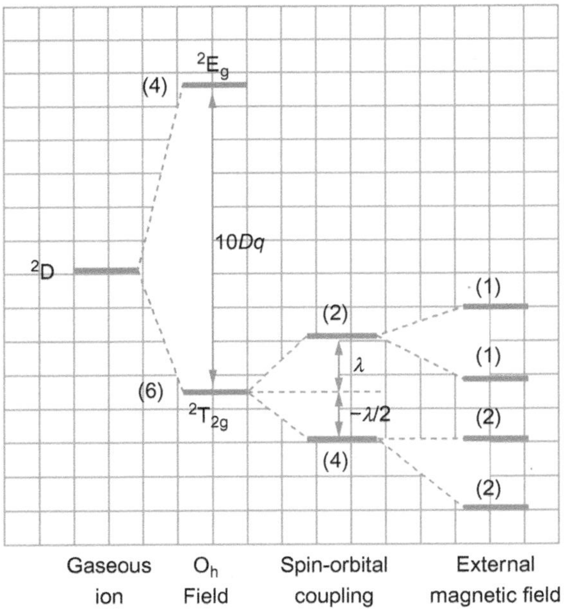

$E^{(0)}$	$E^{(1)}$	$E^{(2)}$
λ	$+\beta H$	$+\frac{4}{3}\frac{\beta^2H^2}{\lambda}$
λ	$-\beta H$	$+\frac{4}{3}\frac{\beta^2H^2}{\lambda}$
$-\frac{\lambda}{2}$	0	0
$-\frac{\lambda}{2}$	0	$-\frac{4}{3}\frac{\beta^2H^2}{\lambda}$

FIG. 11.10 The splitting of gaseous ion ^2D state by O_h-field, by spin-orbital coupling, and by magnetic field. The degeneracy of each level is indicated in parentheses.

- In Ti^{3+} (d^1), the expressions for the energies are obtained from the wave functions resulting from weak crystal field analysis by the operators, $\hat{H}$:

$$\hat{H} = \lambda\hat{L}\cdot\hat{S} + \beta(\hat{L} + g_e\hat{S})\cdot H \tag{11.9.1}$$

- The ^2E_g term can be ignored when calculating the susceptibility, because the separated $10Dq$ is generally large in O_h complexes.
- In the ground state $^2T_2$ level:

$E_n^{(0)} = -\frac{\lambda}{2}, -\frac{\lambda}{2}, \lambda,$ and λ (Fig. 11.10)

$E_n^{(1)} = 0, 0, -\beta H,$ and $+\beta H$ (Fig. 11.10)

$E_n^{(2)} = -\frac{4}{3}\left(\frac{\beta^2 H^2}{\lambda}\right), 0, +\frac{4}{3}\left(\frac{\beta^2 H^2}{\lambda}\right),$ and $\frac{4}{3}\left(\frac{\beta^2 H^2}{\lambda}\right)$ (Fig. 11.10)

By substituting in

$$\chi = N_A \frac{\sum_n \left(\frac{\left(E_n^{(1)}\right)^2}{kT} - 2E_n^{(2)}\right) e^{\frac{-E_n^{(0)}}{kT}}}{\sum_n e^{\frac{-E_n^{(0)}}{kT}}} \tag{11.6.7}$$

and multiplying each term by the corresponding degeneracy,

$$\frac{\chi}{N_A} = \frac{\left\{\begin{array}{l} 2\left[\left(\frac{0^2}{kT}\right) + (2)\left(\frac{4}{3}\right)\left(\frac{\beta^2}{\lambda}\right)e^{\frac{\lambda}{2kT}}\right] + 2\left[\left(\frac{0^2}{kT}\right) - (2)(0)e^{\frac{\lambda}{2kT}}\right] \\ + \left[\left(\frac{\beta^2}{kT}\right) - (2)\left(\frac{4}{3}\right)\left(\frac{\beta^2}{\lambda}\right)\right]e^{\frac{-\lambda}{kT}} + \left[\left(\frac{\beta^2}{kT}\right) - (2)\left(\frac{4}{3}\right)\left(\frac{\beta^2}{\lambda}\right)\right]e^{\frac{-\lambda}{kT}} \end{array}\right\}}{2\left[e^{\frac{\lambda}{2kT}} + e^{\frac{\lambda}{2kT}} + e^{\frac{-\lambda}{kT}}\right]}$$

$$\frac{\chi}{N_A} = \frac{\left\{2\left[\left(\frac{8}{3}\right)\left(\frac{\beta^2}{\lambda}\right)e^{\frac{\lambda}{2kT}}\right] + 2\left[\left(\frac{\beta^2}{kT}\right) - \left(\frac{8}{3}\right)\left(\frac{\beta^2}{\lambda}\right)\right]e^{\frac{-\lambda}{kT}}\right\}}{4e^{\frac{\lambda}{2kT}} + 2e^{\frac{-\lambda}{kT}}}$$

$$\frac{\chi}{N_A} = \frac{\left\{\left[\left(\frac{8}{3}\right)\left(\frac{\beta^2}{\lambda}\right)e^{\frac{\lambda}{2kT}}\right] + \left[\left(\frac{\beta^2}{kT}\right) - \left(\frac{8}{3}\right)\left(\frac{\beta^2}{\lambda}\right)\right]e^{\frac{-\lambda}{kT}}\right\}}{2e^{\frac{\lambda}{2kT}} + e^{\frac{-\lambda}{kT}}}$$

$$\frac{\chi}{N_A} = \frac{\left\{\left(\frac{8}{3}\right)\left(\frac{\beta^2}{\lambda}\right)e^{\frac{\lambda}{2kT}} + \left(\frac{\beta^2}{kT}\right)e^{\frac{-\lambda}{kT}} - \left(\frac{8}{3}\right)\left(\frac{\beta^2}{\lambda}\right)e^{\frac{-\lambda}{kT}}\right\}}{2e^{\frac{\lambda}{2kT}} + e^{\frac{-\lambda}{kT}}}$$

$$\frac{\chi}{N_A} = \frac{\left\{\left(\frac{8}{3}\right)\left(\frac{1}{\lambda}\right)e^{\frac{\lambda}{2kT}} + \left(\frac{1}{kT}\right)e^{\frac{-\lambda}{kT}} - \left(\frac{8}{3}\right)\left(\frac{1}{\lambda}\right)e^{\frac{-\lambda}{kT}}\right\}}{2e^{\frac{\lambda}{2kT}} + e^{\frac{-\lambda}{kT}}} \cdot \beta^2 \cdot \frac{e^{\frac{\lambda}{kT}}}{e^{\frac{\lambda}{kT}}}$$

$$\frac{\chi}{N_A} = \frac{\left\{\left(\frac{8}{3}\right)\left(\frac{1}{\lambda}\right)e^{\left(\frac{\lambda}{2kT} + \frac{\lambda}{kT}\right)} + \left(\frac{1}{kT}\right) - \left(\frac{8}{3}\right)\left(\frac{1}{\lambda}\right)\right\}}{2e^{\left(\frac{\lambda}{2kT} + \frac{\lambda}{kT}\right)} + 1} \cdot \beta^2$$

$$\frac{\chi}{N_A} = \frac{\left(\frac{8}{3\lambda}\right)e^{\left(\frac{3\lambda}{2kT}\right)} + \left(\frac{1}{kT}\right) - \left(\frac{8}{3\lambda}\right)}{2e^{\left(\frac{3\lambda}{2kT}\right)} + 1} \cdot \beta^2 \cdot \frac{e^{\left(\frac{-3\lambda}{2kT}\right)}}{e^{\left(\frac{-3\lambda}{2kT}\right)}}$$

$$\frac{\chi}{N_A} = \frac{\left(\frac{8}{3\lambda}\right) + \left[\left(\frac{1}{kT}\right) - \left(\frac{8}{3\lambda}\right)\right]e^{\left(\frac{-3\lambda}{2kT}\right)}}{2 + e^{\left(\frac{-3\lambda}{2kT}\right)}} \cdot \beta^2$$

If

$$\chi = \frac{N_A\beta^2}{3kT}\mu^2, \quad \text{then: } \mu^2 = \frac{\chi}{N_A} \cdot \frac{3kT}{\beta^2}$$

$$\mu^2 = \frac{\left(\frac{8}{3\lambda}\right) + \left[\left(\frac{1}{kT}\right) - \left(\frac{8}{3\lambda}\right)\right]e^{\left(\frac{-3\lambda}{2kT}\right)}}{2 + e^{\left(\frac{-3\lambda}{2kT}\right)}} \cdot 3kT$$

$$\mu^2 = \frac{\left(\frac{8kT}{\lambda}\right) + \left[(3) - \left(\frac{8kT}{\lambda}\right)\right]e^{\left(\frac{-3\lambda}{2kT}\right)}}{2 + e^{\left(\frac{-3\lambda}{2kT}\right)}} \cdot \frac{\frac{\lambda}{kT}}{\frac{\lambda}{kT}}$$

$$\mu_e^2 = \frac{8 + \left(\frac{3\lambda}{kT} - 8\right)e^{\frac{-3\lambda}{2kT}}}{\frac{\lambda}{kT}\left(2 + e^{\frac{-3\lambda}{2kT}}\right)} \tag{11.9.2}$$

- The dependence of χ or μ on T is expected for molecules that have T ground terms. Results of Boltzmann over these states to the second order in H are summarized in Table 11.6, but not for those with A or E ground terms.
- Most spin-orbital coupling contribute to μ from T terms, less from E or A.

Find the ground term and the magnetic moment, μ_e, of the following ions in Table 11.7.

- The magnetic moments and the ground terms are calculated and listed in Table 11.8.

If:

$3d^5$:Fe^{3+} has $\lambda = -460\ cm^{-1}$, $4d^5$:Ru^{3+} has $\lambda = -1250\ cm^{-1}$, $5d^5$:Os^{3+} has $\lambda = -3000\ cm^{-1}$
$3d^4$:Mn^{3+} has $\lambda = -178\ cm^{-1}$, $4d^4$:Ru^{4+} has $\lambda = -700\ cm^{-1}$, and $5d^4$:Os^{4+} $\lambda = -2000\ cm^{-1}$

Use the above data to model graphically the variation of the magnetic moment and spin-orbital contributions with temperature in a strong octahedral field, using Table 11.6. What do you conclude from these graphs?

- $^2T_{2g}$ is the ground term for nd^5 configuration in strong octahedral field (Table 11.6) the magnetic moment as function of temperature is given by

$$\mu_e^2 = \frac{8 + \left(\frac{3\lambda}{kT} - 8\right)e^{\frac{-3\lambda}{2kT}}}{\frac{\lambda}{kT}\left(2 + e^{\frac{-3\lambda}{2kT}}\right)}$$

TABLE 11.6 Square of the Magnetic Moments of Transition Metal Ions With T Ground Terms in Octahedral and Tetrahedral Fields

Configuration	Ground Term	μ_e^2
Weak field O_h: d^1 T_d: d^9	$^2T_2$	$\dfrac{8+\left(\frac{3\lambda}{kT}-8\right)e^{\frac{-3\lambda}{2kT}}}{\frac{\lambda}{kT}\left(2+e^{\frac{-3\lambda}{2kT}}\right)}$
Weak field O_h: d^2 T_d: d^8	$^3T_1$	$\dfrac{3\left[\frac{0.625\lambda}{kT}+6.8+\left(\frac{0.125\lambda}{kT}+4.09\right)e^{-\frac{3\lambda}{kT}}-10.89e^{\frac{-9\lambda}{2kT}}\right]}{\frac{\lambda}{kT}\left[5+3e^{\frac{-3\lambda}{kT}}+e^{\frac{-9\lambda}{2kT}}\right]}$
Weak field O_h: d^6 T_d: d^4	$^5T_2$	$\dfrac{3\left[\frac{28\lambda}{kT}+9.33+\left(\frac{22.5\lambda}{kT}+4.17\right)e^{\frac{-3\lambda}{kT}}+\left(\frac{24.5\lambda}{kT}-13.5\right)e^{\frac{-5\lambda}{kT}}\right]}{\frac{\lambda}{kT}\left[7+5e^{\frac{-3\lambda}{kT}}+3e^{\frac{-5\lambda}{kT}}\right]}$
Weak field O_h: d^7 T_d: d^3	$^4T_1$	$\dfrac{3\left[\frac{3.15\lambda}{kT}+3.92+\left(\frac{2.84\lambda}{kT}+2.13\right)e^{\frac{-15\lambda}{4kT}}+\left(\frac{4.7\lambda}{kT}-6.05\right)e^{\frac{-6\lambda}{kT}}\right]}{\frac{\lambda}{kT}\left[3+2e^{\frac{-15\lambda}{4kT}}+3e^{\frac{-6\lambda}{kT}}\right]}$
Strong field O_h: d^1 O_h: d^5	$^2T_2$	$\dfrac{8+\left(\frac{3\lambda}{kT}-8\right)e^{\frac{-3\lambda}{2kT}}}{\frac{\lambda}{kT}\left(2+e^{\frac{-3\lambda}{2kT}}\right)}$
Strong field O_h: d^2 O_h: d^4	$^3T_1$	$\dfrac{3\left[\frac{5\lambda}{kT}+15+\left(\frac{\lambda}{kT}+9\right)e^{-\frac{-2\lambda}{kT}}-24e^{\frac{-3\lambda}{2kT}}\right]}{\frac{2\lambda}{kT}\left[5+3e^{\frac{-2\lambda}{kT}}+e^{\frac{-3\lambda}{2kT}}\right]}$

TABLE 11.7 Free Ion Values of Spin-Orbital Coupling Constant, λ, for Some Selected Ions

λ (cm^{-1})	Ion	λ (cm^{-1})	Ion	λ (cm^{-1})	Ion
Weak O_h Field		Weak T_d Field		Strong O_h Field	
155	Ti^{3+}	−830	Cu^{2+}	155	Ti^{3+}
105	V^{3+}	−315	Ni^{2+}	−300	Mn^{2+}
−100	Fe^{2+}	58	Cr^{2+}	−460	Fe^{3+}
−145	Co^{3+}	89	Mn^{3+}	105	V^{3+}
−172	Co^{2+}	57	V^{2+}	−115	Cr^{2+}
−238	Ni^{3+}	92	Cr^{3+}	−178	Mn^{3+}

By using the listed values of λ for Fe^{3+} ($3d^5$), Ru^{3+} ($4d^5$), and Os^{3+} ($5d^5$), the magnetic moments are computed at different of temperatures.

- $^3T_{1g}$ is the ground term for nd^4 configuration in strong octahedral field (Table 11.6) and μ_e^2 is given by

$$\mu_e^2=\frac{3\left[\frac{5\lambda}{kT}+15+\left(\frac{\lambda}{kT}+9\right)e^{-\frac{-2\lambda}{kT}}-24e^{\frac{-3\lambda}{2kT}}\right]}{\frac{2\lambda}{kT}\left[5+3e^{\frac{-2\lambda}{kT}}+e^{\frac{-3\lambda}{2kT}}\right]}$$

TABLE 11.8 Calculated Magnetic Moments and Ground Terms

λ (cm^{-1})		μ^2	μ (BM)	Ion	Ground Term
Weak field					
155	O_h: d^1	3.521	1.876	Ti^{3+}	$^2T_{2g}$
−830	T_d: d^9	4.967	2.229	Cu^{2+}	$^2T_2$
105	O_h: d^2	7.104	2.665	V^{3+}	$^3T_{1g}$
−315	T_d: d^8	15.767	3.971	Ni^{2+}	$^3T_1$
−100	O_h: d^6	31.745	5.634	Fe^{2+}	$^5T_{2g}$
−145	O_h: d^6	32.609	5.710	Co^{3+}	$^5T_{2g}$
58	T_d: d^4	21.705	4.659	Cr^{2+}	$^5T_2$
89	T_d: d^4	19.861	4.457	Mn^{3+}	$^5T_2$
−172	O_h: d^7	10.828	3.291	Co^{2+}	$^4T_{1g}$
−238	O_h: d^7	9.549	3.090	Ni^{3+}	$^4T_{1g}$
57	T_d: d^3	12.487	3.534	V^{2+}	$^4T_1$
92	T_d: d^3	10.704	3.272	Cr^{3+}	$^4T_1$
Strong field					
155	O_h: d^1	3.521	1.876	Ti^{3+}	$^2T_{2g}$
−300	O_h: d^5	6.430	2.536	Mn^{2+}	$^2T_{2g}$
−460	O_h: d^5	6.041	2.458	Fe^{3+}	$^2T_{2g}$
105	O_h: d^2	7.384	2.717	V^{3+}	$^3T_{1g}$
−115	O_h: d^4	12.408	3.522	Cr^{2+}	$^3T_{1g}$
−178	O_h: d^4	13.003	3.606	Mn^{3+}	$^3T_{1g}$

Notice that the orbital contribution depends on the electronic configuration.

This equation is employed for calculating the magnetic moment for Mn^{3+} ($3d^4$), Ru^{4+} ($4d^4$), and Os^{4+} ($5d^4$) as a function of temperature in strong octahedral field.

- The deviation between the calculated values of μ_e and μ_s measures the input of the spin-orbital coupling. The spin-only moment is temperature-independent and given by

$$\mu_s = 2\sqrt{S(S+1)} \qquad (11.4.18)$$

- The magnetic moments-versus-temperature for $^2T_{2g}$ and $^3T_{1g}$ terms are shown in Figs. 11.11 and 11.12. The effect of the temperature upon the spin-orbital coupling and the magnetic moments for nd^4 and nd^5 metal ions are summarized in the following:
  - Ru^{3+} ($4d^5$, $\lambda = -1250$ cm^{-1}, $^2T_{2g}$) and Os^{3+} ($5d^5$, $\lambda = -3000$ cm^{-1}, $^2T_{2g}$) have the highest spin-orbital coupling constants, λ, and less temperature variation (Fig. 11.11).
  - The orbital contribution is more dependent on the electronic configuration than the row number, n, low-spin d^4 (Mn^{3+}, Ru^{4+}, and Os^{3+}) and have large $\partial\mu/\partial T$, while low-spin d^5 (all Fe^{3+}, Ru^{3+} and Os^{3+}) have moderate $\partial\mu/\partial T$:

$$(\partial\mu/\partial T)_{3d^5(Fe^{3+})} > (\partial\mu/\partial T)_{4d^5(Ru^{3+})} > (\partial\mu/\partial T)_{5d^5(Os^{3+})}$$

and

$$(\partial\mu/\partial T)_{3d^4(Mn^{3+})} > (\partial\mu/\partial T)_{4d^4(Ru^{4+})} > (\partial\mu/\partial T)_{5d^4(Os^{4+})}$$

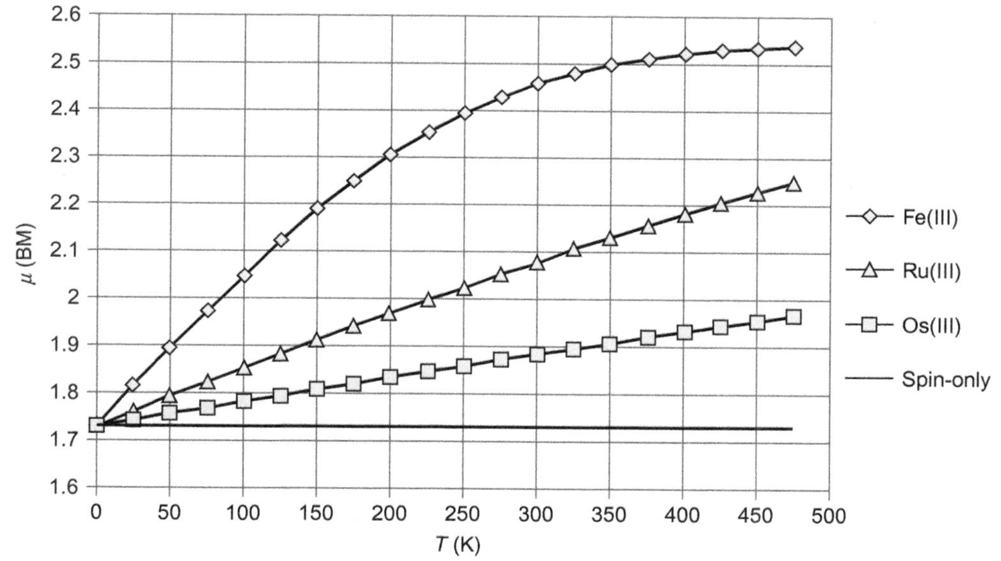

FIG. 11.11 Effect of spin-orbital coupling for $^2T_{2g}$ ground terms of low-spin nd^5 configuration.

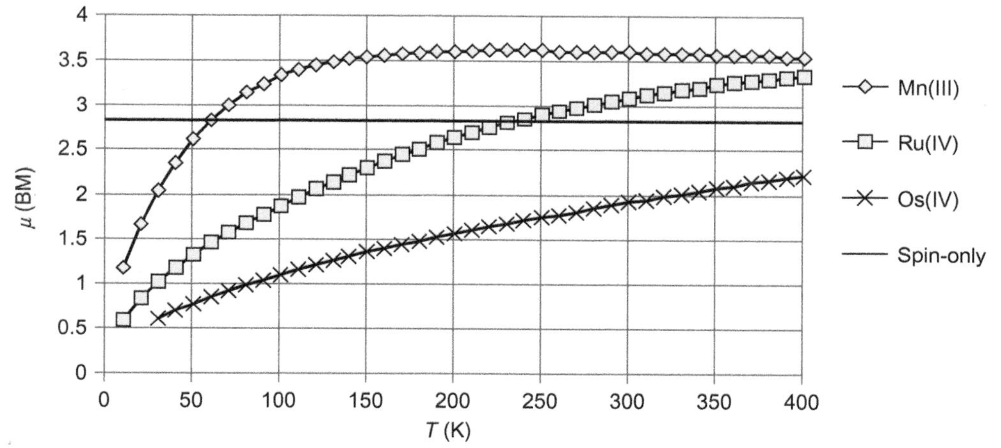

FIG. 11.12 Effect of spin-orbital coupling for $^3T_{1g}$ ground terms of low-spin nd^4 configuration.

TABLE 11.9 Calculated and Observed Magnetic Moments of nd^5 and nd^4 Metal Complexes

Ground Term	Ion	$\mu_{calc.}$	$\mu_{exp.}$ 300 K	Ground Term	Ion	$\mu_{calc.}$	$\mu_{exp.}$ 300 K
$^2T_{2g}$: $3d^5$	Fe^{3+}	2.46	2.3	$^3T_{1g}$: $3d^4$	Mn^{3+}	3.61	3.5
$^2T_{2g}$: $4d^5$	Ru^{3+}	2.08	2.1	$^3T_{1g}$: $3d^4$	Ru^{4+}	3.10	2.7
$^2T_{2g}$: $5d^5$	Os^{3+}	1.89	1.8	$^3T_{1g}$: $3d^4$	Os^{4+}	1.94	1.5

- Fig. 11.11 and Table 11.9 show that
  - the spin-orbital contributions in Ru^{3+}, and Os^{3+} are not so large
  - the magnetic moments are not very different form spin-only value
  - the variation with the temperature is very small. The contribution of the spin-orbital coupling to magnetic moment of Fe^{3+} falls rapidly with temperature

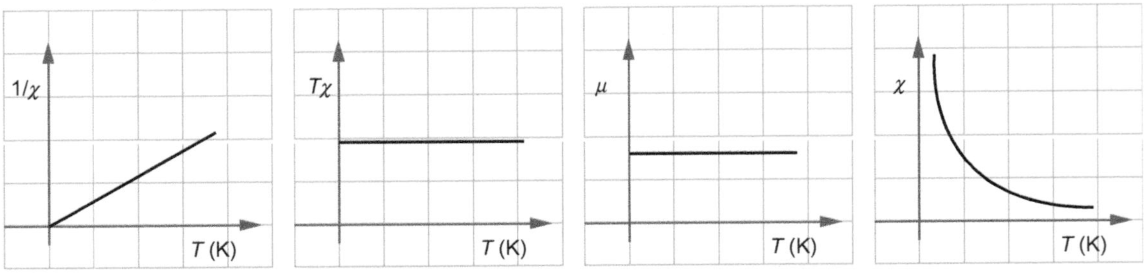

FIG. 11.13 Plots for paramagnets compounds follow Curie law.

- However, Fig. 11.3 and Table 11.9 suggest that
  - the magnetic moment of Mn^{3+}, Ru^{4+}, and Os^{4+} fall rapidly to zero at 0 K
  - the magnetic moment approach to the spin-only value at higher temperatures

11.10 CURIE LAW, DEVIATION, AND DATA REPRESENTATIONS

What is the excepted relationship between the magnetic moment, and susceptibility versus temperature?

- The net angular momentum is taken as a sum over Boltzmann-populated levels in the magnetic field, $H \neq 0$, so χ is temperature dependent.

$$\chi = \frac{N_A \beta^2 \mu_e^2}{3k} \cdot \frac{1}{T} \quad (11.4.19)$$

$$\chi = \frac{C}{T} \text{(Curie Law)} \quad (11.10.1)$$

- A straight-line relationship is predicted between χ and the reciprocal of temperature, given the following:

$$\text{Slope} = C = \frac{N_A \beta^2 \mu_e^2}{3k} \approx \frac{\mu_e^2}{8}, \quad \text{and intercept} = \text{zero}$$

- A considerable number of paramagnetic compounds do not fellow Curie's law (Fig. 11.13). This is caused by the second-order Zeeman effect produced by higher ligand field levels, resulting in paramagnetism that is independent of temperature. Then, the modified form of Curie's equation is

$$\chi = \frac{N_A \beta^2 \mu_e^2}{3k} \cdot \frac{1}{T} + N\alpha, \quad (11.6.1)$$

or

$$\mu_e = 2.84\sqrt{(\chi - N\alpha)T}$$

where $N\alpha$ is the temperature-independent paramagnetic to magnetic susceptibility. The value of $N\alpha$ is very small, and can be neglected in systems containing more than one unpaired electron.

- Various spin-orbital coupling may also contribute to susceptibility, and cause inconsistent with the Curie law in other paramagnetic compounds.

What does a nonzero intercept mean in the Curie law?

- The straight-line relationship is obtained for many cases (Fig. 11.13), but the intercept is nonzero.
- More substances obey the Curie-Weiss law:

$$\chi = \frac{C}{T + \theta} \quad (11.10.2)$$

and is usually expressed as

$$\frac{1}{\chi} = \frac{1}{C} \cdot T + \frac{\theta}{C} \quad (11.10.3)$$

θ corrects the temperature.

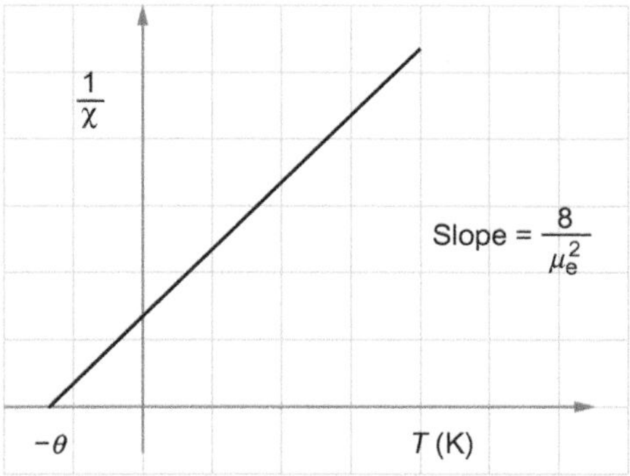

- Curie-Weiss behavior is common in substances that are not pure solid paramagnetic materials. In these materials, interionic or intermolecular interactions align neighboring magnetic moments to contribute to the value of the intercept.
- Many weakly coupled antiferromagnets obey Curie-Weiss, but such obedience is not a reliable indicator of antiferromagnetism.

Calculate the magnetic susceptibility, χ, to model the thermal-magnetic behavior, and plot the change of $1/\chi$ versus temperature (K) for:

i. **Ni(II) in weak octahedral field, $10Dq = 8900\ cm^{-1}$, $\lambda = -315\ cm^{-1}$**
ii. **Fe(III) in strong octahedral field, $\lambda = -460\ cm^{-1}$**
iii. **Mn(III) in strong octahedral field, $\lambda = -178\ cm^{-1}$**

- For Ni((II) in weak octahedral field: $\alpha = 4$ and $^3A_{2g}$ (ground term, $t_{2g}^6e_g^2$),
  Number of unpaired electrons $=2$

$$S = \frac{1}{2} + \frac{1}{2} = 1$$

$$\mu_e = \mu_s\left(1 - \frac{\alpha\lambda}{10Dq}\right) \tag{11.8.1}$$

$$\mu_e = \sqrt{4S(S+1)}\left(1 - \frac{\alpha\lambda}{10Dq}\right) = \sqrt{4(1)(1+1)}\left(1 + \frac{(4)(315)}{8900}\right)$$

$$\mu_e = \sqrt{8}\left(1 + \frac{1260}{8900}\right) = 3.23\ \text{BM, and recall:}$$

$$\chi = \frac{N_A\beta^2\mu_e^2}{3k}\cdot\frac{1}{T} \approx \frac{\mu_e^2}{8T}$$

Curie-Weiss plot of Ni(II) in weak O_h field (Fig. 11.14) shows
  - Straight line goes through zero. The linearity is a clear indication of the trivial contribution of $N\alpha$ (temperature independent paramagnetism).
  - As T approaches infinity, μ_e and χ approach zero, and the Curie law hold for this case.
- For Fe(III) in strong octahedral field has the ground term $^2T_{2g}$, and μ_e can be calculated (Table 11.6) using

$$\mu_e = \left(\frac{8 + \left(\frac{3\lambda}{kT} - 8\right)e^{\frac{-3\lambda}{2kT}}}{\frac{\lambda}{kT}\left(2 + e^{\frac{-3\lambda}{2kT}}\right)}\right)^{1/2}$$

and

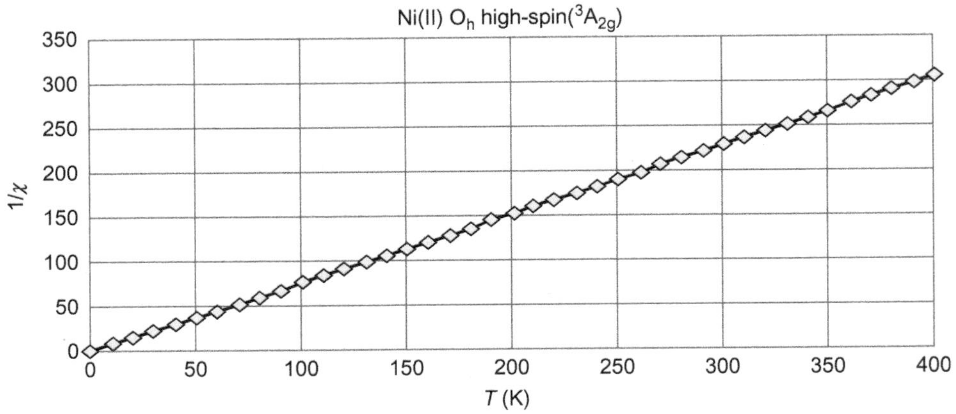

FIG. 11.14 Curie-Weiss plot of Ni(II) in a weak O$_h$ field.

$$\chi = \frac{N_A \beta^2 \mu_e^2}{3k} \cdot \frac{1}{T} \approx \frac{\mu_e^2}{8T}$$

to plot the reciprocal of χ versus temperature.

Curie-Weiss plot of Fe(III) in strong O$_h$ field (Fig. 11.15) indicates

- a straight line relationship at a higher temperature
- Fe(III) becomes diamagnetic at 0°K, which is interesting result that a system with one unpaired electron has zero χ
- as T approach zero the equation no longer apply because $g\beta H \approx kT$. As λ approaches zero, μ_e approaches 1.73 (μ_s)
- this result arises because the spin and orbital contributions cancel

- For Mn(III) in strong octahedral field has the ground term $^3T_{1g}$ (Table 11.6) recalling

$$\mu_e = \left(\frac{3\left[\frac{5\lambda}{kT} + 15 + \left(\frac{\lambda}{kT} + 9\right) e^{-\frac{-2\lambda}{kT}} - 24 e^{\frac{-3\lambda}{2kT}}\right]}{\frac{2\lambda}{kT}\left[5 + 3e^{\frac{-2\lambda}{kT}} + e^{\frac{-3\lambda}{2kT}}\right]} \right)^{1/2}, \text{ and}$$

$$\chi = \frac{N_A \beta^2 \mu_e^2}{3k} \cdot \frac{1}{T} \approx \frac{\mu_e^2}{8T}$$

Curie-Weiss plot of Mn(III) in strong O$_h$ field (Fig. 11.16):

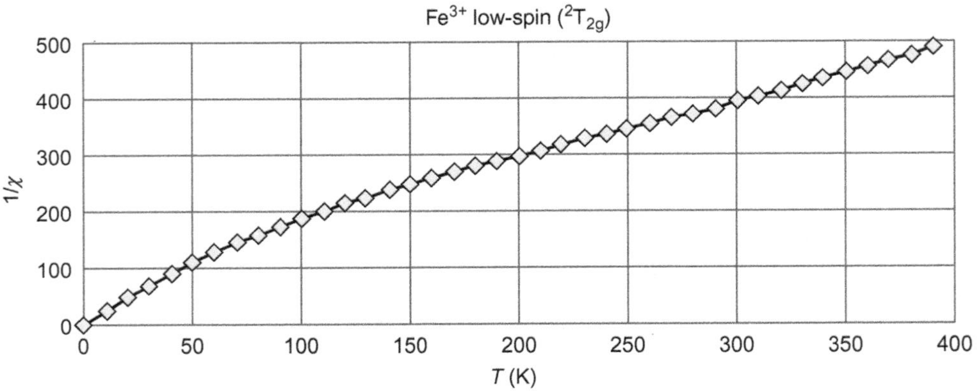

FIG. 11.15 Curie-Weiss plot of Fe(III) in a strong O$_h$ field.

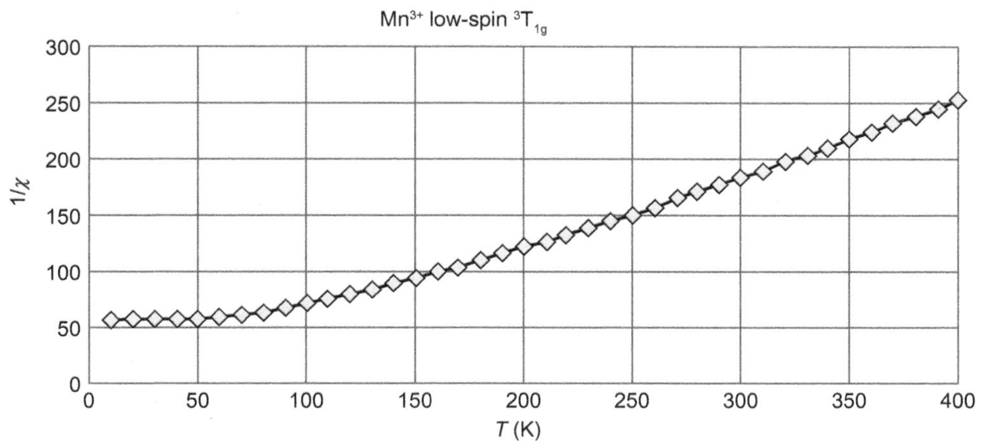

FIG. 11.16 Curie-Weiss plot of Mn(III) in a strong O_h field.

- ○ straight and linear relationship at higher temperature
- ○ paramagnetic behavior at low temperature
- In these analyses, the contribution from ^2E_g excited state has been ignored. However, the above equations are valid for many magnetic applications.

11.11 THE MAGNETIC BEHAVIORS OF COMPOUNDS CONTAIN A UNIQUE MAGNETIC CENTER

Spin Crossover Compounds

Give examples of high-spin, low-spin, and mixed-spin compounds.

- Iron (II) complexes are selected to give examples that cover these spin behaviors (Fig. 11.17).
- High-spin is stable state for: $Fe(H_2O)_6^{2+}$, and $Fe(NCS)_4(OH_2)_2^{2-}$.
- Low-spin is stable state for: $Fe(CN)_6^{4-}$, and $Fe(Bipy)_3^{2+}$.
- Mixed-spin or spin-equilibrium

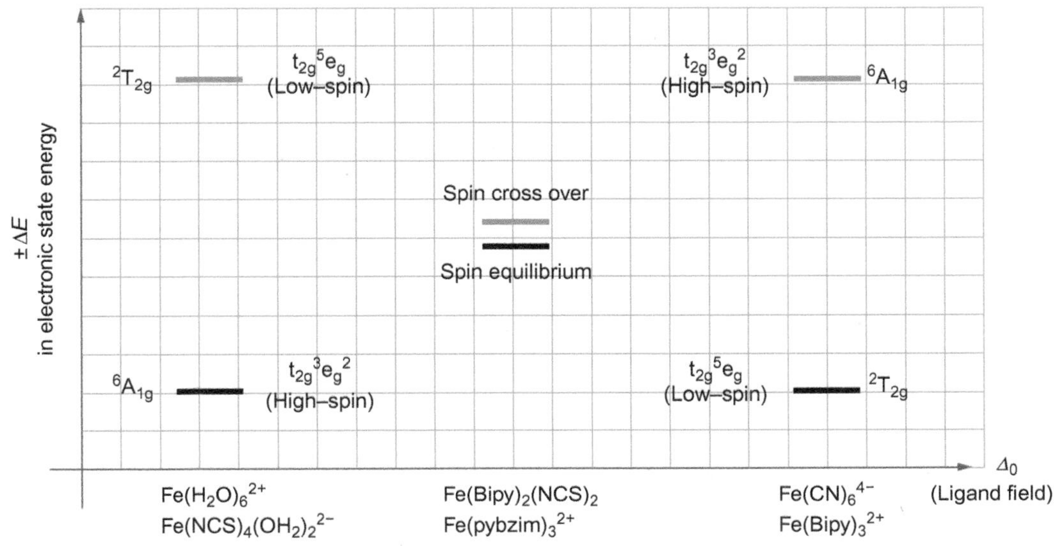

FIG. 11.17 Examples of high-, low-, and mixed-spin iron (II) complexes.

$$\text{in Fe}^{2+}-\text{compounds}:\ \underset{S=0}{\text{Low spin}-\text{Fe}^{2+}} \overset{\Delta E}{\rightleftharpoons} \underset{S=2}{\text{High spin}-\text{Fe}^{2+}}$$

are the states for: $Fe(Bipy)_2(NCS)_2$, and $Fe(Pybzim)_3^{2+}$

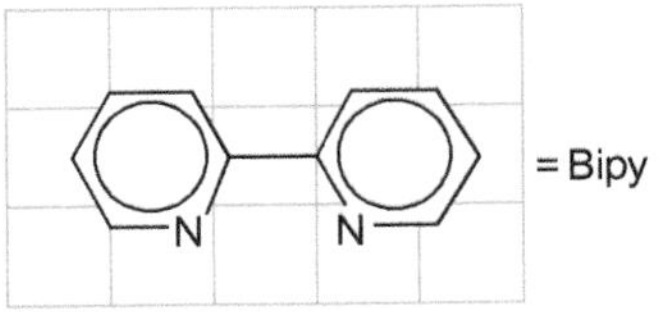

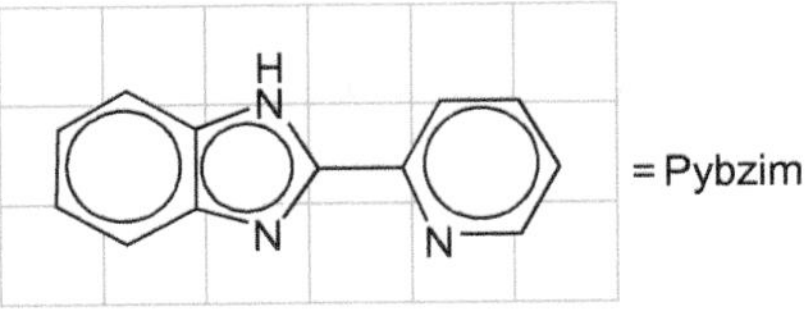

$$\text{in Fe}^{3+}-\text{compounds}:\ \underset{S=\frac{1}{2}}{\text{Low spin}-\text{Fe}^{3+}} \overset{\Delta E}{\rightleftharpoons} \underset{S=\frac{5}{2}}{\text{High spin}-\text{Fe}^{3+}}$$

are the states for: $[Fe^{III}L]^{+}X^{-}$,

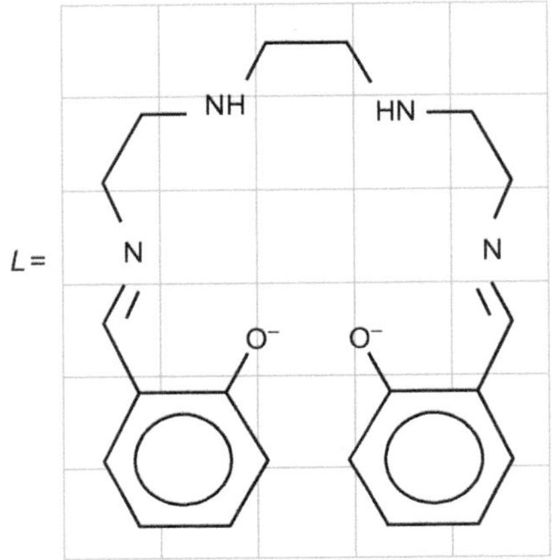

the crossover temperature depends on X^- and solvation.

$$\text{in Mn}^{3+}-\text{compounds }(d^4):\ \underset{S=1}{\text{Low spin}-\text{Mn}^{3+}} \overset{\Delta E}{\rightleftharpoons} \underset{S=2}{\text{High spin}-\text{Mn}^{3+}}$$

are the states for: $[Mn^{III}L]$

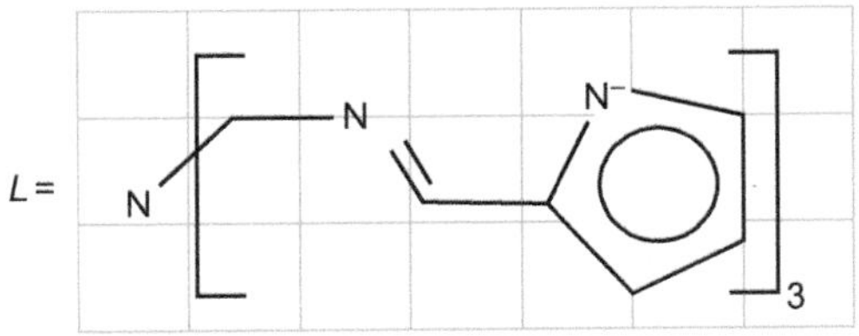

Spin equilibria: one-parameter model, ΔE

How can you calculate the magnetic susceptibility of equilibrium mixture between two spin states that are thermally populated and separated by the energy gap ΔE?

- In Tanabe-Sugano diagrams for d^4, d^5, d^6, and d^7 octahedral complexes, at a certain value of Dq/B (Chapter 8), the ground state changes the spin of the complex.
- When the ligand field is such that the two states are close in energy, the excited state can be thermally populated and the system will consist of an equilibrium mixture of the two forms.
- Boltzmann distribution describes the thermal spreading of molecules between the two states at a given temperature (Fig. 11.18), and ΔE is the thermodynamic barrier. For example:

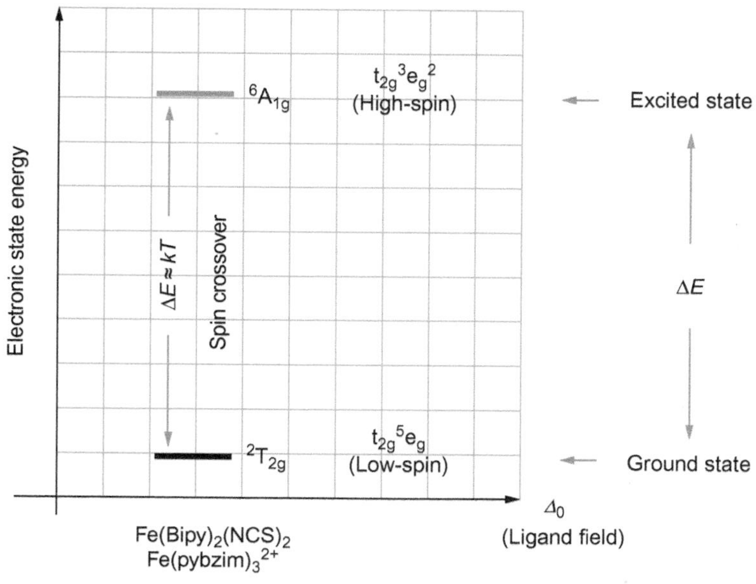

FIG. 11.18 Equilibrium mixture of two spin states separated by energy gap of $\Delta E \approx kT$.

$$\underset{S=0}{\text{Low spin} - Fe^{2+}} \overset{\Delta E}{\rightleftharpoons} \underset{S=2}{\text{High spin} - Fe^{2+}}$$

ΔE is of the same size as kT (200 cm^{-1} at 298 K)

- At low temperatures, no thermal energy is available, and molecules stall in the ground states (e.g., low-spin) (Fig. 11.19).
- At high temperatures, molecules have enough thermal energy to jump the energy gap to high spin state.
- As temperature is raised, fraction in high-spin state increases, and the average μ per molecule increases. For Fe^{2+}-system.
- The Boltzmann distribution at thermal equilibrium:

$$P_n = \frac{N_n}{N} = \frac{e^{\frac{-E_n}{kT}}}{\sum_n e^{\frac{-E_n}{kT}}} \tag{11.4.6}$$

$$P_n = \frac{N_n}{N_{\text{total}}} = \frac{(2S_n+1)e^{\frac{-E_n}{kT}}}{\sum_n (2S_n+1)e^{\frac{-E_n}{kT}}} \tag{11.11.1}$$

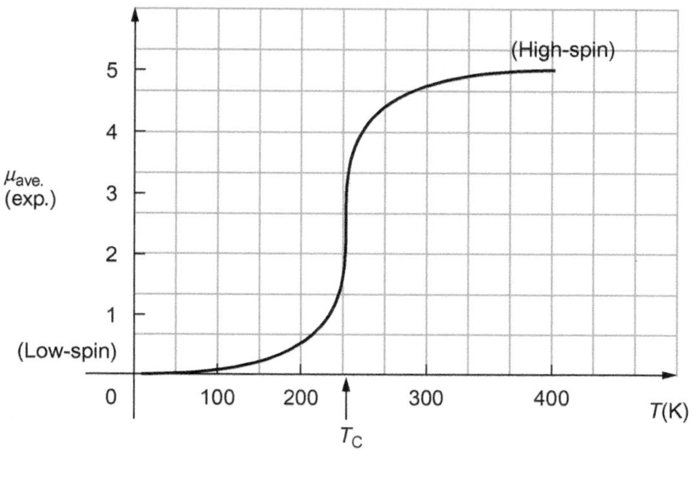

FIG. 11.19 Spin transition occurs at a temperature characteristic of each compound. The steepness is also dependent on compound.

N_n is the population of the states at the energy level E_n, and N_{total} is the total population. $2S_n+1$ is the number of microstates in each energy level.

- The average of macroscopic magnetic susceptibility, χ, is

$$\chi = \frac{\sum_{n=0}^{n}(2S_n+1)\chi_n e^{\frac{-E_n}{kT}}}{\sum_{n=0}^{n}(2S_n+1)e^{\frac{-E_n}{kT}}} \tag{11.11.2}$$

χ_n is the magnetic susceptibility of individual spin at the energy level E_n.

- The ground energy state, $E_0 = E_A$, is taken as zero. The zero of energy is arbitrary for convenience, and $E_B = \Delta E$.

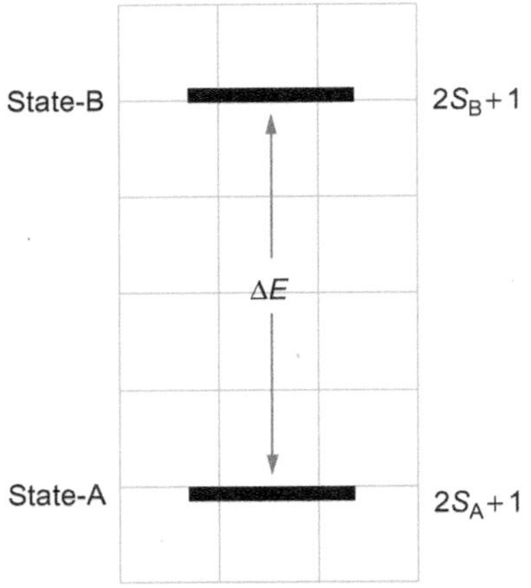

2S+1 is the number of macrostates in each energy level

The value of χ for system contains spin crossover or spin transition between states A and B states:

$$\chi = \frac{(2S_A+1)\chi_A e^{\frac{-E_A}{kT}} + (2S_B+1)\chi_B e^{\frac{-E_B}{kT}}}{(2S_A+1)e^{\frac{-E_A}{kT}} + (2S_B+1)e^{\frac{-E_B}{kT}}}$$

$$\chi = \frac{(2S_A+1)\chi_A + (2S_B+1)\chi_B e^{\frac{-\Delta E}{kT}}}{(2S_A+1) + (2S_B+1)e^{\frac{-\Delta E}{kT}}} \tag{11.11.3}$$

χ is from the weighted (Boltzmann) contributions of the two spin states. This expression is for d^5 and d^6 compounds.

- For d^4 and d^7 compounds, the two spin states mix via spin-orbital coupling, and become complicated.
- The easiest way to deal with T-dependence of χ_A and χ_B is to use spin-only moments, if

$$\chi = \frac{\mu_s^2}{8T}, \text{ and}$$

$$\mu_s = 2\sqrt{S(S+1)}$$

$$\chi_i = \frac{4S_i(S_i+1)}{8T} = \frac{S_i(S_i+1)}{2T}, \text{ so substituting in}$$

$$\chi = \frac{\sum_{n=0}^{n}(2S_n+1)\chi_n e^{\frac{-E_n}{kT}}}{\sum_{n=0}^{n}(2S_n+1)e^{\frac{-E_n}{kT}}} \tag{11.11.2}$$

$$\chi = \frac{S_A(S_A+1)(2S_A+1) + S_B(S_B+1)(2S_B+1)e^{\frac{-\Delta E}{kT}}}{2T\left[(2S_A+1) + (2S_B+1)e^{\frac{-\Delta E}{kT}}\right]} \tag{11.11.4}$$

To see the behavior of Fe(III) spin equilibrium via model, plot:

i. **$1/\chi$ versus temperature when $\Delta E = -100\ \text{cm}^{-1}$ and $k = 0.695\ \text{cm}^{-1}$**
ii. **μ versus temperature when $\Delta E = -100$, 100, 200, and 400 cm^{-1}**

What do you conclude from these graphs?

- In spin crossover:

$$\begin{matrix} LS & \Delta E & HS \\ \text{Fe(III)} & \rightleftharpoons & \text{Fe(III)} \\ S = {}^{1}/_{2} & & S = {}^{5}/_{2} \end{matrix}$$

The values χ at different temperature are computed using

$$\chi = \frac{S_{LS}(S_{LS}+1)(2S_{LS}+1) + S_{HS}(S_{HS}+1)(2S_{HS}+1)e^{\frac{-\Delta E}{kT}}}{2T\left[(2S_{LS}+1) + (2S_{HS}+1)e^{\frac{-\Delta E}{kT}}\right]} \tag{11.11.4}$$

- The appearance of a maximum in the plot of ${}^{1}/_{\chi}$ versus temperature is an index of the spin crossover (Fig. 11.20).
- At low T, no thermal energy is available, and molecules are in a low-spin ground state (Fig. 11.20).
- At high T, and when ΔE is of the same size as kT (200 cm^{-1} at 298 K) or $\Delta E = 100\text{cm}^{-1}$, molecules have enough energy to jump the energy gap to a high-spin state. However, when $\Delta E = 400\ \text{cm}^{-1}$, a partial spin transition is expected (Fig. 11.21).
- However, at $\Delta E = -100\text{cm}^{-1}$, molecules stay in a high-spin form at all temperatures (Fig. 11.21).
- Example: Fe(dtc):

dtc =
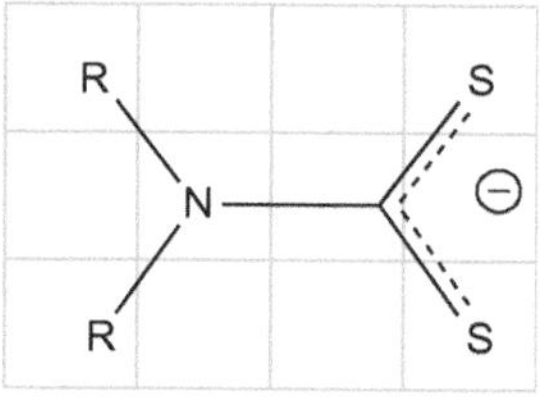

R = isopropyl, crossover at high temperature
R = methyl, crossover at low temperature
RR = —(CH_2)$_4$—, high-spin at all temperatures

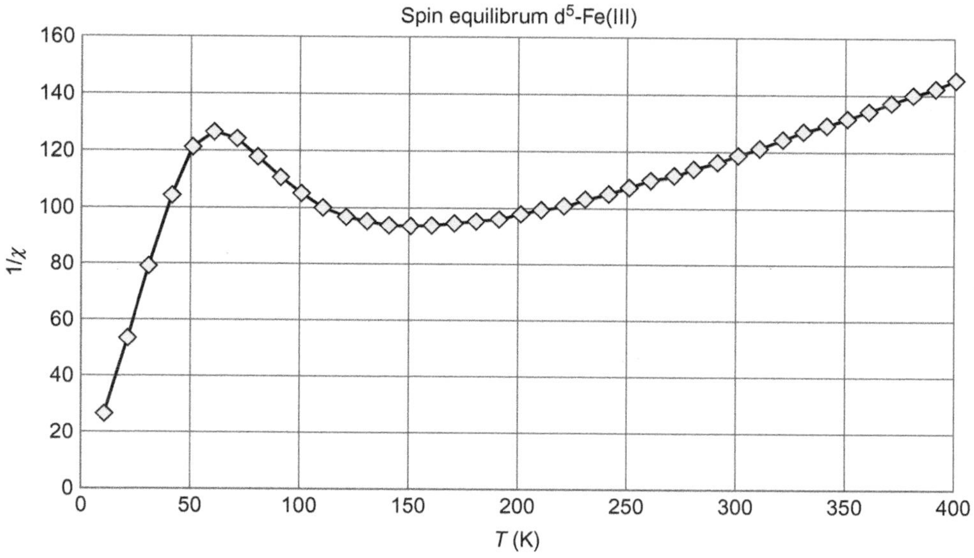

FIG. 11.20 Reciprocal of the magnetic susceptibility versus temperature.

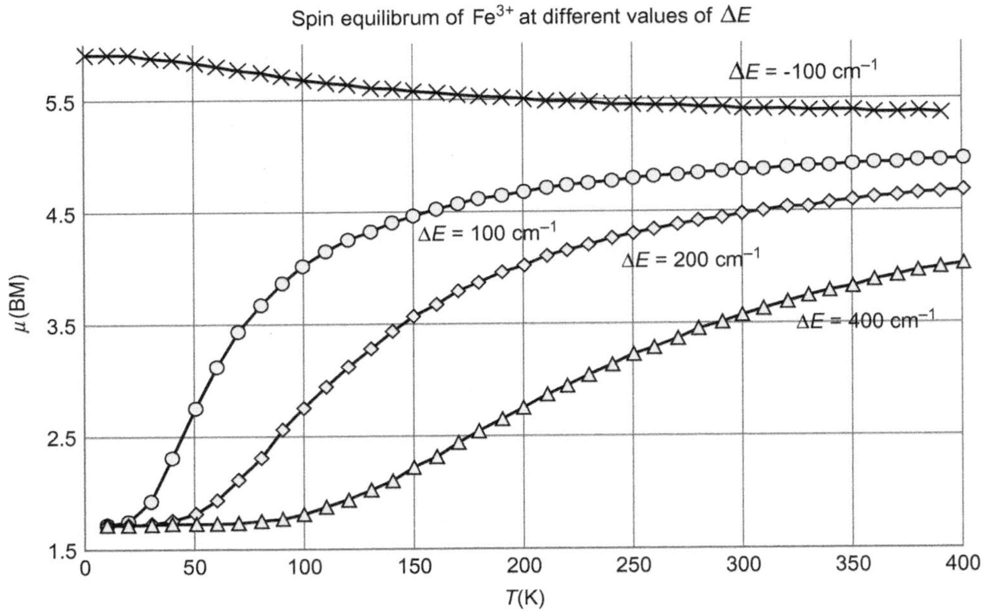

FIG. 11.21 Change in the magnetic moment as a function of temperature.

Spin equilibra: two-parameter model; $\Delta H°, \Delta S°$

For a spin-crossover system:

$$\underset{S=1/2}{\mathrm{Fe(III)}_{LS}} \overset{k}{\rightleftharpoons} \underset{S=5/2}{\mathrm{Fe(III)}_{HS}}$$

Use the following thermodynamic paramters to model the link between μ and temperature:

i. $\Delta S° = 80\,\mathrm{J K^{-1} mol^{-1}}$ and $\Delta H° = 8\,\mathrm{kJ mol^{-1}}$
ii. $\Delta S° = 80\,\mathrm{J K^{-1} mol^{-1}}$ and $\Delta H° = 24\,\mathrm{kJ mol^{-1}}$
iii. $\Delta S° = 160\,\mathrm{J K^{-1} mol^{-1}}$ and $\Delta H° = 20\,\mathrm{kJ mol^{-1}}$
iv. $\Delta S° = 160\,\mathrm{J K^{-1} mol^{-1}}$ and $\Delta H° = 80\,\mathrm{kJ mol^{-1}}$
v. $\Delta S° = 160\,\mathrm{J K^{-1} mol^{-1}}$ and $\Delta H° = 40\,\mathrm{kJ mol^{-1}}$

- For

$$\underset{f}{\mathrm{Fe(III)}_{LS}} \overset{k}{\rightleftharpoons} \underset{1-f}{\mathrm{Fe(III)}_{HS}}$$

- If f is the fraction of $\mathrm{Fe(III)}_{LS}$ and

$$K = \frac{[\mathrm{Fe(III)}_{HS}]}{\mathrm{Fe(III)}_{LS}} = \frac{1-f}{f}$$

$$\therefore f = \frac{1}{K+1}$$

- Let

$$x_{LS} = f, \quad \text{so } x_{HS} = 1-f, \quad \text{and}$$

Using

$$\mu_s = 2\sqrt{S(S+1)}, \quad \text{and } \mu_s = \sqrt{8\chi}, \quad \text{and } \chi = \frac{\mu_s^2}{8T} \text{ for each } \mathrm{Fe(III)}_{LS} \text{ and } \mathrm{Fe(III)}_{HS}$$

$$\chi_m = x_{LS}\chi_{LS} + x_{HS}\chi_{HS}$$

$$\chi_m = \frac{f\mu_{LS}^2 + (1-f)\mu_{HS}^2}{8T}$$

- By introducing $\Delta S^o\,(JK^{-1}mol^{-1})$, $\Delta H^o\,(kJmol^{-1})$, and $R = 8.3\ J\ K^{-1}\ mol^{-1}$, the equilibrium constants, K's, are computed at various temperatures:

$$\Delta G^o = -RT\ \ln\ K = \Delta H^o - T\Delta S^o$$

$$K = \exp\left(\frac{\Delta S^o}{R} - \frac{\Delta H^o}{RT}\right)$$

  then step to calculate f-values, and χ_m's (Figs. 11.22 and 11.23).
  - By increasing ΔH^o, spin-crossover shifts to higher temperature (Fig. 11.22). Therefore, when ΔH^o increases, the thermal energy gap increases, and the low-spin fraction increases.
  - By increasing ΔS^o, spin-crossover shifts to lower temperature (Fig. 11.23). Therefore, when ΔS^o increases, the thermal energy gap decreases, and the low-spin fraction decreases.
- The obtained plots can fit exprimental data by using curve fitting programes.
- Further refinements to obtain more details have involved the following:
  - Introduce term for ΔS^o due to electron multiplicity (small):

$$\pm\Delta S^o_{spin} \cong RT\ \ln\left(\frac{2S_{HS}+1}{2S_{LS}+1}\right)$$

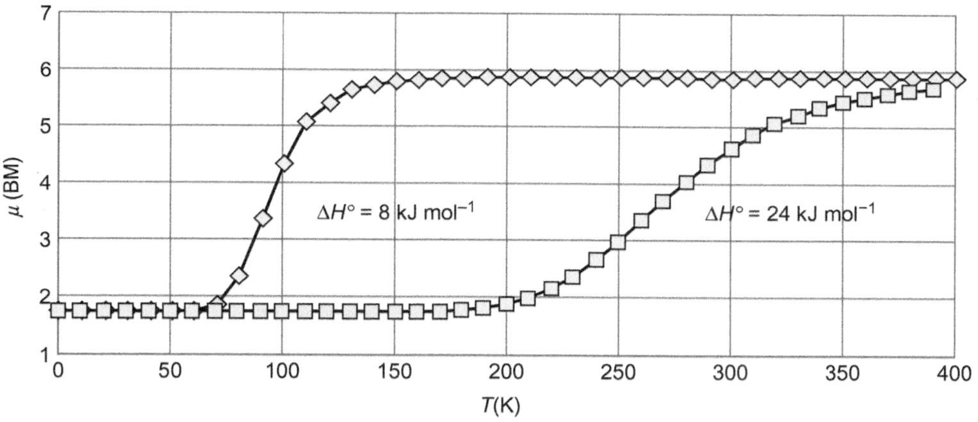

FIG. 11.22 Magnetic susceptibility versus temperature for Fe^{3+}-spin equilibrium at constant $\Delta S^o = +80\ J\ K^{-1}\ mol^{-1}$, and different values of ΔH^o.

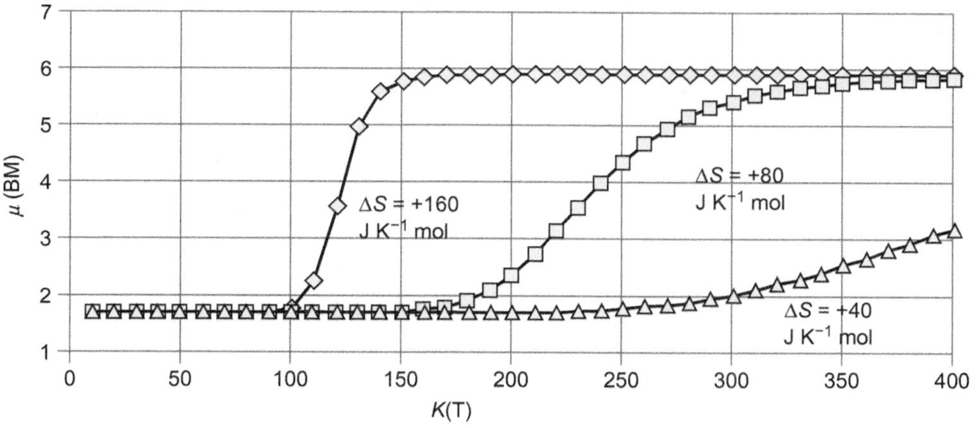

FIG. 11.23 Magnetic susceptibility versus temperature for Fe^{3+} -spin equilibrium at constant $\Delta H^o = +20\ kJ\ mol^{-1}$, and different values of ΔS^o.

- Introduce coopertivity parameter, γ, to account for sharpness of crossover.
- Introduce domain model for coopertivity.
- Attempt to account for hysteresis in μ versus T curves.

11.12 STRUCTURE-LINKED CROSSOVER, THERMAL ISOMERIZATION

In Chromium (III), d^3

i. **Define the ground and excited electronic state in octahedral.**
ii. **What is the effect of the tetragonal distortion on the angular momentum when the energy of the first excited state is about 1800 cm^{-1}?**
iii. **How can you compute the expected magnetic susceptibility, χ?**
iv. **How does the data look when the tetragonal distortion is set to large or small values?**

- Chromium (III) in octahedral field has $^4A_2$ as a ground state and $^4T_2$ and $^4T_1$ as excited levels, as shown for the quartet states in the Tanabe and Sugano diagrams (Chapter 8).
- The electronic state $^4T_2$ is about 1800 cm^{-1} higher in energy than $^4A_2$, so its contribution to χ can be disregarded.
- The ground state is orbitally singlet (A), so has no orbital contribution to χ.
- Therefore, the magnetism is a result of the spin angular momentum only.
- As $S>1/2$, get zero-field splitting spin-spin interaction between/among electron in the ground state. For $S=3/2$ state, there are m_s-components of quartet.
- Fig. 11.24 illustrates splitting of $^4A_2$ term in the tetragonal field by the zero field parameter D:

$$\begin{array}{ccc} \text{Low spin} & D & \text{High spin} \\ Cr^{3+} & \rightleftharpoons & Cr^{3+} \\ \text{Tetragonal} & & \text{Octahedral} \\ S=\pm\frac{1}{2} & & S=\pm\frac{3}{2} \end{array}$$

- The resulting thermal distribution among the spin states is given by the Van Vleck equation:

$$\chi = N_A \frac{\sum_n \left(\frac{\left(E_n^{(1)}\right)^2}{kT} - 2E_n^{(2)} \right) e^{\frac{-E_n^{(0)}}{kT}}}{\sum_n e^{\frac{-E_n^{(0)}}{kT}}} \tag{11.6.7}$$

$$E_n^{(0)} = 0\,(\text{lower level}),\ D\,(\text{upper level})$$

$$E_n^{(1)} = \pm\frac{1}{2}g_z\beta,\ \pm\frac{3}{2}g_z\beta$$

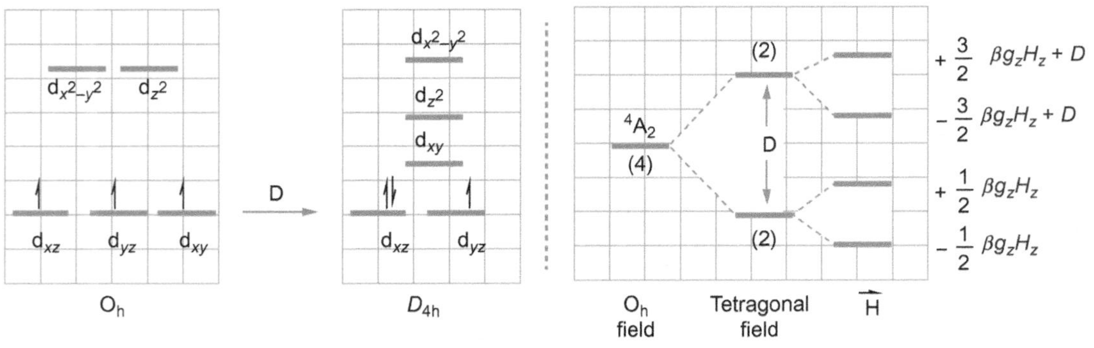

FIG. 11.24 The tetragonal distortion and the splitting of $^4A_2$ term of chromium (III) ion.

$$\frac{\chi}{N_A} = \frac{\frac{2\left(\frac{g_z\beta}{2}\right)^2}{kT}e^{\frac{0}{kT}} + \frac{2\left(\frac{3g_z\beta}{2}\right)^2}{kT}e^{\frac{-D}{kT}}}{2e^{\frac{0}{kT}} + 2e^{\frac{-D}{kT}}} = \frac{\frac{(g_z\beta)^2}{2kT} + \frac{(3g_z\beta)^2}{2kT}e^{\frac{-D}{kT}}}{2 + 2e^{\frac{-D}{kT}}}$$

$$\frac{\chi}{N_A} = \frac{g_z^2\beta^2\left(1 + 9e^{\frac{-D}{kT}}\right)}{4kT\left(1 + 2e^{\frac{-D}{kT}}\right)} \tag{11.12.1}$$

$$\text{when,} \quad D \ll kT, \frac{D}{kT} \ll 1, \quad \therefore e^{\frac{-D}{kT}} = 1 - \frac{D}{kT}$$

which is the situation for very small distortion or at a very high T, χ approaches $\frac{5N_A g_z^2\beta^2}{4kT}$ and μ approaches 3.87 BM, the calculated value for $S = \frac{3}{2}$.

- when D is very large or at T approaches zero: the values of μ approach 1.73 BM (Fig. 11.25), the calculated magnetic moment for $S = \frac{1}{2}$ (Fig. 11.25).

How does the tetragonal distortion affect the magnetism of the nickel (II) complexes?

- An octahedral field gives rise to a $^3A_2$ ground state and $^3T_2$ and $^3A_2$ excited states, as shown for the triplet terms in the Tanabe and Sugano diagrams (Chapter 8).
- The electronic state $^3T_2$ is much higher in energy than $^3A_2$, and does not contribute to χ.
- The ground state $^3A_2$ is orbitally singlet, so no orbital contribution to χ.
- Thus, the magnetism is a result of the spin-only formula.
- Fig. 11.26 shows splitting of $^3A_2$ term in the tetragonal field by the zero field parameter D:

Diamagnetic	D	Low spin
Ni^{2+}	$\rightleftharpoons$	Ni^{2+}
Tetragonal		Octahedral
$S = 0$		$S = \pm 1$

- The thermal distribution among the spin states is given by the Van Vleck equation:

$$\chi = N_A \frac{\sum_n \left(\frac{\left(E_n^{(1)}\right)^2}{kT} - 2E_n^{(2)}\right) e^{\frac{-E_n^{(0)}}{kT}}}{\sum_n e^{\frac{-E_n^{(0)}}{kT}}} \tag{11.6.7}$$

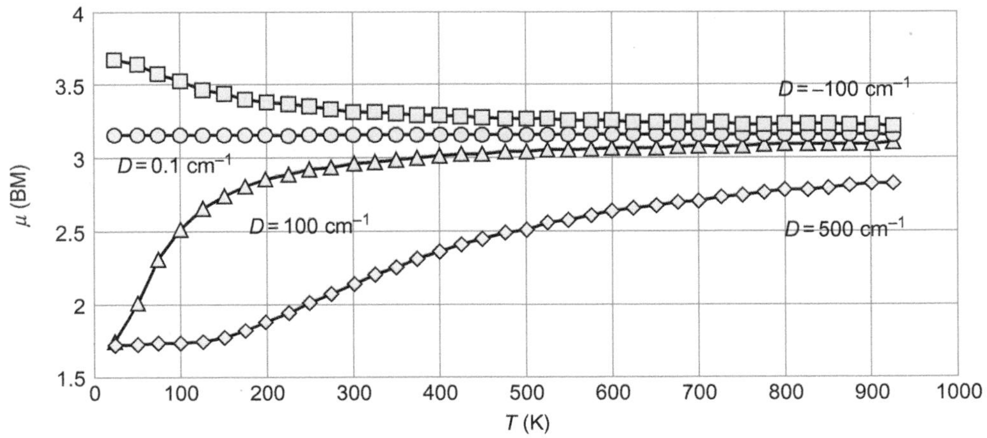

FIG. 11.25 The magnetic moment as a function of temperature at large and small tetragonal distortions of Cr(III) complexes.

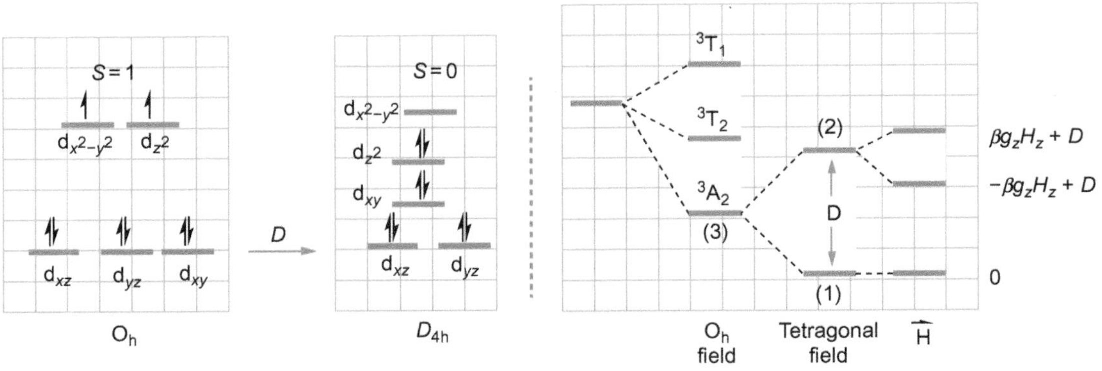

FIG. 11.26 The tetragonal distortion and the splitting of $^3A_2$ term of nickel (II) complexes.

$$E_n^{(0)} = 0\,(\text{lower level}),\, D\,(\text{upper level})$$

$$E_n^{(1)} = 0,\, 0,\, -g_z\beta,\, g_z\beta$$

$$\frac{\chi}{N_A} = \frac{\frac{(0)^2}{kT}e^{\frac{0}{kT}} + \frac{(-g_z\beta)^2}{kT}e^{\frac{-D}{kT}} + \frac{(g_z\beta)^2}{kT}e^{\frac{-D}{kT}}}{e^{\frac{0}{kT}} + 2e^{\frac{-D}{kT}}} = \frac{\frac{g_z^2\beta^2}{kT}e^{\frac{-D}{kT}}}{1 + 2e^{\frac{-D}{kT}}}$$

$$\chi = \frac{N_A g_z^2\beta^2}{kT} \cdot \frac{e^{\frac{-D}{kT}}}{1 + 2e^{\frac{-D}{kT}}} \tag{11.12.2}$$

- Fig. 11.27 shows the magnetic moment $\left(\mu = \sqrt{8\chi}\right)$ as a function of temperature at large and small values of D:
  - For very small distortion or at a very high T:
    μ approaches 2.83 BM, the calculated value for $S = 1$.
  - When D is very large or at T approaches zero:
    The values of μ approach zero BM, and become diamagnetic.

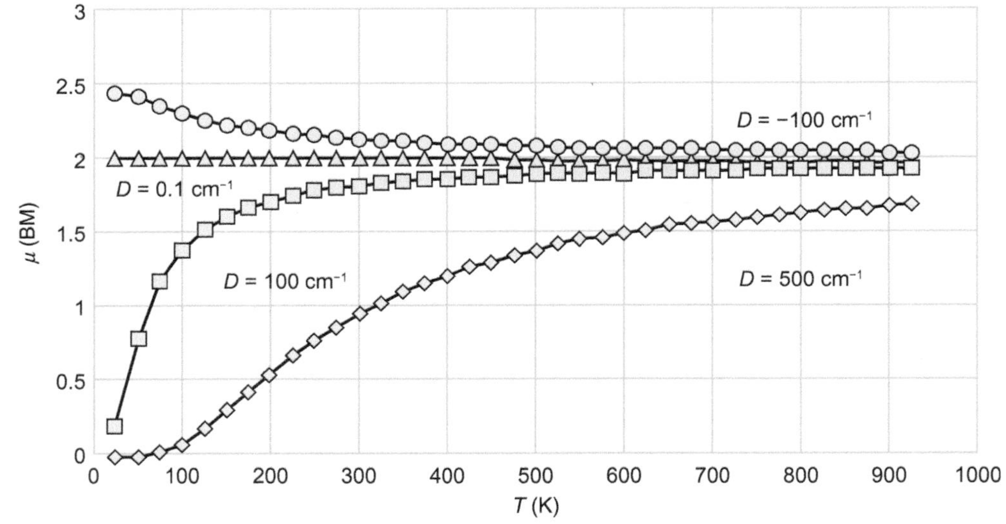

FIG. 11.27 The magnetic moment as a function of temperature at large and small tetragonal distortions of Ni(II) complexes.

11.13 INTERACTIONS BETWEEN MAGNETIC CENTERS

Homobinuclear Interaction

In a homobinuclear complex with two ions of spin S_A and S_B:

How do the two spins interact, and what is the resulting spin angular momentum?
How can the exchange energy be calculated, and derive the equation for χ_M as a function of temperature for antiferromagnetic interaction?

- In general, with two ions of spin S_A and S_B interact, the resulting spin angular momentum is S', with values 0, 1, 2, ...2S, totaling $(2S'+1)$ levels. Each level has spin multiplicity $2S'+1$, and energy $-JS'(S+1)$ above ground level.

$$S' = S_A + S_B \tag{11.13.1}$$

$$S'^2 = S_A^2 + S_B^2 + 2S_AS_B$$

$$S_AS_B = \frac{1}{2}\left[S'^2 - S_A^2 - S_B^2\right]$$

The exchange energy is given by

$$E = -2JS_AS_B = -J\left[S'^2 - S_A^2 - S_B^2\right] \tag{11.13.2}$$

$$E = -J[S'(S'+1) - S_A(S_A+1) - S_B(S_B+1)] \tag{11.13.3}$$

- Let us consider two levels with $S' = 0$ and $S' = S''$, for antiferromagnetic interaction.
The level with $S' = 0$ is the ground state, and

$$\Delta E = E_{S''} - E_0 = -J\left[S''\left(S''+1\right) - S_A(S_A+1) - S_B(S_B+1)\right] + J[-S_A(S_A+1) - S_B(S_B+1)]$$

$$\Delta E = -JS''\left(S''+1\right) \tag{11.13.4}$$

- Therefore, a given S' level will be $-JS'(S'+1)$ above the ground level.
- When the values of $E_n^{(1)}$ term are considered

$$\chi_M = N_A\frac{\sum_n\left(\frac{(E_n^{(1)})^2}{kT} - 2E_n^{(2)}\right)e^{\frac{-E_n^{(0)}}{kT}}}{\sum_n e^{\frac{-E_n^{(0)}}{kT}}} \tag{11.6.7}$$

Each level splits into $2S'+1$ components, ranging in energy from $-g\beta HS'$ to $g\beta HS'$, then

$$E_n^{(1)} = g^2\beta^2\left[S'^2 + (S'+1)^2 + \cdots 0 \cdots + \left(-(S'+1)^2\right) + \left(-S^2\right)\right] = g^2\beta^2\frac{S'(S'+1)(2S'+1)}{3}$$

Since every component of a degenerate set of levels must be counted separately, the denominator of the Van Vleck equation has to include the factor $(2S'+1)$.

$$\chi_M = N_A\frac{\sum_n\left(\frac{(E_n^{(1)})^2}{kT} - 2E_n^{(2)}\right)e^{\frac{-E_n^{(0)}}{kT}}}{\sum_n e^{\frac{-E_n^{(0)}}{kT}}} \tag{11.6.7}$$

If the second-order Zeeman term is considered

$$\chi_M = \frac{N_Ag^2\beta^2}{3kT}\frac{\Sigma_n(S'(S'+1)(2S'+1))e^{\frac{-E_n^{(0)}}{kT}}}{\Sigma_n(S'(S'+1)(2S'+1))e^{\frac{-E_n^{(0)}}{kT}}} + N_\alpha \tag{11.13.5}$$

- E.g., $S_A = S_B = \frac{3}{2}$

$$S'_{max.} = 3, \quad \therefore S' = 0, 1, 2, 3$$

For the antiferromagnetic:

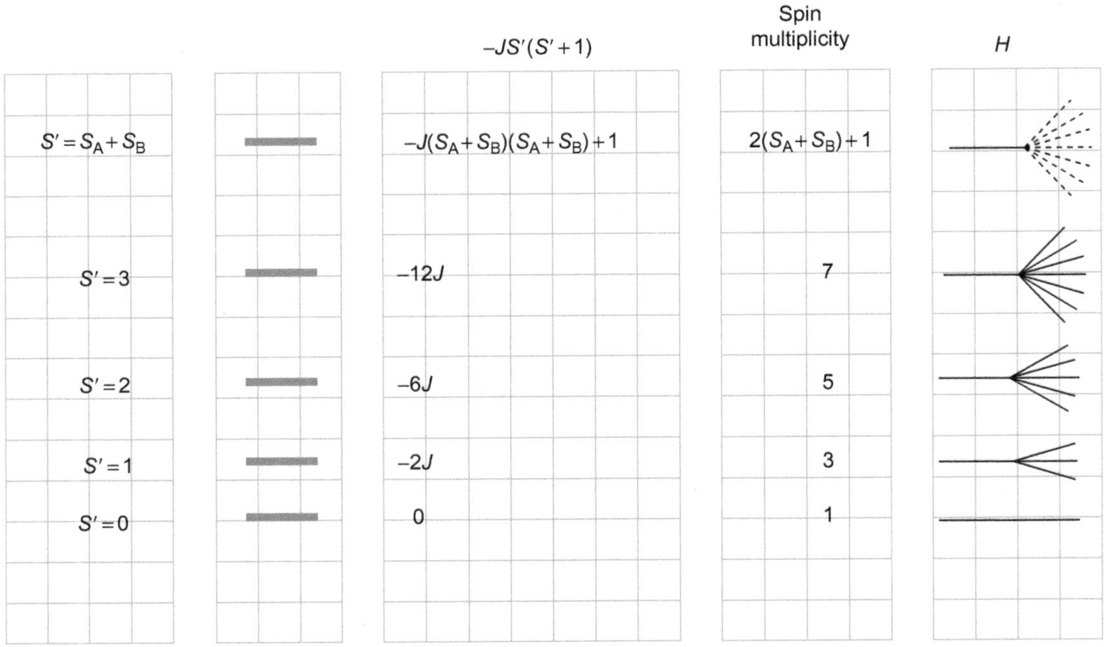

$$\chi_M = \frac{N_A g^2 \beta^2}{3kT} \frac{\left[0 + (1\times 2\times 3)e^{\frac{2J}{kT}} + (2\times 3\times 5)e^{\frac{6J}{kT}} + (3\times 4\times 7)e^{\frac{12J}{kT}}\right]}{\left[e^{\frac{0}{kT}} + 3e^{\frac{2J}{kT}} + 5e^{\frac{6J}{kT}} + 7e^{\frac{12J}{kT}}\right]} + N_\alpha \qquad (11.13.6)$$

$$\chi_M = \frac{N_A g^2 \beta^2}{kT} \frac{\left[2e^{\frac{2J}{kT}} + 10e^{\frac{6J}{kT}} + 28e^{\frac{12J}{kT}}\right]}{\left[1 + 3e^{\frac{2J}{kT}} + 5e^{\frac{6J}{kT}} + 7e^{\frac{12J}{kT}}\right]} + N_\alpha \qquad (11.13.7)$$

- The equations for the other values of S are derived in the same way:

For $S_A = S_B = \frac{1}{2}$

$$C = \frac{N_A g^2 \beta^2}{kT}, \quad x = \frac{J}{kT}$$

$$\chi = C \frac{2e^{2x}}{1 + 3e^{x}}$$

For $S_A = S_B = 1$

$$\chi = C \left[\frac{2e^{2x} + 10e^{6x}}{1 + 3e^{2x} + 5e^{6x}}\right]$$

- For $S_A = S_B = 3/2$

$$\chi_M = C\left[\frac{2e^{2x} + 10e^{6x} + 28e^{12x}}{1+3e^{2x}+5e^{6x}+7e^{12x}}\right]$$

- For $S_A = S_B = 2$

$$\chi_M = C\left[\frac{2e^{2x} + 10e^{6x} + 28e^{12x} + 60e^{20x}}{1+3e^{2x}+5e^{6x}+7e^{12x}+9e^{20x}}\right]$$

- For $S_A = S_B = 5/2$

$$\chi_M = C\left[\frac{2e^{2x} + 10e^{6x} + 28e^{12x} + 602e^{20x} + 110e^{30x}}{1+3e^{2x}+5e^{6x}+7e^{12x}+9e^{20x}+11e^{30x}}\right]$$

How can ligand bridges affect the magnetic-spin coupling in binuclear complexes?

If the splitting energy between the singlet and the triplet states, ΔE, is equal to 9800 cm^{-1}, and the electron pairing, π, is equal to 10,000 cm^{-1}, calculate the exchange coupling, J.

What is the difference between positive and negative exchange coupling? How do J and temperature rule χ_M and μ?

If; $J = -142$ cm^{-1}, and $kT = 200$ cm^{-1} at 298 K, for $Cu_2(CH_3COO)_4(H_2O)_2$, what are the measure and significance of the coupling strength?

Draw the splitting energy levels, and find the expected relationship between χ_M and temperature.

Plot χ_M and μ versus T for $Cu_2(CH_3COO)_4{\cdot}2H_2O$, and write your implications.

- The following omits orbital angular momentum considerations, and applies to ground terms of A or E type in an exact way. For two $S_o = 1/2$ ions, spin-pair is predicted in zero field splitting model.
- Fig. 11.28 describes the electronic properties, using magnetic/spin coupling that is mediated by the ligand bridges. Note the small splitting between levels in the bridge-mediate model.
- The splitting, ΔE, is usually described by magnetic exchange integral J. The size of the splitting between $S = 0$ and $S = 1$ is the measure of the strength of *M-M* interaction (Fig. 11.29), where J in energy units is called the exchange coupling constant or exchange parameter. It is to be distinguished from the J quantum number.
- The interaction between the metal orbitals on the metal centers can be described for a pair i and j by the Hamiltonian:

$$\hat{H} = -2J\,\hat{S}_i\hat{S}_j \qquad (11.13.8)$$

If $\hat{S} = \hat{S}_1 + \hat{S}_2$

$$\hat{S}^2 = \hat{S}_1^2 + \hat{S}_2^2 + 2\hat{S}_1\hat{S}_2$$

$$2\hat{S}_1\hat{S}_2 = \hat{S}^2 - \hat{S}_1^2 - \hat{S}_2^2$$

If $\hat{S}^2\psi = S(S+1)\psi$, then

$$-2J\hat{S}_1\hat{S}_2\psi = -J[S(S+1) - S_1(S_1+1) - S_2(S_2+1)]\psi \qquad (11.13.9)$$

where $S_1 = S_2 = 1/2$ and $S = 0$ or $S = 1$.

- For $S = 0$ (ground state) we obtain
  $-2J\ \hat{S}_1\hat{S}_2 = -J\ [0 - 3/4 - 3/4] = 3/2J$
  While for $S = 1$ (excited state)
  $-2J\ \hat{S}_1\hat{S}_2 = -J\ [2 - 3/4 - 3/4] = -1/2J$
  Then: $-2J = E - \pi$ for $S_1 = S_2 = 1/2$
- If: $\pi = 9800$ cm^{-1}, and $E = 10{,}000$ cm^{-1}:
  $-2J = E - \pi = 200$ cm^{-1}
  $-2J = 200$ cm^{-1}, $J = -100$ cm^{-1}
- J (or $-2J$) is a measure of the coupling strength.
- When $-2J$ is positive:
  $J < 0$
  singlet state ($S = 0$) below triplet ($S = 1$)
  antiferromagnetic coupling/exchange

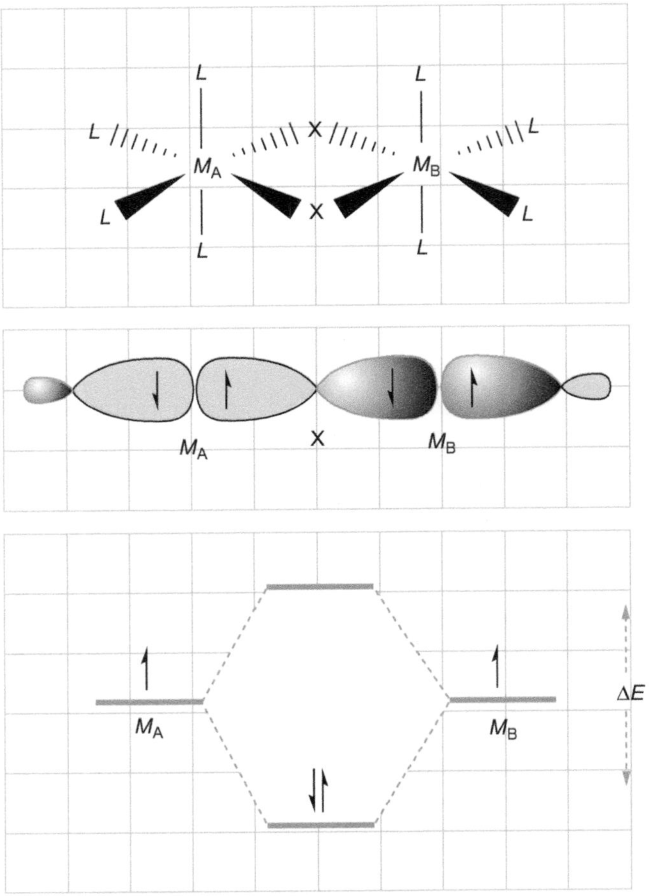

FIG. 11.28 Bridge-mediate: spin-polarization model.

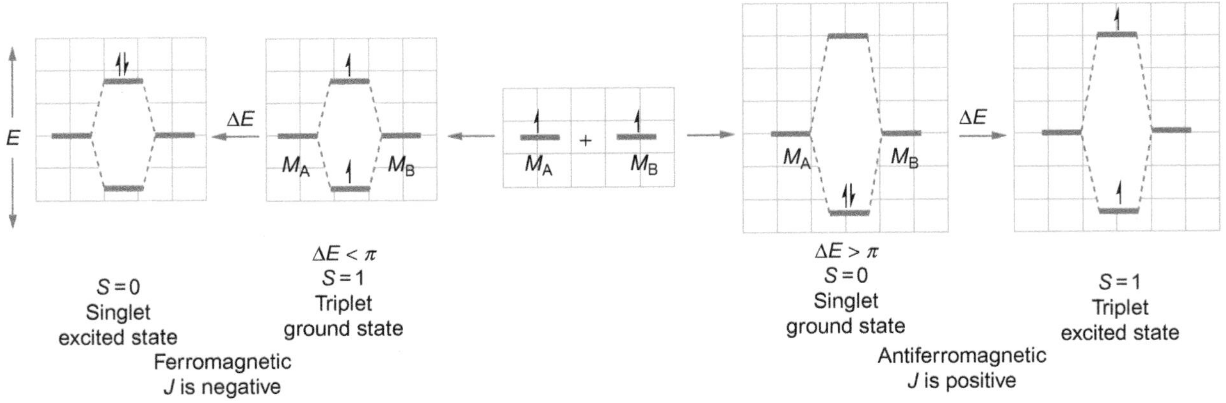

FIG. 11.29 Representation of the interaction of two metal orbitals to produce two nondegenerate levels, where ΔE is the splitting energy, π is the electron pairing, and $2J = \pi - \Delta E$.

- When $-2J$ is negative:
  $J > 0$
  triplet state is the ground state
  ferromagnetic coupling / exchange
- Copper (II) acetate dihydrate is a good example of binuclear interaction, where $\Delta E > \pi$, thus, $S = 0$ is the ground state, and $S = 1$ state is now an excited, and higher by $(\Delta E - \pi)$. Cu^{2+}-acetate (d^9, $S = ½$), $-2J$ of 300 cm^{-1}, $J < 0$, therefore, it is antiferromagnetic.

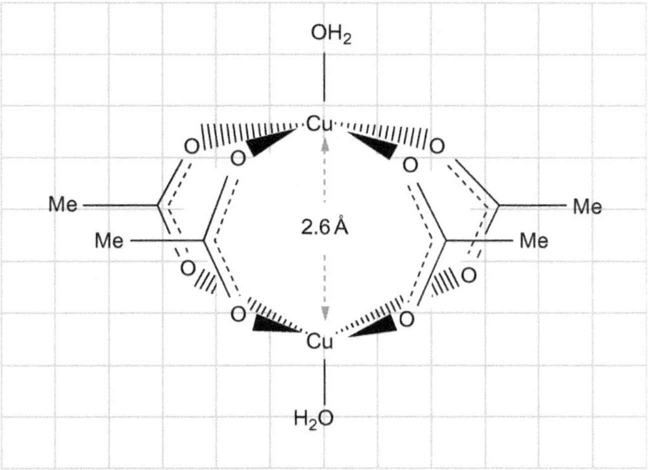

structure of copper (II) acetate dihydrate,
$Cu_2(CH_3COO)_4{\cdot}2H_2O$

- In a magnetic field, and varying the temperature, the Hamiltonian

$$\hat{H} = g\beta\hat{S}_z\hat{H}_z - 2J\hat{S}_1\hat{S}_2 \tag{11.13.3}$$

gives the results shown in Fig. 11.30.
  - If $kT = 200\ cm^{-1}$ at 298 K, molecules have thermally accessible energy levels.
  - The susceptibility of this system when the applied magnetic field is parallel to the principal molecular axis is obtained by the Van Vleck equation:

$$\chi = N_A \frac{\sum_n \left(\frac{(E_n^{(1)})^2}{kT} - 2E_n^{(2)} \right) e^{\frac{-E_n^{(0)}}{kT}}}{\sum_n e^{\frac{-E_n^{(0)}}{kT}}} \tag{11.6.7}$$

By inserting values for
$E_n^{(1)} = 0, -g\beta, 0, +g\beta$

$$E_n^{(0)} = \frac{3}{2}J,\ -\frac{1}{2}J,\ -\frac{1}{2}J,\ -\frac{1}{2}J$$

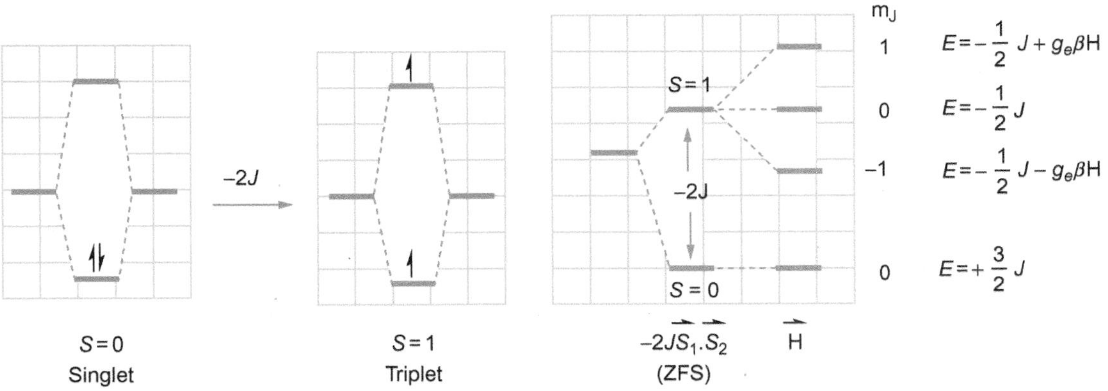

FIG. 11.30 Splitting of the energy levels, ZFS is the zero field splitting.

$$\chi = N_A \frac{\left(\frac{(-g\beta)^2}{kT}\right)e^{\frac{J}{2kT}} + \left(\frac{(+g\beta)^2}{kT}\right)e^{\frac{J}{2kT}}}{e^{\frac{-3J}{2kT}} + e^{\frac{J}{2kT}} + e^{\frac{J}{2kT}} + e^{\frac{J}{2kT}}} = \frac{2\left(\frac{(g\beta)^2}{kT}\right)e^{\frac{J}{2kT}}}{e^{\frac{-3J}{2kT}} + 3e^{\frac{J}{2kT}}}$$

$$\frac{\chi}{N_A} = \frac{\frac{2g^2\beta^2}{kT}}{3 + e^{\left(\frac{-3J}{2kT} - \frac{J}{2kT}\right)}} = \frac{2g^2\beta^2}{3kT}\frac{1}{1 + \frac{e^{\frac{-2J}{kT}}}{3}}$$

$$\chi = \frac{N_A g^2\beta^2}{kT}\frac{2}{3 + e^{\frac{-2J}{kT}}}$$

$$\chi = \frac{N_A g^2\beta^2}{kT}\frac{2e^{\frac{2J}{kT}}}{1 + 3e^{\frac{2J}{kT}}} \tag{11.13.10}$$

$$\text{If}: \frac{N_A\beta^2}{3k} = \frac{1}{8}, \text{ when } g = 2, \text{ then}:$$

$$\chi \cong \frac{2(4)}{8T}\frac{1}{1 + \frac{e^{\frac{-2J}{kT}}}{3}} = \frac{1}{T}\frac{1}{1 + \frac{e^{\frac{-2J}{kT}}}{3}}$$

$$\chi \cong \frac{1}{T} \cdot \frac{3}{3 + e^{\frac{-2J}{kT}}}$$

- By plotting χ versus T for $Cu_2(CH_3COO)_4{\cdot}2H_2O$ ($-2J = 284\ cm^{-1}$), we find that χ becomes small as T becomes small and approaches zero as T approaches infinity. As a result, the magnetic susceptibility rises to a maximum, and then tends to zero when T approaches absolute zero (Fig. 11.31).
- At low T, ion pairs are formed and $S = 0$, the sample is diamagnetic, with $\mu = 0$. On the other hand, at high T, Boltzmann population of the $S = 1$ levels advances μ to become paramagnetic value (Fig. 11.32).

Interactions between magnetic centers in heterobinuclear complexes
Consider the binuclear complex of M_A and M_B, with local spin S_A and S_B:
What are the results of combining S_A and S_B?

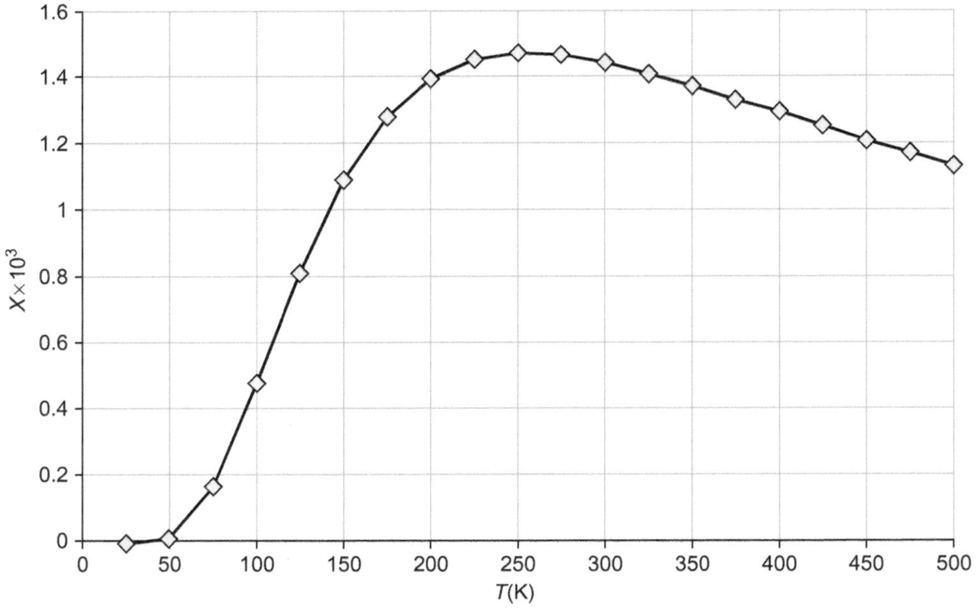

FIG. 11.31 The magnetic susceptibility of $Cu_2(CH_3COO)_4{\cdot}2H_2O$ at different temperatures.

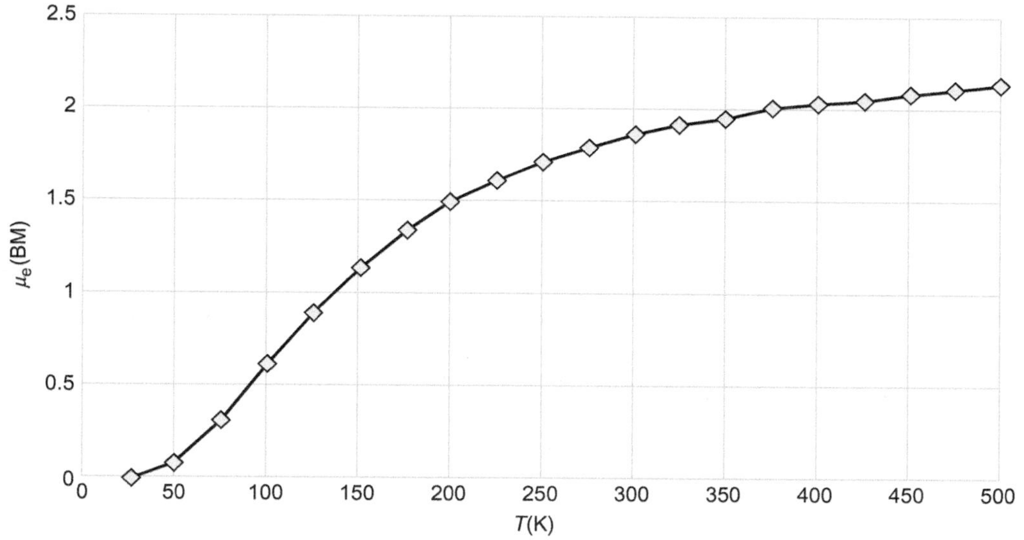

FIG. 11.32 The magnetic moment of $Cu_2(CH_3COO)_4{\cdot}2H_2O$ at different temperatures.

How can you calculate the energy of low-laying states?
When $kT \gg |J|$, the high-T limit of the susceptibility is

$$(\chi T)_{\mathrm{HT}} = \frac{N_A\beta^2}{3k}\left[g_A^2 S_A(S_A+1) + g_B^2 S_B(S_B+1)\right]$$

At 0 K:

$$(\chi T)_{\mathrm{LT}} = \frac{N_A g_s^2\beta^2}{3k}\left[(S_A - S_B)^2 + |S_A - S_B|\right] \quad \text{for } J < 0$$

$$(\chi T)_{\mathrm{LT}} = \frac{N_A g_s^2\beta^2}{3k}\left[(S_A - S_B)(S_A + S_B + 1)\right] \quad \text{for } J > 0$$

$$\text{if}: \boldsymbol{g}_{S'_n} = \frac{(1+C)}{2} g_A + \frac{(1+C)}{2} g_B \text{(relates to } S' = |S_A - S_B|\text{)},$$

$$C = \frac{S_A(S_A+1) - S_B(S_B+1)}{S'(S'+1)}, \quad \text{and}$$

$$\chi = \frac{N_A\beta^2}{4kT}\left(\frac{g_{S'_1}^2 + 10 g_{S'_2}^2 e^{\frac{3J}{kT}}}{1 + 2e^{\frac{3J}{kT}}}\right) - \frac{8\delta^2\beta^2}{3J}\left(\frac{1 - e^{\frac{3J}{kT}}}{1 + 2e^{\frac{3J}{kT}}}\right)$$

$$\text{where, } \delta = \frac{g_A - g_B}{3}$$

In $CuNi(Fsa_2en)(OH_2)_2{\cdot}H_2O$:

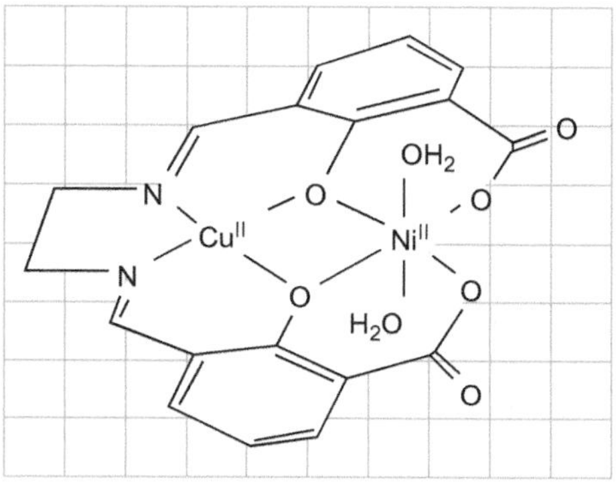

$$g_{Cu^{II}} = 2.1,\ g_{Ni^{II}} = 2.23, \text{ and the energy gap } (3J) = 213\,\text{cm}^{-1}$$

Find $(\chi^T)_{HT}$, $(\chi^T)_{LT}$, and draw χ^T as a function of temperature. How can you rationalize the thermomagnetic behavior of CuNi(Fsa$_2$en)(OH$_2$)$_2$·H$_2$O?

- For any pairs of interacting of magnetic centers with local spins S_A and S_B, the pair states are defined by S', which varies by integer value from $|S_A - S_B|$ to $|S_A + S_B|$, with energies.
- In CuNi(Fsa$_2$en)(OH$_2$)$_2$·H$_2$O:

$$\begin{matrix} Cu^{II} & \cdots & Ni^{II} \\ S_A = \frac{1}{2} & & S_B = 1 \end{matrix}$$

$$S' = |S_A - S_B|, |S_A - S_B + 1|, \ldots, |S_A + S_B| \tag{11.13.11}$$

$$\therefore S' = \frac{1}{2}, \frac{3}{2}$$

- The energy gap between the temperature limits equals $3J$:

		$JS'(S'+1)$	
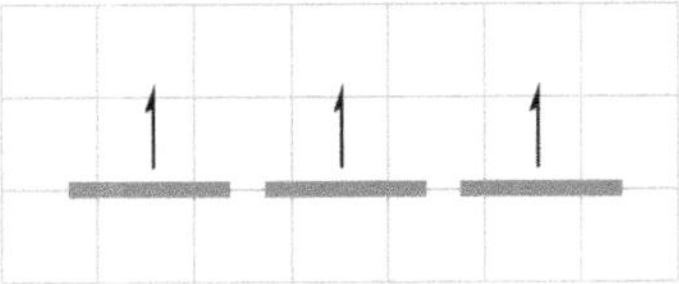	$S' = \frac{3}{2}$	at $\frac{15J}{4}$ (or $3J$)	High T limit
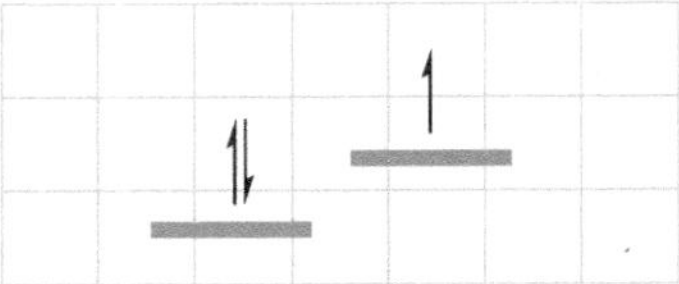	$S' = \frac{1}{2}$	at $\frac{3J}{4}$ (or 0)	Low T limit

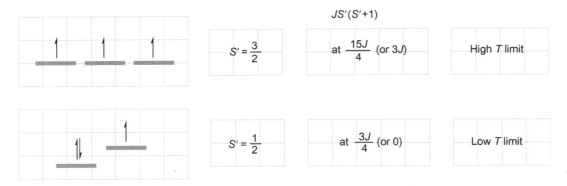

- g_A and g_B are the simple local g-values for Cu^{II} and Ni^{II}:

$$C_{1/2} \text{ for } S' = \frac{1}{2}$$

$$C_{S'} = \frac{S_A(S_A+1) - S_B(S_B+1)}{S'(S'+1)} \tag{11.13.12}$$

$$C_{1/2} = \frac{\frac{1}{2}\left(\frac{1}{2}+1\right) - \frac{3}{2}\left(\frac{3}{2}+1\right)}{\frac{1}{2}\left(\frac{1}{2}+1\right)} = -1.667$$

$$C_{3/2} = \frac{\frac{1}{2}\left(\frac{1}{2}+1\right) - \frac{3}{2}\left(\frac{3}{2}+1\right)}{\frac{3}{2}\left(\frac{3}{2}+1\right)} = -0.333$$

$$g_{s'} = \frac{(1+C_{s'})}{2} g_A + \frac{(1+C_{s'})}{2} g_B \tag{11.13.13}$$

$$g_{1/2} = \frac{(1+C_{1/2})}{2} g_{Cu} + \frac{(1+C_{1/2})}{2} g_{Ni} = \frac{2.1(1-1.667)}{2} + \frac{2.3(1-1.667)}{2} = 2.367$$

$$g_{3/2} = \frac{(1+C_{3/2})}{2} g_{Cu} + \frac{(1+C_{3/2})}{2} g_{Ni} = \frac{2.1(1-0.333)}{2} + \frac{2.3(1-0.333)}{2} = 2.233$$

$$\delta = \frac{g_A - g_B}{3} \tag{11.13.14}$$

$$\delta = \frac{g_{Cu} - g_{Ni}}{3} = \frac{2.1 - 2.3}{3} = -0.0667$$

$$(\chi T)_{HT} = \frac{N_A\beta^2}{3k}\left[g_A^2 S_A(S_A+1) + g_B^2 S_B(S_B+1)\right] \quad (11.13.15)$$

$$(\chi T)_{HT} = 0.1251\left[g_{Cu}^2 S_{Cu}(S_{Cu}+1) + g_{Ni}^2 S_{Ni}(S_{Ni}+1)\right]$$

$$(\chi T)_{HT} = 0.1251\left[(2.1)^2\left(\frac{1}{2}\right)\left(\frac{1}{2}+1\right) + (2.3)^2(1)(1+1)\right] = 1.737$$

$$\mu_{HT} = \sqrt{8(\chi T)_{HT}} = 3.714$$

At low temperatures:

$$(\chi T)_{LT} = \frac{N_A g_s^2 \beta^2}{3k}\left[(S_A - S_B)^2 + |S_A - S_B|\right] \quad \text{for } J<0 \quad (11.13.16)$$

$$(\chi T)_{LT} = (0.1251)(2.367)^2\left[\left(\frac{1}{2}-1\right)^2 + \left|\frac{1}{2}-1\right|\right] = 0.675, \quad \mu = 2.324\text{BM}$$

$$(\chi T)_{LT} = (0.1251)(2.233)^2\left[\left(\frac{1}{2}-1\right)^2 + \left|\frac{1}{2}-1\right|\right] = 0.656, \quad \mu = 2.291\text{BM}$$

- By plotting χT versus T for CuNi(Fsa$_2$en)(H$_2$O)$_2$.H$_2$O ($3J = -213$ cm^{-1}) (Fig. 11.33)

$$\chi T = \frac{N_A\beta^2}{4k}\left(\frac{g_{S_1'}^2 + 10 g_{S_2'}^2 e^{\frac{3J}{kT}}}{1+2e^{\frac{3J}{kT}}}\right) - \frac{8\delta^2\beta^2}{3J}\left(\frac{1-e^{\frac{3J}{kT}}}{1+2e^{\frac{3J}{kT}}}\right) \quad (11.13.17)$$

  where the energy gap $= 3J = -213$ cm^{-1}.
- Fig. 11.33 shows that as T becomes small, μ becomes small and approaches 2.3 BM, which is the average of the spin value of the $S_{Cu(II)}$ and $S_{Ni(II)}$ $\left(\frac{1.73+2.83}{2} = 2.28\right)$. This suggests that as temperature decreases, the spin interaction between the metal centers decreases.
- As T increases, μ approaches the magnetic spin of $S' = 3/2$ (for 3 unpaired electrons, $\mu = 3.87$ BM). This implies that as temperature increases, the spin interaction between the two centers increases.

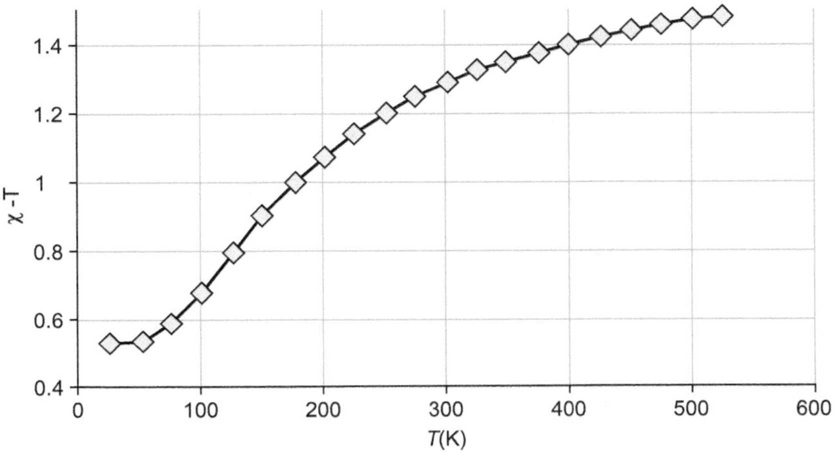

FIG. 11.33 Plot of χT versus temperature for CuNi(Fsa$_2$en)(H$_2$O)$_2$·H$_2$O.

Coupling Mechanisms in Binuclear Compounds

How do two local spin states magnetically couple in binuclear compounds?

- There are two possible coupling mechanisms between the two local centers M_A and M_B:
  a. Dipolar coupling:
     Electron on M_A "feels" the field of M_B-electron through the space. The J values are usually small, and be seen in ESR, but not χ.
  b. Superexchange coupling:
     Spin polarization is mediated by a bridging ligand. The J-values are large and very dependent on geometry. E.g., when unpaired electron occupies $d_{x^2-y^2}$:

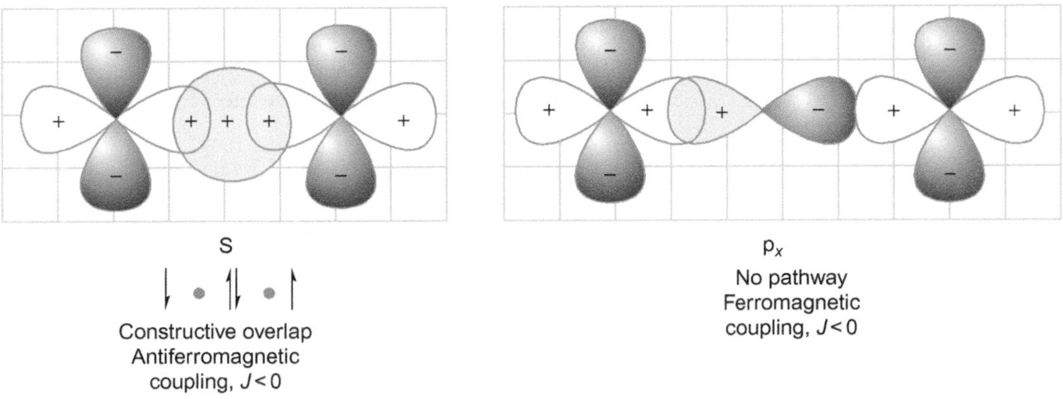

- The spin coupling is accountable for antiferromagnetic behavior, and usually happens through intervening ligands by a superexchange mechanism (Fig. 11.34).
- In the superexchange mechanism:
  - The spin on one metal atom generates a spin polarization on an occupied orbital of a ligand, and this spin polarization results in an antiparallel alignment of the spin of the neighbor metal atom (Fig. 11.34).
  - The produced alternative spin alignments, … ↑↓↑↓…, extents throughout the metal atom.
  - Examples of antiferromagnetic behavior include
    MnO below 122 K
    Cr_2O_3 below 310 K
- Magnetic coupling:

Two or more metal atoms are bridged by ligands that mediate the coupling; for example: copper acetate, which exists as dimer with antiferromagnetic coupling between two d9 centers.

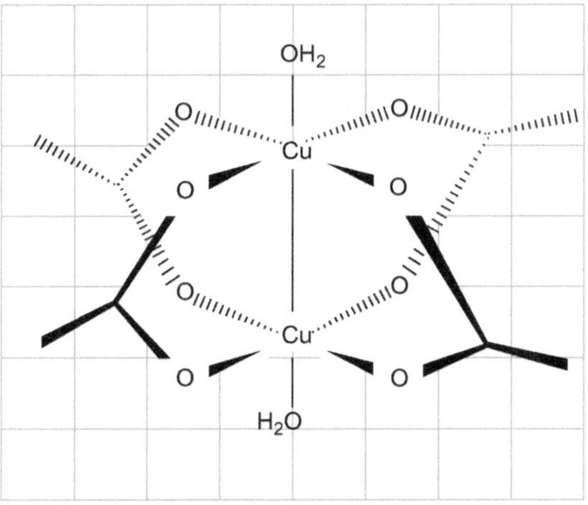

Copper (II) acetate dihydrate, $Cu_2(CH_3COO_4{\cdot}2H_2O$

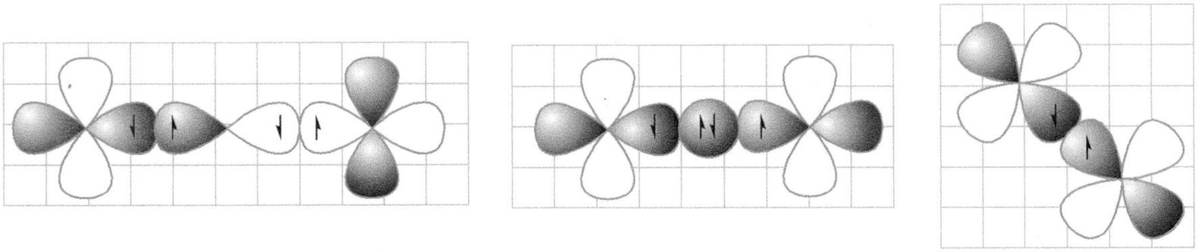

FIG. 11.34 Representations of the exchange mechanism.

Give examples to demonstrate the significance of the proper geometric alignment to the orbital coupling.

- Proper geometric alignment of an orbital has great influence on the coupling strength.
- Examples:
  - Consider that the two local centers of the following binuclear Cu (II)-oxalate isomers have the same geometry, and each center has an unpaired electron in $d_{x^2-y^2}$:

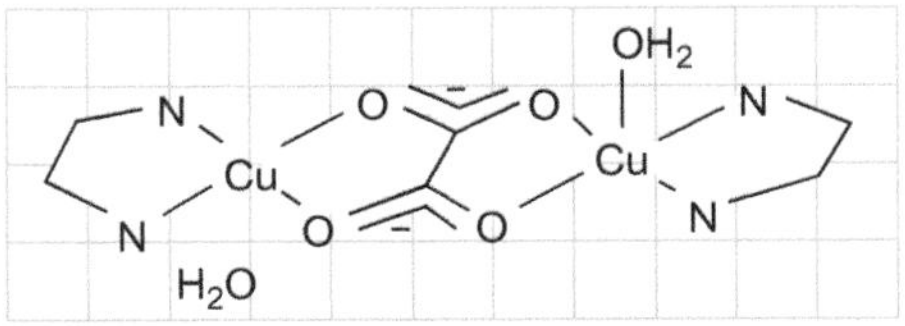

The two $d_{x^2-y^2}$ orbitals are linked by in the same plane, μ-oxalate coupling is strong

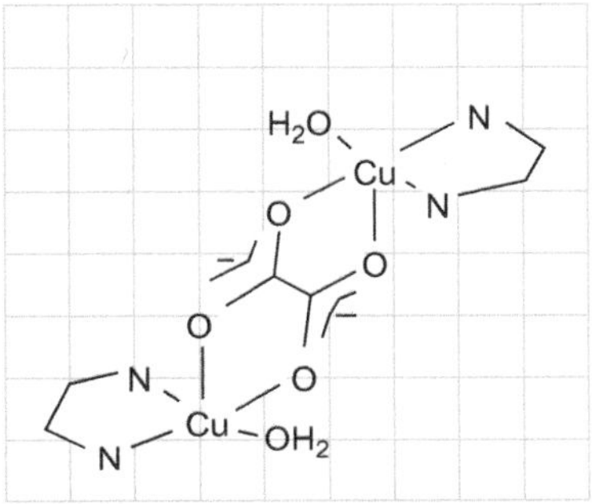

Oxalate is bridging via x and z directions, and coupling is weak

- Examples for coupling between two local centers having different geometry:

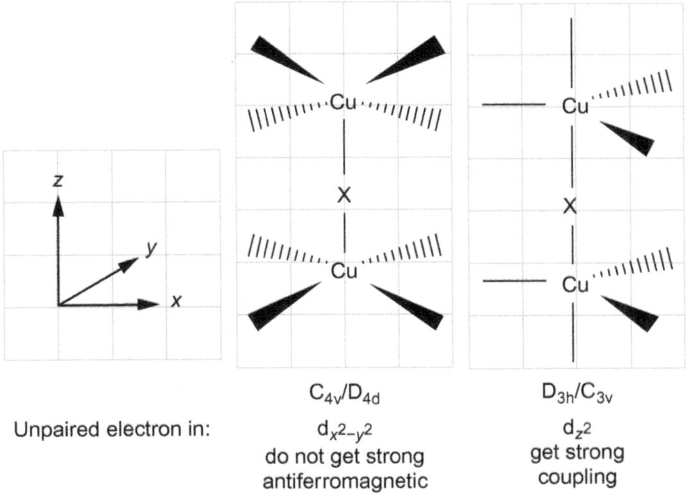

11.14 MEASUREMENT OF THE MAGNETIC SUSCEPTIBILITY

Name the main techniques and the bases that used to determine the magnetic susceptibility.

- Methods for the measurement of the magnetic susceptibility are based on

$$\partial F = H(\kappa_2 - \kappa_1)\partial V \frac{\partial H}{\partial l} \quad (11.14.1)$$

where F is the force acting on the volume element ∂V in the magnetic field of intensity H, with field gradient $\partial H/\partial l$, and κ_2 and κ_1 are the susceptibility per unit volume of specimen and the displaced medium, respectively.

- Methods based on the measurement of F integrated over the field gradient: Gouy's and Quincke's methods.
- Methods based on the direct measurement of F: Faraday's method.

Gouy's Method

How could Gouy determine the magnetic susceptibility, and what are the advantages and the disadvantages of this method?

- A tube is backed to a certain height with powdered sold of the magnetic sample.
- The sample is suspended from one of the arms of a sensitive balance such that the bottom part of the sample tube in the center of the magnetic field, while the top part is outside the magnetic field (Fig. 11.35).
- The whole set of connections is housed in an enclosure that can be flushed with nitrogen or helium.
- The used electromagnet gives a constant magnetic field in the range 5000–20,000 gauss.
- The force, ∂F, exerted by gradient, $\partial H/\partial l$, of intensity, H, is given by

$$\partial F = H(\kappa_2 - \kappa_1)\partial V \frac{\partial H}{\partial l} \tag{11.14.1}$$

κ_2 for nitrogen $\cong -0.0004 \times 10^{-6}$, and the equation reduced to

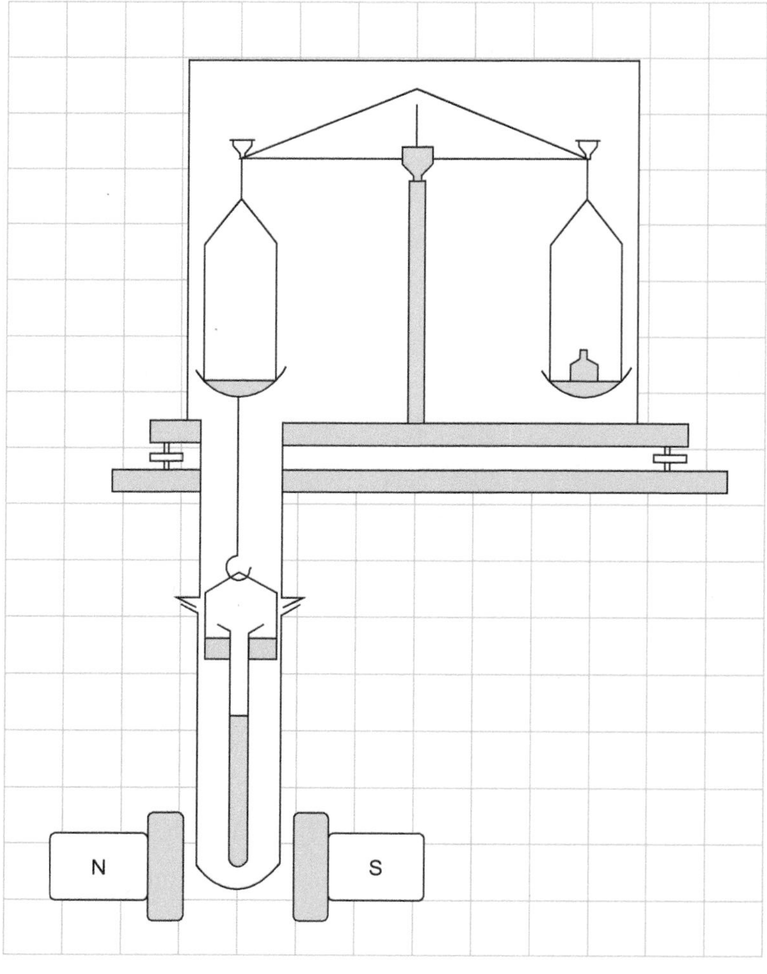

FIG. 11.35 Diagrammatic representation of Gouy's balance.

$$\partial F = H\kappa\partial V\frac{\partial H}{\partial l} \tag{11.14.2}$$

$$\because \partial V = A\partial l$$

$$\therefore \partial F = H\kappa A\partial l\frac{\partial H}{\partial l} = H\kappa A\partial H \tag{11.14.3}$$

where A is the cross-sectional area and ∂l is the height of the sample,

$$F = \int_{H_o}^{H} H\kappa A\partial H$$

where the magnetic field between H_o and H is

$$F = \frac{1}{2}\kappa A\left(H^2 - H_o^2\right) \tag{11.14.4}$$

$$\because \kappa = \rho\chi_g, \quad \rho \text{ is the density}$$

$$\because \rho = \frac{w}{V} = \frac{w}{Al}, \quad l \text{ and } w \text{ are the length and the mass of the sample}$$

$$\therefore F = \frac{1}{2}\kappa A\left(H^2 - H_o^2\right) = \frac{1}{2}\rho\chi_g A\left(H^2 - H_o^2\right) = \frac{1}{2}\left(\frac{w}{Al}\right)\chi_g A\left(H^2 - H_o^2\right) \tag{11.14.5}$$

$$\therefore F = \frac{w\chi_g}{2l}\left(H^2 - H_o^2\right) \tag{11.14.6}$$

$$\text{because } H_o \ll H$$

$$\therefore F = \frac{w\chi_g H^2}{2l} \tag{11.14.7}$$

- If the change in the weight is equal to the force experienced by the sample,

$$\therefore F = \Delta wg$$

where g is the acceleration of gravity ($g = 981$ cm/s^2) and Δw is the change in the weight due to the alteration of the magnetic field. The value of χ_g can be determined directly via the following:

$$\chi_g = \frac{2lF}{wH^2} = \frac{2(981)l\,\Delta w}{(\text{mass of the sample})\,H^2} \tag{11.14.8}$$

- The procedure can be simplified by measuring $\Delta w_{st.}$ for a standard substance of known magnetic susceptibility $\left(\chi_{g_{st.}}\right)$, and measuring Δw of the sample at the same magnetic field H. Then χ_g of the sample is given by

$$\chi_g = \left(\chi_{g_{st.}}\right)\left(\frac{w_{st.}}{\Delta w_{st.}}\right)\left(\frac{\Delta w}{w}\right) \tag{11.14.9}$$

where w and $w_{st.}$ are the weight of the sample and the standard in absence of the magnetic field, respectively.

- The molar susceptibility χ_M is

$$\chi_M = \chi_g(\text{molecular weight of the sample})$$

- The corrected, $\chi_M^{Corr.}$, for diamagnetic of the ligand, anion, solvent of crystallization, and metal ion, and for the temperature independent paramagnetism, TIP, is given by

$$\chi_M^{Corr.} = \chi_M - \text{diamagnetic correction} - \text{TIP}$$

- The advantages of Gouy's method include
  - fair sensitivity
  - convenient solution measurements
  - handy equipment

- The disadvantages include
  - needs a large amount of the sample
  - inaccuracy due to the packing errors

Calculate the magnetic moment, μ_{eff}, of Cu[(biguanide]$_2$Cl$_2$·2H$_2$O, from the following experimental data of Gouy's method using Hg[Co(NCS)$_4$] as standard: $T = 293$ K, $\chi_{Hg[Co(NCS)4]} = 16.44 \times 10^{-6}$cgs.

Diamagnetic correction and atomic mass:

	H	C	N (chain)	O	Cl^-	Cu^{2+}	Structure Correction C=N—R (TIP)
Correction × 10^{-6} g/atom	2.93	6.00	5.57	4.61	26	12.8	8.2
Atomic mass	1.0079	12.011	14.0067	15.999	35.453	63.546	

Weight of empty tube-1 (for standard) without magnetic field = 9.82304 g.
Weight of empty tube-1 with magnetic field = 9.82248 g.
Weight of the tube-1 and Hg[Co(NCS)$_4$] without magnetic field = 10.83980 g.
Weight of the tube-1 and Hg[Co(NCS)$_4$] with the same magnetic field = 10.86280 g.
Weight of empty tube-2 (for the sample) without magnetic field = 9.82285 g.
Weight of empty tube-2 with magnetic field = 9.82232 g.
Weight of the tube-2 and the sample without magnetic field = 10.15926 g.
Weight of the tube-2 and the sample with the same magnetic field = 10.16026 g.

- Answer:
- The change in the weight of tube-1 due to the magnetic field = 9.82248 − 9.82304 = -5.6×10^{-4} g.
- The change in the weight of Hg[Co(NCS)$_4$] and tube-1 due to the magnetic field = 10.86280 − 10.83980 = 0.023 g.
- The change in the weight of Hg[Co(NCS)$_4$] due to the magnetic field, Δw_{st} = 0.023 − (−0.00056) = 0.02356 g.
- Weight of Hg[Co(NCS)$_4$] = 10.83980 − 9.82304 = 1.01676 g.
- The change in the weight of tube-2 due to the magnetic field = 9.82232 − 9.82285 = -5.3×10^{-4} g.
- The change in the weight of the sample and tube-2 due to the magnetic field = 10.16026 − 10.15926 = 0.001 g.
- The change in the weight of the sample due to the magnetic field, Δw = 0.001 − (−0.00053) = 0.00153 g.
- Weight of the sample of Cu[(biguanide]$_2$Cl$_2$·2H$_2$O, w = 10.15926 − 9.82285 = 0.33641 g.
- Using the formula:

$$\chi_g = (\chi_g)_{st}\, x \frac{w_{st}}{\Delta w_{st}}\, x \frac{\Delta w}{w}$$

$$\chi_g = 16.44 \times 10^{-6} \times \frac{1.01676}{0.02356} \times \frac{0.00153}{0.33641} = 3.2268 \times 10^{-6} \text{ cgs units}$$

- Molecular weight of Cu[(biguanide]$_2$Cl$_2$·2H$_2$O = 3.72.5

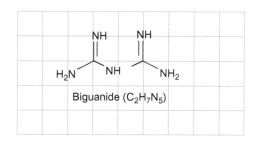

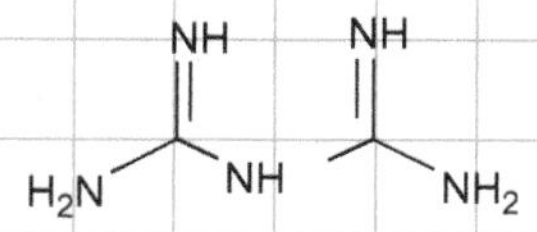

Biguanide ($C_2H_7N_5$)

$$\because \chi_M = \chi_g \times \text{molecular weight}$$

$$\therefore \chi_M = 3.2268 \times 10^{-6} \times 372.5 = 1202 \times 10^{-6} \text{ cgs units}$$

- Diamagnetic correction:

	H	C	N (chain)	O	Cl^-	Cu^{2+}	Structure Correction C=N—R	
Correction $\times 10^{-6}$ g/atom	2.93	6.00	5.57	4.61	26	12.8	8.2	
Number of elements	18	4	10	2	2	1	2	
Diamagnetic correction	−52.74	−24	−55.7	−9.22	−52	−12.8	−16.4	$=-222.86\times 10^{-6}$ g/atom

$$\therefore \chi_{M}^{Corr.} = -1202\times 10^{-6} - \left(-222.86\times 10^{-6}\right) = 1324\times 10^{-6}\,g/atom$$

$$\because \chi_{M}^{Corr.} = \frac{N\mu_{eff}^2}{3KT} = \frac{0.125\mu_{eff}^2}{T}$$

$$\therefore \mu_{eff} = 2.83\sqrt{\chi_{M}^{Corr.}T}$$

$$\therefore \mu_{eff} = 2.83\sqrt{1324\times 10^{-6}\times 293} = 1.78\text{ BM}$$

Faraday's Method

Describe Faraday's method for measuring the magnetic susceptibility. What are the advantages and the disadvantages of this method?

- The whole set of connections is housed in an enclosure that can be flushed with nitrogen or helium, therefore, κ_2 is very small and can be neglected.
- A small amount of a magnetic sample (0.1–10 mg) is packed in a fused quartz ampoule, and suspended from a sensitive balance.
- This method requires special macrobalances; quartz fiber torsion or Sucksmith ring balances are most commonly used (Fig. 11.36).
- The sample is sufficiently small so that the product $H(\partial H/\partial l)$ is constant over the whole volume of the sample, and there is no need to integrate, as was required by Gouy's method:

$$\because \partial F = \kappa \partial V H\frac{\partial H}{\partial l} \tag{11.14.2}$$

$$\therefore \partial F = w\chi_g H\frac{\partial H}{\partial l},\quad \text{or}$$

$$H\frac{\partial H}{\partial l} = \frac{\partial F}{w\chi_g} = \frac{\Delta wg}{w\chi_g} \tag{11.14.10}$$

- The accuracy of this technique is enhanced if the magnet can be moved vertically from bottom to the top of the sample.
- The balance readings are recorded as the magnet is shifted stepwise and sorted from maximum noticeable loss in weight to maximum increase. The difference between these is used to calculate χ_g.
- The section of even $H(\partial H/\partial l)$ is estimated by the packing standard calibrant of mass, $w_{st.}$, of known magnetic susceptibility, $\chi_{g_{st.}}$, and obtaining the value of $H(\partial H/\partial l)$ at different points along the field from

$$H\frac{\partial H}{\partial l} = \frac{\Delta w_{st.}\,g}{w_{st.}\chi_{g_{st.}}} \tag{11.14.10}$$

where $\Delta w_{st.}$ is the deflection in the mass of calibrant and g is the acceleration of gravity.

$$\chi_g = \left(\frac{\Delta w\,g}{w\chi_g}\right)\bigg/\left(H\frac{\partial H}{\partial l}\right) \tag{11.14.10}$$

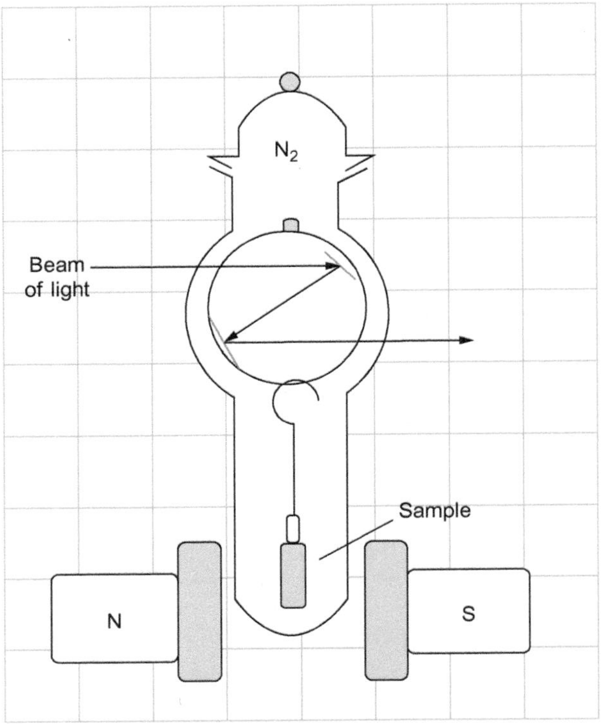

FIG. 11.36 Schematic diagram of a Sucksmith ring balance used in Faraday's method.

- The value of χ_g of a sample can also be obtained by taking the measurement first with the calibrant, then with the sample at constant $H(\partial H/\partial l)$, therefore:

$$\chi_g = \chi_{g_{st.}} \left(\frac{\Delta w}{\Delta w_{st.}}\right)\left(\frac{w_{st.}}{w}\right) \tag{11.14.11}$$

- The advantages of Faraday's method include
  - small amount of sample is required
  - the sample need not to be homogeneous
  - good sensitivity
  - suitable for measurements above room temperature
  - suitable for measuring the magnetic anisotropy
- The disadvantages include
  - requires great manual skill
  - fragile equipment
  - delicate suspension device
  - constructional complexity
  - measurements on solution are difficult
  - small weight changes

Quincke's Method

How is the magnetic susceptibility evaluated in Quencke's method?

- Quincke's method is suitable for liquids and dissolved substance that includes gases.
- The liquid solution to be investigated is placed in a vessel, as shown in Fig. 11.37.
- The vessel is positioned in Quinck's apparatus, such that the meniscus of the liquid in the narrow tube is located in the center or little below of homogeneous field of the electromagnet.

FIG. 11.37 Quincke's vessel.

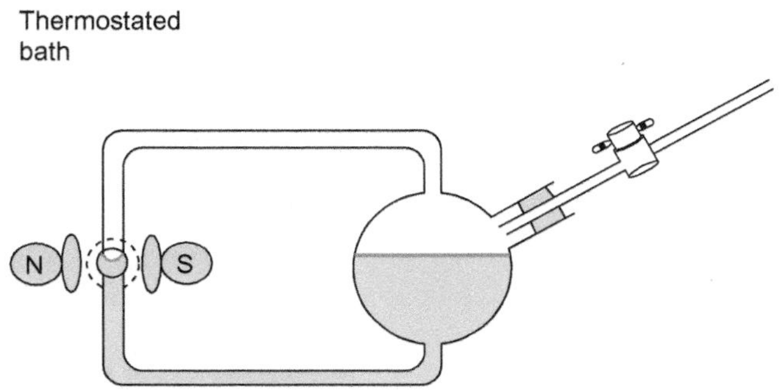

- When applying the magnetic field, the liquid level is shifted to the extent that permitted by the hydrostatic pressure resulting from the difference in the liquid level.
- The level rises if the liquid is paramagnetic and drops when it is diamagnetic.
- Because the reservoir is sufficiently large, the level of its liquid remains almost the same despite of the change in the height of the liquid in the narrow tube.
- The magnetic force is in equilibrium with the hydrostatic pressure.
- The force exerted by the magnetic field is obtained by determining the hydrostatic pressure; therefore, the value of F is obtained from the shift in the height of the liquid in the narrow tube, Δh

$$F = A\left(\rho_{\text{liq.}} - \rho_{\text{gas}}\right) g\,\Delta h = A\rho_1 g\,\Delta h \tag{11.14.12}$$

where $\rho_{\text{liq.}}$ and ρ_{gas} are the densities of the liquid and the gas, respectively.

$$\left(\rho_{\text{liq.}} \gg \rho_{\text{gas}},\ \rho_{N_2} = 0.00125\ \text{g}\,\text{l}^{-1}\ \text{at STP}\right)$$

$$\because \partial F = H\left(\kappa_{\text{liq.}} - \kappa_{\text{gas}}\right)\partial V \frac{\partial H}{\partial h} \tag{11.14.1}$$

$$\because \partial V = A\partial h,\quad A\ \text{is the cross} - \text{section are},\ h\ \text{is the height}$$

$$\therefore \partial F = H\left(\kappa_{\text{liq.}} - \kappa_{\text{gas}}\right)A\partial h \frac{\partial H}{\partial h} = H\left(\kappa_{\text{liq.}} - \kappa_{\text{gas}}\right)A\partial H$$

$$\therefore F = \frac{1}{2l}\left(\kappa_{\text{liq.}} - \kappa_{\text{gas}}\right)A\left(H^2 - H_o^2\right)$$

$$\text{If } H_o \ll H$$

$$\therefore F = \frac{1}{2}\left(\kappa_{\text{liq.}} - \kappa_{\text{gas}}\right)AH^2 \tag{11.14.13}$$

$$\therefore F = \frac{1}{2}\left(\rho_{\text{liq}}\chi_{g_{\text{liq.}}} - \rho_{\text{gas}}\chi_{g_{\text{gas}}}\right)AH^2,\quad \rho\ \text{is the density}$$

$$\because \rho_{\text{liq.}} \gg \rho_{\text{gas}},\ \text{and}\ \chi_{g_{\text{liq.}}} \gg \chi_{g_{\text{gas}}}\left(\chi_{N_2} = -0.0004 \times 10^{-6}\right),\ \text{then}\ \rho_{\text{gas}}\chi_{g_{\text{gas}}}\ \text{is negligible},\ \rho_{gN_2}\chi_{N_2} = 5.004 \times 10^{-13}$$

$$\therefore F = \frac{1}{2}\rho_{\text{liq}}\chi_{g_{\text{liq.}}} AH^2 \tag{11.14.14}$$

$$F = A\rho_1 g\,\Delta h \tag{11.14.12}$$

$$\therefore A\,\rho_{\text{liq.}} g\,\Delta h = \frac{1}{2}A\,\rho_{\text{liq}}\chi_{g_{\text{liq.}}} H^2$$

$$\therefore \chi_g = \frac{2g\,\Delta h}{H^2} \tag{11.14.15}$$

NMR Method

How could Evans determine the magnetic susceptibility with help of NMR spectroscopy?

- The NMR method is based on the principle that the position of a proton resonance frequency of a compound is dependent on the magnetic environment of the proton. Any changes in this surrounding environment induce subsequent changes in the resonance frequency.
- A solution of the paramagnetic sample to be studied, along with 1%–3% of a standard proton containing substance, is placed in an inner tube of the concentric cell (Fig. 11.38), whose length ≫diameter.
- An identical solution without the paramagnetic substance is placed in the annular section of the tube.
- The presence of the paramagnetic substance in the inner tube shifts the proton resonance of the standard; consequently, two-resonance frequencies are obtained for the standard.
- The gram susceptibility, χ_g (cm^3g^{-1}) of the sample is given by

$$\chi_g = \frac{3\Delta f}{4\pi f m} + \chi_o + \frac{\chi_o(\rho_o - \rho_s)}{m} \quad (11.14.16)$$

  where Δf is the observed shift in the frequency Hz, f is the frequency of the proton, m is mass of the paramagnetic substance in g cm^{-3} (concentration in g/mL), χ_o is the gram susceptibility in $cm^3\ g^{-1}$, and ρ_o and ρ_s are the densities of the solvent and the solution, respectively.
- With a dilute solution, the term $\chi_o(\rho_o - \rho_s)/m$ is negligible.
- Identical proton standard in this method: *t*-butyl alcohol for aqueous solution, while benzene or $(CH_3)_4Si$ are suitable for nonaqueous solution.
- The advantages of this method are that it:
  - has good sensitivity
  - requires small amount, as little as 0.02 mL
  - requires dilute solution, using concentrated solutions may cause broadening in the resonance frequency, which reduce the accuracy
  - enables temperature control and variation
- The disadvantage is that it is restricted to the solution.

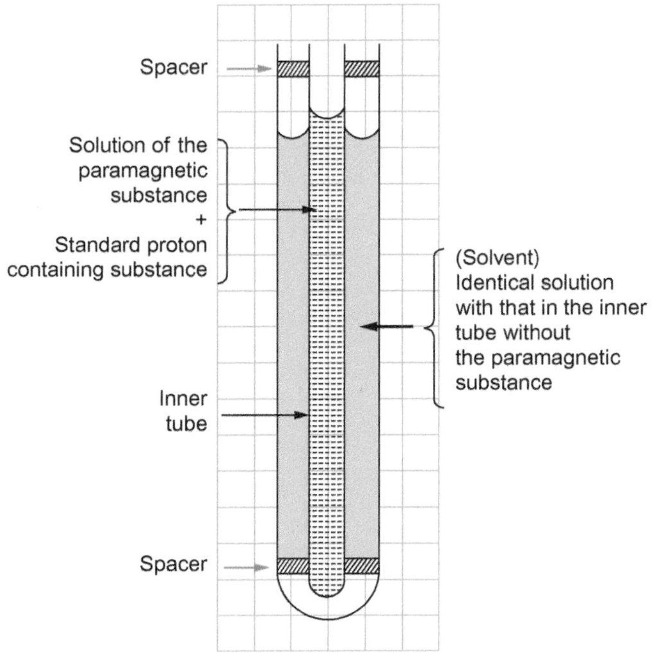

FIG. 11.38 Schematic diagram of the concentric NMR tube for measuring magnetic susceptibility.

SUGGESTIONS FOR FURTHER READING

O. Kahn, Molecular Magnetism, VCH Publishers, New York, NY, ISBN: 1-56081-566-3, 1993.

A. Earnshaw, Introduction to Magnetochemistry, Academic Press, London, 1968a.

R.L. Dutta, A. Syamal, Elements of Magnetochemistry, Affiliated East-West Press PVT Limited, New Delhi, ISBN: 81-85336-92-X, 2004 (Reprint).

R.S. Drago, Physical Methods in Chemistry, W. B. Saunders Company, Philadelphia, PA, 1977 (Chapter 11).

R.W. Jotham, Magnetic moment, in: G.J. Bullen, D.J. Greenslade (Eds.), Problems in Molecular Structure, Pion Limited, London, 1983. Chapter 4 (3).

B.N. Figgis, Introduction to Ligand Fields, Robert E. Krieger Publishing Company, Malabar, FL, 1986 (Chapter 10).

B.N. Figgis, J. Lewis, Magnetic properties of transition metal complexes, in: J. Lewis, R.G. Wilkins (Eds.), Modern Coordination Chemistry, Interscience, 1960a (Chapter 6).

B.N. Figgis, J. Lewis, Technique of Inorganic Chemistry, vol. 4, Interscience Publishers, New York, 1965, p. 137.

L.F. Bates, Modern Magnetism, Cambridge University Press, London, 1961.

J.S. Griffiths, Theory of Transition Metal Ions, Cambridge University Press, London, ISBN: 978-0521051507, 1961.

C.J. Ballhausen, Introduction to Ligand Field Theory, McGraw-Hill, New York, NY, 1962.

A. Earnshaw, Introduction to Magnetochemistry, Academic Press, London, 1968b.

D.F. Evans, The determination of the paramagnetic susceptibility of substances in solution by nuclear magnetic resonance, J. Chem. Soc. 2003 (1959).

B.N. Figgis, M. Gerloch, J. Lewis, R.C. Slade, Paramagnetic anisotropies and covalency of some tetrahedral copper(II) complexes, J. Chem. Soc., A. (1968) 2028.

B.N. Figgis, J. Lewis, The magnetic properties of transition metal complexes, Prog. Inorg. Chem. 6 (37) (1964a) 2028.

B.N. Figgis, R.L. Martin, Magnetic studies with copper(II) salts. Part I. Anomalous paramagnetism and δ-bonding in anhydrous and hydrated copper(II) acetates, J. Chem. Soc. 3837–3846 (1956a).

M. Gerloch, J.R. Miller, Covalence and the orbital reduction factor, *k*, in magnetochemistry, Prog. Inorg. Chem. 10 (1968) 1.

R.M. Golding, Applied Wave Mechanics, Van Nostrand, London, 1969.

F.E. Mabbs, D.J. Machin, Magnetism and Transition Metal Complexes, Chapman and Hall, London, 1973a.

C.G. Shull, W.A. Strauser, E.O. Wollan, Neutron diffraction by paramagnetic and antiferromagnetic substances, Phys. Rev. 83 (1951) 333.

Y.Y. Li, Magnetic moment arrangements and magnetocrystalline deformations in antiferromagnetic compounds, Phys. Rev. 100 (1955) 627.

W.L. Jolly, The Synthesis and Characterization of Inorganic Compounds, Prentice-Hall, Englewood Cliffs, NJ, 1970, pp. 369–384.

T.I. Quickenden, R.C. Marshall, Magnetochemistry in SI units, J. Chem. Ed. 49 (1972) 114.

Diamagnetic Behavior

F.A. Cotton, G. Wilkinson, Advanced Inorganic Chemistry, fourth ed., Wiley-Interscience, New York, NY, 1980, pp. 1359–1365.

Landolt-Börnstein, Magnetic Properties of Coordination and Organometallic Transition metal Compounds, in: E. König (Ed.), Group II Molecules and Radicals, vol. II/2, Springer-Verlag, New York, NY, 1966.

E. König, G. König, Magnetic Properties of Coordination and Organometallic Transition Metal Compounds, vol. II/8, Springer-Verlag, New York, NY, 1976.

Spin-Only Magnetic Moment

A.P. Ginsberg, M.E. Lines, Magnetic exchange in transition metal complexes. VIII. Molecular field theory of intercluster interactions in transition metal cluster complexes, Inorg. Chem. 11 (1972) 2289.

G. Herzberg, Atomic Spectra and Atomic Structure, second ed., Dover Publications, New York, NY, 1944, p. 109.

M. Kotani, On the magnetic moment of complex ions. (I), J. Phys. Soc. Jpn. 4 (1949a) 293.

B. Bleaney, K.W.H. Stevens, Paramagnetic resonance, Rep. Prog. Phys. 16 (1953a) 108.

F.E. Mabbs, D.J. Machin, Magnetism and Transition Metal Complexes, John Wiley, New York, NY, 1973b.

B.N. Figgis, J. Lewis, The magnetic properties of transition metal complexes, in: F.A. Cotton (Ed.), Progress in Inorganic Chemistry, Vol. 6, Interscience, New York, NY, 1964b.

B.N. Figgis, J. Lewis, The magnetochemistry of complex compounds, in: J. Lewis, R.G. Wilkins (Eds.), Modern Coordination Chemistry, Interscience, New York, NY, 1960b.

R.S. Nyholm, Magnetochemistry: introductory lecture, J. Inorg. Nucl. Chem. 8 (1958a) 401.

N.S. Gill, R.S. Nyholm, Complex halides of the transition metals. Part I. Tetrahedral nickel complexes, J. Chem. Soc. 3997 (1959).

R.L. Carlin, Paramagnetic susceptibilities, J. Chem. Educ. 43 (1966) 521.

Spin-Orbital Coupling

J.H. Van Vleck, Theory of Electronic and Magnetic Susceptibilities, Oxford University Press, London, 1932.

N.B. Figgis, J. Lewis, The magnetic properties of transition metal complexes, Prog. Inorg. Chem. 6 (1964c) 37.

J.B. Goodenough, Magnetism and the Chemical Bond, Interscience, New York, NY, 1963.

Spin-Orbital Coupling: In A and E Ground Terms

N.S. Gill, R.S. Nyholm, Complex halides of the transition metals. Part I. Tetrahedral nickel complexes, J. Chem. Soc. 3997 (1959).

R.S. Nyholm, Magnetochemistry: introductory lecture, J. Inorg. Nucl. Chem. 8 (1958b) 401.

J.T. Donoghue, R.S. Drago, Non-aqueous coordination phenomena-complexes of hexamethylphosphoramide. I. Preparation and properties of tetrahedral $[Zn\{PO[N(CH_3)_2]_3\}_4]^{+2}$, $[Co\{PO[N(CH_3)_2]_3\}_4]^{+2}$, and $[Ni\{PO[N(CH_3)_2]_3\}_4]^{+2}$ compounds, Inorg. Chem. 1 (1962) 866.

D.M.L. Goodgame, M. Goodgame, F.A. Cotton, et al., Electronic spectra of some tetrahedral nickel(II) complexes, J. Amer. Chem. Soc. 83 (1961) 4161.

J.T. Donoghue, R.S. Drago, Non-aqueous coordination phenomena-complexes of hexamethyl-phosphoramide. II. Pseudo-tetrahedral complexes of nickel(II) and cobalt(II), Inorg. Chem. 2 (1963) 572.

Spin-Orbital Coupling: In T Ground Terms

M. Kotani, On the magnetic moment of complex ions. (I), J. Phys. Soc. Jpn. 4 (1949b) 293.

B. Bleaney, K.W.H. Stevens, Paramagnetic resonance, Rep. Prog. Phys. 16 (1953b) 108.

F.A. Cotton, E. Bannister, Phosphine oxide complexes. Part I. Preparation and properties of the cation $[(Ph_3PO)_4Ni]^{2+}$, J. Chem. Soc. 1873 (1960).

B.N. Figgis, Introduction to Ligand Field, Interscience, New York, NY, 1966.

Spin and Structure-Linked Crossover

R.L. Martin, A.H. White, Transition Met. Chem 4 (113) (1968a).

H.A. Goodwin, Spin Transitions in six-coordinate iron(II) complexes, Coord. Chem. Rev. 18 (1976) 293.

J.P. Jesson, T. Trofimenko, D.R. Eaton, Spin equilibria in octahedral iron(II) poly((1-pyrazolyl)-borates, J. Am. Chem. Soc. 89 (1967) 3158.

E. König, S. Kremer, Theoret. Chim. Acta. 20 (1971) 143.

E. König, K. Madeja, Equilibriums in some iron(II)-bis(1,10-phenanthroline) complexes, Inorg. Chem. 6 (1967) 48.

M.G. Burnett, V. McKee, S.M. Nelson, $^5T_2 \rightleftharpoons {}^1A_1$ and $^6A_1 \rightleftharpoons {}^2T_2$ spin transitions in iron(II) and iron(III) complexes of 2,2′-bi-2-imidazoline and related ligands, J. Chem. Soc. Dalton Trans. 1492–1497 (1981).

R.L. Martin, A.H. White, Trans. Met. Chem. 4 (1968b) 113.

M.A. Hoselton, L.J. Wilson, R.S. Drago, Substituent effects on the spin equilibrium observed with hexadentate ligands on iron(II), J. Amer. Chem. Soc. 97 (1975) 1722.

B. Bleaney, K.W.H. Stevens, Paramagnetic resonance, Rep. Prog. Phys. 16 (1953c) 108.

Interactions Between Magnetic Centers in Homobinuclear Compounds

B. Bleaney, K.D. Bowers, Anomalous paramagnetism of copper acetate, Proc. Roy. Soc. (London) A214 (1952) 451.

B.C. Guha, Magnetic properties of some paramagnetic crystals at low temperatures, Proc. R. Soc. Lond. A206 (1951) 353.

M. Kato, H.B. Jonassen, J.C. Fanning, Copper(II) complexes with subnormal magnetic moments, Chem. Rev. 64 (1964) 99.

R.L. Martin, in: E.A.V. Ebsworth, A. Maddock, A.G. Sharpe (Eds.), New Pathways in Inorganic Chemistry, Cambridge University Press, Cambridge, 1968.

A.P. Ginsberg, Magnetic exchange in transition metal complexes vi: Aspects of exchange coupling in magnetic cluster complexes, Inorg. Chim. Acta Rev. 5 (1971) 45.

S.J. Lippard, R.J. Doedens, Structure and metal-metal interactions in copper(II) carboxylate complexes, Prog. Inorg. Chem. 21 (2007) 209–231.

O. Kahn, Dinuclear complexes with predictable magnetic properties, Angew. Chem. Int. Ed. Engl. 24 (1985) 834.

M. Kato, Y. Muto, Factors affecting the magnetic properties of dimeric copper(II) complexes, Coord. Chem. Rev. 92 (1988) 45.

S.J. Lippard, C.J. O'Connor, Magnetochemistry—advances in theory and experimentation, Prog. Inorg. Chem. 29 (1982) 203.

G.F. Kokoszka, G. Gordon, Trans. Met. Chem. 5 (1969) 181.

B.N. Figgis, R.L. Martin, Magnetic studies with copper(II) salts. Part I. Anomalous paramagnetism and δ-bonding in anhydrous and hydrated copper(II) acetates, J. Chem. Soc. 3837 (1956b).

E. Kotot, R.L. Martin, Magnetic studies with copper(II) salts. VI. Variable singlet-triplet energies in amine-substituted copper(II) alkanoates, Inorg. Chem. 3 (1964) 1306.

H.J. Schugar, G.R. Rossman, H.B. Gray, Dihydroxo-bridged ferric dimer, J. Amer. Soc. 91 (1969) 4564.

Interactions Between Magnetic Centers in Heterobinuclear Compounds

R.D. Willet, D. Gatteschi, O. Kahn (Eds.), Magneto-Structural Correlation in Exchange Coupled System, NATO ASI Series, Plenum Press, Reidel, New York, 1985.

O. Kahn, M.F. Charlot, in: D.J. Klein, N. Trinajstic (Eds.), Valence Bond Theory and Chemical Structure, Elsevier, Amsterdam, 1990, p. 489.

Y. Pei, Y. Journaux, O. Kahn, Irregular spin state structure in trinuclear species: magnetic and EPR properties of manganese(II)-copper(II)-manganese(II) and nickel(II)-copper(II)-nickel(II) compounds, Inorg. Chem. 27 (1988) 399.

C.C. Chao, On structure and spin Hamiltonian parameters of exchange-coupled ion pairs, J. Mag. Reson 10 (1973) 1.

E. Buluggiu, EPR study on Mn^{2+}-Cu^{2+} and Ni^{2+}-Cu^{2+} mixed pairs, J. Phys. Chem. Solids 41 (1980) 1175.

R.P. Scaringe, D. Hodgson, W.E. Hatfield, The coupled representation matrix of the pair Hamiltonian, Mol. Phys. 35 (1978) 701.

Mathematics Supplement

- Summation formulas
  - Constant series
  - Linear series
- Quadratic series
- Roots of the quadratic equation
- Binomial expansion
- Trigonometric formula
- Logarithms
  - Properties of logarithms
  - Exponential natural logarithm
- Derivative of a function
  - The chain rule
  - The power rule
  - The product rule
  - The quotient rule
  - The exponential function
  - The natural logarithm function
  - Differentiation formulas
- Antiderivatives or indefinite integrals
  - Integration formula
  - The definite integral
- Important mathematical functions
- Kronecker delta function
- Hermite polynomials
- Legendre polynomial
- Laguerre polynomials
- Taylor series
- Matrices
  - Multiplication
  - Terminology
  - The inverse of the matrix
  - Matrix and linear equations
- Determinants

SUMMATION FORMULAS

- **Constant series**:

$$\sum_{k=0}^{S} c = sc$$

- Example:
  If $S=4$

$$\sum_{k=0}^{4} c = 0+c+c+c+c = 4c$$

$$\sum_{k=0}^{4} c = 4c$$

- **Linear series**:

$$\sum_{k=0}^{S} k = \frac{S(S+1)}{2}$$

- Example:
  If $S=4$

$$\sum_{k=0}^{4} k = 0+1+2+3+4 = 10$$

$$\sum_{k=0}^{S} k = \frac{S(S+1)}{2} = 10$$

QUADRATIC SERIES

$$\sum_{k=0}^{S} k^2 = \frac{S(S+1)(2S+1)}{6}$$

Example:
If $S=4$

$$0^2+1^2+2^2+3^2+4^2 = 0+1+4+9+16 = 30$$

$$\sum_{k=0}^{4} k^2 = \frac{4(4+1)(2\times 4+1)}{6} = \frac{4\times 5\times 9}{6} = 30$$

ROOTS OF THE QUADRATIC EQUATION

If

$$ax^2+bx+c=0$$

then

$$x = \frac{-b\pm\sqrt{b^2-4ac}}{2a}$$

BINOMIAL EXPANSION

$$(1+x)^n = 1+nx+\frac{n(n-1)}{2!}x^2+\frac{n(n-1)(n-2)}{3!}x^3+\ldots\ldots\ldots\ldots\ldots\ldots$$

- To compute $(a+b)^n$
  - $(a+b)^n = a^n(1+x)^n$, with $x=\dfrac{b}{a}$

or

- $(a+b)^n = b^n(1+y)^n$, with $y = \frac{a}{b}$

- $(a+x)^n = \sum_{k=0}^{n} \binom{n}{k} x^k a^{n-k}, \binom{n}{k} = \frac{n!}{k!(n-k)!}$
- $(1+r)^{-1} = 1 - r + r^2 - r^3 + \cdots$
- $a\left(\frac{1-r^n}{1-r}\right) = a + ar + ar^2 + ar^3 + \cdots$
- $\sum y^i = 1 + y + y^2 + \cdots\cdots = \frac{1}{1-y}, (y<1)$, and
- $(1-r)^{-1} = 1 + r + r^2 + r^3 + \cdots$
- $\sum iy^i = y(1 + 2y + 3y^2 + \cdots\cdots) = \frac{y}{(1-y)^2}, (y<1)$

TRIGONOMETRIC FORMULA

- $\sin\theta = \cos(90° - \theta)$
- $\cos\theta = \sin(90° - \theta)$
- $\sin^2\theta + \cos^2\theta = 1$
- $\sec^2\theta - \tan^2\theta = 1$
- $\csc^2\theta - \text{ctn}^2\theta = 1$
- $\sin 2\theta = 2 \sin\theta \cos\theta$
- $\cos 2\theta = \cos^2\theta - \sin^2\theta = 2\cos^2\theta - 1 = 1 - 2\sin^2\theta$
- $\tan 2\theta = \frac{2\tan\theta}{1 - \tan^2\theta}$
- $\sin \frac{1}{2}\theta = \sqrt{\frac{1-\cos\theta}{2}}$
- $\cos \frac{1}{2}\theta = \sqrt{\frac{1+\cos\theta}{2}}$
- $\tan \frac{1}{2}\theta = \sqrt{\frac{1-\cos\theta}{1+\cos\theta}}$
- $\sin(A \pm B) = \sin A \cos B \pm \cos A \sin B$
- $\cos(A \pm B) = \cos A \cos B \mp \sin A \sin B$
- $\tan(A \pm B) = \frac{\tan A \pm \tan B}{1 \mp \tan A \tan B}$
- $\sin A \pm \sin B = 2 \sin\left[\frac{1}{2}(A \pm B)\right] \cos\left[\frac{1}{2}(A \mp B)\right]$
- $\cos A + \cos B = 2 \cos\left[\frac{1}{2}(A + B)\right] \cos\left[\frac{1}{2}(A - B)\right]$
- $\cos A - \cos B = 2 \sin\left[\frac{1}{2}(A + B)\right] \sin\left[\frac{1}{2}(A - B)\right]$
- $\tan A \pm \tan B = \frac{\sin(A \pm B)}{\cos A \cos B}$

LOGARITHMS

If $a^b = x$

$$\therefore \log_a x = b$$

a and x are positive real number

Examples

$$2^3 = 8, \quad \therefore \log_2 8 = 3$$

$$10^{-2} = 0.01, \quad \therefore \log_{10} 0.01 = -2$$

$$\log_{10} x = -\frac{1}{2}, \quad \therefore x = \frac{1}{\sqrt{10}}$$

- **Properties of logarithms**:
  - $\log_a x + \log_a y = \log_a (xy)$
  - $\log_a x - \log_a y = \log_a (x/y)$
  - $k \log_a x = \log_a x^k$
  - $a^{\log_a x} = x$
  - $\log_a x = (\log_a b) \log_b x$
- **Exponential natural logarithm**:
  - $e = 2.718282$
  - $e^0 = 1$
  - $\log_e 2.718282 = \ln 2.718282 = 1$
  - $\ln e = 1$
  - $\ln 1 = 0$
  - $y = e^x$

$$\therefore \log_e y = \ln y = x \ldots\ldots\ldots (e = 2.718282)$$

Example:

$$ax = e^{\frac{y}{2}}$$

$$\log_e ax = \log_e e^{\frac{y}{2}}$$

$$\ln e^{\frac{y}{2}} = \frac{y}{2}$$

$$\ln ax = \frac{y}{2}$$

$$y = \ln a^2 x^2$$

  - $e^{\ln x} = x$
  - $e^x e^y = e^{(x+y)}$
  - $(e^x)^y = e^{xy} = (e^y)^x$
  - $\ln xy = \ln x + \ln y$
  - $\ln \frac{x}{y} = \ln x - \ln y$
  - $\ln e^x = x$
  - $e^x = 1 + x + \frac{x^2}{2!} + \frac{x^3}{3!} + \ldots\ldots\ldots$
  - $\ln a^x = x \ln a$
  - $\ln x = (\ln 10) \log x = 2.3026 \log x$
  - $\log x = \log e \ \ln x = 0.43429 \ \ln x$
  - $\ln(1+x) = x - \frac{x^2}{2} + \frac{x^3}{3} - \frac{x^4}{4} + \ldots\ldots\ldots$

DERIVATIVE OF A FUNCTION

- $f'(x) = \lim\limits_{\Delta x \to 0} \frac{f(x+\Delta x) - f(x)}{\Delta x}$

Example:

If $f(x) = 3x^2$

$$f(x+\Delta x) = 3(x+\Delta x)^2 = 3x^2 + 6x\Delta x + 3(\Delta x)^2$$

$$f'(x) = \lim_{\Delta x \to 0} \frac{f(x+\Delta x) - f(x)}{\Delta x} = \lim_{\Delta x \to 0} \frac{3x^2 + 6x\Delta x + 3(\Delta x)^2 - 3x^2}{\Delta x} = 6x$$

$$f'(x) = 6x$$

if

$$y = f(x)$$

$$\frac{\partial y}{\partial x} = \frac{\partial (3x^2)}{\partial x} = 6x$$

- $\frac{\partial c}{\partial x} = 0$
- $\frac{\partial x}{\partial x} = 1$
- $\frac{\partial cu}{\partial x} = c\frac{\partial u}{\partial x}$ c is constant
- $\frac{\partial (u+v)}{\partial x} = \frac{\partial u}{\partial x} + \frac{\partial v}{\partial x}$
- $\frac{\partial x^n}{\partial x} = n x^{n-1}$
- $\frac{\partial \sin \omega\theta}{\partial \theta} = \omega \cos \omega\theta$
- $\frac{\partial \cos \omega\theta}{\partial \theta} = -\omega \sin \omega\theta$

- **The chain rule**: $\frac{\partial y}{\partial x} = \frac{\partial y}{\partial u} \cdot \frac{\partial u}{\partial x}$
  Example:

$$\frac{\partial}{\partial x} = \frac{\partial r}{\partial x}\frac{\partial}{\partial r} + \frac{\partial \theta}{\partial x}\frac{\partial}{\partial \theta} + \frac{\partial \phi}{\partial x}\frac{\partial}{\partial \phi}$$

- **The power rule**: $\frac{\partial (u)^n}{\partial x} = nu^{(n-1)}\frac{\partial u}{\partial x}$
  Example:

$$\frac{\partial (1-2x^2)^3}{\partial x} = 3(1-2x^2)^2(-4x) = -12x(1-2x^2)^2$$

- **The product rule**: $\frac{\partial (uv)}{\partial x} = u\frac{\partial v}{\partial x} + v\frac{\partial u}{\partial x}$
  Example:

$$\begin{aligned}\frac{\partial x^2(1-x^2)}{\partial x} &= x^2\frac{\partial (1-x^2)}{\partial x} + (1-x^2)\frac{\partial x^2}{\partial x} \\ &= x^2(1)(1-x^2)^{1-1}(-2x) + (1-x^2)(2x) \\ &= -2x^3 + 2x - 2x^3 = 2x - 4x^3\end{aligned}$$

- **The quotient rule**: $\dfrac{\partial(u/v)}{\partial x} = \dfrac{v\dfrac{\partial u}{\partial x} - u\dfrac{\partial v}{\partial x}}{v^2}$

  Example:

$$\frac{\partial\left(\dfrac{\sqrt{2x-3}}{x}\right)}{\partial x} = \frac{\partial\left(\dfrac{(2x-3)^{1/2}}{x}\right)}{\partial x}$$

$$= \frac{x\dfrac{\partial(2x-3)^{1/2}}{\partial x} - (2x-3)^{1/2}\dfrac{\partial x}{\partial x}}{x^2}$$

$$= \frac{x\left(\dfrac{1}{2}\right)(2x-3)^{-1/2}(2) - (2x-3)^{1/2}(1)}{x^2}$$

$$= \frac{x(2x-3)^{-1/2} - (2x-3)^{1/2}}{x^2}$$

$$= \frac{(2x-3)^{-1/2}[x-(2x-3)]}{x^2}$$

$$= \frac{3-x}{x^2(2x-3)^{1/2}}$$

- **The exponential function**: $\dfrac{\partial(e^u)}{\partial x} = e^u\dfrac{\partial u}{\partial x}$

  Example:

$$y = ae^{bx}, \text{ then}$$

$$\frac{\partial y}{\partial x} = abe^{bx} = by, \text{ and}$$

$$\frac{\partial^2 y}{\partial x^2} = ab^2e^{bx} = b^2y$$

  Example:

$$\frac{\partial\left(e^{2x^3}\right)}{\partial x} = e^{2x^3}\frac{\partial 2x^3}{\partial x} = 6x^2e^{2x^3}$$

- **The natural logarithm function**: $\dfrac{\partial(\ln u)}{\partial x} = \dfrac{1}{u}\dfrac{\partial u}{\partial x}$

  Example:

$$\frac{\partial[\ln(3-4x)]}{\partial x} = \frac{1}{3-4x}(-4) = \frac{-4}{3-4x} = \frac{4}{4x-3}$$

- **Differentiation formulas are usually used in the differential form**:

$$\partial(uv) = u\partial v + v\partial u \text{ and}$$

$$\partial(\ln u) = \frac{\partial u}{u}$$

ANTIDERIVATIVES OR INDEFINITE INTEGRALS

- If $f'(x) = 3x^2$
  What is $f(x)$?
  By trial and error, we can show that

$$f(x) = x^3,\ f(x) = 3x^2 - 4,\ f(x) = 3x^2 + 10 \ \text{or}$$
$$f(x) = 3x^2 + C$$

where C is any constant.

$$\text{If}: \frac{\partial F(x)}{\partial x} = f(x), \ \text{then}$$

$$\int f(x) = F(x) + C$$

- Suppose that

$$y\partial y = \frac{\partial x}{x} = \frac{1}{x}\partial x, \ \text{we know that}$$

$$y\partial y = \partial\left(\frac{1}{2}y^2\right) \ \text{and}$$

$$\frac{1}{x}\partial x = \partial(\ln x), \ \text{then}$$

$$\partial\left(\frac{1}{2}y^2\right) = \partial(\ln x), \ \text{and}$$

$$\frac{1}{2}y^2 = \ln x + C$$

This is equivalent for

$$\int y\partial y = \int \frac{1}{x}\partial x + C$$

- **Integration formula**:

$$\int a\partial u = a\int \partial u, \ c \text{ is a constant.}$$

$$\int (u+v)\partial x = \int u\partial x + \int v\partial x$$

Example:

$$\int (x^2+3)\partial x = \int x^2\partial x + \int 3\partial x = \frac{x^3}{3} + 3x + C$$

$$\text{If } n \neq -1, \int u^n \partial n = \frac{u^{n+1}}{n+1} + C$$

Example

$$\int \frac{\partial x}{x^3} = x^{-3}\partial x = \frac{x^{-2}}{-2} + C = -\frac{1}{2x^2} + C$$

$$\int u^{-1}\partial n = \int \frac{1}{u}\,\partial u = \ln|u| + C, \quad u \neq 0$$

Example:

$$\int \frac{\partial p}{p-5} = \ln|p-5| + C$$

$$\int e^u\,\partial u = e^u + C$$

$$\int e^{bu}\,\partial u = \frac{1}{b}e^{bu} + C$$

Example:

$$\int e^{2x}\,\partial x = \frac{1}{2}\int e^{2x}\,(2\partial x) = \frac{1}{2}e^{2x} + C$$

$$\int \cos\theta\,\partial\theta = \sin\theta + C$$

$$\int \sin\theta\,\partial\theta = -\cos\theta + C$$

$$\int \cos\omega\theta\,\partial\theta = \frac{1}{\omega}\sin\omega\theta + C$$

$$\int \sin\omega\theta\,\partial\theta = -\frac{1}{\omega}\cos\omega\theta + C$$

- **The definite integral**
Example:

$$\int_a^b x\,\partial x = ?$$

$$\text{since} \int x\partial x = \frac{x^2}{2} + C$$

$$\int_a^b x\,\partial x = \left(\frac{b^2}{2} + C\right) - \left(\frac{a^2}{2} + C\right) = \frac{b^2 - a^2}{2}$$

Example:

$$\int_1^4 x^{1/2}\,\partial x = \left[\frac{x^{3/2}}{3/2}\right]_1^4 = \frac{2}{3}\left[x^{3/2}\right]_1^4 = \frac{2}{3}[8-1] = \frac{14}{3}$$

Example:

$$\text{Evaluate} \int_0^1 \frac{1}{x+1} = ?$$

$$\int_0^1 \frac{1}{x+1} = [\ln(x+1)]_0^1 = \ln 2 - \ln 1 = \ln 2$$

$$\int_0^\infty x^n e^{-ax} \partial r = \frac{n!}{a^{n+1}}, \text{ then}$$

Example:

$$\int_0^\infty r^3 e^{-\frac{r}{a_0}} \partial r = \frac{3!}{\left(\frac{1}{a_0}\right)^4}$$

$$\int_0^\pi \sin^{2n+1}\theta \cdot \cos^m\theta\, \partial\theta = \int_0^\pi \cos^{2n+1}\theta \cdot \sin^m\theta\, \partial\theta = \frac{2^{n+1} \cdot n!}{(m+1)(m+3)\ldots(m+2n+1)}$$

Example:

$$\int_0^\pi \cos^2\theta \sin \partial\theta = \frac{2}{3}$$

$$\int_0^{2\pi} \partial\phi = 2\pi$$

$$\int_0^\pi \sin^{2n}\theta \cdot \cos^{2m}\theta\, \partial\theta = \frac{(2m)!(2n)!\pi}{2^{2(m+n)} m! n! (m+n)!}$$

$$\int_0^\pi \sin^{2n+1}\theta \cdot \cos^m\theta\, \delta\theta = \int_0^\pi \cos^{2n+1}\theta \cdot \sin^m\theta\, \delta\theta = \frac{2^{n+1} \cdot n!}{(m+1)(m+3)\ldots(m+2n+1)}$$

IMPORTANT MATHEMATICAL FUNCTIONS

$$e^x = 1 + x + \frac{x^2}{2!} + \frac{x^3}{3!} + \cdots\cdots\cdots\cdots = \sum_{n=0}^{\infty} \frac{x^n}{n!}$$

$$e^{\pm x} \approx (1 \pm x) \text{ only if } x \ll 1$$

$$e^{\frac{-a-bx-cx^2\ldots}{\alpha}} = \left(1 - \frac{bx}{\alpha}\right) e^{\frac{-a}{\alpha}}$$

$$e^{\pm i\phi} = \cos\phi \pm i \sin\phi$$

$$e^{i\phi} = \cos\phi + i \sin\phi = \varepsilon$$

$$e^{-i\phi} = \cos\phi - i \sin\phi = \varepsilon^*$$

$$e^{\pm in\phi} = \cos n\phi \pm i \sin n\phi$$

$$e^{\pm in\phi} = \cos n\phi \pm i \sin n\phi$$

$$e^{\pm in\phi} = (\cos\phi \pm i \sin\phi)^n$$

$$e^{i\phi} + e^{-i\phi} = 2\cos\phi$$
$$e^{i\phi} - e^{-i\phi} = 2i\sin\phi$$
$$P = \int \psi_1 \hat{M} \psi_2 d\nu = \langle \psi_1 | \hat{M} | \psi_2 \rangle$$

KRONECKER DELTA FUNCTION

$$\langle abc|F|abc\rangle = \langle a|F|a\rangle\langle b|b\rangle\langle c|c\rangle + \langle b|F|b\rangle\langle a|a\rangle\langle c|c\rangle + \langle c|F|c\rangle\langle a|a\rangle\langle b|b\rangle$$
$$\langle (a,b)|F|(c,d)\rangle = \langle a|c\rangle\langle b|F|d\rangle + \langle d|b\rangle\langle (a)|F|c\rangle$$

HERMITE POLYNOMIALS

The Hermite polynomials are generated by

$$S(\alpha,\beta) = e^{\alpha^2-(\beta-\alpha)^2} = e^{-\beta^2+2\alpha\beta}$$
$$= \sum_{n=0}^{\infty} \frac{H_n(\alpha)\beta^n}{n!}$$
$$\frac{\partial S}{\partial \alpha} = 2\beta e^{-\beta^2+2\alpha\beta} = \sum \frac{2\beta^{n+1}}{n!} H_n(\alpha)$$
$$= \sum \frac{\beta^n}{n!} H_n'(\alpha)$$

At equal power of β,

$$\frac{2(n+1)\beta^n}{n!} H_n(\alpha) = \frac{\beta^n}{n!} H_n'(\alpha)$$
$$2(n+1)H_n(\alpha) = H_n'(\alpha), \text{or}$$
$$H_n'(\alpha) = 2nH_{n-1}(\alpha)$$
$$\frac{\partial S}{\partial \beta} = (-2\beta+2\alpha)e^{-\beta^2+2\alpha\beta} = \sum_n \frac{(-2\beta+2\alpha)\beta^n}{n!} H_n(\alpha)$$
$$= \sum_n \frac{\beta^{n-1}}{(n-1)!} H_n(\alpha)$$

At equal power of β,

$$H_{n+1} = 2\alpha H_n - 2nH_{n-1}$$
$$\therefore H_n = 2\alpha H_{n-1} - 2nH_n$$

differentiated with respect to α

$$H_{n+1}'(\alpha) = 2H_n + 2\alpha H_n'(\alpha) - 2nH_{n-1}'(\alpha)$$
$$\because H_n'(\alpha) = 2nH_{n-1}(\alpha)$$
$$\therefore H_n''(\alpha) = 2nH_{n-1}'(\alpha)$$

$$\because H_n'(\alpha) = 2nH_{n-1}(\alpha)$$
$$\therefore H_{n+1}'(\alpha) = 2(n+1)H_n(\alpha)$$

$$\because H'_{n+1}(\alpha) = 2H_n + 2\alpha H'_n(\alpha) - 2nH'_{n-1}(\alpha)$$

$$\therefore 2(n+1)H_n(\alpha) = 2H_n + 2\alpha H'_n(\alpha) - H''_n(\alpha)$$

$$\therefore H''_n(\alpha) - 2\alpha H'_n(\alpha) + 2nH_n = 0$$

Also from the definition:

$$\frac{\partial^n S(\alpha,\beta)}{\partial\beta^n} = e^{\alpha^2}\frac{\partial^n}{\partial\beta^n}e^{-(\beta-\alpha)^2}$$

$$H_n(\alpha) = (-1)^n e^{\alpha^2}\frac{\partial^n}{\partial\alpha^n}e^{-\alpha^2}$$

Example:

$$H_0(\alpha) = 1$$
$$H_1(\alpha) = 2\alpha$$
$$H_2(\alpha) = 4\alpha_2 - 2$$
$$H_3(\alpha) = 8\alpha_3 - 12\alpha$$
$$H_4(\alpha) = 16\alpha_4 - 48\alpha_2 + 12$$
$$H_5(\alpha) = 32\alpha_5 - 160\alpha_3 + 120\alpha$$
$$H_6(\alpha) = 64\alpha_6 - 480\alpha_4 + 720\alpha_2 - 120$$
$$H_7(\alpha) = 128\alpha_7 - 1344\alpha_5 + 3360\alpha_3 - 1680\alpha$$

LEGENDRE POLYNOMIAL

The Legendre polynomial $p_l(\omega)$ is given by ($\omega = \cos\theta$):

$$p_l(\omega) = \frac{1}{2^l l!}\frac{\partial^l}{\partial\omega^l}(\omega^2-1)^l, \quad l = 0, 1, 2, \ldots$$

The Legendre polynomials satisfy the following equation:

$$\frac{1}{\sin\theta}\frac{\partial}{\partial\theta}\left(\sin\theta\frac{\partial p_l}{\partial\theta}\right) + l(l+1)p_l = 0$$

The normalized relation for Legendre polynomials:

$$\int_{-1}^{1}[p_l(\omega)]^2\partial\omega = \frac{2}{2l+1}$$

$$N = \left(\frac{2l+1}{2}\right)^{1/2}$$

Associated Legendre Functions

The associated Legendre polynomial functions $p_l^m(\omega)$ are given by

$$p_l^m(\omega) = (1-\omega^2)^{m/2}\frac{\partial^m}{\partial\omega^m}p_l(\omega) = \frac{(1-\omega^2)^{m/2}}{2^l l!}\frac{\partial^{l+m}}{\partial\omega^{l+m}}(\omega^2-1)^l; \quad -1 \le \omega \le 1$$

The associated Legendre polynomials satisfy the following equation:

$$\frac{1}{\sin\theta}\frac{\partial}{\partial\theta}\left(\sin\theta\frac{\partial p_l^m}{\partial\theta}\right) + \left[l(l+1) - \frac{m^2}{\sin^2\theta}\right]p_l^m = 0$$

The normalized relation for Legendre polynomials:

$$\int_{-1}^{1}\left[p_l^m(\omega)\right]^2\partial\omega=\frac{2}{2l+1}\cdot\frac{(l+m)!}{(l-m)!}$$

$$N=\left(\frac{2l+1}{2}\cdot\frac{(l-m)!}{(l+m)!}\right)^{1/2}$$

LAGUERRE POLYNOMIALS

$$xy''+(1-x)y'+ny=0$$

Associated Laguerre Polynomials

L_n^o: Laguerre polynomials,

$$L_n^o=L_n=e^x\frac{\partial^n}{\partial x^n}(e^{-x}x^n),\quad n=0,1,2,\ldots$$

L_n^k: generalized Laguerre polynomials,

$$L_n^k=(-1)^k e^x\frac{\partial^k}{\partial s^k}L_{n+k}^o;\quad n,k=0,1,2,\ldots$$

The polynomials satisfy the following equation:

$$xy''+(\alpha+1-x)y'+ny=0$$

$$\left[x\frac{\partial^2}{\partial x^2}+(k+1-x)\frac{\partial}{\partial x}+n\right]L_n^k=0$$

The orthonormality relation is

$$\int_0^\infty e^x x^k L_n^k L_m^k\,\partial x=\frac{[(n+k)!]^3}{n!}\partial_{nm}=\begin{cases}0 & m\neq n\\ 1 & m=n\end{cases}$$

The associated Laguerre polynomial of the degree $(q-k)$

$$xY''+[k+1-x]Y'+[q-k]Y=0$$

has a known solution:

$$Y=x^{\frac{(k-1)}{2}}e^{-\frac{x}{2}}L_q^k,\quad\text{where}$$

$$L_q^k=\frac{\partial^k}{\partial x^k}L_q$$

$$L_q=e^\rho\frac{\partial^q}{\partial\rho^q}(e^{-\rho}\rho^q)$$

TAYLOR SERIES

- The Taylor series is a representation of a function as an infinite sum of terms that are calculated from the values of the function's derivatives at a single point.

- The Taylor series of a real or complex-valued function $f(x)$ that is infinitely differentiable at a real or complex number a is the power series

$$f(a)+\frac{f'(a)}{1!}(x-a)+\frac{f''(a)}{2!}(x-a)^2+\frac{f'''(a)}{3!}(x-a)^3+\cdots$$

This can be written in the more compact sigma notation as

$$\sum_{n=0}^{\infty}\frac{f^{(n)}(a)}{n!}(x-a)^n$$

where $n!$ denotes the factorial of n and $f^{(n)}(a)$ denotes the nth derivative of f evaluated at the point a. The derivative of order zero of f is defined to be f itself and $(x-a)^0$ and $0!$ are both defined as 1. When $a=0$, the series is also called a Maclaurin series.

$$\text{If}: \ddot{q}_i + b_{ii}q_i = 0$$

The solution is then given by

$$q_i = q_i^0 \sin\left(\sqrt{b_{ii}}\,t + \delta_i\right)$$

where q_i^0 is the amplitude and δ_i is the phase constant.

MATRICES

- Matrix: is a regular array:

$$a=[a_{ij}]=\begin{bmatrix} a_{11} & a_{12} & \cdots & a_{1j} \\ a_{21} & a_{22} & \cdots & a_{2j} \\ \vdots & \vdots & \ddots & \vdots \\ a_{i1} & a_{i2} & \cdots & a_{ij} \end{bmatrix}$$

Shorthand for representing sets of numbers, which can be combined (+, −, ×, ÷).

- Multiplication

$$\begin{bmatrix} 2 & 0 & 0 \\ 0 & 3 & 0 \\ 0 & 0 & 5 \end{bmatrix}\begin{bmatrix} 7 \\ 2 \\ 4 \end{bmatrix}=\begin{bmatrix} 14 \\ 6 \\ 20 \end{bmatrix} \rightarrow \begin{matrix} (2\times 7)+(0\times 2)+(0\times 4)=14 \\ (0\times 7)+(3\times 2)+(0\times 4)=6 \\ (0\times 7)+(0\times 2)+(5\times 4)=20 \end{matrix}$$

$$A\times B=C$$

The number of columns in A must equal the number of rows in B.
(Matrices are conformable.)
Multiply each element in the first row of the A matrix by the corresponding element in the first column of the B matrix, then sum all these products.

$$\begin{bmatrix} 2 & 3 \\ - & - \\ - & - \end{bmatrix}\begin{bmatrix} 5 & - & - \\ 7 & - & - \end{bmatrix}=\begin{bmatrix} 31 & - & - \\ - & - & - \\ - & - & - \end{bmatrix}$$

$$\begin{bmatrix} a & c \\ - & - \\ - & - \end{bmatrix}\begin{bmatrix} - & b & - \\ - & d & - \end{bmatrix}=\begin{bmatrix} - & (a\cdot b+c\cdot d) & - \\ - & - & - \\ - & - & - \end{bmatrix}$$

$$\begin{bmatrix} a & c \\ - & - \\ - & - \end{bmatrix}\begin{bmatrix} - & - & b \\ - & - & d \end{bmatrix}=\begin{bmatrix} - & - & (a\cdot b+c\cdot d) \\ - & - & - \\ - & - & - \end{bmatrix}$$

$$\begin{bmatrix} - & - \\ a & c \\ - & - \end{bmatrix}\begin{bmatrix} b & - & - \\ d & - & - \end{bmatrix} = \begin{bmatrix} - & - & - \\ (a\cdot b+c\cdot d) & - & - \\ - & - & - \end{bmatrix}$$

$$\begin{bmatrix} - & - \\ a & c \\ - & - \end{bmatrix}\begin{bmatrix} - & b & - \\ - & d & - \end{bmatrix} = \begin{bmatrix} - & - & - \\ - & (a\cdot b+c\cdot d) & - \\ - & - & - \end{bmatrix}$$

$$\begin{bmatrix} - & - \\ a & c \\ - & - \end{bmatrix}\begin{bmatrix} - & - & b \\ - & - & d \end{bmatrix} = \begin{bmatrix} - & - & - \\ - & - & (a\cdot b+c\cdot d) \\ - & - & - \end{bmatrix}$$

$$\begin{bmatrix} - & - \\ - & - \\ a & c \end{bmatrix}\begin{bmatrix} b & - & - \\ d & - & - \end{bmatrix} = \begin{bmatrix} - & - & - \\ - & - & - \\ (a\cdot b+c\cdot d) & - & - \end{bmatrix}$$

$$\begin{bmatrix} - & - \\ - & - \\ a & c \end{bmatrix}\begin{bmatrix} - & b & - \\ - & d & - \end{bmatrix} = \begin{bmatrix} - & - & - \\ - & - & - \\ - & (a\cdot b+c\cdot d) & - \end{bmatrix}$$

$$\begin{bmatrix} - & - \\ - & - \\ a & c \end{bmatrix}\begin{bmatrix} - & - & b \\ - & - & d \end{bmatrix} = \begin{bmatrix} - & - & - \\ - & - & - \\ - & - & (a\cdot b+c\cdot d) \end{bmatrix}$$

- Terminology
  1. If number of rows = number of columns, it is a square matrix.
  2. Note that in a square matrix, the diagonal elements are (a_{11}, a_{22}, a_{33}, … a_{ij}).
  3. If all non-diagonal elements are zero, the matrix is diagonalized, said to be diagonal matrix, or the identity matrix, I.

  If A is a square matrix with n rows and n columns, and I is a matrix with the same dimensions with 1s on the main diagonal and 0s elsewhere, then

$$AI = IA = A$$

$$A \times A^{-1} = I$$

where A^{-1} is the inverse of A, note: $A^{-1} \neq \frac{1}{A}$

Example:

$$\text{Let } A = \begin{bmatrix} 1 & 2 \\ 3 & 5 \end{bmatrix}$$

Then

$$AI = IA = A$$

$$\begin{bmatrix} 1 & 2 \\ 3 & 5 \end{bmatrix}\begin{bmatrix} 1 & 0 \\ 0 & 1 \end{bmatrix} = \begin{bmatrix} 1\times 1+2\times 0 & 1\times 0+2\times 1 \\ 3\times 1+5\times 0 & 3\times 0+5\times 1 \end{bmatrix} = \begin{bmatrix} 1 & 2 \\ 3 & 5 \end{bmatrix}$$

- The inverse of the matrix

To find the inverse of the matrix A, let

$$A^{-1} = \begin{bmatrix} a & b \\ c & d \end{bmatrix}, \text{ then}$$

$$\begin{bmatrix} 1 & 2 \\ 3 & 5 \end{bmatrix}\begin{bmatrix} a & b \\ c & d \end{bmatrix} = \begin{bmatrix} 1 & 0 \\ 0 & 1 \end{bmatrix}$$

$$\begin{bmatrix} a+2c & b+2d \\ 3a+5c & 3b+5 \end{bmatrix} = \begin{bmatrix} 1 & 0 \\ 0 & 1 \end{bmatrix}$$

Therefore,

$$a + 2c = 1$$

$$3a + 5c = 0$$

$$\therefore a = -5, c = 3$$

$$b + 2d = 0$$

$$3b + 5 = 1$$

$$\therefore b = 2, d = -1$$

$$\therefore A^{-1} = \begin{bmatrix} a & b \\ c & d \end{bmatrix} = \begin{bmatrix} -5 & 2 \\ 3 & -1 \end{bmatrix}$$

Example:

$$\text{If}: A = \begin{bmatrix} \cos\theta & -\sin\theta & 0 \\ \sin\theta & \cos\theta & 0 \\ 0 & 0 & 1 \end{bmatrix}, \text{ and } B = \begin{bmatrix} \cos\theta & \sin\theta & 0 \\ -\sin\theta & \cos\theta & 0 \\ 0 & 0 & 1 \end{bmatrix}$$

Show that B is the inverse of A.
Answer:

$$AB = \begin{bmatrix} 1 & 0 & 0 \\ 0 & 1 & 0 \\ 0 & 0 & 1 \end{bmatrix}$$

$$AB = \begin{bmatrix} \cos\theta & -\sin\theta & 0 \\ \sin\theta & \cos\theta & 0 \\ 0 & 0 & 1 \end{bmatrix} \begin{bmatrix} \cos\theta & \sin\theta & 0 \\ -\sin\theta & \cos\theta & 0 \\ 0 & 0 & 1 \end{bmatrix}$$

$$AB = \begin{bmatrix} \cos^2\theta + \sin^2\theta + 0 & \sin\theta\cos\theta - \sin\theta\cos\theta + 0 & 0 \\ \sin\theta\cos\theta - \sin\theta\cos\theta + 0 & \cos^2\theta + \sin^2\theta + 0 & 0 \\ 0 & 0 & 1 \end{bmatrix} = \begin{bmatrix} 1 & 0 & 0 \\ 0 & 1 & 0 \\ 0 & 0 & 1 \end{bmatrix}$$

$$AB = \begin{bmatrix} 1 & 0 & 0 \\ 0 & 1 & 0 \\ 0 & 0 & 1 \end{bmatrix}$$

$$\therefore B = A^{-1}$$

4. The sum of diagonal elements is called the trace or character of the matrix.
5. Block diagonalization: a matrix can be broken into blocks lying along the diagonal.

$$\begin{bmatrix} \begin{bmatrix} 2 & 3 \\ 1 & 2 \end{bmatrix} & \begin{matrix} 0 \\ 0 \end{matrix} & \begin{matrix} 0 & 0 & 0 \\ 0 & 0 & 0 \end{matrix} \\ \begin{matrix} 0 & 0 \end{matrix} & [1] & \begin{matrix} 0 & 0 & 0 \end{matrix} \\ \begin{matrix} 0 & 0 & 0 \\ 0 & 0 & 0 \\ 0 & 0 & 0 \end{matrix} & & \begin{bmatrix} 1 & 2 & 3 \\ 3 & 2 & 1 \\ 0 & 1 & 2 \end{bmatrix} \end{bmatrix}$$

The product of two block diagonal matrices with similar arrays of blocks is another matrix obtained by multiplying the individual blocks:

$$\begin{bmatrix} \begin{bmatrix} 2 & 3 \\ 1 & 2 \end{bmatrix} & \begin{matrix} 0 \\ 0 \end{matrix} & \begin{matrix} 0 & 0 & 0 \\ 0 & 0 & 0 \end{matrix} \\ \begin{matrix} 0 & 0 \end{matrix} & [1] & \begin{matrix} 0 & 0 & 0 \end{matrix} \\ \begin{matrix} 0 & 0 & 0 \\ 0 & 0 & 0 \\ 0 & 0 & 0 \end{matrix} & & \begin{bmatrix} 1 & 2 & 3 \\ 3 & 2 & 1 \\ 0 & 1 & 2 \end{bmatrix} \end{bmatrix} \begin{bmatrix} \begin{bmatrix} 0 & 2 \\ 1 & 2 \end{bmatrix} & \begin{matrix} 0 \\ 0 \end{matrix} & \begin{matrix} 0 & 0 & 0 \\ 0 & 0 & 0 \end{matrix} \\ \begin{matrix} 0 & 0 \end{matrix} & [3] & \begin{matrix} 0 & 0 & 0 \end{matrix} \\ \begin{matrix} 0 & 0 & 0 \\ 0 & 0 & 0 \\ 0 & 0 & 0 \end{matrix} & & \begin{bmatrix} 2 & 0 & 0 \\ 1 & 0 & 2 \\ 0 & 1 & 1 \end{bmatrix} \end{bmatrix} = \begin{bmatrix} \begin{bmatrix} 3 & 10 \\ 2 & 6 \end{bmatrix} & \begin{matrix} 0 \\ 0 \end{matrix} & \begin{matrix} 0 & 0 & 0 \\ 0 & 0 & 0 \end{matrix} \\ \begin{matrix} 0 & 0 \end{matrix} & [3] & \begin{matrix} 0 & 0 & 0 \end{matrix} \\ \begin{matrix} 0 & 0 & 0 \\ 0 & 0 & 0 \\ 0 & 0 & 0 \end{matrix} & & \begin{bmatrix} 4 & 3 & 7 \\ 8 & 1 & 5 \\ 1 & 2 & 4 \end{bmatrix} \end{bmatrix}$$

We can use matrices as representations of symmetry operations.
A matrix also affords a way of representing a vector.

- Matrix and linear equations

  How to go from a set of linear equations to a matrix and solve the matrix equation:

  Example:

  Write the following equations as a matrix equation and then solve the matrix equation:

$$5x+2y-z=-7$$
$$x-2y+2z=0$$
$$3y+z=17$$

Answer:

$$\underbrace{\left.\begin{aligned}5x+2y-z&=-7\\x-2y+2z&=0\\3y+z&=17\end{aligned}\right\}}_{\text{linear equations}}\underbrace{\begin{bmatrix}5&2&-1\\1&-2&2\\0&3&1\end{bmatrix}\begin{bmatrix}x\\y\\z\end{bmatrix}=\begin{bmatrix}-7\\0\\17\end{bmatrix}}_{\text{matrix equation}}$$

If

$$\begin{bmatrix}5&2&-1\\1&-2&2\\0&3&1\end{bmatrix}\begin{bmatrix}x\\y\\z\end{bmatrix}=\begin{bmatrix}-7\\0\\17\end{bmatrix},\text{ then}$$

$$\begin{bmatrix}x\\y\\z\end{bmatrix}=\frac{\begin{bmatrix}-7\\0\\17\end{bmatrix}}{\begin{bmatrix}5&2&-1\\1&-2&2\\0&3&1\end{bmatrix}}=\begin{bmatrix}-7\\0\\17\end{bmatrix}\begin{bmatrix}5&2&-1\\1&-2&2\\0&3&1\end{bmatrix}^{-1}$$

To find $\begin{bmatrix}5&2&-1\\1&-2&2\\0&3&1\end{bmatrix}^{-1}$,

$$\begin{bmatrix}a&b&c\\e&f&g\\h&i&k\end{bmatrix}\begin{bmatrix}5&2&-1\\1&-2&2\\0&3&1\end{bmatrix}=\begin{bmatrix}1&0&0\\0&1&0\\0&0&1\end{bmatrix}$$

$$\begin{bmatrix}5a+b&2a-2b+3c&-a+2b+c\\5e+f&2e-2f+3g&-e+2f+g\\5h+i&2h-2i+3k&-h+2i+k\end{bmatrix}=\begin{bmatrix}1&0&0\\0&1&0\\0&0&1\end{bmatrix}$$

$$5a+b=1$$

$$2a-2b+3c=0$$

$$-a+2b+c=0$$

$A=8/45$, $b=1/9$, $c=-2/45$

$$5e+f=0$$

$$2e-2f+3g=1$$

$$-e+2f+g=0$$

$e=1/45$, $f=-1/9$, $g=11/45$

$$5h+i=0$$

$$2h-2i+3k=0$$

$$-h+2i+k=1$$

$h = -1/15,\ i = 1/3,\ k = 4/15$

$$\begin{bmatrix} x \\ y \\ z \end{bmatrix} = \begin{bmatrix} 8/45 & 1/9 & -2/45 \\ 1/45 & -1/9 & 11/45 \\ -1/15 & 1/3 & 4/15 \end{bmatrix} \begin{bmatrix} -7 \\ 0 \\ 17 \end{bmatrix} = \begin{bmatrix} -2 \\ 4 \\ 5 \end{bmatrix}$$

$$\begin{bmatrix} x \\ y \\ z \end{bmatrix} = \begin{bmatrix} -2 \\ 4 \\ 5 \end{bmatrix}$$

$$\therefore x = -2, y = 4, \text{ and } z = 5$$

DETERMINANTS

- Each square matrix can be assigned a real number, called the determinant of the matrix. A determinant of nth order is a square $n \times n$ array of numbers enclosed by vertical lines.
- The following scheme gives the correct expansions for determinants of order 2, 3, and N, respectively.

If

$$a_1 x + b_1 y = 0,$$

$$a_2 x + b_2 y = 0, \text{ then}$$

$$\begin{vmatrix} a_1 & b_1 \\ a_2 & b_2 \end{vmatrix} = a_1 b_2 - b_1 a_2 = 0$$

If

$$a_1 x + b_1 y + c_1 z = 0,$$

$$a_2 x + b_2 y + c_2 z = 0,$$

$$a_3 x + b_3 y + c_3 z = 0, \text{ then}$$

$$\begin{vmatrix} a_1 & b_1 & c_1 \\ a_2 & b_2 & c_2 \\ a_3 & b_3 & c_3 \end{vmatrix} = a_1 \begin{vmatrix} b_2 & c_2 \\ b_3 & c_3 \end{vmatrix} - b_1 \begin{vmatrix} a_2 & c_2 \\ a_3 & c_3 \end{vmatrix} + c_1 \begin{vmatrix} a_2 & b_2 \\ a_3 & b_3 \end{vmatrix}$$

- When N atoms are combined, there are N combined secular determinants:

$$\begin{vmatrix} x-\alpha & \beta & 0 & 0 & 0 & \cdots & 0 \\ \beta & x-\alpha & \beta & 0 & 0 & \cdots & 0 \\ 0 & \beta & x-\alpha & \beta & 0 & \cdots & 0 \\ 0 & 0 & \beta & x-\alpha & \beta & \cdots & 0 \\ \vdots & \vdots & \vdots & \vdots & \vdots & \cdots & \vdots \\ 0 & 0 & 0 & 0 & 0 & \cdots & E-x \end{vmatrix} = 0$$

where β is the resonance integral, the roots of this symmetric tridiagonal determinant follow the expression

$$x_k = \alpha + 2\beta \cos\left(\frac{k\pi}{N+1}\right), \quad k = 1, 2, \ldots, N$$

If N is infinity large, then $\dfrac{\pi}{N+1} \cong 0$ and $\dfrac{N\pi}{N+1} \cong \pi$:

$$x_1 = \alpha + 2\beta \cos\left(\frac{\pi}{N+1}\right) = \alpha + 2\beta$$

$$x_N = \alpha + 2\beta \cos\left(\frac{N\pi}{N+1}\right) = \alpha - 2\beta$$

Character Tables

NONAXIAL GROUPS

C_1	E
A	1

C_s	E	σ_h		
A'	1	1	x, y, R_z	x^2, y^2, z^2, xy
A''	1	−1	z, R_x, R_y	yz, xz
$\Gamma_{x, y, z}$	3	1		

C_i	E	i		
A_g	1	1	R_x, R_y, R_z	$x^2, y^2, z^2, xy, yz, xz$
A_u	1	−1	x, y, z	
$\Gamma_{x, y, z}$	3	−3		

C_n GROUPS

C_2	E	C_2		
A	1	1	z, R_z	x^2, y^2, z^2, xy
B	1	−1	x, y, R_x, R_y	yz, xz
$\Gamma_{x, y, z}$	3	−1		

C_3	E	C_3	C_3^2		$\varepsilon = e^{2\pi i/3}$
A	1	1	1	z, R_z	x^2+y^2, z^2
E	$\begin{cases}1\\1\end{cases}$	ε ε^*	ε^* ε	$(x, y), (R_x, R_y)$	$(x^2-y^2, xy), (yz, xz)$
$\Gamma_{x, y, z}$	3	0	0		

C_4	E	C_4	C_2	C_4^3		
A	1	1	1	1	z, R_z	x^2+y^2, z^2
B	1	−1	1	−1		x^2-y^2, xy
E	$\begin{cases}1\\1\end{cases}$	i $-i$	−1 −1	$-i$ i	$(x, y), (R_x, R_y)$	(yz, xz)
$\Gamma_{x, y, z}$	3	1	−1	1		

C_5	E	C_5	C_5^2	C_5^3	C_5^4		$\varepsilon = e^{2\pi i/5}$
A	1	1	1	1	1	z, R_z	x^2+y^2, z^2
E_1	$\left\{\begin{matrix}1\\1\end{matrix}\right.$	$\begin{matrix}\varepsilon\\\varepsilon^*\end{matrix}$	$\begin{matrix}\varepsilon^2\\\varepsilon^{2*}\end{matrix}$	$\begin{matrix}\varepsilon^{2*}\\\varepsilon^2\end{matrix}$	$\left.\begin{matrix}\varepsilon^*\\\varepsilon\end{matrix}\right\}$	$(x, y), (R_x, R_y)$	(yz, xz)
E_2	$\left\{\begin{matrix}1\\1\end{matrix}\right.$	$\begin{matrix}\varepsilon^2\\\varepsilon^{2*}\end{matrix}$	$\begin{matrix}\varepsilon^*\\\varepsilon\end{matrix}$	$\begin{matrix}\varepsilon\\\varepsilon^*\end{matrix}$	$\left.\begin{matrix}\varepsilon^{2*}\\\varepsilon^2\end{matrix}\right\}$		(x^2-y^2, xy)
$\Gamma_{x,y,z}$	3	$1+2c\frac{2\pi}{5}$	$1+2c\frac{4\pi}{5}$	$1+2c\frac{4\pi}{5}$	$1+2c\frac{2\pi}{5}$		

C_6	E	C_6	C_3	C_2	C_3^2	C_6^5		$\varepsilon = e^{2\pi i/6}$
A	1	1	1	1	1	1	z, R_z	x^2+y^2, z^2
B	1	-1	1	-1	1	-1		
E_1	$\left\{\begin{matrix}1\\1\end{matrix}\right.$	$\begin{matrix}\varepsilon\\\varepsilon^*\end{matrix}$	$\begin{matrix}-\varepsilon^*\\-\varepsilon\end{matrix}$	$\begin{matrix}-1\\-1\end{matrix}$	$\begin{matrix}-\varepsilon\\-\varepsilon^*\end{matrix}$	$\left.\begin{matrix}\varepsilon^*\\\varepsilon\end{matrix}\right\}$	$(x, y), (R_x, R_y)$	(yz, xz)
E_2	$\left\{\begin{matrix}1\\1\end{matrix}\right.$	$\begin{matrix}-\varepsilon^*\\-\varepsilon\end{matrix}$	$\begin{matrix}-\varepsilon\\-\varepsilon*\end{matrix}$	$\begin{matrix}1\\1\end{matrix}$	$\begin{matrix}-\varepsilon^*\\-\varepsilon\end{matrix}$	$\left.\begin{matrix}-\varepsilon\\-\varepsilon^*\end{matrix}\right\}$		(x^2-y^2, xy)
$\Gamma_{x,y,z}$	3	2	0	-1	0	2		

C_{nv} GROUPS

C_{2v}	E	C_2	$\sigma_{(xz)}$	$\sigma_{(yz)}$		
A_1	1	1	1	1	z	x^2, y^2, z^2
A_2	1	1	-1	-1	R_z	xy
B_1	1	-1	1	-1	x, R_y	xz
B_2	1	-1	-1	1	y, R_x	yz
$\Gamma_{x,y,z}$	3	-1	1	1		

C_{3v}	E	$2C_3$	$3\sigma_v$		
A_1	1	1	1	z	x^2+y^2, z^2
A_2	1	1	-1	R_z	
E	2	-1	0	$(x, y), (R_x, R_y)$	$(x^2-y^2, xy), (xz, yz)$
$\Gamma_{x,y,z}$	3	0	1		

C_{4v}	E	$2C_4$	C_2	$2\sigma_v$	$2\sigma_d$		
A_1	1	1	1	1	1	z	x^2+y^2, z^2
A_2	1	1	1	-1	-1	R_z	
B_1	1	-1	1	1	-1		x^2-y^2
B_2	1	-1	1	-1	1		xy
E	2	0	-2	0	0	$(x, y), (R_x, R_y)$	(xz, yz)
$\Gamma_{x,y,z}$	3	1	-1	1	1		

C_{5v}	E	$2C_5$	$2C_5^2$	$5\sigma_v$		
A_1	1	1	1	1	z	x^2+y^2, z^2
A_2	1	1	1	-1	R_z	
E_1	2	$2c\frac{2\pi}{5}$	$2c\frac{4\pi}{5}$	0	$(x, y), (R_x, R_y)$	(xz, yz)
E_2	2	$2c\frac{4\pi}{5}$	$2c\frac{2\pi}{5}$	0		(x^2-y^2, xy)
$\Gamma_{x,y,z}$	3	$1+2c\frac{2\pi}{5}$	$1+2c\frac{4\pi}{5}$	1		

C_{6v}	E	$2C_6$	$2C_3$	C_2	$3\sigma_v$	$3\sigma_d$		
A_1	1	1	1	1	1	1	z	x^2+y^2, z^2
A_2	1	1	1	1	−1	−1	R_z	
B_1	1	−1	1	−1	1	−1		
B_2	1	−1	1	−1	−1	1		
E_1	2	1	−1	−2	0	0	$(x, y), (R_x, R_y)$	(xz, yz)
E_2	2	−1	−1	2	0	0		(x^2-y^2, xy)
$\Gamma_{x,y,z}$	3	2	0	−1	1	1		

C_{nh} GROUPS

C_{2h}	E	C_2	i	σ_h		
A_g	1	1	1	1	R_z	x^2, y^2, z^2, xy
B_g	1	−1	1	−1	R_x, R_y	xz, yz
A_u	1	1	−1	−1	z	
B_u	1	−1	−1	1	x, y	
$\Gamma_{x,y,z}$	3	−1	−3	1		

C_{3h}	E	C_3	C_3^2	σ_h	S_3	S_3^5		$\varepsilon = e^{2\pi i/3}$
A'	1	1	1	1	1	1	R_z	x^2+y^2, z^2
E'	1	ε	ε^*	1	ε	ε^*	(x, y)	(x^2-y^2, xy)
	1	ε^*	ε	1	ε^*	ε		
A''	1	1	1	−1	−1	−1	z	
E''	1	ε	ε^*	−1	$-\varepsilon$	$-\varepsilon^*$	(R_x, R_y)	(yz, xz)
	1	ε^*	ε	−1	$-\varepsilon^*$	$-\varepsilon$		
$\Gamma_{x,y,z}$	3	0	0	1	−2	−2		

C_{4h}	E	C_4	C_2	C_4^3	i	S_4^3	σ_h	S_4		
A_g	1	1	1	1	1	1	1	1	R_z	x^2+y^2, z^2
B_g	1	−1	1	−1	1	−1	1	−1		x^2-y^2, xy
E_g	1	i	−1	$-i$	1	i	−1	$-i$	(R_x, R_y)	(yz, xz)
	1	$-i$	−1	i	1	$-i$	−1	i		
A_u	1	1	1	1	−1	−1	−1	−1	z	
B_u	1	−1	1	−1	−1	1	−1	1		
E_u	1	i	−1	$-i$	−1	$-i$	1	i	(x, y)	
	1	$-i$	−1	i	−1	i	1	$-i$		
$\Gamma_{x,y,z}$	3	1	−1	1	−3	−1	1	−1		

C_{5h}	E	C_5	C_5^2	C_5^3	C_5^4	σ_h	S_5	S_5^7	S_5^3	S_5^9		$\varepsilon = e^{2\pi i/5}$
A'	1	1	1	1	1	1	1	1	1	1	R_z	x^2+y^2, z^2
E_1'	1	ε	ε^2	ε^{2*}	ε^*	1	ε	ε^2	ε^{2*}	ε^*	(x, y)	
	1	ε^*	ε^{2*}	ε^2	ε	1	ε^*	ε^{2*}	ε^2	ε		
E_2'	1	ε^2	ε^*	ε	ε^{2*}	1	ε^2	ε^*	ε	ε^{2*}		(x^2-y^2, xy)
	1	ε^{2*}	ε	ε^*	ε^2	1	ε^{2*}	ε	ε^*	ε^2		
A''	1	1	1	1	1	−1	−1	−1	−1	−1	z	
E_1''	1	ε	ε^2	ε^{2*}	ε^*	−1	$-\varepsilon$	$-\varepsilon^2$	$-\varepsilon^{2*}$	$-\varepsilon^*$	(R_x, R_y)	(yz, xz)
	1	ε^*	ε^{2*}	ε^2	ε	−1	$-\varepsilon^*$	$-\varepsilon^{2*}$	$-\varepsilon^2$	$-\varepsilon$		
E_2''	1	ε^2	ε^*	ε	ε^{2*}	−1	$-\varepsilon^2$	$-\varepsilon^*$	$-\varepsilon$	$-\varepsilon^{2*}$		
	1	ε^{2*}	ε	ε^*	ε^2	−1	$-\varepsilon^{2*}$	$-\varepsilon$	$-\varepsilon^*$	$-\varepsilon^2$		
$\Gamma_{x,y,z}$	3	α	β	β	α	1	γ	δ	δ	γ		

$$\alpha = 1 + 2c\frac{2\pi}{5} \qquad \gamma = -1 + 2c\frac{2\pi}{5}$$

$$\beta = 1 + 2c\frac{4\pi}{5} \qquad \delta = -1 + 2c\frac{4\pi}{5}$$

C_{6h}	E	C_6	C_3	C_2	C_3^2	C_6^5	i	S_3^5	S_6^5	σ_h	S_6	S_3		$\varepsilon = e^{2\pi i/6}$
A_g	1	1	1	1	1	1	1	1	1	1	1	1	R_z	x^2+y^2, z^2
B_g	1	−1	1	−1	1	−1	1	−1	1	−1	1	−1		
E_{1g}	1	ε	$-\varepsilon^*$	−1	$-\varepsilon$	ε^*	1	ε	$-\varepsilon^*$	−1	$-\varepsilon$	ε^*	(R_x, R_y)	(yz, xz)
	1	ε^*	$-\varepsilon$	−1	$-\varepsilon^*$	ε	1	ε^*	$-\varepsilon$	−1	$-\varepsilon^*$	ε		
E_{2g}	1	$-\varepsilon^*$	$-\varepsilon$	1	$-\varepsilon^*$	$-\varepsilon$	1	$-\varepsilon^*$	$-\varepsilon$	1	$-\varepsilon^*$	$-\varepsilon$		(x^2-y^2, xy)
	1	$-\varepsilon$	$-\varepsilon^*$	1	$-\varepsilon$	$-\varepsilon^*$	1	$-\varepsilon$	$-\varepsilon^*$	1	$-\varepsilon$	$-\varepsilon^*$		
A_u	1	1	1	1	1	1	−1	−1	−1	−1	−1	−1	z	
B_u	1	−1	1	−1	1	−1	−1	1	−1	1	−1	1		
E_{1u}	1	ε	$-\varepsilon^*$	−1	$-\varepsilon$	ε^*	−1	$-\varepsilon$	ε^*	1	ε	$-\varepsilon^*$	(x, y)	
	1	ε^*	$-\varepsilon$	−1	$-\varepsilon^*$	ε	−1	$-\varepsilon^*$	ε	1	ε^*	$-\varepsilon$		
E_{2u}	1	$-\varepsilon^*$	$-\varepsilon$	1	$-\varepsilon^*$	$-\varepsilon$	−1	ε^*	ε	−1	ε^*	ε		
	1	$-\varepsilon$	$-\varepsilon^*$	1	$-\varepsilon$	$-\varepsilon^*$	−1	ε	ε^*	−1	ε	ε^*		
$\Gamma_{x, y, z}$	3	2	0	−1	0	2	−3	−2	0	1	0	−2		

D_n GROUPS

D_2	E	$C_2\ (z)$	$C_2\ (y)$	$C_2\ (x)$		
A	1	1	1	1		x^2, y^2, z^2
B_1	1	1	−1	−1	z, R_z	xy
B_2	1	−1	1	−1	y, R_y	xz
B_3	1	−1	−1	1	x, R_x	yz
$\Gamma_{x, y, z}$	3	−1	−1	−1		

D_3	E	$2C_3$	$3C_2$		
A_1	1	1	1		x^2+y^2, z^2
A_2	1	1	−1	z, R_z	
E	2	−1	0	$(x, y), (R_x, R_y)$	$(x^2-y^2, xy), (xz, yz)$
$\Gamma_{x, y, z}$	3	0	−1		

D_4	E	$2C_4$	$C_2\ (=C_4^2)$	$2C_2'$	$2C_2''$		
A_1	1	1	1	1	1		x^2+y^2, z^2
A_2	1	1	1	−1	−1	z, R_z	
B_1	1	−1	1	1	−1		x^2-y^2
B_2	1	−1	1	−1	1		xy
E	2	0	−2	0	0	$(x, y), (R_x, R_y)$	(yz, xz)
$\Gamma_{x, y, z}$	3	1	−1	−1	−1		

D_5	E	$2C_5$	$2C_5^2$	$5C_2$		
A_1	1	1	1	1		x^2+y^2, z^2
A_2	1	1	1	−1	z, R_z	
E_1	2	$2c\frac{2\pi}{5}$	$2c\frac{4\pi}{5}$	0	$(x, y), (R_x, R_y)$	(xz, yz)
E_2	2	$2c\frac{4\pi}{5}$	$2c\frac{2\pi}{5}$	0		(x^2-y^2, xy)
$\Gamma_{x, y, z}$	3	$1+2c\frac{2\pi}{5}$	$1+2c\frac{4\pi}{5}$	−1		

D_6	E	$2C_6$	$2C_3$	C_2	$3C_2'$	$3C_2''$		
A_1	1	1	1	1	1	1		x^2+y^2, z^2
A_2	1	1	1	1	−1	−1	z, R_z	
B_1	1	−1	1	−1	1	−1		
B_2	1	−1	1	−1	−1	1		
E_1	2	1	−1	−2	0	0	$(x, y), (R_x, R_y)$	(xz, yz)
E_2	2	−1	−1	2	0	0		(x^2-y^2, xy)
$\Gamma_{x, y, z}$	3	2	0	−1	−1	−1		

D_{nd} GROUPS

D_{2d}	E	$2S_4$	C_2	$2C_2'$	$2\sigma_d$		
A_1	1	1	1	1	1		x^2+y^2, z^2
A_2	1	1	1	−1	−1	R_z	
B_1	1	−1	1	1	−1		x^2-y^2
B_2	1	−1	1	−1	1	z	xy
E	2	0	−2	0	0	$(x, y), (R_x, R_y)$	(xz, yz)
$\Gamma_{x, y, z}$	3	−1	−1	−1	1		

D_{3d}	E	$2C_3$	$3C_2$	i	$2S_6$	$3\sigma_d$		
A_{1g}	1	1	1	1	1	1		x^2+y^2, z^2
A_{2g}	1	1	−1	1	1	−1	R_z	
E_g	2	−1	0	2	−1	0	(R_x, R_y)	$(x^2-y^2, xy), (xz, yz)$
A_{1u}	1	1	1	−1	−1	−1		
A_{2u}	1	1	−1	−1	−1	1	z	
E_u	2	−1	0	−2	1	0	(x, y)	
$\Gamma_{x, y, z}$	3	0	−1	−3	0	1		

D_{4d}	E	$2S_8$	$2C_4$	$2S_8^3$	C_2	$4C_2'$	$4\sigma_d$		
A_1	1	1	1	1	1	1	1		x^2+y^2, z^2
A_2	1	1	1	1	1	−1	−1	R_z	
B_1	1	−1	1	−1	1	1	−1		
B_2	1	−1	1	−1	1	−1	1	z	
E_1	2	$\sqrt{2}$	0	$-\sqrt{2}$	−2	0	0	(x, y)	
E_2	2	0	−2	0	2	0	0		(x^2-y^2, xy)
E_3	2	$-\sqrt{2}$	0	$\sqrt{2}$	−2	0	0	(R_x, R_y)	(xz, yz)
$\Gamma_{x, y, z}$	3	$-1+\sqrt{2}$	1	$-1-\sqrt{2}$	−1	−1	1		

D_{5d}	E	$2C_5$	$2C_5^2$	$5C_2$	i	$2S_{10}^3$	$2S_{10}$	$5\sigma_d$		
A_{1g}	1	1	1	1	1	1	1	1		x^2+y^2, z^2
A_{2g}	1	1	1	−1	1	1	1	−1	R_z	
E_{1g}	2	$2c\frac{2\pi}{5}$	$2c\frac{4\pi}{5}$	0	2	$2c\frac{2\pi}{5}$	$2c\frac{4\pi}{5}$	0	(R_x, R_y)	(xz, yz)
E_{2g}	2	$2c\frac{4\pi}{5}$	$2c\frac{2\pi}{5}$	0	2	$2c\frac{4\pi}{5}$	$2c\frac{2\pi}{5}$	0		(x^2-y^2, xy)
A_{1u}	1	1	1	1	−1	−1	−1	−1		
A_{2u}	1	1	1	−1	−1	−1	−1	1	z	
E_{1u}	2	$2c\frac{2\pi}{5}$	$2c\frac{4\pi}{5}$	0	−2	$-2c\frac{2\pi}{5}$	$-2c\frac{4\pi}{5}$	0	(x, y)	
E_{2u}	2	$2c\frac{4\pi}{5}$	$2c\frac{2\pi}{5}$	0	−2	$-2c\frac{4\pi}{5}$	$-2c\frac{2\pi}{5}$	0		
$\Gamma_{x, y, z}$	3	$1+2c\frac{2\pi}{5}$	$1+2c\frac{4\pi}{5}$	−1	−3	$-1-2c\frac{2\pi}{5}$	$-1-2c\frac{4\pi}{5}$	1		

D_{6d}	E	$2S_{12}$	$2C_6$	$2S_4$	$2C_3$	$2S_{12}^5$	C_2	$6C_2'$	$6\sigma_d$		
A_1	1	1	1	1	1	1	1	1	1		x^2+y^2, z^2
A_2	1	1	1	1	1	1	1	−1	−1	R_z	
B_1	1	−1	1	−1	1	−1	1	1	−1		
B_2	1	−1	1	−1	1	−1	1	−1	1	z	
E_1	2	$\sqrt{3}$	1	0	−1	$-\sqrt{3}$	−2	0	0	(x, y)	
E_2	2	1	−1	−2	−1	1	2	0	0		(x^2-y^2, xy)
E_3	2	0	−2	0	2	0	−2	0	0		
E_4	2	−1	−1	2	−1	−1	2	0	0		
E_5	2	$-\sqrt{3}$	1	0	−1	$\sqrt{3}$	−2	0	0	(R_x, R_y)	(xz, yz)
$\Gamma_{x, y, z}$	3	$-1+\sqrt{3}$	2	−1	0	$-1-\sqrt{3}$	−1	−1	1		

D_{nh} GROUPS

D_{2h}	E	$2C_2\ (z)$	$2C_2\ (y)$	$2C_2\ (x)$	i	$\sigma(xy)$	$\sigma(xz)$	$\sigma(yz)$		
A_g	1	1	1	1	1	1	1	1		x^2, y^2, z^2
B_{1g}	1	1	−1	−1	1	1	−1	−1	R_z	xy
B_{2g}	1	−1	1	−1	1	−1	1	−1	R_y	xz
B_{3g}	1	−1	−1	1	1	−1	−1	1	R_x	yz
A_u	1	1	1	1	−1	−1	−1	−1		
B_{1u}	1	1	−1	−1	−1	−1	1	1	z	
B_{2u}	1	−1	1	−1	−1	1	−1	1	y	
B_{3u}	1	−1	−1	1	−1	1	1	−1	x	
$\Gamma_{x, y, z}$	3	−1	−1	−1	−3	1	1	1		

D_{3h}	E	$2C_3$	$3C_2$	σ_h	$2S_3$	$3\sigma_v$		
A'_1	1	1	1	1	1	1		x^2+y^2, z^2
A'_2	1	1	−1	1	1	−1	R_z	
E'	2	−1	0	2	−1	0	(x, y)	x^2-y^2, xy
A''_1	1	1	1	−1	−1	−1		
A''_2	1	1	−1	−1	−1	1	z	
E''	2	−1	0	−2	1	0	(R_x, R_y)	(xz, yz)
$\Gamma_{x, y, z}$	3	0	−1	1	−2	1		

D_{4h}	E	$2C_4$	C_2	$2C'_2$	$2C''_2$	i	$2S_4$	σ_h	$2\sigma_v$	$2\sigma_d$		
A_{1g}	1	1	1	1	1	1	1	1	1	1		x^2+y^2, z^2
A_{2g}	1	1	1	−1	−1	1	1	1	−1	−1	R_z	
B_{1g}	1	−1	1	1	−1	1	−1	1	1	−1		x^2-y^2
B_{2g}	1	−1	1	−1	1	1	−1	1	−1	1		xy
E_g	2	0	−2	0	0	2	0	−2	0	0	(R_x, R_y)	(xz, yz)
A_{1u}	1	1	1	1	1	−1	−1	−1	−1	−1		
A_{2u}	1	1	1	−1	−1	−1	−1	−1	1	1	z	
B_{1u}	1	−1	1	1	−1	−1	1	−1	−1	1		
B_{2u}	1	−1	1	−1	1	−1	1	−1	1	−1		
E_u	2	0	−2	0	0	−2	0	2	0	0	(x, y)	
$\Gamma_{x, y, z}$	3	1	−1	−1	−1	−3	−1	1	1	1		

D_{5h}	E	$2C_5$	$2C_5^2$	$5C_2$	σ_h	$2S_5$	$2S_5^3$	$5\sigma_v$		
A_1'	1	1	1	1	1	1	1	1		x^2+y^2, z^2
A_2'	1	1	1	−1	1	1	1	−1	R_z	
E_1'	2	$2c\frac{2\pi}{5}$	$2c\frac{4\pi}{5}$	0	2	$2c\frac{2\pi}{5}$	$2c\frac{4\pi}{5}$	0	(x, y)	
E_2'	2	$2c\frac{4\pi}{5}$	$2c\frac{2\pi}{5}$	0	2	$2c\frac{4\pi}{5}$	$2c\frac{2\pi}{5}$	0		(x^2-y^2, xy)
A_1''	1	1	1	1	−1	−1	−1	−1		
A_2''	1	1	1	−1	−1	−1	−1	1	z	
E_1''	2	$2c\frac{2\pi}{5}$	$2c\frac{4\pi}{5}$	0	−2	$-2c\frac{2\pi}{5}$	$-2c\frac{4\pi}{5}$	0	(R_x, R_y)	(xz, yz)
E_2''	2	$2c\frac{4\pi}{5}$	$2c\frac{2\pi}{5}$	0	−2	$-2c\frac{4\pi}{5}$	$-2c\frac{2\pi}{5}$	0		
$\Gamma_{x, y, z}$	3	$1+2c\frac{2\pi}{5}$	$1+2c\frac{4\pi}{5}$	−1	1	$-1+2c\frac{2\pi}{5}$	$-1+2c\frac{4\pi}{5}$	1		

D_{6h}	E	$2C_6$	$2C_3$	C_2	$3C_2'$	$3C_2''$	i	$2S_3$	$2S_6$	σ_h	$3\sigma_d$	$3\sigma_v$		
A_{1g}	1	1	1	1	1	1	1	1	1	1	1	1		x^2+y^2, z^2
A_{2g}	1	1	1	1	−1	−1	1	1	1	1	−1	−1	R_z	
B_{1g}	1	−1	1	−1	1	−1	1	−1	1	−1	1	−1		
B_{2g}	1	−1	1	−1	−1	1	1	−1	1	−1	−1	1		
E_{1g}	2	1	−1	−2	0	0	2	1	−1	−2	0	0	(R_x, R_y)	(xz, yz)
E_{2g}	2	−1	−1	2	0	0	2	−1	−1	2	0	0		(x^2-y^2, xy)
A_{1u}	1	1	1	1	1	1	−1	−1	−1	−1	−1	−1		
A_{2u}	1	1	1	1	−1	−1	−1	−1	−1	−1	1	1	z	
B_{1u}	1	−1	1	−1	1	−1	−1	1	−1	1	−1	1		
B_{2u}	1	−1	1	−1	−1	1	−1	1	−1	1	1	−1		
E_{1u}	2	1	−1	−2	0	0	−2	−1	1	2	0	0	(x, y)	
E_{2u}	2	−1	−1	2	0	0	−2	1	1	−2	0	0		
$\Gamma_{x, y, z}$	3	2	0	−1	−1	−1	−3	−2	0	1	1	1		

S_{2n} GROUPS

S_4	E	S_4	C_2	S_4^3		
A	1	1	1	1	R_z	x^2+y^2, z^2
B	1	−1	1	−1	z	x^2-y^2, xy
E	1	i	−1	$-i$	$(x, y), (R_x, R_y)$	(yz, xz)
	1	$-i$	−1	i		
$\Gamma_{x, y, z}$	3	−1	−1	−1		

S_6	E	C_3	C_3^2	i	S_6^5	S_6		$\varepsilon = e^{2\pi i/3}$
A_g	1	1	1	1	1	1	R_z	x^2+y^2, z^2
E_g	1	ε	ε^*	1	ε	ε^*	(R_x, R_y)	$(x^2-y^2, xy), (yz, xz)$
	1	ε^*	ε	1	ε^*	ε		
A_u	1	1	1	−1	−1	−1	z	
E_u	1	ε	ε^*	−1	$-\varepsilon$	$-\varepsilon^*$	(x, y)	
	1	ε^*	ε	−1	$-\varepsilon^*$	$-\varepsilon$		
$\Gamma_{x, y, z}$	3	0	0	−3	0	0		

CUBIC GROUPS

T	E	$4C_3$	$4C_3^2$	$3C_2$		$\varepsilon = e^{2\pi i/3}$
A	1	1	1	1		$x^2+y^2+z^2$
E	$\begin{Bmatrix}1\\1\end{Bmatrix}$	$\begin{matrix}\varepsilon\\\varepsilon^*\end{matrix}$	$\begin{matrix}\varepsilon^*\\\varepsilon\end{matrix}$	$\begin{matrix}1\\1\end{matrix}$		$(2z^2-x^2-y^2,\ x^2-y^2)$
T	3	0	0	−1	$(R_x, R_y, R_z), (x, y, z)$	(xy, xz, yz)
$\Gamma_{x,y,z}$	3	0	0	−1		

T_h	E	$4C_3$	$4C_3^2$	$3C_2$	i	$4S_6$	$4S_6^5$	$3\sigma_h$		$\varepsilon = e^{2\pi i/3}$
A_g	1	1	1	1	1	1	1	1		$x^2+y^2+z^2$
A_u	1	1	1	1	−1	−1	−1	−1		
E_g	$\begin{Bmatrix}1\\1\end{Bmatrix}$	$\begin{matrix}\varepsilon\\\varepsilon^*\end{matrix}$	$\begin{matrix}\varepsilon^*\\\varepsilon\end{matrix}$	$\begin{matrix}1\\1\end{matrix}$	$\begin{matrix}1\\1\end{matrix}$	$\begin{matrix}\varepsilon\\\varepsilon^*\end{matrix}$	$\begin{matrix}\varepsilon^*\\\varepsilon\end{matrix}$	$\begin{matrix}1\\1\end{matrix}$		$(2z^2-x^2-y^2,\ x^2-y^2)$
E_u	$\begin{Bmatrix}1\\1\end{Bmatrix}$	$\begin{matrix}\varepsilon\\\varepsilon^*\end{matrix}$	$\begin{matrix}\varepsilon^*\\\varepsilon\end{matrix}$	$\begin{matrix}1\\1\end{matrix}$	$\begin{matrix}-1\\-1\end{matrix}$	$\begin{matrix}-\varepsilon\\-\varepsilon^*\end{matrix}$	$\begin{matrix}-\varepsilon^*\\-\varepsilon\end{matrix}$	$\begin{matrix}-1\\-1\end{matrix}$		
T_g	3	0	0	−1	3	0	0	−1	(R_x, R_y, R_z)	(xy, xz, yz)
T_u	3	0	0	−1	−3	0	0	1	(x, y, z)	
$\Gamma_{x,y,z}$	3	0	0	−1	−3	0	0	1		

T_d	E	$8C_3$	$3C_2$	$6S_4$	$6\sigma_d$		
A_1	1	1	1	1	1		$x^2+y^2+z^2$
A_2	1	1	1	−1	−1		
E	2	−1	2	0	0		$2z^2-x^2-y^2,\ x^2-y^2$
T_1	3	0	−1	1	−1	(R_x, R_y, R_z)	
T_2	3	0	−1	−1	1	(x, y, z)	(xy, xz, yz)
$\Gamma_{x,y,z}$	3	0	−1	−1	1		

O	E	$6C_4$	$3C_2\ (=C_4^2)$	$8C_3$	$6C_2$		
A_1	1	1	1	1	1		$x^2+y^2+z^2$
A_2	1	−1	1	1	−1		
E	2	0	2	−1	0		$2z^2-x^2-y^2,\ x^2-y^2$
T_1	3	1	−1	0	−1	$(R_x, R_y, R_z), (x, y, z)$	
T_2	3	−1	−1	0	1		(xy, xz, yz)
$\Gamma_{x,y,z}$	3	1	−1	0	−1		

O_h	E	$8C_3$	$6C_2$	$6C_4$	$3C_2\ (=C_4^2)$	i	$6S_4$	$8S_6$	$3\sigma_h$	$6\sigma_d$		
A_{1g}	1	1	1	1	1	1	1	1	1	1		$x^2+y^2+z^2$
A_{2g}	1	1	−1	−1	1	1	−1	1	1	−1		
E_g	2	−1	0	0	2	2	0	−1	2	0		$2z^2-x^2-y^2,\ x^2-y^2$
T_{1g}	3	0	−1	1	−1	3	1	0	−1	−1	(R_x, R_y, R_z)	
T_{2g}	3	0	1	−1	−1	3	−1	0	−1	1		(xz, yz, xy)
A_{1u}	1	1	1	1	1	−1	−1	−1	−1	−1		
A_{2u}	1	1	−1	−1	1	−1	1	−1	−1	1		
E_u	2	−1	0	0	2	−2	0	1	−2	0		
T_{1u}	3	0	−1	1	−1	−3	−1	0	1	1	(x, y, z)	
T_{2u}	3	0	1	−1	−1	−3	1	0	1	−1		
$\Gamma_{x,y,z}$	3	0	−1	1	−1	−3	−1	0	1	1		

LINEAR GROUPS

$C_{\infty v}$	E	$2C_{\infty}^{\phi}$	$2C_{\infty}^{2\phi}$	$2C_{\infty}^{3\phi}$	...	$\infty\sigma_v$		
$A_1 \equiv \Sigma^+$	1	1	1	1	...	1	z	x^2+y^2, z^2
$A_2 \equiv \Sigma^-$	1	1	1	1	...	−1	R_z	
$E_1 \equiv \Pi$	2	2c ϕ	2c 2ϕ	2c 3ϕ	...	0	$(x, y), (R_x, R_y)$	(xz, yz)
$E_2 \equiv \Delta$	2	2c 2ϕ	2c 4ϕ	2c 6ϕ	...	0		(x^2-y^2, xy)
$E_3 \equiv \Phi$	2	2c 3ϕ	2c 6ϕ	2c 9ϕ	...	0		
$E_4 \equiv \Gamma$	2	2c 4ϕ	2c 8ϕ	2c 12ϕ	...	0		
...	...	...	...	...	...	...		
...	...	...	...	...	...	...		
$\Gamma_{x,y,z}$	3	1+2c ϕ	1+2c 2ϕ	1+2c 3ϕ		1		

$D_{\infty h}$	E	$2C_{\infty}^{\phi}$	...	$\infty\sigma_v$	i	$2S_{\infty}^{\phi}$	...	∞C_2		
Σ_g^+	1	1	...	1	1	1	...	1		x^2+y^2, z^2
Σ_g^-	1	1	...	−1	1	1	...	−1	R_z	
Π_g	2	2c ϕ	...	0	2	−2c ϕ	...	0	(R_x, R_y)	(xz, yz)
Δ_g	2	2c 2ϕ	...	0	2	2c 2ϕ	...	0		(x^2-y^2, xy)
...	...	...	...	...	...	...	...	...		
Σ_u^+	1	1	...	1	−1	−1	...	−1	z	
Σ_u^-	1	1	...	−1	−1	−1	...	1		
Π_u	2	2c ϕ	...	0	−2	2c ϕ	...	0	(x, y)	
Δ_u	2	2c 2ϕ	...	0	−2	−2c 2ϕ	...	0		
...	...	...	...	...	...	...	...	...		
$\Gamma_{x,y,z}$	3	1+2c ϕ		1	−3	−1+2c ϕ		−1		

Abbreviation: c, cosine; s, sine.

Index

Note: Page numbers followed by *f* indicate figures, *t* indicate tables, and *s* indicate schemes.

D

E

P

Q

R

S

T

U

V